CURRENT PERSPECTIVES AND NEW DIRECTIONS IN MECHANICS, MODELLING AND DESIGN OF STRUCTURAL SYSTEMS

Current Perspectives and New Directions in Mechanics, Modelling and Design of Structural Systems comprises 330 papers that were presented at the Eighth International Conference on Structural Engineering, Mechanics and Computation (SEMC 2022, Cape Town, South Africa, 5-7 September 2022).

The topics featured may be clustered into six broad categories that span the themes of mechanics, modelling and engineering design:
(i) mechanics of materials (elasticity, plasticity, porous media, fracture, fatigue, damage, delamination, viscosity, creep, shrinkage, etc);
(ii) mechanics of structures (dynamics, vibration, seismic response, soil-structure interaction, fluid-structure interaction, response to blast and impact, response to fire, structural stability, buckling, collapse behaviour);
(iii) numerical modelling and experimental testing (numerical methods, simulation techniques, multi-scale modelling, computational modelling, laboratory testing, field testing, experimental measurements);
(iv) design in traditional engineering materials (steel, concrete, steel-concrete composite, aluminium, masonry, timber);
(v) innovative concepts, sustainable engineering and special structures (nanostructures, adaptive structures, smart structures, composite structures, glass structures, bio-inspired structures, shells, membranes, space structures, lightweight structures, etc);
(vi) the engineering process and life-cycle considerations (conceptualisation, planning, analysis, design, optimization, construction, assembly, manufacture, maintenance, monitoring, assessment, repair, strengthening, retrofitting, decommissioning).

Two versions of the papers are available: full papers of length 6 pages are included in an e-book, while short papers of length 2 pages, intended to be concise but self-contained summaries of the full papers, are in this printed book. This work will be of interest to civil, structural, mechanical, marine and aerospace engineers, as well as planners and architects.

PROCEEDINGS OF THE EIGHTH INTERNATIONAL CONFERENCE ON STRUCTURAL ENGINEERING, MECHANICS AND COMPUTATION, 5-7 SEPTEMBER 2022, CAPE TOWN, SOUTH AFRICA

Current Perspectives and New Directions in Mechanics, Modelling and Design of Structural Systems

Editor

Alphose Zingoni

Department of Civil Engineering, University of Cape Town, South Africa

CRC Press is an imprint of the
Taylor & Francis Group, an **informa** business
A BALKEMA BOOK

Cover image: *Schlaich Bergermann Partner (sbp), Stuttgart, Germany*

First published 2022
by CRC Press/Balkema
4 Park Square, Milton Park, Abingdon, Oxon, OX14 4RN, United Kingdom
e-mail: enquiries@taylorandfrancis.com
www.routledge.com – www.taylorandfrancis.com

CRC Press/Balkema is an imprint of the Taylor & Francis Group, an informa business

© 2022 selection and editorial matter, Alphose Zingoni; individual chapters, the contributors

The right of Alphose Zingoni to be identified as the author of the editorial material, and of the authors for their individual chapters, has been asserted in accordance with sections 77 and 78 of the Copyright, Designs and Patents Act 1988.

All rights reserved. No part of this book may be reprinted or reproduced or utilised in any form or by any electronic, mechanical, or other means, now known or hereafter invented, including photocopying and recording, or in any information storage or retrieval system, without permission in writing from the publishers.

Although all care is taken to ensure integrity and the quality of this publication and the information herein, no responsibility is assumed by the publishers nor the author for any damage to the property or persons as a result of operation or use of this publication and/or the information contained herein.

Library of Congress Cataloging-in-Publication Data

A catalog record has been requested for this book

ISBN: 978-1-032-39114-4 (hbk)
ISBN: 978-1-003-34845-0 (ebk)
DOI: 10.1201/9781003348450

© 2022 Copyright the Editor and Contributors, *ISBN 978-1-032-39114-4*

Table of Contents

Preface xxvii

Committees of the SEMC 2022 International Conference xxxi

1. Keynotes

Metal additive manufacturing in structural engineering: Review, opportunities and outlook 3
L. Gardner

Computational modeling of FRC: From 3D printing to robust design 5
G. Meschke, G. Neu, V. Gudžulić, J. Reinold & T. Iskhakov

FRP strengthening of metallic structures: From research to practice 7
X.L. Zhao

In-test analytical model updating for accurate hybrid dynamic simulations 9
A.S. Elnashai & H.H. El Anwar

Value of vibration-based structural monitoring for bridge emergency management 11
M.P. Limongelli & P.F. Giordano

Understanding tensegrity with an energy function 13
S.D. Guest

2. Dynamic analysis, Vibration response, Vibration control, Environmental vibrations, Fluid-structure interaction

Bridge flutter and how to prevent it at low cost 17
U. Starossek

Early assessment of the vibroacoustic performance of large lightweight buildings 19
L.V. Andersen, L. Mangliar, M.M. Hudert, O. Flodén & P. Persson

Modal analysis of CLT beams: Measurements and predictive simulations 21
B. Bondsman, O. Flodén, H. Danielsson, P. Persson & E. Serrano

Optimal design of viscoelastic tuned mass dampers for structures exposed to coloured excitations 23
M. Argenziano, E. Mele & A. Palmeri

Numerical investigation on train-bridge interaction after ship impact event: Train running safety in different scenarios 25
L. Bernardini, G. Soldavini & A. Collina

Research advancement on submerged floating tunnels: Recent activity at DICA-Politecnico di Milano 27
F. Foti, L. Martinelli, M.G. Mulas & F. Perotti

The influence of near-surface soil layer resonance on vibrations in pile foundations 29
F. Theland, C. Pacoste, J-M. Battini, G. Lombaert, S. François & F. Deckner

Prediction of flutter velocity of long-span bridges using probabilistic approach 31
S. Tinmitonde, X. He & L. Yan

The effect of pile slip on underwater noise emission in vibratory pile driving 33
T. Molenkamp, A. Tsouvalas & A.V. Metrikine

Dynamic analysis of highly symmetric cable dome structures 35
L. Fan, Y. Chen, C. Lu & Y. Sun

Vibroacoustic performance of multi-storey buildings: A comparison of lightweight and heavy structures in the early design phase 37
L.V. Andersen, L. Mangliar, M.M. Hudert, O. Flodén & P. Persson

Effect of modelling the air in rooms on the prediction of vibration transmission in cross-laminated timber buildings 39
O. Flodén, A. Peplow, P. Persson, L. Mangliár & L.V. Andersen

Numerical analysis of transversely isotropic unsaturated railway ground vibrations under a moving load using 2.5D FEM 41
S. Li

Incremental dynamic analysis of a hospital building in Oman 43
M.B. Waris, K. Al-Jabri, W.H. Bhuta & I. El-Hussain

Dynamic response of parabolic reflector antenna subjected to shock load and base excitation considering soil-structure interaction 45
V. Gullapalli, R.R. Kumar & G.R. Reddy

Effect of Cape storms on the flat roof-mounted solar panels 47
Y. Kiama & J. Mahachi

Evaluating static and modal responses of a small scale conventional and bio-inspired wind turbine blade 49
T. Babawarun, H.M. Ngwangwa, A.G. Netshivhulana, T. Pandelani & W.H. Ho

3. Seismic response, Seismic analysis, Earthquake-resistant design

Functional elements exploiting superelasticity and the shape memory effect of Fe-Mn-Al-Ni-X shape-memory-alloys in structural engineering 53
E. Fehling, B. Middendorf, J. Thiemicke, M. Vollmer, A. Bauer, T. Niendorf, V.v. Oertzen & B. Kiefer

Seismic response of curved structures in historical masonry churches 55
C. Chisari, M. Zizi, J. Rouhi & G.De Matteis

Recent advances in tsunami design of coastal structures 57
I.N. Robertson

Load-level isolation system for industrial racks: Evaluations on a case study structure 59
E. Bernardi, M. Donà, A. Zonta, M. Ceresara, S. Mozzon, F. da Porto & P. Tan

Seismic considerations in the dynamic characteristics of masonry infilled RC frames based on experimental tests 61
F.J. Pallarés, A. Agüero & L. Pallarés

Structural plan and design of bridge deck isolation system for seismic redundancy 63
M. Matsumura, K. Annoura & H. Moriyama

Study on seismic performance of fully bolted joints 65
B.-L. Wu, Y.-X. Zhang, S. Shen & Z.-W. Huang

Effects of in-plan irregularity of RC framed buildings on the out-of-plane nonlinear seismic response of masonry infills 67
F. Mazza

Influence of infill walls on lateral load carrying capacity of buildings 69
M.B. Waris, H. Al-Hadi & A. Al-Nuaimi

Simulation of many acceleration time histories with causality from an observed earthquake motion: Modeling of real part and reproduction of imaginary part by Hilbert transform 71
T. Sato

Dynamic characterization of the Tower of Palazzo dei Vicari in Scarperia (Italy) during and after the 2019 Mugello seismic sequence 73
R.M. Azzara, V. Cardinali, M. Girardi, F. Marini, C. Padovani, D. Pellegrini & M. Tanganelli

Push-over tsunami analysis using *ETABS* 75
D.W. Aegerter & I.N. Robertson

Seismic fragility of corroded reinforced concrete highway bridges 77
S. Biswas, G.K. Verma & P. Sengupta

Assessment of the inter-story isolation technique applied to an existing school building 79
M. Donà, E. Bernardi, A. Zonta, E. Saler & F. da Porto

Effects of frequency content of ground motions on seismic response of concrete gravity dams with varying reservoir levels 81
G. Reddy & M. Shrikhande

A proposal for improving the elastic behaviour of dogbone 83
L. Palizzolo, S. Benfratello & S. Vazzano

4. Material modelling, Multi-scale modelling, Micromechanical modelling, Porous media, Biological tissue

Multiscale modelling in materials science and technology 87
D. Rapp, C. Xu, F. Sajadi, X. Xu & S. Schmauder

A multiscale approach to investigate residual stresses due to targeted cooling of hot bulk formed parts 89
S. Hellebrand, D. Brands, J. Schröder & L. Scheunemann

Introduction to lattice discrete particle model for thermoset polymers 91
J. Vorel, J. Vozáb & J. Kruis

Artificial neural networks as material models for finite element analysis 93
D. Sommer, B. Troff & P. Middendorf

Simulations of elasto-plastic soils within the framework of the theory of porous media 95
J. Sunten, A. Schwarz, J. Bluhm & J. Schröder

A coupled multi–field dynamic model for anisotropic porous materials 97
N. De Marchi, M. Ferronato, G. Xotta & V.A. Salomoni

Micromechanical modelling of failure during interactions between ice and subsea structures 99
R.S. Taylor & I. Gribanov

Variational formulation of weak layer stresses in stratified snowpacks 101
F. Rheinschmidt, P. Weißgraeber & P.L. Rosendahl

An exact stiffness matrix method for nanoscale beams 103
R.A.R. Wijesinghe, K.K.V. De Silva, Y. Sapsathiarn & N. Rajapakse

On the deformation of porous spherical shells and other spherical bodies 105
S. De Cicco

Application of quasi-linear viscoelasticity for the characterisation of human buttocks adipose tissue 107
T. Pandelani, S.D. Masouros & J.L. Calvo-Gallego

Nonlocal beam analysis based on the stress-driven two-phase theory 109
F.P. Pinnola, M.S. Vaccaro, R. Barretta & F. Marotti de Sciarra

Modeling the hysteresis behavior of recycled rubber fiber reinforced bearings 111
A. Flora, D. Cardone & A. Calabrese

Numerical permeability prediction of multiscale textile architectures with varying contact angle and surface tension 113
J. Dittmann, P. Seif & P. Middendorf

Data-driven design of high ductile metamaterials under uniaxial tension 115
A.S. Bhuwal, Y. Pang, T. Liu, I. Ashcroft & W. Sun

Modelling of biaxial tensile behaviour of the tracheal tissue using three exponential-based hyperelastic constitutive models 117
H.M. Ngwangwa, L. Semakane, F. Nemavhola, T. Pandelani & D. Modungwa

Evaluating the tensile behaviour of rat myocardium across its three walls from biaxial tensile test data 119
H.M. Ngwangwa, M. Msibi, I. Mabuda, F. Nemavhola & T. Pandelani

Investigation of the performance of different constitutive models on omasum uniaxial tensile behaviour 121
H.M. Ngwangwa, L. Lebea, F. Nemavhola, T. Pandelani & D. Modungwa

Evaluating computational performances of Yeoh, Veronda-Westmann and Humphrey models on supraspinatus tendon uniaxial stress-strain behaviour 123
H.M. Ngwangwa, M. Msibi, I. Mabuda & F. Nemavhola

5. Additive manufacturing, Manufacturing processes, Assembly processes, Kinematics

Robot supported wire arc additive manufacturing and milling of steel columns 127
B. Waldschmitt, J. Lange, C.B. Costanzi, U. Knaack, T. Engel & J. Müller

Global sensitivity analysis of sand-based binder jet 3D printed material 129
L. Del Giudice, M.F. Vassiliou & S. Marelli

Influence of process irregularities in additively manufactured structures 131
M. Erven & J. Lange

Characterization of 3D printed concrete beams after exposure to high temperature 133
O. Shkundalova, T. Molkens, M. Classen & B. Rossi

Experimental studies on the mechanical response for structural design of wire-and-arc additively manufactured stainless steel bars 135
V. Laghi, L. Arrè, M. Palermo, G. Gasparini & T. Trombetti

Tolerance-free assembly of heat-treated reinforced concrete modules with geometric deviations 137
J. Stindt, P. Forman & P. Mark

Dimensional accuracy of casting patterns produced by selected 3D printing technologies 139
P. Zmarzły, T. Kozior & D. Gogolewski

The effect of wind on 3D printed concrete interlayer bond strength based on machine learning algorithms 141
A. Cicione, P.J. Kruger, J.P. Mostert, R. Walls & G.V. Zijl

Tensile test of double-lap GFRP joint applying vacuum infusion wrapping 143
Y. Anan, M. Matsumura, H. Moriyama & S. Inoue

Multiscale analysis of surfaces made with additive technology in medical aspects 145
D. Gogolewski, T. Kozior & P. Zmarzły

Bio-inspired kinematics of reconfigurable linkage structures 147
N. Georgiou & M.C. Phocas

Tribological analysis of selected 3D printing technologies in the context of building foundry models 149
T. Kozior, D. Gogolewski & P. Zmarzły

A case study of structural optimization for additive manufacturing 151
J. Hajnys, M. Pagac, J. Mesicek & P. Krpec

6. Composite structures, Laminated structures, Sandwich structures, Adaptive structures, Numerical modelling, Numerical simulations

Piezoresistive sensor matrix for adaptive structures 155
H. Fritz, C. Walther & M. Kraus

Numerical modelling of the residual burst pressure of thick composite pressure vessels after low-velocity impact loading 157
R. Czichos, P. Middendorf & T. Bergmann

Influence of interfacial debonding on the nonlinear structural response of profiled metal-faced insulating sandwich panels 159
M.N. Tahir & E. Hamed

Accurate numerical integration in 3D meshless peridynamic models 161
U. Galvanetto, F. Scabbia & M. Zaccariotto

Phase field topology optimization of elasto-plastic contact problems with friction 163
A. Myśliński

Computational analysis of magma-driven dike and sill formation mechanisms 165
J.A.L. Napier & E. Detournay

Numerical analysis of a high-rise building near sensitive infrastructures 167
M. Seip, M. Hassan & S. Leppla

Moment support interaction of multi-span sandwich panels 169
A. Engel & J. Lange

Influence of temperature on sandwich panels with PU rigid foam 171
S. Steineck & J. Lange

Minimum weight design of CNT/fibre reinforced laminates subject to a frequency constraint 173
S. Anashpaul, G.A. Drosopoulos & S. Adali

On the application of CFD and MDO to aircraft wingboxes using commercial software 175
X. Duan & K. Behdinan

7. Damage mechanics, Damage modelling, Blast, Impact, Fracture, Fatigue, Progressive collapse, Structural robustness

Research on structural robustness through large-scale testing 179
J.M. Adam, M. Buitrago & N. Makoond

Using optical diagnostics for near field blast testing 181
G.S. Langdon, R.J. Curry, S.E. Rigby, E.G. Pickering, S.D. Clarke, A. Tyas, S. Gabriel, S.C.K. Yuen & C.J. von Klemperer

Experimental evaluation of composite floors under column loss scenarios 183
M. Hadjioannou, E.B. Williamson & M.D. Engelhardt

The role of infills modelling in progressive collapse capacity assessment of RC buildings: A case study 185
M. Scalvenzi, F. Parisi & E. Brunesi

Computational study on the progressive collapse of precast reinforced concrete structures 187
N. Makoond, M. Buitrago & J.M. Adam

Assessment on impact response analysis by chord member fracture of steel truss bridge 189
H. Nagatani, D. Yoshimoto, K. Hashimoto, Y. Kitane & K. Sugiura

Lumped plasticity approach for the robustness assessment of structures 191
T. Molkens

Progressive collapse resistant design of tall buildings under fire load 193
F. Fu

A multi-scale approach for quantifying the robustness of existing bridges 195
V. De Biagi, B. Chiaia & F. Kiakojouri

Interpretation of collapse modalities for a timber roof truss 197
M. Acito, E. Magrinelli, C. Chesi & M.A. Parisi

Blast response of arching masonry walls: Theoretical and experimental investigation 199
I.E. Edri, D.Z. Yankelevsky & O. Rabinovitch

Blast behaviour of natural fibre composites: Response of medium density fibreboard and flax fibre reinforced laminates 201
S. Gabriel, C.K. Yuen, G.S. Langdon & C.J. von Klemperer

Comparison of TNT to PE4 charge using a simplified rigid torso 203
T. Pandelani, D. Modungwa, S. Hamilton & J.D. Reinecke

Blast performance of retrofitted reinforced concrete highway bridges 205
S. Biswas, S. Kulshreshtha & P. Sengupta

Investigation of permanent indentation due to low velocity impact on glass fibre reinforced plastics 207
M. Bastek, L. Merkl & P. Middendorf

Advanced fatigue assessment methods: The future of modern wind turbines 209
M. Rauch, R. Yildirim, H. Bissing & M. Knobloch

Fatigue damage in thick GFRP laminate through the vibration-based bending fatigue experiment 211
A. Sato, Y. Kitane, Y. Goi & K. Sugiura

Decoupling method for fracture modes in an orthotropic elastic material through a three-dimensional vision 213
J. Afoutou, F. Dubois, N. Sauvat & M. Takarli

Investigation on 3D printed dental titanium Ti64ELI and lifetime prediction 215
L. Lebea, H.M. Ngwangwa, D. Desai & F. Nemavhola

Detection of fatigue damage via neural network analysis of surface topography measurements 217
H. Alqahtani, A. Ray

Notch stress analysis on reinforcing steel bars using FE-simulations considering surface topography and structure properties 219
S. Rappl & K. Osterminski

Crashworthiness analysis of conical energy absorber filled with aluminum foam 221
M. Rogala & J. Gajewski

Numerical and experimental analysis of the triggering mechanism of the passive square thin-walled absorber 223
M. Rogala, J. Gajewski & R. Karpiński

Theoretical and experimental investigation into area increase of longitudinal cracks in pressurised pipes 225
D.T. Ilunga, M.O. Dinka & D.M. Madyira

Numerical study of dynamic collapse resistance of an RC flat plate substructure under an interior column removal scenario 227
Z. Yang, H. Guan, Y. Li & X.Z. Lu

8. Buckling of Beams, Columns and Thin-walled sections

Harnessing instabilities within metamaterial structures 231
M.A. Wadee, A. Bekele & A.T.M. Phillips

Plastic buckling of fixed-roof oil storage tanks under blast loads 233
L.A. Godoy & M.P. Ameijeiras

Modelling thin-walled members by means of the dynamic approach within the Generalised Beam Theory (D-GBT) 235
G. Ranzi

On energy based flexural-torsional buckling criteria of double-tee shape elastic thin-walled members 237
M.A. Gizejowski

An individual shear deformation theory of beams with consideration of the Zhuravsky shear stress formula 239
K. Magnucki

Elastic shear buckling of beams with sinusoidal corrugated webs 241
L. Sebastiao & J. Papangelis

Buckling check under biaxial loading considering different launching bearings and eccentric load introduction 243
N. Maier, M. Mensinger & J. Ndogmo

Lateral-torsional buckling of steel double-tee shape beams based on the decomposition of transverse load combinations 245
A.M. Barszcz & M.A. Gizejowski

Flexural-torsional buckling of steel double-tee shape beam-columns with lateral-torsional discrete restraints 247
M.A. Gizejowski, A.M. Barszcz & P. Wiedro

Influence of symmetry on the buckling behaviour of plane frames 249
C. Kaluba & A. Zingoni

Machine learning optimization strategy for the inelastic buckling modelling of uncorroded and corroded reinforcing plain bars 251
F. Pugliese & L. Di Sarno

Numerical study of distortional buckling of single channels restrained by angle cleats 253
G.M. Bukasa & M. Dundu

Experimental investigation of distortional buckling of single channels restrained by angle cleats 255
G.M. Bukasa & M. Dundu

9. Plates, Shells, Membranes, Cable nets, Cable-stayed structures, Lightweight structures, Form-finding, Architectural considerations

Principle of constant stress in analytical form-finding for durable structural design 259
W.J. Lewis

Extraordinary mechanics in slender structural surfaces 261
S. Adriaenssens

Asymmetric multifocal egg-like shells with elastic adaptive frameworks 263
D. Kozlov & Y. Vaserchuk

Analysis of constructional errors of in-service spatial prestressed steel structures 265
A.L. Zhang, J. Wang, X. Zhao & Y.X. Zhang

Examples of application of principle of superposition and rules of symmetry in engineering design 267
J. Rębielak

Full-scale tests of a lightweight footbridge: The Folke Bernadotte Bridge 269
D. Colmenares., G. del Pozo, G. Costa, R. Karoumi & A. Andersson

Numerical analysis of new assembled cable dome ring truss connections 271
A.L. Zhang, M. Zou, Y.X. Zhang, G.H. Shangguan & J. Wang

A numerical displacement-based approach for the structural analysis of cable nets 273
B. Frigo & A.P. Fantilli

Analytical and numerical analysis of anchored flexible steel liquid storage tanks subjected to seismic loads considering soil structure interaction 275
M.S. Pourbehi, J.A.v.B. Strasheim & M. Georgiadis

Parametric study of the catenary dome under gravity load 277
R.A. Bradley & M. Gohnert

10. Structural applications of glass

Structural bonding with hyperelastic adhesives: Material characterization, structural analysis and failure prediction 281
P.L. Rosendahl, F. Rheinschmidt & J. Schneider

Reinforced glass: Structural potential of cast glass beams with embedded metal reinforcement 283
T. Bristogianni & F. Oikonomopoulou

Investigations on the concept of hybrid bonding in glass structures 285
D. Offereins & G. Siebert

Topologically optimized structural glass megaliths: Potential, challenges and guidelines for stretching the mass limits of structural cast glass 287
F. Oikonomopoulou, A.M. Koniari, W. Damen, D. Koopman, I.M. Stefanaki & T. Bristogianni

Rotational stiffness and strength of a two-sided reinforced laminated glass beam-column joint prototype: Experimental investigation 289
M. Pejatovic, R. Caspeele & J. Belis

Transparent smoke barriers 291
A. Haese & B. Siebert

Experimental investigations on EVA interlayers in the regime of large deformations in the context of the examination of the residual load bearing capacity of LSG 293
M. Baric, A. Pauli & G. Siebert

In-plane loaded glass balustrades as structural members for balconies 295
J. Giese-Hinz, F. Nicklisch, B. Weller, M. Baitinger, H. Hoffmann & J. Reichert

Thermal loads in triple glazing: Experimental and numerical case study 297
P. Bukieda & B. Weller

Influence of the interlayer core material in thin glass–plastic composite panels on performance characteristics according to the requirements for laminated safety glass 299
J. Hänig & B. Weller

Load analysis for the development of a bonded edge seal for fluid-filled insulating glazing 301
A. Joachim & B. Weller

The strength of glass with digital printing 303
J. Wünsch, B. Weller, J. Wittwer & A. Rumpf

Numerical simulation of the heat flow through laminated glass beams exposed to fire: A parametric study 305
M. Möckel, T. Juraschitz & C. Louter

11. Steel structures, Steel connections, Steel-concrete composite construction, Cold-formed steel

Full-scale tests of industrial steel storage pallet racks 309
N. Baldassino, M. Bernardi, R. Zandonini & A. di Gioia

Tests of cold-formed steel built-up open section beam-columns subjected to unequal end moments 311
Q.-Y. Li & B. Young

Numerical analysis on semi-rigid pin joint for modular assembled steel reticulated shell 313
A.-L. Zhang & C. Li

Experimental study on shallow H-shaped steel column under combined loading condition 315
S. Nakatsuka & A. Sato

Experimental tests on the rack-to-spine-bracing joints of high-rise steel storage racks 317
Z. Huang, X. Zhao & K.S. Sivakumaran

Experimental and numerical evaluation of various solutions for fire protection of steel structures 319
M.H. Nguyen, S.E. Ouldboukhitine, S. Durif, V. Saulnier & A. Bouchaïr

Microstructure and corrosion resistance of dissimilar welded joints 321
B. Verhoeven, R. Dewil, G. Ferraz, B. Karabulut, S. Sun & B. Rossi

3D models of clad steel structures: Assumptions and validation 323
M.J. Roberts & J.M. Davies

Fatigue resistance of steel arch bridge hanger connection plates due to transverse welding 325
Ph. Van Bogaert

Numerical simulation and design of steel equal-leg angle section beams 327
B. Behzadi-Sofiani, L. Gardner & M.A. Wadee

Behaviour of steel end plate bolted beam-to-column joints 329
J. Qureshi & S. Shrestha

Compression strength of aged built-up column with vanished lacing bars 331
T. Miyoshi, T. Nakakita, K. Iwatsubo, T. Takai & K. Tamada

Machine learning approach for stress analyses of steel members affected by elastic shear lag 333
H. Fritz & M. Kraus

Behaviour and recommended design method for laterally unsupported monosymmetric steel I-section beams 335
S. Suman & A. Samanta

Steel beam upstands as a strengthening approach for hot-rolled I-shaped sections 337
K. Mudenda & A. Zingoni

Bending capacity of single and double-sided welded I-section girders: Part 1: Experimental investigations 339
J. Wang, M. Euler, H. Pasternak, Z. Li, T. Krausche & B. Launert

Bending capacity of single and double-sided welded I-section girders: Part 2: Simplified welding simulation and buckling analysis 341
Z. Li, H. Pasternak, J. Wang, B. Launert & T. Krausche

Issues related to the assessment of an existing reinforcement of a lattice telecommunication tower 343
J. Szafran & J. Telega

Relating stress concentrations in triangular steel bridge piers to simple beam models 345
Ph. Van Bogaert, G. Van Staen & H. De Backer

Experimental investigation of the rack-to-bracing joints between the high-rise steel storage rack frames and the independent bracing towers 347
Z. Huang, X. Zhao & K.S. Sivakumaran

Optimum design of hat cold-formed steel members using direct strength method 349
M.H. Khashaba, I.M. El-Aghoury & A.A. El-Serwi

Performance of light gauge cold-formed steel flexural members subjected to non-uniform fire 351
R. Singh & A. Samanta

Static strength capacity of single-sided fillet welds 353
T. Skriko, A. Ahola, K. Lipiäinen & T. Björk

Influence of web openings on the load bearing behavior of steel beams 355
F. Eyben, S. Schaffrath & M. Feldmann

Effects of elevated temperatures on cold-formed stainless steel double shear bolted connections 357
Y. Cai & B. Young

Simplified approach to calculate welding effect for multi-layer welds of I-girders 359
T. Krausche, H. Pasternak, J. Wang, Z. Li & B. Launert

Resistance of axially and eccentrically loaded steel column at high temperature: A simple expression 361
G. Somma

Large-scale fire testing of an innovative cellular beam and composite flooring structural system 363
J. Claasen, R. Walls & A. Cicione

Design for de-construction of lightweight infill wall systems 365
S. Kitayama & O. Iuorio

Calibration and validation of slender circular CFDST columns subjected to eccentric loading 367
T.N. Haas, A.A. Usongo & J.A. Koen

Investigation of CDP parameters and concrete confinement models for use in circular slender CFDST columns subjected to eccentric loading 369
A.A. Usongo & T.N. Haas

Evaluation of parameters which influence circular slender CFDST columns subjected to eccentric loading 371
A.A. Usongo & T.N. Haas

Evaluating the effect of load eccentricity on slender circular CFDST columns 373
J.P. Rust & T.N. Haas

Shear capacity of the Truedek steel decking system in the formwork 375
N. Singh, O. Mirza, M. Hosseini, A. Wheeler & D. Allan

Numerical modeling for connections of staggered single channel rafters in double-bay cold-formed steel portal frames 377
J. Kafuko & M. Dundu

Numerical modeling of gusseted internal eaves connection of double-bay cold-formed steel portal frames using single channels 379
J. Kafuko & M. Dundu

Stub concrete-filled double-skin circular tubes in compression 381
R. Mahlangu & M. Dundu

12. High strength steel, High performance steel

Welded connections of high-strength steels 385
R. Stroetmann, T. Kästner & B. Rust

A 3D constitutive model of high-strength constructional steels with ductile fracture 387
G.-Q. Li, Y.-B. Wang, Y.-Z. Wang, L.-T. Hai & W.-Y. Cai

Investigation of the effects of an over-elastic preload on the load-bearing behavior of high-strength bolt and nut assemblies 389
J. Reinheimer & J. Lange

Bending effects on the fatigue strength of non-load-carrying transverse attachment joints made of ultra-high-strength steel 391
A. Ahola & T. Björk

Mechanical properties of thin-walled high-strength-steel cold-formed circular hollow sections for crane and scaffolding construction 393
D. Gubetini & M. Mensinger

A basic study on the evaluation of residual axial force of high strength bolts for steel bridges constructed in the same environment at different ages 395
T. Iida, K. Sugiura, T. Yamaguchi & Y. Kitane

Design methodology of grade 12.9 bolts considering hydrogen-induced delayed fracture 397
Y.B. Wang, Z. Sun, B. Tha, G.Q. Li & X.L. Zhao

Influence of high strength steel on the fatigue of welded details 399
H. Bartsch & M. Feldmann

Influence of residual stresses on innovative composite column sections using high strength steel 401
R. Röß, M. Schäfers & M. Mensinger

Experimental and numerical study of butt welded joints made of high strength steel 403
R. Yan, H.E. Bamby & M. Veljkovic

Behaviour of cold-formed high strength steel tubular X-joints with circular braces and rectangular chords 405
M. Pandey & B. Young

Experimental investigation on cold-formed high strength steel tubular T-joints after fire exposures 407
M. Pandey & B. Young

Study on mechanical properties of high-strength steel butt-welded joints 409
W.Y. Cai, Y.B. Wang & G.Q. Li

13. Reinforced concrete structures, Prestressed concrete, Concrete structural elements, Mechanics of concrete

Using nonlinear finite element modeling for punching shear design of concrete slabs 413
P.M. Beaulieu, G.J. Milligan & M.A. Polak

Design of precast concrete framing systems against disproportionate collapse using component-based methods 415
K. Riedel, R. Vollum, B. Izzuddin, G. Rust & D. Scott

Semi-probabilistic nonlinear assessment of post-tensioned concrete bridge made of KT-24 girders 417
B. Šplíchal, D. Lehký & J. Doležel

Punching shear behavior of continuous flat slabs: A new test setup incorporating system influences 419
M. Kalus, J. Ungermann & J. Hegger

Live load test for pedestrian bridge constructed with innovative concrete bridging system 421
H.H. Alghazali, J.J. Myers & K.E. Bloch

Estimating post-punching capacity and progressive collapse resistance of RC flat plates using shell-based FEA 423
A.B. Bahnamiri, T.D. Hrynyk & C.Y. Goh

Long-term deformation of segmented prestress bridges under harsh weather conditions: Case study of Exit 23 Interchange at Riyadh Ring Road 425
H.H. Abbas & M.M. Hassan

Shear capacity of concrete elements with reinforcement made of steel and FRP 427
S. Görtz & R. Haack

Modelling temperature and stress development in large concrete elements under sequential construction conditions: Experience with the Msikaba Bridge foundations 429
Y. Ballim & G. Harli

Characterization of major reinforcing bars for concrete works in Botswana construction industry 431
A.P. Adewuyi, G.B. Eric & O.J. Kanyeto

Experimental assessment of crack width estimations in international design codes 433
I. Ridley, M. Shehzad, J. Forth, N. Nikitas, A. Elwakeel, K. Elkhoury, R. Vollum & B. Izzuddin

RC flooring system recycling plastic bottles: New innovative RC waffle slab 435
C.E. Mankabady, N. Salama, H. Tolba, K. Nassar, M.N. AbouZeid & E.Y. Sayed-Ahmed

Investigations on shear transfer by aggregate interlock with a unique test setup (TorAx) 437
S. Bosbach, M. Schmidt, H. Becks, M. Claßen & J. Hegger

The continuous spalling of reinforced concrete structures of the three bridges on the Lagos lagoon in Nigeria 439
U.T. Igba, J.O. Akinyele, A. Adetiloye & J.O. Labiran

Theoretical analysis and experimental test results calibrated from concrete beams 441
G. Sossou

Behaviour of reinforced concrete beams enhanced with polymer modified ferrocement 443
N. Evbuomwan

Shear capacity of rib and block slab systems 445
Y. Essopjee

14. High strength concrete, High performance concrete, Fibre-reinforced concrete, Fatigue behaviour of HPC and UHPC

Phase-field modeling for damage in high performance concrete at low cycle fatigue 449
J. Schröder, M. Pise, D. Brands, G. Gebuhr & S. Anders

Macroscopic model based on application of representative volume element for steel fiber reinforced high performance concrete 451
M. Pise, D. Brands, J. Schröder, G. Gebuhr & S. Anders

Potential of shape memory alloys in fiber reinforced high performance concrete 453
B. Middendorf, M. Schleiting & E. Fehling

Homogenisation of the material behaviour of UHPFRC under tension 455
L. Gietz, U. Kowalsky, D. Dinkler, J.-P. Lanwer & M. Empelmann

Numerical and experimental analysis of fatigue-induced changes in ultra-high performance concrete 457
S. Rybczyński, F. Schmidt-Döhl, G. Schaan, M. Ritter & M. Dosta

Damage development of steel fibre reinforced high performance concrete in high cycle fatigue tests 459
G. Gebuhr, S. Anders, M. Pise, D. Brands & J. Schröder

Fatigue behavior and crack opening tests under tensile stress on HPSFRC: Experimental and numerical investigations 461
N. Schäfer, V. Gudžulić, R. Breitenbücher & G. Meschke

Influence of concrete compressive strength on fatigue behaviour under cyclic compressive loading 463
M. Markert & M. Deutscher

Influence of loading frequency on the compressive fatigue behaviour of high-strength concrete with different moisture contents 465
M. Abubakar Ali, N. Oneschkow, L. Lohaus & M. Haist

Influence of the concrete moisture content on the strain development due to cyclic loading 467
M. Markert, V. Birtel & H. Garrecht

Influence of moisture content on strain development of concrete subjected to compressive creep and cyclic loading 469
B. Kern, N. Oneschkow, M. Haist & L. Lohaus

Behavior of coupled walls with high performance fiber reinforced concrete coupling beams 471
M. Abdelhafeez & H. Salem

15. Safety and reliability, Design philosophy, Bridge technology, Transport infrastructure, Building performance, Engineering education

Structural reliability estimates by Slepian models 475
M. Grigoriu

Framing of the design base for load bearing structures against climate change 477
J.V. Retief

Mechanical properties of stainless steel coatings formed by build-up spraying 479
E. Horisawa, K. Sugiura, Y. Kitane, Y. Goi, T. Tanimoto & M. Matsumura

Implication of constructing the new Umhlatuzana River bridge deck monolithically with the existing deck 481
A.J. Faure

Field and desktop survey on failures of rail to concrete connections in South African track on concrete ballastless railway systems 483
W.M.P. Makwela & J. Mahachi

Long term energy efficiency of non-conventional building systems: Use of Polyblocks in improving thermal performance 485
D.M. Tshilombo, S.A Alabi & J. Mahachi

Cooling materials that help save lives in the context of Covid-19's economic recession 487
F. Pacheco Torgal

Research for achieving high quality air in architecture and urban planning on the example of the educational and sports center in Józefów near Warsaw, Poland 489
J. Wrana & W. Struzik

Microorganisms and the healthy built environment 491
F. Pacheco Torgal

Do we still need structural engineers? 493
P.M. Debney

Facilitated team-building processes to enhance teamwork skills in engineering education 495
L. Bücking

Peer review of teaching/learning paradigms: A new proposal for engineering education 497
M. Saudy, I. Abotaleb, K. Nassar & E. Sayed-Ahmed

16. Structural applications of FRP composites

Case studies of repurposing FRP wind blades for second-life new infrastructure 501
L.C. Bank, T.R. Gentry, T. Al-Haddad, A. Alshannaq, Z. Zhang, M. Bermek, Y. Henao, A. McDonald, S. Li, A. Poff, J. Respert, C. Woodham, A. Nagle, P. Leahy, K. Ruane, A. Huynh, M. Soutsos, J. McKinley, E. Delaney & C. Graham

Prediction on stress-strain behavior of FRP-confined concrete with passive confinement-based 3D constitutive model 503
J.F. Jiang, P.D. Li & B.B. Li

Development and testing of new precast concrete tunnel segments reinforced with GFRP bars and ties 505
S.M. Hosseini, S. Mousa, H.M. Mohamed & B. Benmokrane

An experimental study on the compression behavior of CFRP-jacketed plastered low-strength RC columns 507
M.N. Yavuzer, S. Kolemenoglu, C. Balcı, M. Ispir & A. Ilki

Parametric study of CFFT systems for small-scale wind turbine towers 509
Y. Gong & M. Noël

Estimating shear strengths of glass fibre reinforced polymer reinforced concrete deep beams with indeterminate strut-and-tie models 511
S. Liu & M.A. Polak

State of the art in research and field applications of post-tensioned structures strengthened with prestressed composites 513
R. Kotynia & M. Staśkiewicz

Reinforcement made of basalt fibre reinforced polymer (BFRP): Load-bearing capacity, durability and applications 515
S. Görtz, K. Lengert, D. Glomb, B. Wolf, A. Kustermann & C. Dauberschmidt

Onsite manufacturing and applications of FRP pipes 517
M. Ehsani

Bolted and hybrid beam-column joints between I-shaped FRP profiles 519
J. Qureshi, Y. Nadir & S.K. John

Performance of self-drilling screw connections for structural pultruded fibre reinforced polymer composites 521
Z. Cai, Y. Bai, L.C. Bank, C. Qiu & X.L. Zhao

Sustainable sandwich composites made of recycled plastics 523
R.A. Kassab & P. Sadeghian

Flexural characteristics of bio-based sandwich beams made of paper honeycomb cores and flax FRP skins 525
Y. Fu & P. Sadeghian

Behavior of NSM FRP flexurally-strengthened RC beams with embedded FRP anchors 527
Y. Ke, S.S. Zhang & X.F. Nie

Standardization, guide development and long-term durability of fiber reinforced polymers (FRP): In situ field results from FRP RC bridge decks after 15+ years of service exposure 529
J.J. Myers & A.F. Al-Khafaji

Strength behaviour of fibre reinforced polymer concrete beams under flexural loading 531
J.T. Senosha & S.D. Ngidi

Properties of a 37 m long FRP wind turbine blade after 11 years in service 533
A.A. Alshannaq, J.A. Respert, L.C. Bank, T.R. Gentry & D.W. Scott

The axial behavior of the low and extremely low strength concrete confined with large-rupture-strain PET-FRPs made of production wastes 535
A.G. Genc, O.F. Eskicumali, N. Aslan, T. Ergin, U. Alparslan, M. Ispir & A. Ilki

Unique FRP solutions for structural repair of piles, seawalls and decks 537
M. Ehsani

Evaluation of using FRP bond equations in alternative types of advanced composite externally bonded to concrete 539
Z. Al-Jaberi, R. Mahdi & J.J. Myers

Sustainable and corrosion-free bridge structures 541
A. Belarbi, D. Vecchio & D. Stefaniuk

17. Construction technology, Construction projects, Construction materials, Properties of concrete

Full-scale experiments on hybrid tunnel lining segments 545
D.N. Petraroia & P. Mark

Environmental inefficiency of the world construction industry 547
M. Kapelko

Natural language processing as work support in project tendering 549
L. Cusumano, R. Rempling, R. Jockwer, R. Saraiva, M. Granath, N. Olsson & S. Okazawa

Improvement of mechanical performance of cement-based composites using a new type of low-cost low-energy demand graphene 551
T.D. Nguyen, M. Su & M. Watson

Relationship between flexural strength and compressive strength in concrete and ice 553
A.P. Fantilli, B. Frigo & F.M. Dehkordi

Comparison and statistical evaluation of Marshall stability and stiffness modulus for asphalt mixtures 555
J. Valentin, P. Vacková & M. Belhaj

Using recycled concrete aggregate from bombed building in new concrete 557
S. Alaud, N.A. Droughi & S. Alaud

Using waste from paint manufacturing in concrete 559
H. Muller & R. Combrinck

Considerations regarding the toxicity of construction and building materials 561
F. Pacheco Torgal

Fire performance of metakolin concrete and mortar blends using Nigerian kaolinite clays 563
J.O. Labiran, B.U. Ezea, U.T. Igba & J.O. Akinyele

Rheological and strength characterisation of limestone calcined clay cement 3D printed concrete 565
K.A. Ibrahim, G.P.A.G. van Zijl & A.J. Babafemi

Fresh properties and strength evolution of slag modified fibre-reinforced metakaolin-based geopolymer composite for 3D concrete printing application 567
M.B. Jaji, G.P.A.G Van Zijl & A.J. Babafemi

Expression for calculating the compressive strength of concrete containing Rice Husk Ash 569
G. Somma

Compatibility issues between dehydrated calcium sulphate cement and plasticiser/superplasticiser 571
J.M.H. Bessinger, M. Meyer & R. Combrinck

18. Timber structures, Timber technology, Properties of wood

Reliability of statically indeterminate timber structures: Modelling approaches and sensitivity study 575
D. Caprio, R. Jockwer & M. Al-Emrani

Future perspectives about timber-hybrid systems: The role of connections 577
G. Di Nunzio, L. Corti & G. Muciaccia

Recent developments in timber-concrete composite construction 579
K. Holschemacher

Improving the modelling of tall timber buildings 581
O. Flamand & M. Manthey

Influence of initial crack width in Mode I fracture tests on timber and adhesive timber bonds 583
S.A. Rahman, M. Ashraf, M. Subhani & J. Reiner

Laminating effect in South African pine glue laminated timber beams 585
F.J. Pretorius & C. Roth

Development of cross-laminated timber composite panels from C16 timber 587
E. McAllister & D. McPolin

Influence of knots and density distribution on compressive strength of wooden foundation piles 589
G. Pagella, M. Mirra, G.J.P. Ravenshorst & J.W.G. van de Kuilen

The effect of wood microstructure on the mechanical properties of some selected tropical hardwood species used in construction 591
J.O. Akinyele, A.B. Folorunsho, U.T. Igba, P.O. Omotainse & J.O. Labiran

Mechano-sorptive behaviour on crack propagation of notched beams of Okume 593
M. Asseko Ella, G. Goli, J. Gril & R. Moutou Pitti

Link between growth strategies and physical-mechanical properties of wood of tropical species from Gabon 595
E. Nkene Mezui, L. Brancheriau, D. Guibal & R. Moutou Pitti

High performance light timber shear walls and dissipative anchors for damage limitation of wooden buildings in seismic areas 597
V. Wilden, G. Balaskas, B. Hoffmeister, L. Rauber & B. Walter

Creep tests on notched beams of silver fir wood (*Abies alba*) 599
A. Bontemps, G. Godi, E. Fournely, J. Gril & R. Moutou Pitti

19. Structural health monitoring, Damage detection, System identification, Maintenance, Durability, Long-term performance

Structural health monitoring in civil engineering: Status and trends 603
G. De Roeck, D. Anastasopoulos & E. Reynders

Development of vibration-based early scour warning system for railway bridge piers 605
C.W. Kim, J. Qi & D. Kawabe

Vibration-based Bayesian anomaly detection of PC bridges 607
D. Kawabe, C.W. Kim, K. Takemura & K. Takase

Vibration-based structural health monitoring of bridges based on a new unsupervised machine learning technique under varying environmental conditions 609
M. Salar, A. Entezami, H. Sarmadi, B. Behkamal, C. De Michele & L. Martinelli

An embedded physics-based modeling concept for wireless structural health monitoring 611
K. Dragos & K. Smarsly

Damage evaluation and tracking using kriging approaches on wave-guide dispersion curves 613
O.A. Bareille, L. Nechak, M. Ichchou & A. Zine

Prompt modal identification with quantified uncertainty of modal properties 615
Y. Goi

Wavelet-based transmissibility for structural damage detection 617
K. Dziedziech, K. Mendrok, T. Uhl & W.J. Staszewski

Explainable framework for Lamb wave-based damage diagnosis 619
L. Lomazzi, M. Giglio & F. Cadini

Application development of distributed fibre optic sensors for monitoring existing bridges 621
B. Novák, F. Stein, A. Dudonu & J. Reinhard

Real-time health monitoring of civil structures by online hybrid learning techniques using remote sensing and small displacement data 623
A. Entezami, C. De Michele & A.N. Arslan

Permanent tunnel lining monitoring system for the purpose of further design optimization 625
M. Jonáš & J. Zatloukal

On the data-driven damage detection of offshore structures using statistical and clustering techniques under various uncertainty sources: An experimental study 627
M. Salar, A. Entezami, H. Sarmadi, C. De Michele & L. Martinelli

Sustainable structural health monitoring using e-waste and recycled materials 629
P. Peralta & K. Smarsly

Corrosion of fasteners in concrete: Literature review and discussion of current test methods 631
M. Cervio & G. Muciaccia

Plastic strain localization behavior of corroded steel plate under tensile loading 633
Y. Kitane, N.K. Gathimba & K. Sugiura

Fundamental study on crack detection method for steel members by thermal image processing 635
S. Watanabe, Y. Goi, Y. Kitane, K. Takase & K. Sugiura

Deflection condition monitoring of a steel bridge via remote sensing techniques 637
M.S. Miah & W. Lienhart

Condition rating for maintenance of existing reinforced concrete bridges in Soweto, South Africa 639
R.K. Maphosa, J Mahachi & S.O. Ekolu

Bridge collapse prediction by small displacement data from satellite images under long-term monitoring 641
A. Entezami, C. De Michele & A.N. Arslan

Structural damage identification using optimization-based FE model updating 643
K. Lamperová, D. Lehký & O. Slowik

The implications of climate change effects on the response parameters of concrete arch dams with respect to anomaly detection 645
T. Tshireletso & P. Moyo

Evaluation of changes in flexural rigidity of cracked concrete railway bridges under high-speed train passages 647
K. Matsuoka

Applying automated damage classification during digital inspection of structures 649
J. Flotzinger, P.J. Rösch, F. Deuser, T. Braml, S. Reim & B. Maradni

Crack intensity closed form solutions by frequency measurements on damaged beams 651
S. Caddemi, I. Caliò, F. Cannizzaro & N. Impollonia

Effect of 2-directional chloride ingress on concrete resistivity and corrosion rate of steel bars at orthogonal edges 653
Z.G. Zakka & M. Otieno

Measurement and simulation of pipeline attached to bridge for vibration-based SHM 655
D. Kobayashi, T. Nakanishi, Y. Sakurada & A. Aratake

Investigation of modulation transfer due to nonlinear shear wave interaction with local source: Numerical and theoretical approach 657
R. Radecki, A. Ziaja-Sujdak, M. Osika & W.J. Staszewski

20. Structural assessment, Historic structures, Masonry structures, Repair, Strengthening, Retrofitting

Intermediate isolation system for the seismic retrofit of existing masonry buildings 661
D. Faiella, M. Argenziano, G. Brandonisio, F. Esposito & E. Mele

A seismic retrofitting design approach for activating dissipative behavior of timber diaphragms in existing unreinforced masonry buildings 663
M. Mirra & G. Ravenshorst

Lifetime assessment for historical cast steel bridge bearings 665
D. Siebert & M. Mensinger

Strengthening of existing infrastructure with concrete screws as post-installed reinforcement 667
J. Feix & J. Lechner

Analytical formulation for plain and retrofitted masonry wall under out-of-plane loading 669
J.A. Dauda, O. Iuorio & F.P. Portioli

Structural behavior of masonry walls with soft-layers: An overview of experimental work 671
N. Mojsilović

Investigating the causes, impact, and repair methods of popping bricks in house construction 673
K. Nongwane, S.A. Alabi & J. Mahachi

Experimental investigation of the influence of embedment depth on the sustained load performance of adhesive anchors 675
A.C.O. Vera, S. Sartipi, W. Botte & R. Wan-Wendner

The review of diagonal compression tests of URM panels strengthened with NSM steel bars 677
S.D. Ngidi

Rehabilitation applications of roadway structures 679
H.H. Abbas & M.M. Hassan

Heritage protection and safety requirements: A difficult relationship 681
U. Quapp & K. Holschemacher

Reconnaissance survey of historic monumental unreinforced masonry building by visual inspection: A case study of senate hall building of Allahabad University, India 683
A. Kumar & K. Pallav

On the characterization of materials and masonry walls of historical buildings: Use of optical system to obtain displacement maps in double-flat jack tests 685
M. Acito, E. Magrinelli, C. Tiraboschi & M. Cucchi

A computational procedure interacting with the Italian "CARTIS" online database to derive residential building portfolios for large scale seismic assessments 687
A. Basaglia, G. Brando, G. Cianchino, G. Cocco, D. Rapone, M. Terrenzi & E. Spacone

Seismic assessment of a colonial adobe building in Cusco, Peru 689
A. Tancredi, G. Cocco, E. Spacone & G. Brando

21. Soil-structure interaction, Foundations, Underground structures, Geotechnical engineering, Rock mechanics

Interaction between the deepest foundation piles in Germany and the superstructure 693
R. Katzenbach, A. Werner & S. Fischer

Derivation of cyclic p-y curves for the design of monopiles in sand 695
M. Achmus & J. Song

Climate-smarter design of soil-steel composite bridges using set-based design 697
J. Lagerkvist, C.G. Berrocal & R. Rempling

Parametric modelling of integral bridge spring reactions 699
N.R. Featherston & C. Viljoen

A parametric study of the vibration of beams resting on elastic foundations with nonlinear cubic stiffness 701
E. Feulefack Songong & A. Zingoni

Degradation of axial friction resistance on buried district heating pipes 703
T. Gerlach & M. Achmus

A hybrid structure to protect infrastructures from high energy rockfall impacts 705
M. Marchelli & V. De Biagi

Author index 707

Current Perspectives and New Directions in Mechanics, Modelling and Design of Structural Systems – Zingoni (ed.)
© 2022 Copyright the Editor and Contributors, ISBN 978-1-032-39114-4

Preface

Six months after the Seventh International Conference on Structural Engineering, Mechanics and Computation, the world went into shutdown in response to the outbreak of the covid-19 pandemic. The next two years were characterised by uncertainty, many engineering research activities were hampered, and conference travel was restricted. Preparations for SEMC 2022 (the Eighth International Conference on Structural Engineering, Mechanics and Computation) began under these circumstances, and to deal with the uncertainty, plans were made for a hybrid event. What was clear as the year 2022 started was that researchers wished to meet face-to-face and interact once again after more than a year of physical isolation. The SEMC 2022 International Conference successfully went ahead in Cape Town from 5 to 7 September 2022, with approximately 250 participants attending physically and 100 attending remotely. These participants came from all over the world, with 45 countries in total being represented. Given the crippling effects of the covid-19 pandemic on research activities over the past two years, that this event was well attended is clear testimony to the resilience of the SEMC series.

The SEMC conferences all have a common aim, namely "to bring together from around the world academics, researchers and practitioners in the broad fields of structural mechanics, associated computation and structural engineering, to review recent achievements in the advancement of knowledge and understanding in these areas, share the latest developments, and address the challenges that the present and the future pose". There is no doubt that SEMC 2022 fulfilled this aim. As I remarked at the last conference in 2019, the SEMC conference continues to attract many of the leading researchers in the field, despite the remoteness of the location from the research hubs of Europe, Asia and America, and the ever-present competition from related events taking place elsewhere around the month of September.

Current Perspectives and New Directions in Mechanics, Modelling and Design of Structural Systems, featuring 330 contributions, are the official Proceedings of the Eighth International Conference on Structural Engineering, Mechanics and Computation. As the title implies, the scope of the Proceedings is wide, covering new theoretical formulations, the mechanics of materials and structures, developments in modelling, simulation and computational techniques, as well as the analysis, design, construction, maintenance, monitoring, repair, strengthening and retrofitting of engineering structures and systems. A significant proportion of the papers were submitted as contributions to special sessions of the conference; the rest of the contributions were submitted independently in response to the general call for papers.

In keeping with the format of past SEMC conferences, the Proceedings have been published in the form of a printed book containing short papers of length 2 pages, and an e-book containing full papers of length 6 pages. The short paper is more than just an extended abstract; it is a self-contained article that is structured in the same way as a normal paper, but shorter in length. The short article in the printed book is intended to give the reader the essence of the paper (aims, results and conclusions); if the reader is interested in the details, they can then consult the full-length paper in the e-book.

A note on where to find which papers is in order. To make this as easy as possible, the contents of the Proceedings have been grouped under 21 sections, the first section featuring keynote papers, and the rest of the sections being clusters of papers that deal with the same issues or related issues. It should be noted that some papers deal with multiple themes, and therefore equally fit into any one of two or more sections. Thus, in viewing the contents of any particular section, readers should be aware that other related papers may lie elsewhere in the Proceedings. The electronic version of the Proceedings features a search tool, which should help in locating all papers on a particular theme or topic, or by a particular author. While most sections feature topics that share some common aspects, occasionally some unrelated topics have been placed in the same section, in order to avoid an excessive number of sections.

In recent years, additive manufacturing has become a subject of intense interest, especially in the mechanical engineering context. The first keynote paper looks at the opportunities within structural engineering. The next contribution considers various aspects of the computational modelling of fibre-reinforced concrete. While fibre-reinforced polymers have been extensively used to strengthen concrete structures, their application to the strengthening of metal structures is not so well-established; the third keynote paper looks at the possibilities. The fourth keynote paper presents a strategy for the economical modelling of the response of structures to earthquakes, while the fifth paper takes a closer look at the role of vibration-based structural health monitoring in bridge management. The concept of tensegrity is still not well-understood among engineers; the last of the keynote papers explains the use of an energy function in better understanding the behaviour of tensegrity structures.

Section 2 of the Proceedings features papers on dynamic analysis, and the response of structures to vibrations caused by fluids, moving people, moving vehicles and trains, the driving of piles into the ground, and various other environmental effects. The response of structures to earthquakes, and how to design structures for earthquake resistance, are specifically dealt with in Section 3. The mechanics and computational modelling of materials is the subject of Section 4, while the growing field of additive manufacturing is given consideration in Section 5, alongside other related topics. Section 6 contains contributions on composite structures, laminated and sandwich structures, alongside contributions on numerical modelling of structures and numerical simulation of processes and phenomena. The mechanics and computational modelling of phenomena specifically related to damage are dealt with in Section 7, which also features contributions on progressive collapse, and how to design against this.

Instability of structural elements under load or environmental effects remains a major concern for structural designers. Papers concerned with the buckling of beams, columns and thin-walled structures appear in Section 8. Interestingly, and as shown in some of the papers, buckling can also be put to good use, instead of being regarded as an adverse condition to be avoided always. Various aspects of lightweight construction, including form-finding for shells, membranes and cable-stayed structures, are reported in Section 9. There is growing interest in using glass not just for aesthetic and cladding purposes, but also as a load-bearing material. Section 10 features results of recent research on structural applications of glass.

Sections 11 to 14 contain papers on various aspects of the behaviour, analysis, design and construction of structures in the two leading construction materials: steel and concrete. Two of these sections are dedicated to special considerations for high-strength steel and high performance steel (Section 12), and high-strength concrete and high performance concrete (Section 14). Issues of safety and reliability, design philosophy, engineering education, building performance and transport infrastructure, while not necessarily related, are conveniently grouped together under Section 15. Research on structural applications of fibre-reinforced polymer composites, and the understanding of associated structural behaviour, continues to increase, as may be seen in the papers of Section 16. Papers on construction technology and construction materials (including properties of concrete) appear in Section 17, while those specifically concerned with construction in timber and mechanical properties of wood appear in Section 18.

In line with the philosophy of reducing life-cycle costs of infrastructure, attention worldwide is increasingly being paid to the development of more effective strategies for condition monitoring, repair and maintenance of existing structures such as buildings, bridges, tunnels and dams. Sections 19 and 20 contain papers on long-term performance of structures, though papers dealing with short-term aspects of the design of masonry structures are also included. A good number of papers report on the latest developments in structural health monitoring and damage detection, while others focus on assessing structures for damage or remaining life, preserving historic buildings, and developing new techniques for the repair, strengthening and retrofitting of existing structures. The last section features papers related to geotechnical engineering, and covering the topics of foundations, soil-structure interaction, buried structures and rock mechanics.

As editor, I had the privilege of reading through an exciting mix of papers reflecting the huge diversity of research that is currently taking place in the field. I am thankful to all authors for their contributions. It is hoped these Proceedings will serve as a valuable resource to students in the early stages of a research career, as well as more seasoned researchers wishing to acquaint themselves with the latest developments in the field, and engineering practitioners looking for specialist information. I am also indebted to the organisers of the twelve special sessions of the conference, who not only invited contributions from known experts on specific topics, but also chaired these sessions during the conference. Members of the International Advisory Board assisted with encouraging participation at the conference. Some served as reviewers. I thank all reviewers for their role in enhancing the quality of the Proceedings. Only those papers that were accepted have been included in the Proceedings.

I would like to thank all those who served as session chairs. The sessions all ran successfully owing to their dedication and professionalism. The enthusiastic support of postgraduate students of the University of Cape Town is gratefully acknowledged, as is the commitment of technical and administrative staff who worked hard to ensure that all aspects of the conference ran smoothly.

Last but not least, I wish to acknowledge the support and encouragement of my family throughout the two years of preparations for this conference. Without this support, it would have been more difficult to complete all the work that needed to be done. I dedicate this edition to the memory of my brother, the late Dr. Tainos Zingoni, who I know would have been proud to receive a copy of these Proceedings.

A. Zingoni

Editor

Current Perspectives and New Directions in Mechanics, Modelling and Design of Structural Systems – Zingoni (ed.)
© 2022 Copyright the Editor and Contributors, ISBN 978-1-032-39114-4

Committees of the SEMC 2022 International Conference

Local Organising Committee
A. Zingoni, *University of Cape Town (Chairman)*
M. Latimer, *Joint Struct. Div. of SAICE & IStructE*
K. Mudenda, *University of Cape Town*
C. Meyer, *Cape Peninsula University of Technology*
C. Kaluba, *University of Cape Town*
L. Zingoni, *University of Cape Town*

International Advisory Board
Prof. M. Achmus, *Leibniz Universität Hannover, Germany*
Prof. S. Adali, *University of KwaZulu-Natal, South Africa*
Prof. J.M. Adam, *Universitat Politècnica de Valencia, Spain*
Prof. H. Adeli, *Ohio State University, USA*
Prof. S. Adriaenssens, *Princeton University, USA*
Prof. J. Ambrosio, *University of Lisbon, Portugal*
Prof. L.V. Andersen, *Aarhus University, Denmark*
Prof. C.J. Anumba, *University of Florida, USA*
Prof. A. Araujo, *University of Lisbon, Portugal*
Prof. F. Armero, *University of California at Berkeley, USA*
Prof. H. Askes, *University of Sheffield, UK*
Prof. A. Astaneh-Asl, *University of California at Berkeley, USA*
Dr. M. Baessler, *Fed. Inst. for Material Research & Testing, Germany*
Prof. B. Baier, *University of Duisburg-Essen, Germany*
Prof. T. Balendra, *National University of Singapore, Singapore*
Prof. Y. Ballim, *University of the Witwatersrand, South Africa*
Prof. J.R. Banerjee, *City University London, UK*
Prof. L.C. Bank, *Georgia Institute of Technology, USA*
Prof. N. Banthia, *University of British Columbia, Canada*
Prof. R.C. Barros, *University of Porto, Portugal*
Prof. K.J. Bathe, *Massachusetts Institute of Technology, USA*
Prof. Z.P. Bazant, *Northwestern University, USA*
Dr. A. Behnejad, *University of Surrey, UK*
Prof. J.L.I.F. Belis, *Ghent University, Belgium*
Prof. K. Bergmeister, *Univ. of Nat. Resources & Life Sciences, Austria*
Prof. D. Beskos, *University of Patras, Greece*
Prof. F.S.K. Bijlaard, *Delft University of Technology, Netherlands*
Prof. L. Bisby, *University of Edinburgh, UK*
Prof. Z. Bittnar, *Czech Technical Univ. in Prague, Czech Republic*
Prof. J. Blachut, *University of Liverpool, UK*
Prof. F. Bontempi, *University of Rome La Sapienza, Italy*
Prof. G. Borino, *University of Palermo, Italy*
Prof. C. Borri, *University of Florence, Italy*
Prof. L.F. Boswell, *City University London, UK*
Prof. A. Bouchair, *Clermont Auvergne University, France*
Prof. M. Bradford, *University of New South Wales, Australia*
Prof. E. Bruehwiler, *Ecole Polytech. Fed. de Lausanne, Switzerland*

Prof. R. Burgueno, *Stony Brook University, USA*
Prof. O. Buyukozturk, *Massachusetts Inst. of Technology, USA*
Prof. S. Caddemi, *University of Catania, Italy*
Prof. D. Camotim, *University of Lisbon, Portugal*
Prof. W. Cantwell, *University of Liverpool, UK*
Dr. C. Caprani, *Monash University, Australia*
Prof. R. Cerny, *Czech Technical Univ. in Prague, Czech Republic*
Prof. J.F. Chen, *Southern Univ. of Science and Tech., China*
Prof. R.L. Chen, *Xiang Tan University, China*
Prof. Y. Chen, *Southeast University, China*
Prof. C.K. Choi, *Korea Advanced Inst. of Science and Tech., Korea*
Prof. A.K. Chopra, *University of California at Berkeley, USA*
Prof. K.F. Chung, *Hong Kong Polytechnic University, China*
Prof. J.G.A. Croll, *University College London, UK*
Prof. G. Cusatis, *Northwestern University, USA*
Prof. A. d'Ambrisi, *University of Florence, Italy*
Prof. R. de Borst, *University of Sheffield, UK*
Prof. G. De Matteis, *University of Campania "Luigi Vanvitelli", Italy*
Prof. V. Denoel, *University of Liege, Belgium*
Prof. G. De Roeck, *Catholic University of Leuven, Belgium*
Prof. A. De Stefano, *Politecnico di Torino, Italy*
Prof. E. Detournay, *University of Minnesota, USA*
Dr. L. Di Sarno, *University of Liverpool, UK*
Prof. G. Domokos, *Budapest Univ. of Tech. and Economics, Hungary*
Prof. D. Dubina, *University of Timisoara, Romania*
Prof. J. Dulinska, *Cracow University of Technology, Poland*
Prof. M. Dundu, *University of Johannesburg, South Africa*
Prof. W. Ehlers, *University of Stuttgart, Germany*
Prof. A.A. El Damatty, *University of Western Ontario, Canada*
Prof. A.Y. Elghazouli, *Imperial College London, UK*
Prof. A. Elnashai, *University of Houston, USA*
Prof. S. El-Tawil, *University of Michigan, USA*
Prof. E. Erdogmus, *Georgia Institute of Technology, USA*
Prof. E. Fehling, *University of Kassel, Germany*
Prof. J. Feix, *University of Innsbruck, Austria*
Prof. G.A. Ferro, *Politecnico di Torino, Italy*
Prof. F.C. Filippou, *University of California at Berkeley, USA*
Prof. D.M. Frangopol, *Lehigh University, USA*
Prof. Y. Fujino, *Yokohama National University, Japan*
Prof. C. Gantes, *National Technical University of Athens, Greece*
Prof. A. Garbacz, *Warsaw University of Technology, Poland*
Prof. L. Gardner, *Imperial College London, UK*
Prof. A. Ghobarah, *McMaster University, Canada*
Prof. A. Ghorbanpoor, *University of Wisconsin-Milwaukee, USA*
Prof. R.I. Gilbert, *University of New South Wales, Australia*
Prof. L. Giuliani, *Technical University of Denmark, Denmark*
Prof. M. Gizejowski, *Warsaw University of Technology, Poland*
Prof. L.A. Godoy, *Universidad Nacional de Cordoba, Argentina*
Prof. M. Gohnert, *University of the Witwatersrand, South Africa*
Prof. P. Gosling, *University of Newcastle, UK*
Prof. P.L. Gould, *Washington University, USA*
Prof. A. Gresnigt, *Technical University of Delft, The Netherlands*
Prof. A.A. Groenwold, *University of Stellenbosch, South Africa*
Prof. S. Guest, *University of Cambridge, UK*
Prof. W. Guggenberger, *Technical University of Graz, Austria*

Prof. G. Hancock, *University of Sydney, Australia*
Prof. R. Harte, *Bergische University Wuppertal, Germany*
Prof. N.M. Hawkins, *Univ. of Illinois at Urbana-Champaign, USA*
Prof. S. Heyns, *University of Pretoria, South Africa*
Prof. M. Holicky, *Czech Tech. Univ. in Prague, Czech Republic*
Prof. K. Holschemacher, *Leipzig Univ. of Appl. Sciences, Germany*
Prof T. Ibell, *University of Bath, UK*
Prof. A. Ibrahimbegovic, *Univ. de Tech. de Compiegne, France*
Prof. A. Ilki, *Istanbul Technical University, Turkey*
Prof. K. Jarmai, *University of Miskolc, Hungary*
Prof. D. Jun, *Hochschule Ruhr West, Germany*
Prof. H.J. Jung, *Korea Advanced Inst. of Science and Tech., Korea*
Prof. A. Kappos, *Khalifa University, United Arab Emirates*
Prof. M. Karmazinova, *Brno Univ. of Technology, Czech Republic*
Prof. R. Karoumi, *KTH Royal Institute of Technology, Sweden*
Prof. A. Kaveh, *Iran Univ. of Science and Technology, Iran*
Prof. K. Kayvani, *HKA, Australia*
Prof. C.W. Kim, *Kyoto University, Japan*
Prof. H.K. Kim, *Seoul National University, Korea*
Prof. S. Kitipornchai, *University of Queensland, Australia*
Prof. U. Kuhlmann, *University of Stuttgart, Germany*
Prof. G. Langdon, *University of Sheffield, UK*
Prof. J. Lange, *Technical University of Darmstadt, Germany*
Prof. D.T. Lau, *Carleton Ottawa University, Canada*
Prof. C. Lazaro, *Universidad Politecnica de Valencia, Spain*
Prof. W. Lewis, *University of Warwick, UK*
Prof. G.Q. Li, *Tongji University, China*
Prof. H.N. Li, *Dalian University of Technology, China*
Prof. M.P. Limongelli, *Politecnico di Milano, Italy*
Prof. A. Long, *Queen's University of Belfast, UK*
Prof. J. Loughlan, *Loughborough University, UK*
Prof. Y. Lu, *University of Edinburgh, UK*
Prof. J. Macdonald, *University of Bristol, UK*
Prof. J.F.A. Madeira, *University of Lisbon, Portugal*
Prof. K. Magnucki, *Institute of Rail Vehicles TABOR, Poland*
Prof. J. Mahachi, *University of Johannesburg, South Africa*
Prof. J. Maljaars, *Eindhoven Univ. of Technology, Netherlands*
Prof. H.A. Mang, *Vienna University of Technology, Austria*
Prof. M. Maslak, *Cracow University of Technology, Poland*
Prof. H. Matsunaga, *Setsunan University, Japan*
Prof. F.M. Mazzolani, *University of Naples Federico II, Italy*
Prof. J. Melcher, *Brno University of Technology, Czech Republic*
Prof. G. Meschke, *Ruhr University Bochum, Germany*
Prof. B. Middendorf, *University of Kassel, Germany*
Prof. F.J. Montans, *Universidad Politecnica de Madrid, Spain*
Prof. G. Morgenthal, *Bauhaus Universität Weimar, Germany*
Prof. C.A. Mota Soares, *University of Lisbon, Portugal*
Prof. M. Motavalli, *EMPA, Switzerland*
Prof. A.L. Mrema, *University of Dar es Salaam, Tanzania*
Prof. A. Muttoni, *Ecole Polytech. Fed. de Lausanne, Switzerland*
Prof. J. Naprstek, *Acad. of Sciences of Czech Rep., Czech Republic*
Prof. R.S. Narayanan, *Consultant to Clark Smith Partnership, UK*
Prof. D.A. Nethercot, *Imperial College London, UK*
Prof. V.T. Nguyen, *Graz University of Technology, Austria*
Prof. Y.Q. Ni, *Hong Kong Polytechnic University, China*

Prof. H.J. Niemann, *Ruhr University Bochum, Germany*
Prof. D. Novak, *Brno University of Technology, Czech Republic*
Prof. G.N. Nurick, *University of Cape Town, South Africa*
Prof. R. Ohayon, *Conservatoire National des Arts et Metiers, France*
Prof. J.P. Ou, *Harbin Institute of Technology, China*
Prof. M. Papadrakakis, *National Technical Univ. of Athens, Greece*
Prof. H. Pasternak, *BTU Cottbus, Germany*
Prof. A. Pavic, *University of Exeter, UK*
Prof. D. Peric, *University of Swansea, UK*
Prof. F. Perotti, *Politecnico di Milano, Italy*
Prof. M. Polak, *University of Waterloo, Canada*
Prof. M. Pulsfort, *Bergische University of Wuppertal, Germany*
Prof. O. Rabinovitch, *Technion - Israel Institute of Technology, Israel*
Prof. E. Ramm, *University of Stuttgart, Germany*
Prof. G. Ranzi, *University of Sydney, Australia*
Prof. K. Rasmussen, *University of Sydney, Australia*
Prof. E. Real, *Universitat Politecnica de Catalunya, Spain*
Prof. B.D. Reddy, *University of Cape Town, South Africa*
Prof. J.V. Retief, *University of Stellenbosch, South Africa*
Prof. T. Ricken, *University of Stuttgart, Germany*
Prof. I. Robertson, *University of Hawaii, USA*
Prof. G. Rombach, *Technical University of Hamburg, Germany*
Prof. B. Rossi, *Catholic University of Leuven, Belgium*
Prof. J.G. Rots, *Technical University of Delft, Netherlands*
Prof. J.M. Rotter, *University of Edinburg, UK*
Prof. C. Sansour, *University of Nottingham, UK*
Prof. B.W. Schafer, *Johns Hopkins University, USA*
Prof. S. Schmauder, *University of Stuttgart, Germany*
Prof. R. Schmidt, *RWTH Aachen, Germany*
Prof. J. Schroeder, *University of Duisburg-Essen, Germany*
Prof. S.A. Sheikh, *University of Toronto, Canada*
Prof. S. Shrivastava, *McGill University, Canada*
Prof. N. Silvestre, *University of Lisbon, Portugal*
Prof. I.F.C. Smith, *Swiss Fed. Inst. of Tech. Lausanne, Switzerland*
Prof. R.A. Smith, *Imperial College London, UK*
Prof. H.H. Snijder, *Eindhoven Univ. of Technology, Netherlands*
Prof. W. Sobek, *University of Stuttgart, Germany*
Prof. U. Starossek, *Hamburg University of Technology, Germany*
Prof. R. Stroetmann, *Technical University of Dresden, Germany*
Prof. K. Sugiura, *Kyoto University, Japan*
Prof. C. Szymczak, *Technical University of Gdansk, Poland*
Prof. T. Tarnai, *Budapest Univ. of Tech. and Economics, Hungary*
Prof. J.G. Teng, *Hong Kong Polytechnic University, China*
Prof. D. Thambiratnam, *Queensland Univ. of Tech., Australia*
Prof. S. Timashev, *Russian Academy of Sciences, Russia*
Prof. J. Torero, *University College London, UK*
Prof. V. Tvergaard, *Technical University of Denmark, Denmark*
Prof. B. Uy, *University of Sydney, Australia*
Prof. P. Van Bogaert, *University of Ghent, Belgium*
Prof. B.W.J. van Rensburg, *University of Pretoria, South Africa*
Prof. G.P.A.G. Van Zijl, *University of Stellenbosch, South Africa*
Prof. R. Vaziri, *University of British Columbia, Canada*
Prof. P. Vellasco, *State University of Rio De Janeiro, Brazil*
Prof. F. Virtuoso, *University of Lisbon, Portugal*
Prof. T. Vogel, *Swiss Fed. Inst. of Tech. Zurich, Switzerland*

Prof. M.A. Wadee, *Imperial College London, UK*
Prof. F. Wald, *Czech Technical Univ. in Prague, Czech Republic*
Prof. J.C. Walraven, *Delft University of Technology, Netherlands*
Prof. Y. Wang, *University of Manchester, UK*
Prof. M.J. Whelan, *University of North Carolina at Charlotte, USA*
Prof. M. Wiercigroch, *University of Aberdeen, UK*
Prof. A.C. Wijeyewickrema, *Tokyo Institute of Technology, Japan*
Prof. W. Witkowski, *Gdansk University of Technology, Poland*
Prof. Y.B. Yang, *National Taiwan University, China*
Prof. D. Yankelevsky, *Technion - Israel Institute of Technology, Israel*
Prof. B. Young, *Hong Kong Polytechnic University, China*
Prof. R. Zandonini, *University of Trento, Italy*
Prof. M.W. Zehn, *Technische Universität Berlin, Germany*
Prof. R.R. Zhang, *Colorado School of Mines, USA*
Prof. W. Zhu, *University of Maryland, USA*
Prof. A. Zingoni, *University of Cape Town, South Africa (Chair)*

1. Keynotes

Current Perspectives and New Directions in Mechanics, Modelling and Design of Structural Systems – Zingoni (ed.)
© 2022 Copyright the Author(s), ISBN 978-1-032-39114-4

Metal additive manufacturing in structural engineering: Review, opportunities and outlook

L. Gardner
Department of Civil and Environmental Engineering, Imperial College London

ABSTRACT: Although still in its infancy, metal Additive Manufacturing (AM) has arrived at construction scale. In this paper, a review of recent developments in metal AM in structural engineering is presented, including the latest research advances, lessons learned from other sectors and applications in practice. Emphasis is placed on Wire-Arc Additive Manufacturing (WAAM) since this is deemed to be the most promising technique for the requirements of the construction sector. Optimization and additive manufacturing go hand in hand, with the latter now enabling the former to be more readily realised in practice. Recent examples of optimized, additively manufactured structural components are presented. Finally, the potential use of WAAM in the strengthening and repair of structures, as well as the economics, sustainability and outlook for WAAM in structural engineering, are presented.

1 INTRODUCTION

Metal AM has already made inroads into a number of engineering disciplines, including in the aerospace, bioengineering and maritime sectors, and is beginning to gain traction in structural engineering (Kanyilmaz et al., 2022). There are many potential uses and benefits, such as reduced material consumption, the ability to readily produce intricate, optimised structural geometries, greater automation, less wastage, ease of production in remote locations, mixed materials, hybrid construction, strengthening and repair. There are of course also numerous challenges, barriers and as-yet unanswered questions surrounding, for example, fundamental mechanical properties, consistency of both short-term and long-term structural performance, geometric accuracy and variability, speed and cost of construction, quality assurance, design, certification and so on. Widespread adoption of metal additive manufacturing in structural engineering will require these issues to be addressed.

2 MATERIAL BEHAVIOUR

The behaviour and design of structures is dominated by the mechanical performance of the constituent material, with strength, stiffness and ductility being among the key properties of interest. Unlike for traditional hot-rolled and cold-formed structural material, where there are established quality assurance protocols and large bodies of experimental data, for additively manufactured material, such protocols and datasets are hitherto somewhat lacking owing to the relatively fresh introduction of the technology. There have however been a number of recent studies into the mechanical properties of WAAM steel and stainless steel plate material, suitable for construction e.g. (Huang et al., 2022) The following conclusions can be drawn from these studies: (1) there is a growing body of data demonstrating the sound mechanical response of WAAM material, (2) cooling rates influence mechanical properties, (3) varying degrees of anisotropy can arise, depending on the material grade and printing procedure, and (4) surface undulations cause a weakening effect on the mechanical properties, emphasising the importance of print quality.

3 ELEMENT BEHAVIOUR

The behaviour and design of metal additively manufactured cross-sections, members, connections and systems have been examined experimentally in a number of recent studies. The optimisation and manufacture of a 2 m stainless steel diagrid column, produced using the dot-by-dot WAAM method, were described by Laghi et al. (2020). The optimisation featured mechanical data obtained from prior tests on individual rods loaded in tension and compression. The study highlighted the importance of taking due account of the particular imperfections associated with the adopted WAAM method and provides insight into the design process and outcomes for optimised WAAM structures. An end-to-end framework for the design, optimisation and manufacture

DOI: 10.1201/9781003348450-1

of WAAM trusses was presented by Ye et al. (2021), in which 2 m span trusses with different boundary conditions were produced. The trusses, an example of which is shown in Figure 1, featured members of varying cross-sectional dimensions to match the demands imposed by the applied loading, but also considering manufacturing constraints. The trusses were shown to provide structural solutions that utilised approximately half the amount of material in comparison to benchmark I-section beams.

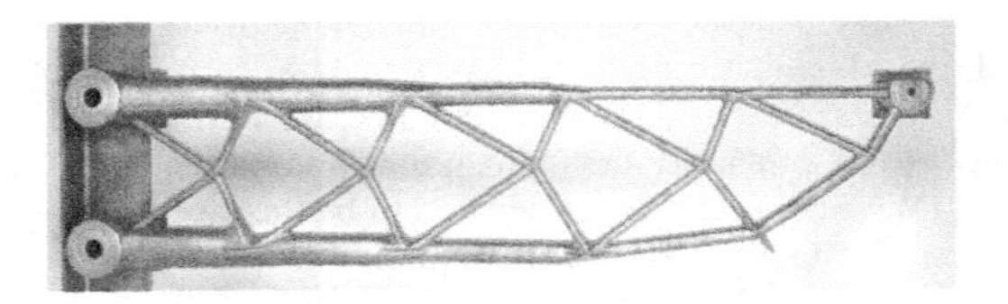

Figure 1. Optimised WAAM steel cantilever truss.

Connections are regarded as the aspect of steel construction that could benefit most immediately from the opportunities offered by additive manufacturing – the connections themselves represent a relatively small portion of the total weight of the structure, so the challenges associated with the relatively slow rate at which material can be deposited is minimised, while the technology and scope for automation can be brought to bear on what is traditionally a rather manual and labour-intensive operation. At the same time, there is significant potential for enhancing the performance of connections through innovation and optimisation, with additive manufacturing enabling geometries to be printed that have thus far, not been possible.

4 STRENGTHENING AND REPAIR

Alongside connections, strengthening and repair represent another aspect of construction where additive manufacturing is considered to offer significant potential in the coming years. The premise is that considerable capacity gains can be achieved through the addition of relatively little WAAM material to existing steel elements but positioned in the optimum location to resist the applied loads.

5 ECONOMICS AND SUSTAINABILITY

There is no straightforward answer to the question of whether additive manufacturing is more economical or sustainable than conventional steel fabrication. The economical and environmental advantage (or disadvantage) will vary from case to case. What is clear though, is that additive manufacturing offers the potential for reduced material use, reduced wastage, greater automation and improved health and safety relative to conventional methods, and that further exploration is warranted.

6 OUTLOOK

Construction-focussed research and applications of metal 3D printing are growing. Although good progress has already been made and early signs of the potential for metal AM in structural engineering are encouraging, it is still relatively early days. Structural optimisation, reduced material usage and wastage, increased automation and the greater reuse of material are all foreseen to be important components of a future, more sustainable construction sector, to which additive manufacturing will no doubt contribute.

7 CONCLUSIONS

Additive manufacturing is a disruptive technology that is emerging as a viable method of construction. A brief review of research and practical advances in metal additive manufacturing in the structural engineering domain has been presented. Wire arc additive manufacturing (WAAM) is regarded as the technique that is best suited to the needs of construction, and there is a growing body of experimental data confirming its suitability for load-bearing applications. The most promising early potential applications in practice are considered to be connections, strengthening and repair and landmark structures, all with the possibility to use optimization to exploit the available geometric freedom offered by AM.

REFERENCES

Huang, C., Kyvelou, P., Zhang, R., Britton, T. B. and Gardner, L. (2022). Mechanical testing and microstructural analysis of wire arc additively manufactured steels. Materials and Design. 216, 110544.

Kanyilmaz, A., Demir, A. G., Chierici, M., Berto, F., Gardner, L., Kandukuri, S. Y., Kassabian, P., Kinoshita, T., Laurenti, A., Paoletti, I., du Plessis, A. and Razavi, S.M.J. (2022). Role of metal 3D printing to increase quality and resource-efficiency in the construction sector. Additive Manufacturing. 50, 102541.

Laghi, V., Palermo, M., Gasparini, G. and Trombetti, T. (2020). Computational design and manufacturing of a half-scaled 3D-printed stainless steel diagrid column. Additive Manufacturing. 36, 101505.

Ye, J., Kyvelou, P., Gilardi, F., Lu, H., Gilbert, M. and Gardner, L. (2021). An end-to-end framework for the additive manufacture of optimized tubular structures. IEEE Access. 9, 165476–165489.

Current Perspectives and New Directions in Mechanics, Modelling and Design of Structural Systems – Zingoni (ed.)
© 2022 Copyright the Author(s), ISBN 978-1-032-39114-4

Computational modeling of FRC: From 3D printing to robust design

G. Meschke, G. Neu, V. Gudžulić, J. Reinold & T. Iskhakov
Institute for Structural Mechanics, Ruhr University Bochum, Bochum, Germany

ABSTRACT: A discrete fiber and a multi-level model for the analysis of SFRC structures are used to assess the influence of the chosen fiber type, content, and orientation on the structural response. Computational simulation of the flow process of fresh FRC, including applications to 3D printing, is performed to analyse the effect of fiber orientation on the the crack bridging behavior of the SFRC in the hardened state. Finally, the prospects of applying a multi-level FRC model in conjunction with methods of optimization to design an SFRC tunnel lining segment are discussed.

1 INTRODUCTION

Concrete is one of the most used building materials globally. Considering its CO2 footprint, sustainability and environmental impact inherently arise as relevant questions. In the past decades, significant progress has been made in improving the performance of concrete structures, in which FRC designs play a substantial role. Enhancing the material and structural performance needs adequate models and design approaches to take full advantage of the potential benefits. This contribution presents a methodology to simulate various stages of the manufacturing process of FRC structures as a supporting tool for the design of efficient concrete structures. The starting point is the simulation of the casting of fresh fiber-reinforced concrete, focussing on 3D printing processes, followed by the analysis of the crack-bridging response of individual fibers, which is upscaled to the structural level in the context of a stochastic multi-level modeling strategy. The performance of this approach is demonstrated by optimizing the design of tunnel lining segments.

2 COMPUTATIONAL MODELS FOR SFRC

The model for fresh concrete flow is based upon the Particle Finite Element Method (PFEM) (Reinold and Meschke 2021, Reinold et al. 2022), with the material considered as a viscous Bingham fluid, and the evolution of the fiber orientations being considered according to the Folgar-Tucker model (Folgar and Tucker 1984). The applicability of the method for simulating the flow of fresh concrete and the fiber orientation evolution is demonstrated in an example of 3D-concrete printing. The fiber orientation state for a single layer of printed concrete is obtained in the form of a fiber orientation tensor (see Figure 1 (top, left)). The fiber orientation tensor contains sufficient information, which can be used to accurately recreate the probability density function of fiber orientations by utilizing the maximum entropy method. This probability density function can subsequently be used to sample the orientations of discrete fibers. The orientation tensor and discrete fiber representations are compared in Figure 1 (top, right).

The crack-bridging actions of fibers are modeled using two approaches: The semi-analytical method (Zhan and Meschke 2014) and the newly developed fully discrete representation. The pullout responses of both models were compared and validated against the experimental data by (Leung and Shapiro 1999), as shown in Figure 1 (bottom, left).

A concrete cracking model is based on cohesive interface elements (Gudzulic and Meschke 2021). Depending on the required level of detail, this approach can be combined with a discrete fiber model as well as an interface law based on a multi-level model for FRC structures (Zhan and Meschke 2016). Both models allow assessing the influence of a chosen fiber type, content, and orientation on the structural response. The multi-level approach for modeling steel-fiber-reinforced structures is based on three levels of analysis: A semi-analytical model for the pullout of steel fibers at the bottom-most level, a crack bridging law accounting for the distribution of fiber orientation and embedment length which finally serves as the traction separation law for zero-thickness interface elements to simulate the structural response (Figure 1,(bottom, right)). For the numerical design of SFRC structures, this multi-level model is employed to assess the influence of relevant SFRC design parameters on the structural response. Inherent material uncertainties, construction tolerances, and accepted failure probabilities are considered as constraints to generate optimized designs. As a representative case, a reliability-based optimization for the design of fiber-reinforced segmental tunnel linings subjected to thrust jack forces is

DOI: 10.1201/9781003348450-2

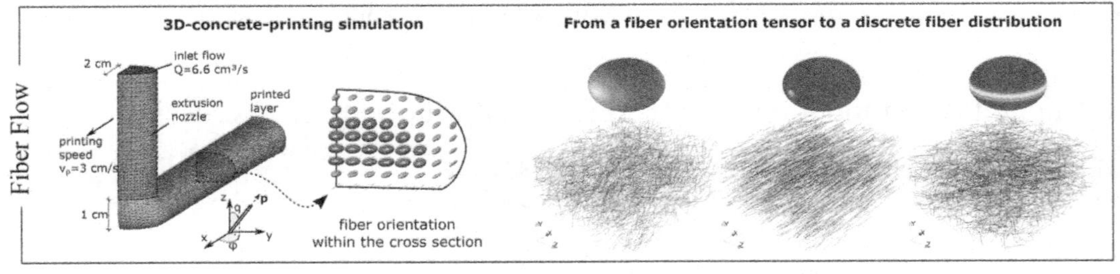

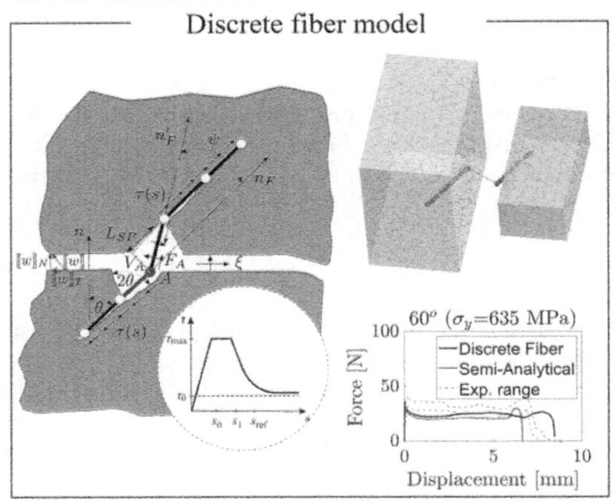

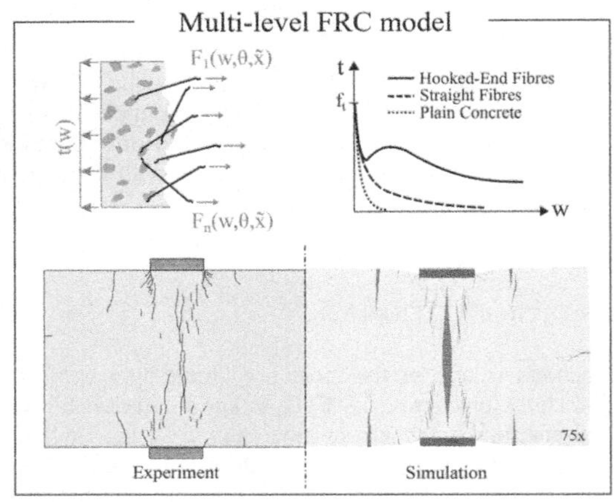

Figure 1. Virtual Lab for fiber-reinforced concrete structures: Simulation of 3D printing of fiber-reinforced concrete and the resulting tensor representation as well as a fully discrete representation of fiber orientations (top). Discrete fiber model with relevant geometrical parameters and simulation of inclined fiber pullout (bottom, left). Multi-level homogenization of fiber crack bridging action and the comparison of experimentally and numerically obtained crack patterns on a tunnel lining segment (bottom, right).

presented. To generate efficient segment designs, while taking material uncertainties into account, the structural reliability is calculated and compared to accepted failure probabilities proposed in (European Comittee for Standardisation 2002). Due to the high computational effort of solving optimization tasks under uncertainty, the structural models are replaced by surrogate models generated by Artificial Neural Networks. It is shown, that this approach leads to a segment design that consumes up to 20% less concrete and reduces the steel content by 59% as compared to a conventional RC design. More details and validations are contained in (Neu et al. 2022).

ACKNOWLEDGEMENT

Financial support was provided by the German Research Foundation (DFG) in the framework of SFB 837 (subproject B2) (Project no.: 77309832) and SPP 2020 (Project no.: 353819637). This support is gratefully acknowledged.

REFERENCES

European Comittee for Standardisation (2002). *EN 1990 - Eurocode: Basis of structural design*. European Comittee for Standardisation.

Folgar, F. & C. Tucker (1984). Orientation behavior of fibers in concentrated suspensions. *J. Reinf. Plast. Compos. 3*, 98–119.

Gudzulic, V. & G. Meschke (2021). Multi-level approach for modelling the post-cracking response of steel fibre reinforced concrete under monotonic and cyclic loading. *Proceedings in Applied Mathematics and Mechanics*.

Leung, C. & N. Shapiro (1999). Optimal steel fiber strength for reinforcement of cementitious materials. *Journal of Materials in Civil Engineering 11*(2), 116–123.

Neu, G., P. Edler, S. Freitag, V. Gudzulic, & G. Meschke (2022). Reliability based optimization of steel-fibre segmental tunnel linings subjected to thrust jack loading. *Engineering Structures*.

Reinold, J. & G. Meschke (2021). A mixed u-p edge-based smoothed particle finite element formulation for viscous flow simulations. *Comput. Mech.* published online.

Reinold, J., V. N. Nerella, V. Mechtcherine, & G. Meschke (2022). Extrusion process simulation and layer shape prediction during 3d-concrete-printing using the particle finite element method. *Autom. Constr. 136*, 104173.

Zhan, Y. & G. Meschke (2014). *Journal of Engineering Mechanics (ASCE) 140 (12)*, 04014091(1–13).

Zhan, Y. & G. Meschke (2016). Multilevel computational model for failure analysis of steel-fiber - reinforced concrete structures. *Journal of Engineering Mechanics (ASCE) 142*(11), 04016090 (1–14).

Current Perspectives and New Directions in Mechanics, Modelling and Design of Structural Systems – Zingoni (ed.)
© 2022 Copyright the Author(s), ISBN 978-1-032-39114-4

FRP strengthening of metallic structures: From research to practice

X.L. Zhao
Department of Civil and Environmental Engineering, The Hong Kong Polytechnic University, Hong Kong, China

ABSTRACT: This paper presents an overview of FRP strengthening of metallic structures from research to practice. The major topics covered include (i) recent applications; (ii) bond behaviour; (iii) flexural strengthening; (iv) compression strengthening; (v) strengthening under dynamic and cyclic loading; (vi) future research. A comprehensive list of publication in this field is made available online to readers.

1 INTRODUCTION

The conventional method of repairing or strengthening aging metallic structures often involves bulky and heavy plates. Fibre reinforced polymer (FRP) offers a great alternative for strengthening metallic structures, while it has been widely used to strengthen concrete structures (Teng et al. 2002). Worldwide research was conducted especially after the establishment of the International Institute for FRP in Construction in 2003, as evidenced by the 700 journal publications shown in Figure 1. The list of publications can be accessed online (Meng et al. 2022).

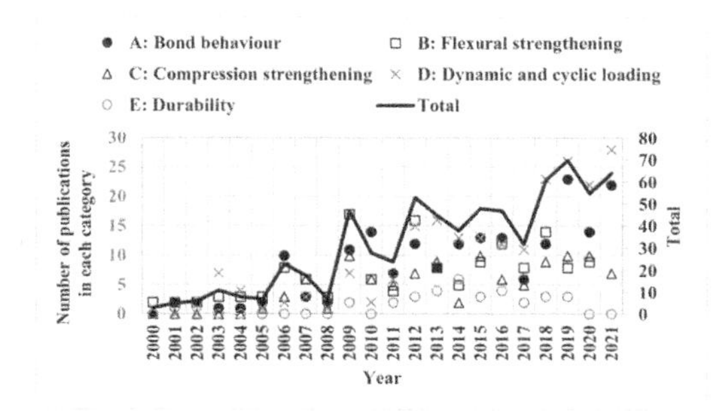

Figure 1. A summary of journal publications on FRP strengthening of metallic structures (by each category (A to E) and total) based on Meng et al. (2022).

The research has turned into field applications including strengthening of cast iron bridges, steel bridge girders, piers and truss members, steel beams in buildings, crane structures, pipelines, and aluminium truss-type highway overhead sign structures, as summarized in Zhao (2013). There are several field applications in the last 10 years, e.g., KY32 Bridge, Kentucky, USA, Münchenstein Bridge, Basel, Switzerland, Diamond Creek Bridge, Melbourne, Australia, Jarama Bridge, Madrid, Spain, Aabach Railway Bridge, Lachen, Switzerland.

The following sections will provide an overview on the research into FRP strengthening of metallic structures in terms of bond behaviour, strengthening under static flexural, compression and dynamic/cyclic loading. Future research areas will be pointed out.

2 BOND BEHAVIOUR

Bond between FRP and metal is vital to ensure a proper strengthening. Six possible failure modes in an FRP bonded steel system subjected to tensile force are illustrated in Zhao & Zhang (2007). The failure modes depend on many parameters, such as steel surface condition, FRP modulus, types of adhesives, adhesive thickness, curing condition, bond length, environmental temperature, and moisture.

Zhao (2013) made a summary of research work before 2013, which dealt with methods of bond test, failure modes, bond-slip model at ambient temperature, effect of temperature, fatigue loading and strain rate on bond strength. Recent studies since 2013 include surface preparation, carbon nano tube modified epoxy, bond-slip model, galvanic corrosion, environmental condition, blast loading, prestressed FRP, non-destructive testing methods and numerical simulation.

3 FLEXURAL STRENGTHENING

Teng and Fernando (2013) presented a systematic treatment (from failure modes to design procedures) of the flexural strengthening of steel and steel-concrete composite beams (in the form of I-section and rectangular hollow section) using FRP laminates. Chapter 6 of Zhao (2013) gave a comprehensive summary of strengthening of web crippling of beams

DOI: 10.1201/9781003348450-3

subjected to end bearing forces. Beams covered include cold-formed rectangular hollow section (RHS), aluminium RHS, LiteSteel beams, channel section and I-section. A summary was given in Zhao & Zhang (2007) for the strengthening of circular hollow section (CHS) and artificially degraded RHS under pure bending.

Here are some recent studies on flexural strengthening since 2013: unified design equations for CFRP strengthened ferritic and lean duplex stainless steel tubular sections against web crippling under four possible load configurations; prestressed FRP with/without anchorage; shear buckling of steel plates and girders; torsion of CHS and square hollow section (SHS), concrete-filled steel tubes (CFST).

4 COMPRESSION STRENGTHENING

FRP can delay, eliminate, or reduce the buckling, leading to a higher load carrying capacity of compression members. A summary of compression strengthening can be found in Zhao (2013), Zhao & Zhang (2007) for CHS, SHS, lipped channel sections, T-sections, and CFST members. It covered methods of strengthening, failure modes, load-displacement curves, and design models.

Recent developments in compression strengthening since 2013 include: a novel prestressed CFRP strengthening system for steel I-section columns; web buckling behaviour of RHS chord in T-joints induced by transverse compression loading of welded brace members; CHS T-joints under axial brace loading; damaged tubular steel T-joints under axial compression; CHS gap K-joints; stability of cylinder and conical shells.

5 DYNAMIC LOADING

Dynamic loading in this paper refers to large deformation cyclic loading, fatigue, impact, and blast loading. Only limited work was covered in Zhao (2013) on FRP strengthening under dynamic loading. They included effect of cyclic loading and impact loading on bond strength, strengthening of CFST columns under earthquake loading, fatigue strengthening of steel plates and I-section beams, prediction of improved fatigue life using boundary element method and fracture mechanics approach for CCT (centre-cracked tensile) steel plates, stress intensity factor (SIF) for CCT steel plates strengthened by CFRP.

Some recent studies in this field include: large deformation cyclic loading of CHS and RHS members, RHS CFST, CHS T-joints and beam-column connections; fatigue strengthening of RHS beams, welded attachments, bolted connections, CHS gap K-joints, web panels, corrugated plates; CHS, SHS and circular CFST members under lateral impact loading, as well as SHS under axial impact loading; spall damage of FRP strengthened metallic structures under blast loading.

6 FUTURE WORK

There is a need to

- Move from accelerated corrosion tests in laboratories to long-term assessment in the field.
- Predict service life of strengthened infrastructure and conduct life cycle analysis of various strengthening techniques.
- Promote more applications through collaboration among universities, industry, and government agencies.
- Include FRP strengthening in university courses and train more engineers and designers.
- Create international standards and guidelines.
- Develop new structural health monitoring techniques, e.g., a carbon nanotube-based sensing layer integrated in a steel/composite adhesive bond (Ahmed et al. 2018).
- Utilise advanced tools such as artificial intelligence, machine learning, machine vision, drones etc.
- Explore new materials in structural strengthening, such as iron-based shape memory alloy (Vůjtěch et al. 2021).

REFERENCES

Ahmed, S., Thostenson, E.T., Schumacher, T., Doshi, S.M, & McConnell, J.R. 2018. Integration of carbon nanotube sensing skins and carbon fiber composites for monitoring and structural repair of fatigue cracked metal structures. *Composite Structures* 203: 182–192.

Meng, Y.R., Zheng, B.T. & Zhao, X.L. 2022. Publication list of FRP strengthened metallic structures. available online from March 2022, https://www.researchgate.net/publication/358795961_FRP-Strengthened_Metallic_Structures_Journal_Paper_Publication_List

Teng, J.G., Chen, J.F., Smith, S.T. & Lam, L. 2002, *FRP-Strengthened RC Structures*. West Sussex: John Wiley & Sons.

Teng, J.G. & Fernando, D. 2013. Flexural strengthening of steel and steel-concrete composite beams with FRP laminates. Chapter 4, *FRP-Strengthened Metallic Structures*. Boca Raton, FL: CRC Press.

Vůjtěch, J., Ryjáček, P., Matos, J.C. & Ghafoori, E. 2021. Iron-based shape memory alloy for strengthening of 113-Year bridge. *Engineering Structures* 248: 113231.

Zhao, X.L. 2013. *FRP-Strengthened Metallic Structures*. Boca Raton, FL: CRC Press.

Zhao, X.L. & Zhang, L. 2007. State-of-the-art review on FRP strengthened steel structures. *Engineering Structures* 29:1808–1823.

Current Perspectives and New Directions in Mechanics, Modelling and Design of Structural Systems – Zingoni (ed.)
© 2022 Copyright the Author(s), ISBN 978-1-032-39114-4

In-test analytical model updating for accurate hybrid dynamic simulations

A.S. Elnashai
University of Houston, USA

H.H. El Anwar
Civil Engineering, Cairo University, Egypt

ABSTRACT: Hybrid simulation is the most reliable assessment approach for structural systems too large for testing in laboratories. It combines the effectiveness of numerical modeling with the accuracy of experimental investigations by sub-dividing the structural system into a computer-simulated part and a physically tested part. The components that exhibit highly inelastic behavior are tested in the laboratory, while the remaining structural system is numerically modelled. Matching of actions and deformations at the boundaries between physical and numerical components takes place at each time interval during the dynamic hybrid simulation process.

In many structural systems, there is an important feature that reduce the effectiveness, efficiency and/or accuracy of hybrid simulation; cases where either it is not possible to predict the critical elements of the structure that will cause failure, or if these critical members are too many to test at full scale. The concept of in-test analytical model updating for hybrid simulations addresses the aforementioned challenge. In the in-test model updating approach, only one component is experimentally investigated. During the test, experimental measurements are utilized to instantaneously update the governing parameters of the numerical model at each time-step. The numerical model therefore exhibits increased accuracy with every increment of the ongoing simulation.

In this paper, an overview of in-test model updating in slow rate hybrid simulation is provided. It is followed by examples of different approaches for tuning the parameters of the numerical model based on the continuous acquisition of experimental measurements. Practical examples from the authors' work and other researchers are presented and critiqued. General conclusions regarding the suitability of various model updating methods for different classes of dynamic testing of structural systems are presented.

1 INTRODUCTION

The assessment of structures subjected to extreme loading has remained a challenging research problem. Hybrid simulation was introduced and verified by researchers to address such problems, particularly when subject to dynamic loading such as earthquakes (Hakuno et al. 1969, Takanashi et al. 1975, Mahin & Shing 1985). Whereby, the parts that show unpredictable behavior are assessed experimentally, while the other parts are numerically modelled. Platforms were developed to establish the communication protocol between the numerical and experimental parts of the structure (Kwon et al. 2008, Schellenberg & Mahin 2006). Such that the restoring forces from the tested specimens and the numerical model are communicated to the equation of motion at each loading increment, which determines the deformations required for the next loading increment. Hybrid simulation has the ability to communicate with tested specimens located in different facilities and to utilize various numerical programs (Elnashai et al. 2016, Kwon et al. 2008).

Although hybrid simulation can be considered as an ideal approach to assess large-scale structure, yet as the number of critical components increase its effectiveness is hindered. The solution is to test a large number of physical specimens overlooking the economical aspects or to sacrifice the outcome accuracy by relying more on numerical models even for some critical components. Such argument motivated the concept of model updating in hybrid simulations. In this concept, the wealth of data acquired from the tested specimen is analyzed to identify the actual parameters that govern its behavior. The corresponding numerical models can be calibrated based on those parameters to enhance their accuracy. Therefore, it is sufficient to only test one sample from each group of elements that share close characteristics, while the remaining are numerically modelled and updated accordingly (Elanwar & Elnashai 2014). The tested component must be selected such that its response is ahead of the other numerically simulated elements in order to identify the selected parameters before they are required to be applied to the numerical model. Figure 1 shows the procedure of the model updating approach.

2 MODEL UPDATING APPLICATIONS

The concept of model application and its applications evolved rapidly as discussed in literature. In 2012, Yang et al. presented a pioneer work that showed that the spring parameters of a simple numerical model can be updated to capture the response of a detailed numerical model using model updating technique (Yang et al. 2012). In 2013, Kwon Kammula proposed

DOI: 10.1201/9781003348450-4

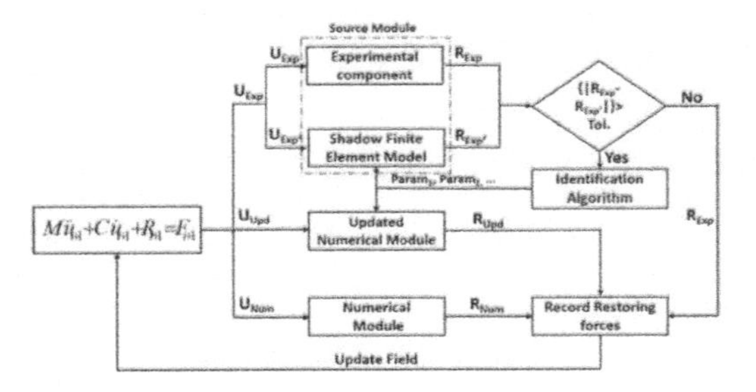

Figure 1. Flowchart describing the main components of model updating in hybrid simulation (Elanwar & Elnashai 2016b).

an approach, where the tested specimen was represented by a group of numerical models each exerts different behavior. Then, weight factors are determined for each numerical models and are added to yield a response close to the tested specimen (Kwon & Kammula 2013). In 2014, Hashemi et al. performed a test on a one bay steel frame, where one of the columns was represented physically and the other was simulated by a rigid element with a spring. The spring parameters defined using Bouc-Wen model were updating and the updated numerical model showed superior performance (Hashemi et al. 2013).

From 2014 through 2016, the authors introduced the concept of updating the parameters that govern the constitutive relationship using optimization and neural networks to capture the response of the tested specimen (Elanwar & Elnashai 2014, Elanwar & Elnashai 2016a & b). The concept was applied to previously performed steel and concrete experiments and the results showed significant improvement after applying the proposed concept. In 2017, Yang et al. aimed to make the model updating approach more practical by implementing an algorithm for constitutive relationship update using Unsent Kalman Filter (UKF) in the existing HyTest platform (Yang et al. 2017). In 2021, Cheng and Becker proposed to use the Weighted Adaptive Constrained Unscented Kalman Filter (WACUKF), which makes the traditional approach less sensitive to measurement noise (Cheng & Becker 2021). In 2021, Aksoylar and Aksoylar used extended Massing rule and showed that no optimization nor knowledge about the model are required (Aksoylar & Aksoylar 2021).

3 CONCLUSIONS

In this paper, an overview of model updating in hybrid simulation of structural response under earthquake loading was discussed. The authors' work and examples of other significant research were presented and critiqued. It is concluded that since its introduction in the last decade, model updating has become a key component in many hybrid simulation applications. The progress of its applications demonstrates significant potential for its utilization in structural earthquake assessment. The discussion highlighted that there are still many components and features of model updating that require further research.

REFERENCES

Aksoylar, C., & Dogramaci Aksoylar, N. 2021. Online Model Identification and Updating in Multi-Platform Pseudo-Dynamic Simulation of Steel Structures–Experimental Applications. *Journal of Earthquake Engineering*, 1–25.

Cheng, M., & Becker, T. C. 2021. Performance of unscented Kalman filter for model updating with experimental data. *Earthquake Engineering & Structural Dynamics*, 50(7), 1948–1966.

Elanwar, H. H., & Elnashai, A. S. 2014. On-line Model Updating in Hybrid Simulation Tests. *Journal of Earthquake Engineering*, 18(3), 350–363.

Elanwar, H. H., & Elnashai, A. S. 2016a. Framework for Online Model Updating in Earthquake Hybrid Simulations. *Journal of Earthquake Engineering*, 20(1), 80–100.

Elanwar, H. H., & Elnashai, A. S. 2016b. Application of In-Test Model Updating to Earthquake Structural Assessment. *Journal of Earthquake Engineering*, 20(1), 62–79.

Elnashai, A. S., Papanikolaou, V., and Lee, D. 2002. Zeus NL – A system for inelastic analysis of structures. *Mid-America Earthquake Center, University of Illinois at Urbana-Champaign*, USA.

Elnashai, A.S., Kwon, O. S, Gencturk, B., Mahmoud, H., Spencer, B., Elanwar, H.H., Kim, S. 2016. Hybrid AnalyticalExperimental Simulation in Earthquake Response Assessment. *Insights and Innovations in Structural Engineering, Mechanics and Computation*, 16–22.

Hakuno, M., Shidawara, M., and Hara, T. 1969. Dynamic destructive test of a cantilever beam, controlled by an analog-computer. *Transactions of the Japan Society of Civil Engineers*, 170, 1–9.

Hashemi, M. J., Masroor, A., and Mosqueda, G. 2014. Implementation of online model updating in hybrid simulation, *Earthquake Engineering & Structural Dynamics* 43(3), 395–412.

Kwon, O. S., Elnashai, A. S., & Spencer, B. F. 2008. A framework for distributed analytical and hybrid simulations. *Structural Engineering and Mechanics*, 30 (3), 331–350.

Kwon, O. S., & Kammula, V. 2013. Model updating method for substructure pseudo-dynamic hybrid simulation. *Earthquake Engineering and Structural Dynamics*, 42(13), 1971–1984.

Mahin, S. A. and Shing, P. B. 1985. Pseudodynamic method for seismic testing. *Journal of Structural Engineering* 111(7), 1482–1503.

Schellenberg A, Mahin S. 2006. Integration of hybrid simulation within the general-purpose computational framework opensees. *In: 8th US National Conf on Earthquake Engineering*, San Francisco, CA, USA.

Takanashi, K., Udagawa, K., Seki, M., Okada, T., and Tanaka, H. 1975. Non-linear earthquake response analysis of structures by a computer-actuator on-line system. *Bulletin of Earthquake Resistant Structure Research Center* 8.

Yang G, Wu B, Ou G, Wang Z, Dyke S. 2017. HyTest: platform for structural hybrid simulations with finite element model updating. *Adv Eng Softw.* 112: 200–210.

Yang, Y. S., Tsai, K. C., Elnashai, A. S., and Hsieh, T. J. 2012. An online optimization method for bridge dynamic hybrid simulations. *Simulation Modelling Practice and Theory* 28, 42–54.

Current Perspectives and New Directions in Mechanics, Modelling and Design of Structural Systems – Zingoni (ed.)
© 2022 Copyright the Author(s), ISBN 978-1-032-39114-4

Value of vibration-based structural monitoring for bridge emergency management

M.P. Limongelli & P.F. Giordano
Politecnico di Milano, Milano, Italy

ABSTRACT: Continuous monitoring of the structural response to vibrations enables to acquire real-time information that can support asset management decisions. Despite the several advantages provided by the availability of continuously updated information, the adoption of vibration-based monitoring systems still encounters difficulties to be implemented at large scale due to their perceived high cost and the difficulty to estimate the return on investment before their implementation. The Value of Information (VoI) analysis from Bayesian decision theory can be used to quantify the benefits associated with vibration-based monitoring information in supporting the selection of optimal asset management actions. In this paper, a framework to quantify the VoI from vibration-based monitoring is presented and the principal open question on this topic are outlined.

1 INTRODUCTION

In the aftermath of a disruptive event, such as a flood or an earthquake, information about the conditions of structures and infrastructures is required to support decision-making. In this context, automatic Vibration-based Monitoring (VBM) systems can be particularly beneficial both for single structures and at the network level. Despite the several advantages they provide, the difficulty to estimate the return over the investment on the VBM system creates some reluctance in the stakeholders (e.g., owners and managers) to invest in this technology. In this paper, a framework for the quantification of the VoI from VBM for emergency management is discussed (Giordano et al., 2022).

2 VALUE OF VBM FOR EMERGENCY MANAGEMENT

The quantification of the benefit from VBM in supporting decisions can be carried out taking basis on the concept of Value of Information (VoI) from Bayesian decision analysis (Raiffa & Schlaifer, 1961). The VoI is a metric that can guide the design of a VBM system as a support tool to address a given decision problem. From a broader perspective, the VoI can support in deciding which type of data should be collected (e.g., data about external actions such as traffic loads, ground acceleration, water flow) to support the decision. The VoI is obtained comparing two situations. In the first, the decision-making is based on the knowledge available without the information whose value must be quantified, for example, the knowledge provided by expert opinions. This analysis is denoted as 'Prior decision analysis' where the term 'prior' refers to the circumstance that the decision is made 'prior' to the acquisition of the information (i.e., not accounting for it). In the second situation, the decision is made assuming that the information is available but before collecting it. This second type of analysis, which requires the modeling of the information before its acquisition, is denoted as 'Pre-Posterior decision analysis'. The term 'pre' refers to the circumstance that the decision is made before the information is available, and the term 'posterior' refers to the fact that the decision is made accounting for the - modeled -information. The analytical formulation of the framework can be found in the extended paper. Herein, a summary of the results obtained for two case studies is reported. The first is relevant to the quantification of the VoI of VBM as a support tool to manage traffic restrictions (close, limit traffic, leave the bridge open) for a bridge during a flood event. Figure 1(a) reported the expected costs of the three emergency management actions computed without the VBM information (Prior decision analysis), as a function of the river flow. For low values of the flow, the optimal action (the action associated with minimum expected costs) is "Open", C_{Open}. "Restrict", C_{Limit}, becomes the optimal choice for intermediate values of the flow whereas "Close", C_{Close}, is optimal action for high values of the flow. Figure 1(b) presents the corresponding VoI as a function of the river flow. The VoI line presents a change of slope in correspondence to the changes of the optimal action due to the different dependency of the expected costs on the direct

DOI: 10.1201/9781003348450-5

consequences. The maximum VoI is reached at the intersection between the expected costs of the actions "Restrict" and "Close" that is in one of the two situations when the decision maker must choose between two cations with the same cost.

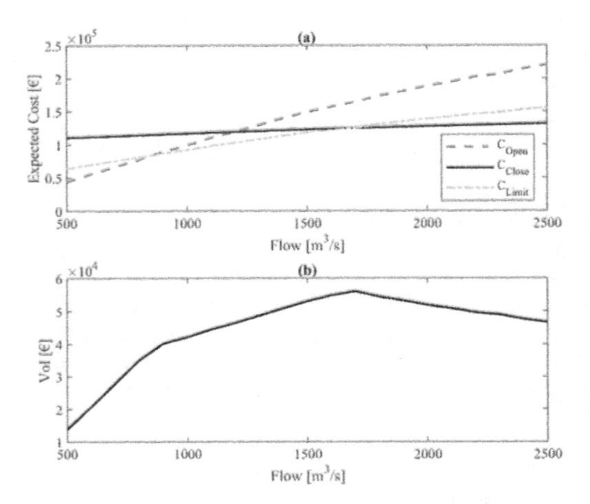

Figure 1. (a) Results of Prior analysis; (b) VoI. [Adapted from (Giordano et al., 2022)].

3 OPEN RESEARCH QUESTIONS

Two main issues in the application of the VoI analysis and the reliability of the results obtained consist in the VBM information modeling and the verification of its quality. To compute the VoI, the possible outcome of the VBM system must be theoretically modeled or estimated based on the VBM observations acquired on the system in its different states. As for full-scale structures, their experimental evaluation in several different damage states is not a feasible procedure. The high costs of such a test, joint with the artificially simulated damage, that may fail to realistically reproduce real damage scenarios, makes this approach rather inefficient. A possible alternative consists in numerical modelling of the structural response in relevant damage states. In this case, all the uncertainties operational and environmental uncertainties that affect the information must be modeled consistently which can prove a very challenging endeavor.

Another aspect that still needs investigation is the quality of the information on which the decision-making process relies to select management actions. This quality depends on several aspects whose impact on the decision-making should be accounted for in strict relation to the relevant decision problem. While the correctness of data, which is related to their precision and accuracy, is usually considered in risk modeling, the impact of other characteristics, such as completeness, timeliness, or relevance, is generally neglected. An additional source of concern dealing with monitoring systems in emergency management is related to the fact that they might be damaged during the disruption and thus provide altered information to decision-makers

4 CONCLUSIONS

This paper describes a framework for the quantification of VBM for emergency management. The VBM system must be designed to efficiently support the decisions the information is meant to support. In the last years, an approach based on VoI from Bayesian decision theory has been developed to this aim. In this paper, the general framework is discussed

An important research question still open for future investigation is relevant to the impact of the quality of the VBM information on the decision process they are meant to support. This entails modeling information quality attributes that go beyond correctness (e.g., precision and accuracy) and account for example the consistency, timeliness, and relevance of the information, in strict connection with the specific decision process they must support.

ACKNOWLEDGMENTS

This study was partially funded by the Italian Civil Protection Department within the WP6 "Structural Health Monitoring and Satellite Data" 2019-21 ReLUIS Project.

REFERENCES

Giordano, P. F., & Limongelli, M. P. (2020). The value of structural health monitoring in seismic emergency management of bridges. *Structure and Infrastructure Engineering*, 1–17. https://doi.org/10.1080/15732479.2020.1862251

Giordano, P. F., Prendergast, L. J., & Limongelli, M. P. (2022). Quantifying the value of SHM information for bridges under flood-induced scour. *Structure and Infrastructure Engineering*. https://doi.org/10.1080/15732479.2022.2048030

Raiffa, H., & Schlaifer, R. (1961). *Applied Statistical Decision Theory*. Division of Research, Graduate School of Business Administration, Harvard University.

Current Perspectives and New Directions in Mechanics, Modelling and Design of Structural Systems – Zingoni (ed.)
© 2022 Copyright the Author(s), ISBN 978-1-032-39114-4

Understanding tensegrity with an energy function

Simon D. Guest
Department of Engineering, University of Cambridge, Cambridge, UK

ABSTRACT: The use of a simple quadratic 'energy function' is explored to show how it can be useful in understanding the behaviour of structures that are stressed. The energy function is quadratic in the coordinates of the structure, and is able to correctly capture the stiffness of inextensional modes of deformation that are found in, for instance, tensegrity structures. The paper shows how the quadratic energy function can be used to derive a 'stress matrix' that is useful both for form-finding, and for checking stability.

1 ENERGY FUNCTIONS FOR A SINGLE TENSION MEMBER

In this paper, we explore how a simple 'energy function' that is a quadratic in nodal coordinates can be useful in understanding the behaviour of structures that are stressed. Figure 1 shows a single bar, and three energy functions for this bar are shown in Figure 2.

- The first energy function E is the 'true' energy function for the bar.
- Traditional structural analysis will assume that there is no variation in length of the bar with lateral deformation — essentially assuming the second 'linearized geometry' energy function L. This correctly gives the stiffness of the structure in the axial direction, but assumes it to be zero in the transverse direction. This is fine for the unstressed configuration, but incorrect when the bar is stressed, as is shown at configurations A and B in the plots. A should be stable in a transverse direction, and B unstable, but the energy function L shows both to be neutrally stable.
- The third energy function $Q_{m=0.25}$ gets the transverse stiffness for A correct, and has a particular simple form, being quadratic in the coordinates x and y (and z for a 3D structure).

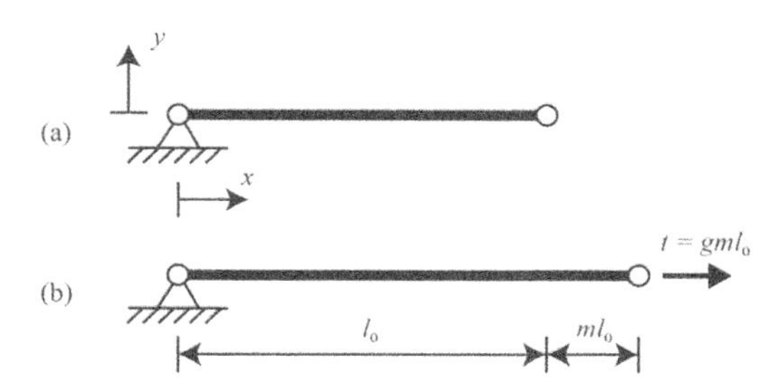

Figure 1. A single bar in two dimensions. One end is constrained at location $(0, 0)$, while the other is at point (x, y). (a) The initial unstressed configuration has $(x, y) = (l_0, 0)$. (b) An initially stressed configuration with $(x, y) = ((1 + m)l_0, 0)$. The bar has stiffness g, and carries a tension $t = gml_0$.

2 THE STRESS MATRIX

For a structure with nodal coordinates written as vectors $\mathbf{x}, \mathbf{y}, \mathbf{z}$, the quadratic stress function for the entire structure can be written as

$$Q = \frac{1}{2}\mathbf{x}^{\mathrm{T}}\mathbf{S}\mathbf{x} + \frac{1}{2}\mathbf{y}^{\mathrm{T}}\mathbf{S}\mathbf{y} + \frac{1}{2}\mathbf{z}^{\mathrm{T}}\mathbf{S}\mathbf{z}$$

where $\mathbf{S}$ is the *stress matrix* $\mathbf{S}$.

The stress matrix derived from the quadratic stress function can do double duty, both for equilibrium as a form-finding tool, and to understand the stiffness of a stressed structure. For equilibrium, the stress matrix must have nullity $d + 1$ to give a d-dimensional structure, with coordinates found by solving, e.g., $\mathbf{Sx} = \mathbf{0}$. For stability, the stress matrix must be positive definite, so that Q is never negative for any deformation. More details of how the stress matrix can be used to design and understand tensegrity structures can be found in Connelly and Guest (2022).

REFERENCES

Connelly, R. & S. D. Guest (2022). *Frameworks, tensegrities, and symmetry.* Cambridge University Press.

DOI: 10.1201/9781003348450-6

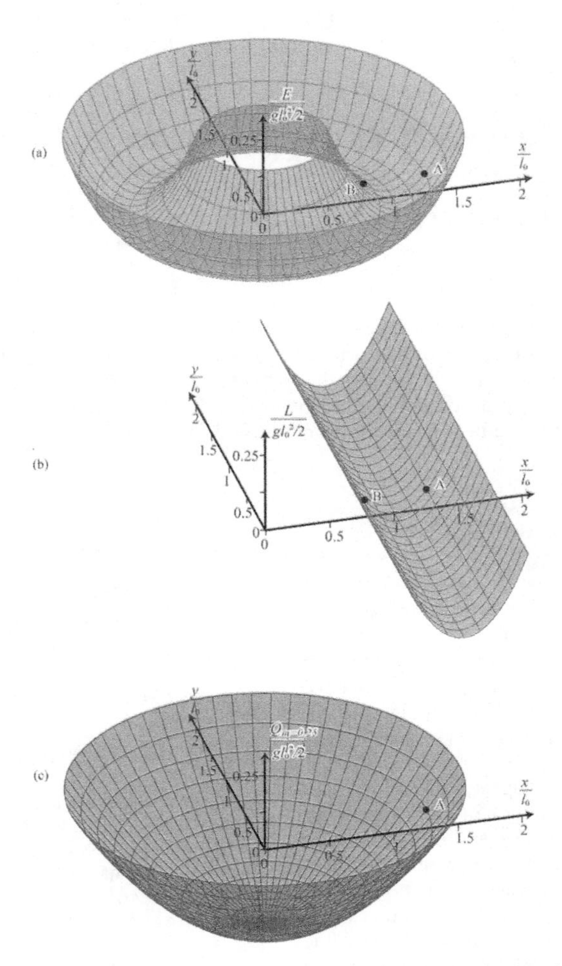

Figure 2. Energy functions for a single bar, non-dimensionalized by a standard energy value $\frac{1}{2}gl_0^2$, plotted varying with two non-dimensional coordinates x/l_0 and y/l_0. Values of energy greater that $0.25 \times \frac{1}{2}gl_0^2$ have been removed for clarity. Three energy functions are shown: (a) the 'exact' energy function E; (b) the 'linearized geometry' energy function L; (c) the 'quadratic' energy function Q_m for $m = 0.25$. All three functions have the same value and the same slope at the point A marked, which is where the bar has the configuration shown in Figure 1(b) with $m = 0.25$. Point B is marked for the first two plots, where $m = -0.25$.

2. Dynamic analysis, Vibration response, Vibration control, Environmental vibrations,Fluid-structure interaction

Current Perspectives and New Directions in Mechanics, Modelling and Design of Structural Systems – Zingoni (ed.)
© 2022 Copyright the Author(s), ISBN 978-1-032-39114-4

Bridge flutter and how to prevent it at low cost

Uwe Starossek
Hamburg University of Technology, Hamburg, Germany

ABSTRACT: The mechanism of flutter and the main features of its mechanical description are explained. Its mathematical treatment as a two-parametric complex eigenvalue problem and a solution procedure for computing the flutter wind speed are outlined. It thus becomes clear why flutter is the all-dominating criterion in the design of very long-span bridges. Various known ways of stabilizing a bridge against flutter, such as the twin deck concept or movable control surfaces, are discussed and it is shown that all are associated with certain disadvantages, such as substantial additional costs or reliability and maintenance concerns. An alternative device is described that avoids these disadvantages: the eccentric-wing flutter stabilizer. In contrast to similar devices proposed in the past, the wings do not move relative to the bridge deck and they are positioned outboard the bridge deck to achieve a greater lateral eccentricity. This enables the wings to produce enough aerodynamic damping to effectively raise the flutter speed. Results of a parametric flutter analysis study are presented in which both the properties of the bridge and the configuration of the wings are varied. Conclusions can thus be drawn about the type of bridges for which the device is particularly effective and cost-efficient.

1 INTRODUCTION

Bridge flutter is a vertical-torsional vibration of the bridge deck that occurs when the structure has certain characteristics and the wind blows at a certain speed (flutter speed). It is shown how the flutter speed of a bridge can be computed and how it can be raised in a reliable and cost-effective manner.

2 FLUTTER ANALYSIS

In the pure form of flutter, the aerodynamic forces on the structure are solely induced by the motion of the structure. When they are modeled as linear functions of the vertical and torsional displacements, frequency-domain analysis is possible and establishing the dynamic force equilibrium leads to the eigen value problem

$$\{(1+ig)\boldsymbol{K} - \omega^2[\boldsymbol{M} + \boldsymbol{A}(k)]\}\hat{\boldsymbol{\Phi}} = \boldsymbol{0} \tag{1}$$

where $\boldsymbol{K}$ =stiffness matrix, $\boldsymbol{M}$ = mass matrix, $\boldsymbol{A}$ = aerodynamic matrix, $\hat{\boldsymbol{\Phi}}$ = displacement amplitude vector, g = damping parameter (assuming damping forces proportional to the elastic restoring forces and in counter-phase to velocity), i = imaginary unit, ω = circular frequency of vibration. The coefficients of $\boldsymbol{A}$ depend on the bridge deck contour. Due to the phase shift between aerodynamic forces and displacements, caused by the unsteadiness of the flow, they are complex quantities, the real parts of which can be interpreted as negative aerodynamic stiffness and the imaginary parts as negative aerodynamic damping. Furthermore, they depend on the degree of unsteadiness of the flow and hence are functions of the reduced frequency

$$k = \frac{\omega b}{u} \tag{2}$$

where u = wind speed and b = half-width of bridge deck. The eigenvalue problem is solved for various (i.e., actually many) fixed values of k with the aim of identifying cases where the imaginary part ω_j'' of one eigenfrequency ω_j is zero (borderline stable state of constant-amplitude vibration). The corresponding wind speed is then obtained from

$$u = \frac{\omega_j' b}{k} \tag{3}$$

The smallest wind speed found in this manner is the flutter speed (Starossek & Starossek 2021b). It can be inferred form Equation (3) that the flutter speed can be raised by increasing the stiffness of the system. However, this would not be cost-efficient for reasons explained in the full paper.

DOI: 10.1201/9781003348450-7

3 ECCENTRIC-WING FLUTTER STABILIZER

Known ways of raising the flutter speed of a bridge, such as the twin deck concept or movable control surfaces, are discussed and it is shown that all are associated with disadvantages, such as substantial costs or reliability and maintenance concerns. An alternative device that avoids these disadvantages is the eccentric-wing flutter stabilizer (Figure 1), given that it consists of light and inexpensive elements and is a passive device with no moving parts. Due to their eccentricity, the wings produce enough aerodynamic damping to effectively raise the flutter speed.

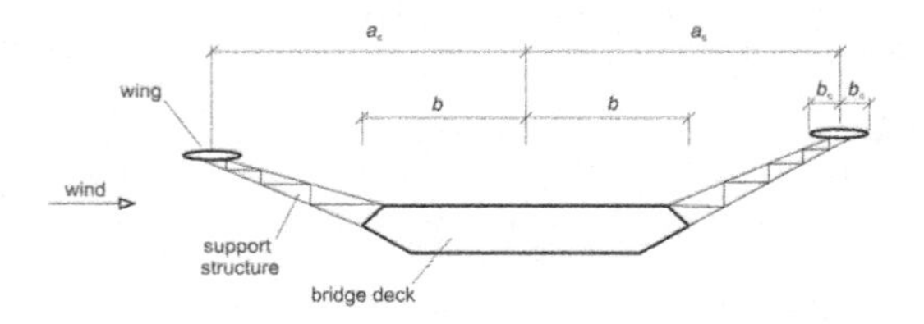

Figure 1. Bridge deck with eccentric-wing flutter stabilizer – cross section.

4 FLUTTER ANALYSIS INCLUDING WINGS

A multi-degree-of-freedom flutter analysis method has been developed based on a specially devised finite aeroelastic beam element that simultaneously models bridge deck and wings (Starossek & Starossek 2021b). Equation (1) and the described procedure for computing the flutter speed continue to apply. The influence of the wings is reflected by respective contributions to system matrices $\boldsymbol{A}$ and $\boldsymbol{M}$.

5 PARAMETRIC FLUTTER ANALYSIS

With this method, a parametric flutter analysis study has been performed in which both the properties of the bridge and the configuration of the wings are varied (Starossek & Starossek 2021a). For generality, a simple generic bridge system is considered, bridge girder and wings are assumed to be streamlined, and all input parameters and results are presented as non-dimensional quantities. As an example, Figure 2 shows the flutter speed increase ratio R, that is, the flutter speed of the system with wings divided by the flutter speed of the system without wings, plotted against frequency ratio ε, that is, the ratio of the lowest frequencies of torsional and vertical vibrations, for selected values of the other bridge and wing parameters. Significant increases of flutter speed are noted. Another example is Figure 3, showing the flutter speed increase ratio R plotted against the relative wing length, $\tilde{L}_c$, that is, the length of the wings divided by the length of the bridge.

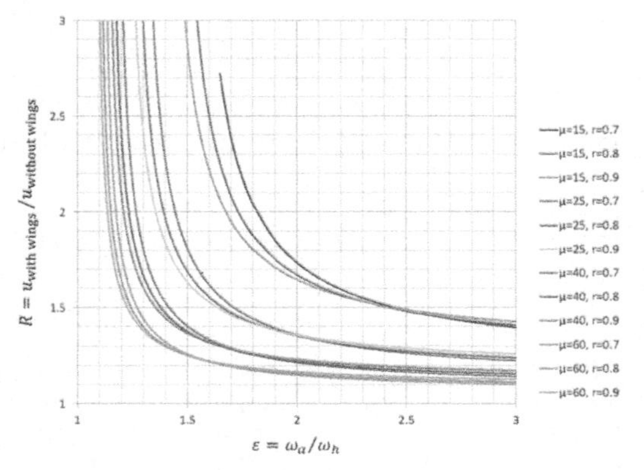

Figure 2. Flutter speed increase ratio, R, against frequency ratio, ε, for selected values of bridge and wing parameters.

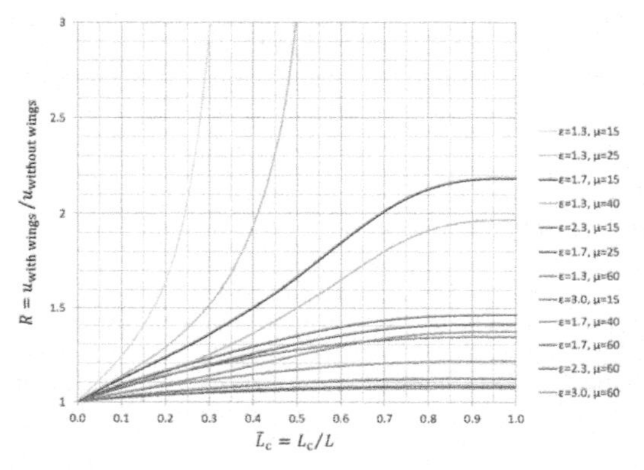

Figure 3. Flutter speed increase ratio, R, against relative wing length, $\tilde{L}_c$, for selected values of bridge and wing parameters.

Accordingly, a flutter speed increase of 64 % is achieved with wings that extend over 20 % of the bridge length, for a frequency ratio of $\varepsilon = 1.3$ and a mass ratio of $\mu = 15$, where $\mu = m/\pi\rho b^2$, m = mass per unit length of bridge girder, ρ = air density. Design studies for the corresponding wing configuration and the respective support structures indicate that the ensuing cost of wings and support structures is on the order of 1.3 % of the cost of the bridge superstructure (Starossek & Starossek 2021a).

REFERENCES

Starossek, U. & Starossek, R. T. 2021a. Parametric flutter analysis of bridges stabilized with eccentric wings. *Wind Engineering & Industrial Aerodynamics* 211, doi: 10.1016/j.jweia.2021.104566.

Starossek, U. & Starossek, R. T. 2021b. Flutter analysis methods for bridges stabilized with eccentric wings. *Wind Engineering & Industrial Aerodynamics* 219, doi: 10.1016/j.jweia.2021.104804.

Current Perspectives and New Directions in Mechanics, Modelling and Design of Structural Systems – Zingoni (ed.)
© 2022 Copyright the Author(s), ISBN 978-1-032-39114-4

Early assessment of the vibroacoustic performance of large lightweight buildings

L.V. Andersen, L. Mangliar & M.M. Hudert
Department of Civil and Architectural Engineering, Aarhus University, Aarhus, Denmark

O. Flodén & P. Persson
Department of Construction Sciences, Lund University, Sweden

ABSTRACT: Driven by the ambition to reduce the embodied energy in buildings, a current trend in the construction sector is to utilize more wood and lightweight composites. In addition to that, the service-life of buildings can be extended by flexible and adaptable floorplans, which typically require large-span structures. However, structures that are both lightweight and large-span are prone to increased vibration levels. This paper presents a computational framework for assessing the vibro-acoustic performance of such buildings at an early stage of the design process. The framework merges advanced digital tools for architectural design with rigorous finite-element models of the building structure. The functionality of the framework is validated by means of analysing a small CLT building, and here specifically the steady-state response to time-harmonic excitation on a floor. It is shown that representative acceleration levels can be achieved with a low computational effort.

1 INTRODUCTION

Using lightweight materials such as cross-laminated timber (CLT) could help to reduce the amount of embodied energy in multi-storey buildings. A drawback of such lightweight structures could be their vibroacoustic performance. If not designed properly, the vibration levels can be expected to be higher than in heavier structures. Existing building codes such as EN12354 are not fully covering the assessment of the vibroacoustic performance of lightweight building structures. The guidance provided by CLT specific design guides is likewise incomplete when it comes to the evaluation of buildings as a whole.

In order to close that gap, this paper presents a computational framework for the vibroacoustic assessment of larger lightweight structures at early design stages. The framework aims to support architects and engineers in making informed decisions based on the combined evaluation of a building's structural and vibroacoustic performance and carbon footprint.

2 FRAMEWORK FOR THE ANALYSIS

The here presented framework aims for computational efficiency. It employs a suite of Python scripts for setting up, running, and extracting the results of a finite-element (FE) model developed in the commercial software Abaqus (*SIMULIA Abaqus version 2021*, 2021). Microsoft Excel is used as a wrapping tool for input and presentation of the main results in a spreadsheet. Further, the model and the results can be transferred to the digital architectural design tool Rhinoceros with the plugin Grasshopper.

The calculations use a high-fidelity finite-element model based on component-mode synthesis, i.e., dynamic sub-structuring, following the Craig-Bampton approach (Craig, 1981). Each floor and wall panel constitutes a substructure attached to a virtual skeleton, as proposed by Andersen et al. (2012) for joisted floors and double-leaf walls, and further developed by Andersen and Kirkegaard (2013) for the analysis of buildings with semi-rigid joints.

The framework allows for different types of dynamic analyses, including the extraction of the global model's eigenfrequencies. Furthermore, the stationary vibration response to time-harmonic sources can be calculated in the frequency domain, and the transient response to time-varying loads can be computed in the time domain.

3 EXAMPLE MODEL OF A CLT BUILDING

In order to validate the functionality of the framework, a small CLT building was analysed. The internal organization of the rooms is shown in Figure 1. The floor panels measure 4.8 m × 6.0 m, have a thickness of 220 mm, and a five-layer layup. The wall panels consist of five layers and have a total thickness of 150 mm. Mean values for European C24 grade timber were used, and edge-glued boards with perfect bonding were

DOI: 10.1201/9781003348450-8

assumed. Based on the approach proposed by Andersen et al. (2012), an FE model was made in Abaqus for each of the panel types. Solid C3D20 elements with 20 nodes and full integration were employed with a maximum mesh size of 100 mm in the local length and width directions, and with one element per layer in the thickness direction (two for the middle layer). Substructures for the three panel types were generated and then assembled to form the case study building.

As part of the global model and the analysis of the building – and based on the component mode synthesis (CMS) – the steady-state response was determined in the frequency domain. A unit-magnitude (i.e., 1 N) time-harmonic point force was applied vertically at the centre of the top surface on the floor panel in the top/left room. In the steady-state analysis, 2.5% structural damping was introduced.

The aim of the analysis was to establish acceleration response levels in the one-third octave bands for the defined load case and in the frequency range that is relevant to vibrations and structure-borne noise, i.e., from 0 to 250 Hz.

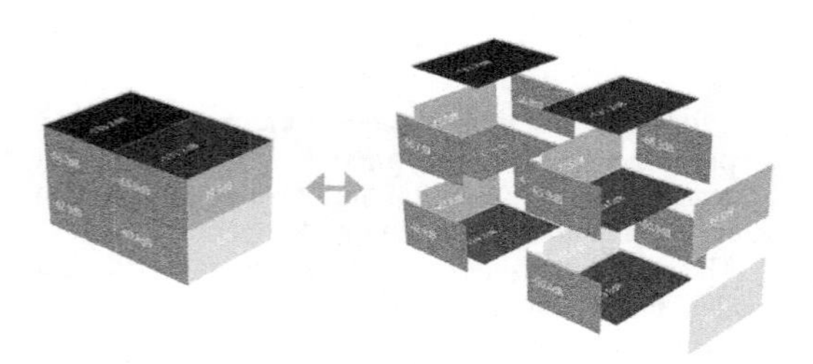

Figure 1. RMS acceleration levels (dB) on the panels of the building due to a unit point force on the floor in Room 1-2-1.

4 RESULTS

The first global modes occurred at the eigenfrequencies 18.3 Hz and 19.2 Hz, which are below the first eigenfrequency of the simply supported floor panel.

For the averaging of the accelerations in each one-third octave band, three approaches were considered:

1. linear spacing with a total of N frequencies;
2. logarithmic spacing with a total of N frequencies;
3. local discretization with n discrete frequencies between two neighbouring eigenfrequencies.

It is assumed that $n \ll N$, e.g., $n = 5$ and $N = 300$ for the frequency range 1–300 Hz. The modal density is relatively constant in the considered frequency range, but slightly higher at high frequencies. At the same time, the one-third octave bands become wider with increasing frequency. Hence, the first approach is judged to provide the better alternative with a fair compromise between the accuracy of the results and the computational effort.

The results obtained by approach 1 with 2×2, 3×3, or 5×5, or 11×11 indicator nodes on each panel are now compared. They indicate that representative RMS levels of a reasonable accuracy can be achieved with the use of 3×3 indicator nodes per panel. On the other hand, 2×2 points are insufficient, especially for higher frequencies. Thus, approach 1 is deemed to provide useful results of a fair accuracy, although the total number of frequencies is 300 compared to about ten times as many frequencies in approach 3.

5 CONCLUSIONS

The paper introduces a computational framework that enables the vibroacoustic assessment of modular timber buildings at early design stages and with a low computational effort. The framework combines simple input definitions, automated modelling, and analysis with impactful visualisations of both the models and the results. The framework is universally valid and can be applied to different dynamic loads, large buildings, and a variety of structural elements and construction materials. Its functionality was demonstrated by analysing the steady-state response to time-harmonic excitation in a small CLT building. The modal analysis and dynamic sub-structuring of the exemplary building provided high accuracy and a significant reduction in computation time compared to conventional FE models. In the frequency range 25–250 Hz, a representative estimate of the RMS acceleration levels was achieved with 3×3 indicator nodes per panel. The latter observation is not general, and more points may be necessary for other building typologies.

ACKNOWLEDGMENTS

The research was carried out in the framework of the project "Reconcile". The authors thank the European Regional Development Fund for the financial support provided via the Interreg V programme.

REFERENCES

Andersen, L. V., Frier, C., Pedersen, L., & Persson, P. 2020. Influence of furniture on the modal properties of wooden floors. *Conference Proceedings of the Society for Experimental Mechanics Series* (pp. 197–204).

Andersen, L. V., & Kirkegaard, P. H. 2013. Vibrations in a multi-storey lightweight building structure: Influence of connections and nonstructural mass. *Research and Applications in Structural Engineering, Mechanics and Computation – Proceedings of the 5th International Conference on Structural Engineering, Mechanics and Computation, SEMC 2013*.

Andersen, L. V., Kirkegaard, P. H., Persson, K., Kiel, N., & Niu, B. 2012. A modular finite element model for analysis of vibration transmission in multi-storey lightweight buildings. *Civil-Comp Proceedings*, 99.

Craig, R. R. 1981. *Structural dynamics: An introduction to computer methods*. New York: John Wiley & Sons, Ltd.

SIMULIA Abaqus version 2021. 2021. Paris, France: Dassault Systemes.

Current Perspectives and New Directions in Mechanics, Modelling and Design of Structural Systems – Zingoni (ed.)
© 2022 Copyright the Author(s), ISBN 978-1-032-39114-4

Modal analysis of CLT beams: Measurements and predictive simulations

B. Bondsman, O. Flodén, H. Danielsson, P. Persson & E. Serrano
Department of Construction Sciences, Lund University, Lund, Sweden

ABSTRACT: Cross-laminated timber (CLT) as a prime example of an innovative product within the field of civil engineering has over the recent years attracted attention of the construction industry around the world. Timber structural systems however are more sensitive to vibrations than conventional concrete based structures. Therefore, a thorough understanding of dynamic characteristics of CLT and accurate predictive models can enhance potential of wooden structures in the construction industry. This work presents a numerical and an experimental modal analysis of a series of CLT beams, cut out from a larger CLT plate, composed of two covering layers of Norway spruce and a mid-layer of Scots pine. The experimental results are utilised for a sensitivity analysis with respect to the mechanical properties of CLT.

1 INTRODUCTION

Timber structural systems have over the recent years increasingly drawn attention of the construction industry in Europe and other parts of the world, primarily due to the commercial launch of the innovative engineered wood product (EWP), Cross-laminated timber (CLT). CLT is constituted by a number of uneven layers of wood boards arranged crosswise at an angle of 90° between one layer and the adjacent ones.

This study presents modal analysis of 24 CLT beams with dimensions $120 \times 120 \times 2000[\text{mm}^3]$, cut out from a larger CLT plate, composed of two covering layers of Norway spruce and a mid-layer of Scots pine. The beams are suspended using bungee cords and excited using an impact hammer within a frequency range $0-2$ [kHz]. The chosen frequency range aims at quantifying deviation in the modal response of CLT as a ground for Finite Element (FE) model validation, which is subsequently utilised for a sensitivity analysis where influence of variation in density and mechanical properties of CLT on the resonant frequencies is studied.

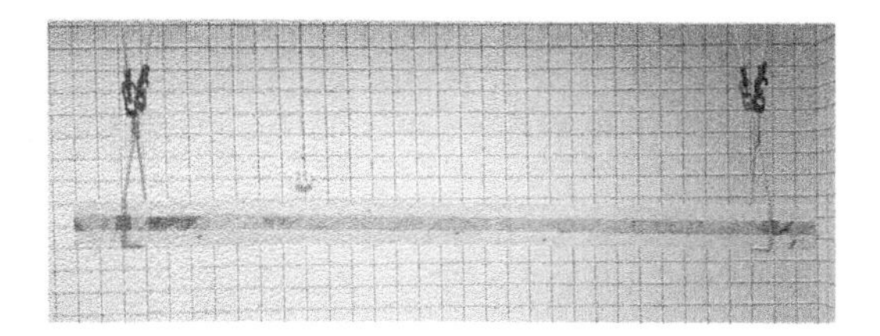

Figure 1. Measurement setup of CLT beam, suspended from the laboratory ceiling using bungee cords.

2 MODAL ANALYSIS

2.1 *Numerical simulation*

The equation of motion for a discretized FE-system in the absence of damping and external forces is given as

$$\mathbf{M}\ddot{\mathbf{a}} + \mathbf{K}\mathbf{a} = \mathbf{0}, \tag{1}$$

where $\mathbf{M}$, $\mathbf{K}$ are mass and stiffness matrices, whilst $\ddot{\mathbf{a}}$, $\mathbf{a}$ are nodal acceleration and displacement vectors, respectively. Natural frequencies of the system and its associated natural modes can be determined by solving the generalised eigenvalue equation as

$$\left(\mathbf{K} - \omega_i^2\mathbf{M}\right)\Phi_i = \mathbf{0}, \tag{2}$$

where ω_i and $\mathbf{\Phi}_i$ are natural frequency and associated natural mode, respectively. Through a numerical convergence study with respect to the equation above, the experimental excitation grid-resolution has been determined and its appropriateness has been verified using Modal Assurance Criterion (MAC)

$$\text{MAC} = \frac{|\mathbf{\Phi}_i^{\text{T}}\mathbf{\Phi}_j|^2}{\left(\mathbf{\Phi}_i^{\text{T}}\mathbf{\Phi}_i\right)\left(\mathbf{\Phi}_j^{\text{T}}\mathbf{\Phi}_j\right)}, \tag{3}$$

yielding a matrix with values between 0 and 1, where 1 denotes exact match between two mode shapes, i.e. when $i = j$. An AutoMAC has been computed for six in-plane, six out-of-plane bending and six torsional simulated modes within a frequency $0-2$ [kHz] with the aim of ensuring

DOI: 10.1201/9781003348450-9

dissimilarity between the vibration modes so they can be separated from each other.

To investigate influence of material properties on the resonant frequencies response, a Normalised Relative Frequency Difference (NRFD) is utilised in a sensitivity analysis

$$\text{NRFD} = \frac{f_i - f_j}{f_j}, \quad (4)$$

where i and j are order of modal resonant frequencies to be compared.

2.2 *Condition and properties of the beams*

All the 24 beams have mutually two covering layers of Norway Spruce of strength class C24. Ten of those have Scots Pine with strength class C16 in the mid-layer whilst the other fourteen have Scots Pine with strength class C24 in the mid layer.

The mean density and moisture content of each specimen were 473 [kg/m^3] and 10 [%], discretely.

2.3 *Experimental Modal Analysis*

Experimental Modal Analysis (EMA) has been employed, where impulse hammer excitation was chosen. The beams were suspended using bungee cords and the vibration response was measured using three single-axis accelerometers in axial, transversal and lateral direction, see Figure 1.

2.4 *Finite element model*

In the FE model, mechanical properties of wood strength class C35, corresponding to a mean value of the measured mass density in EN338 (SIS 2016), was chosen. It was assumed that $G_{LT} = G_{LR}$, based on which a relationship between the mentioned quantities and rolling shear modulus results in $G_{LT} = 14G_{RT}$. In the same manner, a relationship between tangential and radial stiffness modulus results in $E_T = 0.61E_R$, (Dahl 2009). To investigate influence of uncertainty in mechanical properties of wood on the resonant frequencies, minimum and maximum value of the mass density was chosen to govern the lower and upper limit of the corresponding mechanical properties, i.e. strength classes C24 and C50 in EN338 (SIS 2016), respectively.

3 RESULTS

Throughout the measurement campaign one axial, six out-of-plane bending, six in-plane bending, six torsional modes have been determined for all the beams.

The results indicate a deviation of the resonant frequencies from a normalised mean value up to 14 [%] for the out-of-plane bending modes, whilst less than ± 10 [%] for the in-plane, torsional and axial modes individually. The deviation was greater among the beams having Scots Pine of strength class C24 in the mid-layer. This was clearly visible in the out-of-plane bending and torsional modes. The deviation was also greater among low order vibration modes, and decreased in higher order modes of vibration.

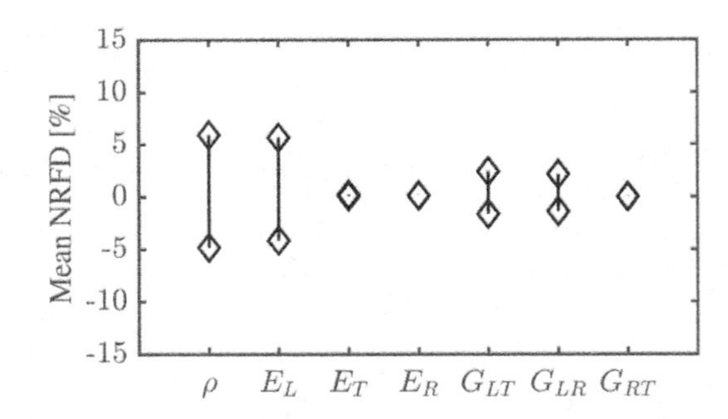

Figure 2. Average influence of variation in mechanical properties on resonant frequencies on averaged resonant frequencies associated with six out-of-plane, in-plane, torsional and one axial modes.

The sensitivity analysis and in particular the average influence approach used herein, discloses greatest influence from the mass density and axial stiffness on the natural frequencies, see Figure 2.

4 CONCLUSIONS

In the present study, the variability in resonant frequencies of beams cut out from a CLT panel has been studied using a parametric numerical model and experiments. The deviation was greater among the beams having Scots Pine of strength class C24 than C16 in the mid-layer.

The sensitivity analysis indicate visible influence of axial stiffness and density on the resonant frequencies. Thus, it can be concluded that the use of mass density is relevant for selecting elastic constants of wood and capturing variation in the modal response.

Further development of the present study can comprise analyses of the Frequency Response Functions of vibration with respect to stochastic propagation of variation in material parameters.

ACKNOWLEDGEMENT

The authors acknowledge financial support from Södra Foundation of Research, Development and Education, Ref. no. 2020-196, Swedish Innovation Agency Vinnova Ref. no. 2018-04159, and the Crafoord Foundation Ref. no. 20210941.

REFERENCES

Dahl, K. (2009, 12). *Mechanical properties of clear wood from Norway spruce*. Ph. D. thesis, Norwegian University of Science and Technology's.

SIS (2016). Structural timber – strength classes. Standard SS-EN 338:2016, Swedish Institute for Standards, Stockholm, Sweden.

Current Perspectives and New Directions in Mechanics, Modelling and Design of Structural Systems – Zingoni (ed.)
© 2022 Copyright the Author(s), ISBN 978-1-032-39114-4

Optimal design of viscoelastic tuned mass dampers for structures exposed to coloured excitations

M. Argenziano & E. Mele
Department of Structures for Engineering and Architecture, University of Naples Federico II, Naples, Italy

A. Palmeri
School of Architecture, Building & Civil Engineering, Loughborough University, Loughborough, England
Department of Structural, Geotechnical and Building Engineering, Politecnico di Torino, Turin, Italy

ABSTRACT: Dynamic interaction between primary and secondary structures can alter the response of buildings, bridges and other civil engineering structures to external stressors such as earthquakes and windstorms. TMDs (tuned mass dampers) are a well-known example of passive control devices that exploit this concept. A TMD consists of a secondary mass attached to the primary structure through a linear or nonlinear link. Various formulations exist to optimize the performance of TMDs, depending on the chosen criterion. Typically, the steady-state amplitude of motion of the primary structure is minimized, e.g., when subjected to monochromatic harmonic excitation (H_∞ criterion) or white noise input (H_2 criterion). Several closed-form analytical solutions have been formulated for linear TMDs with elastic and viscoelastic damping. However, the available expressions only cover the case of constitutive laws described by two or three parameters, e.g., the Kelvin-Voight model (an elastic spring in parallel with a viscous dashpot) and the standard linear solid (SLS) model (an elastic spring in parallel with a single Maxwell element). Furthermore, the inherent damping of the primary system is usually neglected, even though it can drastically affect the performance of the TMD. Similarly, the effects of coloured excitations, e.g., the Kanai-Tajimi model of ground shaking, are often disregarded, leading to sub-optimal designs. Aimed at overcoming these limitations, a new stochastic approach is proposed. The only assumptions are: *i*) the salient dynamic features of the TMD-controlled structure can be captured with a linear 2-DoF (degree of freedom) system; *ii*) the dynamic action can be represented as a stationary Gaussian process.

1 INTRODUCTION

In classical applications of TMDs (e.g., Den Hartog 1985), a 2-DoF lumped mass model is utilized and the primary-to-secondary link is modelled by means of a KV model.

The SLS model is able to capture the frequency-dependent behaviour of several passive control devices, e.g., elastomeric rubber bearings and fluid viscous dampers. The adoption of the SLS model leads to the introduction of an additional internal variable (AIV) in the dynamic problem, which represents the internal deformation of the viscoelastic model. Further, more refined linear models exist that require multiples AIVs (e.g., Palmeri, Ricciardelli, De Luca, & Muscolino 2003).

In this paper, the H_2 optimization of SLS-type TMDs is implemented, with the aim to overcome some limitations of previous optimization studies (e.g., Asami & Nishihara 2002, Batou & Adhikari 2019) that have not addressed the influence of inherent damping in the primary structure and coloured excitations.

2 STATEMENT OF THE PROBLEM

The key dynamic features of SLS-type TMDs can investigated through a simplified 2-DoF lumped-mass system in which a SLS model describes the primary-to-secondary link. The application of the classical Buckingham π theorem allows identifying three dimensionless design parameters measuring the rigidity coefficients in the link (named ρ_0 and ρ_1) and its relaxation time $\bar{\tau}_1 = \omega_1 \tau_1$, ω_1 being the undamped natural circular frequency of the primary system. The primary-to-secondary

DOI: 10.1201/9781003348450-10

mass ratio μ and the equivalent viscous damping ratio of the main structure ξ_1 are further governing parameters.

3 NUMERICAL RESULTS

By adopting a random vibration approach, the proposed optimization procedure seeks to minimize the root mean square (RMS) of the displacement of the main mass when subjected to a stationary ground motion excitation, modelled either as a white noise or a coloured input. For the latter case, the Kanai-Tajimi filter (KT) (Kanai 1957, Tajimi 1960) is adopted. Genetic and particle swarm algorithms are utilized for numerically deriving optimal parameters as a function of μ and ξ_1 (see Figure 1).

4 CONCLUSIONS AND FUTURE RESEARCH

The results show that the optimal parameters for the problem in hand are sensitive to relatively low variations of the primary damping and strongly depend on the frequency content of the dynamic action. Thus, these aspects should be carefully considered and managed when designing viscoelastically damped TMDs, either reducing the relevant epistemic uncertainties or increasing the robustness of the seismic protection offered by this type of vibration absorbers.

Research is currently underway to apply probabilistic and possibilistic methods, i.e., random, interval and fuzzy variables, to quantify the effects of different sources of uncertainty on the optimal design of the SLS-type TMDs for M-DoF (multiple degrees of freedom) structures.

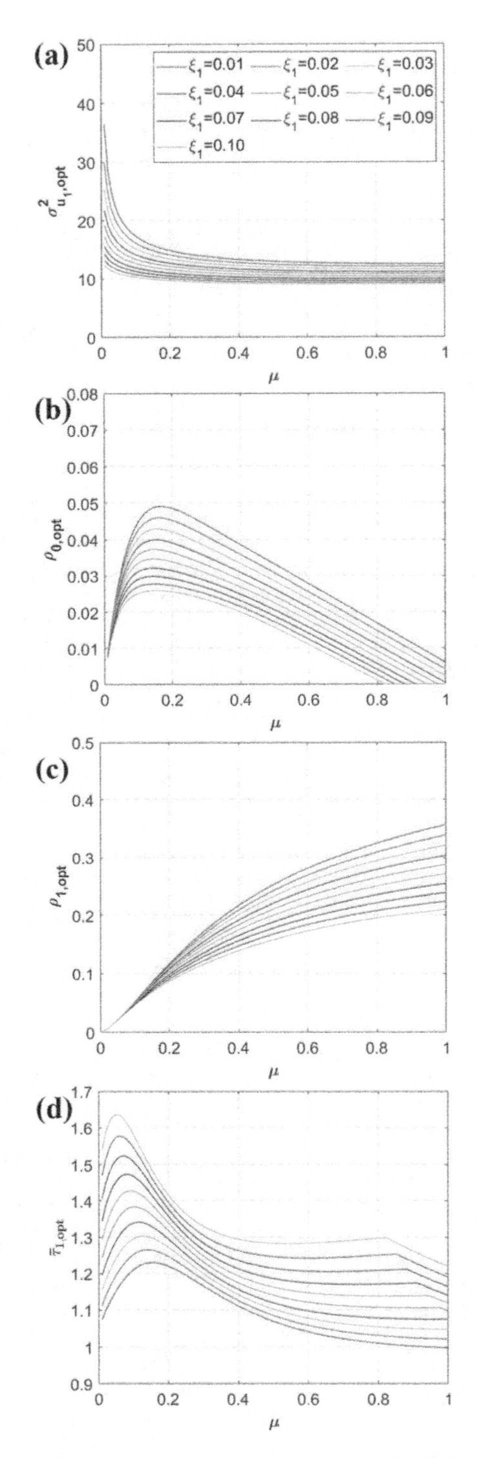

Figure 1. Dimensionless optimal parameters of the SLS-type TMD as a function of the mass ratio μ for several values of the primary damping ξ_1 under filtered white noise excitation: (a) optimal variance of U_1; (b) optimal modulus of equilibrium ρ_0; (c) optimal ρ_1; (d) optimal relaxation time $\bar{\tau}_1$.

REFERENCES

Asami, T. & O. Nishihara (2002). H 2 optimization of the three-element type dynamic vibration absorbers. *J. Vib. Acoust. 124*(4), 583–592.

Batou, A. & S. Adhikari (2019). Optimal parameters of viscoelastic tuned-mass dampers. *Journal of Sound and Vibration 445*, 17–28.

Den Hartog, J. P. (1985). *Mechanical vibrations*. Courier Corporation.

Kanai, K. (1957). Semi-empirical formula for the seismic characteristics of the ground. *Bulletin of the earthquake research institute 35*, 309–325.

Palmeri, A., F. Ricciardelli, A. De Luca, & G. Muscolino (2003). State space formulation for linear viscoelastic dynamic systems with memory. *Journal of Engineering Mechanics 129*(7), 715–724.

Tajimi, H. (1960). A statistical method of determining the maximum response of a building during earthquake. PROCEEDING OF WORLD CONFERENCE ON EARTHQUAKE ENGINEERING.

Current Perspectives and New Directions in Mechanics, Modelling and Design of Structural Systems – Zingoni (ed.)
© 2022 Copyright the Author(s), *ISBN 978-1-032-39114-4*

Numerical investigation on train-bridge interaction after ship impact event: Train running safety in different scenarios

L. Bernardini
PhD Candidate, Department of Mechanical Engineering, Politecnico di Milano, Milano, Italy

G. Soldavini
Master Thesis Student in Mechanical Engineering, Politecnico di Milano, Milano, Italy

A. Collina
Full Professor, Department of Mechanical Engineering, Politecnico di Milano, Milano, Italy

ABSTRACT: This work presents a numerical study on the vehicle-bridge dynamic interaction in the catastrophic event of a ship impact. In particular, the objective is to study the behavior of train running safety coefficients as a function of bridge and vehicle mechanical properties. Two representative FE models of a continuous beam bridge were built up, featured by different pier heights. Critical running scenarios have been deepened, studying wheel-rail interaction and eventual detachment to physically interpret the results provided by the computation of unloading and derailment coefficients.

1 INTRODUCTION

1.1 *Runability analysis*

Bridges and viaducts enhance daily transportation of passengers and goods. Their design phase requires the implementation of an accurate runability analysis able to represent the effect of both environmental (Olmos & Astiz, 2018) and operational actions on the rail vehicle dynamics, and consequently allow for the evaluation of its running safety. Since runnability analysis involves, at the same time, global and local aspects of the structure dynamics, while the vehicle dynamics depends on the local wheel and rail interaction (Diana, Cheli, & Bruni, 2000), a sophisticated wheel-rail contact model is required that considers wheel profile and geometry, local deformability and allows for multiple contacts.

2 MODELS AND METHODS

In this paper, ship impacts occurring on central pier are simulated, due to barge vessels. The resulting force, is featured by an impulsive shape (Gholipour, Zhang, & Mousavi, 2019) with a maximum peak value slightly below 40 MN, with a time duration of about 1.2 s. This force has been scaled from a peak value of 5 MN up to 40 MN. The impact is always set to occur when the first train car reaches the impacted pier (i.e., pier 3 in Figure 1). Two bridge FE models have been considered in this work (see Figure 1), featured by linear foundations. The lateral motion of the deck due to the ship impact excites train lateral motion. In the case of proximity between bridge and rail carbody lateral mode frequencies, a resonance-like phenomenon may originate after the collision.

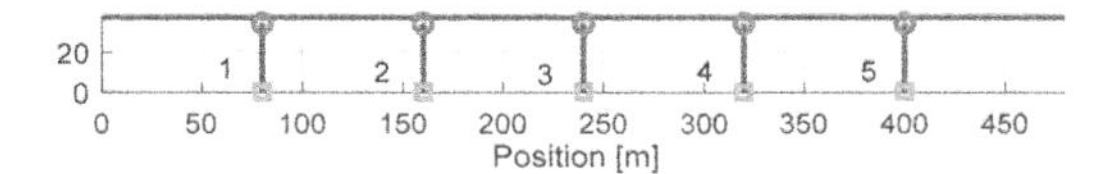

Figure 1. Bridge FE model, lateral view.

Two different rail vehicles have been considered, a Twindexx and an ICE train. The two trains are modelled through 3D multibody models, and they travel at constant speed.

The track has been considered as rigidly connected to bridge deck, modelling a direct fastening between track and deck slab, with the addition of track irregularity profile, computed referring to ORE B176. The wheel-rail contact geometry is taken into account considering the possibility of more than one contact point. The normal forces are computed through a multi-Hertzian approach, while the tangential ones according to the creepage dependence formulation of Shen-Hedrick-Elkins. The contact forces exchanged between wheel and rail in correspondence of each active contact point therefore depend on wheel-rail relative motion. As a result, the contact forces depend on structure and train motion. The procedure for integrating in time domain considers the two subsystems, structure and train as separated, interacting through the contact forces.

DOI: 10.1201/9781003348450-11

For all the investigated scenarios, train running safety has been evaluated through derailment (EN14363, 2018) and unloading (EN14067-6, 2010) coefficients computed for each wheel.

3 RESULTS

As an example, extracted from the simulations pattern performed on the lower bridge, Figure 2 shows that train coaches featured by lateral car frequencies closer to bridge first lateral frequency are characterized by the highest running coefficients. With the lower bridge, some of the treated impact scenarios led to critical results, in terms of safety coefficients, as shown in Figure 3. The most critical outcomes concern ICE train: in particular, the unloading threshold is exceeded for any travelling speed once the impact force peak overcomes 25 MN, as shown in Figure 3.

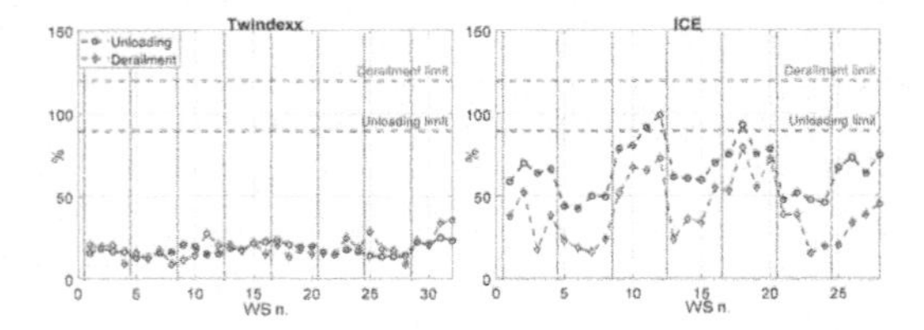

Figure 2. Derailment and unloading coefficients for the two trains, travelling over the lower bridge at 200 km/h, with an impact peak value of 30 MN.

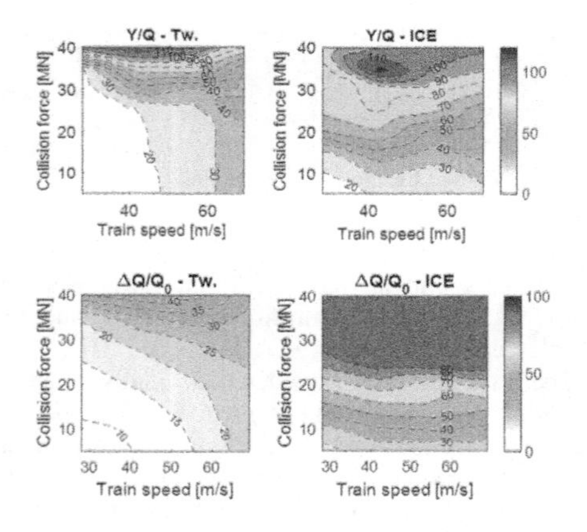

Figure 3. Distribution of running safety coefficients vs train speed and collision intensity. Lower bridge.

Figure 4 refers, as an example, to the unloading and derailment coefficients behaviour for the 15th wheelset of ICE train running at 150 km/h, with a peak collision force of 40 MN.

After the ship impact occurred, unloading values over the threshold were obtained. It was observed that when the unloading coefficients reach the value of 100% and that value is not immediately left, a wheel-rail detachment can be observed. Moreover, it was observed that overcoming the unloading

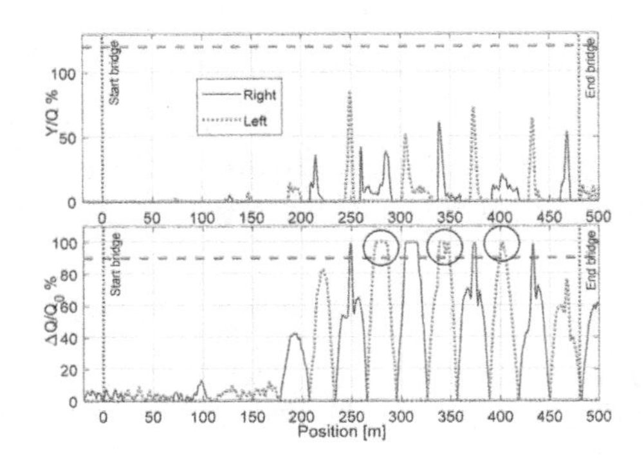

Figure 4. Derailment and unloading coefficients. ICE train, 150 km/h, 15th wheelset, maximum impulsive force.

threshold of 90%, that is considered as a critical scenario according to (EN14067-6, 2010), does not imply that the wheel detaches from the rail.

4 CONCLUSIONS

This work presents a numerical analysis on the dynamic train-track-bridge interaction after ship collision events. As a result of the simulation pattern, coaches featured by lateral mode frequencies closer to structural ones, present higher safety coefficients: therefore, the effects of resonance-like phenomena exert significant influence on train running safety. Adopting a complex modelling of the wheel-rail contact allows to consider non-linearities and to physically interpret a situation in which the unloading coefficients overcome the threshold, reaching their maximum value (i.e., 100%). Through this approach it was possible to show that overcoming the unloading threshold (i.e., 90%) does not imply a wheel uplift. The latter is rather caused by reaching the limit of 100 % without readily leaving it.

REFERENCES

Diana, G., Cheli, F., & Bruni, S. (2000). Railway runnability and train track interaction in long span cable supported bridges. *Advances in Structural Dynamics*, 43–54.

EN14067-6. (2010). *Railway applications - Aerodynamics - Part 6: Requirements and test procedures for cross wind assessment*. Brussels: European Committe for Standardisation (CEN).

EN14363. (2018). *Railway applications - Testing and Simulation for the acceptance of running characteristics of railway vehicles - Running Behvaiour and stationary tests.*

Gholipour, G., Zhang, C., & Mousavi, A. A. (2019). Analysis of girder bridge pier subjected to barge collision considering the superstructure interactions: the case study of a multiple-pier bridge system. *Structure and Infrastructure Engineering*, 392–412.

Olmos, J. M., & Astiz, M. Á. (2018). Improvement of the lateral dynamic response of high pier viaduct under turbulent wind during the high-speed train travel. *Engineering Structures*, 368–385.

Current Perspectives and New Directions in Mechanics, Modelling and Design of Structural Systems – Zingoni (ed.)
© 2022 Copyright the Author(s), *ISBN 978-1-032-39114-4*

Research advancement on Submerged Floating Tunnels: Recent activity at DICA-Politecnico di Milano

F. Foti, L. Martinelli, M.G. Mulas & F. Perotti
Department of Civil and Environmental Engineering, DICA, Politecnico di Milano, Milano, Italy

ABSTRACT: A few topics, selected from the recent research activities performed at DICA on Submerged Floating Tunnels (SFTs), are presented, concerning load modelling, design procedures and dynamic interaction with travelling vehicles. A model for generating seaquake effects taking into account the variability of the seabed ground motion over the length-scale of a few kilometers is described. A semi-analytical procedure is proposed to perform a preliminary design, without using a complete finite element model. The bridge-vehicle dynamic interaction is analyzed, to assess the serviceability conditions under the transit of heavy vehicles and high- and medium-speed trains. The research effort aims to the development of a complete analysis and simulation toolbox for the design of SFTs in medium-to-high sea-depth conditions.

1 INTRODUCTION

The interest of the engineering community about the Submerged Floating Tunnel (SFT) solution has recently increased due to infrastructural planning activities, such as for route E39 in Norway and Messina Strait in Italy. This offers the chance of assessing the state-of-the-art of analysis tools, design procedures, and supporting experimental activity, capable of handling the complex task of drafting, detailing, building and maintaining a construction of such relevance and innovative character.

The research group on Structural Dynamics based at DICA – Politecnico di Milano, engaged on the topic since the 90's, has actively participated to this flourishing trend of studies and applications. Present and future activity will also benefit from the contribution of the engineers of the Offshore Division of SAIPEM, with whom a cooperation agreement has been recently established.

This paper is devoted to concisely reporting the main research aspects recently addressed by the authors' research group, involving: (a) the treatment of the "seaquake" seismic excitation, (b) the development of a semi-analytic continuous model of the SFT for preliminary design calculations, and (c) the modeling of the dynamic interaction between the tunnel and a passing train/vehicle.

Relevant results are presented with reference to a prototype case study originating from a design proposal for the Messina Strait crossing. The crossing is 4680 m long and the maximum depth of the sea is about 350 m. The tunnel is located about 40 m below the water surface. A side view of the tunnel is depicted in Figure 1.

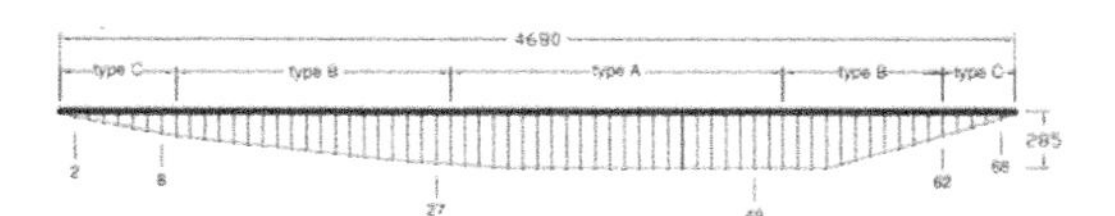

Figure 1. Lateral view of the SFT at study.

2 SEAQUAKE EXCITATION MODEL

SFTs can be restrained at several positions along their length and, in the case of an earthquake, will have to withstand the effects of ground motion at the anchoring points, as well as compressive waves generated by the seismic seabed motion (i.e. the so-called"seaquake" loading).

A simplified model based on a linearized one-dimensional wave propagation equation and free surface boundary conditions, has been developed by the research group. This model is formulated in frequency domain, and only accounts for the vertical water motion. The scattered waves due the structure presence are neglected and water is assumed as inviscid, incompressible and with an irrotational velocity field. The model, initially developed under the assumption of rigid seabed, was subsequently generalized to incorporate the compressibility of the last soil strata beneath the seabed (Martinelli 2018).

DOI: 10.1201/9781003348450-12

Seaquake effects are computed starting from the generation in time domain of a realization of the seismic vertical acceleration at the bedrock beneath the nodes of a SFT finite element model for which seaquake forces are needed. The vertical acceleration is generated starting from its Power Spectral Density accounting for spatial variability of the free-field ground motion through a stochastic model. Current and future studies concern an upgrade of the seaquake model to include the effects of wave reflections at the water free surface and at the seabed and extensions to 2D and 3D models.

3 SEMI-ANALYTICAL PROCEDURE FOR DYNAMIC ANALYSIS OF THE SFT

Conceptual design and structural optimization of SFTs require simple but reliable analytical tools to perform parametric analyses with an affordable computational cost. To this aim, a semi-analytical continuous model has been recently developed and applied to the analysis of the seismic response of SFTs (Foti et al., 2022).

The main assumptions of this model are: (a) the vertical and transverse response to seismic excitation can be decoupled; (b) the seismic excitation can be assumed as perfectly correlated along the tunnel; and (c) the hydrodynamic effects can be modeled in an equivalent linearized form. The anchoring system is modeled by introducing a "smeared" stiffness distribution.

Most literature works dealing with continuous equivalent beam models for SFTs are based on the assumption of a constant distributed stiffness of the mooring devices. Conversely, a variable stiffness of the mooring system along the length of the tunnel is here considered, to properly account for the possible variability of the depth of the seabed, the spacing between anchored sections and the inclination and the cross-section of the anchoring tethers.

4 DYNAMIC INTERACTION WITH MOVING VEHICLES

In addition to the loads generated by natural phenomena, the interaction with travelling vehicles, including trains, deserves attention in the assessment of the SFT operability.

Previous studies on vehicle-bridge dynamic interaction provided a reliable approach to account for the problem, adopting discrete models, with a finite number of DOFS, for both bridge and vehicle. The same FE model of the structure set up at the design phase is adopted. A system of rigid masses, springs and dampers models the vehicle.

The coupled equations of motion of the vehicle-bridge system are derived under the assumption of point-wise and perfect wheel-bridge (or rail) contact. Contact points are massless, and the vehicle, which travels with a constant horizontal velocity, transmits only vertical forces. The derivation accounts for the roughness of the contact surface, either the road pavement or the rail irregularity.

The integration of the coupled equations of motion is performed through an ad-hoc iterative strategy that allows to handle computational problems due to the time-varying properties of the structural matrices. Forced uncoupling of the coupled equations, with separate integration of the two systems, is associated to an iteration process to restore equilibrium and compatibility at the systems interface. The two systems are separately analyzed, the bridge subjected to the vertical forces transmitted by the vehicle, and the vehicle excited by the motion of the contact point, including the road roughness or the rail irregularities.

The proposed formulation was recently applied (Mulas et al., 2018, 2022) to investigate the tunnel response due to the transit of both heavy vehicles and medium- to high-speed passenger trains in a double tube SFT configuration with two straight independent tunnels. The results highlighted that the passage of any type of vehicle in a tube causes a significant variation of the stress state only in that tube, with a local effect around the vehicle position. The loss of tension in the anchoring tethers, less than 10% of the initial tension, is largely within an acceptable range.

5 CONCLUSIONS

Research by the Structural Dynamics group based at DICA – Politecnico di Milano is still ongoing, focusing on several aspects of the structural analysis of SFTs. Aim of the activity is contributing to the development of a simulation environment supporting the SFT design in its various phases. Within this context both safety issues related to extreme environmental conditions (eg earthquakes and sea waves) and operability issues (eg response to passing trains) have been successfully addressed and are still under further development.

REFERENCES

Foti, F. & Martinelli, L. & Perotti, F., 2022. A semi-analytical model for the design and optimization of SFTs under seismic loading, Submitted to 11th Intern. Conference on Bridge Maintenance, Safety and Management IABMAS 2022.

Martinelli, L. 2018. Submerged floating tunnels under seismic and seaquake loadings. Proc. of IABMAS 2018, 9th Intern Conf. on Bridge Maintenance, Safety and Management, 877–884.

Mulas, M.G. & Martinelli, L. & Palamà, G. 2018. Dynamic interaction with travelling vehicles in a submerged floating tunnel. Proc of 9th Int Conf on Bridge Maintenance, Safety and Management IABMAS 2018, 838–844.

Mulas, M.G. & Martinelli, L. & Zambon, S. 2022. The coupled dynamic response of a prototype SFT to high speed trains. Accepted, Proc. of 11th Intern. Conf. on Bridge Maintenance, Safety and Management IABMAS 2022.

Current Perspectives and New Directions in Mechanics, Modelling and Design of Structural Systems – Zingoni (ed.)
© 2022 Copyright the Author(s), *ISBN 978-1-032-39114-4*

The influence of near-surface soil layer resonance on vibrations in pile foundations

F. Theland, C. Pacoste & J-M. Battini
Department of Civil and Architectural Engineering, KTH Royal Institute of Technology, Stockholm, Sweden

G. Lombaert & S. François
Department of Civil Engineering, KU Leuven, Leuven, Belgium

F. Deckner
GeoMind KB, Nacka, Sweden

ABSTRACT: This paper investigates the influence of an unsaturated layer in a soft soil on the dynamic response of piles subjected to an incident wave field caused by a vertical surface load. The free field response is compared to the response of single floating and end-bearing piles, and different configurations of square end-bearing pile groups. Simulations are made using a finite element model with perfectly matched layers where the incident wave field is computed from a separate source model employing a subdomain formulation. The presence of the unsaturated layer results in a resonance phenomenon in the top layer, amplifying critically refracted waves as they reach the surface. The associated response is shown to be significantly lower for concrete piles subjected to the same incident loading and practically independent of the pile end condition, in contrast to the response associated with the incident surface waves. For small end-bearing pile groups, the response caused by surface waves are further reduced while the response due to layer resonance show no further reduction. As a consequence, the frequency content associated with the layer resonance might constitute the dominating part of the vertical velocity response for end-bearing pile groups.

1 INTRODUCTION

The prediction of vibrations in the built environment involves multiple sources of uncertainty, making it difficult to estimate expected levels of vibration in buildings with accuracy prior to construction. Therefore, site measurements in the free field can be used to estimate the vibrations transmitted into buildings using numerical models or empirical adjustment factors (Kuo et al. 2016). However, the transmission of vibrations both depends on the properties of the subsoil and the character of the incident wave field impinging on the foundation. It is therefore important to understand the influence of the soil conditions on the incident wave field and its interaction with the foundation.

Soft clays underlain by a stiff bedrock are soil conditions encountered in urban areas in Sweden. Under these soil conditions, end-bearing piles are often used, in which the vibration response can differ substantially from the free field vibrations (Makris and Badoni 1995). On the other hand, the topmost unsaturated soil layer may cause significant amplification of reflected and refracted waves due to layer resonances which are situated within the frequency range of interest for environmental vibrations (Theland et al. 2021).

This paper investigates the response of piles and small pile groups in a layered soil exhibiting resonance of the top soil layer. The objective is to compare the response in the free field to the response of the pile foundations subjected to an incident wave field caused by a load applied at the soil's surface.

2 NUMERICAL MODEL

A finite element model with perfectly matched layers (FE-PML) is implemented where a subdomain formulation is utilized to prescribe an equivalent traction field at the FE-PML boundary to propagate an incident wave field computed from a separate source model (Papadopoulos et al. 2018). The considered source model consists of a horizontally layered half-space and the required displacements and tractions are obtained semi-analytically using the MATLAB toolbox EDT (Schevenels et al. 2009). In this paper, a vertical point load is applied at the soil's surface outside the FE-PML domain at a distance of 100 m.

DOI: 10.1201/9781003348450-13

3 CASE STUDY

3.1 *Soil profile*

Table 1 presents the small-strain soil properties assumed for the layered soil of the present analysis.

Table 1. Small-strain soil properties of a horizontally layered soil in terms of layer thickness h, S- and P-wave speeds C with their corresponding hysteretic damping ratios β and the material density ρ.

h [m]	C_s [m/s]	C_p [m/s]	β_s [-]	β_p [-]	ρ [kg/m^3]
1	70	125	0.03	0.03	1700
7	90	1200	0.03	0.03	1700
∞	2000	3600	0.03	0.03	2500

The large contrast in P-wave speed between the unsaturated top layer and the underlying saturated soil cause amplification of reflected and refracted waves in the top layer at the layer resonance frequencies for vertically propagating P-waves $f_{p1n} = C_{p1}/(4h_1)(2n-1)$. The P-waves that are critically refracted in both the saturated soil and in the elastic bedrock are of significantly longer wavelength than the surface waves propagating at the same frequency, which leads to slow attenuation with distance.

3.2 *Pile response due to incident loading*

Presents a comparison of the response in the free field, at the top of an end-bearing and a floating concrete pile with a vertical pile-bedrock spacing of 0.5 m. The end condition has a significant influence on the response at the lower frequencies. This part of the response is

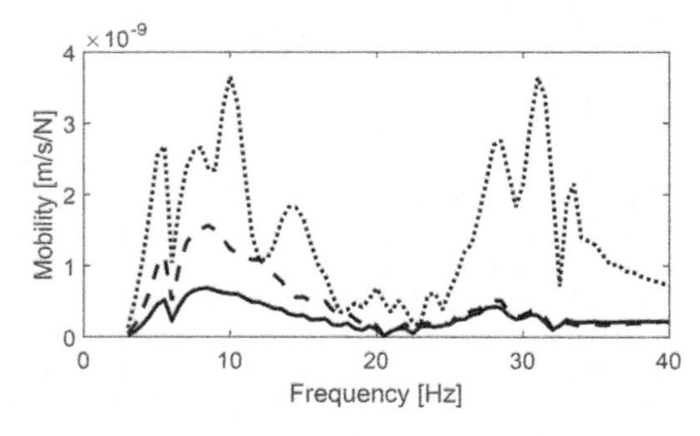

Figure 1. Vertical mobility of the end-bearing pile (solid line), floating pile (dashed line) and the free field (dotted line).

related to the incident surface waves. On the contrary, the response associated with the layer resonance is reduced by essentially the same factor regardless of the end condition.

3.3 *Vertical response of pile groups*

Pile groups are also affected by the phase differences between the response of the individual piles. This further reduces the response in the lower frequency range of the different pile groups considered in Figure 2. Due to the long wavelengths involved in the response around the layer resonance frequency, the piles move almost in phase and the response is not further reduced compared to the single pile and the differences observed are mainly attributed to pile-soil-pile interaction.

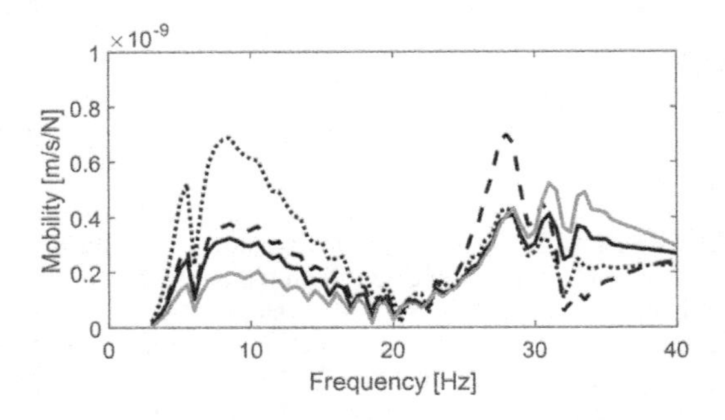

Figure 2. Vertical mobility of end-bearing pile groups with the configurations 2×2 with spacing $s = 0.8$ m (solid black line), $3 \times 3, s = 0.8$ m (solid gray line) and $2 \times 2, s = 1.3$ m (dashed line) compared to a single end-bearing pile (dotted line).

4 CONCLUSIONS

An analysis of the response of pile foundations in a soil with an unsaturated top layer exhibiting layer resonance is presented. The responses of the different pile foundations are computed using a FE-PML model. The amplitude of the pile response associated with layer resonance is found to be (1) significantly lower than the free field response, (2) independent of the pile end condition and (3) not significantly affected by kinematic interaction in small pile groups.

REFERENCES

Kuo, K., H. Verbraken, G. Degrande, & G. Lombaert (2016). Hybrid predictions of railway induced ground vibration using a combination of experimental measurements and numerical modelling. *Journal of Sound and Vibration 373*, 263–284.

Makris, N. & D. Badoni (1995). Seismic response of pile groups under oblique-shear and Rayleigh waves. *Earthquake Engineering & Structural Dynamics 24*(4), 517–532.

Papadopoulos, M., S. François, G. Degrande, & G. Lombaert (2018). The influence of uncertain local subsoil conditions on the response of buildings to ground vibration. *Journal of Sound and Vibration 418*, 200–220.

Schevenels, M., S. François, & G. Degrande (2009). EDT: An ElastoDynamics Toolbox for MATLAB. *Computers & Geosciences 35*(8), 1752–1754.

Theland, F., G. Lombaert, S. François, C. Pacoste, F. Deckner, & J.-M. Battini (2021). Assessment of small-strain characteristics for vibration predictions in a Swedish clay deposit. *Soil Dynamics and Earthquake Engineering 150*, 106804.

Current Perspectives and New Directions in Mechanics, Modelling and Design of Structural Systems – Zingoni (ed.)
© 2022 Copyright the Author(s), ISBN 978-1-032-39114-4

Prediction of flutter velocity of long-span bridges using probabilistic approach

S. Tinmitonde, X. He & L. Yan
School of Civil Engineering, Central South University, Changsha, People Republic of China
National Engineering Laboratory for High-Speed Railway Construction, Changsha, People Republic of China

ABSTRACT: Flutter has been prove to be the most dangerous wind-induced excitation phenomena on long-span bridges such structures. Until now, the flutter velocity can be computed after conducting an experimental wind tunnel test or by the Computational Fluid Dynamics (CFD) method. Despite the robustness of the approaches mentioned above, they are cost-prohibitive and time-consuming. Therefore, it is crucial to develop a cost-effective strategy for evaluating the critical flutter velocity of such structures. To accommodate to this issue, scholars have proposed Machine Learning (ML). However, ML techniques are subjected to bias and do not present accurate results for small datasets. To tackle this weakness, the present study proposes a probabilistic approach. The dataset used to train the model was obtained from a fully numerical approach that combined 2D-CFD simulations, and multimode approach. The finding indicates that the proposed probabilistic model provides accurate prediction of the flutter velocity while using tiny dataset.

1 INTRODUCTION

The spectacular urbanization worldwide has resulted in an emerging tendency toward constructing super long-span cables-supported bridges. However, wind excitation poses a significant risk to such megastructures. Researchers have revealed that flutter instability is the most harmful wind-induced vibration effect on long-span bridges that need to be addressed (Argentini et al., 2022).

The flutter velocity can be calculated using either an experimental wind tunnel test or computational fluid dynamics. Despite the robustness of the techniques outlined above, they are expensive and time-consuming (use of heavy equipment during wind tunnel testing, electricity, models manufacture, computation time). As a result, it is necessary to develop an approach for accelerating the assessment of the critical flutter velocity of such structures at an earlier design stage. To address these shortcomings; researchers have suggested artificial intelligence (AI) models.

Chen and co-authors (Chen et al., 2008) proposed an artificial neural network to accurately predict the flutter derivatives of a rectangular plate based on a dataset acquired from an experimental wind tunnel test. Recently, (Liao et al., 2021) utilized machine learning algorithms to forecast the flutter velocity of a streamlined bridge deck cross-section subjected to shape modifications and found that the ANN model exhibited the best predictive performance.

However, machine learning algorithms are prone to bias in the absence of big data(Zhou et al., 2020). To overcome this drawback, (Rizzo & Caracoglia, 2020) suggested data augmentation methods which included Nataf-model Monte Carlo and Polynomial Chaos Expansion model. However, the primary consequence of data augmentation is data bias, reducing the accuracy of the prediction (Xu et al., 2020). To accommodate this issue, we introduced a probabilistic machine learning approach in this article to forecast the critical flutter velocity of long-span bridges.

2 GENERAL WORKFLOW OF THE STUDY

Firstly, we generated the design of experiments (DoE) and computed the geometrics parameters of each DoE. Secondly, we conducted a series of 2D-URANS (Unsteady Reynolds-averaged Navier-Stokes) CFD simulations on each DoE to calculate the integral parameters of the aerodynamic coefficients and their derivatives. After that, an example of a long-span Bridge whose dynamics and structural characteristics are known to compute the critical flutter velocity of each DoE based on a multimode approach.

Finally, the probabilistic approach was introduced to predict the flutter velocity using five design parameters as inputs variables and the critical flutter velocity as target (dependent variable). Figure 1 depicts the entire workflow adopted in the present study.

DOI: 10.1201/9781003348450-14

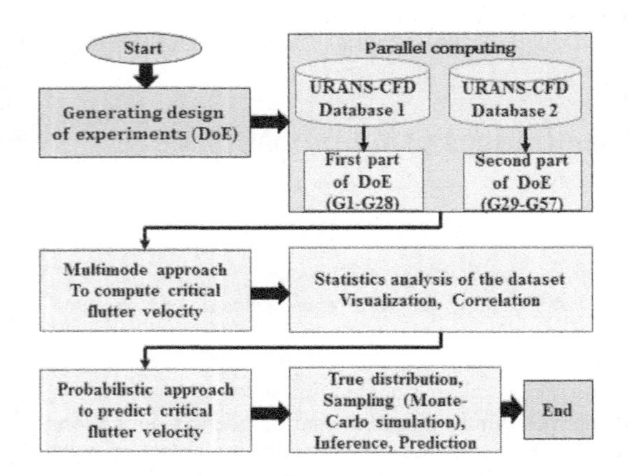

Figure 1. General roadmap of the present study.

3 PROBABILISTIC MODELING

The probability distribution of the data can be expressed using the following equation (Eq.(1)):

$$f(y|\alpha,\beta,\sigma_\varepsilon) = \rho\left(\frac{y-\alpha-\beta X}{\sigma_\varepsilon}\right) \quad (1)$$

y = response;
$X = (x_{ij})$ = matrix [x_{ij} denotes the value of the $j^{th}(j = 0, 1, \ldots, 4)$ variable for the $\mathrm{i}^{\mathrm{th}}(\mathrm{i} = 1, 2, \ldots, \mathrm{n})$ observations];

$\alpha, \beta_0, \beta_1, \beta_2, \beta_3, \beta_4$ = unknow coefficients parameters ρ = probability density function(PDF).

Assuming that all data points are statistically independent identically distribution (i.i.d), then the equation (0.6) can be rewritten as follows (Eq.(2)) (Gelman et al., 2014):

$$f(D|\alpha,\beta,\sigma_\varepsilon) = \prod_i^n f(x_i, y_i|\alpha,\beta,\sigma_\varepsilon) \quad (2)$$

$\theta = \left\{\mu_\alpha, \mu_\beta, \sigma_\beta, \sigma_\varepsilon, \alpha, \beta\right\}$ = random variable to be estimated

$\pi(\mu_\alpha), \pi(\mu_\beta), \pi(\sigma_\alpha), \pi(\sigma_\beta), \pi(\sigma_\varepsilon)$ are the prior distribution.

The posterior distribution was then computed based on the Baye theorem(Gelman et al., 2014) using the *Python programming* package called *PyMC3* (Salvatier et al., 2016). To conduct the probabilistic model, a finite number of 500 samples were drawn from the flutter velocity dataset based on the normal distribution assumption previously demonstrated. Figure 2 is the results obtained from the simulations.

Figure 2, indicates that a slight variation of the sample was observed for the four chains involved during the 500 samplings using MCMC simulation.

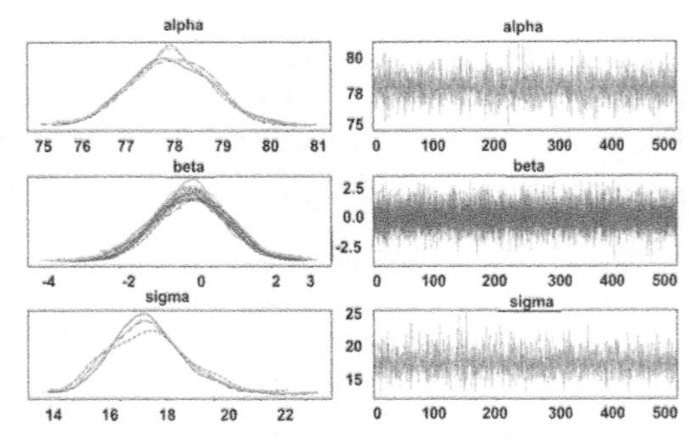

Figure 2. Posterior density of the prior parameters generated with four chains.

This result means that the number of chains did not affect the sampling accuracy.

4 CONCLUSION

The findings indicate that the proposed probabilistic approach provides an accurate estimation of the flutter velocity with negligible error. The results substantiate that; the proposed probabilistic model can effectively be used to forecast the flutter velocity of long-span bridges at a lower cost.

REFERENCES

Argentini, T., Rocchi, D., Somaschini, C., Spinelli, U., Zanelli, F., & Larsen, A. (2022). Aeroelastic stability of a twin-box deck: Comparison of different procedures to assess the effect of geometric details. *Journal of Wind Engineering and Industrial Aerodynamics, 220* (May 2021), 104878.

Chen, C.-H., Wu, J.-C., & Chen, J.-H. (2008). Prediction of flutter derivatives by artificial neural networks. *Journal of Wind Engineering and Industrial Aerodynamics, 96* (10–11), 1925–1937.

Gelman, A., Carlin, J. B., Stern, H. S., Dunson, D. B., Vehtari, A., & Rubin, D. B. (2014). Bayesian Data Analysis. In Taylor& Francis Group (Ed.), *The Statistician* (Third, Vol. 45, Issue 2). Chapman and Hall/CRC.

Liao, H., Mei, H., Hu, G., Wu, B., & Wang, Q. (2021). Machine learning strategy for predicting flutter performance of streamlined box girders. *Journal of Wind Engineering and Industrial Aerodynamics, 209*(January), 104493.

Rizzo, F., & Caracoglia, L. (2020). Artificial neural network model to predict the flutter velocity of suspension bridges. *Computers & Structures, 233*, 106236.

Salvatier, J., Wiecki, T. V., & Fonnesbeck, C. (2016). Probabilistic programming in Python using PyMC3. *PeerJ Computer Science, 2*(4), e55.

Xu, Y., Noy, A., Lin, M., Qian, Q., Li, H., & Jin, R. (2020). *WeMix : How to Better Utilize Data Augmentation.*

Zhou, J., Shen, J., & Xuan, Q. (2020). Data Augmentation for Graph Classification. *International Conference on Information and Knowledge Management, Proceedings,*

Current Perspectives and New Directions in Mechanics, Modelling and Design of Structural Systems – Zingoni (ed.)
© 2022 Copyright the Author(s), ISBN 978-1-032-39114-4

The effect of pile slip on underwater noise emission in vibratory pile driving

T. Molenkamp, A. Tsouvalas & A.V. Metrikine
Faculty of Civil Engineering, Delft University of Technology, Delft, The Netherlands

ABSTRACT: Due to the growing demand in offshore wind, increasing numbers of foundation piles are planned to be installed in the coming decades. Monopiles driven by impact hammers have a large environmental impact on aquatic life. Vibratory pile driving is a promising alternative that generates less noise nuisance. Despite the lower levels of noise expected, modeling of noise radiation from vibratory piling is still required due to the large size of the foundation piles used nowadays and the changes in the radiated spectrum of the noise. The existing models used to assess the noise emission are calibrated against impact piling and are not accurate when it comes to noise radiation from vibratory installation. Existing models either represent the sediment as an acoustic fluid or, when the seabed is modelled as elastic medium, they couple the soil and pile displacements fully at their interface. The effect of both these assumptions on the radiated noise still needs to be verified in vibratory pile installation. Additionally, the effect of the secondary noise path, i.e. noise channeling into the seawater via the soil, is expected to play a more significant role in vibratory pile driving because more energy is concentrated at the lower frequencies. In this paper, a pile-water-soil model to predict the noise emission due to vibratory pile driving is developed which describes the soil as an elastic medium and allows the pile to move relative to the soil during the pile driving process. To maintain a computationally efficient solution method, the effect of friction is linearized via a spring connection between soil and pile. Finally, a study is conducted and the effect of the slip on the noise emission is studied in detail for the first time.

1 MODEL DESCRIPTION

The model consists of the pile and the soil-fluid domain, coupled via interface conditions as shown in Figure 1. The problem is solved fully in the frequency domain. The pile is modelled as a cylindrical shell described by Flügge's shell theory (Leissa 1973) with a load applied on the top boundary. The governing equations for the pile read:

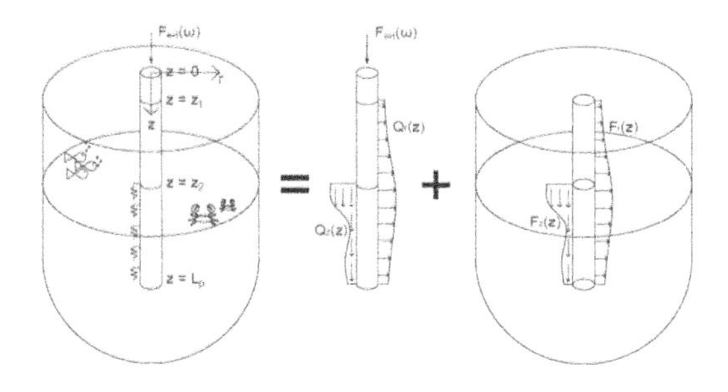

Figure 1. Model schematization and the decoupling in substructures.

$$\boldsymbol{L}_{\mathrm{p}}\boldsymbol{u}_{\mathrm{p}}(z)-\rho_{\mathrm{p}}t_{\mathrm{p}}\omega^{2}\boldsymbol{u}_{\mathrm{p}}(z)=\boldsymbol{Q}_{\mathrm{p}}(z)+F_{\mathrm{ext}}\delta(z)\hat{e}_{z} \quad (1)$$

The fluid and soil domains are described by the acoustic and the elastic wave equations, respectively, i.e.

$$\left(\nabla^{2}+\frac{\omega^{2}}{c_{\mathrm{f}}^{2}}\right)\phi_{\mathrm{f}}(r,z)=-S_{\mathrm{f}}(z)\delta(r-r_{\mathrm{p}}) \quad (2)$$

$$\begin{aligned}(\lambda_{\mathrm{s}}+2\mu_{\mathrm{s}})\nabla\nabla\cdot\mathbf{u}_{\mathrm{s}}(r,z)-\mu_{\mathrm{s}}\nabla\times\nabla\times\mathbf{u}_{\mathrm{s}}(r,z)\\-\rho_{\mathrm{s}}\omega^{2}\mathbf{u}_{\mathrm{s}}(r,z)\\=F_{\mathrm{s}}(z)\delta(r-r_{\mathrm{p}})\end{aligned} \quad (3)$$

The displacement field of the fluid and the pressure are described in terms of the fluid potential as: $\mathbf{u}_{\mathrm{f}}(r,z)=\nabla\phi_{\mathrm{f}}(r,z)$ $p_{\mathrm{f}}(r,z)=\rho_{\mathrm{f}}\omega^{2}\phi_{\mathrm{f}}(r,z)$, respectively. The interface conditions coupling both pile and soil-fluid substructures are given as:

$$\begin{aligned}&u_{\mathrm{pr}(z)}=u_{\mathrm{fr}(r_{\mathrm{p}},z)},\\&Q_{\mathrm{pr}}(z)+F_{\mathrm{fr}}(z)=0,\quad z_{1}<z<z_{2}\end{aligned} \quad (4)$$

$$\begin{aligned}&u_{\mathrm{pr}}(z)=u_{\mathrm{sr}}(r_{\mathrm{p}},z),\\&Q_{\mathrm{pr}}(z)+F_{\mathrm{fr}}(z)=0,\\&Q_{\mathrm{pr}}(z)=\tilde{k}_{Fr}(u_{\mathrm{pr}}(z)-u_{\mathrm{sz}}(r_{\mathrm{p}},z)),\\&Q_{\mathrm{pr}}(z)+F_{\mathrm{fr}}(z)=0,\quad z_{2}<z<L_{\mathrm{p}}\end{aligned} \quad (5)$$

DOI: 10.1201/9781003348450-15

where $\tilde{k}_{Fr}$ is the dynamic linearized friction stiffness of the spring connecting the pile and soil displacements along the vertical coordinate. The limit cases of $\tilde{k}_{Fr} \to \infty$ and $\tilde{k}_{Fr} \to 0$ correspond to the cases of the fully kinematic constraint (pile and soil are not allowed to slip) and of no vertical constraint (no reaction force to the pile in the vertical direction). These two cases represent two extrema of the real case. The response of the fluid-soil domain is described by Green's theorem.

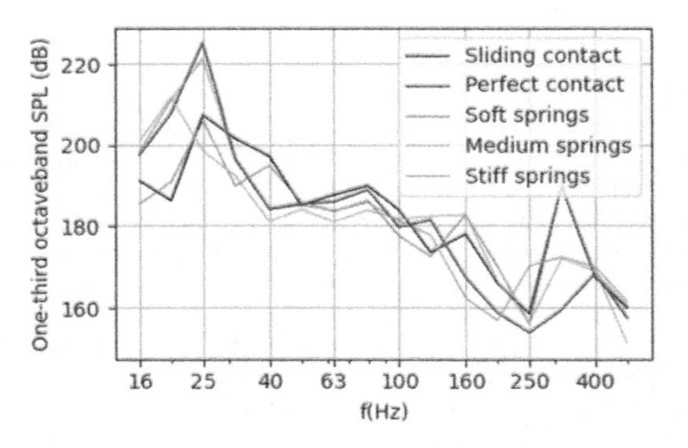

Figure 2. Sound pressure levels 2 meter above the seabed.

2 RESULTS

To evaluate the impact of the pile-soil interface conditions on the noise emission a case study is set up. The case study considers a pile of 76,9 meter length and 8 meter diameter. An initial penetration depth of 20 meters is considered and a fluid depth of 40 meters as starting points for the analysis. The Fourier amplitude spectrum of the vibratory hammer force has a main driving frequency at 22 Hz but super harmonics are also clearly present.

Figure 2 shows the SPL computed for the various configurations at 2 meter above the seabed. It is clear that the way the pile-soil interface is modelled has a strong influence on the sound levels. Moreover, the influence of the interface conditions shows a strong frequency dependence.

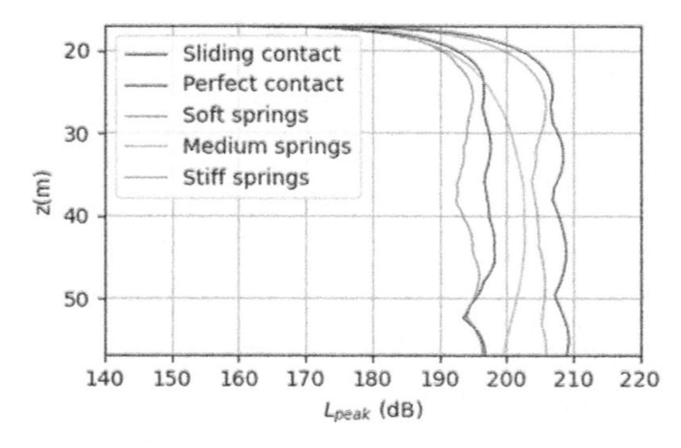

Figure 3. L_{peak} levels over the depth of the water column at r=10 meter for varying interface conditions.

The peak pressure levels in Figure 3 show that in a case in which the pile cannot move relative to the soil (perfect contact), theL_{peak} is the highest. Varying spring stiffness results in peak pressure levels in between both ultimate cases except for the least stiff friction springs. The variation in the L_{peak} values are more than 10 dB.

3 CONCLUSIONS

This study shows that the pile-soil interface conditions do have a significant impact on the noise levels generated. A linearized frictional model has been developed which introduces a linear pile-soil slip condition and allows thus the systematic study of the effect of this slip on the predicted noise underwater. We conclude that the sound levels are strongly influenced by the assumed interface conditions, i.e. stiff or soft pile-soil connection. The mechanism that contributes to the observed differences is primarily the alteration of the vibrations of the pile itself which results in different energy transfers into the elastic and acoustic waves. Differences of about 20 dB in the one-third octave band SPL are computed at some frequencies which justifies the need to examine this more carefully in the future.

It seems that models which assume a perfect pile-soil contact (no sliding), most likely, overestimate the peak pressure levels, whereas models that do neglect the shear-rigidity of the seabed underestimate this. There is a significant difference in peak pressure levels in this study, i.e. order of 10 dB in L_{peak} at 10-meter from the pile centre.

There are two main reasons why the noise levels vary based on the interface conditions. First, models that do assume perfect contact between pile and soil tend to overestimate the energy transfer into the elastic waves in the soil and in particular the Scholte and shear waves which in turn can attribute to the so-called second noise path. Second, the interface conditions strongly influence the vertical resistance of the pile and therefore the vibrations of the pile, i.e. the acoustic source in the examined cases. The contribution of both physical mechanisms and their interrelation, together with the influence of the pile dimensions, penetration depth, soil properties, etc. needs further research and validation with measured data.

ACKNOWLEDGEMENTS

This research is associated with the GDP project in the framework of the GROW joint research program. Funding from "Topsector Energiesubsidie van het Ministerie van Economische Zaken" under grant number TE-HE117100 and financial/technical support from the following partners is gratefully acknowledged: Royal Boskalis Westminster N.V., CAPE Holland B.V., Deltares, Delft Offshore Turbine B.V., Delft University of Technology, ECN, EnecoWind B.V., IHC IQIP B.V., SHL Offshore Contractors B.V., Shell Global Solutions International B.V., Sif Netherlands B.V., TNO, and Van Oord OffshoreWind Projects B.V.

Current Perspectives and New Directions in Mechanics, Modelling and Design of Structural Systems – Zingoni (ed.)
© 2022 Copyright the Author(s), ISBN 978-1-032-39114-4

Dynamic analysis of highly symmetric cable dome structures

Linzi Fan
School of Civil Engineering, Sanjiang University, Nanjing, China

Yao Chen*, Chenhao Lu & Yue Sun
School of Civil Engineering, Southeast University, Nanjing, China

ABSTRACT: Cable dome structures belong to a new type of efficient and robust prestressed cable-strut struc-tures. Similar to crystals, a highly symmetric cable dome is the regular special arrangement of the tension cables, compression struts, and pin-joints. Recent investigation on such kind of regular and symmetric structures is extensive. However, there is limited dynamic research on cable dome structures. Here, frequency analysis is firstly carried out and compared by Abaqus between highly symmetric dome structures with Geiger type and those with Levy type. Further, the influencing factors of initial prestress and loading states are studied. Subsequently, time history analysis is performed in different directions to study dynamic behavior of cable dome structures. The traveling wave effect and changes of cross-sectional areas of all the members are considered. The results show that dynamic behavior of Levy dome structure is superior to that of Geiger dome structure, whereas the natural frequencies are relatively low and intensively distributed. The increase of initial prestress and cross-sectional areas of the members is conducive to improvement of structural stiffness. Moreover, the stiffness will be weakened by enlargement of loads. Levy dome structure is sensitive to traveling wave effect.

1 INTRODUCTION

In recent years, many scholars have focused on morphological analysis (Chen et al., 2018; Koohestani, 2017; Zhang et al., 2014), stiffness analysis (Chen et al., 2020; Qiu et al., 2013; Sultan, 2013; Zhang & Feng, 2017), and optimization design (Chen et al., 2021; Feng, 2017; Quagliaroli et al., 2015; Xu et al., 2018) on different cable dome structures. However, there is limited research on the dynamic analysis of cable domes. In fact, dynamic analysis plays an important role in understanding and evaluating the ultimate bearing capacity, failure mode and seismic performance of these prestressed cable-strut structures (Chen et al., 2019; Fan et al., 2020).

In this study, the nonlinear finite element method using Abaqus is utilized to carry out frequency analysis on symmetric cable domes. Besides, the existing El-centro waves would be introduced to perform multi-dimensional seismic response of these cable domes. To further consider the influence of initial prestress level, load state, traveling wave effect and cross-sectional areas of the members on the dynamic behavior of these structures, detailed time history analysis on cable domes with different parameters is presented.

2 NUMERICAL MODEL OF CABLE DOME STRUCTURES

Numerical models are based on Geiger and Levy cable domes with inner pull ring, as shown in Figure 1.

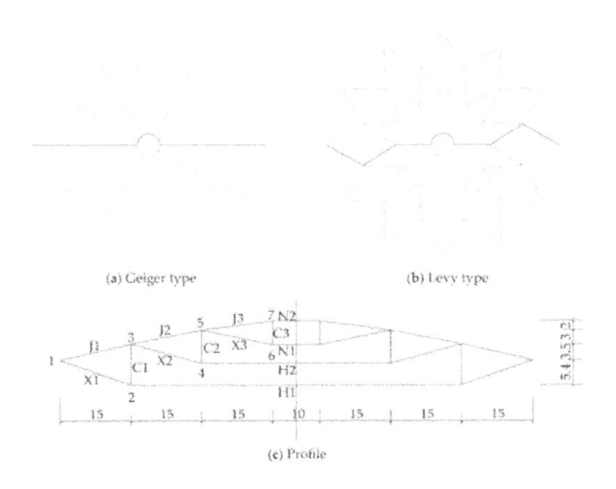

Figure 1. Two basic cable dome models.

*Corresponding author: chenyao@seu.edu.cn

DOI: 10.1201/9781003348450-16

3 NATURAL FREQUENCY ANALYSIS ON CABLE DOMES

The first 100 natural frequencies of the cable dome structures are extracted by the Subspace method. The results have been shown in Figures 2.

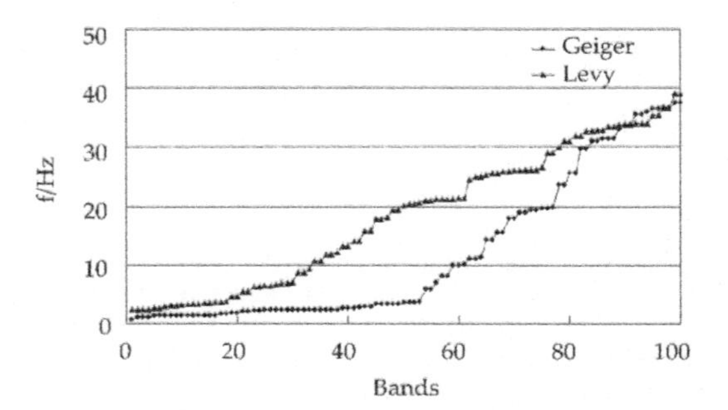

Figure 2. First 100 modes of natural frequencies of two types of cable domes.

4 TIME HISTORY ANALYSIS

We introduce the classical El-centro wave (1940) within 20*s*, and the time interval is 0.02*s*. The peak acceleration along the horizontal direction is 0.3*g*, and that of the vertical direction is 0.15*g*. Both the structural self-weight and the dead load of $0.35kN/m^2$ are considered during time history analysis. Figure 3 shows the calculated dynamic displacements of reference node 7.

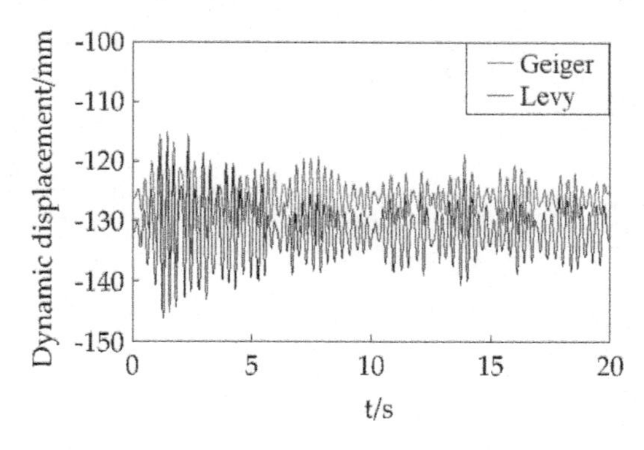

Figure 3. Dynamic displacements of node 7 along the vertical direction.

5 CONCLUSIONS

After calculating the integral self-stress states and considering the contributions of the initial prestresses, we have performed nonlinear frequency analysis and time history analysis on two kinds of symmetric cable dome structures. The results can be mainly concluded as follows:

1. Cable dome structures are flexible and prestressable. The Geiger cable dome structure exhibits mainly torsional vibrations, while the Levy cable dome structure shows mainly vertical vibrations. The structural stiffness of the former dome is more significantly affected by the prestress level and external load state.
2. Tension cables for the two kinds of cable dome structures work effectively under earthquake actions. Levy cable dome is more affected by the traveling wave effect. Those members parallel to the wave propagation direction is very sensitive to the traveling wave.

Admittedly, the span of the cable domes and the heights of the vertical struts also have great influence on the static and dynamic behavior of the cable dome structures. Besides, the dynamic performance of the structures with the interaction of the cables and the membranes needs to be further studied in future work.

REFERENCES

Chen, Y., Sun, Q., & Feng, J. 2018. Improved form-finding of tensegrity structures using blocks of symmetry-adapted force density matrix. *Journal of Structural Engineering*, *144* (10), 04018174.

Chen, Y., Yan, J., & Feng, J. 2020. Nonlinear form-finding of symmetric cable–strut structures using stiffness submatrices associated with full symmetry subspace. *Archive of Applied Mechanics*, *90*(8), 1783–1794.

Chen, Y., Yan, J., Feng, J., & Sareh, P. 2021. Particle swarm optimization-based metaheuristic design generation of non-trivial flat-foldable origami tessellations with degree-4 vertices. *Journal of Mechanical Design*, *143*(1), 011703.

Chen, Y., Yan, J., Sareh, P., & Feng, J. 2019. Nodal flexibility and kinematic indeterminacy analyses of symmetric tensegrity structures using orbits of nodes. *International Journal of Mechanical Sciences*, *155*, 41–49.

Fan, L., Sun, Y., Fan, W., Chen, Y., & Feng, J. 2020. Determination of active members and zero-stress states for symmetric prestressed cable–strut structures. *Acta Mechanica*, *231*(9), 3607–3620.

Feng, X. 2017. The optimal initial self-stress design for tensegrity grid structures. *Computers & Structures*, *193*, 21–30.

Koohestani, K. 2017. On the analytical form-finding of tensegrities. *Composite Structures*, *166*, 114–119.

Qiu, C., Aminzadeh, V., & Dai, J. S. 2013. Kinematic analysis and stiffness validation of origami cartons. *Journal of Mechanical Design, 135*(11).

Quagliaroli, M., Malerba, P. G., Albertin, A., & Pollini, N. 2015. The role of prestress and its optimization in cable domes design. *Computers & Structures*, *161*, 17–30.

Sultan, C. 2013. Stiffness formulations and necessary and sufficient conditions for exponential stability of prestressable structures. *International Journal of Solids and Structures, 50*(14-15), 2180–2195.

Xu, X., Wang, Y., & Luo, Y. 2018. Finding member connectivities and nodal positions of tensegrity structures based on force density method and mixed integer nonlinear programming. *Engineering structures*, *166*, 240–250.

Zhang, L.Y., Li, Y., Cao, Y.P., & Feng, X.Q. 2014. Stiffness matrix based form-finding method of tensegrity structures. *Engineering structures*, *58*, 36–48.

Zhang, P., & Feng, J. 2017. Initial prestress design and optimization of tensegrity systems based on symmetry and stiffness. *International Journal of Solids and Structures*, *106*, 68–90.

Current Perspectives and New Directions in Mechanics, Modelling and Design of Structural Systems – Zingoni (ed.)
© 2022 Copyright the Author(s), ISBN 978-1-032-39114-4

Vibroacoustic performance of multi-storey buildings: A comparison of lightweight and heavy structures in the early design phase

L.V. Andersen, L. Mangliar & M.M. Hudert
Department of Civil and Architectural Engineering, Aarhus University, Aarhus, Denmark

O. Flodén & P. Persson
Department of Construction Sciences, Lund University, Sweden

ABSTRACT: With the construction sector being responsible for about 40% of the energy and material use, the sector has a great responsibility for lowering the consumption if we are to succeed in our global pursuit for the green transition. However, buildings must still comply with the demands of users. For large-span, open-space lightweight, multi-storey buildings, there is a potential risk related to annoyance caused by vibrations and structure borne noise. This paper addresses the manifold effects of building typology in terms of the vibroacoustic performance and the environmental impact. Using an automized digital framework, multistorey buildings made of different materials, including cross-laminated timber and concrete, are modelled and compared. Finite-element analysis is used for the dynamic structural analysis. The design software Rhino, together with Grasshopper, is used for material take-offs and visualisation. The aim of the paper is to provide insight into the pros and cons of different building typologies, in order to support informed decision making in the early stages of design.

1 INTRODUCTION

The typology of the loadbearing structure has a great impact on the performance of a building and should therefore be determined already in the early phase of the design workflow. Lightweight structures have little embodied energy. However, compared to well-known, heavy structures, i.e., reinforced concrete, relatively little is known about their vibroacoustic performance. Our aim is to help to find the optimum in the design decision, balancing between the pros and cons of lightweight structures.

Studies on finding the right balance between the structural performance and environmental impact, with a focus on CLT and concrete buildings, were recently carried out by Hassan et al. (2019). Moreover, hybrid CLT/concrete slabs have gained attention within the last few years. The dynamic response of such panels was examined by Barbosa et al. (2021).

In this study, we compare three versions of a multi-storey building structure, with the following typologies: cross-laminated timber (CLT) wall panels and floor slabs, timber–concrete composite (TCC) structures using CLT floor panels with a concrete top layer and CLT walls, and reinforced concrete (RC) wall panels and hollow-core concrete slab (HCS) floors. We discuss the multilateral effects of the choice of the building structure, highlighting the importance of considering the vibroacoustic performance as well as the environmental impact of the loadbearing structures already in the early design stage.

2 METHODOLOGY

The companion paper "Early assessment of the vibroacoustic performance of large lightweight buildings" describes the computational framework applied for the present analyses. Based on a finite-element (FE) approach, substructures were generated for the various floor and wall panels defined in Section 3. These were then used to construct a global model of a multi-storey building, and the steady-state response to time-harmonic loading was determined in the frequency range relevant for structure-borne noise. Based on 5×5 uniformly distributed indicator nodes, the root-mean-square (RMS) value of the accelerations normal to a panel due to point-force excitation was computed. The results were achieved for the one-third octave bands with centre frequencies in the range 25–250 Hz, and the RMS acceleration levels were converted to dB relative to a reference acceleration of $1/s^2$. This has then been used to compare the vibration performance of the three versions.

In parallel to this, a digital model of each building has been constructed in the graphical architectural design environment Rhino/Grasshopper. Based on material quantity take-offs from this model, the global warming potential has then been assessed. The Grasshopper plugin Bombyx (Basic, Hollberg, Galimshina, & Habert, 2019) and the cloud-based tool CAALA (CAALA GmbH, 2018) were employed, and the estimated environmental impact of the building versions compared.

3 BUILDING MODELS AND RESULTS

Overall, the CLT, TCC/CLT, and HCS/RC buildings share the geometry illustrated in Figure 1. The spans in the length, height, and width directions are 4.8 m, 3.6 m, and 6.0 m, respectively. The material properties and cross-sections, and the dynamic properties of the individual panels, can be found in the full paper.

DOI: 10.1201/9781003348450-17

Table 1 shows the material use and CO_2-e emission of the three buildings, while Figure 2 shows the mode count for the buildings in the frequency range up to 450 Hz. The HCS/RC building has about half the modal density of the other buildings for frequencies above 100 Hz. Finally, Table 2 shows the RMS acceleration for the loads indicated in Figure 1 and for selected combinations of panels and 1/3 octave bands.

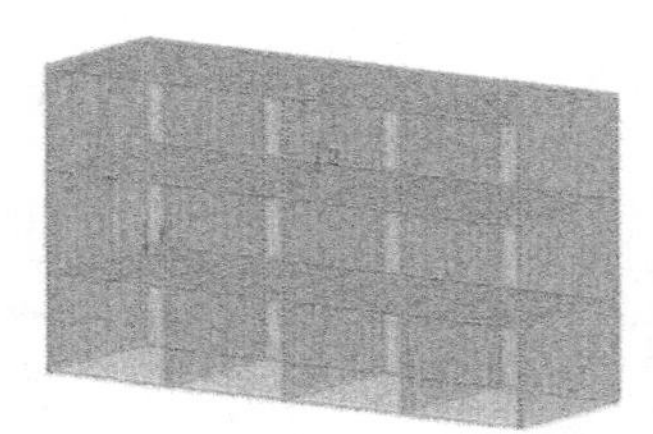

Figure 1. Overall layout of the buildings constructed in Abaqus (*SIMULIA Abaqus version 2021*, 2021). The red arrows and numbers indicate the applied loads.

Table 1. Materials and global warming potential of the buildings.

Quantity	CLT	TCC/CLT	HCS/RC
t wood	78,1	78,1	0
t concrete	0	42,5	353
t CO_2-e (Bombyx)	–0,15	–0,15	1,67
t CO_2-e (CAALA)	–0,184	–0,175	1,9

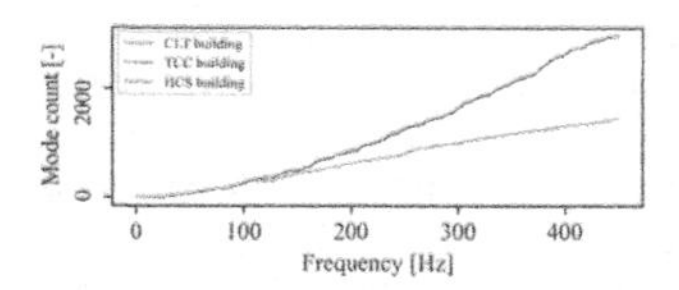

Figure 2. Mode counts for the three buildings.

4 DISCUSSION AND CONCLUSION

Based on the comparison of the three studied building models, the following main points can be stated:

1. The CLT building has a significantly lower global warming potential than the HCS/RC building.
2. The RMS acceleration levels of the floors and walls in the CLT building, caused by the loads applied on two selected floors, are about 10–35 dB higher than for the HCS/RC building, depending on the frequency.
3. For the walls, the differences are generally smaller for the considered building models.

While the conclusions may not be truly general, the study clearly indicates shortcomings of a lightweight structure, in this case a CLT building, compared to

Table 2. RMS acceleration levels (in dB) of selected panels and one-third octave bands. Shaded cells mark the loaded floors.

		Load 1 (Room 1-2-1)			Load 2 (Room 2-3-1)		
Panel	Band (Hz)	CLT	TCC/ CLT	HCS/ RC	CLT	TCC/ CLT	HCS/ RC
Floor	50	74.1	68.1	49.1	56.9	54.9	27.7
1-2-1	100	83.1	78.9	62.0	64.5	60.2	52.9
	200	89.0	82.0	70.4	67.7	63.6	51.2
Floor	50	56.9	53.8	21.6	74.9	68.1	53.4
2-3-1	100	63.0	58.5	48.0	81.4	77.3	63.4
	200	68.8	64.0	54.2	89.1	82.8	68.9
Wall[1]	50	54.1	42.3	31.1	51.2	44.9	29.1
3-1-2	100	51.4	50.1	41.4	51.1	49.7	46.9
	200	50.6	50.1	47.6	51.5	51.8	48.7
Wall[2]	50	56.7	57.8	21.4	55.8	52.1	28.1
1-3-1	100	66.3	61.0	44.5	65.1	57.4	46.9
	200	60.7	58.5	57.2	61.2	55.4	53.2

[1]Wall oriented in the longitudinal direction of the building.
[2]Wall oriented in the transverse direction of the building.

a traditional heavy concrete structure. Also, adding a concrete layer on top of the CLT floor panels only provides an insertion loss of about 0–5 dB compared to the pure CLT building.

ACKNOWLEDGMENTS

The research was carried out in the framework of the project "Reconcile". The authors thank the European Regional Development Fund for the financial support provided via the Interreg V programme. Further, the authors would like to thank former MSc students Mette S. F. Bak and Jonatan Vethanayagam for their contributions to the panel model development.

REFERENCES

Barbosa, A. R., Rodrigues, L. G., Sinha, A., Higgins, C., Zimmerman, R. B., Breneman, S., Pei, S., van de Lindt, J. W., Berman, J., & McDonnell, E. 2021. Shake-Table Experimental Testing and Performance of Topped and Untopped Cross-Laminated Timber Diaphragms. *Journal of Structural Engineering.*

Basic, S., Hollberg, A., Galimshina, A., & Habert, G. 2019. A design integrated parametric tool for real-time Life Cycle Assessment – Bombyx project. *IOP Conference Series: Earth and Environmental Science*, 323(1): 1–10.

CAALA GmbH. 2018. Computer-Aided Architectural Life cycle Assessment (CAALA). Retrieved April 6, 2022, from https://caala.de/

Hassan, O. A. B., Öberg, F., & Gezelius, E. 2019. Cross-laminated timber flooring and concrete slab flooring: A comparative study of structural design, economic and environmental consequences. *Journal of Building Engineering.*

SIMULIA Abaqus version 2021. 2021. Paris, France: Dassault Systemes.

Current Perspectives and New Directions in Mechanics, Modelling and Design of Structural Systems – Zingoni (ed.)
© 2022 Copyright the Author(s), *ISBN 978-1-032-39114-4*

Effect of modelling the air in rooms on the prediction of vibration transmission in cross-laminated timber buildings

O. Flodén, A. Peplow & P. Persson
Department of Construction Sciences, Lund University, Sweden

L. Mangliár & L.V. Andersen
Department of Civil and Architectural Engineering, Aarhus University, Denmark

ABSTRACT: In this paper, the necessity of modelling the enclosed air in room volumes of buildings made of cross-laminated timber, with respect to the prediction of vibration transmission, is investigated. Coupled structural-acoustic finite element analysis is employed for an example building structure. It is found that modelling the air in the receiver room, i.e. in the room where the vibration response is evaluated, has a marked effect on the vibration transmission between panels.

1 INTRODUCTION

The development of new and improved technical solutions for mitigating noise and vibrations in buildings of cross-laminated timber (CLT) will facilitate increased competitiveness for the buildings. Accurate numerical models are considered to be of great value for the development; the models should take into account all physical phenomena that are relevant for achieving adequate accuracy in predictions while not being overly detailed so that the computational efficiency is impaired. In this paper, finite element (FE) modelling is employed to investigate the effect of an enclosed acoustic medium in rooms within CLT buildings on the vibration transmission. In previous work by Andersen et al. (2012), the authors studied the modelling of air in room volumes of mass-timber buildings using 2D FE-modelling, and concluded that the acoustic media in room volumes has a small effect on the vibration transmission. Flodén et al. (2015) then studied the modelling of acoustic media in thin cavities located between floor and ceiling structures in modular-based lightweight timber buildings using 3D FE-modelling, and concluded that the acoustic media has appreciable effect on the transmission of vibrations between adjacent storeys.

The present paper utilizes 3D FE-modelling of CLT buildings to study the effect of modelling the enclosed air in the receiver room, i.e. the room where the vibration field is evaluated, on the vibration transmission between adjacent stories. An example building structure representing part of a CLT building is employed to compare between the two cases of including or excluding the air in the receiver room of the model. Frequencies in the one-third octave bands with centre frequencies between 20 and 250 Hz are considered.

2 EXAMPLE CASE

As example case in the paper, a CLT building structure representing part of a multi-storey building is used; see Figure 1 for an illustration (in the figure, the two walls closest to the observer are hidden at each storey). The building structure includes three storeys, the middle room being the receiver room for the vibration transmission. The floor panels have dimensions 7000 mm × 4000 mm and the walls are 3000 mm high. The panels consist of five layers of laminae that are 40 and 20 mm thick, respectively, for the floor and wall panels. The outmost layers are oriented parallel to the short side of the panels.

Only the bare CLT panels are included. Adjacent storeys are separated by one single panel, i.e. the floor panels are also the ceiling panels of the storey below. Elastomer strips (100 mm wide and 12 mm thick) are placed between adjacent storeys so that the strips separate the walls of one storey from the floor and walls of the storey above.

A coupled structural-acoustic FE model of the example structure was established. The CLT panels were modelled using shell finite elements, assuming orthotropic material with rectilinear orientation for the layers of timber boards. Full interaction between individual boards is assumed. The elastomer strips were modelled using frequency-dependent spring–dashpot elements. The connections wall–wall and floor–wall within each storey was modelled by assuming full interaction for translational degrees of freedom (dofs) while introducing springs for the rotational dofs. The spring stiffness was chosen to represent the midpoint between clamped and pinned conditions. The enclosed air in the receiver room was meshed with

DOI: 10.1201/9781003348450-18

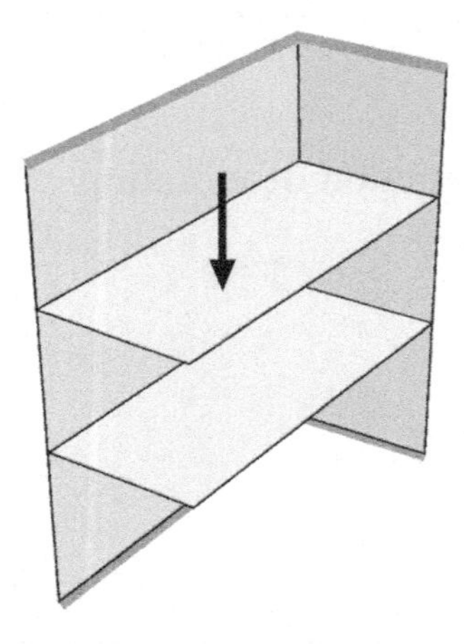

Figure 1. The example case considered in the paper, comprising three storeys of a CLT building. The positions of the load (black arrow at the topmost floor) and the displacement boundary conditions (grey thick lines at the topmost and the bottommost storeys) are indicated. The two walls closest to the observer are hidden at each storey, i.e. there are in total four walls at each storey.

acoustic finite elements. Structural damping and volumetric acoustic damping was employed; the exception being the elastomer strips.

As shown in Figure 3, displacement boundary conditions were applied at the elastomer strips located above the topmost walls and below the bottommost walls; the spring–dashpot elements were constrained in all six dofs at their outer nodes. The dynamic loading was modelled as a unit amplitude harmonic point load located at the floor of the topmost storey as shown in Figure 3. The load was placed at an offset of 600 mm and 1100 mm in the width and length directions, respectively, from the centre of the floor panel. Direct

Table 2. Results for the example case in terms of relative difference in vibration response (given in percent) when including the air in the receiver room, as compared to excluding the air.

One-third octave band centre freq.	Rel. diff. in vib. resp. (%)		
	Ceiling	Floor	Walls
20 Hz	–0.88	74	115
25 Hz	–8.2	47	64
31.5 Hz	–8.6	109	107
40 Hz	–0.68	104	112
50 Hz	–1.1	51	78
63 Hz	–4.3	–5.7	38
80 Hz	–2.1	–3.5	23
100 Hz	–6.3	19.4	23
125 Hz	–3.6	–4.2	46
160 Hz	–4.6	–2.3	57
200 Hz	–3.6	–6.9	81
250 Hz	–4.9	–5.8	90

steady-state analyses were performed with a step size of 0.5 Hz. The vibration response in the receiver room was evaluated using the vibration response metric defined by Flodén & Persson (2021); a grid of 5×5 evaluation points was used at each of the six panels in the receiver room.

3 EFFECT OF MODELLING THE AIR IN THE RECEIVER ROOM

Table 2 presents the relative difference in vibration response when including the air in the receiver room as compared to the reference case of excluding the air.

4 CONCLUSIONS

Based on the presented results, the following conclusions can be made:

- The effect of modelling the air in the receiver room on the vibration levels in the ceiling panel is small.
- The vibration transmission from the ceiling panel to the floor and wall panels is affected substantially by modelling the enclosed air; the vibration levels are increased by a factor of two in some of the evaluated one-third octave bands.

It should be noted that the absolute vibration levels are approximately an order of magnitude higher in the ceiling than in the walls and the floor. The conclusions imply that it may be important to consider the effect of air in the receiver room of CLT buildings when evaluating the vibration transmission between panels.

ACKNOWLEDGEMENTS

The European Regional Development Fund via the Interreg V project Reconcile; the Swedish Innovation Agency Vinnova (grant number 2018-04159); Södra Foundation of Research, Development and Education; and the J. Gust. Richert foundation are gratefully acknowledged for financial support.

REFERENCES

Andersen, L.V., Kirkegaard, P.H., Dickow, K.A., Kiel, N. & Persson, K. 2012. Influence of wall surface and air modelling in finite-element analysis of sound transmission between rooms in lightweight buildings. In *Proceedings of the Internoise 2012/ASME NCAD Meeting, New York City*, August 19–22, *Vol. 45325, pp.* 393–404.

Flodén, O., Negreira, J., Persson, K. & Sandberg, G. 2015. The effect of modelling acoustic media in cavities of lightweight buildings on the transmission of structural vibrations. *Engineering Structures* 83: 27–41.

Flodén, O. & Persson, P. 2021. Robust prediction metrics for structure-borne noise in timber buildings. In *INTER-NOISE and NOISE-CON Congress and Conference Proceedings, Washington, DC*, August 1–5, *Vol. 263(4), pp.* 2348–2359.

Current Perspectives and New Directions in Mechanics, Modelling and Design of Structural Systems – Zingoni (ed.)
© 2022 Copyright the Author(s), ISBN 978-1-032-39114-4

Numerical analysis of transversely isotropic unsaturated railway ground vibrations under a moving load using 2.5D FEM

Shaoyi Li
School of Civil Engineering and Architecture, Jishou University, Zhangjiajie Hunan, P.R. China

ABSTRACT: The dynamic responses of unsaturated transversely isotropic railway ground under a moving load were studied by 2.5D finite element method (FEM). The unsaturated ground was modeled by unsaturated poroviscoelastic medium with consideration of the transversely isotropic constitutive relationship. A track model laying on the ground surface including the rail, sleeper and ballast was introduced to build the coupled track-ground equations. A moving load was applied on the track to study the dynamic responses of the railway ground for different load speeds, soil saturation and ratios of horizontal and vertical elastic modulus. This research found that: the horizontal shear wave velocity has large influences on the vibration resonance of the transversely isotropic unsaturated ground.

1 INTRODUCTION

The ground vibrations are important problems for the transportation engineering. The mechanical properties of ground for the horizontal and vertical directions are different by the geological sedimentation. It is more accurate to analyze the layered ground vibration considering the transversely isotropic characteristics. Ba et al. (2019) used semi-analytical method to study the transversely isotropic railway ground vibrations. The soil actually contains solid, fluid and gas phases and is an unsaturated medium. The results of unsaturated railway ground vibration analysis would be more instructive for practical engineering problems. Lu et al. (2018) used the analytical solution of unsaturated poroviscoelastic medium to analyze the unsaturated ground vibrations under moving load. Ai & Ye (2021) used analytical method to studied the dynamic responses of unsaturated transversely isotropic ground.

The dynamic responses of the unsaturated transversely isotropic railway ground are less discussed. In this paper, the governing equations for the unsaturated transversely isotropic ground are derived. The 2.5D FE method, which uses the Fourier transform for the longitudinal coordinate and time variables to solve the 3D problem using 2D model in the wavenumber and frequency domains, is applied to derive the weak form governing equations of the unsaturated medium. By coupling a track model, the numerical model for unsaturated railway ground is built to analyze the influences of load speeds, soil saturation and transversely isotropic ratio on the dynamic responses of the unsaturated ground.

2 GOVERNING EQUATIONS

Considering the transversely isotropy for the medium, the effective stress theory can be formulated as:

$$\sigma_{ij} = \mathbf{C}\varepsilon_{ij} - \delta_{ij}\alpha_i p \qquad (1)$$

where $\mathbf{C}$ is the tensor for the constitutive relationship of transversely isotropic medium, that

$$\mathrm{C} = \begin{bmatrix} c_{11} & c_{13} & c_{12} & 0 & 0 & 0 \\ c_{13} & c_{33} & c_{13} & 0 & 0 & 0 \\ c_{12} & c_{13} & c_{11} & 0 & 0 & 0 \\ 0 & 0 & 0 & c_{44} & 0 & 0 \\ 0 & 0 & 0 & 0 & c_{44} & 0 \\ 0 & 0 & 0 & 0 & 0 & c_{66} \end{bmatrix}; \text{ in which}$$

$c_{11}\sim c_{66}$ are the coefficients, and can be expressed using the engineering elastic parameters as:

$c_{11} = \frac{E_h(1-(E_h/E_v)v_{vh}^2}{(1+v_{hh})\Delta}$, $c_{33} = \frac{E_v(1-v_{hh})}{\Delta}$,

$c_{12} = \frac{E_h(v_{hh}+(E_h/E_v)v_{vh}^2}{(1+v_{hh})\Delta}$, $c_{13} = \frac{E_h v_{vh}}{\Delta}$, $c_{44} = G_{vh}$, $\Delta = 1 - v_{hh} - 2(E_h/E_v)v_{vh}^2$, $c_{66} = G_{hh} = \frac{1}{2}c_{11} - c_{12}) = \frac{E_h}{2(1+v_{hh})}$; the E_h, E_v, v_{hh}, v_{vh}, G_{vh}, G_{hh} are the horizontal elastic modulus, vertical elastic modulus, horizontal poisson ratio, vertical-horizontal poisson ratio, vertical shear modulus and horizontal shear modulus respectively. The $\alpha_i = 1 - K_{ib}/K_s$ is Biot coefficient, in which K_{ib} is the compressional modulus for the soil skeleton and $K_{1b} = K_{2b} = \frac{c_{11}+c_{12}+c_{13}}{3}$, $K_{3b} = \frac{2c_{13}+c_{33}}{3}$, K_s is the compressional modulus for the grain; p is the pore pressure, that $p = \chi p^w + (1-\chi)p^a$, in which p^w and p^a are the pore water pressure and gas pressure, χ is the effective stress coefficient and assigned equal to the soil saturation S_r; considering the viscosity of the soil η_s, the coefficients for the elastic constitutive relationship are replaced as $c_{ij} = c_{ij}(1 + i\eta_s sign(\omega))$, ω is the circular frequency.

By utilizing the Darcy law for the fluid and gas phases to eliminate the relative displacement vectors for the fluid and gas, and using the mass conservative equations for the solid, fluid and gas phases, the governing equations for the unsaturated poroviscoelastic medium can be obtained regarding the pore pressures for the water and air as:

DOI: 10.1201/9781003348450-19

$$\mathbf{L}^{\mathrm{T}}\overline{\sigma} - \delta_{ij}\alpha[S_r\overline{p}^w + (1 - S_r)\overline{p}^a] + \omega^2[\rho\overline{\mathbf{u}} + \rho_w(A_{11}\overline{\mathbf{u}} + A_{12}\nabla(\overline{p}^w))\rho_a(A_{21}\overline{\mathbf{u}} + A_{22}\nabla(\overline{p}^a))] = 0 \quad (2)$$

$$\nabla\mathbf{B}_{11}\overline{u} + A_{11}\nabla\overline{\mathbf{u}} + A_{12}\nabla^2\overline{p}^w + B_{12}\overline{p}^w + B_{13}\overline{p}^a = 0 \quad (3)$$

$$\nabla\mathbf{B}_{21}\overline{u} + A_{21}\nabla\overline{\mathbf{u}} + A_{22}\nabla^2\overline{p}^a + B_{22}\overline{p}^w + B_{23}\overline{p}^a = 0 \quad (4)$$

where ρ is the soil density, $\rho = (1 - n)\rho_s$, $nS_r\rho_w + n(1 - S_r)\rho_a$,

$\mathbf{B}_{11} = S_r\left(\mathbf{I} - \frac{\mathbf{K}_b}{K_s}\right)$,

$B_{12} = \left[\frac{nS_r}{K_w} - nA_{ss} + \frac{S_r}{K_s}(\alpha - n)S_r\right]$,

$B_{13} = \left[nA_{ss} + \frac{S_r}{K_s}(\alpha - n)(1 - S_r)\right]$,

$\mathbf{B}_{21} = (1 - S_r)\left(\mathbf{I} - \frac{\mathbf{K}_b}{K_s}\right)$,

$B_{22} = \left[nA_{ss} + \frac{1-S_r}{K_s}(\alpha - n)S_r\right]$,

$B_{23} = \left[\frac{n(1-S_r)}{K_a} - nA_{ss} + \frac{1-S_r}{K_s}(\alpha - n)(1 - S_r)\right]$,

$\mathbf{I} = [1, 1, 1]^{\mathbf{T}}$; K_w and K_a are the compressional modulus for the fluid and gas phases; $\alpha = \frac{2\alpha_1+\alpha_3}{3}$, $\mathbf{K}_b = [K_{1b}, K_{3b}, K_{1b}]^{\mathrm{T}}$; A_{ss} is the partial derivative of the soil saturation S_r regarding the matric suction $s = p^a - p^w$; The relationship of the matric suction and the soil saturation can be described by the soil water characteristic curve (SWCC).

The perfectly match layer (PML) is deployed at the edges of the FE model to simulate the infinite half space. The PML uses the coordinate transform to reduce the reflected waves from the boundary, the transformed coordinate can be formulated as:

$$\hat{s} = \int_0^s \lambda_s(s)ds = s + \int_{s_0}^{s_t} \lambda_s(s)ds \quad (5)$$

where s_0 and s_t are the coordinates for the PML; $\lambda_s(s)$ is the complex transform coefficient.

The strain vector ϵ can be expressed as:

$$\epsilon = \mathbf{L}_{11}\mathbf{N}\overline{\mathbf{u}} + \mathbf{L}_{12}\mathbf{N}\frac{\partial\overline{\mathbf{u}}}{\partial z} = \mathbf{B}_1\overline{\mathbf{u}} + \mathbf{B}_2\frac{\partial\overline{\mathbf{u}}}{\partial z} \quad (6)$$

where $\mathbf{L}_{11}^{\mathrm{T}} = \begin{bmatrix} \frac{1}{\lambda_x}\frac{\partial}{\partial x} & 0 & 0 & \frac{1}{\lambda_y}\frac{\partial}{\partial y} & 0 & 0 \\ 0 & \frac{1}{\lambda_y}\frac{\partial}{\partial y} & 0 & \frac{1}{\lambda_x}\frac{\partial}{\partial x} & 0 & 0 \\ 0 & 0 & 0 & 0 & \frac{1}{\lambda_y}\frac{\partial}{\partial y} & \frac{1}{\lambda_x}\frac{\partial}{\partial x} \end{bmatrix}$,

$\mathbf{L}_{12}^{\mathrm{T}} = \begin{bmatrix} 0 & 0 & 0 & 0 & 0 & 1 \\ 0 & 0 & 0 & 0 & 1 & 0 \\ 0 & 0 & 1 & 0 & 0 & 0 \end{bmatrix}$, $\mathbf{N}$ is the shape function.

The differential operator $\nabla = [\frac{1}{\lambda_x}\frac{\partial}{\partial x}, \frac{1}{\lambda_y}\frac{\partial}{\partial y}, \frac{\partial}{\partial z}]$ can be rewritten as$\nabla = \mathbf{L}_{21} + \mathbf{L}_{22}\frac{\partial}{\partial z}$, in which $\mathbf{L}_{21} = [\frac{1}{\lambda_x}\frac{\partial}{\partial x}, \frac{1}{\lambda_y}\frac{\partial}{\partial y}, 0]$ and $\mathbf{L}_{22} = [0, 0, 1]$. Considering the shape functions, and applying the Fourier transform regarding the longitudinal coordinate z, the weak form governing equations in the wavenumber domain can be obtained as:

$$\begin{bmatrix} K_{11} & Q_{12} & Q_{13} \\ K_{21} & Q_{22} & Q_{23} \\ K_{31} & Q_{32} & Q_{33} \end{bmatrix} \begin{bmatrix} \tilde{\overline{\mathbf{u}}} \\ \tilde{\overline{p}}^w \\ \tilde{\overline{p}}^a \end{bmatrix} = \begin{bmatrix} \tilde{\overline{\mathrm{f}}} \\ 0 \\ 0 \end{bmatrix} \quad (7)$$

3 NUMERICAL ANALYSIS

When the S_r=0.5 and the load speed increases from 90m/s to 120m/s, the ground displacement is increased. With the v=120m/s and a modulus ratio of 0.5, the ground vibration is large and an oscillation occurs after the moving load. When the modulus ratio is 0.5, the vertical shear velocity of the ground is qSV=140m/ss and the horizontal one is qSH=117m/s. The load speed 120m/s approaches the qSH, which causes the ground vibration amplification.

When the v=120m/s and S_r=0.7, the decreasing modulus ratio from 2.0 to 1.0 increases the vibration amplitude extensively. The shear wave velocities for the horizontal and vertical direction are qSV=qSH=127m/s for S_r=0.7 and E_h/E_v=1.0, which makes the load speed close the shear wave velocity causing the vibration increment.

When the S_r=0.9, load speed is 120m/s and modulus ratio is 2.0, the shear wave velocities for the vertical and horizontal directions are qSV=115m/s and qSH=155m/s, the velocity of qSH is larger than the load speed, which does not cause resonant vibration; when the modulus ratio is 1.0, the two shear wave velocities are qSV=qSH=115m/s, which are close to the load speed and cause the increasing of vibration amplitude. So, the velocity of the horizontal shear wave qSH has large influences on the vertical ground vibrations.

4 CONCLUSIONS

Several conclusions can be drawn as:

(1) The displacement amplitude of the unsaturated ground is increased by the increasing soil saturation and the reduction of the horizontal elastic modulus.
(2) The horizontal shear wave qSH has large influences on the vertical ground vibrations.
(3) The dominant frequency of the ground vibration is controlled by the soil saturation and elastic modulus ratio.

REFERENCES

Ai, Z. Y., & Ye, Z. (2021). Extended precise integration solution to layered transversely isotropic unsaturated poroelastic media under harmonically dynamic loads. *Engineering Analysis with Boundary Elements, 122*, 21–34.

BA, Zhenning., J, Liang., Lee, VW., et al. (2019). A semi-analytical method for vibrations of a layered transversely isotropic ground-track system due to moving train loads. *Soil Dynamics and Earthquake Engineering, 121*, 25–39.

Lu, Z., Fang, R., Yao, H., et al. (2018). Dynamic responses of unsaturated half-space soil to a moving harmonic rectangular load. *International Journal for Numerical and Analytical Methods in Geomechanics, 42*(9), 1057–1077.

Current Perspectives and New Directions in Mechanics, Modelling and Design of Structural Systems – Zingoni (ed.)
© 2022 Copyright the Author(s), ISBN 978-1-032-39114-4

Incremental dynamic analysis of a hospital building in Oman

M.B. Waris, K. Al-Jabri & W.H. Bhuta
Department of Civil and Architectural Engineering, Sultan Qaboos University, Muscat, Oman

I. El-Hussain
Earthquake Monitoring Center, Sultan Qaboos University, Muscat, Oman

ABSTRACT: This study evaluates the seismic performance of a six-story hospital building in Muscat, Oman using Incremental dynamic analysis. The building is a reinforced concrete flat-slab system with peripheral beams having a height of 25.2 m. The building has footprint of 65 m x 74 m that is divided into four portions through expansion joints with four shear walls. ETABS is used for the numerical modelling of the building considering all possible geometric details. The natural period of the building is estimated as 0.87 sec, which agrees with estimates suggested by the Omani Seismic Design Code. The nonlinear structural behavior is modeled using plastic hinges as per ASCE 41-13. Natural earthquake record matching the elastic response spectrum for the location is selected from the Pacific Earthquake Engineering Research Center (PEER) database. First-mode spectral acceleration is chosen as the intensity measure and the damage is measured using the maximum inter-story drift. 10 simulations are run using the record with varying scales until the collapse is observed for the Incremental dynamic analysis (IDA). Each simulation produced a single discrete Intensity vs Damage point. The building is assessed for Immediate Occupancy and Global Instability limit states. The immediate occupancy-damage control was identified as 0.084 g, which is well above the site-specific design spectral acceleration of 0.067. The results indicate that building satisfies drift limits and immediate occupancy criteria.

1 INTRODUCTION

Incremental Dynamic Analysis (IDA) can adequately describe the performance of the structure in various performance states and its capacity to resist seismic loadings assessed using a single ground acceleration record in multiple incremental nonlinear time-history analyses that can define the capacity curve of a building. The varying earthquake intensity covers a range of linear to nonlinear behavior and finally to the global dynamic instability of structure (FEMA 2012).

The current study used the IDA analysis to evaluate the seismic performance of a hospital building in Oman, to verify its performance under probable seismic activity in the region. The building was modeled using ETABS incorporating all structural details. Spectral acceleration was used as the intensity measure and maximum inter-story drift was used as the damage measure.

2 METHODOLOGY

Figure 1 shows the detailed 3D numerical model of the building generated as per the structural and architectural drawings of the hospital building. Incremental dynamic analysis was employed in this study, considering spectral acceleration was used as intensity measure (IM), and maximum inter-story drift as damage indicator (DM). The accelerogram was selected using target spectrum matching as per the location of the hospital. These results were then compared with the seismic requirement for the location to access the building performance.

The model was verified using the calculated natural period of the building, which was found to be 0.87 sec, which matched the estimates by the Omani Seismic Code (2013). Hilber-Hughes-Taylor (HHT) integration with γ=0.833, and β=0.444, and α= -1/3 was used to perform the non-linear time history analysis with a time step of 0.05 sec. Considering the seismology of the region, this study used spectral matching to select the Lytle Creek 1970 earthquake. The acceleration record in X and Y directions were simultaneously applied to the 3D-model of the hospital building. Spectral acceleration was considered as the Intensity Measure (IM). Joint displacements were used as output and absolute maximum displacement of each story in both lateral directions was extracted using the displacement histories. The inter-story drift was calculated using the displacement and its maximum value was used as Damage Measure (DM) for the respective time history analysis.

DOI: 10.1201/9781003348450-20

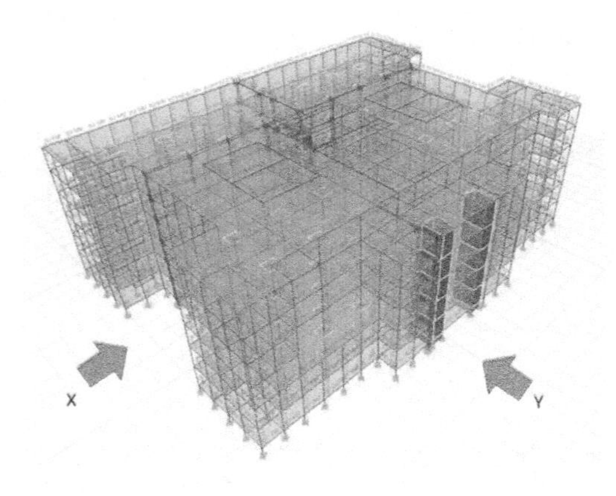

Figure 1. 3D Numerical model of the hospital building.

3 RESULTS AND DISCUSSION

Ten numerical simulations were carried out using different Scale Factors (SF) ranging between 0.1 to 1.4. Intensity measure was estimated using SRSS for the two directions in each run and plotted against the maximum storey drift to develop the IDA curve as shown in Figure 2.

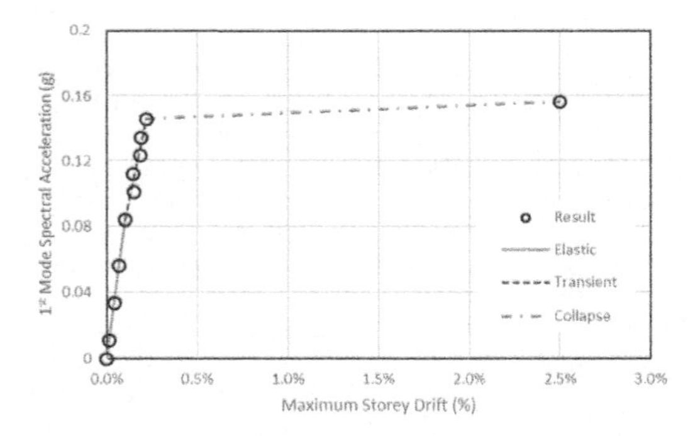

Figure 2. The IDA Curve developed in this study.

The maximum inter-story drift was observed at Level-2 for all the scenarios. The IDA curve remained perfectly elastic till spectral acceleration of 0.084g that produced maximum storey drift of 0.10 %. A transient region was observed between IM of 0.084g to 0.146g, which was indicated by fluctuations in IDA gradient with the drift value reaching 0.22%. This is still below the drift limit of 2.0% specified by the Omani Seismic Code. When the IM was increased from 0.146 g to 0.157 g, it resulted in a near tenfold increase in drift to 2.5%. The curve did not show any significant decrease in the slope before reaching the flat-line, which is an indicator of global collapse. From the IDA curve, the immediate occupancy- minimum damage performance level is identified as 0.084g, while the life safety-collapse prevention limit is 0.146g. Considering the building is situated in Zone-1 as per the Omani Seismic Code resting on soil type-C with. The design spectral acceleration based on the natural period of the hospital as per the Omani Seismic Code is 0.067g as indicated in Figure 3. This indicates that the building satisfies the immediate occupancy performance criteria considering the seismic demand.

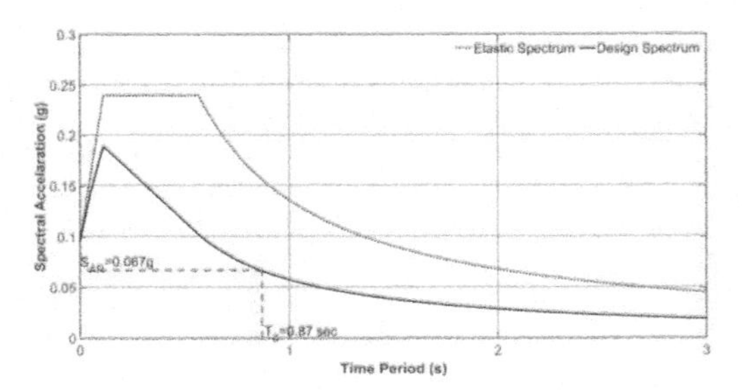

Figure 3. Elastic and Design Response Spectrum for OSC.

4 CONCLUSIONS

The study employed a detailed 3D model and nonlinear time history analysis using ETABS to develop an IDA curve for a hospital building in Oman. Spectral acceleration was used as intensity measure, while maximum storey drift was considered as damage measure. The IDA curve was developed using ten time-history simulations and presented typical regions of behavior. The immediate occupancy-damage control limit was identified as 0.084g, which is well above the site-specific design spectral acceleration of 0.067g.

REFERENCES

Earthquake Monitoring Center. 2013. "Seismic Design Code for Buildings, Sultanate of Oman." Oman.

FEMA, P. 2012. "58-1 (2012) Seismic Performance Assessment of Buildings (Volume 1-Methodology)." Federal Emergency Management Agency, Washington.

Current Perspectives and New Directions in Mechanics, Modelling and Design of Structural Systems – Zingoni (ed.)
© 2022 Copyright the Author(s), *ISBN 978-1-032-39114-4*

Dynamic response of parabolic reflector antenna subjected to shock load and base excitation considering soil structure interaction

VenkataLakshmi Gullapalli
Department of Civil Engineering, Christ University, Bangalore, India DGM, ECIL, Hyderabad, India

R. RaghuNandanKumar
Department of Civil Engineering, Christ University, Bangalore, India

G.R. Reddy
SED, VJTI, India BARC, India
DOC, NITK, India
Formerly: HBNI, Mumbai, India

ABSTRACT: Parabolic reflector antenna structures are subjected to dynamic loads along with normal loads. Finding of dynamic response of the antenna structure subjected to short duration loads such as earthquake loads and shock loads considering soil structure interaction is very important to ensure safety and functionality of the antenna system resting on soft soil. 7.2m diameter parabolic reflector antenna with 90 degree elevation orientation is considered for the study. A triangular pulse of shock load is applied on antenna at different locations and responses are estimated to understand the coupling effect of soil and structure on frequencies, damping and response. Transient response analysis is carried out. Earthquake analysis is also carried out as per IS 1893 part 4:2016 considering Zone V site location. The foundation soil below antenna is considered as homogeneous with shear wave velocity (V_s) 100m/sec. Direct method of analysis considering soil structure interaction as per ASCE 4-16 is carried out. FEM software MSC NASTRAN is used for analysis. The absorbing boundary conditions are used to reflect radiation damping. The depth wise stress variation in foundation soil is evaluated. The results of free vibration analysis, transient response analysis with fixed base and SSI are compared.

Keywords: Dynamic response, Soil Structure Interaction, Transient Response, base excitation, Response Spectrum

1 INTRODUCTION

Of late, a large number of satellite communication antennas across the world for effective communication in all fields such as education, weather forecasting, global mobile communication, military, soil study, etc., are being installed. For fixed satellite services, parabolic reflector antennas are required. Antennas rest on the ground operating with prescribed speeds and accelerations. During unforeseen circumstances, say, extreme wind loads, followed by wind-generated dated missiles, and earthquake loads, the antenna structure may be subjected to short-duration loads. Shock loads are considered as per IS 4991, and earthquake loads are considered as per IS 1893. To ensure the safety and functionality of the system after such incidents, transient response analysis for shock loads and response spectrum analysis for earthquake loads are to be carried out for finding the dynamic response. Finding of dynamic response considering soil-structure interaction is essential for antennas resting on soft soils.

7.2m diameter parabolic reflector antenna structure with 90 degrees elevation orientation (looking to sky) is considered for finding of dynamic response of structure subjected to shock load and base excitation due to earthquake load.

2 METHODOLOGY

In this study, shock load excitation in x, y and z directions is given at three locations of antenna i.e., i. Reflector edge, ii. Yoke base, iii. RCC pedestal and responses are studied in four locations that are Quadripod tip, Reflector edge, yoke base, RCC pedestal for fixed base, Soil Structure Interaction with ASCE 4-16 soil springs. In addition to these four responses studies,

DOI: 10.1201/9781003348450-21

the response in soil is separately studied for SSI 3D soil model.

Shock load as triangular pulse of 10000 N at 100milli seconds as load y-axis, 0 to 200 milli seconds as time in x- axis is applied on antenna. Transient response analysis carried out 10 sec with time step 25 milli seconds. Earthquake load is applied as base excitation. Earthquake load is calculated as per IS: 1893 (2016) for Zone V as frequency vs acceleration and time vs acceleration. Finite Element Method (FEM) is used for analysis. Fixed base analysis and Soil Structure Interaction Analysis with ASCE 4-16 Soil Springs and Direct method are carried out for free vibration, shock load and Earthquake load.

2.1 *Configuration of Antenna*

Antenna system consists of parabolic reflector, subreflector, quadripod, hub with central opening for mounting antenna feed, yoke, screw jack for elevation rotation, mount with azimuth bearing at top and RCC foundation. Reflector consists of 16 numbers radial panels rest on backup structure with 16 radial trusses connected with circumferential members. All panels along with backup structure are connected to cylindrical hub. Sub-reflector is hyperboloid rests on quadripod. Reflector assembly rests on yoke connected to mount. Mount consists of two parts, bottom is cylindrical, top one is circular tapered cylinder and is fixed on RCC foundation with anchor bolts. Reflector assembly is made up of aluminium. Yoke arm, mount are made up of Steel. The foundation considered is of square in plan with dimensions of 4m x 4m and 0.6m thick raft and a square RCC pedestal with a plan dimensions of 2.1m x 2.1m and 1.7m height.

2.2 *Analysis*

Analysis is carried out using FEM software MSC PATRAN pre and post processor and NASTRAN solver. For direct method of SSI analysis, to minimize the boundary effects, soil boundary is considered at 2.5 times the size of footing in plan and depth is considered at 1.5 times. Numerical analysis is carried out for antenna with 90 deg elevation orientation. Founding medium is considered as soil, with shear wave velocity 100m/sec.

Material damping 5% considered for aluminium, steel and RCC as per IS 1893 (2016). Damping of Soil is calculated as per ASCE 4-16 (2016).

2.3 *Boundary conditions*

For SSI direct method, 3 translations fixed at soil bottom. For each layer of soil, master slave element created for shear boundary, soil dynamic properties shear modulus, elastic modulus and poisons ratio are only defined and soil density not considered. Dashpots calculated using shear wave velocity (Lysmer and Kuhelemeyer, 2014), damping constant, $c = \rho * Vs * Area$ of node element and applied as absorbing boundaries.

3 RESULTS

The response in soil domain is reviews, in terms of stresses in the elements. It is observed that, is the stresses are decreasing from raft bottom to down with clear representation of pressure bulb. Frequencies and mass participations of free vibration analysis, transient response analysis with shock load, and Frequency response analysis with base excitation are same.

4 CONCLUSIONS

Depth wise stresses are decreasing from raft bottom to down the soil for both shock load and base excitation. Displacement and acceleration patterns are same for both shock load and base excitation. Mode shape of the frequency of the system is same for both shock load and base excitation.

REFERENCES

Banerjee R, Senugupta A and G R Reddy, Effect of Radiation Damping on Dynamic Soil Structure Interaction, 16th Symposium on Earthquake Engineering, 20-22 December, 2018, I.I.T. Roorkee, India

ASCE 4-16 (2016), Seismic Analysis of Safety Related Nuclear Structures, ASCE, Reston, Virginia.

Indian standard IS 1893, Criteria for Earthquake Resistant Design of Structures, Part 1-General Provisions and Buildings, 2015, Part 4, Industrial Structures including Stack like Structures, 2016.

Sanaz Mahmoudpour, Reza Attarnejad, and Cambyse Behnia, Dynamic Analysis of Partially Embedded Structures Considering Soil-Structure Interaction in Time Domain, Hindawi Publishing Corporation, Mathematical Problems in Engineering Volume 2011, Article ID 534968, 23 pages, doi:10.1155/2011/534968

Shehata E.Abdel Raheem, Mohamed M. Ahmed,Tarek M. A. Alazrak, Soil-Raft Foundation-Structure Interaction Effects on Seismic Performance of Multi-Story MRF Buildings, Engineering Structures and Technologies, Taylor & Francis, 2014 6(2):43–61, doi:10.3846/2029882X.2014.972656

Hee Seok Kim, A Study on the Performance of Absorbing Boundaries Using Dashpot, Engineering, 2014, 6, 593–600, September 2014, Scientific Research, dx.doi.org/10.4236/eng.2014.610060

MSC/Patran User's Manual, MacNeal-Schwendler Corporation, USA, 2017

MSC/ Nastran Dynamic Analysis User's Guide, MacNeal-Schwendler Corporation, USA, 2017

MSC/ PATRAN 322 Exercise Workbook, Shock Analysis of a 3-Story Structure, Lesson 14, MacNeal-Schwendler Corporation, USA, 2017

Indian standard IS 4991-1968, Reaffirmed 2003, Criteria for Blast Resistant Design of Structures for Explosions above Ground

Current Perspectives and New Directions in Mechanics, Modelling and Design of Structural Systems – Zingoni (ed.)
© 2022 Copyright the Author(s), ISBN 978-1-032-39114-4

Effect of Cape storms on the flat roof-mounted solar panels

Y Kiama & J Mahachi
University of Johannesburg, Johannesburg, South Africa

ABSTRACT: The City energy goal for 2020 included significant small scale embedded generation primarily in the form of a photovoltaic (PV) system on rooftops in the commercial sector. However, severe storms subject the photovoltaic panels to harsh wind conditions, lift forces and drag. This study simulates the wind effects on roof-mounted photovoltaic panels at various panel tilt angles (25° to 35°). Furthermore, different wind speed magnitudes were used (35 m/s), and building heights (30 m) were also used. The CFD simulation of wind flow around the building is performed with simulation software. The net pressure coefficients of the PV increased with the higher panel tilt angle. It is recommended that attention be paid to roofs with lower heights as they cause remarkable variation of wind pressure on the PV.

1 INTRODUCTION

Cape Town is a coastal city in the Western Cape Province of South Africa. During the winter months (May to September), strong cold fronts pass the coast from the Atlantic Ocean and result in high precipitation and strong north-westerly winds. These systems are also associated with significant ocean swell waves (Troch et al., 2021). According to (Troch et al., 2021) The extreme storm event produced extremely high offshore waves and high westerly winds. Considerable swell wave heights of up to 11.5m were recorded with a maximum peak period of 18s, maximum wind speeds of approximately 120 km/h were measured at the Cape Town port.

Previous research has reported the effect of wind pressures on solar panels; these experiment were performed on downscale PV prototypes. Wood et al., (2001) assessed the effect of various PV frame mounting heights. The wind pressure results were mostly constant across all solar panels. In addition to the mounting heights Kopp (2014) examined the effect of two tilts angles namely 2^0 and 20^0. It was noted that the pressure flow around the 20^0 tilt angle PV significantly increased comparatively to the pressure equalisation effects observed at a 2^0 Tilts.

2 RESEARCH METHODOLOGY

2.1 *Computational Fluid Dynamic (CFD)*

Figure 2 indicates the CFD framework used within the context of assessing the roof-mounted PV panels system subjected to the cape storms. The CFD and Finite Element Analyses were conjointly used to deduce the PV panels' stress coefficients based on the different tilt angles. The Finite element Analysis sand computational fluid Dynamic analysis consisted of building a model of the Reinforced concrete building and the solar panels mounted on the roof top.

2.2 *Geometrical modeling*

The domains were initially drawn using Caddie commercial software in 2D. After that, it was imported to Rhinoceros 3D before being integrated into ANSYS. The benchmark building height used throughout the simulation is a 30m building; this is because the wind velocity is expected to be higher between 23-30m of height. the computation domain size is conformed to the guideline suggested by Enteria et al., (2015).

A 1m width by 2m length by 17 mm thickness module mounted on an inclined (at an angle of 25, 30 and 35°) support structure was modeled on ANSYS Space Claim 2022R1. An enclosure of air flow domain with inlet and outlet length of 5H and 15H (with H being the maximum height of the mounting and the module) was modeled as a separate body. The enclosure is also extended 5H times on both sides.

2.3 *Mesh generation*

The mesh generation for both the domain and the panels where set to medium with a smoothing increasing to high. A tetrahedral mesh of 429398 elements and 83384 nodes was created using ANSYS Meshing 2022R1. The element size for panel is finer in such a way that can capture the proximity with the flow domain. Tetrahedral mesh where not converted to polydromic elements. The rationale for using the tetrahedral is accuracy in the output.

DOI: 10.1201/9781003348450-22

2.4 *Boundary conditions*

A velocity inlet boundary conditions of 35 m/s at 27 °C temperature and -30 Pa pressure (to accommodate for the height of the building from the sea level) with 1% turbulent intensity and 5 turbulent viscosity ratio was applied at inlet wall of the enclosure. A pressure outlet boundary conditions of 0 Pa was used at the outlet wall with 1% turbulent intensity and 5 turbulent viscosity ratio. A symmetry boundary condition was applied to the remaining walls except the bottom wall which was assigned as rough wall.

3 RESULTS AND DISCUSSION

This section presents the results and the findings of the research focusing on four key themes namely the pressure contours, the velocity contours, imported pressure loads and panel deformations.

3.1 *Pressure contours*

Pressure is much higher on the underside of the panel in such a way that the panel will feel a lift force. A flow separation was observed right after at the end of the panel because of the blockage. Both observations are more pronounced as the tilt angle of the panel increases from 25° to 35°.

3.2 *Velocity contours*

As shown velocity is higher where pressure is the lower which basically at the back of the panel, specifically near the region of flow separation. A perfect correspondence of pressure and velocity increase and decrease respectively was observed which complies with the Bernoulli's principle.

3.3 *Imported pressure load*

Coupling the ANSYS Fluent result to the ANSYS Structural will directly transfer the pressure load on to the surface of the panel and the mounting structure. The imported pressure are shown in the figure below and it increases with tilt angle.

3.4 *Deformations*

Coupling the Deformation is higher for tilt angle of 25° than 30°. But the for tilt angle of 35°, it was observed to be slightly higher than that of at 30°. That can be attributed to the fact that the angle of tilt reaches to a limit after which the air flow will no longer be blocked rather a phenomenon related to a stall was created. From the contour plot the bottom most regions of the panel are happened to be the most deformed.

4 CONCLUSION

The study has successfully demonstrated the impact that different tilt angles have on the surface generated pressures coefficients and strains induced as a result of the Cape storm. The PV panels are considerably affected in the wind ward side, the angle tilts greater than 30 degrees have demonstrated higher stress concentration. Conical vortices effects are noted with the largest peak net pressures located near the ends of the panels. The most critical peak show a reduction of 10 % to 20 % from the 25 degree module induced pressure.

REFERENCES

Barnes, M. A., Turner, K., Ndarana, T., & Landman, W. A. (2021). Cape storm: A dynamical study of a cut-off low and its impact on South Africa. *Atmospheric Research*, *249*(May 2020), 105290. https://doi.org/10.1016/j.atmosres.2020.105290

Enteria, N., Awbi, H., & Yoshino, H. (2015). Application of renewable energy sources and new building technologies for the Philippine single family detached house. *International Journal of Energy and Environmental Engineering*, *6*(3), 267–294. https://doi.org/10.1007/s40095-015-0174-0

Kopp, G. A. (2014). Wind loads on low profile tilted solar arrays placed on large flat lowrise building roofs.pdf. *Journal of Structural Engineering*, *140*(2), 04013057.

Kopp, G. A., & Banks, D. (2013). Use of the wind tunnel method for obtaining design wind load on roof mounted solar array.pdf. *Journal of Strutural Engineering*, *139* (2), 284–287.

Stathopoulos, T., Zisis, I., & Xypnitou, E. (2014). Local and overall wind pressure and force coefficients for solar panels. *Journal of Wind Engineering and Industrial Aerodynamics*, *125*, 195–206. https://doi.org/10.1016/j.jweia.2013.12.007

Troch, C., Terblanche, L., & Henning, H. (2021). Modelling and measurement of low-frequency surge motion associated with extreme storm conditions in the Port of Cape Town. *Applied Ocean Research*, *106*(July 2020), 102452. https://doi.org/10.1016/j.apor.2020.102452

Wood, G. S., Denoon, R. O., & Kwok, K. C. (2001). Wind loads on industrial solar panel arrays and supportin.pdf. *Wind and Structures*, *4*(6), 481–494.

Current Perspectives and New Directions in Mechanics, Modelling and Design of Structural Systems – Zingoni (ed.)
© 2022 Copyright the Author(s),
ISBN 978-1-032-39114-4

Evaluating static and modal responses of a small scale conventional and bio-inspired wind turbine blade

T. Babawarun, H.M. Ngwangwa & A.G Netshivhulana*
Unisa Biomedical Engineering Research Group, Department of Mechanical Engineering, University of South Africa, Florida 1710, Johannesburg, South Africa

T. Pandelani
Defence and Security, Council for Scientific and Industrial Research (CSIR), Pretoria, South Africa.

W.H. Ho
School of Mechanical, Industrial and Aeronautical Engineering, University of the Witwatersrand, Johannesburg, South Africa

ABSTRACT: Wind turbine blade is considered as one of the most vital component of the wind turbine system meant to produce energy from the power of the wind. A finite element model of a three bladed small-scale Kestrel e230i horizontal axis wind turbine will be used in this study. Firstly, the three-dimensional solid geometry of the conventional wind turbine was generated using a customized aerofoil measurement technique in the laboratory. The generated geometry was then cleaned to remove all abrupt changes in surface profile. Then the actual conventional wind turbine blade was statically tested and resulting strains measured in different positions along its blade length. The blade was rigidly fixed at its root by bolting it into a laboratory test bench. The three-dimensional model was subsequently loaded and supported at its root in a similar manner. Finite element analysis and model updating was performed in ANSYS v 19.0 by iteratively correlating stress values between the finite element model and the test. A bio-inspired version of the updated finite element model was generated by adding undulations/corrugations to the bottom surface of the conventional blade in a longitudinal direction. Modal analyses were then performed for the two models to compare their natural frequencies in first two bending and first twist modes. These two modes are reported to be the most important modes in wind turbine dynamic response and may also be very influential in energy production of wind turbines. The numeral results show that the inclusion of undulations/corrugations increases the natural frequencies, which typically means that for low-speed inland applications, the risk of wind turbine blade structural failure due to resonant or near-resonant excitations is heavily averted due to increased potential operating frequency ranges.

1 INTRODUCTION

While most of the wind energy market is dominated by megawatt-size wind turbines, the increasing importance of distributed electricity generation gives way to small, personal-size installations [1]. Therefore, mechanical analysis of wind turbine blades is also important from the point of view of their inertia.[1]. Wind turbine is a device that converts kinetic energy from the wind, also called wind energy, into mechanical energy a process known as wind power [2].

In structural dynamics analysis, it is important that the operating frequencies do not lie close to any of the structure's natural frequencies [3]. In the design of a wind turbine it is important that resonance be avoided.

In this paper, two different blade models are considered, a conventional NACA airfoil and a bio-inspired airfoil are analyzed using modal analysis and the changes in mode shapes and first three natural frequencies are observed to examine if the bio-inspired shape provides a much wider operating range. We further wish we wish to use the static results that was obtained from earlier validation of the model, the model was then updated by imputing the bio-inspired corrugations and modal analysis was evaluated.

*Corresponding author

DOI: 10.1201/9781003348450-23

2 MATERIALS AND METHODS

In this paper, two blade models are analyzed and named conventional and bio-inspired. The assembled model is a horizontal axis wind turbine (HAWT). As seen in Figure 1, the structure is a three bladed wind turbine. The conventional blade is modelled as a prototype of a Kestrel e230i small-scale wind turbine blade. The bio-inspired blade second is modelled with the corrugations introduced into the blade to give it a bio structure, this blade is then modelled into the three-bladed structural system. The modelling of both systems is done in ANSYS Space Claim Direct Modeler (SCDM) v 19.

Figure 1. 3D isometric view of the coupled HAWT wind turbine.

3 RESULTS

From Table 1, for the conventional blade, the first and the third mode show frequency decreases as the rotational speed increases (Backward whirl), while for the second mode there is an increase in the frequency as the rotational speed increases (Forward whirl).

Table 1. Natural frequencies of the first three mode for the original and modified blade.

	Conventional blade			Bio-inspired blade		
	Mode (Hz)			Mode (Hz)		
Rotational speed	1	2	3	1	2	3
0 rpm (Stationary)	3.3	3.3	19.9	6.9	9.1	12.6
261 rpm	1.3	8.4	19.9	5.6	10.6	12.7
347 rpm	1.0	10.4	19.9	5.1	11.1	12.8
414 rpm	0.9	11.9	19.9	4.7	11.3	13.0
521 rpm	0.7	14.2	19.9	4.2	11.6	13.1
608 rpm	0.6	15.8	19.9	3.8	11.7	13.2

4 DISCUSSION

In the case of the bio-inspired blade, only the first mode shows a decrease in frequency as the rotational speed increases while for the second and third mode there is an increase in frequency as the rotational speed increases. These trends are expected for backward mode where the frequencies decrease with increase in rotational speed, and forward modes, where the frequencies increase with rotational speeds. It can be seen that for the conventional blade, the first natural frequency of the rotational speed within the range with which we are considering has a maximum frequency of 3.3 Hz while for the bio-inspired blade, the maximum natural frequency is at 6.8 Hz. This is a substantial shift in frequency from the conventional blade to the bio-inspired blade; hence, there is an increase in the operating range of the wind turbine.

In the case of the bio-inspired blade, only the first mode shows a decrease in frequency as the rotational speed increases while for the second and third mode there is an increase in frequency as the rotational speed increases. These trends are expected for backward mode where the frequencies decrease with increase in rotational speed, and forward modes, where the frequencies increase with rotational speeds. It can be seen that for the conventional blade, the first natural frequency of the rotational speed within the range with which we are considering has a maximum frequency of 3.3 Hz while for the bio-inspired blade, the maximum natural frequency is at 6.8 Hz.

5 CONCLUSIONS

Two wind turbine models were considered, model 1, which was a conventional shaped blade, modelled after a small-scale kestrel e230i blade, and model 2 which is a blade with bio-inspired corrugation added at one side (base) of the blade span. Results showed that the model 2 had higher natural frequencies than model 1 in the operating wind speed range of Johannesburg, which was the area of analysis. The change in the bio-inspired blade is found to be better for a changing wind speed. The higher natural frequency in model 2 implies that there is an increase in the blade stiffness hence an overall improvement in the structure. The new blade model was also found to possess better vibrational characteristics because of the modification.

REFERENCES

1. Michal Lipian, Pawel Czapski and Damian Obidowski, 2020, Fluid–Structure Interaction Numerical Analysis of a Small Urban Wind Turbine Blade, pp 1
2. Ramesh J, Dr P. Rathma Kumar, Md Umar, Dr M VMallikarjuna, 2017, Static and Dynamic Analysis of 1KW small wind turbine blades by various materials.
3. Thresher RW. 1982. Structural dynamic analysis of wind turbine systems. J. Solar Energy Engineering. 104:89–95.

3. Seismic response, Seismic analysis, Earthquake-resistant design

Current Perspectives and New Directions in Mechanics, Modelling and Design of Structural Systems – Zingoni (ed.)
© 2022 Copyright the Author(s), ISBN 978-1-032-39114-4

Functional elements exploiting superelasticity and the shape memory effect of Fe-Mn-Al-Ni-X shape-memory-alloys in structural engineering

E. Fehling, B. Middendorf & J. Thiemicke
Institute for Structural Engineering, University of Kassel, Kassel, Germany

M. Vollmer, A. Bauer & T. Niendorf
Institute for Materials Engineering, University of Kassel, Kassel, Germany

V. v. Oertzen & B. Kiefer
Institute of Mechanics and Fluid Dynamics, TU Bergakademie Freiberg, Freiberg, Germany

ABSTRACT: Innovative iron-based shape memory alloys, e.g. Fe-Mn-Al-Ni-based SMAs, provide superelastic behavior as well as shape memory effects at much lower material costs than conventional SMA. By exploiting the superelasticity of Fe-Mn-Al-Ni-based alloys, functional elements can be designed to limit the forces during seismic events as well as to dissipate energy. In comparison with plastic material behavior, e.g. of conventional structural steel, this offers a huge gain of resilience to structures since residual deformations are suppressed due to the inherent capability of recentering a structure after seismic loading. Temperature driven activation of the newly developed alloys leads to the recovery of the original shape of a prestrained SMA sample and makes it possible to prestress any 2D- or 3D-shaped reinforcement, not only tendons but, e.g. also stirrups or spiral reinforcement, given the essential deformation restraint.

1 INTRODUCTION

Shape memory alloys (SMAs) are used in many different fields. However, due to high costs, their applications are restricted to niche applications where small parts can offer significant advantages, e.g. when used as flexible dental braces, as actuators in airplanes. First examples to integrate shape memory alloys in structural engineering are, for example, application of superelastic Ni-Ti SMA as seismic dampers or of Fe-Mn-Si SMA for additional strengthening of damaged structures induced by prestressing (Cladera 2014). The characteristics of Fe-Mn-Al-Ni-based SMA enable the use of the one-way effect and superelasticity, as well as their combination. They promise, furthermore, to be able to be produced at much lower costs than the widespread NiTi SMAs. This opens up new methodological and constructive design approaches in civil engineering to extend the primary properties and to integrate new functions as well as to give some answers on how to reduce the ecological impact due to the high demand for the construction of industrial and residential buildings and infrastructure facilities.

2 EXPERIMENTAL INVESTIGATIONS

Innovative Fe-Mn-Al-Ni SMAs show the one-way shape memory effect, similar to, e.g. Ni-Ti alloys. This effect is activated by temperature and leads to the recovery of the original shape (or length, respectively) of a prestrained SMA sample. Using this effect in combination with deformation restraint after prestraining and unloading, prestressing of reinforcement of concrete structures can be achieved easily.

It has been shown (Vollmer 2021), that for Fe-Mn-Al-Ni SMA even multiple cycles of increased prestressing are possible by increasing the maximum temperature T_{max} as well as the dwell time at T_{max} in the heating and cooling cycles (see Figure 1).

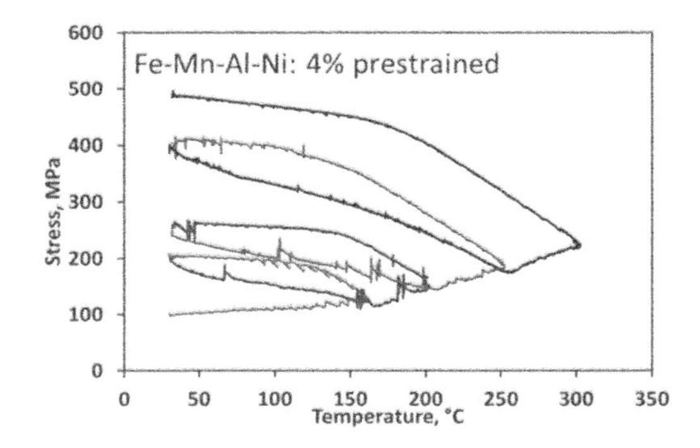

Figure 1. Prestressing of an Fe-Mn-Al-Ni SMA by multiple thermal activation cycles with different values of maximum temperature T_{max} (Vollmer 2021).

DOI: 10.1201/9781003348450-24

Moreover, the superelastic hysteresis of SMAs provides the possibility to withstand large strains without significant increase of the stress level, similar to materials that exhibit plastic behavior, such as steel.

In that respect, the advantage of SMAs in comparison to conventional steel is their capability of recentering after unloading due to the superelastic effect. First results for the superelastic hysteresis observed in Fe-Mn-Al-Ni-Ti SMA specimens are depicted in Figure 2. The results reveal the potential of these alloys for force limitation, e.g. during seismic events, as well as for energy dissipation and recentering (Vollmer 2019).

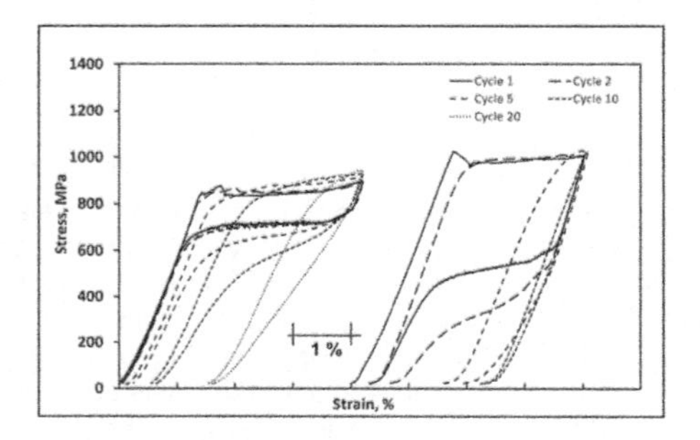

Figure 2. Superelastic behavior of two Fe-Mn-Al-Ni-Ti conditions in compression with different widths of the stress-strain hysteresis (Vollmer 2019).

3 STRUCTURAL APPLICATIONS

For prestressing, e.g. of concrete structures, SMAs can be used by exploiting the one-way shape memory effect with deformation restraint. This offers several advantages:

- no friction losses during posttensioning,
- no (hydraulic) actuators required,
- the possibility of posttensioning of practically all conceivable shapes of reinforcement, eventually leading to effectively 2D and 3 D prestressing states.

Lastly, also prestressing of SMA fibers in fiber reinforced concrete members as outlined by Middendorf et al. (Middendorf 2022) becomes possible.

With respect to seismically loaded structures, it is possible to design improved types of bracings, e.g. cross bracings with superelastic behavior or seismic shear links for eccentrically braced frames (EBFs) that do not suffer from persistent plastic deformations after a seismic event.

4 CONSTITUTIVE MATERIAL MODELS

Based on some first experimental results for Fe-Mn-Al-Ni alloys, an established constitutive model for SMA (Lagoudas 2012) has been calibrated based on experimental data available for (cyclic) tension-compression loading tests (Vollmer 2019). For shear loading, Figure 3 shows a comparison of ordinary steel with Ni-Ti SMA and an example for the newly developed Fe-based SMA. For the SMAs, the force limitation due to the superelastic hysteresis becomes evident as well as the recentering capability.

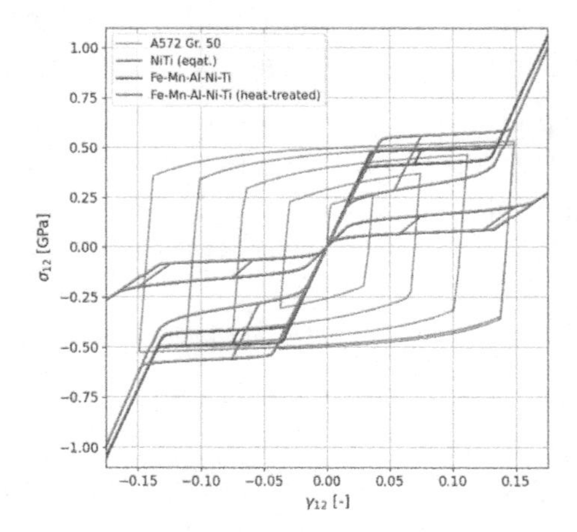

Figure 3. Simulation of behavior for alternating shear loading - comparison between different SMA (superelasticity) and an A572 grade 50 conventional steel (plastic hysteresis).

5 CONCLUSION

This paper discusses the characteristics and possible applications of functional elements constructed of Fe-Mn-Al-Ni shape memory alloys as well as experimental and numerical investigations of their characteristics. It was demonstrated that applications in structural engineering become possible which exploit the capacity of SMA to prestress concrete members as well as to provide force limitation, energy dissipation, and recentering of structures after unloading, when subjected to seismic loading.

REFERENCES

Cladera, A., Weber, B., Leinenbach, C., Czaderski, Ch., Shahverdi, M. & Motavalli, M. 2014. Iron-based shape memory alloys for civil engineering structures: An overview. *Constr. Build. Mater. 63: 281–293.*

Lagoudas, D. C., Hartl, D. J., Chemisky, Y., Machado, L. & Popov, P., 2012. Constitutive Model for the Numerical Analysis of Phase Transformation in Polycrystalline Shape Memory Alloys, *Int. J. Plast.* 32-33, 155–183.

Middendorf, B., Schleiting, M. & Fehling, E., 2022, Potential of Shape Memory Alloys in fiber reinforced High Performance Concrete, submitted for publication, *SEMC 2022*

Vollmer, M., Arold, T., Kriegel, M. J., Klemm, V., Degener, S., Freudenberger, J. & Niendorf, T. 2019. Promoting abnormal Grain Growth in Fe-based Shape Memory Alloys through Compositional Adjustments. *Nature communications* (2019), 10:2337.

Vollmer, M., Bauer, A., Frenck, J.M., Krooss, P., Wetzel, A., Middendorf, B., Fehling, E. & Niendorf, T. 2021. Novel Prestressing Applications in Civil Engineering Structures enabled by Fe-Mn-Al-Ni Shape Memory Alloys. *Eng. Struct. 241: 112430.*

Current Perspectives and New Directions in Mechanics, Modelling and Design of Structural Systems – Zingoni (ed.)
© 2022 Copyright the Author(s), ISBN 978-1-032-39114-4

Seismic response of curved structures in historical masonry churches

Corrado Chisari, Mattia Zizi, Jafar Rouhi & Gianfranco De Matteis
Department of Architecture and Industrial Design (DADI), University of Campania "Luigi Vanvitelli", Aversa (CE), Italy

ABSTRACT: Curved structures (i.e. arches and vaults) are recognized as distinctive elements of the masonry built heritage, and in particular of churches. If on the one hand they represent structural and architectural elements of significant values, on the other they can experience important damage in case of seismic actions, being characterized by significant vulnerability. The structural assessment of masonry curved structures can be based on several simplified geometrical representations and performed at different detail levels, including limit analysis (with a kinematic or static approach) and finite or discrete element modelling. Therefore, the present paper focuses on the seismic response of curved structures in historical masonry, investigated by means of different approaches. In particular, a comparative assessment of these approaches is carried out considering multiple configurations including i. stand-alone arches; ii. arches supported by lateral piers; iii. triumphal arches (diaphragm arches). Strengths and limits of each method are compared, highlighting possible differences in the collapse predictions. Also, by means of parametric analyses, the main sources of vulnerability and failure modes of curved structures are investigated and critically assessed.

1 INTRODUCTION

Masonry churches constitutes a significant part of the historical built heritage in Europe. Among local collapse modes typical of churches, those regarding curved structures, i.e., arches and vaults, are very important, as they may substantially govern the vulnerability of the entire church.

For this reason, methodologies for assessing the seismic capacity of masonry curved structures are discussed in this paper, by considering three different configurations: i. stand-alone arches; ii. arches supported by lateral piers; iii. triumphal arches (diaphragm arches).

2 METHODS FOR VULNERABILITY ASSESSMENT OF ARCHES

Two main approaches can be identified for the seismic vulnerability analysis of curved structures: Limit Analysis (LA), providing the load multiplier bringing the structure to collapse, and thus the collapse PGA, and Finite Element (FE) analysis, for which by tracking the entire force-displacement plot it is possible to perform a seismic check in terms of displacements. Both methods have advantages and drawbacks. The first is conceptually simple, robust, does not require any information about material properties and usually provides realistic results in case of simple arches. FE modelling provides significantly more accurate results when the hypothesis at the base of LA become unrealistic, but at expenses of higher computational demand, potential sensitivity to several material parameters, not always easy to calibrate, and reduced robustness, i.e., possibility of not obtaining any results due to lack of convergence.

In this paper, limit analysis is used to parametrically investigate the vulnerability of stand-alone and buttressed arches; FE modelling are conversely exploited to study the response of diaphragm arches, where the presence of the diaphragm complicates the characterization of the collapse mechanism.

3 PARAMETRIC STUDY

Three configurations of arches are considered, parameterized by the ratio e/s, where e is the eccentricity of the arch and s the half-span: e/s=0.0, 0.5, 1.0. The effect of the thickness, applied load, dimensions of the buttresses on the PGA at collapse are investigated (Zizi, et al., 2021; Chisari, et al., 2021). The analysis of the results shows that the thickness of the arch and the superimposed load have a generalized positive effect on the PGA at collapse.

When it is necessary to account for the effect of buttresses on the mechanism, three types of collapse can be identified (Como, 2013): (a) local, involving the arch only; (b) global, with two hinges at the base of the buttresses and two in the arch; (c) semi-global, with one hinge in at the base

DOI: 10.1201/9781003348450-25

of a buttress and three within the arch. The effect of the geometrical characteristics of the buttresses was investigated, and the results are summarized in Figure 1. Along with the value of collapse acceleration, the type of collapse is identified by letters L (local), S (semi-global) and G (global). It is possible to see that global collapses never occur for the investigated geometries, and the activation of semi-global mechanisms reduces the capacity of the system.

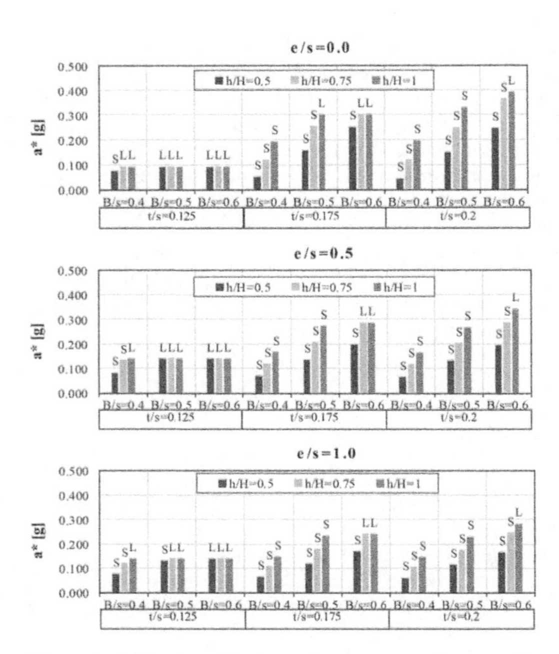

Figure 1. Influence of buttress parameters on the capacity of the system, from (Chisari, et al., 2021).

4 SEISMIC RESPONSE OF DIAPHRAGM ARCHES

The collapse mechanism of diaphragm arches also requires four hinges to be activated. Since in homogenized FE models a separation of the volume cannot be obtained, a hinge is conventionally defined as a region where plastic strain concentration is observed. In addition to the collapse modes already described, a new collapse mode, of type strong diaphragm-weak piers was also observed (Figure 2).

Focusing on a reference case, the influence of the diaphragm height h_d, measure from the impost, on the maximum horizontal force of the structure F_{max} and the collapse weight multiplier $\lambda = F_{max}/W$ (where W is the self-weight) was investigated. The results show that stronger diaphragms increase the vulnerability of the structure, which is governed by the strength of the buttresses. The results obtained in terms of safety factor show a strong yet qualitative consistency with LA.

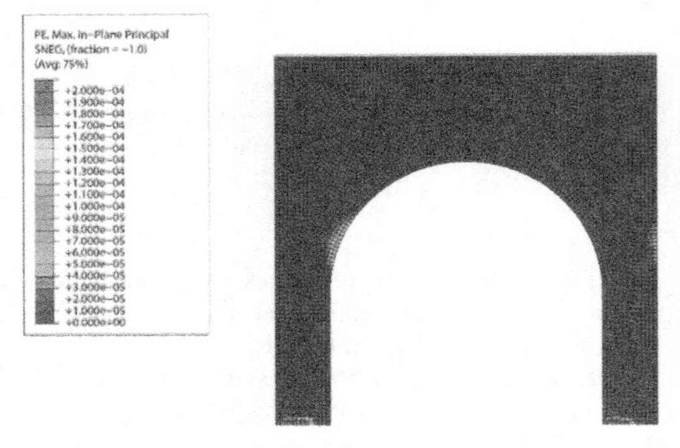

Figure 2. An example of strong diaphragm-weak piers collapse mechanism in a diaphragm arch with e/s=0.0, B/s= 0.4, H/s=1.0 and h/y_{max}= 1.5.

5 CONCLUSIONS

In the present paper, methodologies for assessing the seismic vulnerability of curved structures were discussed and the suitability of LA and FE analyses for this purpose was investigated. Ongoing research is investigating further the relationship between different levels of analysis, with the aim of setting up a general methodology for vulnerability assessment of curved structures.

ACKNOWLEDGEMENTS

Part of this work has been carried out within the activities of the research project "PREVENT - Integrated PRocedure for assEssing and improVing the resiliENce of existing masonry bell Towers on a territorial scale", funded under the VALERE 2019 program by the University of Campania. C.C. is funded by MUR (Ministry of University and Research) through PON FSE 2014-2020 program (project AIM1879349-2). M. Z. is funded by MUR (Ministry of University and Research) through PON FSE 2014-2020 program (project CUP: B61B21005470007).

REFERENCES

Chisari, C., Cacace, D. & De Matteis, G., 2021. Parametric investigation on the effectiveness of frm-retrofitting in masonry buttressed arches. *Buildings*, 11(9), p. 406.

Como, M., 2013. *Statics of historic masonry constructions*. Berlin: Springer.

Zizi, M. et al., 2021. Automatic Procedures for the Safety Assessment of Stand-alone Masonry Arches. *International Journal of Architectural Heritage*, Volume (in press).

Current Perspectives and New Directions in Mechanics, Modelling and Design of Structural Systems – Zingoni (ed.)
© 2022 Copyright the Author(s), *ISBN 978-1-032-39114-4*

Recent advances in tsunami design of coastal structures

I.N. Robertson
University of Hawaii at Manoa, Honolulu, USA

ABSTRACT: The 2004 Indian Ocean Tsunami initiated a rapid increase in tsunami research, particularly as it relates to the performance of coastal structures during tsunami inundation. The subsequent Chile tsunami in 2010 and Great Japan Earthquake and Tsunami (or Tohoku Tsunami) in 2011 re-invigorated the urgency of developing design provisions for tsunami loading on coastal structures. The culmination of this experimental and theoretical research, and field reconnaissance after damaging tsunamis, resulted in the development of a new Chapter 6 "Tsunami Loads and Effects" in the ASCE7-16 Standard "Minimum Design Loads and Associate Criteria for Buildings and Other Structures". This paper briefly reviews the ASCE7-16 tsunami design provisions. The application of these provisions to the design of new tsunami vertical evacuation refuge structures in Oregon and Washington States, and new multistory residential buildings in Waikiki, Hawaii, will be presented. This paper also introduces recent modifications to the tsunami design provisions approved for the ASCE7-22 Standard published in December 2021. These modifications were prompted by recent laboratory research, observations after the earthquake and tsunami in Palu, Indonesia, and updates to numerical modeling procedures for tsunami inundation.

1 INTRODUCTION

For five years, the ASCE Tsunami Loads and Effects subcommittee worked to develop a new chapter in the ASCE/SEI 7-16 Standard, Minimum Design Loads and Associated Criteria for Buildings and Other Structures. This new Chapter 6, Tsunami Loads and Effects, provides comprehensive provisions for design of coastal structures for tsunami loads, scour and related considerations. These tsunami design provisions now apply to all coastal communities in California, Oregon, Washington State, Alaska and Hawaii. A companion design manual by the author explains the new provisions and demonstrate their application to prototypical reinforced concrete buildings in coastal communities in the Western USA.

2 OVERVIEW OF ASCE7-16 TSUNAMI LOADS AND EFFECTS

2.1 *Tsunami hazard*

Tsunamis can be generated by a number of natural phenomena, but the most common damaging trans-oceanic tsunamis are caused by subduction zone earthquakes. Local damaging tsunamis can also be caused by thrust or strike-slip faults, aerial or submarine landslides, volcanic explosions, and island flank failures.

2.2 *Tsunami flow parameters*

To design structures for tsunami loading, it is necessary to determine the maximum flow depth and velocity at the project site during the 2500-year Maximum Considered Tsunami. These flow characteristics can be determined either by site specific tsunami inundation modeling, or by use of the Energy Grade Line Analysis (EGLA).

2.3 *Structural design provisions*

2.3.1 *Tsunami risk categories*

ASCE7 defines four Tsunami Risk Categories (TRC). TRC IV includes all essential buildings. TRC III includes large occupancy buildings and critical facilities. TRC I includes buildings that are generally not occupied continuously, and TRC II includes all other buildings, which is the vast majority, including all residential, office, educational buildings, etc. All TRC IV and TRC III buildings and structures, located within the Tsunami Design Zone (TDZ), must consider the effects of tsunami loading in their design. The local jurisdiction is encouraged to require tsunami design for TRC II buildings with sufficient height to provide emergency refuge for people stranded within the TDZ.

DOI: 10.1201/9781003348450-26

2.3.2 *Tsunami load cases*

Three tsunami load cases must be considered, namely a buoyancy check, the maximum flow velocity assumed to occur when the flow depth is 2/3 of the maximum flow depth, and the maximum flow depth with 1/3 of the maximum velocity. Design is required for both incoming and outgoing flow.

2.3.3 *Hydrostatic forces*

Hydrostatic forces that must be considered in the structural design include buoyancy, unbalanced lateral forces when fluid loads are applied to only one side of a structural element such as a wall, and residual water surcharge loads that can apply to elevated floors during drawdown. Design equations are provided for each of these loading conditions.

2.3.4 *Hydrodynamic forces*

Hydrodynamic loads include traditional drag forces on both the overall structure and on individual structural elements, lateral impulsive forces due to the initial impact from a tsunami bore on structural walls or other broad structural elements, pressurization of enclosed spaces due to flow stagnation, and shock pressure effects below piers and elevated floors due to entrapped bore conditions. Design equations and procedures are given for each of these conditions.

2.3.5 *Waterborne debris impact forces*

Tsunamis can generate a large quantity of debris, some of which can cause substantial forces when impacting a structural member. ASCE7-16 provisions consider impact from utility poles/logs, shipping containers, passenger vehicles, tumbling boulders/concrete debris, and ships. Impact force equations for all these conditions, except ships, are specified in the provisions.

2.3.6 *Scour effects*

Sediment transport and scour around building foundations can result in localized and overall structural failure. The ASCE7-16 provisions are based on an empirical expression derived from field observations after numerous past tsunamis.

2.3.7 *Tsunami vertical evacuation refuge structures*

The design of structures that are specifically designated as tsunami vertical evacuation refuge structures must, of necessity, be more conservative than typical construction. This is particularly important when considering the height of the refuge levels to minimize the potential for overtopping during an extreme event.

3 TSUNAMI VERTICAL EVACUATION STRUCTURES

ASCE7-16 tsunami design provisions have already been applied to several tsunami vertical evacuation structures and multi-story residential buildings. In Westport, Washington, the Ocosta Elementary School constructed a tsunami evacuation area on the roof of a new gymnasium building. In Newport, Oregon, the Oregon State University (OSU) Hatfield Marine Science Center is designed as a three-story research building with a tsunami vertical evacuation refuge on the roof easily accessible via a wide ramp.

4 ASCE7-22 UPDATES

Several significant updates have been made in the 2022 edition of ASCE7. High-resolution Tsunami Design Zone maps have been developed for highly populated coastlines of California, for the Islands of Oahu and Hawaii, and for the Salish Sea region of Washington State. ASCE7-22 now includes clarification as to the use of push-over analysis for tsunami design and assessment of existing structures.

Design expressions are included in ASCE7-22 for the horizontal drag and vertical uplift on horizontal pipelines. Based on field reconnaissance after the Palu Earthquake and Tsunami in 2018 debris damming and impact loads must now be applied to interior columns for certain structures.

Recent laboratory research results have been included in ASCE7-22 with application to scour around vertical elements such as piles. In addition, a method for estimating the depth of soil affected by pore pressure softening has been included in ASCE7-22 for designers to include this effect in their estimates of sediment transport and scour.

5 SUMMARY AND CONCLUSIONS

This paper presents an overview of the Tsunami Loads and Effects Chapter of ASCE7-16 and several examples of its application to the design of tsunami vertical evacuation structures. Enhancements made to these design provisions in the 2022 edition of ASCE7 are introduced with background information.

Current Perspectives and New Directions in Mechanics, Modelling and Design of Structural Systems – Zingoni (ed.)
© 2022 Copyright the Author(s), ISBN 978-1-032-39114-4

Load-level isolation system for industrial racks: Evaluations on a case study structure

E. Bernardi & M. Donà
Earthquake Engineering Research & Test Center, Guangzhou University, Guangzhou, China
Department of Geosciences, University of Padova, Padova, Italy

A. Zonta, M. Ceresara, S. Mozzon & F. da Porto
Department of Geosciences, University of Padova, Padova, Italy

P. Tan
School of Civil Engineering, Guangzhou University, Guangzhou, China

ABSTRACT: Recent Italian earthquakes have dramatically demonstrated the high seismic vulnerability of industrial racks, raising awareness of the need to increase their seismic resilience in industry, academia, and society in general. Originally, these structures were conceived and designed to meet load-bearing requirements for static loads only. Although recent racking regulations contain clear requirements against seismic actions, most of the racks still in use today do not comply with them. To reduce their vulnerability, but also to increase the resilience of the newly designed racks, specific base isolation systems have recently been studied as an alternative to conventional strengthening solutions (e.g. thicker profiles, bracing systems). In this context, the Load-Level Isolation System (LLIS), i.e. the isolation system directly applied to some load levels of the rack, represents a new and interesting seismic mitigation technique, which deserves to be further explored. This paper aims to evaluate the effectiveness of the LLIS by analyzing a 5-span, 7-level case study rack. The structure is analyzed through 3D finite element modeling and time-history analysis, using a set of bidirectional natural events. The effects of the LLIS are evaluated in terms of maximum displacement and base forces.

1 INTRODUCTION

Industrial racks, especially pallet racks, are becoming increasingly popular thanks to globalization and the increase in e-commerce.

The racks systems, characterized by high repeatability of the elements and simplicity of assembly, are made using cold-formed open profiles. In the longitudinal direction, the uprights are connected to each other by beams which, in addition to supporting the pallet load, create with the uprights a system resistant to lateral actions. Instead, in the transverse direction, the horizontal actions are absorbed by the diagonals and horizontals frame elements that connect the uprights creating a truss system.

Most of the steel racks used today resulting highly inadequate with respect to seismic risk. Various strategies are available to improve the seismic behavior of racks, for example the insertion of bracing system or the base seismic isolation (Simoncelli et al. 2020). Furthermore, an innovative strategy, which requires further numerical and experimental investigations, consists in applying specific isolation systems directly to the load levels. This seismic mitigation technique is therefore based on the concept of mass damping, exploiting the masses of the pallet which are often very large. This paper aims to assess the effectiveness of the Load-Level Isolation System (LLIS) by analyzing a real case study rack. The chosen isolation devices are based on the Rolling-Ball Rubber-Layer (RBRL) isolation system (Donà et al. 2017) and are applied to the top two load levels. The 3D finite element model, which includes the main structural non-linearities, is analyzed in time-history (TH) under seven bidirectional natural events. The effects of the LLIS are evaluated and, for comparison purposes, the same rack was analyzed without LLIS (i.e. existing reference structure), and also by inserting a longitudinal bracing system.

2 CASE STUDY

The 5-span, 7-level case study rack is shown in Figure 1. The structure consists of double entry and the plan dimensions are 13.5 m x 2.40 m, with a height of 9.25 m. Each span is 2.7 m long, and each load level supports the weight of 2.1 t.

The frames are composed of two uprights connected to each other through diagonal and horizontal

DOI: 10.1201/9781003348450-27

frame elements. The latter are connected to the uprights through bolts, creating a hinge connection. The frames are connected to each other by beams with semi-rigid joints. The connections at the base of the uprights are bolted in such a way that they can be considered semi-rigid around the y direction and hinged around the x direction.

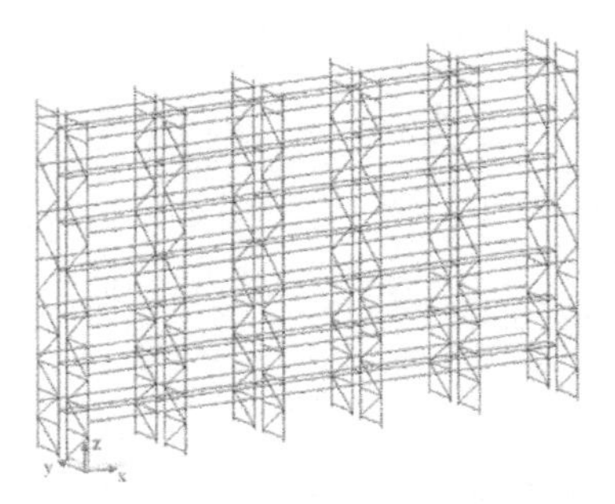

Figure 1. Finite element model of the case study rack.

3 RBRL SYSTEM

The RBRL system (Donà et al. 2017) was tailored to be used as a Load-Level Isolation System (LLIS), allowing the decoupling of motion between the pallet and its load level in the y direction.

The RBRL is composed of a steel rolling ball system, two rubber-layer tracks and recentering springs in the shape of rubber cylinders, which define the stiffness of the system in the steady-state rolling phase. The design of the LLIS was based on the evaluation of the recentering force offered by the cylindrical springs and the rolling friction provided by the rubber tracks. Therefore, the isolation system behavior is represented by an elastic-plastic model. The tailored RBRL device is shown in Figure 2 and its isolation period T_{IS} was assumed equal to 1 s.

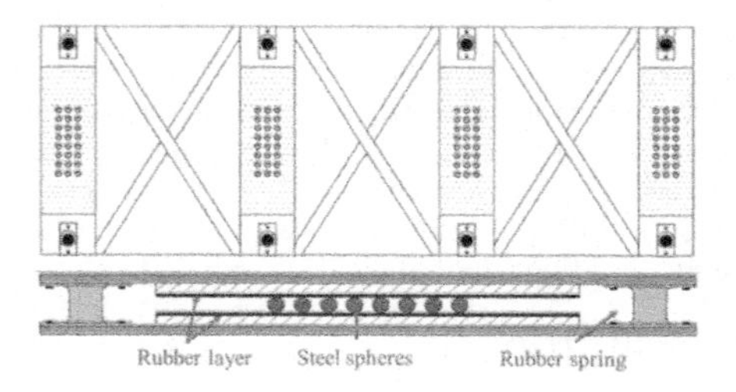

Figure 2. RBRL system tailored for the LLIS application.

4 EFFECTIVENESS OF THE LLIS AND CONCLUSIONS

The effectiveness of the LLIS was investigated considering three different configurations: existing structure, rack with LLIS and rack strengthened with a bracing system.

Seven bidirectional Time-history analysis were performed using natural records selected from the SIMBAD database, and scaled to be compatible with the EC8 Type 1 spectrum, with PGA=0.3 g and soil type B.

The maximum TH responses, averaged among seismic events, were evaluated in terms of displacement as well as axial and shear forces ratio at the base.

The graphs in Figure 3 shows that, in the y direction, the isolation of the load levels allows to reduce the maximum displacement up to of 30%, while it does not show any influence in the x direction.

Figure 4 shows the ratios between the maximum axial and shear forces at the base, obtained in the rack with bracing or LLI system, and those obtained in the existing structure. The presence of the bracing increases the axial forces, keeping the shear force unchanged compared to the existing structure; on the other hand, the LLIS technique allowed a reduction of 21% and 13% respectively of the maximum axial and shear forces.

Considering the results obtained and the possible margins for improvement of the isolator parameters, the LLIS shows interesting potential and deserves to be further investigate.

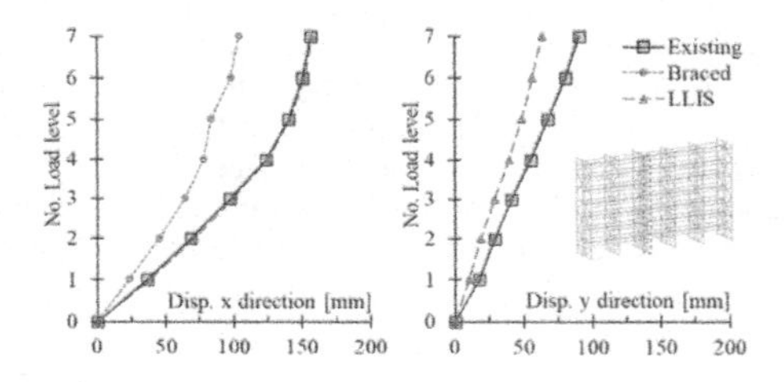

Figure 3. Displacement profiles of the external upright in the center of the rack, for the three rack configurations.

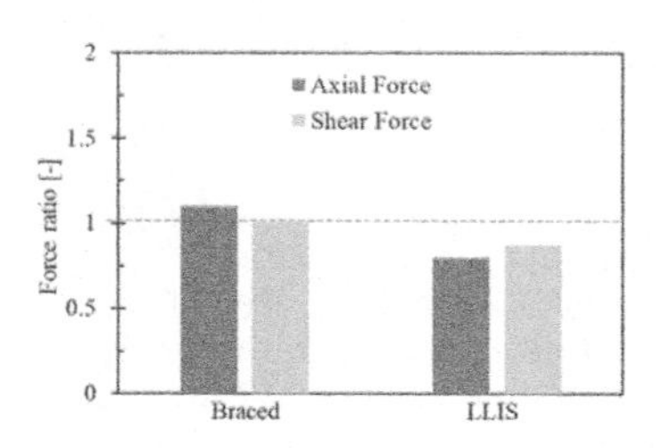

Figure 4. Ratios between the maximum forces of rack with intervention (braced or with LLIS) and those of the existing rack.

REFERENCES

Donà, M., Muhr, A.H., Tecchio, G. & da Porto., F. 2017. Experimental characterization, design and modelling of the RBRL seismic-isolation system for lightweight structures. Earthquake Engineering and Structural Dynamics 46: 831–853.

Simoncelli, M., Tagliaferro, B. & Montuori, R. 2020. Recent development on the seismic devices for steel storage structures. *Thin-Walled Structures* 155: 106827.

Current Perspectives and New Directions in Mechanics, Modelling and Design of Structural Systems – Zingoni (ed.)
© 2022 Copyright the Author(s), ISBN 978-1-032-39114-4

Seismic considerations in the dynamic characteristics of masonry infilled RC frames based on experimental tests

F.J. Pallarés, A. Agüero & L. Pallarés
Universitat Politècnica de València, Valencia, Spain

ABSTRACT: In seismic design of buildings, partitions and enclosures are usually treated as non-structural elements, although, experiences such as the one that occurred in the Lorca earthquake in Spain (2011), among others, emphasize the importance of a global conception of all the elements that make up the design of a building. Many international Standards indicate that non-structural elements that can alter the conditions of the structure will be taken into account for the structural analysis model, and they will be checked accordingly; or, alternatively, constructive solutions should be adopted that guarantee the resistant non-participation of these elements. The interaction between the brick walls and the structural skeleton modifies vibration frequency of the buildings, introducing significant changes in the dynamic characteristics of the frames, as stiffness. This paper presents an experimental study where accelerations have been registered in full-scale reinforced concrete frames infilled with masonry, which are excited using a hammer to know their dynamic characteristics. The obtained results are compared with the results registered for the bare frame specimen, and both tested under cyclic loading until failure. The increase in the stiffness becomes clear, showing important changes in the response of the frames that should be considered in the designing phase, either taking the walls into consideration in the calculations or adopting isolation measures as commented in this text.

1 INTRODUCTION

The very important role played by the non-structural elements made with brick (interior partitions and façades or exterior enclosures) is critical in seismic behavior. The stiffness and strength of a masonry layer in its own plane are in many occasions greater than that of the structure in which it is placed, determining its behavior.

The present contribution aims to show an experimental test where the dynamic properties of the frames are evaluated in terms of natural frequencies and stiffness for infilled and unfilled frames, showing the clear disagreement between design and construction phase, where nonstructural elements are not taken into account.

2 STATE OF THE ART

As commented, in seismic design of buildings, partitions and enclosures are usually treated as non-structural elements, although, experience shows the importance of a global conception of all the elements that make up the design of a building and the role played by these non-structural elements in the seismic response.

The interaction between the brick walls and the structural skeleton modifies the vibration frequency of the building, as demonstrated by Smith and Carter (1967) back in the 60s, or Bertero and Brokken (1983), who indicated that the infills introduce significant changes in the dynamic characteristics of the frames, decreasing the fundamental period of vibration by 54% due to the effect of the infills, while the mass hardly represents an increase of 10%. The presence of masonry infills leads to the stiffening of the structure, what is translated in a decrease of the natural period of vibration of the structure and the consequent increase of the seismic forces.

Penava et al. (2018) state that ignoring contribution of masonry infill wall to the shear resistance of the infilled frame structural system can lead to inaccurate predictions of the system's behavior.

3 EXPERIMENTAL TEST

An experimental laboratory test has been carried out to compare a reinforced concrete bare frame and a masonry infilled RC frame with double masonry panel as it is typically constructed in the facades of real buildings. The bare frame is usually accounted for the calculations during design phase in routine

DOI: 10.1201/9781003348450-28

seismic design of buildings, while the infilled frame is the usual as-built wall with a conventional façade masonry infill type made up of two different layers of bricks with mortar. Both frames have been instrumented with 4 PCB high sensitivity seismic accelerometers according to the locations shown in Figure 1 for the infilled frame. Furthermore, the frames have also been tested under cyclic loading to compared the change in stiffness as a consequence of the infill.

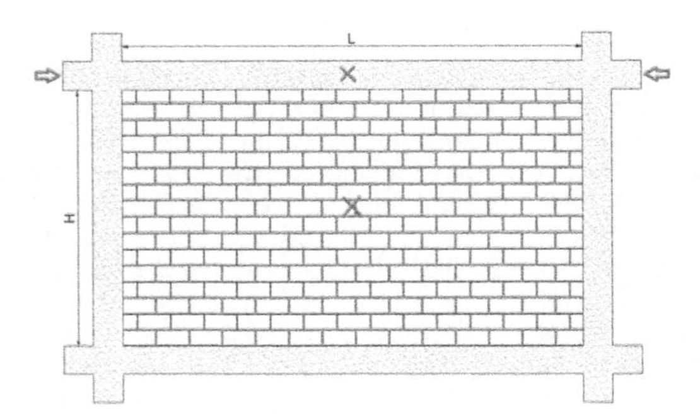

Figure 1. Accelerometers locations in infilled frame.

4 RESULTS

Only in plane accelerations are shown in this paper with the aim to show the change in dynamic parameters for in plane calculations. The recorded accelerations have been treated to get modal parameters, mainly the natural frequencies through an FFT analysis. Figure 2 shows the in plane first natural frequency detected for the infilled frame. The first natural frequency is increased from 9.2 Hz up to 38.6 Hz when compared to the bare frame, four times the initial value, which is in accordance with the significant changes mentioned in the Introduction section.

The stiffness of frames is a key parameter in seismic design and the amount of attracted seismic force. The effective secant stiffness of positive and negative displacement cycles is combined and plotted in Figure 3 for each drift cycle.

5 CONCLUSIONS

This contribution presents the influence of masonry infills in the dynamic characteristics (mainly natural frequency and stiffness) of reinforced concrete frames through an experimental test using accelerometers to record free damping vibration, and using an actuator to induce cyclic loading.

The initial stiffness of the infilled frame that would be used in elastic calculations in which the effect of the façade partition is taken into account, is 8 times higher than that of the bare frame for designs not considering the infills.

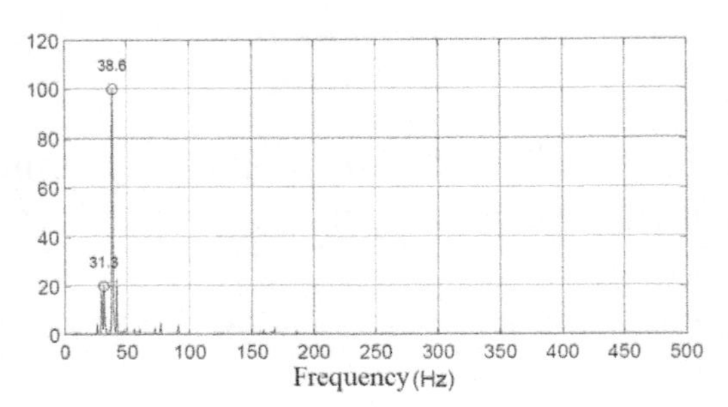

Figure 2. FFT analysis for the infilled frame.

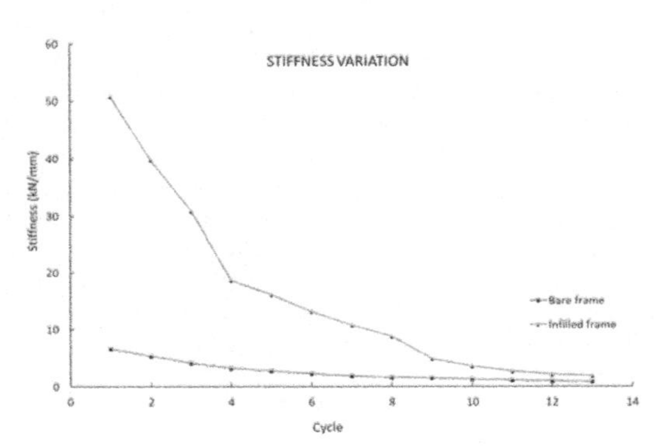

Figure 3. FFT analysis for the bare frame.

The consideration of masonry infills in design phase for building concrete frames significantly increases the stiffness of the structure, decreasing the fundamental periods and increasing seismic forces.

ACKNOWLEDGMENTS

The authors would like to express their sincere gratitude to the AGENCIA ESTATAL DE INVESTIGACION (Spanish Ministry of Science and Innovation) supporting and funding through the Project PID2019-110460RB-I00/AEI/10.13039/501100011033.

REFERENCES

Smith, S. & Carter, B. 1967. Methods for predicting the lateral stifness and strength of mutli-storey infilled frames. *Building SCI*. p. 247–257.

Bertero, V. & Broken S. 1983. Infills in Seismic Resistant Building. *Journal of Structural Engineering*. 109(6).

Penava D., Sarhosis V., Kožar I., Guljaš I. 2018. Contribution of RC columns and masonry wall to the shear resistance of masonry infilled RC frames containing different in size window and door openings. *Engineering Structures*, 172, 105–130.

Current Perspectives and New Directions in Mechanics, Modelling and Design of Structural Systems – Zingoni (ed.)
© 2022 Copyright the Author(s), ISBN 978-1-032-39114-4

Structural plan and design of bridge deck isolation system for seismic redundancy

M. Matsumura, K. Annoura & H. Moriyama
Kumamoto University, Kumamoto, Japan

ABSTRACT: Bridge isolation using laminated rubber bearing has been widely used for seismic safety and reduces response displacement of substructures during earthquake. But in recent large earthquakes in Japan, damages at girder ends of steel girder bridges prevented an easy repair and an early recovery of traffic. Then the authors focused on a bridge deck isolation of the steel girder bridge having deck system for damage mitigation and seismic redundancy and have conducted feasibility studies of the system. This paper presents application examples of the bridge deck isolation into a steel I-shaped girder bridge and its seismic responses through dynamic response analysis. The analytical results indicate that the bridge deck isolated by using a low frictional sliding surface over the steel girder and supported laterally by buffers effectively reduces reaction forces of bearings. Structural plans to enhance repairability and maintainability are also taken into the application examples.

1 INTRODUCTION

Bridge isolation using laminated rubber bearing has been widely used for seismic safely of bridges after the Hyogo-ken Nambu Earthquake. The isolation bearings are set at the connections between the superstructure and the substructure. The seismic isolation of the superstructure is expected to reduce the seismic force delivered to the substructure and their displacement response by making their natural period long and adding damping. However, the displacement response of the superstructure tends to increase. Figure 1 show damages of girder ends of steel girder bridges. Breaks of rubber bearing and collisions of superstructure and etc. were observed in recent large earthquakes in Japan and such damages prevented an easy repair and an early recovery.

Here, deteriorations of RC deck due to increasing traffic are problems in bridge maintenance. Then in restoring work of existing steel bridges having deteriorated RC deck, the RC deck are replaced to precast PC deck, providing a short construction period and high durability, and existing steel bearings are also replaced to the seismic isolation rubber bearings. Then the size of the seismic isolation rubber bearing tends to become larger to support the dead load, especially when the PC deck of heavier weight is adopted.

As these result, inertia force (horizontal seismic force) in direct proportion to mass distributions of a superstructure having the PC deck of a heavy weight increases and the increased horizontal force is subjected to the steel girder of a less shear resistance. Hence, steel girder and bridge bearing in bridge isolation using the rubber bearings will be vulnerable against seismic accelerations exceeding the maximum acceleration assumed in seismic design.

Another problem of the bridge isolation using the rubber bearings is a narrow space around the bearings at the connections between the superstructure and the substructure.

Then, the authors are to propose a new type bridge isolation system, isolating the heavy RC deck supported by steel girders of small stiffness, to mitigate damages of steel girder and to improve maintainability.

This paper presents application of the steel girder bridge having the isolated deck system and feasibility investigations of the system.

Figure 1. Breaks of rubber bearing.

DOI: 10.1201/9781003348450-29

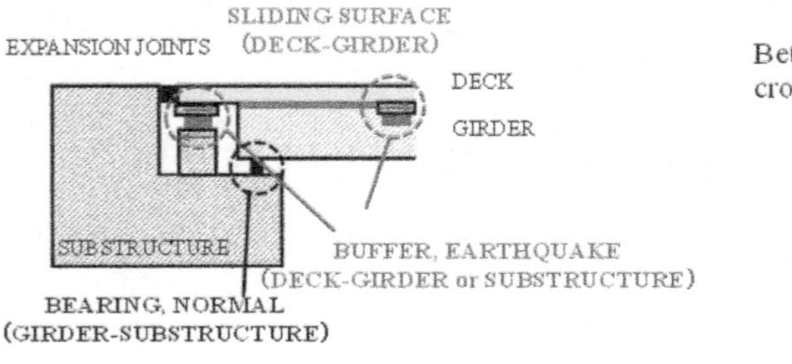

Figure 2. Function separation and deck isolation.

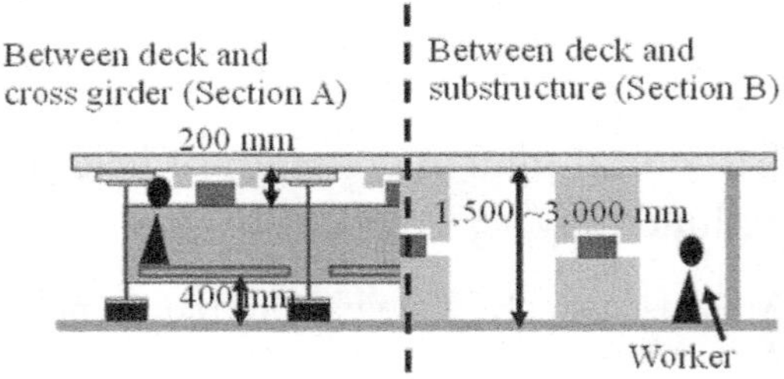

Figure 4. Maintenance space in deck isolation.

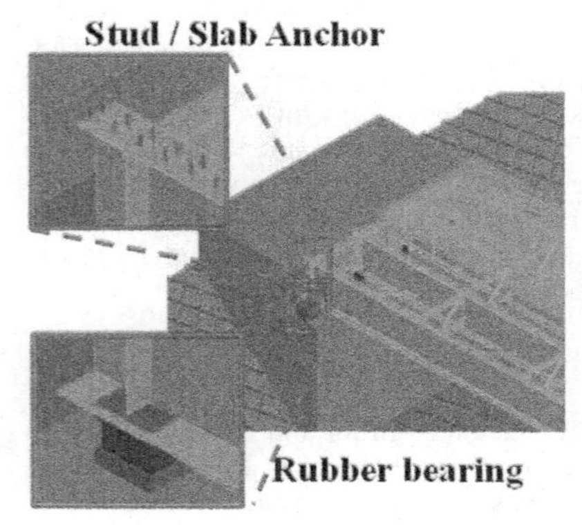

(a) Isolation using rubber bearing

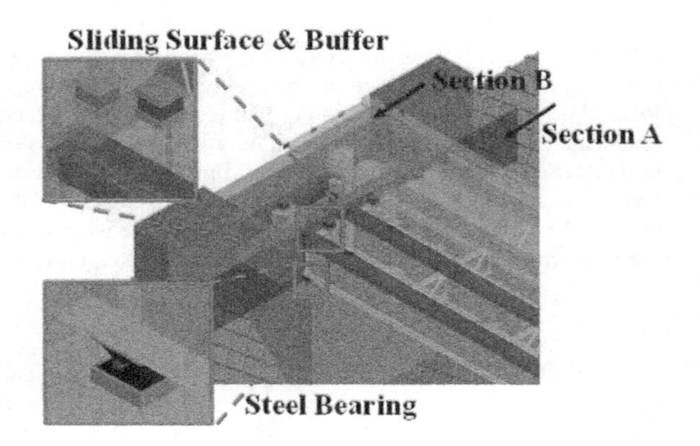

(b) Deck isolation

Figure 3. Conventional isolation vs deck isolation.

2 CONCEPT OF DECK ISOLATION

2.1 *Features of deck isolation*

Effective way to mitigate damage of steel girder of less shear resistance is to reduce the seismic force subjected to it by isolating the bridge deck of the heavy weight. Then installation a sliding surface between the upper part of steel girder and the lower surface of the concrete deck is considered as illustrated in Figure 2. Figure 3 compares (a) conventional type of bridge isolation using the rubber isolation bearings and (b) bridge deck isolation.

Figures 3 (b) and 4 illustrate installations of the bearings at the girder end, considering space for maintenance. As the deck isolation is required only during a severe earthquake, the figure explains function separation of the bearings; ones for normal loading conditions and ones during earthquake. The bearings for normal loading conditions in survive are located at the connecting part between the superstructure and the substructure and the other bearings horizontally support the bridge deck only during earthquake. These structures may be adoptable when replacing the RC deck to the PC deck in restoring work and reduce the seismic force subjected not only to the superstructure but to the substructure.

2.2 *Performances of deck isolation*

Figure 4 shows installations of the buffers to support the isolated deck during earthquake in the horizontal direction and to control seismic response of the deck. These buffers will be installed on the cross girder between the deck and the cross girder and/or the substructure to secure a wide space so that workers can access them for normal maintenance work and for emergent inspection, including repair work.

3 CONCLUSIONS

In this study, the bridge deck isolation to mitigate damage of steel girders and substructures and to reduce the risk of bridge fall are introduced and the application example are discussed based on the dynamic response analysis.

The differences between the conventional bridge isolation using the rubber isolating bearings and the deck isolation using the sliding of the deck are compared. The performance requirements of the deck isolation are also summarized.

REFERENCE

Annoura, K. Matsumura, M. Kasai, A. and Moriyama, H. 2020. Fundamental consideration damage mitigation of superstructure by deck isolation. *Proceeding. of Construction Steel*. Vol.28: pp.629–635. (in Japanese)

Current Perspectives and New Directions in Mechanics, Modelling and Design of Structural Systems – Zingoni (ed.)
© 2022 Copyright the Author(s), ISBN 978-1-032-39114-4

Study on seismic performance of fully bolted joints

Bing-Long Wu
School of Civil and Transportation Engineering, Beijing University of Civil Engineering and Architecture, Beijing, PR China

Yan-Xia Zhang
School of Civil and Transportation Engineering, Beijing University of Civil Engineering and Architecture, Beijing, PR China
Beijing Advanced Innovation Center for Future Urban Design, Beijing, PR China

Sen Shen & Zhe-Wen Huang
School of Civil and Transportation Engineering, Beijing University of Civil Engineering and Architecture, Beijing, PR China

ABSTRACT: Conventional steel box-section column connections are welded onsite, high labor cost, poor seismic performance, and environmental air pollutions. In view of the above questions,the core tube flange connection structure is proposed in this paper.A core tube is arranged inside the column and is connected between the upper and lower columns by flange plates and high-strength bolts .The technique is efficiently assembled and environmentally friendly. Based on the previous research, tapping bolts are added to improve the joint stiffness and bearing capacity. High-fidelity finite-element models are established. The feasibility and correctness of the finite element method are verified by comparison with the experimental structure.A comparative study is carried out on the TFBC with respect to the traditional box-section column connections utilizing finite element models. The research effectively demonstrates that the proposed TFBC has sufficient stiffness, desired loading capacity, and favourable rigid performance and TFBC has similar mechanical properties to welding joint.

Keywords: Box-Section Column Connection, Core-Tube, Flanges, Steel Structure, Performance Analysis

1 INTRODUCTION

Since 2016, the research group has proposed a prefabricated box-column flange-bolt column connection joint, and successively investgated an ample amount of experiments and numerical analysis of the prefabricated flange-bolt connection joint with double flanges and single flanges [1–4], and proposed a joint design method. The research showed that the flange-bolt column connection had the same structural properties as the welded joint. When the inter-story displacement is large, the flange plate will deform to a certain extent, and the deformation cannot be repaired generally. Furthermore, the setting of the prestressed steel strands can improve the toughness of the structure and increase the stiffness of the joints.

However, the research of prestressed steel strands in multi-story steel structures mainly focuses on the connection of horizontal members, yet there are few studies on the connection of vertical members. Based on the above researches, the research group proposed a prefabricated vertical short strand and tapping high-strength bolt-core sleeve flange column connection joint (TFBC). Through parametric analysis, the influence of whether to set tapping high-strength bolts and tapping high-strength bolts of different performance rating on the joint performance is further explored.

2 JOINT STRUCTURE

The prefabricated vertical short strand and tapping high-strength bolt-core sleeve flange column connection joint is mainly composed of column, flange plate, core sleeve, vertical short strand, high-strength bolts and tapping high-strength bolts. The structure of the joint is shown in Figure 1.

DOI: 10.1201/9781003348450-30

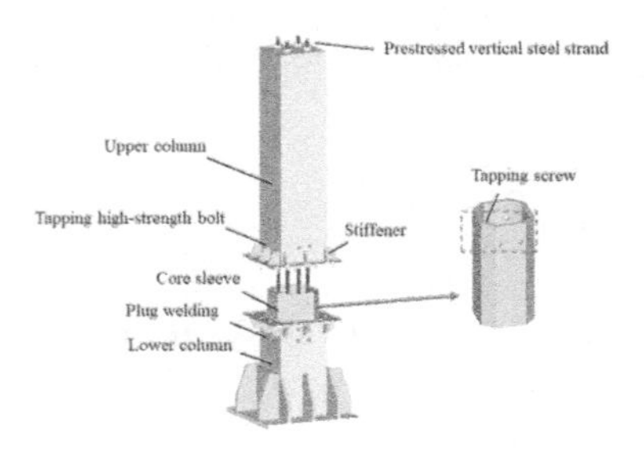

Figure 1. Prefabricated vertical short strand and tapping high-strength bolt-core sleeve flange column connection joint.

3 TEST AND FINITE ELEMENT MODEL

3.1 *Test description*

The test adopts the pseudo-static loading method, and the loading is controlled by the horizontal displacement. In the horizontal direction, the 200t servo actuator is connected to the connecting piece fixed on the top of the column, and the vertical axial force is applied by 100t center-hole jacks.

3.2 *Material properties*

The constitutive relations of steel and bolts adopt the bilinear model, and simulate the prestress of the vertical short strand by setting the linear expansion coefficient of the vertical short strand.

3.3 *Geometry and finite element mesh*

The main part of the joint finite element model adopts the solid element C3D8R eight-node hexahedron element, and the prestressed vertical short strand adopts the T3D3 three-dimensional three-node truss element.

4 VALIDATION OF THE FINITE ELEMENT MODEL

4.1 *Hysteretic curves*

The finite element model is in good agreement with the hysteretic curve obtained from the test.

4.2 *Skeleton curves*

The skeleton curves of the numerical simulation match the one of tests.

4.3 *Flange opening*

Under each loading level, the flange opening obtained by the finite element model is similar to the test.

4.4 *Failure mode*

The failure mode of the test component is the tearing of the steel at the column foot, and the test results are in good agreement with the phenomenon of the numerical simulation.

5 PARAMETRIC STUDY

5.1 *Flange opening*

Setting tapping high-strength bolts can limit the relative deformation at the flange plate of the joints. The higher the rating of tapping high-strength bolts, the smaller the flange opening.

5.2 *Joint's bear capacity*

Setting tapping high-strength bolts can improve the bearing capacity of the joints. The higher the rating of tapping high-strength bolts, the higher the joint's bear capacity.

6 CONCLUSIONS

The main conclusions are as follows:

(1) The overall shape of the hysteresis curve of the TFBC is relatively full, and it has a high flexural bearing capacity, realizing the seismic design goal of "strong joint, weak member".
(2) Comparing the hysteretic curves and skeleton curves obtained by the finite element model with the experiment, the curves are in good agreement. The flange opening and the tearing of the steel at the column foot in the finite element model are basically consistent with the experimental phenomena.
(3) Setting tapping high-strength bolts can limit the relative deformation at the flange plate and improve the bearing capacity of the joints. The higher the rating of tapping high-strength bolts, the better the mechanical properties.

ACKNOWLEDGEMENTS

This study is supported by Beijing Energy Conservation & Sustainable Urban and Rural Development Provincial and Ministry Co-construction Collaboration Innovation Center and Joint Program of Beijing Natural Science Foundation and Education Commission (KZ201910016018).

REFERENCES

[1] Y.X. Zhang, W.Z. Huang, M.Z. Zheng, Y. Wang, G. Ning, Quasi-static experimental research on box-section all-bolted column connection with inner sleeve[J]. Industrial Construction, 48(05):37–44, 2018
[2] Y.X. Zhang, M.Z. Zheng, W.Z. Huang, K. Jiang, G. Ning, Analytical study on fully-bolted box section column-column connection with inner sleeve[J]. Progress in Steel Building Structures, 20(04):34–46, 2018
[3] Y.X. Zhang, B.L. Wu, A.L. Zhang, A.R. Liu, Experimental study on seismic performance of plane frame with box column connection by core tube and double flange[J]. Journal of Building Structures, 43(02):29–42, 2022
[4] A.L. Zhang, J. Wang, Y.X. Zhang, A.R. Liu, Quasi-static testing of core-tube box-column steel frames with double flanged rigid connections. J. Engineering Mechanics, 38(09):146–160, 2021.

Current Perspectives and New Directions in Mechanics, Modelling and Design of Structural Systems – Zingoni (ed.)
© 2022 Copyright the Author(s), *ISBN 978-1-032-39114-4*

Effects of in-plan irregularity of RC framed buildings on the out-of-plane nonlinear seismic response of masonry infills

F. Mazza
Dipartimento di Ingegneria Civile, Università della Calabria, Rende (Cosenza), Italy

ABSTRACT: Masonry infills (MIs) are typically found in reinforced concrete (RC) framed buildings, but they are commonly neglected in seismic design practice. As expected, the increase of lateral stiffness induced by the in-plane (IP) behaviour of MIs may emphasize or mitigate torsional effects in a building with asymmetric plan, depending on their distribution in the perimeter frames and potential damage related to the out-of-plane (OOP) failure modes. The present work is aimed at identifying the effects of lower and upper bound IP and OOP nonlinear modelling assumptions of MIs on their seismic damage. To this end, a five-storey RC framed structure, characterized by an L-shaped plan and bays of different length, is chosen as representative of a spread typology of residential buildings in Italy. A simulated design of the fixed-base structure is preliminarily carried out in line with a former Italian code, for medium-risk seismic region and typical subsoil class. Then, seismic retrofitting with elastomeric and sliding bearings is carried out, in order to attain performance levels imposed by the current Italian code in a high-risk seismic zone. Four structural models are considered: i.e. the bare structure, with nonstructural MIs; three infilled structures, with MIs arranged in order to reduce (elastic) torsional effects of the fixed-base structure along one or both the in-plan principal directions. Nonlinear dynamic analyses are carried out by employing a self-built C++ code for RC infilled framed structures. Records of recent earthquakes in central Italy are considered, matching on average the design response spectrum of acceleration for the geographical coordinates at the selected site.

1 INTRODUCTION

The first aim of this work is to identify the effects of lower and upper bound assumptions for IP and OOP stiffness and strength properties of MIs on their nonlinear seismic response. To this end, a fixed-base RC framed building, with L-shape in plan and bays of different length along both principal directions, is simulated as built in L'Aquila (Italy), assuming three alternative arrangements of MIs made with two leaves of clay hollow bricks. Then, the effectiveness of elastomeric (e.g. high damping rubber) and sliding (e.g. curved surface sliding) bearings as retrofitting technique of the original structure able to improve the IP and OOP behaviour of MIs is investigated. Nonlinear seismic analyses are carried out by employing a self-built C++ computer code for RC framed structures, where a non-structural module allows nonlinear modelling of MIs through a five-element macro-model. It predicts the IP and OOP force-displacement laws of MIs and describes the effects of simultaneous or prior IP damage on the OOP seismic response. Attention is focused on seismic sequences recorded in the near-fault area during recent earthquakes in central Italy (i.e. L'Aquila in 2009 and Rieti in 2016).

2 IN-PLAN IRREGULAR TEST STRUCTURES

A five-storey RC framed building with asymmetric plan (Figure 1), the first storey of 4 m and the other ones of 3.3 m height each, is designed as bare frame in line with provisions of a former Italian code for a medium-risk zone. Three in-plan configurations of structural MIs are assumed in order to reduce (elastic) torsional effects of

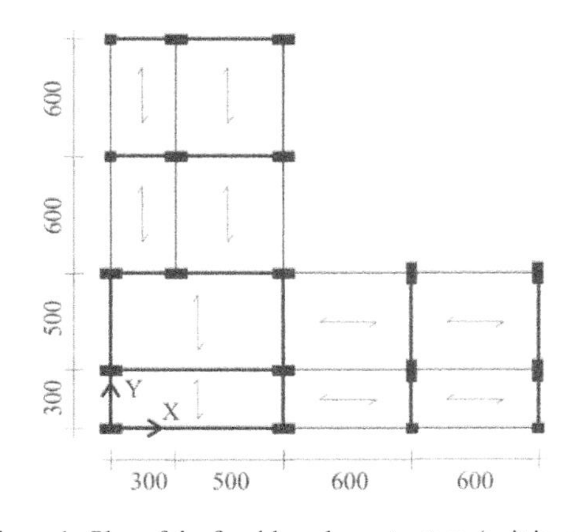

Figure 1. Plan of the fixed-base bare structure (unit in cm).

DOI: 10.1201/9781003348450-31

the fixed-base infilled structure along X (C1) and Y(C2) or both (C3) the in-plan principal directions.

Base-isolation with elastomeric (i.e. high damping rubber, HDR) and sliding (i.e. curved surface sliding, CSS) bearings is considered for the seismic retrofitting of the above mentioned fixed-base bare structure in a high-risk seismic zone. Nineteen HDRBs with the same dimensions are designed assuming an equivalent viscous damping ratio in the horizontal direction $\xi_{I,H} = 15\%$ and a fundamental vibration period $T_{I,H}$=2.5 s. Nineteen CSSBs with constant values of the radius of curvature (R=450 cm) and dynamic friction coefficient (μ =4.5%) are assumed in order to obtain an equivalent viscous damping ratio in the horizontal direction $\xi_{I,H} = 31.6\%$ and a fundamental vibration period $T_{I,H}$=3.1 s.

3 NUMERICAL RESULTS

Nonlinear dynamic analysis of the L-shaped fixed base and base-isolated RC framed structures is carried out to evaluate IP and OOP seismic response of MIs. Specifically, eighteen structural solutions are tested, combining three in-plan configurations of MIs together with lower and upper bound IP and OOP force-displacement laws of MIs. Eleven near-fault earthquakes (EQs) in central Italy are selected from the Italian Accelerometric Archive (ITACA): i.e. AQA, AQK, AQV and AQG stations, for L'Aquila EQ (6/4/2009); AMT, NOR, NRC and FEMA stations for Accumoli (Rieti) EQ (24/8/2016); NOR station for Ussita (Rieti) EQ (26/10/2016); FCC and NOR stations for Norcia (Rieti) EQ (30/10/2016).

Firstly, the global percentage of IP and OOP collapse mechanisms of MIs are plotted in Figure 2 for three (i.e. C1, C2 and C3) in-plan configurations of MIs of the fixed-base infilled structure. As can be observed, all configurations are not able to prevent collapse mechanisms, with similar IP behaviour. The C3 configuration is found to be more effective at improving the OOP behaviour, while OOP collapses about twice than those observed in the IP direction are resulted for the C1 and C2 configurations.

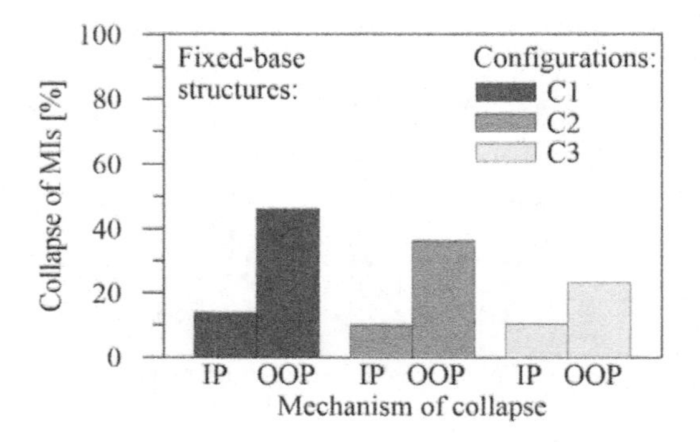

Figure 2. Percentage of IP and OOP collapse mechanisms for three configurations of MIs of the fixed-base structure.

Extremely low percentages of IP collapses is highlighted for the C1 configuration opposed to high OOP risk of collapse, especially for MIs with the lowest value of the aspect ratio. On the other hand, the C2 configuration is prone to both IP and OOP collapses of all typologies of MIs, thus proving the worst solution among those examined. Note that infill panels with the highest values of the aspect ratio are particularly inclined to premature IP collapse, while the opposite (OOP) behaviour is observed for the typology characterized by the lowest aspect ratio.

The maximum IP drift ratio of the original fixed-base (FB) and retrofitted base-isolated (BI.HDRB and BI.CSSB) infilled structures is shown in Figure 3. As shown, BI.HDRB are sound to be more effective than BI.CSSB at reducing the IP drift ratio, so avoiding predominant IP collapse expected for panels with the highest value of the aspect ratio. The overview of MIs collapsed in facades, omitted for the sake of brevity, confirms that IP collapses are totally removed in base-isolated structures, while only limited OOP phenomena happen at first floor.

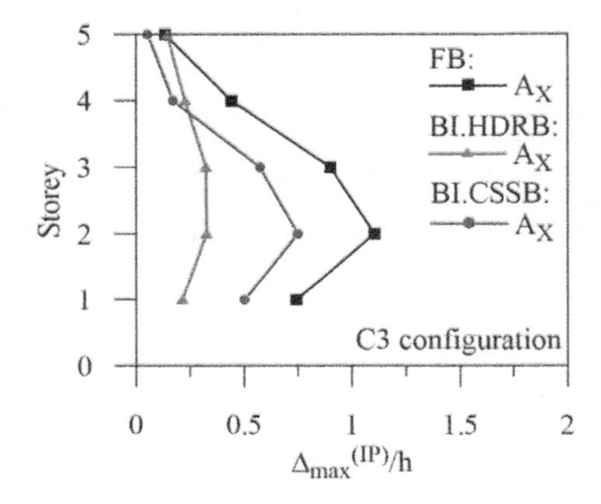

Figure 3. Maximum IP drift ratio for fixed-base and base-isolated structures: MIs for the C3 configuration.

4 CONCLUSIONS

The seismic protection of MIs against their IP and OOP collapse is investigated herein with reference to in-plan irregular fixed-base framed structures retrofitted by means of elastomeric and sliding base-isolation systems. Variability of IP and OOP mechanical properties and aspect ratio of MIs are also studied considering three in-plan distributions aimed to reduce torsional effects. Nonlinear seismic analysis of eighteen test structures is carried out by means of eleven near-fault earthquakes recently occurred in central Italy. Fixed-base structures with different in-plan configurations of MIs are not able to prevent IP collapse, albeit the C3 solution is resulted more effective than C1 and C2 at reducing the OOP collapse of MIs. Infill panels with the highest values of the aspect ratio are particularly inclined to premature IP collapse, while the typology with the lowest aspect ratio is prone to OOP collapse. Both base-isolation retrofitting solutions are sound to be effective at limiting the IP drift ratio and OOP drift and acceleration ratios, totally removing IP collapses of MIs and allowing only few OOP collapses.

Current Perspectives and New Directions in Mechanics, Modelling and Design of Structural Systems – Zingoni (ed.)
© 2022 Copyright the Author(s), ISBN 978-1-032-39114-4

Influence of infill walls on lateral load carrying capacity of buildings

M.B. Waris, H. Al-Hadi & A. Al-Nuaimi
Department of Civil and Architectural Engineering, Sultan Qaboos University, Muscat, Oman

ABSTRACT: This study investigates the contribution of infill walls to the capacity and behavior of reinforced concrete structures against lateral loads. A 27.2 m high 8-storey building with a footprint of 20m x 40m is modelled using ETABS. Cases with bare frame, fully infilled frames, and partially infilled with openings (25%, 50%, 75%) are considered. The study considered infill thicknesses of 90, 140, 200 and 230 mm. Single equivalent nonlinear strut is used to model the infill and non-linear pushover analysis is used to investigate the model behavior. The consideration of infill walls had a significant effect on the natural period that dropped from 1.45s to 0.77s due to infill. It is found that the presence of infill wall increases the lateral stiffness of the structure. The increase in stiffness is directly proportional to the infill thickness. The initial stiffness was 4.4 times the bare frame for the case with 230 mm thick infill walls. Presence of openings significantly reduced the contribution of infill walls. For cases of infill walls with 50 % opening, the additional stiffness is reduced to just 10 % while for 75% opening it is less than 1 %. Therefore, the presence infill with openings of 50% or higher will not play a significant role in the analysis and design process.

1 INTRODUCTION

This research investigates the effect of infill walls on an eight-story building resisting the lateral loads applied on the structure design as per Omani Seismic Code without consideration for infill walls. The infill is modeled using the single strut model to representation of the infill. Pushover analysis is used to compare the seismic capacity of the building to investigate the effect of infill thickness and opening in infill. Infill thickness of 90 mm, 140 mm and 200 mm is considered, while opening of 25%, 50% and 75% are investigated. Also the study considered the case of using dual infill considering 90+140 mm vs 230 mm infill.

2 METHODOLOGY

The study considered an 8-story building modelled using ETABS. The structure has been loaded using a live load of 1 kN/m^2 for floors on the roof, and 2.5 kN/m2 for other floors. The dead load applied was 2 kN/m^2 on all floors. Concrete strength used for the structure is 25 MPa while the reinforcement steel yield strength is 460 MPa. The plan view is shown in Figure 1. The multi-response spectrum analysis is used to design the building in Zone-1 considering the soil type as Type C considering the Oman Seismic Code (OSC). The structure as a moment resisting system used for residential purpose. The base shear estimated as 1338 kN and 1267 kN in the X and Y-directions respectively. The design was carried out as per EC-2 considering the provisions of Omani Seismic Code considering infill load on the peripheral beams based on thickness of 230 mm. Constant dimension of 250 x 450 mm and 450 mm x 450 mm were selected for beams and columns respectively. This is to ensure that the design is safe for all infill thicknesses and to maintain symmetry in plane and elevation of the structural system. The study used the equivalent single strut to model infill with different thickness and used reduction factor to the strut stiffness to account for the openings.

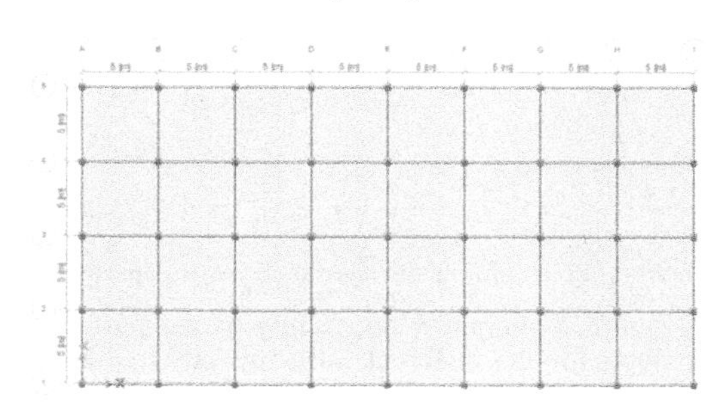

Figure 1. Plan of the 8-story building used in the study.

3 RESULTS AND DISCUSSION

Since with the change in infill thickness the weight and therefore, the base shear of each case will change the discussion is made considering the ratio

DOI: 10.1201/9781003348450-32

of base shear-to-weight. Figure 2 compares the base shear-to-weight ratio (V_b/W) obtained from the push-over analysis with the design base shear to weight ratio from OSC. There is an increase in the relative design base shear of the building due to the drop in natural period. However, the change in base shear capacity due to the infill is more significant. The base shear/weight ratio of a 90 mm thick wall is approximately 73.4% greater than the base shear of the bare frame. Therefore, having a very thin wall is still improvement to the lateral resistance of the building. The thickness of the infill therefore does affect the design as well as the capacity of the building.

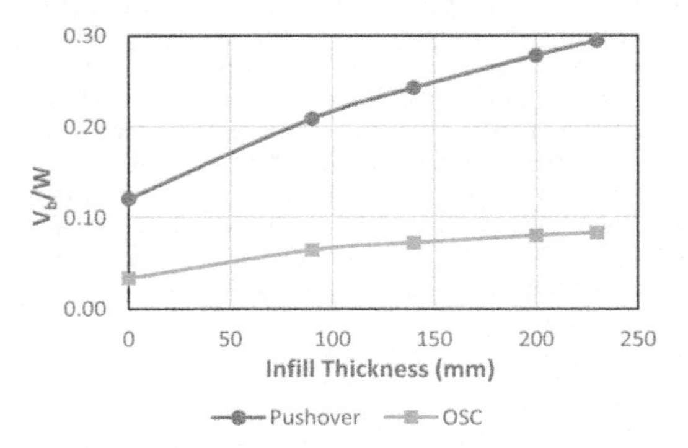

Figure 2. Effect of Infill Thickness on Base Shear.

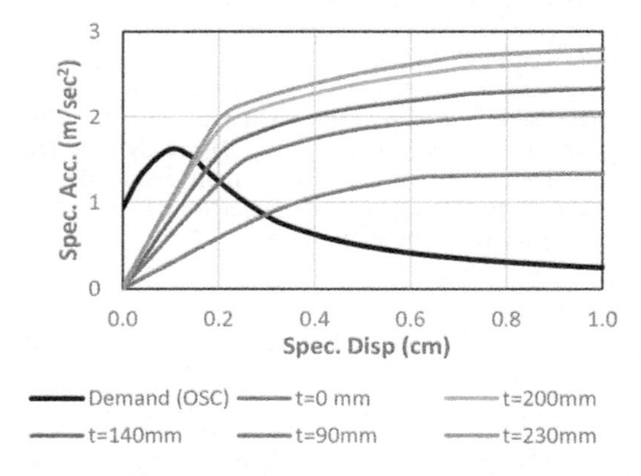

Figure 3. Effect of Infill thickness on Seismic Capacity.

Figure 3 compares the results of the pushover analysis for the different infill thickness with the demand based on OSC. As the thickness increases, the wall provides more assistance to resist lateral loading. The performance point obtained by the intersection between the demand and the capacity represents the model performance under the seismic demand. It can be observed that the performance point for the bare frame model is located approximately on the yield point. This indicates that the bare frame is progressively damaged under the subjected demand. However, it can be noticed that the performance point is located at earlier stages for the infill models compared to the bare frame. Also, the models with higher thickness have their performance point located at within the elastic range of behavior.

Figure 4 shows the base shear/weight ratio follow inverse relationship to the opening percentage. It is noticed that the behavior of 50% & 75% opening models is approximately identical to the bare frame. For the 25% opening model, the base shear/weight ratio is reduced to 56%, while that for 50%.75% and 100% is around 44%. This indicates that if the infill has more than 50% opening, it will not provide any considerable resistance to the lateral load applied on the structure.

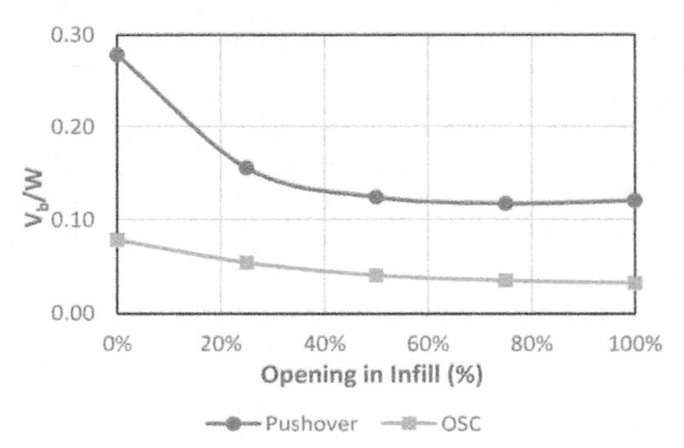

Figure 4. Effect of openings in infill on base shear.

4 CONCLUSIONS

The study ETABS is used to model 8-storey buildings to investigate the effect of infill thickness and openings on the performance of an 8-story building design as per Oman Seismic Code (OSC). The infill walls is modelled using the single equivalent diagonal strut method. The presence of infill wall increased the lateral stiffness of the structure significantly even for the small wall thickness of 90 mm. With increase in infill thickness of the wall, the more lateral stiffness gained by the structure. The presence of infill effects the natural period making it stiffer. Infill also causes a shift in the performance point of the building, therefore, should be considered in the analysis and design process even for smaller thickness. For infill walls with openings equal to or exceeding 50%, the effect on stiffness, natural period, and performance in insignificant. Therefore, such walls may be ignored during analysis and design.

REFERENCES

Oman Seismic Code for Buildings (2013), Earthquake Monitoring Center, Sultan Qaboos University Press.

Current Perspectives and New Directions in Mechanics, Modelling and Design of Structural Systems – Zingoni (ed.)
© 2022 Copyright the Author(s), ISBN 978-1-032-39114-4

Simulation of many acceleration time histories with causality from an observed earthquake motion: Modeling of real part and reproduction of imaginary part by Hilbert transform

Tadanobu Sato
Emeritus Professor of Kyoto University and Research Associate of Kobe-Gakuin University

1 INTRODUCTION

To conduct the performance design we need many acceleration time histories to evaluate the dynamic nonlinear behavior of concerned structural systems. One practical method is to apply the stochastic process for simulating acceleration time histories. For this purpose we extract the stochastic characteristic of acceleration time history and its uncertainty from an observed earthquake motion. In this research we make a model to simulate a sample real part of Fourier transform of an observed acceleration time history. The imaginary parts of the Fourier transform of a causal acceleration time history can be reproduced from the real part by using the Hilbert transformation. Using this well known fact we develop a method to simulate a causal acceleration time history. First we extract an average trend from the root square of the real part. Dividing the real part by this average trend we obtain a new process, which is named as the standardized real part. After proving that this process have had a stationary characteristic we extract the probabilistic characteristics from this standardized real part process. We then develop a method to simulate a sample process of the standardized real part by using the concept of the autoregressive process. Multiplying the average trend to this sample standardized real part process we can obtain a sample real part process. Appling the concept of the Hilbert transform to this sample real part process we can obtain the imaginary part.

In this paper we assume that an observed acceleration time history is given at the design site, from which we can extract its stochastic characteristics. For that purpose, in the following analysis we use an acceleration time history that is the EW component observed at Suttus Observatory during 1993 Off Hokkaido southwest earthquake, which is shown in Figure 1. The total data points of an offered acceleration time history was 16384 then adding zero to the later part of time history until total data points bcome 2^{16}, by which the causality of acceleration time history is assumed to be realized. We model the real part of the Fourier transform of an observed acceleration time history.

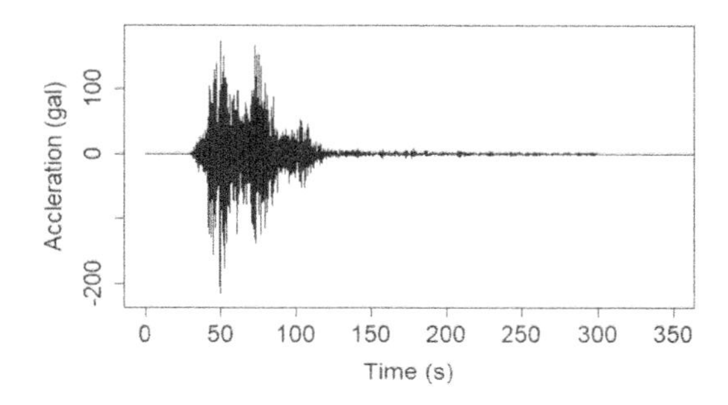

Figure 1. Observed acceleration time history.

By the inverse Fourier transform of a sample real part process with the causal imaginary part we can simulate sample causal acceleration time histories.

2 HILBERT TRANSFORM

The real part $R(\omega)$ and imaginary part $i\Im(\omega)$ of a causal acceleration time history has the following Hilbert transform relation

$$i\Im(\omega) = -\frac{i}{\pi}\int_{-\infty}^{\infty}\frac{R(y)}{\omega - y}dy = \frac{1}{2\pi}R(\omega)\otimes\frac{2}{i\omega} \quad (1)$$

We have the following Fourier transform relations

$$\mathrm{sgn}(t) \Leftrightarrow \frac{2}{i\omega} \quad (2)$$

Considering Eq.2 and taking the inverse Fourier transform both sides of Eq.1 we have

$$h_i(t) = \mathrm{sgn}(t)h_e(t) \quad (3)$$

where $h_i(t)$ is the time function, which is obtained by the inverse Fourier transform of

DOI: 10.1201/9781003348450-33

$i\Im(\omega)$. $h_e(t)$ is the time function, which is obtained by the inverse Fourier transform of $R(\omega)$.

Eq.3 give us the easy way to calculate the Hilbert transform, which is realized by taking the inverse Fourier transform of the sample real part that gives the time function $h_e(t)$ and multiplying $\text{sgn}(t)$ to $h_e(t)$ we can obtain the time function $h_i(t)$. Then $i\Im(\omega)$ can be obtained by the Fourier transform of $h_i(t)$. Taking the inverse Fourier transform of $(R(\omega) - i\Im(\omega))$ we can obtain a sample causal time history.

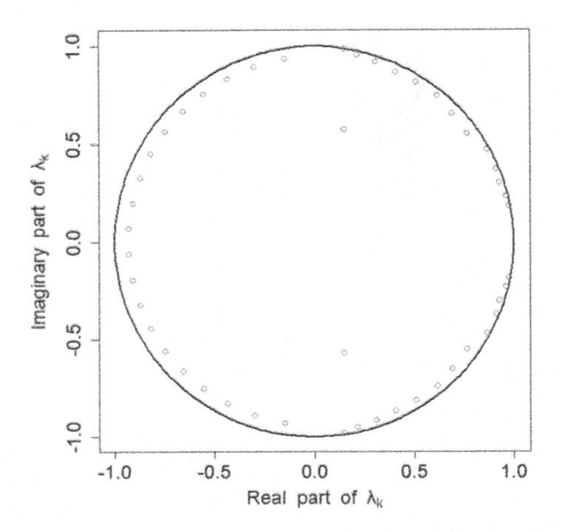

Figure 2. The complex roots distribution of $P(\lambda) = 0$. We can conclude that the discretized standardized real part process is a stationary process.

3 STATIONARITY OF STANDARDIZED REAL PART

The optimal order of autoregressive process is $q = 48$, which is obtained at the minimum AIC, at where the $a_k$values are given in Table 1. The higher order equation composed of a_k values is given by

$$P(\lambda) = \lambda^q - a_1\lambda^{q-1} - \cdots - a_{q-1}\lambda - a_q \quad (4)$$

There are 48 complex roots of the equation, which satisfies $P(\lambda) = 0$. These roots are expressed by $\lambda_k(k = 1, 2, \cdots, q)$. If the distribution of these roots are within the unit circle, which means $|\lambda_k| < 1$, it is proved that the discrete standardized real part process is a stationary process. Figure 2 shows the complex roots distribution, which are expressed by the red open circles. All roots are within the unit circle that is expressed by a black circle. This means that the discrete standardized real part process is a stationary process.

Table 1. The identified values of a_k at $q = 48$.

4.1342	7.7851	7.951	3.4007	1.6519
2.3005	0.3816	1.5658	-0.051	1.1339
0.0538	-0.891	0.0576	0.7118	-0.065
0.6167	0.0313	0.5227	0.0237	0.4756
0.0106	0.4132	0.0127	0.3912	0.0434
0.3435	0.0413	0.3333	0.0624	0.3005
0.054	0.2966	0.0624	0.278	0.0448
0.2848	0.0367	0.2791	0.0047	0.2929
0.0649	0.29	0.2015	0.2293	0.4657
0.3695	0.1672	0.0433		

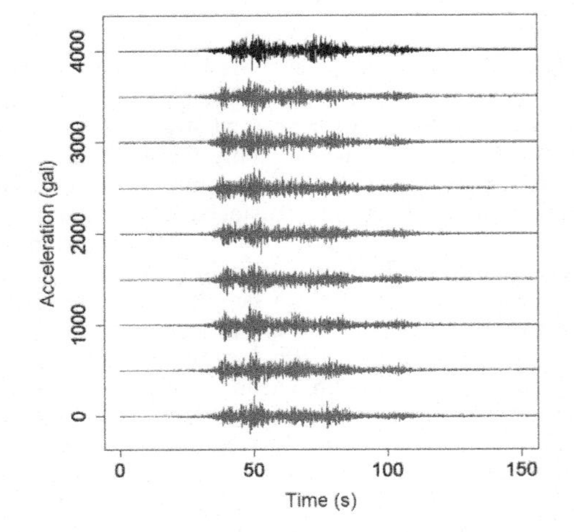

Figure 3. Simulated sample acceleration time histories (blue lines) and the acceleration time history observed at Suttsu Observatory (black line).

4 CONCLUDING REMARKS

The main results obtained through this research are as follows:

- Using root square of real part obtained by Fourier transform of an observed acceleration time history we extract the average trend of the real part. The process divided the real part by this average trend was named as the standardized real part process.
- We use the autoregressive process to simulate a sample standardized real part process. The order of the autoregressive process and the coefficients of the autoregressive process are determined, which are depended on the circular frequency interval.
- Based on the higher order equation composed of the autoregressive coefficients we prove that the standardized real part process has a stationary characteristic.
- A sample real part is obtained by multiplying the average trend of the real part to simulated standardized real parts. We used the Hilbert transform to obtain the imaginary part from the simulated real part. Using the inverse Fourier transform we could simulate many sample acceleration time histories, which are shown in Figure 3.

Current Perspectives and New Directions in Mechanics, Modelling and Design of Structural Systems – Zingoni (ed.)
© 2022 Copyright the Author(s), ISBN 978-1-032-39114-4

Dynamic characterization of the Tower of Palazzo dei Vicari in Scarperia (Italy) during and after the 2019 mugello seismic sequence

R.M. Azzara
Istituto Nazionale di Geofisica e Vulcanologia (INGV), Osservatorio Sismologico di Arezzo, Arezzo

V. Cardinali
Università degli Studi di Firenze, Dipartimento di Architettura (DiDA), Firenze

M. Girardi, F. Marini, C. Padovani & D. Pellegrini
Institute of Information Science and Technologies "Alessandro Faedo", Pisa

M. Tanganelli
Università degli Studi di Firenze, Dipartimento di Architettura (DiDA), Firenze

ABSTRACT: The paper presents the first results of an experimental campaign conducted on the Tower of Palazzo dei Vicari, in Scarperia, Italy. The Tower is a slender medieval structure located at the corner of the main façade of Palazzo dei Vicari. Over the centuries, both Tower and Palace have undergone several severe earthquakes; therefore, they have been reinforced with steel tie rods. In 2019 the structure was hit by the Mugello seismic sequence, which occurred between 9th December 2019 and the first half of January 2020. The mainshock was registered as 4.5 M_L, with an epicenter 5 km far from the city of Scarperia. The earthquake was felt in Toscana and Emilia Romagna Regions and caused some damage in the area's cities. During the seismic sequence, two seismometers were installed on the Tower to record earthquakes and natural vibrations and evaluate the behavior of the structure under seismic loads. A second monitoring experiment has been more recently performed, in June 2021. Eight seismic stations were deployed along the height of the Tower, to achieve a complete dynamic identification of the structure. The ambient vibration tests before and after the seismic sequence made it possible to exclude the presence of damage and to calibrate the numerical model of the Tower via model updating techniques.

1 INTRODUCTION

This paper is devoted to the dynamic identification of the medieval Tower of the Palazzo dei Vicari. The building is in the historical centre of Scarperia, in the Mugello area, a hilly territory 20 km North of Florence characterized by a significant seismicity. On December 9th, 2019 (02:37 UTC), an M_L 4.5 earthquake hit the territory; the seismic sequence lasted over a month, producing about 300 events with a magnitude from 0.5 to 4.5 M_L. After the M_L 4.5 mainshock, two seismic stations were installed in the Tower The instruments recorded 22 earthquakes with a magnitude M_L between 1.8 and 3.1 and allowed evaluating the Tower's structural response under seismic stress. In the following months a further experimental campaign was performed to improve the Tower's dynamic characterization, and eight seismometers were installed at different heights along the structure. The ambient vibration tests were adopted to calibrate a finite-element model of the Tower via sensitivity analysis.

1.1 *The Tower of Palazzo dei Vicari: Dynamic test*

Palazzo dei Vicari represents the most iconic landmark in the city of Scarperia. Its Tower, located in the northern corner of the building in adjacence with the rest of the Palace, is strengthened by several layers of tie rods at different heights, and the different finishes of the bolted endplates indicate a continuous insertion of the reinforces over the centuries. The seismometers position in the Tower is shown in Figure 1. The analysis of the data recorded during the seismic events made it possible to exclude any damage caused by the seismic sequence.

DOI: 10.1201/9781003348450-34

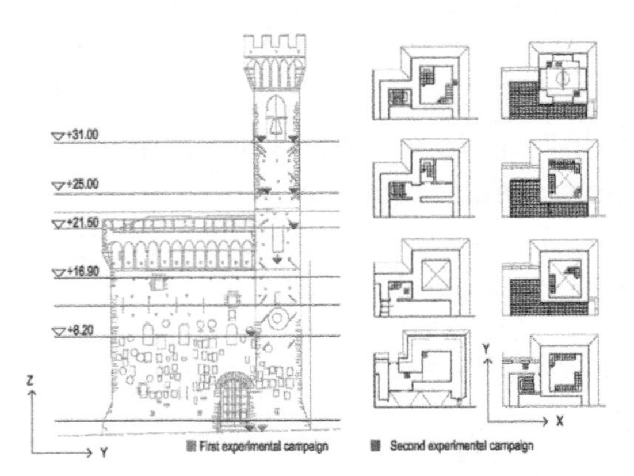

Figure 1. Experimental setup of the monitoring campaign. Red: position of the seismometers during the first campaign; blue: position during the second campaign.

In the following month a further experimental campaign based on 8 seismometers allowed the complete characterization of the structure. The signals were processed through different system identification techniques: the stochastic subspace identification (SSI) and the complex mode identification function (CMIF) algorithm (Brincker and Ventura 2015). Table 1 shows the results obtained. Both algorithms identify the first four frequencies of the Tower. The CMIF procedure identifies two additional mode shapes confirmed by finite-element calibration.

Table 1. Mode shapes, natural frequencies and relative Modal Phase Collinearity (MPC) of the Tower of the Palazzo dei Vicari.

Mode shape	SSI		CMIF	
	f [Hz]	MPC	f [Hz]	MPC
1 (bending x)	1.22±0.002	0.95	1.21	0.97
2 (bending y)	1.25±0.002	0.91	1.25	0.95
3 (torsional)	3.01±0.011	0.90	3.18	0.45
4 (bending x)	3.42±0.121	0.97	3.38	0.97
5 (bending -x+y)			4.08	0.96
6 (bending y)			6.15	0.86

1.2 *Numerical modelling and sensitivity analysis*

A finite-element model of the Tower has been created in "Nastran" format using the Abaqus code and imported in NOSA-ITACA (www.nosaitaca.it) for analysis and calibration. The resulting model is made up of 66221 4-node iso-parametric tetrahedrons, for a total number of 16562 nodes and 49686 degrees of freedom. The Tower is restrained at the base and the presence of the adjacent buildings is taken into account via fixed constraints. A first calibration of the model was obtained by considering four sets of materials with different Young's moduli, then, a sensitivity analysis was performed via the SAFE Matlab toolbox (Pianosi et al., 2015) coupled with NOSA-ITACA (Girardi et al., 2021). A first calibration of the finite-element model has been obtained by using a homogenous value of the Young's modulus and the Poisson's ratio: E = 1.863 MPa; $\upsilon = 0.2$.

2 CONCLUSIONS

The paper describes the monitoring and the dynamic identification of the Tower of Palazzo dei Vicari in Scarperia (Tuscany, IT). The study considered data recorded during two distinct experimental campaigns. The first setup was installed during the 2019 Mugello seismic sequence, recording 22 earthquakes with magnitude greater than M_L 2. Although only two seismic stations were installed on the Tower, it was possible to identify the first frequencies of the structure. Coherently with the low energetic content of the seismic sequence, the FTT of the signals suggests excluding the presence of damage in the Tower. After this first experimental test, a new setup consisting of eight different seismometers targeted at the complete characterization of the structure was executed. SSI and CMIF algorithms allowed to identify the torsional and higher bending modes. Finally, a finite-element model of the Tower was built, and the presence of the adjacent structures was taken into account via fixed constraints. Sensitivity analysis showed that the Young's moduli of the Tower's frame are the most influencing parameters, while the stiffness of the vaults does not influence the dynamic properties of the Tower. A first calibration was obtained using a homogeneous Young's modulus value. Further numerical investigations are planned to model the dynamic behaviour of the Tower subjected to the recorded seismic load, while taking into account the interaction with the Palace complex.

REFERENCES

Brincker R., Ventura C., Introduction to Operational Modal Analysis, 2015, Wiley.

E. Reynders, M. Schevenels and G. De Roeck, "Macec 3.3. A Matlab toolbox for experimental and operational modal analysis.," 2014. http://bwk.kuleuven.be/bwm/macec/.

Pianosi, Francesca, Fanny Sarrazin, and Thorsten Wagener, A Matlab toolbox for global sensitivity analysis. Environmental Modelling & Software, 70: 80-85, 2015.

Girardi, M., Padovani, C., Pellegrini, D., Robol, L., A finite element model updating method based on global optimization. Mechanical Systems and Signal Processing, 152, 107372, 2021.

Current Perspectives and New Directions in Mechanics, Modelling and Design of Structural Systems – Zingoni (ed.)
© 2022 Copyright the Author(s), ISBN 978-1-032-39114-4

Push-over tsunami analysis using *ETABS*

D.W. Aegerter
University of Hawai'i at Mānoa, Honolulu, Hawaii, USA
Martin, Chock & Carden, Inc., Honolulu, Hawaii, USA

I.N. Robertson
University of Hawai'i at Mānoa, Honolulu, Hawaii, USA

ABSTRACT: ASCE 7-16 Chapter 6 *Tsunami Loads and Effects* offers a conservative prescriptive approach while also allowing for alternative performance-based analysis. This study presents a procedure for performing a non-linear static pushover analysis for tsunami loading in accordance with ASCE 7-16 using *ETABS* (CSI, 2018). This paper establishes that the use of the procedure described can assist in targeted strengthening of a building, reducing construction costs while adhering to the strength capacity requirements of ASCE 7-16.

1 INTRODUCTION

The addition of Chapter 6 in ASCE 7-16 seeks to address the lack of design guidance for tsunami prone areas.

Selection of the same building and locations used by Robertson (2020) and Baiguera et al. (2022) allow results of this analysis to be compared with the procedures described in those studies.

The prototype building is a 6-story reinforced concrete structure. The lateral force resisting system (LRFS) consists of Special Moment Resisting Frames in both orthogonal directions. The building is considered to be Tsunami Risk Category (TRC) II and is 254 feet wide by 86 feet deep and 74 feet tall. The building satisfies the seismic and wind demands for both geographic locations, providing a reasonable example of a building in these areas.

2 METHODS

2.1 *Design requirements*

ASCE 7-16 prescribes the criteria which must be met for the Maximum Considered Tsunami (MCT) based on the building's TRC.

Given that the buildings at both the Hilo and Seaside sites are classified as Seismic Design Category D, the following equation for determining Life Safety Structural Performance of the LFRS can be used: $F_{TSU} < 0.75\ \Omega_0 E_{mh}$ where Ω_0 is the seismic overstrength factor and E_{mh} is the horizontal seismic load effect, determined by Chapter 12, and F_{TSU} is the tsunami load effect determined by Energy Gradeline Analysis (EGLA) in Chapter 6.

F_{TSU} is maximum at the point of maximum flow velocity, defined as Load Case 2 (LC2). The net maximum F_{TSU} transmitted to the LFRS will be referred to as $F_{T,net,LC2}$, in keeping with the convention followed by Baiguera et al. (2022). Inundation depth and flow velocity for LC2 and LC3 for the Hilo Site as determined by EGLA are shown in Table 1.

Table 1. Results of EGLA for Hilo Site and Associated Load Cases.

	Inundation depth	Flow velocity
Case	ft./s	ft./s
Load Case 2	38.0	50.2
Load Case 3	57.0	16.7

2.2 *Loading assumptions*

To ensure plastic hinges form in the order most likely occur under actual tsunami conditions, the load distributions and sequences need to be properly determined. For all inundation depths the total F_{TSU} is assumed to be equally distributed across the entire face of the building up to the inundation depth. By analysis it is evident that for $0 < t < t_{LC2}$, F_{TSU} is increasing because both flow velocity and inundation depth are increasing for that time interval.

The method used for analyzing the LFRS applies a tributary-equivalent monotonically increasing point load at each storey diaphragm concentrically about the center of rigidity. Loads are applied sequentially up the building, only after the previous storey has been loaded for the relative tsunami depth. In *ETABS*, this is modeled using a series of nonlinear static pushover (NLSP) Load Cases which add upon previous loads. To avoid confusion between *ETABS* and ASCE "Load Case," "*ETABS* Load Case" is referred to as "ELC". Each NLSP ELC begins with referenced Load Patterns having a magnitude of 0. The ELC then incrementally increases the magnitudes of all selected Load Patterns proportionally, checking for nonlinearity, until full loading is reached or convergence issues arise, with stresses, deflections, and hinges all carried over to the beginning of the next ELC.

Loads are discretized into one ELC per inundated storey, and each storey is loaded with its maximum

DOI: 10.1201/9781003348450-35

force prior to the next higher storey experiencing load, which occurs when the inundation depth is halfway between the two storeys. It is therefore sufficient to model only the conditions where the load at the highest inundated storey is a maximum prior to the next storey experiencing load. For the Hilo, site this results in three ELCs. The ELC maximum storey loads for the Hilo site are listed in Table 2.

Table 2. Total Loads for each ELC.

	ELC 1	ELC 2	ELC 3
Storey	kips	kips	kips
Fourth Floor	0	0	3694
Third Floor	0	6961	7403
Second Floor	5912	7541	8020

2.3 *Computational Analysis using ETABS*

After MCT forces have been determined for LC2, a building model can be analyzed using a two-phase approach in *ETABS* which is similar to the ASCE-VDPO2 structural analysis method presented in Baiguera et al. (2022)

The building model is constructed in *ETABS* in a manner consistent with traditional pushover analysis. In this analysis, the buildings have already been designed for the local seismic forces so the reinforced concrete beams and columns have been detailed in the *ETABS* model as such. Because only the LFRS is being analyzed, storey diaphragms can be modeled as rigid and massless. Axial load are added on the columns due to the appropriate gravity loads, which contribute to both the column strengths as well as P-delta effects. Plastic hinges are assigned to all column and beam segments up to the maximum inundation depth.

The loads for each ELC sum to the total expected force for each storey. Each NLSP ELC continues from the previous ELC with loads carrying over, therefore, the loads applied in each ELC need to be decreased by the total amount previously applied to that storey. Continuing with the Hilo site example, the loads for each ELC are shown in Table 3.

Table 3. *ETABS* Loads for each ELC, kips.

	ELC 1	ELC 2	ELC 3
Storey	kips	kips	kips
Fourth Floor	0	0	3694
Third Floor	0	6961	442
Second Floor	5912	1629	479

3 RESULTS AND DISCUSSION

Results for the Seaside, Oregon site are presented from this analysis as it affords the most direct comparison with Baiguera et al. (2022). These results are shown in Figure 1.

Strength of the building at Seaside exceeds the requirements of ASCE 7-16, even when using the prescriptive approach, that is: $0.75\Omega_0\ E_{mh} > F_{T,net,LC2}$. However, analysis by Baiguera et al. shows that the building is ~25% stronger than the reduced seismic strength would suggest. The *ETABS* analysis similarly shows a ~20% increase in strength over the ASCE 7-16 prescriptive approach.

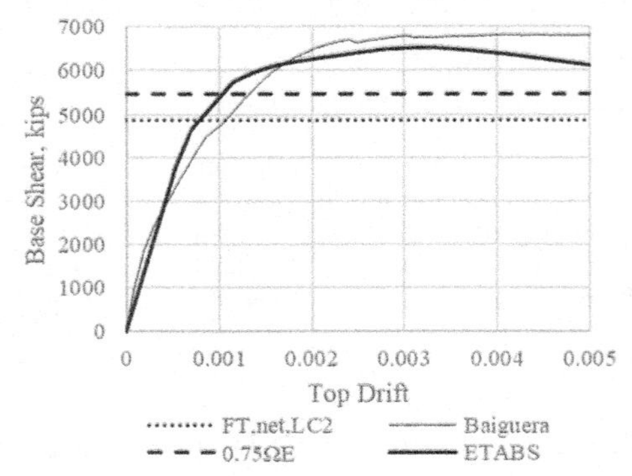

Figure 1. *ETABS* Analysis Results.

4 CONCLUSION

The procedure described in this analysis provides increased accuracy when determining building strength against tsunami forces. Comparing the results of this analysis with that of other researchers, it is evident that this methodology provides a framework for engineers to perform tsunami analysis using a relatively-common software *ETABS*.

Additional research is needed to assess the requirements of Chapter 6 of ASCE 7-16 as it applies to component-based strength, foundational tsunami effects, and member shear strength. It is expected that a convenient addendum to this procedure can be developed to account for these effects.

REFERENCES

ASCE 2017a. *Minimum design loads and associated criteria for buildings and other structures (ASCE 7-16)*. Reston: ASCE/SEI.

Baiguera, M., Rossetto, T., Robertson, I., & Petrone, C. 2019a. A nonlinear static procedure for the tsunami design of a reinforced concrete building to the ASCE7 standard. *Proceedings of the SECED 2019 Conference*. London: SECED

Baiguera, M., Rossetto, T., Robertson, I., & Petrone, C. 2019b. Towards a tsunami nonlinear static analysis procedure for the ASCE 7 standard. *Proceedings of the 2nd International Conference on Natural Hazards & Infrastructure 2019*. Chania: National Technical University of Athens.

Baiguera, M., Rossetto, T., Robertson, I., & Petrone, C. 2022. A procedure for performing nonlinear pushover analysis for tsunami loading to ASCE 7. *ASCE Journal of Structural Engineering* 148(2).

CSI. 2018. *ETABS 2018*. Computers and Structures, Inc.

Robertson, I. 2020. *Tsunami loads and effects: guide to the tsunami design provisions of ASCE 7-16*. Reston: ASCE.

Current Perspectives and New Directions in Mechanics, Modelling and Design of Structural Systems – Zingoni (ed.)
© 2022 Copyright the Author(s), ISBN 978-1-032-39114-4

Seismic fragility of corroded reinforced concrete highway bridges

S. Biswas, G.K. Verma & P. Sengupta
Department of Civil Engineering, Indian Institute of Technology (ISM), Dhanbad, India

ABSTRACT: Highway bridges play a crucial role in ensuring smooth and fast transportation system across the regions. During their service life, highway bridges experience structural deterioration due to fatigue as well as harsh environmental conditions, like chloride penetration, acid and sulphate attacks, carbonation, freezing and thawing, etc. Reinforcement corrosion in concrete bridges, leads to reduction in the effective cross-sectional area of the steel, changes in material properties, lower bond strength between reinforcement bars and the surrounding concrete, spalling of concrete cover, loss of confinement, etc. Consequently, this study aims at seismic vulnerability evaluation of corroded reinforced concrete highway bridges using three-dimensional nonlinear finite element modelling philosophy. The numerical simulation results of corroded reinforced concrete structural components are successfully validated with respect to the experimental results from the literature. Fragility functions represent the conditional probability of a structure surpassing a particular damage state under seismic excitations. Hence, time-dependent seismic fragility curves are derived for corroded reinforced concrete highway bridges using incremental dynamic analysis philosophy. Excerpts from the study would enlighten the researchers and practicing engineers about the seismic vulnerability of corroded reinforced concrete highway bridges across the world in a realistic manner.

1 INTRODUCTION

Corrosion of steel reinforcement in reinforced concrete (RC) structures is a common, natural phenomenon that can accelerate the rate of deterioration of a structure. Specifically, Vu and Stewart (2000) mentioned corrosion as one of the primary causes of deterioration in bridge decks and piers.

Corrosion leads to loss of mass in reinforcement, cracking of concrete cover, degradation of the bond capacity on the interface of concrete and reinforcement, deterioration of mechanical properties of steel bars, cross sectional damage of concrete members, decreasing of bearing capacity and deformation capacity of structures which will even cause the complete destruction of bridges and endanger people safety. For bridge piers, they dissipate earthquake energy through the formation of plastic hinges. Ou et al. (2013) developed a seismic evaluation methodology for RC bridges undergoing chloride attack based on a nonlinear static pushover analysis approach.

This paper presents a numerical study of an RC bridge pier with varying degrees of corrosion, subjected to repeated cyclic loading to analyse its seismic response. The FEA package, ABAQUS, has been used to develop the finite element model of the RC bridge pier and simulate the seismic response. Numerical models adopted from Guo et al. (2015) were successfully validated using ABAQUS.

2 NUMERICAL SIMULATION

Finite element analyses of numerical models of the bridge pier were performed by using the finite element software ABAQUS (Dassault Systèmes 2019). The concrete was modelled using solid C3D8R elements and the steel rebars, both longitudinal reinforcements and stirrups, were modelled using T3D2 elements. The base of the foundation was fixed with "ENCASTRE" boundary condition. To account for the nonlinear behaviour of concrete, the Concrete Damaged Plasticity Model (CDPM) available in ABAQUS, was used. For the steel reinforcement bars, an elasto-plastic material model was used. An increasing lateral displacement was applied on top of the FE model to simulate seismic loading conditions.

3 FRAGILITY ANALYSIS

Fragility analysis is a very effective technique to study the extent of vulnerability of damage in the structures. Fragility functions represent the conditional probability of a particular structure surpassing a certain damage state when subjected to external loading. Intensity measure (IM) quantifies the extent of input external excitations. Damage state may be referred to as the level of damage that a structure suffered when subjected to some external excitations. Engineering demand parameter (EDP) is a parameter which can predict the damaged stage properly.

DOI: 10.1201/9781003348450-36

In this study, Spectral Acceleration (Sa/g) is considered as Intensity Measure (IM) and drift ratio at the pier top is taken as the Engineering demand parameter (EDP). Seismic fragility curves for the uncorroded, corroded at 50 years age and corroded at 100 years age highway bridge piers for three damage states low, moderate and severe are shown in Figures 1, 2 and 3 respectively.

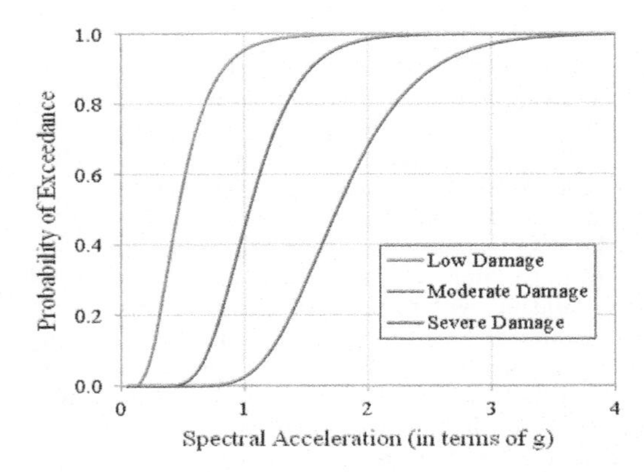

Figure 1. Seismic fragility curves of un-corroded highway bridge pier.

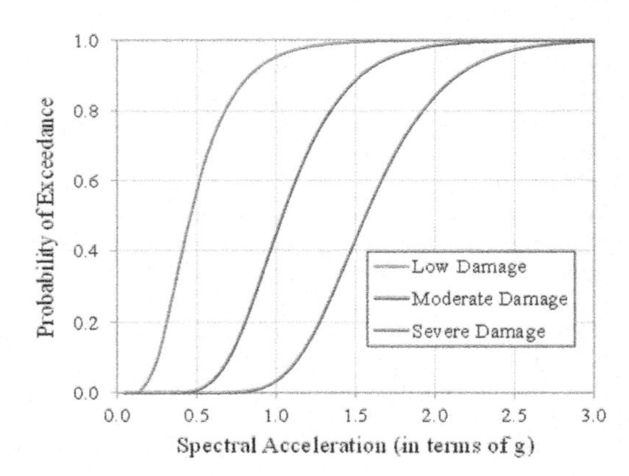

Figure 2. Seismic fragility curves of corroded highway bridge pier at 50 years age.

4 CONCLUSIONS

Reinforcement corrosion has been observed as a major factor contributing to the performance degradation of aged highway bridges. Reinforcement corrosion not only causes reduction in area of steel reinforcement, but also affects the bond strength between reinforcement bars and surrounding concrete. Under repeated cyclic as well as seismic loading, the corroded bridge piers exhibit spalling of concrete cover, loss of confinement and subsequently

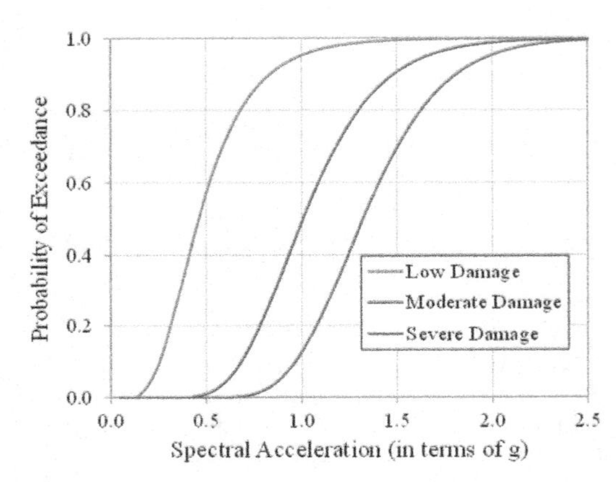

Figure 3. Seismic fragility curves of corroded highway bridge pier at 10 years age.

reduced load carrying and deformation capacity. This study presents the seismic vulnerability of corroded reinforced concrete highway bridge piers. A numerical simulation approach is adopted to capture the seismic response of corroded bridge piers. The numerical simulation results of bridge piers at zero, low, moderate and severe levels of corrosion damage are successfully validated with respect to the experimental results from the literature.

Incremental dynamic analysis is performed in this study so as to construct the time-dependent fragility functions of the un-corroded and corroded highway bridge piers. In this study, Spectral Acceleration (Sa/g) is considered as Intensity Measure (IM) and drift ratio at the pier top is taken as the Engineering demand parameter (EDP). Subsequently, time-dependent seismic fragility curves for the uncorroded, corroded at 50 years age and corroded at 100 years age highway bridge piers for three damage states, i.e. low, moderate and severe are derived.

REFERENCES

Dassault Systèmes, U. S. (2019). Abaqus analysis user's manual. *ABAQUS Academic-Research Edition Software 2019*.

Guo, A., H. Li, X. Ba, X. Guan, & H. Li (2015). Experimental investigation on the cyclic performance of reinforced concrete piers with chloride-induced corrosion in marine environment. *Engineering Structures 105*, 1–11.

Ou, Y.-C., H.-D. Fan, & N. D. Nguyen (2013). Long-term seismic performance of reinforced concrete bridges under steel reinforcement corrosion due to chloride attack. *Earthquake Engineering & Structural Dynamics 42*(14), 2113–2127.

Vu, K. A. T. & M. G. Stewart (2000). Structural reliability of concrete bridges including improved chloride-induced corrosion models. *Structural safety 22*(4), 313–333.

Current Perspectives and New Directions in Mechanics, Modelling and Design of Structural Systems – Zingoni (ed.)
© 2022 Copyright the Author(s), ISBN 978-1-032-39114-4

Assessment of the inter-story isolation technique applied to an existing school building

M. Donà & E. Bernardi
Department of Geosciences, University of Padova, Padova, Italy
Earthquake Engineering Research & Test Center, Guangzhou University, Guangzhou, China

A. Zonta, E. Saler & F. da Porto
Department of Geosciences, University of Padova, Padova, Italy

ABSTRACT: Inter-story isolation system (IIS) is increasingly attractive for the seismic risk mitigation of both new and existing buildings. An interesting but still poorly studied use of the IIS is to add isolated stories on the top of an existing building to increase its functionality while reducing its base shear forces. To this end, some design methods are available in the literature. However, there are still few studies demonstrating the potential of IIS through real case studies. Therefore, this paper presents an ideal application of the IIS to an existing two-story RC school building, with the aim of raising it by one story while improving its seismic performance. The optimization study of the IIS parameters (frequency and damping ratios) is presented and the effectiveness of the optimal IIS is assessed by comparing the TH responses of the FE models, with and without IIS, under a set of bidirectional records. The results confirm the beneficial effects of the IIS.

1 INTRODUCTION

The inter-storey isolation system (IIS) consists of inserting the isolation layer between the storeys rather than at the base of the building, combining the functions of seismic isolation and mass damping. In addition to promoting architectural freedom in the conception of skyscrapers and multipurpose buildings, it represents a sustainable solution for densely populated areas (Liu et al. 2018).

Furthermore, when it is required to raise an existing building, the IIS is very useful by acting as a non-conventional tuned mass damper (TMD), thus allowing control of the substructure response (Faiella et al. 2020, Bernardi et al. 2021).

Although various research is available in the literature on the IIS behavior and its optimization (mainly through TMD methods), only a few studies show its effectiveness through real case studies, especially as regard its use as a seismic retrofit technique (Faiella and Mele 2020).

Therefore, this paper presents an ideal application of the IIS to an existing two-story RC school building. First, the structure and the associated 3D finite element (FE) models, with and without IIS, are presented. Then, the method proposed by Donà et al. (2021) and based on surrogate response models is used to optimize the IIS parameters. In this study the only goal was to minimize the inter-story drift of the substructure. Lastly, the effectiveness of the IIS is demonstrated by comparing the TH responses of the two FE models, considering a set of natural records.

2 CASE STUDY

The case study structure, representative of the macro-class of two-story RC school buildings built in Italy in the 1960s, has bidirectional frames designed for gravitation loads only and characterized by a weak column-strong beam configuration.

The school has an irregular configuration in plan (L-shape, see Figure 1), with an area of about 800 m^2. The school is 52 m long in the main direction and 22 m in the opposite direction. The façades have a significant portion of openings, common in this type of buildings; only five bays are fully filled with masonry panels. Regarding the material properties, the reinforcement class is AQ50 (i.e. smooth rebars), while a mean strength of 20 MPa was assumed for the concrete. The infill panels are 13 mm thick and made of hollow clay bricks with 35% void area.

The IIS technique is used to add one story to the building, in order to increase the space available for school activities while also improving the overall seismic performance of the structure.

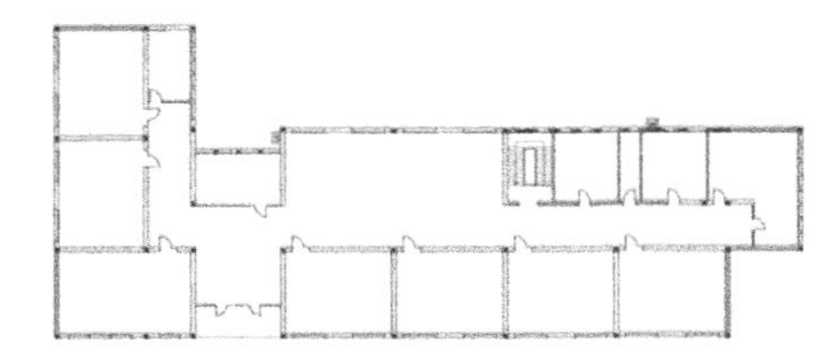

Figure 1. Plan of the RC school building.

DOI: 10.1201/9781003348450-37

3 OPTIMIZATION OF THE IIS PARAMETERS

To maintain an adequate safety index of the columns to vertical loads, the maximum weight of the isolated superstructure (including the isolation layer) was about 50% of the weight of the existing structure; therefore, the superstructure was designed with steel frames, to be as light as possible.

To derive the optimal values of the IIS parameters, i.e. frequency (ν) and damping (ζ) ratios, a parametric TH analysis was performed using an equivalent 3-DOF structural model (Figure 2a), calibrated for each direction of analysis, evaluating combinations of ν-ζ within their ranges of interest, i.e. 0.01 to 1.

Regarding the seismic input, seven bidirectional natural records were selected from the SIMBAD database and scaled to be compatible with the EC8 Type 1 spectrum, with PGA=0.25 g and soil type B.

The inter-story drift of the first story was assumed as the main performance parameter and its minimization as the only objective function, corresponding to minimizing the base shear force.

The inter-story drift results were therefore used to calibrate the surrogate response models (SRMs) in both directions. These are complete fourth degree polynomials in the variables ν and ζ, normalized to the case without IIS (see Figure 2b). The minimization of these SRMs allowed to derive the values of ν and ζ which, for each direction of analysis, optimize the seismic response of the substructure. These values are reported in Table 1.

4 IIS EFFECTIVENESS AND CONCLUSIONS

Figure 3 compares the maximum inter-story drift profiles, averaged among the seven seismic events, in the cases with and without IIS. The beneficial effect of the IIS is significant, especially in the y direction, where the building is more flexible, and in the POS.2 position, where the torque effects are greater.

Figure 4 compares the maximum stress state of the structural elements at the end of one TH analysis. The original structure shows a widespread yielding of these elements, and the use of the IIS significantly reduce their stress state. With the IIS, only a crack state is generally observed at the base of the first-story columns; moreover, an improvement in the stress state of the beams is observed, especially in the frames in the y direction (more deformable).

Ultimately, the results confirm the effectiveness of the IIS, supporting its use as a seismic retrofit strategy, even if specific case-by-case assessments are required.

Table 1. Optimal values of the IIS parameters.

	$\zeta_{,\text{Opt}}$ [-]	$\nu_{,\text{Opt}}$ [-]
x direction	0.41	0.37
y direction	0.34	0.41

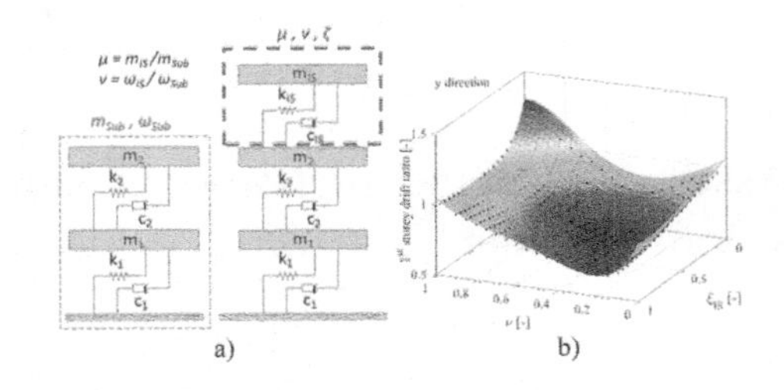

Figure 2. (a) 2-DOF (without IIS) and 3-DOF (with IIS) systems. (b) SRM of the first inter-story drift ratio, y direction.

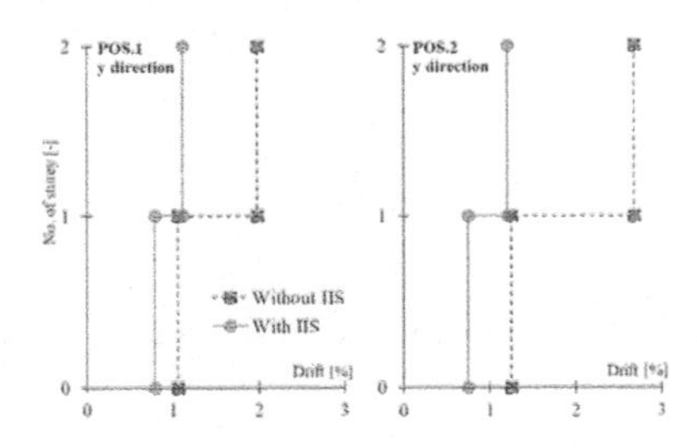

Figure 3. Maximum inter-story drift profiles, with and without IIS, in positions POS.1 and POS.2 and for the y direction.

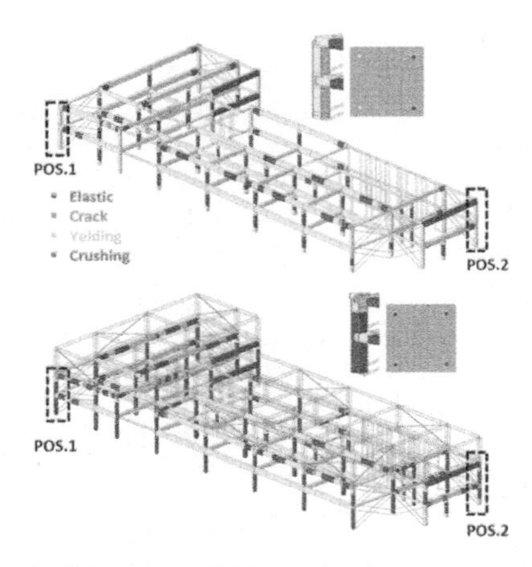

Figure 4. Comparison of the structural stress state: without (above) and with (below) IIS, for the record no. 1.

REFERENCES

Bernardi et al. 2021. Investigations on inter-storey seismic isolation as a technique for adding upper storeys. *Proc. of 8th COMPDYN*, Athens, Greece.

Donà et al. 2021. Evaluation of optimal FVDs for inter-storey isolation systems based on surrogate performance models. *Bulletin of Earthquake Engineering* 19: 4587–4621.

Faiella et al. 2020. Seismic Retrofit of Existing Masonry Buildings through Inter-story Isolation System: A Case Study and General Design Criteria. *J. of Earthquake Engineering*.

Faiella & Mele. 2020. Insights into inter-story isolation design through the analysis of two case studies. *Engineering Structures* 215:110660.

Liu et al. 2018. Effectiveness of fluid-viscous dampers for improved seismic performance of inter-storey isolated buildings. *Engineering Structures* 169: 276–292.

Current Perspectives and New Directions in Mechanics, Modelling and Design of Structural Systems – Zingoni (ed.)
© 2022 Copyright the Author(s), ISBN 978-1-032-39114-4

Effects of frequency content of ground motions on seismic response of concrete gravity dams with varying reservoir levels

Gautham Reddy & Manish Shrikhande
Department of Earthquake Engineering, Indian Institute of Technology Roorkee, India

ABSTRACT: The seismic response of the dam-foundation-reservoir systems involves complex dynamics of interacting systems and varying reservoir water levels are known to influence the extent of damage sustained by the dam body during an earthquake. In this preliminary study, we investigate the effects of the frequency content of ground motion on the seismic behaviour of concrete gravity dams with varying reservoir levels. Crest displacements, maximum tensile stresses, damage area and energy dissipated are employed to establish the ultimate damage state of the dam. The results demonstrate the significant effects of the frequency content of ground motion and reservoir water levels on accumulated damage and damage distribution in the dam body with implications on the seismic design criteria.

1 INTRODUCTION

1.1 *Seismic response of concrete gravity dams*

The seismic response of a concrete gravity dam involves a complex dynamic interaction of dam-foundation-reservoir systems. Seismic response of a concrete arch dams are influenced by varying reservoir levels (Akkose, Bayraktar, & Dumanoglu 2008), strong motion duration, the frequency content and phase content of ground motions (Soysal, Binici, & Arici 2016). This preliminary work investigates the effect of frequency content of ground motion and varying reservoir water levels on seismic response and the accumulated damage in a concrete gravity dam.

2 CHARACTERIZATION OF FREQUENCY CONTENT

Stochastic indicator of frequency content ε, as shown Equation (1) of a ground motion is based on modeling the strong phase of an acceleration record as a stationary process is computed from the power-spectral density (PSD) of an acceleration record.

$$0 \leq \varepsilon = \sqrt{\left(1 - \frac{m_2^2}{m_0 m_4}\right)} \leq 1 \qquad (1)$$

where m_n is the n-th moment of the PSD, $S_x(\omega)$ The deterministic indicator of frequency content of a ground motion as shown in Equation (2) is computed from the response spectrum of a ground motion proposed by (Rathje, Abrahamson, & Bray 1998)

$$f_M = \frac{\sum C_i^2 f_i}{\sum C_i^2} \qquad (2)$$

where C_i is the Fourier amplitude and f_i are the discrete Fourier transform frequencies between 0.25 - 20 Hz.

3 GROUND MOTION RECORDS

The target spectrum for the Pine Flat Dam site is obtained for a return period of 10,000 years.

The ground motion record bin is populated with ground motions with mean square frequency $2\omega_n \geq f_M \geq \omega_n$, where ω_n is the fundamental frequency of dam-foundation-reservoir system with frequency indicator ε, $0.7 \leq \varepsilon \leq 1$.

4 NON-LINEAR ANALYSIS OF CONCRETE GRAVITY DAM

The geometrical details of the tallest monolith of the Pine Flat Dam is shown in Figure 1. A 2D model of dam-foundation-reservoir system each with a reservoir water level of 116 m, 58 m and 29 is prepared using the general purpose finite element software, ABAQUS 2017. Material damping is incorporated using Rayleigh damping with viscous damping value of $\zeta = 5\%$. Fluid-structure interaction of dam and reservoir is modeled using coupled

DOI: 10.1201/9781003348450-38

Table 1. Ground Motion Data.

# RSN	Earthquake (Year)	Station Name	M_W	Rrup (km)	$V_{s,30}$ (m/sec)	f_M	ε	$t_{sig}(sec)$
765	Loma Prieta (1989)	Gilroy Array #1	6.93	8.84	1428.14	3.73	0.95	3.69
1011	Northridge (1994)	LA - Wonderland Ave	6.69	15.11	1222.52	6.41	0.80	8.7
455	Morgan Hill (1984)	Gilroy Array #1	6.19	14.9	1428.14	7.90	0.77	9.53

Table 2. Damage and drift ratios for different reservoir levels.

				Water Level: 116 m		Water Level: 58 m		Water Level: 29 m	
#RSN	P_D	f_M	ε	CR (%)	DR (%)	CR (%)	DR (%)	CR (%)	DR (%)
765	0.008	3.73	0.95	0.05	0.03	7.3	0.04	7.8	0.05
1011	0.01	6.41	0.80	0.14	0.03	6.7	0.05	7.2	0.05
455	0.07	7.90	0.77	6.4	0.05	17.7	0.06	17.9	0.03

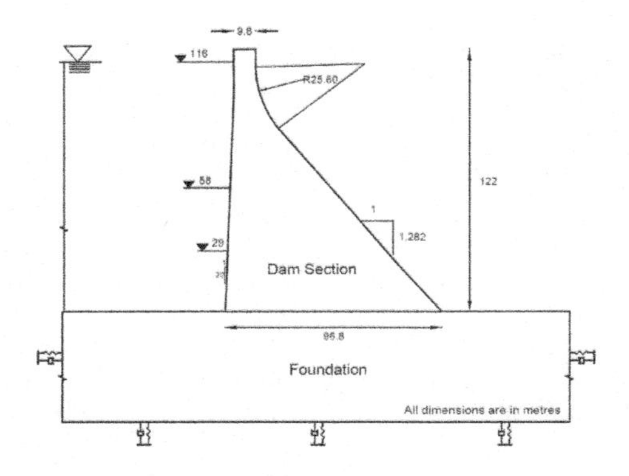

Figure 1. Dam-foundation-reservoir system.

acoustic-structural formulation. Massed foundation with viscous spring-boundary conditions are provided at far ends of foundation rock. Concrete damage plasticity (CDP), a continuum Damage Model is used to model material non-linearity of dam concrete.

5 RESULTS AND DISCUSSION

Dam-foundation-reservoir system with three reservoir levels 116 m, 58 m & 29 m each are subjected to set of selected ground motions scaled to match the target spectrum their spectral values at each fundamental time periods. The results are presented in Table 2.

Across the dam-foundation-reservoir system at each constant reservoir water level, the cracked ratio of dam increases with increase in f_M and decrease in ε, decreases As reservoir water level decreases from 116 m to 58 m and 29 m, the downstream slope of the dam body accumulated increasingly higher damage, see 3.

The Destructiveness Potential, P_D which accounts for the frequency characteristics of the ground motion and Significant Duration t_{sig} correlates well with the damage measure cracked ratio. Crest displacement or drift ratio do not correlate well with the damage state of the dam evaluated using cracked ratio, see Table 2.

6 CONCLUSIONS

1. Although the ultimate damage states (crack ratio) does not vary significantly between reservoir level of 58 m and 29 m, it is however important to note that changing reservoir levels contribute significantly to the seismic response of concrete gravity dams.
2. The frequency content of ground motion characterised either using ε, f_M or P_D correlate well with the ultimate damage state of the gravity dam. Ground motions with wide frequency band process $(2/3 \leq \varepsilon \leq 0.85)$ and high mean square frequency, f_M may cause extensive damage in the dam body.
3. Crest displacement or drift ratio do not correlate well with the damage state of the dam evaluated using cracked ratio.

REFERENCES

Akkose, M., A. Bayraktar, & A. Dumanoglu (2008). Reservoir water level effects on nonlinear dynamic response of arch dams. *Journal of Fluids and Structures 24*(3), 418–435.

Lubliner, J., J. Oliver, S. Oller, & E. Oñate (1989). A plastic-damage model for concrete. *International Journal of solids and structures 25*(3), 299–326.

Rathje, E. M., N. A. Abrahamson, & J. D. Bray (1998). Simplified frequency content estimates of earthquake ground motions. *Journal of geotechnical and geoenvironmental engineering 124*(2), 150–159.

Soysal, B., B. Binici, & Y. Arici (2016). Investigation of the relationship of seismic intensity measures and the accumulation of damage on concrete gravity dams using incremental dynamic analysis. *Earthquake Engineering & Structural Dynamics 45*(5), 719–737.

Current Perspectives and New Directions in Mechanics, Modelling and Design of Structural Systems – Zingoni (ed.)
© 2022 Copyright the Author(s), ISBN 978-1-032-39114-4

A proposal for improving the elastic behaviour of dogbone

L. Palizzolo, S. Benfratello & S. Vazzano
Department of Engineering, University of Palermo, Palermo, Italy

ABSTRACT: The Reduced Beam Section (RBS) connections are prequalified connections approved by international standards and adopted to improve the seismic behaviour of steel Moment Resisting Frames (MRF). This goal is reached by locally reducing the resistance of a selected portion of the beam by suitably trimming the flanges and promoting in that portion the onset of plastic deformations. Among RBS, the most adopted typology is the so-called dogbone (DB). Advantages of DB are related to the protection of the welded cross section between beam and column, to the increasing of the total energy dissipation and to the high rotational capacity. Drawbacks are represented by the local stiffness decrease, the reduced lateral strength and the risk of web buckling. The paper is devoted to proposing a stiffness improvement of beam equipped with DB. Specifically, suitably steel elements are added to the beam with DB to avoid any global reduction for the beam stiffness. Numerical examples are carried out for different flange reductions and the related stiffness interventions are evaluated by solving a simple optimization problem. The goodness of the results is verified by analyzing both mechanical and kinematical response of frames equipped with both classical DB and improved beam.

1 INTRODUCTION

The brittle damages observed at the beam-to-column flange welds during the 1994 Northridge and 1995 Kobe earthquakes stimulated studies aimed to improve the performance of connections. One of the developed approaches is the Reduced Beam Section (RBS) connection, where the protection is obtained ensuring a plastic hinge to take form outside the connection. The main approach leads to the so-called dogbone (DB) connection by suitably trimming the flanges of I-shaped the cross-section beams. Nowadays the trimming profile commonly adopted in international codes (see, e.g., FEMA 2000) is the radius one. The main drawback of DB is the reduction in the beam stiffness, an increment of the frame drift and of the beam deflection. Recently, to overcome this drawback, the authors proposed a new connection device, called Limited Resistance Plastic Device (LRPD) (Benfratello et al., 2022).

Aim of the present paper is to propose a new approach to improve the bending stiffness of a beam equipped with dogbones. The goal is reached by adding to the beam suitable steel elements able to take back the beam stiffness to that of the original beam. The stiffness intervention is evaluated as the solution to simple optimization problem. The results are verified by analyzing the response of frames equipped with both classical dogbones and improved beam.

2 REINFORCEMENT DESIGN

In Figure 1 the geometrical scheme of the dogbone and of the proposed reinforcement interventions are reported. To remedy to the greater deformability caused by dogbone, it is possible to improve the elastic behaviour of the beam adding appropriate steel bars (Fig. 2) of suitable radius to restore the original global elastic stiffness of the beam.

The continuous reinforcement steel bars are disposed just along the central ℓ_r portion of the beam and their design is performed by imposing that the relative rotation between the end sections A and B of the central portion of the reinforced beam, subjected just to unitary constant bending moment, equals the analogous quantity related to the original beam.

The following equation is obtained

$$2\pi r^4 + 2\pi(2t_p - h_p)r^3 + \pi\left(h_p^2 + 4t_p^2 - 4h_p t_p\right)r^2 + I_p - \frac{\ell_r}{\frac{\ell_p}{I_p} - \frac{2\ell_d}{I_d}} = 0 \tag{1}$$

where $\ell_p = 2\ell_d + \ell_r$, h_p, t_p and I_p are the height, the flange thickness and the moment of inertia of the original beam cross section, respectively while I_d is the moment of inertia of the dogbone cross section.

DOI: 10.1201/9781003348450-39

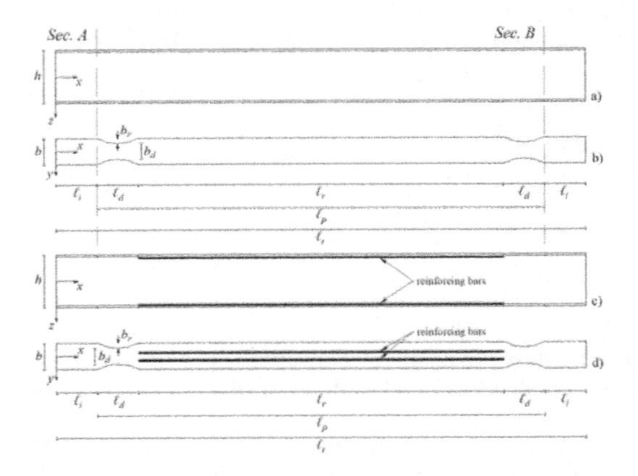

Figure 1. Geometrical scheme for the dogbone and for the reinforcement: a) lateral view; b) upper view; c) lateral view with reinforcement bars; d) upper view with reinforcement bars.

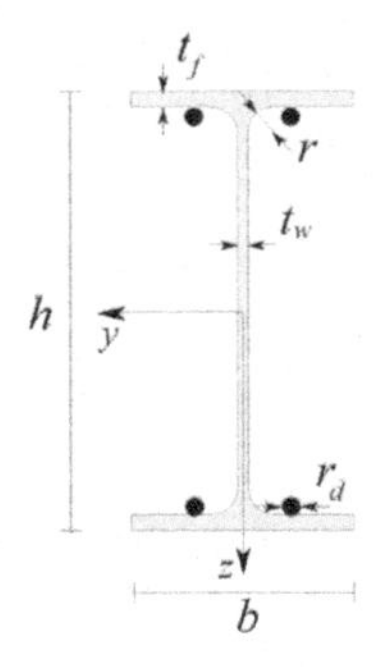

Figure 2. Proposed position of the reinforcement bars.

3 APPLICATION

The reinforcement approach is applied to the frame sketched in Figure 3. The frame is constituted by S275 IPE300 steel profiles ($E = 210\,\text{GPa}$) with span length B=6 m, height H=3.5 m, $F_H = 50\,\text{kN}$ and $q_V = 45\text{kN/m}$. The frame is studied in three different configurations: a) no dogbones; b) classical dogbones; c) dogbones with reinforced beam. The case b) is studied assuming values for ℓ_i ranging between $h/4$ and $3h/4$, $\ell_d = 3h/4$ and for three values of base reduction (40%, 50%, 60%). The nine models are studied by means of a linear analysis by SAP2000 software.

The beam midspan deflection, chosen as control parameter, is calculated for all the models. In Table 1 the results are reported, being δ, δ_{db} and δ_{imp} the control parameters related to the case of no-dogbone configuration, dogbone configuration and reinforced beam, respectively, and where Δ_{db} and Δ_{imp} are the percentage variation of control parameter for dogbone and for the reinforced beam, respectively, calculated with respect to the no-dogbone configuration.

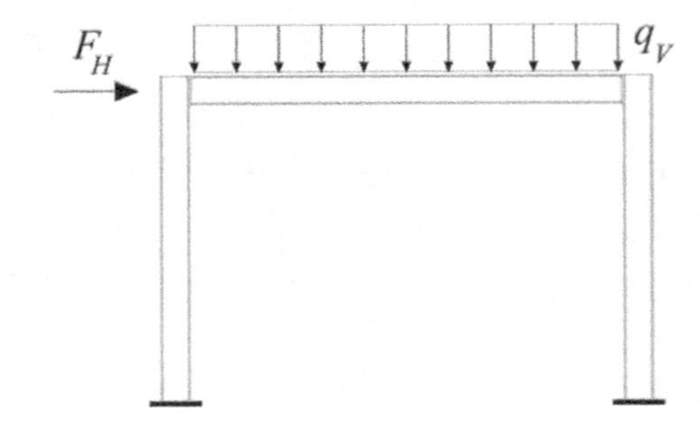

Figure 3. Loads scheme for the frame samples.

Table 1. Dogbone and improved beam models' output.

n°	δ [cm]	δ_{db} [cm]	δ_{imp} [cm]	Δ_{db} [%]	Δ_{imp} [%]
1	1.908	1.955	1.914	2.41%	0.34%
2	1.908	1.946	1.905	1.96%	-0.13%
3	1.908	1.938	1.897	1.57%	-0.57%
4	1.908	1.974	1.915	3.34%	0.38%
5	1.908	1.961	1.910	2.73%	0.13%
6	1.908	1.950	1.891	2.19%	-0.89%
7	1.908	1.998	1.915	4.51%	0.35%
8	1.908	1.981	1.897	3.71%	-0.56%
9	1.908	1.966	1.882	2.97%	-1.39%

4 CONCLUSIONS

In the paper a simple procedure has been proposed to design the reinforcement for steel beam elements interested by dogbone intervention. The reliability of the procedure is verified by an application related to a simple steel frame. The obtained results showed that the designed reinforcement guarantees an elastic behaviour very close to that of the original beam.

REFERENCES

FEMA, Recommended Seismic Design Criteria for new-Steel Moment Frame Buildings, FEMA 350, Federal Emergency Management Agency, Washington, DC, 2000.

Benfratello, S.; Palizzolo, L.; Vazzano, S. 2022. A New Design Problem in the Formulation of a Special Moment Resisting Connection Device for Preventing Local Buckling. *Appl. Sci.*, 12:202. https://doi.org/10.3390/app12010202

4. Material modelling, Multi-scale modelling, Micromechanical modelling, Porous media, Biological tissue

Current Perspectives and New Directions in Mechanics, Modelling and Design of Structural Systems – Zingoni (ed.)
© 2022 Copyright the Author(s), ISBN 978-1-032-39114-4

Multiscale modelling in materials science and technology

D. Rapp, C. Xu, F. Sajadi, X. Xu & S. Schmauder
Institute for Materials Testing, Materials Science and Strength of Materials, Stuttgart, Germany

ABSTRACT: In the recent past, multiscale materials modelling became a central idea in understanding present day's complex composites and in making progress in the development of advanced materials. Applications range from structural materials used in construction, predicting microstructures and mechanical properties of additively manufactured lightweight designs as well as functional materials such as electrically conductive polymers. This paper aims at giving a recent example from our work on a multiscale material model for additively manufactured AlSi10Mg alloy. Further examples on hydrogen, Ni_3Al and polymers are in the long version.

1 INTRODUCTION

This brief version of our contribution to the SEMC2022 proceedings only discusses the microstructure prediction for AlSi10Mg using statistical physics. The further works on electrically conductive polymers, hydrogen diffusion and MD potential from machine learning can be found in the long version.

2 MICROSTRUCTURE EVOLUTION IN SELECTIVE LASER MELTING (SLM)

Modeling and simulation also enable modeling of processes such as additive manufacturing where source of energy melts layers of metal layer by layer in order to fabricate a bulk material. One aspect in which modeling can deepen our understanding is the thermal history prediction and its effect on the microstructural evolution.

There has been different attempt in order to simulate the thermal history, e.g., Rosenthal's analytical solution of the heat equation (Rosenthal 1941), or solving it through finite element methods (Romano 2016) or even taken fluid dynamics of the melt pool into account (Chen 2020). The thermal history then is being considered in order to model the resulting microstructural features.

In the following, we have made use of Rosenthal's analytical solution of heat source to calculate the shape and size of the melt pool which is produced during the laser-based powder bed fusion of an aluminum alloy and simulated the resulting grain structure using Potts Monte Carlo method. The resulting microstructure has also been validated by the experimental results.

2.1 *Materials and method*

To fabricate the material, an aluminum powder of AlSi10Mg was used and fabricated by a Trumpf TruPrint3000 machine. The process conditions were kept constant and are indicated in Table 1. To obtain detailed information about the grain morphology and crystallographic orientations, EBSD measurements were carried out at an acceleration voltage of 20 kV and a working distance of 14 mm. Orientation maps were acquired for an area of $550 \times 400\ \mu m^2$ with a step size of 2 μm at 200× magnification (cf. Figure 2).

To model the melt pool, The Rosenthal (Rosenthal 1941) general analytical solution for the partial differential equation of heat flow due to a moving heat source in one direction can be written as follows

$$T = T_0 + \frac{\eta P}{2\pi k R} \exp \frac{-\rho c v}{2k} (R + \xi)$$

where P is the energy source, η is the absorptivity, k is the heat conductivity, R is the distance between the moving heat source and the point of interest, ρ is material density, c is the specific heat, v is the speed of heat source and ξ is the moving coordinate parallel to the energy source trajectory. To find the melting temperature line, CALPHAD calculations have been performed at Scheil's condition to simulate non-equilibrium transformation and averaging the solidus and liquidus temperatures. The parameters listed in Table and Table have been used for the aluminum alloy.

For simulating the grain structure, the Potts Monte Carlo method on the basis of curvature driven grain growth with temperature driven mobility (Rodgers 2021) was used. Therein grain growth is driven by minimizing the total grain boundary energy via probabilistic switching of lattice spin, i.e., grain orientation. The total energy of the system is calculated by

Table 1. Process parameters used for the fabrication of the samples using the TruPrint3000 SLM machine.

Laser power	Scanning speed	Beam diameter	Hatch space	Slice thickness	Scanning pattern
420 W	1300 mm/s	0.1 mm	0.21 mm	0.06 mm	Chess

DOI: 10.1201/9781003348450-40

$$E = \frac{1}{2}\sum_{i=1}^{N}\sum_{j=1}^{L} 1 - \delta_{ij}$$

where N is the total number of lattice sites, L is the number of neighboring sites (L=26 for a cubic 3x3 lattice) and δ_{ij} equals one if the lattice spins of sites i and j are identical and is zero otherwise. Initially the N lattice sites are assigned 256 spin values randomly. For the evolution by the Metropolis algorithm, at each time step the probability of accepting a new spin value is given by

$$P = \begin{cases} M(T) & \Delta E \leq 0 \\ M(T)e^{\frac{-\Delta E}{k_B T_S}} & \Delta E > 0 \end{cases}$$

meaning that new spins that reduce the overall energy are accepted with the temperature dependent grain boundary mobility $M(T)$, whereas spins that increase the energy are thermally suppressed. The temperature T_S used for this is an artificial simulation temperature above the physical T to accelerate the evolution.

2.2 *Results*

A snapshot of the heat flux simulation for the melt process is shown in Figure 3. From this temperature profile, the melt pool parameters for the Potts Monte Carlo model have been derived and used to predict the microstructure evolution depicted in Figure 1. The predicted grain morphology exhibits visible weld bead borders and contains equiaxial and columnar grains as the experimental observation shown in Figure 2.

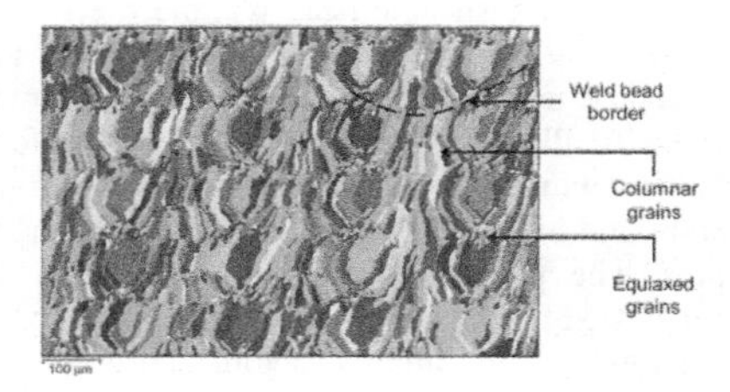

Figure 1. Predicted grain morphology from Potts Monte Carlo simulation after several laser scanning passes.

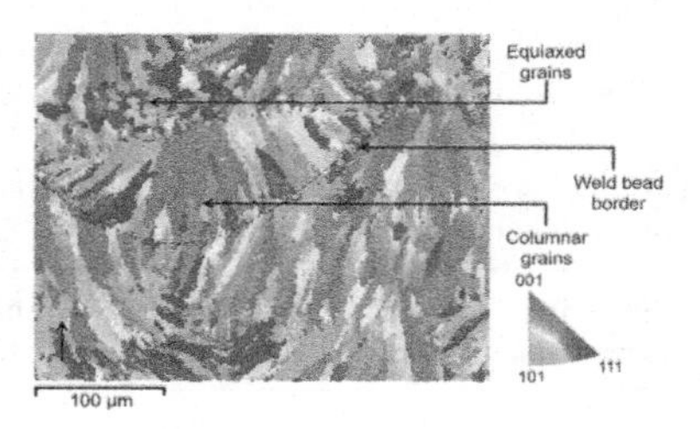

Figure 2. EBSD micrograph of a SLM manufactured sample.

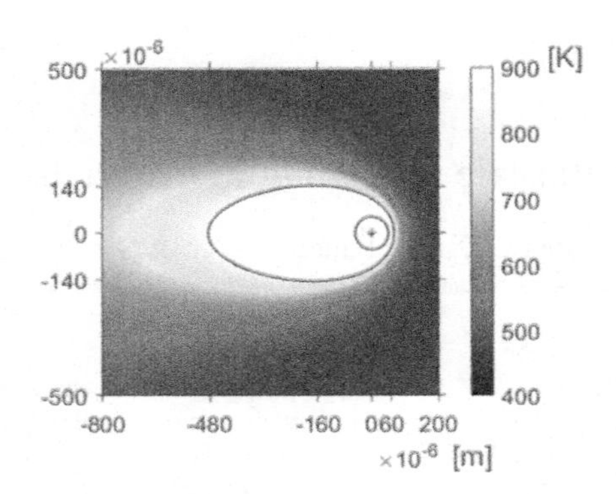

Figure 3. Temperature profile during simulated SLM process obtained using thermophysical parameters from Table 1 and 2.

Table 2. Thermophysical parameters for melt pool analysis.

T_0	η	k	ρ	c	T_M
473 K	30 %	110 J/smK	2670 kg/m^3	915 J/kgK	849 K

3 SUMMARY

The multiscale material simulation setup can be used to obtain the grain morphology (sizes, aspect ratio, borders, texture) prior to performing an actual experiment, aiding in the search for ideal process parameters for a desired microstructure.

ACKNOWLEDGEMENT

We thank the Deutsche Forschungsgemeinschaft (DFG, German Research Foundation) (GSC 260 – 49619325) for financial support and acknowledge the support by the Graduate School for advanced Manufacturing Engineering (GSaME).

REFERENCES

Rosenthal, D. 1941. The Welding Journal 20:220–234.

Romano, J. and Ladani, L. and Sadowski, M. 2016. The Journal of The Minerals, Metals & Materials Society 68, 967–977.

Chen, F. and Yan W. 2020. Materials & Design 196, 109185.

Yang, M. and Wang, L. and Yan, W. 2021. Additive Manufacturing 47, 102286.

Rodgers, T.M. et al. 2021. Additive Manufacturing 41, 101953.

Current Perspectives and New Directions in Mechanics, Modelling and Design of Structural Systems – Zingoni (ed.)
© 2022 Copyright the Author(s), *ISBN 978-1-032-39114-4*

A multiscale approach to investigate residual stresses due to targeted cooling of hot bulk formed parts

S. Hellebrand, D. Brands & J. Schröder
Institute of Mechanics, Faculty of Engineering
University Duisburg-Essen, Essen, Germany

L. Scheunemann
Chair of Applied Mechanics
Department of Mechanical and Process Engineering
Technical University Kaiserslautern, Kaiserslautern, Germany

ABSTRACT: In order to produce components with desired properties, current research focuses on the induction of targeted residual stress states inside the material. The main goal is the evocation of compressive residual stresses in regions near the surface. The so-called FE^2 approach provides an insight on microscopic level and thereby enables the analysis of macro- and microscopic residual stresses.

1 INTRODUCTION

Hot bulk forming offers great opportunities to evoke and adapt residual stress states inside a component during its manufacturing process. This contribution investigates residual stress distributions as a consequence of cooling and related phase transformation. Starting with a two-dimensional single-scale boundary value problem (bvp), an approach is presented to reduce the computational domain and thereby the numerical costs. Thus, it is possible to utilize a two-scale finite element analysis, c.f. (Schröder 2014), to compute residual stresses on macro- and microscale.

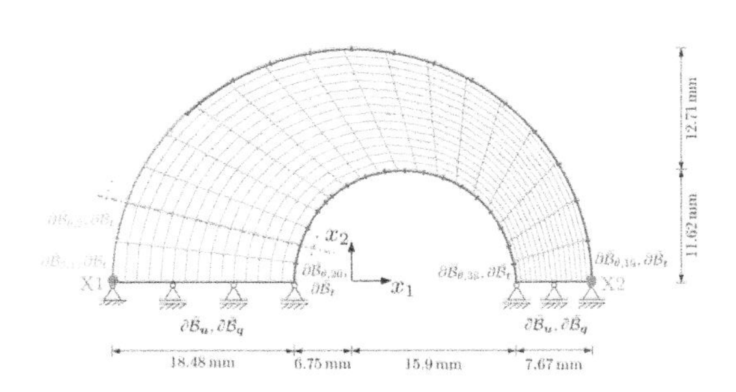

Figure 1. Single-scale boundary value problem for the cooling of a two-dimensional section of a cylindrical specimen with eccentric hole. Thermal boundary conditions are applied on the lateral surface, with spray applied in light colored region. The evolution over time of the displacement are stored along the dashed line.

2 FORMULATION OF THE PROBLEM

The bvp at hand is a cylindrical specimen with an eccentric hole made from Cr-alloyed steel 1.3505 (100Cr6), see Figure 1. After heating to above 1000°C, it is cooled down to room temperature by application of a spray, which is a mixture of water and air, see Figure 2. The fast cooling results in a phase transformation from initially 100% austenite to approximately 87% martensite and 13% retained austenite. Experimental and numerical investigations, which are carried out at Institute of Forming Technologies and Machines (IFUM), Leibniz University Hannover, serve as basis for necessary input data.

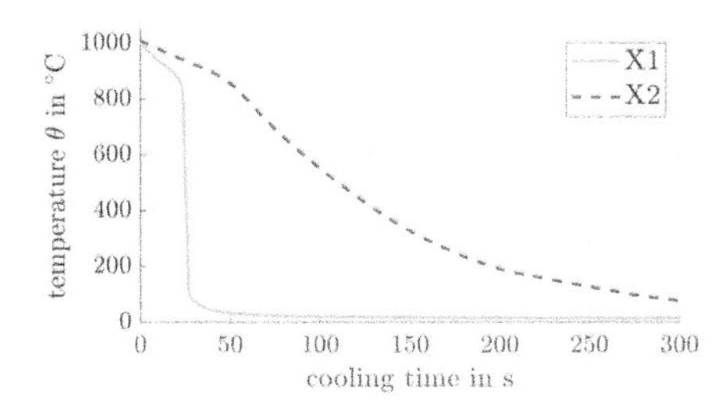

Figure 2. Thermal boundary conditions applied in measuring points X1 and X2, which are defined in Figure 1.

The geometry is discretized using 100 × 50 (tangential×radial direction) 9-noded quadrilateral elements. An adaptive time stepping scheme is applied with refinement in case of phase transformation. In absence of outer forces or moments, the resulting

DOI: 10.1201/9781003348450-41

stresses inside the material can be interpreted as residual stresses in consequence of the fast cooling and the phase transformation.

In order to reduce the numerical costs of the bvp, only a subsection of the bvp should be computed with a single-scale FE- and a two-scale FE^2-method. For this purpose, the evolution of the vertical displacement along a radial line is stored over time during the single-scale cooling simulation of the complete bvp, see Figure 1. This information is used to calculate the cooling process on a reduced geometry with 300 elements, depicted in Figure 3. For this purpose an additional displacement boundary condition is applied to the cutting edge. These additional boundary conditions are intended to represent the constraints of the geometry parts that are cut off. On microscale, 900 linear quadrilateral elements are taken into account.

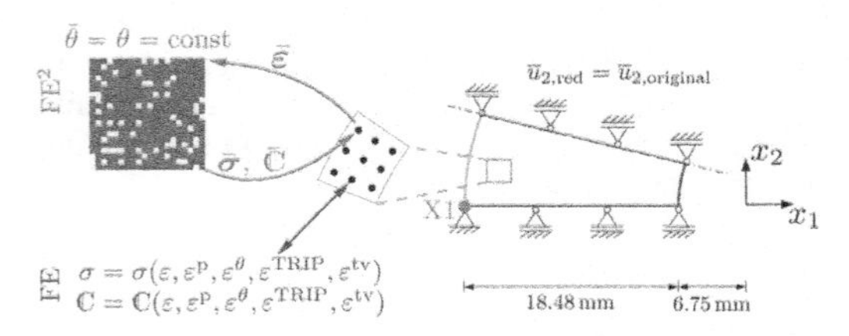

Figure 3. Reduced geometry as boundary value problem in a two-scale finite element simulation.

For the material description in the framework of a small strain theory, an additive split of the total strains is taken into account. The basic approach has been presented in (Mahnken, Schneidt, & Antretter 2009). Thus, the strains are decomposed into elastic (e), plastic (p), thermal (θ), transformation induced plasticity (TRIP) and transformational volumetric (tv) strains, i.e., $\varepsilon = \varepsilon^e + \varepsilon^p + \varepsilon^\theta + \varepsilon^{TRIP} + \varepsilon^{tv}$, respectively. The initially face-centered cubic (fcc) unit cells of austenite are transformed to body-centered tetragonal (bct) unit cells of martensite. This comes along with a volumetric expansion, taken as $K_{tv} = 1\%$ in each direction uniformly. The actual martensitic volume fraction c^M depends on the temperature and is computed by the classical approach of (Koistinen & Marburger 1959).

In this contribution, focus lies on the simulation of the cooling process. In order to take into account the plastic deformation of the hot bulk forming process taking place before the cooling, the computed accumulated plastic strains (by IFUM) are applied as initial values to the bvp at hand. The incorporation of phase transformation on microscopic level for the FE^2 approach follows (Uebing, Brands, Scheunemann, & Schröder 2021).

3 NUMERICAL RESULT

For the residual stress analysis, the tangential stresses are in focus, since these are known to be most relevant regarding service-life or strength. In Figure 4, the tangential stress evolution in measuring point X1 is given. In the beginning of the cooling process, before the onset of phase transformation, thermal contraction leads to tensile stresses in the region, in which the spray cooling is applied. The onset of martensitic evolution, which is related to the volumetric expansion of the atomic lattice, leads to a superposition and, thus, compressive stresses. As the cooling progresses, the phase transformation also takes places in the bulk material. As a result, the material at the lateral surface experiences a reduction in compressive stresses or even tension as shown in the final stress state.

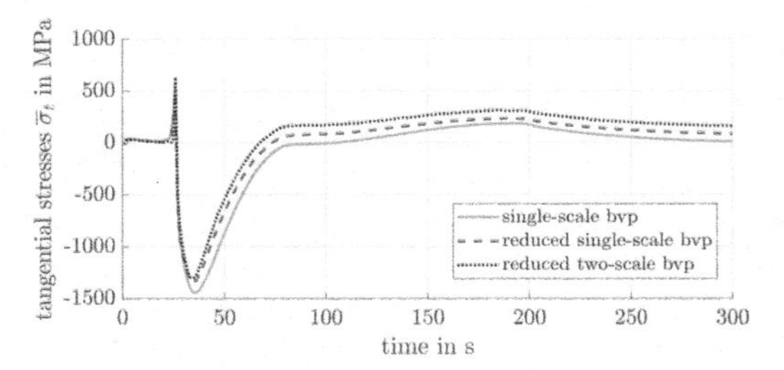

Figure 4. Tangential stress evolution in measuring point X1.

Comparing the results of all three bvp, only small deviation on macroscopic level is visible. These differences can be traced back to some simplifications in the material modeling. One example are the simplified boundary conditions, which have been applied to the cutting edge of the reduced geometry, leading to an overestimation of the stresses. In general, the comparison of the computing time of one time step shows that by reducing the bvp, the numerical cost is significantly reduced by factor of approximately 16.

4 CONCLUSIONS

In this contribution the hot bulk forming of a cylindrical specimen with eccentric hole is analyzed. A method is presented, to reduce the geometry and thereby the numerical costs to enable a two-scale analysis. It can be concluded, that the reduced formulation is already close to the full solution of the problem, but also that there is room for improvement.

REFERENCES

Koistinen, D. & R. Marburger (1959). A general equation prescribing the extent of the austenite-martensite transformation in pure iron-carbon and plain carbon steels. *Acta metallurgica 7*(1), 59–60.

Mahnken, R., A. Schneidt, & T. Antretter (2009). Macro modelling and homogenization for transformation induced plasticity of a low-alloy steel. *International Journal of Plasticity 25*(2), 183–204.

Schröder, J. (2014). A numerical two-scale homogenization scheme: the FE–method. In J. Schröder and K. Hackl (Eds.), *Plasticity and Beyond - Microstructures, Crystal-Plasticity and Phase Transitions*, Volume 550 of *CISM Courses and Lectures*, pp. 1–64. Springer.

Uebing, S., D. Brands, L. Scheunemann, & J. Schröder (2021). Residual stresses in hot formed bulk parts - Two-scale approach for austenite-to-martensite phase transformation. *Archive of Applied Mechanics 91*, 542–562.

Current Perspectives and New Directions in Mechanics, Modelling and Design of Structural Systems – Zingoni (ed.)
© 2022 Copyright the Author(s), ISBN 978-1-032-39114-4

Introduction to lattice discrete particle model for thermoset polymers

J. Vorel, J. Vozáb & J. Kruis
Faculty of Civil Engineering, Czech Technical University in Prague, Prague, Czech Republic

ABSTRACT: Thermosetting polymers are widely used as adhesives or matrices in fibre reinforced, or particulate filled composites, with broad applications in many industries involving, but not limited to, building, automotive, aerospace, marine industries. Thermosets and their fibre-reinforced composites have been progressively replacing traditional construction materials in civil engineering. The main applications of particle-filled thermosets and thermoset-matrix composites in building and construction include adhesives, repair and rehabilitation of civil structures. Such applications require new approaches and reliable computational models that allow the accurate yet computationally efficient prediction of structures or structural elements. This paper aims to introduce a Lattice Discrete Particle Model suitable for simulating the failure behaviour of particle-filled polymers (LDPM-P) used in civil engineering. LDPM-P can simulate the polymer composites of interest at the particle scale, considering their size and distribution. The constitutive relations are based on the phenomenological approach. The proposed approach is intended for a reliable design of a broad group of structural elements such as adhesive anchors and rebar connections. The proposed model is implemented into the MARS software.

1 INTRODUCTION

Thermosetting polymers are widely used as adhesives or matrices in fibre reinforced, or particulate filled composites, with broad applications in many industries involving, but not limited to, building, automotive, aerospace, marine industries. Thermosets and their fibre-reinforced composites have been progressively replacing traditional construction materials in civil engineering. The main applications of particle-filled thermosets and thermoset-matrix composites in building and construction include adhesives, repair and rehabilitation of civil structures. Nowadays, continuum-based finite elements numerical models are usually utilized to capture and predict the behaviour of particulate polymers (Clarijs, Leo, Kanters, van Breemen, & Govaert 2019, Krop, Meijer, & van Breemen 2019, Müller, Rozo Lopez, Klein, & Hopmann 2020). Lower scale models were also proposed by several authors (Lieou, Elbanna, & Carlson 2013, Kothari, Hu, Gupta, & Elbanna 2018). However, to keep the reasonable computational time and still consider the material internal structure, the Lattice Discrete Particle Model (LDPM), usually used for concrete and similar quasi-brittle materials, seems to be a good candidate. Therefore, the paper introduces the modification of the standard LDPM (Cusatis, Pelessone, & Mencarelli 2011, Cusatis, Mencarelli, Pelessone, & Baylot 2011) to fit the needs of polymer-based composites. Moreover, the Poisson ratio exceeding the limiting value of 0.25 for the current LDPM formulations is desirable. The standard polymers can experience a value between 0.3-0.4 based on the material type.

2 LATTICE DISCRETE PARTICLE MODEL

The LDPM simulates the material as a collection of rigid bodies (cells) interacting over the facets defined between them. These facets are assumed to be in the matrix phase between the adjacent cells and are interpreted as potential crack surfaces. First, particles with the assumed spherical shape are introduced into the studied volume. A Delaunay tetrahedralization of the particle centres and the nodes used to describe the external surface of the volume is used to define the lattice system that represents the mesostructure topology. The system of polyhedral cells is then created based on the 3D tessellation. Note that there are different options utilized for the tessellation, e.g., described in (Cusatis, Pelessone, & Mencarelli 2011) or (Eliáš 2017). Each cell consists of the aggregate and surrounding matrix phase found between the particles (Figure 1). Contrary to the original LDPM formulation, no mix design is needed, and only the distribution of filler sizes are assigned.

The rigid body kinetics is utilized to describe the deformations associated with the facets (Cusatis, Pelessone, & Mencarelli 2011).

$$u(x) = u_i + \theta_i \times (x - x_i), \qquad (1)$$

where u_i and θ_i are the translational and rotational degrees of freedom of node i. For the given displacements and rotations of the associated particles, the relative displacement at the centroid of facet k can be determined as

DOI: 10.1201/9781003348450-42

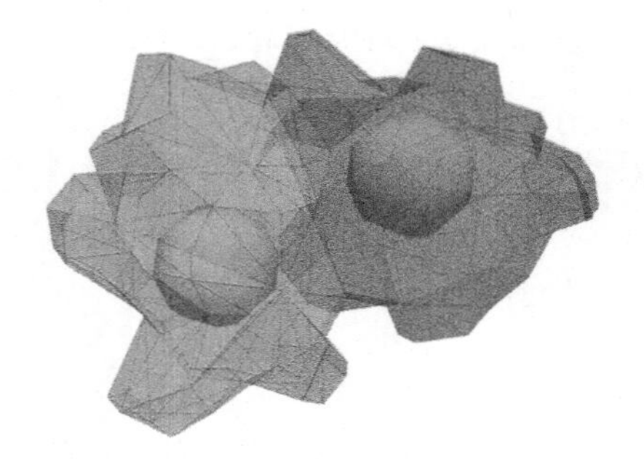

Figure 1. Cells of two adjacent particles.

$$u_{Ck} = u_{Cj} - u_{Ci}, \tag{2}$$

where u_{Ci} and u_{Cj} are the displacements at the facet centroid caused by the translations and rotations of the adjacent nodes *i* and *j*, respectively. Displacement vector u_{Ck} is then employed to define the strain measures and discrete compatibility equations as follows:

$$\varepsilon_{Nk} = \frac{n_k{}^{\mathrm{T}} u_{Ck}}{l_{ij}}, \quad \varepsilon_{Mk} = \frac{m_k{}^{\mathrm{T}} u_{Ck}}{l_{ij}},$$

$$\varepsilon_{Lk} = \frac{l_k{}^{\mathrm{T}} u_{Ck}}{l_{ij}}, \tag{3}$$

where $n = (x_j - x_i)/l_{ij}$, m and l are two mutually orthogonal vectors in the plane of the projected facet and $l_{ij} = \| x_j - x_i \| = [(x_j - x_i)^{\mathrm{T}}(x_j - x_i)]^{1/2}$. x_i and x_j stand for the positions of node *i* and *j*, respectively. Because of the restriction on Poisson's ratio $(-1 < \nu < 0.25)$ caused by the aforementioned split into normal and shear components, the volumetric-deviatoric split introduced in the microplane models (Carol & Bažant 1997, Bažant & Zi 2003) is considered. The volumetric-deviatoric split allows to recover the full Poisson ratio range $(-1 < \nu < 0.5)$ needed for polymers.

The constitutive material law defined on the facets is described in the full version of the paper. By imposing the equilibrium through the Principle of virtual work, the internal work and nodal forces associated with the facet can be calculated (Cusatis, Pelessone, & Mencarelli 2011).

3 NUMERICAL IMPLEMENTATION

The proposed lattice discrete particle model for polymers is implemented in the MARS software[1] which is a special purpose computational software for simulating the mechanical response of structures under various loading conditions. A detailed description of the similar model implementation can be found in (Cusatis, Pelessone, & Mencarelli 2011). Moreover, the definition based on the angle ω can be also extended to hydrostatic compressionx, see (Cusatis,Bažant, & Cedolin 2003) for similar approach.

4 CONCLUSIONS

The material model based on the lattice particle model has been formulated. It has various attractive features, which still need to be verified in numerical simulations, including various loading scenarios. With the rigorous formulation of conclusions, this task is left for further study.

ACKNOWLEDGEMENT

The financial support provided by the GAČR grant No. 21-28525S is gratefully acknowledged.

REFERENCES

Bažant, Z. & G. Zi (2003). Microplane constitutive model for porous isotropic rocks. *International journal for numerical and analytical methods in geomechanics 27*(1), 25–47.

Carol, I. & Z. Bažant (1997). Damage and plasticity in microplane theory. *International Journal of Solids and Structures 34*(29), 3807–3835.

Clarijs, C., V. Leo, M. Kanters, L. van Breemen, & L. Govaert (2019). Predicting embrittlement of polymer glasses using a hydrostatic stress criterion. *Journal of Applied Polymer Science 136*(17), 47373.

Cusatis, G., Z. Bažant, & L. Cedolin (2003). Confinement-shear lattice model for concrete damage in tension and compression: I. theory. *Journal of engineering mechanics 129*(12), 1439–1448.

Cusatis, G., A. Mencarelli, D. Pelessone, & J. Baylot (2011). Lattice discrete particle model (LDPM) for failure behavior of concrete. II: Calibration and validation. *Cement and Concrete Composites 33* (9), 891–905.

Cusatis, G., D. Pelessone, & A. Mencarelli (2011). Lattice discrete particle model (LDPM) for failure behavior of concrete. I: Theory. *Cement and Concrete Composites 33*(9), 881–890.

Eliáš, J. (2017). Boundary layer effect on behavior of discrete models. *Materials 10*(2), 157.

Kothari, K., Y. Hu, S. Gupta, & A. Elbanna (2018). Mechanical response of two-dimensional polymer networks: role of topology, rate dependence, and damage accumulation. *Journal of Applied Mechanics 85*(3).

Krop, S., H. Meijer, & L. van Breemen (2019). Sliding friction on particle filled epoxy: Developing a quantitative model for complex coatings. *Wear 418*, 111–122.

Lieou, C., A. Elbanna, & J. Carlson (2013). Sacrificial bonds and hidden length in biomaterials: A kinetic constitutive description of strength and toughness in bone. *Physical Review E 88*(1), 012703.

Müller, J., N. Rozo Lopez, E. Klein, & C. Hopmann (2020). Predicting the damage development in epoxy resins using an anisotropic damage model. *Polymer Engineering & Science 60*(6), 1324–1332.

1 https://www.es3inc.com/mars-capabilities/

Current Perspectives and New Directions in Mechanics, Modelling and Design of Structural Systems – Zingoni (ed.)
© 2022 Copyright the Author(s), *ISBN 978-1-032-39114-4*

Artificial neural networks as material models for finite element analysis

D. Sommer, B. Troff & P. Middendorf
Insitute of Aircraft Design, University of Stuttgart, Germany

ABSTRACT: An obstacle for development of structural components is often the adequate description of materials for computational methods, especially in finite element analysis. The substitution of classical constitutive material models with data-driven models using machine learning techniques could be a new strategy, providing a faster description, faster computation and efficient parameter identification for the modelling of materials. For this purpose, an artificial neural network (ANN) is trained, which learns the relationship between strains and stresses and is capable of modelling the hypo-elastic behaviour from a constitutive model. To train the ANN, artificial stress-strain curves are generated using a programming interface for LS-DYNA's material model driver. A hyperparameter study finds the optimal set of hyperparameters for training of the neural network. Finally, the trained ANN is implemented in a user-defined subroutine, which provides a link between modern machine learning frameworks and commercial FEA codes. Explicit simulations using the ANN model run stable and show good agreement in global response as well as on local element stress level.

1 INTRODUCTION

A material description for finite element analysis (FEA) used for component design or complete car-crash analysis is only possible after going through a complex, costly and mostly manual testing and calibration process. Also, a high-level of experience is required to choose the appropriate material model. An automated material model description process by means of data driven methods is aimed to be used to create a material description inside FEA, evaluated for robustness and its predictive capability. The motivation of this work is to construct a machine learning model (MLM), which can substitute classical analytical models and which is then transferable to describe a new material without the drawbacks outlined before.

Using a surrogate neural network model was proposed by Ghaboussi et al. in 1998 with several publications from different researchers since. Recently Bonatti and Mohr (2021) formulated a custom neural network architecture for the specific needs of materials modelling.

In this paper, the primary focus is the generation of virtual training data and on finding the best parameters for a feed-forward neural network, which is then shown running in a commercial code and evaluated.

2 TRAINING DATA ACQUISITION

The Material Model Driver (MMD) allows generating stress-strain data for any constitutive material model implemented in LS-DYNA for a given strain path. It calculates the course of the stress along a predefine-able strain path from the chosen material model. The calculation of the stress is carried out without the need for an element formulation or other FEA settings such as hourglass control. This is the advantage of this method, compared to the use of stress-strain paths from direct FEA simulations. In order to cover the entire strain space, it is parametrized (for the two-dimensional case) as shown in Figure 1a. This is done by generating straight lines from the origin to the points uniformly distributed on a spherical boundary surface, defined by an effective maximum total strain. A Python framework was established to automate the MMD. A resulting data set of stress-strain curves, is shown in Figure 1a. As the material model, an elasto-plastic model with J2-plasticity representing a generic aluminium alloy was selected.

After the data is generated, it has to be normalised by transforming the different inputs to the ANN into the same order of magnitude. This brings advantages for the training of the neural network, such as increasing the learnability of the data and acceleration of training time (Sola & Sevilla 1997). The input vector x is linearly transformed to the interval $[a, b]$, giving the normalised input $\tilde{x}$. For each input variable, $max(x_i)$ or $min(x_i)$ of an input variable in the training dataset x are determined and mapped to the

DOI: 10.1201/9781003348450-43

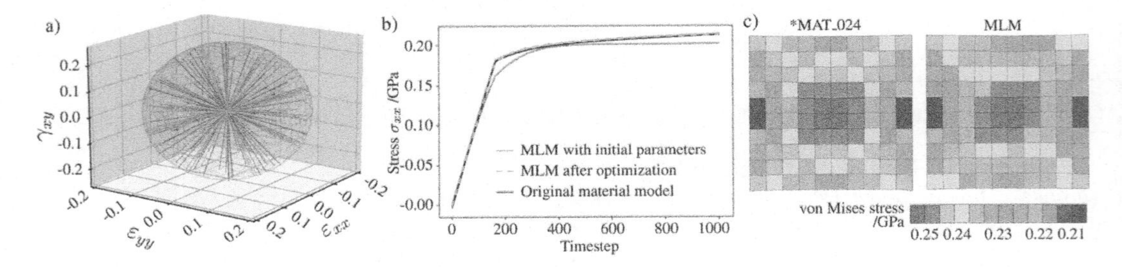

Figure 1. A) Generic linear strain paths with boundary sphere at effective strain of 0.15; b) Performance of MLM before and after optimization on a single strain path; c) Von Mises stress of original *MAT_024 vs MLM on a tensile patch of elements.

interval. The normalised output $\widetilde{y}$ is calculated analogously before training the ANN. This requires a back-transformation of the values of the output vector returned by the neural network to the physical values.

3 THE MACHINE LEARNING MATERIAL MODEL

For the machine-learning material model (MLM) we consider the mapping in the form of $g : \widetilde{\varepsilon}_{ij}(t) \mapsto \widetilde{\sigma}_{ij}(t)$, where g is the neural network, predicting the material's normalized stress response $\widetilde{\sigma}$ from the normalized strain $\widetilde{\varepsilon}$, the latter as given from the explicit FEA cycle. To find the best fitting hyperparameters to learn the original material behaviour with a feed forward neural network (FFNN) a study of parameters is conducted. Hypoelasticity is assumed for simplicity, as the FFNN cannot capture unloading correctly, since it contains no state- or history-variables. The network is trained with the dataset which was created using the MMD.

An extensive hyperparameter study is performed, training the MLM iteratively with different settings. Found parameters on presented training data are summarized in Table 1. The optimal MLM is trained to a MSE loss of $2.8 \cdot 10^{-4}$ over 1000 epochs with a mean absolute prediction error over all components is found as $3.1 \cdot 10^{-3}$ MPa.

Table 1. Material parameters for a generic aluminium alloy.

Hyperparameter	Initital setting	Optimal setting
Number of Layers	3	4
Neurons per Layer	10	15
Activation function φ	sigmoid	tanh
Batch size	500	200
Loss function	MSE	MSE
Optimizer	Adam	Adam
Learningrate	0.1	decay to 10^{-6}

Using the model in a finite element analysis with explicit timestepping with our own user subroutine, it is possible to evaluate the model on non-predefined strain-paths. Inside FEA the next strain increment in each element will be determined by the current stress response. A misprediction of stress can result in a different progression of strain within live FEA. A patch of 10x10 Belytschko-Tsay shell elements with nodes fixed on the left side while on the right side, nodes are displaced outwards linearly. Figure 1b shows the progression of the MLM with initial parameters before optimization in blue. The MLM using the optimum parameters predicts the stress correctly. The von Mises stress (Figure 1c) is slightly mispredicted, showing the state of maximum error at high plastic strains.

4 CONCLUSION

This study presented a machine-learning-based material model in the commercial code LS-DYNA. The MLM as a surrogate model was shown to be precise, given hyperparameters of the neural network are tuned to training data. This set of parameters is determined once for a class of materials and given scheme of training-data generation. A methodology of obtaining stress-strain paths from classical models for pre-training MLM was also given and evaluated. Important future work has to capture unloading, but mainly focus on transfer learning of MLM onto new materials with only data obtained from real-world tests.

Funding: This work was supported by Federal Ministry for Economic Affairs and Climate Action of Germany for project AIMM (Artificial Intelligence for Materials Modelling; identification 19I20024).

REFERENCES

Bonatti, C. & D. Mohr (2021). One for all: Universal material model based on minimal state-space neural networks. *Science Advances 7* (26),eabf3658.

Ghaboussi, J., D. A. Pecknold, M. Zhang, & R. M. Haj-Ali (1998). Autoprogressive training of neural network constitutive models. *International Journal for Numerical Methods in Engineering 42*(1), 105–126.

Sola, J. & J. Sevilla (1997). Importance of input data normalization for the application of neural networks to complex industrial problems. *IEEE Transactions on Nuclear Science 44*(3), 1464–1468.

Current Perspectives and New Directions in Mechanics, Modelling and Design of Structural Systems – Zingoni (ed.)
© 2022 Copyright the Author(s), ISBN 978-1-032-39114-4

Simulations of elasto-plastic soils within the framework of the Theory of Porous Media

J. Sunten, A. Schwarz, J. Bluhm & J. Schröder
Institute of Mechanics, Department of Civil Engineering, University of Duisburg-Essen, Essen, Germany

ABSTRACT: The Theory of Porous Media (TPM) is an established tool in the examination of elasto-plastic deformations of soils. Therein the incorporation of dynamic effects into a simplified binary model grants realistic simulations in which the material behavior is described by distinct yield criteria. The application of TPM and plasticity can lead to diverse material models but this paper concentrates on one specific combination.

1 INTRODUCTION

Going back to the work of (1) and the references therein, the TPM, as an extension of the mixture theory, see (2), by the concept of volume fractions, is an appreciated tool for the mechanical description of soils. The paper at hand is limited to a binary, dynamic, elasto-plastic models dealing with dynamic simulations in coalescence with a plasticity model for granular porous solids. All following simulations have been done with AceGen and AceFEM packages (version 7.114), see (3),(4) and (5) of Mathematica (version 12.3.1.0).

2 THEORETICAL BACKGROUND

The here presented equations and simulations are based on the well known simplified binary model (solid and liquid) with the exception of the inclusion of dynamic effects. As an additional assumption, the body forces are considered as zero and all calculations are restricted to small strain theory. The two governing equations (the balance of momentum of the mixture and the material time derivative of the saturation condition) are therefore directly given in their dynamic forms as

$$\mathrm{Div}(\boldsymbol{\sigma}_E^S - \mathcal{P}\mathbf{I}) - \rho^S \mathbf{x}_S'' - \rho^L \mathbf{x}_L'' = \mathbf{0} \qquad (1)$$

$$\mathrm{Div}\left(\mathbf{x}'_S - \frac{k^L}{\gamma^{LR}}\mathrm{Grad}\,\mathcal{P} - \frac{k^L\rho^{LR}}{\gamma^{LR}}\mathbf{x}''_L\right) = 0 \qquad (2)$$

with $\boldsymbol{\sigma}_E^S$ being the effective Cauchy stress of solid, $\mathcal{P}$ the pore water pressure, k^L the Darcy parameter, ρ^α and $\rho^{\alpha R}$ (with $\alpha = S, L$) the partial and real densities and $\gamma^{LR} = 9.81\,\rho^{LR}\,kg/(\mathrm{s}^2\mathrm{m}^2)$ denoting the specific weight. The weighted seepage (difference) velocity $\mathrm{n}^L\mathbf{w}_{LS} = \mathrm{n}^L(\mathbf{x}'_L - \mathbf{x}'_S)$ can be expressed with the help of an extended Darcy law as $\mathrm{n}^L\mathbf{w}_{LS} = \frac{k^L}{\gamma^{LR}}\mathrm{Grad}\,\mathcal{P} - \frac{k^L\rho^{LR}}{\gamma^{LR}}\mathbf{x}_L''$ where $\mathbf{x}_L''$ is the acceleration of the fluid phase. As there are three unknowns ($\mathcal{P}$, $\mathbf{x}_S$ and $\mathbf{x}_L$) in these equations, at least one additional equation is needed to solve the equational set up. Several approaches to include dynamic effects into the TPM are known and the weak forms of (6) are utilized in the following simulations. This work is limited to an elasto-plastic behavior without hardening of the solid phase, so that

$$\boldsymbol{\sigma}_E^S = 2\mu^S\varepsilon_e^S + \lambda^S \mathrm{tr}\,\varepsilon_e^S\,\mathbf{I} \qquad (3)$$

denotes the Cauchy stress of the solid phase arising from the elastic part of the strains $\varepsilon_e = \varepsilon - \varepsilon_p$. To determine ε_e and ε_p a yield criterion called Ehlers 7 (E 7) as presented in (7) is applied. It was especially developed for the use in saturated soils making it the preferred choice for accurate simulations, where its fitting behavior is based on seven carefully chosen material parameters. In its formulation J_2^d denotes the second and J_3^d the third invariant of the deviatoric part of the Cauchy stress of the solid phase and J_1 the first invariant of the solid Cauchy stress. The expression $\sqrt{-J_2^d}$ can also be transformed into the first Reuss variable, the yield radius $\hat{r} = \sqrt{-2J_2^d}$. In this work associative flow rules are used.

3 NUMERICAL EXAMPLES

It was chosen to use so called Taylor-Hood elements and an implicit backward Euler time integration scheme. The considered body is depicted in Figure 1 and all given material parameters are taken from (6) (TPM parameters, 2D example) and (7) (plasticity parameters), respectively. The body is exposed to two load cases (LC). For LC1 $q_{1_{max}} = 200,000$ N/m2) is linearly increased in the first 0.5 s and is constant thereafter. The liquid phase can leave the body only at

DOI: 10.1201/9781003348450-44

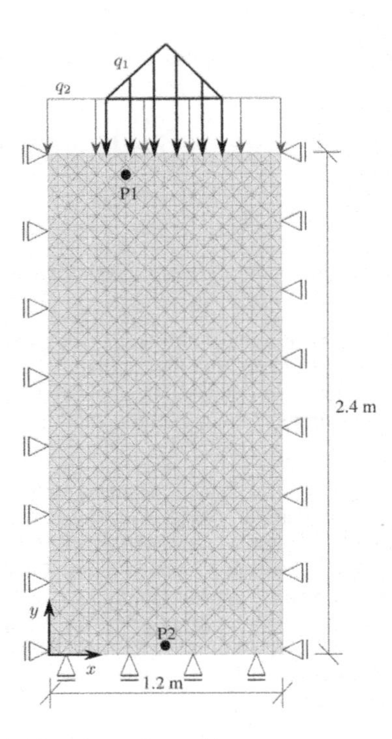

Figure 1. Boundary value problem.

the upper edge next to the load. The classical yield curves ($\hat{r}$ over J_1) are interpreted at the point P1 (0.4|2.3) and the boundary value problem (BVP) is calculated until complete consolidation ($\mathcal{P} = 0$ in the complete body) which takes about 200 s. Contrary LC2 is a load with ($q_{LC2} = 280,000$ N/m2) linearly increased in 0.01 s, with the out stream of the liquid at the lower edge and the vertical displacement being interpreted at point P2 (0.6|0.05). The goal of example one is to sketch the influence of the TPM model in comparison to a one phase (without liquid) simulation, whereas example two illustrates the effect of the dynamic simulation. In all figures, the abbreviation "Q-S" stands for "quasi-static" referring to simulations without acceleration. The graphs in Figure 2a belong to LC1 and indicate that the liquid phase does influence the material behavior in its final response hinting more noticeable differences for more complex BVPs. Figure 2b represents LC2 and it can be seen, that the oscillating response of the material can not be depicted by the quasi-static approach, rendering a dynamic simulation inevitable for these BVPs. The dynamic and quasi-static simulations lead to the same results in the consolidated state.

4 CONCLUSION

Conclusively it can be said, that the use of the TPM does influence more complex BVPs and the dynamic effects can influence the responding behavior at the

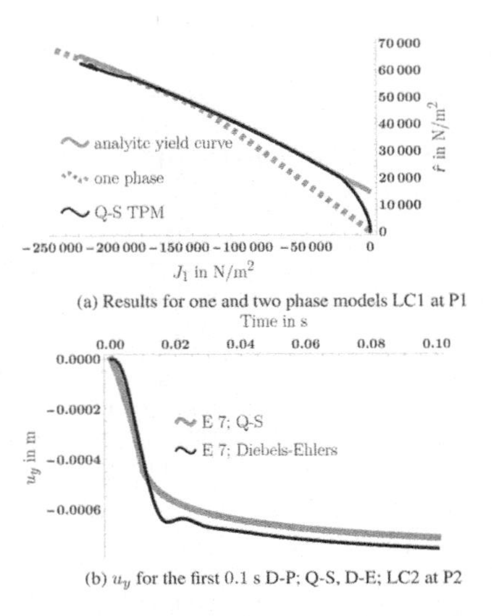

(a) Results for one and two phase models LC1 at P1

(b) u_y for the first 0.1 s D-P; Q-S, D-E; LC2 at P2

Figure 2. Results for rectangle body ($dx = 1.2$ m & $dy = 2.4$ m) with 2304 triangular elements.

beginning, but not in the consolidated state. That is why, in a first step the BVP under examination has to be specified to determine the question whether to use the TPM at all. Then the target of interest has to be clarified as an observation over time needs other characteristics as a pure consolidation simulation. Further research in this field shall among others focus on the implementation of different dynamic approaches, adaptive time integration schemes and parallel simulations for elasto-plastic, dynamic TPM problems.

REFERENCES

[1] W. Ehlers. *Poröse Medien - ein kontinuummechanisches Modell auf der Basis der Mischungstheorie*. Habilitationsschrift, Universität - Gesamthochschule - Essen, 1989.

[2] R. M. Bowen. Incompressible porous media models by use of the theory of mixtures. *International Journal of Engineering Science*, 18:1129–1148, 1980.

[3] J. Korelc. Automatic generation of finite-element code by simultaneous optimization of expressions. *Theoretical Computer Science*, 187(1):231–248, 1997.

[4] J. Korelc. Multi-language and Multi-environment Generation of Nonlinear Finite Element Codes. 18:312–327, 2002.

[5] J. Korelc. Automation of primal and sensitivity analysis of transient coupled problems. *Computational Mechanics*, 44:631–649, 2009.

[6] S. Diebels and W. Ehlers. Dynamic analysis of a fully saturated porous medium accounting for geometrical and material non-linearities. *International Journal for Numerical Methods in Engineering*, 39:81–97, 1996.

[7] W. Ehlers. A single-surface yield function for geomaterials. *Archive of applied mechanics*, pages 246–259, 1995.

Current Perspectives and New Directions in Mechanics, Modelling and Design of Structural Systems – Zingoni (ed.)
© 2022 Copyright the Author(s), ISBN 978-1-032-39114-4

A coupled multi–field dynamic model for anisotropic porous materials

N. De Marchi, M. Ferronato, G. Xotta & V.A. Salomoni
Department of Civil, Environmental and Architectural Engineering - ICEA & Department of Management and Engineering – DTG, University of Padova, Padova, Italy

ABSTRACT: A fully coupled multi-field model for the dynamic simulation of anisotropic porous materials is here presented. The related discrete problem is addressed in a fully-implicit way and the Bi-Conjugate Gradient Stabilized (Bi-CGStab) algorithm, properly accelerated by an ad-hoc preconditioner, was used. A series of dynamic analyses has been performed, in order to test the potential and computational efficiency of the developed numerical tool.

1 INTRODUCTION

Predicting the mechanical behavior of anisotropic porous media, subject to different types of dynamic loads, is still a challenging task. At the same time, also having an efficient, stable and robust numerical model for these problems, is a necessary condition to obtain good results in a relatively short time.

For this reason a fully coupled multi-field model for the dynamic simulation of anisotropic porous materials, implemented in the MATLAB research code *GeoMatFEM* (De Marchi et al. 2019), is presented here. Three-dimensional dynamic analyses of seismic waves propagation in a fully saturated anisotropic soil were conducted to validate and test the computational efficiency of the developed code.

2 THEORETICAL FRAMEWORK

2.1 *Governing equations*

The mass balance and the linear momentum equations for a fully saturated porous medium in the dynamic regime, by assuming that neither thermal effects nor mass exchanges between phases are possible, are:

$$(\dot{\rho}^{\alpha})_{\alpha} + \rho^{\alpha}\nabla^{x} \cdot v_{\alpha} = 0; \tag{1}$$

$$\rho^{\alpha}(\dot{v}_{\alpha})_{\alpha} = \nabla^{x} \cdot \sigma^{\alpha} + \rho^{\alpha} g + h^{\alpha}, \tag{2}$$

where $\rho^{\alpha} = n^{\alpha}\rho_{\alpha}$ is the partial mass density of α-phase, computed as the product of the volume fraction n_{α} and the intrinsic mass density of α constituent. v_{α} is the velocity vector, σ^{α} is the partial Cauchy stress tensor of the α phase and h^{α} is the volume specific local interaction force between the phases.

2.2 *Constitutive equations*

A homogeneous and anisotropic linear elastic porous medium is here assumed, so that the constitutive relations for the solid matrix and for the fluid are provided by an extension of the Terzaghi's principle and the Hooke's law:

$$\sigma' = \sigma + Bp = C : \varepsilon; \tag{3}$$

$$B = (I - CS_s) : \mathbf{1}, \tag{4}$$

where σ' σ are the effective and the total stress tensors, B is the Biot effective stress coefficient tensor, p is the pore fluid pressure, C is the elastic constitutive tensor of the mixture, ε is the small strain tensor of the solid matrix and S_s is the elastic compliance tensor of the solid matrix.

Regarding the fluid phase, by assuming laminar flow, the generalized form of Darcy's law, that relates the fluid mass flow rate to the pore pressure gradient, is considered:

$$n^{F} w_{F} = -K^{F}(\nabla^{x} p - \rho_{F} g), \tag{5}$$

where $w_F = v_F - v_S$ is the relative velocity, $w^F = n^F w_F$ is the Darcy velocity, $n^F = \varphi$ is the porosity and K^F is the second-order specific permeability tensor.

3 NUMERICAL SOLUTION

The numerical solution of the initial-boundary value coupled problem in space was obtained by using the FE Method with inf-sup stable discretizations, while the θ-Method was adopted in time.

DOI: 10.1201/9781003348450-45

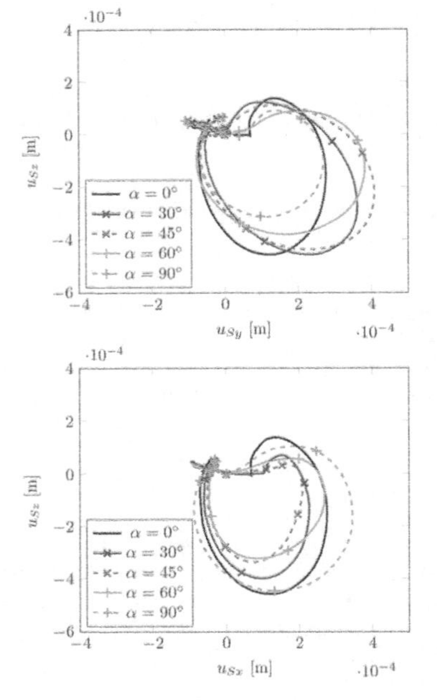

Figure 1. Raileigh waves, for 2 nodes, considering different rotations α of the isotropic axes of symmetry.

In order to improve the linear solver efficiency and robustness and allow for the simulation of larger size problems, the Bi-CGStab iterative algorithm was used to solve the time-sequence of large sparse non-symmetric linear systems, with convergence accelerated by an ad-hoc preconditioner P^{-1} (Castelletto et al. 2016, Ferronato et al. 2019).

4 NUMERICAL RESULTS

A square soil prism, subject to an impulsive pressure load over the top surface, was considered. A transversely isotropic elastic constitutive model was adopted for both the solid (intrinsic) material and the porous (structural) matrix.

A series of analyses were performed by rotating by α (from 0 to 90°) the isotropy plane and, thus, giving rise to different soil responses.

An impulsive load, impacting a transversal isotropic medium, generates two different shear waves with different velocities and orientations; this phenomenon is known as shear wave splitting.

Rayleigh waves, arising from the body waves interaction with the upper free surface, are depicted in Figure 1 for 2 nodes located on two orthogonal directions of wave propagation. It can be appreciated how, considering no rotation of the isotropy plane (α=0°), nodes motion is the same (black continuous line) while, as α increases, two different motions appear. Splitting phenomenon is thus caught.

Figure 2 shows the convergence profiles of the Left-preconditioned Block Bi-CGStab for four different time steps at the last Newton-Raphson iteration, by considering the anisotropic model with α=90° and by using a 21×42×20 FE discretization (dof equal to 295552). For all time steps, the convergence profiles show a steep slope and are quite stable, requiring only 36 iterations on average to obtain the linear system solution. Finally, in Table 1, the solver performance is reported for numerical examples of different dimension and mesh discretizations and compared, in terms of solution times, with the direct Matlab solver "\".

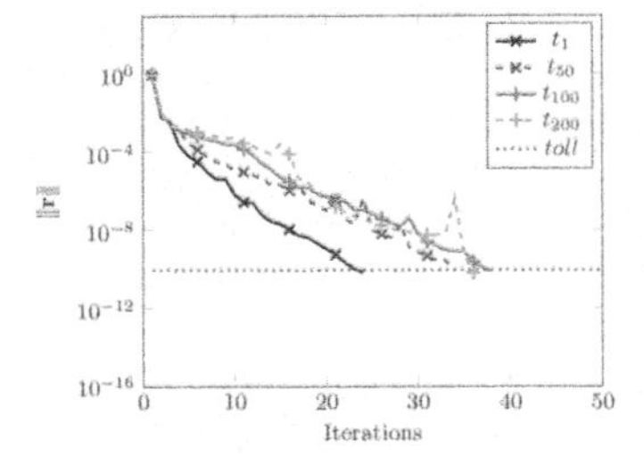

Figure 2. Convergence profile.

Table 1. Solver performance.

			CPU Time [s]	
Case	Mesh	dof	Bi-CGStab	Matlab
2D	1×42×40	33969	5.34	3.42
2D	1×84×80	142028	36.21	63.82
2D	1×168×160	566384	628.55	964.501
3D	21×21×20	148804	74.29	1303.28
3D	21×42×20	295552	325.86	–
3D	42×42×20	586948	956.24	–
3D	42×42×40	1159788	3367.95	–

5 CONCLUSIONS

A fully-coupled fully implicit FE model, for the solution of the equations representing the dynamics of fully saturated anisotropic porous media, was presented. For the efficiency and robustness of the inner linear solver, the Bi-CGStab algorithm was used, accelerated by an ad-hoc preconditioning technique. The developed numerical tool proved to be efficient and reliable for modeling the wave propagation in anisotropic soils.

REFERENCES

Castelletto, N., J. White, & M. Ferronato (2016). Scalable algorithms for three-field mixed finite element coupled poromechanics. *Journal of Computationa Physics 327*, 894–918.

De Marchi, N., V. Salomoni, & N. Spiezia (2019). Effects of finite strains in fully coupled 3d geomechanical simulations. *International Journal of Geomechanics 19*(4), 04019008.

Ferronato, M., A. Franceschini, C. Janna, N. Castelletto, & H. A. Tchelepi (2019). A general preconditioning framework for coupled multiphysics problems with application to contact-and poro-mechanics. *Journal of Computational Physics 398*, 108887.

Current Perspectives and New Directions in Mechanics, Modelling and Design of Structural Systems – Zingoni (ed.)
© 2022 Copyright the Author(s), ISBN 978-1-032-39114-4

Micromechanical modelling of failure during interactions between ice and subsea structures

R.S. Taylor & I. Gribanov
Memorial University of Newfoundland, St. John's, NL, Canada

ABSTRACT: For coastal, offshore and subsea infrastructure in ice prone regions, the potential for interactions between engineered structures and ice features must be accounted for during the design of such systems. As a natural material, ice contains many flaws and is typically at a high homologous temperature, resulting in material behavior similar to other brittle solids under high-temperature conditions. The particularly applies to its susceptibility to fracture, creep and damage under typical conditions of interest for engineering applications. Ice features contain many material defects, large grain boundaries, pre-existing cracks and other flaws that play a critical role in the initiation and propagation of fractures during interactions. The distribution of such flaws within the material contributes to statistical scale effects, wherein larger ice features are expected to contain larger flaws resulting in observed decreasing average failure pressures as interaction areas increase. In addition, since ice in the ocean is at a high homologous temperature, it is prone to rate and temperature effects. Micromechanical modelling of ice failure processes is further complicated by the fact that ice expands upon freezing, which is a thermodynamically reversible process that causes ice to melt when subjected to sufficiently high hydrostatic pressure. This pressure melting phenomena is itself a function of temperature, as is embodied in the relationship of Clapeyron and Clausius. In this paper, micromechanical models of local damage and fracture processes in ice under compression, which limit loads transmitted during interactions with engineered structures are presented, including recent experimental and computational developments to support micromechanical modelling of ice material behaviour. From this work it is concluded that since statistical distributions of flaws are ultimately responsible for the underpinning mechanisms that limit ice loads during interactions, it is essential that micromechanical models adequately capture these scale dependent aspects of the material behaviour over the relevant range of interaction scales.

1 INTRODUCTION

In ice prone regions, the potential for interactions between drifting ice features and subsea structures presents a significant challenge for designers of such systems. Traditional approaches typically have assumed that such systems always fail during interactions with any ice. Consequently risk mitigation approaches have historically focused on avoidance strategies through route modification or excavation/ trenching infrastructure below the seabed to reduce the likelihood on an ice impact. Recent work has focused on reexamining the assumption of contact failure to explore avenues for improving the ice resistance of such engineered systems. Fundamental to such evaluations is the modelling of ice failure and associated loads.

During interactions between ice features and engineered structures, localized ice failure processes are dominated by spalling and crushing (Sanderson, 1988). Spalling fractures reduce contact areas which concentrate the transmission of force into small regions of high pressure, termed high pressure zones (*hpzs*). These *hpzs* are important features that result in highly non-uniform contact through which intense local pressures are transmitted. The conditions under which *hpzs* form and evolve must be accounted for in ice modelling.

For ice interactions involving subsea infrastructure, contact between ice and seabed soils introduces additional modelling complexities. In particular, there is potential for embedment of gravel particles in the ice. Recent work has examined aspects of ice-particle interactions in the context of subsea engineering applications. One of the main conclusions of that work, which is on the scale of an individual rock particle (see Figure 1) is that observed ice failure processes very closely reflect those observed during steel spherical indentation tests.

Based on the above rationale, it assumed that recent developments of micromechanical models for ice compressive failure based on steel spherical indentation may be applied to the case of ice-rock indentation at the individual particle scale.

DOI: 10.1201/9781003348450-46

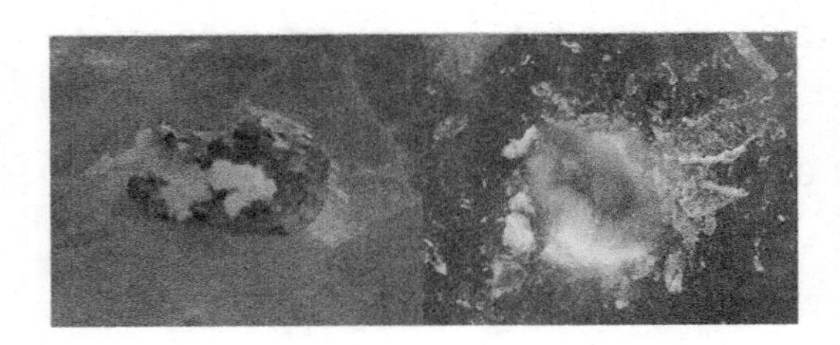

Figure 1. Ice-rock particle indentation test showing: (left) rock embedded in ice with sintered, crushed ice adhering to the rock; (right) spalls around periphery, with white crushed ice and grey recrystallized ice in the centre of the rock indentation site.

2 METHODOLOGY

In the present paper, ice compressive failure processes are examined to allow for analysis of individual physical mechanisms and support the development of modelling strategies that are informed by experimental observations.

The formation of a damaged layer is characterized by two aspects. The first characteristic of a damaged layer is the presence of white zones around the periphery of the *hpz* corresponding to a layer of microfractured, crushed and sintered ice particles, which correspond to lower pressure regions. The second characteristic is the presence of a central dark zone through which the majority of loads are transmitted. This central zone corresponds to high-confinement, high-shear conditions in an *hpz*. Pressure softening in this part of the damage layer is associated with dynamic recrystallization and localized pressure-melting processes in this layer. These processes are modelled using two Schapery damage measures, S_1 and S_2, to account for these two distinct damage sources. S_1 is a state variable that accounts for the low confinement processes (primarily microcracking) and S_2 is a state variable that models pressure softening processes in high confinement regions (Turner, 2018).

Spalls are characterized by a large asymmetric reduction in area and an associated drop in load. Since all ice will have some distribution of grain sizes, links between flaw and grain size distributions are important for modelling random fracture events (Taylor and Jordaan, 2015). Gribanov et al. (2018) developed a cohesive zone model that was applied to simulate random fracture in polycrystalline ice. This model is based on damaging cohesive elements in which energy associated with shear and tensile loading are taken into account. In addition to capturing the effects of grain size (and hence flaw size) distribution, as well as being able to capture fracture in both ductile and brittle domains, this approach also explicitly models microcracking and crack coalescence which triggers spalls and macrocracks. Overall this model gave very good agreement with data from ice fracture experiments.

While this cohesive zone model has been implemented initially only for elastic materials, extension of this approach to link with Schapery-based damage constitutive models for simulating pressure-softening behavior is planned.

3 RESULTS AND CONCLUSIONS

Ice compressive failure processes are complex and present significant challenges for the development of micromechanical models of ice behavior. The implementation of a two-term Shapery damage model in a conventional FE software can simulate dominant processes, but is limited to a single loading cycle, since severe element distortion arises in highly softened elements. The use of element deletion techniques to remove highly damaged elements has been demonstrated to be suitable for modelling the extrusion of extensively damaged material. However, such techniques are not well suited for modelling the effects of discontinuous, randomly formed microcracks and spalls. To better capture microcracking damage, the thermodynamically consistent cohesive zone model of Gribanov et al. (2018) provides a suitable framework for modelling micro and macrocracks linked to distributions of flaws (e.g. grain boundaries or pre-existing flaws). Linking these recent developments in modelling discrete fracture processes with earlier developments of ice continuum damage processes represents an important next step in micromechanical model development for ice.

REFERENCES

Gribanov, I., Taylor, R., & Sarracino, R. (2018). Cohesive zone micromechanical model for compressive and tensile failure of polycrystalline ice. Engineering Fracture Mechanics, 196, 142–156.

Sanderson, T.J.O., 1988. Ice Mechanics: Risks to Offshore Structures. Graham and Trotman Ltd., London.

Taylor, R.S., & Jordaan, I.J., (2015). Probabilistic fracture mechanics analysis of spalling during edge indentation in ice. Engineering Fracture Mechanics 134 (2015) 242–266

Turner, J. (2018). Constitutive behaviour of ice under compressive states of stress and its application to ice-structure interactions. Doctoral dissertation, Memorial University.

Current Perspectives and New Directions in Mechanics, Modelling and Design of Structural Systems – Zingoni (ed.)
© 2022 Copyright the Author(s), ISBN 978-1-032-39114-4

Variational formulation of weak layer stresses in stratified snowpacks

F. Rheinschmidt
Institute of Structural Mechanics and Design, Technical University of Darmstadt, Germany

P. Weißgraeber
Chair of Lightweight Design, University of Rostock, Germany

P.L. Rosendahl
Institute of Structural Mechanics and Design, Technical University of Darmstadt, Germany

ABSTRACT: Dry snow slab avalanches are a severe hazard for infrastructure and back country and free ride skier. Avalanche release strongly depends on the stratification of the snow cover and the mechanical properties of the individual snow layers. The condition of so-called weak layers in the snow pack is primarily critical. Overloading of these layers results in anticracks, which propagate as collapse until an avalanche is triggered.

To provide an efficient stability assessment of stratified snowpacks, we present an analytical model for snow cover deformations and stresses within the weak layer for arbitrarily layered snowpacks. In particular, the model covers the effect of the layering order on both the extensional and bending stiffness of the slab. It can be used for externally-loaded slopes and for stability tests such as the propagation saw test. The model is highly efficient and can easily be used for parameter studies and implementation into other toolchains.

1 INTRODUCTION

With the work of Rosendahl and Weißgraeber (2020) a closed-form analytical framework for the prediction of weak layer collapse is available. Modelling weak layers with normal and shear springs represents a simplified kinematic of the crucial weak layer. For bedded segments this study proposes a model improving the predictions for thick weak layers within the framework of the previously mentioned model.

2 MODELLING APPROACH

For bedded segments Timoshenko kinematics are employed for the slab. Weak layer kinematics are enhanced such that continuous displacements at the interface and clamping at the bottom end of the weak layer are provided,

$$u_{\text{weak}}(x,z) = u(x)\left(\frac{1}{2}-\frac{z}{t}\right) + \cos\left(\frac{z}{t}\pi\right)\phi_u(x), \quad (1)$$

$$w_{\text{weak}}(x,z) = w(x)\left(\frac{1}{2}-\frac{z}{t}\right) + \cos\left(\frac{z}{t}\pi\right)\phi_w(x). \quad (2)$$

From this displacement field, stresses in slab and weak layer are computed with a layer-wise isotropic constitutive law at plane strain conditions. In order to apply the principle of minimum potential energy, the strain energy as the inner potential of the bedded segment is defined by

$$\Pi^{\mathrm{i}} = \frac{1}{2}\int_V \sigma : \varepsilon \mathrm{d}V. \quad (3)$$

The contribution of the weight of the slab is considered as external work,

$$\Pi^{\mathrm{ext}} = -\int_V \rho\vec{g}\cdot\vec{u}\mathrm{d}V. \quad (4)$$

The total potential is now given as the sum of inner potential and external work,

$$\Pi = \Pi^{\mathrm{i}} + \Pi^{\mathrm{ext}}. \quad (5)$$

Applying the principle of the minimum total potential energy in form of the method of calculus of variations,

DOI: 10.1201/9781003348450-47

$$\delta\Pi \stackrel{!}{=} 0, \quad (6)$$

yields the Euler-equations. Those form a system of five differential equations of order two with constant coefficients. Integration by parts yields boundary conditions which are identified as the work conducted by internal forces at the slab and stresses at the weak layer's free edges. In case of a bedded segment bordering an unbedded one, transitional conditions are formulated. Equilibrium of forces is applied for the internal forces of the slab besides continuous displacements. Simultaneously, stress free boundary conditions can be applied to the weak layer's stress-free edge. With the calculus of variations stress free boundary conditions at weak layers can be enforced in integral average. A fully compatible stress solution is obtained with a more complex weak layer kinematic. Vanishing shear stresses at the vertical edges of the weak layer, enable one to predict failure loads more exactly. This results in a crucial advancement compared to previous studies where no stress-free boundary conditions at the weak layer's free end could be applied.

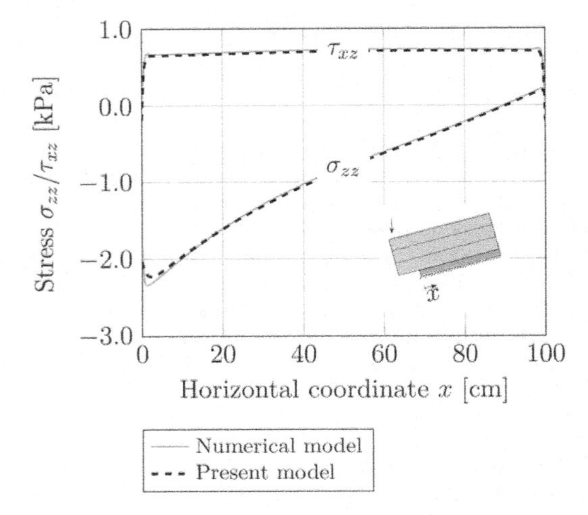

Figure 1. Comparison of the stresses in the weak layers mid-plane between present model and finite element computations for a PST-configuration.

3 RESULTS

The present model and its basic assumptions are validated with finite element computations (FEA) based on the numerical model of Rosendahl and Weißgraeber (2020). Additionally, a layering of the snowpack is considered. In Figure 1 and Figure 2 the normal and shear stresses are evaluated in the mid-plane of the weak layer for two standard configurations.

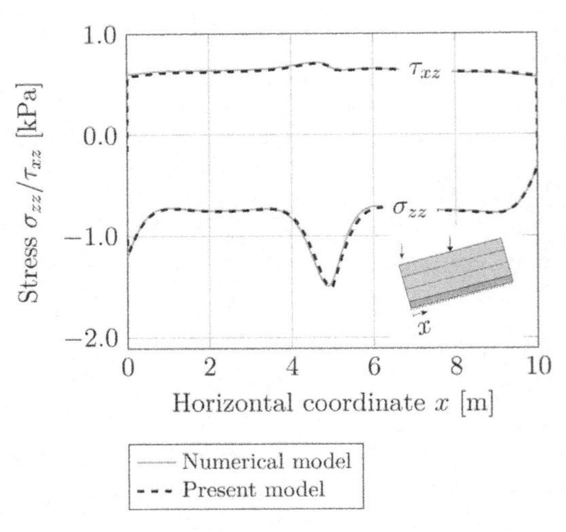

Figure 2. Comparison of the stresses in the weak layers mid-plane between present model and finite element computations for a snow pack loaded with weight and skier.

4 CONCLUSION

The introduced model provides a simple and comprehensible closed form analytical model for the computation of stresses within the weak layer of stratified snow packs. The agreement with numerical computations is exceptional. The absence of shear stresses at the free edge of the weak layer in the PST is a great improvement compared to previous models. Effects like the influence of layering, the thickness of the weak layer or superposition of different load cases can be rendered with the present model. The model shows very good results for thick weak layers due to its improved kinematics. The assumptions on the distribution of displacements over the weak layer's thickness are consistent. The enhanced kinematics of the weak layer in combination with the powerful method of calculus of variations improves the rendering of shear and normal stresses in the case of thick weak layers.

REFERENCES

Rosendahl, P. L. & P. Weißgraeber (2020). Modeling snow slab avalanches caused by weak-layer failure – part 1: Slabs on compliant and collapsible weak layers. The Cryosphere 14(1),115–130.

Current Perspectives and New Directions in Mechanics, Modelling and Design of Structural Systems – Zingoni (ed.)
© 2022 Copyright the Author(s), ISBN 978-1-032-39114-4

An exact stiffness matrix method for nanoscale beams

R.A.R. Wijesinghe
Faculty of Engineering, General Sir John Kotelawala Defence University, Sri Lanka

K.K.V. De Silva
Faculty of Applied Science and Engineering, University of Toronto, Canada

Y. Sapsathiarn
Department of Civil Engineering, Mahidol University, Thailand

N. Rajapakse
Sri Lanka Institute of Information Technology, Malabe, Sri Lanka

ABSTRACT: Conventional continuum theories are inapplicable to nanoscale structures due to their high surface-to-volume ratios and the effects of inter-atomic forces. Although atomistic simulations are more accurate for nanostructures, their use in practical situations is often constrained by the high computational cost. Therefore, simplified continuum methods accounting for the surface energy are considered computationally efficient engineering approximations for nanostructures. The modified continuum theory of Gurtin and Murdoch accounting for the surface energy effects has received considerable attention in the literature. This paper focuses on developing an exact stiffness matrix for nanoscale beams based on the Gurtin-Murdoch theory. The proposed approach is based on the analytical solutions to the governing partial differential equations established in the Gurtin-Murdoch continuum theory for defining nanoscale beams. For the first time, an exact stiffness matrix and a mass matrix were derived based on the general analytical solution of a nanobeam. The study examines the response of thin nanoscale beams under both static and dynamic loading conditions. Normalized displacements are obtained from the finite element analysis for thin nanoscale beams under point and distributed loading for different boundary conditions. Natural frequencies and mode shapes are also computed. The final results agree with the solutions in the literature, demonstrating the accuracy and efficiency of the presented method.

1 INTRODUCTION

Nanoscale beams are a subject of increasing interest with the rise of nanoscale technology. The analysis of nanoscale beams is considerably more complex than classical beams due to the surface effects that become prevalent at this scale because of the very high surface area to volume ratio.

Methods such as atomisation (Shenoy, 2005) and finite element analysis (Liu et al., 2011) may be used to analyse more complicated beams. Atomisation results in a very accurate formulation but is computationally expensive, thus not always ideal for nanoscale beam analysis. There are many different formulations for finite element analysis for nanoscale beams. The shortcomings identified in the literature were the lack of accuracy due to approximations in shape functions and stiffness and mass matrices, leading to the need for many elements to converge towards accurate results.

This paper pr`esents an exact stiffness matrix method for accurate finite element analysis of thin nanoscale beams by using the general analytical solution of a nanobeam.

2 FORMULATION

Consider a nanoscale beam with its cross-section on the z-y plane, its length parallel to the x-direction, as shown in Figure 1.

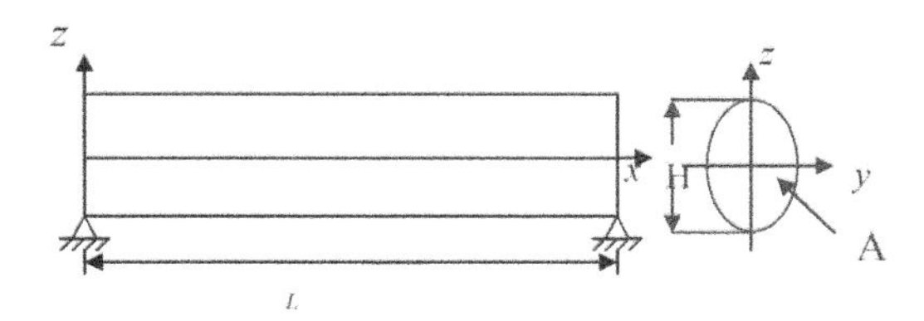

Figure 1. Beam geometry with coordinate system.

Consider the following differential equation that governs the behaviour of thin nanobeams (Liu and Rajapakse, 2009).

$$\left[EI + I^*(2\mu_0 + \lambda_0) - \frac{2\nu I\tau_0}{H}\right]\frac{\partial^4 w}{\partial x^4} - \tau_0 s^* \frac{\partial^2 w}{\partial x^2} + q(x) = -(\rho A + \rho_0 s^*)\frac{\partial^2 w}{\partial t^2} - \frac{2\nu I\rho_0}{H}\frac{\partial^4 w}{\partial x^2 \partial t^2} \quad (1)$$

DOI: 10.1201/9781003348450-48

where I, I^* are the moment of inertia of the cross-section and perimeter of the beam respectively; E is Young's modulus of the beam; μ_0 and λ_0 are the surface Lame constants; v, τ_0, H, A, ρ, ρ_0 are Poisson's ratio, residual surface stress, the height of the beam, area of the cross-section, the density of the material, and the density of the surface layer respectively; $q(x)$ is the transverse distributed load; $w(x)$ is the transverse displacement; and $\theta(x)$ is the rotation of the beam (Liu and Rajapakse, 2009).

The general solution of Equation (1) has been presented by Liu and Rajapakse (2009) in terms of four arbitrary coefficients. We consider a 2-node beam finite element with vertical displacement and rotation at the nodes as the degrees of freedom. Using the general solution of Equation (1), the arbitrary functions in the general solution can be solved in terms of the nodal displacements and rotations. After that, the general solution for beam displacement can be expressed in terms of the nodal degrees of freedom. The generalised forces (bending moments and shear forces) at the beam nodes can be determined in terms of the nodal degrees of freedom by using the general solution. This relationship between the generalised forces at the nodes and the nodal degrees of freedom can be expressed as,

$$\boldsymbol{F} = \boldsymbol{K}\boldsymbol{w} \tag{2}$$

Where $\boldsymbol{F}$ is a column vector containing the shear forces and bending moments at the nodes, $\boldsymbol{K}$ is the exact stiffness matrix, and $\boldsymbol{w}$ is the nodal degrees of freedom vector.

The Equation (2) forms the basis for applying the classical finite element method to analyse nanobeams. A beam can be discretised using several finite elements, and the global stiffness matrix can be assembled based on Equation (2). After that, the relevant boundary conditions can be applied to solve simply supported, cantilever and clamped-clamped beams.

To obtain the natural frequencies and mode shapes under time-harmonic vibrations, we also derive the exact element mass matrix by utilising the exact general solution of Equation (1). An eigenvalue problem is established using traditional methods, and its solution yields the natural frequencies and mode shapes.

3 NUMERICAL RESULTS AND DISCUSSION

In this section, the results obtained by utilising the exact finite element method is presented in Figure 2. The dimensions of the uniform rectangular Si nanobeams considered are as follows, length(L) is 120 nm, width(b) is 3 nm, and height(H) is 6 nm. Twelve identical finite elements were used.

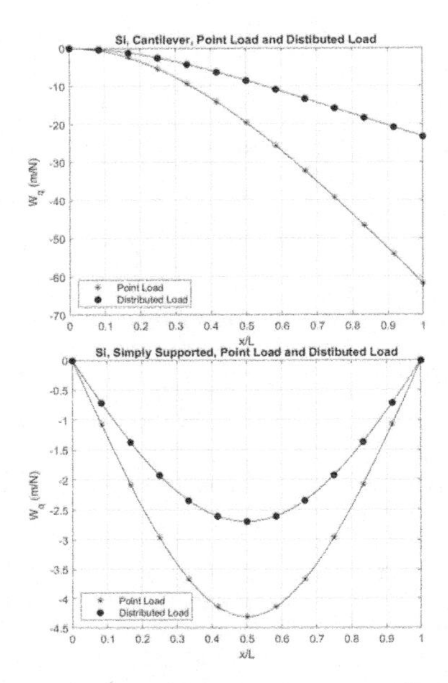

Figure 2. Deflection under static loading conditions (central point load for simply supported, and end point load for cantilever).

Table 1 contains the natural frequencies of Si beams for the different boundary conditions.

Table 1. Natural Frequencies.

	Natural Frequencies				
Beam Type	1st (GHz)	2nd (GHz)	3rd (GHz)	4th (GHz)	5th (GHz)
Si CC	3.3647	9.2530	18.127	29.976	44.839
Si C	0.5645	3.3418	9.2722	18.138	29.984
Si SS	1.5070	5.9384	13.326	23.682	37.031

4 CONCLUSION

An exact finite element method for the analysis of thin nanoscale beams has been derived, which shows good agreement with the analytical results for both static and free vibration problems.

REFERENCES

Liu, C. & Rajapakse, R. K. N. D. 2010. Surface Energy Incorporated Continuum Models for Static and Dynamic Response of Nanoscale Beams. *IEEE Trans. Nanotechnol.* pp. 422–431.

Liu, C., Rajapakse, R. K. N. D. & Phani, A. S. 2011. Finite element modeling of beams with surface energy effects. *Journal of applied mechanics* 78(3).

Shenoy, V. B. 2005. Atomistic calculations of elastic properties of metallic fcc crystal surfaces. *Physical Review* vol. 71.

Current Perspectives and New Directions in Mechanics, Modelling and Design of Structural Systems – Zingoni (ed.)
© 2022 Copyright the Author(s), ISBN 978-1-032-39114-4

On the deformation of porous spherical shells and other spherical bodies

S. De Cicco
Department of Structures for Engineering and Architecture, University of Naples Federico II, Naples, Italy

ABSTRACT: In this paper we consider the linear theory of porous materials introduced by Nunziato and Cowin. In this model the interstitial pores are empty. The deformation is measured by four kinematic variables. We study the elastic deformation of porous spherical bodies under the action of radial pressure. As application, we study the case of a spherical shell and that of a sphere with a rigid nucleus.

1 INTRODUCTION

The deformation and the motion of spherical bodies are classical problems of the theory of elasticity. The attention to this topic is due to its practical applications in engineering, geomechanical science and geophysics.

In this paper we address the equilibrium problem of spherical bodies under the action of given surface tractions in the context of the linear theory of elastic materials with voids. This theory was introduced by Cowin & Nunziato (1983). The theory differs from the well-known porous theory. Whereas in the porous theory the pores are filled with fluids, in the N-C model are empty. Recent applications are given in the works by De Cicco & De Angelis (2019) and De Cicco & Iesan (2013).

The Nunziato-Cowin model is based on the concept of volume fraction We consider an element of volume dv_0 in a point X_0 in the reference configuration. Let dv_0 P_0 be the volume of the skeleton matrix in, we define the volume fraction field the ratio $\sigma_0 = dv_0/dV_0$. If we denote by σ the volume fraction field in a generic deformed configuration, the difference $\psi = \sigma - \sigma_0$is a scalar function measuring the change in volume fraction the kinematic variables are four: the three components of the displacement $u_i(i = 1, 2, 3)$ and the change in volume fraction ψ. When $\psi = 0$ the theory reduces to the classical theory of elasticity. The model is suitable to describe the behaviour of rocks, ceramics, pressed powders, bones as well as concrete.

First, we present the basic equations of the equilibrium theory of elastic materials with voids and then derive the solution in closed form for a class of problems for which the kinematic functions depend only upon the radial variable. As applications we investigate the case of a porous spherical shell loaded with internal and external pressure and the case of a porous sphere with a rigid core. All the results are explicitly determined and generalize the solutions of analogous problems treated in the classical theory of elasticity.

2 BASIC EQUATIONS

The basic equations of the equilibrium theory of isotropic elastic solids whit voids are

$$
\begin{aligned}
& E = \frac{1}{2}\left(\nabla u + \nabla u^T\right) \\
& div\ T + \boldsymbol{f} = \boldsymbol{0}, \quad \text{div}\ \mathbf{h} - p + \text{q} = \mathbf{0} \\
& \mathbf{T} = \mathbf{2}\mu E + \lambda\ \text{tr}EI + \ \beta\psi I \\
& \mathbf{h} = \alpha\nabla\psi \\
& p = \beta\ \text{div}\ \mathbf{u} + \zeta\ \psi
\end{aligned}
$$

where $\mathbf{u}$ is the displacement field, T denotes the stress tensor, $\boldsymbol{f}$ the body force, $\boldsymbol{h}$ the equilibrated stress vector, p the intrinsic equilibrated body force and q the extrinsic equilibrated body force, ψ is the volume fraction function, I is the identity tensor, and $\mu, \lambda, \beta, \alpha$ and ζ are constitutive coefficients.

The boundary conditions are expressed by

$$
\boldsymbol{t} = T\boldsymbol{n}, \quad h = \boldsymbol{h} \cdot \boldsymbol{n}
$$

where $\boldsymbol{t}$ is the surface traction and h the equilibrated surface force.

3 SPHERICAL SHELLS

We consider the equilibrium of a porous body bounded by a spherical surface under the action of given surface tractions. The displacement vector and the volume fraction function will be expressed in

DOI: 10.1201/9781003348450-49

spherical coordinates (r, φ, ϑ). As consequence the basic equations are expressed in spherical coordinates. Then we restrict our attention to the class of problems whose solutions are expressed by functions that depend only upon the radial variable r. Precisely we suppose that

$$u_r = u(r), \quad u_\varphi = u_\vartheta 0, \quad \psi = \psi(r)$$

The problem has solution

$$u = \frac{\zeta}{3\alpha\xi^2} B_1 r - B_2 \frac{1}{r^2} - \nu C_1 i_1(\xi r) + \nu C_2 k_1(\xi r)$$

$$\psi = C_1\ i_0(\xi_r) + C_2\ k_0(\xi_r) - \frac{\beta}{\alpha\xi^2} B_1$$

where C_1,C_2 B_1 and B_2 are arbitrary constants and i_n and k_n are spherical modified Bessel functions of the first and second kind, respectively.

4 APPLICATIONS

1 *A shell bounded by concentric spherical surfaces*

We consider a body bounded by concentric spherical surfaces under the action of internal and external pressure. We denote by r_1 and r_2 the radius of the external and internal boundaries respectively. Let p_1 be the pressure on the external spherical surface and p_2 the pressure on the internal spherical surface. The following expressions are the explicit formulas of the solution

$$u = \frac{1}{r_1^3 - r_2^3}\left[\frac{1}{3K}(p_1 r_1^3 - p_2 r_2^3) r - \frac{1}{4\mu}(p_1 - p_2)\frac{r_1^3 r_2^3}{r^2}\right]$$

$$\psi = \frac{1}{K^*}\frac{p_1 r_1^3 - p_2 r_2^3}{r_1^3 - r_2^3}$$

2 *A sphere with a rigid core*

Now, we consider a spherical porous body B of radius r_1 in which has been inserted a concentric rigid sphere of radius r_2. The body is in equilibrium under the action of uniform pressure $-t$ $(t>0)$.

the solution can be rewritten in explicit form. The kinematic variables are

$$u = -\frac{t}{D}\left\{\frac{\zeta}{3\beta} Q_1(\xi r_2)\left(r - \frac{r_2^3}{r^2}\right)\right.$$
$$\left. +\nu\left[Q_2(\xi r_2)\frac{r_2^2}{r^2} - -Q_2(\xi r)\right]\right\}$$
$$\psi = -\frac{t}{D}[Q_1(\xi r) - Q_1(\xi r_2)]$$

Where

$$Q_1(\xi r) = \frac{i_0(\xi r) k_1(\xi r_1) + k_0(\xi r) i_1(\xi r_1)}{i_1(\xi r_1)}$$

$$Q_2(\xi r) = \frac{i_1(\xi r) k_1(\xi r_1) - i_1(\xi r_1) k_1(\xi r)}{i_1(\xi r_1)}$$

The maximum value of σ_{rr} occurs at $r = r_2$

$$\sigma_{rr}^{max} = -\frac{t}{D} Q_1(\xi r_2)\left(K^* + \frac{4\mu\zeta}{3\beta}\right)$$

REFERENCES

Cowin S.C. & Nunziato J.W., 1983. Linear elastic materials with voids, *J. Elasticity*, 13, Issue 2: 125–147.

De Cicco S. & Iesan D., 2013. Thermal effects in anisotropic porous elastic rods, *Journal of Thermal Stresses*, 36 (4): 364–377.

De Cicco S. & De Angelis F., 2019. A plane strain problem in the theory of elastic materials with voids, *Math. and Mech. of Solids*, 25, n. 1: 46–59.

De Cicco S. & Iesan D., 2021. On the theory of thermoelastic materials with double porosity structure, *Journal of Thermal Stresses*, 44 (12): 364–377.

Goodman M.A. & Cowin S.C., 1972. A continuum theory for granular materials, *Arch. Rational Mech. Anal.*,. 44, 249–266.

Current Perspectives and New Directions in Mechanics, Modelling and Design of Structural Systems – Zingoni (ed.)
© 2022 Copyright the Author(s), *ISBN 978-1-032-39114-4*

Application of quasi-linear viscoelasticity for the characterisation of human buttocks adipose tissue

T. Pandelani
Council for Scientific and Industrial Research, Pretoria, RSA

S.D. Masouros
Imperial College London, London, UK

J.L. Calvo-Gallego
University of Seville, Spain

ABSTRACT: The mechanical properties of the human buttocks adipose tissue were investigated theoretically. The material properties obtained from the stress relaxation test were fitted to with a new algorithm using Fung's quasilinear viscoelastic (QLV) model with Prony series in conjunction with a hyperelastic model. The model was applied to fit the stress-time data during both the ramp and relaxation phases. By using the material properties obtained from the stress relaxation test, a numerical model was developed and the data from the model was fitted to Polynomial, Ogden and Exponential hyperelastic strain energy density (SED) function. It was found that, the QVL with the Polynomial SED function could predict the peak stress better than the other SED functions.

1 INTRODUCTION

The mechanical properties of biological materials such as soft tissue are critical in tissue and biomedical engineering as well as to simulate surgeries. Quasi-static characterisation of various biological tissues has been previously conducted [1]. However, human buttocks adipose tissues are subjected not only to low-rate but also to high-rate loading in impact loading from falls and blasts from under body explosions been sparsely characterised [2].

Due to the scarce information that can be found in the literature about the mechanical properties of the adipose tissue in the buttocks, material properties that can be used to build a FE model of the pelvis for UBB must be developed.

In a more recent study [3], an algorithm was proposed to obtain the material parameters for Quasi linear viscoelastic theory. This algorithm is intended for fitting the Prony series coefficients and the hyperelastic constants of the quasilinear viscoelastic model by considering that the relaxation test is performed with an initial ramp loading at a certain rate.

In the current study, first, the QVL model using the algorithm was applied to fit the stress-time data [4]. Second, in order verify the predictive ability of the model, a numerical model of the specimen was developed to simulate the actual experiment. The predicted stress-time was compared to the experimental data of the relaxation test.

2 METHODS AND MATERIALS

2.1 *Quasi linear viscoelastic model*

Fung (1981) developed the quasi-linear viscoelastic (QLV) theory that combines non-linear elasticity and viscoelasticity [1]. Fung's approach is a creative use of the well-developed theory, linear viscoelasticity, in a new and different way [1].

2.2 *FE of experimental data*

A FE model of the sample representing the physical specimen dimensions under uniaxial compression was analysed to verify the algorithm (See Figure 1).

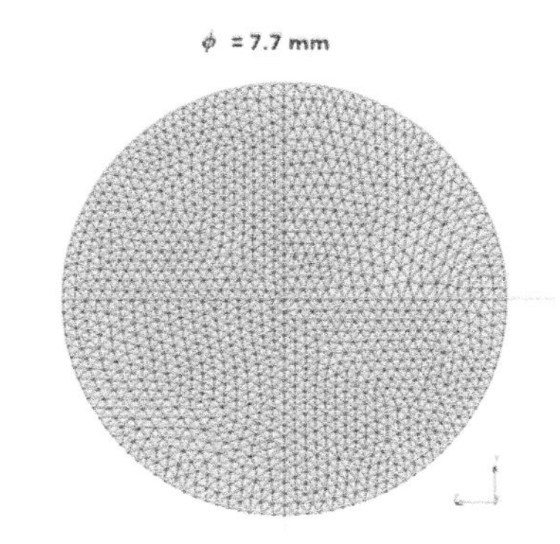

Figure 1. Top view of the specimen used in the finite element model.

DOI: 10.1201/9781003348450-50

Since the solution of the FE model with ideal boundary conditions during uniaxial compression is only affected by the numerical approximation, which is, in particular, the time integration of the viscoelastic effect.

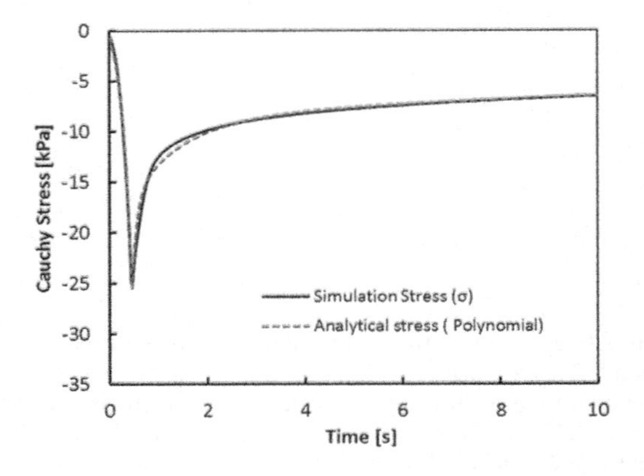

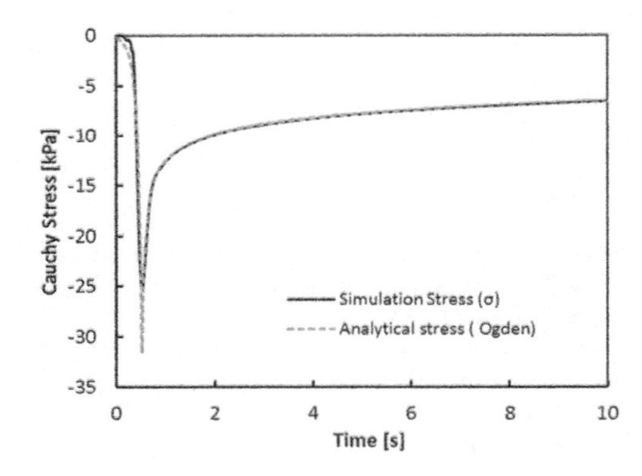

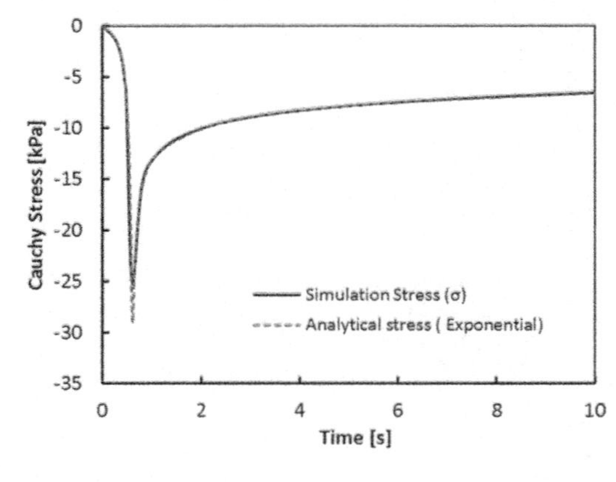

Figure 2. Comparison between the stress-time curves expressed by the derived hyperelastic material formulations (Polynomial, Ogden and Exponential) and data from the FE model relaxation test simulated.

3 RESULTS

The stress record obtained from the FE simulation compared to curves expressed by the derived QVL material formulations from the Polynomial, Ogden and Exponential hyperelastic model are shown in Figure 2. The polynomial material fit predicted better than the other two fitting methods the behaviour of tissues.

4 DISCUSSION

It can be seen in Figure 2, that the fit with the QLV models was quite accurate compared with the simulation response. In fact, it was equally accurate for all the three SED functions. It can be also noticed that the polynomial models fit the simulation curve slightly better than the Ogden and exponential functions, although not too much.

Using a simple model of uniaxial compression, it was possible to verify the developed the algorithm. Because the curves developed from the constants fitted from the uniaxial compression FE test are almost identical to the curve from the algorithm constants, the proposed algorithm appears to be valid.

5 CONCLUSIONS

The QLV theory has been used to analyse a relaxation test in the articular disc of the TMJ of pigs. Using a

simple model of uniaxial compression, it was possible to validate the developed algorithm and, with this algorithm, to analyse the different boundary conditions that simulate the actual experimental setup. The objective was to propose a testing procedure to characterize the QLV behaviour of the articular disc.

In the present work, the non-linear elasticity and the viscoelasticity have been reviewed and the important characteristics of each model have been identified. These models can be adopted with various material formulations. The SED functions formulations offer a robust way to model the non-linearly elastic response of soft tissue such as the adipose tissue.

REFERENCES

[1] Fung, Y. C. (1993). Biomechanics: Mechanical Properties of Living Tissues. Springer.

[2] Then C, Vogl TJ, Silber G. Method for characterizing viscoelasticity of human gluteal tissue. J Biomech. 2012 Apr 30;45(7):1252–8. doi: 10.1016/j.jbiomech.2012.01.037. Epub 2012 Feb 22. PMID: 22360834.

[3] Commisso, M. S. et al. (2016) 'Quasi-Linear Viscoelastic Model of the Articular Disc of the Temporomandibular Joint', Experimental Mechanics, 56(7), pp. 1169–1177. doi: 10.1007/s11340-016-0161-2.

[4] Pandelani, T., Ngwangwa, H.M., Nemavhola F.J, Calvo-Gallego J.L.,(2022). Characterisation of viscoelasticity of human buttocks gluteal muscles, 9th World Congress of Biomechanics, 10–14 July 2022.

Current Perspectives and New Directions in Mechanics, Modelling and Design of Structural Systems – Zingoni (ed.)
© 2022 Copyright the Author(s), ISBN 978-1-032-39114-4

Nonlocal beam analysis based on the stress-driven two-phase theory

F.P. Pinnola, M.S. Vaccaro, R. Barretta & F. Marotti de Sciarra*
Department of Structures for Engineering and Architecture, Naples, Italy

ABSTRACT: The size-dependent behaviour of elastic beams is investigated using Bernoulli-Euler kinematics. The two-phase stress-driven integral elasticity is adopted to model size effects. Biaxial bending is considered and an effective coordinate-free solution procedure is proposed. The corresponding governing equations of nonlocal elasticity are established and discussed. The contributed theoretical results could be useful for the implementation of procedure oriented to design and optimization of modern sensors and actuators.

Keywords: Biaxial bending of nano-beams, Integral elasticity, Nonlocal stress-driven model

1 INTRODUCTION

Analysis of small-scale phenomena in micro- and nano-structures is a subject of large investigation in literature (see e.g. Farajpour et al. 2018, Ghayesh & Farokhi 2020, Basutkar 2019). It is well-known that classical continuum models are not able to capture size-dependent behaviors and thus a key point is to take into account small-scale effects which are technically noteworthy and cannot be overlooked (Rafii-Tabar et al. 2016). Motivation of the present paper is to generalize the stress-driven two-phase formulation (Barretta et al. 2018, Scorza et al. 2022, Behdad & Arefi 2022) to biaxial bending for functionally graded (FG) nano-beams. An effective coordinate-free formulation is proposed and the obtained theoretical outcomes can be effectively exploited to model and design of small-scale continua.

2 BIAXIAL BENDING FOR FUNCTIONALLY GRADED NANO-BEAMS

Let us consider a functionally graded (FG) straight nano-beam subjected to biaxial bending moment undergoing small displacements. The length of the nano-beam is L, the x-coordinate is taken along the length of the nano-beam with the y-coordinate along the thickness (the height) and the z-coordinate along the width of the nano-beam originating at the cross-section geometric centre O. The position vector $\mathbf{p} = [y, z]^T$ collects the cartesian coordinates of a point P of the cross-section. The material is assumed to be elastically isotropic and the effective Euler-Young modulus E is functionally graded along the thickness.

Position vector $\mathbf{p}_G = [y_G, z_G]^T$ about the geometric centre O of the elastic centre G is provided by the formula $\mathbf{p}_G = \frac{\mathbf{S}_O(E)}{A(E)}$, where $A(E)$ is the elastic area and $\mathbf{S}_O(E)$ is the first-order moment vector of elastic area about the z axis. The new elastic reference system has the origin in the elastic centre G and the axes (η, ς) in the cross-section plane are the principal axes of the symmetric and positive definite elastic inertia moment tensor given by $\mathbf{I}_G(E) = \int_\Omega E(\mathbf{r})\mathbf{r} \otimes \mathbf{r} dA$, where vector $\mathbf{r}$ is the position of the point P with respect to the elastic centre, id est $\mathbf{r} = \mathbf{p} - \mathbf{p}_G$. The expression of the effective Euler-Young modulus E in the elastic reference system with the origin in the elastic centre G is provided as

$$E(\eta, \zeta) = E_1 + (E_2 - E_1)\left(\frac{y_G + \eta}{h} + \frac{1}{2}\right)^k, \quad (1)$$

where the elastic moduli of axial fibers at the bottom and top sides of the beam are provided by the parameters E_1 and E_2, respectively. Let us define $\mathbf{R}$ as the linear operator performing the rotation in the cross-section Ω by $\pi/2$ counterclockwise. In the framework of the linearised Bernoulli-Euler theory, the kinematic compatibility condition provides the expression of the axial strain field as function of the transverse displacement vector, i.e. $\varepsilon(\mathbf{r}, x) = -\partial_x{}^2\mathbf{v}(x) \cdot \mathbf{r} = -\chi(x) \cdot \mathbf{r}$ being $\chi(x) = [\chi_\eta(x), \chi_\zeta(x)] = [\partial_x^2 v_\eta(x), \partial_x^2 v_\zeta(x)]^T$ the bending curvature vector. The applied distributed loading is collected in the vector $\mathbf{q}(x) = [q_\eta(x), q_\zeta(x)]^T$ and the bending moment is given by the vector $\mathbf{M}(x) = [M_\eta(x), M_\zeta(x)]^T$. The differential condition of equilibrium in $[0, L]$ prescribes that

*Corresponding author: marotti@unina.it

DOI: 10.1201/9781003348450-51

$$\mathbf{R}\partial_x^2\mathbf{M}(x) = -\mathbf{q}(x)\,, \tag{2}$$

and the shear force vector is defined as $\mathbf{T}(x) := \mathbf{R}\partial_x\mathbf{M}(x)$, with components $[T_\eta(x), T_\zeta(x)]^T$. Boundary conditions at $x_1 = 0$ and $x_2 = L$, having normals $n(x_1) = -1$ and $n(x_2) = 1$ respectively, are given by $\mathbf{M}(x_i)n(x_i) = \mathcal{M}(x_i)$ and $\mathbf{T}(x)n(x_i) = \mathcal{F}(x_i)$, with $i = \{1,2\}$, being $\mathcal{F}(x_i) = [\mathcal{F}_\eta(x_i), \mathcal{F}_\zeta(x_i)]^T$ and $\mathcal{M}(x_i) = [\mathcal{M}_\eta(x_i), \mathcal{M}_\zeta(x_i)]^T$ concentrated shearing force and bending couple vectors.

3 STRESS-DRIVEN NONLOCAL INTEGRAL MODEL FOR FG NANO-BEAMS

The stress-driven (SD) two-phase model of elasticity for nano-beams subject to biaxial bending is here formulated. Accordingly, the constitutive law expressing the bending curvature vector $\boldsymbol{\chi}$ in terms of the bending moment vector $\mathbf{M}$ is given as

$$\boldsymbol{\chi} = -\alpha\mathbf{I}_G^{-1}(E)\mathbf{RM} + (\alpha - 1)\left(\phi * \left(\mathbf{I}_G^{-1}(E)\mathbf{RM}\right)\right) \tag{3}$$

where $0 \le \alpha \le 1$ is the mixture parameter and ϕ is a scalar averaging kernel described by a nonlocal non-dimensional parameter $\lambda := L_c/L$, that is $\phi := \phi_\lambda$. For simplicity, in the sequel we will investigate the purely nonlocal case for biaxial bending, i.e. we set $\alpha = 0$ in Eq. ((3). It can be proven that by adopting the Helmholtz's bi-exponential averaging kernel, the nonlocal SD integral relation Eq. (4) for biaxial bending of nano-beams, defined on the bounded interval $[0, L]$, is equivalent to a differential vector law with constitutive boundary conditions (Romano et al. 2017).

Proposition 1 - Constitutive equivalence property for biaxial bending of nano-beams.

The fully nonlocal vectorial constitutive law Eq. (4)

$$\boldsymbol{\chi}(x) = -\int_0^L \phi(x-y,\lambda)\left(\mathbf{I}_G^{-1}(E)\mathbf{RM}\right)(y)dy \tag{4}$$

equipped with the Helmholtz's bi-exponential kernel

$$\phi(x, L_c) = \frac{1}{2L_c}\exp\left(-\frac{|x|}{L_c}\right), \tag{5}$$

is equivalent to the vectorial differential relation

$$\boldsymbol{\chi}(x) - L_c^2\partial_x^2\boldsymbol{\chi}(x) = -\mathbf{I}_G^{-1}(E)\mathbf{RM}(x)\,, \tag{6}$$

with the constitutive boundary conditions (CBCs)

$$\begin{cases} \partial_x\chi(0) = \chi(0)/L_c\,, \\ \partial_x\chi(L) = -\chi(L)/L_c. \end{cases} \tag{7}$$

The exact solution pertaining to the SD model for biaxial bending of nano-beams can be performed as explained in the sequel. Solve the differential equilibrium condition Eq. (2) to get the expression of the bending moment $\mathbf{RM}(x)$ in terms of the integration constants $(A_{j\eta}, A_{j\varsigma})$ with $j = 1,2$. Solve the second-order differential equation Eq. (6) to get the expression of the bending curvature $\boldsymbol{\chi}$ of the nano-beam in terms of integration constants $(A_{j\eta}, A_{j\varsigma})$ with $j = 1,\ldots,4$ to be determined. Solve the second-order differential equation $\boldsymbol{\chi}(x) = \partial_x^2\mathbf{v}(x)$ involving the curvature vector in terms of the transverse displacement vector $\mathbf{v}$ of the FG nano-beam to get the expression of $\mathbf{v}$ in terms of integration constants $(A_{j\eta}, A_{j\varsigma})$ with $j = 1,\ldots,6$ to be determined. Determine the integration constants by imposing the two CBCs given by Eq. (7) in terms of $\mathbf{v}(\mathbf{x})$, that is

$$\begin{cases} \mathbf{v}^{(3)}(0) = \mathbf{v}^{(2)}(0)/L_c\,, \\ \mathbf{v}^{(3)}(L) = -\mathbf{v}^{(2)}(L)/L_c\,, \end{cases} \tag{8}$$

and standard boundary conditions at the FG nano-beam end points $x \in \{0, L\}$.

REFERENCES

Farajpour A, Ghayesh MH & Farokhi H. 2018. A review on the mechanics of nanostructures. *International Journal of Engineering Science* 133: 231–263.

Ghayesh MH & Farokhi H. 2020. Nonlinear broadband performance of energy harvesters. *International Journal of Engineering Science* 147: 103202.

Basutkar R. 2019. Analytical modelling of a nanoscale series-connected bimorph piezoelectric energy harvester incorporating the flexoelectric effect. *International Journal of Engineering Science* 139: 42–61.

Rafii-Tabar H, Ghavanloo E & Fazelzadeh SA. 2016. Nonlocal continuum-based modeling of mechanical characteristics of nanoscopic structures. *Phys. Rep.* 638: 1–97.

Barretta R, Fabbrocino F, Luciano R & Marotti de Sciarra F. 2018. Closed-form solutions in stress-driven two-phase integral elasticity for bending of functionally graded nano-beams. *Physica E: Low-dimensional Systems and Nanostructures* 97: 13–30.

Scorza D, Luciano R & Vantadori S. 2022 Fracture behaviour of nanobeams through Two-Phase Local/Nonlocal Stress-Driven model, *Composite Structures* 280: 114957.

Behdad S & Arefi M. 2022. A mixed two-phase stress/strain driven elasticity: In applications on static bending, vibration analysis and wave propagation, *European Journal of Mechanics - A/Solids* 94:104558.

Romano G, Barretta R, Diaco M & Marotti de Sciarra F. 2017. Constitutive boundary conditions and paradoxes in nonlocal elastic nano-beams. *International Journal of Mechanical Sciences* 121:151–156.

Current Perspectives and New Directions in Mechanics, Modelling and Design of Structural Systems – Zingoni (ed.)
© 2022 Copyright the Author(s), ISBN 978-1-032-39114-4

Modeling the hysteresis behavior of recycled rubber fiber reinforced bearings

A. Flora & D. Cardone
School of Engineering, University of Basilicata, Italy

A. Calabrese
CSULB CECEM Department, California State University Long Beach, USA

ABSTRACT: Recycled Rubber Fiber-Reinforced Bearings (RR-FRBs) represent a suitable low-cost solution of isolation system especially for structures in developing countries. The numerical modelling of the cyclic behavior of RR-FRBs appears more challenging than for traditional High Damping Rubber Bearings (HDRBs), due to a significant softening response, associated with the axial-shear interaction. In this paper, an advanced hysteresis model, capable of capturing the nonlinear cyclic response of RR-FRBs is adopted and properly calibrated. Results of preliminary numerical analyses performed in OpenSees are shown and compared with the results of experimental tests on low-cost rubber bearings. The findings of this study could provide a new impulse to the application of low-cost rubber bases devices in current practice.

1 INTRODUCTION

In the attempt of extending the range of application of seismic isolation to low-rise residential buildings in developing countries, Kelly (1999) proposed to use Fiber Reinforced rubber Bearings (FRBs) as strip-type base isolation devices directly under the foundation of masonry buildings. To further reduce the construction costs, Calabrese et. al (2015) verified the feasibility of replacing natural rubber with Recycled Rubber-like materials obtained from the reuse of scrap tires and industrial leftovers. The resulting device was named Recycled Rubber-Fiber Reinforced Bearing (RR-FRB).

In RR-FRBs, the bonding between the rubber and the reinforcement layers is achieved by gluing, with a polyurethane binder, the layers of reinforcement and the layers of elastomer, with no need for vulcanization of the entire device. While many hysteresis models have been developed to describe the nonlinear response of conventional rubber-based, the applicability of these models to describe the cyclic response of RR-FRBs has to be verified.

In this paper, advanced hysteresis model, capable of capturing the axial-shear interaction and the coupled (3D) response of RR-FRBs are examined and properly calibrated, using consolidated procedures. In particular, three different "element objects", implemented in the OpenSees framework library (McKenna et al 2000) are considered: (i) the Elastomeric Bearing (Bouc-Wen) Element, (ii) the HDR element (Kumar et al 2014) and, (iii) the Kikuchi Bearing Element (Kikuchi et al 2012).

2 METHODOLOGY AND RESULTS

The experimental results reported in Calabrese et al. (2019) have been used for the parametric identification and calibration of the aforesaid hysteresis models. The bearings were tested under displacement control by applying constant axial load (corresponding to 3.85 MPa vertical pressure) and increasing lateral displacements (from 10% to 60% shear strain) at 0.5 Hz frequency of loading.

Figure 1 compares experimental (black line) and numerical (red line) results, for each cyclic hysteretic model. As can be seen, the HDR element is able to capture the nonlinear cyclic response at low levels of deformation, while it is not able to capture the softening of the bearing at large deformation amplitudes. On the other hand, the Bouc-Wen element shows a satisfactory fitting at large deformations while it completely fails in capturing the experimental stiffness in the low deformation range. Finally, a very good fit, both at low and large deformations, is observed using the optimized Kikuchi Bearing element.

DOI: 10.1201/9781003348450-52

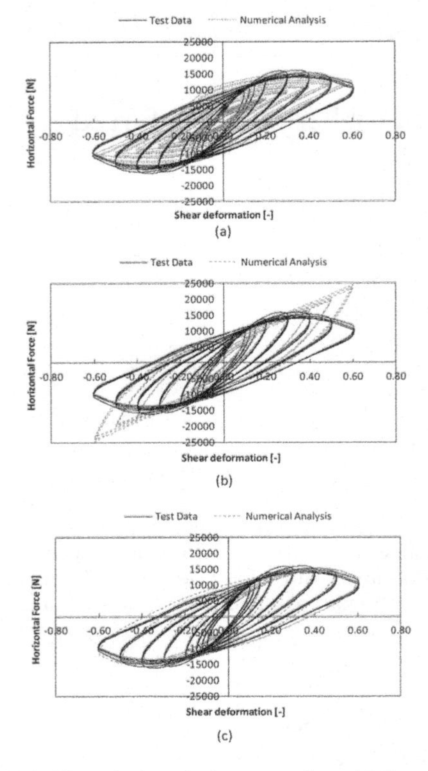

Figure 1. Numerical analysis vs test data, (a) Bouc-Wen element, (b) HDR element and (c) Kikuchi bearing element.

3 SUMMARY AND CONCLUSIONS

Recycled Rubber-Fiber Reinforced Bearings (RR-FRBs) are a suitable base isolation system for ordinary low-rise buildings. They can be adopted to mitigate the consequences of seismic events for both residential and commercial structures. The dynamic behavior of RR-FRBs, however, is significantly influenced by P-delta effects, which cannot be described using simple bilinear or trilinear hysteretic models, as done for conventional elastomeric bearings. This paper discusses the applicability and calibration of advanced 3D hysteresis models, already included in the OpenSees framework library, to capture the cyclic response of RR-FRBs. In particular, the Elastomeric Bearing (Bouc-Wen) Element, the HDR element and the Kikuchi Bearing Element have been examined. An iterative procedure, based on the fitting of experimental data has been followed to calibrate the parameters of each model. A very good fitting has been observed for the Kikuchi Bearing element. In particular (see Table 1), a mean error lower than 20%, with respect to the experimental data, has been computed considering effective lateral stiffness, dissipated energy and equivalent viscous damping ratio. The findings of this study could provide a new impulse to the application of low-cost rubber bases devices in current practice. Additional studies are needed to compare results of response time-history analyses and results of experimental tests for real structures using RR-FRBs.

REFERENCES

Calabrese, A., Spizzuoco, M., Strano, S. and Terzo, M., 2019. Hysteresis models for response history analyses of recycled rubber–fiber reinforced bearings (RR-FRBs) base isolated buildings. Engineering Structures, 178, pp.635–644.

Calabrese, A., Spizzuoco, M., Serino, G., Della Corte, G. and Maddaloni, G., 2015. Shaking table investigation of a novel, low-cost, base isolation technology using recycled rubber. Structural Control and Health Monitoring, 22(1), pp.107–122.

Kelly, J.M., 1999. Analysis of fiber-reinforced elastomeric isolators. Journal of Seismology and Earthquake Engineering, 2(1), pp.19–34.

Kikuchi, M., Aiken, I.D. and Kasalanati, A., 2012. Simulation analysis for the ultimate behavior of full-scale lead-rubber seismic isolation bearings. In 15th world conference on earthquake engineering (No. 1688, p. 24).

Kumar, M., Whittaker, A., and Constantinou, M. (2014). "An advanced numerical model of elastomeric seismic isolation bearings." Earthquake Engineering & Structural Dynamics, Published online, DOI: 10.1002/eqe.2431.

McKenna, F., Fenves. G.L., Scott, M.H. and Jeremic, B. Open system for earthquake engineering simulation (OpenSees). Pacific Earthquake Engineering Research Center, University of California 2000, Berkeley, CA.

Table 1. Comparison of numerical predictions (Kikuchi Bearing element) with experimental data.

Shear strain	0.10	0.13	0.15	0.20	0.30	0.40	0.50	0.60
	Effective lateral stiffness [N/mm]							
Experimental	*476.5*	*421.6*	*381.3*	*332.5*	*223.7*	*181.1*	*121.0*	*84.8*
Kikuchi	339.3	319.7	314.7	273.8	222.3	182.9	144.4	100.3
	Dissipated Energy [KNmm]							
Experimental	*197.7*	*288.4*	*383.0*	*638.6*	*1170.2*	*1872.4*	*2515.3*	*3207.8*
Kikuchi	174.6	279.9	385.1	615.2	1191.5	1906.7	2750.0	3720.9
	Equivalent viscous damping ratio [-]							
Experimental	*0.17*	*0.17*	*0.18*	*0.19*	*0.23*	*0.26*	*0.33*	*0.42*
Kikuchi	0.19	0.21	0.22	0.22	0.24	0.26	0.30	0.41

Current Perspectives and New Directions in Mechanics, Modelling and Design of Structural Systems – Zingoni (ed.)
© 2022 Copyright the Author(s), ISBN 978-1-032-39114-4

Numerical permeability prediction of multiscale textile architectures with varying contact angle and surface tension

J. Dittmann*, P. Seif & P. Middendorf
Institute of Aircraft Design, University of Stuttgart, Stuttgart, Germany

ABSTRACT: Experimental permeability measurements of continuous fiber reinforced plastics are state of the art since many years. Nevertheless, obtained permeability values still scatter up to 8-12 % with no identifiable reason, although strict test conditions are specified [1]. Simulation models are getting more complex and are capable of representing realistic fiber yarn architectures, fiber yarn surfaces and dual scale flow conditions, which have not been investigated in the past. This study will show effects and flow phenomena which are not yet considered in experimental permeability measurements. Here, the effect of varying contact angles and surface tension values will be shown in simulations, as resulting effects on permeability values are more reproducible compared to experimental tests. Therefore, the open source CFD program OpenFOAM will be used. A mesoscopic unit cell model of a braided textile will be modelled to show the already addressed influences. Results will be discussed and compared, deviations in the forecasts will be attributed to their causes.

1 INTRODUCTION

Contact angle measurements and surface tension determination are state of the art processes. It is therefore surprising that in the field of fiber composite materials, permeability determinations with different fluids are carried out without these basic measurements.

In recent years, various international benchmark studies have been carried out to determine the permeability values of endless fiber materials [1, 2, 3, 4].

As a first step towards more fluid comparability, the viscosity of the fluid systems of each charge was tested by the same organization and was included in the comparison [1]. However, even in this study, a scatter in permeability measurement of 8-12 % could not be undercut.

The hypothesis in this study is, that the contact angle and the surface tension of the fluid should also be determined, since these parameters could also have a large influence on the resulting permeability values.

2 MATERIAL & METHODS

2.1 *Measuring the contact angle*

To measure the static contact angle we use the method of fluid drops on a single filament. We measure the contact angle of Glycerol 85 % (Hedinger GmbH) on a carbon fiber filament optically (SGL SIGRATEX KDK 8042). The filament diameter is 7 µm. The dynamic contact angle was not measured in this study, but we can refer to T. D. Blake and Y. D. Shikhmurzaev [5], who tested Water-Glycerol solutions at different fluid concentrations and different velocities.

2.2 *Measuring the surface tension*

To measure the surface tension we use the Du-Noüy method [6]. With the Wilhelmy equation [7], the pull-out force F, the wetting length P and the contact angle Θ ($\Theta = 0°$), we calculate the surface tension σ of the fluid.

2.3 *Use of contact angle and surface tension in simulation models*

OpenFOAM defines the contact angle values as a starting condition and the surface tension values as a constant value. If the velocity of the flow is low, the constant contact angle value Θ_0 becomes dominant. If flow velocities are increasing the dynamic contact angle becomes more important. Θ_A and Θ_R are the advancing angle and the receding angle of the fluid, which are limiting the resulting dynamic contact angle [8].

*Corresponding author: dittmann@ifb.uni-stuttgart.de

DOI: 10.1201/9781003348450-53

For the mesoscopic model we simulate a triaxially braided unit cell (+45°, -45°, 0°) with a fiber volume content of 35% in OpenFOAM.

2.4 *Permeability determination*

To simulate the flow, we use a transient dual-scale, two-phase solver called interFoam. Afterwards, we will calculate the permeability K based on Darcy's equation [9].

3 RESULTS

We analyzed fluid drops on filaments using the method by X. Wu et al. [10]. For the constant contact angle Θ_0 this results in a value of $\Theta_0 = 52.96°$. X. Wu et al. obtain an average value of $\Theta_0 = 50°$.

For the surface tension determination we performed two separate measurements for each fluid concentration, each with 30 repetitions. Results are summarized in Table 1.

Table 1. Surface tension measurements.

	σ [N/m]
Glycerol 85 %	0.0652
Glycerol 90 %	0.0586
Glycerol 95 %	0.0565
Glycerol 99.5 %	0.0542

To test our hypothesis, we simulate four different contact angle settings and three different surface tension values (cf. Table 2).

Table 2. Contact angle and surface tension settings.

Θ_0 *[°]*	45	52.9	60	70
σ [N/m]	0.0652	0.0326 0.0652 0.1304	0.0326 0.0652 0.1304	0.0652

Simulation results show an inhomogeneously propagating flow front. In areas with little distance between the fibers and the model wall, areas with remaining gas phase content are formed.

The calculated permeability values are summarized in Table 3. As one can see the varying contact angles and the varying surface tension values are influencing the resulting permeability values.

Table 3. Simulated permeability values.

	Θ_0*[°]*			
σ [N/m]	45	52.96	60	70
0.0326		3.018*	1.506*	
0.0652	2.530*	2.024*	1.245*	2.531*
0.1304		1.354*	1.172*	

* Permeability values K in 10^{-9}

4 SUMMARY AND CONCLUSION

It was shown that the contact angle and the surface tension of fluids in permeability measurements have a strong influence. In this study, deviating permeability predictions for the simulated textile of up to 55 % were observed.

ACKNOWLEDGEMENT

The present work is funded by the German Research Foundation (DFG) - Project number: 432847151, which is gratefully acknowledged by the authors.

REFERENCES

[1] D. May, A. Aktas, S. G. Advani, et al. In-plane permeability characterization of engineering textiles based on radial flow experiments: A benchmark exercise. Composites: Part A, 21:100–114, 2019.

[2] A. X. H. Yong, A. Aktas, D. May, et al. Out-of-plane permeability measurement for reinforcement textiles: A benchmark exercise. Composites: Part A, 148, 2021.

[3] N. Vernet, E. Ruiz, S. Advani, et al. Experimental determination of the permeability of engineering textiles: Benchmark ii. Composites: Part A, 61:172–184, 2014.

[4] R. Arbter, J. M. Beraud, C. Binetruy, et al. Experimental determination of the permeability of textiles: A benchmark exercise. Composites: Part A, 42:1157–1168, 2011.

[5] T. D. Blake and Y. D. Shikhmurzaev. Dynamic wetting by liquids of different viscosity. Journal of Colloid and Interface Science, 253:196–202, 2002.

[6] P. L. Du Noüy. A new apparatus for measuring surface tension. The Journal of general physiology, 1(5), 521, 1919.

[7] C. D. Volpe and S. Siboni. The Wilhelmy method: a critical and practical review. Surface Innovations, 6(3), 120–132, 2018.

[8] M. Isoz. Dynamics of rivulets and other multiphase flows. Phd thesis, Faculty of Chemical Engineering, Prague, 2018

[9] J. Dittmann, S. Hügle, P. Seif, L. Kauffmann and P. Middendorf. Permeability Prediction using Porous Yarns in a Dual-Scale Simulation with OpenFOAM. ICCM21, Xi'an, China, 2017.

[10] X. Wu and Y. A. Dzenis. Droplet on a fiber: geometrical shape and contact angle. Acta Mechanica, 185 (3):215–225, 2006.

Current Perspectives and New Directions in Mechanics, Modelling and Design of Structural Systems – Zingoni (ed.)
© 2022 Copyright the Author(s), ISBN 978-1-032-39114-4

Data-driven design of high ductile metamaterials under uniaxial tension

A.S. Bhuwal, Y. Pang & T. Liu
Composite Research Group, Faculty of Engineering, University of Nottingham, University Park, Nottingham, UK

I. Ashcroft
Center of Additive Manufacturing, Faculty of Engineering, University of Nottingham, University Park, Nottingham, UK

W. Sun
Gas Turbine and Transmissions Research Centre, Faculty of Engineering, University of Nottingham, University Park, Nottingham, UK

ABSTRACT: A data-driven Face Cubic Centre (FCC) metamaterial design for improved ductility, based on a deep learning framework has been developed. The method relies on Artificial Neural Network (ANN) fundamentals to train a transfer function between input and output variables. A range of finite element (FE) calculations is used to generate macroscopic stress-strain responses of FCC metamaterials under uniaxial tension. It is found that FCC metamaterial lattices can enhance the ductility of the structure compared to FCC periodic lattices. The FCC metamaterial design is achieved through distortion of coordinates and varying strut diameters in an FCC periodic lattice. The disorderliness is taken as input variables to an ANN which is trained to predict the stress responses of metamaterials and to optimize the lattice structure for high ductility. In this work, two functions are used to optimize the structure. The FCC metamaterials after optimization are validated by FE simulation which shows an increase in ductility of the structure of ~30-40%.

1 INTRODUCTION

Modifying the microstructure or microgeometry of a material to change its mechanical characteristics is a well-known approach, including examples in composite materials, sandwich structures, and cellular materials. Laser powder bed fusion (L-PBF) and similar manufacturing processes have enabled the cost-effective manufacture of highly complex geometries (Gibson et al. 2021) and made it possible to design for engineering architected materials or periodic lattices with optimized mechanical properties. With the advancement of the manufacturing processes, we have developed this methodology to approach the optimum design of Face Cubic Centre (FCC) metamaterials using a machine learning neural network.

Machine learning approaches are designed to capture highly nonlinear relations and do not require an auxiliary database for their implementations (Chollet 2018). Artificial Neural Network (ANN) models have been explored for predicting the capabilities of feed-forward and gated recurrent neural networks for predicting stress states based on applied strain (Gorji et al. 2020). An ANN framework is demonstrated in a recent study to predict the evolution of local strain distribution, plastic anisotropy, and failure during tensile deformation of a 3D-printed aluminium alloy (Muhammad et al. 2021). In this, small dataset to map the functional relationship between disorderliness and non-monotonic stress. The mapped equation is used in a non-gradient based optimization algorithm to achieve the desired FCC metamaterials.

2 METHODOLOGY

In this work, to the determination of the optimal FCC metamaterial for the desired structural property, we follow 3-steps in this framework: 1) FE analysis to create a database of outputs corresponding to each input configuration; 2) ANN to find the functional relationship that links inputs and outputs, and 3) non-gradient optimization to determine the optimal FCC metamaterial. This is followed by FE validation and comparing ANN and FEA results.

2.1 *Data generation and collection*

In this work, the input database of spatial coordinate perturbations and strut diameter variations is obtained by perturbating nodes and varying strut diameter by a uniform random distribution function defined as (Bhuwal et al. 2021):

DOI: 10.1201/9781003348450-54

2.2 *Material and finite element model*

The Ramberg-Osgood model is used to represent the true stress-strain relationship of the material. The struts of the FCC lattices are modelled as 2-node Timoshenko-beam elements (B21 in ABAQUS notation) with rigid connections between struts with appropriate constitutive damage model. The relative density $\bar{\rho}$ of the FCC unit cell lattice material is taken as 0.1.

2.3 *Artificial neural network*

The cost function C is calculated as an average of the individual losses over the entire dataset, as follows

$$C = \frac{1}{m}\sum_{i=1}^{m} L_i \tag{1}$$

where L_i represents the loss value for the ith training sample and m is total number of training samples. A mean-squared error function (MSE) is used to compute the losses and evaluate the network's training performance. The error function L is written as:

$$L = \frac{1}{n}\sum_{i=1}^{n} (y_i - f(x_i))^2 \tag{2}$$

where n is a batch size in an output vector.

2.4 *Simulated annealing optimization*

This method randomizes the iterative improvement phase (Monte-Carlo) and allows for occasional uphill moves (moves that do not enhance the solution) to lessen the likelihood of slipping into a local minimum.

3 RESULTS

3.1 *Distortion of coordinates*

We trained an ANN model with a customized loss function for 70% of the database using this dataset. We validated the model efficiency with 15% of the database, and the final 15% database was used to test the ANN model. The objective functions were maximized with the constraint of allowable nodal perturbation $\alpha = 0.2$ and iterated for 10000 solutions.

3.1.1 *Ductility optimization*

The mapped ANN prediction function was used to maximize the ductility of the macroscopic stress. The ANN model showed a 30-40% increase in ductility. The output from the simulated annealing optimization in nodal coordinates was used to create a corresponding FE model for validation.

3.1.2 *Strain energy density optimization*

In this case, the ANN prediction function maximizes the strain energy density. After optimization, the ANN model showed a 26% increase in strain energy density. As before, the output from the simulated annealing optimization was used to create a corresponding FE model for validation.

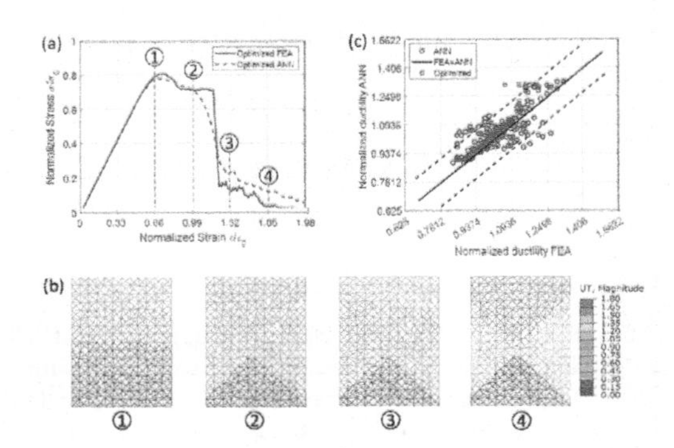

Figure 1. Design of FCC metamaterials based on spatial coordinate perturbations using ductility function for optimization (a) optimized FCC metamaterial's stress versus strain plot comparing the results obtained from ANN and FE validation (b) detailed lattice displacement plot (c) correlation plot between the ANN ductility and FE ductility.

4 CONCLUSION

This study shows that both methods of disorderliness investigated yielded promising results for improved ductile fracture. With this data-driven methodology, we can increase the ductility of lattice structures metamaterials by 30-40%. We also show that determining the metamaterials with optimally improved ductility is possible with unprecedented accuracy and efficiency.

REFERENCES

Bhuwal, A. S. *et al.* (2021) 'Localization and coalescence of imperfect planar FCC truss lattice metamaterials under multiaxial loadings', *Mechanics of Materials*, 160, p. 103996. doi: 10.1016/j.mechmat.2021.103996.

Chollet, F. (2018) *Deep Learning mit Python und Keras: Das Praxis-Handbuch vom Entwickler der Keras-Bibliothek, MITP-Verlags GmbH & Co. KG.*

Gibson, I. *et al.* (2021) *Additive Manufacturing Technologies, Additive Manufacturing Technologies.* doi: 10.1007/978-3-030-56127-7.

Gorji, M. B. *et al.* (2020) 'On the potential of recurrent neural networks for modeling path dependent plasticity', *Journal of the Mechanics and Physics of Solids.* doi: 10.1016/j.jmps.2020.103972.

Muhammad, W. *et al.* (2021) 'A machine learning framework to predict local strain distribution and the evolution of plastic anisotropy & fracture in additively manufactured alloys', *International Journal of Plasticity.* doi: 10.1016/j.ijplas.2020.102867.

Current Perspectives and New Directions in Mechanics, Modelling and Design of Structural Systems – Zingoni (ed.)
© 2022 Copyright the Author(s), ISBN 978-1-032-39114-4

Modelling of biaxial tensile behaviour of the tracheal tissue using three exponential-based hyperelastic constitutive models

H.M. Ngwangwa, L. Semakane & F. Nemavhola
Unisa Biomedical Engineering Research Group, Department of Mechanical Engineering, School of Engineering, College of Science Engineering and Technology, University of South Africa, Science Campus, Private Bag X6, Florida, 1710.

T. Pandelani & D. Modungwa
Defence and security, Council for Scientific and Industrial Research (CSIR), Pretoria, South Africa

ABSTRACT: The trachea experiences some mechanical loading during its physiological functioning when breathing in and out. It is reported that an understanding of its mechanical behaviour is essential in studying the biological function of the whole respiratory system. Therefore, we study the biaxial tensile properties of sheep tracheal tissue by fitting three different constitutive models namely the Choi-Vito, Holzapfel 2000, and Four-fiber family. The results show that the Choi-Vito and Holzapfel models perform very well especially along the longitudinal direction while the Four-fiber family performs better than in the circumferential direction.

1 INTRODUCTION

The trachea is the long tube that connects the larynx to the bronchi and delivers air to the lungs. It is made of cartilaginous rings. In its normal physiological function, the trachea experiences some mechanical loading (Safshekan et al., 2020).

In this paper, we study the mechanical behaviour of the trachea in two directions under biaxial tensile testing: along the circumferential direction and along the longitudinal direction. The stress-strain behaviour of the tracheal muscle is studied by examination of the fitting performances of three different constitutive hyperelastic models: the Choi-Vito, Holzapfel 2000, and the four-fiber family models.

2 MATERIALS AND METHODS

A sheep (Vlein Merino) weighing between 40 and 42 kg was slaughtered and then delivered to the University of South Africa's Biomechanics Laboratory.

The CellScale Biaxial testing system (BioTester 5000 CellScale, Waterloo, ON, Canada®) was used to test the mechanical properties of the sheep trachea tissue.

A preload of 10% at the strain rate of 0.001 per second was applied to remove tissue slack at the beginning of the tensile test.

3 THEORETICAL FORMULATION

The passive response of a biological soft tissue is more complex than the response of elastic solids due to the fact that they undergo finite deformations under mechanical loads.

3.1 *Choi-Vito model*

Choi-Vito model is hyperelastic anisotropic material model developed for canine pericardium whose strain-energy function is expressed as (Choi & Vito, 1990)

$$W = b_0\left[exp\left(b_1 E_{11}^2\right) + exp\left(b_2 E_{22}^2\right) + exp(2b_3 E_{11} E_{22}) - 3\right] \quad (1)$$

where b_i are the material parameters.

3.2 *The Holzapfel (2000) model*

This model is hyperelastic anisotropic material model for stress-strain description of arterial layers with strain energy function given by (Holzapfel et al., 2000).

$$W = \frac{c_1}{2c_2}\left[exp\left(c_2(I_4 - 1)^2\right) - 1\right] \quad (2)$$

DOI: 10.1201/9781003348450-55

where c_i are the material parameters. The model is implemented in an exponential format.

3.3 *The four-fiber family model*

This model is hyperelastic anisotropic material model for stress-strain description of aortas and aneurysms. Its strain energy function is expressed as (Baek et al., 2007; Ferruzzi et al., 2011)

$$W = \frac{c}{2}(I_1 - 3) + \sum_{i=1}^{4} \frac{c_{1i}}{4c_{2i}} \left\{ exp\left[c_{2i}(I_{4i} - 1)^2 \right] - 1 \right\} \quad (3)$$

This model implements a hybrid polynomial and exponential format where c, c_{1i}, c_{2i} are material parameters.

4 RESULTS AND DISCUSSION

Stress-strain curves for the tracheal muscle subjected to equi-biaxial tensile loading show that the longitudinal direction attains 30 % strain below 50 kPa whereas the circumferential direction attains the same level of strain at twice as much as stress (100 kPa). When the three models were fitted to the stress-strain curves all of the models yielded qualitatively similar results (see Figure 1).

However, when the evaluation index was calculated it showed that the Choi-Vito yielded and Holzapfel 2000 models yielded 70 % EI while the Four-fiber family yielded 20 % EI on average. However, the Four-fiber family model performed much better on the circumferential stress than on the longitudinal stress.

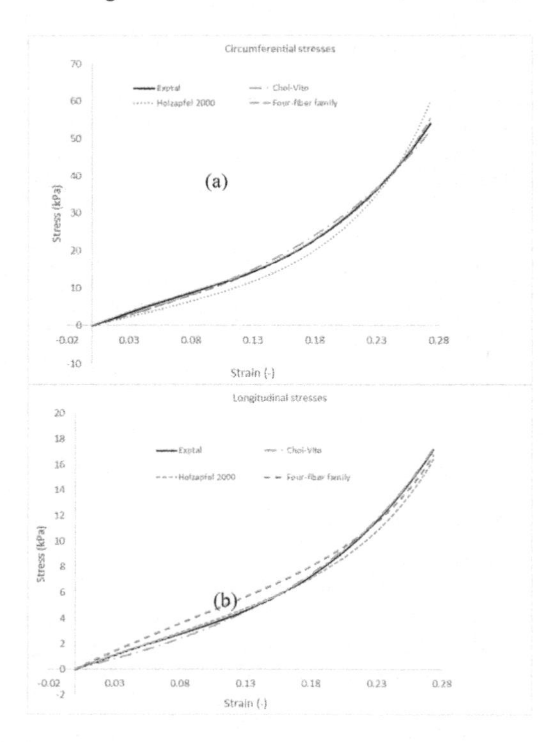

Figure 1. The model calculated results plotted over the measured results for the tracheal tissue with the graphs showing in (a) the circumferential stresses (b) longitudinal stresses. The concerned hyperelastic models are the Choi-Vito, Holzapfel and Four-fiber family as displayed in the figure legends.

5 CONCLUSIONS

The trachea tissue tensile response is studied in the light of three different constitutive models: Choi-Vito, Holzapfel 2000, and Four-fiber family. All these parameters are based on exponential frameworks. However, the EI shows that on average the Choi-Vito and Holzapfel 2000 perform better than the Four-fiber family. This is a very encouraging result since both of these modules have fewer numbers of material parameters than the Four-fiber family. It is observed that the Four-fiber family model performed better in the circumferential than the longitudinal direction. The circumferential direction is the direction in which the cartilaginous rings have the most influence.

REFERENCES

Baek, S., Gleason, R. L., Rajagopal, K. R., & Humphrey, J. D. (2007). Theory of small on large: Potential utility in computations of fluid-solid interactions in arteries. *Computer Methods in Applied Mechanics and Engineering, 196* (31–32),3070–3078. 10.1016/j.cma.2006.06.018

Choi, H. S., & Vito, R. P. (1990). Two-dimensional stress-strain relationship for canine pericardium. *Journal of Biomechanical Engineering, 112*(2), 153–159. https://doi.org/10.1115/1.2891166

Ferruzzi, J., Vorp, D. A., & Humphrey, J. D. (2011). On constitutive descriptors of the biaxial mechanical behaviour of human abdominal aorta and aneurysms. *Journal of the Royal Society Interface, 8*(56), 435–450. https://doi.org/10.1098/rsif.2010.0299

Holzapfel, G. A., Gasser, T. C., & Ogden, R. W. (2000). A new constitutive framework for arterial wall mechanics and a comparative study of material models. *Journal of Elasticity, 61*(1–3), 1–48. https://doi.org/10.1023/A:1010835316564

Safshekan, F., Tafazzoli-Shadpour, M., Abdouss, M., Shadmehr, M. B., & Ghorbani, F. (2020). Finite element simulation of human trachea: Normal vs. surgically treated and scaffold implanted cases. *International Journal of Solids and Structures, 190*, 35–46. 10.1016/j.ijsolstr.2019.10.021

Current Perspectives and New Directions in Mechanics, Modelling and Design of Structural Systems – Zingoni (ed.)
© 2022 Copyright the Author(s), ISBN 978-1-032-39114-4

Evaluating the tensile behaviour of rat myocardium across its three walls from biaxial tensile test data

H.M. Ngwangwa, M. Msibi, I. Mabuda & F. Nemavhola
Unisa Biomedical Engineering Research Group, Department of Mechanical Engineering, School of Engineering, College of Science Engineering and Technology, University of South Africa, Science Campus, Private Bag X6, Florida, 1710.

T. Pandelani
Defence and security, Council for Scientific and Industrial Research (CSIR), Pretoria, South Africa

ABSTRACT: Though studies in mechanical behaviour of myocardial tissues have been dominated by left ventricle research, recent findings on certain pathological conditions caused by COVID-19 have changed that outlook and have led to studying other walls of the heart. This paper reports findings of biaxial tensile tests conducted across the three walls of the Wistar rats' hearts namely left ventricle, septum, and right ventricle. The tensile loads were applied along the directions of the sagittal plane (longitudinal direction), and along the transverse plane (circumferential direction). The atria, pulmonary trunk and all unwanted soft tissues were dissected away. The tests were conducted using a BioTester 5000 CellScalle soft tissue testing machine. Before collecting data, the specimens were preconditioned by applying a 10% strain at a strain rate of 0.001/s. A preload of 5 mN was applied for 0.53 seconds to remove tissue slackness. In order to maintain hydration and mimic the body temperature, saline 0.91% w/v of NaCl was placed in the bath and heated to 37°C and maintained for the duration of testing. The results show that rat myocardium has different tensile properties in the two directions as well as across its three walls.

1 INTRODUCTION

Passive response of soft tissue to mechanical loading is widely used to understand the behavior of the tissue without the consideration of effects of humoral factors (Emig et al., 2021). The passive biaxial responses of can also be used to understand the physiological conditions that are necessary for construction of computational models for growth and remodeling.

The myocardial tissue is heterogenous and anisotropic material. Besides, it possesses a hierarchical and helical structure (Antúnez Montes, 2020) that results in complex three-dimensional (3D) mechanical behaviour which forms a critical component of ventricular function in health and disease. Its helical structure is essential to achieve 60-80 % blood ejection since the myofiber only shortens by approximately 15 % or less. Mechanical test combined with microstructural examinations assist in understanding the components of its structure that are responsible for certain functions. Furthermore, histological evaluations reveal that growth factors are beneficial in treatment of ischemic myocardium.

In this paper, we study the three main walls of the rat heart, namely, left ventricle (LV), the septum (SPT), and right ventricle (RV). Blair et al. (2020) showed that these walls respond different to certain pathological conditions for example it is reported that the right ventricle is more severely affected by heart failure than the left ventricle.

2 MATERIALS AND METHODS

Fifteen (N = 15) Wistar rats (200-250 g) aged between 8 and 10 weeks, were sacrificed in accordance with animal welfare regulations in South Africa. A total of 15 samples were harvested from the LV, SPT and RV and subjected to equi-biaxial tensile testing.

3 RESULTS AND DISCUSSION

In Fig. 1(a) & (b), the results show that the myocardial tissue is stiffer along the circumferential direction up to approximately 30 % strain for the LV wall and up to approximately 15 % strain for the SPT wall. Voigt & Cvijic (2019) show that the circumferential strain is higher than longitudinal strain for the LV.

DOI: 10.1201/9781003348450-56

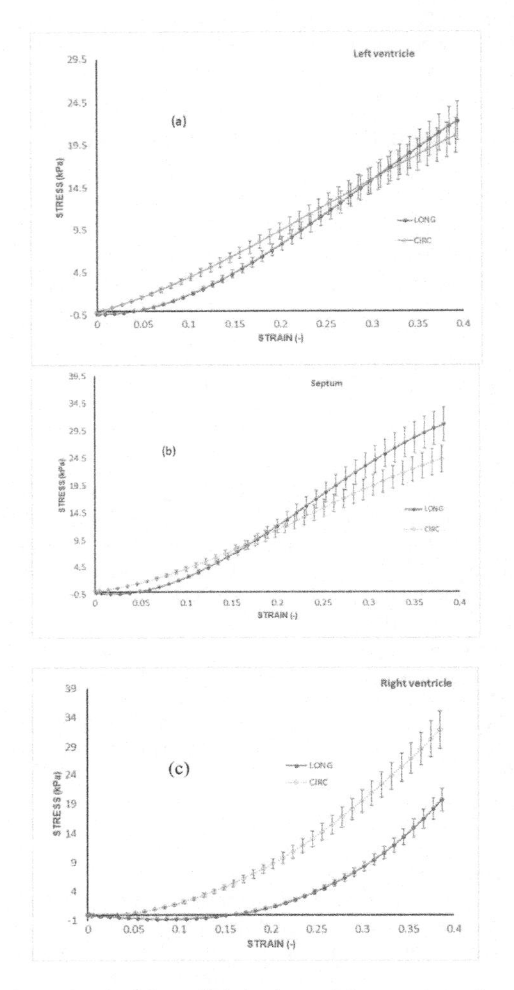

Figure 1. Biaxial tensile behaviour of the rat myocardium showing the stress-strain curves in longitudinal and circumferential direction.

In Fig. 1(c) for RV, the circumferential stress was consistently higher up to 40 % strain than the longitudinal stress. In terms of wall-to-wall variations of the tensile properties for the rat myocardia, the results in Fig. 2 show that the walls are different along the longitudinal directions. In the circumferential direction, the RV is stiffest only beyond 30 % strain but the lowest below 20 % strain.

The study found that the myocardial tissues were stiffer along the circumferential direction for the initial phase of the tensile loading, specifically up to 30 % in the LV and 15 % in the SPT, after which the tissues were stiffer along the longitudinal direction. However, the RV showed that the tissue was consistently stiffer along the circumferential direction in the strain range of interest. For variations of the tissue stiffness across the walls, the difference was only clear along the longitudinal direction where the SPT is the stiffest wall, and the RV is the most compliant wall.

This paper shows that the myocardium tissue has both cross-directional and cross-wall variations. Thus, it is necessary to consider such differences in computational modeling of myocardium tissue.

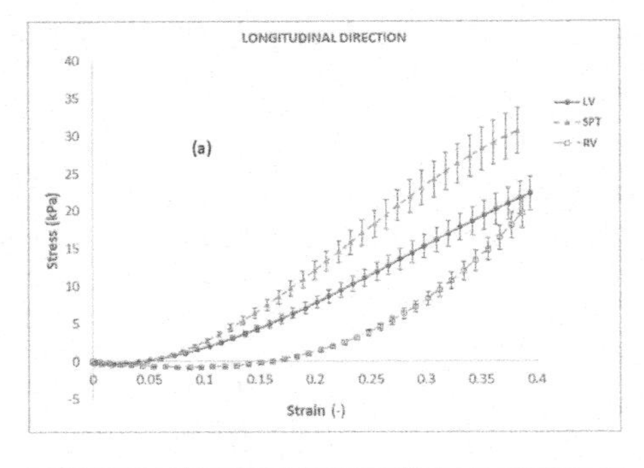

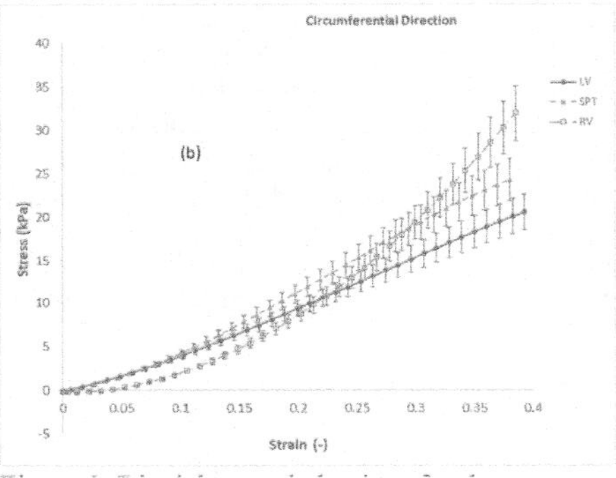

Figure 2. Biaxial stress behaviour for the rat myocardium showing the stress-strain results for the three different walls.

REFERENCES

Antúnez Montes, O. Y. (2020). Anatomical Correlation of the Helical Structure of the Ventricular Myocardium Through Echocardiography. *Revista Española de Cardiología (English Edition)*, *73*(2), 153–160. https://doi.org/10.1016/j.rec.2018.10.016.

Blair, C. A., Brundage, E. A., Thompson, K. L., Stromberg, A., Guglin, M., Biesiadecki, B. J., & Campbell, K. S. (2020). Heart Failure in Humans Reduces Contractile Force in Myocardium From Both Ventricles. *JACC: Basic to Translational Science*, *5*(8), 786–798. https://doi.org/10.1016/j.jacbts.2020.05.014.

Emig, R., Zgierski-Johnston, C. M., Timmermann, V., Taberner, A. J., Nash, M. P., Kohl, P., & Peyronnet, R. (2021). Passive myocardial mechanical properties: meaning, measurement, models. *Biophysical Reviews*, *13*(5), 587–610. https://doi.org/10.1007/s12551-021-00838-1.

Voigt, J. U., & Cvijic, M. (2019). 2- and 3-Dimensional Myocardial Strain in Cardiac Health and Disease. *JACC: Cardiovascular Imaging*, *12*(9), 1849–1863. https://doi.org/10.1016/j.jcmg.2019.01.044.

Current Perspectives and New Directions in Mechanics, Modelling and Design of Structural Systems – Zingoni (ed.)
© 2022 Copyright the Author(s), *ISBN 978-1-032-39114-4*

Investigation of the performance of different constitutive models on omasum uniaxial tensile behaviour

H.M. Ngwangwa, L. Lebea & F. Nemavhola
Unisa Biomedical Engineering Research Group, Department of Mechanical Engineering, School of Engineering, College of Science Engineering and Technology, University of South Africa, Science Campus, Florida 1710, South Africa

T. Pandelani & D. Modungwa
Defence and security, Council for Scientific and Industrial Research (CSIR), Pretoria, South Africa

ABSTRACT: There remains a challenge with continued research on how to achieve functionality to mimic the biomechanical environment, induce bioactivity, and support critical size tissue reintegration. Furthermore, the persistent use of animals for regenerative medicine or drug discovery is ethically challenging. This dictates/highlights the need for further development and exploration of mathematical models that can be used to study the properties of biological tissues. As such this research study investigates the ability of three constitutive material models in representing the tensile behaviour of the omasum tissue for its potential attractive features/advantages towards biomimetics and material development. The results from the study indicate a tissue with high levels of compliance and that it is very hard to reproduce the test results for the omasum, which will present a challenge in verification/validation towards high fidelity models.

1 INTRODUCTION

In this paper the uniaxial tensile behaviour of a sheep's omasum is studied. One of the most attractive features of the omasal tissue is that its mechanical properties would offer opportunities for biomimetics and material development. Lun et al. (2010) identified ovine forestomach matrix as a potential biomaterial for wound healing and tissue regeneration.

Therefore, the current research investigates the ability of five constitutive material models in representing the tensile behaviour of the omasum namely, the Fung, Holzapfel 2000, Holzapfel 2005, Polynomial (anisotropic), and Four-fiber family model.

1.1 *The Fung model*

The Fung model is a hyperelastic anisotropic material model for stress-strain description of arterial wall whos strain energy density function is given by (Chuong & Fung, 1983)

$$W = \frac{c}{2}\left(e^{Q} - 1\right) \tag{1}$$

Where $Q = b_1 E_{\theta\theta}^2 + b_2 E_{ZZ}^2 + b_3 E_{RR}^2 + 2b_4 E_{\theta\theta} E_{ZZ} + 2b_5 E_{ZZ} E_{RR} + 2b_6 E_{RR} E_{\theta\theta}$; and b_i are the material parameters.

1.2 *The Holzapfel (2000) model*

This model is constituted by forms of the strain energy function that represent isotropy and anisotropy given by (Holzapfel & Ogden, 2009)

$$W = \frac{c_1}{2c_2}\left[exp\left(c_2(I_4 - 1)^2\right) - 1\right] \tag{2}$$

Where c_i are the material parameters. The model is implemented in an exponential format.

1.3 *The Holzapfel (2005) model*

This model has a strain energy function which is expressed as (Holzapfel et al., 2005)

$$W = \frac{c_1}{2c_2}\left\{exp\left[c_2\left((1-\kappa)(I_1 - 3)^2 + \kappa(I_4 - 1)^2\right) - 1\right]\right\} \tag{3}$$

Where c_i are the material parameters and κ is a parameter that modulates the convergence rate.

1.4 *The four-fiber family model*

This model is hyperelastic anisotropic material model for stress-strain description of aortas and aneurysms with strain energy function (Ferruzzi et al., 2011)

DOI: 10.1201/9781003348450-57

$$W = \frac{c}{2}(I_1 - 3) + \sum_{i=1}^{4} \frac{c_{1i}}{4c_{2i}} \left\{ exp\left[c_{2i}(I_{4i} - 1)^2 \right] - 1 \right\} \tag{4}$$

This model implements a hybrid polynomial and exponential format where c, c_{1i}, c_{2i} are material parameters.

1.5 *The polynomial (anisotropic) model*

This model is hyperelastic anisotropic material model whose strain energy function is expressed by (Bursa et al., 2008)

$$W = \sum_{i=1}^{3} a_i(I_1 - 3)^i + \sum_{j=1}^{3} b_j(I_2 - 3)^j + \sum_{k=2}^{6} c_k(I_4 - 1)^k + \sum_{m=2}^{6} e_m(I_6 - 1)^m \tag{5}$$

Where $a_i, b_j, c_k,$ and e_m are material parameters.

2 RESULTS AND DISCUSSION

The uni-axial stress-strain results for the omasum show a tissue with high levels of compliance. For the initial 20 % strain most of the tissues had stresses below 10 kPa. At 50 % strain, over half of the tissues have stresses of less than 30 kPa.

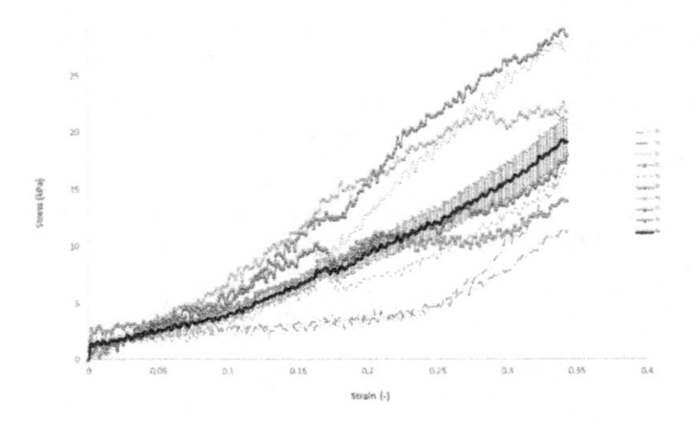

Figure 1. Measured stress curves with an average curve plotted over them. The mean curve shows a 10 % error bound on both sides.

The results in Figure 1 also show that it is very hard to reproduce the test results for the omasum. There are wide deviations in the measured results.

Figure 2 shows that the polynomial (anisotropic) model and four-fiber family model are the best performing models. The performance of the models on the omasum's tensile behaviour shows that its biomechanics is influenced by fiber orientation as all the best performing material models incorporate fiber orientation in their architecture. The Fung model performed very poorly on this tissue. The versatility of the polynomial (anisotropic) model enabled it to perform very well on Omasum in terms of mean stress.

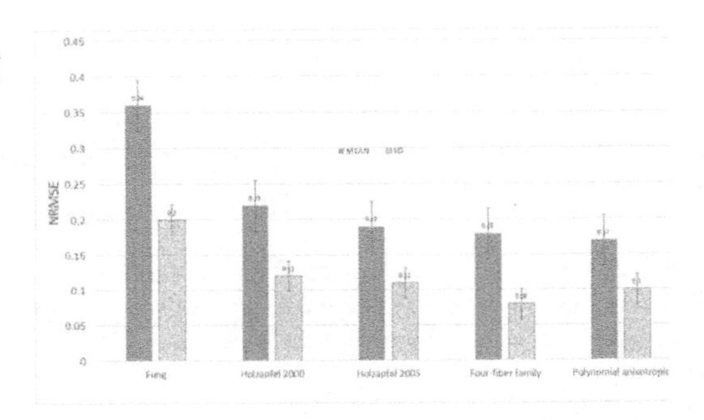

Figure 2. The mean normalised RMS error plotted with their corresponding standard deviation in the result. All the five material models are shown: Fung, Holzapfel 2000, Holzapfel 2005, Four-fiber family and Polynomial anisotropic models.

3 CONCLUSION

In this paper, the mechanical behavior of Omasum was evaluated, and it was found that four-fiber family and the Holzapfel 2005 may be a better choice for computational efficiency. However, the Fung model performed poorly.

REFERENCES

Bursa, J., Skacel, P., Zemanek, M., & Kreuter, D. (2008). *Implementation of hyperelastic models for soft tissues in FE program and identification of their parameters*. https://doi.org/10.2316/Journal.216.2008.6.601-143

Chuong, C. J., & Fung, Y. C. (1983). Three–dimensional stress distribution in arteries. *Journal of Biomechanical Engineering, 105*(3), 268–274. https://doi.org/10.1115/1.3138417

Ferruzzi, J., Vorp, D. A., & Humphrey, J. D. (2011). On constitutive descriptors of the biaxial mechanical behaviour of human abdominal aorta and aneurysms. *Journal of the Royal Society Interface, 8*(56), 435–450. https://doi.org/10.1098/rsif.2010.0299

Holzapfel, G. A., & Ogden, R. W. (2009). Constitutive modelling of passive myocardium: A structurally based framework for material characterization. *Philosophical Transactions of the Royal Society A: Mathematical, Physical and Engineering Sciences, 367*(1902), 3445–3475. https://doi.org/10.1098/rsta.2009.0091

Holzapfel, G. A., Sommer, G., Gasser, C. T., & Regitnig, P. (2005). Determination of layer-specific mechanical properties of human coronary arteries with nonatherosclerotic intimal thickening and related constitutive modeling. *American Journal of Physiology - Heart and Circulatory Physiology, 289* (5 58–5), 2048–2058. https://doi.org/10.1152/ajpheart.00934.2004

Lun, S., Irvine, S. M., Johnson, K. D., Fisher, N. J., Floden, E. W., Negron, L., Dempsey, S. G., McLaughlin, R. J., Vasudevamurthy, M., Ward, B. R., & May, B. C. H. (2010). A functional extracellular matrix biomaterial derived from ovine forestomach. *Biomaterials, 31*(16), 4517–4529. https://doi.org/10.1016/j.biomaterials.2010.02.0

Current Perspectives and New Directions in Mechanics, Modelling and Design of Structural Systems – Zingoni (ed.)
© 2022 Copyright the Author(s), ISBN 978-1-032-39114-4

Evaluating computational performances of Yeoh, Veronda-Westmann and Humphrey models on supraspinatus tendon uniaxial stress-strain behaviour

H.M. Ngwangwa, M. Msibi, I. Mabuda & F. Nemavhola
Unisa Biomedical Engineering Research Group, Department of Mechanical Engineering, School of Engineering, College of Science Engineering and Technology, University of South Africa, Science Campus, Private Bag X6, Florida, 1710.

ABSTRACT: It is understood that accurate modelling of the mechanical behaviour of tendon tissues is vital due to the tendon's essential role in the facilitation of joint mobility in biological mechanisms. The supraspinatus tendon helps to maintain dynamic stability at the glenohumeral joint in conjunction with other rotator-cuff tendons, namely infraspinatus, subscapularis and teres minor. These tendons are thought to support the action of the more powerful forces that are innovated at that joint by the deltoid, pectoralis major, biceps and triceps muscles. The supraspinatus tendon is often prone to injury especially in careers or sporting activities that involve frequent arm abduction. It is therefore important to understand the mechanical behaviour of this tendon in order to devise better rehabilitation therapies in case of injury. This paper evaluates the relative modelling capabilities of three hyperelastic models, namely the Yeoh, Humphrey and Veronda-Westmann material models on the tensile behaviour of three tendon specimens. We compare their fitting accuracies, convergence rates during optimisation, and the different forms of sensitivities to data-related features and initial parameter estimates. The results show that the Yeoh model outperforms the other two hyperelastic models and therefore makes it relatively better choice for further investigation of the mechanical behaviour of tendons.

1 INTRODUCTION

Tendons are connective tissues that connect muscles to bones and in so doing they transmit forces from the muscles to the bones across joints.

In this study, we examine the performance of the following three hyperelastic models in modelling the tendon:

Yeoh model with Cauchy stress given by (Yeoh, 1993)

$$\sigma = 2\left(\lambda^2 - \frac{1}{\lambda}\right)\left(c_{10} + 2c_{20}(I_1 - 3) + 3c_{30}(I_1 - 3)^2\right) \quad (1)$$

Humphrey model with Cauchy stress given by (Humphrey & Yin, 1987)

$$\sigma = 2\left(\lambda^2 - \frac{1}{\lambda}\right)c_1 c_2 e^{c_2(I_1-3)} \quad (2)$$

and Veronda-Westmann model whose Cauchy stress is given by (Veronda & Westmann, 1970)

$$\sigma = 2\left(\lambda^2 - \frac{1}{\lambda}\right)c_1 c_2\left(e^{c_2(I_1-3)} - \frac{1}{2\lambda}\right) \quad (3)$$

Where the c_i or c_{i0} $(i = 1, 2, 3)$ are the material parameters to be determined through a nonlinear least squares routine. Although these are well-known hyperelastic models and widely used for the modelling of rubberlike soft materials, their accuracies and applicability are dependent on a lot of factors which remain a subject of research. In this study, we show that the Yeoh model outperforms the other two models.

2 MATERIALS AND METHODS

The details of the experimental procedure can be read from (Ngwangwa & Nemavhola, 2021). The MTS EM Tensile tester was operated in displacement control at an average strain rate of 1.72 % s^{-1}.

3 RESULTS AND DISCUSSIONS

The results in Figures 1 & 2 show that the Yeoh material model yields the best performance at correlation percentages of greater than 99 % over all the tests in both scenarios. However, when the test range was taken slightly beyond the linear region, the correlation percentages dropped. The Humphrey and Veronda-Westmann were more affected by this change as shown in Figure 2.

DOI: 10.1201/9781003348450-58

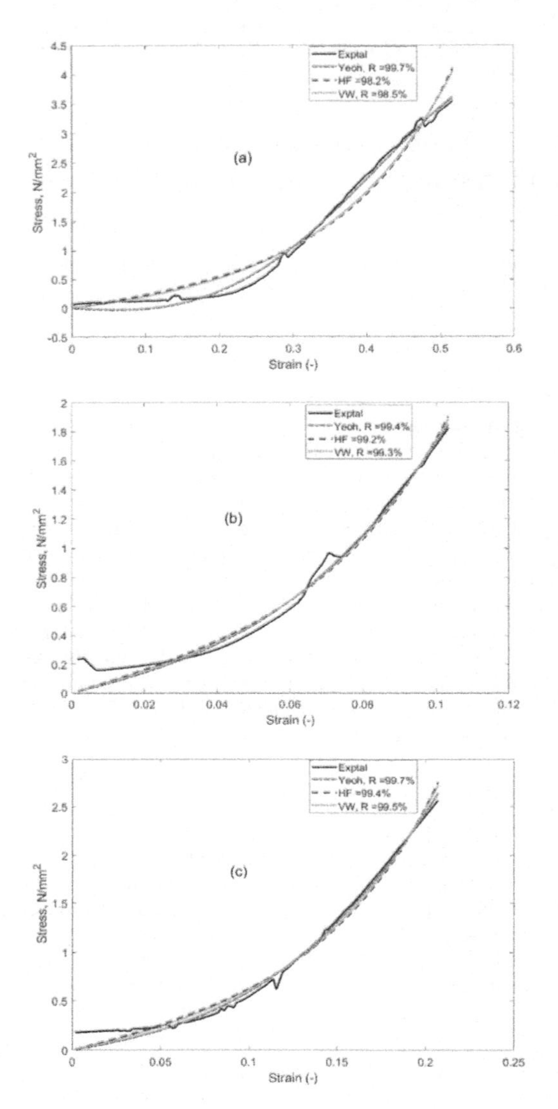

Figure 1. The hyperelastic model results correlated with (a) Test 1, (b) Test 2, and (c) Test 3 data with correlation percentages.

The Yeoh model also yielded much faster convergence rates and better sensitivity in the stress results due to changes in its material parameters than the other two materials, which qualified it to be the best choice for modeling of tendons tensile behaviour.

REFERENCES

Humphrey, J. D., & Yin, F. C. P. (1987). On constitutive relations and finite deformations of passive cardiac tissue: I. A pseudostrain-energy function. *Journal of Biomechanical Engineering*, *109*(4), 298–304. https://doi.org/10.1115/1.3138684

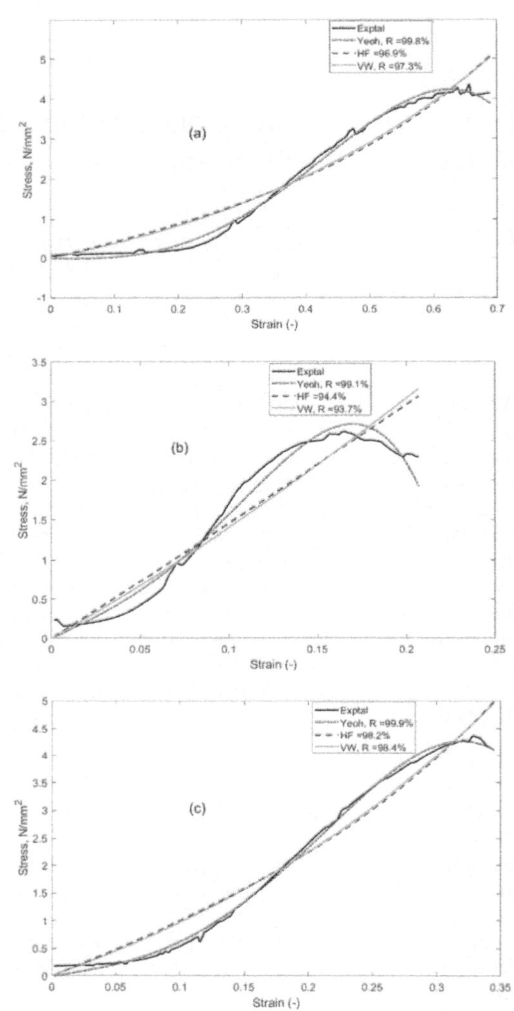

Figure 2. The hyperelastic model results correlated with (a) Test 1, (b) Test 2, and (c) Test 3 data showing the correlation percentages.

Ngwangwa, H. M., & Nemavhola, F. (2021). Evaluating computational performances of hyperelastic models on supraspinatus tendon uniaxial tensile test data. *Journal of Computational Applied Mechanics*, *2021*(1), 27–43. https://doi.org/10.22059/jcamech.2020.310491.559

Veronda, D. R., & Westmann, R. A. (1970). Mechanical characterization of skin-Finite deformations. *Journal of Biomechanics*, *3*(1). https://doi.org/10.1016/0021-9290(70)90055-2

Yeoh, O. H. (1993). Some forms of the strain energy function for rubber. *Rubber Chemistry and Technology*, *66* (5), 754–771. https://doi.org/10.5254/1.3538343

5. Additive manufacturing, Manufacturing processes, Assembly processes, Kinematics

Current Perspectives and New Directions in Mechanics, Modelling and Design of Structural Systems – Zingoni (ed.)
© 2022 Copyright the Author(s), ISBN 978-1-032-39114-4

Robot supported wire arc additive manufacturing and milling of steel columns

B. Waldschmitt & J. Lange
Institute for Steel Construction and Materials Mechanics, Technical University of Darmstadt, Darmstadt, Germany

C.Borg Costanzi & U. Knaack
Institute for Structural Mechanics and Design, Technical University of Darmstadt, Darmstadt, Germany

T. Engel & J. Müller
spannverbund GmbH, Waldems, Germany

ABSTRACT: Wire Arc Additive Manufacturing (WAAM) is a robot-controlled welding process used to build up three-dimensional structures in steel, which may be unfeasible to manufacture using conventional methods. This article presents the design-to-manufacturing process of a steel column node of a varying cross section and high complexity using partial parametric robot programming and multiple process-control checks. To achieve the desired geometries and high level of dimensional accuracies, the printed structure will undergo post processing by means of machining. Therefore the milling of printed structure's areas which are over-printed as planned is presented.

1 INTRODUCTION

Within the steel construction industry the research in Wire Arc Additive Manufacturing (WAAM) has raised attention for manufacturing non-standardized steel members in addition to standard elements. WAAM is an additive manufacturing process based on gas shielded metal arc welding (GMAW), in which a welding wire serves as the printing material to achieve desired three dimensional target geometries and material properties. Compared to other metallic additive manufacturing processes, WAAM is characterized by a 30 to 40-fold higher deposition rate (Bergmann et al. 2018).

Figure 1. Additively manufactured steel node.

2 STEEL MEMBERS

The article focuses on presenting a steel node to join different standard profiles of closed cross-section (Figure 1) or additively manufactured elements. Thereby cross-sectional profiles of different dimensions or geometric shapes (here: square and circle) can be joined together. In addition, the combined process of additively manufactured components and post-processing by removing material to a planned wall thickness using various milling and grinding processes is explained.

3 DESIGN-TO-MANUFACTURING WORKFLOW

The design-to-manufacturing workflow was applied to all components. A digital model contains information about process and input parameters, weld-seam geometries and material properties. The weld paths are generated using partial parametric robot programming (PRP), a coordinate determination method based on mathematical functions (Feucht et al. 2020). By regularly checking the current height among other things and comparing it with the digital model by the

DOI: 10.1201/9781003348450-59

robot controller, the coordinates of the weld path are adaptively updated. The final evaluation of the recorded data, supplemented by 3D-scanning for a target-actual-comparison, is used to generate milling trajectories for a surface finish and to determine, for example, optimized motion sequences or cooling and measuring processes. All stored information results in a rudimentary digital twin.

4 EVALUATION

The node was fabricated by 379 layers (297 vertical; 82 cantilevered). The total manufacturing time was 28.51 h. Of this, only 43.5% were spent on welding directly, 31.9% on tactile measurement, and 20.3% on cooling the structure. The total weight of the node amounts to 20.12 kg. The deposition rate related to the welding time results in 1.65 kg/h.

The recorded data are supplemented by a final 3D-scan model. A point cloud was generated by using a laser scanner and converted into a mesh geometry, which is necessary for performing a target-actual-comparison between the planned and manufactured object as well as generating milling trajectories This was achieved by slicing the mesh model into layers (Figure 2 right) with spacing equal to the milling trajectories.

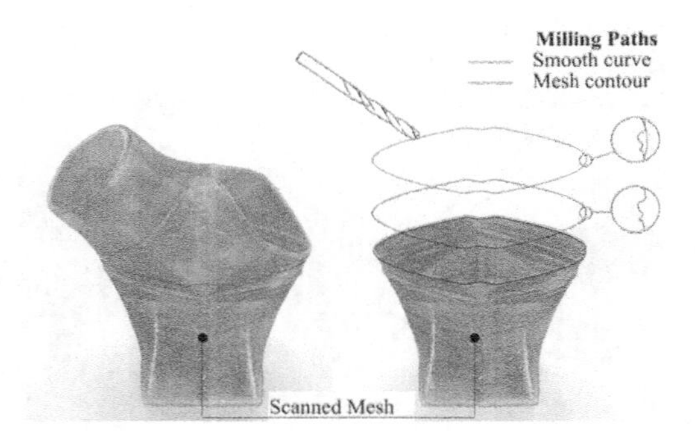

Figure 2. Generating milling trajectories.

5 POST PROCESSING - MILLING

For the machining, a rotational speed of approx. 5000 rpm at a feed rate of approx. 3-4 mm/s and an ablation rate of 2-3 mm turned out to be reasonable parameters. A test specimen was created that included different surface shapes, in order to test the limits of the machining process. The six axes of the robot offer advantages in the machinability of complex shaped workpieces compared to conventional CNC machines.

The milling of the specified paths is executed with the cutter aligned orthogonally to the workpiece's surface (Figure 3 right) and an overlap of the paths of 2 mm to prevent the formation of burrs as much as possible. Nevertheless, small burrs appear between the milling paths, which could be traced back to the non-perfect rigidity, the repetition accuracy of the robot and the setup's possible tilting of a few tenths of a degree. To achieve the desired, aesthetically pleasing surface finish, the milled surfaces were finished with flap sanders (Figure 3 left). A final finish was applied using sisal polishing wheels (d=150 mm, w=20 mm) and appropriate polishing paste.

Figure 3. Milling and grinding of a test specimen.

6 CONCLUSION

Robot supported additive manufacturing (WAAM) and post processing of a steel node was presented. It is another implementation option for the steel industry. The design-to-manufacturing process followed illustrates a reasonable manufacturing flow. The determination of weld path coordinates using partial PRP needs further improvement to reduce manual adjustment. The recorded data also indicate the difficulties in manufacturing very complex free forms with overhang. The presented post-processing methods will be applied in a next step to adjust the outer part surfaces.

ACKNOWLEDGMENTS

We would like to thank the GMSS of TU Darmstadt for their kind support. This research was funded by the Bundesministerium für Wirtschaft und Energie (Grant Number: 16KN076133).

REFERENCES

Bergmann, Jean Pierre/Henckell, Philipp/Reimann, Jan/Ali, Yarop/Hildebrand, Jörg (2018). Grundlegende wissenschaftliche Konzepterstellung zu bestehenden Herausforderungen und Perspektiven für die Additive Fertigung mit Lichtbogen. Studie im Auftrag der Forschungsvereinigung Schweißen und verwandte Verfahren e.V. des DVS. Düsseldorf, DVS Media GmbH.

Feucht, Thilo/Lange, Jörg/Erven, Maren/Costanzi, Christopher Borg/Knaack, Ulrich/Waldschmitt, Benedikt (2020). Additive manufacturing by means of parametric robot programming. Construction Robotics 4 (1-2), 31–48. https://doi.org/10.1007/s41693-020-00033-w.

Current Perspectives and New Directions in Mechanics, Modelling and Design of Structural Systems – Zingoni (ed.)
© 2022 Copyright the Author(s), ISBN 978-1-032-39114-4

Global sensitivity analysis of sand-based binder jet 3D printed material

L. Del Giudice, M.F. Vassiliou & S. Marelli
Institute of Structural Engineering (IBK), Swiss Federal Institute of Technology (ETH), Zürich, Switzerland

ABSTRACT: Sand based binder jet 3D printing (BJ3DP) is one of several additive manufacturing processes that have been growing in the last years. This particular technology was developed for the foundry industry, to create complicated sand molds in a consistent and rapid way. However, in recent years this technology has been used in different research fields to manufacture small-scale physical models of brittle materials, e.g. sedimentary rocks. The mechanical properties of the printed material depend heavily on the printing parameters. In this work, the influence of the printing parameters is investigated and a correlation-based global sensitivity analysis is performed. Uniaxial compression tests are performed to measure compressive strength, and Young's modulus of samples printed with varying printing parameters. Four parameters are identified as the main contributors of the variation of mechanical properties, namely printing speed, droplet mass, printing resolution, and activator percentage. An experimental design with 18 combinations of the mentioned process parameters is devised and the influence of each parameter is evaluated. It is concluded that only the printing resolution is important.

1 INTRODUCTION

In the last decade, the use of Additive Manufacturing (AM) technologies, also referred to as three-dimensional printing (3DP), has been steadily growing in both research and industry. Since its first application in the 1980s, AM technologies have evolved from a tool for rapid prototyping to a technology used for producing parts made out of a wide variety of materials. Besides the typical use cases, i.e. rapid prototyping and manufacturing of plastic and steel parts, AM has been used in the field of Civil Engineering using concrete 3D printers. It has also been used in in Civil and Seismic engineering as a tool to create small-scale physical models of reinforced concrete (Del Giudice et al. 2021).

Regardless of the technologies employed (fused-deposited-material, binder jetting, selective laser sintering, etc.) and of the specific application for the 3D printer, all 3D printers have a plethora of parameters that can be tuned, which affect the characteristics of the printed material. It is therefore important to quantify the effect of each input parameter to the desired output of interest (a specific property of characteristic of the printed material/part). This can be achieved through a sensitivity analysis on the input/output of the process.

In this paper, a sensitivity analysis on the input/output relationship of a Voxeljet VX500 sand based binder jetting 3D printer (BJ3DP) is carried out. "Input" represents the printing parameters, while "output" relates to the physical properties of the resulting specimens. Four input parameters are identified as relevant for this study, namely the activator percentage, the resolution, the print speed and the droplet mass (see Del Giudice et al. 2020 for an in-depth description of the printing process and parameters). The output of interest is the compressive strength and Young's modulus of the printed material. Latin-hypercube-sampling (LHS) is used to determine the experimental design, which resulted in 18 print-jobs with different combinations of the input parameters. Finally, the necessary mechanical tests are performed and the correlation coefficients are calculated.

2 EXPERIMENTAL DESIGN

The impact of each input variable on the compression properties is studied following these steps: first, the 3D printing process is studied and inputs that are relevant with respect to the output of interest are identified. Second, the experiments are designed by sampling the input parameters using a LHS. Third, the samples are printed according to the printing parameter combinations defined in the experimental design, and the tests necessary to quantify the output of interest are performed. Finally, the correlation between inputs and outputs is calculated with Pearson's coefficient and Spearman's rank correlation coefficient. The relevant printing parameters to be studied depend on the printed material mechanical properties of interest. This study aims at evaluating some mechanical properties of the printed material, namely its compressive strength and Young's modulus. Those are determined by testing cylindrical specimens with 50mm diameter and height of 100mm. Due to the inherent orthotropic behavior of the material (Del Giudice & Vassiliou, 2020) each test is performed along X, Y, and Z direction, i.e. $\sigma_{c,x}$, $\sigma_{c,y}$, $\sigma_{c,z}$, E_x, E_y, E_z, which correspond to the compressive strength and the Young's modulus in X, Y, and Z direction. Four parameters are identified as the most

DOI: 10.1201/9781003348450-60

relevant to the mechanical properties of the printed material, i.e. activator percentage, voxel resolution, droplet mass, and print speed. The activator percentage ranges form 0.30-0.33%, the voxel resolution ranges from 80-150 μm, the droplet mass ranges from 105-115 ng, and the print speed ranges from 0.3-0.5mm/s. Once the input parameters are identified and the relative ranges are defined, the experimental design (the combinations of the printing parameters) is sampled with LHS using UQLab (Marelli& Sudret, 2014). A total of 18 experimental points were selected, resulting in 18 print-jobs each one corresponding to the relevant printing parameters.

3 RESULTS

In this study, the correlation between input and output variables is evaluated with Pearson's correlation coefficient (ρ) and Spearman's rank correlation coefficient (ρ_s). The results from the uniaxial compression tests and reported in Table 1 and the correlation coefficients are reported in Table 2.

Table 1. Compressive strength and Young's modulus of the tested specimens.

Data Point	$\sigma_{c,x}$ MPa	$\sigma_{c,y}$ MPa	$\sigma_{c,z}$ MPa	E_x GPa	E_y GPa	E_z GPa
1	8.3	6.7	8.8	3.89	3.75	3.87
2	12.6	11.0	14.4	5.27	5.33	5.07
3	11.1	9.6	12.1	4.96	4.80	4.71
4	9.4	7.3	9.4	4.35	4.72	3.60
5	8.0	6.2	7.8	3.83	3.87	3.62
6	11.6	9.7	11.7	4.61	4.60	4.37
7	9.9	8.3	10.3	4.41	4.14	4.05
8	11.9	11.0	12.1	4.74	5.08	4.74
9	13.6	12.5	13.1	5.31	5.08	4.78
10	12.7	11.0	12.4	4.99	5.24	4.56
11	14.9	13.3	14.7	6.08	5.73	5.18
12	8.1	6.5	8.4	4.12	4.16	3.58
13	11.5	10.3	11.5	4.43	4.51	4.09
14	10.1	9.5	10.2	4.40	4.47	4.04
15	13.2	11.8	12.8	4.88	5.13	4.56
16	12.0	11.4	12.2	5.07	4.66	4.48
17	15.3	14.3	15.7	5.41	5.60	4.72
18	10.5	9.8	11.0	4.58	4.64	4.32

Based on the correlation coefficients, the compressive strength and the Young's modulus in all directions have a strong linear correlation to the dX resolution. In fact, both correlation coefficients vary between -.78 and -.91, showing that not only there is a strong correlation between dX resolution and the mechanical properties of the printed material, but also that this correlation is close to linear. On the contrary, the correlation coefficients for both print speed and droplet mass are close to zero. Therefore, the mechanical properties are not affected by these two input

Table 2. Linear (ρ) and rank (ρ_s) correlation coefficients between input/output variables.

	Activator		dX		Print Speed		Droplet Mass	
Output	ρ	ρ_s	ρ	ρ_s	ρ	ρ_s	ρ	ρ_s
$\sigma_{c,x}$	.33	.36	-.87	-.84	.07	.05	.03	.06
$\sigma_{c,y}$	.38	.36	-.80	-.78	.03	.04	.00	.07
$\sigma_{c,z}$	.34	.32	-.91	-.91	.15	.11	.03	.05
E_x	.40	.39	-.86	-.89	.12	.05	-.08	-.02
E_y	.26	.20	-.85	-.86	.18	.19	-.16	-.17
E_z	.42	.36	-.85	-.86	.01	.03	-.07	-.05

parameters. Finally, the correlation coefficients related to the activator percentage are around 0.4, meaning that there is probably no clearly linear or monotonic correlation between the mechanical properties and the activator percentage. Based on the simple coefficients used in the present study, the only variable that affects the compressive strength and Young's modulus of the material in all three principal directions (X, Y, and Z) is the dX resolution.

4 CONCLUSIONS

The correlation between the printing parameters of sand based binder jetting 3D printer and the compressive properties of the printed material is evaluated. Four input parameters were considered in this study, i.e. activator percentage, voxel size, print speed, and droplet mass. The mechanical properties of interest investigated were the compressive strength and Young's modulus along X, Y, and Z direction. The correlation between the inputs (printing parameters) and the outputs (mechanical properties examined) was quantified using linear and rank correlation coefficients. Activator percentage, print speed and droplet mass do not have significant correlation with the mechanical properties examined. However, the results have shown a clear correlation between the voxel size, defined by the dX resolution, and both mechanical properties examined.

REFERENCES

Del Giudice, L., & Vassiliou, M. F. (2020). Mechanical properties of 3D printed material with binder jet technology and potential applications of additive manufacturing in seismic testing of structures. *Additive Manufacturing*, *36*, 101714.

Del Giudice, L., Wróbel, R., Katsamakas, A. A., Leinenbach, C., & Vassiliou, M. F. (2022). Physical modelling of reinforced concrete at a 1: 40 scale using additively manufactured reinforcement cages. *Earthquake Engineering & Structural Dynamics*, *51*(3), 537–551.

Marelli, S., & Sudret, B. (2014). UQLab: A framework for uncertainty quantification in Matlab. *Proceedinds 2nd International Conference on Vulnerability, Risk Analysis and Management (ICVRAM2014)*, 2554–2563. Liverpool, United Kindom.

Current Perspectives and New Directions in Mechanics, Modelling and Design of Structural Systems – Zingoni (ed.)
© 2022 Copyright the Author(s), ISBN 978-1-032-39114-4

Influence of process irregularities in additively manufactured structures

M. Erven & J. Lange
Institute for Steel Construction and Materials Mechanics, Technical University of Darmstadt, Germany

ABSTRACT: For steel construction, Wire Arc Additive Manufacturing (WAAM) appears very promising. Here, complicated 3D structures made of steel can be printed comparatively quick by welding multiple layers on top of each other. In contrast to conventional manufacturing this allows a structure to be individually adapted to its requirements. However, unforeseen process instabilities or irregularity make each additively manufactured structure unique. In this paper, the material properties of process irregularities (missing shielding gas and voids) are investigated and their consequences are implemented in the finite element analysis. While voids lead to a reduction of the bearable tensile force, missing shielding gas completely changes the behavior of the printed structure and causes a brittle behavior.

1 INTRODUCTION

Additive manufacturing offers the possibility to produce individual shapes comparatively easily. Especially WAAM (Wire Arc Additive Manufacturing) seems promising for the steel construction industry (Buchanan & Gardner 2019) due to its comparatively fast build rate and good material properties (Silvestru *et al.* 2021), (Tarus *et al.* 2021). For complicated structures it seems reasonable since the time-consuming preparation of the construction elements is not necessary (Bergmann *et al.* 2020).

Although the manufacturing process runs automatically, the large number of welds results in an error-prone process, especially in the nowadays initial development stage. Here it is advantageous to know what kind of defect has what kind of effect on the mechanical behavior of a structure or the material and if this is neglectable. Since a structure which needs to be completely remanufactured due to an irregularity doubles the costs.

2 APPROACH

Various possible sources of process irregularities were identified (see Figure 1) and the effects of two important types are investigated. Tensile specimens are used to investigate the material behavior of the real material. These findings are used to simulate the irregularities in finite elements and show their effect.

Therefore, steel plates were additively printed. Failures were introduced at defined places. Welding current and voltage were observed with a rate of 10 Hz. After printing, the specimens were milled or turned respectively and tensile test were made. In addition to the specimens with failures, plates and specimens were made with a "normal" process and the same welding parameters to have a possibility for comparison and evaluate to consequences of the prearranged process irregularities.

Figure 1. Investigated failures: over-welded void (top left), over-welded oil (bottom left), welding process without gas (top right), weld without shielding gas (bottom right).

3 INVESTIGATED WELDING FAILURES

To get a first insight on how the mechanical behavior is influenced by an interrupted welding process, two specimens (3_2, 3_3) were examined which contained voids. A reference specimen (3_1) was also created. All tensile specimens had a thickness of 4 mm and a width of 16 mm in the decisive cross-section.

DOI: 10.1201/9781003348450-61

Specimen 3_1 shows the normal specimen without a defect. While no influence can be seen in specimen 3_3, specimen 3_2 fails early without prior notice due to a load drop. Until that point all three specimens show a rather uniform behavior, with a high Young's Modulus, a distinctive yielding plateau and a clear solidification.

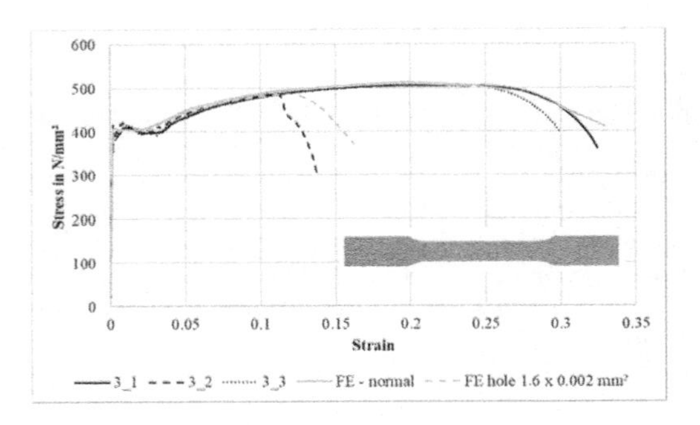

Figure 2. Comparison of real testing and simulation – void.

A multilinear material model was created using the normal specimen 3_1. With this material and the insertion of a void in the geometry, specimen 3_2 was simulated. A minimal imperfection in the FE program illustrates a non-closed joint between the seams. It can be seen that both ultimate load and associated strain can be predicted well, dashed lines in Figure 2.

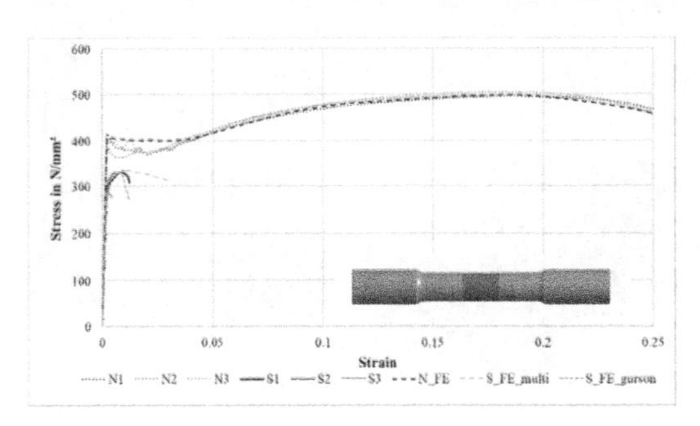

Figure 3. Stress-Strain-Curve - investigation on shielding gas.

Similarly, the effect of welding without shielding gas was investigated. For this purpose, three "normal" specimens "N" and three specimens "S" with layers welded without shielding gas (20 mm) were examined. The tested specimens were cylindrical specimens of the geometry DIN 50125 – A 10 × 50.

Looking at the results of the tensile tests in Figure 3, it can be seen that the material of the "S" specimens behaves significantly different than the "N" specimens. After leaving the region of linearity at about two third of the normal yield strength, the specimens fail abruptly a short time later.

Two multilinear material models were created. One for the standard "N" specimen and one for the area without inert gas. Using these a combined "S" specimen was simulated and it was possible to predict the behavior until the ultimate load and also the associated elongation is reached (Figure 3 S_FE_multi). However, the strong load drop and the almost direct brittle failure cannot be represented with this. The Gurson model was implemented (Figure 3 S_FE_gurson), which fits well with the material welded without inert gas.

4 SUMMARY AND OUTLOOK

A first attempt was made for the identification of process irregularities in WAAM. Here, two critical, but also possible and relevant failures were investigated. Both failures result in an abrupt failure, which can be classified as very critical. However, while a small notch primarily leads to a reduction in tensile strength while otherwise maintaining the elastic characteristic values and the yielding plateau, a weld printed without shielding gas harbors the risk of abrupt collapse, still in the elastic range.

REFERENCES

Bergmann, J.P., Lange, J., Hildebrand, J., Eiber, M., Erven, M. & Gaßmann, C., *et al.* 2020, 'Herstellung von 3D-gedruckten Stahlknoten', *Stahlbau* 89(12), 956–969.

Buchanan, C. & Gardner, L., 2019 'Metal 3D printing in construction: A review of methods, research, applications, opportunities and challenges', *Engineering Structures* 180, 332–348.

Silvestru, V.-A., Ariza, I., Vienne, J., Michel, L., Aguilar Sanchez, A.M. & Angst, U., *et al.* 2021, 'Performance under tensile loading of point-by-point wire and arc additively manufactured steel bars for structural components', *Materials & Design* 205, 109740

Tarus, I., Xin, H., Veljkovic, M., Persem, N. & Lorich, L. 2021, 'Evaluation of material properties of 3D printed carbon steel for material modelling', *ce/papers* 4(2-4), 1650–1656.

Current Perspectives and New Directions in Mechanics, Modelling and Design of Structural Systems – Zingoni (ed.)
© 2022 Copyright the Author(s), ISBN 978-1-032-39114-4

Characterization of 3D printed concrete beams after exposure to high temperature

O. Shkundalova & T. Molkens
Department of Civil Engineering, KU Leuven, Sint-Katelijne-Waver, Belgium

M. Classen
Institute of Structural Concrete, RWTH University, Aachen, Germany

B. Rossi
Department of Civil Engineering, KU Leuven, Sint-Katelijne-Waver, Belgium
University of Oxford, New College, Oxford, UK

ABSTRACT: 3D printing of concrete (3DPC) shows wide applicability and brings significant benefits to the construction sector, such as fast and cost-efficient production and geometrical freedom of structural elements, limited need in heavy equipment and program-controlled erection process. Naturally, when it comes to safety and structural stability, fire resistance is of paramount importance. The post-fire behaviour of 3D printed concrete is a new and nearly unexplored field of knowledge. Thereby, the thermal response of small-scale 3DPC beams to high temperature was investigated and their post-heating behaviour characterized. The beams were subjected to several ranges of elevated temperatures and then tested to measure their flexural and compressive strengths. The influence of various parameters, such as the presence of geometrical irregularities or higher porosity due to 3D printing and cooling ways was also studied and described in this paper.

1 INTRODUCTION

1.1 *State-of-the-art*

3D printing of concrete (3DPC) is gaining wide use in the construction sector due to several advantages, such as fast program-controlled erection process, freedom of geometric shapes of the produced elements, limited use of heavy construction technique. Due to the layer-by-layer extrusion process, 3DPC has many features distinguishing it from conventional construction techniques:

- the mortar used for 3D printing is fast curing, and non-uniform shrinkage can take place;
- the produced specimens are not compacted, as it would be the case for regular concrete production, which results in larger number of pores and voids in the printed material;
- the mechanical properties of 3D printed concrete, as well as the interlayer bond quality between subsequently printed layers, are highly affected by environmental conditions (temperature, humidity) during printing, speed of concrete extrusion, and interlayer time (Babafemi et al., 2021);
- anisotropic mechanical properties of 3DPC are usually observed (Liu et al., 2021; Meurer & Classen, 2021).

It was found that 3D printed concrete has different failure modes compared to conventional concrete, such as interfacial delamination, interface shear-slip, intralayer tensile cracking, intralayer cracking under compression-shear, and crushing (van den Heever et al., 2021).

There is a very limited number of studies shedding light on mechanical properties of 3D printed concrete after fire. To better understand the strength retention and failure modes of 3DPC after exposure to high temperature, small-scale 3DPC beams were tested in this paper.

Figure 1. Samples cut out of large walls with the printed layer orientation parallel, perpendicular and at an angle of 45°.

DOI: 10.1201/9781003348450-62

2 EXPERIMENTAL STUDY

2.1 *Preparation of the samples*

The wall panels were produced from the 3D printing mortar Weber 3D 145-2. Small beams were cut out of straight sections of the wall with dimensions 40 mm x 40 mm x 160 mm. The concrete layer directions in the cut samples were either parallel (along the print-line), perpendicular (perpendicular to the print-line), or at an angle of 45°. The samples can be seen in Figure 1.

2.1.1 *Heating and cooling conditions*

To study the post fire behaviour of 3D printed concrete, the samples were firstly heated up to the following temperatures: 150°C, 400°C, 800°C, 1000°C. Material strengths at ambient temperature of 20°C were also measured as per reference.

To achieve high temperatures when heating 3DPC beams, a Naber L47T 33kW muffle furnace was used. To ensure uniform distribution of heat in the beams, holes were drilled in several samples, in which K-type thermocouples were placed for continuous temperature monitoring. An example of one sample during heating can be seen in Figure 2 (left).

Two types of cooling ways were used: air cooling at 20°C and water cooling with a water temperature of 10°C. The samples were placed in a large bath of cold water in order to get as close as possible to real conditions when the fire brigade extinguishes the structure in the event of a fire. The water bath can be seen in Figure 2 (right).

Figure 2. Furnace heating the sample with the thermocouple installed (left); water cooling bath (right).

2.1.2 *Test types*

After heating and cooling, 3-point bending and compression tests were carried out according to EN 196, ISO 679 and EN 12390-5:2009 to determine the residual flexural and compressive strengths for each tested direction and temperature (see Figures 3 and 4). The samples were tested at a displacement-controlled speed of 0.25mm/min during flexural tests. During the compressive tests, two small cubical elements of 40 mm x 40 mm x 40 mm were used and tested at a speed of 2.4kN/s.

3 CONCLUSIONS

In this preliminary research on post-fire properties of 3D printed concrete, samples made of 3D printing mortar Weber 3D 145-2 were heated up to 150°C, 400°C, 800°C and 1000°C and cooled down in air or in water.

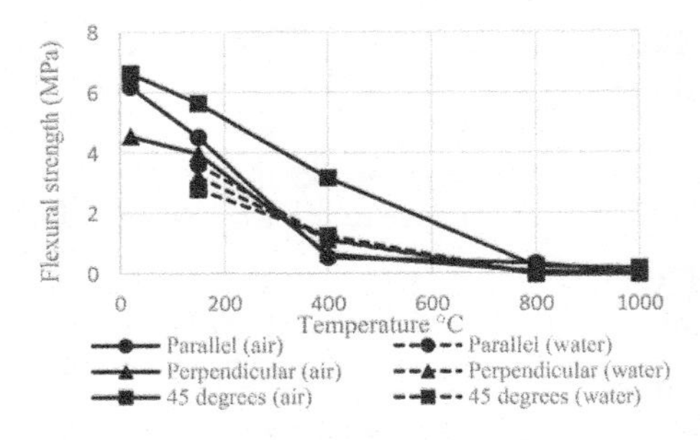

Figure 3. Post fire flexural strength of 3DPC.

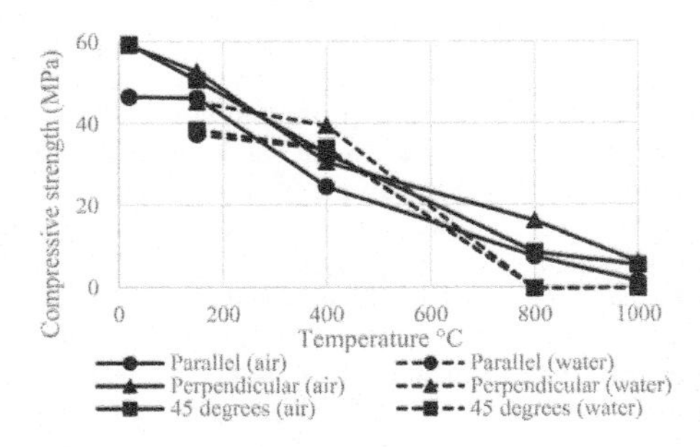

Figure 4. Post fire compressive strength of 3DPC.

Their flexural and compressive strengths were then measured. Based on the study performed, several conclusions could be drawn:

- 3D printed concrete clearly shows orthotropic material properties, with strengths of the samples with parallel concrete layer orientation generally higher than that of the samples with perpendicular layer orientation.
- For samples heated to 400 °C and above, after cooling in air, delamination and formation of cracks at the interlayer interfaces between concrete layers occurred before any tests took place. This was observed both for samples with parallel and perpendicular orientations of 3D printed concrete layers.
- After the 3D printed concrete samples were heated above 400°C, the flexural and compressive strengths decrease substantially with mostly linear trends. In addition, the test results showed high variability.
- Cooling conditions after heating 3D printed concrete have a huge impact on the strengths. When the samples were cooled in air, the strength decreased linearly, and a residual capacity could be observed. When the samples were cooled in water after heating to 400°C, the measured strengths amounted to only about 25% of the initial values. After heating to 800°C and 1000°C, the 3D-printed concrete samples immediately collapsed in water and thus had no residual strength.

Current Perspectives and New Directions in Mechanics, Modelling and Design of Structural Systems – Zingoni (ed.)
© 2022 Copyright the Author(s), ISBN 978-1-032-39114-4

Experimental studies on the mechanical response for structural design of wire-and-arc additively manufactured stainless steel bars

V. Laghi, L. Arrè, M. Palermo, G. Gasparini & T. Trombetti
DICAM – Department of Civil, Chemical, Environmental and Materials Engineering, University of Bologna, Italy

ABSTRACT: Advancements of metal-based Additive Manufacturing lead to the first applications in structural engineering field. In detail, Wire-and-Arc Additive Manufacturing (WAAM) technique allows to realize large-scale structural elements by means of off-the-shelf welding equipment. WAAM-produced stainless steel exhibits non-negligible anisotropic behavior and geometrical irregularities influenced by the build direction and printing orientation. With reference to dot-by-dot printing strategy, used to fabricate bars and lattice structures, the research group at University of Bologna carried out experimental tests to assess the mechanical response of WAAM stainless steel bars for structural design purposes. In detail, tensile, bending and compressive tests have been carried out to assess the main mechanical parameters of bars printed at different build angles and with different levels of slenderness. Additional geometrical characterization was performed to assess both the inherent and the specimen-to-specimen geometrical variability. These results will then be used to calibrate design values, partial safety factors and buckling curves for structural design purposes.

1 WIRE-AND-ARC ADDITIVE MANUFACTURING (WAAM)

Metal Additive Manufacturing (AM) technologies are catalogued in: (i) Powder Bed Fusion (PBF); (ii) Directed Energy Deposition (DED) and (iii) sheet lamination (Buchanan & Gardner 2019). While PBF process has limited building space size, DED process has ideally no geometrical constraint in size. In particular, Wire-and-Arc Additive Manufacturing (WAAM) is a DED technique which uses arc welding as a power source, welding wire as feedstock and robotic arm as motion. Key features of WAAM-produced elements are: (i) the inherent surface roughness, (ii) the marked mechanical anisotropy, (iii) the influence of process parameters (Dinovitzer et al. 2019, Kyvelou et al. 2020).

Among different printing deposition strategies, current research is based on the "dot-by-dot" technique, a spot-like deposition strategy to realize bars and lattice structures (Müller et al. 2019, Laghi et al. 2020). The investigated bars have a constant nominal diameter of 6 mm (as governed by the welding dot), manufactured by MX3D by means of a stainless steel welding wire with grade ER308LSi. Details on the process parameters are reported in (Laghi et al. 2022a).

The deposition of metal droplets is along the main axis (z-axis), developing 3D elements (bars) with a certain build direction and constant input nominal diameter. However, the real diameter varies along the height of the bar (with a variability of around 0.5 mm) due to the successive deposition of metal droplets, resulting in non-negligible lack of straightness (non-straight longitudinal axis connecting the centroids of each circular cross-section) and a visible surface roughness (similar to the one of continuously-printed plates, see e.g. Laghi et al. 2021).

The geometrical irregularities and different mechanical response, with respect to the feedstock, evidenced the need of ad-hoc guidelines and standards to design structures realized with WAAM technique (Laghi et al. 2022a). Geometrical irregularities and material behavior are properly taken into account for structural design purposes.

2 EXPERIMENTAL CHARACTERIZATION OF WAAM STAINLESS STEEL BARS

The mechanical characterization has been performed on different specimens through tensile, three- point bending and compression tests. The mechanical properties of the printed elements are estimated in terms of their effective values, assuming an effective cross-sectional area from volume equivalency (Laghi et al. 2022a). On the other hand, detailed information on the geometrical irregularities from 3D scan (cross-section variability, lack of straightness) are studied for advanced modelling, to quantify the variability of the local geometrical parameters.

The tensile tests have been performed in displacement control on specimens printed at 0°, 10° and 45° build angles (e.g. dot-0, dot-10 and dot-45 specimens) for a total of 29 tests. Overall, for all three inclinations

DOI: 10.1201/9781003348450-63

considered, a significant inherent variability in the tensile response can be detected for each single orientation. Further details are provided in (Laghi et al. 2022a).

The three-point bending tests have been performed on 10 dot-0 specimens of 200-mm length then tested under compression. The aim of the tests is to measure the elastic flexural stiffness of the bars within the elastic limit of the material, hence before reaching yielding (corresponding to around 200 MPa of stress). The results confirm an effective flexural elastic modulus of around 110 GPa for bars printed with 0° build angle, thus having 85% reduction from the elastic modulus value registered from tensile tests. This result is then used to interpret the compression tests from which ad-hoc buckling curves will be calibrated. Further details can be found in (Laghi et al. 2022b).

The compression tests have been performed in displacement-control (with an initial velocity of 0.2 mm/min, then unloading and reloading until 12-mm displacement) on dot-0 specimens only (with constrained bars), with different lengths to estimate the buckling strength for different levels of slenderness. Figure 1 reports the ultimate compression force vs. specimen length. For each level of slenderness, also the mean force and the characteristic value (mean minus two standard deviations) are reported.

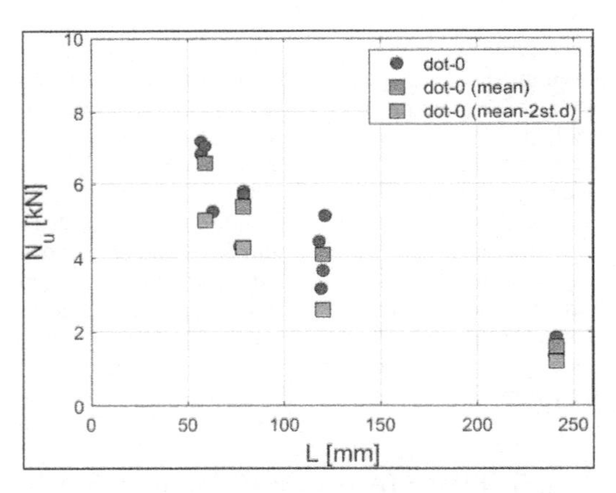

Figure 1. Compression results for WAAM-produced stainless steel bars.

Table 1 . Results of the compression tests on WAAM dot-by-dot bars.

Mean slenderness	$N_{u,m}$ [kN]	$N_{u,k}$ [kN]	$N_{cr,E}$ [kN]	N_{pl} [kN]
Very stub (λ=0.57)	6.57	5.04	21.51	
Stub (λ=0.74)	5.39	4.30	12.99	
Slender (λ=1.12)	4.11	2.59	5.69	14.14
Very slender (λ=2.27)	1.62	1.22	1.38	

3 CONCLUSIONS AND FUTURE PERSPECTIVES

The current work has the intent to characterize WAAM-produced stainless steel bars, pointing up the following key features.

- The geometrical characterization of the bars was performed by means of detailed measurements on the diameter distribution and lack of straightness.
- The results provided by the tensile tests revealed a marked influence of the build angle on the overall mechanical response.
- The compression tests were carried out on bars with different levels of slenderness to estimate their compression capacity.

On-going research is devoted to draw new buckling curves for WAAM-produced bars from additional experimental tests, parametric studies and Advanced Finite Element Modelling to account also the influence of different geometrical and mechanical parameters on the buckling behavior.

REFERENCES

Buchanan, C. & Gardner L., 2019. Metal 3D printing in construction: A review of methods, research, applications, opportunities and challenges. *Eng. Struct.*, Volume 180, p. 332–348.

Dinovitzer, M., Chen X., Laliberte J., Huang X. & Frei H., 2019. Effect of wire and arc additive manufacturing (WAAM) process parameters on bead geometry and microstructure. *Addit. Manuf.*, Volume 26, p. 138–146.

Kyvelou, P., Slack H., Daskalaki Mountanou D., Wadee M. A., Ben Britton T., Buchanan C. & Gardner L., 2020. Mechanical and microstructural testing of wire and arc additively manufactured sheet material. *Mater. Des.*, Volume 192.

Laghi, V., Palermo M., Gasparini G. & Trombetti T., 2020. Computational design and manufacturing of a half-scaled 3D-printed stainless steel diagrid column. *Addit. Manuf.*, Volume 36, pp. 1–13.

Laghi, V., Palermo M., Gasparini G., Girelli V.A. & Trombetti T., 2021. On the influence of the geometrical irregularities in the mechanical response of Wire-and-Arc Additively Manufactured planar elements. *J. Constr. Steel Res.*, Volume 178, pp. 1–23.

Laghi, V., Palermo M., Tonelli L., Gasparini G., Girelli V.A., Ceschini L. & Trombetti T., 2022. Mechanical response of dot-by-dot wire-and-arc additively manufactured 304L stainless steel bars under tensile loading. *Constr. Build. Mater.*, Volume 318, p. 125925.

Laghi, V., Palermo M., Gasparini G., & Trombetti T., 2022. Flexural mechanical response of wire-and-arc additively manufactured stainless steel bars. *Journal of Structural Engineering* (under review).

MX3D Web page, www.mx3d.com

Müller, J., Grabowski M., Müller C., Hensel J., Unglaub J., Thiele K., Kloft H. & Dilger K., 2019. Design and Parameter Identification of Wire and Arc Additively Manufactured (WAAM) Steel Bars for Use in Construction. *Metals (Basel)*, 9(7), p. 725

Current Perspectives and New Directions in Mechanics, Modelling and Design of Structural Systems – Zingoni (ed.)
© 2022 Copyright the Author(s), ISBN 978-1-032-39114-4

Tolerance-free assembly of heat-treated reinforced concrete modules with geometric deviations

J. Stindt, P. Forman & P. Mark
Ruhr University Bochum, Bochum, Germany

ABSTRACT: Modular construction has economic advantages since the modules are prefabricated with quality assurance and must only be assembled on site. To enable tolerance-free assembly of the modules using screw connections, the hole clearance of the screws must be sufficiently large. In the present work, a method is developed that models modular deep beam-like honeycomb structures made from Y-shaped precast concrete components as direct kinematics. This method considers both translational and rotational positional deviations that arise from dimensional deviations of the modules due to shrinkage deformations. For selected scenarios concerning the size of the structure and the concrete used, the required hole clearances for connecting the modules are derived. These hole clearances increase with increasing structure size so that bearing structure with horizontal dimensions of approx. 59 m and up to 30 m in height can be assembled with standard holes $\Delta d \leq 3$ mm.

1 INTRODUCTION

Monolithic structures with in-situ concrete are primarily handcrafted in the open field under changeable weather conditions and thus strongly scattering material properties (Mark et al. 2021). In contrast, precast concrete elements can be prefabricated under controlled conditions and assembled on-site since production and assembly are spatially separated. Truss structures are suitable for modular construction because they consist of almost equal components, such as tension ties and compression struts. This results in slender structures with high load-bearing when assembled (Busse & Empelmann 2014) The number of contact points increases with increasing segmentation of the structures. Contact points compensate geometric deviations, for example, due to shrinkage (Tiltmann 1977). For bolted connections, the hole clearance Δd limits compensation in the shear direction (Polini 2014). For this purpose, the translation and rotation of the segments due to shrinkage is described by a direct kinematic to derive the necessary hole clearances for tolerance-free assembly.

2 KINEMATIC DESCRIPTION OF THE POSITIONAL DIMENSIONS OF MODULAR HONEYCOMB STRUCTURES DUE TO SHRINKAGE

The direct kinematic is used to describe the changes in position of the segments within a honeycomb structure. The honeycomb structure consists of articulated Y-modules with columns and fitting pieces at the edges, which are assembled via bolted connections. The structure size is defined by N_{max} Y-modules in the first layer as well as in M_{max} layers above each other which is exemplarily shown for $N_{max} = M_{max} = 2$ in Figure 1.

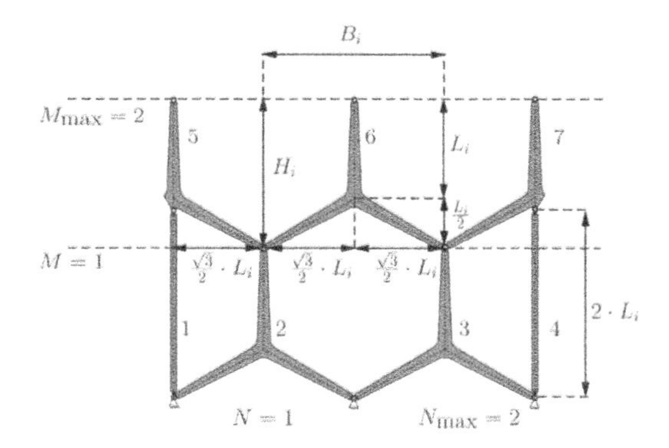

Figure 1. Example of a honeycomb structure consisting of two rows and layers of Y-modules and detail of the different types of segments.

The dimensions of the Y- modules, the columns, and the fitting pieces (half Y- modules) are described as a function of the arm length L_i (Figure 1, segment 6+4+5). L_i depends on the nominal length L_0 of 2 m and the shrinkage strain $\varepsilon_{cs,\infty}$ acc to Eq. (1).

$$L_i = L_0 \cdot \left(1 - \varepsilon_{cs,\infty}\right) \tag{1}$$

In this paper, different $\varepsilon_{cs,\infty}$ from an normal strength concrete (NC) and high-performance concrete (HPC) are investigated. Shrinkage strains of 0.525 mm/m and 0.401 mm/m acc. to (CEN 2020) result for the NC and HPC with compressive strengths of 38 MPa (NC) and 108 (HPC). In addition, an ultra-high-performance concrete (UHPC) based on the binder Nanodur compound 5941 and a rebar ratio of 1.8 % with and without a heat treatment (HT) is investigated (Sagmeister, 2017). The empirical shrinkage strains of the UHPC are 0.257 mm/m (without

DOI: 10.1201/9781003348450-64

HT) und 0.077 mm/m (with HT) (Stindt et al. 2021). The segments shift due to rotations and displacements resulting from geometric deviations due to shrinkage. These deformations are described by means of a direct kinematic (Woernle 2016). The basics of direct kinematics is the idealization of the bodies as local coordinate systems (LCS). The relative position changes between the LCS of the bodies A and B are summarized by homogeneous transformation matrix **T** (Weitz 2018). In direct kinematics, transformation matrices are used according to the Denavit-Hartenberg convention (Spong et al. 2006). These combine multiple spatial motions such as rotation θ_i around the z-axis, and translation a_i around the x-axis.

3 REQUIRED HOLE CLEARANCE DUE TO SHRINKAGE DEFORMATION

The prediction of the required hole clearance is investigated for different structure sizes with N_{max} = 2 to 17 and M_{max} = 2 to 20. This results to dimensions of H x W = 60 m x 59 m for L_0 of 2 m. A Monte Carlo simulation with 10^4 data sets shows that the 95% quantile of the required hole clearance $\Delta d_{req,95}$ depends on the number of layers. $\Delta d_{ref,95}$ is shown in Figure 2 for NC (circle, white), HPC (circle, grey), UHPC without (square, white) and with heat treatment (square, grey).

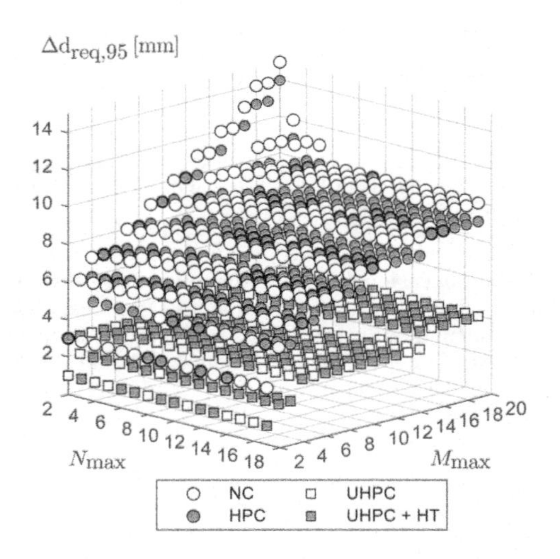

Figure 2. 95 % quantile of the required hole clearance $\Delta d_{req,95}$ as a function of N_{max} and M_{max}.

For flat structures (M_{max} = 2), minimum hole clearances of 3 mm for NC and HPC and of 1 mm for the UHPC result. For very slender structures ($M_{max} \gg N_{max}$) $\Delta d_{req,95}$ increases to 15 mm (NC/HPC) and 6 mm (UHPC). The reason for this is the disproportionate share of columns in the total number of segments within the honeycomb structure. Due to their length, the columns exhibit more significant absolute shrinkage deformations than the Y-modules. As a result, honeycomb structure made from NC and HPC can be assembled with standard hole clearance of 3 mm for M_{max} up to 2, which corresponds to a height of 6 m. For UHPC, it is possible to assemble structure of a minimum of 30 m with horizontal dimensions of approx. 59 m.

4 CONCLOUSIONS

The presented model describes the rotational and translational position changes of segments within a honeycomb structure to predict the required hole clearances for assembly. The main findings are:

- The required hole clearance essentially compensates for geometric deviations from rotations of the segments. Unequal shrinkage deformations, such as different nominal dimensions of the segments and high dispersions of the shrinkage strains, increase the rotation.
- With standard holes of up to 3 mm, honeycomb structures made of UHPC based on the binder Nanodur Compound 5941 can be installed up to a height of 30 m and up to approx. 6 m for ordinary concrete and high-performance concrete.

REFERENCES

Busse, D., & Empelmann, M. 2014. Concrete truss structures – an untapped potential? *Bautechnik, 91*(6), 438–447.

CEN 2020. *Eurocode 2: Design of concrete structures – Part 1-1: General rules, rules for buildings, bridges and civil engineering structures.* (prEN 1992-1-1:2020-11). CEN, Brüssel.

Mark, P., Lanza, G., Lordick, D., Albers, A., König, M., Borrmann, A., Stempniewski, L., Forman, P., Frey, A. M., Renz, R., Manny, A., & Stindt, J. 2021. Industrializing precast productions. *Civil Engineering Design, 3*(3), 87–98.

Polini, W. 2014. To model joints with clearance for tolerance analysis. *Proceedings of the Institution of Mechanical Engineers, Part B: Journal of Engineering Manufacture, 228*(12), 1689–1700.

Sagmeister, B. 2017. *Maschinenteile aus zementgebundenem Beton.* (1st ed.). Beuth Praxis. Berlin: Beuth Verlag.

Spong, M. W., Hutchinson, S., & Vidyasagar, M. 2006. *Robot modeling and control.* Hoboken NJ: John Wiley & Sons.

Stindt, J., Forman, P., & Mark, P. 2021. Influence of Rapid Heat Treatment on the Shrinkage and Strength of High-Performance Concrete. *Materials (Basel, Switzerland), 14*(15).

Tiltmann, K. O. 1977. *Toleranzen bei Stahlbetonfertigteilen: Normen, Größen und Kosten; mit 16 Tabellen.* Köln-Braunsfeld: R. Müller.

Weitz, E. 2018. *Konkrete Mathematik (nicht nur) für Informatiker.* Wiesbaden: Springer Fachmedien Wiesbaden.

Woernle, C. 2016. *Mehrkörpersysteme.* Berlin, Heidelberg: Springer Berlin Heidelberg.

Current Perspectives and New Directions in Mechanics, Modelling and Design of Structural Systems – Zingoni (ed.)
© 2022 Copyright the Author(s), ISBN 978-1-032-39114-4

Dimensional accuracy of casting patterns produced by selected 3D printing technologies

P. Zmarzły, T. Kozior & D. Gogolewski
Department of Manufacturing Technology and Metrology, Kielce University of Technology, Kielce, Poland

ABSTRACT: The traditional production process of casting pattern, e.g. using machining methods, takes up to several days depending on the complexity of the cast. Therefore, alternative manufacturing methods should be developed that will significantly shorten the production time of casting patterns and thus improve the casting process. Because the dimensional accuracy of casting patterns is of key importance, the article presents the results of metrological measurements of linear dimensions of patterns manufactured using three different 3D printing technologies, i.e. Fused Deposition Modeling (FDM), PolyJet Matrix (PJM) and Selective Laser Sintering (SLS).

1 INTRODUCTION

In the foundry industry, new technological solutions are increasingly used to improve the process of creating casting patterns and moulds, which allows for the improvement of the casting process. Therefore, additive technologies commonly known as 3D printing are successfully used in the foundry industry.

Additive technologies are used in many sectors of industry, e.g. in the automotive industry (Romero et al. 2021), medicine (Gogolewski et al. 2021) robotics (Blasiak et al. 2021) but also in the foundry industry (Sithole et al. 2019). Jandyal et al. (2022) reported that 3D printing is gaining more and more popularity especially now, in the era of industry 4.0. Adamczak et al. (2017) show that the most significant parameters affecting the dimensional accuracy of components produced using are 3D printing are printing technical parameters, e.g. printing direction, laser power and applied material layer thickness. These parameters also affect the mechanical properties of the printed components (Kozior et al. 2022). Therefore, when planning the printing process, one should select such values of 3D printing parameters to obtain the final product with the desired physicochemical properties.

Despite many works related to the assessment of the possibility of using additive technologies in the foundry industry, there is still a lack of comprehensive metrological studies that would allow assessing the dimensional accuracy of casting patterns manufactured with the most popular additive technologies. Therefore, the subject of the paper concerns the examination of the dimensional accuracy of samples with a shape that corresponds to typical casting patterns.

2 MATERIALS AND METHOD

In order to evaluate the dimensional accuracy of the casting patterns, a special sample for research purposes was designed. The research pattern has geometric features that are critical in terms of casting technology, i.e. edge rounding and draft. The overall dimensions (length, width and height of the pattern) and the diameter of the cone measured at the height were selected to evaluate the dimensional accuracy. The samples in the form of casting patterns were made with the use of three additive technologies, i.e. Fused Deposition patterning (FDM), PolyJet Matrix (PJM) and Selective Laser Sintering (SLS). Measurements of the linear casting patterns dimensions were taken using a Prismo Navigator coordinate measuring machine by Zeiss.

3 RESULTS

Due to the fact that the linear dimensions of the casting patterns are of critical importance, they were subjected to analysis. Three casting patterns were made using each of the technologies. Once the metrological measurements were taken, the average values of the measurements were determined for each series of samples. Additionally, in order to better visualize the results, the relative errors of the obtained measurement results were calculated. Table 1 presents the measurement results, where d1-d4 denote the measured dimensions x_n - nominal

DOI: 10.1201/9781003348450-65

values, $\bar{x}$ - average value of the measurement results, δ - relative errors.

Table 1. Measurement results of patterns made using additive technologies.

		FDM		PJM		SLS	
	x_n	$\bar{x}$	δ	$\bar{x}$	δ	$\bar{x}$	Δ
d1	100.00	99.97	0.03%	100.04	-0.04%	99.88	.12%
d2	100.00	99.96	0.04%	99.96	0.04%	99.83	0.17%
d3	34.50	34.74	-0.70%	34.82	-0.93%	34.62	-0.35%
d4	55.00	55.36	-0.65%	54.85	0.27%	54.79	0.38%

By analysing the results of the tests presented in Table 1, it can be concluded that the measured values of the external dimensions of the casting patterns made using the FDM and PJM technologies with 0° printing angle are similar to each other. These results were in most cases close to the adopted nominal values. Whereas in the case of the samples made using the SLS technology, the differences between the measured values of the d1 and d2 patterns' sides were greater than for the patterns made using other additive technologies. Different results were obtained for the cylinder's diameter. Based on the evaluation of this geometrical parameter, the SLS technology turned out to be the most accurate one. By evaluating the height of the casting pattern described by the d4 feature, it can be concluded that the PJM technology turned out to be the most accurate one. Considering the overall measurement results presented in Table 1, it can be noticed that, in most cases, the obtained dimensions are smaller than the nominal dimensions. This is due to the shrinkage of plastics during the printing process.

4 SUMMARY

The test results showed that the dimensional accuracy of the produced casting patterns depends on the type of the additive technology utilised. The best dimensional accuracy of the selected geometrical features of the casting pattern obtained using the FDM and PJM technologies. While the greatest deviations from the nominal dimensions were shown for the SLS technology. It should be added that there were clear differences in accuracy between the dimensions related to the length and the width of the pattern and the dimensions related to its height. This is the result of a specific arrangement of the casting patterns on the printer's working platform. When 0° printing direction was used, successive layers were applied in the direction perpendicular to the base of the pattern, which resulted in less accuracy of d4 area printing. Hence the conclusion that when developing the technological process of printing casting patterns, appropriate technological parameters of 3D printing, (e.g. printing direction), should be carefully selected to obtain satisfactory accuracy of printing specific geometric features. Moreover, visible material shrinkage was noticed in the case of printed and wooden patterns. This should also be taken into consideration when creating the technological process.

Therefore, a compromise seems to be the use of FDM technology to make prototype casting patterns. This technology is relatively cheap in commercial and research applications (low cost of printers and raw material). It should be noticed that despite obtaining a satisfactory dimensional accuracy, the shape accuracy and the surface quality of the printed patterns should also be evaluated. This will allow obtaining comprehensive knowledge facilitating the selection of appropriate additive technologies, materials and technological parameters for making the designed casting patterns. Such comprehensive research will be further continued under the Lider project.

ACKNOWLEDGEMENTS

\The research presented in this paper was financed by the National Centre for Research and Developments part of the Lider XI project, number LIDER/44/0146/L-11/19/NCBR/2020, under the title "An analysis of application possibilities of the additive technologies to rapid fabrication of casting patterns".

REFERENCES

Adamczak, S., Zmarzly, P., Kozior, T., & Gogolewski, D. 2017. Analysis of the Dimensional Accuracy of Casting Models Manufactured By Fused Deposition Modeling Technology. *Engineering Mechanics 2017*, (May): 66–69.

Blasiak, S., Laski, P. A., & Takosoglu, J. E. 2021. Rapid prototyping of pneumatic directional control valves. *Polymers* 13(9)

Gogolewski, D., Kozior, T., Zmarzły, P., & Mathia, T. G. 2021. Morphology of models manufactured by slm technology and the ti6al4v titanium alloy designed for medical applications. *Materials* 14(21).

Jandyal, A., Chaturvedi, I., Wazir, I., Raina, A., & Ul Haq, M. I. 2022. 3D printing – A review of processes, materials and applications in industry 4.0. *Sustainable Operations and Computers*: 33–42.

Kozior, T., Bochnia, J., Gogolewski, D., Zmarzły, P., Rudnik, M., Szot, W., Musiałek, M. 2022. Analysis of Metrological Quality and Mechanical Properties of Models Manufactured with Photo-Curing PolyJet Matrix Technology for Medical Applications. *Polymers* 14(3).

Current Perspectives and New Directions in Mechanics, Modelling and Design of Structural Systems – Zingoni (ed.)
© 2022 Copyright the Author(s), ISBN 978-1-032-39114-4

The effect of wind on 3D printed concrete interlayer bond strength based on machine learning algorithms

A. Cicione, P.J. Kruger, J.P. Mostert, R. Walls & G.Van Zijl
Department of Civil Engineering, University of the Stellenbosch, Stellenbosch, South Africa

ABSTRACT: Over the past 5 years, research focused on 3D printed concrete has increased significantly. One specific issue that attracts significant attention, is the reduction in mechanical strength as a result of weak bond between filament layers. Deriving analytical expressions for this reduction in flexural capacity has proven to be extremely challenging, since there are multiple variables that affect the interlayer bond strength. For example, pass time (time between layer depositions), ambient temperature, wind, humidity, and concrete mixture impact the bond. Hence, a more sophisticated approach, such as Machine Learning (ML) models, could be beneficial to determine the effect of a large number of variables on, for example, the interlayer bond strength. It is with this background, that this paper seeks to investigate the effect of wind on the interlayer bond strength based on supervised ML algorithms, as a basis for future research. In this work, 29 3D printed concrete samples were printed under different wind conditions. The input data (wind speed) and output data (tensile capacity) were collected and cleaned using Python. The data was split into two categories, namely training data (data used to develop the model using a regression algorithm with the built-in Scikit package in Python) and test data (data used to test the accuracy of the model). It was found that an increase in wind speed up to approximately 25 km/h, correlates to a decrease in interlayer bond.

Keywords: 3D printed concrete, tensile capacity, machine learning, supervised learning, wind, regression modelling

1 INTRODUCTION

Three-dimensional concrete printing (3DCP) is an automated layer-wise construction technique that requires no formwork (Paul, van Zijl, and Gibson 2018). A number of researchers investigated the behaviour of 3D printed concrete (3DPC) at ambient conditions (Paul, van Zijl, and Gibson 2018). A current area of study is the influence that the interlayer bond has on the mechanical properties of 3DPC. Some studies find that the interlayer bond can cause the flexural capacity of the sample to reduce by up to 36% (Le et al. 2012). At elevated temperatures, Cicione et al. (2020) found that samples tend to delaminate between layers, which further highlights the need to understand the factors affecting the interlayer bond strength.

A number of factors influence the bond strength of 3DPC, such as pass time, wind, temperature, humidity. The relationships between these factors and the interlayer strength are highly non-linear and complex, implying that the derivation of analytical expressions to calculate the interlayer bond strength become extremely challenging. This paper seeks to develop a simple ML model that can predict interlayer bond strength given a specific wind condition. Although, there are a number of other factors influencing the bond strength, making the problem a multi-dimensional problem, this paper serves as a building block for future work.

2 EXPERIMENTAL SETUP

A standard high-performance 3D printable material was used throughout the experiments (Moelich et al. 2021). A gantry-type 3D concrete printer along with a wind tunnel, as shown in Figure 1, was used to conduct the experiments. Five wind velocities, (0, 15, 25, 35 and 40 km/h), were selected for the experiments. Each wind configuration was printed with a 5 min pass time and at a printing speed of 35 mm/s. All samples were left to cure in a climate-controlled room for 7 days at 23±1°C and 60±2% relative humidity. On the third day of curing, five 50x40 mm sections were saw cut from each wind configuration and steel T-shaped brackets epoxy glued to both sides of the samples and left to cure for the remaining period. Thereafter, the samples were subjected to a direct tensile test on the MTS testing machine at a crosshead displacement-controlled test rate of 0.250 mm/min until failure occurred.

DOI: 10.1201/9781003348450-66

Figure 1. 3D concrete printer and wind tunnel setup.

3 MACHINE LEARNING METHODOLOGY AND RESULTS

Although there are many ML algorithms available within Python v3.10.2, in this work a Linear Regression ML algorithm was used to train the computer using experimental data.

Before a linear regression ML model can be applied to the training data, the Polynomial Features

function from sklearn.preprocessing was used to transform independent variable X_train (the wind speed input data in this case) into a new matrix (X_train_poly), where values are calculated by setting X_train to the power of two, three, etc., to account for nonlinearities within the input data.

The test_train_split function from sklearn, was used to randomly split the data into training (80%) and testing (20%) data and the model trained using the training data. The model was then tested by plotting the model's predicted values against the actual measured values, as depicted in Figure 2. The model predicted the results (y_test) relatively well. We can further evaluate the model using a regression evaluation metrics. In this case, Root Mean Squared Error (RMSE) calculates to 466. It should also be noted here that only 29 data points were used in this preliminary study, which is not ideal. Once more data points become available for training, one should see a decrease in the RMSE. However, the purpose of this work is to show the methodology and to lay the foundation for future research.

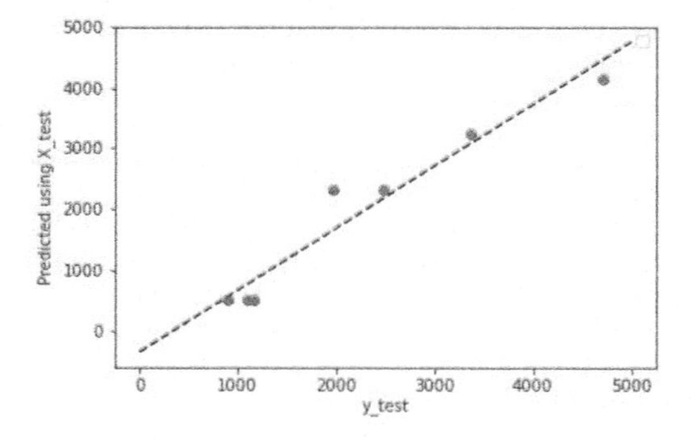

Figure 2. Scatter plot of the linear regression model's predictions versus the measured values.

4 CONCLUSION

In this paper, 3DPC samples printed at different wind speeds, ranging from 0 – 40 km/h, at constant pass time and other environmental conditions, were tested under direct tension. The results showed that an increase in wind speed, resulted in a decrease in tensile capacity, up to 25 km/h, whereafter the tensile capacity starts to increase again. It is hypothesized that the wind speeds affect the surface moisture condition of the substrate filament, which, as a result, affect the tensile capacity. Although further testing is needed before any assertive conclusions can be made.

The work then continued by investigating the feasibility of using Machine Learning algorithms to predict interlayer tensile capacity under different wind speed conditions. The input data was split into training and testing data (80/20 split) and a Linear Regression ML algorithm used to train the model. The predictions were relatively accurate, with a Root Mean Squared Error of 466 N. It was noted that more samples are needed to improve the model accuracy.

This paper served as the foundation for future research, which should focus on incorporating more variables such as wind speed, ambient temperature, humidity, pass time, etc. into ML models. Additionally, the use of different, more sophisticated, ML models such as Artificial Neural Network should be investigated.

REFERENCES

Cicione, Antonio, Pienaar Jacques Kruger, Richard Shaun Walls, and Gideon van Zijl. 2020. "An Experimental Study of the Behavior of 3D Printed Concrete at Elevated Temperatures." *Fire Safety Journal* 120. https://doi.org/https://doi.org/10.1016/j.firesaf.2020.103075.

Le, T.T., S.A. Austin, S. Lim, R.A. Buswell, R. Law, A.G.F. Gibb, and T. Thorpe. 2012. "Hardened Properties of High-Performance Printing Concrete." *Cement and Concrete Research* 42 (3): 558–66. https://doi.org/10.1016/j.cemconres.2011.12.003.

Paul, Suvash Chandra, Gideon P.A.G. van Zijl, and Ian Gibson. 2018. "A Review of 3D Concrete Printing Systems and Materials Properties: Current Status and Future Research Prospects." *Rapid Prototyping Journal*. Emerald Group Publishing Ltd. https://doi.org/10.1108/RPJ-09-2016-0154.

Current Perspectives and New Directions in Mechanics, Modelling and Design of Structural Systems – Zingoni (ed.)
© 2022 Copyright the Author(s), ISBN 978-1-032-39114-4

Tensile test of double-lap GFRP joint applying Vacuum Infusion wrapping

Y. Anan, M. Matsumura & H. Moriyama
Kumamoto University, Kumamoto, Japan

S. Inoue
COMTEC Inc., Kumamoto, Japan

ABSTRACT: Bolted or riveted joints are used to connect FRP members in FRP bridge construction. But corrosion deteriorations of such metallic parts are to decrease structural performance of the connections. Double-lap joint using FPR connecting plates, one of typical adhesive bonded connections between FRP plates, will be an alternative when peeling at the tips of the connecting plates is prevented. Then the authors focus on the Vacuum Infusion molding technique into the double-lap joint of GFRPs to decrease shear stress and peeling stress at the tips of the connecting plates and install additional GFRP sheet wrapping through a whole joint length. Carried out in this study are uniaxial tensile tests of the double-lap joint of GFRPs with/without one layer of the GFRP sheet by VI wrapping. Test results indicate that one layer of GFRP by VI wrapping increases the maximum load and axial stiffness and reduces peeling stress at the tips.

1 INTRODUCTION

Bolted and riveted connections are generally used to connect the plates in FRP footbridges in Japan. But corrosion damages of such metallic parts are to decrease structural performance of the connections.

Adhesive bonded connections are considered a possible method to connect the FRP segments and to expand the span of the GFRP footbridge of less maintenance. Double-lap joint using GFRP connecting plates is one of the typical adhesive bonded connections, but shear stress and peeling stress are concentrated at the tips of the connecting plates. Then, the authors focus on an additional wrapping GFRP sheet over the joint length. Also Vacuum Infusion technique, which is an effective molding technique providing stable volume fraction for FRP members.

In this study uniaxial tensile tests of double-lap joints of GFRPs with/without one additional layer of the GFRP sheet wrapped by the VI molding technique.

2 APPLICATION EXAMPLES OF VI MOLDING TECHNIQUE

COMTEC Inc. manufactured some of GFRP made box girder footbridges by using the VI molding technique. Figure 1 shows a GFRP box girder footbridge manufactured and erected by COMTEC Inc. A light weight feature of the GFRP made bridge brought advantages in a shorter construction period and so on.

Now that the bridge length is limited to transport it by land, development of some jointing technique applicable at the construction site will make the span long for more wider applications of the GFRP footbridges.

Figure 1. GFRP box girder bridge construction.

3 TEST PROGRAM

Material tests and joint tests subjected to tensile force were conducted. In the Joint tests, 2 types of the specimens are prepared; "ad" where the connecting plates are bonded only with adhesive and "ad-VI" where they are bonded with adhesive and with one layer of the GFRP sheet by the VI wrapping.

Both the mother plate and the connecting plates were also made by the VI molding technique. 4-axial roving cloth and 1-axial mat of the glass fiber and vinyl ester resin as the matrix resin are used for the plates. As a tapered shape can reduce stress concentrations at the tip of the connecting plate, then connecting plate in the joint specimen has a tapered shape by 20 mm steps introduced between the upper two layers and the lower

DOI: 10.1201/9781003348450-67

ones. The thickness of the adhesive was 1 mm. The connecting plate are shown in Figure 2.

A universal testing machine of 2 MN are used in the test and the loading speed was set at 20 μ/s for both the material and the joint tests.

4 TEST RESULTS

Figure 3 plots the load-displacement relationship of "ad" and "ad-VI". The averaged value of the maximum load among the VI wrapping "ad-VI" specimens take 1.17 times the specimens without the VI wrapping, and the axial stiffness of the "ad-VI" specimens become larger, too.

Figure 4 shows load-strain relationship of specimens "ad" and "ad-VI".

In the "ad" specimen, as strains of the steps at the locations 4 to 6, of the connecting plate take smaller values, but peeling was initiated from the tip. On the other hand, in the "ad-VI" specimen, strains of the connecting plate at the locations 6 & 7, suddenly decreased at about 100 kN, because some damages occurred at the corner parts at the steps of the outer additional layer. However, strain at the location 8 at the tips of the additional VI wrapping layer increases lineally and no significant changes can be seen. Therefore, the adhesive between the connecting plates and the mother plate and the adhesive between the VI wrapping and the mother plate were kept sound conditions, and the maximum load takes a larger value than that of the "ad" specimen.

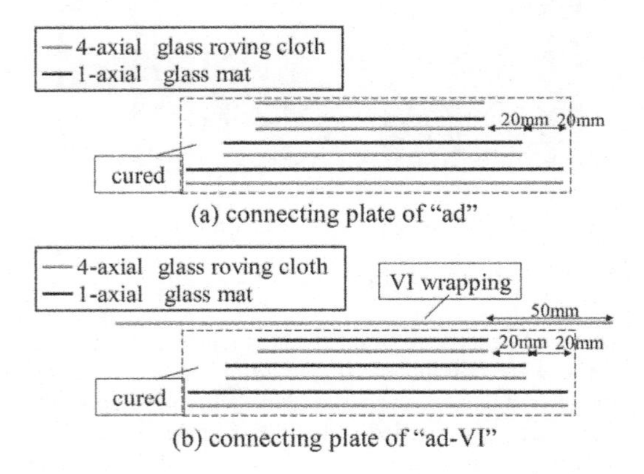

Figure 2. Layer constitution of connecting plates.

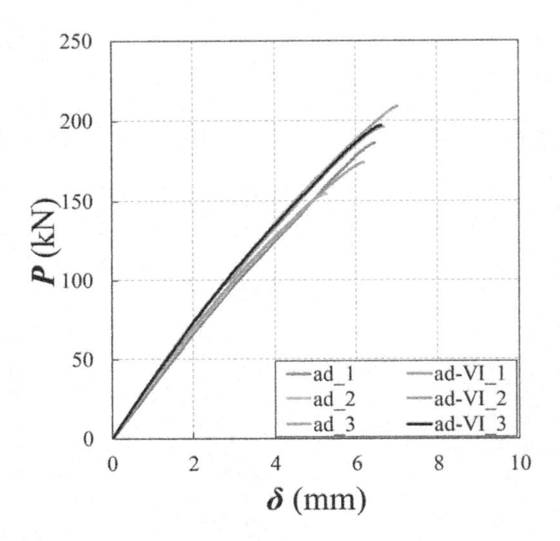

Figure 3. Load-Displacement relationship.

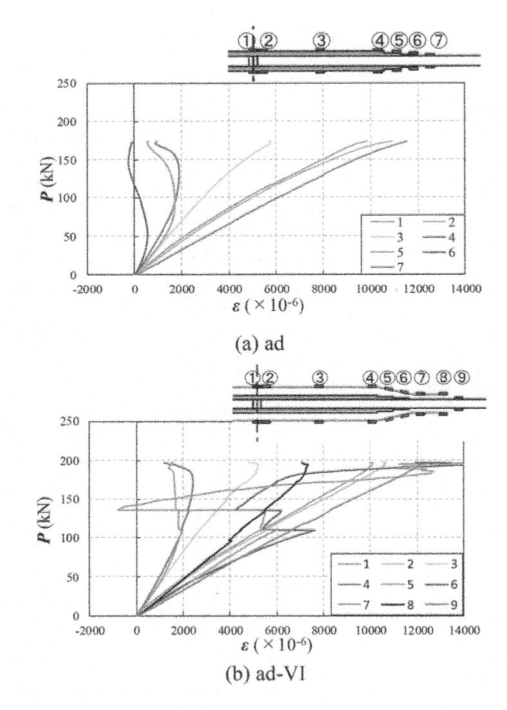

Figure 4. Load-Strain relationship.

5 CONCLUSIONS

Focused on in this study is GFRP double-lap joints. Uniaxial tensile test of the plates and the joints are carried out. Both the plates are molded by Vacuum Infusion molding technique. In the joint test, one layer of GFRP sheet wrapping by the VI technique is adopted to prevent breaks and peelings of adhesive between the connecting plates and the mother plate.

Test results indicate that the maximum load and axial stiffness increase when the one additional layer of GFRP by the VI wrapping is adopted, as the outer wrapping layer make working stress at the tips smaller.

REFERENCES

COMTEC Inc. 2022. *GRP BRIDGE*. (in Japanese)

Current Perspectives and New Directions in Mechanics, Modelling and Design of Structural Systems – Zingoni (ed.)
© 2022 Copyright the Author(s), ISBN 978-1-032-39114-4

Multiscale analysis of surfaces made with additive technology in medical aspects

D. Gogolewski, T. Kozior & P. Zmarzły
Department of Manufacturing Technology and Metrology, Kielce University of Technology, Kielce, Poland

ABSTRACT: The paper presents the results of the assessment of surface texture irregularities of elements manufactured with the use of the SLM additive technology. The research was carried for a hip implant made of Ti6Al4V material. The surface texture analysis was carried out using the modern, multiscale approach, which is the one-dimensional continuous wavelet transformation. In addition, the areas on the samples on which the morphological features in the form of dimples and bumps corresponding to the example distribution of porosity had been modelled were subjected to similar analyses. Research carried out on a wide range of scales showed differences in both the distribution and size of individual surface irregularities. The research results will allow for better understanding of possibilities to produce fully functional, user-dedicated elements, which is particularly important, among others, in production of human implants. Positive verification of the concept of adaptation of wavelets in the analysis of the surface texture will certainly allow expanding the applicability of modern multiscale methods, which are part of the Fourth Industrial Revolution, Metrology 4.0.

1 INTRODUCTION

The dynamic development of additive manufacturing determines the possibility of their use in an ever wider range of applications. The possibility of fast and precise production of personalized elements with specific properties and dimensions, which are dedicated to a user, results in a symbiosis of modern manufacturing techniques and the medical industry. In addition, it is crucial to assess the surface texture mapping specified in CAD software, among other this directly affects the implants ability to nest human tissue. However, the individual surface irregularities are visible only in a certain range of the scale. It is often assessed that the application of traditional methods to analyse irregularities on the surface of elements produced using additive technologies is insufficient.

The analysis of the current state-of-the-art showed that assessment of surfaces with a characteristic distribution of irregularities using the multiscale method, can still be considered a research gap and requires further analysis. In the paper, the surfaces of the titanium femoral stem were examined to evaluate surface irregularities, as well as the surfaces on which the characteristic distribution of irregularities(porosity) based on dimples and bumps of various sizes was modelled.

The samples were manufactured using selective laser melting (SLM) of titanium-powder-based material (Ti6Al4V), produced by EOS. Three different orientations in towards the building platform were adopted: 0°, 45° and 90°. An EOS M290 machine was used to prepare the samples. Measurements of the surface texture were carried out using Talysurf CCI Lite optical profilometer. TalyMap Platinum 6 and Matlab software were used to analyse the surface texture. Measurements were carried out both on the surface where dimples and bumps were modelled, as well as in the area where they did not exist.

A multiscale analysis of the obtained surfaces was carried out with the use of a number of wavelet forms. The diversity of the mother wavelet with different properties made it possible to perform a broader analysis of individual surface irregularities. For this purpose, a one-dimensional continuous wavelet transform with a wide-scale range was used, i.e. $0.01 \div 6$, which corresponds, depending on the mother wavelet used, to the individual effective support width, to over 54 μm. The surface profiles were analysed using: db2, db10, db20, sym8, coif3, morlet, mexican hat (Ricker wavelet), Meyer, bior1.5, bior2.4, bior3.9, bior5.5, shan1-1.5 and shan2.3.

The first stage of the research was a multiscale assessment of the surface profiles in areas without modelled morphological features. The *Rq* parameter was selected for the quantitative assessment in particular scales. Figure 1a shows the averaged results of the wavelet analysis, taking into account the values obtained for thirty profiles from each surface. It should be mentioned that the differences in the value of the parameter for individual scales, wavelets and building angle for samples, were relatively insignificant.

DOI: 10.1201/9781003348450-68

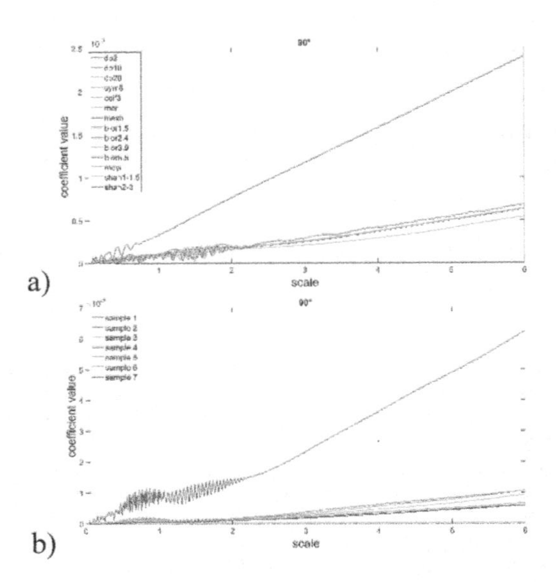

Figure 1. Evolution of root mean square height (*Rq*) with scale for sample a) without b) with; morphological features, db20.

It should be noted that the value of the parameter depends on both the mother wavelet and the scale. For the Mexican hat wavelet, the obtained values differed significantly from the values observed for other wavelets. It was proven, that in the case of the assessed surfaces, there is a relatively large number of peaks and pits, which have already been visualised in the range greater than 0.73 μm. Up to this value of this scale size, the values of the obtained coefficient for the individual wavelets are close to zero. From this value, however, the occurrence of even greater differences in the parameter in comparison with other forms of the wavelet was observed. For the remaining types of transformation, the relative difference in the parameter values remains at the level of a few per cent for each scale value. Moreover, in the band up to 21.7 μm, the occurrence of periodic changes in the parameter value was observed. The length of the period increases with the increase of the scale value and, at the end of the range, it reaches the value close to the effective support width of individual wavelets. These changes provide diagnostic clues about the surface, indicating the size of individual features and their irregularities. The research also showed that the building angle has a significant impact on the number of surface irregularities. Even the smallest scales assessed showed differences in the parameter values for individual wavelets. It proves the presence of a much larger number of relatively small irregularities on the assessed surfaces as compared to the surfaces produced at other angles.

The second stage of the research was a multiscale assessment of the surface profiles in areas on which morphological features in the form of dimples and bumps were modelled in CAD software. The tests were carried out for convex (No. 1-4) and concave (No. 5-8) hemispheres with diameters of 0.3, 0.22, 0.15 and 0.08 mm. The obtained values for the db20 mother wavelet are shown in Figure 1b.

On the basis of the obtained results and by assessing the surface quality of the produced samples, it was found that not all areas have features modelled in the CAD program. This is due to the problems with mapping individual dimples and bumps, which are affected, among others, by technical parameters. Diagnosis, verification of the surface with the use of the wavelet transformation showed that for sample No. 1 made at an angle of 90°, much higher values of the parameter were obtained. The assessment showed that the phenomenon is related to the application and sintering of an additional layer of material during the process of building. For the remaining samples and material building angles, it should be noted that there are relatively small differences in the obtained values. For small ranges of scales, no significant changes and differences in the topography of the individual surface irregularities were observed. Thus, its size and nature are similar for a scale smaller than 27.98 μm. For larger scales, for all three angles and the mother wavelet, a much greater variation in the parameter values was observed. This may be due to flaws in the production process, the inability to accurately reproduce the actual hemisphere, material flow or the staircase effect. It was also observed that the type of hemispheres (convex or concave) did not cause any significant differences in the value of the coefficient.

The research conducted provide hints on producing bone parts and implants, but also on modelling porosity. The research allowed to assess the influence of the mother wavelet and the character of morphological features depending on the building angle. The type and size of individual irregularities were determined, both for surfaces without and with morphological features modelled in the CAD program. It was proven that the type of wavelet transformation allows for the differentiation of the surface texture irregularities, and the individual properties of the wavelets affect the results, in particular for small scales, which correspond to the high-frequency change in the height of irregularities - roughness and microroughness.

The research presented in this paper was supported by the National Science Centre of Poland under the scientific work No. 2020/04/X/ST2/00352 "Multiscale analysis of free-form and functional surfaces manufactured by additive technology".

REFERENCES

Brown, C. A. et al. 2018 Multiscale analyses and characterizations of surface topographies, CIRP Annals, 67(2), 839–862.

Gogolewski, D. 2021 Fractional spline wavelets within the surface texture analysis, Measurement 179, 109435.

Gogolewski, D. et al. 2021 Morphology of Models Manufactured by SLM Technology and the Ti6Al4V Titanium Alloy Designed for Medical Applications, Materials, 14 (21), 6249.

Current Perspectives and New Directions in Mechanics, Modelling and Design of Structural Systems – Zingoni (ed.)
© 2022 Copyright the Author(s), ISBN 978-1-032-39114-4

Bio-inspired kinematics of reconfigurable linkage structures

N. Georgiou & M.C. Phocas
Department of Architecture Faculty of Engineering, University of Cyprus

ABSTRACT: The present study refers to a linkage structure typology development and kinematics mechanism application based on the particular function and kinematics of the human fingers. The linkage structure consists of cross hybrid rigid bars interconnected in series that are additionally integrated with a secondary system of struts and two continuous diagonal cables. The kinematics mechanism corresponds to the effective crank–slider approach that stepwise reduces a planar linkage system to an externally actuated 1-DOF system, in order to adjust the joint angles from the initial to the target values for its complete reconfiguration. The two ends of the linkage structure are supported on the ground, through a pivot joint on one end and a linear sliding block on the other end. Each intermediate joint is equipped with brakes. Two linear motion actuators placed on the ground, are associated to the cables. A reconfiguration of the linkage structure can be accomplished through different control sequences and an optimal one can be selected based on specific criteria, such as lowest maximum required brake torques and actuators motion. A simulation study refers to the preliminary kinematics and the comparative Finite-Element Analysis of a group of planar linkage systems of eight cross hybrid rigid aluminum bars of constant and variable length.

1 INTRODUCTION

In architectural context, the exploration of kinematics and responsiveness brings significant scientific challenges related to the properties of structures and materials, in order to facilitate reconfigurations or real transformations of systems. In this framework, the kinematics mechanism of the human hand can be associated with the kinematics of articulated bar linkage structures. The latter constitute promising systems for the development of deployable and reconfigurable modular structures with high flexibility and controllability. Actively controlled bar linkage structures with a sliding block have been proposed, in order to succeed different configurations based on the effective crank–slider (ECS) kinematics approach (Konatzii et al. 2021; Phocas et al. 2020).

Along these lines, the present study focuses on the kinematics analysis, decoding and evaluation of re-configurable cross hybrid linkage structures based on the effective crank–slider approach. The study presented refers to the preliminary kinematics and the comparative Finite-Element Analysis (FEA) of a group of planar cross hybrid linkage systems of eight rigid aluminum bars of variable length.

2 BIO-INSPIRED DESIGN

The emerging scientific discipline of biomimetics can bring new insights into the field of architecture. An analysis of both architectural and biological method-ologies can demonstrate valuable aspects connecting the disciplines. In the field of engineering research termed 'bio-inspired engineering' the aim is to generate innovative engineering designs by abstracting a mechanism of interest from a biological system. The kinematics of the human hand is of great interest, as it refers to articulated serial chains comprised of links and joints. The structural design synthesis of each hand finger, except the thumb, can be paralleled with a single degree-of-freedom (SDOF) closed loop linkage based the ECS approach for a reconfiguration.

3 EFFECTIVE CRANK-SLIDER APPROACH

The basic structural and kinematics element of the proposed system is a planar linkage structure that consists of n-serially connected rigid links with pivot joints between them. The structure is connected to its ground supports by a pivot joint on one end and a linear sliding block on the other. Each intermediate joint is equipped with brakes, while a linear actuator is fixed to the sliding block on the ground. The reconfiguration process of the structure is accomplished by a stepwise adjustment of the joints, where in each step, the brake of one intermediate joint is selectively released and the pin joints at the supports always remain unlocked. Through this procedure, every angle of the n-bar linkage system is transmitted and adjusted to its target position in a step-by-step procedure, reducing the system

DOI: 10.1201/9781003348450-69

to a 1-DOF mechanism, and in particular, in an ECS system. In the case of a system with n bodies (including the ground and the slider block), a complete reconfiguration will require a number of $(n-3)$ intermediate steps, given that during the final step the four remaining joint variables will be adjusted simultaneously.

3.1 *Simulation study*

The simulation study refers to the preliminary kinematics and FEA of five 8-bar cross hybrid linkage systems (CHS). The systems consist of the joint connected rigid bars and a secondary system of struts and two diagonal continuous cables. Each system has identical initial length, target configuration span and number of bars, but different length of bars. The basic planar hybrid linkage system of eight rigid aluminum bars of same length, CHS0, is presented in Figure 1.

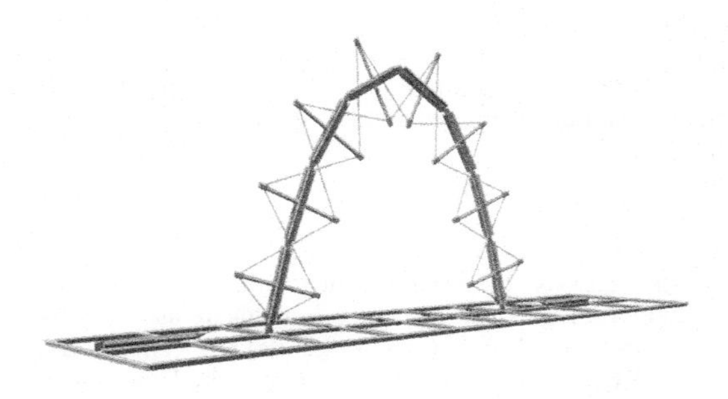

Figure 1. Perspective view of CHS0.

The kinematics approach has been examined from an initial, almost flat configuration of the linkages to a respective span of half their initial length in their symmetric arch-like target configuration with span of 6.0 m.

3.2 *Kinematics analysis*

The kinematics analysis of the planar bar linkages has been conducted with the software program Working Model 2D. The kinematics analysis was based on numerical integration of the linkage model, based on the Kutta-Merson method. The respective time step is automatically adjusted during the course of the simulation.

4 STRUCTURAL ANALYSIS

The structural behavior of the systems in each recon-figuration step has been examined with the FEA software program SAP2000. The linkage bars consist of aluminum of 69.6 GPa elastic modulus and 241.3 MPa yield strength. The cables are assigned to steel S450 of 24.82 GPa elastic modulus, and a pretension of 2 kN. Throughout the reconfiguration, the static structural analysis considers only the geometrical nonlinearity of the systems and their self-weight. In all cases, the response of the systems was within the elastic range. Furthermore, no geometric imperfections, nor any initial deformation of the systems in each configuration step, have been considered.

5 CONCLUSIONS

The current paper investigates the kinematics of the stepwise deployment and reconfiguration process of cross hybrid bar linkages based on the ECS. The simulation study conducted refers to the preliminary kinematics analysis and FEA of five planar cross hybrid linkages with eight rigid bars, which have the same initial overall length of 12.0 m in their almost flat position, and the same span of 6.0 m in the target configuration of a symmetric arch, but different bars length. The numerical analysis shows that the CHS4 with relatively longer bars placed symmetrically on both sides of the system, registers the lowest bending moment in the beams and system deformation. The analysis results demonstrate the feasibility and potential of the ECS approach that needs to be further examined through prototyping and testing of system typologies.

REFERENCES

Konatzii, P., Matheou, M., Christoforou, E.G. & Phocas, M.C. 2021. Versatile reconfiguration approach applied to articulated linkage structures. *Architectural Engineering*, 27(4): 04021039-1–04021039-12.

Phocas, M.C., Christoforou, E.G. & Dimitriou, P. 2020. Kinematics and control approach for deployable and reconfigurable rigid bar linkage structures. *Engineering Structures*, 208: 110310-1–110310-8.

Current Perspectives and New Directions in Mechanics, Modelling and Design of Structural Systems – Zingoni (ed.)
© 2022 Copyright the Author(s), ISBN 978-1-032-39114-4

Tribological analysis of selected 3D printing technologies in the context of building foundry models

T. Kozior, D. Gogolewski & P. Zmarzły
Department of Manufacturing Technology and Metrology, Kielce University of Technology, Kielce, Poland

ABSTRACT: The article presents the results of tribological research of samples models manufactured with the use of three additive technologies: Selective Laser Sintering - SLS, Fused Deposition Modeling - FDM and Photo-Curing Polymer Resin - PJM. The tests were carried out with the use of a ring on disc tribological tester (T-15), where the ring-shaped samples were manufactured of commercial polymer materials, and the counter-sample (disc-shaped) was made of C45 steel. During the tests, the temperature change in the friction zone, the wear measured using the linear method and the value of the friction force were determined. The test results showed large differences in abrasion resistance depending on the technology and material used. Large differences were also observed in the nature of the friction process, in particular, it concerns the friction force.

The ongoing fourth industrial revolution 4.0 (IR 4.0) is one of the most technologically advanced economic changes. The technologically advanced tools used in this transformation require very large investments in the purchase of new technologies (mainly digitization), machines, materials, and their research. In the case of production technology, such tools can certainly include additive technologies, otherwise known as 3D printing technologies. The development of 3D printing technology over the last several years has been very dynamic. It is particularly visible in the case of material chemistry, where composite materials based on glass and carbon fibers are increasingly used with very good rheological properties (Bochnia and Blasiak, 2019). Due to the large development in the field of accuracy and technological possibilities, 3D printing technologies are increasingly used to build foundry models (Upadhyay, Sivarupan and El Mansori, 2017).

Tribological research in the field of 3D printing technology (FDM, PJM and SLS) are rarely used compared to many tests of mechanical and metrological properties or other unconventional manufacturing technologies described in many research works (Madej *et al.*, 2016; Adamczak *et al.*, 2017). Selected tribological aspects of models produced with 3D printing are described in the works (Rudnik *et al.*, no date; Pawlak, 2018). The work (Rudnik *et al.*, no date) entitled: *Tribological properties of medical material (MED610) used in 3D printing PJM technology* extensively describes tribological research of the biocompatible material MED 610 (3D printing technology - PJM), taking into account the analysis of variable friction parameters in the PV system (pressures - rotational speed). The maximum friction parameters and tribological properties, such as wear resistance of ring-shaped samples and the value of the friction force, were determined. The work by *Pawlak* entitled *Wear and coefficient of friction of PLA - Graphite composite in 3D printing technology* presents the research results of one of the most popular materials in 3D printing - PLA (3D printing technology - FDM). Creating composite structure of 50% graphite and 50% polylactide allowed to decrease coefficient of friction and also linear wear.

In the research three 3D printing technologies were used to build the sample models - selective laser sintering - SLS, photo-curing of liquid polymer resins - PJM and fused deposition modeling - FDM. In the case of the selective laser sintering technology, a well-known construction material with the trade name PA 2200, based on the well-known polyamide PA 12, was used to build the samples. Based on liquid polymer resins under the trade name FullCure 720 - FC 720. For models made in the FDM technology, the known construction material ABS was used. The test specimens were made using 3D printers: Formiga P 100 – SLS (EOS), Connex 350 – PJM (Stratasys), Dimension 1200es – FDM (Stratasys). All samples were made in one building plane, i.e. lying with a flat tested surface of the sample on the printing platform of the machine. The samples were made with the given thicknesses of the layer: SLS - 0.1 mm, PJM - 0.016 mm and FDM - 0.254 mm. Moreover, in the case of SLS technology, the energy density delivered to the sintered powder layer was 0.056 J/mm^2. In all the above-mentioned cases, the models were made as SOLID with 100% filling. Each sample was adjusted to the friction

DOI: 10.1201/9781003348450-70

process conditions among others, the roughness parameter Ra was below 1 μm. 3 samples were made for each technology, the total number of both samples and tests was 9.

The test samples were made in the shape of a ring. Counter-sample for the disc-shaped tests was made of C45 steel.

The tribological tests were performed with the use of a T-15 tribological tester of the ring on disc type (company Łukasiewicz Research Network - Institute for Sustainable Technologies, Radom, Poland). The tribological tests were performed with the same friction parameters for all samples, that is: the pressure force was always 45 N, and the rotational speed of the sample was 150 rpm. The friction process was a process without the use of lubricants. Example of friction results is presented in Figure 1.

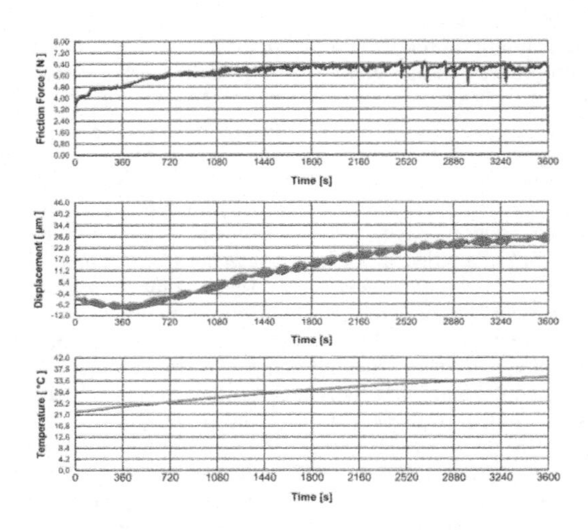

Figure 1. Friction results, FDM – material ABS.

Based on the results of previously published own works and on the basis of the knowledge of literature, it can be concluded that the PJM technology is the least optimal choice in terms of the application of models in the foundry industry. It happened that the test samples (in PJM) used to select the friction parameters were damaged, for example, at the end of the test cycle, which was not recorded for other models in SLS and FDM technologies.

The best tribological properties (the lowest coefficient of friction and the lowest linear wear) were determined for the samples made in the Selective Laser Sintering technology and the PA 2200 polyamide. It seems, however, that the measurement of ABS P 430 material wear shows the interesting data due to the slightly higher wear and values of the friction force at a much lower cost of sample production (3D printers in FDM technology are several times cheaper than for SLS technology, as well as ABS material compared to PA 2200).

Summing up, it should be stated that the optimal choice in the case of the production of casting models for short casting series is the FDM technology with satisfactory linear wear, not too high friction force and low cost of model production, including the possibility of the production of large-size models.

In the future, it is planned to conduct extensive research for the three 3D printing technologies, including the analysis of the influence of technological parameters of the process on the tribological properties of the models produced in terms of the application of the test results in the foundry industry.

The research presented in this paper was financed by the National Centre for Research and Developments part of the Lider XI project, number LIDER/44/0146/L-11/19/NCBR/2020, under the title "An analysis of application possibilities of the additive technologies to rapid fabrication of casting patterns".

REFERENCES

Adamczak, S. Zmarzły, P. Kozior, T. and Gogolewski D. 2017b, Assessment of roundness and waviness deviations of elements produced by selective laser sintering technology, *Engineering Mechanics 2017*.

Bochnia, J. and Blasiak, S. 2019, Fractional relaxation model of materials obtained with selective laser sintering technology, *Rapid Prototyping Journal*, 25, 76–86. doi: 10.1108/RPJ-11-2017-0236.

Madej, M. Ozimina, D. Kurzydłowski, K. Płociński, T. Wiecinski, P. Baranowicz, P. 2016, Diamond-like carbon coatings in biotribological applications, *Kovove Materialy*. doi: 10.4149/km20163185.

Pawlak, W. 2018, Wear and coefficient of friction of pla-graphite composite in 3D printing technology, *Engineering Mechanics 2018*. doi: 10.21495/91-8-649.

Rudnik, M. Hanon, M. Szot, W. Beck, K. Gogolewski, D. Zmarzly, P. Kozior, T. 2022, Tribological properties of medical material (MED610) used in 3D printing PJM technology, *Tehnički vjesnik/Technical Gazette*.

Upadhyay, M., Sivarupan, T. and El Mansori, M. 2017, 3D printing for rapid sand casting—A review, *Journal of Manufacturing Processes*, 29, 211–220. doi: 10.1016/j.jmapro.2017.07.017.

Current Perspectives and New Directions in Mechanics, Modelling and Design of Structural Systems – Zingoni (ed.)
© 2022 Copyright the Author(s), ISBN 978-1-032-39114-4

A case study of structural optimization for additive manufacturing

J. Hajnys, M. Pagac & J. Mesicek
Department of Machining, Assembly and Engineering Metrology, Faculty of Mechanical Engineering, VSB - Technical University of Ostrava, Ostrava, Czech Republic

P. Krpec
V-NASS , Ostrava , Czech Republic

ABSTRACT: The article deals with structural optimization and their simulation. The paper is divided into two sections, where the first is a simplified review of the current state of structural optimization with a focus on topological optimization and lattice structures. In the second part, a simulation of a selected component used in the automotive industry is performed. It is clear from the results that it is appropriate to use lattice structure to reduce weight and to use topological optimization to improve stiffness. The results are discussed and summarized in table.

1 INTRODUCTION

3D printing or additive manufacturing (AM) has been dynamically dominated in recent years and found its irreplaceable place not only for rapid prototyping, but also as an alternative to conventional production. Thanks to the freedom of modelling, designers are not constrained by conventional modeling rules as when they work with subtractive methods. Based on this fact, the design can be easily optimized in many ways, which is in line with the recent trend of reducing weight and saving material. In general, weight reduction can be realized in two ways, i.e., by topological optimization (TO) or by replacing the solid material with lattice structures. For these purposes, it is necessary to know what specific properties are to be achieved - stiffness, strength, dynamic stress or visual appearance.

2 CASE STUDY

For this article, the component which serves as a bracket in the automotive industry was chosen to demonstrate the optimization works, see Figure 1. This bracket was originally manufactured by using a 3-axis milling machine, where chips were gradually removed from the block of material until the final form is accomplished. Used material is AISI 316L stainless steel.

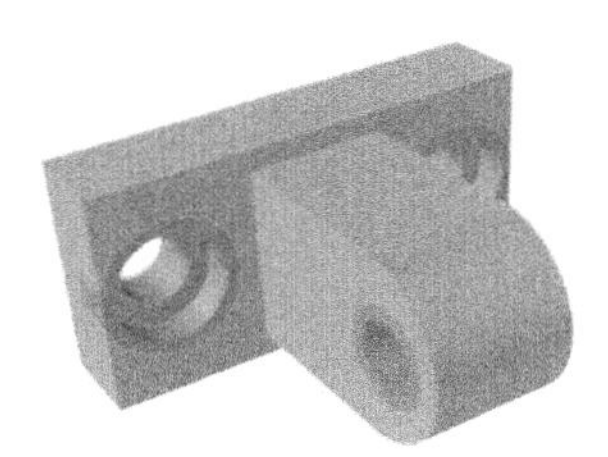

Figure 1. Bracket before optimization - default state.

2.1 *Topology optimization of bracket*

The commercial software SolidThinking Inspire, which uses the SIMP method, was chosen for the TO study. Firstly, the Design Area was created, which artificially created the maximum allowable volume. Then, to determine the areas that must remain as they are in this case the holes, Non-design Areas were created, see Figure 2, dark purple parts.

Figure 2. Topologically optimized bracket.

Followingly, corresponding boundary conditions had to be defined. Boundary conditions were defined based on forces ratio, torque moments and their distribution, which load the bracket. The minimum wall thickness was set at 10 mm. However, it was in a state where the model was not smoothed and contained various sharp edges. At this stage, it is possible to produce the part using AM, but it may happen that the printed part would fail during

DOI: 10.1201/9781003348450-71

operation due to numerous small peaks on the surfaces, which acts as stress concentrators. The Polynurbs function is used for smoothing, which tangentially connects individual peaks. After smoothing the surface, a stiffness analysis was performed. The results and TO data are shown in Table 1. Optimized bracket, with tangentially rounded shapes, and reduced number of yield stress concentrators is showed in Figure 2.

Table 1. Results of optimalizations.

Phase of shape	TO	LS	LS
Mass [kg]	1.196	0.936	0.187
Displacement [mm]	0.077	0.017	0.1
Von Mises Stress [MPa]	13	42	133

2.2 *Lattice structure optimization of bracket*

Solid-Thinking software was again used to create the cellular structure. This lattice structure (LS) software uses a hybrid lattice mesh representation. The result is the structure shown in Figure 3. As with this TO case, the goal was to maximize the stiffness. The setting of individual parameters was Target Length: 5 mm, Minimum Diameter: 0.5 mm and Maximum Diameter: 1 mm. When creating a lattice structure, it is not necessary to smooth the surface, as it depends on the setting of the primary cell, which is tetrahedral cell in this case. The only refinement which can be done with the resulting structure is remesh. Remeshing would yield individual struts with higher resolution in the final model. This is beneficial for printing and functionality of the part because smoother surfaces reduce or eliminate the risks of stress concentrators. However, it is necessary to take into account that the size of the final file computational operation in the case of editing (merging, adding supports, etc.) or slicing before printing in other software.

Figure 3. Bracket with tetrahedral lattice structure filling.

2.3 *Discussion and results of optimization bracket*

The optimization was performed for the AISI 316L corrosion-resistant steel material. In terms of weight reduction, the lattice structure with the lowest value of 0.187 kg is the best result, see Table 1. Furthermore, FEM analysis was performed for both the optimized parts. Considering both the Von Mises Stress and maximum displacement result, the TO model is better. In terms of computational time, the TO was relatively shorter because the mathematical operations for creating the lattice structures are more sophisticated. In general, it should be noted that the computational time depends on the power of the station and also the chosen optimization method.

3 CONCLUSION

This paper dealt with structural optimization for production using AM. In the first part of the article, a simplified review of TO and cellular structures is performed. For demonstration, an automotive part was optimized using both the TO method and with lattice structures. The objectives of the optimization for the selected part were to reduce the weight and increase the stiffness. The key knowledges of this paper can be summarized as follows:

- Structural optimizations are divided into size optimization, shape optimization and topological optimization.
- The goal of TO is to reduce the weight, and in particular load cases, increase the strength of the given parts.
- In comparison with TO, optimization with cellular (lattice) optimization is more efficient in weight reduction. However, the stiffness reduction of the optimized structure must be considered.
- The performed simulations confirmed that TO is preferred for stiffness improvement while LS is for weight reduction.
- The mentioned optimizations are highly applicable for production using AM.

The achieved results are from simulation, which could serve as inspiration for designers to Design for Additive Manufacturing (DfAM). Part fabrication and strength tests are the subjects of further research.

ACKNOWLEDGMENTS

The research was funded in association with project Innovative and additive manufacturing technology—new technological solutions for 3D printing of metals and composite materials, reg. no. CZ.02.1.01/0.0/0.0/17_049/0008407 financed by Structural Funds of the European Union and project.

6. Composite structures, Laminated structures, Sandwich structures, Adaptive structures, Numerical modelling, Numerical simulations

Current Perspectives and New Directions in Mechanics,
Modelling and Design of Structural Systems – Zingoni (ed.)
© 2022 Copyright the Author(s), *ISBN 978-1-032-39114-4*

Piezoresistive sensor matrix for adaptive structures

H. Fritz, C. Walther & M. Kraus
Bauhaus University of Weimar, Germany

ABSTRACT: Adaptability in structural engineering refers to adjusting the behavior and functionality of structural components with respect to changing conditions, such as varying loads. Corresponding optimization processes can be actively controlled by adaptive systems, whereby main components comprise actuators for manipulating the structural behavior and sensors for regulating the actuators. To date, such sensor measurements are mostly carried out locally. However, in adaption processes caused by time-varying loads, local measurements may lead to inaccuracies due to individual load phenomena in specific parts of structures. Limited attention has been paid on developing holistic, global monitoring systems for detecting and quantifying loads since a multitude of sensors is cost intensive, and the integration into the building envelope challenging. Based on previous work, this paper investigates slender piezoresistive pressure sensor matrices to quantify wind and snow loads. After a brief introduction into adaptive structures, a framework for large-scale monitoring systems based on sensor matrices is proposed. Subsequently, working principles and low-cost fabrication of piezoresistive sensor matrices are presented. Finally, the sensors are validated and applications as well as future work for practical implementation are discussed. In summary, the low-cost pressure sensor matrices appear promising to quantify and localize actions on structures globally and may offer a wide application range for adaptive structures.

1 INTRODUCTION

The importance of adaptive load-bearing structures increases as they represent a decisive step towards more sustainable buildings (Sobek et al., 2000). Amongst others, adaptive structures may be designed for time-varying, non-scheduled actions, such as wind or snow loads. To control adaptive structures reliably based on sensors and actuators, measuring time-varying actions for obtaining estimates of the real load situation is of great importance. To date, such sensors are usually integrated locally in or on structures, as large-scale integration is expensive and elaborate. However, individual local measurements may not adequately capture individual effects of actions, such as snow accumulations, snow drifts or wind turbulence distributions on surfaces. Nevertheless, imprecise load assumptions may be hazardous for reliably interpreting the current structural condition and controlling adaptive structures. This paper investigates force sensors to be applied in holistic monitoring schemes of adaptive structures for quantifying time-dependent loads resulting from wind or snow actions. Based on arrays of individual piezoresistive pressure sensors presented in a previous work (Fritz et al., 2022), flexible and slender sensor matrices are investigated.

2 PIEZORESISTIVE SENSOR MATRICES

The sensor matrices in the present work consist of copper strips as electrodes, arranged as active elements in i rows and j columns, see Figure 1.

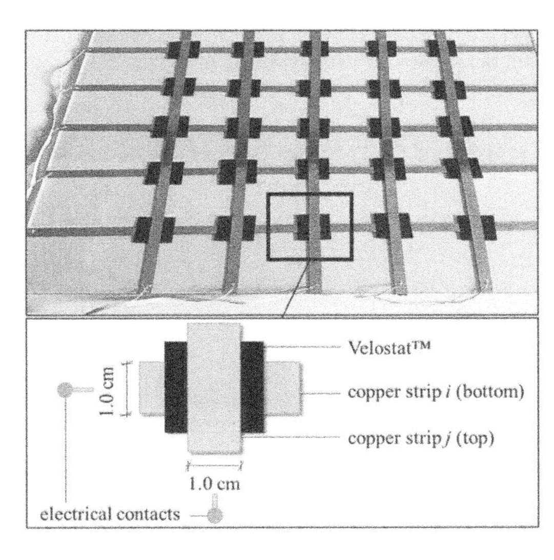

Figure 1. Prototype and principle structure of a sensor matrix containing 25 sensors.

DOI: 10.1201/9781003348450-72

The piezoresistive sensor matrices consist of the semiconducting polymer composite material "Velostat™", whose electrical resistance changes when pressure is applied (Ruschau et al., 1992). This change in resistance is measured by applying electrodes, attached on Velostat. The columns and rows are orthogonally assembled in matrix arrangement and Velostat is interposed between the copper strips at points of intersections, resulting in *i* x *j* sensors sensitive to pressure. For fabricating the prototype sensor matrix, squared active sensor areas of 1.0 cm² are obtained using copper strip widths of 1 cm. In total, the prototype fabricated consists of five rows and five columns of copper strips, resulting in 25 sensors (see Figure 1). Top and bottom (non-conductive) polystyrene plates frame the sensor matrix.

3 CALIBRATION AND VALIDATION

A mass-resistance relationship (i.e., transfer function, describing the change of the sensor resistance $R_{\mathrm{Vel},i,j}$ in relation to the analog (unknown) mass $m_{i,j}$) is derived during calibration by applying different masses. The best fitting transfer function is obtained using a logarithmic function of 6th order, yielding in the following relationship for a sensor at position *i* and *j*:

$$m_{i,j} = 10^{\sum_{n=0}^{6} 1000\, a_n \log_{10}\left(R_{\mathrm{Vel},i,j}\right)^n}, \qquad (1)$$

with $a_0 = 49.8$, $a_1 = -28.7$, $a_2 = 68.4$, $a_3 = -86.3$, $a_4 = 60.8$, $a_5 = -22.7$, and $a_6 = 3.5$. Applying the transfer function in Equation (1), the best approximations are obtained in the ranges of approx. 100 g to 1000 g (max. error of 11% – 8.0%). Outside this range, the measurements may be imprecise, i.e., small changes in the resistance $R_{\mathrm{Vel},i,j}$ lead to great changes in mass and vice versa. For validation, a mass of 4.8 kg is applied over six sensors. The load is introduced over 14 cm x 21 cm = 294 cm², resulting in pressures of 16.32 g/cm². The mass measured by all 25 sensors of the matrix with a sum of 4950 g agrees well to the applied mass of 4800 g.

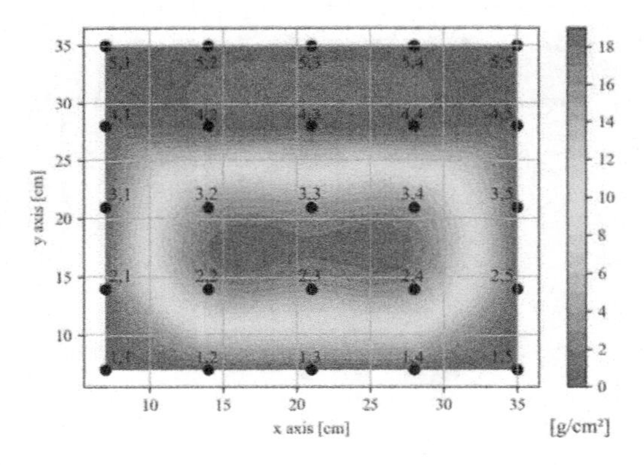

Figure 2. Pressure distribution obtained by the prototype sensor matrix under 4.8 kg load introduced over 294 cm².

In Figure 2, the measurements are divided by the load catchment area of one sensor (49 cm²) and interpolated between the sensors. As can be seen, the pressure is localized and quantified well.

4 SUMMARY AND CONCLUSIONS

For precisely controlling adaptive structures, this paper aims to quantify wind and snow loads through holistic large-scale monitoring systems. To this end, piezoresistive pressure sensors developed in a previous study are advanced to piezoresistive sensor matrices. In summary, the piezoresistive sensor matrices are suitable for localization and quantification of wind and snow loads. Due to the fabrication of sensors in matrix arrangement, the effort for multiple sensors and technical components are reduced, while allowing high spatial sensor resolution. Furthermore, through low-cost fabrication and slender geometries, large-scale integration into the building envelope is possible. The custom-made sensor matrix investigated is not yet feasible for practical applications due to not fully quantified time-dependent effects and the necessity of recalibrating individual sensors as a result of fabrication imperfections. Presupposing corresponding corrections of measurement uncertainties will offer to control structural components based on pressure sensor matrices integrated in building envelopes. For example, the sensor matrix can be employed as large-scale snow load monitoring system on roofs with heating elements or for adaptive membranes as pressure-sensitive skins with active bracing.

ACKNOWLEDGEMENTS

This research is supported by the European Social Fund (ESF) of the European Union and by the Thuringian Ministry for Economic Affairs, Science and Digital Society (TMWWDG) under Grant 2019 FGR 0096. This work is also supported by Carl Zeiss Foundation. The financial support is gratefully appreciated. Any opinions, findings, conclusions, or recommendations expressed in this paper do not necessarily reflect the views of the aforementioned institutions.

REFERENCES

Fritz, H., Walther, C. & Kraus, M. 2022. Piezoresistive sensors for monitoring actions on structures. In *Proceedings of the European Workshop on Structural Health Monitoring (EWSHM)*. Palermo, Italy, 07/07/2022 (submitted).

Ruschau, G. R., Yoshikawa, S. & Newnham R. E. 1992. Resistivities of conductive composites. *Journal of Applied Physics* 72(3): 953–959.

Sobek, W., Haase, W. & Teuffel, P. 2000. Adaptive Systeme. *Stahlbau* 69(7): 544–555.

Current Perspectives and New Directions in Mechanics, Modelling and Design of Structural Systems – Zingoni (ed.)
© 2022 Copyright the Author(s), ISBN 978-1-032-39114-4

Numerical modelling of the residual burst pressure of thick composite pressure vessels after low-velocity impact loading

R. Czichos & P. Middendorf
Institute of Aircraft Design, Stuttgart, Germany
University of Stuttgart, Stuttgart, Germany

T. Bergmann
Audi AG, Development Lightweight Construction/Body Structure, Neckarsulm, Germany

ABSTRACT: In order to reduce greenhouse gas emissions, sustainable drive systems with batteries or fuel cells are increasingly being used in electric vehicles. Especially with the latter, the energy is provided by compressed hydrogen, which is usually stored at 700 bar in type IV composite overwrapped pressure vessels (COPV). These structures are prone to damage (e.g. delamination) caused by dynamic loading such as low-velocity impact, which might occur during manufacturing, operation or maintenance. Since this type of damage has an influence on the residual burst pressure, a numerical method is presented to estimate this influence. Models with different levels of complexity are compared to show the advantages and disadvantages of the respective approaches for different impact energies and locations. The numerical methods are validated with experimental results, where a general good agreement can be seen especially for the detailed models.

1 INTRODUCTION

The climate change leads to an increased use of applications with sustainable energy sources. One of these is fuel cell technology, using hydrogen as an energy carrier, stored at high pressure or cryogenic temperature. Especially for the former, hydrogen is compressed up to 700 bar nominal working pressure (NWP) and stored in type IV composite overwrapped pressure vessels (COPV), which offer the advantage of lower weight and higher fatigue strength, resulting in higher operational lifetime compared to conventional pressure vessels of types I-III (Kim, Lee, & Hwang 2012). This is of particular interest for mobile applications, but composites are prone to damage caused by dynamic loading such as low-velocity impact (LVI), which might occur during manufacturing, operation or maintenance. Such impact events can have a critical influence on the residual burst pressure, which depends on several factors such as mass, velocity and shape of the impactor, impact position as well as internal pressure of the vessel during impact (Weerts, Cousigné, Kunze, Geers, & Remmers 2021, Choi 2017). The understanding of damage and failure during LVI loading is therefore of high interest and investigated within this work. From literature, a wide variety of modelling approaches are known and the present study will compare common approaches.

2 PRESSURE AND IMPACT LOADING

A type IV pressure vessel is investigated, whose structure usually consist of a plastic liner, two metal bosses and a composite laminate to withstand the NWP of 700 bar. The laminate is made of carbon fibre reinforced plastics (CFRP) with a thickness of approx. 25 mm. Since this no longer fulfils the assumption of a thin-walled structure, the influence of normal and inter-laminar stresses through the thickness cannot be neglected and three-dimensional finite elements are used. The simulations are performed in *Pam-Crash 14.5* and the vessel structure is built with *μWind v0.76b* from *MeFeX GmbH* (MeFex GmbH 2022), which uses a layer-by-layer calculation of the roving position and thus allows the stiffness in the cylindrical and dome region to be calculated with high precision. The material is a *T700* carbon fibre with an *Epon826* epoxy resin, which is suited for filament winding and the constitutive behaviour and progressive damage is modelled by a modified LADEVÈZE model. Fibre failure (FF) is modelled by a stress-based YAMADA SUN criterion, whereas inter-fibre failure (IFF) is calculated by Puck's failure criterion. Delamination is considered by an additional criterion between individual layers and described by a cohesive contact. In addition to the 3D solid model with around 3.5 mio. elements and delamination interfaces, a simple layered shell model is derived. Since it is less computationally

DOI: 10.1201/9781003348450-73

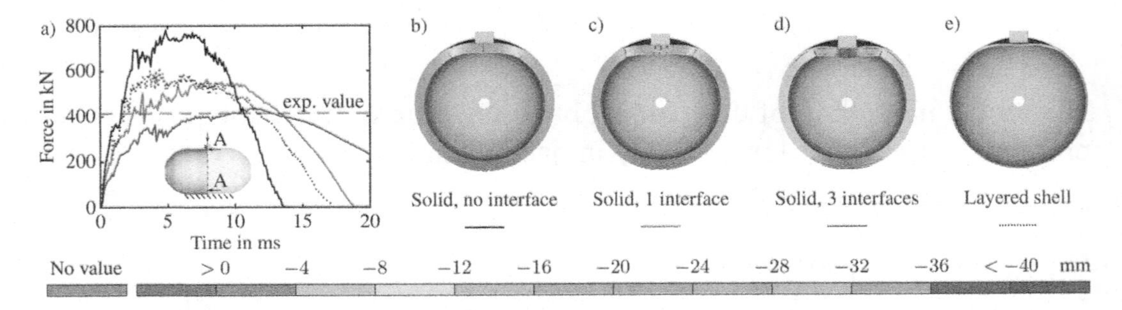

Figure 1. Results of a 12 kJ impact with internal pressure of 20 bar. a) Contact force of the stamp over time; b) - e): Displacement contour plots in downward stamp direction for section A-A: b) Solid with no interface; c) Solid with 1 interface; d) Solid with 3 interfaces and e) layered shell.

intensive than the solid model, it is well suited for usage in an overall system. The criterion to estimate the burst pressure is set equal to FF, as IFF might occur, but normally does not lead to final failure. The results are compared with experimental tests, where two identical vessels were tested and the results are

Table 1. Comparison of the burst pressure w/ and w/o impact.

	Experiment	Shell	Solid
w/o impact	1788 ± 38bar	1234bar	*1685bar
w/ impact	560 ± 53bar	N/A	**505bar

* w/o delamination interface; ** 3 delamination interfaces

listed in Table 1. It can be seen that the shell model underestimates the burst pressure, which can be attributed to the limited applicability of this model for thick laminates, as it assumes a plane stress state and leads to a stiffer behaviour. In contrast, the solid model gives quite accurate results with a minor underestimation of only 6%. In addition to the burst pressure, the vessel is also subjected to an impact load of 12 kJ, which can be seen in Figure 1. Due to confidentiality, the experimental force vs. time graph is not shown, instead the maximum value is used and indicated by the dashed line. The stiffest behaviour is seen for the solid model without interface, overestimating the reaction force by 90%. One delamination interface reduces the deviation to around 36% and similar results can be observed for the layered shell model, which is not capable to model delamination. The best results are obtained by the solid model with 3 delamination interfaces, overestimating the maximum force by only 5%. In Figure 1 b) - e) the y-displacement from section A-A are shown. Element deletion can be seen for the solid models under the impactor and in close proximity. These elements are predominantly found in the hoop layers, reaching the FF criterion due to the local bending stresses. From Figure 1 c) and d), it can be seen that delamination occurs in the model with only 1 interface, but this is predominantly found in the area around the impactor. Somewhat different results are seen for the model with 3 interfaces, where the greatest delamination is observed between the outer plies and decreases towards the inside. This is also consistent with the experimental observations. The residual burst pressure is estimated by applying temporally different loads and boundary conditions. The results are shown in Table 1, where the solid model shows a good agreement with the tests and underestimates the final burst pressure by only 10%. For the shell model excessive element deletion was observed, leading to an unfolding during pressurising, so no evaluation could be made for this approach.

3 CONCLUSION AND OUTLOOK

Different approaches for numerical modelling of burst and residual burst pressure of COPV related to LVI were compared. Models of varying complexity from layered shell to solid models with multiple delamination interfaces were investigated. The best results were achieved with solid models and 3 interfaces but with high computing times. In contrast, the shell approach showed a stiffer behaviour but still reasonable results with less computation time, which suits the use in an overall system. Upcoming work needs to modify the shell model behaviour to better match the solids deformation with delamination interfaces.

REFERENCES

Choi, I. H. (2017). Low-velocity impact response analysis of composite pressure vessel considering stiffness change due to cylinder stress. *Composite Structures 160*, 491–502.

Kim, E.-H., I. Lee, & T.-K. Hwang (2012). Low-velocity impact and residual burst-pressure analysis of cylindrical composite pressure vessels. *AIAA journal 50*, 2180–2193.

MeFex GmbH (2022). Konstuktion und Berechnung von Faser-Kunststoff-Verbund-Bauteilen - μWind. https://www.mefex.de/software/winding. [Online; accessed 07-January-2022].

Weerts, R. A., O. Cousigné, K. Kunze, M. G. Geers, & J. J. Remmers (2021). The initiation and progression of damage in composite overwrapped pressure vessels subjected to contact loads. *Journal of Reinforced Plastics and Composites 40*, 594–605.

Current Perspectives and New Directions in Mechanics, Modelling and Design of Structural Systems – Zingoni (ed.)
© 2022 Copyright the Author(s), ISBN 978-1-032-39114-4

Influence of interfacial debonding on the nonlinear structural response of profiled metal-faced insulating sandwich panels

M.N. Tahir & E. Hamed
Centre for Infrastructure Engineering and Safety, School of Civil and Environmental Engineering, The University of New South Wales, Sydney, Australia

ABSTRACT: This study aims to investigate the effects of debonding at the sheet-foam interface on the local buckling (wrinkling) and failure load of profiled metal-faced insulating sandwich panels. Two modeling approaches are adopted and compared in the analysis including full 3D finite element analysis and a simplified finite element approach. The results show that the debonding of the top face sheet has a substantial influence on the local buckling capacity of the panel. Both approaches agreed well with a maximum difference of around 15.5 %. The full 3D Finite element analysis results show a significant reduction in the failure load of the panel due to debonding.

Keywords: Elastic foundation, Finite element modeling, Interfacial debonding, Local buckling, Profiled sandwich panels

1 INTRODUCTION

Metal-faced insulating sandwich panels (MFISPs) typically consist of cold-formed thin steel faces which may be flat, lightly, or heavily profiled and an intermediate thick layer of the foam core like expanded polystyrene (EPS) and others. The debonding regions in MFISPs can be created at the interfaces due to imperfections in the manufacturing process, or damage induced during handling or shipping of the panels. This phenomenon can also occur due to thermal degradation of the bond between the sheet-core interface or by the core itself when they are used as a roof panel (Tahir & Hamed 2021). Interfacial debonding is very critical in such panels because it increases the local bending stresses and deformations in the face sheets and reduces the overall stiffness of the panel. This might lead to local buckling (wrinkling) of the face sheet at the debonded region under a smaller load magnitude as compared to fully bonded panels. The local buckling or wrinkling can be referred to as a rippling/dimpling of the compressed face sheet that usually occurs well below the yield stress of the face sheet (Carlsson 2011). It is evident from the previous studies that it is important to investigate the influence of debonding in profiled metal-faced sandwich panels.

2 NUMERICAL MODELLING

To demonstrate the influence of debonding at the top face-core interface on the nonlinear structural response in terms of local buckling and strength of the panel, a simply supported sandwich panel under a downward uniformly distributed load is considered here. Two debonded regions of length L_{deb} are considered near the central span at 200 mm apart.

In the simplified approach, only the top profiled face sheet is modelled as a folded plate resting on a two-parameter elastic foundation (closely spaced horizontal and vertical springs) that is loaded by interfacial shear at the bottom surface as shown in Figure 1 and can be approximated by (for details, see Tahir & Hamed (submitted) study);

$$\tau(x) = N_0 \frac{\pi}{bL} \cos\left(\frac{\pi x}{L}\right) \tag{1}$$

where b and L are the width and span of the panel, respectively and N_o is the peak axial force in the face sheet.

Following the modified nonlinear Vlasov & Leont'ev (1966) and Vallabhan & Das (1991) elastic foundation model, the vertical (k_w) and horizontal (k_u) spring constants take this form;

$$k_w = \frac{1.077E_c(1-\vartheta_c)}{(1+\vartheta_c)(1-2\vartheta_c)H} * S \tag{2}$$

$$k_u = \frac{0.516E_c}{(1+\vartheta_c)H} * S \tag{3}$$

where H is the thickness of the elastic foundation and it is taken as a center to center distance between the face sheets here; S is the distance between the springs in the FE mesh.

The modeling is performed using the finite element software ABAQUS. A strip of the folded plate of width 250 mm is modelled using a four-node shell element with reduced integration (S4R). The horizontal and vertical springs are modelled using element type SPRING1 which are discretely distributed along the bottom side of

DOI: 10.1201/9781003348450-74

the flat regions of the plate only. Accordingly, the interfacial shear is only applied at the flat region of the folded plate, see Figure 1. For the debonding regions, only vertical nonlinear springs are attached to simulate the contact effect between the face sheet and the core due to local bending and buckling. Their stiffness is taken as k_w for compression and as zero for tension. Linear eigenvalue analysis is performed first to obtain the critical buckling mode which is subsequently used in the geometric nonlinear analysis as an imperfection.

Also for the full 3D FE modelling, the finite element software ABAQUS is used. Four-node shell elements (S4R) are used to model the face sheets, whereas the foam core is modelled using 8-node linear brick elements (C3D8R). Interfaces of the foam core and face sheets are fully bonded using tie contact interaction type except the debonded regions, where the surface to surface contact is provided.

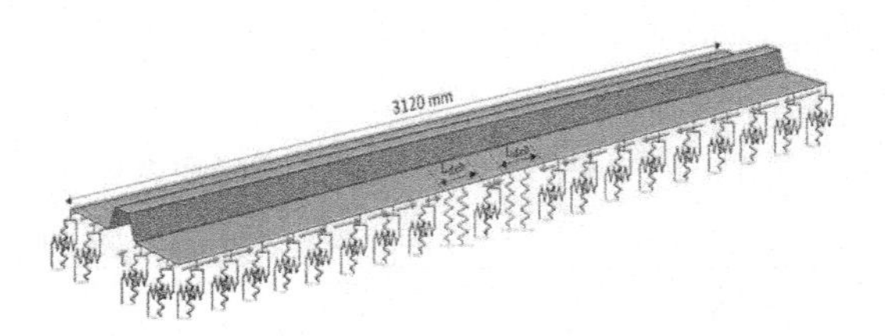

Figure 1. Conceptual simulation of simplified FE modelling approach.

3 RESULTS AND DISCUSSION

3.1 *Influence of debonding on local buckling pressure*

Three different cases are examined here, including no debonding regions and L_{deb} of 30 mm and 60 mm, to investigate the influence of debonding on local buckling pressure. Table 1 represents the results obtained from both approaches. It can be seen that the buckling pressure reduced appreciably (around 27%) with increasing the L_{deb}. The reasons for such substantial reductions in local buckling pressure are the decrease in stiffness of the face sheet and the increase in local bending stresses at the debonding regions. Both the approaches show a good correlation with a maximum absolute percentage difference of around 15.5 %.

Table 1. Effects of debonding at the profiled face-core interface on local buckling pressure.

	Local buckling pressure (KPa)		
Cases	Simplified Modelling	Full 3D FE Modelling	%Diff
No debonding	2.38	2.01	15.5
L_{deb}= 30 mm	1.8	1.85	2.16
L_{deb}= 60 mm	1.26	1.46	13.69

3.2 *Influence of debonding on the strength of the panel*

In order to determine the strength of the panel, both material and geometric nonlinearities are incorporated in the full 3D FE analysis. The elastic-perfectly-plastic response is assumed for the face sheet. Figure 2 represents the applied pressure versus displacement curves which show a consistent reduction of the failure pressure with increasing the debonding length. It can be seen that there is a negligible effect on the initial stiffness of the panel up to the debonding length of 60 mm. However, a substantial reduction in initial stiffness and failure pressure of the panel is observed at a higher value of debonding length i.e., 90 mm.

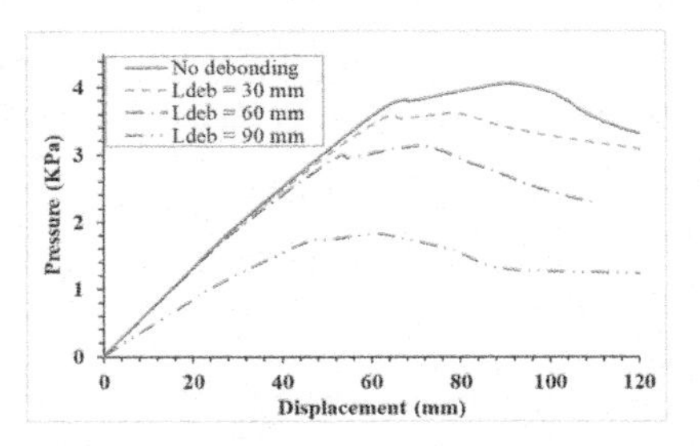

Figure 2. Effect of debonding at the face-core interface; Applied pressure versus displacement.

4 CONCLUSIONS

In this study, a nonlinear analysis is performed using two modeling approaches including 3D FE modeling and simplified FE approach, to investigate the influence of interfacial debonding on the local buckling of profiled MFISPs. The results show that debonding of the face sheet can significantly reduce the local buckling pressure and strength of the panel, which highlights the importance of considering this imperfection scenario in the analysis and design of MFISPs. In this investigation, a reduction of about 27 % and 23 % in the local buckling and failure load respectively, is predicted for a debonding of about 4 % of the panel.

REFERENCES

Carlsson, L.A. & Kardomateas, G.A. 2011. *Structural and failure mechanics of sandwich composites*. Dordrecht, USA: Springer.

Tahir, M.N. & Hamed, E. 2021. Effects of temperature and thermal cycles on the mechanical properties of expanded polystyrene foam. *Journal of Sandwich Structures & Materials* 10996362211063152.

Tahir, M.N. & Hamed, E. 2022 (Submitted). A simplified approach for local buckling (wrinkling) in metal-faced profiled sandwich panels.

Vlasov, V.Z. & Leonti'ev, N.N. 1966. *Beams, Plates, and Shells on Elastic Foundation*, Israel Program for Scientific Translation.

Vallabhan, G.C.V. & Das, Y.C. 1991. Modified Vlasov Model for Beams on Elastic Foundations. *Journal of Geotechnical Engineering* 117(6): 956–966.

Current Perspectives and New Directions in Mechanics, Modelling and Design of Structural Systems – Zingoni (ed.)
© 2022 Copyright the Author(s), *ISBN 978-1-032-39114-4*

Accurate numerical integration in 3D meshless peridynamic models

U. Galvanetto
Department of Industrial Engineering, University of Padova, Padova, Italy

F. Scabbia
Center of Studies and Activities for Space "G. Colombo" (CISAS), University of Padova, Padova, Italy

M. Zaccariotto
Department of Industrial Engineering, University of Padova, Padova, Italy

ABSTRACT: In mathematical terms, physical problems are often described by equations involving the computation of integrals, either because they appear in the weak form of differential equations or for the nature itself of the strong formulation. The recent theory of Peridynamics expresses the internal forces of mechanical problems as the integral of the nonlocal interactions of the material points. The numerical evaluation of those integrals strongly affects the accuracy of the solution of discretized peridynamic problems. The most common discretization of a peridynamic body is a meshless regular grid of points such that a cube is the volume associated to every point of the grid. Since the neighborhood of every (source) node is a sphere containing a large number of (family) nodes the intersection between the neighborhood and some of the cubic volumes associated to the family nodes is partial, and it is a complex function of the ratio between the horizon of the neighborhood (δ) and the grid spacing (h). Such a problem has affected the numerical solution of peridynamic problems since the first time Peridynamics was proposed. The paper presents the following two main results:

- the exact solution to the geometrical problem of computing the partial volumes generated by all possible combinations of δ/h;
- the application of the above exact evaluation to the numerical integration of peridynamic problems and the evaluation of the impact of the exact integration on the solution of structural problems

Several examples will illustrate various aspects of the newly proposed numerical integration techniques.

1 INTRODUCTION

Peridynamics is a non-local continuum theory devised to naturally model fracture in solid bodies (Silling 2000). The equations in the peridynamic formulation are based on integrals over spheric domains, named "neighborhoods", with a radius δ. Near the boundaries of the body some neighborhoods are not complete, which leads to undesired stiffness fluctuations, the so-called surface effect, and problems in imposing the boundary conditions. These issues can be solved as shown in the works of Scabbia et al. (2021) and Scabbia et al. (2022a).

The meshfree method with a uniform grid spacing h is the most commonly used for the discretization procedure: the body is subdivided into cubic cells with edges of length h. The quadrature weight β of a family node is defined as the fraction of cell volume which lies within the spheric neighborhood, as shown in Figure 1. The quadrature weights are commonly computed with approximated algorithms. However, Seleson (2014) showed that approximated algorithms may lead to inaccurate results and poor convergence behavior. Therefore, he proposed the first algorithm for the analytical computation of the quadrature weights in 2D peridynamic problems, which improves the accuracy of the numerical results and the convergence behavior. We propose an innovative algorithm capable of computing exactly the quadrature weights in a 3D peridynamic problem.

2 EXACT COMPUTATION OF THE QUADRATURE WEIGHTS

For reasons of cell-neighborhood symmetries, the number of quadrature weights to be computed can be reduced. We developed an integral formula for the computation of the quadrature weights β which is valid for each position of the cell in the neighborhood and for each ratio δ/h between the horizon size and the grid spacing (Scabbia et al. 2022b):

$$\beta = \frac{1}{h^3}\int_{a_x}^{b_x}\int_{a_y}^{b_y}\int_{a_z}^{b_z} \mathrm{d}z\,\mathrm{d}y\,\mathrm{d}x, \tag{1}$$

where the limits of the integrals are defined as follows:

$$a_x = \max(0, x_1),\, b_x = \min\left(x_2, \sqrt{\delta^2 - a_y{}^2 - z_1{}^2}\right),$$
$$a_y = \max(0, y_1),\ \ b_y = \min\left(y_2, \sqrt{\delta^2 - x^2 - z_1{}^2}\right),$$
$$a_z = z_1, \qquad b_z = \min\left(z_2, \sqrt{\delta^2 - x^2 - y^2}\right),$$

where x_1, x_2, y_1, y_2, z_1 and z_2 are the coordinates of the faces of the cubic cell. The vertices of the cubic

DOI: 10.1201/9781003348450-75

cell that lie within the neighborhood determine the different possible cases, as shown in Figure 2.

The integrals in Equation 1 should be split into intervals which have the same integrand. Some of those integrands do not have an explicit solution, but we can perform a Taylor series expansion of them and then integrate the polynomials. In this way, we can compute exactly the quadrature weights when the truncation order of the Taylor expansions is chosen to be 20 or higher.

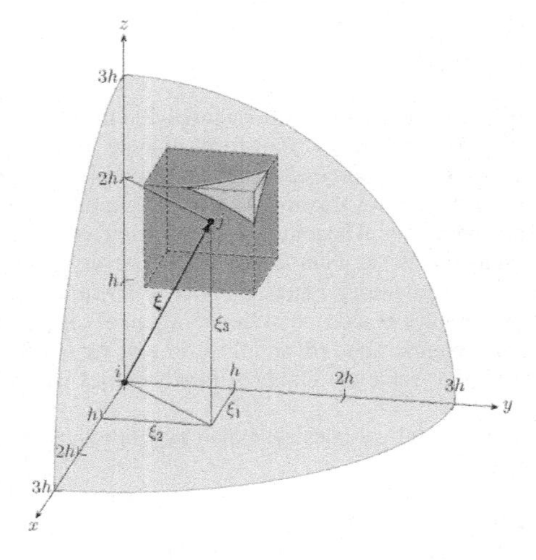

Figure 1. The quadrature weight of node j is the fraction of cell volume (in green) lying within the spheric neighborhood of node i (in gray).

3 CONCLUSIONS

The different possible cases of intersections between the cubic cells and the spheric neighborhood in 3D Peridynamics make the computation of the quadrature weights rather complex. We propose a novel algorithm to compute exactly the quadrature weights. This improves by a great extent the accuracy of the peridynamic numerical results and the convergence behavior under grid refinement.

ACKNOWLEDGEMENTS

The authors would like to acknowledge the support they received from the Italian Ministry of University and Research under the PRIN 2017 research project "DEVISU" (2017ZX9X4K) and from University of Padova under the research projects BIRD2018 NR.183703/18 and BIRD2020 NR.202824/20.

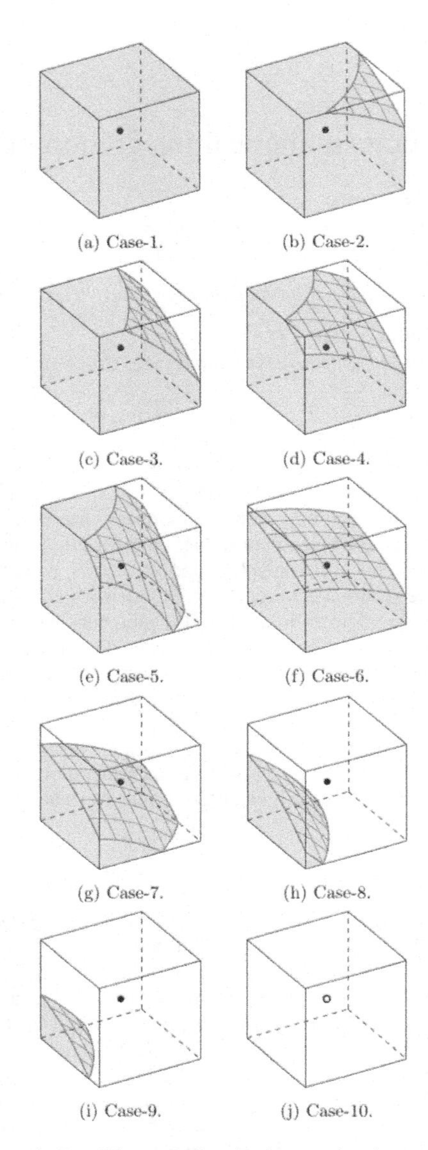

Figure 2. Possible cases for the intersection between the spheric neighborhood and the cubic sphere. The same cases can be found also when considering only one half or one fourth of the cell for symmetry reasons.

REFERENCES

Scabbia, F., Zaccariotto, M. & Galvanetto, U. 2021. A novel and effective way to impose boundary conditions and to mitigate the surface effect in state-based Peridynamics. *International Journal of Numerical Methods in Engineering*.

Scabbia, F., Zaccariotto, M. & Galvanetto, U. 2022a. A new method based on Taylor expansion and nearest-node strategy to impose Dirichlet and Neumann boundary conditions in ordinary state-based Peridynamics. *Submitted*.

Scabbia, F., Zaccariotto, M. & Galvanetto, U. 2022b. Exact computation of the quadrature weights in 3D peridynamics. *Submitted*.

Seleson, P. 2014. Improved one-point quadrature algorithms for two-dimensional peridynamic models based on analytical computations. *Computer Methods in Applied Mechanics and Engineering* 282: 184–217.

Silling, S.A. 2000. Reformulation of elasticity theory for discontinuities and long-range forces. *Journal of the Mechanics and the Physics of Solids* 48(1): 175–209.

Current Perspectives and New Directions in Mechanics, Modelling and Design of Structural Systems – Zingoni (ed.)
© 2022 Copyright the Author(s), ISBN 978-1-032-39114-4

Phase field topology optimization of elasto-plastic contact problems with friction

A. Myśliński
Systems Research Institute, Warsaw, Poland

ABSTRACT: This work aims to solve numerically the topology optimization problem for two bodies in bilateral frictional contact. The incremental elasto-plastic material model with linear kinematic hardening rather than pure elastic material model is assumed. The contact phenomenon is governed by the system of the coupled variational inequalities in terms of the displacement and the generalized stress. The structural optimization problem consists in finding such material distribution of the body in contact in terms of phase field function to minimize maximal contact stress and to ensure the uniform distribution of this stress. The Lagrange multiplier technique is used to formulate the set of necessary optimality conditions. Generalized Newton method combined with the phase field method are used to solve numerically this optimization problem. The results of computation are provided and discussed. The obtained results indicate that structures designed with accounting for plastic deformation have a reinforced area where plastic deformation occurs.

1 INTRODUCTION

Contact problems appear in many different fields of engineering including earthquake or civil engineering or machine dynamics. Therefore modeling of contact processes is an important topic which is currently still under investigation. Mathematical models describing the frictional or frictionless contact phenomenon between a deformable body and an obstacle, the so-called foundation, have been considered in many works.

In many applications the areas where the contact occurs are very small It implies that the transmitted contact force densities are usually rather big in the contact zone and it leads to plastic deformations. Therefore it is reasonable to consider an elasto-plastic rather than elastic model for the material. The high contact stress may also lead to undesired vibrations or the degradation of surfaces of the contacting bodies as well as the deterioration of the working conditions for employees. The reduction of high contact stress or the obtaining the uniform distribution of this stress is the main aim of the topology or shape optimization problems for bodies in contact. The topology optimization consists in such distribution of the material filling the structure within the design domain to minimize the given cost functional describing the required features of the structure. Most research related to topology optimization has been concerned with linear elastic structures. The amount of papers dealing with non-linear elastic structures or nonlinear mechanical behavior is limited.

The aim of this work is to analyze and to solve numerically the shape and topology optimization problem for two bodies in contact assuming static elasto-plastic material model with linear kinematic hardening rather than elastic material model. The optimization problem consists in finding such material distribution inside the domain occupied by the body in contact to minimize the contact stress. In the paper penalization approach will be used to regularize the structural optimization problem. Moreover the phase field approach is used to solve it numerically. The small strain plasticity material model is used. Numerical results are provided and discussed.

2 STRUCTURAL OPTIMIZATION PROBLEM

Consider the bilateral contact with friction between the elasto-plastic body occupying two dimensional domain and the rigid foundation. The body is loaded by body and boundary forces. The Tresca friction condition is imposed on the contact boundary. The elasto-plastic deformation of the body in contact is described by the additive small strain plasticity model with the linear kinematic hardening. The yield function is given as Von Mises function. The maximal plastic work principle for the generalized stresses is assumed to hold. The displacement and the generalized stress of the body are governed by the system of the coupled variational inequalities.

This nonlinear system is difficult for numerical modelling due to its non-differentiability. For the

DOI: 10.1201/9781003348450-76

sake of its analysis and numerical solution we shall regularize and transform it into the system of two nonlinear coupled equations depending on the regularization parameter. The regularization is based on the penalization of the plasticity condition and smoothing of the projection operator on the set of the generalized stresses. Moreover the friction functional has been regularized using smooth function approximating tangent displacement.

The material density function at any generic point in a design domain has been chosen as a design variable. It is a phase field variable taking value close to 1 in the presence of solid material, while value 0 corresponds to regions of the design domain where there is a void, i.e. the material is absent. The cost functional consists from two terms. The first term is dependent on the stress and approximates the normal contact stress. The second term is Ginzburg-Landau free energy functional depending on the phase field variable only. This term bounds the perimeter of the design domain and penalizes intermediary values of the material density inside the design domain. The volume constraint is imposed. The topology optimization problem consists in finding such material density function determining the material distribution in the domain occupied by the body in contact to minimize the contact stress.

3 GRADIENT FLOW EQUATION

The Lagrangian approach is used to formulate the necessary optimality condition for the structural optimization problem. The gradient of the cost functional has been calculated. This gradient depends on the design variable as well as on the solutions to the state and the adjoint equations.

In order to find the optimal material density function and topology domain we shall use the gradient flow equation rather than the necessary optimality condition. Assuming that the material density function is dependent on the artificial time variable the time evolution of this function is governed by the minus gradient of the cost functional with respect to the design variable. Gradient flow equation takes the form of the fourth order modified Cahn-Hilliard equation. The stationary solutions to this equation are being calculated, i.e., such that the time derivative of the material density function is equal to zero. By using the gradient flow method, we can accurately capture the trajectory leading to the optimal solution satisfying the necessary optimality condition. Moreover the solutions calculated iteratively possesses the property that the cost functional value is decreasing with time.

The four order gradient flow equation is splitted into two second order equations. The bilinear quadrangular finite element method is used. Generalized Newton method is used to solve numerically state and adjoint equations.

4 NUMERICAL RESULTS

Figure 1 presents the optimal topology domain obtained by solving structural optimization problem. In a case of elasto-plastic materials the mass of the structure is larger than in a case of elastic model. The algorithm tries to avoid the generation of plastic zones which are less rigid and induce larger displacements.

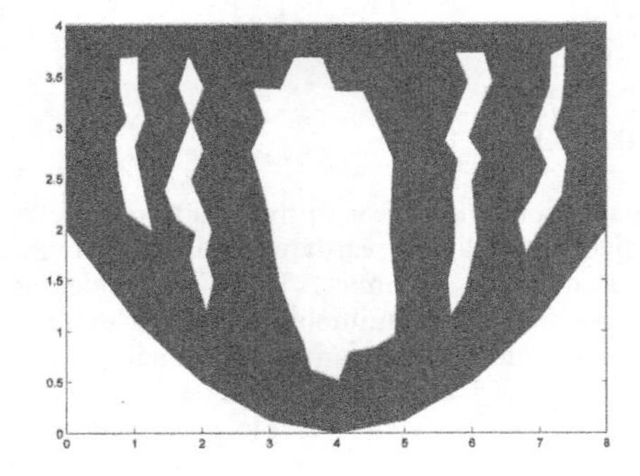

Figure 1. Optimal topology domain.

5 CONCLUSIONS

The obtained results indicate that presented approach based on the application of the phase field technique can be applied to solve numerically a topology optimization problem for bodies in bilateral frictional contact where nonlinear small strain elasto-plastic with linear kinematic hardening material model rather than elastic material model is used. It is capable of finding topologies that generates minimum contact stress. The obtained normal contact stress is almost constant along the optimal shape boundary and has been significantly reduced comparing to the initial one.

Current Perspectives and New Directions in Mechanics, Modelling and Design of Structural Systems – Zingoni (ed.)
© 2022 Copyright the Author(s), *ISBN 978-1-032-39114-4*

Computational analysis of magma-driven dike and sill formation mechanisms

J.A.L. Napier
Department of Mining Engineering, University of Pretoria, South Africa

E. Detournay
Department of Civil, Environmental, and Geo- Engineering, University of Minnesota, USA

ABSTRACT: The paper describes the numerical simulation of a vertically propagating dike sheet and the subsequent intersection and fluid branching of magma into a sill. This study represents a preliminary exploration of the intersection problem using a uniform fixed grid mesh, which allows some insights to be gained into the field stress conditions that control the dike propagation through the plane of weakness or which can lead to the deflection of the magma flow and subsequent sill structure formation. An approximate method is introduced to represent the crack tip opening shape assuming zero toughness at the dike edges.

1 INTRODUCTION

The present paper describes an initial attempt to simulate the intersection of a magma-filled vertical dike with a horizontal sill structure. This problem imposes an immediate requirement to construct a generalised edge tracking procedure that can accommodate the intersection line between multiple flow branches. A second requirement is to introduce a numerical scheme to evaluate the flow splitting partition at the intersection line.

2 NUMERICAL METHOD

In order to address these considerations the flow simulations are carried out using a square element solution mesh with an adjustment to the self-effect stiffness values in the displacement discontinuity elements that are adjacent or close to the fluid front. This provides an approximate method to account for the tip asymptotic. In particular for the viscosity-dominated regime, the correction of the self-effect stiffness avoids having to impose a velocity-dependent crack tip shape expansion.

3 DIKE PROPAGATION WITH NO SILL INTERSECTION

The basic flow source is assumed to be driven from an approximate representation of a magma chamber which is modelled as a vertical fluid-filled rectangular crack having a height of 200 m and a width of 440 m. It is assumed in all cases that the source region has a net fluid pressure of 5 MPa and that the simulations are carried out using regular square element meshes having a grid size of 40 m. The primitive (virgin) horizontal field stress is $\sigma_{zz}^{V} = -70 + 0.030y$ MPa.

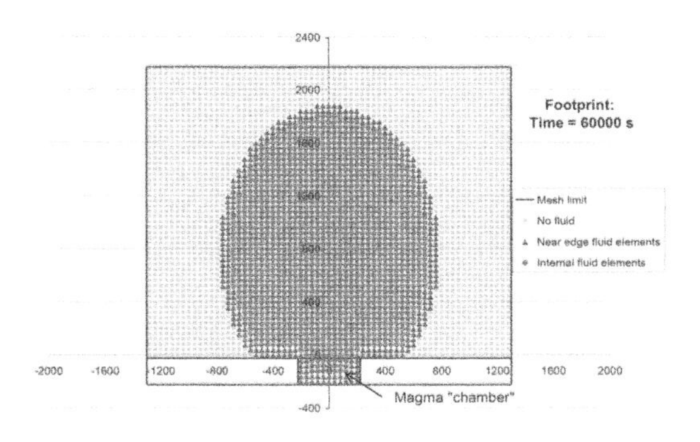

Figure 1. Dike extent after a flow period of 60000 s with a constant net pressure flow source of 5.0 MPa.

Figure 1 shows the evolved pattern of fluid-filled elements in the vertical dike structure after 600 time steps representing a total flow time of 60000 s. The fluid edge elements with adjusted self-effect stiffness values which are adjacent to the unfilled region are highlighted. The calculated flow width profiles along the vertical centreline are plotted as a function of the distance from the flow source at the top edge of the magma "chamber" in Figure 2 at time intervals of 10000s. The flow width profiles resemble the plane strain shape results, characterized by a leading edge "bulge".

DOI: 10.1201/9781003348450-77

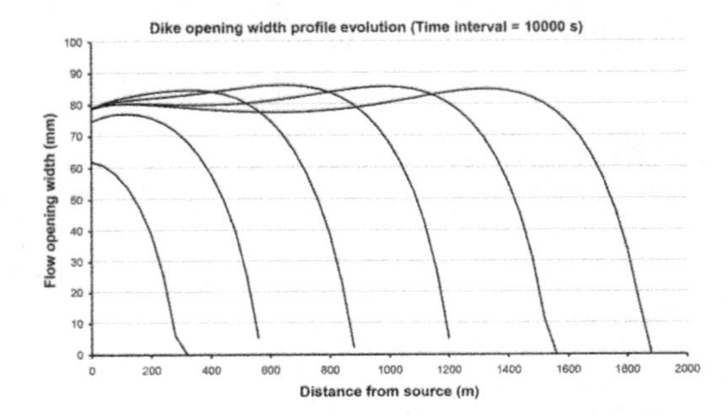

Figure 2. Flow width evolution.

4 SILL INTERSECTION

The sill structure is modelled as a horizontal plane of weakness that is located at the y-coordinate value of 1020 m. Three cases were considered with different magnitudes of the vertical field stress. In the first case (a) the vertical stress variation σ_{yy}^V is assumed to be equal to the horizontal stress σ_{zz}^V used in the previous problem. In case (b) the vertical field stress is $\sigma_{yy}^V = \sigma_{zz}^V - 5\text{MPa}$ and in case (c) $\sigma_{yy}^V = \sigma_{zz}^V - 10\text{MPa}$.

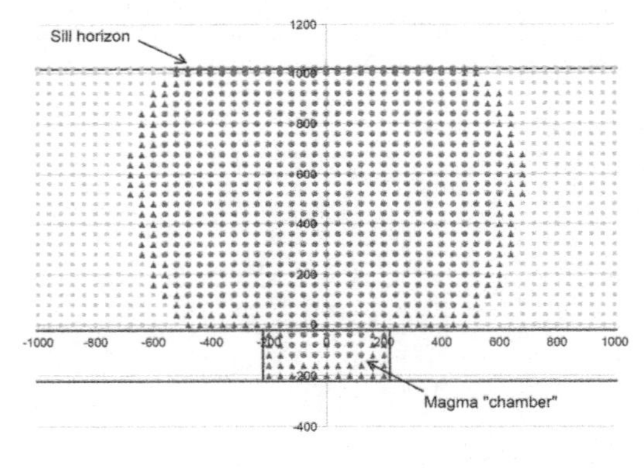

Figure 3. Fluid-filled elements on a vertical dyke intersection of a sill structure nominally 1000 m above the flow source. (The vertical field stress is 5 MPa lower than the horizontal confining stress).

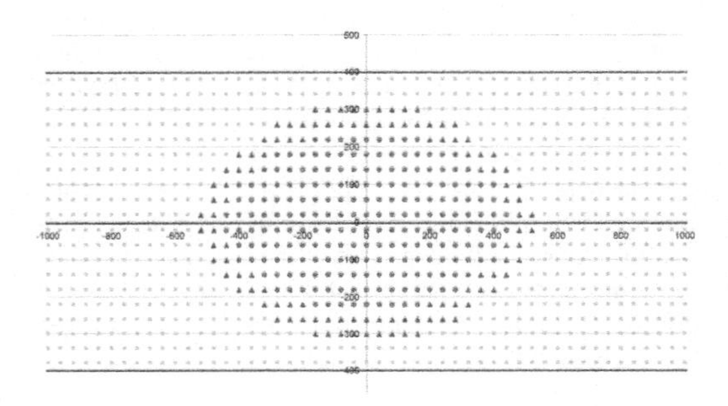

Figure 4. Fluid-filled elements on sill plane at time 50000 s.

The evolved pattern of fluid-filled elements on the dike plane at the elapsed time of 50000 s is shown in Figure 3 for case (b) where the vertical field stress is 5 MPa lower than the horizontal field stress. The intersection front with the sill plane is 1080 m wide in this case. The corresponding fluid footprint on the sill plane is shown in Figure 4.

The effect of varying the vertical stress on the fluid opening width profile is shown in Figure 5. The opening width is plotted along the vertical dike centreline and across one half of the minor axis centreline on the sill plane after an elapsed fluid propagation time of 50000 s. The gaps in the curves correspond to the transition from the vertical path on the dike centreline to one half of the sill minor axis.

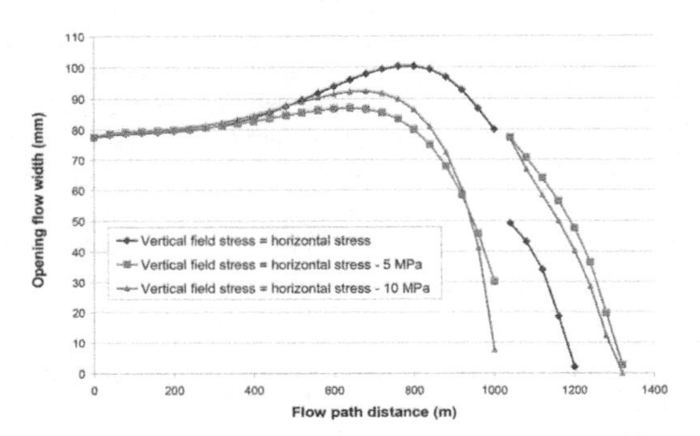

Figure 5. Opening flow width profiles for different vertical stress conditions.

It is interesting to note the discontinuous variation in the fluid opening width at the dike-sill intersection point as the vertical field stress is varied. In case (a) where the field stress is uniform it is observed that the opening width is reduced as the fluid flow is split into each half of the sill plane on each side of the dike plane. Conversely, when the vertical stress is lower than the horizontal stress, the fluid opening width increases discontinuously across the intersection line and exhibits a "pinching" effect in the dike opening width immediately below the sill plane.

5 CONCLUSIONS

A computational scheme using displacement discontinuity boundary elements has been devised to simulate flow branching in a fixed element grid mesh. A simple approximation has been implemented to represent the element self-effect stiffness that can be used to simulate viscous controlled fluid-driven crack propagation.

The flow algorithm has been applied to the case of vertical dike propagation and intersection with a horizontal sill plane. Some initial explorations of the effect of varying the stress field on the dike-sill intersection flow characteristics have been carried out indicating that this can affect the volume flow partitioning between the dike structure and the sill significantly.

Current Perspectives and New Directions in Mechanics, Modelling and Design of Structural Systems – Zingoni (ed.)
© 2022 Copyright the Author(s), ISBN 978-1-032-39114-4

Numerical analysis of a high-rise building near sensitive infrastructures

Matthias Seip & Mohamed Hassan
Ingenieursozietaet Professor Dr.-Ing. Katzenbach GmbH, Germany

Steffen Leppla
Frankfurt University of Applied Sciences, Germany

ABSTRACT: A new high-rise building is planned in the east of Frankfurt am Main, Germany. This new building is located directly between the river Main and the eastern inner harbour. The high-rise building will have a height of about 75 m and will be founded on a Combined Pile-Raft Foundation (CPRF) in the tertiary, soft marl. The new structure will be constructed next to important, sensitive traffic infrastructure and adjoins directly existing historic quay structures, which are more than 100 years old. For the analysis of the stability and the serviceability of the heavily eccentric loaded foundation of the new high-rise building and the existing structures and for the investigation of the soil-structure interaction complex three-dimensional, non-linear numerical simulations have been carried out. These numerical simulations take into account all geometric and material specific boundary conditions, including the history of the construction of the whole area. The paper introduces a very complex construction project in which the realistic consideration of the soil-structure interaction is very important for the serviceability of the foundation of the planned high-rise building as well as for the surrounding infrastructures. The three-dimensional numerical simulations are explained, and the results are shown.

1 INTRODUCTION

A new high-rise building is planned in the east of Frankfurt am Main, Germany. This new building is located directly between the river Main and the eastern harbour. The construction site is shown in Figure 1. The high-rise building will be a 75 m high hotel with 22 floors (Figure 2). Under the roundabout intersection two levels will be constructed for building equipment and appliances, for the entrance and for parking.

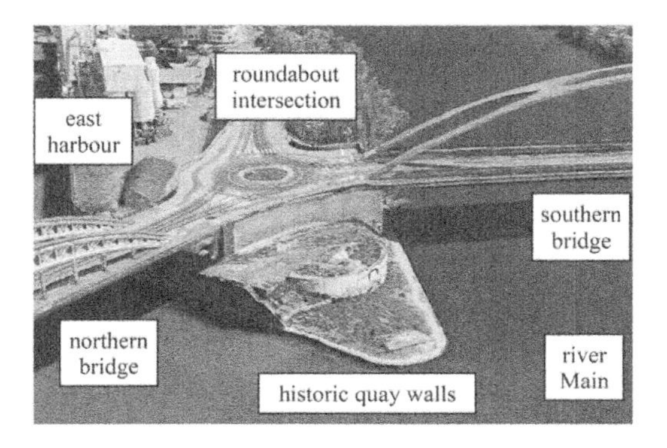

Figure 1. Construction site (google earth).

The vehicle access will be facilitated under the southern bridge along the river Main by a ramp.

Figure 2. Visualisation of the planned high-rise building (B&L Group, Hamburg, Germany).

The high-rise building will be founded on a Combined Pile-Raft Foundation (CPRF). The load of the superstructure is transferred into the CPRF via walls and columns and via a large single column on the tip of the triangular raft. The total load of the high-rise building including the foundation raft is about 507 MN. The load has no symmetric distribution and is explicitly eccentric. According to the regulations the whole project is classified into the Geotechnical Category GC 3, which is the category for the most complex construction projects.

DOI: 10.1201/9781003348450-78

2 SUBSOIL CONDITIONS

Mainly by two core drillings the subsoil was investigated with the following results:

- artificial filling (0.7 m to 16 m thick)
- quaternary alluvial clay } 0.6 m to 3.1 m thick
- quaternary sand and gravel }
- tertiary marl (minimum 80 m thick)

3 FOUNDATION OF HIGH-RISE BUILDING

The foundation of the high-rise building is planned as a CPRF (ISSMGE 2013). Due to the interaction between the foundation elements and the subsoil CPRFs have a very complex bearing and deformation behaviour. Hence CPRFs have to be classified into the Geotechnical Category GC3 according to EC 7. More explanations about CPRF and several examples from engineering practice are given in Katzenbach et al. (2016).

As a result of the comprehensive numerical analysis using the Finite-Element-Method (FEM) the CPRF consists of the 2 m thick raft and of 34 piles. The piles have a diameter of 1.5 m. The length of the piles is 40 m to 45 m (Figure 3).

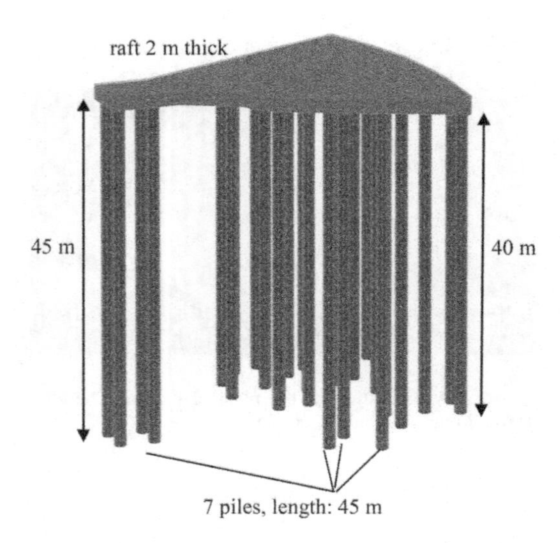

Figure 3. FE-model of the CPRF.

4 NUMERICAL CALCULATIONS

The 3-dimensional FE-model (cutout) of the project is shown in Figure 4. The calculated settlements are shown in Figure 5.

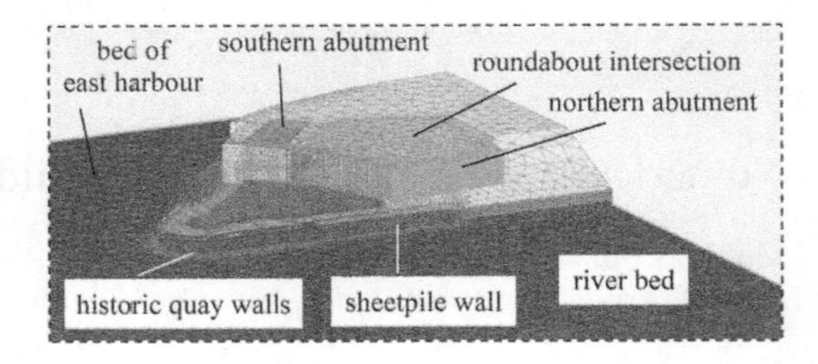

Figure 4. FE-model of the project (cutout).

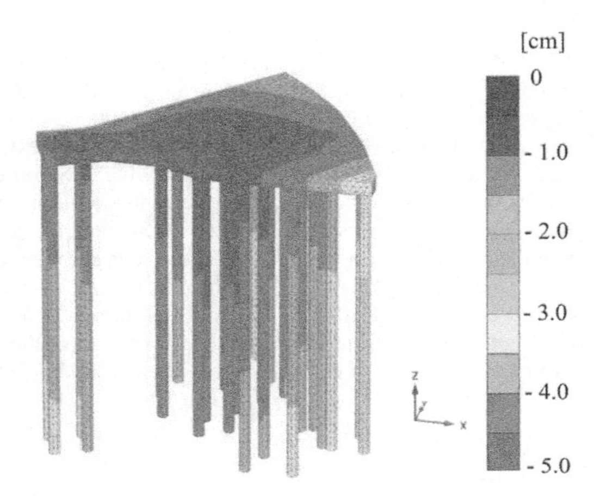

Figure 5. Calculated settlements of the CPRF for ($G_{tot,k}$ + Q_{tot}).

5 CONCLUSIONS

For analysis of the stability and the serviceability of the heavily eccentric loaded foundation of the new high-rise building and the existing structures and for the investigation of the soil-structure interaction complex three dimensional, non-linear numerical simulations are very important. These numerical simulations are to take into account all geometric and material specific boundary conditions, including the history of the construction of the whole area.

REFERENCES

ISSMGE International Society of Soil Mechanics and Geotechnical Engineering. 2013. *Combined Pile-Raft Foundation Guideline*. R. Katzenbach, D. Choudhury (ed). Darmstadt, Germany.

Katzenbach, R., Leppla, S., Choudhury, D. 2016. *Foundation systems for high-rise structures*. CRC Press Taylor & Francis Group, New York, USA.

Current Perspectives and New Directions in Mechanics, Modelling and Design of Structural Systems – Zingoni (ed.)
© 2022 Copyright the Author(s), *ISBN 978-1-032-39114-4*

Moment support interaction of multi-span sandwich panels

A. Engel & J. Lange
Institute for Steel Construction and Materials Mechanics, Technical University Darmstadt, Germany

ABSTRACT: Self-supporting double skin metal faced sandwich panels are usually designed and installed as multi-span beams. In this case, there is always an interaction at the intermediate supports between the compressional normal force in the face sheet and the support reaction orthogonally to it, which can reduce the bending capacity. Several effects influencing the reduced load-bearing capacity are not considered in current European approval practice (and design). Within a research project funded by the DFG, experimental, numerical and analytical investigations were therefore carried out on sandwich panels. The main influences on the reduced wrinkling stress under an applied transverse compressive load were quantified and a possible solution was identified in order to take them into account in the future, e.g. within the scope of the approval tests.

1 INTRODUCTION

The sandwich panels used in the construction industry for roof and wall cladding are predominantly installed as multi-span beams. The local indentations in the face sheet resulting from the support reaction at the intermediate supports can cause a reduction of the load bearing capacity (wrinkling stress) there. Several effects influencing the reduced wrinkling stress are not considered in current approval practice according to EN 14509 (see Nelke 2018). There, the behaviour at the intermediate support of the multi-span system is approximated by a single-span beam. Its length L_{SB} (SB-length or M/R-ratio between the support moment M and support reaction R) is decisive for the result.

Within a research project, various influences on the reduced wrinkling stress, such as the SB-length and load width, were to be quantified and transferred to an interaction model. The investigations were limited to the load case "transverse compression" and to panels with PU foam core.

Material and full-scale tests were carried out on wall and roof panels and supplemented by numerical parameter studies with FE models (also see Engel 2021).

2 MATERIAL & COMPONENT TESTS

Material tests according to EN 14509 were carried out to determine the properties of the PU foam core in the specific directions. The core showed a strong orthotropic behaviour. Complementary tests with a system for optical strain measurement were performed to investigate the inhomogeneous stiffness distribution over the core depth due to the continuous production process. The results showed a distinctive inhomogeneity of the foam core, depending on both panel depth and manufacturer.

Simulated central support tests were carried out with different spans and load application widths L_S. The wrinkling stress increases nonlinearly with increasing M/R-ratio (Figure 1) - for large SB-lengths, the normal force in the face sheet is already high at comparably small transverse loads and indentations - and converges towards the undisturbed wrinkling stress. The stability failure of the embedded face sheet dominates. At low M/R-ratios, the normal force is rather small and the core indentation correspondingly big. Here, the load-bearing capacity is limited by the plasticization of the core material or the face sheet. Furthermore, it can be seen that the wrinkling stress increases with increasing load width. Investigations on the local indentation under the load showed that it is reduced with increasing load width, but is partly compensated by local edge compression appearing at wide supports.

3 NUMERICAL INVESTIGATIONS

Ultimate load calculations were performed on a 3-D beam model considering geometric and physical nonlinearities, as well as a local imperfection of the system with the first wrinkling eigenmode.

The investigations showed that the reduced wrinkling stress is mainly influenced by the N/R-ratio between the compressive normal force in the face sheet and the transverse load R. The effect of the load width can be big, but is limited because of the edge compression that occurs with increased load

DOI: 10.1201/9781003348450-79

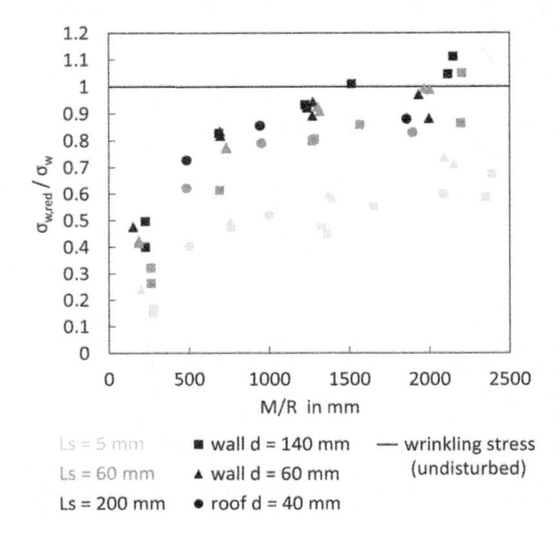

Figure 1. Relative reduced wrinkling stress (mean value of about 3 tests each).

width. This effect depends on the ratio L_S/d_C and should be considered for $L_S/d_C \geq 1.5$.

Further investigations on the core strength and the applied imperfection allowed conclusions about the failure mechanism. For big N/R-ratios a stability failure of the compressed face sheet dominates, while for small N/R-ratios stress failure of the core or face sheet occurs due to plasticization.

4 INTERACTION RELATIONSHIP AND SUMMARY

A practical experimental approach for determining the reduced wrinkling stress will be presented in the following.

If the reduced wrinkling stress is plotted against the ratio M/R, the curves differ regarding the component depth and the load width. These differences can be approximately calculated by normalizing the abscissa with Equation 1.

$$\theta = \frac{N}{R}\sqrt{L_S^*} = \frac{N}{R}\sqrt{\min\begin{cases} L_S \\ 1,5\ d_C \end{cases}} \quad (1)$$

In order to take the edge compression into account, the equation contains a limitation of L_S/d_C to 1.5.

Between the relative reduced wrinkling stress $\sigma_{w,red}/\sigma_w$ and the parameter $1/\theta$, an almost linear relationship can be assumed using a logarithmic representation, as shown in Figure 2. For better manageability, the function values of $1/\theta$ were multiplied by a factor of 100. The results are taken from the numerical parameter study on sandwich panels with flat face sheets and an imperfection of e_0 = 0,03 mm.

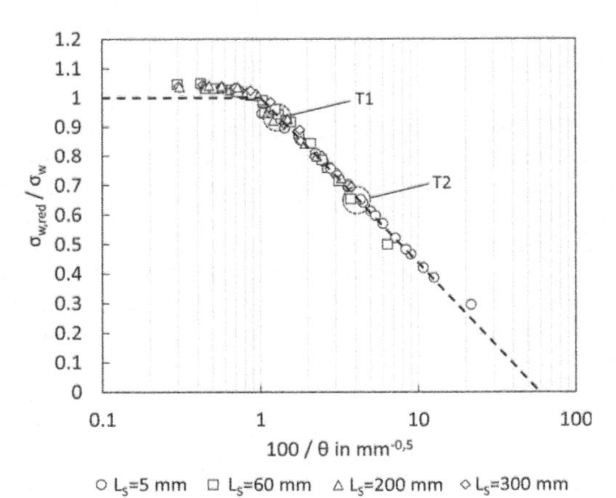

Figure 2. Interaction between $\sigma_{w,red}$ and θ.

The interaction relationship provides accurate results for most of the parameter range. The interaction curve can be determined, for example, by two simulated central support tests (T1 and T2, Figure 2) with different spans, whereby it must be ensured that both tests lie within the linear range shown. As a span for the first point (T1) of the interaction, the length suggested in EN 14509 for the wrinkling stress tests provides a first reference, for the second point (T2) half the length - both tests with L_S = 60 mm load width each.

In summary, the main influences on the reduced wrinkle stress were identified and a practical approach was proposed to take them into account.

REFERENCES

EN 14509:2013. Self-supporting double skin metal faced insulating panels – Factory made products – Specifications. CEN European Committee for Standardization. Brussels: CEN-CENELEC Management Centre.

Engel, A.; Lange, J. 2021. Momenten-Auflager-Interaktion von mehrfeldrig gespannten Sandwichelementen. In *Stahlbau 90, H. 11, S.* 798–806. Berlin: Ernst & Sohn. https://doi.org/10.1002/stab.202100050

Nelke, H. 2018. Tragfähigkeit von Sandwichelementen unter Biegung und Querdruck. Dissertation. Darmstadt: Technical University Darmstadt. https://tuprints.ulb.tu-darmstadt.de/7498/

Current Perspectives and New Directions in Mechanics, Modelling and Design of Structural Systems – Zingoni (ed.)
© 2022 Copyright the Author(s), ISBN 978-1-032-39114-4

Influence of temperature on sandwich panels with PU rigid foam

S. Steineck & J. Lange
Institute for Steel Construction and Materials Mechanics, Technical University of Darmstadt, Germany

ABSTRACT: Sandwich panels are subject to a variety of different temperature loads. The continuously fabricated PU core changes its material properties in the temperature range relevant to construction practice. It is striking that the material behavior under temperature conditions of the currently predominantly used PIR is in some cases contrary to that of the previously mainly used PUR. This has not yet been considered in the currently valid verification formats for sandwich panels.

1 INTRODUCTION

Sandwich panels used in the building industry are three-layer structures consisting of two thin outer face sheets and a thicker core of PU rigid foam or mineral wool. They are used in the building industry primarily in hall construction. Sandwich panels experience different loads. In addition to the loads from their dead weight, they also are exposed to climatic loads such as wind, snow and temperature due to their function as the outer shell of the building.

There are different types of failure of sandwich panels. The wrinkling failure is usually the critical failure mode for design. The wrinkling stress σ_w can be described simplified by the following formula (EN 14509:2013):

$$\sigma_w = 0.5 \cdot \sqrt[3]{G_C \cdot E_C \cdot E_F} \quad (1)$$

The Equation 1 shows that the wrinkling stress is mainly dependent on the stiffness of the steel layer E_F, the stiffness of the core material E_C and the shear modulus of the core layer G_C. The material parameter of the polymeric interlayer are not constant over the temperature. Therefore, it is important to consider the load-bearing behavior under different temperatures.

2 STATE OF THE ART

Sandwich panels are regulated in the European area in the standard EN 14509:2013, which mainly comprises test-based approval procedures. The standard describes a test to determine the wrinkling stress, which is performed under room temperature. To account for an increased temperature in the design, the wrinkling stress σ_w is multiplied by the correction factor k_1 (Equation 2). E_{Ct} is the stiffness of a sandwich cube in a tensile test at 80 °C or 20 °C. The cubes must be stored for 24 h at this temperature beforehand.

$$\sigma_w^{80^\circ C} = k_1 \cdot \sigma_{cr} \text{ with } k_1 = \sqrt[3]{\left(\frac{E_{Ct,+80^\circ C}}{E_{Ct,+20^\circ C}}\right)^2} \quad (2)$$

The PU core material is constantly being developed to improve its mechanical and environmental properties. The core material polyurethane (PUR) used in the past has a balanced ratio of the chemical components polyol to isocyanates. Currently, it is no longer used in sandwich panels, especially in Germany, due to the improved properties of the further development of the material from PUR to PIR. The core material used in most PU sandwich panels today is called polyisocyanurate (PIR). The designation arises from the excessive presence of isocyanates in the chemical composition. The result is a chemically and thermally more stable rigid foam. (Briehl 2021). Up to the present time, the standard has not been adapted to the design requirements due to the further development of the core material.

3 EXPERIMENTAL STUDIES

3.1 *Different behavior between PUR and PIR*

Current test results indicate a change in the material behavior of today's foam in the high temperature range. Figure 1 shows the ratio of the Young's moduli at 80 °C and 20 °C. The values marked as circles are from the year 2010. Experiments from the year 2020 are marked as triangles. The values show reduction of the Young's modulus of about 5 % to 30 % for all specimens. The test data from 2020 show an increase in the

DOI: 10.1201/9781003348450-80

Young's modulus at 80 °C of up to 46 % compared with the Young's modulus at 20 °C. It can be seen that the correction factor k_1 does not lead to a reduction in the wrinkling stress.

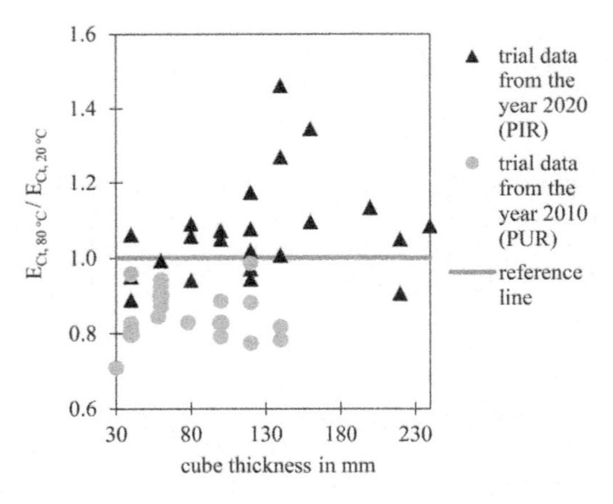

Figure 1. Ratio of tensile young`s modulus (left) and tensile strength (right) at 80 °C and 20 °C.

3.2 *Temperature range for building practice*

The temperature loads relevant for the practical construction condition are currently recorded in a long-term measurement. Temperature differences of up to 50 °C occur within a half-day cycle (12 h). Sudden sharp drops in the temperature of the outer surface layer and the foam layer underneath are striking. This rapid cooling is caused by rain events. Within the evaluation period, temperature gradients of up to 40 °C occur in the core material. The temperature recording shows the importance of considering temperature in the design of sandwich panels. In addition to elevated temperatures, temperature gradients and sudden temperature jumps must also be taken into account. Constant low temperatures and high temperature gradients also occur when sandwich panels are used in cold stores. When storing frozen goods, temperatures of -20 °C are common on the inside of the elements.

3.3 *Material parameter under different temperatures*

In the standard for sandwich testing, the influence of elevated temperature on core material parameters is determined on cube tensile tests, which are first tempered at 80 °C for 20 h to 24 h and then tested under this temperature. Thus, two effects, the storage temperature and the test temperature, are included in the result obtained from this test. For an accurate analysis, it makes sense to consider these two effects separately. Tests on specimens with an age of more than one year show that the effect of the increase in Young's modulus from Figure 1 up to 46 % is less pronounced in these specimens shown in Figure 2. It seems that some material change (e.g. post-reaction, diffusion) has already taken place even without a large temperature effect. The effect of tempering was recorded separately here. For this purpose, half of the samples were stored at 80 °C for 24 h before the test (grey). Figure 1 confirms that the increase in Young's modulus from Figure 2 is due to the warm storage temperature and is imprinted in the material. This can indicate a post-reaction due to storage time and temperature. The fracture stress is primarily dependent on the prevailing test temperature during tensile loading.

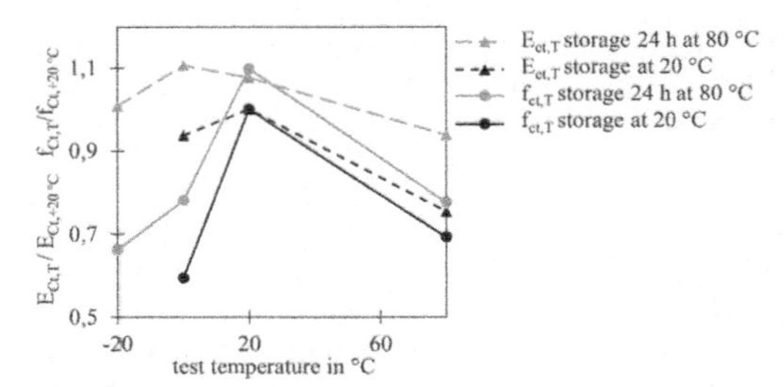

Figure 2. Ratio of Young´s moduli and tensile strength at different test temperatures related to the test at 20 °C.

4 CONCLUSION AND OUTLOOK

Particularly due to the constantly developing technologies with regard to core production and its chemical composition the test methods and the design approaches of sandwich panels must be critically examined. The calculation of the wrinkling stress in the current design specifications is based on an outdated material behavior. Initial proposals for modifying the k_1-factor also take into account the relationship between the tensile strengths. In further tests, it must be investigated whether this approach is realistic and on the safe side. Tests show that even short-term temperatures of 80 °C change the stiffness in the long term. The change in stiffness is due on the one hand to the storage under elevated temperature and on the other hand to the elevated temperature at the time of the tensile test. Initial results show that the stiffness increases due to the elevated storage temperature. Increased test temperature leads to a decrease in tensile strength. In addition, the transfer of the small part tests to entire components must be investigated. These results should then provide information on the overall load-bearing behavior of sandwich panels so that they can be designed economically for the climatic effects.

REFERENCES

EN 14509:2013. Self-supporting double skin metal faces insulating panels – Factory made products – Specifications CEN European Committee for Standardization. Brussels: CEN-CENELEC Management Center

Briehl, H. 2021. Chemie der Werkstoffe. Springer Nature. Wiesbaden

Current Perspectives and New Directions in Mechanics, Modelling and Design of Structural Systems – Zingoni (ed.)
© 2022 Copyright the Author(s), ISBN 978-1-032-39114-4

Minimum weight design of CNT/fibre reinforced laminates subject to a frequency constraint

S. Anashpaul
Discipline of Civil Engineering, University of KwaZulu-Natal, Durban, South Africa

G.A. Drosopoulos
Discipline of Civil Engineering, University of Central Lancashire, Preston, UK
Discipline of Civil Engineering, University of KwaZulu-Natal, Durban, South Africa

S. Adali
Discipline of Mechanical Engineering, University of KwaZulu-Natal, Durban, South Africa

ABSTRACT: Multiscale polymer/fibre/carbon nanotube (CNT) nanocomposites are being used to obtain laminates with improved properties as compared to fibre composites. The properties of the nanocomposite laminates can be further improved by optimizing the distribution of the reinforcements. In the present study, minimum weight design of a nanocomposite laminate is investigated subject to a frequency constraint. Halpin-Tsai micromechanical equations are used to determine the material properties of the three-phase composite.

1 INTRODUCTION

The need for lightweight structures in many fields of engineering led to minimum weight designs and composites are often the preferred material for this purpose. Development of composite materials with nano size reinforcements such as CNTs improved the weight advantage of composites further. One problem with CNT reinforced composites is the random distribution of CNTs in the matrix leading to composites without stiffness and strength in a particular direction. Development of three-phase nanocomposites solved this problem (Jeawon et al 2021), (Radebe et al 2019). Weight of three-phase composites can be further reduced by optimally determining the distribution of reinforcing components (fibres and CNTs). In the present article the weight of the laminate is minimized by optimally distributing the reinforcing components subject to a frequency constraint.

2 BASIC EQUATIONS

We study a symmetrically laminated plate of rectangular shape with lengths a in the x direction and b in the y direction. The thickness of the laminate is H. The number of layers is denoted by N. The laminated plate undergoes free vibrations and the equation governing the vibrations of laminate is

$$D_{11}\frac{\partial^4 w_0}{\partial x^4} + 2(D_{12} + 2D_{66})\frac{\partial^4 w_0}{\partial x^2 \partial y^2} + D_{22}\frac{\partial^4 w_0}{\partial y^4} + I_0\frac{\partial^2 w_0}{\partial t^2} = 0 \quad (1)$$

where I_0 is the mass per unit area. The solution of equation (1) is given by

$$w_0(x,y) = C_{mn}\, sin\frac{m\pi x}{a} sin\frac{n\pi y}{b} \quad (2)$$

for simply supported laminates. The expression for the fundamental frequency is given by

$$\omega_{11}^2 = \frac{\pi^4}{I_0 b^4}\left[D_{11}r^4 + 2(D_{12} + 2D_{66})r^2 + D_{22}\right] \quad (3)$$

with the aspect ratio $r = b/a$. Micromechanical expression for the Young's modulus of the CNT reinforced matrix is given in reference [3] as

$$E_{CNT} = \frac{E_M}{8}\left[5\left(\frac{1 + 2\beta_{dd}V_{CNT}}{1 - \beta_{dd}V_{CNT}}\right) + 3\left(\frac{1 + 2(l_{CNT}/d_{CNT})\beta_{dl}V_{CNT}}{1 - \beta_{dl}V_{CNT}}\right)\right] \quad (4)$$

In Eq. (4), the subscripts CNT and M. Micromechanical equations are given in [3, 4].

3 WEIGHT MINIMIZATION PROBLEMS

In the first optimal design problem, only the fibre distributions across the thickness are taken as the design variables with the distribution of the CNTs being the same for all layers. In the second design problem distributions of both the fibres and the CNTs are taken as design variables and these distributions are determined optimally. In the third design problem, fibre and CNT distributions and the fibre angles are taken as the design variables. Results are given for a glass fibre reinforced polymer matrix and the material properties are given in Table 1.

Problem 1: Minimum weight laminate with non-uniform fibre distribution and uniform CNT distribution (Design variable: V_{Fk})

Design parameters for Problem 1 are the fibre contents of layers, that is, V_{Fk}, $k = 1, 2, \ldots, N$ and CNTs are uniformly distributed. The minimum weight problem can be expressed as

DOI: 10.1201/9781003348450-81

Table 1. Material properties.

Material	E (GPa)	G (GPa)	ν	Density (kg/m³)
CNTs	640	$E/(2(1+\nu))$	0.27	1350
Matrix	3.5	$E/(2(1+\nu))$	0.35	1200
Glass fibers	72.4	$E/(2(1+\nu))$	0.20	2400

$$\min W_L(V_{Fk}, V_{CNTk}) \text{wrt } V_{Fk} \tag{5}$$

$$\text{subject to } \frac{\omega_{11}}{\omega_{IP}} \Omega_O \tag{6}$$

$$0 \leq V_{Fk} \leq V_{Fmax} \text{ for } k = 1, 2, \ldots, N \tag{7}$$

where V_{CNTk} is the CNT volume content of the kth layer and V_{Fmax} is the maximum fibre content of the kth layer. Equation (6) is the non-dimensional design constraint on the frequency. In the ratio ω_{11}/ω_{IP}, ω_{11} is the fundamental frequency of the minimum weight laminate and ω_{IP} is the frequency of the isotropic plate. The design efficiency factor provides a comparison between the weights of optimal laminate and an isotropic plate with the same fundamental frequency. Results are shown in Table with the CNT content specified as 1% by weight for each layer.

Table 2. Optimal fibre volumes V_{fk} and the fibre orientations for Problem 1 with the laminate thickness $H = 0.01m$, $V_{fk} \leq 0.6$, frequency constraint $\Omega_O = 1.3$ and $W_{CNT} = 1\%$.

Aspect ratio b/a	Optimal V_f per layer	Optimal fibre angle	Weight W_L	Efficiency factor
1.00	[0.15/0/0/0]$_s$	45	12.06	0.773
1.50	[0.31/0/0/0]$_s$	60	19.42	0.599
2.00	[0.45/0/0/0]$_s$	90	26.73	0.535

Problem 2: Three-phase laminate with non-uniform (optimal) distributions of fibres and CNTs (Design variables: V_{Fk} and V_{CNTk})

Design parameters for this problem are the distributions of the fibres and CNTs across the laminate thickness. The design problem can be stated as

$$\min W_L(V_{Fk}, V_{CNTk}) \text{ wrt } V_{Fk} \text{ and } V_{CNTk} \tag{8}$$

subject to the constraints (6) and (7) and the constraint

$$0 \leq V_{CNTk} \leq V_{CNTmax} \text{ for } k = 1, 2, \ldots, N \tag{9}$$

where V_{CNTmax} is the maximum volume content of CNTs. Results for this case is given in Table 3.

Problem 3: Three-phase laminate with optimal) distributions of fibers and CNTs and optimal fiber orientations (Design variables: V_{Fk}, V_{CNTk} and θ_k)

In the last design problem, in addition to fibre and CNT distributions (Problem 2), fibre orientations are also included in the set of design parameters. As such, ply angles, the volume content of fibres and

Table 3. Three-phase square laminates with 8-layers and with $H = 0.01m$, $V_{fk} \leq 0.6$, $W_{CNTk} \leq 0.05$, $W_{CNTmax} = 1.25\%$, frequency constraint $\Omega_O = 1.75$.

Stacking sequence	V_f per layer	W_{CNT} per layer	Weight W_L	Efficiency factor
[90/0/90/0]$_s$	[0.1/0/0/0]$_s$	[0.05/0/0/0]$_s$	12.33	0.587
[45/45/4545]$_s$	[0.06/0/0/0]$_s$	[0.05/0/0/0]$_s$	12.21	0.581

Table 4. Optimal fibre orientations and volume contents and optimal CNT weight contents with $H = 0.01m$, $W_{CNTk} \leq 0.05$, $W_{CNTmax} = 1.25\%$, $b/a = 1.0$.

Freq. const.	Optimal fibre angles	Volume content of fibres	Weight content of CNTs	Efficiency factor
1.5	[45/45/30/45]$_s$	[0/0/0/0]$_s$	[0.036/0/0/0.014]$_s$	0.667
2.0	[45/60/0/30]$_s$	[0.26/0/0/0]$_s$	[0.05/0/0/0]$_s$	0.533
2.5	[45/45/90/45]$_s$	[0.6/0.47/0/0]$_s$	[0.05/0/0/0]$_s$	0.507

the weight content CNTs are determined optimally. The statement of the problem is given by

$$\min W_L(V_{Fk}, W_{CNTk}, \theta_k) \text{ } wrt \text{ } V_{Fk}, V_{CNTk} \text{ } and \text{ } \theta_k \tag{10}$$

The results are given in Table 4 showing the values of the optimal design parameters.

4 CONCLUSIONS

In the present study, weight of a multiscale nanocomposite laminate was minimized subject to a frequency constraint using the fibre distribution (Problem 1), fibre and CNT distributions (Problem 2) and fibre/CNT distributions and ply angles (Problem 3) as the design variables. Effectiveness of the minimum weight designs was assessed by introducing a design efficiency factor showing the reduction in the weight of the laminate. It was shown that fibres and CNTs are more effective in minimizing the weight if they are placed close to the outer layers.

REFERENCES

Jeawon, Y., Drosopoulos, G.A., Foutsitzi, G., Stavroulakis, G. E. & Adali S., Optimization and analysis of frequencies of multi-scale graphene/fibre reinforced nanocomposite laminates with non-uniform distributions of reinforcements, Engineering Structures, 2021, 228, 111525.

Radebe, I.S., Drosopoulos, G.A. & Adali, S. Buckling of non-uniformly distributed graphene and fibre reinforced multiscale angle-ply laminates. Meccanica, 2019, 54 (14), 2263–2279.

Gholami, R., Ansari, R. & Gholami, Y. Numerical study on the nonlinear resonant dynamics of carbon nanotube/fiber/polymer multiscale laminated composite rectangular plates with various boundary conditions. Aerospace Science and Technology, 2018, 78, 118–129.

Rafiee, M., Nitzsche, F. & Labrosse, M. Cross-sectional design and analysis of multiscale carbon nanotubes-reinforced composite beams and blades. International Journal of Applied Mechanics, 2018, 10, 1850032.

Current Perspectives and New Directions in Mechanics, Modelling and Design of Structural Systems – Zingoni (ed.)
© 2022 Copyright the Author(s), *ISBN 978-1-032-39114-4*

On the application of CFD and MDO to aircraft wingboxes using commercial software

Xu Duan & Kamran Behdinan
Department of Mechanical and Industrial Engineering, University of Toronto, Toronto, Canada

ABSTRACT: When designing for aircraft structures, it is common to begin with a previous design, and adapt it to new applications. Using multidisciplinary design optimization, refinement can be automatically performed on initial designs using high-fidelity tools. In this paper, multidisciplinary design optimization is applied to the design of an aircraft wingbox using four design variables. The Bruguet range equation was chosen to be the objective function due to its consideration of lift, drag, and weight of the aircraft. Industry standard software such as ANSYS and MATLAB are used for analysis and optimization. Using the operating conditions of the Boeing 747-400 as a basis, the design parameters were compared to the initial design and showed variations below 15%.

1 INTRODUCTION

To generate a design for an aircraft wingbox, there are many design requirements that need to be satisfied such as weight, stress, and aerodynamic performance. In a traditional design process, an initial design is often generated using past experience and previous designs as a basis to be modified. Multidisciplinary Design Optimization (MDO) automates this process by optimizing based on cost functions and constraints. This method can also allow the generation of designs that are vastly different from the initial given enough iterations.

Cramer et al. (Cramer *et al.*, 1994) is often credited for formalizing many simple MDO methods and are well documented in literature. A summary of multidisciplinary approaches has been compiled by Martins et al. (Martins and Lambe, 2013) An early method of design optimization was done by Gumbert et al. (Gumbert, Hou and Newman, 2005), used Computational Fluid Dynamics (CFD) and Finite Element Method (FEM) to optimize a wing. One method used to reduce the computational cost is to not completely reaching a converged solution at each optimization step.

2 SET-UP

2.1 *Optimization*

The objective function chosen was the Breguet range equation as shown in Equation 1 where k is a constant chosen to represent the fuel consumption rate of the engines,C_L is the coefficient of lift,C_D is the coefficient of drag,W_T is the total weight of the aircraft, andW_F is the weight of the fuel. Since k is a constant, it can be removed from the optimization process.

$$R = k\frac{C_L}{C_D}\ln\left(\frac{W_T}{W_T - W_F}\right) \tag{1}$$

Several constraints were also applied to the optimization process. First, upper and lower bounds were placed on the design variables. An inequality constraint was placed on the stress within the structure to ensure it does not exceed a factor of safety of 4.5. This maximum stress constraint is applied on the maximum Von-Mises stress in the FEM model. The material chosen for the structure of the wing was 7075-T6 aluminum alloy.

2.2 *Geometry and mesh*

The design variables chosen to represent the shape of the wing are shown in Figure 1. To provide rigidity, ten ribs were chosen to be placed equally spaced from the root to the chord. Two spars were also placed with distances with respect to the leading and trailing edge of the wing. The skin thickness for wing was chosen to be 2mm and the airfoil shape chosen for the wing is the NACA 2412. The wing was also set to have an angle of attack of 1 degree.

ANSYS Meshing was used to generate the mesh to be used in the both the CFD and FEM simulations. Due to the automatic generation of geometry, 3D tetrahedral unstructured mesh was used to minimize the chance of meshing failures during the optimization process. To accurately measure the aerodynamic performance of the wing, a very fine mesh is required but due to computational limits, the mesh was increased until the variation in the Breguet range equation was equal to 1%.

MATLAB was chosen as the software to program the optimizer due to its many advanced built-in optimization functions. It was found that the patternsearch

DOI: 10.1201/9781003348450-82

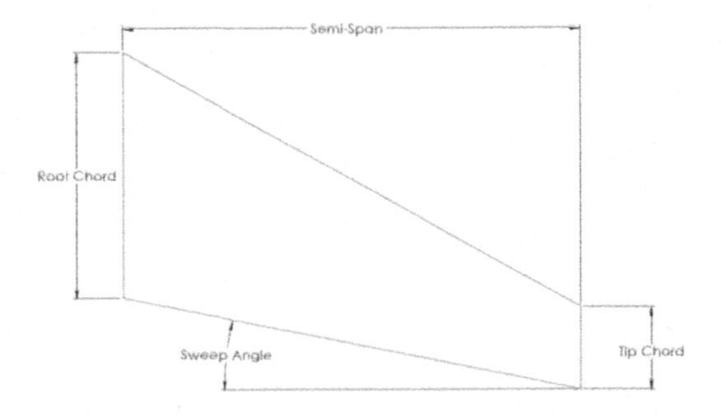

Figure 1. Geometry of Wing.

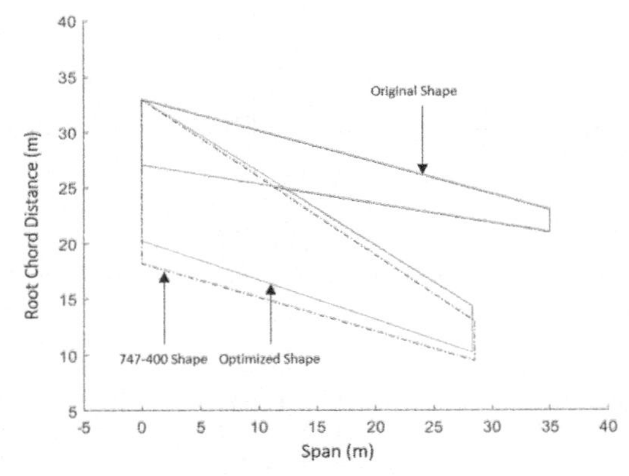

Figure 2. Optimized Shape Compared to Original and Boeing 747-400.

optimizer did not get stuck in local minima as often as other local optimizers and required significantly fewer iterations to optimize than global optimizers. Output data from ANSYS was transferred to MATLAB via scripting, then the geometry is generated based off the outputs of the optimizer.

2.3 *Solver*

ANSYS Fluent was chosen to be the CFD solver, and ANSYS Mechanical was chosen to be the FEM solver. The CFD program solves the Reynolds-Averaged Navier-Stokes Equation with the SST k-omega turbulence model. Air was the working fluid and was modeled as an ideal gas at the operating conditions of a 747-400. Air enters the fluid domain using a velocity inlet boundary condition at 255m/s and exits the domain with a pressure outlet boundary condition. All plane surfaces are set with the no-slip boundary condition. From the CFD analysis, the lift and drag coefficients are obtained, as well as the pressure acting on the surface of the wing. The pressure can then be mapped onto the FEM mesh which is done through ANSYS Workbench. One major benefit of using ANSYS Workbench is that it does not require the data to be processed by an external program which decreases computation time and ensures better accuracy of the mapping. Once the pressure has been mapped onto the aircraft wing, ANSYS Mechanical can then calculate the Von Mises stress to be used for the constraints of the optimization.

3 RESULTS

The model was run for a maximum duration of 24 hours to limit the computational cost as well as a convergence target of 3 significant digits. Figure 2 shows a comparison between the original shape, the optimized shape, as well as a comparison to the Boeing 747-400. The initial parameter values were chosen at random to demonstrate the capabilities of MDO on an initial design that is significantly different from the final optimized shape. Parameters of the Boeing 747-400 are estimated from drawings obtained via Boeing's website

Based on the Breguet range equation, the optimized shape performs better than the Boeing 747-400. The variation between the sweep and span are within 5% of the Boeing reference. However, the root chord is approximately 16% smaller than the Boeing reference which also causes the taper to be larger. The smaller root chord causes an increased variation in the taper between the optimized and reference geometry. When looking at the tip chord separately, the optimized geometry is only 5% larger than the reference. One reason for the increased variation is due to a higher stress constraint than what may be required.

4 CONCLUSIONS

In this paper, it has been demonstrated that high fidelity simulation tools can be applied to a crude initial design to optimize the wingbox of an aircraft. The simulation was constrained to a maximum of 24 hours to limit the computational cost associated with the analysis. The optimization was constrained based on publicly available data on the Boeing 747-400, and results varied between 0% for the span, to 16% for the root chord. Since the stress constraint is based on factor of safety for the yield strength and load, it can be increased to force the wing to have a larger root chord to give the optimized result slower. Reviewing the pressure distribution of the wing shows that it follows an expected distribution.

REFERENCES

Cramer, E. J. *et al.* (1994) 'Problem Formulation for Multidisciplinary Optimization', *SIAM Journal on Optimization* 4(4), 754–776

Gumbert, C. R., Hou, G. J. W. and Newman, P. A. (2005) 'High-fidelity computational optimization for 3-D flexible wings: Part I - Simultaneous Aero-Structural Design Optimization (SASDO)', *Optimization and Engineering* 6(1), 117–138

Martins, J. R. R. A. and Lambe, A. B. (2013) 'Multidisciplinary design optimization: A survey of architectures', *AIAA Journal*

7. Damage mechanics, Damage modelling, Blast, Impact, Fracture, Fatigue, Progressive collapse, Structural robustness

Current Perspectives and New Directions in Mechanics, Modelling and Design of Structural Systems – Zingoni (ed.)
© 2022 Copyright the Author(s), ISBN 978-1-032-39114-4

Research on structural robustness through large-scale testing

J.M. Adam, M. Buitrago & N. Makoond
ICITECH, Universitat Politècnica de València, Valencia, Spain

ABSTRACT: As a result of the persistent occurrence and apparently increasing frequency of catastrophic structural failures, recent years have been marked by a growing body of literature on progressive collapse and structural robustness. At present, the vast majority of these studies focus on computational simulations and laboratory testing of reduced-scale sub-assemblages. Although many vital aspects of structural behaviour under extreme conditions have been uncovered through such research, these strategies are characterised by significant limitations which can only be overcome through full-scale testing of real structures. This article presents some of the major research works performed in this regard by the *Building Resilient* research group from the *ICITECH* institute of the *Universitat Politècnica de València*. The most important works related to temporary shoring of buildings, cast-in-place and precast reinforced concrete building structures, steel truss bridges, and fuse-segmented buildings are presented together with the most significant results achieved so far.

1 INTRODUCTION

In recent years, several structural failures leading to catastrophic consequences for society have occurred. Some of impactful recent examples of bridge and building failures include the collapse of the I-35W bridge in Minneapolis in 2007 and of the Champlain Towers in Miami in 2021. Progressive collapse is often found to be at the heart of such failures. This phenomenon, through which an initial localised failure propagates to other parts of a structural system, often results in the collapse of the entire structure or of a disproportionate part of it (Adam et al. 2018).

At present, buildings and bridges are increasingly more exposed to the devastating consequences of extreme events caused by climate change, terrorist threats, their own ageing, or inadequate maintenance.

This article presents the research currently being carried out at the *Building Resilient* research group aiming to contribute to improving the resilience of buildings and bridges, by avoiding possible collapses or, at least, by minimising their consequences. The research being carried out typically relies on ambitious experimental campaigns involving tests on full-scale structures. Specifically, the following areas of work are presented in this article: 1) Temporary shoring of buildings; 2) Flat-slab reinforced concrete (RC) building structures; 3) Precast RC building structures; 4) Steel truss bridges; 5) Development of a new design philosophy based on connecting building segments with structural fuses.

2 STRUCTURAL ROBUSTNESS RESEARCH

2.1 *Temporary shoring*

The research performed by the group on this topic was pioneering for being the first study analysing how the failure of a few elements of a temporary shoring system can lead to the progressive collapse of the entire system. One of the most innovative outcomes achieved relates to the patented design of the first example of a structural fuse specifically intended to prevent progressive collapse (Buitrago et al. 2018).

2.2 *Flat-slab RC building structures*

The most notable research performed on this structural typology involves an experimental campaign in which a full-scale building was subjected to different corner-column failure scenarios (Figure 1).

Figure 1. Test building prepared for sudden column removal.

DOI: 10.1201/9781003348450-83

This research enabled the analysis and evaluation of alternative load paths that can be activated after the failure of corner columns considering dynamic effects caused by the sudden loss (Adam et al. 2020).

2.3 *Precast RC building structures*

Precast concrete elements are definitely gaining importance in the field of building construction. However, certain characteristics of these constructions make them, a priori, more vulnerable when exposed to extreme events. With the *PREBUST* project, *Building Resilient* is working to improve the robustness of building structures made with precast RC components. The most ambitious part of *PREBUST* is an experimental campaign involving a purposely built real-scale test building of two storeys subjected to three different sudden column removal scenarios.

2.4 *Steel truss bridges*

Steel truss bridges are particularly sensitive to progressive collapse, which can often result in catastrophic consequences. In collaboration with the *Calsens* spin-off company, *Building Resilient* is investigating how alternative load paths are activated after local-initial failures in this typology of bridges.

The work carried out has included the laboratory testing of a 21 m railway bridge span (Figure 2). This experimental campaign has allowed the robustness of the bridge to be evaluated (Buitrago et al. 2021) and monitoring guidelines to be defined for the early detection of local failures with propagation potential.

Figure 2. Complete span of a railway bridge inside the ICITECH laboratories.

2.5 *Endure project*

Current building design codes are based on providing structures with a high degree of continuity. Thus, when one element fails, the load it supported can be redistributed among the other elements of the structure. Although this design philosophy has been effective on many occasions, there are certain scenarios in which it is not, and can in fact even increase the risk of progressive collapse. Therefore, it is necessary to define new design approaches to remedy these limitations to mitigate the risk of disaster.

The aim of the *Endure* project is to develop a new building design philosophy based on fuse-segmentation to prevent failure propagation. This philosophy aims to protect buildings against progressive collapse by connecting different segments of a building with structural fuses. These fuses will give continuity to the structure for scenarios considered by current design codes but will separate the segments when failure propagation is inevitable during exceptional scenarios for which design codes are not effective.

3 CONCLUSIONS

The knowledge acquired as a result of the presented research is allowing: 1) the definition of design and construction strategies to achieve resilient structures; 2) the definition of preventive and remedial actions for reducing the vulnerability of buildings and bridges to extreme events; and 3) the proposal of monitoring guidelines for the early detection of local failures with a high potential for causing progressive collapse.

ACKNOWLEDGMENTS

The work presented would not have been possible without the funding and support received from: *Calsens, LIC - Levantina, Ingeniería y Construcción, Alsina, FGV (Ferrocarrils de la Generalitat Valenciana), BBVA Foundation, Generalitat Valenciana, Ministerio de Educación y Formación Profesional, Ministerio de Ciencia e Innovación* and the *European Research Council.*

This work is the result of the dedication and enthusiasm of many people. The following deserve special thanks: Pedro A. Calderón, Juan J. Moragues, Elisa Bertolesi.

REFERENCES

Adam, J. M., Parisi, F., Sagaseta, J., & Lu, X. 2018. Research and practice on progressive collapse and robustness of building structures in the 21st century. *Engineering Structures*, *173*, 122–149. https://doi.org/10.1016/j.engstruct.2018.06.082

Adam, J. M., Buitrago, M., Bertolesi, E., Sagaseta, J., & Moragues, J. J. 2020. Dynamic performance of a real-scale reinforced concrete building test under a corner-column failure scenario. *Engineering Structures*, *210*, 110414. https://doi.org/10.1016/j.engstruct.2020.110414

Buitrago, M., Adam, J. M., Calderón, P. A., & Moragues, J. J. 2018. Load limiters on shores: Design and experimental research. *Engineering Structures*, *173*, 1029–1038. https://doi.org/10.1016/j.engstruct.2018.07.063

Buitrago, M., Bertolesi, E., Calderón, P. A., & Adam, J. M. 2021. Robustness of steel truss bridges: Laboratory testing of a full-scale 21-metre bridge span. *Structures*, *29*, 691–700. https://doi.org/10.1016/j.istruc.2020.12.005

Current Perspectives and New Directions in Mechanics, Modelling and Design of Structural Systems – Zingoni (ed.)
© 2022 Copyright the Author(s), ISBN 978-1-032-39114-4

Using optical diagnostics for near field blast testing

G.S. Langdon, R.J. Curry, S.E. Rigby, E.G. Pickering, S.D. Clarke & A. Tyas
Department of Civil and Structural Engineering, University of Sheffield, Sheffield, UK

S. Gabriel & S. Chung Kim Yuen
BISRU, Department of Mechanical Engineering, University of Cape Town, Rondebosch, South Africa

C.J. von Klemperer
Department of Mechanical Engineering, University of Cape Town, Rondebosch, South Africa

ABSTRACT: The ability to measure the structural and material response to air-blast loading is vital to developing a proper understanding of near-field blast loading and response. Computational modelling has advanced significantly but, until recently, experimental techniques lagged behind. This paper discusses recent advances in these experimental techniques. The first part describes a bilateral test programme between the UK and South Africa. The high-speed imaging and digital image correlation system at Cape Town gives repeatable and accurate impulse distributions across a central strip of a panel, useful for model validation. Flexural wave behaviour was observed from the transient velocity and displacement profiles, giving good insights into the mechanics of plate response from blast loads. The second part demonstrates the value of high-speed stereo-imaging for measuring the transient response of blast loaded fibre reinforced polymer panels and sandwich structures. The peak displacements. elastic rebounds and transient oscillations provide valuable insights into the damage propagation within these types of structures. The final part of the paper describes some of the continued developments since the success of those early trials, resulting in a new optical diagnostics for blast capability at the University of Sheffield. The imaging system operates at higher frame rates and can cover a wider region of interest on the structure. Ultra-high speed imaging is also shown to be a useful tool for visualising detonations fronts in explosive charges and the expanding fireball.

1 INTRODUCTION

Optical diagnostics offer exceptional potential for measuring the dynamic structural response and loading characteristics arising from explosive detonations in the near and far field. However, there are considerable challenges involved in this. In recent years, improvements in instrumentation have provided renewed impetus to the goal of better experimental measurements from laboratory scale explosion tests (Curry & Langdon 2017, Clarke et al. 2020). These approaches offer potential for measurements in the near-field, especially of the loading (specific impulse distribution). In this paper we describe the use of optical diagnostics for transient blast measurements.

2 USING OPTICAL DIAGNOSTICS FOR SPECIFIC IMPULSE MEASUREMENT

The Characterisation of Blast Loading (CoBL) test facility (Clarke et al. 2020) was developed at the University of Sheffield to measure pressure at discrete locations across a rigid target plate, in order to characterize the load arising from an explosive detonation in the near-field. Spatial and temporal pressure histories are obtaining by analysing the Hopkinson bar signals, allowing a map of specific impulse distribution across the target face, at discrete locations. Carefully controlled explosion detonations were performed at Sheffield using the CoBL test facility. Spherical and cylindrical PE4 charges were located at predetermined stand-off distances from the charge surface. The data recorded from the Hopkinson pressure bar array was used to determine the specific impulse distribution at each bar location.

Experiments at BISRU involved detonating spherical and cylindrical PE4 charges. The blast loading was directed at deformable Domex 355MC steel plates mounted via a clamp frame to a pendulum. The transient response was obtained from high-speed stereo images of the rear face motion of the deforming plates following Curry & Langdon (2017).

The initial velocity profiles were used to infer a continuous specific impulse distribution which, after suitable scaling, was compared to CoBL specific impulse distributions. Good agreement was demonstrated (Figure 1).

DOI: 10.1201/9781003348450-84

The spherical detonations produced repeatable specific impulse distributions, while there was greater spread for the cylindrical tests. This work shows that the two measurement techniques (CoBL and high speed video/DIC) give excellent agreement and are both suitable methods for obtaining the spatial impulse distribution across a target structure (Rigby et al., 2019).

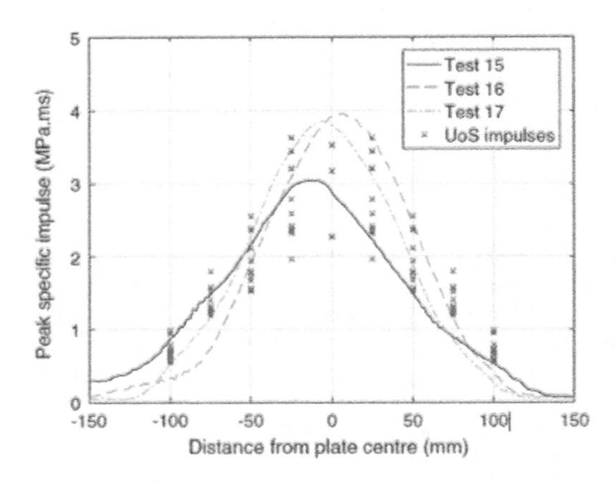

Figure 1. Specific impulse distributions obtained from cylindrical charge detonations at Sheffield and BISRU.

3 USING OPTICAL DIAGNOSTICS TO MEASURE STRUCTURAL RESPONSE

Recent blast tests on fibre reinforced composite structures using high-speed stereo-imaging and DIC revealed highly transient effects in these elastically dominated materials (see Figure 2), and led to the development of a proposed sequence of damage within blast-loaded sandwich panels (Gabriel et al., 2021). These observations would not have been possible without the use of high-speed stereo-imaging.

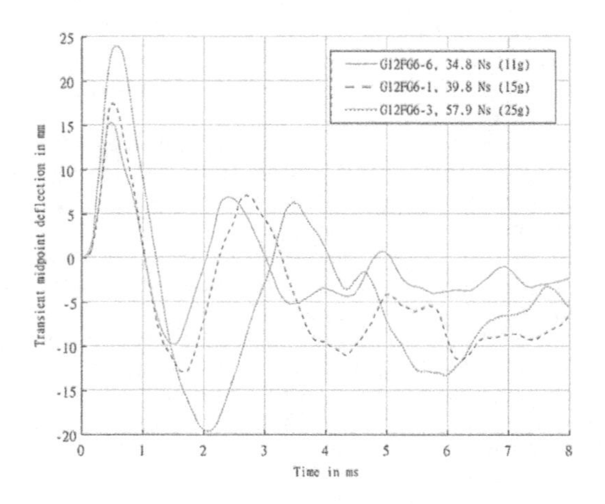

Figure 2. Graph showing rear transient mid-point displacement-time histories for sandwich panels.

4 NEXT GENERATION OF OPTICAL DIAGNOSTICS

A new experimental arrangement, comprising a clamping frame attached to a rigid steel test-frame has been developed, instrumented with two Shimadzu HPV-X2 cameras capable of filming at rates up to 5 Mfps and illuminated by a Luminys 30k high speed lab light. The increased frame rate allows our optical measurement techniques to be extended into new time domains showing, in much finer detail, the initial velocity fields resulting from an explosive detonation (and hence the spatial impulse distribution) and the transient evolution of out-of-plane displacement across these structures. This is being commissioned

The higher frame rates are also useful for examining highly transient phenomena such as propagation of detonation fronts through an explosive charge during the first few microseconds. So far, we have filmed the propagation of the detonation front through the explosive charge during the first few microseconds, including the lateral expansion of the fireball from the point of detonation (Figure 3). It could be used alongside the Photron SA-Z camera, creating the potential to capture images of the detonation wave early-time fireball development, as well as the later fireball and shock wave features and target interaction, from the same experiment, in the near-field.

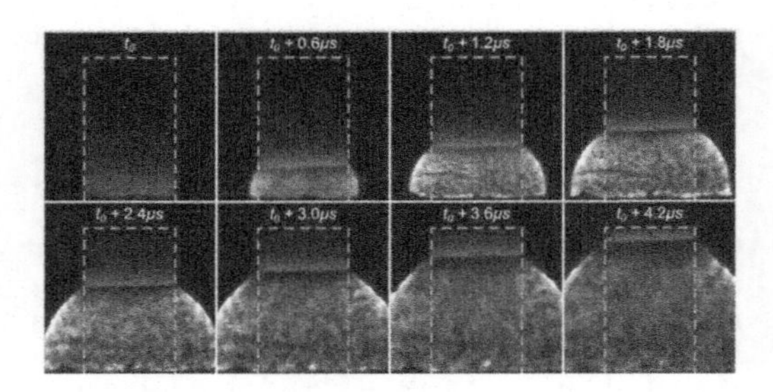

Figure 3. Camera footage showing detonation front (dotted line = initial charge shape outline overlaid).

REFERENCES

Clarke S.D, Rigby S.E, Fay S., Barr A., Tyas A., Gant M., Elgy I. 2020. Characterisation of buried blast loading. *Proc. R. Soc. A*, 476(2236):20190791.

Curry R.J. & Langdon G.S. 2017. Transient response of steel plates subjected to close proximity explosive detonations in air, *Int J Impact Eng*, 102:102–116.

Gabriel S, von Klemperer CJ, Chung Kim Yuen S, Langdon GS. 2021. Towards an understanding of the effect of adding a foam core on the blast performance of glass fibre reinforced epoxy laminate panels, *Materials* 14 (23):7118.

Rigby S.E, Tyas A., Curry R.J., Langdon G.S. 2019. Experimental measurement of specific impulse distribution and transient deformation of plates subjected to near-field explosive blasts, *Expt Mechanics*, 59:163–178.

Current Perspectives and New Directions in Mechanics, Modelling and Design of Structural Systems – Zingoni (ed.)
© 2022 Copyright the Author(s), ISBN 978-1-032-39114-4

Experimental evaluation of composite floors under column loss scenarios

M. Hadjioannou
Protection Engineering Consultants, Dripping Springs, Austin, USA

E.B. Williamson & M.D. Engelhardt
University of Texas at Austin, Civil, Architectural & Environmental Engineering Department, Cockrell School of Engineering, Austin, USA

ABSTRACT: To better understand the performance of steel-concrete floor slabs under a critical column removal, two large-scale composite floor systems were tested at The University of Texas at Austin under two different column removal scenarios. Early analytical research indicates that floor slabs play an essential role in mitigating collapse. Nonetheless, the limited experimental data currently available provide inconsistent results. The main objective of the experimental program was to identify basic behaviors of composite floor systems and to estimate their ultimate capacity under the absence of a critical column. The specimens were representative of isolated regions of the gravity-load resisting system of a typical steel-framed building, with simple shear connections and no provisions to mitigate disproportionate collapse. Both specimens were tested under quasi-static loading conditions. Detailed finite element models were developed and validated against the collected experimental data. Using the validated numerical models, the collapse load under sudden column loss was estimated.

1 INTRODUCTION

A commonly used approach to assess the collapse potential of structures under column loss scenarios is the alternate load path (ALP) method. The ALP method is considered threat-independent by evaluating structural response following local damage of critical load-bearing members that can be overloaded and damaged by beyond design-basis loads, such as blast or impact. The goal of the analysis is to evaluate structural response under service-level gravity loads following instantaneous removal of load-bearing members (e.g., columns and wall sections) to ensure that damage does not propagate and lead to progressive collapse. The ALP methodology is outlined in design documents such as the Department of Defense (DoD) (2016) guidelines.

Early numerical studies that followed the ALP framework highlighted the important contributions of floor slabs in mitigating progressive collapse, such as the work by Alashker and El-Tawil (2011). The study presented in this paper is a continuation of the effort to develop a better understanding of the behavior of steel-concrete composite floor slabs under column removal scenarios. Specifically, two large-scale experimental tests were performed at the Ferguson Structural Engineering Laboratory (FSEL) at the University of Texas at Austin (Hadjioannou *et al.*, 2018). These tests demonstrated the appreciable capacity of such floor systems. They also provided quality experimental data that were used to develop and validate high-fidelity finite element (FE) models. These models were shown to predict the observed collapse behavior with good accuracy (Hadjioannou, Williamson and Engelhardt, 2020). This paper presents an overview of the collapse tests, the validated high-fidelity FE models, and an estimate of dynamic load effects due to sudden column removal.

2 COLLAPSE TESTS OF FLOOR SLABS

2.1 *Description of test specimens*

Two full-scale steel-concrete floor slabs were tested at FSEL with a typical bay width of 4.60 m. The two specimens were representative of isolated sections of the gravity-load resisting system of a typical steel-frame building with simple shear connections and no provisions for progressive collapse. The first specimen was a 2-bay × 2-bay floor slab, representative of an interior column loss (ICL) scenario. The second specimen was a 2-bay × 1-bay floor slab, representative of an exterior column loss (ECL) scenario. The two specimens were built within a specially designed test frame to account for the boundary conditions provided by the adjacent bays in an actual building. A plan view of the ECL specimen is shown in Figure 1.

DOI: 10.1201/9781003348450-85

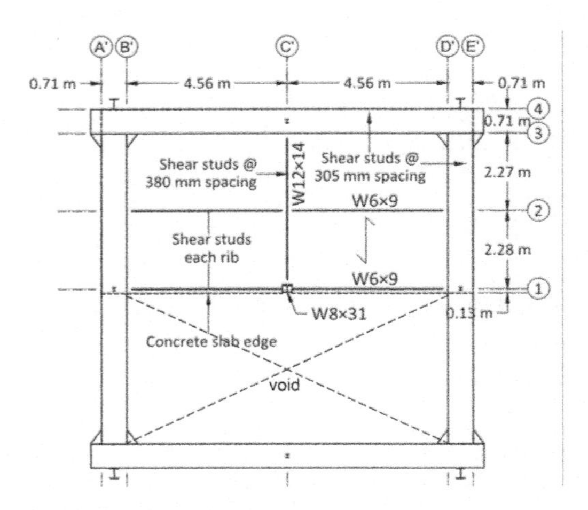

Figure 1. Plan view of ECL specimen.

2.2 *Test procedure and response*

To simulate column removal, the W8×31 (Figure 1) column was discontinued 150 mm below the bottom flange of the W12×14 girder in both specimens, leaving a 4-m clearance above the ground. This "stub" column was supported underneath by a 140-kN long-stroke hydraulic actuator. Initially, the slab was loaded with a uniformly distributed load, consistent with the DoD-recommended (2016) service-level design load for progressive collapse. This load combination resulted in a total uniform load of 5.1 kN/m^2. Water vessels were placed over the slab area and filled with enough water to achieve the target load level. For the ECL specimen, a series of concrete blocks were placed along the free edge of the specimen, to simulate the presence of a façade load, resulting in a line load of 4.76 kN/m.

Each test was performed in two distinct stages. During the first stage, the hydraulic actuator was gradually released until the measured reaction on the actuator load cell reached zero. Subsequently, during the second stage, water was remotely added to the water vessels using a pump system that was designed to evenly fill them. The added water volume was monitored with flowmeters to determine the added load on the slab. The additional water within the vessels was added over a period of a few hours, ensuring pseudo-static loading. The test was terminated when the floor slab collapsed.

The ICL specimen collapsed at a uniform load of 8.65 kN/m^2, after pre-damaging all steel connections by removing all the bolts. The ECL specimen collapsed at a uniform load of 9.1 kN/m^2 and a line load along the free edge of the slab of 4.76 kN/m^2.

3 HIGH-FIDELITY NUMERICAL SIMULATIONS

Detailed numerical models of both specimens were developed using the multi-physics finite element code LS-DYNA. The FE models were able to compute the response of the two tests up to total collapse. Nearly all specimen components were modeled using brick elements, including the concrete slab, steel beams, bolts, and connections. The corrugated metal decking was represented with shell elements, while beam elements were used for the shear studs and the light reinforcing steel within the concrete slab. A view of the ICL FE model is shown in Figure 2.

A series of dynamic analyses was performed to estimate the collapse load of the ECL specimen under sudden column loss. Due to the dynamic effects associated with instant column removal, the collapse load was approximately 1.6 times lower than the computed static collapse load.

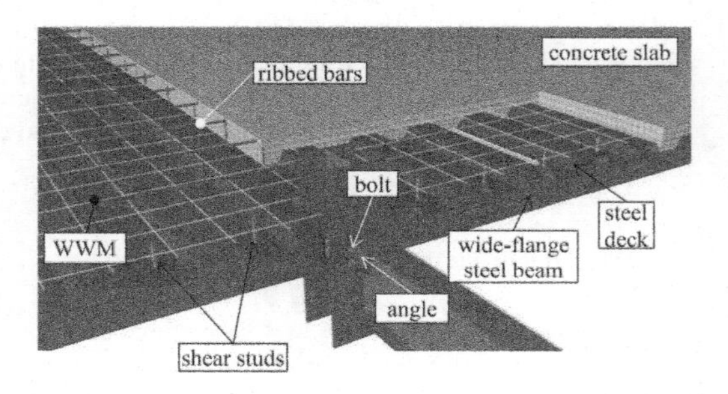

Figure 2. View of ICL FE model near center (removed) column (some parts omitted for clarity).

REFERENCES

Alashker, Y. and El-Tawil, S. (2011) 'A design-oriented model for the collapse resistance of composite floors subjected to column loss', *Journal of Constructional Steel Research*, 67 (1), pp. 84–92. doi:10.1016/j.jcsr.2010.07.008.

DoD (2016) *Design of buildings to resist progressive collapse*. Washington, DC: United Facilities Criteria (UFC) 4-023-03.

Hadjioannou, M. *et al.* (2018) 'Large-scale experimental tests of composite steel floor systems subjected to column loss scenarios', *Journal of Structural Engineering*, 144(2), p. 04017184. doi:10.1061/(ASCE)ST.1943-541X.0001929.

Hadjioannou, M., Williamson, E.B. and Engelhardt, M.D. (2020) 'Collapse Simulations of Steel-Concrete Composite Floors under Column Loss Scenarios', *Journal of Structural Engineering*, 146(12), p. 04020275. doi:10.1061/(ASCE)ST.1943-541X.0002841.

Current Perspectives and New Directions in Mechanics, Modelling and Design of Structural Systems – Zingoni (ed.)
© 2022 Copyright the Author(s), *ISBN 978-1-032-39114-4*

The role of infills modelling in progressive collapse capacity assessment of RC buildings: A case study

M. Scalvenzi & F. Parisi
Department of Structures for Engineering and Architecture, University of Naples Federico II, Naples, Italy

E. Brunesi
EUCENTRE, European Centre for Training and Research in Earthquake Engineering, Pavia, Italy

ABSTRACT: The robustness of reinforced concrete (RC) frame buildings can be influenced by several factors, as highlighted by several experimental tests and numerical simulations. This study wants to investigate the role of infill walls in the progressive collapse resistance of a real RC frame building, which was constructed in the 1950s and suffered a partial progressive collapse during retrofitting works. The structural model, analyzed in a previous study, is further refined by integrating nonlinear fiber-based finite element modelling of frame members with nonlinear macro-element modelling of infill walls. Nonlinear pushdown analysis were performed to assess the effects of infill walls on a realistic structural model that is representative of existing RC frame buildings.

1 INTRODUCTION

In recent years, the interest in progressive collapse has significantly increased. Very few studies evaluated the gravity load-bearing capacity of existing buildings during retrofitting operations, particularly in transient stages during which the structure may be temporarily weakened or subjected to loads different from those assumed at the time of its structural design.

This study want to investigate the role of infill walls, which could provide extra vertical resistance to the frame system, and the effects related to different modelling strategies (i.e. implementation of either infill loads only or infill loads plus their stiffness and resistance). Specifically, the case study is a real, masonry-infilled, reinforced concrete (RC) frame building that was constructed in the 1950s and suffered a partial progressive collapse during retrofitting works.

2 CHARACTERISTICS AND MODELLING OF THE CASE-STUDY STRUCTURE

The building under study is a multi-story RC frame structure that was located in Naples, Italy, and suffered the collapse of a corner during retrofitting works in 2001 (Figure 1a). The structural retrofitting consisted in a simultaneous removal of concrete cover from ten columns at the ground floor and soil excavation around column bases. Soil excavation was done to allow a complete removal of concrete cover along column height. As columns were founded on isolated footings, soil excavation was assumed to cause the loss of load-bearing capacity in the soil-foundation system. The structure was designed only to gravity loads according to a past Italian code (Regio Decreto 2229/1939) and practice rules, using the permissible stress method and nominal values of loads and material properties.

Figure 1. Building corner: (a) after collapse and (b) fiber-based model.

As such, a simulated design procedure was integrated with forensic investigation data to derive

DOI: 10.1201/9781003348450-86

sectional size and steel reinforcement arrangement of structural elements. Details of the simulated design procedure are described in Scalvenzi & Parisi (2021). In order to assess the effects of infill walls on progressive collapse resistance of the structure, this study is based on pushdown analyses that were carried out on partial capacity model of the structure, that was coincident with the corner of the building that suffered collapse.

The structure was modelled through the finite element (FE) code SeismoStruct (2019), as shown in Figure 1b. The nonlinear capacity modelling of the structure was based on a spread plasticity approach with force-based fiber formulation. Each fiber was assigned a uniaxial stress–strain relationship. In this study, each cross section was subdivided in 200 fibers.

Gravity loads were directly assigned to the beams as distributed loads, according to the following combination rule recommended by UFC guidelines (2013) for progressive collapse assessment via nonlinear static (pushdown) analysis:

$$Q_{bd} = \Omega_N(1.2DL + 0.5LL) \quad (1)$$

The model of the infill walls implemented in SeismoStruct (2019) was developed and initially programmed by Crisafulli (1997) for seismic loads, and later validated for gravity-induced progressive collapse applications against experimental pushdown testing. Masonry infill walls were considered to be of a 'weak' type consisting of two hollow clay block masonry leaves with 60% volumetric percentage of holes and an internal cavity for thermal isolation.

3 RESULTS OF PUSHDOWN ANALYSIS

As a continuation of previous work, pushdown analysis was carried out by changing the intensity of gravity loads on beams under a monotonically increasing downward displacement D_v imposed to a control node of the structure. The beam-column joint on top of the most stressed column was assumed to be the control point. In this study, at first, the structural behavior was assessed under the soil excavation (SE) and the concrete cover removal (CCR) scenarios involving one column: the soil was excavated around a central and a corner column, while the concrete cover was removed around the central column. In another analysis case, the number of columns involved in SE and CCR scenarios increased to 3 and 10 (in the CCR scenario only). Three different modelling options were considered for the structure (and its infills): (i) Bare frame (BF), (ii) Bare frame with infill loads (BF+IL) (iii) and Infilled frame (IF).

Analysis results are listed in Table 1 in terms of α_{max} and its percentage variation (in round parentheses) with respect to the maximum load multiplier of the intact structure for the three models. The IF model was able to withstand 3.18 times the design load, with an increase from 13% to 20% compared to BF and BF+IL models, respectively. Among the three models under study, the worst case turned out to be the BF+IL model, in which the infills were modelled as loads producing a demand increase on the structure.

Table 1. Maximum load multiplier and its variations related to retrofitting scenarios on bare frame (BF), bare frame with infill loads (BF+IL) and infilled frame (IF) models.

	α_{max}		
Model/Scenario	BF	BF+IL	IF
Intact structure	2.78	2.55	3.18
SE @ central column (C9-16)	0.13 (-96%)	1.28 (-50%)	2.46 (-23%)
SE @ corner column (C11-19)	2.30 (-17%)	1.61 (-37%)	2.14 (-33%)
SE @ 3 columns	0.03 (-99%)	0.74 (-71%)	1.96 (-39%)
CCR @ central column (C9-16)	2.27 (-18%)	2.20 (-14%)	2.95 (-7%)
CCR @ 3 columns	2.27 (-18%)	2.20 (-13%)	2.90 (-9%)
CCR @ 10 columns	2.22 (-20%)	1.99 (-22%)	2.35 (-26%)

4 CONCLUSIONS

In this study, the progressive collapse capacity of a real masonry-infilled RC frame building that suffered partial collapse during structural retrofitting has been discussed. Pushdown analysis with displacement control was carried out on a partial model of the structure, consisting in the building corner directly involved in retrofitting and progressive collapse. Analysis results have outlined that the modeling of infill walls had a beneficial impact, since the maximum load withstood by the intact structure increased from 2.78 to 3.18 times design loads.

REFERENCES

CNR-DT 212/2013. Istruzioni per la valutazione affidabilistica della sicurezza sismica di edifici esistenti. Rome: Italian National Research Council; 2013 [in Italian].

Crisafulli, F.J. 1997. Seismic behaviour of reinforced concrete structures with masonry infills. PhD thesis, University of Canterbury.

Royal Decree 16.11.1939,n. 2229, Norme per la esecuzione delle opere in conglomerato cementizio semplice ed armato, G.U. n. 92 issued on 18 April 1940 (in Italian).

Scalvenzi, M. & Parisi, F. 2021. Progressive collapse capacity of a gravity-load designed RC building partially collapsed during structural retrofitting. *Engineering Failure Analysis* 121: 105164.

Seismosoft. SeismoStruct — A computer program for static and dynamic nonlinear analysis of framed structures; 2019. Available from: www.seismosoft.com.

Current Perspectives and New Directions in Mechanics, Modelling and Design of Structural Systems – Zingoni (ed.)
© 2022 Copyright the Author(s), ISBN 978-1-032-39114-4

Computational study on the progressive collapse of precast reinforced concrete structures

N. Makoond, M. Buitrago & J.M. Adam
ICITECH, Universitat Politècnica de València, Valencia, Spain

ABSTRACT: Although recent years have been marked by a substantial research effort on the progressive collapse of frame structures, precast reinforced concrete (RC) structures have been the subject of fewer studies when compared to cast-in-place RC or steel ones. Given that precast RC is being increasingly used and that it can be particularly vulnerable under accidental loading conditions, a better understanding of how secondary resisting mechanisms can be activated in such structural systems is a necessary requirement towards building more robust structures in the future. This paper presents computational simulations of a two-floor precast RC frame structure used to predict and extrapolate results from an ambitious experimental campaign involving the sudden removal of edge and corner columns from a purposely built real-scale building.

1 INTRODUCTION

Progressive collapse can be defined as the phenomenon through which an initial localised failure propagates to other parts of a structural system, often leading to the collapse of the entire structure or to a disproportionate part of it. Such events usually occur when structures are exposed to abnormal loading conditions and typically result in significant negative consequences for society. Some classic examples of progressive collapse of building structures include that of the Ronan Point tower (London, 1968) and of the A.P. Murrah Federal Building (Oklahoma, 1995), while more recent occurrences include the collapse of the Hard Rock hotel (New Orleans, 2019) and of the Champlain towers (Miami, 2021). The occurrence of such events over the years and the huge losses they entail have undoubtedly contributed to increased awareness on the need for robust structures that are insensitive to initial local damage. This is clearly evidenced by the growing number of publications on progressive collapse and structural robustness (Adam et al. 2018).

Precast reinforced concrete (RC) components are being increasingly used nowadays due to noteworthy advantages in terms of cost-effectiveness, quality assurance, and durability. However, this structural typology can be characterised by a greater vulnerability to progressive collapse due to the clear lines of weakness it exhibits at joints between precast components (Van Acker et al. 2012), which can contribute to limiting the available alternative load paths (ALPs) in the event of a partial collapse. Despite this fact, the vast majority of research on structural robustness that has been carried out up to now has focused on cast-in-place RC or steel/composite frame structures.

To this end, the research presented in this paper aims to contribute to better understanding the ALPs that may be activated in precast RC structures after the sudden loss of key columns. This is to be achieved through an experimental campaign in which a full-scale precast RC building will be subjected to different sudden column removal scenarios. The two-storey 15×12 m^2 test structure will be heavily monitored during the tests and the acquired results will be employed to calibrate suitable numerical models. These will then be used to extrapolate the experimentally observed response to other situations that are relevant for robustness considerations. In this paper, results of simulations performed prior to testing the actual structure are presented.

2 EXPERIMENTAL CAMPAIGN

The test building that will be subjected to sudden column removal has two floors and a rectangular shape in plan (Figure 1). The structure's skeleton consists of precast RC beams resting on corbels of precast RC columns. Each floor is made up of hollow-core slabs and a cast-in-place RC topping, with the precast slabs placed as indicated by the dotted lines in Figure 1. The planned experimental program involves three individual tests, each intended to simulate the sudden loss of a specific edge or corner column of the first floor (Figure 1).

DOI: 10.1201/9781003348450-87

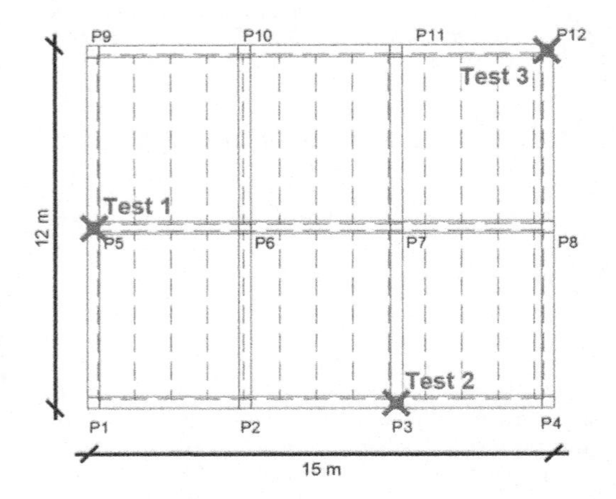

Figure 1. Schematic plan view of test building and location of columns to be removed for each test scenario.

3 COMPUTATIONAL SIMULATION

The Applied Element Method (AEM) has been used to perform nonlinear dynamic computational simulations of the sudden column removal scenarios due to its ability to accurately represent different stages of failure including cracking, separation, and collision (Tagel-Din & Meguro 2000). The chosen simulation strategy was deemed as being adequate following a validation exercise performed before any predictive simulations (Makoond et al. 2021).

For each planned test scenario, two sets of dynamic simulations were performed over an analysis duration of 2 s after sudden column removal. The first involved evaluating the structural response under the effect of a uniformly distributed load of 4 kN/m^2 imposed on bays adjacent to the removed column. This load will be reproduced during the experimental tests. These simulations are therefore useful for predicting the ALPs expected to develop during testing (Figure 2).

Figure 2. Prediction of alternative load paths that will be activated during Test 1.

The second series of dynamic simulations involved gradually increasing the distributed load until collapse occurred (Figure 3) in order to estimate the residual capacity of the test structure after column loss.

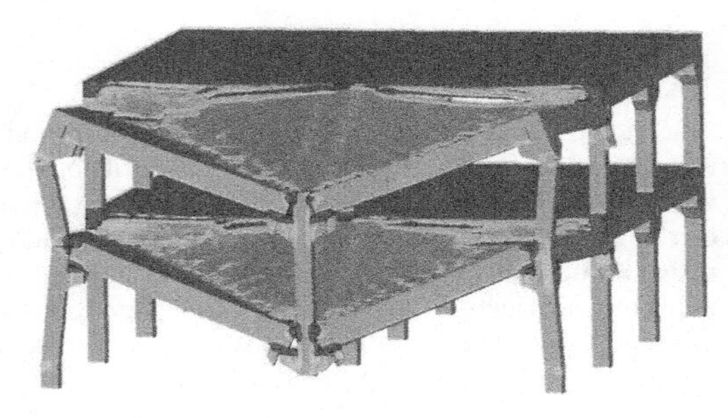

Figure 3. Collapse simulation for the sudden column removal scenario corresponding to Test 1.

4 CONCLUSIONS

The results presented in this article have been used for supporting key decisions on the final design of the test structure and on the loading and monitoring strategy to be employed for testing. In addition, preliminary conclusions on the effectiveness of possible measures for improving structural robustness are also described.

ACKNOWLEDGMENTS

The work presented in this article would not have been possible without the funding received from the *Ministerio de Ciencia e Innovación*.

The authors would like to thank Pedro A. Calderón and Juan J. Moragues for their invaluable help and support. Thanks are also addressed to Ayman El-Fouly (Applied Science International).

REFERENCES

Adam, J. M., Parisi, F., Sagaseta, J., & Lu, X. 2018. Research and practice on progressive collapse and robustness of building structures in the 21st century. *Engineering Structures*, *173*, 122–149. https://doi.org/10.1016/j.engstruct.2018.06.082

Makoond, N., Buitrago, M., & Adam, J. 2021. Progressive collapse assessment of precast reinforced concrete structures using the Applied Element Method (AEM). *6th International Conference on Mechanical Models in Structural Engineering (CMMoST 2021)*.

Tagel-Din, H., & Meguro, K. 2000b. Applied element method for dynamic large deformation analysis of structures. *Doboku Gakkai Ronbunshu*, *2000*(661), 1–10. https://doi.org/10.2208/jscej.2000.661_1

Current Perspectives and New Directions in Mechanics, Modelling and Design of Structural Systems – Zingoni (ed.)
© 2022 Copyright the Author(s), *ISBN 978-1-032-39114-4*

Assessment on impact response analysis by chord member fracture of steel truss bridge

H. Nagatani & D. Yoshimoto
Miyaji Engineering Co., Ltd, Chiba, Japan

K. Hashimoto
Kobe University, Hyogo, Japan

Y. Kitane & K. Sugiura
Kyoto University, Kyoto, Japan

ABSTRACT: In this paper, the impact response due to member fracture was evaluated by dynamic 3D FEM analysis using a frame model considering the composite effect of the steel members with the deck. Furthermore, the experimental results and the analysis results of fracturing the members of the existing steel truss bridge were compared. Therefore, it was confirmed that the FEM analysis result of the impact coefficient has an evaluation accuracy within an error of about 10% compared with the experimental result in the practical stress fluctuation range. Furthermore, it was analytically verified that the impact coefficient decreases to about 1.0 when the fracture time from the start to the completion of the member fracture increases to about the natural period. In addition, it was analytically verified that the influence of the load bending moment and the change in deck rigidity on the impact coefficient was small.

1 INTRODUCTION

In Japan, the significant number of cases resulted in member fracture of steel truss bridges due to insufficient maintenance have been reported, and the safety evaluation on such accidents is very important for aging bridges.

In the safety evaluation of the entire bridge subjected to the member fracture, it is important to assess whether the impact response may cause the successive member breakage or not. In the existing evaluation, the impact coefficient of 1.84 was simply used[1)]. In addition, an analysis study of the impact coefficient focusing on the impact coefficient, which influences the magnitude of dynamic stress reallocation due to member fracture, is being conducted[2)]. But this value has not been verified by the experimental observations.

In this paper, the impact response due to member fracture is evaluated by dynamic 3D FEM analysis using a frame model that considers the composite effect of steel members and deck slab. Then, the experimental result of breaking the existing steel truss bridge member and the analysis result are compared, and the accuracy of the analysis for the evaluation of the impact coefficient is evaluated.

Furthermore, the effect on the impact coefficient will be analytically verified using the fracture time from the start to the completion of fracture of the member, the bending moment to be loaded, and the rigidity of the deck as parameters.

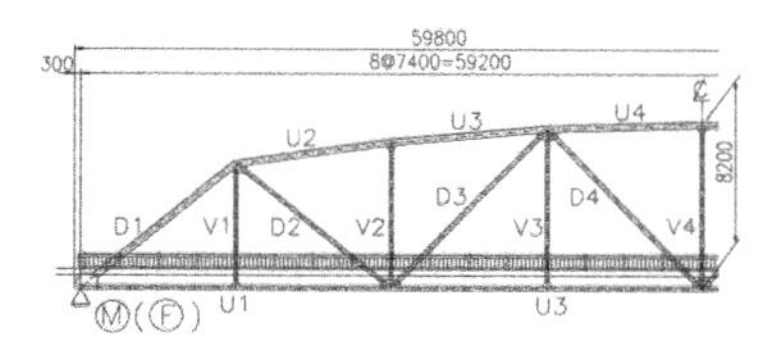

Figure 1. Overview of the targeted truss bridge.

2 MEMBER FRACTURE EXPERIMENT USING EXISTING STEEL TRUSS BRIDGE[3)]

The target bridge is the through steel truss bridge shown in Figure 1 (started operation in 1959, removed in 2012). This bridge is a simple steel truss type with a total length of 80 m, a maximum span length of 59.2 m, a main structure height of 8.2 m, and a width of 3.6 m (1 lane).In this experiment, V4 member and V3 member were cut at a position of about 2.55 m from the upper surface of the lower chord.

The impact coefficient Ii was calculated by the following Eq. (1) with reference to Reference5).

$$I_i = \frac{\sigma_{idm} - \sigma_{is}^{(0)}}{\sigma_{is} - \sigma_{is}^{(0)}} \tag{1}$$

where $\boldsymbol{\sigma_{idm}}$ = maximum value of the absolute value of the dynamic response value of the axial stress at the member stress evaluation point *i* at the time of member fracture; $\boldsymbol{\sigma_{is}}$ = static axial stress after fracture of

DOI: 10.1201/9781003348450-88

member at the same position as σ_{idm}; and $\sigma_{is}^{(0)}$ =axial stress at the same position as σ_{idm} before member fracture.

From Figure 2, 3 the smaller the static stress fluctuation due to member fracture, the larger the impact coefficient. It can be seen that as the static stress fluctuation increases, it converges to a certain value.

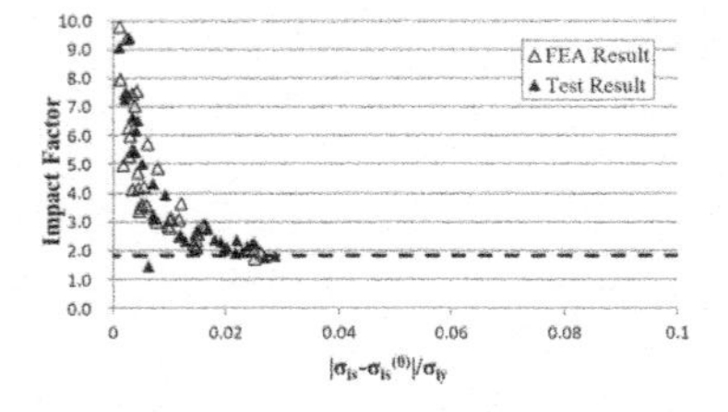

Figure 2. Relationship between impact coefficient and static stress fluctuation (V4 member Fracture).

Table 1. Impact factor by FEA and Test result.

	Fracture member V4			Fracture member V3		
	①Test	②FEA	②/①	①Test	②FEA	②/①
I_i	2.09	1.91	0.91	1.44	1.55	1.07

3 VERIFICATION OF IMPACT COEFFICIENT EVALUATION BY FEM ANALYSIS

In this analysis, the general-purpose structural analysis code Abaqus/Standard 6.17 was used. The steel member was modeled with beam elements and the RC deck was modeled with shell elements1).The stringer and the deck were rigidly connected in consideration of the composite effect. Statically and dynamically loaded on the fractured member with the acting section force as an external force.

Figure 2, 3 and Table1 show a comparison of the analysis results and experimental values for the impact coefficients when the V4 and V3members fracture. From this results, when the static stress fluctuation is 2% or less of the yield stress, the difference between the two is large, but in the range beyond that, the two are in good agreement. The difference between the two is within 10%.

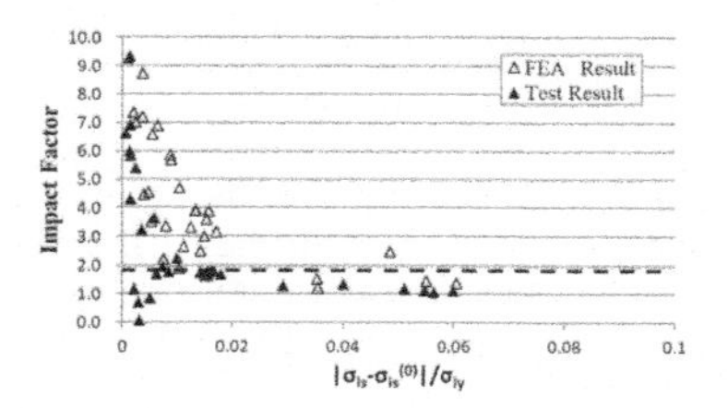

Figure 3. Relationship between impact coefficient and static stress fluctuation (V3 member Fracture).

Furthermore, the table shows the analysis results of the fracture time Δt from the start to the completion of fracture of the member, the decrease in deck rigidity, and the impact coefficient when the acting bending moment is 0. Where, T is the natural period of the dominant vibration mode.

Table 2. Impact coefficient analysis results.

	Fractre time Δt (seec)				Δt=0.010 sec	
	T/32 0.010	T/8 0.042	T/4 0.083	T 0.332	M=0	E=1/2$E0$
I_i	1.82 (1.00)	1.66 (0.91)	1.37 (0.75)	1.03 (0.57)	1.82 (1.00)	1.82 (1.00)

4 CONCLUSIONS

In this study, the impact coefficient used to evaluate the redundancy of a steel truss bridge was evaluated by static and dynamic FEM analysis, and the applicability was evaluated by comparing it with the actual bridge loading test results. In addition, analysis was performed by changing the loading methodand deck rigidity, and their effects were verified. The results obtained in this study are summarized below.

(1) The FEM analysis result of the impact coefficient has an evaluation accuracy within an error of about 10% compared with the experimental result in the practical stress fluctuation range. Therefore, the FEM analysis shown here is effective in evaluating the impact coefficient.
(2) The impact of the fracture time Δt from the start to the completion of the member fracture when loading the member force released by the member fracture is large, and if the time exceeds the natural period of the dominant vibration mode, the impact The coefficient approaches 1.0.
(3) The influence of the change in the rigidity of the deck and the bending moment of the loaded load is small on the evaluation of the impact coefficient.

REFERENCES

1) Nagatani,H., Akashi,N., Matsuda,T., Yasuda,M., Ishii, H., Miyamoto,M., Obata,Y., Hirayama,H. & Okui,Y. 2008.Structural redundancy analysis for steel truss-bridges in Japan. *Journal of Japan Society of Civil Engineers, Division A* 65(2):410–425
2) Goto,Y.,Kawanishi,N. & Honda,I. 2011. Dynamic stress amplification caused by sudden failure of tension members in steel truss bridges, *J. Struct. Eng. ASCE* 137 (8): 850–861
3) Hashimoto,K.,Nakamura,E. & Sugiura,K. 2013.Measurement of impact coefficient and stress redribution due to facture of mambers of the existing steel trass, *Journal of Structural Engineering, Japan Society of Civil Engineers* 59(A):180–189

Current Perspectives and New Directions in Mechanics, Modelling and Design of Structural Systems – Zingoni (ed.)
© 2022 Copyright the Author(s), ISBN 978-1-032-39114-4

Lumped plasticity approach for the robustness assessment of structures

T. Molkens
Department of Civil Engineering, KU Leuven, campus De Nayer, Sint-Katelijne Waver, Belgium
Sweco Belgium bv, Berchem, Belgium

ABSTRACT: Most codes specify that a structural design should meet minimum requirements for robustness. However, simple design rules that allow to comply with these requirements are scarce and can be improved. In this contribution, a method is set out which can be directly implemented in commercial software available at most engineering offices. The ductility is hereby included by a combination of a definable permissible rotation angle that can be implemented by the user as a material-dependent parameter and a lumped plasticity function in the nodes. By doing so, all plasticity is thus concentrated in the nodes instead of smeared out over the elements. The method is illustrated by a real case-study and presented in flow-chart format. It permits an easy at hand verification of robustness using the same models and software used for the ordinary design of framed structures.

1 INTRODUCTION

Unfortunately, the robustness of structures (or lack of) is often only assessed after a dramatic event. In September 2021, a forklift in the port of Antwerp (B) hit a column, resulting in the collapse of 1600 m² of warehouse space, Figure 1 (Van den Buijs, 2021). Human error (Linssen, 2019) resulted in a partial collapse of a parking building in Eindhoven (NL), Figure 2. Nevertheless, it is essential to ensure minimum structural integrity in the design phase.

Figure 1. Collapse of a warehouse in the port of Antwerp (B) (Van den Buijs, 2021).

When developing alternative load paths, catenary action in beams or membrane action in plates are often used, but this can only be substantiated if the stiffness of the surrounding structure is also known in the event of such an accidental action. Moreover, there is no doubt that the material behaviour needs to be known because only by exploiting these alternative load paths can be developed (e.g. via rotational ductility). Finally, and possibly most importantly, designers must carry out such simulations using their known easy-at-hand software tools. The proposed lumped plasticity methodology answers this problem for framework-like constructions. It is an easily applicable method that avoids ignoring or using overly simple models to assess the robustness criterion.

2 ENHANCED REDUNDANCY

A major obstacle contemporary practising engineers face is that their software is usually based on an elastic force distribution. Usually, commercial software allows for the implementation of a function description. It has a significant advantage in composing load combinations because of the superposition principle. However, it has the initially insurmountable disadvantage of a staggering amount of plastic calculation capacity necessary to generate alternative load paths. The most critical internal force effects are usually, if not always, located at the joints between elements. Therefore, it is that plastic hinges will develop at these nodes. An elegant solution is obtained by concentrating all the plasticity at the level of the joints/nodes.

In the first step, we assume a monolithic construction in which the connections are as strong as the constituent elements. The derivation for calculating the rotational constant C_{el} in the elastic domain of an entire rectangular section and modulus of resistance W_{el} can be found in Figure 2 and equations (1) to (3), with the section h's height elongation ε_x.

It may be noted that the final calculation of the rotational constant depends only on Young's

DOI: 10.1201/9781003348450-89

modulus E, the second moment of area I of the section and an equivalent length L_{eq} over which the strains are built up.

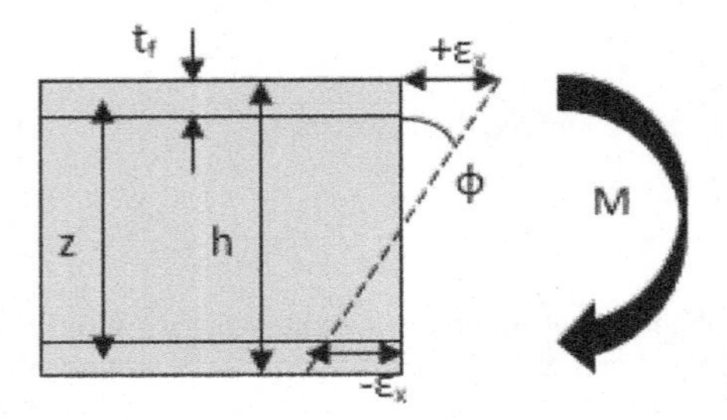

Figure 2. Derivation of rotational ductility function.

$$C_{el} = \frac{M_{el}}{\varphi_y} = \frac{E \cdot W_{el} \cdot h}{2L_{eq}} = \frac{E \cdot I}{L_{eq}} \tag{1}$$

$$M_{el} = f_y W_{el} = \varepsilon_Y \cdot E \cdot W_{el} \tag{2}$$

$$\varphi_y = \frac{\varepsilon_y L_{eq}}{h/2} = \frac{2\varepsilon_y L_{eq}}{h} \tag{3}$$

Because of the displacement ductility μ_δ, defined in this contribution as the ratio between the elongation at ultimate tensile strength ε_u and that at the yield strength ε_y, a second horizontal branch should be added (or for numerical reasons a quasi-horizontal branch). The value C_{el} calculated above is the gradient of the first branch in Figure 5 taken from (EN 1993-1-8, 2003). For a fully plastic solid section (class 1 sections according to (EN 1993-1-1, 2003)), it can be calculated via the equations (4) to (6). The end of this plateau is determined by the virtual spring constant value C_{pl}.

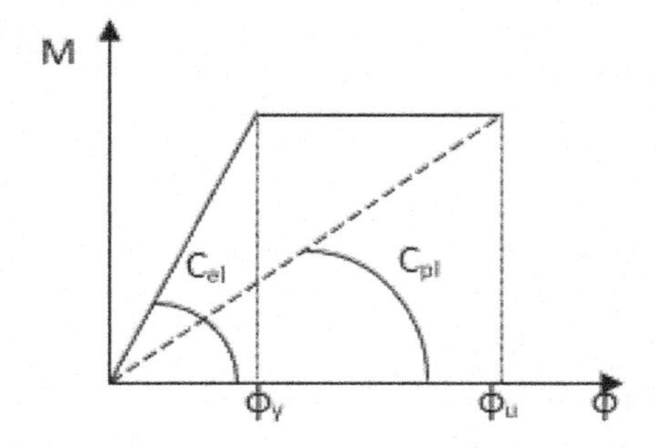

Figure 3. Moment-rotation relation, lumped plasticity.

$$C_{pl} = \frac{M_{pl}}{\varphi_u} = \frac{E \cdot W_{pl} \cdot h}{2 \cdot \mu_\delta \cdot L_{eq}} = \frac{2}{3} \frac{E \cdot I}{\mu_\delta \cdot L_{eq}} \tag{4}$$

$$M_{pl} = f_y W_{pl} = \varepsilon_Y \cdot E \cdot W_{pl} \tag{5}$$

$$\varphi_u = \frac{\varepsilon_u L_{eq}}{h/2} = \frac{2\mu_\delta \varepsilon_y L_{eq}}{h} \tag{6}$$

3 CASE-STUDY

In 2012, a power plant with a capacity of 300 MW was commissioned in the port of Ghent. Prefabrication was used for the cooling tower unit to achieve a short construction period. Therefore all beam-column connections can be assumed as hinged. Afterwards, a question arises about the robustness criterium. To illustrate; the loss of one column in the facade is considered. Because of the prefab, damping of only 2% is considered.

Normal forces of up to 34 kN are calculated, which are far below the capacity of joint details. No risk for a progressive collapse was found, and the structure, although not designed for such events, can be considered robust. Despite significant horizontal deformations, the remaining columns also proved capable of ensuring the stability of the structure as a whole.

4 CONCLUSIONS

The proposed method uses simple commercial design software that practising engineers are familiar with. Concentrating all plasticity in the nodes (connections) via a moment-rotation function elaborated in this contribution answers shortcomings of quasi-static approaches. The time expenditure remains limited without essential additions because the starting point is an already existing model. Therefore, this method can be considered a steppingstone for verifying robustness criteria in design practice.

REFERENCES

EN 1990, 2015. Eurocode 0 Basis of structural design (consolidated version including A1:2005 and AC:2010). Brussels, Belgium: CEN250.

EN 1991-1-7, 2006. Eurocode 1 - Actions on structures - Part 1-7: General actions - Accidental actions. Brussels: CEN250.

EN 1992-1-1, 2005. Design of concrete structures - Partl 1-1: General rules and rules for buildings (+AC 2008). Brussels: CEN250.

EN 1993-1-1, 2003. Eurocode 3 - Design of steel structures: Part 1-1: General rules and rules for buildings. Brussels: CEN.

EN 1993-1-8, 2003. Eurocode 3: Design of steel structures - Part 1.8: Design of joints. Brussels: CEN250.

EN 1999-1-1, 2007. Design of aluminium structures - part 1-1: General structural rules. Brussels: CEN.

Hoffmann, N., Kuhlmann, U. & Skarmoutsos, G., 2019. Design of steel and composite structures for robustness. Ghent, Belgium, Taylor and Francis Group, pp. 2193–2199.

Linssen, J., 2019. Oorzaken instorting parkeergarage Eindhoven. Cementonline.

Van den Buijs, D., 2021. Loods ingestort in de Antwerpse haven: "Heftruck reed tegen steunpilaar". VRT NWS, 21 september.

Current Perspectives and New Directions in Mechanics, Modelling and Design of Structural Systems – Zingoni (ed.)
© 2022 Copyright the Author(s), ISBN 978-1-032-39114-4

Progressive collapse resistant design of tall buildings under fire load

Feng Fu
Department of Civil Engineering, School of Mathematics, Computer Science and Engineering. City, University of London, London, UK

ABSTRACT: Fire is one of the key causes of the progressive collapse of a tall building. One of the remarkable fire induced collapse incidents is the world trade center collapse. Therefore, the progressive collapse resistant design of tall buildings under fire is a pressing issue for structural engineers. In this paper, Based on the existing fire incidents as well as available experimental tests and numerical models developed by the author and other researchers, the local and whole building behavior of a tall building in fire will be first investigated, followed by detailed discussions of the collapse mechanism of a tall building under fire. The key factors affecting the collapse of a tall building in fire will be identified. Based on these investigations, the effective design measures to prevent the collapse of a tall building will be recommended.

1 INTRODUCTION

Fire can cause major social and economic loss in a community. Fire damage is a major factor for steel tall building collapse. Steel framed structures are vulnerable to fire hazard. They affect either the lateral buckling of a structural column, or the longitudinal axis of a structural beam. In time, the damaged steel members fail, leading to the collapse of the structure. This will be either by partial damage to the structure or the structure as a whole will collapse. The fire damages cause unprotected or protected steel frame being directly exposure to different levels of temperatures– which makes the structural systems more vulnerable. Damages due to fire in protected steel building is also substantially significant.

The standard approach in fire design is mainly based on the component levels using prescriptive approaches, however, the uncertainties in variables are not incorporated to this approach. There are large uncertainties in the values of the parameters that affects the fire behavior of the structures. This includes thermal conductivity of insulation material, fire load and material mechanical properties at elevated temperatures.

To fully understand the behaviour of tall buildings in fire and the effective method to design progressive collapse resistant of tall buildings under fire load, in this paper based on the existing fire incidents as well as available experimental tests and numerical models developed by the author and other researchers, the whole building behavior of a tall building in fire will be first investigated, followed by detailed discussions of the collapse mechanism of a tall building under fire. The key factors affecting the collapse of a tall building in fire will be identified. Based on these investigations, the effective design measures to prevent the collapse of a tall building will be recommended.

2 FE ANALYSIS OF MULTI-STORY STEEL BUILDINGS IN FIRE

2.1 *Geometry and design*

In the first exercise, Six prototype steel buildings with 3,4 & 5 storey heights are considered with single damage state. This approach can also be applied in different types of building and damages states. All modelled buildings for this analysis have similar 15 m by 15 m plan area consisting of 3 bays of 5 m in x direction and 2 bays in y-direction (Table 1).

Table 1. Steel building models information and geometry.

Model number	Model Type	Geometry
Model 1	3-storey building frame with 3-bays-without bracing	12m height & 15m width
Model 2	3-storey building frame with 3-bays-with bracing	12m height & 15m width
Model 3	4-storey building frame with 3-bays-without bracing	16m height & 15m width
Model 4	4-storey building frame with 3-bays-with bracing	16m height & 15m width
Model 5	5-storey building frame with 3-bays-without bracing	20m height & 15m width
Model 6	5-storey building frame with 3-bays-with bracing	20m height & 15m width

DOI: 10.1201/9781003348450-90

2.2 *Heat transferring analysis using parametric fire*

The 6 multi-storey buildings were replicated using ANSYS . The fire is applied in different locations as shown in Figure 2, to simulate different fire scenarios.

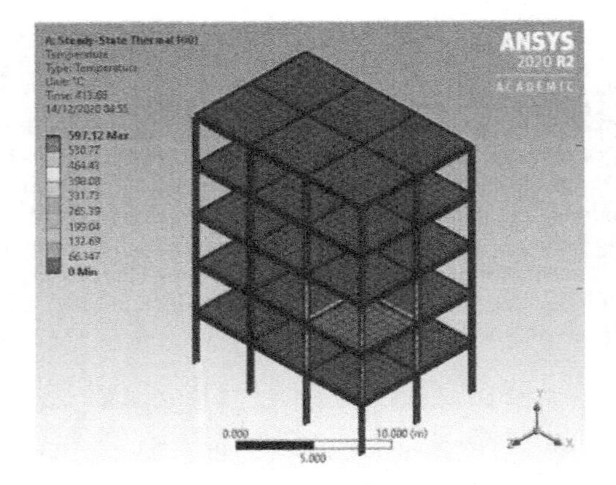

Figure 2. Fire applied in conner location.

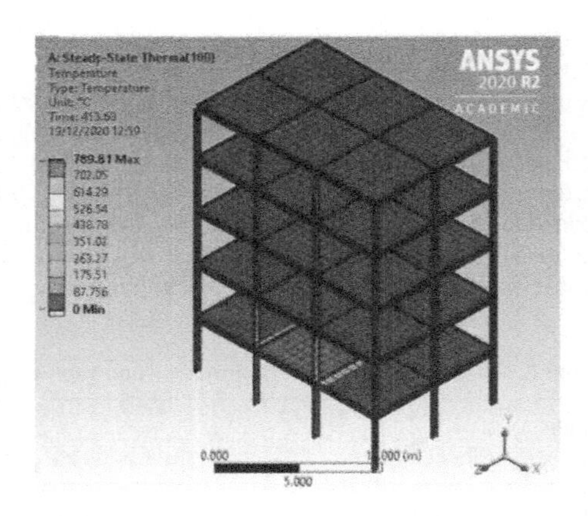

Figure 3. Fire applied in middle location.

3 FE SIMULATION OF CARDINGTON FIRE TEST

In this exercise the Cardington fire test is simulated using ANSYS (as shown in Figure 4). The model replicate the building used in Cardington test. It is also a steel frame with cross bracing system. Three fire scenarios was simulated, as it shown in Figures 5, 6, 7.

Figure 4. Fire analysis of Cardington fire test.

4 CONCLUSION

In this paper, based on the existing fire incidents as well as available experimental tests and numerical models developed by the author and other researchers, the local and whole building behavior of a tall building in fire will be first investigated, followed by detailed discussions of the collapse mechanism of a tall building under fire. The key factors affecting the collapse of a tall building in fire is identified. Based on these investigations, the effective design measures to prevent the collapse of a tall building is recommended.

ACKNOWLEDGEMENT

The author shows his gratitude for my two students, Marifhe C. Mindanao and Wing Sing Tsang for their work in the fire models in section 2 and 3 when they were doing their project under my supervision.

Current Perspectives and New Directions in Mechanics, Modelling and Design of Structural Systems – Zingoni (ed.)
© 2022 Copyright the Author(s), *ISBN 978-1-032-39114-4*

A multi-scale approach for quantifying the robustness of existing bridges

V. De Biagi, B. Chiaia & F. Kiakojouri
Department of Structural, Geotechnical and Building Engineering, Politecnico di Torino, Torino, Italy

ABSTRACT: Bridges are among the most relevant structural engineering works in transport and mobility infrastructure. Depending on a wide range of needs and constraints, various types of structures are found: simply supported beams on piers, box girders, Gerber decks, arches, balanced systems, etc. European infrastructural heritage has now more than 50 years of working life, with increasing traffic loads and continuous ageing and needs maintenance. Recent cases of existing bridge failures have opened the problem of the robustness of such systems. To this aim, a multilevel framework is formulated. This approach is needed for studying the propagation of damage from the element level to the whole structure. In the proposed multilevel approach each single part is studied and its damage tolerance is assessed. The effects of the damage on the single part on the overall bridge structural scheme are then assessed. This multilevel analysis allows to define a member consequence factor, i.e., a measure of the overall effects of the local damage.

1 INTRODUCTION

Bridges are among the most important structural engineering works in the transportation and mobility infrastructure, allowing high-speed and regular tracks to be built in densely populated places or in topographically difficult areas. Recent bridge disasters have raised concerns about the damage tolerance of such systems (Morgese et al. 2020). Although robustness is a key aspect in assessing the safety of the structures subjected to threads, the large part of the studies focuses on specific typologies of existing bridges, rather than on general approaches to quantify the robustness for a wide set of structures. A multi-scale approach is herein proposed to address this goal.

2 A MULTI-SCALE APPROACH FOR BRIDGE ROBUSTNESS ASSESSMENT

Although various metrics for quantifying the robustness already exist (Starossek and Haberland 2011), to define a profitable approach to quantify the robustness of bridges, as they present different static schemes, basic properties should be first reported. First, (i) a statically determinate structure is not robust since every possible damage can have consequences. (ii) Load transfer is achieved through specific paths and it is governed by the ways the elements are arranged and the stiffnesses are distributed across the structure (De Biagi & Chiaia 2013). (iii) Each element of the structure has its own robustness (intrinsic robustness). (iv) The failure of a single element has an effect on the whole structure. Elements arrangement and single-component properties (e.g., ductility) influence collapse propagation. The previously reported concepts should be kept in mind when formulating a method for assessing the robustness of existing bridges. Differently from new constructions, where the project information is sufficient to draw considerations on the geometry of the system, the weights and the load distribution, the capacity of the single elements and their ductility, in existing constructions hypotheses should be made. Meanwhile, the existing bridges exhibit different structural conception, depending on where they are located, who designed them, when they have been designed. In the Sixties, large span arch bridges were built, while nowadays viaducts are preferred.

To this aim, the proposed multi-scale approach herein proposed intends to highlight a framework for assessment of the robustness of such a variable structural item. Essentially, the robustness is the cross-result of two separate evaluations: the robustness of each component of the bridge and the robustness of the static scheme of the bridge, as detailed in the following.

2.1 *Robustness of the single component*

Each component of the bridge is made of various elements. For example, a concrete deck might consist in a grillage of main beams connected by transverse beams over which a slab lays. Similarly, bridge substructure is composed of a cab beam and the piles. The foundations are usually made of separate piles connected in the top by a slab or a beam. Other examples can be traced, considering that the technologies in bridge design and construction are various and

DOI: 10.1201/9781003348450-91

sometimes tailored to specific site constraints (material, geotechnical problems, construction phases…).

Although a large variety of components can be traced, it should be noted that a certain amount of robustness can be associated to each. For example, considering prestressed beams, the number of tendons is usually larger than one, with the possibility of redistributing the forces among the remaining parts if degradation phenomena, like corrosion, act on one of the tendons. This allows a certain amount of robustness with respect to environmental phenomena and degradation. Considering a concrete bridge deck, the arrangement of the elements foster the robustness of the component. In this sense, although is difficult to correctly quantify the contribution, the slab redistribution transfers the loads from the elements that might reduce their capacity (due to a damage) to the elements that are still "safe". It results that, in general, a certain degree of robustness exist withing each component of the bridge. This results in a sort of extra-capacity for providing alternative load paths and redistribution.

2.2 *Robustness of the arrangement of the components*

Depending on the arrangement of the elements, a general theory on the robustness of the bridge can be traced. This serves for understanding how a damage on an element can progress into a local or total collapse of the structure. The large variety of static schemes that are present in bridges implicitly requires a general approach for dealing with progressive damage and global failure. To this aim, there are points that must be considered for understanding the role of each component in the general structural setup. The analysis can be generally performed considering the statics. In detail, there is a sort of hierarchy in the load transfer, with elements that are carried by others. This is the case, for example, in Gerber support, with an element that is carried by another one. The typology of the support, the possibility of working both in compression and in traction, or in compression, only, must be considered in the analysis of the robustness of the bridge. Depending on the arrangement of the elements, the typology of bridge, some considerations on the overall robustness can be formulated. Statically determinate schemes, cannot tolerate local damage, since a hinge in the beam produces a mechanism with the consequent failure of the span. Meanwhile, progressive collapse is prevented by the inherent compartmentalization of the deck. Balanced systems, on the contrary, are prone to progressive collapse when an element is removed. Finally, the arrangement of components that presents the larger robustness is the one in which statically indeterminacy holds. For example, in continuous decks over supports, the formation of a hinge would not cause a mechanism to be formed. To generalize, the use of statics allows to understand the potential effects of a variation in the static scheme of the bridge. If the system is turned into a mechanism after the failure of the component, the robustness is null.

3 CONCLUSIONS

The present paper details an approach for defining the robustness of existing bridges. The method accounts for the mutual dependence between the single components and their arrangement. Although the approach is still at a preliminary, it contains all the ingredients to develop a more detailed framework. Differently from common approaches that tend to define a thread and, then, to compute the outcomes on the structure, the present approach deals with a multi-stage analysis, focusing on the mutual interaction between each part of the structure. Although simple, this method can be adopted in any sort of bridge since a hierarchy can always be traced. Future developments will account for the quantification of the robustness.

REFERENCES

De Biagi V., Chiaia B. 2013. Complexity and robustness of frame structures. International Journal of Solids and Structures 50(22-23): 3723–3741.

Morgese M., Ansari F., Domaneschi M., Cimellaro G.P. 2020. *Post-collapse analysis of Morandi's Polcevera viaduct in Genoa Italy.* Journal of Civil Structural Health Monitoring 10(1):69–85.

Starossek U. 2018. *Progressive collapse of structures*, Second Editon. ICE Publishing, London, 234 pp.

Starossek U., Haberland M. 2011. *Approaches to measures of.*

Current Perspectives and New Directions in Mechanics, Modelling and Design of Structural Systems – Zingoni (ed.)
© 2022 Copyright the Author(s), ISBN 978-1-032-39114-4

Interpretation of collapse modalities for a timber roof truss

M. Acito, E. Magrinelli, C. Chesi & M.A. Parisi
Politecnico di Milano, Milan, Italy

ABSTRACT: Roof structures in historical buildings are often based on complex timber trusses; this is typically the situation of theatre buildings, where a wide room is present and large spans have to be covered. Several studies have been done on such structural systems showing that, in the case of complex roof systems, connections between timber elements were conceived in the past mainly to resist vertical loads corresponding to symmetrical patterns, with no consideration of eccentric loads and horizontal forces. In the paper, this problem is investigated with reference to the specific case study offered by a theatre, where a queen post truss was adopted for the roof system and the sudden collapse of one truss has occurred. The theatre, which is located in northern Italy, was built at the beginning of the 19th century. Although some modifications were made through time, the roof structure has preserved the original configuration. A precise identification of the real structural conditions has been obtained on the basis of the results of laser scanning and photogrammetric surveys. On this basis, numerical models have been developed in order to analyse a variety of load scenarios compatible with the observed structural collapse, with the final purpose of highlighting possible deficiencies in the original structural configuration, in line with the collection of similar cases which are reported in the literature.

1 INTRODUCTION

The performance of traditional timber roof systems under static loads is discussed with reference to the queen post truss in relation to possible decay conditions of its structural elements. The research was originated by the survey and inspection activities required for the identification of the collapse modalities of a portion of the roof system of an historical building in northern Italy, originally conceived as a theatre. The roof portion affected by the collapse, occurred in 2021, corresponds to the entire tributary area of a truss. The interpretation of the static-geometric configuration of the collapsed elements has led to the identification of two independent structural systems: a main one, consisting of the timber trusses providing support to the roofing, and a secondary one, constituted by a rib structure, to which the suspended ceiling was hanging. After commenting the methodological approach adopted for the survey and technical analysis of the collapsed elements, historical and construction properties of the queen post truss system are recalled. Specifically, the structural performance has been investigated through static and kinematic analyses, in relation to realistic damage conditions affecting the timber elements.

2 METHODOLOGICAL APPROACH TO THE SURVEY OF POST-COLLAPSE SCENARIO

In order to preserve the safety of the technicians involved in the post-collapse phase, photogrammetric analyses were performed by the use of drones and the geometric survey was done by laser-scanner techniques. Complementary to this, historical critical analysis of the structure, in-situ and laboratory testing were performed for the characterization of the geometry of structural elements and material decay conditions as well. By means of images and data coming from the on-site inspection, the global configuration of the roof system has been interpreted as constituted by a primary supporting system (queen post truss) and a secondary rib structure, supporting the ceiling. The collapse involved: a full queen post truss, a couple of rib structures inside the collapsed area, a couple of portions of rib systems, three transversal beams running from the collapsed truss to the next one, a system of secondary beams providing support to the roofing layer, the suspended ceiling portions inside the collapsed area, the entire roof area between the non-collapsed trusses. In the paper, attention is focused only on the queen post truss.

DOI: 10.1201/9781003348450-92

3 HISTORICAL HINTS ON THE QUEEN POST TRUSS SYSTEM

The collapsed truss, commonly identified as a of the "queen post" type, is normally used when a span larger than 10 m has to be covered. Queen post truss was extensively used by Andrea Palladio in the roof systems of buildings he designed.

4 KINEMATIC AND STATIC PROPERTIES OF DIFFERENT TRUSS SCHEMES

In the case of the theatre which is here discussed, the lower chord is composed by three parts, overlapping to each other for some length, where transversal nails and hooping iron strips are present. This can be considered a full connection, able to transfer bending, normal and shear actions. Hinges can be considered at the base and at the top of the lateral posts, where they are connected to the chord and the rafter respectively. Hinge connection can also be assumed for the upper chord. Since rafters are well connected to the chord, it can be assumed that the truss system is simply supported by masonry walls. From the kinematic point of view, one degree of redundancy can be recognized in the above scheme. The kinematic analysis shows that, even in the presence of redundancy, the collapse of a rafter results into the global collapse of the structure, due to the collapse of the chord in bending and shear at the connection with the lateral post. If the lower chord fails, a rigid mechanism would be activated, leading the structure to collapse. In view of interpreting the reasons of collapse, the static response to permanent loads has to be analyzed with reference to different levels of material decay. In the assumed geometrical configuration, a truss spacing of 2.30 m has been adopted. The permanent load condition is characterized by the self-weight of the roof. The stress distribution following the collapse of a rafter demonstrates the impossibility of resisting this configuration, even in the presence of permanent loads only. Indeed, maximum tension stress has a value of 24 MPa, higher than the strength of fir wood, even without material decay.

5 MATERIAL DECAY IN TIMBER ELEMENTS: CAUSES AND STRUCTURAL IMPLICATIONS

Timber elements may be attacked by fungi and insects. Damage to structural elements by insect attack simply consists in the cross-section reduction. The attack by fungi can develop with a humidity level higher than 20% and produces a drastic reduction of the mechanical properties. The effects of material decay have been clearly highlighted through Resistograph® experimental testing on the chord and the rafter, comparing results with the case of intact material. The structural effects of the material decay in the rafter and the chord can be interpreted with reference to the stress increase generated by the cross-section reduction. Assuming that the cross-section reduction due to material decay is by 20% or 30% of the original size, it can be observed that maximum stresses increase by a factor of 2 or 3. In detail, maximum compression in the rafter is increased from 5 to 14 MPa, while maximum traction in the chord from 5 to 13 MPa. Considering that the material is fir wood, belonging to S3 category (C18, see UNI 11035 and EN 1912), a value of 25 MPa can be assumed for the average strength in bending of the intact material. Such value must be reduced in relation to both the long-term duration of the applied load (60%) and a humidity level higher than 20% produced by water seepage, which leads to a further strength reduction by 50%. Therefore, the long-term strength in bending of water saturated fir wood can be estimated to be around 7-8 MPa. Even in the case of a cross section reduction by 10-15%, such strength value would not be sufficient to resist permanent loads.

6 DISCUSSION AND CONCLUSIONS

The need to interpret the collapse condition of a queen post truss in a historical theater building has provided the opportunity to discuss its structural performance. With reference to this structural scheme, kinematic and static considerations clearly show that, even in the presence of structural redundancy, the collapse of the most stressed element, i.e., the rafter, would result in the collapse of the entire truss. Under the effect of permanent loads, however, the strength properties of fir wood, which was here employed, are totally compatible with the imposed stress level. The occurred collapse condition, therefore, could be associated to material strength decay, which has been clearly enhanced by experimental testing: a consistent strength reduction was present at some locations in both the rafter and the chord due to insects and fungi attack. From the mechanical point of view, this is equivalent to a reduction of the resisting cross-section. In the specific case, a reduction of 10-15% in the resisting cross-section of the rafter and the chord would be sufficient to produce stress levels not compatible with material strength. The observed collapse can therefore be explained on the basis of the material decay occurred at critical locations of the timber structural elements.

Current Perspectives and New Directions in Mechanics, Modelling and Design of Structural Systems – Zingoni (ed.)
© 2022 Copyright the Author(s), ISBN 978-1-032-39114-4

Blast response of arching masonry walls: Theoretical and experimental investigation

Idan E. Edri, David Z. Yankelevsky & Oded Rabinovitch
Faculty of Civil & Environmental Engineering, National Building Research Institute, Technion-Israel Institute of Technology, Haifa, Israel

ABSTRACT: The out-of-plane behavior of unreinforced masonry walls is typically characterized by low flexural resistance and brittle failure. Still, providing suitable boundary conditions that enable an arching action enhances the out-of-plane flexural strength and ductility, even under blast loading. The present study, which is reviewed in this paper, combines experimental and theoretical methods. The experimental phase starts with monotonic and cyclic tests on small-size masonry specimens, with focus on the interaction of the masonry unit and the mortar layer in the arching wall. The test tracks the nonlinear behavior of a single mortar joint during complex loading and unloading scenarios. The review of the experimental phase then continues to a laboratory blast test on a full-size wall. In the theoretical part, four analytical models for the dynamic analysis with different structural resolution levels are reviewed. The first model assumes a single-degree-of-freedom response, while the other three account for the multi-degree-of-freedom nature of the response. The physical modeling assumptions in each model are discussed, and their effect on the assessment of the wall response is highlighted. The static and dynamic tests are used as benchmarks for validation of the theoretical models, and physical features of the response are discussed through a series of numerical studies. The combined experimental and theoretical efforts reviewed in this paper provide insight into the complex behavior of such walls, sets quantitative tools for its investigation and analysis, and demonstrates the unique aspects of the blast response of arching masonry walls.

1 INTRODUCTION

The out-of-plane behavior of unreinforced masonry (URM) walls is characterized by relatively low flexural resistance and ductility. The blast response of such walls is of great interest, and the quantification of the nonlinear and dynamic mechanisms that evolve during the blast response is important for understanding the behavior of URM walls under such extreme conditions.

This paper reviews an ongoing research effort that combines theoretical and experimental methodologies. The theoretical part presents an overview of four analytical models for the analysis of the blast response of one-way arching masonry walls. Each model represents the wall at a different structural modeling resolution. The experimental part presents monotonic and cyclic static tests, and a blast test using a blast simulator.

2 THEORETICAL APPROACHES

This section presents a review of four analytical approaches to the analysis of the wall.

The first model (the ARCH model) simplifies the arching masonry wall by an equivalent SDOF system. The model, which is described in detail in Edri & Yankelevsky (2018), is based on the arching mechanism that is enabled due to the axial restraints at the supports. Due to the oscillatory response, two different thrust lines referring to the inbound and rebound resistance mechanisms are considered. The masonry units are assumed as rigid blocks, and all the axial deformations are attributed to the axial strains of the mortar joints. The static resistance term in the equation of motion adopts the resistance function calculated through the analytical procedure presented in Edri & Yankelevsky (2017). Along with the capabilities of the ARCH model, using such approach still limits the assessed dynamic response to the three-hinged-arch mechanism.

This motivated the development of the SIM model in Edri et al. (2019) and its dynamic version in Edri et al. (2020). Both models extend the analysis to consider the multi-degree-of-freedom (MDOF) nature of the wall. This model considers rigid blocks connected by arrays of normal and shear springs that represent the mortar bed joints. The model formulation considers geometrical and material nonlinearities, and the governing equations are derived in the mathematical framework of variational principles. The dynamic analysis presented in Edri et al. (2020)

DOI: 10.1201/9781003348450-93

reveals that due to the complex cracking and arching effects, the qualitative deformation profile of the wall changes during the dynamic response. This observation manifests the necessity of a MDOF modeling that can capture the variation of the deformation profile and its impact on the dynamic behavior of the wall.

In an attempt to take a deeper look into the stress field within each mortar joint, a richer physical modeling technique was proposed in Edri et al. (2021a) that develops the dynamic BIM model. The physical modeling in this case also accounts for rigid masonry units, but now they are connected by nonlinear beam members representing the mortar material. The formulation of the model is based on variational calculus, and it is established in the mathematical framework of the principle of virtual work and Newtonian dynamics.

In the above three models, the physical modeling refers to rigid blocks. Edri et al. (2021b) developed a static continuous beam-type model (CBM model) where the masonry units and the mortar joints are modeled as beam members with nonlinear kinematic and constitutive relations. The model augments the above models and generalizes the structural modeling of masonry walls. The model is derived through variational principles, employing the finite element method for the numerical solution of the governing equations.

3 EXPERIMENTS

This section reviews the static and dynamic tests carried out in the course of this research.

The static tests included small-size arching masonry specimens with two and four blocks subjected to monotonic and cyclic loading. The experimental setup, the geometrical and mechanical properties, and the monitoring system are described in Edri et al. (2022). The 2-block specimen focuses on the complex behavior of the single central joint. The 4-block specimen studies and verifies the local behavior of the joint and the global response of the assembled specimen. Comparisons of the CBM model results with the experimental results showed good agreements.

The dynamic test included a full-scale one-way arching masonry wall loaded by a blast simulator. This experimental study is presented in details in Edri et al. (in prep.). The experiment was designed to explore global and local dynamic measures of the response with emphasis on ones that are not available in the open literature. The dynamic CBM model is used to analyze the blast response of the tested wall. Good agreement is obtained between the model and the test in terms of the global displacement vs. time curves, the temporal variation of the local strains at the blocks, and the dominant frequency of the global and local responses. The dynamic in-plane arching force was evaluated by utilizing the dynamic strain measurements at both sides of the wall.

4 CONCLUDING REMARKS

This paper has reviewed an ongoing effort to investigate the dynamic response of arching masonry walls to blast. The theoretical part of the study has included the development of four analytical models. Starting with SDOF and MDOF representations and ending with a continuous modeling approach has allowed to enhance the understanding of the influential parameters and to quantify their effect on the response. The spectrum of the analytical models, with different structural resolution levels, has set theoretical platforms for the static and dynamic analyses of arching masonry walls, clarified their response, and provided inclusive insight into their nonlinear behavior.

The experimental part of this study has included monotonic and cyclic static tests on small-size masonry specimens, and a controlled laboratory dynamic blast test on a full-size arching wall using a blast simulator. Overall, the static and dynamic experiments have provided new data for the examination and evaluation of theoretical and numerical models, quantified the static and dynamic response, and illuminated the arching mechanism.

The findings of the analyses, the experimental work, and the numerical studies have indicated that the blast response of arching URM walls is characterized by unique nonlinear phenomena that affect the behavior and the analysis of such walls.

REFERENCES

Edri, I. E., and Yankelevsky, D. Z. (2017). An analytical model for the out-of-plane response of URM walls to different lateral static loads. *Engineering Structures*, 136.

Edri, I. E., and Yankelevsky, D. Z. (2018). Analytical model for the dynamic response of blast-loaded arching masonry walls. *Engineering Structures*, 176.

Edri, I. E., Yankelevsky, D. Z., and Rabinovitch, O. (2019). Out-of-plane response of arching masonry walls to static loads. *Engineering Structures*, 201.

Edri, I. E., Yankelevsky, D. Z., and Rabinovitch, O. (2020). Blast response of one-way arching masonry walls. *International Journal of Impact Engineering*, 141, 103568.

Edri, I. E., Yankelevsky, D. Z., and Rabinovitch, O. (2021a). Nonlinear Rigid–Flexible Multibody Modeling of Arching Masonry Walls Subjected to Blast Loading. *Journal of Engineering Mechanics*, American Society of Civil Engineers, 147 (3),04021002.

Edri, I. E., Yankelevsky, D. Z., and Rabinovitch, O. (2021b). Continuous beam-type model for the static analysis of arching masonry walls. *European Journal of Mechanics - A/Solids*, 104387.

Edri, I. E., Yankelevsky, D. Z., and Rabinovitch, O. (2022). Experimental study on one-way arching masonry specimens under monotonic and cyclic loads. *Structures*, 37, 1142–1156.

Edri, I. E., Yankelevsky, D. Z., Remennikov, A. M., and Rabinovitch, O. Dynamic response of arching masonry walls under blast load. To be submitted.

Current Perspectives and New Directions in Mechanics, Modelling and Design of Structural Systems – Zingoni (ed.)
© 2022 Copyright the Author(s), ISBN 978-1-032-39114-4

Blast behaviour of natural fibre composites: Response of medium density fibreboard and flax fibre reinforced laminates

S. Gabriel & S. Chung Kim Yuen
Blast and Impact Survivability Research Unit (BISRU), Department of Mechanical Engineering, University of Cape Town, Rondebosch, South Africa

G.S. Langdon
Department of Civil and Structural Engineering, University of Sheffield, Sheffield, UK

C.J. von Klemperer
Department Of Mechanical Engineering, University Of Cape Town, Rondebosch, South Africa

ABSTRACT: There is currently a poor understanding of blast behaviour of natural fibre composites, yet these materials are extensively used in various other applications. An experimental study was carried out to examine the behaviour of flax fibre reinforced epoxy composites and a South African locally produced medium density fibreboard subjected to uniform blast loading. The transient response was similar for both materials and showed that the evolution of the mid-point displacement was viscous elastic and damped. A similar cracking pattern was observed on the surfaces; however, a different cracking phenomenon were found along the cross-section.

1 INTRODUCTION

Numerous studies have been investigated to determine the blast related properties of synthetic composites, yet little is known on the response of natural fibre composites to an explosive load (Mouritz, 2019). These materials are utilised in various applications in automotive, construction and other industries (Mohammed et al., 2015). As such, there is a possibility that natural fibre composites could be exposed to a deliberate or accidental blast event. Therefore, understanding their response to blast is important and could help mitigate the detrimental consequences associated with explosive loading.

The purpose of this paper is to describe the preliminary blast response of two natural fibre composites, namely flax fibre reinforced epoxy composite and medium density fibreboard (MDF).

2 MATERIALS AND PROPERTIES

Nine layers of 440gsm flax fabric was infused with Prime 20LV epoxy resin to yield ~10 mm thick laminates for the blast tests using VARTM manufacturing process. A 16mm thick MDF, manufactured in South Africa, was purchased and specimens were cut for the blast tests.

Quasi-static tests were performed to determine material characteristics. These included flexural and tensile tests conducted according to applicable ASTM standards for composite materials. The internal properties of MDF were determined based on SANS standards. The results of the quasi-static tests are shown in Table 1.

Table 1. Properties of the composites based on the quasi-static tests performed.

Property	Flax fibre reinforced epoxy	Supawood MDF
Flexural strength (MPa)	112	30
Flexural strain at failure (mm/mm)	0.030	0.017
Tensile strength (MPa)	65 (0/90°); 40 (45°)	17
Tensile strain at failure (mm/mm)	0.025 (0/90°); 0.022 (45°)	0.010
Internal properties	~	0.38 MPa (tensile) 115.5 MPa (comp)

3 EXPERIMENTAL BLAST WORK

Prior to any experimental work, numerical simulations were carried out to determine the experimental design parameters. Based on the simulation results, the experiments were conducted under what was considered a uniform loading condition. Carefully controlled, scaled explosion experiments were conducted in air. A schematic of the testing arrangement is shown in Figure 1. A horizontal central portion of the back face of a panel was recorded for some of

DOI: 10.1201/9781003348450-94

the low charge tests to obtain the transient response of the panels.

The charge mass range tested was between 5 g to 11 g, and 2 g to 6 g for the flax fibre reinforced polymer (FRP) and MDF respectively. A total number of 14 blast tests were carried out in this series

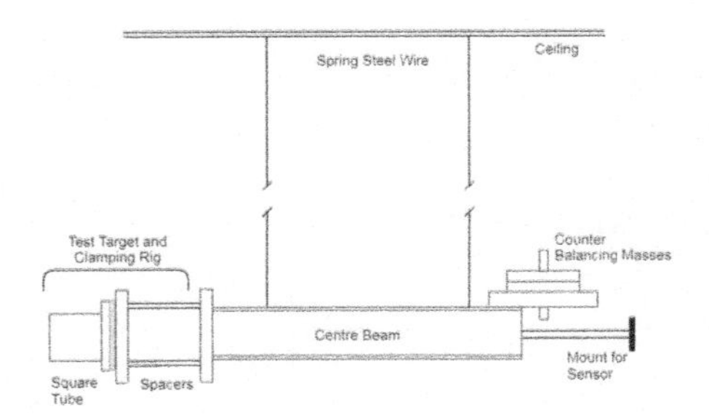

Figure 1. Schematic of the blast test arrangement.

4 RESULTS AND DISCUSSION

The transient behaviour appeared to be similar for both materials. Viscously damped harmonic oscillations were observed on transient midpoint time histories as shown in Figure 2.

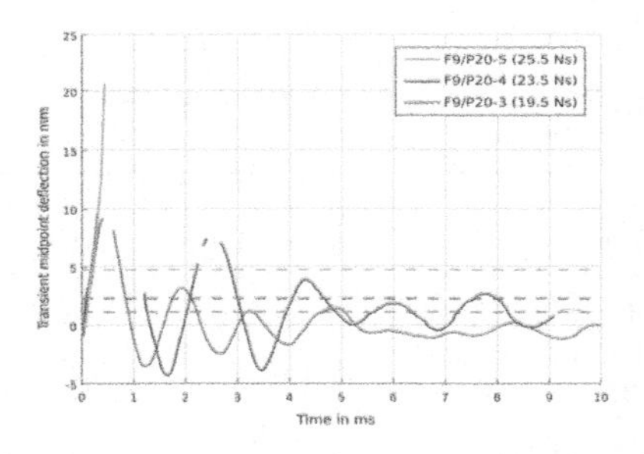

Figure 2. Midpoint deflection time history for flax fibre reinforced composite panels (dotted lines represent permanent deflection).

The transient midline displacement profiles for the panels also appeared to have a similar behaviour. Dome-shaped profiles were initially observed before any cracking occurred. Once cracking occurred, the profile changed to a conical shape. Upon rebound as the crack closes, the profile returned to the dome shape. This phenomenon is shown in Figure 3.

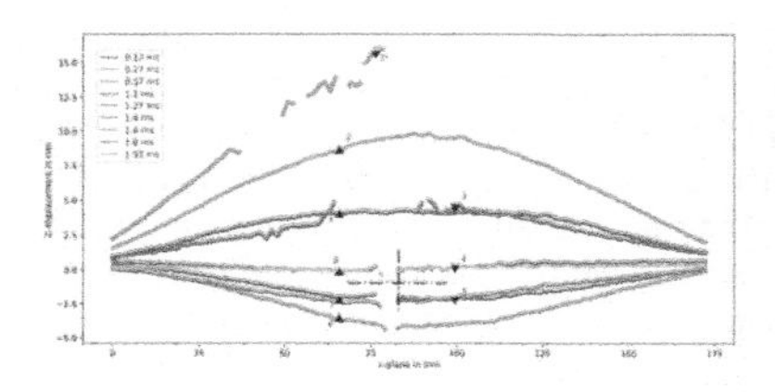

Figure 3. Midline displacement profile for a flax FRP.

Cracking on the front and back faces, and permanent deflection were observed on the panels. A result of the cracks traced on the panels is shown on Figure 4. Fragmentation occurred on the MDF at an impulse of 20 Ns (6 g). In the flax FRP, 39.5 Ns (11 g) was the limit where fragmentation would start to occur.

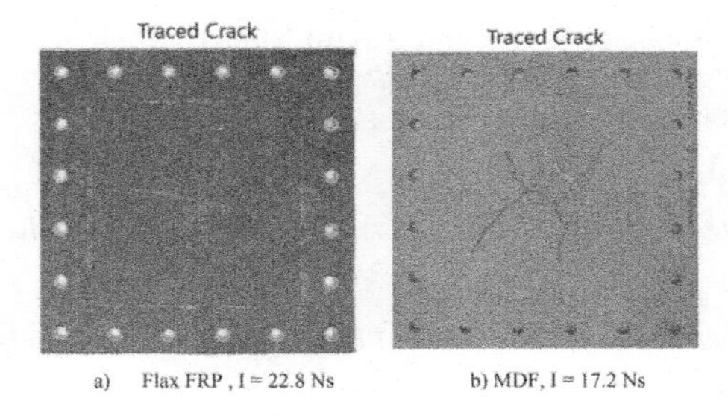

Figure 4. Photographs of the back surface of blast tested panels.

In-plane cracking was observed on the flax fibre reinforced composites as shown in Figure 5a. In comparison, out of plane cracking was observed on the MDF as shown in Figure 5b.

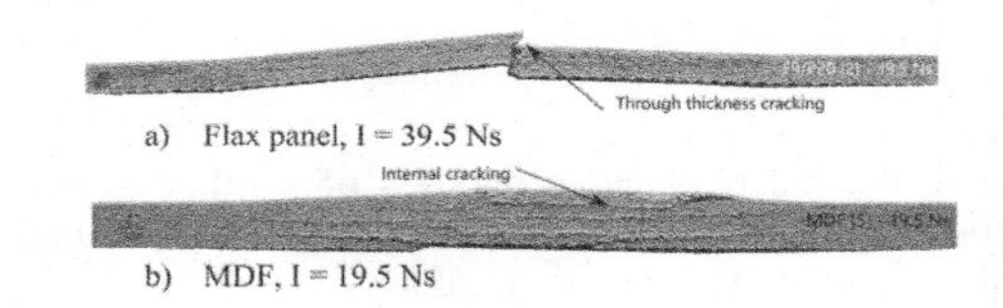

Figure 5. Photographs of the midline cross section of a) flax panel and b) MDF.

5 CONCLUDING REMARKS

A series of small-scale uniformly loaded blast experiments were conducted. The blast performance and transient response was evaluated. It was found that the blast resistance of the flax fibre reinforced composites was superior compared to the MDF. Similar observations were made for the transient behaviour for both materials where viscous elastic damped harmonic vibrations were observed. Cracking appeared to be the main damage observed on both panels. In comparison to synthetic materials, both materials have weak blast resistance properties, however it may be possible to create a more sustainable blast resistant material if hybrid composites are considered.

REFERENCES

Mohammed, L., Ansari, M. N. M., Pua, G., Jawaid, M. & Islam, M. S. 2015. A Review on Natural Fiber Reinforced Polymer Composite and Its Applications. *International Journal of Polymer Science*, 2015, 1–15.

Mouritz, A. P. 2019. Advances in understanding the response of fibre-based polymer composites to shock waves and explosive blasts. *Composites Part A: Applied Science and Manufacturing*, 125, 105502–105502.

Current Perspectives and New Directions in Mechanics, Modelling and Design of Structural Systems – Zingoni (ed.)
© 2022 Copyright the Author(s), ISBN 978-1-032-39114-4

Comparison of TNT to PE4 charge using a simplified rigid torso

T. Pandelani, D. Modungwa, S. Hamilton & J.D Reinecke
Defence and security, Council for Scientific and Industrial Research (CSIR), Pretoria, South Africa

ABSTRACT: The energetic output and the blast load associated with the detonation of a mass of high explosive will differ depending on the chemical composition of the explosive itself. It is therefore convenient to equate the effects of an explosive to TNT. The "TNT equivalent (TNTeq) ratio" refers to the mass ratio of the explosive in question that will produce equal peak overpressure to that equivalent mass of TNT. There is vast range of equivalency ratios for PE4 in the literature. However, there appears to be no research that determines effect of TNT equivalency of spherical shape of PE4 charge detonated on rigid ground surface

This paper presents experimental measurements of reflected pressure-time histories from a series of well-controlled small scale blast tests. An investigation of TNT equivalent ratio of spherical PE4 charge detonation on rigid ground was performed using the Blast Test Device (BTD). Our results presented here clearly demonstrates that the TNTeq ratio of 1.37 is higher in the near field environment and TNTeq of 1.2 will be the best to produce blast wave by the hemispherical PE4 detonations on rigid surface.

1 INTRODUCTION

Explosions, whether accidental or intentional, can cause severe injuries and even death. The severity and type of blast injury depends on many factors such as: the type of explosives used, distance of the victim from the explosion, protective clothing and surrounding objects and walls [1].

There are four categories of blast related injuries: Primary, Secondary, Tertiary and Quaternary, as illustrated in Figure 1. Primary blast injuries (PBI) are caused by the propagation of the blast wave from the centre of detonation through the victim, resulting in damage to organs predominantly filled with air [2].

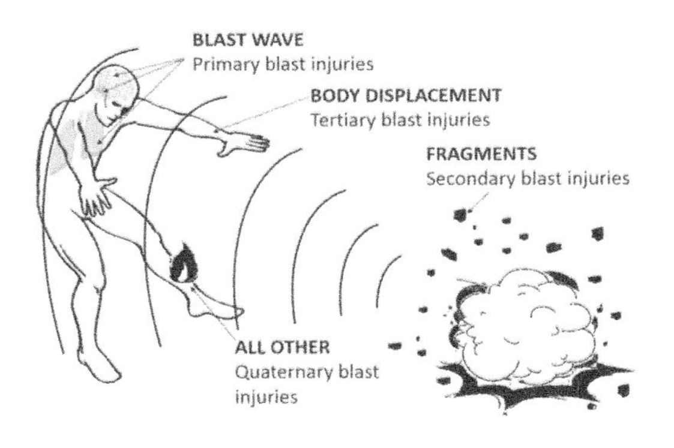

Figure 1. Blast injury categories [1, 2].

The influence of different explosives such as plastic explosives (PE4) and trinitrotoluene (TNT) which have been used in Improvise Explosives Devices (IEDs) should be investigated to begin to assess the effectiveness of the performance of materials and personal protective equipment used in this environment. It is therefore convenient to equate the effects of an explosive to a common material, such as trinitrotoluene (TNT). There is the need, therefore, to better quantify TNT equivalence for commercial explosives.

This paper presents an experimental investigation of TNT equivalence to hemispherical PE4 charges in far-field blast events. Experimental results are presented and conclusions are drawn in order to provide a more informed value of TNT equivalence based on measured pressure.

2 MATERIALS AND METHODS

The test scenario considered in this study was the detonation of a spherical charge over a flat solid concrete surface using TNT and PE4 explosives. This configuration generates complex blast wave behaviour, depending on the mass and height of the charge, such as a reflection wave or Mach reflection. The BTDs and all other measurement sensors were located at a standoff distance (SOD) of 2 m from the charge as in Figure 2. In the vertical direction, the charge was positioned at

DOI: 10.1201/9781003348450-95

three different heights of burst (HoB), which is defined as the distance from the bottom of the charge to the ground. Each HoB formed a distinct test series.

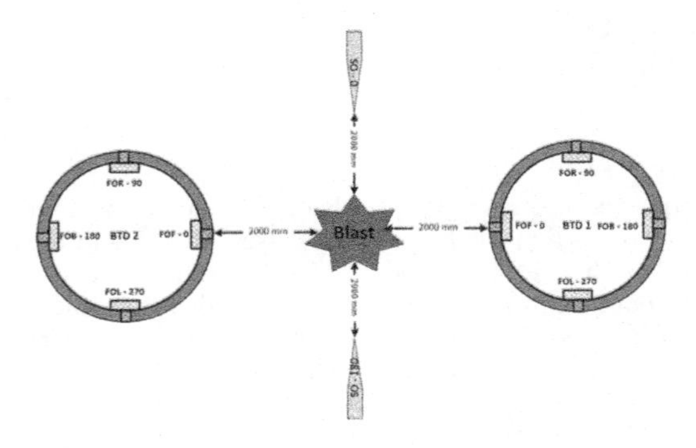

Figure 2. Top view of experimental setup.

3 RESULTS

Figure 3 shows the typical average incident pressure measured from the Face-On (FO) pressure sensor for the 1 kg PE4 and the 1.37 kg TNTeq from 220mm HOB. The average FO pressure was calculated based on the pressure measurements of the two BTDs sensors. The averaged peak pressures for the 1 kg PE4 tests at HoB's of 220 mm was 1209 kPa and 1309 kPa for the PE4 and TNTeq charges respectively.

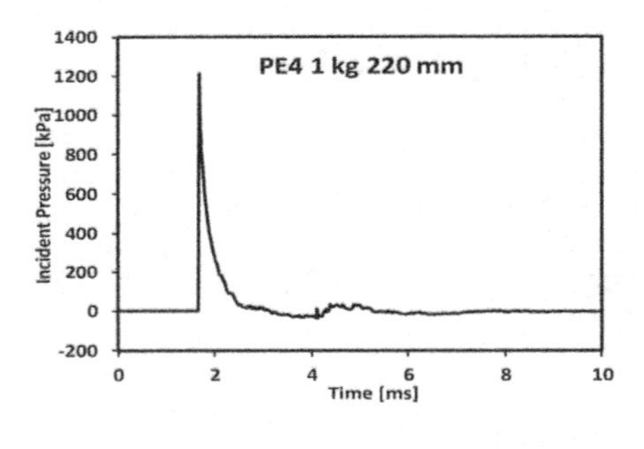

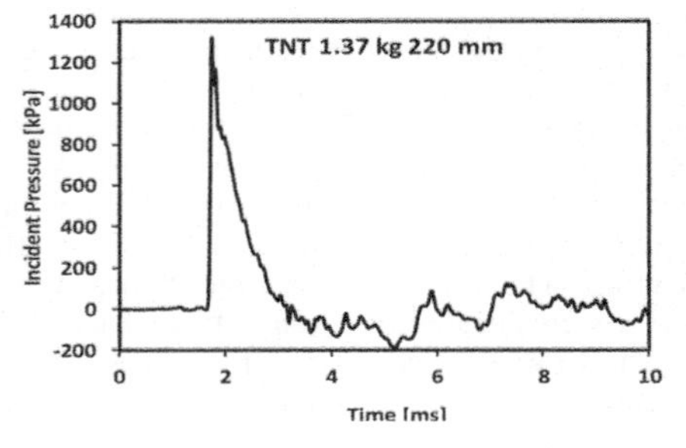

Figure 3. Average FO Pressure for PE4 and TNT Tests at HoB of 220 mm.

4 DISCUSSION

A series of experiments were undertaken using the BTDs to investigate the reflected peak overpressure, positive and negative phase impulse and primary shock were measured from PE4 and TNTeq charges.

The resulting pressure blast profiles were very reproducible.

The PE4 explosives were compared to the TNT eq and the peak overpressures were measured. The results shows that the PE4 1kg 220mm HOB pressure profiles were less by 7% compared to the TNTeq pressure profiles.

5 CONCLUSIONS

The use of explosives can cause severe injuries and even death. To reduce the severity of human injury from these explosions research is being carried out in the realm of blast protection. It is important to understand the complexity, paths and pressure range of the blast waves associated with different explosives used in IEDs to design protection that will attenuate those effects. IEDs are made of various explosives, and it will be better to compare them using TNT.

The TNT equivalence of an explosive is given as the equivalent mass of TNT required to produce a blast wave of equal magnitude to that produced by a unit weight of the explosive in question. There are currently different numbers used by various studies on the TNT equivalence (TNTeq) of PE4 with the ratio of 1.37 being common [2].

The PE4 charge was shown to produce less peak pressure as compared to the TNT eq. Thus, the equivalency ratio required will be much less than the 1.37 used. It appears most researchers use the peak pressure based equivalency ratio of 1.37 when deformation effects are being studied [2-3], however an equivalency ratio of 1.09 - 1.21 maybe more appropriate for pressure measurement.

It is suggested that an equivalency ratio of 1.2 may be appropriate, work conducted by Rigby [4] also indicated an overall TNT to PE4 equivalency of approximately 1.2.

REFERENCES

1. Jorolemon M.R., and D.M. Krywko, 2019. Blast Injuries. *In: Stat Pearls [Internet]. Treasure Island (FL): StatPearls Publishing*; Available from: https://www.ncbi.nlm.nih.gov/books/NBK430914/
2. Weckert, S. and Anderson, C. (2006). A Preliminary Comparison between TNT and PE4 Landmines. Australian Government, Department of Defence.
3. Hyde D.W., Conventional Weapons Program (ConWep), U.S Army Waterways Experimental Station, Vicksburg, MS, USA, 1991
4. Rigby, S. E.; Sielicki, P.W. An Investigation of TNT Equivalence of Hemispherical PE4 Charges. Engineering Transactions, [S.l.], v. 62, n. 4, p. 423–435, 2015. ISSN 2450–8071.

Current Perspectives and New Directions in Mechanics, Modelling and Design of Structural Systems – Zingoni (ed.)
© 2022 Copyright the Author(s), ISBN 978-1-032-39114-4

Blast performance of retrofitted reinforced concrete highway bridges

S. Biswas, S. Kulshreshtha & P. Sengupta
Department of Civil Engineering, Indian Institute of Technology (ISM), Dhanbad, India

ABSTRACT: With numerous incidents involving malicious and accidental bomb attacks across the globe, blast performance evaluation of existing structures has become of utmost significance. In general, the enormously high peak blast pressure and the extremely short loading duration make blast loading distinct from other conventional loading. Hence, the analysis and design principles adopted for conventional loading can-not be applicable while dealing with blast loading acting on the structures. Since highway bridges play a pivotal role in infrastructural and economic development of any country, this research aims at blast performance assessment of reinforced concrete highway bridges before and after application of various retrofit schemes. Three-dimensional nonlinear finite element modelling philosophy is employed to assess the efficiency of different retrofit schemes, like steel jacketing and Fibre Reinforced Polymer (FRP) wrapping. The numerical simulation results of retrofitted reinforced concrete structural components are successfully validated with respect to the experimental results from the literature. Subsequently, an extensive parametric study is conducted so as to explore the effects of various structural and loading parameters on the blast performance of retrofitted reinforced concrete highway bridges. Substantive findings from this research would enlighten the engineering community on selection of a cost-effective but efficient retrofitting strategy applicable for reinforced concrete highway bridges under blast loading environment.

1 INTRODUCTION

Bridges are part of civil infrastructure that are very crucial in the interconnectivity of a place and also require heavy investment be it in terms of money, labour or time during construction. The increase in global terrorism attacks with use of explosives results in the damage of critical infrastructure such as bridges, buildings etc. Due to this the study of impact of blast loading on these structures becomes very important. Retrofitting comes into the picture to protect the old structures that were not adequately designed for blast loading. Retrofitting can be done in many ways such as concrete jacketing, steel jacketing and fibre reinforced polymer (FRP) wrapping.

In this study, we have considered a normally reinforced concrete slab and a CFRP-retrofitted reinforced concrete slab similar in configuration to the experimental models considered by Wu et al. (2009) and analysed it for the impact of blast loading using finite element modelling software ABAQUS. Further, parametric study and fragility analysis have also been performed on the considered specimens.

2 NUMERICAL MODELLING

Finite element analyses of numerical models of the slabs (Figure 1) were performed by using the finite element software ABAQUS (Dassault Systèmes 2019).

The Concrete Damage Plasticity Model (CDPM) was incorporated in the model to take into account the nonlinear behaviour of concrete. The nonlinear behaviour of FRP was incorporated using Hashin Damage parameters in ABAQUS. The reinforcements were modelled by assuming elasto-plastic behaviour. A perfect bond was assumed between the concrete and reinforcement.The blast loading was defined using the CONWEP model already present in ABAQUS.

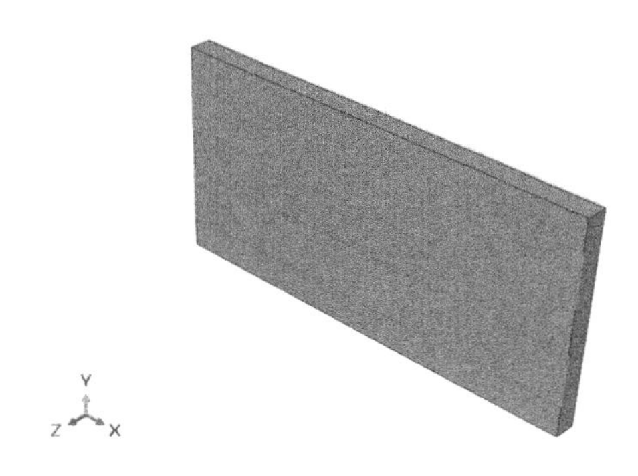

Figure 1. CFRP retrofitted slab model.

The numerical models were successfully validated with the experimental models obtained from literature (Wu et al. 2009).

Various parametric studies were performed on the normally reinforced concrete slabs by varying the slab

DOI: 10.1201/9781003348450-96

thickness as 100 mm, 200 mm and 300 mm, as well as varying the thickness of the CFRP and Steel plates.

3 FRAGILITY ANALYSIS

Fragility functions represent the conditional probability of a particular structure surpassing a certain damage state when subjected to external loading. Intensity measure (IM) quantifies the extent of input external excitations. Damage state may be referred to as the level of damage that a structure suffered when subjected to some external excitations. Engineering demand parameter (EDP) is a parameter which can predict the damaged stage properly.

In this study, the effectiveness of steel jacketing and CFRP wrapping is investigated by comparing the fragility curves of RC highway bridge decks, before and after retrofit under blast loading environment. Inverse of Scaled Stand-off Distance is considered as the intensity measure (IM) while the ratio of peak deflection (d) and span length (L), a dimensionless parameter is taken as the engineering demand parameter (EDP). The damage states are classified as Low, Moderate and Severe on the basis of the different ranges of the EDP. The fragility functions of unretrofitted reinforced concrete slab of 100 mm thickness is shown in Figure 2. Thereafter, the fragility curves of retrofitted NRC-3 slabs of 100 mm thickness with 1.4 mm thick CFRP wrapping and 1.4 mm thick steel jacketing are presented in Figures 3 and 4 respectively.

4 CONCLUSIONS

With increasing incidents involving accidental and deliberate explosions across the world, it is crucial to assess the vulnerability of various infrastructural systems against blast loading. This study presents an extensive numerical investigation of blast performance of Reinforced Concrete Highway Bridges.

Nonlinear finite element models of unretrofitted and retrofitted highway bridge deck slabs developed in ABAQUS were successfully validated with the data from literature (Wu et al. 2009). An extensive parametric study is conducted by varying the structural parameters like slab depth, retrofit sheet thickness, retrofit material and loading parameters, like charge weight and stand-off distances of explosives. Conventional Weapons Effects Programme (CONWEP) algorithm was used to analyse the deflections of the slab models caused by blast charges.

Bassed on the results of the parametric study, the effectiveness of CFRP wrapping and steel jacketing in blast resistance of RC highway bridge decks were subsequently established by deriving the fragility functions. This research serves as a stepping stone for the engineering community across the world to assess the effectiveness of different retrofit schemes in blast resistance of reinforced concrete highway bridges.

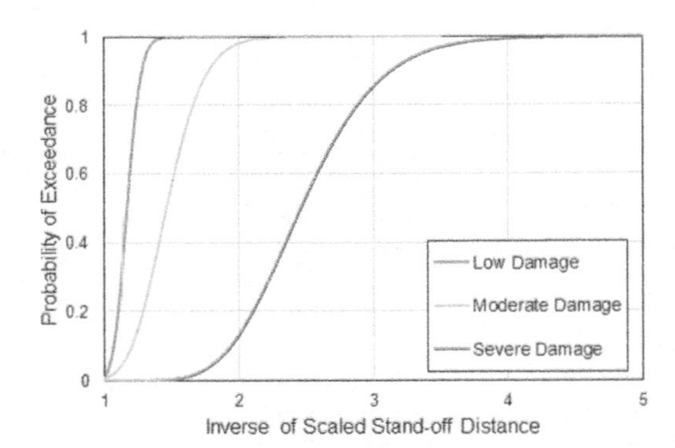

Figure 2. Fragility Curves of Unretrofitted Bridge Deck under Explosion.

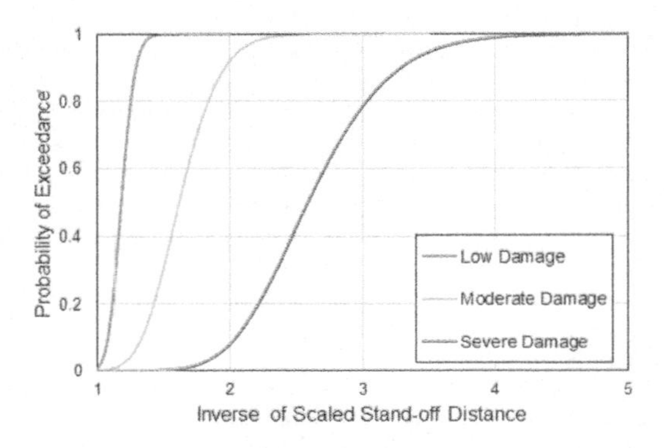

Figure 3. Fragility Curves of Bridge Deck after CFRP Wrapping under Explosion.

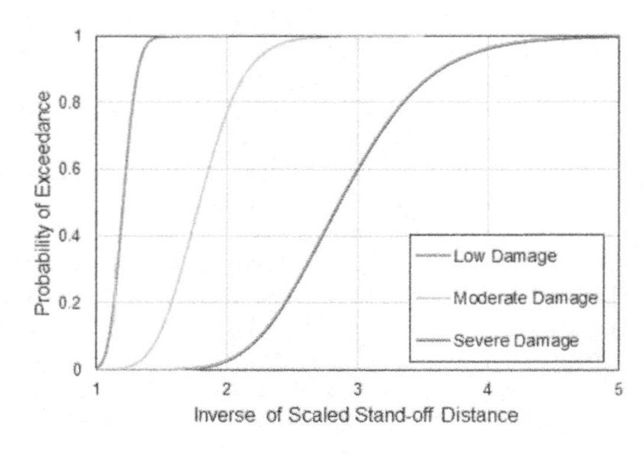

Figure 4. Fragility Curves of Bridge Deck after Steel Jacketing under Explosion.

REFERENCES

Dassault Systèmes, U. S. (2019). Abaqus analysis user's manual. *ABAQUS Academic-Research Edition Software 2019*.

Wu, C., D. Oehlers, M. Rebentrost, J. Leach, & A. Whittaker (2009). Blast testing of ultra-high performance fibre and frp-retrofitted concrete slabs. *Engineering structures 31*(9), 2060–2069.

Current Perspectives and New Directions in Mechanics, Modelling and Design of Structural Systems – Zingoni (ed.)
© 2022 Copyright the Author(s), ISBN 978-1-032-39114-4

Investigation of permanent indentation due to low velocity impact on glass fibre reinforced plastics

M. Bastek, L. Merkl & P. Middendorf
Institute of Aircraft Design, University of Stuttgart, Germany

ABSTRACT: To investigate the damage mechanisms under transverse loadings in glass fibre reinforced plastic (GFRP) materials, low velocity impact (LVI) tests and quasi-static (QS) indentation tests were conducted. Three energy levels were chosen for the impact tests and the maximum indenter displacement of these tests served as termination criterion for the QS indentation tests. The load-displacements curves of both testing configurations showed strong deviations as the QS indentations tests formed a plateau at significantly lower loads where the reaction force of impact tests further increased. Microsections of the specimen revealed excessive fibre breakages in the QS specimen as the main reason for this behaviour. This failure further led to higher indentation depth values compared to the impacted specimen. The strain-rate dependency of glass fibres is assumed to be the main reason for the differences.

1 INTRODUCTION

Fibre reinforced plastics (FRP) are prone to transversal impact loadings which lead to internal damages reducing the residual strength of a structure. Fibre failure, matrix cracks and delamination can occur as damage mechanisms. Low velocity impacts (LVI) caused for example by tool drops, are critical as they lead to barely visible impact damage (BVID). This type of damage is hardly visible, but can reduce the in-plane residual compressive strength up to 55%. Compression after impact (CAI) tests provide experimental evidence of the damage behaviour of a laminate. Here, a specimen is impacted with a specific impact energy level followed by the measurement of the permanent indentation. Finally, the residual in-plane compressive strength is determined. Many studies focused on the damage and residual strength at low impact energy levels ranging from 2 J to 40 J. Other authors use quasi-static (QS) indentation tests to investigate the sequence of damage mechanisms during a transversal loading. The comparability of GFRP impact tests and QS indentation tests has hardly been investigated and the influence of dynamic effects has not yet been sufficiently analysed. This study investigates the influence of different damage mechanisms in GFRP specimen on the development of the permanent indentation by comparing the results of LVI tests with medium impact energies and QS indentation tests.

2 EXPERIMENTAL INVESTIGATIONS

2.1 *Materials and specimen manufacturing*

The fabric used for the study is a pseudo-unidirectional E-glass fabric with 8% of the fibres woven in 90° direction and an epoxy resin as matrix. The specimens (150 mm x 100 mm) are manufactured by a VAP (Vacuum Assisted Process) method. The stacking sequence of the specimens is [0/90/0/90]s with an average laminate thickness of 4.248 mm and a fibre volume fraction of $\varphi_F = 46.86\%$.

2.2 *Testing set-up*

The impact tests are performed in accordance with AITM1-0010. The indenter hits a composite specimen with a defined impact energy level. Three energy levels (25 J, 50 J, 75 J) are tested and during the tests, the acceleration of the indenter and its displacement d are recorded. The boundary conditions of the QS indentation tests are analogous to those of the impact tests. The QS tests are displacement-controlled and the maximum displacement d_{max} of the impact tests serves as the termination criterion. The indenter is moved downwards with 1 mm/min and its reaction force is recorded. Directly after each test the permanent indentation depth i_p of the specimens is measured. This procedure is repeated after 24 h, 72 h and one week to track the relaxation behaviour.

3 RESULTS AND DISCUSSION

3.1 *Load and displacement history*

The load-displacement curves from three impact tests and QS indentation tests are shown in Figure 1. The load increases with an increasing indenter displacement d. All impact tests reach higher maximum loads compared to the QS indentation tests with the same indenter displacement. For the 25 J tests, an average maximum load of 7.51 kN is reached before the indenter rebounds. The maximum load of the

DOI: 10.1201/9781003348450-97

Table 1. Comparison of total energy E_{tot}, absorbed energy E_{abs} and the relative absorbed energy E^{rel}_{abs} as well as the indentation depths i_p and their ratio i^{QS}_p / i^{impact}_p.

Impact					QS indentation					
E_i [J]	E_{tot} [J]	E_{abs} [J]	E^{rel}_{abs} [-]	i^{impact}_p	d_{max} [mm]	E_{tot} [J]	E_{abs} [J]	E^{rel}_{abs} [-]	i^{QS}_p	i^{QS}_p / i^{impact}_p
-	-	-	-	-	3	4.16	1.21	0.29	0.148	-
25	24.72	8.03	0.32	0.291	6.8	21.79	8.93	0.41	0.574	1.97
50	50.26	21.86	0.43	0.454	9.3	40.21	23.30	0.58	1.284	2.83
75	75.58	41.89	0.55	0.829	11.2	56.07	41.96	0.75	2.731	3.29

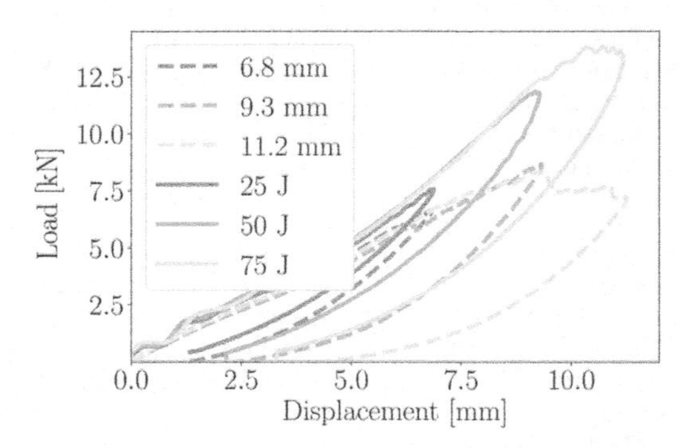

Figure 1. Load-displacement curves of QS indentation tests with 3 mm, 6.8 mm, 9.3 mm and 11.2 mm indenter displacement.

corresponding QS test is 6.54 kN. The maximum loads for 50 J and 75 J are 11.71 kN and 13.99 kN respectively. In contrast, the maximum load of the QS tests with $d = 9.3$ mm and $d = 11.2$ mm is approximately 8.5 kN. After this maximum, the tests with 11.2 mm displacement form a distinctive load plateau which is an indicator for progressive fibre failure.

The relative absorbed energy E^{rel}_{abs} is also evaluated. E^{rel}_{abs} is defined as E_{abs}/E_{tot} where E_{tot} is the total energy and E_{abs} the absorbed energy during an impact. E_{tot} is defined as the integral from P over d_{max} and E_{abs} is determined by the integral of P over d. Table 1 compares the averaged E^{rel}_{abs} of all tests. It can be seen that despite an equal displacement of the indenter, relatively more energy is absorbed in the QS indentation tests, whereas the absorbed energy E_{abs} does not show any significant differences in the two experimental set-ups. From this and the differences in the maximum loads, it can be concluded that the damage mechanisms of the impact specimen and the QS indentation are quite different.

3.2 *Indentation and internal damage*

To analyse and compare the damage mechanisms, microsections are taken directly from the indented area. An increase in fibre failure with increasing indenter displacement is observed for both testing configurations. The extent of fibre failure in the QS indentation tests increases significantly more than in the impacted specimen with the same indenter displacement. For all microsections, it can be observed that the permanent indentation depth i_p after the tests is higher for the QS indentation tests compared to impacted specimen with the same indenter displacement. The averaged values of the permanent indentation i_p measured directly after the tests are given in Table 1.

3.3 *Summary and conclusion*

To investigate the damage mechanisms during an impact, LVI tests and QS indentation tests with identical indenter displacements were performed. With increasing impact energies and indenter displacements, increasing deviations in the load-displacement curves were observed. While the reaction force of the impacted specimen increased, the loads of the QS tests formed a load plateau. In addition, under QS loadings, a significant deeper indentation depth developed, which is a result of extensive fibre failure. This was observed by analysing microsections of the QS specimen. Distinctive fibre failure was identified, whereas the extent of fibre failure was lower for the impact tests. This failure is a reason for the higher relative absorbed energy E^{rel}_{abs} of the QS specimen. Besides the dominant fibre breakage which strongly contributes to the formation of the permanent indentation i_p more pronounced plastic deformations of the matrix are assumed to be the cause. Strain-rate dependent material behaviour is a possible reason for the different damage evolution in the two test set-ups. Despite relatively low impact velocities, the dynamic loading can lead to higher stiffness and strength values of glass fibres. Since no strain measurements were carried out during the investigations of this study, no assertion can be made about the occurred strain-rates during the tests. Based on the experimental results, a limited comparability between the QS indentation and LVI tests is observed due to higher extent of damage in the QS tests.

Current Perspectives and New Directions in Mechanics, Modelling and Design of Structural Systems – Zingoni (ed.)
© 2022 Copyright the Author(s), *ISBN 978-1-032-39114-4*

Advanced fatigue assessment methods: The future of modern wind turbines

M. Rauch & R. Yildirim
Hochschule Kaiserslautern, Kaiserslautern, Germany

H. Bissing & M. Knobloch
Ruhr-Universität Bochum, Bochum, Germany

ABSTRACT: Affordable and clean energy is one of the Sustainable Development Goals of the UN and part of the Green Deal established by the European Union. To reach these ambitious goals the contribution of wind energy is already indispensable among the renewable energies. For the realization of future wind projects, economic aspects will gain more importance. The requirement for efficient use of materials will be essential for economical wind turbine towers. Advanced fatigue assessment methods, such as the Two Stage Model, enable the consideration of structural and material input parameters explicitly. The required characterization of cyclic material properties is derived by different methods, e.g. the incremental step test. In this paper, an alternative way to determine cyclic material properties of the base material, the heat affected zone and weld metal using the easy-to-use Hardness method is evaluated by a proof-of-concept study. The study proves the potential for advanced fatigue assessment methods.

1 INTRODUCTION

Affordable and clean energy is one of the most important societal needs and therefore also one of the Sustainable Development Goals of the UN and part of the Green Deal established by the European Union. To reach these ambitious goals the contribution of wind energy is already indispensable among the renewable energies. For the realization of future wind projects, economic aspects will gain even more importance. The demand for efficient use of materials and resources as well as energy-saving constructional details will be essential for wind turbine towers. Depending on the site conditions, wind turbines consist of either an entire steel tower or a hybrid tower, where at least a significant part of the structure is made of steel. To ensure safe and continuous operation throughout the entire service life of the wind turbine, fatigue design is paramount and often crucial to the design of wind turbine towers. Regarding fatigue design, various assessment methods can be applied for steel towers differing in effort, complexity and accuracy. Compared to the nominal stress method, local concepts increase the effort and complexity, but also the accuracy of the prediction and offer new opportunities concerning the application of advanced manufacturing concepts as well as lifetime extensions.

This paper first reviews the different frameworks for determining the cyclic material properties. Next, a brief overview of the results of incremental step tests of a butt weld connection are presented. The investigations carried out on Vickers hardness are then presented, more specifically the Vickers hardness of the base material, heat affected zone and weld metal of a butt weld. The application of simplified methods from the literature is then demonstrated to determine the cyclic material properties based on Vickers hardness. By a proof-of-concept study, the applicability of the easy-to-use Hardness method is shown as an alternative approach to the incremental step test.

2 DETERMINATION OF CYCLIC MATERIAL PROPERTIES

The strain-life approach is used for the first stage of the Two Stage Model (Röscher & Knobloch 2019) to determine the crack initiation life. Bissing et al. (2022) proves the applicability of this assessment method for welded connections of wind turbine towers in the fatigue life. In the present German national research project "FutureWind" (IGF-No. 20987 N), a welded connection typically used for wind turbine towers made of steel is investigated in detail by applying the Two Stage Model. Among others, the Two Stage Model can consider the material characterisation not only of the base material but also of the heat affected zone and weld metal. The required cyclic material properties to determine the crack initiation life with the Two Stage Model are derived based on incremental step tests (Bissing et al. 2021).

DOI: 10.1201/9781003348450-98

3 VICKERS HARDNESS OF BUTT WELDS

The Vickers hardness of a butt weld connecting two heavy plates of grade S 355 J2+N has been systematically investigated in order to determine the cyclic material properties as material input for advanced fatigue assessments. In a first step, the Vickers hardness of HV0,1, HV1 and HV10 has been determined for a butt weld. Based on the microstructure the Vickers hardness of HV10 of the base material exhibit the smallest scatter and a suitable correspondence with the ultimate strength. The Vickers hardness of HV10 was used for the further investigations and as basis of the determination of the cyclic material properties.

4 CYCLIC MATERIAL PROPERTIES BY SIMPLIFIED METHODS

Wächter & Esderts (2018a, b) evaluate the simplified methods available in the literature for determining the cyclic material properties For structural steels, the Hardness Method according to Roessle & Fatemi (2000), the FKM-Method developed by Wächter & Esderts (2018 b) and the uniform material law (Bäumel et al. 1990) are found to be the most suitable simplified methods (Wächter & Esderts 2018a, b).

In this paper, the Hardness Method and the FKM-Method are evaluated for a butt weld connecting two 35 mm thick heavy plates of grade S 355 J2+N to determine the cyclic material properties separately for base material, heat affected zone and weld metal. The investigations demonstrated that the mean values of the fatigue strength coefficient σ'_f and the cyclic strength coefficient K' according to the Hardness method and the FKM-Method differ only slightly for all three areas, but the fatigue ductility coefficient ε'_f showed a larger difference. Taking into account the results of the incremental step test, the results of the Hardness method appeared to be more reliable.

The simplified methods, i.e., Hardness method, FKM-method and uniform material law, demonstrated an increase of the cyclic strength coefficient K' of the heat affected zone and weld metal compared to the base material. Bissing et al. (2022) demonstrated the applicability of the Two Stage Model for steel towers and analysed the influence of scattering influence variables. Based on these investigations, the scatter of the cyclic hardening coefficient K' shown in the present study seems appropriate when simplified methods, such as the Hardness method, are applied. Simplified methods therefore appear to be suitable for determining the cyclic material properties for reliable service life prediction of wind turbines.

5 CONCLUSIONS

The paper presented two procedures, i.e., incremental step tests and simplified methods, to determine cyclic material properties required for advanced fatigue assessment methods, such as the Two Stage Model. The applicability of the Hardness method, the FKM-Method and the uniform material law as simplified methods for steels have been evaluated for a butt weld using the Vickers hardness. As demonstrated by the hardness tests for a steel of grade S 355 J2+N, the hardness and ultimate strength can be derived by Vickers hardness of HV10 for the base material, the heat affected zone and weld metal. For the presented simplified methods, the mean values of the cyclic strength coefficient K' differ only slightly and demonstrate an increase of K' of the heat affected zone and weld metal compared to the base material. The Hardness method is an easy-to-use test method with sufficient accuracy to obtain relevant cyclic material properties for the strain-life approach. The study has proven the potential of the easy-to-use method for advanced fatigue assessments.

ACKNOWLEDGEMENTS

The research project IGF-No. 20987 N /P1398/04/ 2020 "Prediction models for the remaining service life and further operation of wind turbines (FutureWind)" from the Research Association for Steel Application (FOSTA), Düsseldorf, was supported by the Federal Ministry of Economic Affairs and Climate Action through the German Federation of Industrial Research Associations (AiF) as part of the programme for promoting industrial cooperative research (IGF) on the basis of a decision by the German Bundestag. The project was carried out at Ruhr-Universität Bochum and Hochschule Kaiserslautern. The authors also thank SEH Engineering for providing the test specimens.

REFERENCES

Bäumel, A. & Seeger, T. & Boller, C. 1990. *Materials data for cyclic loading, Supplement 1 (Materials science monographs, vol. 61)*. Amsterdam: Elsevier.

Bissing, H. & Knobloch, M. & Rauch, M. 2021. Improving economic efficiency of wind energy using data-based fatigue assessment methods. In H.H. Snijder, Bart De Pauw, Sander van Alphen, Pierre Mengeot (eds), *IABSE Congress Ghent 2021, Structural Engineering for Future Societal Needs, Congress Proceedings*. Zurich: IABSE.

Bissing, H. & Knobloch, M. & Rauch, M. 2022. Advanced fatigue assessment - the future of wind turbine towers. *Structural Integrity Procedia* 38: 372–381.

Röscher, S. & Knobloch, M. 2019. Towards a prognosis of fatigue life using a Two-Stage-Model: Application to butt welds. *Steel Construction* 12(3): 198–208.

Roessle, M.L. & Fatemi, A. 2000. Strain-controlled fatigue properties of steels and some simple approximations. *International Journal of Fatigue* 22: 495–511.

Wächter, M. & Esderts, A. 2018a. On the estimation of cyclic material properties: Part 1. *Materials Testing* 60: 945–952.

Wächter, M. & Esderts, A. 2018b. On the estimation of cyclic material properties: Part 2. *Materials Testing* 60: 953–959.

Current Perspectives and New Directions in Mechanics, Modelling and Design of Structural Systems – Zingoni (ed.)
© 2022 Copyright the Author(s), ISBN 978-1-032-39114-4

Fatigue damage in thick GFRP laminate through the vibration-based bending fatigue experiment

A. Sato, Y. Kitane, Y. Goi & K. Sugiura
Department of civil and Earth Resources Engineering, Kyoto university, Kyoto, Japan

ABSTRACT: This research assessed the bending fatigue life and the change in mechanical property under the fatigue loading of thick GFRP laminate. The specimen has 32 layers of glass woven roving $(0/90)_{32}$, unsaturated polyester resin as matrix, a thickness of 15.5 mm and a width of 30 mm at the test section. The bending fatigue test was conducted with the vibration-based fatigue testing machine. As a result, when the maximum applied stress was equal to 20% of the static bending strength, the fatigue limit for 10 million cycles was observed. The visual inspection revealed the expansion of the fatigue damage. It was also confirmed that residual stiffness of the specimen decreased due to cyclic loading.

1 INTRODUCTION

Many previous researches about fatigue of FRP revealed the fatigue strength expressed by S-N diagram. It is also known that fatigue damage of composite is accumulated as the matrix cracks, the debondings and the delamination.

However, most of these researches was conducted with the thin composite which thickness is less than 10 mm, not focusing on thick laminates. INEEL bridge which is the demonstration GFRP bridge in United States uses about 19 mm thick GFRP laminate in its slabs and girders (Rodriguez et.al. 1997). The fatigue behavior of such a thick laminate has not been revealed. Thus, it is necessary to conduct material revel fatigue test using thick GFRP laminate.

This study aims to develop the high speed fatigue testing method and examine fatigue behavior of thick GFRP laminate. The testing method is developed in reference to the vibration-based bending fatigue testing machines which have been used for the steel welded joints.

2 EXPERIMENTAL OUTLINE

This study used Hand lay-up molded GFRP laminate as the specimens. The laminate which was 15.5 mm in thickness consisted of woven roving glass fiber and unsaturated polyester resin. The material properties are summarized in Table 1. The specimen was cut out from the laminate as shown in Figure 1. The test section was introduced in the specimen and the width at the section was decreased compared with normal section so as to induce the fatigue failure there. The simple theoretical calculation determined the width of the test section as 30 mm to apply twice larger stress than normal section.

Table 1. Material properties.

Fiber	Woven roving glass fiber
Resin	Unsaturated Polyester
Molding	Hand lay-up
Structure	$[0/90]_{32}$
Thickness	15.5 mm
V_f	46.6%
Bending strength:σ_u	369.7MPa
Bending modulus:E_b	46.6 GPa

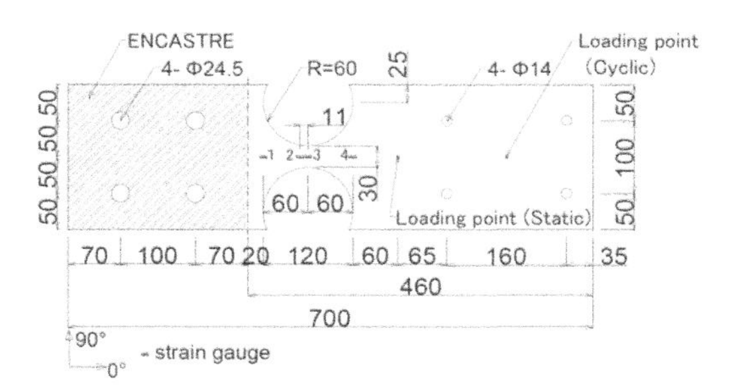

Figure 1. Top view of the specimen.

The bending fatigue test was conducted under R = 0.1 with the testing machine shown in Figure 2. In the machine, the specimen is rigidly connected at the left end and sup-ported by coil springs at the right end via steel plate. During cyclic loading, the eccentric mass in the vibrator rotates to generate a centrifugal force in the vertical direction, and a bending moment is applied to the specimen.

DOI: 10.1201/9781003348450-99

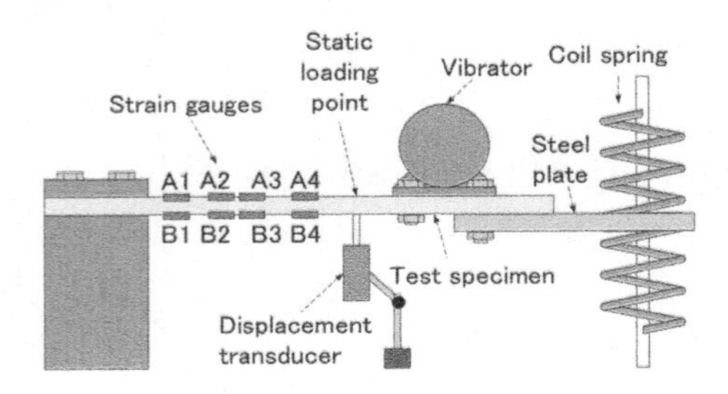

Figure 2. Schematic view of experiment set up.

3 RESULT AND DISCUSSION

Figure 3 shows the comparison of S-N diagram between the bending and the tensile fatigue strength. The bending fatigue strength shown in Figure 3 is not only the experimental result in this study but also the tensile fatigue strength quoted from the previous study which authors conducted (Sato et. al. 2021). The specimens used for tensile fatigue test was $[0/90]_4$ of GFRP which fiber volume fraction was 42.7%, and the reinforcing fiber, matrix resin, molding method and stress ratio were the same between bending and tensile fatigue test. First of all, the bending fatigue limit of thick GFRP laminate was 75 MPa which was 20% of static bending strength. It can be also seen from Figure 3 that the bending fatigue strength is less than the tensile. The bending fatigue strength was thought to be decreased due to shear stress. GFRP is orthotropic material so material properties varies in its direction. In the specimens, only matrix resin contributed the inter-lamina shear strength. Thus, the bending fatigue strength may be affected by the shear stress. This is verified from the matrix cracks observed near the neutral axis as shown in Figure 4. Figure 4 was captured at early cycles of fatigue life and this implies that not only tensile but shear contribute fatigue damage in the early region. The effect of shear stress needs further discussion in future works.

The effect of thickness on the fatigue strength is discussed using Figure 5. Figure 5 summarizes the bending fatigue limit (σ_f). The scattered data was obtained from not only this study but also previous researches working on bending fatigue test of GFRP, and are organized by the thickness and types of reinforcing fiber. Comparing the data in woven roving, most data shows that fatigue limit was about 70 MPa. Therefore, Figure 5 implies that the thickness doesn't affect the fatigue limit between 2 mm and 15.5 mm.

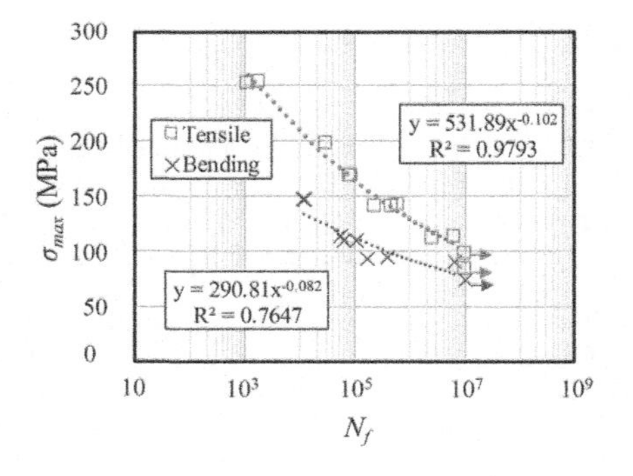

Figure 3. Comparison of S-N diagram under Tensile & Bending fatigue test.

Figure 4. Matrix cracks observed from side view of specimen.

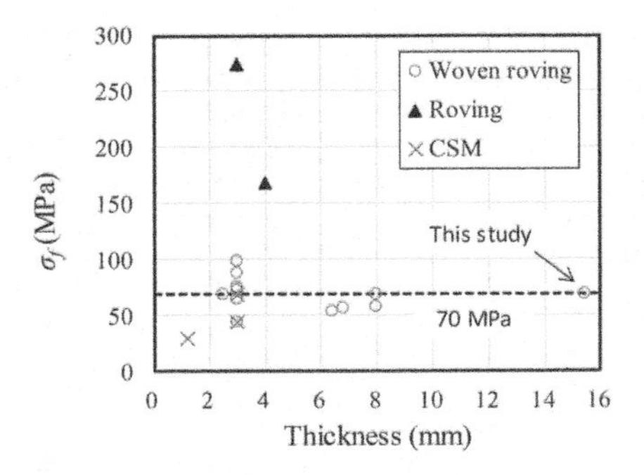

Figure 5. Effect of thickness on fatigue limit.

4 CONCLUSION

The fatigue limit of thick GFRP laminate (15.5 mm in thickness) was 20% of static bending strength. This is the same level with the other thin laminates and no effect of thickness was recognized in the comparison performed here.

The initial fatigue damage was observed as micro cracks and result in sharp stiffness degradation. The matrix cracks were induced by not only tensile stress but also shear stress. The area of matrix cracks expanded as the number of cycles increased, and then fatigue failure occurred following to the partial delamination in the compression side.

REFERENCES

Rodriguez, G, J. Dumlao, C. Ciolko, T, A. Pfister, J, T and Schemeckpeper, R, E. 1997. Test and Evaluation Plan for the Composite Bridge, Idaho National Engineering Laboratory.

Sato, A. Kitane, Y. Hibi, H. Goi, Y and Sugiura, K. 2021. Fatigue Strength and Stiffness Degradation of Woven Roving GFRP, Proceedings of the 14th Symposium on Research and Application of Hybrid and Composite Structures, JSCE, No.29-8. (In Japanese)

Current Perspectives and New Directions in Mechanics, Modelling and Design of Structural Systems – Zingoni (ed.)
© 2022 Copyright the Author(s), ISBN 978-1-032-39114-4

Decoupling method for fracture modes in an orthotropic elastic material through a three-dimensional vision

Jérôme Afoutou, Frédéric Dubois, Nicolas Sauvat & Mokhfi Takarli
Laboratory of Civil Engineering, Durability and Diagnostic, University of Limoges, Egletons, France

ABSTRACT: This work deals with the generalization of the M integral initially developed by Chen and Shield. This paper presents its adaptation for the decoupling of fracture modes for orthotropic elastic medium and written for a three-dimensional vision. The use of this integral is based on the definition of the virtual mechanical fields (displacement and stress) in the crack tip vicinity. This is the main difficulty especially for orthotropic medias. So, it proposed a transformation of virtual displacement field proposed by Irwin for isotropic media in order to take into account the orthotropic effects. The analytical developments are followed by a numerical implementation in the finite element method. The application is limited to the crack initiation phase of a double cantilever beam specimen including a variable inertia (DCBIV) requested in opening mode. The results are presented in terms of an energy release rate. It is shown that $M_\theta{}^{3D}$ is path independent.

Keywords: 3D fracture, analytical formulation, energy release rate, orthotropy, $M_\theta{}^{3D}$ integral

1 INTRODUCTION

The effects of temperature and moisture in the process of cracking mechanisms of timber structures are studied in the literature. But the analysis tools are generally established for two-dimensional configurations (Hamdi et al., 2017). The objective of this work is to propose an independent integral path allowing to decouple the fracture modes for orthotropic materials in a three-dimensional configuration. To this end, after having generalized the M-integral for a three-dimensional configuration, we propose a virtual displacement field having a form like the singular fields generally defined in the vicinity of the crack tip. This field is based on an isotropic form proposed by Irwin(Irwin, 1957). The transformation performed allows to consider the orthotropy effects without taking into account the usual fields because its three-dimensional format is not defined. The energy release rates are calculated as a function of the crack length. In addition, to study the three-dimensional effects, a parametric study is proposed according to the thickness of the specimen. Finally, the distribution along the crack front line of the effects of each mode is presented.

2 BACKGROUND ON THE PATH-INDEPENDENT INTEGRAL AND DECOUPLING PROCEDURE

This section presents the invariant integrals used to evaluate the fracture parameter of the energy approach to fracture mechanics.

$$G_\theta^{3D} = -\int_V \left(W \cdot \theta_{k,1} - \left(\sigma_{ij} \cdot u_{i,1}\right) \cdot \theta_{k,j}\right) \cdot dV \quad (1)$$

Where W, σ_{ij} and u_{ij} denote the elastic energy density, stress field and displacement field respectively. The domain of integration is the volume V. $G_\theta{}^{3D}$ is proposed by (El Kabir et al., 2018).

For the decoupling modes of fractures, the general form of $M_\theta{}^{3D}$ is given by the expression below:

$$M_\theta{}^{3D} = \frac{1}{2}\int_V \left(\sigma_{ij}^u \cdot v_{i,1} - \sigma_{ij,1}^v \cdot u_i\right) \cdot \theta_{k,j} \cdot dV \quad (2)$$

It allows mixed mode decoupling for a stationary crack. (u) and (v) represent kinematically admissible real and virtual displacement fields. In the same spirit, $\sigma_{ij}{}^u$ and $\sigma_{ij}{}^v$ are the real and virtual stress fields, respectively. The integration domain is common to the integral $G_\theta{}^{3D}$.

In the absence of a three-dimensional orthotropic displacement field, we propose to use the isotropic displacement field. The originality of this approach is to replace the isotropic reduced elastic compliances by orthotropic elastic compliances. C_{e1} and C_{e2} are the orthotropic reduced elastic compliances in modes I and II.

In a mixed configuration and thanks to the principle of superposition, we write:

DOI: 10.1201/9781003348450-100

$$M_{\theta}^{3D}(u,v) = C_{e1}\frac{K_I^u.K_I^v}{8} + C_{e2}\frac{K_{II}^u.K_{II}^v}{8} + C_{e3}\frac{K_{III}^u.K_{III}^v}{8} \quad (3)$$

K_I, K_{II}, K_{III}, are the stress intensity factors for open, planar shear and torsion modes, respectively. C_{e1}, C_{e2} and C_{e3} represent the corresponding reduced elastic compliances.

To quantify the share of each fracture mode through the stress intensity factors, separate integral calculations are performed in order to have only one pure mode. In parallel, a calculation of the global energy release rate G_{θ}^{3D} is performed.

3 FINITE ELEMENT IMPLEMENTATION

The modelling is done on the Finite Element code Castem18. The objective is to show that k_i^* is a constant depending only on the orthotropy rate. The loading applied allows for a mode I crack opening by applying a unit force of 1kN. Half of the geometry (Figure 1) is modelled to take into account the geometric and loading symmetries. We consider a straight through crack front.

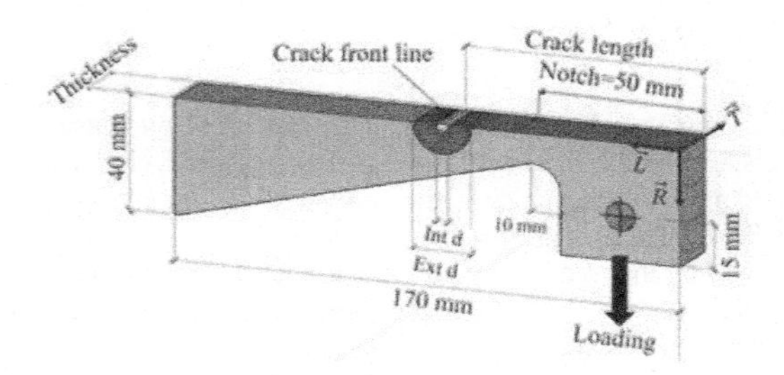

Figure 1. Model DCBIV half specimen.

4 OPTIMIZATION OF VIRTUAL FIELDS

We first validated the stability of the integral by substituting the virtual displacement field with the real displacement field.

The results of the parametric study show that:

- The assumption of a straight crack front is valid for thicknesses up to 20 m.
- The parameter k_i^* hardly varies whatever the thickness and length of the crack. It can therefore be assumed to be constant.
- The proposed virtual displacement field is relevant in view of these first validations.

Figure 2 deals with the invariance of the integral. A fluctuation of 5.8% and 2.6% for the mode I share is observed for the 10mm and 100mm thicknesses respectively. The proportion of mode II is negligible since the specimen is loaded only in tensile mode.

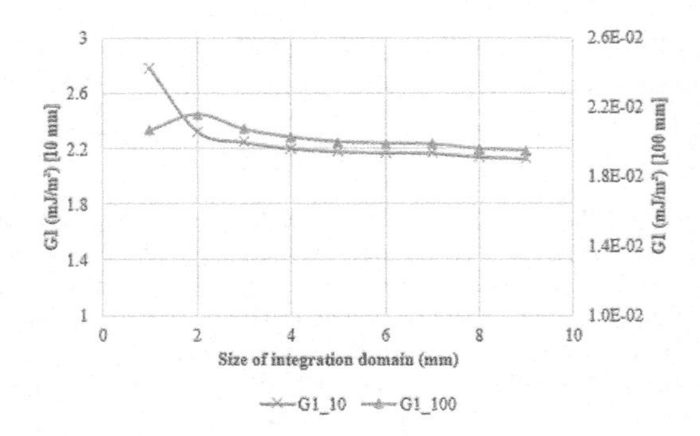

Figure 2. Invariance of the integration domain.

5 CONCLUSION AND OUTLOOK

In this work, we have presented an approach to decoupling fracture modes in an orthotropic material for a three-dimensional configuration. The results obtained are a first validation step for the generalization of M_{θ}^{3D} integral.

In view of the distribution of the release rate in thick specimens, it is relevant to develop a model approach integrating a non-rectilinear crack front in order to have a homogeneity of the release rate along the crack front by including a multi-criteria including the part of each mode. Finally, a last outlook is to implement the integral M_{θ}^{3D} in a specimen geometry allowing to associate the three cracking modes.

REFERENCES

El Kabir, S., Dubois, F., Moutou Pitti, R., Recho, N., Lapusta, Y., 2018. A new analytical generalization of the J and G-theta integrals for planar cracks in a three-dimensional medium. Theoretical and Applied Fracture Mechanics 94, 101–109. https://doi.org/10.1016/j.tafmec.2018.01.004

Hamdi, S.E., Pitti, R.M., Dubois, F., 2017. Temperature variation effect on crack growth in orthotropic medium: Finite element formulation for the viscoelastic behavior in thermal cracked wood-based materials. International Journal of Solids and Structures 115–116, 1–13. https://doi.org/10.1016/j.ijsolstr.2016.09.019

Irwin, G., 1957. Analysis of stresses and strains near the end of crack traversing a plate. J. Appl. Mech 24, 361–364.

Current Perspectives and New Directions in Mechanics, Modelling and Design of Structural Systems – Zingoni (ed.)
© 2022 Copyright the Author(s), *ISBN 978-1-032-39114-4*

Investigation on 3D printed dental titanium Ti64ELI and lifetime prediction

Lebogang Lebea*
Department of Mechanical and Mechatronic Engineering, Tshwane University of Technology
Unisa Biomechanics Research Group, Department of Mechanical and Industrial Engineering, University of South Africa

Harry M. Ngwangwa
Unisa Biomechanics Research Group, Department of Mechanical and Industrial Engineering, University of South Africa

Dawood Desai
Department of Mechanical and Mechatronic Engineering, Tshwane University of Technology

Fulufhelo Nemavhola
Unisa Biomechanics Research Group, Department of Mechanical and Industrial Engineering, University of South Africa

ABSTRACT: Fatigue analysis plays a vital role in determining the structural integrity and life of a dental implant. With the use of such implants on the rise, there is a corresponding increase in the number of implant failures. As such, the aim of this research paper is to investigate the life of 3D-printed dental implants. The dental implants considered in this study were 3D printed according to the direct metal laser sintering (DMLS) method. Additionally, a finite element model was developed to study their performance, while fatigue lifetime was predicted using Fe-Safe software®. The model was validated experimentally by performing fatigue tests. The life of the dental implants was analysed based on Normal strain and the Brown-Miller with Morrow mean correction factor algorithm. The model revealed that there was a strong correlation between the FEA and the experimental results. The clinical success of 3D-printed dental implant experimentally is 20.51 years and computationally under Normal strain is 19.89 years and Brown-Miller with Morrow mean correction factor is 26.82 years.

Keywords: Fatigue, oblique loading, axial loading, finite element analysis, 3D-printed dental implant

1 INTRODUCTION

Dental diseases have a considerable impact not only on people's self-esteem but also on their eating ability, nutrition and health (Isabel and Ribeiro, 2016). Problems concerning various aspects related to dental implants always remain relevant and aesthetics is the main reason why people often choose implants instead of other methods (Agnieszka and Adam, 2018). The aim of this study is to numerically investigate the life of 3D printed dental implants under Normal strain and Brown-Miller with Morrow mean correction factor algorithm. The objective of this study was to validate experimentally by performing fatigue tests using an MTS Acumen fatigue testing machine 3D-printed dental implants as specimens.

2 METHODOLOGY

The design of dental implants was carried out by the authors at the University of South Africa using Abaqus CAE software and then saved as a STL file format, which is one of the most common file formats used in 3D printing process. The dental implants were 3D printed according to the direct metal laser sintering (DMLS) method. Dental implants with a diameter of 3.4 mm were 3D printed and the crown and one-piece dental implant were printed as one unit. The study considered a single body dental implant due to the difficulties in 3D-printing of dental implants. Additionally, a finite element model was developed to study their performance, while fatigue lifetime was predicted using Fe-Safe software® (Figure 1).

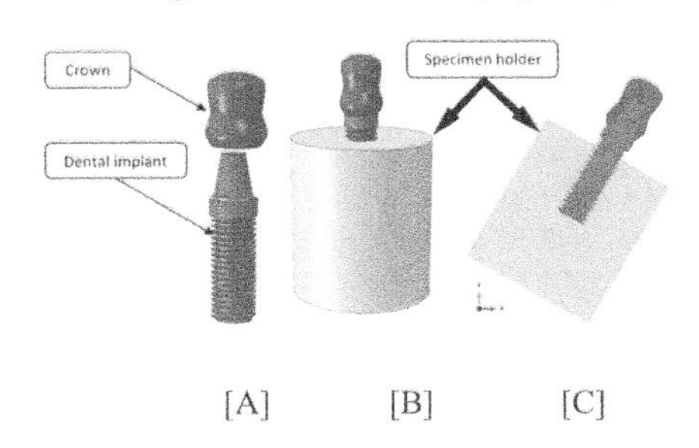

Figure 1. [A] Crown and one piece dental implant, [B] 3D model of specimen holder, dental implant and crown assembly, [C] Section view of the 3D model assembly.

The lifetime of the dental implants was analysed based on Normal strain and the Brown-Miller theory with

*Corresponding author: ebealc@unisa.ac.za

DOI: 10.1201/9781003348450-101

Morrow mean correction factor algorithm. The model was validated experimentally by performing fatigue tests. The specimens were fixed according to the International Standard of dynamic testing of single-post endosseous dental implants, which were replicated in the FEA model. The current study consider air as testing environment. The experimental models were tested at 80 % of the maximum load, as recommended by ISO 14801 (Figure 2), and this maximum load was derived from the FEA model (BSI Standards Publication dentistry, 2016; Nagy and Griggs, 2018) (BSI Standards Publication dentistry, 2016; Nagy and Griggs, 2018).

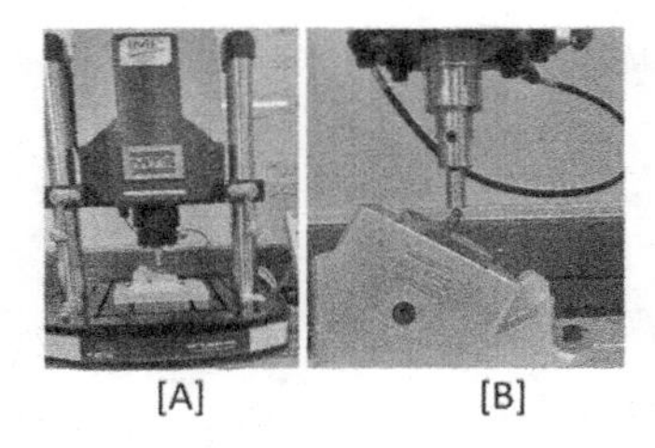

Figure 2. [A] Full view of MTS Acumen fatigue testing machine, [B] Detailed setup of 3D printed dental implant, specimen holder inclined at 30° and load cell in vertical direction (axial loading).

3 RESULTS AND DISCUSSION

The failure of the 3D printed dental implant occurred in the second and third threads, which is in agreement with FEA results (Figure 3).

Figure 3. Fractured 3D-printed dental implants after experimental study according to ISO 14801 standard's fatigue test.

The cycle target was set for 5×10^6 cycles but, with a loading of 80 %, it was observed that a minimum of 137 433 and maximum of 262 142 cycles respectively could be reached before failure occurs. Computational statistical estimators, F-numbers, and R^2 values were used to determine that the model was statistically valid.

The results shows that the model F-value of 4.6326, which means the model for the Normal strain response to the experimental results. Once the p-value is less than 0.05, the terms of the model have a significant effect on the response output. The Model's p-value for the Normal strain, which is 0.0027, implies that the model terms are also significant. Similarly, the Brown-miller has an F-value of 0.6859 and a p-value of 0.5163, implying that the model is insignificant. Calculating the correlation coefficient between the experiment and the finite element analysis model, it was found that R^2 of the experiment vs. Normal strain was 0.8256, experiment vs. Brown-miller was 0.9973, and Normal vs. Brown-miller was 0.8484. Thus, this confirms that there are no issues with the data or the model.

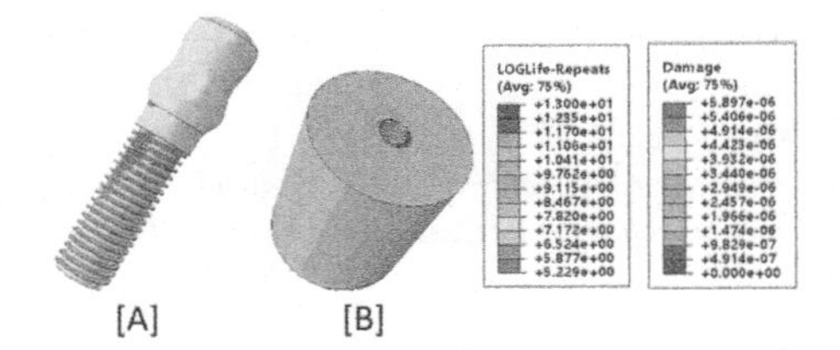

Figure 4. [A-B] Fatigue analysis results for Normal strain algorithm.

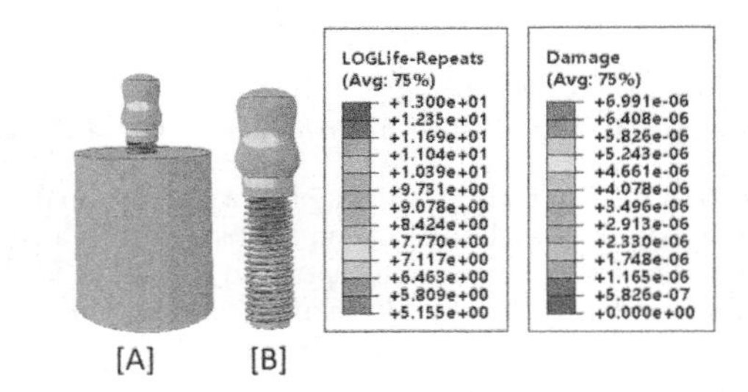

Figure 5. [A-B] Fatigue analysis for Brown-Miller with Morrow mean correction factor.

The results shows that the Normal strain is 19.89 years, Brown-Miller with Morrow mean correction factor is 26.82 years, and experimental is 20.51 years (Figures 4 and 5).

4 CONCLUSIONS

The life prediction differences between two algorithms were reported and it was found that the 3D-printed dental implants in the current study exhibit their highest performance under the Normal strain method, as opposed to the Brown-Miller with Morrow mean correction factor. It was observed that there is a strong correlation between the projected FEA results and the results achieved during experimental testing.

REFERENCES

Agnieszka, A. and Adam, O. N. Y. (2018) 'Stress analysis of dental implant inserted in the mandible', 68(1), pp. 25–32.

BSI Standards Publication dentistry (2016) 'BSI Standards Publication: Dentistry Implants: Dynamic loading test for endosseous dental implants'.

Isabel, S. and Ribeiro, C. (2016) 'Dental Implants Resistance : Computational Analysis', (November).

Nagy, W. W. and Griggs, J. A. (2018) 'Fatigue lifetime prediction of a reduced-diameter dental implant system : Numerical and experimental study', *Dental Materials*, pp. 1–11.

Current Perspectives and New Directions in Mechanics, Modelling and Design of Structural Systems – Zingoni (ed.)
© 2022 Copyright the Author(s), ISBN 978-1-032-39114-4

Detection of fatigue damage via neural network analysis of surface topography measurements

Hassan Alqahtani
Department of Mechanical Engineering, Taibah University, Medina, KSA

Asok Ray
Department of Mechanical Engineering & Department of Mathematics, Pennsylvania State University, University Park, PA, USA

ABSTRACT: The fatigue strength of the machined structure is significantly affected by the surface roughness because the surface topography creates the stress concentration in valleys of the surface topography. The surface roughness usually illustrates the micro-geometric appearance variation of the machined structure. The investigated surface indices are the average surface roughness, root-mean-square height, maximum peak height, maximum valley depth, and surface flatness. This paper applies the neural network to determine the most significant surface texture that is affected by the fatigue damage. The results show that the best NN model is the surface flatness model, with (up to) 87.5% accuracy.

1 INTRODUCTION

The deterioration of a component or material caused by a large number of applied stress cycles, maximum tensile stress, and fluctuations in the applied stress is generally known as "fatigue damage". In general, the fatigue life of a component is influenced by the surface finish. One of the most important surface textures that could describe surface micro topography is surface roughness. The fatigue strength decreases as the surface roughness increases (Anderson & Anderson 2005, Campbell 2008).

The most common technique that is used to detect fatigue damage is visual inspection (VT). The VT can only detect large flaws. Also, it is possible to have misinterpretation flaws using VT. Thus, automated crack detection is deemed very necessary. Recent years have witnessed a growing academic interest in using machine learning techniques to detect fatigue damage. This paper investigates the evolving fatigue damage in machinery structures using surface topography measurements. The reason for using the surface topography measurements to detect fatigue damage (e.g. the crack) is that the surface textures describe generally the deviations of a real surface from its ideal form, and the crack initiation has a significant effect on the deviations of a real surface; hence, the crack initiation can be detected using the surface topography measurements. All surface textures were measured by an optical meteorology device (Alicona). The investigated surface textures are the average surface roughness, root-mean-square height, maximum peak height, maximum valley depth, and the surface flatness. In addition, this paper applies the neural network to determine the most significate surface texture indices that are affected by the fatigue damage, which can be determined by obtaining the best NN model performance for fatigue damage detection and classification.

The major contributions of this paper are delineated as follows:-

1. *Fatigue damage detection & classification*
2. *Best surface index selection based NN-based pattern classification*

2 DESCRIPTION OF THE EXPERIMENTAL APPARATUS

In this paper, the experimental apparatus is built upon a computer-instrumented and computer-controlled fatigue testing machine, equipped with a confocal microscope. The objective of this experiment is to obtain the measurements of the surface textures for evaluation of the surface state. Ten typical experiments have been conducted to build a strategy of fatigue damage detection based on the surface textures data of polycrystalline alloys. The type of cyclic loading of these experiments is tension-tension load cycles.The optical metrology device (Infinite-Focus Alicona) is used to measure the surface textures such as the average surface roughness (S_a), root-mean-square height

DOI: 10.1201/9781003348450-102

(S_q), maximum peak height (S_p), maximum valley depth (S_v), and the surface flatness . In the optical system, the surface roughness is usually obtained by estimating the variations in the focus, where the small depth of the focus in an optical system is combined with vertical scanning.

3 METHODOLOGY OF DAMAGE ANALYSIS

The artificial Neural network (ANN) was the methodology of damage analysis in this investigation. The artificial Neural network (ANN) presents a set of nodes that are associated in a syntactic way that is similar to the association of neurons in the human brain; thus, ANN follows the system of a human brain. ANN has different forms of nodes connections, and each form creates a new type of ANN. The Feed-Forward Neural is commonly used in ANN applications. The mechanism process of the Feed-Forward Neural starts from input nodes and ends at the output nodes; hence, the direction of the sequence in this type is a forward direction. This paper presents an ANN structure called the Shallow or Vanilla multi-layer neural network. The Shallow ANN architecture contains an input layer, one hidden layer, and an output layer. The computational process of each node is that the input data are multiplied by weights and added by a bias. practically, When the input data has valuable information, a higher weight is applied and vice versa. Thus, weight has played an important role in the ANN computational process. The applied activation function in this paper is known as the Sigmoid function, and it allows a neural network model to learn complex problems by finding non-linear relationships between data features. The Sigmoid function generates an analogue output that is limited between 0 and 1.

Adjusting ANN structure weights is a vital process; where these weights are adjusted based on the value of the error between the ANN model outputs and desired outputs. This technique is defined as the back-propagation technique. In this technique, a set of input data is passed through ANN structure generating outputs. Based on the measured error, back-propagation technique is applied in the reverse path of ANN architecture (from the output layer to the input layer (the reverse path of ANN architecture) to re-adjusted the weights. This cycle is defined as an epoch, and the number of epochs is determined by achieving the best model performance (the minimum error). Generally, the model performance improves gradually as the model error gradually decreases. In this paper, the scaled conjugate gradient back propagation algorithm (SCG) was applied to re-adjust weights and biases of the ANN by using the Deep Learning ToolboxTM of MATLABTM 2020a. The NN architecture that is applied in this study is constructed with ten Sigmoid hidden neurons and two Softmax output neurons. The splitting ratio of the input data, which is measurements of the surface roughness, is 70% for training, 15% for validation, and 15% for testing.

4 RESULTS AND DISCUSSION

In this paper, the classifier performance is illustrated in the confusion matrix; where the columns of the matrix refer to the target class (true class); while the rows of the matrix refer to the output of the classifier (the predicted class).The accuracy of the classifiers of S_a, S_q, S_p, and S_v is similar, which is between almost 69% & 70%. On the other hand, the accuracy of the surface flatness classifier is 87%, which exceeds the accuracy of other classifiers by almost 25%. The reason for having the best accuracy in the flatness classifier is that the data of the undamaged class is separable from the data of the damaged class because the surface flatness texture measures the amount of variation up and down the y-axis over the entire surface, and definitely, these variations in the y-axis of the damaged surface differ from the undamaged surface, as shown in the figure. In contrast, the data of undamaged surfaces and damaged surfaces using SS_a, S_q, S_p, and S_v show a similarity of almost 30%, and that is the reason for the lack of accuracy in these surface textures.

5 CONCLUSION

The presented research has examined the effects of surface topography on prediction of the fatigue damage in polycrystalline alloys. A neural network model has been built to classify the undamaged and damaged classes of surface textures in test specimens. It is concluded that The surface flatness provides the best classier with an accuracy 87%, while the lowest classifier performance is the average surface roughness, with (up to) 67.9% accuracy.

REFERENCES

Anderson, T. L. & T. L. Anderson (2005). *Fracture mechanics: fundamentals and applications*. CRC press, Boca Raton, FL, USA.

Campbell, F. C. (2008). *Elements of metallurgy and engineering alloys*. ASM International.

Current Perspectives and New Directions in Mechanics, Modelling and Design of Structural Systems – Zingoni (ed.)
© 2022 Copyright the Author(s), *ISBN 978-1-032-39114-4*

Notch stress analysis on reinforcing steel bars using FE-simulations considering surface topography and structure properties

S. Rappl & K. Osterminski
Technical University of Munich, Munich, Germany

ABSTRACT: The fatigue behavior of reinforcing steel bars (rebars) is significantly influenced by its surface geometry and the manufacturing process. As a result of loading, stress concentrations occur at the feet of transverse ribs. Here, the fillet radius has a significant influence on the resulting notch stresses. A laser-based line scan system was used to generate three-dimensional digital geometric twins of the rebars. The resulting surface models were implemented in a Finite Element Method (FEM) program for linear-elastic simulations. These results were compared with the fracture locations of the fatigue tests and the local strength.

1 INTRODUCTION

1.1 *Motivation*

Many buildings suffer a high number of dynamic loads throughout their service lifetime. On the one hand, the design of such structures often proves to be very complex. On the other hand, there is also a large scatter of the materials resistance against fatigue.

1.2 *Production and common properties of rebars*

The most common manufacturing process for reinforcing steel bars (rebars) is the Tempcore®-process. Herein, the hot rolled rebar is locally quenched on its surface, resulting in a variation of the structural setup distributed over the cross section of the rebar. On the surface a tempered martensite can be found which possesses good strength ($R_{p0,2}$ = 700 N/mm², R_m = 800 N/mm²). The rebar core is composed of a ferrite-pearlite structure, providing a lower strength but a very good ductility (Rappl et al., 2022).

1.3 *Fatigue behavior of reinforcing steel bars*

At notches or cross section transitions the local stresses are increased. Investigations detected the fillet radius r as key influencing factor (Jhamb, 1972). Three-dimensional FE-Simulations on idealized rebar geometries (Rappl et al., 2022) confirmed the previous results. Due to the simplified geometries, impacts of e.g. surface roughness or notches were neglected. However, to predict the fatigue crack initiation precisely, these effects must also be taken into account. This emphasizes the need of a three-dimensional FE-Simulation on a digital twin.

2 EXPERIMENTAL

2.1 *Preliminary investigations*

In this study 29 rebars of diameter d = 12 mm and grade B500B (Tempcore®) containing two transverse rib rows were used. All specimens were taken from one manufacturing charge. In a first step, the rebars were scanned with a laser-based line scan (LLS) system (Osterminski & Gehlen, 2019). A program was developed to convert the point clouds into a surface model (Wich, 2017).

2.2 *Fatigue tests*

The fatigue behavior of the rebars was investigated in a uniaxial fatigue test using a resonance pulsator. The test parameters were selected according to the standard (DIN 488-1, 2009). Three different termination criteria were given: runouts after 10 million load cycles, a failure in the free length or a failure in the clamping area, when the requirement of the free length could no longer be fulfilled.

2.3 *FEM program*

The surface models were implemented in ANSYS Workbench and converted into a volume body. The material parameters of the standard (DIN 488-1, 2009) were used for a linear-elastic simulation. In a next step, an adaptive mesh was generated, consisting of tetrahedral elements. On the end faces, a loading force and a fixed support were applied, respectively.

DOI: 10.1201/9781003348450-103

3 RESULTS AND DISCUSSION

3.1 *Preliminary investigations*

Surface models of the failure specimens were generated with the crack origin in the middle.

3.2 *Fatigue tests*

Besides eleven failures in the clamping area and ten runouts eight valid failures were observed.

3.3 *FEM program*

Figure 1 shows the von Mises stresses σ_{vM} near the crack origin. The highest notch stresses were obtained on the surface in or near the fillet radius. On the alternated rib side mutual effects of converging ribs could be observed. The rib itself and the valley show much lower stresses. This is in good agreement with the literature (Jhamb, 1972, Rappl et al. 2022).

The von Mises stresses at the crack origins $\sigma_{vM,crack}$ are between 623 N/mm² and 740 N/mm². The notch stresses of the surface model are approx. 15% higher than the one of the simplified geometry (Rappl et al., 2022). A variation of the fillet radius over the rib foot leads to three-dimensional notches, where the stress trajectories are also deflected in circumferential direction and therefore lead to the higher notch stresses.

In (Rappl et al., 2022) a yield strength on the surface was calculated to approx. 690 N/mm². Assuming the same material properties, all $\sigma_{vM,crack}$ are below or in the scattering area of the surface yield strength. However, also residual stresses have to be taken into account. A deformation measurement during and after slicing a rebar into segments results in compressive residual stresses up to -40 N/mm² on the surface but with a decreasing trend (Volkwein et al., 2020). Tensile residual stresses occur at 2 mm beneath the surface and raise up to 40 N/mm². A high slope of von Mises stresses to the core could be observed in the surface region. After 0.5 mm the nominal stress of 300 N/mm² was reached. Based on these results an unfavorable effect of mechanical and tensile residual stresses is not assumed. However, it has to be mentioned that in (Volkwein et al., 2020) a locally differentiated point of view is not taken into account.

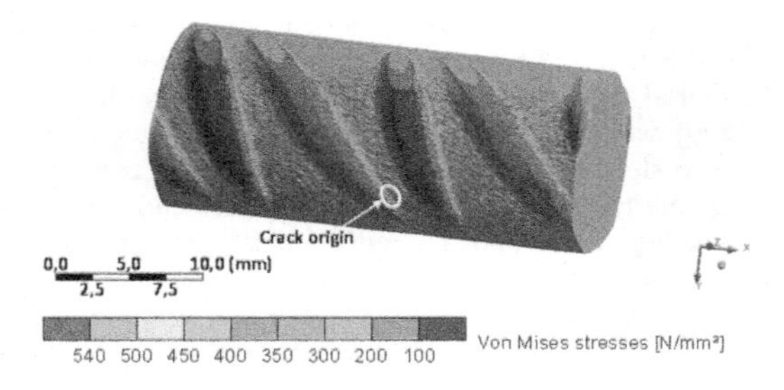

Figure 1. Von Mises stresses near the fatigue failure.

4 CONCLUSION AND OUTLOOK

The current paper presents the investigation of FE-Simulations carried out on digital geometric twins of rebars. Resulting notch stresses at the crack origin (observed in fatigue tests) were compared with local microstructure strengths.

- The highest notch stresses were observed on the surface in the fillet radius. On the alternated rib side mutual effects were detected between converging ribs.
- The maximum von Mises stresses at the crack origin were between 623 N/mm² and 740 N/mm² and approx. 15% higher than the results of idealized geometries. This was attributed to three-dimensional notches of a digital geometric twin.
- The calculated notch stresses at the crack origins were all in the scatter of the surface yield strength.

FE-Simulations carried out over the whole length of the digital rebars allow a statistical evaluation. Here, it can be observed whether the maximum notch stress itself or a local stress concentration is the origin of fatigue failure. Moreover, residual stresses will be included in future studies.

ACKNOWLEDGEMENTS

The authors wish to thank the German Research Foundation (DFG) for funding the project (GE 1973/33-1). We would also like to thank Prof. Gehlen for the opportunity to work on the research project. Finally yet importantly, our thanks are dedicated to the student assistants helping with the data processing.

LITERATURE

DIN 488–1:2009-08. 2009. Betonstahl - Teil 1: Stahlsorten, Eigenschaften, Kennzeichnung, Berlin: Beuth Verlag GmbH.

Jhamb, I. C. 1972. Fatigue of reinforcing bars, University of Alberta Libraries.

Osterminski, K., Gehlen, C. 2019. Development of a laser-based line scan measurement system for the surface characterization of reinforcing steel, Materials Testing, vol. 61, no. 11, pp. 1051–1055.

Rappl, S., Hameed, M. Z. S., Osterminski, K. Experimental and numerical analyses on the importance of the fillet radius for the fatigue behavior of rebars. Structural Concrete (in review)

Volkwein, A., Osterminski, K., Meyer, F., Gehlen, C. 2020. Distribution of residual stresses in reinforcing steel bars, Engineering Structures, vol. 223, p. 111140.

Wich, M. 2017. Entwicklung einer Schnittstelle zur grafischen Weiterverarbeitung von Laserscanergebnissen von korrodierten und nicht-korrodierten Betonstählen. Master's Thesis, Technical University of Munich

Current Perspectives and New Directions in Mechanics, Modelling and Design of Structural Systems – Zingoni (ed.)
© 2022 Copyright the Author(s), ISBN 978-1-032-39114-4

Crashworthiness analysis of conical energy absorber filled with aluminum foam

Michał Rogala & Jakub Gajewski
Faculty of Mechanical Engineering, Lublin University of Technology

ABSTRACT: This article presents an analysis of dynamic crushing of thin-walled conical aluminum profiles. The profiles were subjected to the energy of 1700J. The cones had a variable upper diameter specified in the paper as D2. Samples were tested in three configurations, empty, half-filled and completely filled with aluminum porous material. The analysis was performed using the finite element method. On the basis of the data obtained, the crush effectiveness indicators were determined, which were used to determine the effect of wall inclination and foam material on the energy absorber performance.

1 INTRODUCTION

Safety is nowadays a critical issue, especially considering that due to technological, social and economic progress, vehicles have become a daily solution in transport. The popularity of this mode of transport carries with it a certain danger of accidents, which entails overloading that can lead to fatal injuries to people. Crush zones at the front and back are responsible for vehicle safety.

2 AIM OF THE RESEARCH

The subject of research was a conical aluminum profile. The cone had a variable diameter D2 and was filled with aluminum foam. The length of the foam filling was 100mm and 200mm from the bottom base.

The numerical analysis was performed in a dynamic way, which means that a mass of 70kg was assigned to the top plate and the initial velocity was defined as 7m/s. The boundary conditions were made with the use of reference points. The upper plate has an unlocked possibility to move in the Y direction, the lower plate serving as a base has been fixed by blocking all translational and rotational degrees of freedom.

The plates are connected to the profile using the *Tie* function. The type of element used to model a non-deformable disc is R3D4 i.e., three-dimensional, 4-node quadrilateral element, while for the cone S4R i.e., 4-node quadrilateral shell element with reduced integration.

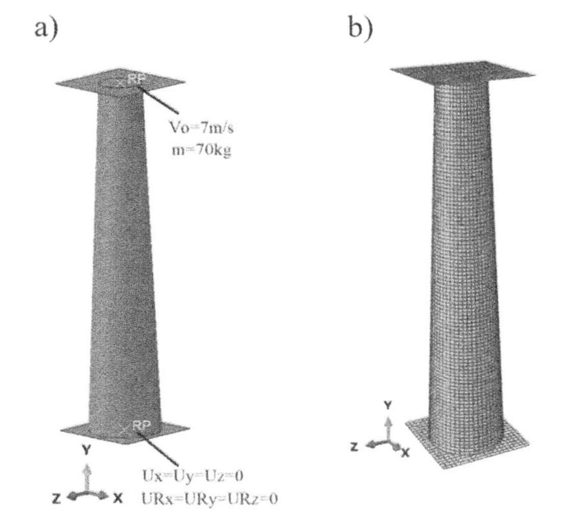

Figure 1. A) boundary conditions of performed analysis b) discretized conical energy absorber.

3 NUMERICAL ANALYSIS

The finite element method analysis set out to detect the effect of two variable dimension on the energy-absorption properties of the models, specifically: the upper diameter (D2) and the foam filling. The cross-sectional diameter of profile changed gradually from 24mm to 40mm at 4-mm steps. The analysis used aluminum foams with two lengths of 100mm and 200mm measured from the base. The dynamic tests were carried out until the tup lost its entire impact velocity.

DOI: 10.1201/9781003348450-104

The Figure 1 below shows a C40-32 profile crushing model with a thickness of 1 mm in the upper position.

4 RESULTS OF NUMERICAL ANALYSIS

The results of the numerical analysis were presented using the energy-absorption indicators described in first section. All the results were obtained under the same boundary conditions and mesh size, so that the results are the most reliable. The first quantity describes the mean force detected during the dynamic analysis (Figure 2), its size is reflected in the amount of absorbed energy. The higher the mean crushing force, the higher the efficiency of the energy absorber.

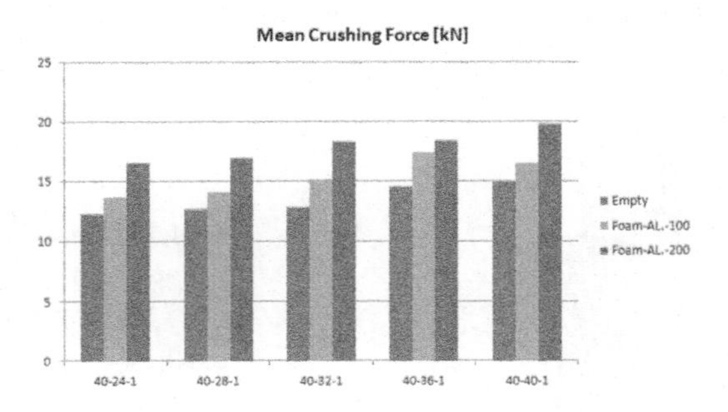

Figure 2. MCF indicator for conical energy absorber.

The highest mean force can be observed for models filled with 200mm long aluminum foam. There is a clear correlation between the length of the filling and the MCF value. The relationship between the length of the filling and the value of the MCF is also growing for the variable diameter D2 in the upper position.

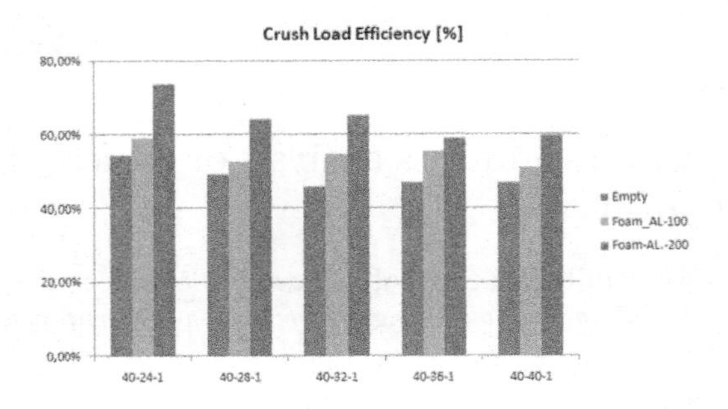

Figure 3. CLE indicator for conical energy absorber.

The performance of a thin-walled energy absorber changes clearly both by changing the D2 diameter and the length of the foam. The samples show a decreasing tendency with an increase of the diameter in the upper position, which is caused by a change of peak crushing force. The CLE index increases with a change in the length of the porous filling (Figure 3). The efficiency changes in the range from 45% to 75%, which indicates a significant influence of geometrical parameters.

5 CONCLUSIONS

The numerical tests carried out show that the use of foam inside the energy absorber has a positive effect on the design properties. Research shows that it is beneficial to fill with foam along the entire height of the cone. Has also been shown because of the easy formation of a plastic hinge, it is preferable that the ratio of the cone diameters is as large as possible. Confirmation of the above conclusions concerns both the MCF and CLE coefficients.

Current Perspectives and New Directions in Mechanics, Modelling and Design of Structural Systems – Zingoni (ed.)
© 2022 Copyright the Author(s), ISBN 978-1-032-39114-4

Numerical and experimental analysis of the triggering mechanism of the passive square thin-walled absorber

Michał Rogala, Jakub Gajewski & Robert Karpiński
Faculty of Mechanical Engineering, Lublin University of Technology

ABSTRACT: The subject of this study is the analysis of thin-walled aluminum profiles with triggers in the form of embossing. Samples with square cross-section loaded dynamically were studied until the complete loss of velocity by the tup. The numerical analyses were based on an elastic-plastic material model. The material properties of AA-6063 aluminum were derived from own tests performed on a tensile machine. The analyses were performed using the finite element method (Abaqus CAE). Based on the results obtained, the crush efficiency indicators were determined.

1 INTRODUCTION

Today, passive safety is a particularly important design aspect of mechanical structures. Vehicles have specific crumple zones designed to protect vehicle components during a road crash. Particularly important is the crash-box, which protects the passengers in a frontal collision at a maximum speed of 15-20km/h without damage to the vehicle stringers. A passive energy absorber is characterized by the fact that, with certain design assumptions, the structure folds like an accordion to absorb mechanical energy without generating large overloads. The field of energy absorption by vehicle protective components began in the early 1960s. Before that, vehicles had rigid steel bumpers to protect the occupants, but because of the overloads generated during an accident, they could be fatal to the passengers.

2 NUMERICAL ANALYSIS

Numerical analyses were performed using Abaqus CAE software. Each simulation was carried out in two steps, in the first one the buckling form characteristic for the respective geometry was obtained, then the model with implemented geometric disturbance was subjected to dynamic crushing. The profile was between two non-deformable plates serving as a tuple and a base. It consisted in loading a non-deformable plate of mass 70 kg with an initial velocity of 7 m/s, which defined a mechanical energy of 1700 J. All the profiles tested were loaded with the same energy and the analysis continued until the tup lost velocity. The aim of the study was to determine the effect of the triggering position in the form of embossing on the behavior of the thin-walled profile and the obtained crush efficiency indices.

The variable parameter was the position of the center of the trigger relative to the base denoted by the parameter X, cases for the range 20-100mm were studied. The initial size was due to geometrical constraints as the trigger had to fit all the way through, the maximum initiator position of 100mm was due to the fact that the authors wanted the column to start the initiation of the crushing process at its bottom.

3 RESULTS

Based on the numerical analyses, the crush efficiency indicators described in more detail in section one were determined. The basic values are based on the course of force characteristics. Figure 1 presents a compilation of characteristics for the trigger center position in the range 20-100 mm. Already on the basis of the graphs the variation in the formation of folds as a function of displacement is apparent. Moreover, it is possible to see the variation of the value of the total shortening of the profile for the same initial energy, i.e. 1700 J

The first indicator determined is PCF, which is the force generated at the beginning of the crush and is related to the shape and position of the trigger. For the case studied, the force spread is in the range of 37.5 to 40.5 kN. A large force peak generated over a long period of time can cause fatalities.

DOI: 10.1201/9781003348450-105

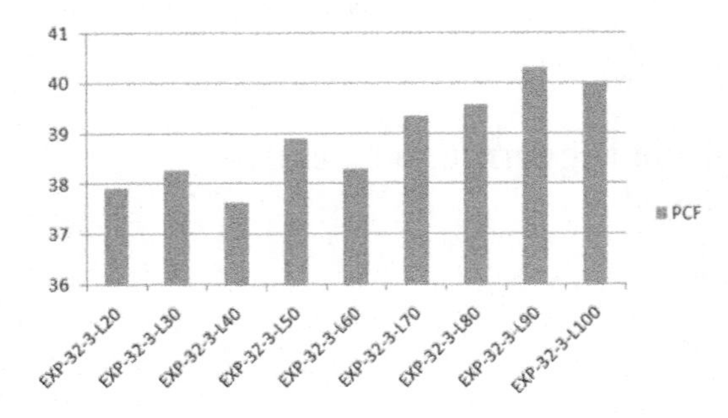

Figure 1. PCF value for different trigger location.

Figure 2 shows the mean crushing force, which reached values in the range of 15.25 to 16.75 kN. The value of the average force in the large miter is responsible for the overall performance of the thin-walled structure. Due to the crush characteristics, thin-walled structures often show a large decrease in force after fold formation which is equivalent to a decrease in MCF. By using the shape of the trigger and its position it is possible to change its value significantly.

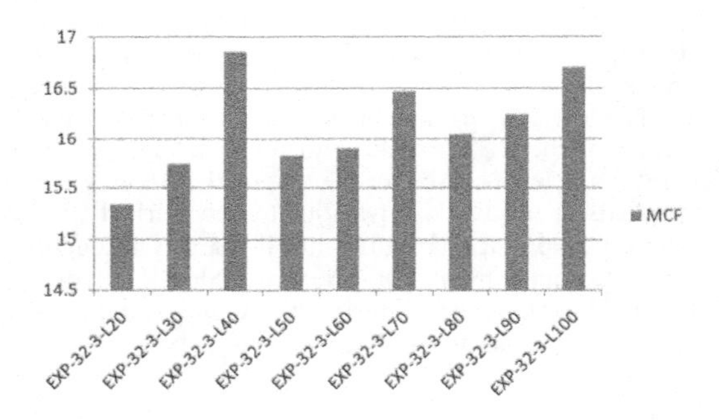

Figure 2. MCF value for different trigger location.

The last indicator determined is CLE which represents the crushing force efficiency. It significantly represents the thin-walled profile with the best force performance. The best effect is shown by the profile with the position of the trigger center at a distance of 40 mm from the bottom edge of the profile.

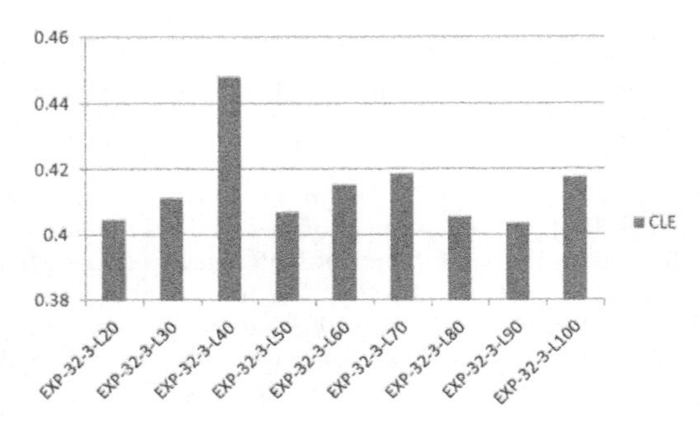

Figure 3. CLE value for different trigger location.

4 CONCLUSIONS

The conducted numerical tests show that the best location of the trigger is a height of 40 mm from the base of the column. Determining the optimal height is related to the diameter of the press forming and the formation of the first plastic hinge. The value of CLE was identified as the most important optimization criterion. Figure 3 shows that the Crush Load Coefficient for a height of 40 mm increases the value of the CLE by about 15%. Optimal value of the trigger execution height assumed it also positively influences the values of PCF and MCF coefficients.

It should be also considered that the trigger height is related to the formation of the first plastic joint. The numerical tests require verification on a test stand. However, from the research carried out so far, it can be concluded that the height of the trigger execution is of key importance for the design of energy absorption of column.

Current Perspectives and New Directions in Mechanics, Modelling and Design of Structural Systems – Zingoni (ed.)
© 2022 Copyright the Author(s), ISBN 978-1-032-39114-4

Theoretical and experimental investigation into area increase of longitudinal cracks in pressurised pipes

D.T. Ilunga
Department of Civil Engineering Science, University of Johannesburg, South Africa

M.O. Dinka
Department of Civil Engineering, University of Johannesburg, South Africa

D.M. Madyira
Department of Mechanical Engineering, University of Johannesburg, South Africa

ABSTRACT: The purpose of this study was to investigate the effect of the pipe material stiffness on the behaviour of water leakage through a longitudinal crack in pressurised pipe due to the expansion of the crack area. Two theoretical equations were derived for water leakage flow rate through longitudinal cracks opening in a uniaxial and biaxial stress. The two equations were derived as a function of pressure, fluid properties, pipe geometry and pipe mechanical properties, with special emphasis on the pipe material stiffness. The stiffness of the pipe was incorporated in the actual derived equations as a mechanical property to characterise the resistance offered by the elastic behaviour of the material around the crack to deformation due to pressure variations. This is justified, because the material properties were not taken into consideration in the derivation of the orifice equation for flow through openings. The two derived equations were used as mathematical models to predict flow leakage through longitudinal cracks in pressurised water pipes. Results show that the material around the longitudinal crack exhibits elastic expansion behaviour due to hoop stresses induced by internal pressure within the pipe. The leakage exponent values resulting from a Class 6 uPVC pipe, using the two derived equations are much higher than the theoretical value of 0.5: ranging from 1.01 to 2.68 for uniaxial stress state and biaxial stress state conditions respectively. The pipe material stiffness plays a major role in the leakage flow rate behaviour.

1 AIM OF THE INVESTIGATION

The purpose of this study is firstly to investigate the effect of the pipe material stiffness on the behaviour of a longitudinal crack area present on a pressurised pipe-wall. Secondly, investigate how the material behaves around a longitudinal crack in sidewall of pressurised pipes; and thirdly, to provide two mathematical models which explain the increase of leakage exponent related to the increase of longitudinal crack area due to pressure increase.

2 STRUCTURAL BEHAVIOUR OF LONGITUDINAL CRACK IN PRESSURISED WATER PIPES

A pressurised pipe is subjected to hoop and longitudinal stresses. The normal circumferential (or hoop) stresses are twice the size of the longitudinal stresses. Due to pressure increase, there is appearance of stress concentration around the crack flanks as well as at the tips of the longitudinal crack resulting in the crack area increase in circumferential direction and may lead to crack propagation in longitudinal direction.

2.1 *Experimental verification of the behaviour of longitudinal crack in pipes subjected to internal pressure*

The behaviour of crack subjected to internal pressure was verified by applying an internal pressure with the aid of two rubeer bladders placed inside the sample and measuring how much stresses were generated and required to open the crack area until it starts to propagate.

3 RESULTS OF THE INVESTIGATION

3.1 *Determination of actual longitudinal crack opening area in pressurised pipes under uniaxial stress state conditions*

Equation 1 was derived as actual orifice area of the strained longitudinal crack due to the expansion of the material around the crack in pressurised pipe material under uniaxial stress state as follows:

$$A_{Actual} = a_{cr}.u_{y\max}\left(1 + \frac{vK\rho gh_{cr}R}{tE}\right) \quad (1)$$

Where A_{Actual} is the actual strained crack area, a_{cr} the critical crack length, u_{ymax} the maximum crack faces opening, K the stress intensity factor, ρ the fluid

DOI: 10.1201/9781003348450-106

density, g the gravitational acceleration, h_{cr} the critical pressure head of the fluid, R the internal radius of the pipe, t the thickness of the pipe wall, v the Poisson's ratio of the pipe material and E Young's modulus of the pipe material.

3.2 *Derivation of the equation for leakage flow rate through longitudinal crack opening area in pressurised pipes under uniaxial stress state conditions (Ilunga's Equation)*

Equation 2 was derived using equation 1, as a mathematical model to predict leakage flowrate through longitudinal crack incorporating mechanical properties of the pipe material under increasing pressure in uniaxial stress state as follows:

$$Q_{Actual} = C_d\sqrt{2g}\left[a..u_y\left(h^{\frac{1}{2}} + \frac{\lambda v K\rho g h^{\frac{3}{2}} R}{tE}\right)\right] \quad (2)$$

Where Q_{Actual} is the leakage flowrate, C_d the discharge coefficient, λ the shell curvature factor due to bulging strain of the crack.

Figure 1 was obtained by applying equation 2 as follows:

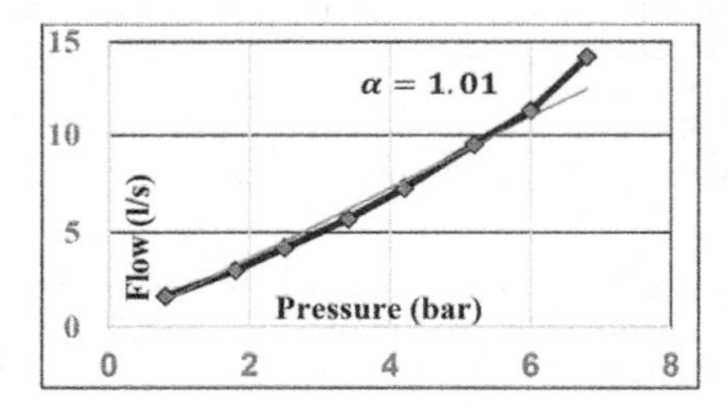

Figure 1. Leakage flow behaviour through a 90mm Crack within a 6 Class uPVC pipe using the derived flow equation under uniaxial stress state conditions.

3.3 *Determination of actual longitudinal crack opening area in pressurised pipes under biaxial stress state conditions*

Equation 3 was derived as actual orifice area of the strained longitudinal crack due to the expansion of the material around the crack in pressurised pipe in biaxial stress state as follows:

$$A_{Actual} = (u_y.a)\left(1 + \frac{K\rho g h R}{2tE}\right)^2[(2-v)(1-2v)] \quad (3)$$

3.4 *Derivation of the equation for leakage flow rate through longitudinal crack opening area in pressurised pipes under biaxial stress state conditions (Ilunga's Equation)*

Equation 4 was derived as a mathematical model to predict leakage flowrate through longitudinal crack incorporating mechanical properties of the pipe material under increasing pressure in biaxial stress state as follows:

$$Q_{Actual} = Cd\sqrt{2g}\left[h^{\frac{1}{2}}(a..u_y)\left(h^{\frac{1}{2}} + \frac{\lambda K\rho g h^{\frac{3}{2}} R}{2tE}\right)^2 h^{\frac{1}{2}}(2-v)(1-2v)\right] \quad (4)$$

Figure 2 was obtained by applying equation 4 as follows:

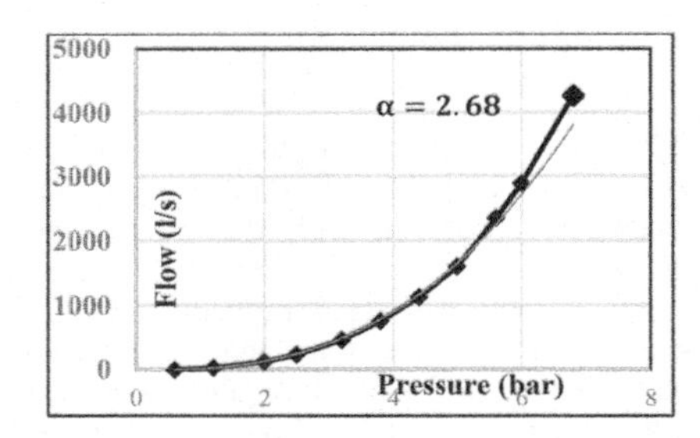

Figure 2. Leakage flow through a 90mm Crack within a 6 Class uPVC pipe using the derived flow equation under biaxial stress state conditions.

4 CONCLUSIONS

This paper investigated the theoretical and experimental behaviour of an area increase of a longitudinal crack in pressurised water pipes. Two theoretical equations for crack opening area in pressurised pipe were derived, resulting in stress-strain due to elastic property of the pipe material, in uniaxial stress state and biaxial stress state conditions. Subsequently, two theoretical equations for increased flow through a longitudinal crack incorporating pipe material properties (stiffness) were derived as mathematical models to predict leakage rate under increasing pressure in uniaxial stress state conditions as well as in biaxial stress state conditions. This is justified, because the material properties were not taken into consideration in the derivation of the orifice equation for flow through openings.

The experimental investigation results have indicated that '*materials exhibit expansion properties which alter the size of longitudinal cracks as a leak on pressurized cylinders, increasing the area therefore experiencing higher leakage flow rates than the theoretical value of 0.5 (Torricelli Theorem)*'.

Finally, the two derived equations proposed in this study to predict the leakage exponent values resulting from a Class 6 uPVC pipe, have given exponents of leakage which are much higher than the theoretical value of 0.5, ranging from 1.01 to 2.68 under uniaxial stress state and biaxial stress state conditions respectively. The effect of the pipe material stiffness plays a major role in the leakage flow rate behaviour.

Current Perspectives and New Directions in Mechanics, Modelling and Design of Structural Systems – Zingoni (ed.)
© 2022 Copyright the Author(s), ISBN 978-1-032-39114-4

Numerical study of dynamic collapse resistance of an RC flat plate substructure under an interior column removal scenario

Z. Yang & H. Guan
Griffith University Gold Coast Campus, Queensland, Australia

Y. Li
Beijing University of Technology, Beijing, China

X.Z. Lu
Tsinghua University, Beijing, China

ABSTRACT: If one or more columns of a flat plate structure are damaged in an accidental event, the load redistribution process may trigger punching shear failures at the adjacent slab-column joints. Progressive collapse of a large portion of the structure, or even its entirety, is like to occur following punching failure propagations. Given that progressive collapse is a dynamic phenomenon, investigating the dynamic responses of the structure under sudden column removal scenarios are imperative. In this study, a high-fidelity numerical model is developed to simulate the dynamic behaviour of a multi-panel flat plate substructure under a sudden removal of an interior column. The model is validated by three sequential experimental tests on the same substructure, by which the load responses of the interior column stub, the slab deflections and the crack patterns are verified. Based on the validated model, the dynamic load redistribution patterns of the three tests are evaluated. Two series of parametric studies are also conducted to investigate the influence of the load level on the slab overhangs and the load release speed on the slab deflections.

1 INTRODUCTION

Abnormal loading events (overloading, fire, explosion, etc.) can trigger punching shear failure initiated at one or several slab-column joints of a reinforced concrete (RC) flat plate structure. Following subsequent load redistribution, punching shear failure may progressively spread to adjacent joints. If these joints have insufficient strength to resist additional loads, the entire structure or a significant part of it may eventually collapse. A very recent collapse of a flat plate building happened in June 2021 in Miami, USA (Lu et al. 2021), causing mass casualties and unmeasurable economic loss. In this study, a numerical modelling approach was proposed and validated. The load redistribution patterns of the three tests were numerically studied. Finally, parametric studies were conducted to investigate the influence of overhang loads and the load release speed on the slab deflections.

2 NUMERICAL MODELS

2.1 *Brief of experimental program*

The numerical analysis in the present study was conducted based on a 1/3-scale 2×2-bay flat plate substructure specimen, which was tested by suddenly removing the interior column under three different levels of the gravity loads. The test specimen was designed in accordance with both Chinese Building Codes (MHURDPRC 2010) and Australia Standard of Concrete Structures (Australian Standard 2009). The slab had an overall dimension of 5000 mm × 5000 mm × 90 mm, and it was supported by one interior column and eight surrounding columns at a space of 2000 mm. In addition, since the test specimen was extracted from the centre region of the first floor of a 4×4-bay multi-level flat plate structure, a 500 mm slab overhang was extended from the peripheral column grid to simulate the continuity of the structure.

2.2 *Details of numerical model*

A high-fidelity finite element (FE) model was established in this study to replicate the dynamic behaviour of the flat plate substructure subjected to the sudden removal of the interior column. Considering the accuracy, efficiency, and stability during the numerical analyses, a FE software LS-DYNA (LSTC 2016) was adopted in this study. To avoid any computational issue during the large structural deformation, the explicit method was used for the model analyses.

DOI: 10.1201/9781003348450-107

2.2.1 *Geometrical model*

The concrete slab and column stubs were modelled by 8-node solid elements. The steel rebars were simulated by Hughes-Liu beam elements. The steel rebars were bonded to the concrete through a keyword card named beam_in_solid, and a perfect bond condition was defined for this model.

2.2.2 *Material model*

The continuous surface cap model (CSCM) was adopted in this study to simulate the concrete material due to its effectiveness in the modelling of concrete softening, dilation, and compaction. The default configurations of the CSCM with 32.6 MPa unconfined compressive strength and 19 mm aggregate size were adopted, which exhibited a satisfactory accuracy in the model validations. The piecewise linear plasticity material model was used for the material modelling of the steel rebars.

2.2.3 *Load applications and support sudden release*

The gravity loads applied to the slab during the experimental tests were simulated by increasing the densities of the slab elements, the keyword card load_body was employed to assign the gravity acceleration. In addition, the specially designed connector, which was used to support and release the interior column stub in the experimental tests, was simplified as a steel plate underneath the column stub in the model development. The sudden load release in the experimental tests was simulated by defining the downward accelerations to the bottom steel plate to ensure that the numerically obtained load release durations matched the experimentally observed ones.

2.2.4 *Model validations*

Table 1 compares the typical results obtained from the tests and models, which include the reaction forces of the interior column and the peak vertical displacement of the slab. Good correlations can be seen for both types of results.

Table 1. Results comparisons between tests and models.

Test No.	Reaction forces of interior column		Peak vertical displacement	
	kN		mm	
	Tests	Models	Tests	Models
T1	98.2	105.4	52.1	50.7
T2	114.8	117.2	52.3	54.4
T3	172.2	162.3	123.3	121.3

In addition, similar crack patterns were also seen from experimental and numerical results for both bottom and top surfaces, which demonstrated the effectiveness of the proposed modelling method in predicting the structural behaviour of the specimen.

3 NUMERICAL STUDIES

Based on the validated numerical models, the dynamic load redistribution patterns of the three experimental tests were obtained and discussed herein. At the peak load states of T1 to T3, the total reaction forces increased to 116.4%, 109.6%, and 117.8% of the slab loads within the column grid due to the inertia effect. At the stable load state, most of the loads (no less than 89%) previously carried by the interior column were transferred to the edge columns.

Due to safety reasons, the slab loads on the overhangs were different in the three tests, which may influence the resistance capacities of the test specimen. To quantify such an influence, T2 and T3 were numerically repeated with the overhang loads identical to that of T1. Fairly small differences were found in the slab deformation responses between the two pairs of models. For T3, which had a larger load difference on the overhangs than T2, the peak vertical displacement of the original model was only 3.8 mm larger than that of the repeated model, indicating that little influence can be made by the load amount on the overhangs.

The effect of the load release speed was evaluated based on the T1 model. Almost no influence on the peak vertical displacement was exhibited when the load release duration is no larger than 0.05 s. However, the peak vertical displacement may decrease by 10% when the load release duration equals 0.15s.

4 CONCLUSION

In this study, a high-fidelity numerical model was developed based on a series of experimental tests, which were conducted on an RC flat plate substructure. The model was validated by a set of experimental results, including the load responses of the interior column stub, the slab deflections, and the crack patterns. Thereafter, the dynamic load redistributions were numerically discussed. It was indicated that more than 89% of the interior column load was transferred to the edge columns. Furthermore, the validated models were also used to evaluate the influences of the overhang loads and the load release speed, and both of them were found to be small on the dynamic resistance capacities of the specimen.

REFERENCES

Australian Standard (AS). 2009. *Concrete Structures*. AS3600, Sydney, NSW, Australia.

LSTC (Livermore Software Technology Corporation). 2016. *LS-DYNA keyword user's manual*. Livermore, CA.

Lu, X., H. Guan, H. Sun, Y. Li, Z. Zheng, Y. Fei, Z. Yang, and L. Zuo. 2021. "A preliminary analysis and discussion of the condominium building collapse in surfside, Florida, US, June 24, 2021." *Front. Struct. Civ. Eng.*

MHURDPRC. 2010. *Code for Design of Concrete Structures*. Ministry of Housing and Urban Rural Development of the People's Republic of China.

8. Buckling of Beams, Columns and Thin-walled sections

Current Perspectives and New Directions in Mechanics, Modelling and Design of Structural Systems – Zingoni (ed.)
© 2022 Copyright the Author(s), ISBN 978-1-032-39114-4

Harnessing instabilities within metamaterial structures

M. Ahmer Wadee, Adam Bekele & Andrew.T.M. Phillips
Department of Civil & Environmental Engineering, Imperial College London, London, UK

ABSTRACT: Advances in additive manufacturing are increasingly allowing bespoke, carefully designed, metamaterial lattice structures to be constructed for enhanced mechanical performance. For instance, in structural elements that are designed to absorb energy and shield a more valuable structure, the properties combining a high initial stiffness followed by a practically zero underlying stiffness, ensure that a desired energy quantity may be absorbed within a limited displacement and that any stress transfer to the valuable structure is minimized. Presently, a lattice structure is studied that is deliberately designed to switch locally under compression between conventional material behaviour to that exhibiting auxetic (negative Poisson's ratio) behaviour through a sequence of snap-through instabilities within layers of individual lattice cells. The sequence of buckling instabilities can potentially be controlled to maintain the loading level alongside the low structural stiffness while the required energy quantity is absorbed, without necessarily damaging the material in the process. This departs from the usual paradigm where such structures are presently designed to be sacrificial and opens up the intriguing possibility of introducing such structural elements that are repairable and therefore reusable.

1 INTRODUCTION

There has been a growing requirement for buildings and infrastructure to be shielded with protective layers, especially those that may be vulnerable to impact or blast from external agencies. The search for energy absorbing structures has therefore obtained increased importance. The current work introduces a methodology for enhancing energy absorption and structural isolation performance of a protective cellular structure. For such a combination of applications, the desired load–displacement response is an initially high stiffness that generates a large load carrying capacity, followed by a quasi-zero post-buckling stiffness response (Reis, 2015; Champneys et al., 2019; Wadee et al., 2020). Whilst the latter (stable) low stiffness post-buckling response restricts a transfer of large impact forces to the protected structure, the high initial stiffness and high load carrying capacity maximizes the energy absorbency of the system.

2 NONLINEAR ANALYTICAL MODELLING

A unit cell configuration, referred to presently as a 'flint arrowhead', as shown in Figure 1, is selected. This bistable geometry can snap from the flint arrowhead, whereby under compression the cell structure expands in the transverse direction, to a conventional (auxetic) arrowhead with effectively a negative Poisson's ratio. Figure 1(c) depicts a series of n cells with the lengths of the rigid links all being equivalent to those of the unit cell. The model equations are formulated by applying the principle of stationary total potential energy V, which comprises the total strain energy stored within the individual longitudinal and rotational springs minus the work done by the external load. The resulting equilibrium equations are solved within the numerical continuation software AUTO-07P since it can pinpoint bifurcations and compute the emerging solution branches.

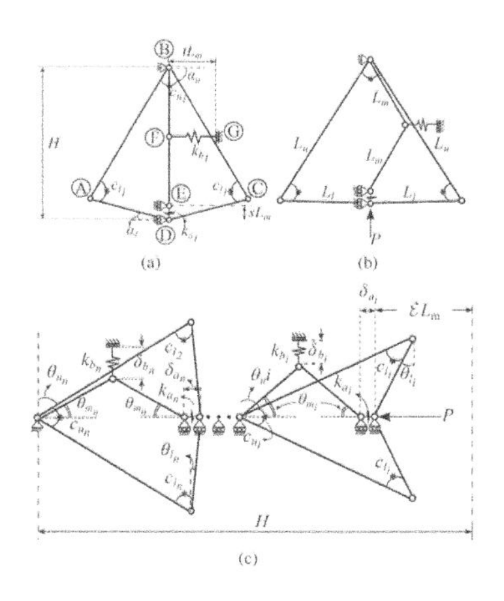

Figure 1. Flint arrowhead model configuration. (a)–(b) represent the unit cell at various stages of axial compression; (d) represents a stack of n cells and $i = \{1, 2, \ldots, n\}$.

DOI: 10.1201/9781003348450-108

The behaviour of five flint arrowhead cells in series is examined primarily to understand the overall mechanical response of this cellular architecture. Figure 2 shows the normalized load versus axial end-shortening. Owing to the cellular arrangement, there are five separate primary bifurcating paths branching from the fundamental path. The first branch relates to the buckling of the first cell that requires the lowest buckling load. The solutions that branch from the primary equilibrium path signify the subsequent buckling of adjoining cells. Figure 2 also shows deformations with corresponding tags on the branches. The model behaviour also shows each cell collapsing before the subsequent cell buckles, as desired.

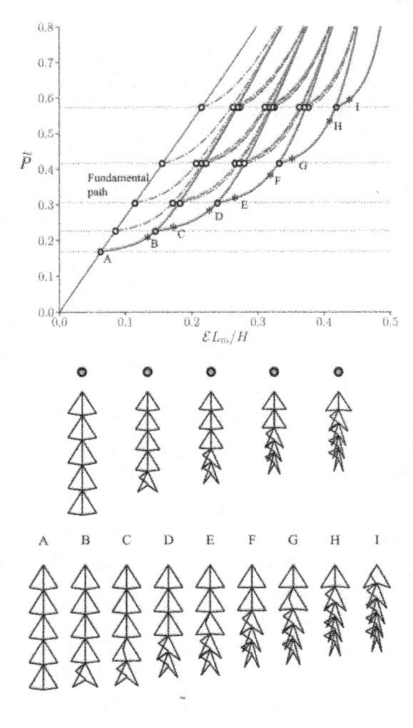

Figure 2. Normalized load $\tilde{P} = P/(k_{a_1} L_m)$ versus normalized axial end-shortening $\Delta L_m/H$ for a stack of five flint arrowhead cells.

3 SIMULATION OF REAL BEHAVIOUR AND CONCLUSIONS

Studies were conducted within ABAQUS on an entire lattice comprising a network of 3D flint arrowhead unit cells under compression, a 3D extension of the planar arrowhead geometry examined in the analytical model. Rigidly connected 3D quadratic three-noded Timoshenko beam elements (B32) are utilized with circular solid sections with Young's modulus $E = 2\text{GPa}$ alongside the material density of 1.15g/cm^3. These represent the properties of Nylon which was used in the physical specimens that were produced and tested presently. Figure 3 shows comparisons between the experimental and numerical results for a physically tested lattice panel. The curves presented are for where minimal internal structural damage was evident and good agreement is shown. However, the panels were tested to destruction and exhibited considerable densification subsequently, as is usual in such arrangements when under compression.

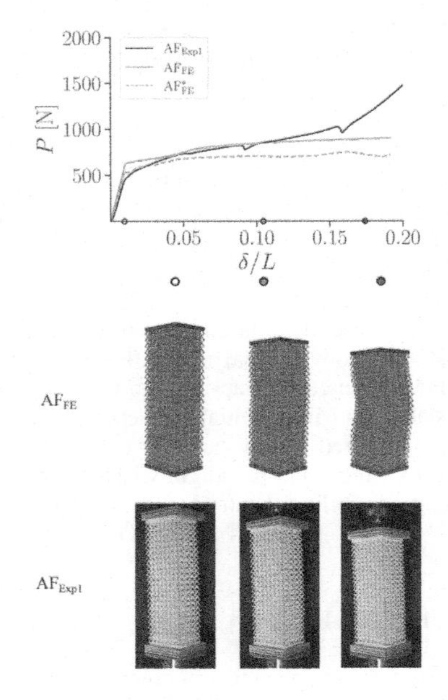

Figure 3. Numerical and experimental mechanical behaviour comparison for a 3D printed lattice.

It is concluded that the nonlinear analytical model for elastic sequential buckling predicts the observed behaviour where softening of the mechanical response occurs owing to internal instability within the lattice cells. The effects of varying different stiffnesses within the model were identified and allow the mechanical response to be altered both qualitatively and quantitatively such that favourable behaviour is observed for energy absorption purposes. Since the driving parameters have been identified, the present work provides a theoretical framework for engineering and optimizing such lattices.

REFERENCES

Champneys, A. R., Dodwell, T. J., Groh, R. M. J., Hunt, G. W., Neville, R. M., Pirrera, A., Sakhaei, A. H., Schenk, M., & Wadee, M. A. 2019. Happy catastrophe: Recent progress in analysis and exploitation of elastic instability. *Front. Appl. Math. Stat.*, **5**. Article 34.

Reis, P. M. 2015. A perspective on the revival of structural (in)stability with novel opportunities for function: from buckliphobia to buckliphilia. *J. Appl. Mech. – Trans. ASME*, **82** (11),111001–1–111001–4.

Wadee, M. A., Phillips, A. T. M., & Bekele, A. 2020. Effects of disruptive inclusions in sandwich core lattices to enhance energy absorbency and structural isolation performance. *Front. Mater.*, **7**. Article 34.

Current Perspectives and New Directions in Mechanics, Modelling and Design of Structural Systems – Zingoni (ed.)
© 2022 Copyright the Author(s), ISBN 978-1-032-39114-4

Plastic buckling of fixed-roof oil storage tanks under blast loads

L.A. Godoy
Institute for Advanced Studies in Engineering and Technology, IDIT/CONICET/UNC, Argentina

M.P. Ameijeiras
FCEFyN, Universidad Nacional de Cordoba, and Institute for Advanced Studies in Engineering and Technology, IDIT/CONICET/UNC, Argentina

ABSTRACT: This paper deals with the dynamic behavior of cylindrical shells under blast loads, with special reference to plasticity and plastic buckling. Interest in this problem focuses on the response of steel vertical oil storage tanks with a flat roof under a nearby explosion. The problem is modeled using a finite element discretization of the structure, with the loads represented by an impulsive pressure which is variable around the circumference. The results for tanks with flat roof show that there is an elastic behavior of the shell at low pressure levels, and first yield occurs at the center of the tank as the peak blast pressure is increased. For subsequent increases of peak pressure, the dimple extends in the area affected by the pressures, so that vertical stripes with plastic material behavior are formed from top to bottom of the shell. For even higher pressures, a global mode develops with large amplitude displacements. The total plastic energy is investigated under increasing peak pressure levels.

1 INTRODUCTION

Problems of large plastic deformations and buckling of shells under blast loads have been reported in recent years with reference to oil storage tanks. Flammable fuels are stored in vertical tanks, leading to an industrial process with high probability of occurrence of oil spills and explosions (Godoy et al. 2022).

The identification of the onset of buckling requires the use of controlled experiments, such as those carried out by means of computational or analytical studies. To understand this buckling behavior, this paper explores thresholds at which the shell changes the way of providing equilibrium along the dynamic process.

2 METHODOLOGY

It is assumed that an explosion occurs at ground level, and a vertical steel tank receives time-dependent pressures. The impulsive reflected pressure is here given in the convenient form:

$$p(\theta, t) = p_0 e^{-(k_1\theta)^2}\left(1 - \frac{t}{t_0}\right)e^{-\frac{k_2 t}{t_0}} \tag{1}$$

where p_0 is the peak reflected overpressure, t_0 is the blast duration, and θ is the central angle. Values of $k_2 = 1.43$ and $k_1 = 1.1$ were adopted in this work to approximate results of small-scale models tested by Duong et al. (2012). Elastic-perfectly plastic material behavior was assumed using the von Mises yield criterion and an associated flow rule. Material properties for A-36 steel were assumed in the computations.

The computations reported in this paper were carried out using the general-purpose finite element software ABAQUS (2010). The equivalent plastic strain $\bar{\varepsilon}^y$ is computed in the form

$$\bar{\varepsilon}^y = \int_0^t \sqrt{\frac{2}{3}\dot{\underline{\varepsilon}}^y : \dot{\underline{\varepsilon}}^y}\, dt \tag{2}$$

where $E_P = \int_0^t \left(\int_V {}^c : {}^y dV\right) dt$ is the plastic deformation rate tensor, computed at the yield condition, and the integration extends from t = 0 to a time t at which the equivalent plastic strain is computed. Another variable considered is the Total Plastic Energy, defined as:

$$E_P = \int_0^t \left(\int_V \underline{\sigma}^c : \dot{\underline{\varepsilon}}^y dV\right) dt \tag{3}$$

where the stress tensor $\underline{\sigma}^c$ does not account for viscous or time-dependent effects. The volume integral is dominated by the area with dynamic pressures.

3 CASE STUDY

The vertical cylindrical shell considered here has a fixed flat roof and clamped boundary conditions at the bottom. This shell has diameter $D = 15$m, height

DOI: 10.1201/9781003348450-109

H = 12m, and thickness h = 7.5mm. A geometric imperfection with the shape of the static critical eigenvector, and amplitude equal to the shell thickness was assumed. An explicit time integration technique was used with an approximate stable increment of 0.03ms. The duration of the load was 30ms.

The plastic behavior was computed at a time for which the load already ceased to act, and transient vibrations would tend to vanish. If the peak pressure reaches p_0 = 50KPa, a wavy pattern develops with a maximum radial displacement u_r = $7h$. Coupled with this change in shape, a small plastic dimple is seen to develop; this state may be considered as the onset of plasticity and is clearly non-uniform in the vertical direction. For further increases in p_0, plasticity extends to regions towards the top and bottom of the shell, and around the circumference.

The buckling mode in Figure 1 shows several waves around the circumference for p_0 = 175KPa, and a global mode develops due to lateral buckling. The radial displacements are in the order of u_r = $34h$.

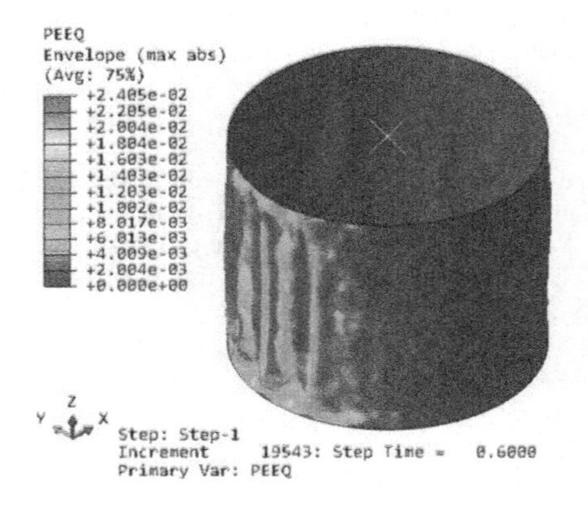

Figure 1. PEEQ and deflected shape of a cylindrical shell with flat roof under blast pressure; t_0 = 30ms and p_0 = 175KPa.

The results of total plastic energy for increasing values of p_0 were computed using ABAQUS (2010) and are shown in Figure 2. At p_0 = 70KPa, for which a plastic dimple first occurs, the plastic energy is very small, i.e., E_p = 1.5KJ. At p_0 = 105KPa there is a clear increase to E_p = 27.6KJ, and for higher values of p_0 there is a non-linear increase in the plot, with a slope that gradually tends to zero. The load at 175KPa may be considered as an estimate of shell collapse.

As a tentative criterion to establish a limit state, one could take a peak load equal to a factor of the value of the pressure at first yield. If one adopts a factor of 1.5, then at p_0 = 105KPa the total plastic energy becomes 27.6KJ. The ratio of energy with respect to first yield in this case is 27.6/1.5 = 18.4.

4 CONCLUSIONS

Finite element analyses were reported in this paper to investigate plastic buckling of vertical oil storage

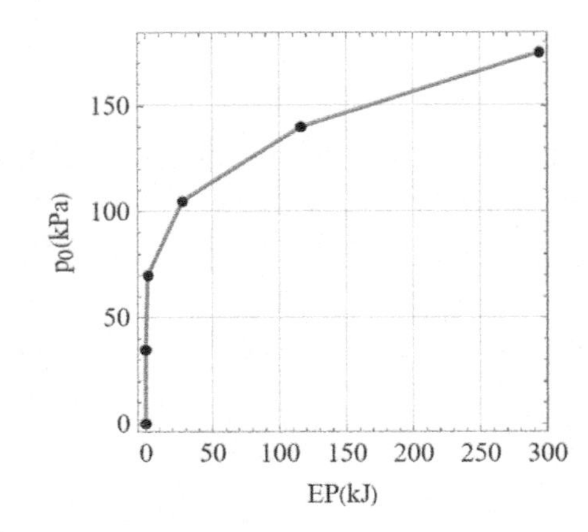

Figure 2. Total plastic energy versus peak pressure; t_0 = 30ms.

tanks under blast loads due to an explosion. For increasing values of blast pressure, the plastic zone extends to reach the top and bottom boundaries of the shell. At higher blast pressure levels, a global mode develops with large lateral displacements. To follow the progress of plasticity with increasing values of blast pressures, the total plastic energy E_p has been computed. A nonlinear increase in E_p is found that tends to stabilize at approximately three times the pressure required to first yield. It is here proposed to take a pressure equal to 1.5 times the pressure at first yield, as an estimate of limit state under blast loads.

Regarding post-buckling modes, one may find plastic buckling with a large number of wrinkles in the circumferential direction (short-wave modes), but an overall (global mode) with large displacements towards the inside of the shell is obtained in other cases. The paper shows that both modes are part of the same process, but the localized mode seems to occur at the onset of plastic buckling, whereas the global mode is observed at advanced post-buckling states.

ACKNOWLEDGEMENTS

The authors thank the support of a grant from the Science and Technology Research Council of Argentina (CONICET) during the present research.

REFERENCES

ABAQUS, 2006. Simulia: Unified FEA. Dassault Systems. Warwick, RI, USA.

Duong D. H., Hanus J. L., Bouazaoui L., Pennetier O., Moriceau, J., Prod'homme G. & Reimeringer M. 2012. Response of a tank under blast loading – Part I: Experimental characterization, *European Journal of Environmental and Civil Engineering* 16(9): 1023–1041.

Godoy L. A., Jaca R. C. & Ameijeiras M. P. 2022. On buckling of oil storage tanks under nearby explosions and fire, Chapter 7 in: Above Ground Storage Tank Spills, Ed. M. Fingas, Elsevier.

Current Perspectives and New Directions in Mechanics, Modelling and Design of Structural Systems – Zingoni (ed.)
© 2022 Copyright the Author(s), ISBN 978-1-032-39114-4

Modelling thin-walled members by means of the dynamic approach within the Generalised Beam Theory (D-GBT)

G. Ranzi
School of Civil Engineering, The University of Sydney, Sydney, Australia

ABSTRACT: The Generalized Beam Theory (GBT) is a powerful tool for the analysis of thin-walled members. The use of the GBT requires the identification of a suitable set of deformation modes that define the possible behaviours that the structural cross-section can undergo and this is then used as the basis to model the member response of the thin-walled elements. This paper provides an overview of the dynamic approach within the broader context of GBT formulations where the set of cross-sectional modes is obtained by means of dynamic analyses of the member cross-sections. This approach is applicable to open, closed and partially-closed cross-section and it is applicable to members formed by steel, composite and multi-component sections.

1 INTRODUCTION

The Generalized Beam Theory, commonly referred to as GBT, is a powerful tool for the analysis of thin-walled members. The basic idea of the method consists in describing the displacement field of the thin-walled element as a linear combination of assumed 'deformation modes' of the cross-section (including in-plane and warping components), and 'amplitude modes', which are unknown functions depending on the axial coordinate. The use of the GBT requires the execution of two main steps: (i) the identification of a suitable set of deformation modes capable of describing the cross-sectional behaviour, referred to as 'cross-sectional analysis', and (ii) the 'member analysis' that determines the amplitude values defining the intensity of the deformation modes along the member axis. The initial formulation of the GBT was presented by Schardt (1994) and further refined and extended by several researchers, e.g. (Schardt 1994, Silvestre & Camotim 2002, Camotim et al. 2010). A wide range of methods of analysis have been developed and applied to predict the response of thin-walled members. The particularity of these elements relies on the fact that they undergo significant deformations both in-the-plane and out-of-the-plane of the cross-section. In this context, this paper presents an overview of the use of the Generalized Beam Theory developed with the D-GBT formulation to highlight the ease of use of this approach. With this method, the dynamic analysis is performed on a planar frame that represents the cross-section and this procedure enables the evaluation of the conventional, extension and shear modes. This approach has been applied to different structural typologies that involve steel, composite and multi-component cross-sections.

2 D-GBT FORMULATION FOR STEEL MEMBERS

The first paper considering the D-GBT (Ranzi and Luongo, 2011) focused on the analysis of thin-walled members with the use of the GBT conventional modes. The model was derived to analyse the response of thin-walled members formed with open, closed and partially-closed cross-sections.

A similar formulation was presented in (Ranzi and Luongo, 2014) in which the cross-sectional analysis was performed using the direct stiffness method and the element formulation was obtained as the analytical solution of the dynamic problem. Because of this, the terms considered in the solution procedure were expressed in terms of cosh and sinh, and the identification of the dynamic modes required an iterative process. The approach was validated against specific open and partially-closed cross-sections. A direct approach for the evaluation of the conventional modes in a simple analysis process was subsequently presented in (Piccardo et al., 2014a).

In later work, the D-GBT was extended to enable the evaluation of both conventional and non-conventional modes. This was an important step to enable the D-GBT formulation to find wide applicability to the broad range of steel, composite and multi-component sections (Piccardo et al., 2014b). The ability in determining the entire set of deformation modes relied on the performance of two dynamic analyses, i.e. one to capture the in-plane

DOI: 10.1201/9781003348450-110

oscillations and one the out-of-plane oscillations of the cross-sections.

An alternative strategy for the evaluation of the cross-sectional modes was presented in (Taig et al., 2015a) in which an unrestrained planar frame was used for the evaluation of the conventional and extension modes. This provided an alternative procedure to the dynamic GBT approaches available at that time, in which the conventional modes were obtained from a planar dynamic analysis enforcing inextensibility to the members of the frame, and the extension modes were then evaluated from the conventional modes enforcing particular constraint conditions.

3 D-GBT FORMULATION FOR COMPOSITE AND MULTI-COMPONENT MEMBERS

The use of the D-GBT approach applied to composite structures was first proposed in references (Taig & Ranzi 2015, Taig et al. 2015b) by considering the longitudinal partial interaction that can occur in layered composite members. This model was then further extended to consider both transverse and longitudinal partial interaction in (Taig & Ranzi 2016). These formulations relied on a dynamic cross-sectional analysis for the identification of the deformation modes and were applicable for the analysis of multi-component sections formed with thin-walled open, closed and partially-closed sections.

In (Ferrarotti et al., 2017, Ferrarotti et al., 2018), the applicability of the partial shear interaction model was extended to the analysis of multi-component members within the framework of the GBT. The partial interaction is included in the analysis by means of rectilinear lines of shear-deformable linear elastic springs assumed to be uniformly distributed along the member length. With this approach it is possible to evaluate the complete and suitable set of deformation modes, including conventional, extension and shear modes, to account for both transverse and longitudinal shear interaction that occurs between the various elements composing the multi-component cross-section.

4 CONCLUSIONS

The Generalized Beam Theory (GBT) has been used for decades for the modelling of structural typologies. In this context, this paper has presented the use of a dynamic approach within the GBT framework for the derivation of the cross-sectional deformations modes required as the basis for the member analysis. For this purpose, the formulation has been applied to different structural typologies, such as steel and composite members. In the case of composite members (for two-layered and multi-component cross-sections), the influence of the partial interaction that occurs between adjacent components forming the cross-section has been accounted for. The D-GBT has been shown to represent an effective and efficient method of analysis within the family of GBT procedures.

REFERENCES

Schardt, R. 1994. Generalised Beam Theory – An adequate method for coupled stability problems. *Thin-Walled Structures* 19: 161–180.

Silvestre, N. & Camotim, D. 2002, First-order generalised beam theory for arbitrary orthotropic materials. *Thin-Walled Structures*. 40: 755–789.

Camotim, D. & Basaglia, C. & Silvestre, N. 2010. GBT buckling analysis of thin-walled steel frames: A state-of-art report. *Thin-Walled Structures* 48: 726–743.

Ranzi, G. & Luongo, A. 2011. A new approach for thin-walled member analysis in the framework of GBT. *Thin-Walled Structures* 49: 1404–1414.

Ranzi, G. & Luongo, A. 2014. An analytical approach for the cross-sectional analysis of generalized beam theory, *Proceedings of the Institution of Civil Engineers, Structures and Buildings*, 167: 414–425.

Piccardo, G. & Ranzi, G. & Luongo, A. 2014a. A direct approach for the evaluation of the conventional modes within the GBT formulation. *Thin-Walled Structures* 74: 133–145.

Piccardo, G. & Ranzi, G. & Luongo, A. 2014b. A complete dynamic approach to the GBT cross-section analysis including extension and shear modes. *Mathematics and Mechanics of Solids* 19(8): 900–24.

Taig, G. & Ranzi, G. & D'Annibale, F. 2015a. An unconstrained dynamic approach for the Generalised Beam Theory. *Continuum Mechanics and Thermodynamics* 27: 879–904.

Taig, G. & Ranzi, G. & Dias-da-Costa, D. & Piccardo, G. & Luongo, A. 2015b. A GBT Model for the Analysis of Composite Steel-Concrete Beams with Partial Shear Interaction. *Structures* 4: 27–37.

Taig, G. & Ranzi, G. 2015b. Generalized Beam Theory (GBT) for composite beams with partial shear interaction. *Engineering Structures* 99: 582–602.

Taig, G. & Ranzi, G. 2016. Generalized Beam Theory (GBT) for composite beams with longitudinal and transverse partial shear interaction. *Mathematics and Mechanics of Solids*, DOI: 10.1177/1081286516653799.

Taig, G. & Ranzi, G. & Dias-da-Costa, D. & Piccardo, G. & Luongo, A. 2015. A GBT Model for the Analysis of Composite Steel-Concrete Beams with Partial Shear Interaction. *Structures* 4: 27–37.

Ferrarotti, A., Ranzi, G., Taig, G., Piccardo, G. (2017). Partial interaction analysis of multi-component members within the GBT. *Steel and Composite Structures*, 25(5), 625–638

Ferrarotti, A., Ranzi, G., Piccardo, G. (2018). Partial interaction analysis of multi-component members with the D-GBT approach. In Amin Heidarpour, Xiao-Ling Zhao (Eds.), *Tubular Structures XVI: Proceedings of the 16th International Symposium for Tubular Structures (ISTS 2017), Melbourne, Australia, 4–6 December 2017*, (pp. 459–466). Leiden: CRC Press.

Current Perspectives and New Directions in Mechanics, Modelling and Design of Structural Systems – Zingoni (ed.)
© 2022 Copyright the Author(s), ISBN 978-1-032-39114-4

On energy based flexural-torsional buckling criteria of double-tee shape elastic thin-walled members

M.A. Gizejowski

Division of Concrete & MetalStructures, Faculty of Civil Engineering, Warsaw University of Technology, Warsaw, Poland

ABSTRACT: Double-tee shaped steel beam-columns under compression and major axis bending (about the section stronger principal axis) are generally sensitive to flexural-torsional buckling when insufficiently restrained against lateral-torsional deformations. Depending upon the section inertia properties about principal axes, out-of-plane restraining conditions and a ratio of the dimensionless compressive axial force and the maximum bending moment, the influence of prebuckling in-plane displacements and the in-plane buckling effect on the flexural-torsional resistance become more or less important, or even negligible. The energy formulations dedicated to the elastic flexural-torsional buckling analysis of I- or H-section steel members are usually based on the Vlasov theory of thin-walled structures in the form of linear buckling analysis. In such a classical approach, only the effect of prebuckling stress resultants is taken into account in the energy formulation, and the buckling analysis may be converted to the linear eigenproblem analysis. When nonlinear rotation terms of the displacement field components, and consequently the curvature strain components, maintain rigorously the trigonometric functions of the twist rotation, a more accurate buckling energy equations and corresponding differential equilibrium equations may be obtained. As a result, the buckling analysis may be converted to the nonlinear eigenproblem analysis. In this paper, formulations of the displacement field and the energy equation, leading to the elastic flexural-torsonal buckling criteria are widely discussed. Conclusions are drawn with regard to the utilization of obtained solutions in engineering practice.

In memory of the late Professor Nicholas Snowden Trahair

1 INTRODUCTION

The classical energy formulation dedicated to the flexural-torsional buckling of double-tee steel members is based on the so-called linear buckling analysis (LBA) and inclusion the prebuckling stress resultants in the energy equation. As a result, the buckling analysis may be converted to the linear eigenproblem analysis (LEA). More accurate theoretical models deal with the modified energy formulations leading to the nonlinear eigenproblem analysis (NEA). Using such methods of analysis, practical approximate solutions may be obtained for any asymmetric transverse loading conditions that produce a moment gradient.

The Classical Energy Method (CEM) for the evaluation of elastic lateral-torsional buckling (LTB) problems of beams and flexural-torsional buckling (FTB) of beam-columns was studied by many authors. The summary in relation to LTB problems was given by Pi et al. (1992). Giżejowski et al. (2021) used CEM for solving a general buckling problem of asymmetric transverse loading conditions. The stability criterion was formulated by treating the general loading pattern and corresponding moment diagram as a superposition of symmetric and antisymmetric components. The results were presented in a table format for the equivalent uniform moment factors for single loading cases and in the form of charts for combined loading cases.

A way for the improvement of CEM was shown by Timoshenko & Gere (1961). In case of LTB problems, the energy based solution may be improved by making use of the minor axis bending differential equilibrium equation. Trahair (1993) have referred this method to TEM (Timoshenko Energy Method). TEM was subsequently used by many authors to solve different LTB problems of double-tee section beams, both bisymmetric and monosymmetric.

In this study, different issues related to the formulation of displacement field equations and stability criteria for the elastic FTB problems of beam-columns, and also the elastic LTB problems of beams (when the axial force equals to zero) are presented and discussed.

DOI: 10.1201/9781003348450-111

2 ENERGY FORMULATION

The non-classical energy equation at the buckling state, based on the theory of thin-walled members (Vlasov 1961) is derived. It involves the field pre-buckling major axis moment M_y and the field buckling minor axis moment M_z, together with the effect of prebuckling displacements expressed by the factor $k_1 = 1 - I_z/I_y$:

$$\frac{1}{2}\left\{\int_0^L \left[EI_z\delta(v_0'')^2 + EI_w\delta(\theta_x'')^2 + GI_T\delta(\theta_x')^2 + k_1M_y\delta\left[\theta_x\left(2v_0'' + \frac{M_z}{EI_z}\right)\right]\right]dx + \int_0^L N\delta\left[(v_0')^2 + i_0^2(\theta_x')^2\right]dx + \sum_i \int_{x_{qi,1}}^{x_{qi,2}} q_{z,i}z_{q,i}\delta\theta_x^2 dx + \sum_j Q_{z,j}z_{Q,j}\delta\theta_x^2\right\} = 0 \quad (1)$$

where δ = variation; u_0, v_0, w_0 = member axis displacements dependent upon x-coordinate (x is the member axis while y,z are the section axes); θ_x = twist angle of rotation; N = prebuckling axial force (tension is positive); i_0 = polar radius of gyration; E,G = Young modulus and Kirchhoff modulus; I_y, I_z, I_T, I_w = section moment of inertia about the principal axes, and section torsion as well as warping constants; $q_{z,i}$ = distributed and $Q_{z,j}$ = concentrated conservative load components; $z_{q,i}$, $z_{Q,j}$ = load heights in the plane of bending.

3 GENERAL SOLUTION OF FTB PROBLEMS

Simply supported beam-columns with end sections free to warp are considered. The general solution of FTB problems of such a member static scheme is developed and based on the energy Eq. (1). It may be written down as follows:

$$\left(\frac{M_{y,max}}{C_{bc}M_{cr,0}}\right)^2 = \left(1-\frac{N}{N_y}\right)\left(1-\frac{N}{N_z}\right)\left(1-\frac{N}{N_T}\right) \quad (2)$$

where $M_{cr,0}$ = critical moment under uniform bending and compression; C_{bc} = equivalent uniform moment factor; $M_{y,max}$ = maximum in-plane moment; N_y, N_z, N_T = lowest major axis flexural, minor axis flexural and torsional bifurcation loads.

The structure of equivalent uniform moment factor depends upon the different approximations used for M_y and M_z. The said moments are decomposed into symmetric and antisymmetric components. As an example, adopting the second order amplified moment components, symmetric and antisymmetric, and the minor axis curvature v_0'' delivered from the second order differential equilibrium equation, yields the following relationship:

$$\frac{1}{C_{bc}} = \sqrt{\frac{k_1}{\zeta}}\left[\left(\frac{M_{ys,\max}}{M_{y,\max}}\frac{1}{C_{bs,rem}}\right)^2 + \frac{\left(1-\frac{N}{N_y}\right)\left(1-\frac{N}{N_z}\right)}{\left(1-\frac{N}{N_{ya}}\right)\left(1-\frac{N}{N_{za}}\right)}\left(\frac{M_{ya,\max}}{M_{y,\max}}\frac{1}{C_{ba,rem}}\right)^2\right]^{0.5} \quad (3)$$

where

$$\frac{1}{C_{bs,rem}} = \left(2\int_0^1\left[\frac{M_{ys}}{M_{ys,\max}}\right]^2 sin^2(\pi\xi)d\xi\right)^{0.5} \quad (4)$$

$$\frac{1}{C_{ba,rem}} = \left(2\int_0^1\left[\frac{M_{ya}}{M_{ya,\max}}\right]^2 sin^2(\pi\xi)d\xi\right)^{0.5} \quad (5)$$

and $\xi = x/L$; ζ = factor representing the load height effect on the buckling state; N_{ya}, N_{za} = second lowest major axis and minor axis flexural bifurcation loads; M_{ys}, M_{ya} = symmetric and antisymmetric filed moment components; $M_{ya,\max}$, $M_{ya,\max}$ = maximum moments of symmetric and antisymmetric components.

4 CONCLUSIONS

The solution presented in the previous section coincides with that obtained by Mohri et al. (2008) in special cases of the loads symmetrically distributed with regard to the member half-length. Eq. (2) and equivalent uniform moment factors of different analytical forms presented in the full length paper, and corresponding to different approximations of M_y and M_z moments, allow for more detailed investigations into their accuracy verification using FEM. Such investigations are underway in relation to the elastic FTB of beam-columns under arbitrary transverse load combinations.

REFERENCES

Giżejowski, M.A., Barszcz, A. M. & Stachura, Z. 2021, Elastic Flexural-Torsional Buckling of Steel I-Section Members Unrestrained Between End Supports. *Archives of Civil Engineering* 67(1): 635–656.

Mohri, F., Bouzerira, Ch. & Potier-Ferry, M. 2008. Lateral buckling of thin-walled beam-column elements under combined axial and bending loads. *Thin-Walled Structures* 46(3): 290–302.

Pi, Y-L, Trahair, N.S. & Rajasekaran, S. 1992. Energy equation for beam lateral buckling. *Journal of Structural Engineering* 118(6):1462–1479.

Timoshenko, S.P. & Gere, J.M. 1991. *Theory of Elastic Stability*, 2nd edition. New York: McGraw-Hill.

Trahair, N.S. 1993. *Flexural-Torsional Buckling of Structures*. Boca Raton: CRC Press.

Vlasov, V.Z. 1961. *Thin Walled Elastic Beams*. Second ed., Jeruzalem: Israel Program for Scientific Translations.

Current Perspectives and New Directions in Mechanics, Modelling and Design of Structural Systems – Zingoni (ed.)
© 2022 Copyright the Author(s), ISBN 978-1-032-39114-4

An individual shear deformation theory of beams with consideration of the Zhuravsky shear stress formula

K. Magnucki
Łukasiewicz Research Network – Poznan Institute of Technology, Poznan, Poland

ABSTRACT: The paper is devoted to simply supported beams under three-point bending. The homogeneous beams with bisymmetrical cross-sections, sandwich beams and beams with symmetrically varying mechanical properties in the depth direction are studied. Analytical models of these beams are presented in detail. The shear deformation theories of the planar cross-sections of these beams are developed with consideration of the Zhuravsky shear stress formula. Based on the principle of stationary total potential energy differential equations of equilibrium of these beams are obtained. These equations are analytically solved. The deformation of the planar cross-sections of exemplary beams are analytically determined and graphically presented in figures. Moreover, the maximum deflections of these beams are derived.

1 INTRODUCTION

The shearing effect existing in main elements of engineering structures was recognized in 19th century by D.I. Zhuravsky. The shear stresses of this effect are described in many engineering books and monographs, e.g. Gere & Timoshenko (1984). The Bernoulli-Euler beam theory was generalized by Timoshenko in 1921. (Wang at al. 2000) described the shear theories of beams and plates developed in the 20th century. The intensive development of this theories is evident in the 21st century. Selected papers on this topic are as follows: Shen (2009), Reddy (2010), (Carrera et al. 2011), Xiang (2014), Polizzotto (2015), (Chen et al. 2016), Genovese & Elishakoff (2019), (Magnucki et al. 2020a, b), Magnucki & Magnucka-Blandzi (2021).

The main purpose of this work is to present an individual shear deformation theory of the homogeneous and nonhomogeneous beams with consideration on the Zhuravsky shear stress formula. The effectiveness of this theory is demonstrated for bending of the example beams. This work is a continuation of the studies presented in the paper (Magnucki et al. 2020a).

2 HOMOGENEOUS BEAMS WITH BISYMMETRICAL CROSS SECTIONS

The bisymmetrical cross section of the homogeneous beam of depth h is shown in Figure 1.

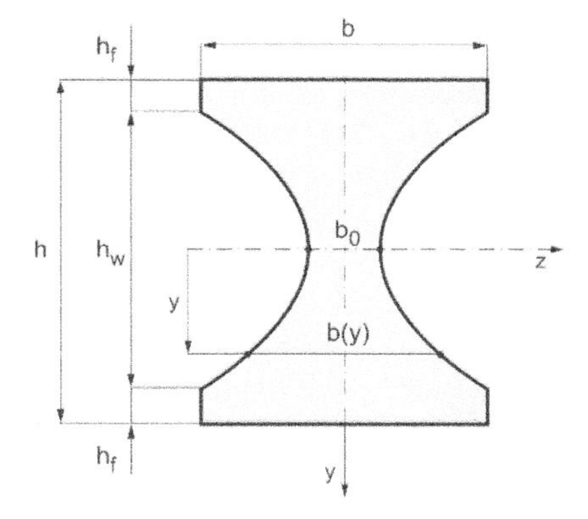

Figure 1. Scheme of the bisymmetrical cross section.

The symmetrically varying width of this cross section in the depth direction is as follows

$$b(\eta) = b\bar{b}(\eta), \tag{1}$$

where $\eta = y/h$ – dimensionless coordinate, and dimensionless width in the middle part ($-\chi_w/2 \le \eta \le \chi_w/2$) is assumed in the form

$$\bar{b}(\eta) = \beta_0 + (1 - \beta_0)\left[\eta^2 + 4(4 - \chi_w^2)\left(\frac{\eta}{\chi_w}\right)^4\right]^{k_c}, \tag{2}$$

where $\chi_w = h_w/h$ – relative depth of the middle part, $\beta_0 = b_0/b$ – dimensionless coefficient, k_c – exponent.

The deformation of a planar cross section and maximum deflections of exemplary beams for three-point bending are analytical determined.

3 SANDWICH BEAMS

The rectangular cross section of the sandwich beam of depth h and width b is shown in Figure 2.

The symmetrically varying mechanical properties in the depth direction of this beam are assumed with consideration of the paper Magnucki & Magnucka-Blandzi (2021).

DOI: 10.1201/9781003348450-112

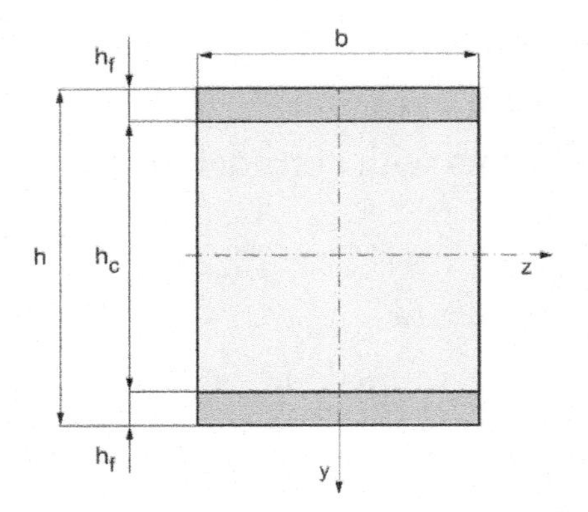

Figure 2. Scheme of the cross section of the sandwich beam.

The form of this function is identical to the function of the symmetrically varying width of the homogeneous beam (1).

The deformation of a planar cross section and maximum deflections of exemplary beams for three-point bending are analytical determined.

4 BEAMS WITH SYMMETRICALLY VARYING MECHANICAL PROPERTIES

The rectangular cross section of the beam with symmetrically varying mechanical properties is shown in Figure 3.

The symmetrically varying mechanical properties in the depth direction of this beam, with consideration of the paper (Magnucki et al. 2020b), are assumed in the following form:

$$E(\eta) = E_1 f_e(\eta), \tag{3}$$

where dimensionless function

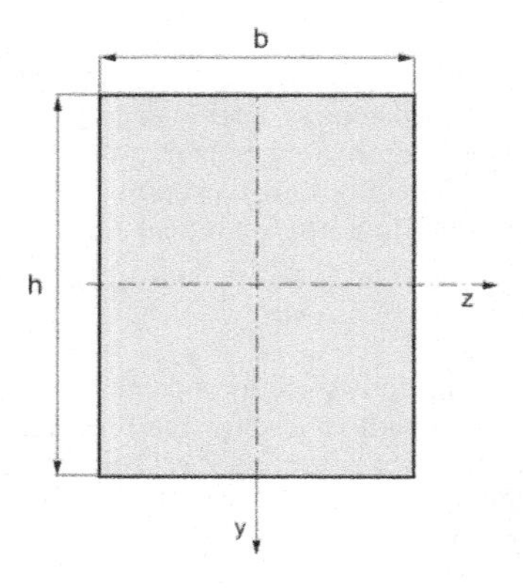

Figure 3. Scheme of the cross section of the beam with symmetrically varying mechanical properties.

$$f_e(\eta) = e_0 + (1 - e_0)\left(5\eta^2 - 256\eta^{10}\right)^{k_e}, \tag{4}$$

and k_e – exponent ($0 \leq k_e$), $\eta = y/h$ – dimensionless coordinate ($-1/2 \leq \eta \leq 1/2$), $e_0 = E_0/E_1$ – dimensionless parameter, $E_0 = E(0)$, $E_1 = E(\mp 1/2)$ – Young's modules.

The deformation of a planar cross section and maximum deflections of exemplary beams for three-point bending are analytical determined.

5 CONCLUSIONS

The proposed nonlinear theory of deformation of the planar cross section of beams with consideration the Zhuravsky shear stress formula is a generalization of previously known beam theories and exactly described the shear effect in the bent beams. This theory may be applicated to the analytical modelling of the plates and shells.

REFERENCES

Carrera, E., Giunta, G. & Petrolo, M. 2011. *Beam structures. Classical and advanced theories*. John Wiley & Sons, Ltd.

Chen, D., Yang, J. & Kitipornchai, S. 2016. Free and forced vibrations of shear deformable functionally graded porous beams. *International Journal of Mechanical Sciences* 108-109: 14–22.

Genovese, D. & Elishakoff, I. 2019. Shear deformable rod theories and fundamental principles of mechanics. *Archive of Applied Mechanics* 89(10): 1995–2003.

Gere, J.M. & Timoshenko, S.P. 1984. *Mechanics of materials*. Second Ed. Boston: PWS-KENT Publishing Company.

Magnucki, K., Lewinski, J. & Magnucka-Blandzi, E. 2020a. A shear deformation theory of beams with bisymmetrical cross sections based on the Zhuravsky shear stress formula. *Engineering Transactions* 68(4): 353–370.

Magnucki, K., Lewinski, J., & Magnucka-Blandzi, E. 2020b. An improved shear deformation theory for bending beams with symmetrically varying mechanical properties in the depth direction. *Acta Mechanica* 231: 4381–4395.

Magnucki, K. & Magnucka-Blandzi, E. 2021. Generalization of a sandwich structure model: Analytical studies of bending and buckling problems of rectangular plates. *Composite Structures* 255: 112944.

Polizzotto, C. 2015. From the Euler-Bernoulli beam to the Timoshenko one through a sequence of Reddy-type shear deformable beam models of increasing order. *European Journal of Mechanics A/Solids* 53: 62–74.

Reddy, J.N. 2010. Nonlocal nonlinear formulations for bending of classical and shear deformation theories of beams and plates. *International Journal of Engineering Science* 48: 1507–1518.

Shen, H-S. 2009. *Functionally graded materials: nonlinear analysis of plates and shells*. Boca Raton, London, New York: Taylor & Francis Group.

Wang, C.M., Reddy, J.N. & Lee, K.H. 2000. *Shear deformable beams and plates: Relationships with classical solutions*. Amsterdam, Lausanne, New York, Oxford, Shannon, Singapore, Tokyo: Elsevier.

Xiang, S. 2014. A New shear deformation theory for free vibration of functionally graded beams. *Applied Mechanics and Materials* 455: 198–201.

Current Perspectives and New Directions in Mechanics, Modelling and Design of Structural Systems – Zingoni (ed.)
© 2022 Copyright the Author(s), ISBN 978-1-032-39114-4

Elastic shear buckling of beams with sinusoidal corrugated webs

L. Sebastiao & J. Papangelis
School of Civil Engineering, University of Sydney, NSW, Australia

ABSTRACT: In this paper numerous beams with sinusoidal corrugated webs are investigated by finite element analysis to understand the elastic local shear buckling phenomena. Over 600 models with different geometric parameters were analysed to investigate the behaviour by considering a cantilever acted upon by an edge shear stress at the free end. This method of loading induces the beam entirely being under uniform shear force, therefore shear buckling modes may occur. For better characterisation of this behaviour a sensitivity analysis of beams with distinct geometric parameters was examined under this load condition, considering the effect of variation of the web sizes. Furthermore, considering the data generated, it was possible to propose an analytical equation to predict the critical stress. The key factor was the local buckling coefficient calculated for different aspect ratio values from the relationship of geometric parameters. This paper might add significant value to the structural engineering industry by providing unique data, which has not been done before extensively for such beams.

1 INTRODUCTION

Corrugated steel web beams are largely used in buildings and bridges. Its first application was seen in the aeronautic field in 1924, and later used as an ordinary steel structure from the 1960s (Basiński, 2018 and Shon et al., 2017). The configuration of this type of beam enhances the stability against shear buckling failures, allowing the beam to resist higher shear forces when compared to traditional flat web beams. There are different types of corrugated web shapes and the most commonly used are the trapezoidal and sinusoidal. The trapezoidal corrugated web beam is the most researched, developed (design and production matters), and intensively used. However, in previous research good performance has been observed for sinusoidal corrugated webs (Lee et al., 2020) due to the development of automation and welding technology which has been crucial in expecting a new trend for its fabrication and applicability in bridges and building (Kiymaz et al., 2010, Shon et al., 2017 and Basiński, 2018).

2 FINITE ELEMENT MODEL

In this paper, a finite element elastic buckling analysis was deployed to investigate the shear behaviour of beams with sinusoidal corrugated webs by using the software program Strand7 (Strand7, 2022). In order to understand the shear buckling behaviour of the sinusoidal web beam, the finite element method gives wide flexibility to perform parametric analyses with different variable dimensions. The local shear buckling coefficient k_L was determined using the results from the finite element analysis.

The sinusoidal web beam model was modelled with a web and two flanges. The beam was constructed as a cantilever with one end built-in and the other end free, as shown in Figure 1. An edge shear stress was applied to the web at the free edge. As a result, a uniform shear force is propagated along the full length of the beam. At the support edge, vertical, lateral and rotational restraints were considered in order to generate a reaction force and simulate a cantilever beam. In addition, to prevent flexural-torsional buckling of the beam, a lateral restrain was applied at the free edge.

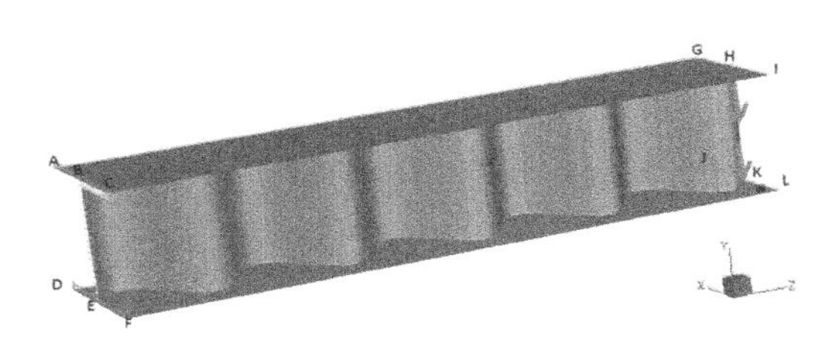

Figure 1. Finite element model.

DOI: 10.1201/9781003348450-113

3 ANALYSIS

This study considers an extended range over the wave amplitude A from 50 to 175 mm in 25 mm increments, with the wavelength B ranging between 400 to 800 mm in 100 mm increments. The web height h varied between 400 to 1200 mm in 200 mm increments while the web thickness t_w ranged from 1 to 3 mm in 1 mm increments.

4 RESULTS

A local shear buckling mode from the finite element analysis is shown in Figure 2.

Figure 2. Local shear buckling.

Figure 3 shows the value of the local shear buckling coefficient k_L calculated for all the finite element results plotted with the expression $(ht_w/AB)^{1/4}$. It can be seen in Figure 3 that the results for k_L show a general decreasing exponential trend as the expression $(ht_w/AB)^{1/4}$ increases. From the line of best fit, the following equation can be derived

$$k_L = 6.31\left(\frac{AB}{ht_w}\right)^{0.41} \quad (1)$$

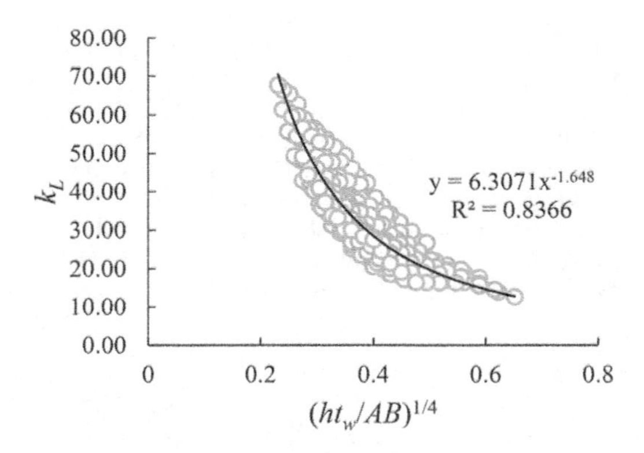

Figure 3. Variation of local shear buckling coefficient with expression $(ht_w/AB)^{1/4}$.

Thus, based on this investigation and the finite element results, a final equation for the local shear buckling stress τ_L can be derived as follows

$$\tau_L = 6.31\left(\frac{AB}{ht_w}\right)^{0.41}\frac{\pi^2 E}{12(1-\nu^2)}\left(\frac{t_w}{w}\right)^2 \quad (2)$$

5 CONCLUSIONS

This paper presents unique results for the elastic shear buckling of beams with sinusoidal corrugated webs. The aim of this work was to derive an equation for the local shear buckling stress for these types of beams.

A finite element analysis has been described to calculate the local shear buckling stress for beams with sinusoidal corrugated webs. The analysis was conducted on numerous beams by modelling the beam as a cantilever subjected to an edge shear stress. This type of loading places the beam in a state of uniform shear along the full length of the beam.

The finite element analysis was conducted by varying the geometric parameters of the sinusoidal corrugated web which affect local shear buckling, namely the wave amplitude, wavelength, web height and web thickness.

It was found that the local shear buckling stress decreases as the wave amplitude, wavelength, and web height all increase. On the other hand, the local shear buckling stress increases significantly as the web thickness is increased.

From the finite element results, an equation for the local shear buckling stress was derived. This was achieved by curve fitting all the finite element results. This equation will be useful for design engineers who wish to determine the elastic local shear buckling of beams with sinusoidal corrugated webs.

REFERENCES

BASIŃSKI, W. 2018. Shear Buckling of Plate Girders with Corrugated Web Restrained by End Stiffeners. *Periodica Polytechnica Civil Engineering*, 62, 757–771.

KIYMAZ, G., COSKUN, E., COSGUN, C. and SECKIN, E. 2010. Transverse load carrying capacity of sinusoidally corrugated steel web beams with web openings. *Steel and Composite Structures*, 10, 69–85.

LEE, S., PARK, G. and YOO, J. 2020. Analytical Study of Shear Buckling Behavior of Trapezoidal and Sinusoidal Corrugated Web Girders. *International Journal of Steel Structures*, 20, 525–537.

SHON, S., JIN, S. and LEE, S. 2017. Minimum Weight Design of Sinusoidal Corrugated Web Beam Using Real-Coded Genetic Algorithms. *Mathematical Problems in Engineering*, 9184292.

STRAND7 2022. *Finite Element Analysis*. Sydney, Australia

Current Perspectives and New Directions in Mechanics, Modelling and Design of Structural Systems – Zingoni (ed.)
© 2022 Copyright the Author(s), ISBN 978-1-032-39114-4

Buckling check under biaxial loading considering different launching bearings and eccentric load introduction

N. Maier, M. Mensinger & J. Ndogmo
Technical University Munich, Chair of Metal Structures, Munich, Germany

ABSTRACT: In general, two types of launching bearings are used for launching: sliding rockers and systems with hydraulic bearings. During incremental launching, the centre of the webs of the superstructure is not perfectly in line with the centre of the launching bearings due to unavoidable tolerances. These eccentricities are not considered in the current design against plate buckling according to EN 1993-1-5 [1]. Furthermore, there is a significant difference between the different types of launching bearings due to the boundary conditions. At the Technical University Munich, large-scale buckling tests on longitudinally stiffened plates under biaxial stresses with different types of launching bearings and eccentric load introduction were carried out. The results from the validated numerical model and parameter study of the influence due to different types of launching bearings on the buckling behavior are shown. The results are compared with the buckling verification acording to the reduced stress method proposed in EN 1993-1-5:2019 [1].

1 INTRODUCTION

During incremental launching biaxial stresses occur. The buckling verification must be performed considering longitudinal and transverse stresses. Additionally, the load introduction is not in centre line with the webs or launching bearings. The eccentric load introduction can be caused by imperfections in the production process, misalignment of the launching bearings, the necessary air gap to the lateral guide as well as the different thicknesses of the web along the superstructure. The load introduction from the bearings into the web depends strongly on the type of launching bearing used. When using a rocker bearing, the rotation in launching direction of the bridge is restrained since the rocker bearing system is linearly supported along its length. This leads to a better absorption of eccentricities from the bearing system. In contrast, the hydraulic bearing system is not linearly supported, due to the using of a spherical bearing instead of the rocker. This leads to the assumption that the eccentricity from load introduction has an influence on the buckling behavior.

2 TEST PROCEDURE

To introduce the biaxial stress state into the specimens as it occurs during incremental launching, a special test stand is developed. During the whole test program, 12 specimens are investigated for various influences on the buckling behaviour, such as the torsional and bending stiffness of the longitudinal stiffeners and their position, the stress ratio σ_z/σ_x, and different bearing systems with eccentric load introduction. The webs have external dimensions of 3 meters x 4 meters in order to obtain information about imperfections from the production process in a realistic scale.

3 NUMERICAL MODEL AND PARAMETERS

The numerical model is created with the software ANSYS [2] and validated. Only the specimen and the load introduction beams are modelled (see Figure 1).

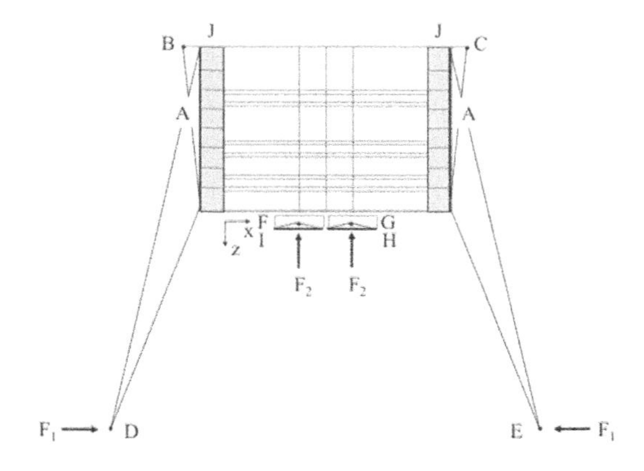

Figure 1. Numerical model ANSYS [2].

The boundary conditions of the test setup and the load introduction are modelled using external points, which are coupled with the respective surfaces of the specimen by using definition "beam" with a defined stiffness of the test setup. The boundary conditions of the external points F and G and their position in z-direction depends on the launching bearing and is different for the both types of bearings. The name of the models and parameters are shown in Table 1.

$$M1/M2 - c - EF/GF - s - b_1 - i(+/-)$$

DOI: 10.1201/9781003348450-114

Table 1. Parameter and geometry definition.

M1	φ_x free
M2	φ_x fix
c [a]	Load introduction length
EF	Local buckling
GF	Global buckling
s	position of external points G and F in z direction
b1	Distance to the first stiffener
i (+/-)	Equivalent geometric imperfection

[a]c = 2 · 950 mm with distance 100 mm for M1

4 BUCKLING VERIFICATION ACCORDING TO EN 1993-1-5 SECTION 1

In order to verify, if the influence of the eccentric load introduction, is included in the current buckling verification using the reduced stress method according to [1], the numerical results are recalculated with this method. For this purpose, stresses are calculated from the load capacity of the FEM results and compared with the design value of stresses according to [2]. The design value of stresses can be calculated by a defined stress ratio β due to the reformulation of the following equation.

$$\sqrt{\left(\frac{\sigma_{x,Ed}}{\rho_{c,x}}\right)^2 + \left(\frac{\sigma_{z,Ed}}{\rho_{c,z}}\right)^2 - V\left(\frac{\sigma_{x,Ed}}{\rho_{c,x}}\right)\left(\frac{\sigma_{z,Ed}}{\rho_{c,z}}\right)} \leq \frac{f_y}{\gamma_{M1}} \tag{2}$$

The results from FEM include the flange. The results therefore have been compared with different approaches.

Approach 1: MRS(1) B
- α_{cr} from FEM (is considered)
- Reduction factor ρ according to [1] Annex B

Approach 2: MRS(3) B
- α_{cr} of the hinged plate
- $\alpha_{cr,p}$ and $\alpha_{cr,c}$ are determined separately for each loading direction
- Reduction factor ρ according to [1] Annex B

Approach 2.1: MRS(3) Th. II. O.
- Verification of longitudinal stiffener according to Th. II. O.

For all approaches:
- $\gamma_{M,1} = 1.0$
- The longitudinal stiffeners are closed

The results are shown in Figures 2 for the Model M1-950-GF-310-400-8(+)

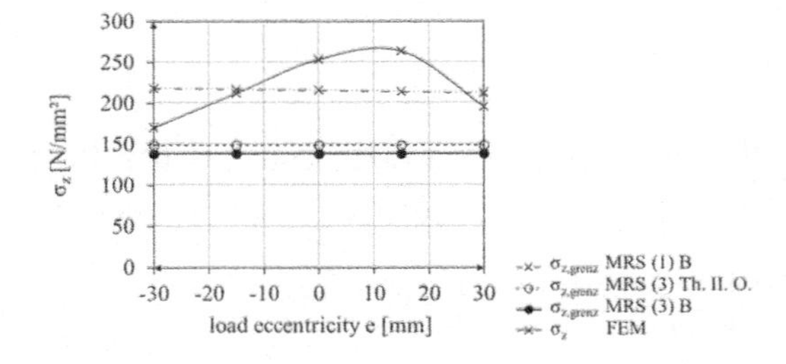

Figure 2. Comparison M1 global buckling.

The design value of stresses using MRS (1) B are larger for cases were eccentricities are larger than ± 15 mm. However, since the common practices of the buckling check according to EN 1993-1-5:2019 [2] assume Navier's boundary conditions (MRS (3) B), the allowable stresses are limited more by this regulation. Eccentricities of up to 30mm can easily be accommodated by using this approach.

5 CONCLUSION

The parameter study shows that an eccentric load introduction with a possible rotation φ_x has an influence on the load capacity during buckling behaviour. A higher load capacity can be achieved by positioning the rotation point closer to the flange when using the hydraulic launching bearing system (M1). Eccentric load introduction don't lead to problems with the existing standard [2] when using hydraulic bearings with eccentricities up to ±30 mm. Larger eccentricities then 30 mm were not part of the investigations. This conclusion of the results is valid for systems with rotatable flange. The reaction to eccentric load introduction is expected to be less sensitive in systems with a closed hollow cross section due to the higher rotational and additional bending stiffness of the bottom plate. The investigations are not yet final.

All presented results are part of the dissertation [3] and will be published there first.

REFERENCES

[1] EN 1993-1-5, Eurocode 3: Design of steel structures –Part 1–5: Plated structural elements; German version EN 1993-1-5:2006 + AC:2009 + A1:2017 + A2:2019 (October 2019).

[2] ANSYS workbench and mechanical, 2019.

[3] Maier, N. „Einfluss auf das Beulverhalten längsversteifter Platten unter biaxialer Beanspruchung aus exzentrischer Lasteinleitung unterschiedlicher Verschublagertypen"Dissertation - In Bearbeitung (Bearbeitungsstand: 22.02.2022; Abgabe Voraussichtlich 2022). Technische Universität München. Lehrstuhl für Metallbau

Current Perspectives and New Directions in Mechanics, Modelling and Design of Structural Systems – Zingoni (ed.)
© 2022 Copyright the Author(s), *ISBN 978-1-032-39114-4*

Lateral-torsional buckling of steel double-tee shape beams based on the decomposition of transverse load combinations

A. M. Barszcz & M.A. Gizejowski
Division of Concrete and Metal Structures, Faculty of Civil Engineering, Warsaw University of Technology, Warsaw, Poland

ABSTRACT: Double-tee shape steel beams under major axis bending (bent about the section stronger principal axis) are generally sensitive to lateral-torsional buckling (LTB), when insufficiently restrained against lateral-torsional deformations between supports. The formulation of LTB problems presented in this paper is based on the Vlasov theory of thin-walled structures, within the concept of linear buckling analysis (LBA). In the classical approach, only the effect of prebuckling bending moment is taken into account in the energy formulation, therefore the buckling analysis is converted to the linear eignenproblem analysis (LEA). When the energy term being a product of twist rotation and minor axis curvature is replaced by the term dependent upon the twist rotation (in the same way as in the Timishenko Energy Method), LTB solutions are based on the quadratic eigenproblem analysis (QEA). In this paper, the effect of prebuckling displacements is taken into account and more accurate normal strain relationship is formulated. The energy term being dependent upon the minor axis curvature and the twist rotation is decoupled and the resultant energy equation becomes dependent only upon the twist rotation and its derivatives. Moreover, any complex loading case is treated as a superposition of two components, symmetric and antisymmetric ones. Thus, the final solution for the critical moment is workable and easy to apply for any loading condition.

1 INTRODUCTION

An accurate closed form solution of lateral-torsional buckling problem may only be obtained for the case of uniform bending. Moment gradient cases need approximate analytical methods or numerical methods to be used. The Classical Energy Method (CEM) for the evaluation of elastic lateral-torsional buckling problems of beams was studied by many authors, the summary of which has been given by Pi et al. (1992). A way for the improvement of CEM was shown by making use of the minor axis bending differential equilibrium equation. Trahair (1993) refers this method to TEM (Timoshenko Energy Method). This method was used by many authors to solve different LTB problems of double-tee section beams. Barszcz et al. (2021) used TEM for approximate solutions to be obtained for any asymmetric transverse loading combinations that produce a moment gradient.

This study discusses different issues related to the energy formulation of LTB of beams leading to the stability criterion formulated for a general case of transverse loading, the solution of which is obtained by decomposing the general loading pattern, and corresponding moment diagram, into two components, namely symmetric and antisymmetric. The superposition rule of symmetric and antisymmetric components makes a suitable starting point for the analytical closed form solution of elastic LTB problems to be developed.

2 GENERAL SOLUTION OF LTB PROBLEMS

The beam combined loading scheme of an arbitrary asymmetric pattern is shown in Figure 1.

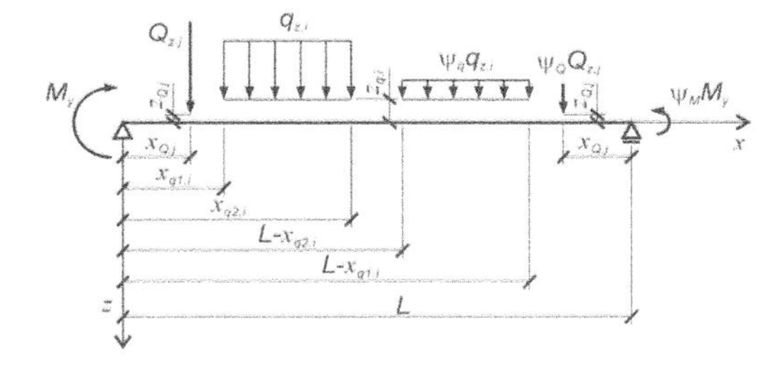

Figure 1. General asymmetric loading pattern.

The solution of LTB bifurcation criterion may be conveniently sought for by decomposing the general, asymmetric loading pattern shown in Figure 1 into symmetric and antisymmetric components, presented in Figure 2 and Figure 3, respectively.

DOI: 10.1201/9781003348450-115

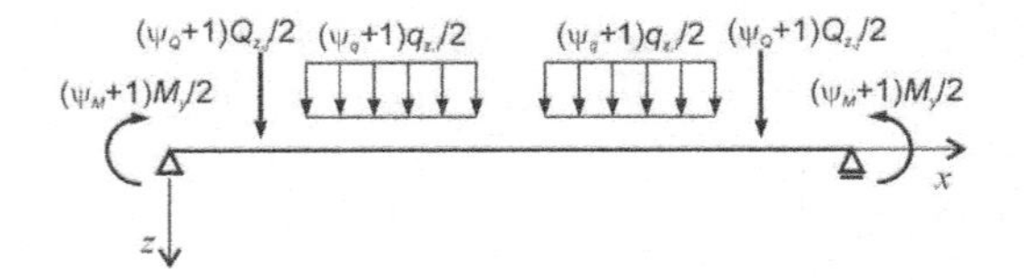

Figure 2. Symmetric component of the loading pattern given in Figure 1.

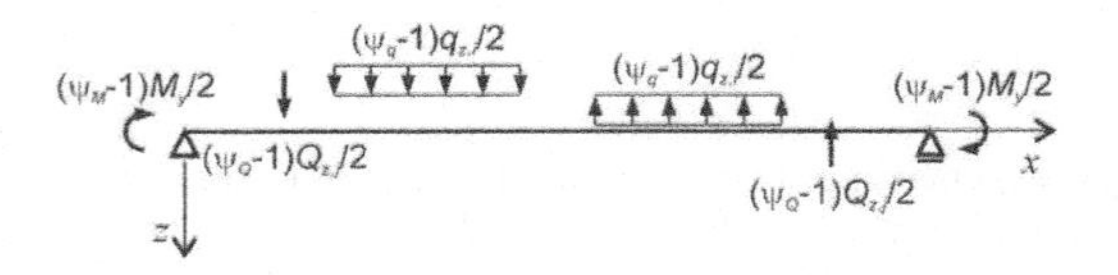

Figure 3. Antisymmetric component of the loading pattern given in Figure 1.

Carrying out the variation calculations for the stability energy equations presented in the full length paper, and considering symmetric and antisymmetric moment field components M_{ys} and M_{ya} (produced by loads shown in Figure 2 and Figure 3, respectively), the following closed form solution is obtained:

$$\left(\frac{M_{y,max}}{C_{bc}M_{cr,0}}\right)^2 = 1 \qquad (1)$$

where

$$\frac{1}{C_{bc}} = \frac{1}{\sqrt{\zeta}} \sqrt{\left(\frac{M_{ys,\max}}{M_{y,\max}}\frac{1}{C_{bs}}\right)^2 + \left(\frac{M_{ya,\max}}{M_{y,\max}}\frac{1}{C_{ba/}}\right)^2} \qquad (2)$$

and, $M_{cr,0} = i_0\sqrt{N_z N_T}$ = critical moment under uniform bending; C_{bs}, C_{ba} = equivalent uniform moment factors for the symmetric and antisymmetric loading components dependent upon the form of energy equation used, namely CEM (Classical Energy Method) or REM (Refined Energy Method); $M_{y,max}, M_{ys,\max}, M_{ya,\max}$ = maximum in-plane moments, produced respectively by total applied loads, symmetric loading pattern component and antisymmetric loading pattern component; i_0 = radius of gyration; N_z, N_T = lowest minor axis flexural and torsional bifurcation loads; ζ = off-shear centre load effect parameter.

3 COMPARISON OF LTB SOLUTIONS

Table 1 presents the comparison of equivalent uniform moment factors for three different loading cases, namely Case 1: unequal end moments, Case 2: uniformly distributed loads unequal in two half-lengths, Case 3: concentrated loads at equal distance x_Q from the supports but of an unequal magnitude in both half-lengths. In the latter case, three distances are adopted, namely $L/2$, $L/3$ and $L/4$. The load factors ψ_k (where k=M, q,Q) are used for three load cases specified above, giving symmetric and antisymmetric load components.

Table 1. Comparison of equivalent uniform moment factors from LEA and QEA.

		LEA*		QEA**	
Load type	x_Q	$C_{bs,cem}$	$C_{ba,cem}$	$C_{bs,rem}$	$C_{ba,rem}$
Case 1	-	1.00	2.78	1.00	2.77
Case 2	-	1.15	1.43	1.13	1.37
Case 3	$L/2$	1.42	0.00	1.37	0.00
	$L/3$	1.12	1.74	1.10	1.56
	$L/4$	1.05	1.81	1.04	1.73

* Classical Energy Method (CEM).
** Refined Energy Method (REM).

4 CONCLUDING REMARKS

General concluding remarks are as follows:

1. For beams in case of a single loading type, CEM and TEM results are close to each other, especially for the symmetric loading pattern.
2. For combined loading cases there might be a noticeable difference between the moment conversion factors. REM moment onversion factors $C_{bs,rem}$ and $C_{ba,rem}$ are of smaller values than their CEM counterparts $C_{bs,cem}$ and $C_{ba,cem}$, and the difference increases with the increase of ψ_M.
3. Verification of TEM/REM results made for selected combined loading cases showed that TEM/REM results of present study are closer to the results from the computer code LTBeam than those from the Access Steel design equations and charts, especially in the range of ψ_M between 0.5 and 1.0.
4. Results based on the Access Steel design charts are placed well below those from the computer code LTBeam and this study, especially in the range mentioned above.

REFERENCES

Barszcz, A.M., Giżejowski, M.A. & Stachura, Z. 2021. On elastic lateral-torsional buckling analysis of simply supported I-shape beams using Timoshenko's energy method. In M.A. Giżejowski et al. (eds), *Modern Trends in Research on Steel, Aluminium and Composite Structures*: 92–98. London: Routledge.

Pi, Y-L, Trahair, N.S. & Rajasekaran, S. 1992. Energy equation for beam lateral buckling. *Journal of Structural Engineering* 118(6):1462–1479.

Trahair, N.S., (1993) Flexural-torsional buckling of structures, E & FN Spon, London.

Current Perspectives and New Directions in Mechanics, Modelling and Design of Structural Systems – Zingoni (ed.)
© 2022 Copyright the Author(s), ISBN 978-1-032-39114-4

Flexural-torsional buckling of steel double-tee shape beam-columns with lateral-torsional discrete restraints

M.A. Gizejowski & A. M. Barszcz
Department of Concrete and Metal Structures, Faculty of Civil Engineering, Warsaw University of Technology, Warsaw, Poland

P. Wiedro
Faculty of Civil Engineering, Warsaw University of Technology, Warsaw, Poland

ABSTRACT: The classical energy formulation is dedicated to the flexural-torsional buckling analysis of I-section steel members and referred to the isolated members without intermediate lateral-torsional restraints. The second order effects of in-plane bending are neglected in this formulation. In design situations met in practice, the effect of flexural-torsional buckling of double-tee shape beam-columns is mitigated by using lateral-torsional restraints. In such a case, the load bearing system becomes a multi-segment system in which segments are spanning either between the support and the internal lateral-torsional restraint being the closest one to that support or spanning between the lateral-torsional restraints. In this case, there are two important effects that contribute to the system failure mode. The first one deals with the in-plane buckling while the second one with the continuity of deformations at the points of lateral-torsional restraints. In this paper, a finite element numerical model is presented for the detailed analysis of a slender simply supported beam-column, unrestrained between the supports and with the centrally placed lateral-torsional restraint. Numerical simulations are carried out with use of the computer software LTBeamN and Abaqus. Numerical results are used for the verification of the analytical solution based on the Vlasov theory of thin-walled members. Conclusions are drawn with regard to engineering practice.

1 INTRODUCTION

1.1 *Introductory remarks*

Torsionally and/or laterally braced structural elements are frequently used in real framed structures in order to increase the lateral-torsional buckling (LTB) resistance of beams or flexural-torsional buckling (FTB) resistance of beam-columns. Stability problems of such elements in relation to elastic/inelastic restrained out-of-plane buckling are therefore studied together with investigations into the brace stiffness requirements for treating the systems as rigidly braced, Nethercot & Trahair (1976), among others. A summary of LTB and FTB investigations has been presented in many textbooks, e.g. Trahair (1993).

Finite element numerical simulations are widely used nowadays in studying different problems of instability of simple and continuous structures, Papangelis et al. (1998).

1.2 *Scope of investigations*

Present study is concerned with a steel bisymmetric double-tee section of the flange and web geometrical dimenstions equivalent to that of rolled HEB 300 made of steel grade S235. The simply supported beam-column of such a section is loaded with a half-length uniformly distributed load. Two design situations are considered, namely with no intermediate discrete restraints and with the central rigid LT restraint. Numerical finite element models are built according to the linear buckling analysis (LBA), the version based on the linear eigenproblem analysis (LEA) and implemented into the LTBeamN. Numerical simulations are also carried out according to the nonlinear buckling analysis (NBA), the version based on GMNA+ (for the definition of analysis types refer to Gizejowski et al. 2019) and implemented into the Abaqus software. In the latter, the effect of prebuckling deformations and the continuity of beam-column buckling deformations over the discrete restraints of a quasi-perfect member is included. An analytical solution for FTB of beam-columns is presented in the way consistent with that of Salvadori (1951), proposed for beams with intermediate lateral-torsional restraints. The obtained results may create a basis for a generalization of such approach in order to facilitate the buckling analysis of multi-segment beam-column systems. The proposed analytical model is verified by comparing its results with those from LEA numerical simulations carried out with use of computer software LTBeamN and Abaqus.

DOI: 10.1201/9781003348450-116

2 NUMERICAL SIMULATIONS

2.1 *Description of the modeling technique*

The beam modeling technique conforming the Vlasov theory with 7 degrees of freedom per node is adopted in the LTBeamN software. The shell model of S4R is adopted in the Abaqus software with a size of approximately 30 mm by 30 mm for flanges and the web of considered double-tee section. Simulations based on LBA are carried out by assuming the elastic behavior in both LTBeam and Abaqus codes for the ideally straight and residual stress free member. Simulations based on NBA are carried out in Abaqus environment for a quasi-perfect member made of elastic-plastic material model, residual stress free, the initial geometry of which corresponds to the lowest out-of-plane overall instability mode with the amplitude as small as possible, i.e. scaled from unity to that ensuring the numerical convergence. The NBA numerical results are presented in the full length paper.

Figure 1 concerns with the comparison of LBA finite element results and the analytical based solution presented in the full length paper for the LT unrestrained beam-column. The numerical results are marked by circles and squares.

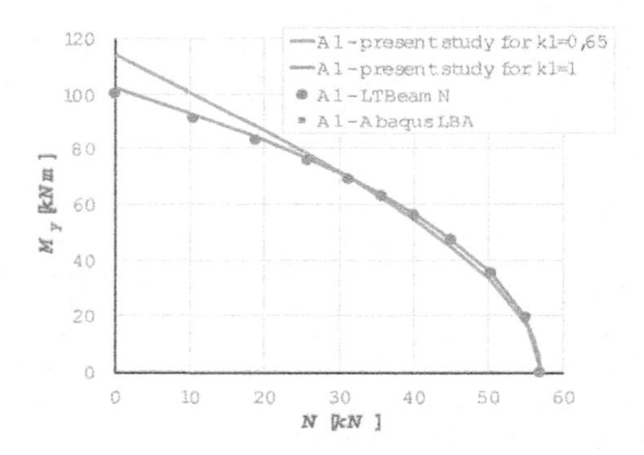

Figure 1. Comparison of LEA finite element results with those from the proposed analytical formulation for the unrestrained beam-column.

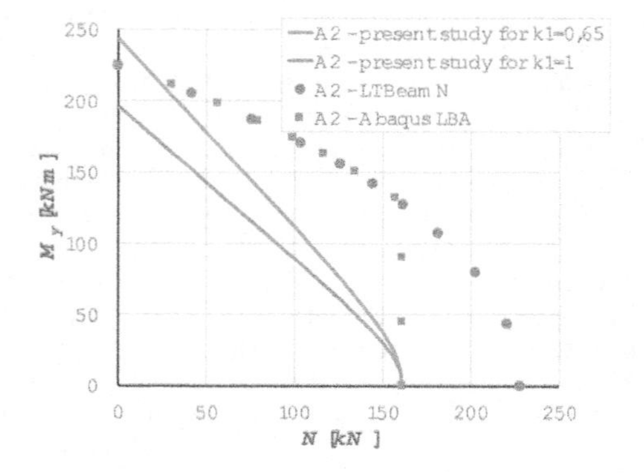

Figure 2. Comparison of LEA finite element results with those from the proposed analytical formulation for the restrained beam-column.

Figure 2 concerns with the comparison of LBA finite element results and the analytical Salvadori's based solution presented in the full length paper for the beam-column being LT restrained at the mid-length section. The curve representing analytical results is based on the solution for the critical segment being that spanning between the left support and the mid-length discrete restraint.

3 CONCLUDING REMARKS

Conclusions from the investigations are as follows:

1. The comparison shown in Figure 1 suggests that the analytical solution with k_1 factor equal to unity agrees well with the LBA numerical results based on LEA. This is not a surprise since LEA does include the effect of prebuckling displacements on the bifurcation state.
2. The curve representing the LBA analytical solution with the actual value of k_1 is placed higher than that representing the results of numerical simulations. This is a well-known fact resulting from the concave curvature effect of the deflected member profile, increasing the buckling resistance (Trahair 1993).
3. The comparison presented in Figure 2 shows that the analytical solution with k_1 factor equal to unity considerably underestimates the LBA numerical results based on LEA. The curve representing the LBA analytical solution with the actual value of k_1 is closer to numerical results.
4. There is a stronger interaction of the critical loads in compression and in-plane bending for discretely restrained members than that observed for unrestrained members.

REFERENCES

Gizejowski, M.A., Stachura, Z., Szczerba, R.B. & Gajewski, M.D. 2019. Buckling resistance of steel I-section beam-columns: in-plane buckling resistance. *Journal of Constructional Steel Research* 157: 347–358

Nethercot, D.A., Trahair, N.S. 1976. Lateral buckling approximations for elastic beams. *The Structural Engineer*, 54(6): 197–204.

Papangelis, J.P., Trahair, N.S. & Hancock, G.J. 1998. Elastic flexural-torsional buckling of structures by computer, *Computers and Structures* 68: 125–137.

Salvadori, M.G. 1951. Lateral buckling of beams of rectangular cross-section under bending and shear. *Proceeings of the First US National Congress of applied Mechanics*: 403–405.

Trahair, N.S., (1993) Flexural-torsional buckling of structures, E & FN Spon, London

Current Perspectives and New Directions in Mechanics, Modelling and Design of Structural Systems – Zingoni (ed.)
© 2022 Copyright the Author(s), ISBN 978-1-032-39114-4

Influence of symmetry on the buckling behaviour of plane frames

C. Kaluba & A. Zingoni
University of Cape Town, Rondebosch, Cape Town, South Africa

ABSTRACT: This paper presents the findings of a study on buckling behaviour of symmetric plane frames. The buckling behaviour of the frames were studied using finite element modelling and group-theoretic adapted slope deflection method. It was found that for plane frames symmetric to a C_{nv} group, the lowest buckling load had an eigenmode with the symmetry of a subgroup of C_{nv} with the highest order of elements; and the order of emergence of symmetries for eigenmodes of higher eigenvalues was from the subgroups of C_{nv} with the highest order of elements to those with the lowest order of elements.

1 INTRODUCTION

Symmetry is known to induce complex bifurcation behaviour in symmetric structures. In particular, double critical points at which multiple eigenvalues of the stiffness matrix simultaneously vanish and appear inherently on the equilibrium paths of symmetric structures such as lattice domes (Ikeda & Murota 1991). The occurrence of repeating eigenvalues is similar to that noted in the vibration behavior of symmetric systems, where group theory has been employed to study the physical system (Zingoni 1996, 2005, 2019). Through the results of heuristic case studies, Ikeda & Murota (1991) designed a framework to describe the bifurcation hierarchy of lattice domes symmetric to the dihedral D_n symmetry groups in terms of geometry, loading and stiffness. The equilibrium paths of the lattice domes were traced, and it was found that on the fundamental equilibrium paths the deformation patterns of the domes were D_n symmetric, while the deformation patterns on the bifurcated paths were symmetric to subgroups of the particular D_n symmetry group of the problem and this is described as the bifurcation hierarchy (Ikeda et al. 1991). In these studies, the numerical properties of the problem in question, which would help predict which subgroups of D_n emerge in the bifurcation hierarchy were not investigated.

Unlike lattice domes, research on the influence of symmetry on the stability behaviour of plane frames appears to be very limited. Mises & Ratzersdorfer (1926), investigated the lowest buckling load of plane frames that were symmetric to the C_{nv} symmetry group in terms of geometry, loading and stiffness. They found that the critical force in the members at the lowest buckling load is given as (Timoshenko & Gere 1963):

$$P_{cr} = \left(\frac{4\pi}{n}\right)^2 \frac{EI}{l^2} \tag{1}$$

where n= number of sides of polygon, with $n>3$; EI= flexural rigidity of members; and l=length of members. For $n=3$, the critical load was given by:

$$P_{cr} = (1.23\pi)^2 \frac{EI}{l^2} \tag{2}$$

The symmetry of the buckling mode for each n-sided plane frame was not reported. Further, other combinations of symmetry for the loading, geometry and stiffness were not investigated. A survey of literature on the influence of symmetry on the buckling behaviour of symmetric plane frames, failed to find other studies addressing this subject.

2 METHOD

In this study, each frame was first studied using the slope deflection method to obtain analytical results that were then used to validate finite element models (FEM) created in the software Abaqus. Since symmetric structures display repeating eigenvalues, which result in numerical ill conditioning when computing eigenvalues, the group-theoretic approach was applied to the conventional slope deflection method when computing eigenvalues. A similar group-theoretic approach to the buckling analysis of symmetric plane frames using the matrix stiffness method was presented in an earlier publication by the authors of this paper (Kaluba & Zingoni 2021).

DOI: 10.1201/9781003348450-117

The results obtained from the analytical model were used to validate the FEM created in Abaqus. These FEM were then used to study the effects of different symmetric configurations on buckling behaviour. The models created in Abaqus were created from a circular hollow section of outer diameter of *406 mm*, thickness of *6 mm* and length of *2 m*. The material properties of the cross-section were as follows: $E=200$ MPa and Poisson's ratio $\nu=0.3$. The beam elements used for the model were *B21* two-node linear elements. The mesh refinement was validated using results obtained from the analytical analysis.

3 PLANE FRAMES WITH C_{NV} SYMMETRY

The buckling behaviour of C_{2v}, C_{3v}, C_{4v}, C_{6v} and C_{8v} frames was investigated. A sketch of one such frame is shown in Figure 1. The frame in Figure 1 was C_{8v} symmetric in terms of stiffness. However, the symmetry of the loading is varied as shown in the Figure 1.

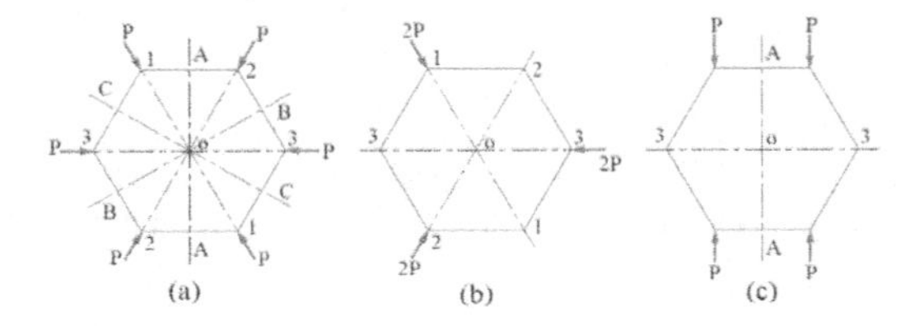

Figure 1. C_{6v} symmetric frame (a) C_{6v} loading, (b) C_{3v} loading (c) C_{2v} loading.

From the results obtained from the analytical model and finite element models, it was found that Equation 1 did not predict the correct lowest buckling load for the C_{5v}, C_{6v} and C_{8v} frames with C_{5v}, C_{6v} and C_{8v} symmetric stiffness and loading. The predicted lowest buckling loads from Equation 1 are 0.8π, 0.667π, and 0.5π respectively. However, this study found the lowest buckling loads as 1.092π, π and 0.873π respectively.

4 INFLUENCE OF SYMMETRY ON STABILITY BEHAVIOUR OF PLANE FRAMES

From the analysis of the results obtained from the buckling analysis of the frames studied, the following was concluded for plane frames that are C_{nv} symmetric in terms of stiffness and loading and with $n \geq 3$:

i. The lowest buckling load will have an eigenmode with the symmetry of a subgroup of C_{nv} with the highest order of elements;
ii. the order of emergence of symmetries, for eigenmodes of higher eigenvalues will be from the subgroups of C_{nv} with the highest order of elements, to those with the lowest order;
iii. eigenvalues whose eigenmodes have the symmetries of subgroup of C_{nv} form the lower bound eigenvalues;
iv. eigenvalues whose eigenmodes have C_n symmetries form the higher bound eigenvalues;
v. plane frames that are C_{nv} symmetric in terms of stiffness and subjected to a loading arrangement symmetric to a subgroup of C_{nv}, display the buckling behaviour of a frame that is symmetric to that particular subgroup in terms of stiffness and loading;
vi. plane frames that are C_{nv} symmetric in terms of stiffness and subjected to $C_{nv/2}$ loading, display the buckling behaviour of a plane frame symmetric to C_{nv} with respect to loading and stiffness for the case where $n/2$ is even.

REFERENCES

Ikeda, K., and Murota, K. (1991). "Bifurcation analysis of symmetric structures using block-diagonalization." *Computer Methods in Appl. Mech. and Engineering*, 86 (2), 215–243.

Ikeda, K., Murota, K., and Fujii, H. (1991). "Bifurcation hierarchy of symmetric structures." *International Journal of Solids and Structures*, 27(12), 1551–1573.

Kaluba, C., and Zingoni, A. (2021). "Group-Theoretic Buckling Analysis of Symmetric Plane Frames." *J. Struct.Eng.*, 147(10), 04021153.

Mises, R. V., and Ratzersdorfer, J. (1926). "Hauptaufsätze: Die Knicksicherheit von Rahmentragwerken." *Z.Angew. Math.Mech.*, 6(3), 181–199.

Zingoni, A. (1996). "An efficient computational scheme for the vibration analysis of high-tension cable nets." *Journal of Sound and Vibration*, 189(1), 55–79.

Zingoni A. (2005). "On the symmetries and vibration modes of layered space grids." *Engineering Structures*, 27(4), 629–638.

Zingoni, A. (2019). "Group-theoretic vibration analysis of double-layer cable nets of D4h symmetry." *International Journal of Solids and Structures*, 176/177, 68–85.

Current Perspectives and New Directions in Mechanics, Modelling and Design of Structural Systems – Zingoni (ed.)
© 2022 Copyright the Author(s), *ISBN 978-1-032-39114-4*

Machine learning optimization strategy for the inelastic buckling modelling of un-corroded and corroded reinforcing plain bars

F. Pugliese
Institute for Risk and Uncertainty and Department of Civil Engineering and Industrial Design, School of Engineering, University of Liverpool, Liverpool, UK

L. Di Sarno
Department of Civil Engineering and Industrial Design, School of Engineering, University of Liverpool, Liverpool, UK
Department of Engineering, University of Sannio, Benevento, Italy

ABSTRACT: This paper presents an optimization strategy based on genetic algorithms to simulate the cyclic response of steel plain bars. A refined finite element model of the steel bar is adopted for the numerical analysis. Thus, the genetic algorithm optimizes the main parameters of the most adopted constitutive model for steel rebars. The optimization procedure compares the available experimental tests with the numerical results by reducing a pre-defined objective function. Regression analyses are then performed for each calibrated model parameter. Therefore, taking advantage of the comprehensive numerical procedure, a parametric analysis is conducted to include the effects of corrosion on the inelastic buckling of plain rebars. The parametric study aims to develop accurate and adequate constitutive models for steel reinforcing bars in robust seismic analyses for RC structures. A case study of a typical old RC column with longitudinal plain bars under cyclic loading is presented.

1 INTRODUCTION

This study presents a nonlinear optimization strategy, through the evolutionary algorithm named genetic algorithm, to investigate the model parameters of the most adopted constitutive material for smooth steel bars. Corrosion is applied to reproduce its effects on the inelastic buckling of smooth rebars as a sort of virtual laboratory testing. A case study of a typical RC column with un-corroded and corroded plain reinforcing bars is presented and the results are discussed.

2 FINITE ELEMENT MODEL FOR THE INELASTIC BUCKLING

The force-based element with distributed plasticity is utilized to simulate the nonlinear behavior of the steel bar (Figure 1).

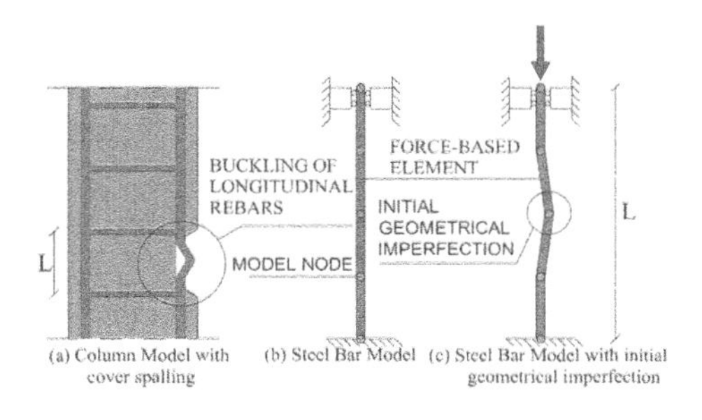

Figure 1. Finite Element Model of the inelastic buckling.

The constitutive material named STEEL02 in OpenSees [MKenna et al. (2000)], is employed in the force-based elements.

3 OPTIMIZATION PROCEDURE

The Genetic Algorithm (GA) is used to calibrate the parameters of the Steel02 constitutive material. The objective function is as follows:

$$error = \frac{\sqrt{\sum_{j=1}^{m}\left(F_{exp,i} - F_{num,i}\right)^2}}{\sqrt{\sum_{j=1}^{m}\left(F_{exp,i}\right)^2}} \quad (1)$$

where m is the total number of experimental points and, F_{exp} and F_{num} are the forces from the experimental and the numerical results. An example of the optimization procedure is presented in Figure 2.

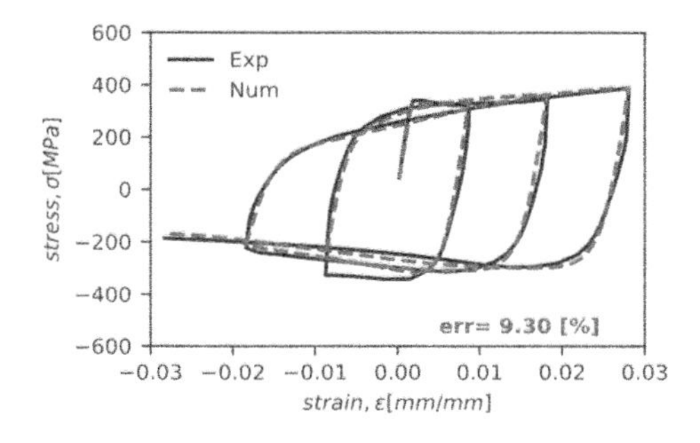

Figure 2. An example of the GA application for L/D = 10.

DOI: 10.1201/9781003348450-118

It can be observed that the proposed approach (GA plus FE model) can adequately and accurately predict the cyclic response of smooth rebars with various slenderness ratios.

4 CASE STUDY

A case study of a typical RC column reinforced with plain rebars is herein investigated. The experimental test results of RC columns under cycling loading from Di Ludovico et al. (2014) are used.

Using the FE model of the steel bar, the buckling response with the slenderness ratio equal to 12.5 is simulated, including the effects of corrosion. The results are presented in Figure 3.

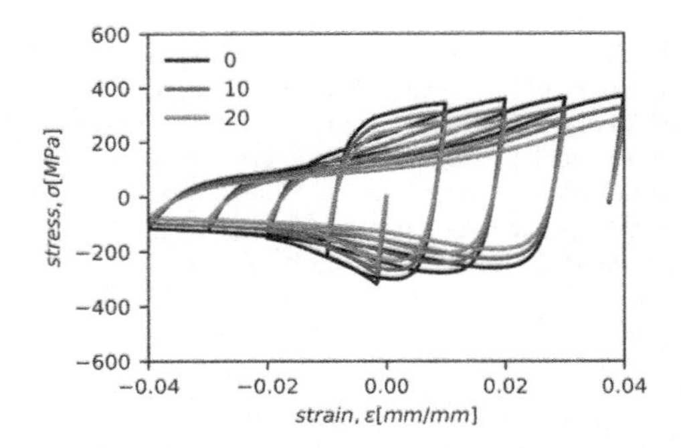

Figure 3. Inelastic buckling considering the corrosion effects. (Corrosion equal to 0%, 10% and 20%).

A hysteretic material available in OpenSees is used to simulate the inelastic buckling according to the numerical simulation presented in Figure 3. The GA is used in the FE model of the un-corroded RC column with smooth rebars to calibrate the pinching parameters and the length for the displacement-based element to adequately match the experimental results.

The outcomes of the GA are shown in Figure 4.

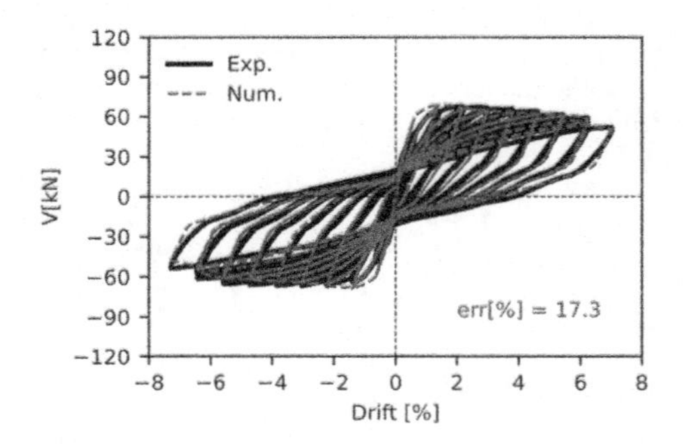

Figure 4. Experimental vs numerical results for the RC column with smooth rebars.

The FE model can reasonably capture the cyclic response of the experiment with an error of 17.3%. The optimized values for the pinching of the hysteretic material are 0.5 and 0.8 along the x-axis and the y-axis, respectively; instead, the height of the displacement-based element including the inelastic buckling is equal to 450mm.

Applying the corrosion effects according to Pugliese et al. (2022), the cyclic responses of the RC columns with aged smooth rebars are graphically illustrated in Figure 5.

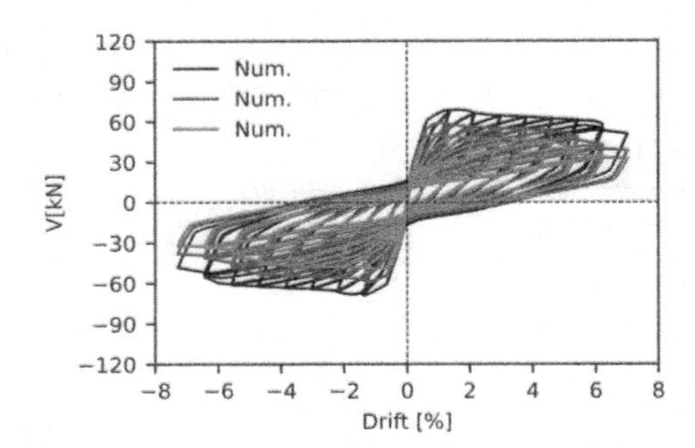

Figure 5. Cyclic response of the RC column with un-corroded and corroded smooth rebars.

The numerical results in Figure 5 show a substantial reduction of the maximum lateral strength of the RC column. Specifically, the un-corroded RC column presents a positive and negative peak equal to 68.6kN and -68.7kN, compared with 59.1kN and -58.5kN for CR equal to 10%, and 50.1kN and -49.1kN for CR equal to 20%. Besides, the energy dissipated for each cycle appears to reduce because of the corrosion effects.

5 CONCLUSIONS

The following conclusion can be drawn:

- The results of the FE model combined with the GA adequately minimize the objective function while matching the experimental tests;
- The effects of corrosion are significant for the inelastic buckling of smooth rebars;
- The effects of corrosion on RC columns can be detrimental over the lifetime of structural components.

REFERENCES

Di Ludovico, Verderame, Prota, Manfredi & Cosenza (2014). Cyclic behavior of nonconforming full-scale RC columns, *J. Struct. Eng.*, 140 (2014).

McKenna, Fenves, & Scott (2000). Open system for earthquake engineering simulation. Berkeley, CA: *Univ. of California*.

Pugliese & Di Sarno (2022). Probabilistic structural performance of RC frames with corroded smooth bars subjected to near- and far-field ground motions, *Journal of Building Engineering*, Volume 49, 2022, 104008, ISSN 2352-7102.

Current Perspectives and New Directions in Mechanics, Modelling and Design of Structural Systems – Zingoni (ed.)
© 2022 Copyright the Author(s), ISBN 978-1-032-39114-4

Numerical study of distortional buckling of single channels restrained by angle cleats

G.M. Bukasa & M. Dundu
University of Johannesburg, Johannesburg, South Africa

ABSTRACT: The paper presents a developed finite element model and its details used for investigating distortional buckling behaviour of lipped cold-formed steel beam. The latter is part of structural framing system where cold-formed purlin section is connected to it through angle-cleat. Both beam and purlin are connected to angle-cleat through their webs. All the lengths attributed to the investigated beams were determined using Constrained and Unconstrained finite strip method, CUFSM. Two concentrated point loads were applied to the beam in order to initiate uniform bending moment under the investigated span. The model included the actual material properties obtained from the tensile tests, the geometric imperfections, loading arrangement and boundary conditions. The model was verified against the experimental results published by GM Bukasa and M Dundu (2016). The results from the model were compared with the direct strength method predictions.

Keywords: single channels, restrained, purlin–angle cleat, connection, distortional buckling

1 INTRODUCTION

Cold-formed lipped steel members have different usage in building construction, amongst them they are used as rafter through which purlin is connected in the framing system. These members are generally thin walled material with cross-section classified among the unsymmetrical sections. Hence their susceptibility to local, distortional, lateral-torsional and interaction buckling behaviour is significantly higher and remarkable in comparison to other type of sections. Previous research studies reviewed in this paper have shown great interest on the distortional behaviour of steel sections flexural members, especially lipped cold-formed steel purlin, lipped cold-formed stainless steel and aluminum. Among the published numerical studies on distortional buckling behaviour are the work of Camotim et al. (2018) conducted a review: interactive behaviour, failure and DSM design of cold-formed steel members prone to distortional buckling, while Yu and Schafer (2007) performed a numerical study on simulation of cold-formed steel beams in local and distortional buckling with applications to direct strength method.

This paper presents details of the finite element model study on distortional buckling and its results, which are compared to the codified Direct Strength Method.

2 BEAM GEOMETRY AND LENGTH SELECTION

Lipped cold-formed cross-sections selected for numerical study consist of the following nominal dimensions: 300mm depth, 20mm lips and flange width that varied from 50mm, 65mm and 75mm, respectively. All the cross-section profiles were made of 3mm nominal thickness. The beam lengths were selected according to the approach based on CUFSM elastic analysis so that the beams would experience distortional buckling.

3 DESCRIPTION OF THE FINITE ELEMENT MODEL

3.1 *Material properties*

A 100kN capacity displacement controlled testing Intron was used to test all the coupons according to the guidelines provided by the British Standard, BS EN ISO 6892-1 (2009).

Table 1. Average material properties.

Channel sections	f_y (MPa)	f_u (MPa)	f_y/f_u	ε_f (%)	E (GPa)
300x75x20x3.0	265.70	351.55	1.32	30.29	205.58
300x65x20x3.0	265.00	340.89	1.29	28.57	201.50
00x50x20x3.0	267.60	360.00	1.35	27.01	203.60

DOI: 10.1201/9781003348450-119

The average of coupon test results including yield stress, f_y, ultimate strength, f_u and the elastic modulus, E, are summarized in Table 1.

4 ANALYSIS METHOD

Two types of analysis were conducted: elastic finite element analysis and nonlinear finite element analysis.

5 VERIFICATION OF THE FINITE ELEMENT MODEL

The validation of the developed finite element model in this study, was made possible by using the previous experimental results published by GM Bukasa and M Dundu (2016).

6 FAILURE MODES

Distortional buckling failure mode of the specimens with Lu =1335mm and 1625mm were observed in the middle span as shown in Figure 1. Distortional buckling occurred first at the compression portion of the web and followed by distortion of the compression flange.

Figure 1. Distortional buckling.

7 RESULTS AND COMPARISON WITH DESIGN METHOD

The finite element analysis results and the Direct Strength Method (DSM) predictions of the single channel are given in Table 2. In this table, M_{FE} is the failure moment of the simulated specimens, M_{nd} is the distortional buckling moment, determined using the Direct Strength Method (AISI S100-12).

Table 2. Comparison of FEM and design.

Channel sections	L_u (m)	f_y (MPa)	M_{FE} (kNm)	M_{nd} (kNm)	M_{FE}/M_{nd}
300x 75x 20x3.0	1.33	265.7	43.76	31.10	1.4
300x 75x 20x3.0	1.62	265.7	35.42	28.68	1.2
300x 65x 20x3.0	1.62	265.0	34.88	28.63	1.2
300x 50x 20x3.0	1.62	267.6	24.68	24.19	1.0

8 CONCLUSIONS

A large number of numerical simulations were performed with a finite element programme. Some of the results have been presented in this paper. The direct strength method was used for comparison with the finite element model results. From the numerical analysis, the following conclusions are made:

As expected, all the cold-formed steel channel beams failed by distortional buckling mode of the compression web followed by compression flange distortion. This caused the mid-span to deform, while both shear spans remained straight. The distortional failure of the beam with 1335mm unbraced length was more pronounced compared to other beams. Finally, distortional buckling was the main failure mode observed in the collapse of the four cold-formed steel channel beams considered in the investigation.

ACKNOWLEDGEMENT

The authors wish to acknowledge the University of Johannesburg Research Committee (URC) for sponsoring this research.

REFERENCES

BS EN ISO 6892-1. 2009. Tensile Testing of Metallic Materials-Part 1- Method of Test at Room Temperature, British Standard Institution.

Camotim, D., Dinis, B., Martins, D., Young, B. 2018. Review: Interactive behaviour, failure and DSM design of cold-formed steel members prone to distortional buckling. Thin walled structures 128: 12–42

CUFSM v2.6. 2004. Elastic buckling analysis of thin-walled members by finite strip analysis.

Dundu, M. and Kemp, A. R. (2006b) "Plastic and flexural behaviour of single cold-formed channels connected back-to-back." Journal of Structural Engineering, ASCE, 132(8), pp. 1223–1233.

Current Perspectives and New Directions in Mechanics, Modelling and Design of Structural Systems – Zingoni (ed.)
© 2022 Copyright the Author(s), *ISBN 978-1-032-39114-4*

Experimental investigation of distortional buckling of single channels restrained by angle cleats

G.M. Bukasa & M. Dundu
University of Johannesburg, Johannesburg, South Africa

ABSTRACT: This paper experimentally investigates distortional buckling behaviour and strength of cold-formed steel lipped channels, restrained at interval points by purlin-angle cleats, and subjected to pure bending tests. Angle cleats with cross-section nominal dimension of 100x75x3.0mm and height of 300mm were attached to the web of the channel, and connected to the web of the purlin section. The main aim of this experimental study is to investigate distortional buckling between two unbraced cleat-purlin connections of three different cross-sections, viz; 300x75x20x3.0mm, 300x65x20x3.0mm, and 300x50x20x3.0mm. All the experimental span lengths of the test specimens were determined using the Constrained and Unconstrained Finite Strip Method (CUFSM). Distortional buckling strength was evaluated using the Direct Strength Method (DSM), as specified in the North American design standard (AISI S100-12). The experimental results showed that by using this type of restraining system, the DSM for distortional buckling predictions were conservative.

Keywords: Distortional buckling, single cold-formed channels, restrained, purlin–angle cleat, connection

1 INTRODUCTION

This paper investigates the strength and behaviour of the lipped cold-formed steel beam rafter-purlin sys-tem of a four-point loaded frame. The purpose of the tests was to examine the strength and distortional failure modes of the internal unbraced length of single channel beams. Distortional buckling mode causes premature failure, which results in a decrease of the ultimate strength of the beam. The experimental lengths used to study this span have been selected from signature curves of the cross section, obtained using CUFSM software.

Research on distortional buckling on cold-formed steel channel sections have been increasing significantly in recent years. Previous experimental studies related to the present investigation were carried out by Yu and Schafer (2006), Martins et al. (2017), Pharm and Hancock (2013), G. J. Hancock (1996. This paper contains an experimental investigation on distortional buckling of three different cross-sections (300×75×20×3.0mm, 300×65×20×3.0mm, and 300×50×20×3.0mm) between two unbraced angle cleat-purlin connections.

2 MATERIAL PROPERTIES

In total 24 coupons were machined and tested according to the specifications in BS EN ISO 6892-1. The average values of f_y and E for each cross-section are presented in Table 1.

Table 1. Average material properties.

Channel	f_y (MPa)	f_u (MPa)	f_u/f_y	ε_f (%)	E (GPa)
300×75×20×3.0	265.70	351.55	1.32	30.29	205.58
300×65×20×3.0	265.0	340.89	1.29	28.57	201.50
300×50×20×3.0	267.6	360.00	1.35	27.01	203.60

3 EXPERIMENTAL PROGRAMME

The full experimental test set-up is shown in Figure 1. Four-point loading was applied to the lipped cold-formed steel beam, to generate pure bending created within the middle unbraced length.

A pair of identical lipped cold-formed channel beams with the same span lengths and geometry were

DOI: 10.1201/9781003348450-120

Figure 1. Test set-up.

positioned in parallel and oriented in the same direction (Dundu and Kemp, 2006a,b). Beams were separated by a distance of 1.84m measured from their webs (Dundu and Kemp, 2006a,b). All the simply supported beam tests were performed using 250kN capacity hydraulic Instron testing machine. All the beam tests were conducted at a displacement rate of 0.5mm/min until the load significantly dropped so that a good failure pattern could be observed.

4 TEST RESULTS

In Table 2, M_t is the maximum moment applied to the tested specimens, M_y is the yield moment and M_{DSM} is the distortional buckling moment, determined using the North American Specification for the Design of Cold-Formed Steel Structural Members, AISI S100-12.

Table 2. Test results.

Channel sections	L_u (m)	f_y (MPa	M_t (kNm)	M_y (kNm)	M_{DSM}	M_t/M_{DSM}
300×75×20×3.0	1.33	265.7	31.10	31.51	29.85	1.04
300×75×20×3.0	1.62	265.7	28.68*	31.51	29.85	0.96
300×65×20×3.0	1.62	265.0	28.63	28.69	28.02	1.02
300×50×20×3.0	1.62	267.6	24.19	27.46	27.07	0.89

(Note: * indicates the test was repeated.)

It can be seen from the table that the large flange size of 75mm exhibits a higher capacity than the 65mm and 50mm, but the difference between them is not so large. Section with 75mm flange developed a 0,17% more strength than the 65mm flange and 15,65% more than the section with 50mm flange width, respectively.

5 CONCLUSIONS

A series of tests on lipped cold formed steel channels are reported in this paper. The main conclusions are summarized as follows:

- Distortional buckling of the web and flange was observed in the tested specimens. This buckling mode generally affect the strength of the specimens. In all the tested beams, failure was observed in the internal span with no localised failure or deformation in the shear spans as well as in the connection cleat-purlin. The flange size had more influence on the strength of the beam as demonstrated by the experimental results presented in this paper. In comparison with the lateral-torsional buckling tests conducted by GM Bukasa and M Dundu (2014), which the thickness was the governing parameter, it was observed that varying the thickness influenced the buckling mode of the beams having the same unbraced length and same flange width size.
- Comparison of the experimental results with the codified DSM, the North American Specification for the Design of Cold-Formed Steel Structural Members, AISI S100-12 provide good agreement for the members failing by distortional buckling. The effect of this buckling has to be accounted for, especially in design.

ACKNOWLEDGEMENT

The authors wish to acknowledge the University of Johannesburg Research Committee (URC) for sponsoring this research.

REFERENCES

AISI S100. 2013. North American Specification for the Design of Cold-Formed Steel Structural Members, Washington D.C.

BS EN ISO 6892-1. 2009. Tensile Testing of Metal-lic Materials-Part 1- Method of Test at Room Tem-perature, British Standard Institution.

CUFSM v2.6. 2004. Elastic buckling analysis of thinwalled members by finite strip analysis.

Dundu, M. and Kemp, A. R. (2006a) "Strength requirements of single cold-formed channels connect-ed back-to-back. " Journal of Construction Steel Research, 62, pp. 250–261.

Martins et al. 2017. Distortional failure of cold-formed beams under uniform bending: Behaviour, Strength and DSM design.

Pham, CH., Hancock, GJ. 2013. Experimental investigation and Direct Strength Design on high strength complex C-Sections in pure bending. Jour-nal of Structural Engineering, ASCE, Vol. 139.

9. Plates, Shells, Membranes, Cable nets, Cable-stayed structures, Lightweight structures, Form-finding, Architectural considerations

Current Perspectives and New Directions in Mechanics, Modelling and Design of Structural Systems – Zingoni (ed.)
© 2022 Copyright the Author(s), ISBN 978-1-032-39114-4

Principle of constant stress in analytical form-finding for durable structural design

W.J. Lewis
School of Engineering, University of Warwick, UK

ABSTRACT: The paper presents examples of nature-inspired structural forms modelled using a form-finding approach that preserves the principle of constant stress in structures under statistically prevalent load. It is a principle observed in highly optimized natural objects, such as: trees, bones, or shells, and one that can be used to model shapes of a variety of engineering structures, ranging from roofing forms to arches. Its main advantage is that it produces structures characterised by a minimal stress response to loading. While constant stress roofing forms (minimal surfaces) have a limited design space determining their existence, it is found that the same applies to arches of constant axial stress. Despite that, the design space is very large, allowing an enormous number of natural designs to be created. It is suggested here that the principle of constant stress should be used as the main optimisation criterion in generating durable and sustainable design solutions.

1 INTRODUCTION

1.1 *Design methodology: Form-finding implementing the principle of constant stress*

In general, form-finding is a process of shaping structures using, or controlling, forces developing in them.

Many methods are possible, depending on the type of structure being modelled; they can employ computational and analytical models, supported by physical experiments (Lewis, 2018).

Membrane structures, in the form tensioned roofing forms, can be form-found using statistically prevalent load, such as pre-stress, and chosen boundaries/supports. When the chosen pre-stress is constant, the resulting structures take on natural forms, known as minimal surfaces (Otto & Rasch, 1995) that can be modelled using soap-film models (Hildebrand & Tromba, 1985).

Two-pin arch structures featured here are shaped by analytical form-finding process implementing the principle of constant axial stress that developsg when the arch is subjected to statistically prevalent loads, such as their own weight (Lewis, 2022).

Analytical form-finding differs from computational models by providing a continuous, not discretised, description of structure's geometry.

Form-found structures modelled on the principle of constant stress exhibit maximum strength/mass ratio, as observed in natural objects (plants, shells, trees).

2 CONSTANT STRESS ROOFING FORMS

Shown below is one of the oldest examples of a constant stress (minimal surface) roofs.

Figure 1. IL now ILEK) Tent, Stuttgart, 1965, designed by Frei Otto using cable net and soap-film models. The height, h_c, is limited to that of an attainable soap-film surface.

3 CONSTANT STRESS ARCH FORMS

Presented here are arch structures that are not only moment-less under statistically prevalent load (such as theirs own weight), but also characterised by a constant value of axial stress along their entire length. Their centre-line profile and material distribution along the length are determined by the deck and arch loads, as well as the basic input parameters, such as the span, span/rise ratio, and a chosen value of constant stress. Full details of the theory of

DOI: 10.1201/9781003348450-121

analytical form-finding are given in Lewis, 2022. It is recommended that the chosen stress is low enough to ensure that, under ultimate load (combining permanent and variable load), the material design strength is not exceeded. Further research is needed to establish an optimum value of constant stress, so that the structure does not become too slender (when the chosen stress is too high), or too bulky (when the chosen stress is too low). As a rule of thumb, the value to be used should be at least ten times lower than the characteristic strength of the material.

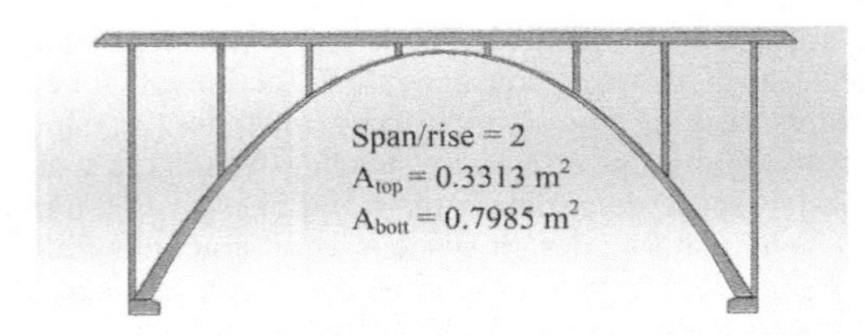

Figure 2. 1:20 Model of a constant stress arch. Input: span: 50 m, deck load density: 50 kN/m, arch weight density: 25 kN/m^3 stress: 2.4 MPa. Output: centre line profile and cross-section area, *A*.

3.1 *Material usage: Constant stress and other arch forms*

Apart from constant stress arches, there are exist also moment-less arches of constant cross-section, described in Lewis, 2016. Material efficiency of both types has been studied in relation to the conventional arch forms, typically of parabolic and circular configurations. Results, for span/rise ratios of 2 and 4, are shown in Figure 3.

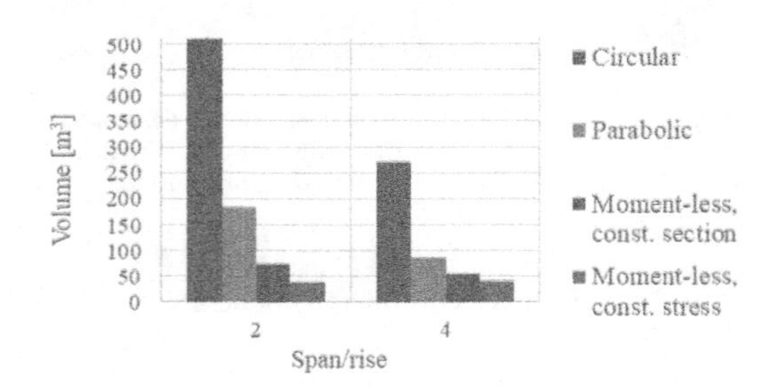

Figure 3. Volume of material in moment-less and other forms of arches.

4 DISCUSSION

With reference to Figure 3, it can be seen that moment-less arches use much less material than the conventional arch forms, with constant stress arches being most materially efficient. The latter are not necessarily structures of least weight. This is demonstrated in Lewis, 2022, where it is shown that least weight structures are just a subset of constant stress arches, and where it is argued that the criterion of minimum weight restricts design options, and may raise questions regarding durability of such structures.

5 CONCEPT OF THE DESIGN SPACE

It is known that constant stress (minimal) surface structures do not exist between any boundaries; in the case of a catenoid surface, its attainable height depends on the ratio of the larger to smaller radii of the boundaries. This has been demonstrated analytically by Alexander & Macho, 2020, and earlier found by Frei Otto through experimentation with soap-film models, helping him to determine the maximum attainable height of the IL Tentg (Figure 1). The structure is now a listed building.

There exists a parallel between constant surface stress structures and arches of constant axial stress, as both of have a limit imposed on their existence (Lewis, 2022). In the latter case, the limits are defined by a combination of two dimensionless parameters.

6 CONCLUSIONS

The principle of constant stress observed in Nature produces highly optimised objects. Incorporating it into form-finding design of engineering structures will lead to durable and sustainable design solutions demanded by our future infrastructure.

ACKNOWLEDGEMENT

The work has been supported by the Leverhulme Trust (grant no. EM-2020-59/9).

REFERENCES

Alexander, G. P. & Macho T. 2020. A Björling representation for Jacobi fields on minimal surfaces and soap film instabilities. *Proc. R. Soc.* A (doi.org/10.1098/rspa.2019.0903).

Hildebrandt, S. & Tromba, A. 1985. *Mathematics and optimal form*. Scientific American Library, USA.

Lewis, W. J. 2016. Mathematical model of a moment-less arch. *Proc. R. Soc. A* 472. (doi:10.1098/rspa.2016.0019).

Lewis W. J. 2018. *Tension structures: form and behaviour*, 2nd edn. London, UK: ICE Publishing.

Lewis, W. J. 2022. Constant stress arches and their design space. *Proc. R. Soc. A* (doi.org/10.1098/rspa.2021.0428).

Otto F. & Rasch B. 1995. *Finding form: towards an architecture of the minimal*. Deutscher Wergbund Bayern. Edition Axel Menges.

Current Perspectives and New Directions in Mechanics, Modelling and Design of Structural Systems – Zingoni (ed.)
© 2022 Copyright the Author(s), ISBN 978-1-032-39114-4

Extraordinary Mechanics in Slender Structural Surfaces

S. Adriaenssens
Form Finding Lab, Department of Civil and Environmental Engineering, Princeton University

ABSTRACT: Our research focuses on the extraordinary mechanics of slender large-span shells, membranes, and elastic rod networks under extreme loading and during construction. Our work concerns shape generation and analysis approaches for innovative lightweight structural systems that enable a resilient, and sustainable built environment. we focus on the advancement of theoretical and computational approaches to predict and design the overall properties, stability, and failure of these systems. In this paper, I outline how we discovered, studied, designed and even built large-scale structural surfaces that can efficiently carry extreme loading, self-assemble, adjust their stiffnesses, elastically shift from one shape to another, or amplify motion.

1 INTRODUCTION

The construction industry is one of most resource-intensive sectors, and yet our urban infrastructure continues to be built in the massive tradition in which strength is pursued through material mass. In contrast, we, at the Form Finding Lab at Princeton University, have focused our research on structural systems that derive their performance from their curved shape, dictated by the flow of forces. Our core research question is 'What is the relationship between form and efficiency in civil-scale structures?'.

2 RIGID SHELLS WITH ENHANCED LOADBEARING CAPACITY

Anecdotal evidence of the survival of the San Francisco City Hall Cupola (1906) and the Los Manantiales Shells (Mexico City, 1985) suggests that shell roofs remain intact and safe during and after extreme earthquake events. We demonstrated that shells owe this seismic resistance to their large geometric stiffness and low mass, which lead to high fundamental frequencies well above the driving frequencies of earthquakes. For the shape generation of arches and shells in seismic regions, we presented novel optimization and form finding approaches as well as the first geometries for compression-only shells under seismic loading. One such approach relies on the construction of funicular polygons and generates shells with corrugations. Using a non-linear pushover analysis, we discovered that by tailoring the corrugation wavelength and the depth, the shell's lateral loading capacity could be increased by as much as 80% before it collapses. The proof that corrugations can substantially increase the loading capacity in large-span shells opens up new avenues for the design of robust and resilient infrastructure.

3 SELF-LOCKING AND GRAVITY-ACTUATED CONSTRUCTION PRACTICES

Because of gravity, large-span structures demand wasteful temporary support during constructionwhich adds to the economic and environmental cost of the construction. We explored the feasibility of gravity-actuated assemblies in composite bending active/rigid shell systems, reciprocal shell systems, and plate-bande systems. In our latest work on plate-bande systems, we investigated the static equilibrium of a self-locking octagonal shell, constructed in a herringbone pattern and closed-loop sequencing, to gain insight into the local interactions between the more than 10,000 bricks. A discrete element analysis exposed the existence of plate-bande resistance within the herringbone pattern. Even during construction, these plate-bande resistance systems prevented the sliding and overturning of the shell. These studies formed the basis for the construction of a scaffold-free glass brick vault with collaborative robots.

4 PRESSURIZED MEMBRANES WITH TUNABLE STIFFNESS

Motivated by the need to combat coastal flooding, we re-invented pressurized thin-wall structures as an alternative to existing large-scale rigid solid seawalls. The core concept is that under extreme storm surge loading, the pressurized semi-cylindrical

DOI: 10.1201/9781003348450-122

clamped membrane undergoes large displacements and geometrically stiffens because of the membrane's inc reasing internal tension. This adaptive stiffening ensures that the structure has an adequate height to prevent water overtopping.To optimize the design of the pressurized membrane, we coupled an adjoint-based method based on automatic differentiation of a finite-element model to a shape parametrization approach. At the large-scale, we were able to show that the stiffness of pressurized membranes (diameter 8m) adjusts and can be optimally designed as a function of internal pressure (20-40 kPa), under changing flood conditions found in Jamaica Bay (NY, USA), to prevent water overtopping while avoiding material rupture and resonance with the waves. The originality of this work lies in harnessing and designing the geometric stiffening effect, due to the changing tension in the membrane, as a strategy to design novel resilient structural systems with loading-dependent stiffness characteristics

5 SHAPE-MORPHING COMPLIANT SHELLS

We are interested in understanding and controlling the mechanics and geometry of elastic surface structures to imbue them with advanced functionality useful for green architecture. Adaptive building skins act as active climate filters and moderate the influence of variable weather conditions on the interior of the building. However, current adaptive systems rely on rigid body motions and complex hinges that increase the complexity of control and construction costs. We sought to simplify and tailor this macro-scale shape-shifting through the application of elastic botanic principles. Plants utilize the elastic properties of their organs to move with minimum energy and maximum effect. We identified 5 main botanic strategies to create and amplify movement and showed that many plant movements are rooted in compliance and multi-functionality. Our attention was drawn by two shape-morphing mechanisms. The first one couples geometry and shell mechanics and is found in the *Stylidium graminifolium* flower. The second one exhibits bending, torsion, and warping action through the shrinking and swelling of cells in bilayers of the *Triticum turgidum* wheat. The pollination column in the *Stylidium* flower consists of an active anticlastic shell part that through progressive curvature inversion, amplifies the motion of a passive plate part. Using an optimization framework, we interpreted the scale limits of these (and other) shells and tailored this dynamic amplification mechanism to an efficient adaptive shading shell that can be installed on a building skin of any orientation. In the *Triticum turgidum* wheat seed, the amount of stress related to the hygroscopic shrinking or swelling of its cells ranges from almost zero to high values. Being able to design bilayer wood devices with limited stresses under humidity changes is structurally desirable for adaptive building skins when trying to avoid fatigue. To describe the large displacements of such devices, we presented a mechanical model, based on the non-linear theory of hyper-elastic solids that simulates to a high degree of accuracy the reversible hygroscopic orthotropic (de-)swelling phenomenon. We combined this model with physical explorations to imagine, optimize and prototype adaptive bilayer wood veneer shell devices . On the exterior of any glass building skin, the *Stylidium* and *Triticum* shell devices change shape as a function of the solar radiation intensity, humidity, and desired views to the outside. This research provided a novel paradigm for building skin technology and promoted biomimicry as a driver for green architecture.

6 VISION FOR THE FUTURE

Other leading large-scale manufacturing industries (like the automotive industry) are 50 years ahead of the construction industry in terms of digitization, yet robots can revolutionize construction through automation, precision, customization, differentiation, and the absence of external support. Today, advances in robotic construction directly map from robotic approaches in other manufacturing industries that use robots solely for automation and the execution of pre-determined tasks. In other words, the robot's role is reduced to that of a super muscle: it only improves and speeds up the construction process. We want to challenge this dogma by equipping the human-robotic construction framework not only with super muscles, but also with senses and a super brain that can plan and make structural decisions autonomously. Therefore, our ultimate research goal is to leverage principles of self-assembly and collective human/artificial intelligence for the construction of large span structural systems. In practice, this means that when starting construction, neither the global geometry nor the exact assembly program are known. If successful, this will allow for automated large-scale construction in unknown contexts (such as disaster areas, extraterrestrial, or submarine environments). This concept is large in scope, bold and innovative in character. However, the transformative concept of self-assembly at the architectural and civil scale has yet to be computationally programmed and physically realized. If successful, this paradigm shift will allow for waste-free construction practices, increase construction productivity and economic growth.

Current Perspectives and New Directions in Mechanics, Modelling and Design of Structural Systems – Zingoni (ed.)
© 2022 Copyright the Author(s), *ISBN 978-1-032-39114-4*

Asymmetric multifocal egg-like shells with elastic adaptive frameworks

D. Kozlov
Research Institute of Theory and History of Architecture and Urban Planning, Moscow, Russia

Y. Vaserchuk
Institute of Modern Art, Moscow, Russia

ABSTRACT: Geometry of natural asymmetric eggshells often forms by rotating of multifocal plane curves around their axes of symmetry. The curves can be described by an equation of the order of 2^n, where n is the number of foci of the curves. Models of the curves were taken as the basis for experiments with complex asymmetric structures with elastic frame and rigid coverings. In these experiments, adaptive frame structures made in the form of cyclic closed knots from elastic rods of composite materials simulated the bearing principle of the inner woven structures of an eggshell.

1 INTRODUCTION

Mechanical "work of shape" of the shells of living nature often manifests in their asymmetry. The asymmetry of natural forms is the result of adaptation of living organisms to the directed forces of the external environment, such as wind loads, water pressure, solar radiation, the action of gravitation, etc. The experience of nature is applicable to create asymmetric shells in the field of structural engineering and architecture (Lebedev et al. 1990).

2 MATHEMATICAL DESCRIPTION OF ASYMMETRIC SHELLS

Geometric models of many asymmetric shells of living nature can be formed by multifocal plane curves – a generalization of the second-order curves: ellipse, parabola and hyperbola, which rotate around their axes of symmetry.

When the number of foci of the parabola increases to two or more and its directrix property preserves, the curve becomes closed and acquires an asymmetric egg-like shape (Brandt 1979). The curve with two foci corresponds to a fourth-order equation. Its real number solution gives the formula

$$y = \frac{1}{2}\sqrt{3x(2-x)\left(1 - \frac{\beta^2}{(x+1)^2}\right)} \qquad (1)$$

where $0 \leq x \leq 2$, $0 \leq \beta \leq 1$, $\beta = c/a$, and a – half of the big axis; c – distance between foci. If the β ratio changes within 0.75 – 0.86, a number of meridional sections of bird eggs fall into this range (Figure 1).

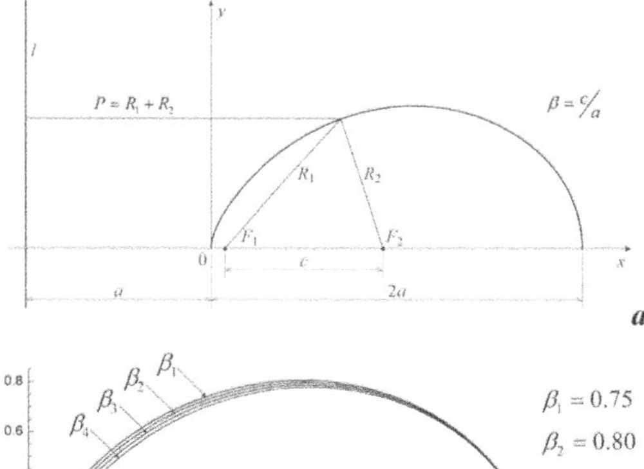

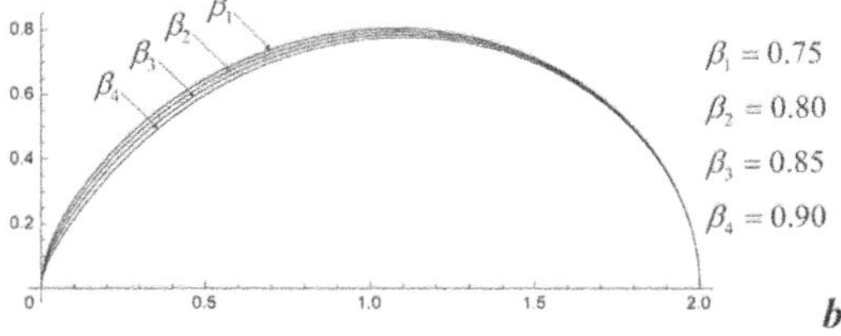

Figure 1. Parabolic two-focus curve (a), the range of β parameter values typical for the longitudinal section of bird eggs (b).

The ratio from 0.75 to 0.80 corresponds to the shapes of chicken eggs, and from 0.81 to 0.84 to the shapes of quail eggs (Figure 2) (Lebedev 1983). A surface formed by rotation of the curve (1) around its big axis can serve as the basis for computer design and the creation of physical models made on numerically controlled machines.

DOI: 10.1201/9781003348450-123

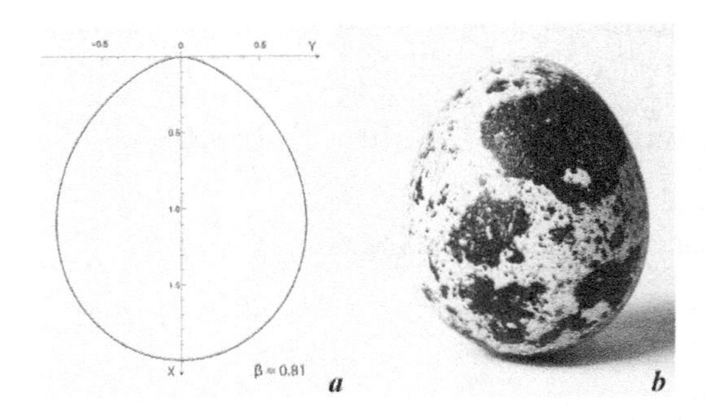

Figure 2. Parabolic bifocal curve with parameter $\beta = 0.81$ (a), quail egg (b).

3 LOAD-BEARING PROPERTIES OF NATURAL SHELLS

Further bionic studies demonstrated that the living nature shells are not homogeneous in their inner structure, but consist of several layers that perform different physiological and structural functions. An eggshell has a three-part structure, in which the cuticle is an external protective layer for the main inner woven support layer that is relatively soft in terms of mechanical properties. The inner woven shells perform mechanical functions, working in tension and, at the same time, serve as functional biological membranes (Voznesenskiy and Khanuhov 1970).

4 ASYMMETRIC COMPLEX SHELLS WITH ELASTIC FRAME

The research of the load-bearing properties of asymmetric natural shells served as the basis for experiments with complex structures composed of elastic frame and rigid coverings. In these experiments, adaptive frame structures made in the form of complicated cyclic knots from elastic and closed rods simulated the bearing principle of the inner woven structures of eggshells (Kozlov 2018).

These structures have great adaptivity of their forms. The same structure made on a plane can take different spatial forms including egg-shaped multifocal shells of different curvatures. Any of these shapes by rigidly fixing with additional fastening elements turn into a geometrically unchangeable structure.

In complex structures, a rigid outer covering, similar to the structure of an arthropod shell, attaches to an elastic inner shell. The resulted structure is a multi-layer one, and its outer shell is more rigid than the inner load-bearing core. In this case, the supporting structure, made in the form of a multifocal surface of revolution from elastic continuous closed rod, acts as a reinforcing structure for the complex shell structure.

Freeform rigid outer plates attach to the elastic inner structure to stabilize its shape and transfer forces to it (Figure 3). Such an integrated design makes it possible to make the rigid outer shell as thin as possible like in similar structures of living nature.

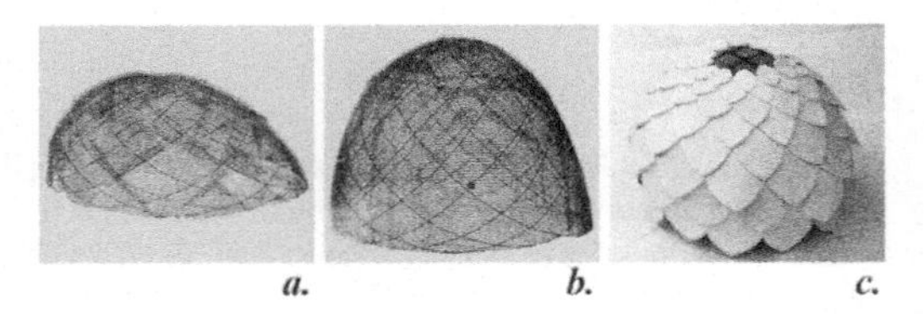

a. *b.* *c.*

Figure 3. Elastic complex structures of egg-like shells with rigid coatings: horizontal asymmetric shell with net coating (a), asymmetric vertical shell with net coating (b), shell with elastic frame and rigid lamellar coating (c).

5 CONCLUSION

In the classical structural engineering, structures with a rigid frame are mainly considered. This provides the building with stability and durability and allows the direct transfer of loads from the frame elements to the foundation. Elastic complex structures offer a completely different principle. They consist of an elastic frame with the ability to change geometric parameters and a rigid coating that adapts to the frame. This approach would make it possible to reduce the cost of building structures due to the adaptive capabilities of the frame and the relatively small thickness of the coating.

REFERENCES

Brandt G.V. 1979. *Geometry of Multifocal Surfaces in Application to Architectural and Constructional Bionics*, Moscow: USSR Acad. Sci. (in Russian).

Kozlov D. 2018. Bionic type structures based upon resilient cyclic knots. In Adriaenssens S., Mueller C. (eds.), *IASS 2018. Creativity in Structural Design*, International Association for Shell and Spatial Structures (IASS).

Lebedev Yu. S. et al. 1990. *Architectural Bionics*, Moscow: Stroyizdat (in Russian).

Lebedev, Yu. S. 1983. Architectural Bionics. In F. Otto, H. Pollig & B. Burkhardt (eds), *Information of the Institute for Lightweight Structures, IL-32*: 78–106. Stuttgart: University of Stuttgart.

Voznesenskiy S., Khanuhov H. 1970. New type of shells. In L.I. Kirillova (ed.), *Architectural Composition. Contemporary Problems*: 147–149. Moscow: Stroyizdat (in Russian).

Current Perspectives and New Directions in Mechanics, Modelling and Design of Structural Systems – Zingoni (ed.)
© 2022 Copyright the Author(s), ISBN 978-1-032-39114-4

Analysis of constructional errors of in-service spatial prestressed steel structures

A.L. Zhang
Beijing University of Civil Engineering and Architecture, Beijing Advanced Innovation Center for Future Urban Design, Beijing Engineering Research Center of High-rise and Long-span Pre-stressed Steel Structures, Beijing, China

J. Wang, X. Zhao & Y.X. Zhang
Beijing University of Civil Engineering and Architecture, Beijing Advanced Innovation Center for Future Urban Design, Beijing, China

ABSTRACT: The project conducts sensitivity study on constructional errors including deviations of joints, prestress loss of prestressed cables, and deformation of struts. The research data are carefully extracted from a high-fidelity laser measurement point-cloud model of a typical spatial prestressed steel structure. Data light-weight process is conducted to the point-cloud model where key nodes are automatically marked and stored according to the design BIM model. Constructional errors are efficiently characterized and compared with the design model through adapted pattern recognition algorithm. The statistical analysis is conducted to the constructional errors where the statistical models are obtained for future reliability study.

1 INTRODUCTION

The long-span prestressed steel structure system is widely used in the world because of its own advantages. Due to the existence of prestress and construction factors, the spatial position of cable-strut joints often deviates greatly from the design, which will have a certain impact on the overall structural stability (Yujie et al. 2019). For this reason, it is extremely important to measure construction deviations with high accuracy. Using 3D laser scanning technology to monitor the construction process of long-span steel structures and using the point cloud data 3D model to realize the real visualization of the steel structure (Dongling et al. 2014), but the relationship between the measured deviation data and the intelligent evaluation of structural performance is missing.

In this paper, the point cloud model of the suspend-dome structure is obtained by 3D laser scanning. The actual position coordinates of the cable-strut joints are obtained, and the joint position deviation is obtained after comparing with the BIM model. Statistical data were analyzed for structural bias sensitivity analysis.

2 RESEARCH OBJECTS AND TECHNICAL SOLUTIONS

The laser scanning data in this paper comes from the badminton stadium of Beijing 2008 Olympic Games. The Beijing Olympic Badminton Gymnasium occupy 24383 m^2 in overall structural area (Figure 1). The geometric point cloud of the suspend-dome structure was collected by the mid-to-long-range laser scanner RIGEL. Iterative Closest Point (ICP) was applied to accurately reconstruct the suspend-dome structure model (Figure 2). The registered point cloud error is within 5mm. Then it is compared with the BIM model to get construction errors of joints.

Figure 1. The badminton arena for 2008 Olympic Games.

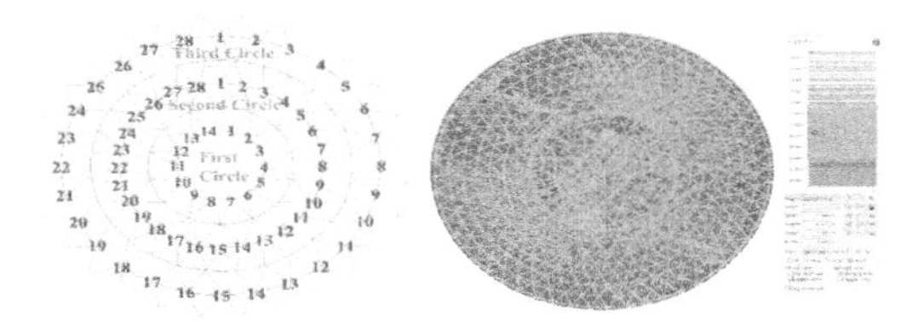

Figure 2. Number of cable-strut joints and point cloud model.

DOI: 10.1201/9781003348450-124

3 RESULTS

We obtained the deviation data of each loop cable-strut joint in the x, y, z direction (Figure 3), and calculated the relative spatial position deviation. We can find that the maximum deviation in the x-direction occurred at the second loop cable-strut joint, which was 0.087m. While the maximum y deviation was 0.081 m at the second loop joint, and the maximum z deviation was 0.061m at the third loop joint.

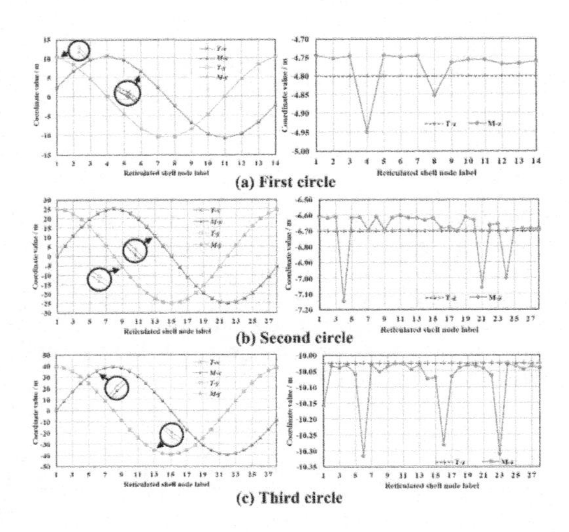

Figure 3. Comparison of coordinate design value and measured value of cable-strut joints.

According to the Standard for Acceptance of Construction Quality of Steel Structures, maximum longitudinal deviations are required to be within ± l/ 2000, and ± 40 mm. The deflection of joints should be within 1.15 from the designed values. Therefore, it is determined whether the standard requirements are met by calculating the root mean square of the spatial deviation of the cable-strut joints in all directions. The maximum deviations in the x and y directions are 38.7mm and 39.8mm within the standards. The maximum deflection d of the structure under constant load is 80mm, and $1.15d$ is 92mm as the deflection limit in the z direction. The maximum deviation in the z direction is calculated to be 46.4mm, which also satisfies the standard.

The circumferential deviation data were analyzed in the frequency domain through Fourier transform. The power spectrum densities of 3-loop cable-strut joints' deviations demonstrate 1st eigen-mode energy (Figure 4). Through the power spectrum, the mathematical expression of the deviation of each loop cable-strut joint is obtained, such as Equation (1)–(3).

$$f(x) = \frac{-801.1x^2 + 3025x - 5810}{24.67x^2 - 117x + 239.7} \quad (1)$$

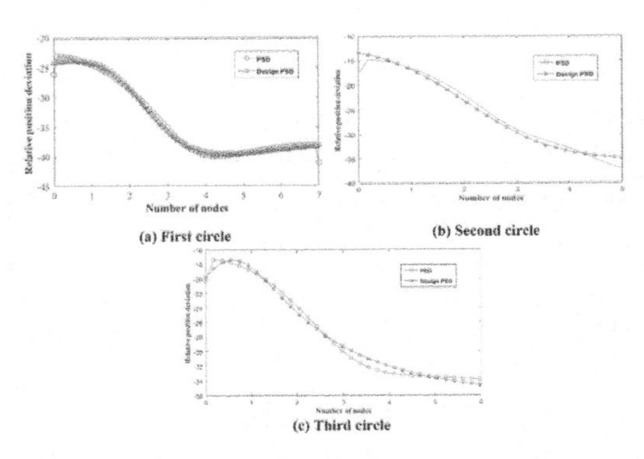

Figure 4. Comparison among measured PSD and Design PSD.

$$f(x) = \frac{-418.9x^2 + 629.4x - 3025}{17.03x^2 - 72.51x + 228.5} \quad (2)$$

$$f(x) = \frac{709.9x^2 - 661.1x + 1083}{-18.66x^2 + 11.95x - 54.79} \quad (3)$$

4 CONCLUSIONS

In this research, the construction deviations of key cable-strut joints of in-service long-span prestressed steel structures were obtained by comparing the point cloud model with the BIM model. Among them, the application of the algorithm makes the deviation calculation faster. Maximum deviations are 38.7 mm, 39.8 mm, and 46.4 mm with respect to x, y, and z directions. The deviation data is Fourier transformed to the frequency domain to obtain the power spectral density and the mathematical expression of the fitting. The deviation data will be used for finite element model and deviation sensitivity analysis. The structural reliability for the spatial prestressed-steel structure can be thoroughly analyzed in the future, and the structural stability of this type of in-service structure can be well evaluated. In turn, it provides a pivotal technical scheme with high precision and high efficiency for in-situ health monitoring of structures.

REFERENCES

Dongling, MA., Jian, CUI. & Shiyan, WANG. 2014. Application and Research of 3D Laser Scanning Technology in Steel Structure Installation and Deformation Monitoring. *Applied mechanics and materials* 580-583: 2838–2841.

Yujie, YU., Zhihua, CHEN. & Renzhang, YAN. 2019. Finite element modeling of cable sliding and its effect on dynamic response of cable-supported truss. *Frontiers of structural and civil engineering* 13(5): 1227–1242.

Current Perspectives and New Directions in Mechanics, Modelling and Design of Structural Systems – Zingoni (ed.)
© 2022 Copyright the Author(s), ISBN 978-1-032-39114-4

Examples of application of principle of superposition and rules of symmetry in engineering design

J. Rębielak
Professor Emeritus of the Tadeusz Kosciuszko Cracow University of Technology, Cracow, Poland
Chairman of the Committee for Architecture and Town Planning of Wroclaw Branch of the Polish Academy of Sciences, Wroclaw, Poland

ABSTRACT: Architectural and engineering spatial structures are most often built of modular repetitive elements. Therefore considering the principles of symmetry is very helpful in design processes of them. The paper presents examples of shaping of the innovative structural systems of buildings and their foundation structures together with the tension-strut structures of various geometric shapes as well as the structures of special types of aircrafts. The paper discusses the basic assumptions adopted in the design processes of system of the composite foundation and an example of its application in the design of a tall building. There are spoken selected aspects of shaping of some lightweight structures mostly applied in the roof covers. Moreover there are presented some proposals of design of the structural systems of e.g. a vertical take-off and landing aircraft.

1 INTRODUCTION

Architectural objects and engineering structures subjected to high-value loading forces, which forces can be applied at any angle to such structures, are quite often symmetrical in the shape. The required analyses of them can be performed in several steps during which the application of the superposition principle and the rules of symmetry are very useful and effective in engineering and in architecture (Nagy 2009).

2 TALL BUILDING STRUCTURES

In designing of the tall building a particularly important issue is the stability of such an object some-times situated on a ground with very low bearing capacity and/or in the seismically active area. The support structure of the tall building has to have at the same time two contradictory features. It has to be very rigid but on the other hand it should be to some degree flexible (Kowalczyk et al. 1993). The general rules of symmetry as well as the principle of superposition have been applied in the shaping process of system of the composite foundation. The composite foundation system was applied in the design of a building complex called the GeoDome SkyTowers, Figure 1 (Rębielak 2012). The four towers, each approximately 380 m high, are based on a composite foundation system in the shape of a square with dimensions of approximately 250 m x 250 m, which is placed directly on the ground level.

Figure 1. Perspective view of the GeoDome Sky Towers complex.

3 LIGHTWEIGHT SPATIAL STRUCTURES

Systems composed only of tensile elements are the lightest supporting structures for roofs, especially those with large spans. These include certain types of pneumatic structures and membrane structures, which, however, cannot be used for most forms of the roof covers. Tension-strut systems composed of a relatively small number of compression members and a considerably larger number of tensile members have a wider range of practical applications.

A certain group of the tension-strut structures is shaped on basis of the triangular-hexagonal grid of

DOI: 10.1201/9781003348450-125

members. The grid is centrally located in the space of a given structure while other structural components are arranged around this grid. In space of structure called the I(TH) tension-strut system the additional single members are perpendicularly directed to the surface of the triangular-hexagonal grid of members and they are connected to chosen nodes of this grid, Figure 2.

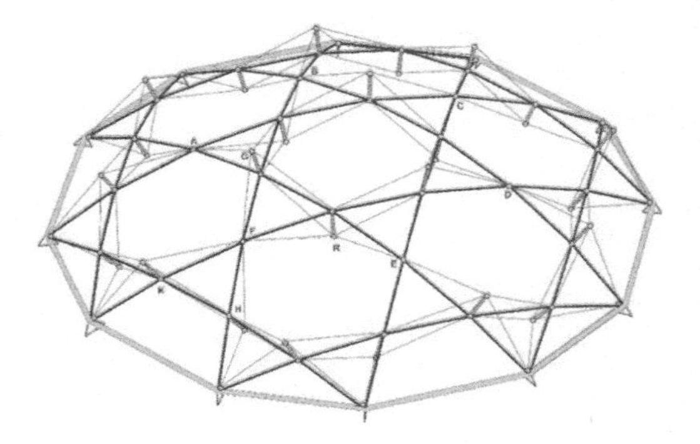

Figure 2. A spherical form of the I(TH) tension-strut structure.

4 PROPOSALS OF AIRCRAFT STRUCTURES

The basic tasks of an aircraft's structural system are to transmit and resist safely all types of the applied loads. The airplane structure has to be lightweight and to provide an aerodynamic shape and to protect passengers and technical systems, etc. from the environ-mental conditions during the flight especially on the very high altitude and with the very big cruse speed. (Megson 2007).The main purpose of the first proposed structural system was striving after getting an lightweight airplane having a large range of the flight speed, in particular of high cruising speed, by an airplane equipped with propulsion of relatively low power, Figure 3.

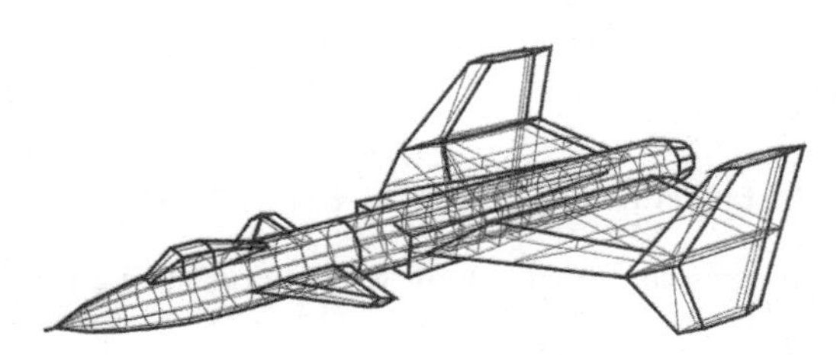

Figure 3. Take-off configurations of an airplane with a rotating fuselage.

The nose part of the fuselage, containing the cockpit, is an integral part of the fuselage structure and is connected with it so that it can rotate along the longitudinal axis of the entire fuselage. Only the nose part of the fuselage is rotating in this concept of structural system of the aircraft and in case of emergency it can be suitably and very fast separated from the main body of the aircraft.

Recently, new types of turbojet and jet-propelled airplanes have appeared with the characteristics of a short take-off and landing aircraft and also the possibility of vertical take-off and landing. The design of a vertical take-off and landing aircraft with a dorsal stringer (Rębielak 2021) enables the processes of safe and precise vertical take-off and landing as well as performance of horizontal task flights at various heights and with a relatively broad range of the cruising speeds. During take-off and landing the axes of the engines are directed vertically, see Figure 4, and after reaching the appropriate height and the appropriate rotation of the engines by 90 degrees, the plane continues the flight in the horizontal configuration.

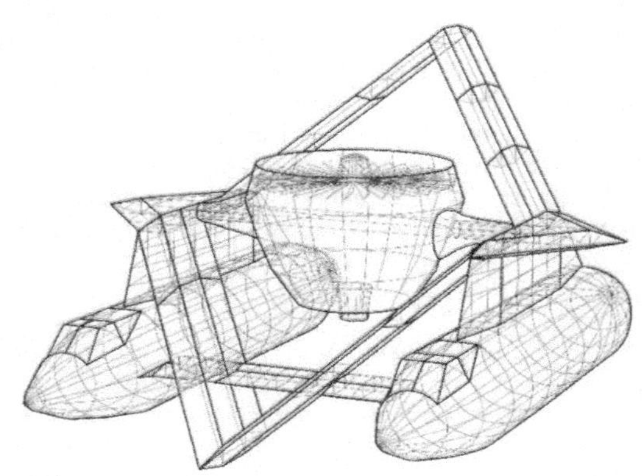

Figure 4. Take-off configurations of an aircraft with a dorsal stringer.

5 CONCLUSIONS

The presented examples of shaping spatial structural systems prove the great usefulness of applications the principle of superposition and the rules of symmetry in designing effective forms of such systems. In addition this principle and these rules can be used in the processes of comprehensive static and dynamic analyses of various types of structures carried out in order to obtain optimal design solutions.

REFERENCES

Megson, T. H. G. 2007. *Aircraft structures for engineering students*, Amsterdam, Boston, Heidelberg, London, New York, Oxford Paris, San Diego, San Francisco, Singapore, Sydney, Tokyo: Elsevier.

Nagy, D. 2009. Symmetry in architecture: saving the Greek symmetria and helping the birth of the modern concept. *The Journal of the International Society for the Interdisciplinary Study of Symmetry*, Nos 1-4: 184–187.

Rębielak, J. 2012. System of combined foundation for tall buildings, *Journal of Civil Engineering and Architecture*, Vol. 6 (12): 1627–1634.

Rębielak, J. 2021. *Vertical take-off and landing airplane*. Patent Application No 221971. Warsaw: Patent Office of the Republic of Poland.

*Current Perspectives and New Directions in Mechanics,
Modelling and Design of Structural Systems – Zingoni (ed.)*
© 2022 Copyright the Author(s), *ISBN 978-1-032-39114-4*

Full-scale tests of a lightweight footbridge: The Folke Bernadotte Bridge

D. Colmenares
Royal Institute of Technology, Stockholm, Sweden

G. del Pozo
Norwegian University of Science and Technology, Trondheim, Norway

G. Costa
Politecnico Di Milano, Milan, Italy

R. Karoumi & A. Andersson
Royal Institute of Technology, Stockholm, Sweden

ABSTRACT: This work presents the results from an experimental test on a 97 m long lightweight footbridge in Stockholm, Sweden. The modal properties were estimated based on impact hammer testing. The experimental results show that the lightweight steel footbridge presents damping ratios in accordance with the Sétra Technical Guide for footbridges. Furthermore, the work presents a 3D Finite Element (FE) model of the footbridge. The bridge model is first tested by performing a sensitivity analysis of the boundary conditions with respect to the modal properties of the system. Next, the model is adjusted using a frequency-based modal updating technique with a Genetic Algorithm (GA) to solve the optimization problem. Finally, the model results are compared with the experimental test, and the vicinity of the obtained solution is evaluated. The FE model shows a good agreement with respect to the experimental results in terms of bending natural frequencies and corresponding mode shapes.

1 INTRODUCTION

The dynamic characterization of slender and lightly damped footbridges can be a difficult challenge due to the uncertainties associated to these systems, as shown in Hallak Neilson, J (2020). This is particularly true when determining the inherent damping of such structures. One way to overcome this problem is to use a modal updating technique to minimize the errors caused by measurements and the assumptions commonly used in design. The aim of this work is to characterize the dynamic behavior of the Folke Bernadotte Bridge. The investigation is focused on the first three bending modes of vibration of the footbridge which are in the frequency range between 1 Hz and 6 Hz. Firstly, a Finite Element (FE) model of the structure is built, and the nature of its boundary conditions is studied to maximize the consistency between the model and the measurements. Secondly, field measurements are performed to characterize the footbridge by performing hammer tests. The frequency response functions (FRFs) are obtained, and modal properties of the footbridge are estimated focusing on the bending behavior of the structure. Finally, a modal updating technique is applied to the FE-model to identify the modal properties of the structure.

2 CASE STUDY

The Folke Bernadotte Bridge is a footbridge across Djurgårdsbrunn Bay in Stockholm, Sweden. A picture of the footbridge is shown in Figure 1.

Figure 1. Folke Bernadotte Bridge.

The footbridge has a span of 97 meters forming a single shallow arch bridge made of stainless steel. The system conforms a 3D truss like structure in which all elements are welded together. The estimated total mass of the bridge is 97.3 tons.

DOI: 10.1201/9781003348450-126

2.1 *FE model*

The FE-model of the Folke Bernadotte Bridge was built in ABAQUS (Abaqus/CAE Copyright 2002-2020 Dassault Systèmes Simulia Corp). The model was constructed using cubic Euler-Bernoulli B33 beam elements. The masses of the timber deck and the railings are considered by adjusting the linear mass of the steel from 7850 kg/m^3 to 8800 kg/m^3. The lateral view of the FE-model is shown in Figure 2.

Figure 2. FE-model lateral view. Units are in mm.

2.2 *Field measurements*

The field measurements were performed on March 23rd 2021. The temperature on the testing day was 19 °C with moderate wind conditions. The measurement plan is shown in Figure 3. A total of 11 hits were made at each excitation point. The sampling frequency for the experiment was 1200 Hz and the record time for each hit was 50 s obtaining a frequency resolution of 0.02 Hz. Using the H_1 estimator and averaging, the FRFs were obtained and evaluated through the coherence function. The exponential window was applied to improve the SNR with an end factor of 0.01 %.

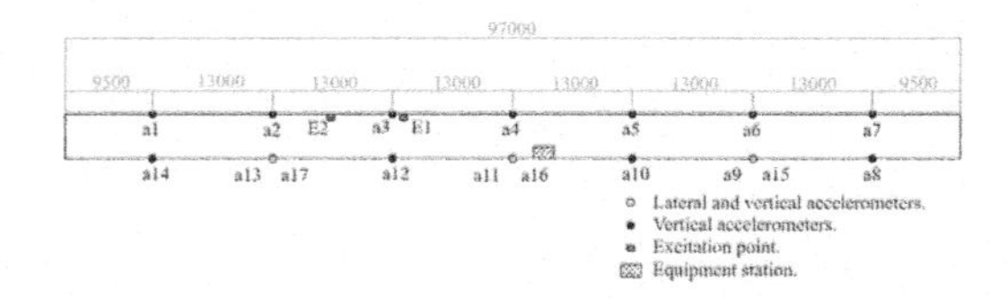

Figure 3. Measurement plan. Units are in mm.

3 MODAL UPDATING

In order to calibrate the FE-model a FRF-based modal updating procedure has been used to find the model parameters that best fit with respect to the measurements made. A detail description of the procedure can be found in Grafe (1999). The updating procedure is made by solving a non-linear optimization problem using the Frequency Response Assurance Criterion (FRAC) and the Frequency Amplitude Assurance Criterion (FAAC). The modal updating procedure is treated as a non-linear optimization problem solved using GA. Elastic springs in the longitudinal direction of the footbridge at the boundary conditions, elastic modulus of the stainless steel, damping ratios and the density of the model are taken as the decision variables of the modal updating procedure. Suitable lower and upper bounds for variable are assumed.

4 RESULTS

The comparison between the theoretical FRFs (black line) of the best solution with respect to the experimental FRFs (gray line) is shown in Figure 4 corresponding to the pairs of accelerometers 6-9. It can be seen that there is a small frequency shift between the FE-model FRFs and the experimental FRFs around the third mode of vibration. However, the general trend and the peak values of the experimental FRFs were well captured by the calibrated FE-model. The differences between the model and the measurements may be due to either the disturbances in the measurements, the uncertainties associated to the boundary conditions as well as damping mechanisms and that in regards to the railings, only its corresponding mass was taken into account in the model.

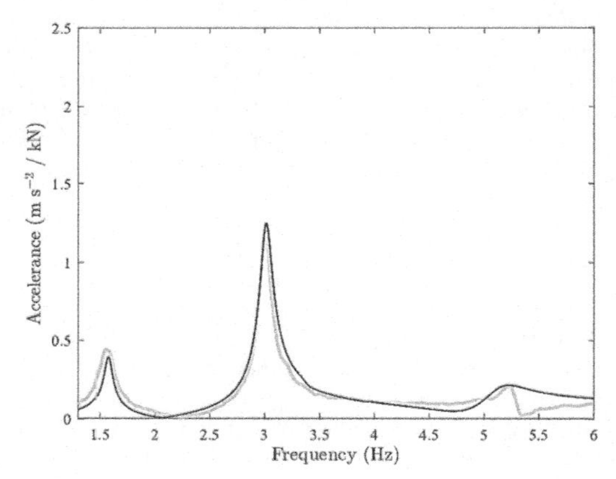

Figure 4. Theoretical (black line) vs. experimental (gray line) FRFs at accelerometers 6-9.

5 CONCLUSIONS

In this work, the results of a successful modal updating procedure applied to the Folke Bernadotte Bridge have been presented resulting in a calibrated FE-model. The work presented a slender and lightly damped footbridge in which the difficulties of the inherent uncertainties of the system were overcome.

REFERENCES

Grafe, H. 1999. Model updating of large structural dynamics models using measured response functions (Doctoral dissertation, University of London).

Current Perspectives and New Directions in Mechanics, Modelling and Design of Structural Systems – Zingoni (ed.)
© 2022 Copyright the Author(s), ISBN 978-1-032-39114-4

Numerical analysis of new assembled cable dome ring truss connections

A.L. Zhang
Beijing University of Civil Engineering and Architecture, Beijing Advanced Innovation Center for Future Urban Design, Beijing Engineering Research Center of High-rise and Long-span Pre-stressed Steel Structures, Beijing, China

M. Zou, Y.X. Zhang, G.H. Shangguan & J. Wang
Beijing University of Civil Engineering and Architecture, Beijing Advanced Innovation Center for Future Urban Design, Beijing, China

ABSTRACT: The ring trusses of the existing cable dome projects are welded. A new type of prefabricated ring truss connection is proposed, which can realize that all the ring trusses can be assembled with high-strength bolts on the construction site, and provides a promising solution for the rapid and environmental protection construction of the ring truss. The joint can realize the modular assembly of ring truss. The purpose of this study is to investigate the flexural properties of the new joints. The finite element model of the new joint is established by ABAQUS and the initial stiffness, ultimate bending moment and failure mode of the new joint were analyzed in detail by considering the, welded cruciform joints thickness and high-strength bolt number. The results show that the new joint has good mechanical properties, and the influence mechanism of various parameters on the mechanical properties of the new joint are obtained.

1 INTRODUCTION

Long-span cable dome structure is widely used in buildings such as stadiums, and the construction method of ring truss of cable dome is welding, which exists some problems such as slow construction speed and environmental pollution. The full-bolt assembled ring truss can improve the construction speed, reduce energy consumption and environmental pollution in the construction process, which is of great significance to the development of industrialized buildings and carbon neutralization peaks. For spatial truss joints, many scholars have done a lot of research. Zhang (Ai-lin Z et al 2019, Ai-lin Z et al 2020) conducted experiments and finite element analysis on welded joints with different parameters, and obtained the design value of bearing capacity and the increase coefficient η of ultimate bearing capacity of DKT-shaped joints with internal stiffeners, as well as the design value of bearing capacity and the calculation formula of ultimate bearing capacity of welded joints with internal stiffeners. However, there is little research about the assembled ring truss of cable dome. So, an assembly scheme of ring truss is proposed in this paper, and the joints suitable for assembled ring truss are mainly studied. The finite element model of the new joint is established by ABAQUS, which is compared with the welded joint. At the same time, the thickness of welded cruciform joints and the number of high-strength bolts are considered. The initial stiffness, ultimate bending moment and failure mode of the novel connection joint are analyzed in detail, and the influence of various parameters on the mechanical properties of the new joint is obtained, which verifies the reliability of the joint.

2 RING TRUSS MODULAR MODE AND JOINT CONSTRUCTION

A gymnasium is taken as the prototype structure. And the ring truss is divided into modules of the same size. The chord and web members are cut in the middle at the block and assembled with new joints. The details of the connection joints are shown in Figures 1.

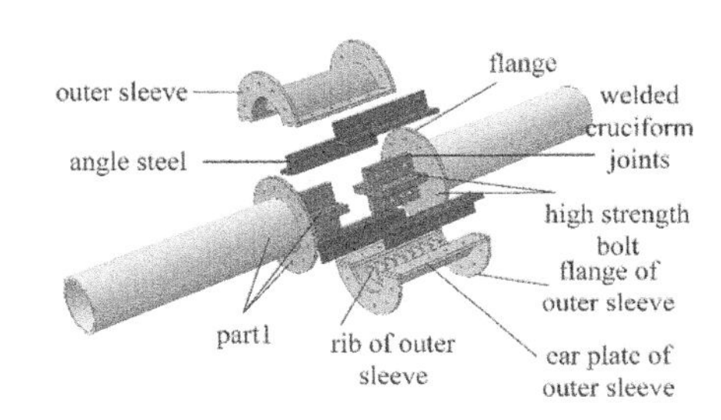

Figure 1. Details of the new assembled cable dome ring truss connections.

DOI: 10.1201/9781003348450-127

3 FINITE ELEMENT MODEL

ABAQUS finite element software is used to establish the splicing joint model, and the model components are all made of solid element C3D8R8 joint hexahedron linear reduced integration element. Material nonlinearity, geometric nonlinearity and contact nonlinearity are fully considered during the analysis. The mesh density is adjusted according to the component size and analysis requirements. The steel members' property is Q355B in the model. In addition, the Von Mises yield criterion and the follow-up strengthening criterion are adopted. The elastic part of the material is defined by the elastic modulus and Poisson's ratio. According to the average value of the material property test results, the elastic modulus is taken as E=2.06×105MPa, One side of the finite element model is fixed and 6 degrees of freedom are restrained; on the other side, the axial force and horizontal displacement are applied according to the axial compression ratio of 0.2, the displacement angle of the member, and the loading history is: 0.00375rad, 0.005 rad, 0.0075 rad, 0.01 rad, 0.02 rad, 0.03 rad, 0.04 rad, 0.05 rad, 0.06 rad, 0.07 rad.

4 COMPARATIVE ANALYSIS OF NEW JOINTS AND WELDED JOINTS

The bearing capacity and stiffness of the new joint (NJ) and welded specimen (WJ). By comparison, it is found that the bearing capacity and stiffness of the new joints are close to those of the welded joints. The yield load of the welded joint is 1.6% higher than that of the new joint, the ultimate load is 5.3% higher, and the initial stiffness is 0.8% higher, and the difference is within 6%. The new joint can be approximately considered as a rigid joint.

5 ANALYSIS OF THE INFLUENCE OF THE THICKNESS OF THE WELDED CRUCIFORM JOINTS

The thickness of the welded cruciform joints increases the bearing capacity and stiffness of the joint, the yield load of WCJ50 is increased by 7.4%, the ultimate load is increased by 4.3%, and the initial stiffness is increased by 3.8% compared with WCJ40. The yield load of WCJ60 is 1% less than that of WCJ50, the ultimate load is increased by 5.8%, and the initial stiffness is decreased by 0.4%; from the comparison results, when the thickness of the welded cruciform joints is less than 50mm, increasing the thickness can improve the bearing capacity and stiffness. When the steel is larger than 50mm, increasing the thickness of the welded cruciform joints has little effect on the bearing capacity and stiffness.

6 ANALYSIS OF THE INFLUENCE OF THE THICKNESS OF THE WELDED CRUCIFORM JOINTS

Comparing specimens with different numbers of high-strength bolts, the number of bolts increases the bearing capacity and stiffness of the joints. The increase of flange bolts will increases the yield load by 0.6%, the ultimate load by 2%, and the initial stiffness by 0.8%; the increase of internal bolts and external bolts makes the yield load is increased by 1.6%, the ultimate load is increased by 6.3%, and the initial stiffness is increased by 1.3%. With the increase of high-strength bolts, the bearing capacity and stiffness of the new joint also increase, but the increase phenomenon is not obvious, indicating that the effect of welded cruciform joints is greater than that of bolts effect.

7 CONCLUSION

By comparing the established finite element model with the welded joints, and considering the variable parameters of the welded cruciform joints thickness, the number of bolts and other factors, the finite element calculation is carried out, and the following conclusions are drawn:

(1) The stiffness and bearing capacity of the new joint and the welded joint are not much different, the new joint has strong energy dissipation capacity, and the new joint can realize rigid connection.
(2) Increasing the thickness of the welded cruciform joints and increasing the flange bolts can improve the stiffness and bearing capacity of the new joint.
(3) The connection between the welded cruciform joints and the flange is easy to be damaged. When the welded cruciform joints and the flange connection are damaged, the bending moment shared by the flange bolts will increase, resulting in the failure of the bolts. Adding bolts has limited effect on slowing down the damage. So an effective method is to increase the thickness of the welded cruciform joints. At the same time, it also shows that the internal welded cruciform joints is the main stress part of the component in the case of bending, and it should be actively strengthened in the design of the frame.

REFERENCES

Ailin Z, Dinan S, Yanxia Z. Research on mechanical performance of complex tubular joints of Beijing New Airport Terminal C type column[J]. Journal of Building Structures, 2019, 40(3): 210–220.

Ai-lin Z, Wen-chao C A I, Yan-xia Z, et al. Bearing capacity formulas of multi-planar DKT-joint containing internal stiffeners under axial compression[J]. Engineering Mechanics, 2020, 37(9): 50–62.

Current Perspectives and New Directions in Mechanics, Modelling and Design of Structural Systems – Zingoni (ed.)
© 2022 Copyright the Author(s), *ISBN 978-1-032-39114-4*

A numerical displacement-based approach for the structural analysis of cable nets

B. Frigo & A.P. Fantilli
Politecnico di Torino, Torino, Italy

ABSTRACT: A new displacement-based approach is proposed herein to predict the behaviour of pre-tensioned cable nets subjected to vertical loads. The cables are contained in horizontal plane and have a single degree of freedom in each node, where loads are applied. The model is based on the equilibrium equations of an infinitesimal cable-element, which are solved by considering the catenary equation under the hypothesis of (a) zero bending stiffness, (b) linear elastic behaviour of materials, (c) small deformation, and (d) the existence of perfect hinges in each node. More precisely, a finite-difference numerical procedure is introduced in order to evaluate nodal displacements related to a set of applied vertical forces. The effectiveness of the proposed approach is then assessed by comparing the numerical results with those obtained by other models found in the current literature. Finally, the proposed approach is used to design new and more efficient insulating panels for the green house technology.

1 INTRODUCTION

A numerical model able to analyze the non-linear static response of pre-tensioned cable nets, to be used in the feasibility study of an ultra-insulating panel with vacuum technique (Figure 1), is herein introduced. For this purpose, the model should have the following characteristics:

- easy to use.
- requires few input data.
- reliable.

In other words, a tool for the conceptual design of the panel has to be introduced with the aim of defining the main structural part: the net.

Unfortunately, in the current literature these models cannot be easily found. Indeed, the numerical model proposed by Lewis (1989) concerns the analysis of pre-stressed nets and pit-jointed frame structures. It was based on the principle of minimum energy and was applied to steel trusses using the linear load - displacement constitutive relationship. To simplify the approach, a new numerical procedure, to be used for the analysis of pre-tensioned 2D nets, and based on the displacement method, is herein presented.

The proposed model is then applied to cable nets, loaded in parallel planes, having a single degree of freedom in each node. As results, the nodal displacements, and the induced stress of each cable, are calculated when a set of nodal forces are applied.

To validate the proposed approach, the results of a case study are compared with those obtained with the Lewis' model (Lewis 1989).

2 ANALYSIS OF PRE-STRESSED CABLE NETS USING THE DISPLACEMENT-BASED APPROACH

The structural analysis of a generic net structure composed by m cable elements connected in n nodes is based on following assumptions:

1. the bending stiffness of each cable can be neglected.
2. cable shows a linear-elastic behavior: instability, slackening phenomena, and viscous plastic deformations are not considered.
3. only small deformations of the cable are allowed.
4. nodes are assumed to be perfect hinges.
5. cross-section of each cable is constant throughout its length.
6. self-weight and variable loads are assumed to be concentrated load acting on the nodes.

The initial equilibrium configuration (*status 0*) represents the undeformed state of the cable net and the pure prestress. After applying external loads, to define the final state, the equilibrium and compatibility equations are included in the matrix relationship:

DOI: 10.1201/9781003348450-128

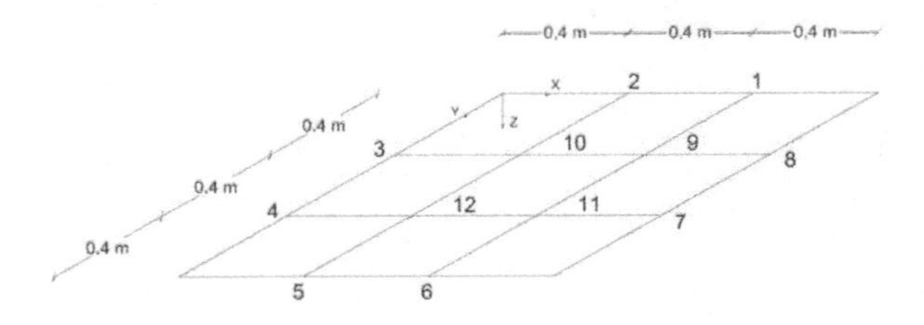

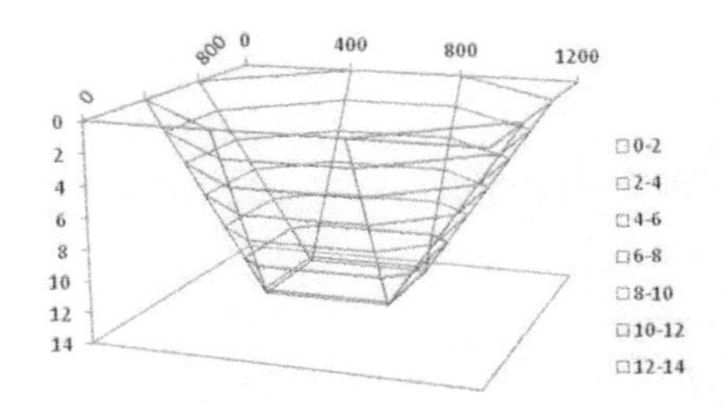

Figure 1. (a) Geometric scheme of the 2 × 2 cable net with total dimension equal to 120 cm × 120 cm and cable length of 400 mm (Example 3 - Lewis 1989) and (b) the vertical displacements (along the z axis) of the nodes.

$$\{P\} = [K] \cdot \{X\} \tag{1}$$

where:

$\{P\}$ = vector ($2n \times 1$) of the load on the net node.
$[K]$ = global stiffness matrix ($2n \times 2n$).
$\{X\}$ = vector ($2n \times 1$) of the displacements in a net node.

It is solved by using the linear load-displacement relationship, which results in a stiffness matrix algorithm.

From this analytical model, a numerical algorithm has been developed to calculate the displacements and stresses induced by deformation of the cable net.

3 VALIDATION OF THE NUMERICAL MODEL

To validate the proposed algorithm, it is applied to an ideal net of simple geometry (i.e., that reported in Figure 1 by Lewis 1989). To give a quantitative rate of the model, we also computed:

To give a quantitative rate of the model, the following parameters are computed:

- percentage of the maximum deflection (at the central nodes of the net) of the cable:
- the deviation of the calculated deflection from that of Lewis (1989):

Table 1. Comparison between the results obtained with the proposed model and those computed by Lewis (1989).

	Deflection (mm)	$\Delta l_{\%}$ (%)	$Dev_{\%}$ (%)
Proposed model	12.78	+3.20	-0.56
Lewis (1989)	12.22	+3.06	+5.75

- percentage of deviation from Lewis' (1989) displacement:

The comparison shows that the deflections given by the numerical model and those computed by Lewis (1989) are coincided (see Table 1).

4 CONCLUSIONS

Based on the results of a simplified nonlinear procedure for the analysis of cable nets, the following conclusion can be drawn:

- the equations of the dynamic equilibrium of a body in motion (the D'Alembert principle), leads to the formulation of the dynamic relaxation algorithm, as obtained by Lewis (1989).
- in the same way, in the proposed model the linear load- displacement relationship can be obtained by means of a stiffness matrix algorithm.
- both the models calculate approximately the same maximum vertical displacement of the nets made with cables.
- Thus, the proposed simplified can be effectively used to design the nets made with steel cables.

Future paper will be devoted to use the proposed model for the feasibility analysis of prototype of a vacuum ultra-high insulating panel.

REFERENCES

Buchholdt, H. A. 1985. *An Introduction to Cable Roof Structures*. Cambridge: Cambridge University Press.

Leonard, J. W.1988. *Tension Structures, Behavior & Analysis*. New York: McGraw - Hill.

Lewis, W. J. 1989. The Efficiency of Numerical Methods for the Analysis of Prestressed Nets and Pin-jointed Frame Structures. *Computers and Structures*, Vol. 33, No 3, pp. 791–800.

Current Perspectives and New Directions in Mechanics, Modelling and Design of Structural Systems – Zingoni (ed.)
© 2022 Copyright the Author(s),
ISBN 978-1-032-39114-4

Analytical and numerical analysis of anchored flexible steel liquid storage tanks subjected to seismic loads considering soil structure interaction

M.S. Pourbehi
Cape Peninsula University of Technology, Cape Town, South Africa

JAvB. Strasheim
Stellenbosch University, Stellenbosch, South Africa

M. Georgiadis
VPE Engineering Services, Durban, South Africa

ABSTRACT: This paper deals with the seismic design of modular steel water storage tanks designed and manufactured with Zincalume© steel panels. The tanks have a novel design and do not have bottom plates. They are fitted with a synthetic liner, are stiffened with vertical wind girts and are anchored to a concrete ring beam at the bottom of each wind girt using holding down bolts. Due to the high radius to wall thickness ratio (R/t_w) of these tanks, the structural performance, interaction with the stored fluid and the foundation soil under dynamic loads, have not been well understood. In the research reported, the effect of seismic loads on the global structural behaviour of these tanks such as base shear, overturning moment, circumferential stress as well as buckling behaviour of the thin metal panels is studied using both analytical and numerical Finite Element Method (FEM) analyses. Parametric studies using spring-mounted mass analogy techniques specified by the American Petroleum Institute code (API650) and the New Zealand Society for Earthquake Engineering recommendation (NZSEE) are conducted for three tanks with different fluid height to tank radius ratios (Hw/R).

1 INTRODUCTION

Steel storage tanks are considered a significant part of the industry and are widely used for water supply, industrial chemical storage, and nuclear power plant systems. Due to their importance, the storage tanks should, where required, be designed to withstand severe earthquakes. The seismic response of steel storage tanks which store fluids differs from that of typical buildings due to the hydrodynamic fluid-structure interaction of the tank and stored fluid. The hydrodynamic pressure acts on the tank wall, roof, and foundation causing fluid sloshing motion and dynamic hoop stresses. This action can cause structural damage to tanks during earthquakes such as elephant's foot buckling, diamond shape buckling, damage to the tank roof and failure of the anchor bolts.

The dynamic behaviour of the steel storage tanks including fluid structure interaction has been studied by many researchers. The work of George Housner (Housner G.W, 1955) is fundamental to the development of the quantification of the impulsive and convective components of the hydrodynamic pressure on the fluid container. This original paper presented an analysis of the hydrodynamic pressures developed when a fluid container is subjected to horizontal accelerations. Other follow on papers by Housner Housner (Haroun M.A. & Housner G.W, 1981; Housner G.W, 1963) as well as a paper by Howard Epstein (Howard I. E., 1976) also deal with the analytical solutions of this problem.

2 ANALYTICAL CODE BASED STRUCTURAL MODELS AND ANALYSIS METHODS

The mass-spring tank structural models on which the API 650 and NZSEE approaches are based are shown in Figure 1. The API method models the convective mass mc and the rigid impulsive mass m_i. The NZSEE method models the convective mass mc and the flexible impulsive mass m_f as well as the rigid body mass m_r. Based on this code, a tank with H_w/R (aspect ratio) larger than 1 can be defined as a flexible tank.

As far as local elastic and/or plastic shell buckling is concerned the critical flexural compression stress induced in the wall of pressurised cylindrical tank needs to be determined. It can be seen that for membrane circumferential stress parameter, pR/t (p - the pressure at the bottom of the tank) less than 100 MPa for a tank with aspect ratio R/t larger than 2000, elastic diamond buckling (elastic plastic collapse) governs as a failure mode. When internal pressure parameter is more than 100 MPa, the stress in the wall is dominated by elephant's foot buckling (Teng & Rotter, 2004). Equation (1) was developed by Rotter to compute the elastic-plastic buckling capacity f_{pb}.

DOI: 10.1201/9781003348450-129

Table 3. Comparison between, NZSEE, API650 and FEM for different flexible tanks.

	NZSEE			API650			FEM		
Variable	T1	T2	T3	T1	T2	T3	T1	T2	T3
Hydrodynamic pressure (kPa)	37	15.8	12.5	21.6	9.1	5.9	39	18.5	13.2
Base shear (MN)	4.96	0.34	0.08	4.1	0.27	0.053	4.56	0.356	0.11
Over-turning moment (MN.m)	19.1	0.606	0.117	15.46	0.48	0.069	19	0.67	0.15
Maximum hoop stress (MPa)	345	307	106	259	244.3	80.7	300	318	116
Tension force in anchors (kN)	111	28.4	18	94	23.2	16.1	103	32.2	19.5

$$f_{pb} = \sigma_{cl} \cdot \left[1 - \left(\frac{p.R}{t_w.f_y}\right)^2\right] \cdot \left(1 - \frac{1}{1.12 + r^{1.5}}\right) \cdot \left(\frac{r + f_y/250}{r + 1}\right) \quad (1)$$

3 FEM MODELLING OF TANKS WITH THREE HW/R RATIOS

3.1 *Tank layout, geometry and material parameters*

Three steel storage tanks with aspect ratios H_w/R ranging from 0.87, 1.07 and 1.77 was considered for this research. Only one soil type is considered in the analytical code-based method. The usual material parameters for steel, water, concrete and soil was used in the studies and analyses.

3.2 *Finite element models of tanks*

The tank geometry and elements for both panels and water are shown in Figure 2. The mesh size for shell elements is 0.8 x 0.8 m and the Abaqus element S4R types are used in this model. Acoustic elements AC8DR with 8 nodes and one degree of freedom of pressure is utilised in modelling of water behaviour due to seismic excitation. The size of acoustic elements in this model is 0.5 x 0.5 m.

3.3 *Dynamic seismic loading and analysis procedure*

The Endurance Time (ET) technique, developed by Estekanchi (Estekanchi & Alembagheri, 2012), is used to overcome the difficulties with regard to nonlinear time history dynamic analysis in typical engineering problems, such as water storage tanks. To be compatible with the peak ground acceleration (PGA) used in the analytical analysis of 0.2 g, only the part of the signal up to 3 s is used. Figure 3 shows the comparison between design elastic response spectrum for ground type 3 (SANS code) [(SANS10160-4, 2011) used in the analytical analysis (dashed line) and the spectrum derived from the ET signal (solid line)

4 ANALYSIS RESULTS AND DISCUSSION

Good comparison between the two analytical techniques and the FEM method were found for the Hydrodynamic pressure, base shear, over-turning moments, maximum hoop stress and tension force in the wind girt anchors. Details can be found in the full paper. Table 3 shows the comparative values obtained for the, NZSEE, API650 and FEM analyses for the set of flexible tanks considered.

4.1 *Discussion of the FEM modelling procedure and results*

The nonlinear geometrical behaviour of structure is considered in the dynamic FEM analysis. The analysis steps include static analysis for gravity and hydrostatic loading which provides a pre-stress condition before the dynamic analysis of 3.0 seconds commences.

Figures showing the typical hydrodynamic pressure distribution in the water at the time of maximum hoop stress, time history values of typical hoop stress at bottom of the shell elements, base shear and moment at bottom of the tank and the vertical force in the anchors are included in the full paper.

5 SUMMARY AND CONCLUSIONS

Parametric studies using spring-mounted mass analogy techniques specified by the American Petroleum Institute code (API650) and the New Zealand Society for Earthquake Engineering recommendation (NZSEE) using South African seismic standards were conducted for three tanks with different fluid height to tank radius ratios (H_w/R). The results were compared with nonlinear time history dynamic analyses performed using Abaqus 3D finite element software.

REFERENCES

American Petroleum Institute (API). Welded steel tanks for oil storage, 2007

New Zealand Society for Earthquake Engineering. *Seismic Design of Storage Tanks: 2009 Recommendations of a NZSEE Study Group on Seismic Design of Storage Tanks*. New Zealand Society for Earthquake Engineering, November 2009.

H.E. Estekanchi and M. Alembagheri. *Seismic analysis of steel liquid storage tanks by endurance time method.* Thin-Walled structures, 50:14–23, August 2012.

Teng, J. G., & Rotter, J. M. (2004). Buckling of Thin Metal Shells. In *Buckling of Thin Metal Shells*. https://doi.org/10.4324/9780203301609

Current Perspectives and New Directions in Mechanics, Modelling and Design of Structural Systems – Zingoni (ed.)
© 2022 Copyright the Author(s), ISBN 978-1-032-39114-4

Parametric study of the catenary dome under gravity load

R.A. Bradley & M. Gohnert
School of Civil and Environmental Engineering, University of the Witwatersrand, Johannesburg, RSA

ABSTRACT: The catenary form is optimal for arches and singly curved (non-gaussian) shells, i.e. the catenary arch is a momentless and tensionless structure under self-weight. This advantage is especially appealing for the construction of low-rise structures built from materials which perform poorly in tension, e.g. masonry and concrete. Though the catenary is particularly important in arch and vault design, it also engenders a form for domes which, although not momentless, is in pure compression under gravity load. In this paper, the structural efficiency of the catenary dome is presented by finite element analysis (FEA) and membrane stress solutions. These analyses confirm that catenary domes experience only compressive hoop and meridian stresses under gravitational (self-weight) load. The influence of materials and geometric characteristics (e.g. thickness and height) as well as support type for several concrete domes are also investigated. These analyses revealed that Poisson's ratio and shell thickness had the largest impact on the hoop stress and bending moment, respectively, toward the dome base. Nevertheless, all catenary shells considered remained in pure compression, irrespective of the parameter changes. Finally, the catenary dome is compared with circular, elliptical, and parabolic profiles to highlight its structural efficacy over these more conventional forms.

1 INTRODUCTION

Engineers and architects of the past have realized the importance of keeping arches, vaults, and domes free of tension. The catenary is a well-established form which offers such a benefit when applied to the design of arches and vaults (Bradley et al. 2017). The catenary profile is physically defined by the shape of a hanging chain – between two supports at the same level – under self-weight. Freeze and invert the profile and you achieve the theoretical momentless, and tensionless, arch under the given load. Although the catenary is physically defined in two-dimensions (2D) by a hanging chain under load, a similar shape may be obtained for doubly curved shells of revolution using physical form-finding. For example, a thermoformed dome under self-weight is shown, alongside a hanging chain, in Figure 1. The material used in the form-finding was acrylic plate, which becomes semi-fluid when heated (Bradley & Gohnert 2016). This aforesaid form-finding technique demonstrates the structural advantage of the catenary dome, and this is further highlighted mathematically by Gohnert & Bradley (2021,2022), i.e., the authors showed that the catenary dome is pure compression under self-weight. The abovementioned study was, however, based on steeper profiles. In this paper we extend the investigation by considering the impact of geometrical, material, and support parameters on the stress patterns in the catenary dome.

2 RESULTS AND DISCUSSION

A stress analysis of several concrete domes was implemented in the software package Abaqus/CAE. The resultant FE stresses were compared with membrane solutions presented in the literature (Gohnert & Bradley 2021). Table 1 shows the material and geometric characteristics utilized in the analyses.

The analysis of four catenary domes, with aspect ratio ranging from H/L = 0.25 to 1.00, revealed that inner and outer stresses were compression in the

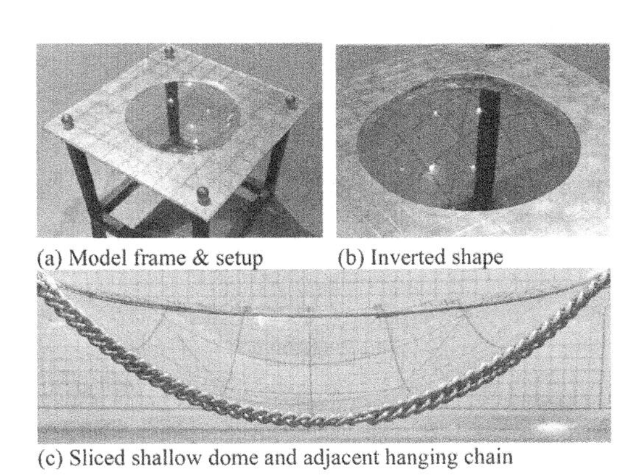

(a) Model frame & setup (b) Inverted shape

(c) Sliced shallow dome and adjacent hanging chain

Figure 1. Thermoformed catenary dome.

DOI: 10.1201/9781003348450-130

Table 1. Geometric and material characteristics.

	Geometry			Materials*		
H/L	$L(m)$	$H(m)$	$t(m)$	E(kN/m^2)	$Y(kN/m3)$	v
0.25-1.0	8	2 -8	0.15	30E6	23.5	0.15

* E – Young's Modulus, Y – Unit weight, v – Poisson's Ratio.

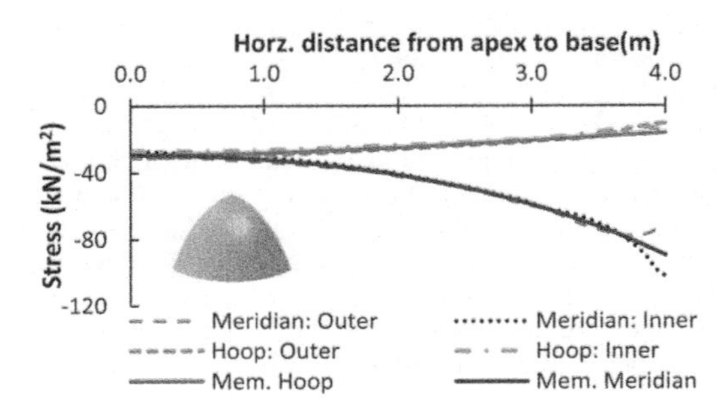

Figure 2. Catenary dome (H/L= 0.50).

hoop and meridian directions. The stresses for one of the catenary domes, with H/L = 0.5, is shown in Figure 2. Although all domes were in compression, there were varying degrees of discrepancy between inner and outer stresses for each shell.

The greatest difference between the inner and outer stresses was observed for the shallowest dome (H/L = 0.25), i.e. the dome experienced the greatest bending moment. Furthermore, the discrepancy between the mid-plane FE and membrane stresses was largest for the shallow dome. In contrast, the influence of the boundary, through bending and in-plane forces, was far less prominent in the steeper domes considered in the study. The minimal impact of the boundary for these profiles emphasizes the structural efficiency of the steeper catenary dome, i.e. gravity load is carried primarily through membrane action.

A sensitivity analysis was implemented to assess the influence of the geometric and material characteristics for a catenary dome with H/L = 0.50. Poisson's Ratio (v) had a noticeable influence on the disparity between inner and outer stresses in the hoop and meridian directions. Specifically, bending moment increased as v decreased. This was similarly apparent for the section thickness, where bending moment increased as thickness increased. In contrast, Young's Modulus had no impact on either stress distribution. Although the FE stresses were influenced by certain parameter changes, their disparity with the membrane solution remained largely limited and localized toward the base.

The stresses for two catenary domes, having an H/L ratio of 0.50 and 0.75, were compared with those for several conventional dome shapes, e.g., circular, paraboloid, and ellipsoid. Both catenary domes were pure compression, which was alike the paraboloid domes. In contrast, the circular and ellipsoidal domes experienced hoop tension as well as significant meridional moment toward the base support.

3 CONCLUSIONS

Several structural characteristics of the catenary dome were reported in this paper; the key findings are summarized below.

- Principal stresses in the catenary dome were compression only under a uniform gravitational load, regardless of the geometrical and material characteristics of the shell. This conclusion may differ for very shallow domes not considered in this study.
- Self-weight was predominantly carried through membrane action in the steeper catenary domes. Bending action was significantly lower in the steeper profiles, and this effect was also localized toward the base support. This was not the case for the shallowest dome (H/L = 0.25), where bending action extended along much of the dome profile.
- In-plane meridional boundary forces were insignificant in all domes considered. In contrast, the in-plane hoop boundary forces were comparable in magnitude to the membrane forces.
- The influence of the boundary – through in-plane forces and/or bending – was minimal in the steeper profiles, which emphasizes the structural efficiency of the form under gravitational load.
- The membrane solution proves to be a reasonable approximation of the stresses in steep domes under uniform gravity load, irrespective of material characteristics and boundary fixity.

REFERENCES

Bradley, R.A. & Gohnert, M. 2016. Three lessons from the Mapungubwe shells. *Journal of the South African Institution of Civil Engineering* 58(3): 2–12.

Bradley, R.A., Gohnert, M., Bulovic, I., Goliger, A.M. & Surat D.B. 2017. Steep catenary earth-brick shells as a low-cost housing solution. *Journal of Architectural Engineering* 23(2).

Dassault Systemes 2021. Abaqus/CAE [Software]. Dassault Systemes Simulia Corp.

Gohnert, M. & Bradley, R. 2021. Membrane solution for a catenary dome. *J. Int. Assoc. Shell Spatial Struct.*62(1): 37–49.

Gohnert, M. & Bradley, R. 2022. Membrane stress equations for a catenary dome with a variation in wall thickness. *Engineering Structures* 253.

10. Structural applications of glass

Current Perspectives and New Directions in Mechanics, Modelling and Design of Structural Systems – Zingoni (ed.)
© 2022 Copyright the Author(s), ISBN 978-1-032-39114-4

Structural bonding with hyperelastic adhesives: Material characterization, structural analysis and failure prediction

P.L. Rosendahl, F. Rheinschmidt & J. Schneider
Institute of Structural Mechanics and Design, Department of Civil and Environmental Engineering, Technical University of Darmstadt, Germany

ABSTRACT: Ever since (Leguillon 2002) proposed the coupled stress and energy criterion within the framework of finite fracture mechanics for the assessment of crack nucleation, many authors proved its capabilities in a multitude of structural situations. Requiring both stress and energy conditions to be met simultaneously proved key to modeling brittle crack formation at singular and nonsingular stress concentrations. However, only very few studies explore the potential of this so-called coupled stress and energy criterion beyond linear elasticity. The present work aims at extending finite fracture mechanics to brittle crack nucleation in hyperelastic media using the example of silicone adhesives. For this purpose, we use the comprehensive constitutive as well as fracture mechanical characterization of DOWSIL™ TSSA to propose a mixed-mode failure model for crack initiation in nonlinear elastic materials. Characterized in independent experiments, the model is used to determine critical loads of hyperelastic adhesive bonds of volumetric expansion dominated samples. For the examined joints the model predicts and explains size effects and agrees well with experimental findings.

1 INTRODUCTION

The unique molecular structure of silicones provides mechanical properties that render them excellent structural adhesives. However, at bi-material corners between adherends and adhesive stress singularities owing to geometrical and material discontinuities are present. In order to capture crack onset at these singularities, classical approaches such as stress-based criteria or fracture mechanics can only be applied using an additional length parameter.

Different nonlocal approaches were successfully applied to hyperelastic materials (Clift et al. 2014, Hagl 2016, Berto 2015). Yet, the length parameter involved in all of these approaches is not known a priori and lacks definite physical meaning.

Assuming the sudden nucleation of a finite sized crack introduces a length scale with clear physical meaning - the finite size of the initiated crack. The concept is known as finite fracture mechanics (FFM). In order to determine the finite crack size, Leguillon (2002) proposed requiring the simultaneous satisfaction of both a stress and an energy criterion as necessary and sufficient condition for crack nucleation. This so-called coupled stress and energy criterion involves two equations that allow for computing two unknowns: the critical loading and the size of the initiating crack. The coupled criterion requires only the fundamental material properties strength and fracture toughness as inputs. It provides excellent predictions for many structures with stress concentrations (Felger et al. 2019, Rosendahl et al. 2017, Dölling et al. 2020, Rosendahl and Weißgraeber 2020a, Rosendahl and Weißgraeber 2020b). A comprehensive review is given by Weißgraeber et al. (2016).

Common to all of the above FFM studies are the assumptions of brittleness and linear-elastic material behavior. The present work proposes a formulation of finite fracture mechanics for hyperelastic materials.

2 PANCAKE TEST

Pancake tests are designed to cause cavitational failure in nearly incompressible materials in order to study cavitation. However, since cavitation is a phenomenon of elastic instability, it is not necessarily linked to ultimate failure. Indeed, surfaces of fractured DOWSIL™ TSSA specimens have smooth fracture planes and show no signs of bubble formation or coalescence. Instead, fracture appears to originate from the bi-material point where DOWSIL™ TSSA and substrate meet. Therefore, we use the strain energy density function proposed by Drass et al. (2018) accounting for the elastic growth of cavities to simulate DOWSIL™ TSSA pancake tests. The onset of cavitation is assessed using a critical volume change criterion. Crack nucleation from the bi-material point is assessed using the coupled criterion.

DOI: 10.1201/9781003348450-131

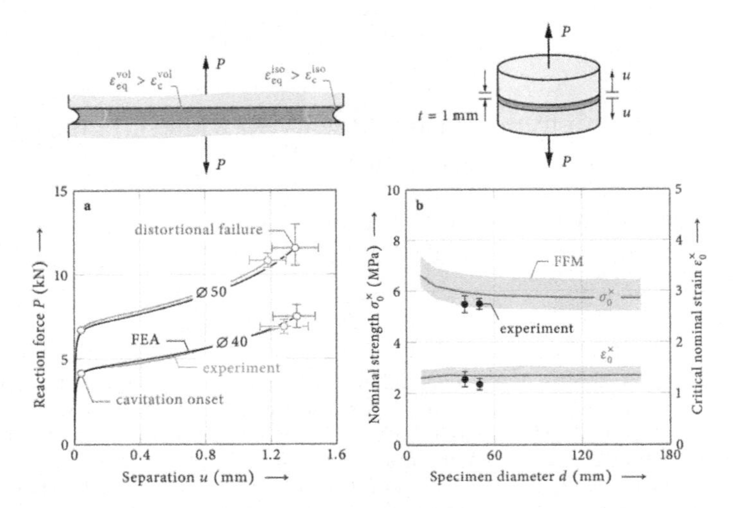

Figure 1. Finite fracture mechanics (FFM) predictions for DOWSIL™ TSSA pancake experiments. **a** Load displacement response with failure modes. **b** Influence of specimen diameter on ultimate nominal strength and strain.

We measured ultimate nominal stresses and strains of 18 DOWSIL™ TSSA pancake specimens. The measurements are compared against model predictions of the force-displacement response and predictions of critical loads for crack nucleation from the bi-material point in Figure 1. The material model captures the load-displacement response accurately (Figure 1a). Using the fracture parameters of DOWSIL™ TSSA reported by Rosendahl et al. (2019) and Rosendahl et al. (2021), cavitation onset is predicted at the experimentally observed onset of softening in the force-displacement response. It first appears in the specimen center and then expands across the adhesive layer with increasing loading. Dilatational failure in the specimen center causes load redistribution towards the specimen perimeter. Ultimate loading is associated with crack nucleation from the bi-material point and is predicted well by FFM. The pictogram sketches regions of overloaded material shortly before ultimate failure. The influence of the specimen diameter on the nominal ultimate strength $\sigma_0^\times$ is only observed for small samples (Figure 1b).

REFERENCES

Berto, F. (2015). A criterion based on the local strain energy density for the fracture assessment of cracked and V-notched components made of incompressible hyperelastic materials. *Theoretical and Applied Fracture Mechanics 76*, 17–26.

Clift, C. D., L. D. Carbary, P. Hutley, & J. H. Kimberlain (2014). Next generation structural silicone glazing. *Journal of Facade Design and Engineering 2*(3–4), 137–161.

Dölling, S., J. Hahn, J. Felger, S. Bremm, & W. Becker (2020). A scaled boundary finite element method model for interlaminar failure in composite laminates. *Composite Structures* (under review).

Drass, M., J. Schneider, & S. Kolling (2018). Novel volumetric Helmholtz free energy function accounting for isotropic cavitation at finite strains. *Materials & Design 138*, 71–89.

Felger, J., P. L. Rosendahl, D. Leguillon, & W. Becker (2019). Predicting crack patterns at bi-material junctions: A coupled stress and energy approach. *International Journal of Solids and Structures 164*, 191–201.

Hagl, A. (2016). Development and test logics for structural silicone bonding design and sizing. *Glass Structures & Engineering 1*(1), 131–151.

Leguillon, D. (2002). Strength or toughness? A criterion for crack onset at a notch. *European Journal of Mechanics A/Solids 21*(1), 61–72.

Rosendahl, P. L., M. Drass, & J. Schneider (2021). Distortional-dilatational strain failure mode concept for hyperelastic materials. *ce/papers 4*(6), 301–322.

Rosendahl, P. L., Y. Staudt, C. Odenbreit, J. Schneider, & W. Becker (2019). Measuring mode I fracture properties of thick-layered structural silicone sealants. *International Journal of Adhesion and Adhesives 91*, 64–71.

Rosendahl, P. L. & P. Weißgraeber (2020a). Modeling snow slab avalanches caused by weak-layer failure – Part 1: Slabs on compliant and collapsible weak layers. *The Cryosphere 14*(1), 115–130.

Rosendahl, P. L. & P. Weißgraeber (2020b). Modeling snow slab avalanches caused by weak-layer failure – Part 2: Coupled mixed-mode criterion for skier-triggered anticracks. *The Cryosphere 14*(1), 131–145.

Rosendahl, P. L., P. Weißgraeber, N. Stein, & W. Becker (2017). Asymmetric crack onset at open-holes under tensile and in-plane bending loading. *International Journal of Solids and Structures 113-114*, 10–23.

Weißgraeber, P., D. Leguillon, & W. Becker (2016). A review of Finite Fracture Mechanics: crack initiation at singular and non-singular stress raisers. *Archive of Applied Mechanics 86*(1-2), 375–401.

Current Perspectives and New Directions in Mechanics, Modelling and Design of Structural Systems – Zingoni (ed.)
© 2022 Copyright the Author(s), *ISBN 978-1-032-39114-4*

Reinforced glass: Structural potential of cast glass beams with embedded metal reinforcement

T. Bristogianni & F. Oikonomopoulou
Department of Architectural Engineering + Technology, Faculty of Architecture, TU Delft, The Netherlands

ABSTRACT: The shaping freedom of cast glass in combination with the robustness of the resulting voluminous components opens up new, exciting directions in the field of structural glass. Yet, cast glass components remain brittle, limiting their structural applications in hyper-static compressive structures designed with conservative safety factors. Stretching these limits, this work investigates the reinforcement of cast glass by incorporating metal bars during the casting process, in a similar principle to reinforced concrete. Aim is to increase the ductility of the composite glass component, provide a warning mechanism prior to ultimate fracture and secure a post-failure load-bearing capacity. The development of hybrid glass components involves kiln-casting experiments using different metal-glass combinations, of similar thermal expansion coefficients. The method of introducing the metal bar in the glass during casting, and the effect of the selected forming temperature are investigated. The resulting metal-glass interfaces are examined for micro-cracks using a digital microscope, and for internal stresses using cross-polarized light. Two material combinations are found successful; soda lime silica with titanium and alkali borosilicate with Kovar. A hybrid borosilicate-Kovar 30*30*240mm beam is further tested in 4-point bending until failure, while its displacement is measured by Digital Image Correlation (DIC). The flexural response of the composite component is compared to the performance of unreinforced cast glass beams of similar composition. Although reinforced and unreinforced specimens show a comparable flexural strength, the reinforced specimen exhibits a warning mechanism well before failure, a gradual fracture and a post-failure load-bearing capacity. These attributes encourage the further exploration of cast glass reinforcement.

1 INTRODUCTION

Cast glass can play a promising role in structural applications, however, due to its brittleness and unpredictable failure behavior, it is considered an unsafe material. In the case of float glass, several safety strategies have been developed to reduce the inherent risk in the event of collapse, among which the mechanical reinforcement of laminated float components can be of interest in the field of cast glass. In specific, the direct incorporation of the reinforcement in the glass during casting can allow for a high composite action, where the metal reinforcement works in tension and the glass in compression; in essence, an approach similar to that of reinforced concrete. To achieve such a direct bond without the implementation of an adhesive medium, it is crucial that the chosen glass recipe and metal composition present an almost identical thermal expansion coefficient, in order to prevent cracking occurring due to thermal stresses.

2 EXPERIMENTAL WORK AND RESULTS

Three different metal-glass combinations are evaluated by kiln-casting hybrid 50mm cubic samples at various forming temperatures (870-1120°C). The experiments show compatibility for two material combinations of similar thermal expansion coefficient (CTE): Alkali Borosilicate with Kovar and Soda Lime Silica with Titanium (Figure 1). The combination of Soda Borosilicate and Kovar is unsuccessful due to the $\approx 2*10^{-6}$ K^{-1} difference in the CTE of the two materials. Attention should be given to the selected forming temperature, to avoid the partial diffusion of the metal's surface into the surrounding glass, and to the orientation of the reinforcement within the investment mould prior to casting.

Figure 1. Titanium reinforced soda-lime glass (left) and Kovar reinforced alkali borosilicate (right).

DOI: 10.1201/9781003348450-132

As a proof of concept, a 30*30*240mm alkali borosilicate beam reinforced with a 240mm long ø6mm Kovar bar at the middle of its bottom surface is tested in 4-point bending until failure (100:200mm span). The specimen fails in a progressive manner, initially presenting a series of perpendicular flexural cracks at the bottom surface, within the maximum tensile zone. Thereafter, angular shear cracks appear at the zones between the support and the loading rollers. The analysis of the DIC images taken during the experiment (Figure 2) reveal the arrest of the created cracks as they propagate upwards towards the compression zone of the beam, and their slow growth as the force increases. The first crack is observed at 16.6MPa, upon which the stiffness of the hybrid component is reduced. Complete glass failure occurs at 40.8MPa. The experiment is stopped at this point, and no permanent deformation is detected in the bar. The glass at the bottom part is shuttered in several small pieces, yet the majority of the glass mass still remains connected to the metal bar, and as such the component remains in place. At the end of life of the hybrid component, the glass can be easily mechanically separated.

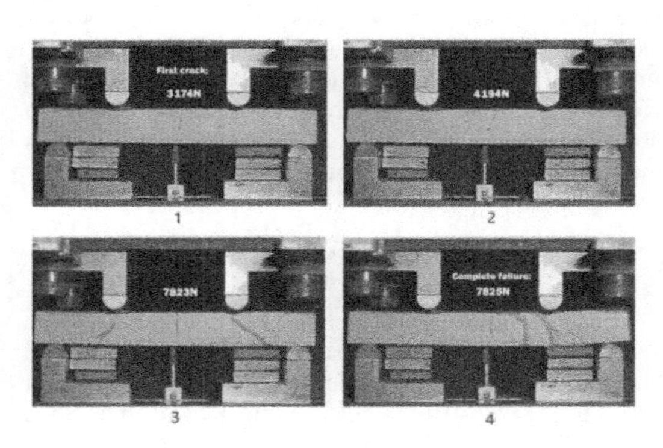

Figure 2. Major strain in the reinforced beam during loading, as obtained by the DIC analysis. The appearance of different cracks is associated with the corresponding load (in Newton).

3 DISCUSSION

The flexural behaviour and fracture pattern of the reinforced beam is compared to the performance of unreinforced Soda Borosilicate specimens, previously tested by the authors (Figures 3–4). The unreinforced beams show an elastic behaviour followed by sudden catastrophic fracture at an average flexural strength of 43.3MPa. Upon failure, the beam separates in two or more distinct pieces without any post-fracture load-bearing capacity. The reinforced beam on the other hand, fails progressively. Although the reinforced beam exhibits the first crack at approximately one third of the unreinforced beam's strength (16.6MPa), it continues to perform, sustaining up to 1.44 times more load prior to failure, at a similar stress level (40.8MPa). The emerging cracks at an early loading stage are speculated to be linked to peak shear stresses developing at the metal-glass interface. The progressive failure offers a valuable warning mechanism, and allows the use of a lower safety factor, provided that the design strength used corresponds to the elastic stage of the reinforced beam. Overall, the described experiment successfully exhibits the behaviour of the hybrid cast beam, encouraging the further development and systematic testing of the concept.

Figure 3. Side view of the reinforced (top) and unreinforced beam (bottom).

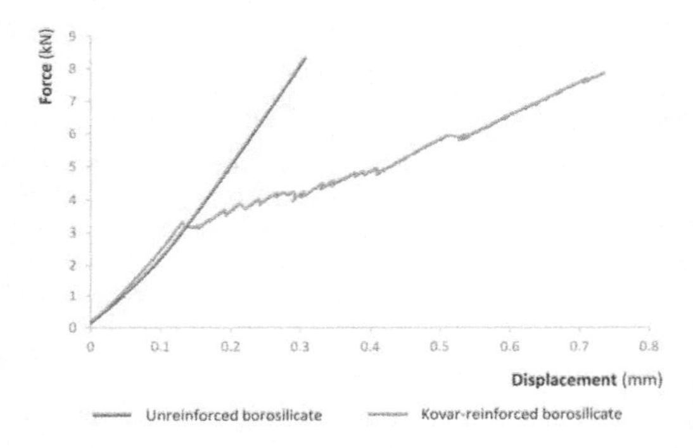

Figure 4. Force-displacement graph of the tested beams.

4 CONCLUSION

Kiln-cast experiments conclude to the possibility of embedding a metal reinforcement in a glass component during the casting process, provided that the thermal expansion coefficient of the two parts is (almost) identical. In this study, the combinations of soda-lime-silica with titanium, and alkali borosilicate with Kovar, are proven successful, yet the forming temperature requires adjustment to prevent the occurrence of colour streaks and large bubbles. The proof of concept flexural testing of a Kovar reinforced alkali borosilicate beam shows that the hybrid glass component fails progressively, in a similar manner to reinforced concrete. Although the failure strength of the component is comparable to the strength of a non-reinforced beam, a warning mechanism in the form of small arrested cracks appears well before failure, at 1/3 of the ultimate strength. During cracking and upon glass failure, the reinforced beam remains connected through the metal bar and exhibits post-fracture load-bearing capacity. The experiment suggests that further development of the concept is meaningful, and can lead to safer cast glass components for structural applications.

Current Perspectives and New Directions in Mechanics, Modelling and Design of Structural Systems – Zingoni (ed.)
© 2022 Copyright the Author(s), *ISBN 978-1-032-39114-4*

Investigations on the concept of hybrid bonding in glass structures

D. Offereins & G. Siebert
University of the Bundeswehr Munich, Germany

ABSTRACT: The concept of hybrid bonding refers to point-fixings in glass facades, that connect the glazing and the steel substructure by adhesion. Since adhesives have weaknesses with regard to use in facades, the aim of the concept is to compensate for these by combining to different types of adhesive systems: stiff adhesives, whose weaknesses are particularly found in insufficient temperature resp. environmental resistances and soft adhesives that provide comparatively low load-bearing capacities. The soft (secondary) adhesive can also provide additional redundancy in the load-bearing system in the case of non-load-induced damage to the stiff (primary) adhesive. To investigate the basic feasibility, uniaxial tensile tests in different configurations are carried out on aged and non-aged specimens of the bulk material of one epoxy resin adhesive and one silicone. The experimental results are used to calibrate common material models for each material.

1 INTRODUCTION

The stiff adhesive is applied in a circular pattern in the center of the point-fixing. These adhesives are usually used in thicknesses between 0.1 and 2.0 mm. The soft adhesive is applied as a circular ring around the primary adhesive.

With regard to the load-bearing behavior, the stiff adhesive is expected to carry the entire load in the undamaged state. The task of the silicone hereby is to protect the primary adhesive from harmful influences. Furthermore, the silicone shall be able to withstand a certain level of stress in case of damage to the primary adhesive to provide additional redundancy in the load-bearing system.

Stiff adhesives usually can be classified into the group of acrylates, epoxy adhesives, or polyurethanes. Structural silicones are recommended for soft adhesives, especially because of their resistance against environmental influences. In this paper, the epoxy adhesive Scotch-Weld 9323 B/A and the structural silicone DowCorning 993 are investigated.

2 EXPERIMENTAL AND NUMERICAL WORK

Uniaxial tensile tests on aged and non-aged specimens on both, the epoxy and the silicone adhesive, were carried out. The aging process consists of conditioning at 80 °C and 100 % RH for three weeks in a Binder MKF 720 climate chamber.

While for the epoxy adhesive only simple tensile tests up to breakage were carried out, the experimental program for the silicone consisted of so-called staircase tests, cyclic tests, and investigations on the Mullins effect, the latter two not being discussed in the short paper. At this point, it should be mentioned that the epoxy adhesive had to be mixed by hand, which is why air inclusions could not be completely avoided. The applied strain rate, regarding the initial cross-section of the specimens, was 0.05 1/s for the soft and 0.001 1/s for the stiff adhesive, respectively.

2.1 *Structural silicone*

The specimens were conditioned for at least 24 h at room temperature after the curing process (see manufacturer's technical data sheets) and also after the aging process was done, respectively.

In Figure 1, the typical behavior of silicone appears, which is characterized by large strains at low stresses. The graph shows the relaxation behavior of the silicone, which starts immediately but ends within the applied holding period of 2 min. The maximum loss of stiffness is approximately 19 %, whereby it can be determined that the higher the applied strain and stress, the higher the loss.

After aging, a considerable loss of stiffness can be identified. In addition, the maximum strain decreases to a similar extent. It must be noted, that the specimens were so soft that the extensometers cut into the material, causing local failure.

For means of numerical simulation, the lowest points of each holding period of the staircase tests were determined and the material simulated with a Yeoh 3-Parameter model in the finite element software ANSYS.

DOI: 10.1201/9781003348450-133

The Yeoh model is a hyperelastic material model that accurately describes the typical behavior of silicone, as illustrated in Figure 1. Hyperelastic material models are usually described by the strain energy density function W, which in this case reads

$$W = C_{10}(I_1 - 3) + C_{20}(I_2 - 3)^2 + C_{30}(I_1 - 3)^3, \quad (1)$$

where I_1 is the first strain invariant of the Cauchy-Green deformation tensor and C_{10}, C_{20} and C_{30} are material constants. The model represents the material behavior sufficiently for both, the aged and non-aged case, as can be seen in Figure 1.

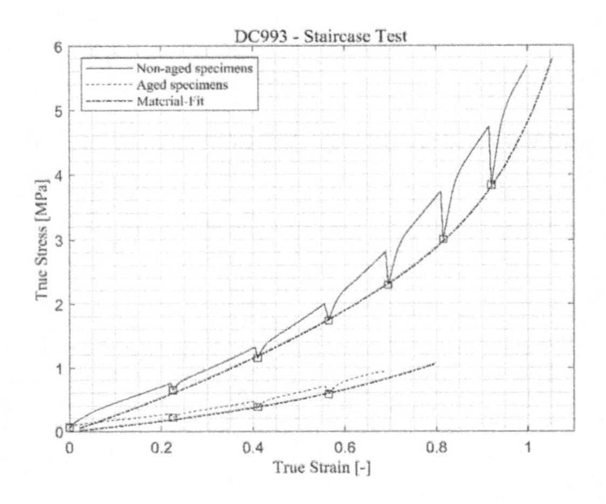

Figure 1. Staircase Test on DC993 in aged and non-aged condition.

2.2 *Epoxy resin adhesive*

The adhesive shows a substantially stiffer behavior, with stresses that are higher by a factor of about 10 than with the structural silicone.

After the aging process, the adhesive's color changed from pink to dark red. Although there is a certain level of loss in stiffness, the material still reaches a high stress level. In addition, the formation of a plateau area immediately before failure can be observed.

A common approach to numerically represent stiff adhesives such as epoxy or acrylates is by using a simple linear-elastic model. The elastic modulus was determined to be 1,678 MPa for the non-aged state and 1,190 MPa for the aged state, respectively. The Poisson's ratio was taken from the literature and amounts to 0.392, cf. Dispersyn (2016).

As Figure 2 illustrates, the model accurately reproduces the material behavior at small strains. However, the deviations become larger at a level of approximately 50 % of the maximum strains and the stresses at the failure of the specimens are thus significantly overestimated. In further simulations, a more suitable material model should therefore be elaborated.

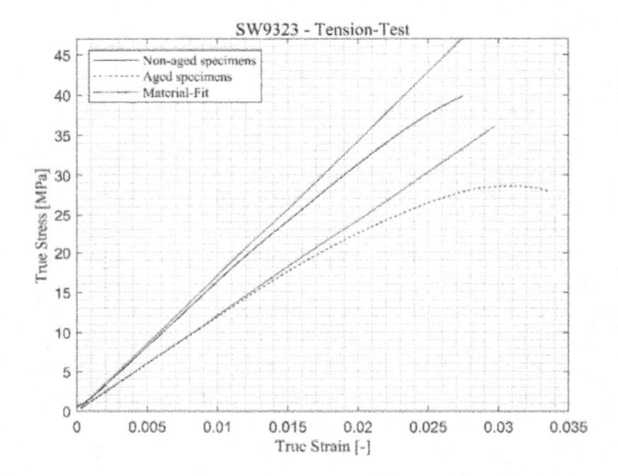

Figure 2. Tensile Test on SW9323 in aged and non-aged condition.

3 CONCLUSION AND OUTLOOK

In the non-aged state, both adhesives exhibited appealing material behavior, which qualifies them for further investigations. For the stiff adhesive, this also applies to the aged condition. The silicone performed worse than the results in the literature suggested.

On the other hand, the aging conditioning was significantly stricter than required by guidelines and standards, which usually prescribe a temperature of 45 °C. Moreover, other adhesives are planned to be investigated with regard to chemical reciprocal effects between the adhesives, their mechanical behavior, and their resistance against aging and environmental influences.

REFERENCES

Bues, M. D. 2021. *Ein Beitrag zur Auslegung tragender Klebverbindungen im Fassadenbau*. Karlsruhe.

Dispersyn, J. 2016. *Evaluation and optimisation of adhesive point-fixings in structural glass*. Gent.

Santarsiero, M. & Louter, C. 2015. The mechanical behavior of SentryGlas ionomer and TSSA silicon bulk materials at different temperatures and strain rates under uniaxial tensile stress state. *Glass Struct. Eng. 1*: 395–415.

Current Perspectives and New Directions in Mechanics, Modelling and Design of Structural Systems – Zingoni (ed.)
© 2022 Copyright the Author(s), ISBN 978-1-032-39114-4

Topologically optimized structural glass megaliths: Potential, challenges and guidelines for stretching the mass limits of structural cast glass

F. Oikonomopoulou, A.M. Koniari, W. Damen, D. Koopman, I.M. Stefanaki & T. Bristogianni
Department of Architectural Engineering + Technology, Faculty of Architecture, TU Delft, Delft, The Netherlands

ABSTRACT: This paper introduces the use of structural topology optimization (TO) as a new design approach that enables the creation of monolithic load-bearing cast glass components of substantial dimensions with significantly reduced annealing times, rendering such components viable in terms of manufacturing. Using TO, the glass mass can be optimized to match design loads whilst maintaining high stiffness and a homogeneous mass for even cooling. Initially, the two main TO approaches are discussed in terms of suitability for cast glass. A strain-based optimization is eventually preferred over Von Mises optimization in the specific study. To explore the potential of TO for optimizing structural cast glass components, three distinct studies are analyzed in ANSYS workbench: (i) a structural glass node, (ii) a cast glass floor and (iii) a pedestrian bridge. These lead to the establishment of a set of design/input criteria, taking into account glass as a material, casting as a manufacturing method, addressing also the safety of the structure. The design studies also reveal the inherent challenges of using TO for load-bearing glass components, which, in turn, lead to the establishment of design guidelines for developing a TO tool specifically for glass. Towards the real-life applicability of such complex-shaped, customized components, possible manufacturing methods are also discussed.

1 INTRODUCTION

In this paper, structural topology optimization (TO) is investigated as a promising approach for designing monolithic, structural cast-glass elements of substantial mass in all three dimensions. Using TO the glass mass can be designed to match design loads whilst maintaining high stiffness and a homogeneous mass for even cooling. The possibility to generate cast glass forms of reduced mass and complex geometry, results in interesting structures of reduced material and most importantly, of the annealing time involved, rendering them viable in terms of manufacturing.

There are currently two main TO approaches: strain-based and Von Mises (stress-based) optimization. Common stress-based TO aims to minimise the stresses in an object for a given set of boundary conditions. Stress-based TO typically employ a Von Mises stress criteria, which is an abstraction that does not distinguish between tension and compression and allows in turn for a simplified and faster optimisation progress. However, if the Von Mises stress criteria is applied to glass, the tensile strength becomes the main restrain, leaving underutilized the considerably higher compressive strength of glass. This will have to be evaluated in a later stage increasing, therefore, the amount of manual work needed. Strain-or compliance based TO is an alternative approach that aims to maximize stiffness of an object by setting as objective the minimization of the compliance energy of the structure. It provides better efficiency in terms of stiffness and uses the compressive capacity of glass more beneficially. Though this should result in more reliable geometries, this approach also does not distinguish between tensile- and compressive stresses. In addition, as stress is not directly taken into account, post-analysis is necessary here as well to check for possible peak stresses in the glass structure, increasing in this regard the manual work and the time needed for each operation. It is evident that none of the existing TO algorithms is fully fitting to glass. In this research, strain-or compliance based TO has been chosen, as it is more suitable for a 3D element and thus allows for a better exploration of the thickness reduction, which, in turn, has a major influence on the annealing behaviour of cast glass.

To explore the potential of TO for optimizing structural cast glass components, 3 distinct studies are analyzed in ANSYS workbench: (i) a cast grid-shell node (Figure 1), (ii) a glass floor (Figure 2) and (iii) a pedestrian bridge (Figure 3).

2 DISCUSSION

The examples highlight the need for a TO tool customized to the brittle behaviour of glass and, most importantly, to the differentiation between the allowable limits regarding tension and compression. If this is achieved, it is expected to reduce drastically the amount of manual post-processing work and, thus, make the overall design process less time-consuming.

DOI: 10.1201/9781003348450-134

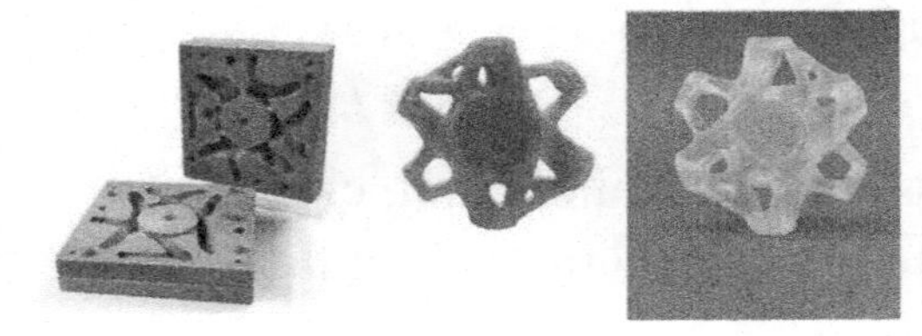

Figure 1. Left: 3D-printed sand moulds produced by ExOne; middle: 3D-printed wax positive model; right: cast glass prototype of the grid-shell node made by the wax model and a silica-plaster mould.

Figure 2. The final mass-optimized cast glass slab.

Figure 3. Model used for linear (only float glass) and non-linear (float glass and PU layer), indicating as well the splitting of the structure for redundancy.

Regarding the TO methodologies, although the Level-Set method generally results in a clearer form, its strong dependence on the initial guess design is questioning its ability to result in a global optimum solution. In contrary, SIMP is well-proven to have good convergence and end in robust solutions. Hence, despite the need for translation of the final pseudo density result, it should be selected for the algorithm implementation. Structural evaluation of principal stresses and deformation can be held through its combination with FEM equations. In this regard, since the evaluation of the structural criteria will be ensured (as distinct tension and compression values will be set as limits in the algorithm), minimization of volume can be set as the objective, since it is the most crucial aspect for the reduction of the annealing time and thus, the viable production of the component.

Other criteria linked to manufacturing, such as the cross section size, can be applied with the adaptation of methods that have already been introduced, such as the max. length scale constraint. Given that the annealing rate of glass can be described as a function of the cross section's size, the value for the max. element size constraint can combine both the limit for the range of allowable cross sections, in terms of homogeneous mass distribution, and the max. allowable annealing time so that the result is timeand cost-efficient. However, constraints, such as eliminating the sharp edges, should still be done in a post-processing phase.

Regarding the designs' voids, apart from checking their min. and max. size, in terms of fabrication, connectivity between them should be assured. This can be evaluated through applying the Virtual Temperature Method; converting the connectivity problem to a temperature problem and performing a heat flow analysis.

Research in the direction of integrating these criteria into an algorithm, special for cast glass but potentially applicable also to other brittle materials, will be able to greatly facilitate the overall process for achieving viable structural cast glass megalithic components.

3 CONCLUSIONS

Overall, the discussed case-studies showcase the potential of using TO for engineering massive load-bearing cast glass components and demonstrate the large potential of this new architectural vocabulary for the built environment. In all 3 case-studies, a minimum compliance objective is followed; this is particularly meaningful for examples (ii) and (iii) where deformation is anticipated to be more critical than stresses. This decision is strongly related to the fact that it is not possible to apply different constraints for tension and compression during the optimization process. In this regard, compliance-based optimization is selected since it uses the compressive capacity of glass more beneficially. The stress constraint in this case refers to the allowable limit of tensile strength which, for glass, is more critical. However, a major drawback in the process is that heavy approximations have to be done and it is necessary to manually evaluate the peak stresses after every TO iteration. This increases considerably the amount of manual work and renders the whole process very time consuming.

Regarding the TO software, all the projects make use of gradient-based algorithmic methodologies. Particularly (i) and (ii) use SIMP, which is the most widely popular method, but case study (iii) follows the Level-Set Method aiming to have a clearer final shape that will need less post-processing.

Safety is also taken into consideration in all examples either by redundancy, e.g. by splitting the monolithic structures into different parts, and/or by placing float glass layers on the top in order to protect the monolithic elements from accidental impact. Lastly, the need for having a fixed governing load case that determines the necessary stiffness is highlighted in order to ensure that the varying loads, e.g. wind, will have little influence on the structure.

In terms of manufacturing, initial prototype work at TU Delft has indicated kiln-casting with 3d-printed sand moulds (Figure 1) as the most promising approach for the manufacturing of these components; whereas 3d printing glass may also be an option in the future.

Current Perspectives and New Directions in Mechanics, Modelling and Design of Structural Systems – Zingoni (ed.)
© 2022 Copyright the Author(s), ISBN 978-1-032-39114-4

Rotational stiffness and strength of a two-sided reinforced laminated glass beam-column joint prototype: Experimental investigation

M. Pejatovic, R. Caspeele & J. Belis
Department of Structural Engineering and Building Materials, Ghent University, Ghent, Belgium

ABSTRACT: Glass is a fragile material that experiences only elastic deformations under common loading conditions during its service life. However, occurrence of an unforeseen event may lead to breakage of glass and hence redistribution capacity at both component and system level may be necessary to prevent collapse. Previous research illustrated load bearing mechanisms that can be developed in reinforced glass multi-span systems and horizontally restrained beams prior to collapse. Moment resisting connections are of crucial importance for a structural system subjected to a combination of vertical and horizontal loads. The main post-fracture assumption is that a well-designed connection enables the full plastic capacity of one of the adjoining members to be reached. Strength and stiffness prediction of such joints should rely on experimental data acquired by means of a customized testing setup and adequate measurement techniques. In this paper, special attention is given to the testing procedure and acquisition of data necessary for the assessment of rotational stiffness and strength of reinforced glass beam-column joints. Main features of the testing setup, specimens manufacture process and measuring equipment are presented.

1 INTRODUCTION

Utilization of structural glass as a building material is a trend which has been increasingly present in various architectural contexts. Glass load-bearing members in a combination with bonded reinforcement (e.g. stainless steel) possess additional safety upon glass fracture in case of extreme events (Louter, 2007). In other words, a high capacity and ductility in the post-fracture domain is achieved in case the well-bonded reinforcement can be exploited (Martens, et al., 2016). To expand the knowledge of the system actions in hybrid glass frames with the additional post-fracture capacity under multiple loading scenarios, one has to develop a testing strategy and a database of experiments related to beam-column joints. However, no data in the literature is present about rotational characteristic of reinforced moment-resistance joints. For the purpose of performing a test of a robust moment-resistant joint prototype that will be presented in this paper, the hybrid specimens were built by means of lamination with SentryGlas® interlayer, whose mechanical properties have been investigated by many authors (e.g. (Callewaert, et al., 2012)).

2 PRODUCTION OF SPECIMENS AND EXPERIMENTAL TESTING

For the production of the specimen soda-lime silica glass panels, welded stainless steel bars and a SentryGlas® interlayer are used. Prior to the final placement in the autoclave, such a specimen is put in a vacuum bag and sealed (shown in Figure 1) so that the air does not influence formation of bubbles in the interlayer.

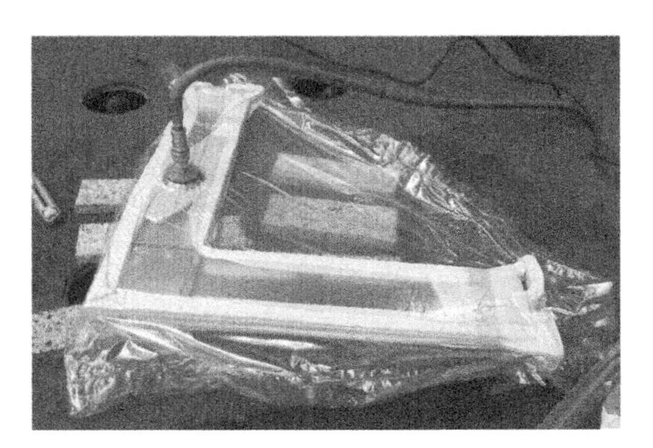

Figure 1. (a) Arrangement of the constitutive layers; (b) Sealed specimen in a vacuum bag ready for lamination.

Figure 2 shows the testing setup and the measurement system. The setup consists of steel attachments that enable transfer of forces from the testing machine through hinges to the joint (Figure 2(2)). The forces are transferred by means of a combined utilization of

DOI: 10.1201/9781003348450-135

bolted steel parts and welding of additional perforated stainless steel plates on the reinforcement.

A DIC system (see Figure 2(3)) is used to capture kinematics during the test from one side of the specimen, whereas from another side the crack pattern evolution is followed.

(1) The specimen
(2) The steel attachment
(3) DIC cameras
(4) Strain gauges

Figure 2. Testing setup and the measurement system.

3 RESULTS

Some experimental results are summarized in Figure 3. The behavior obtained in the test could be divided into 3 characteristic phases: a linear-elastic, post-fracture and residual phase. The M-$\Delta\phi$ curve in Figure 3(a) suggests that the joint prototype exhibits a significant post-fracture moment and displacement capacity. The appearance of new cracks is followed by load drops, which is followed by a deterioration of the rotational stiffness. The stainless steel bar in tension does not reach the yielding strain which explains the absence of the plastic phase. Finally, the failure mode is governed by breakage of glass in the joint core which can also be observed in the M-$\Delta\phi$ curve (final drop of the moment) and its confirmation can be found in the crack pattern at failure in Figure 3(b). The observation related to the application of DIC for testing such a member is that a high percentage of area of the initially determined region of interest has been preserved after the specimen failed. This gives a possibility to get information about the performance in the residual phase too.

4 CONCLUSIONS

The goal of the test was to experimentally assess the rotational stiffness and the post-fracture performance of such joint. DIC system was used for measurement of kinematics and performance of glass. The results

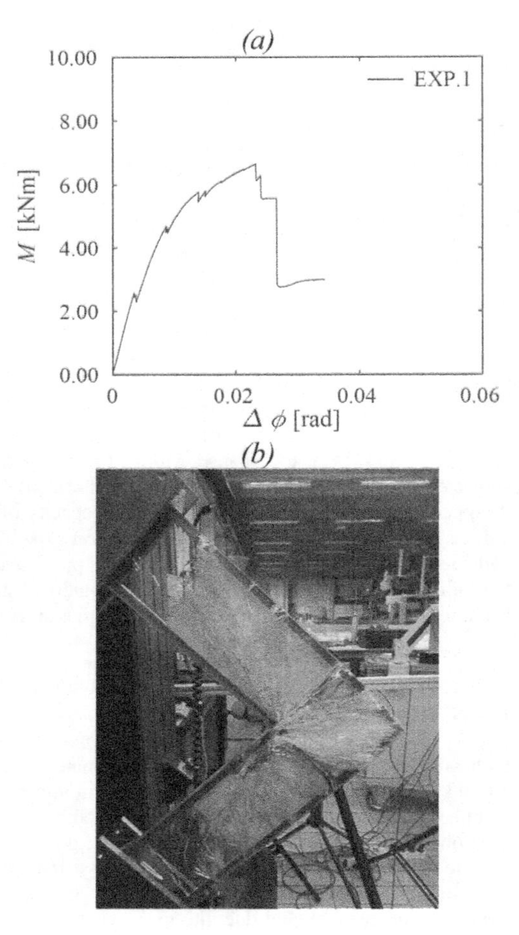

Figure 3. (a) Experimental M-$\Delta\phi$ curve, (b) Crack pattern at failure.

showed that the joint prototype exhibited significant additional load carrying and deformation capacity prior to a complete failure. For the chosen configuration, brittle failure of the joint core occurred. In the next steps of the research the influence of a different technology, steel ratios and glass panes thickness will be investigated.

REFERENCES

Callewaert D. [et al.] Experimental stiffness characterisation of glass/ionomer laminates for structural applications [Journal] // Construction and Building Materials. - 2012. - 37. - pp. 685–692.

Louter C. Adhesively bonded reinforced glass beams [Journal] // HERON Vol. 52 (2007) No. 1/2. - 2007. - Vol. 52.

Martens K., Caspeele R. and Belis J. Development of Reinforced and Post-tensioned Glass Beams: Review of Experimental Research [Journal] // Structural Engineering, 2016, 142(5). - 2016.

Current Perspectives and New Directions in Mechanics, Modelling and Design of Structural Systems – Zingoni (ed.)
© 2022 Copyright the Author(s), ISBN 978-1-032-39114-4

Transparent smoke barriers

Andreas Haese & Barbara Siebert
Ingenieurbüro Dr. Siebert, München, Germany

ABSTRACT: Glass smoke barriers are part of many infrastructure projects. They combine the architect's idea of low visibility with the function of effectively deviating the smoke to keep certain areas free of smoke for a certain time. Often standard fire-safe framed glass set-ups are not wanted due to the frame necessary to keep the fire-resistant glass in place. Frameless glass elements on the other side do have a higher risk of glass breakage due to the unprotected and accessible glass edges. Additionally, the free edge is exposed directly to the (usually) applied temperature of 600°C, a temperature standard glass interlayers cannot withstand. Fire-resistant glass set-ups, typically with hydrogel or soluble materials between glass layers, on the other side do have a very poor remaining load capacity in case of glass breakage. For several subway stations in Munich different approaches to find the best compromise of fire resistance, remaining load capacity and the optical effects were studied, some tests have already been carried out. The paper will discuss different glass types and interlayers in different installation situations and show the effects on fire resistance and remaining load carrying capacity in regard to the actual requirements and depending on the chosen system and detailing.

1 INTRODUCTION

Although transparent smoke barriers do already exist or are even common and most of them look quite similar, the actual setup can be very different, depending on the requirements they must fulfill.

The combination of requirements that had to be fulfilled in a Munich underground station led to the situation that no product was available on the market, that would fulfill those requirements, making it necessary to search for a possible solution.

2 SITUATION AND REQUIREMENTS

In the course of the refurbishment of an underground station, smoke barriers will be placed around the openings of the floor slabs, where the stairs or escalators are connecting the train level to an intermediate level. Additionally, some will also be placed along the platform edge.

Given the local situation, the following requirements must be considered.

- smoke protection
- sufficient fire resistance
- durability during lifetime
- residual capacity if glass breaks
- wind load resistance
- enhanced residual load capacity for 2'000 wind load cycles with glass (partly) broken
- lower glass edge free (no profile wanted)

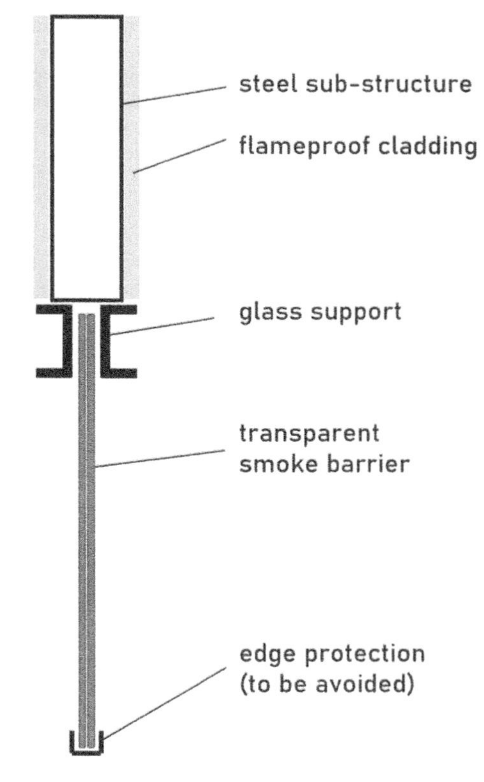

Figure 1. Glass smoke barrier system.

DOI: 10.1201/9781003348450-136

3 INNOVATIVE APPROACH

As within the materials and systems available on the market such as

- borosilicate glass
- laminated (safety) glass
- fire protection glazing
- coated tempered glass

No solution has been found that covered all requirements, a new approach was considered, using a "cold liquid interlayer" to laminate borosilicate glass panes.

"Cold liquid interlayer" (although it's only liquid when applied but then cures to be solid), initially developed to glue glass on the uneven surface of natural stone under the name "protectstone" showed a completely different behavior when tested for fire safety than common film interlayers or the interlayers in fire protection glazing. Instead of liquifying and potentially burning when draining out between the glass panes like PVB or EVA or foaming up and destroying the laminate like the interlayers of fire protection glazing, the bonding material just carbonized without burning.

It was decided to test a triple build-up of borosilicate glass with the middle pane just 5 mm shorter at the lower edge. This makes it highly unlikely that the middle pane will break when hitting the edge of the glass. As a consequence, the middle glass can be assumed to be intact in every situation (except vandalism).

4 TESTING

The testing program was designed to address the fire safety, the durability and the structural performance.

Figure 2. Three pane build-up (middle) almost black after 15 minutes.

The testing of durability and structural performance is still in progress with the results yet received showing sufficient durability and sufficient structural performance to pass the testing procedure for laminated safety glass.

The fire testing was done with two other elements with "best-practice" build-ups as comparison.

Although the innovative build-up became untransparent very quickly, till the end of the test no major changes appear and only little interlayer material has fallen down in light flakes. The filter paper did not catch fire and had no burn marks.

Figure 3. Burnt interlayer material on filter paper.

5 CONCLUSION

The tests have shown, that the innovative approach of using cold liquid interlayers that up to now have been mostly used for decorative purposes for smoke barriers is promising. Not only did this glass build-up pass the fire test, it also passed it with an unprotected free lower glass edge, directly exposed to the furnace temperature of up to 660°C.

Although the tests regarding composite behavior have not yet been completed, the first results indicate sufficient impact resistance to pass the LSG criteria. The durability has been tested successfully on similar products.

When the tests are completed, the glass build-up might not only be a solution to the very specific requirements of transparent smoke barriers, but also find a wider field of applications. Although glass itself is undoubtedly non-combustible, this is not completely true for laminated (safety) glass with its standard PVB-interlayer. Especially after the Grenfell disaster, materials used in facades have come under close scrutiny. For fire critical situations, a glass build-up with a carbonizing interlayer might be preferable to a standard LSG with a liquifying and flammable interlayer.

Current Perspectives and New Directions in Mechanics,
Modelling and Design of Structural Systems – Zingoni (ed.)
© 2022 Copyright the Author(s), *ISBN 978-1-032-39114-4*

Experimental investigations on EVA interlayers in the regime of large deformations in the context of the examination of the residual load bearing capacity of LSG

Milica Baric, Alexander Pauli & Geralt Siebert
Institut für Konstruktiven Ingenieurbau, Universität der Bundeswehr München, Neubiberg, Deutschland

ABSTRACT: Recently, the investigation of the residual load-bearing behaviour of LSG is a very important topic within the research on structural glass design. The most important component regarding residual load-bearing capacity in the context of LSG is the interlayer. First tests on LSG with EVA led to promising results regarding residual load-bearing capacity. Therefore, within this work, several tests within the regime of large deformations, are carried out on the pure interlayer, in order to characterize the material behaviour. The respective tests concern uniaxial tension until breakage, in order to examine the failure of the material, cyclic loading with different strain rates, in order to examine the material for rate-dependent behaviour, and cyclic tests with several relaxation steps, in order to investigate the equilibrium behaviour. Subsequently, the test results are examined and the material is assigned to a respective behaviour as well as a matching model.

1 INTRODUCTION

This paper is concerned with the experimental investigation of the material behavior of Ethylene-Vinyl-Acetate (EVA) in the context of laminated safety glass. The main focus lies on the residual load-bearing capacity within the fully broken state of laminated safety glass. As first tests on EVA led to promising results, this work focuses on the material behavior within the regime of finite strains. The overall goal is a numerical model capable of calculating the load resistance of laminated safety glass within the fully broken state.

2 EXPERIMENTAL INVESTIGATION

The respective experimental program involves strain-driven uniaxial tension tests with different loading procedures: (i) tension until breakage, (ii) one loading cycle at two different strain rates, and (iii) one loading cycle with several relaxation steps. All tests were conducted at a temperature of 21.5 +/- 1°C and relative humidity of 40 +/- 10%.

2.1 *Tension until breakage*

In the beginning, up to true strains of 1 [-], the material shows a moderately stiff, almost linear behavior. At higher strains from 1.0 [-], a large increase in stiffness together with a highly nonlinear relationship between stress and strain can be observed. The elongation at breakage lies between true strains of 1.7 – 1.8 [-] at true stresses of 150 up to 200 [MPa].

2.2 *Cyclic loading with different strain rates*

The tests with loading and unloading at a high and a low strain rate were carried out to examine the rate-dependent behavior of the material. To ensure, that the specimens do not break during the loading cycle, the maximum strain was set to a true strain of 1.6 [-] (compare results of tension until breakage). The tests were performed with a technical strain rate of 0.01 1/s and a strain rate of 0.0018 1/s. The unloading of the specimens was stopped at a force of 5 N.

After testing the specimens were laid flat and the shrinkage was qualitatively measured over time. It could be observed, that even after several days a considerable amount of deformation about a true strain of 1.4 (300 % technical strain) was still left.

2.3 *Cyclic loading with several relaxation steps 'staircase test'*

To investigate the equilibrium behavior of the material, a cyclic loading with several relaxation steps was performed. The test procedure was comprised of 10 loading steps, followed by a relaxation of 300 s, and two unloading steps, followed by a relaxation of 300 s. The loading steps from 0 % technical strain until 100 % technical strain were increased by steps of 25 %, the rest of the steps were increased or decreased by steps of 50% technical strain.

DOI: 10.1201/9781003348450-137

The material relaxes until a certain point if time tends towards infinity. The steps of the loading and unloading cycle seem to head towards each other. Nevertheless, having in mind that there are irreversible deformations left within the specimen after unloading, the conclusion is drawn that the material tends towards an equilibrium hysteresis.

3 MATERIAL CHARACTERIZATION

The observations made during the evaluation of the test results lead to the assumption, that EVA within the regime of finite strains could be described by a material model of viscoplasticity.

The material behavior of viscoplasticity can be mathematically described by the sum of two functionals (cf. Equation (1)). A rate-dependent functional, describing the viscoelastic part and a rate-independent functional, describing the plastic part (Haupt 1993; Haupt und Lion 1995; Bergstrom 2015).

$$T(t) = \underset{s\geq 0}{\mathfrak{F}_{eq}}[E(t-s)] + \underset{s\geq 0}{\mathfrak{E}_{ov}}[E(t-s)] \quad (1)$$

T ... current stress
E ... current strain
$\mathfrak{F}_{eq}$ ... functional of equilibrium stress
$\mathfrak{E}_{ov}$ ... functional of over stress

Considering the purpose of numerical modeling of laminated safety glass, a decisive simplification is made. The material is simplified by a curve, represented by the hyperelastic Yeoh model (Yeoh 1993). Under the assumption of incompressibility, for the stress due to uniaxial deformation, it holds (Yeoh 1993; Bergstrom 2015):

$$T_{11} = 2\left[c_{10} + 2c_{20}(I_1 - 3)^2 + 3c_{30}(I_1 - 3)^2\right]\left(\lambda^2 - \frac{1}{\lambda}\right) \quad (2)$$

where T_{11} equals the uniaxial Cauchy stress, I_1 equals the first invariant of the right Cauchy Green stretch tensor and $\lambda = (\varepsilon_{tech} + 1)$ is the stretch in case of uniaxial deformation.

'Figure 1' shows the course of the Yeoh-Model within the result of the staircase test. As it can be seen the stress stays inside the hysteresis.

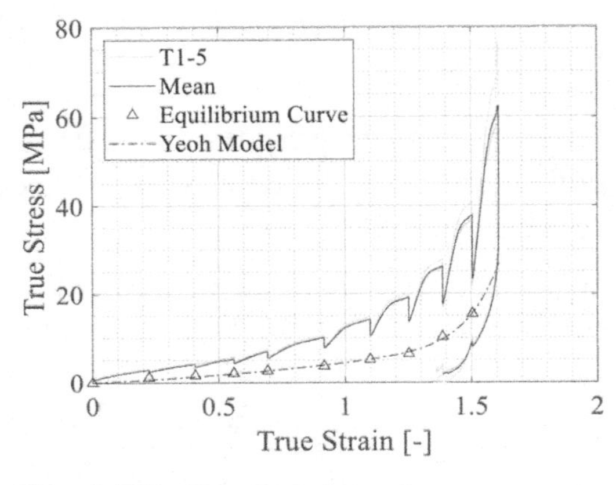

Figure 1. EVA - Hyperelastic Approach.

4 CONCLUSIONS

The goal of this paper was the investigation of the behavior of EVA in the regime of finite strains in the context of the numerical modeling of the residual load-bearing capacity of laminated safety glass. The material behavior was observed to be viscoplastic, To create a simple numerical model, a decisive simplification was made. With respect to that, the material is described by the hyperelastic Yeoh Model. As it is ensured, that the stresses never exceed critical stresses, this assumption is valid. The disadvantage of this approach is a quite big loss of material reserves.

REFERENCES

Bergstrom, Jörgen (2015): Mechanics of solid polymers. Theory and computational modeling. First ed. Amsterdam: William Andrew (imprint of Elsevier).

Haupt, P. (1993): On the mathematical modelling of material behavior in continuum mechanics. In: *Acta Mechanica* 100 (3-4), S. 129–154. DOI: 10.1007/BF01174786.

Haupt, P.; Lion, A. (1995): Experimental identification and mathematical modeling of viscoplastic material behavior. In: *Continuum Mech.* Thermodyn 7 (1), S. 73–96. DOI: 10.1007/BF01175770.

Yeoh, O. H. (1993): Some Forms of the Strain Energy Function for Rubber. In: *Rubber Chemistry and Technology* 66 (5), S. 754–771. DOI: 10.5254/1.3538343.

Current Perspectives and New Directions in Mechanics, Modelling and Design of Structural Systems – Zingoni (ed.)
© 2022 Copyright the Author(s), ISBN 978-1-032-39114-4

In-plane loaded glass balustrades as structural members for balconies

J. Giese-Hinz, F. Nicklisch & B. Weller
Technische Universität Dresden, Institute of Building Construction, Dresden, Germany

M. Baitinger, H. Hoffmann & J. Reichert
VERROTEC GmbH, Mainz, Germany

ABSTRACT: Many residential buildings have projecting balconies with glass balustrades. However, the architectural quality of glass balustrades is often impaired by the diagonal tensile rods that are used wherever cantilevered balconies are impractical. The project team explored the load-bearing potential of glass balustrades employed as bracing elements at the sides of a balcony. A linear adhesive bond connects the glass at the bottom to the balcony platform, while another bond line at the top connects the handrail. The design is subjected to a complex series of tests as a proof of concept. This paper focuses on the testing of life-size specimens under soft body impact. The test results are compared to FEA using a simplified spring model for the adhesive. The numerical analysis determines the relevant stresses and deformations of the glass and adhesive joints. Both component tests and FEA confirm the usability of the construction as a barrier glazing.

1 INTRODUCTION

In new buildings, balconies are often realised as projecting slabs in conjunction with various balustrades; this is particularly the case with reinforced concrete constructions. Pre-existing buildings or certain other types of construction, such as timber structures, frequently do not support a balcony platform construction; in such situations, balconies have to be supported on columns in front of the building. The use of support-free construction would be architecturally more attractive (Figure 1).

Figure 1. Balcony with back-anchored platform.

When employed in conjunction with a transparent glass balustrade, the required diagonal tensile rods disturb the overall appearance as well as reducing the floor space of the balcony; they also represent a potential safety risk, as children may be tempted to climb on them. However, using modern technologies and measuring tools, it is now possible for the protective balustrade glazing to additionally function as the primary load-bearing element of the balcony.

2 GLASSBRACE ELEMENT

2.1 *Objective and requirements*

The goal of this research project is to harness the load-bearing ability of the glass side balustrade and so to dispense with the previously required tensile rods. The glazing is now no longer merely a fall guard but also has a bracing function (Figure 2). This enables both constant loads, such as the dead load of the platform and balustrade, and variable loads from wind and general usage to be transferred into the building. As the glazing is now a component of the balcony's primary support construction, a suitable glazing structure is required to ensure a sufficient residual load-bearing capacity in the event of failure of individual panes. A further focus of the development are the glued joints between the glass elements and the handrail and platform, since excessively high concentrations of tensile stresses at these points could lead to premature glass failure.

DOI: 10.1201/9781003348450-138

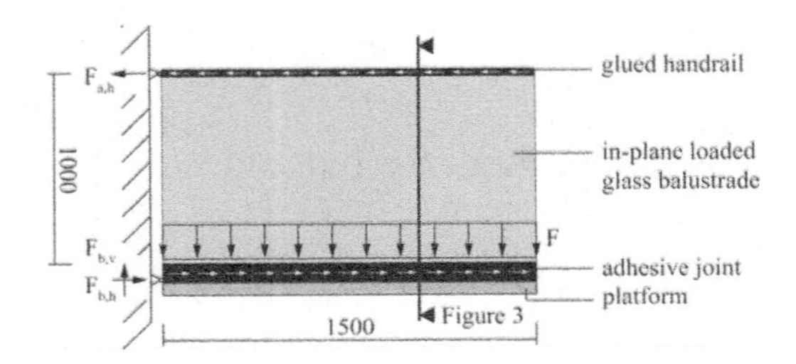

Figure 2. Construction concept of the bracing balustrade element.

2.2 *Construction and mode of action*

The design is based on load-bearing adhesive joints between the glass pane and the balcony platform and between the glass pane and the U-shaped handrail. Bonding by an elastic silicone sealant enables a secure load transfer with no increased tensile stress peaks in the glass. In this way, the glass panel functions as a bracing for the balcony construction. Serial manufacture is facilitated by the modular structure of the element and the use of an adapter profile at the foot, with very high quality and rapid installation at the construction site.

3 COMPONENT TESTS

3.1 *Investigation programme*

The development of the load-bearing adhesive bracing element requires a detailed verification concept for the adhesive and glass. For this purpose, a comprehensive test programme is performed along with numerical calculations. The experimental investigations are divided into component and bench scale tests. The small part tests are concerned with adhesive properties and creep behaviour, taking into consideration the effects of ageing (Giese-Hinz *et al.*, 2021). The bench tests focus on the stress resistance of the element to static and dynamic loads, as well as the component behaviour in the extraordinary event of glass breakage.

3.2 *Setup and verification of impact resistance*

The component tests assume a balcony with a cantilever of 1.50 m and a width of 3.00 m. The experimental verification of impact resistance is performed from a drop height of 900 mm. In addition, the balcony is loaded with the live load and dead weight in the limit state of the load capacity using equivalent steel weights. All specimens passed the tests. Neither the glass nor the adhesive joints failed at any of the impact points.

4 NUMERICAL SIMULATION

The aim of numerical simulation is to create a model for calculating the system with static stress at pane level in conjunction with a dynamic stress from a soft impact. The model is used to verify the stresses and deformations in all components and the connections to the existing structure. To determine the maximum stresses and deformations of the glass pane and the adhesive joint due to the impacts, the results of the static and the dynamic calculations are superimposed at the decisive points.

The calculations show that the vertical static loads from dead weight and live load only place the glass under low stress. The decisive stress on the glass panes comes from the pendulum impact. The impact load-bearing capability of the glass panes can be verified by calculation and agreement with the component tests. The maximum load on the adhesive joint at the bottom edge of the glass (shear stress in the adhesive joint) is, on the other hand, very high. The load on the adhesive due to the pendulum impact is smaller. The verification is performed successfully and confirms the results of the experiments.

5 SUMMARY

The tests and numerical calculations show that it is possible to replace previously necessary tensile anchors on balconies by a glued all-glass bracing balustrade and to verify its protection against falling. Moreover, it appears that sufficient residual load-bearing capacity can be attained for relevant damage scenarios.

ACKNOWLEDGEMENTS

This study arose out of the 'GLASSBRACE' research project and was funded by the German Federal Ministry for Economic Affairs and Energy (BMWi) through the Central Innovation Program (ZIM) within the KLEBTECH network. Special thanks go to the other project partners BONDA Balkon- und Glasbau GmbH and Thiele Glas Werk GmbH for their helpful cooperation and technical support.

REFERENCES

Giese-Hinz, J., Kothe, C., Louter, C. & Weller, B. 2021. Mechanical and chemical analysis of structural silicone adhesives with the influence of artificial aging, Int. J. Adhes. Adhes.doi:10.1016/j.ijadhadh.2021.103019.

Current Perspectives and New Directions in Mechanics, Modelling and Design of Structural Systems – Zingoni (ed.)
© 2022 Copyright the Author(s), ISBN 978-1-032-39114-4

Thermal loads in triple glazing: Experimental and numerical case study

P. Bukieda & B. Weller
Institute of Building Construction, Technische Universität Dresden, Dresden, Germany

ABSTRACT: Increasing demands with regard to size, energy efficiency and user comfort result in complex façade elements, often consisting of multi-pane insulating glass units with thermal or solar control glass and shading elements. This leads to numerous structural and environmental factors, which influence the development of temperatures in glass panes and thus thermal stress. Glass breakage due to thermal stress is already experienced in practice. Nevertheless, there are no generally valid and introduced engineering approaches for determining thermal loads under affecting all significant key influence factors for complex façade elements or triple glazing. In this contribution to the subject, the temperatures of triple glazing with a functional coating and various shading elements was investigated under outdoor exposure. The environmental parameters and temperatures of the glass panes were recorded and validated by transient simulation. This paper outlines the experimental monitoring and presents the validation of the simulation results. The heat transfer coefficient has proven to be one of the most influential factor in the simulation of glass temperatures. Since the heat transfer coefficient cannot be directly determined by measurement, a number of calculation approaches were evaluated, leading to dimensioning assumptions for the simulation of thermal loads in triple glazing.

1 INTRODUCTION

The absence of any specifications of thermal stress on the action and resistance side often results in conservative design assumptions that necessitate the use of thermally toughened or heat strengthened glass. Specifications for calculating thermal loads and their proofs might enable a safe and more economic use of annealed glass panes in window and façade constructions. In order to investigate the use of an existing software tool for the determination of temperatures in triple glazing, experimental monitoring and numerical simulation is implemented. The heat transfer coefficient has a considerable impact on the simulation of thermal loads. This paper examines a number of heat transfer coefficient approaches and assesses them with regard to the simulation of thermal loads.

2 EXAMINATION AND RESULTS

2.1 *Heat transfer coefficient*

A significant heat transfer is described by stating the heat transfer coefficient h W/(m²·K). In civil engineering the heat transfer coefficient comprises a component resulting from radiation h_r and another due to convection h_c. As the heat transfer coefficient cannot be directly determined by measurement, four different calculation approaches were examined to quantify their extent of the impacts on evolving thermal stress. To create a data basis for the simulation, a number of experiments were conducted with flow velocity measurements on the façade level.

2.2 *Monitoring and measuring results*

Monitoring took place in the period between June and August 2021 on two identical triple glazing with a sun protection coating and wooden frame under outdoor exposure. The triple glazing was measuring 1000 mm x 1000 mm x 38 mm. The ambient climatic conditions were measured, comprising the irradiation intensity at pane level, the interior and exterior air temperatures and the direction-independent flow velocity n the vicinity of the triple glazing. Also, three shading scenarios were examined: no shading (scenario 1), external (scenario 2) and one internal (scenario 3) shading. In the investigations, the maximum pane temperature differences and, in turn, the resulting thermal stress, were determined from different measuring points of one pane. Figure 1 shows the maximum temperature differences determined for selected monitored days in dependence of the shading scenario. The maximum temperature difference of 17.7 K was measured on 7 August 2021 in the outer pane with an external shading (scenario 2). This corresponds to a thermal stress of 11.2 N/mm².

DOI: 10.1201/9781003348450-139

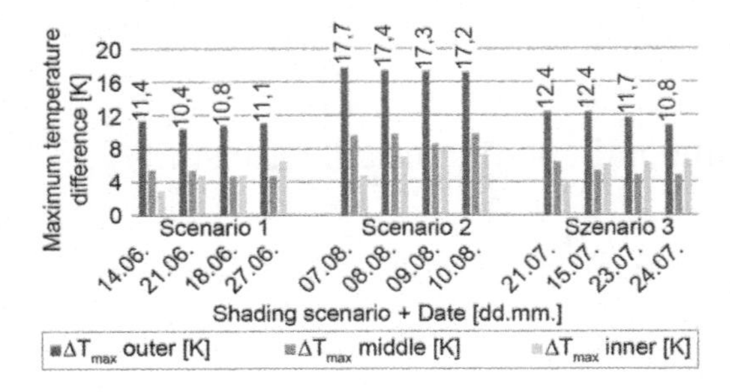

Figure 1. Maximum temperature differences determined in all three panes of the triple-glazed unit on all reference days.

2.3 *Numerical simulation*

The DELPHIN simulation program made by Bauklimatik Dresden Software GmbH was used to calculate transient glass temperatures. First, a two-dimensional model of the triple glazing was created for each scenario. The frame and seal area was greatly simplified. Figure 2 shows the model with external shading (scenario 2). All structural factors relating to the glazing and its material properties were defined in the model.

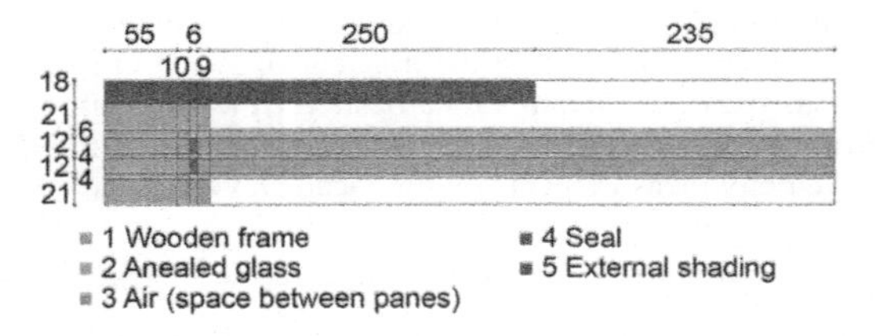

Figure 2. 2D model of the triple-glazed unit with external shading (Scenario 2).

Marginal and field conditions were designated on the basis of the model to enable all essential heat transfers to be included in the calculation. The following are defined as structural processes for the calculation:

- Heat transfer and internal and external air temperatures
- Convection within the pane cavities
- Absorption of short-wave radiation for each pane
- A long-wave radiation exchange between the panes

The simulations were performed with different simulation durations (1 hour, 2 hours, 24 hours).

2.4 *Simulation results*

In the validation procedure, the calculated pane temperatures were compared with the measured ones in the measured positions of each pane (middle area, shadowed area, edge area behind the frame). A deviation of 10 % was set between the measured and the calculated pane temperature. In addition, it was taken into account whether the pane temperatures were on the safe side with regard to the calculation of the maximum temperature difference.

To determine the heat transfer coefficient, the approach pursuant to DIN EN ISO 6946 was determined in the course of the validation to be suitable for the simulation of thermal loads. Moreover, simulation durations of 2 and 24 hours were validated. For the edge areas, the simulated temperatures were too high in the scenarios with no external shading (scenarios 1 and 3). Accordingly, the maximum temperature difference works out too low. Further adjustments in the implementation of the frame and shading materials are needed. Within the simulation, the thermal stress in an amount of about 3 N/mm² was overestimated with the current boundary conditions.

3 RECOMMENDATION AND OUTLOOK

An determination of the heat transfer coefficient pursuant to DIN EN ISO 6946 is recommended for the simulation of thermal loads but requires assumptions to be made regarding flow velocities on the façades.

The examination of thermal loads in triple glazing lead to maximal thermal stresses of 11.2 N/mm². The amount of stresses is in the area of the current allowable stresses in glass edges. However, no critical scenario or safety factor was taken into account in this study. Current design differences will be harmonised within the framework of the new Eurocode 10, which is expected to come into force in 2024 (Feldmann & Di Biase 2018). The current draft of Eurocode 10 (CN/TS 250) provides for the assessment of thermal loads. Until a full calculation concept is available, the risk continues to be borne by the planners. To choose limit values and scenarios, planners can refer to current research studies (Gulappi 2021) or more conservative standards for guidance.

REFERENCES

ISO 6946: 2017. Building components and building elements - Thermal resistance and thermal transmittance - Calculation methods. German version. Beuth Verlag GmbH.

Feldmann, M. & Di Biase, P. 2018. The CEN-TS “Structural Glass - Design and Construction Rules” as pre-standard for the Eurocode. ce/papers, Wiley, 2, 71–80.

Galuppi, L., Maffeis, M. & Royer-Carfagni, G (2021). Enhanced engineered calculation of the temperature distribution in architectural glazing exposed to solar radiation. In: Glass Struct Eng, Vol. 6, pp. 425–448.

Current Perspectives and New Directions in Mechanics, Modelling and Design of Structural Systems – Zingoni (ed.)
© 2022 Copyright the Author(s), ISBN 978-1-032-39114-4

Influence of the interlayer core material in glass–plastic composite panels on performance characteristics according to the requirements for laminated safety glass

Julian Hänig & Bernhard Weller
Institute of Building Construction, Technische Universität Dresden, Dresden, Germany

ABSTRACT: There is an increasing need for dematerialization and to reduce the consumption of resources in the building industry. This can be addressed by the use of innovative lightweight designs in conjunction with new material combinations. The architectural trend towards maximum transparency in glass façades is an indication of the demand for new glass products that enable lightweight designs, which, in turn, drives on the development of new type of lightweight glass–plastic-composite panel that consists of a stiff, thick, transparent polymeric interlayer core covered by thin layers of glass. With outstanding structural performance, full transparency, and reduced self-weight, these composite panels are conceived as a lightweight substitute for conventional glass that saves resources by virtue of its reduced self-weight and innovative lightweight design. However, the influence of the interlayer core material on the performance characteristics – durability and impact resistance under hard and soft body impact – that are used to categorize glass–plastic-composite panels as either laminated glass or laminated safety glass is unknown, even though they are an essential aspect of safety applications in the building industry.

Composite panels were investigated in line with the requirements for laminated safety glass using two different polymeric interlayer cores: 1) a stiff and brittle polymethylmethacrylate and 2) polyurethane, which is more flexible. The experimental study incorporated a durability test pursuant to EN ISO 12543-4, a ball drop test for laminated glass pursuant to EN 14449 and a pendulum test pursuant to EN 12600. In line with the performance characteristics relating to durability, impact resistance and fracture behavior, the two panel configurations were classified according to laminated safety glass requirements. This enables the potential applications of glass–plastic-composite panels in the building industry to be derived.

1 INTRODUCTION

The ongoing trend towards large glass structures and the need to keep the use of material to a minimum is being addressed by developments towards enhancing the performance and weight aspects of transparent building materials. As glass becomes thinner and its availability increases, the building industry demands more and more new applications. However, as thin glass is highly flexible, it requires enhanced stiffness to ensure sufficient load-bearing capacity. This has led to the development of a new material combination in the form of a composite panel comprising a thick, lightweight, transparent polymeric interlayer core bonded directly to covering layers of thin glass (Figure 1). The composite panel should serve as a glass substitute characterized by low self-weight, good ductility and high structural performance and has now been jointly developed by the Technische Universität Dresden and KRD Coatings GmbH.

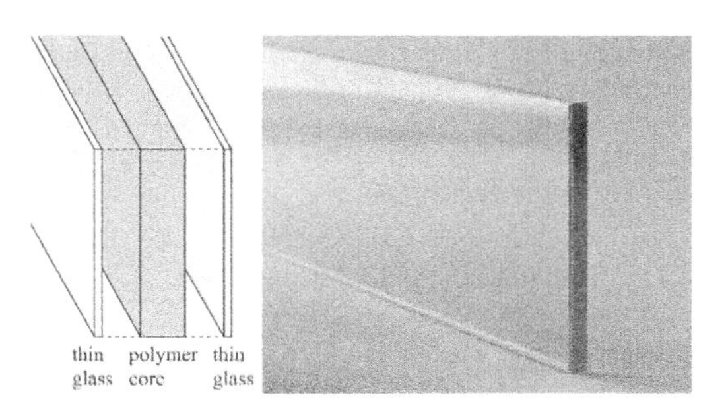

Figure 1. Glass–plastic-composite panel structure (schematic assembly and view).

Two quite different transparent polymeric materials, a stiff and relatively brittle PMMA and a high-flexibility polyurethane (PU), were considered for the interlayer core (Hänig & Weller 2019, Wittwer & Schwarz 2016).

To satisfy the safety needs as a substitute for safety glass, the durability and safe mode of failure

DOI: 10.1201/9781003348450-140

requirements for laminated safety glass (LSG) must be fulfilled. The novel composites were tested according to the requirements set out in EN ISO 12543 'Glass in building – Laminated glass and laminated safety glass'.

These incorporate durability tests pursuant to EN ISO 12543-4, ball drop tests pursuant to EN 14449 and pendulum tests pursuant to EN 12600 for both the PMMA and the PU composite configurations. Glass–plastic-composite panels are classified in accordance with the requirements for LSG, corresponding to the performance characteristics for durability, impact resistance and fracture behavior.

2 TEST METHODS

The experimental study method (Figure 2) covers the general requirements for LSG.

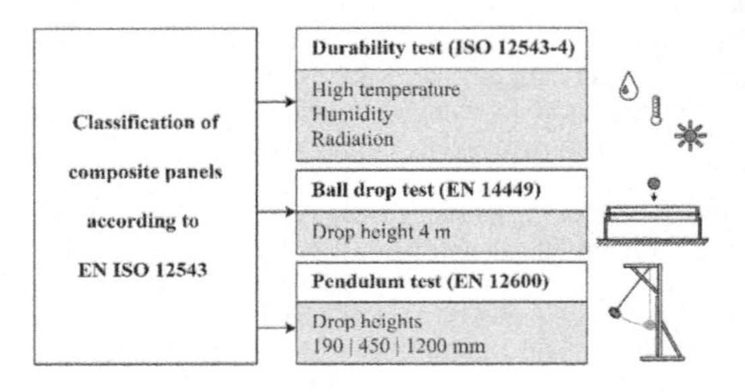

Figure 2. Experimental study method for the assessment of glass–plastic-composite panels.

3 RESULTS

The full paper presents the test results in detail. The influence of the interlayer material in glass–plastic-composite panels on performance characteristics is discussed along with the results. This enables a concluding classification and reveals the potential of these composite panels.

4 SUMMARY, CONCLUSIONS AND OUTLOOK

This research incorporates a summary of the findings regarding the influence of the interlayer core material in glass–plastic-composite panels on the performance characteristics according to the requirements for LSG. Glass–plastic-composite panels with a PMMA interlayer core are classified as LG, since no safe mode of breakage or residual capacity was determined. However, the panels still display great potential for applications in which safety is not a requirement. Their high levels of stiffness and durability as well as their complete transparency indicate their suitability for indoor applications, such as all-glass systems, partition walls and glass furniture, in which low self-weight is crucial to the design. Another approach is pursued in (D'Ettore et al. 2021). Small and unobtrusive connection details that are directly integrated into the PMMA interlayer core (Figure 3) are developed to generate maximum transparency by taking advantage of the panels' low self-weight, high composite stiffness, and the reduced stress concentrations in the glass resulting from direct joint in the PMMA. Moreover, its use in insulating glass units to reduce self-weight, in particular for load-sensitive rehabilitation, enables an extensive range of applications.

Figure 3. Prototypes of integrated connections in the PMMA interlayer core of glass–plastic-composite panels.

The PU interlayer is already being successfully employed in building industry applications. The safe mode of breakage combined with the reduction in self-weight mean that composite panels with a PU interlayer core are suitable for a wide range of applications in the building industry.

REFERENCES

D'Ettore, E., Hänig, J., Weller, B. 2021. Connection Joints for Glass–plastic-Composite Panels - Experimental Research. In: *engineered transparency 2021*: 237–250. Berlin: Ernst & Sohn. https://doi.org/10.1002/cepa.1635

EN 12600 2003. Glass in building - Pendulum tests - Impact test method and classification for flat glass.

EN 14449 2017. Glass in building - Laminated glass and laminated safety glass - Product standard.

EN ISO 12543 2011. Glass in building - Laminated glass and laminated safety glass part 1-4.

Hänig, J., Weller, B. 2019. Load-bearing behaviour of innovative lightweight glass–plastic-composite panels. *Glass Structures & Engineering* 5: 83–97. https://doi.org/10.1007/s40940-019-00106-5

Wittwer, W. & Schwarz, T. 2016. A deterministic mechanical model based on a physical material law for glass laminates. In: *engineered transparency 2016*: 489–500. Berlin: Ernst & Sohn.

Current Perspectives and New Directions in Mechanics, Modelling and Design of Structural Systems – Zingoni (ed.)
© 2022 Copyright the Author(s), ISBN 978-1-032-39114-4

Load analysis for the development of a bonded edge seal for fluid-filled insulating glazing

A. Joachim & B. Weller
Institute of Building Construction, Technische Universität Dresden, Dresden, Germany

ABSTRACT: Aesthetic and energy efficiency considerations play a decisive role in façade construction. In recent years, the use of fluids in pane cavities has been the subject of increased study, the aim being to satisfy the requirements of energy saving legislation. Such fluids are useful for transforming façades into multifunctional building envelopes. The combination of hydrostatic pressure and constant fluid contact places the edge seal of such glazing under considerable loads. For this reason, initial pilot projects could only be realized using mechanical clamping aids. However, this goes against the aesthetic requirements of modern glass façades, which strive to achieve a maximum proportion of glass combined with a minimum amount of edging. The goal, therefore, is to develop a bonded edge seal for use in fluid-filled insulation glazing.

This paper presents a brief introduction to the subject and explains the research goals. Followed by a potential edge seal design and a presentation of the acting loads. Adhesives used in the edge seal are exposed to physical, mechanical and chemical loads. This often leads to degradation processes in the polymers. A load analysis forms the basis of both a list of requirements and a suitable test program.

1 INTRODUCTION

More than any other part of a building, it is the glass façade that typifies modern architecture. The aim is for maximum transparency and natural light, particularly in prestigious office and administrative buildings. However, one effect of such a large glass surface area is that a relatively large proportion of energy is lost. For this reason, much research is being devoted to façade optimization. In recent years, studies have focused on the use of fluids in façades. They enable the thermal control of façade elements while counteracting both unwanted energy entry and energy loss. However, the aesthetic demands of a façade are also high. The demand is for structural glazing façades in which horizontal forces are borne by a load-transferring adhesive and the proportions of the frame are kept to a minimum.

However, the pane cavity is filled with a fluid rather than with an air-gas mixture, the hydrostatic pressure and the ageing processes caused by the fluid can lead to high stress levels acting on the edge seal. To satisfy the aesthetic demands, the aim is to develop a bonded edge seal for fluid-filled multi-pane insulating glass units (IGUs). The aim is for the bonded edge seal to transfer the mechanical loads resulting from hydrostatic pressure, wind, and payload without the need for external clamping. The constant fluid exposure represents a further essential load, as the adhesive must be constantly able to bear a load while retaining its sealing effect.

2 FLUID-FILLED IGUS AND BONDED EDGE SEAL

For an overview of the state of the art regarding fluid-filled IGUs, please refer to the publications by Joachim et al. (2021) and Joachim et al. (2022).

The development of a bonded edge seal for fluid-filled façade elements requires a number of geometrical and structural boundary conditions to be defined. A room-high façade element is assumed with a height of h = 3000 mm and a width of b = 1350 mm (Figure 2, left). The pane cavity measures approximately d = 24 mm. The pane cavity contains a fluid mixture of water and ethylene glycol with a mixing ratio of 70:30, which is used for heating and cooling.

The fluid causes a hydrostatic pressure to build up in the pane cavity, which exerts a tensile load on the bonded edge seal. To reduce this and limit pane deformation, a vacuum is generated by technical means in the façade element. This causes the pressure zero line to shift from the upper edge to the middle of the element. The load configuration assumes an antisymmetric form (Figure 2, right).

DOI: 10.1201/9781003348450-141

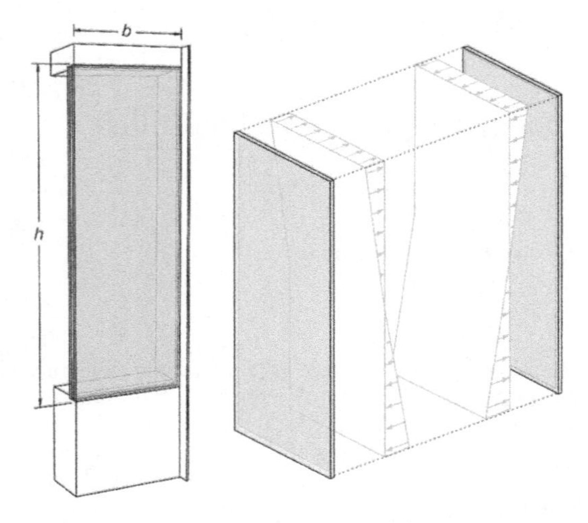

Figure 1. Example construction and antisymmetric load from hydrostatic pressure and technically generated vacuum.

In order to divert the mechanical load from the hydrostatic pressure and moreover to withstand the anticipated strong ageing load from the constant contact with the fluid mixture, a two zone edge seal is to be developed. This is similar to a conventional gas-filled multi pane insulating glass unit. The purpose of combining the two adhesives is to distribute the sealing and load transfer functions to suitable materials.

To enable such an edge seal to be realised, it is necessary to locate suitable adhesives. An analysis of the resulting loads is essential for enabling a requirements catalogue to be compiled and a suitable test program to be devised.

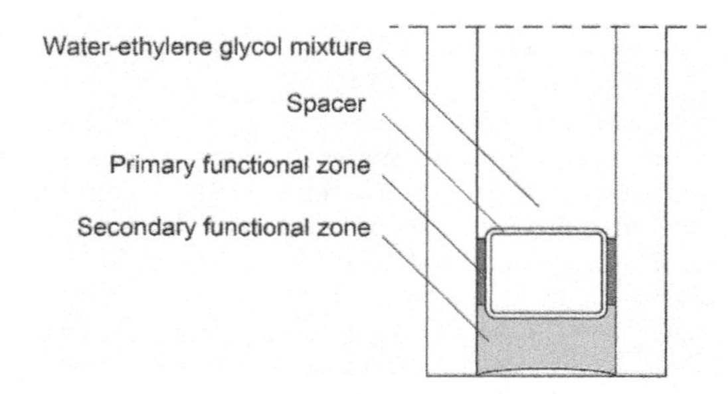

Figure 2. Approach for a two zone edge seal.

3 LOAD ANALYSIS

The load types are subdivided into physical, mechanical and chemical types. Table 1 presents an overview of load types.

The load analysis is presented in the full paper. For each load type, an explanation is given as to why it might be present in the edge seal, and, if the load is quantifiable, what load level can be anticipated. In addition, the load is assigned to the functional zone. The effects of physical and chemical ageing are then discussed.

Table 1. Overview of load types.

Physical	Mechanical	Chemical
Temperature	Hydraulic pressure	Water
Solar radiation	Wind load	Ethylene glycol
	Horizontal live load	

4 SUMMARY AND OUTLOOK

The load analysis reveals how diverse and extreme the loads are that must be taken into account in the development of a bonded edge seal for fluid-filled multi pane insulating glass.

The fluid in the cavity represents a particular challenge to the development of a bonded edge seal, both in terms of mechanical and chemical load. Even with the aid of vacuum technology, the hydraulic pressure of $p_h(h) \approx 15$ kN/m^2 is still more than ten times higher than the usually assumed distributed load due to wind. In addition, this is a constant load, which is far more critical for an adhesive than a transient load. But even chemical load from the fluid in the pane cavity represents a considerable obstacle to the development of a bonded edge seal. The water-ethylene glycol mixture penetrates the edge seal and causes a change to its mechanical properties.

ACKNOWLEDGEMENTS

This study arose out of the research project “fluidIGU” and was funded by the German Federal Ministry for Economic Affairs and Energy (BMWi) through the Central Innovation Program (ZIM) within the KLEBTECH network. Special thanks go to the project partners Bollinger + Grohmann Ingenieure and ADCO Technik GmbH for their pleasant cooperation and technical support.

REFERENCES

Joachim, A. et al. 2021. Selection of adhesives for the realisation of a bonded edge seal for liquid-filled insulating glazing. In: Weller, B.; Tasche, S. (eds.) *Glasbau 2021*, pp. 197–214. Ernst & Sohn, Berlin (2021). https://doi.org/10.1002/cepa.1606

Joachim, A. et al. 2022 Leak test for the material selection of a bonded edge seal for fluid-filled façade elements. In: *International Journal of Adhesion and Adhesives (IJAA)*. Volume 113. https://doi.org/10.1016/j.ijadhadh.2021.103082

Current Perspectives and New Directions in Mechanics, Modelling and Design of Structural Systems – Zingoni (ed.)
© 2022 Copyright the Author(s), ISBN 978-1-032-39114-4

The strength of glass with digital printing

J. Wünsch & B. Weller
Institute of Building Construction, Technische Universität Dresden, Dresden, Germany

J. Wittwer & A. Rumpf
Polartherm Flachglas GmbH, Großenhain, Germany

ABSTRACT: Enamelled glass has been used in opaque or semi-transparent elements of the building envelope for a considerable time. Once the ceramic ink or paste is applied, it is burned in at a high temperature during manufacture, and the glass panes are subjected to thermal pre-stressing at the same time. In the case of enamelled glass, the pre-stressing process does not run undisturbed, meaning that the bending strength of the glass is sometimes significantly lower than that of standard panes. Digital printing is a method that can be used to apply the ink to the enamelled glass. Ink is applied with a large-scale inkjet printer at a high resolution, using a digital template. Multi-colour printing takes place continuously, and any number of primary colour inks can be mixed during printing. The coating thickness is initially about 15-20 μm. Overprinting can be carried out to improve the opacity of translucent colours. The colour composition differs from that of other application methods (such as screen printing). The ratio of glass flow to pigment is altered in favour of the pigments. This is necessary to achieve sufficient opacity with thinner coatings. This article presents the results of four-point bending tests pursuant to the EN 1288-3 standard. This article compares the experimental results of digitally printed glazing with conventional printing methods and outlines the influence of the digital printing ink's colour on the bending strength.

1 INTRODUCTION

Enamelled glass is used on a regular basis in the building envelope. Depending on the method used to apply the ink, the enamel can take the form of a single-colour, full or partial coating, ranging to a high resolution, multi-colour photorealistic image. The technically important step of burning in takes place at the same time as thermal pre-stressing. The design value of the bending strength achieved for enamelled glass is generally considerably lower than for standard glass. This is addressed in the standard by a reduction in the minimum strength. For toughened, enamelled glass, the design value is reduced from 120 N/mm^2 to 75 N/mm^2 (EN 12150-1). The lower strength levels apply irrespective of the printing method used. Several academic works consider the influence of the printing method on the bending strength. The results are in part w ell above the minimum strengths stipulated in the standard.

2 MATERIALS

This study investigates flat glass made from soda-lime-silica with a nominal thickness of 6 mm. All the glass elements were coated in enamel over one full side using a variety of application methods, each with their respective ink compositions. For this article, the author has chosen digital printing with four different colours, as well as screen printing and spraying with black screen printing ink. Drying takes place in a kiln at around 180 °C, to ensure a fast and reproducible process. The duration of the drying process depends on the solvent content and the film thickness of the applied ink. The actual burning in of the ceramic inks takes place during pre-stressing at temperatures of between 600 °C and 650 °C (Wörner 2001).

3 EXPERIMENTAL

The dimensions of the samples are prescribed in the standard as 1100±5 mm for the length and 360±5 mm for the width, (EN 1288-3). The specimens are installed in the test assembly in such a way that the printed glass surfaces are oriented downwards and lie in the flexural tension zone under load. The tests are performed using a test stand for four-point bending (Figure 1), whose dimensions and structure correspond with the specifications given in EN 1288-3. A load is applied continuously to the glass plane using a hydraulic cylinder until breakage occurs.

DOI: 10.1201/9781003348450-142

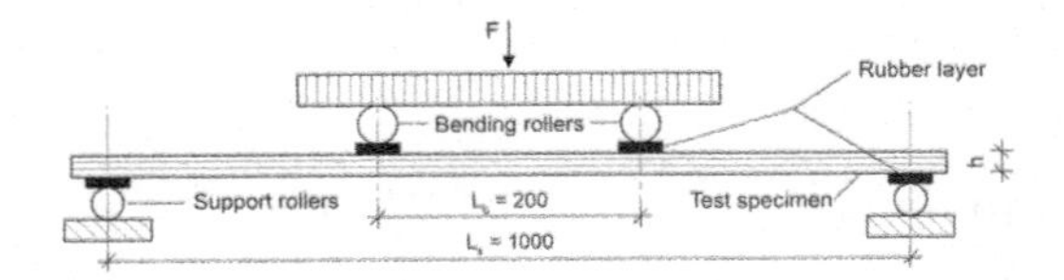

Figure 1. Four-point bending test setup according to EN 1288-3.

4 RESULTS

Following the toughened glass test, the test specimens in all series displayed a typical post-fracture structure. The fracture pattern was characterised by small cube-shaped fragments that present a low risk of injury from cutting.

The 5 % fractile of bending strength of the glass elements with the sprayed-on screen print ink attained a value of 89.2 N/mm². The bending strength of the panes with the screen print was 111.3 N/mm². The highest value of 124.7 N/mm² attained among the enamelled panes was calculated for the digital black print. By far the highest bending strength of 148.4 N/mm² was attained by the reference series of toughened glass panes.

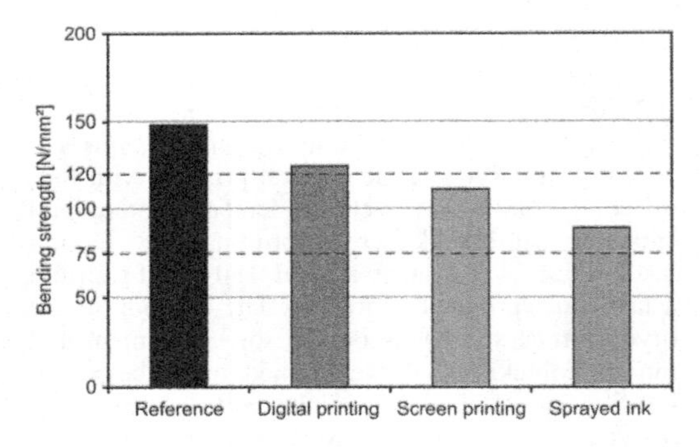

Figure 2. 5 % fractile of the bending strength tensile strength obtained with various printing methods in comparison with the design values.

All results obtained with digitally printed glazing are of a similarly high level. The bending strength of the panes with the black digital print attained a value of 124.7 N/mm^2 stated above. The other bending strength values were calculated as 133.3 N/mm^2 for the red ink, 132.3 N/mm^2 for the blue ink, and 135.0 N/mm^2 for the white ink.

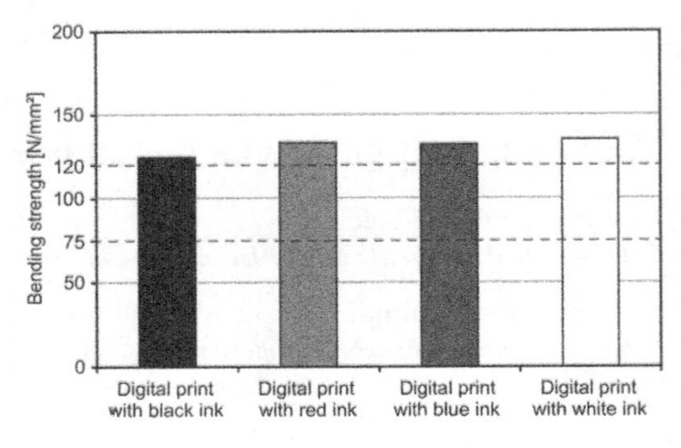

Figure 3. 5 % fractile of the bending strength for digitally printed glazing obtained by varying the ink colour in comparison with the design values.

5 CONCLUSION

For all three application procedures, significant reductions in strength were observed compared to the reference. The results attained here are comparable with those obtained by other research studies (Maniatis & Elstner 2016, Krampe 2013, Krohn 2002). If a sufficient amount of pre-stressing is attained, the design values of the bending strength of 120 N/mm² required by the EN 12150-1 standard can be achieved with all the digitally printed glazing elements investigated.

REFERENCES

Maniatis, I. & Elstner, M. 2016. Investigations on the mechanical strength of enamelled glass. *Glass Struct. Eng.* (2016) Volume 1: 277–288.

Krampe, P. 2013. Zur Festigkeit emaillierter Gläser. Dissertation, Technische Universität Dresden.

Krohn, M.H. et al. 2002. Effect of Enamelling on the Strength and Dynamic Fatigue of Soda-Lime-Silica Float Glass. *Journal of the American Ceramic Society* 85, Volume 10: 2507–2514.

Wörner, J.D. et al. 2001. Glasbau – Grundlagen, Berechnung, Konstruktion, Berlin, Heidelberg: Springer.

Wünsch, J. et al. 2021. Biegezugfestigkeit von emailliertem Flachglas unter Variation des Farbauftrages. In Weller, B. & Tasche, S. (eds), *Glasbau 2021*: 345–354. Berlin: Ernst & Sohn.

EN 1288–3:2000-09. Glass in Building – Determination of the Tensile Strength of Glass – Part 3: Test with Sample Supported on Two Sides (Four-Point Bending Procedure). Berlin: Beuth.

EN 12150–1:2000-11. Glass in Building – Thermally-Toughened Soda-Lime-Silicate Safety Glass – Part 1: Definition and Description. Berlin, Beuth.

Current Perspectives and New Directions in Mechanics, Modelling and Design of Structural Systems – Zingoni (ed.)
© 2022 Copyright the Author(s), ISBN 978-1-032-39114-4

Numerical simulation of the heat flow through laminated glass beams exposed to fire: A parametric study

M. Möckel & T. Juraschitz
Institute of Building Construction, Technische Universität Dresden, Dresden, Germany

C. Louter
Faculty of Civil Engineering & Geoscience, Structural Design & Building Engineering, TU Delft, Delft, The Netherlands

ABSTRACT: Glass surfaces are getting larger and more impressive to obtain maximum transparency for buildings. To increase the transparency of the whole structure, load-bearing glass structures are getting more and more important. However, their use is currently still limited regarding concerns about its performance in case of fire. To reduce high costs and the high environment pollution caused by fire tests, preliminary examinations in form of simulations could be a solution to examine the fire performance. This paper describes the design of a numerical model to calculate the heat flow through glass beams under fire load. The model was validated using the results of experimental examinations.

1 INTRODUCTION

Load-bearing glass beams or columns become increasingly popular. Planning a building includes fire protection requirements, which must meet a high standard in public buildings. Previous research studies showed that the load-bearing capacity of glass beams exposed to fire load is low. Consequently, to use structural glass, alternative fire protection methods such as sprinkler systems or alternative escape routes must be used in public buildings.

The content of this study is the numerical simulation of the heat flow through a glass beam under fire load. To validate these simulations, the results of experimental fire resistance tests of previous research studies will be used. Even though fire tests are both expensive and environmentally harmful they need to be carried out to examine the fire resistance of building components.

2 STATE OF RESEARCH

Several research studies examined the effects of high temperatures on the load-bearing behaviour of structural glass. For example, Sturkenboom (2018), Louter et al. (2021) and Möckel et al. (2022) examined the fire behaviour of laminated glass beams under static load. The explicit description of the test set-up is defined in Möckel et al. (2022). The results of these three research studies showed that laminated glass beams without any heat protection resists the heat load from 5 min to 28 min. In addition to the experimental examinations, Louter et al. (2021) carried out numerical simulations as well. Due to the good representation of the reality, the results of the simulations will be compared to the experimental set-up of Sturkenboom (2018). Additionally, the numerical results from Louter et al. (2021) will be used for comparison.

The fire tests to examine the fire resistance of glass beams base on the set-up of the four point bending test according to EN 1288-3. Figure 1 shows the set-up of the fire tests of Möckel et al. (2022).

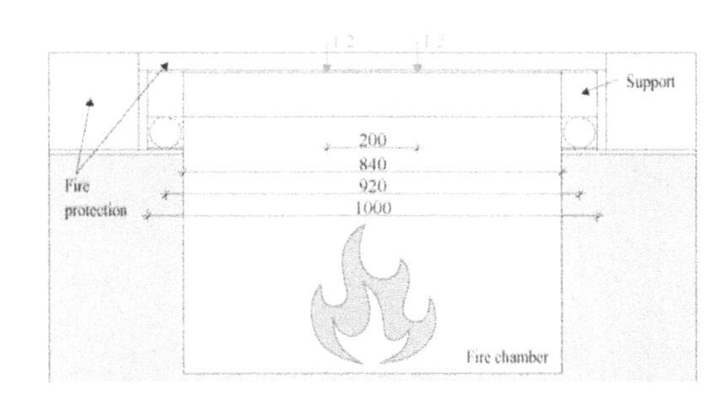

Figure 1. Test set-up according to Möckel et al. (2022).

The beam stands vertically on two hinged supports and the load was applied on two points on top of the beam. Additionally, the specimens were supported vertically at both ends to prevent them from buckling. In these tests, the whole beam except the support areas was loaded with the heat. The tests were carried out with the standard temperature-time (stt) curve according

DOI: 10.1201/9781003348450-143

to EN 1363-1. Thermocouple wires were embedded in the interlayers of the glass beams to measure the temperature (temp.) in dependence on time.

3 FE-MODEL SET-UP

The numerical simulations in this study were carried out with Ansys 2021 R2. The glass structure in the FE model was chosen as a laminated fully tempered glass beam with three glass panes with 10 mm thickness each and two PVB interlayers with a thickness of 1.52 mm each, like the specimens in the references.

Due to the symmetry of the beam and the support situation, the beam was split within the use of two symmetry planes in y- and x-direction. In the experimental tests, the support areas were covered completely by the steel supports to prevent the beam from buckling during the test. Therefore, in the FE model these areas were chosen as unexposed surfaces. The heat load was applied on the outer glass pane and the edges from below as heat radiation. To simplify the FE model, the temp. load was applied consistently over the height of the beam. The input temp. curve was the stt curve according to EN 1363-1.

4 RESULTS & DISCUSSION

The initial simulation with the reference model was carried out to examine how the heat flows through the single layers of the glass beam and for the validation with the experimental results of Sturkenboom (2018). Figure 2 shows the experimental studies in comparison with the simulation results. The input temp. curve "Furn. EN 1363-1" is presented in light grey dashed line. The black line defines the temp. of the reference model "FE-REF". All curves were calculated at one point inside the interlayer of the beam.

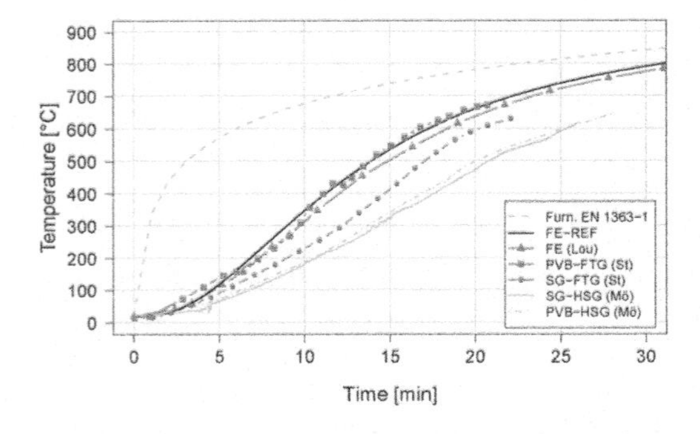

Figure 2. Comparison of the experimental results of Sturkenboom (2018) (St), Möckel et al. (2022) (Mö) and the results of the FE simulation of this study and Louter et al. (2021) (Lou).

The evaluated temp. curve has not increased that much as the input curve. The maximum temp. difference between the input curve and the result equals about 450 K after about 5 min. The high difference between the temp. curves of the FE simulation and the experimental results of Möckel et al. (2022) stands out at first sight. The maximum gap equals up to 200 K. In difference, the results of Sturkenboom (2018) are closer with a maximum gap of 100 K. Especially the temp. curve of the specimen "PVB-FTG (St)" fits almost completely to the temp. curves of the FE simulations. The diagram shows also the temp. curve of the FE simulation of Louter et al. (2021), named "FE (Lou)". This curve has the same shape as the simulation curves of this study. There is only a small gap to the results of "FE-REF" of maximum 40 K. Due to the well-matching experimental results of Sturkenboom (2018) and the FE results of Louter et al. (2021), the simulation results of the reference model "FE-REF" were validated.

5 SUMMARY & OUTLOOK

The numerical evaluation of the fire tests on laminated glass beams showed that the results are comparable with the results of the experimental tests. The simulations fit well with the experimental results of Sturkenboom (2018).

However, these results cannot be generalised as they based on the set-up of the experimental tests by Sturkenboom (2018). To achieve secure results, which comply the current fire resistance standards, further research projects should focus on validate FE simulations with different test set-ups with experimental results. Materials and glass beam compositions, static load level and load introduction points should be included into the FE model and pre-selected before experimental testing. By this, the number of fire tests and the associated pollution as well as high costs could be reduced.

Forward-looking, the TU Dresden will investigate systems to reduce the heat load on the load-bearing centre of the laminated glass beam in further research projects in order to define a fire resistance for laminated glass beams.

REFERENCES

DIN Deutsches Institut für Normung e.V., 2000. Bestimmung der Biegefestigkeit von Glas: Teil 3: Prüfung von Proben bei zweiseitiger Auflagerung (Vierschneiden-Verfahren), 81.040.20(DIN EN 1288-3). Berlin: Beuth Verlag GmbH.

DIN Deutsches Institut für Normung e.V., 2020. Feuerwiderstandsprüfungen: Teil 1: Allgemeine Anforderungen, 13.220.40; 13.220.50(DIN EN 1363-1). Berlin: Beuth Verlag GmbH.

Louter, C.; Bedon, C.; Kozłowski, M., and Nussbaumer, A., 2021. Structural response of fire-exposed laminated glass beams under sustained loads; exploratory experiments and FE-Simulations. Fire Safety Journal, 123. doi:10.1016/j.firesaf.2021.103353.

Möckel, M. and Louter, C., 2022. Study on the load-bearing behaviour of laminated glass beams exposed to fire. Challenging Glass Conference Proceedings, Vol. 8 (2022). Challenging Glass Conference Proceedings, 8.

Sturkenboom, J., 2018. Fire Resistant Structural Glass Beams: Delft University of Technology, master thesis.

11. Steel structures, Steel connections, Steel-concrete composite construction, Cold-formed steel

Current Perspectives and New Directions in Mechanics, Modelling and Design of Structural Systems – Zingoni (ed.)
© 2022 Copyright the Author(s), ISBN 978-1-032-39114-4

Full-scale tests of industrial steel storage pallet racks

N. Baldassino, M. Bernardi & R. Zandonini
Università di Trento, Trento, Italy

A. di Gioia
Metalsistem S.p.A., Rovereto, Italy

ABSTRACT: Industrial steel storage pallet racks are framed structures typically made of cold-formed steel profiles characterised by a quite complex behaviour. As part of an extensive research on the racks' static and seismic behaviour, eight full-scale tests on four-level two-bay commercial pallet racks were carried out: two vertical loads preliminary tests and six monotonic push-over tests with three different levels of vertical loads and an inverse triangular pattern of horizontal forces in the down-aisle (longitudinal) direction.

1 INTRODUCTION

Industrial steel storage pallet racks, worldwide used in the logistic field to store goods and products, are typically framed light systems made of cold-formed steel profiles, with open mono-symmetric perforated members connected through non-linear semi-rigid joints with a non-symmetric response. The racks' behavior is quite complex to be predicted and the so-called "design by testing" approach is commonly adopted in their design. A number of experimental studies on racks' components and sub-assemblies can be found in literature. In addition, various experimental investigations focused on their overall structural performance were carried out starting from the mid-70s, although the variability of the racking systems and of the testing set-ups adopted in the different studies calls for additional studies on this topic.

An extensive research has been recently completed at the University of Trento, focusing on the static and seismic racks' performance at both the components and the global level (Bernardi 2021). To investigate the racks' global response eight full-scale tests were accomplished, using an innovative full-scale testing set-up: two vertical load tests and six monotonic push-over tests. The main features and results of the full-scale tests are presented and discussed.

2 THE EXPERIMENTAL PROGRAMME

2.1 *The testing set-up and the reference structure*

An innovative testing set-up was developed and adopted for the tests (Baldassino et al. 2021). A 'rigid' steel trussed testing tower acts as a reaction frame. The use of eighteen independent dynamic actuators permits to apply different vertical forces on each bay of the racking specimen, as well as different horizontal loads at each storey level. The key point of this testing set-up is its capability to maintain the verticality of the loads during all the phases of the tests, while the specimen sways even significantly. In fact, the vertical actuators can horizontally translate on rail beams, as a function of the horizontal displacement of the rack measured at the corresponding level.

A typical pallet rack made of cold-formed steel profiles was selected as case study for the research. Four-level two-bay racks braced only in the cross-aisle direction with irregular "D" bracings were tested. The structures had a nominal height of 8000 mm, with an inter-storey height of 2000 mm, and the bays were 1927 mm long. For all the tests the racks were provided with ideal hinges at the bases.

2.2 *The vertical loads and push-over tests*

Two preliminary tests were first performed on specimens with an initial out-of-plumb in down-aisle direction of 1/126, monotonically increasing the vertical loads applied to the racks up to their collapse. The tests allowed evaluating the load bearing capacity of the racks. Both tested racks collapsed in the same sway mode, with plastic deformations of the nodal zones of the first load level.

Push-over tests were then carried out in the down-aisle direction, considering three levels of vertical loads and an inverse triangular horizontal loading pattern. Two tests were performed for each level of vertical load P, selected as a fraction of an "experimental service load" (ESL) obtained from the two vertical loads tests. The tests were performed with vertical loads equal to

DOI: 10.1201/9781003348450-144

ESL, *2/3ESL* and *1/3ESL*. The vertical loads were first applied to the racks with no applied out-of-plumb and then maintained constant during the monotonically horizontal loading, up to the collapse. The structures' global capacity curves for the three levels of vertical loads are plotted in Figure 1, in which the normalised total base shear *V* vs. top drift curves are reported. The base shear is normalised by the maximum value of *V* obtained for *P* = *1/3ESL*. The comparison of the curves shows that the initial lateral stiffness of the rack is not affected by the level of *P*. When the lateral displacement increases, the influence of *P* becomes more apparent, in terms of both lateral stiffness and maximum shear load. The higher deformability of the frames in case of lower levels of *P* is probably due to the flexibility of the racks, whose global behaviour is highly influenced by the beam-to-column joints performance. The key role played by the joints can be appreciated also by considering the frames' failure mode (Figure 2), characterised by the localization of the plastic deformations of the frames at the joints level. All the racks showed the same collapse mode, with the sway of their first level and the plastic deformation of the nodal zones of the first and second level.

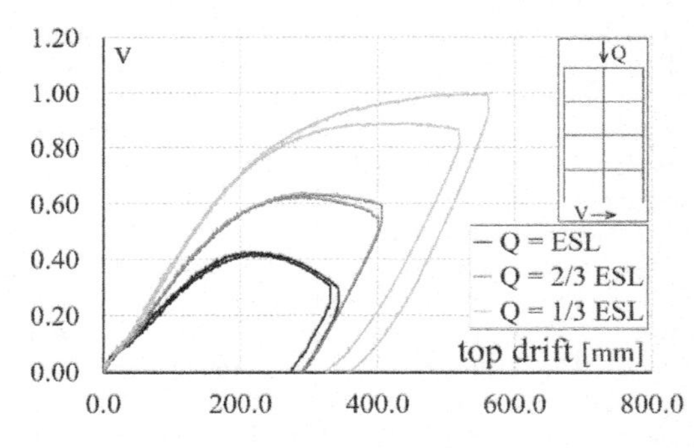

Figure 1. Normalised push-over curves for the six push-over tests, with three levels of vertical loads *P*.

The results of the push-over tests allowed evaluating the behaviour factor q of the analysed racks. Mean values of q equal to 1.44, 1.61 and 2.31 were respectively obtained for *P* = *ESL*, *P* = *2/3ESL* and *P* = *1/3ESL*. Higher values of q are associated with lower levels of *P*, with a non-negligible difference between the results obtained for the three levels of gravity loads. However, all the values of q are quite low, as typical values for low dissipative structures. This is probably associated with the localisation of the plasticity at the first levels of the frames only. For the two higher values of *P*, the behaviour factors are approximately equal to 1.5, which is the minimum value that can be assumed (EN 16681 2016). The obtained results are consistent with the values recommended by the standards and comparable with the values calculated by Kanyilmaz et al. (2016), from similar push-over tests' results.

Figure 2. Collapse of a full-scale rack subjected to push-over loading with *P* = *1/3 ESL*.

3 CONCLUDING REMARKS

As part of a wider research on pallet racks, eight full-scale tests were carried out: two vertical loads tests on specimens with an applied initial out-of-plumb of 1/126 and six push-over tests in down-aisle direction. In both the vertical loads and the push-over tests, the racks exhibited the same collapse mode, with the sway of the structures and the plastic deformation of the first levels of the frames. The push-over tests allowed evaluating the behaviour factor q of the racks, ranging from 1.44 to 2.31. These values are consistent with the design standard recommendations and with the findings of a similar research project.

REFERENCES

Baldassino, N., Bernardi, M., Bernuzzi, C., di Gioia, A. & Simoncelli, M. 2021. Seismic Performance Monitoring and Identification of Steel Storage Pallet Racks. In P. Rizzo & A. Milazzo (eds), *Vol. 2. Lecture Notes in Civil Engineering - Volume 128*: 447–455. Cham, Switzerland: Springer. Online ISBN: 978-3-030-64908-1. DOI: https://doi.org/10.1007/978-3-030-64908-1_42.

Bernardi, M. 2021. *Industrial steel storage racks subjected to static and seismic actions: an experimental and numerical study*. PhD thesis. Trento, Italy: University of Trento.

Kanyilmaz, A., Castiglioni, C.A., Brambilla, G. & Chiarelli G.P. (2016). Experimental assessment of the seismic behavior of unbraced steel storage pallet racks. *Thin-Walled Structures* 108: 391–405. DOI: http://dx.doi.org/10.1016/j.tws.2016.09.001.

EN16681 2016. *Steel static storage systems - Adjustable pallet racking systems - Principles for seismic design*. Brussels, Belgium: European Committee for Standardization.

Current Perspectives and New Directions in Mechanics, Modelling and Design of Structural Systems – Zingoni (ed.)
© 2022 Copyright the Author(s), *ISBN 978-1-032-39114-4*

Tests of cold-formed steel built-up open section beam-columns subjected to unequal end moments

Qiu-Yun Li & Ben Young
Department of Civil and Environmental Engineering, The Hong Kong Polytechnic University, Hong Kong, China

ABSTRACT: A test program is presented in this paper to investigate the behaviour of cold-formed steel built-up open section members under combined compression and non-uniform minor axis bending. A total of 17 specimens were eccentrically loaded under pinned supports. The end moment ratio varied for the specimens of each test series to examine the influence of non-uniform bending on beam-column members. Moreover, the comparison of experimental loading capacities with nominal strengths was performed to assess the applicability of the current design rules for the thin-walled section beam-columns under moment gradients

1 INTRODUCTION

In practice, non-uniform moment distributions along the member length are found for beam-columns. The current design rules to consider the effect of non-uniform bending on member strengths were initially put forward for non-slender section members, which should be evaluated for slender section components. To this end, the experimental study is described in this paper, which has been reported by Li & Young (2022), to explore the instability of cold-formed steel (CFS) built-up open section beam-columns subjected to non-uniform minor axis bending. Four series of eccentric compression tests with varying end moment ratios were conducted. The test results were obtained and used to assess the predictions determined by the AISI S100 (2016), ANSI/AISC 360 (2016) and AS/NZS 4600 (2018).

2 TEST PROGRAM

2.1 *Test specimens and material properties*

The built-up open sections were devised in this study and formed by assembling two identical channels with self-tapping screws. The folded-flange channels were longitudinally stiffened and press-braked from zinc-coated steel sheets that had the nominal 0.2% proof stresses of 500 and 550 MPa. A general screw spacing of 100 mm along the member length and a small spacing of 20 mm at the ends were employed to compose the built-up section specimens. A total of 17 specimens with nominal plate thicknesses of 0.75 and 1.2 mm and nominal member lengths of 900 and 1500 mm were categorized into four test series. The dimensions of test specimens were measured for individual channels.

2.2 *Test rig and procedure*

The beam-column specimens were eccentrically loaded under pin-ended boundary condition. A hydraulic testing machine with a loading rate of 0.2 mm/min was utilized to conduct the tests. Two LVDTs were horizontally set at quarter- and mid-heights of the members to obtain the lateral deflections. Three LVDTs were vertically installed at each end of the specimens to measure the end rotations. Moreover, four strain gauges were pasted onto each member at quarter- and mid-heights to capture the longitudinal strains.

2.3 Determination of e_0 and ψ

In this study, the beam-column specimens experienced synchronous compression and non-uniform minor axis bending. For the specimens of each test series, the nominal initial loading eccentricities at the bottom ends (e_0) were identical, while those at the top ends changed to generate varying end moment ratios (ψ). The actual values of e_0 and ψ were carefully determined for the test specimens, as detailed in Li & Young (2022).

2.4 *Experimental results*

The interaction between local, distortional and flexural buckling was observed for the built-up open section members in the tests. The experimental loading capacities (P_{Exp}) were obtained and are listed in Table 1. Compared to the specimen with the greatest end moment ratio in the same test series, the loading capacity of the specimen with the smallest end ratio increased by 32% on average, which indicated that the non-uniform bending had a beneficial effect on the strengths of the built-up section beam-columns.

DOI: 10.1201/9781003348450-145

Table 1. Experimental results and comparison with strength predictions (Li & Young 2022).

Specimen	P_{Exp} (kN)	$\frac{P_{Exp}}{P_{AISI}}$	$\frac{P_{Exp}}{P_{AISC}}$	$\frac{P_{Exp}}{P_{AS/NZS}}$
OI-T0.75L900A	70.1	1.07	1.03	1.16
OI-T0.75L900B	69.0	1.04	1.00	1.12
OI-T0.75L900C	79.9	1.10	1.07	1.19
OI-T0.75L900D	86.0	1.07	1.04	1.16
OI-T0.75L900E	92.2	1.02	1.00	1.11
OI-T1.2L900A	142.4	1.15	1.11	1.15
OI-T1.2L900B	160.1	1.15	1.11	1.15
OI-T1.2L900C	161.5	1.07	1.04	1.07
OI-T1.2L900D	171.0	1.05	1.03	1.05
OI-T0.75L1500A	49.6	1.05	1.02	1.05
OI-T0.75L1500B	54.2	1.09	1.06	1.10
OI-T0.75L1500C	62.9	1.16	1.13	1.16
OI-T0.75L1500D	71.3	1.19	1.16	1.19
OI-T1.2L1500A	90.7	1.13	1.10	1.13
OI-T1.2L1500B	98.3	1.17	1.14	1.17
OI-T1.2L1500C	108.6	1.18	1.15	1.18
OI-T1.2L1500D	120.2	1.21	1.18	1.21
Mean (P_m)	-	1.11	1.08	1.14
COV (V_P)	-	0.052	0.054	0.042
ϕ	-	0.85	0.90	0.85
β	-	3.01	2.65	2.91

3 DESIGN OF BEAM-COLUMNS SUBJECTED TO NON-UNIFORM BENDING

The reliability analysis and comparison of experimental loading capacities with nominal strengths were performed to assess the applicability of the current design rules for the CFS built-up open section beam-columns under moment gradients. The reliability index (β) with the prescribed minimum value taken as 2.5 was calculated from Eq. K2.1.1-2 of the AISI S100 (2016). In this study, the linear, bi-linear and two-stage interaction formulae, as specified in Section H1.2 of the AISI S100 (2016), Section H1 of the ANSI/AISC 360 (2016) and Section 3.5.1 of the AS/NZS 4600 (2018), respectively, were adopted to predict the nominal strengths of the built-up section beam-columns experiencing non-uniform bending. The nominal resistances against pure compression and pure bending were obtained according to the codified direct strength method equations. In addition, the influence of non-uniform bending is considered in the current beam-column design by introducing the equivalent uniform moment factor (C_m). Eq. (1) as stipulated in the aforementioned design codes was utilized in this study to calculate C_m.

$$C_m = 0.6 + 0.4\psi \quad (1)$$

The strength predictions of P_{AISI}, P_{AISC} and $P_{AS/NZS}$ were determined by the provisions specified in the AISI S100 (2016), ANSI/AISC 360 (2016) and AS/NZS 4600 (2018), respectively, which were compared with the experimental loading capacities (P_{Exp}). As presented in Table 1, the mean values of P_{Exp}/P_{AISI}, P_{Exp}/P_{AISC} and $P_{Exp}/P_{AS/NZS}$ were 1.11, 1.08 and 1.14 for the 17 built-up section members with the corresponding coefficients of variation (COVs) of 0.052, 0.054 and 0.042, respectively. Based on the resistance factors (ϕ) specified in the aforementioned design codes, the values of β equal to 3.01, 2.65 and 2.91 for the predictions of P_{AISI}, P_{AISC} and $P_{AS/NZS}$, respectively, which were all greater than the lower limit of 2.5. Therefore, it was revealed that the AISI S100 (2016), ANSI/AISC 360 (2016) and AS/NZS 4600 (2018) were generally conservative and reliable for the design of the CFS built-up open section members experiencing synchronous compression and non-uniform minor axis bending.

4 CONCLUSIONS

An experimental study is presented in this paper on cold-formed steel (CFS) built-up open section beam-columns subjected to non-uniform minor axis bending. A total of 17 specimens were loaded under eccentric compression. The interaction between local, distortional and flexural buckling was observed for the built-up section members in the tests. The results manifested that the non-uniform bending had a beneficial effect on the strengths of the built-up open section beam-columns. Moreover, the experimental loading capacities were compared with nominal strengths to assess the appropriateness of the current design rules for the thin-walled section beam-columns under moment gradients. It was demonstrated that the strength predictions determined by the provisions of the AISI S100 (2016), ANSI/AISC 360 (2016) and AS/NZS 4600 (2018) were generally conservative and reliable for the CFS built-up open section members undergoing combined compression and non-uniform minor axis bending.

REFERENCES

AISI. (2016). North American specification for the design of cold-formed steel structural members. *AISI S100-16, American Iron and Steel Institute*, Washington, DC.

ANSI/AISC. (2016). Specification for structural steel buildings *ANSI/AISC 360–16, American National Standards Institute and American Institute of Steel Construction*, Chicago.

AS/NZS. (2018). Design of cold-formed steel structures. *AS/NZS 4600: 2018, Standards Australia and Standards New Zealand*, Sydney, Australia.

Li, Q.-Y., & Young, B. (2022). Experimental study on cold-formed steel built-up section beam-columns experiencing non-uniform bending. *Engineering Structures*, (In press).

Current Perspectives and New Directions in Mechanics,
Modelling and Design of Structural Systems – Zingoni (ed.)
© 2022 Copyright the Author(s), ISBN 978-1-032-39114-4

Numerical analysis on semi-rigid pin joint for modular assembled steel reticulated shell

Ai-Lin Zhang
College of Architecture and Civil Engineering, Beijing University of Technology, Beijing, PR China
Beijing Engineering Research Center of High-rise and Large-Span Pre-stressed Steel Structures, Beijing, PR China

Chong Li
College of Architecture and Civil Engineering, Beijing University of Technology, Beijing, PR China

ABSTRACT: In this article, a new type of semi-rigid pin joint was proposed. High-strength bolts were not selected as connectors, which avoided the complicated procedures as well as achieved high efficient assembly. The assembly procedure of the joint is simple, and the new joint can adapt to the complex curvature changes of the reticulated shell. Arranged in the middle of the shell members, the novel joint can not only achieve the modular assembly of the reticulated shell but also reduce the number of assembled joints. First, a finite element model of the semi-rigid pin joint was established by ABAQUS, and several parameters of the joint were analyzed. Besides, the axial tensile and compressive properties of semi-rigid pin joints were studied, and the influence law of each parameter on the mechanical properties of the new joint was obtained. The results showed that the semi-rigid pin joint inherits the good axial bearing performance of the traditional pin joint.

1 INTRODUCTION

In recent years, to change the current situation of labor-intensive, high energy consumption, and high pollution in the construction industry [1], assembly building has become the key direction for the development of the construction industry. Therefore, it is necessary to vigorously develop prefabricated space reticulated shells, and the core is to develop fabricated joints suitable for fabricated spatial reticulated shells.

Several new fabricated joints for space reticulated shell structures have appeared. For example, Temcor joint, Mero joint, Inlay hub joint, Socket joint, and so on.

However, current research is aimed at the scattered assembly of bars and joints, and a large number of fabricated joints affect the overall rigidity of fabricated reticulated shells.

A new modular assembled reticulated shell system proposed in this study is shown in Figure 1.

By arranging the semi-rigid pin joints in the middle of the bars, it can adapt to the complex curvature changes of the reticulated shell. On the other hand, the number of semi-rigid pin joints in the fabricated reticulated shell is reduced.

2 SEMI-RIGID PIN JOINT

The components of the semi-rigid pin joint are shown in Figure 2. It includes steel pipe 1, end plate 1, middle ear plate, wing plate 1, stiffening plate 1, wing plate 2, screw rod, side ear plate, wing plate pin, ear plate pin, welding nut, end plate 2, and steel pipe 2.

Figure 1. Modular assembled reticulated shell.

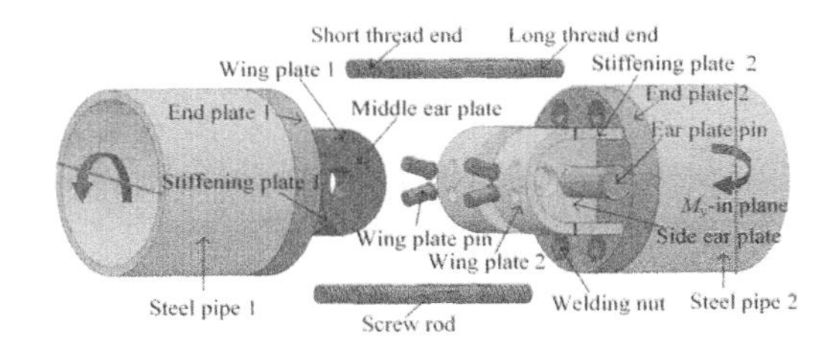

Figure 2. Components of semi-rigid pin joint.

DOI: 10.1201/9781003348450-146

3 THE FINITE ELEMENT MODEL

The finite element software ABAQUS was used to establish the three-dimensional model of the semi-rigid pin joint and further analyze the influence of different parameters on the mechanical properties of the joint. To obtain more accurate analysis results, the wing plate, stiffening plate, middle ear plate, side ear plate, wing plate pin, and ear plate pin have been finely meshed (mesh size 5mm). The finite element model of the semi-rigid pin joint is shown in Figure 3.

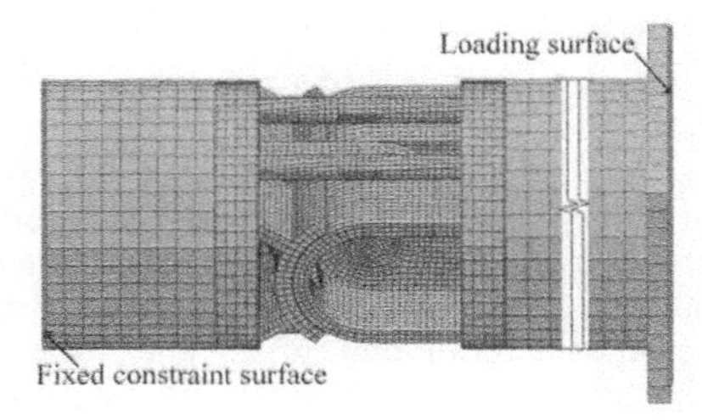

Figure 3. Finite element model of the semi-rigid pin joint.

The parameters considered in the model included the diameter of the wing plate pin (d_{wpp}), the number of the wing plate pin (N_{wpp}), the thickness of the wing plate (t_{wp}), and the diameter of the screw rod (d_{sr}).

In the FE models, 45# steel was selected as the material of wing plate pin, ear plate pin, and screw rod, and the material of all other components was selected as Q345 grade steel. The detailed dimensions of the finite element models are shown in Table 1.

Table 1. Geometric parameters of semi-rigid pin joints.

FEM ID	Steel pipe	d_{wpp} (mm)	t_{wp} (mm)	d_{sr} (mm)	N_{wpp}
FEM-1	245×12	12	15	20	4
FEM-2	245×12	16	15	20	4
FEM-3	245×12	20	15	20	4
FEM-4	245×12	20	10	20	4
FEM-5	245×12	20	15	24	4
FEM-6	245×12	20	15	27	4
FEM-7	245×12	16	15	20	8

4 COMPARISON OF TENSILE AND COMPRESSIVE BEARING CAPACITY OF THE JOINTS

Based on the defined curve parameters, the initial stiffness of each joint under axial tension and compression was calculated respectively, as shown in Table 2.

It can be clearly seen from Table 2 that the stiffness of the seven joints under the axial compression of the members is greater than that under the axial tension. Taking joint FEM-3 as an example, the stress nephograms of the whole joint FEM-3 under axial tension and compression ares shown in Figure 4.

Table 2. Comparison of joint stiffness under axial tension and compression.

FEM ID	1	2	3	4	5	6	7
Tensile	668	675	695	691	766	825	705
Compress	947	956	966	889	1000	1018	989

Note: unit (kN/mm)

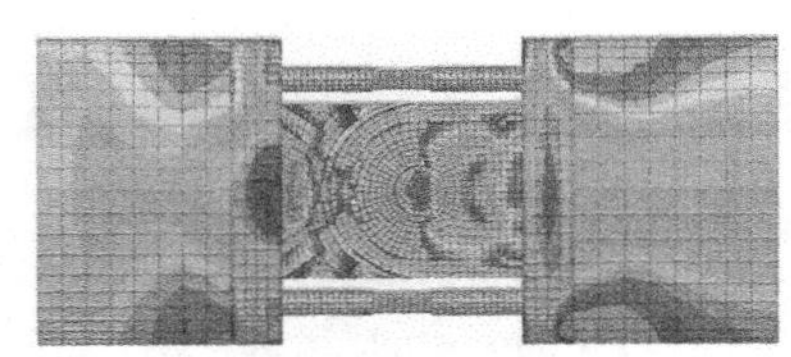

(a) Joint FEM-3 under axial tension.

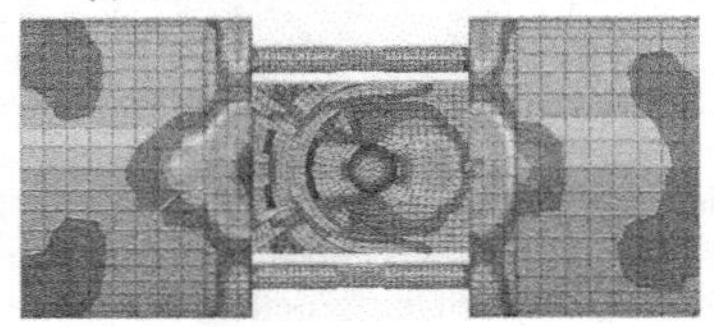

(b) Joint FEM-3 under axial compression.

Figure 4. Comparison of stress nephograms of the joint FEM-3 under axial tension and compression.

By comparison, it can be found that the stress of steel pipe 1 and steel pipe 2 under axial compression was significantly greater than that under axial tension.

5 CONCLUSIONS

The improved semi-rigid pin joint can inherit the excellent tensile and compressive performance of the traditional pin joint. The diameter and number of the wing plate pins and the diameter of screw rods were important factors affecting the axial tensile mechanical properties of the joints. However, these three factors had little influence on the axial compression performance of the joint. In addition, the stiffness of the seven joints under the axial compression of the bars was greater than that under the axial tension.

ACKNOWLEDGEMENTS

This research is supported by the Key projects of the National Natural Science Foundation of China (Grant number 52130809) and the National Natural Science Foundation of China (Grant number 51878025).

REFERENCES

[1] A.l. Zhang, T.Y. Liu, Y.X. Zhang, et al. Research prospect of rapid and fully assembled long-span prestressed spatial steel structure system innovation based on intelligent construction [J]. Journal of Beijing University of Technology, 2020(6):591–603 [in Chinese].

Current Perspectives and New Directions in Mechanics, Modelling and Design of Structural Systems – Zingoni (ed.)
© 2022 Copyright the Author(s), ISBN 978-1-032-39114-4

Experimental study on shallow H-shaped steel column under combined loading condition

S. Nakatsuka & A. Sato
Nagoya Institute of Technology, Japan

ABSTRACT: When moment-resisting frames resist the horizontal forces, the columns are subjected to axial force with bending moment (i.e., under combined loading conditions). The lateral-torsional buckling shall be considered when the H-shaped steel columns are subjected to major axis bending. The collapse modes that will dominate the H-shaped steel column capacity are in-plane instability (caused by $P\delta$ effects), out-of-plane instability (i.e., lateral-torsional buckling: LTB), and local buckling. If the column is expected to form a plastic hinge, that column shall guarantee sufficient ductility to prevent the building collapse. Moreover, the collapse mode that will determine the ultimate limit state should be informed. In the previous study, structural performances of the steel column were evaluated; however, the classification of the collapse modes was not studied sufficiently due to the limit of image recording equipment. Moreover, the structural behaviour reported in the previous study was based on monotonic loading, and the observed results from the cyclic loading were not sufficiently gathered. Cyclic loading may significantly impact the in-elastic behaviour of the H-section section column due to the potential of LTB. The difference in loading sequence must be clarified when the design rules are compiled. In this paper, the test results of different axial force ratios and lengths of members are reported. Sets of specimens are prepared for direct comparison of the loading sequence differences. The difference of collapse modes and the influence on the structural performances (such as max bending moment and plastic deformation capacity) is shown. Finally, comparing the monotonic and cyclic loading condition results, the effect of loading condition on the structural performances are also shown with the derived evaluation formula.

1 INTRODUCTION

In a moment-resisting frame (MRF), column will be subjected to the bending moment around the major axis combined with compressive axial force (so-called beam-column). Generally, the strong-column weak-beam concept is adopted for the MRF direction. The two-phase design procedure is the default for building design, and the strong-column weak-beam concept is allowed to be checked at the floor level instead of at each joint level in the ultimate limit state (ULS) design. In this situation, the columns may have a chance to form the plastic hinge. Moreover, the column may yield due to the strain hardening at the beam hinge under cyclic loading. Therefore, it is crucial to understand the structural behaviour of the column in the in-elastic range.

It is well known that the H-shaped section has three failure modes that will dominate the ULS: Local buckling, out-of-plane instability (so-called lateral-torsional buckling; LTB), and in-plane instability (derived from $P\delta$ effects). Moreover, combined failure mode might happen in the actual members.

In the current design regulations, the difference of failure mode is not properly considered in the design formulae. In the previous study, structural performances of the steel column were evaluated; however, the classification of the collapse modes was not studied sufficiently due to the limit of image recording equipment. Moreover, the structural behaviour reported in the previous study was based on monotonic loading, and the observed results from the cyclic loading were not sufficiently gathered. Cyclic loading may significantly impact the in-elastic behaviour of the H-section section column due to the potential of LTB. The difference in loading sequence must be clarified when the design rules are compiled. In this paper, the test results of different axial force ratios and lengths of members are reported. Sets of specimens are prepared for direct comparison of the loading sequence differences. The difference of collapse modes and the influence on the structural performances is shown. Finally, comparing the monotonic and cyclic loading condition results, the effect of loading condition on the structural performances are also shown with the derived evaluation formula.

2 PLASTIC DEFORMATION CAPACITY

In previous research [1], the index considering lateral torsional buckling $n_y \cdot \lambda_{c0}{}^2 \lambda_{b,N}$ is suggested and the evaluation of plastic deformation capacity considering the effect of lateral torsional buckling is indicated as

DOI: 10.1201/9781003348450-147

$$R = \frac{-0.59(Z_{wp}/Z_p) + 0.18}{n_y \cdot \lambda_{c0}^2 \cdot \lambda_{b,N}} + [-8.2(Z_{wp}/Z_p) + 0.65] \quad (1)$$

3 TESTING AND RESULTS

In order to figure out the behaviour of H-shaped beam-columns under constant compressive axial force combined with the bending moment, the full-scale test was conducted. The testing under cyclic and monotonic bending moment was conducted respecitvely. Figure1 shows the example of relation between applied end moment M_1 and end rotation θ. There are three ultimate limit states observed in the specimens under Monotonic loading such as local buckling, in-plane instability, and lateral torsional buckling (Figure 1). The failure modes change from local buckling to lateral torsional buckling as the axial compressive force increases. And the specimens which undergo higher axial compressive force level collapsed with in-plane instability ($P\delta$ effect). In every specimen, in-plane and out-plane deformation at the maximum strength point or M_{pc} point were relatively small, and deformation become more significant as the end rotation increased.

The failure modes of all specimens under cyclic loading were lateral torsional buckling. In every specimen, the out-plane deformation increased as the loading cycle progresses. The in-plane deformation didn't get large as much as out-plane deformation.

The skelton curve located closely to the monotonic curve. Therefore, monotonic curve can infer the skelton curve obtained from cyclic loading.

4 STRUCTURAL PERFORMANCE

Figure 2 shows the comparison between the evaluation value $n_y{\cdot}\lambda_{c0}^2\lambda_{b,N}$ and plastic deformation capacity R. In this figure, the results obtained from the previous study which treated the same section as this paper are also plotted. As shown in this figure, inversely proportional relation between the evaluation value and R. Therefore, it can be said that R can be evaluated by the value $n_y{\cdot}\lambda_{c0}^2 \lambda_{b,N}$. The solid curve shown in the figure indicates Eq(1) and the vertical solid line shows the limitation

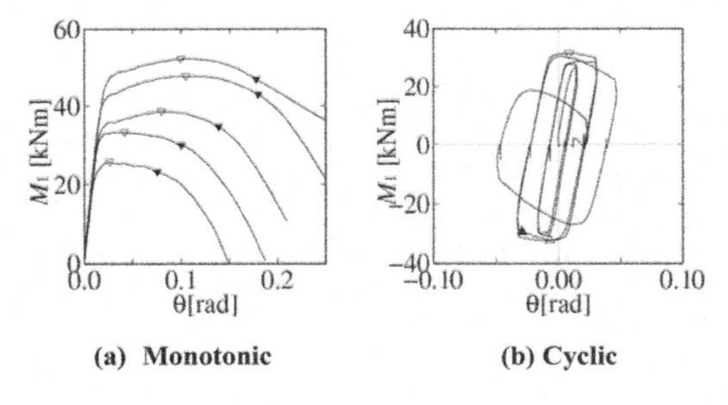

Figure 1. Example of M-θ. (a) Monotonic (b) Cyclic.

Figure 2. Failure modes of Monotonic Loading. (From left; Local buckling, LTB, $P\delta$)

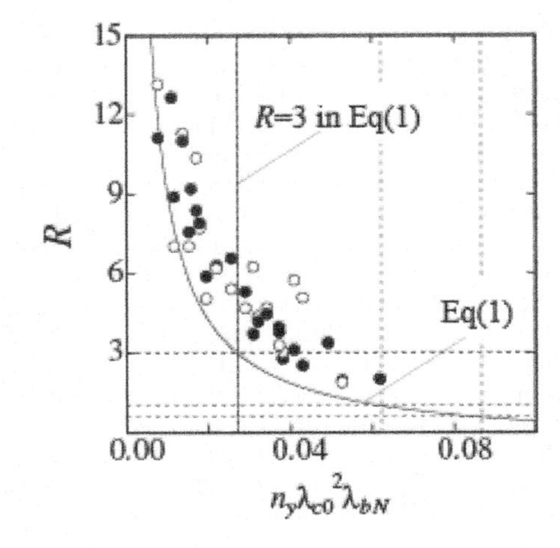

Figure.3. Evaluation of R.

of the H-shaped columns to satisfy R>3 (based on eq (1)). As can be seen from the figure, the R of the specimens which fulfil the limitation was larger than 3. Moreover, Eq(1) showed conservative and reasonable result to the test result.

5 CONCLUSIONS

In this study, the result of the full-scale testing of wide flange beam-columns was reported. In the specimens under monotonic loading, there were three ultimate limit states: local buckling, in-plane instability, and lateral torsional buckling. The failure modes change into local buckling, lateral torsional buckling and in-plane instability as the axial compressive force increases. On the other hand, lateral torsional buckling was dominant in all specimens under cyclic loading. The monotonic curve and the skelton curve fitted well and the evaluation of R considering the effect of lateral torsional buckling proposed in the previous study showed conservative and reasonable result to the test result.

REFERENCES

[1] Yoshioka, S., Sato, A., "Analytical Study on H-shaped Columns subjected to Compressive Axial Force with One End Monotonic Bending Moment", Summaries of Technical Papers of Annual Meeteing, Architecrural Institute of Japan, Structures III, pp.813–814, 2021.09 (in Japanese)

Current Perspectives and New Directions in Mechanics, Modelling and Design of Structural Systems – Zingoni (ed.)
© 2022 Copyright the Author(s), ISBN 978-1-032-39114-4

Experimental tests on the rack-to-spine-bracing joints of high-rise steel storage racks

Z. Huang & X. Zhao
Tongji University, Shanghai, China

K.S. Sivakumaran
McMaster University, Hamilton, Ontario, Canada

ABSTRACT: High-rise steel storage racks are extensively used in the warehouses. In order to avoid obstruction of the placement of the pallet goods, a spine bracing system, locates at a short horizontal distance away from the rack frame, maybe used to stabilize such high-rise racks. The spine bracing members, which are designed to be working both in tension and in compression, are commonly connected to the main rack frame through a bracket assembly, forming a rack-to-bracing joint. Recently the behaviors of four groups of three nominal identical rack-to-spine-bracing joints subjected to lateral load were experimentally established. The influences of the key design factors on the shear behavior of the joints were explicitly considered and discussed. The experimental observations indicated that the local deformation of the flange and web panels of the upright components, as well as the flexural bending of the endplate components, contributed to the overall shear deformation of the rack-to-bracing joints.

1 INTRODUCTION

Compared with regular steel storage racks, the high-rise racks provide much higher storage capacities while taking only limited footprints and thus, such high-rise storage systems are widely used in the warehouses in recent years. The high-rise storage racking systems require additional bracing systems, namely the spine bracing systems, in the down-aisle direction to stabilize the main racks. It must be pointed out that the spine bracing systems are usually located at a short horizontal distance away from the main rack frames so that the placement of the pallet goods adjacent to the spine bracings would not be obstructed. Consequently, the bracing members are connected to the main racks through a custom configured bracket assembly, forming a rack-to-spine-bracing joints.

Past studies indicated that the performance of the global high-rise racks was strongly intertwined with the behavior of the rack-to-spine-bracing joints (Huang & Zhao 2019, Kanyilmaz et al. 2016, Teh et al. 2004). However, the behavior of such joints is not yet fully understood. As such, the experimental project reported in this paper aims to provide comprehensive and reliable experimental data on the behavior of the rack-to-spine-bracing joints subjected to the lateral load. The load-deformation relations and the failure modes of the rack-to-spine-bracing joints, as well as the effects of the influencing design factors on the behavior of such joints are summarized and discussed in this paper.

2 EXPERIMENTAL INVESTIGATION

The experimental test program involved four rack-to-spine-bracing joint specimen groups with each specimen group consisting of three nominally identical test specimens. The design factors associated with such rack-to-spine-bracing joints, including the thickness of the upright and the thickness of the endplate, were explicitly considered in the tests. Specimens in the groups RSPJ-A1, RSPJ-A2 and RSPJ-A3 used 3 mm thick, 2.5 mm thick and 2 mm thick upright pieces, respectively, and all the specimens in these groups used 6 mm thick endplates. The upright piece and the endplate associated with the RSPJ-A4 group specimens were fabricated out of 3 mm thick and 4 mm thick material, respectively.

Figure 1 shows the test set-up used in the experimental program. The rack-to-spine-bracing joint specimen was mounted in a rigid frame with both ends of the upright piece fixed connected to the frame using four bolts. An actuator was used to apply lateral load to the test specimen. The fixing end of the actuator was mounted to a sliding support which allowed the actuator to move freely in the horizontal direction, so that the direction of the lateral load applied by the actuator would remain

DOI: 10.1201/9781003348450-148

unchanged throughout the test. The sliding support also avoided the actuator being subjected to undesired shear force during the loading process. The loading end of the actuator was attached to the connecting plate through a pin connection. The lateral load applied by the actuator, F, was recorded by a load cell. The shear deformation of the considered rack-to-spine-bracing joint, Δ, was measured using three transducers.

Figure 1. The test set-up.

3 EXPERIMENTAL RESULTS

Figure 2 exhibits the representative load-deformation relations associated with the joint specimens in each test group. Even though the load-deformation curves associated with all the joint specimens are not explicitly shown here, it is stated that the three individual test specimens within the same group show consistent load-deformation relations. In general, the rack-to-spine-bracing joints under consideration may fail by the fracture of the connecting bolts (RSPJ-A1), the yielding of the upright flanges (RSPJ-A2 and RSPJ-A3), and the yielding and tearing of the endplates (RSPJ-A4).

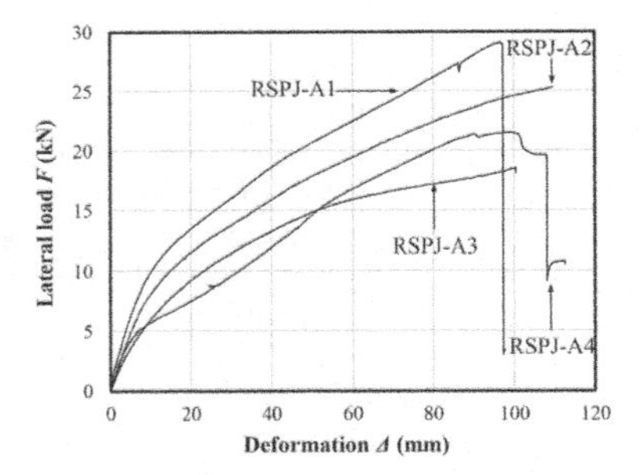

Figure 2. The representative load-deformation curves.

4 DISCUSSIONS OF THE TEST RESULTS

Comparing the load-deformation relations associated with the specimens in the groups RSPJ-A1 thru RSPJ-A3 in Figure 2 we can noticed that, when the thickness of the upright was reduced from 3 mm (RSPJ-A1) to 2.5 mm (RSPJ-A2), the initial shear stiffness of the entire rack-to-spine-bracing joint reduced by 32.7%. When the upright thickness was further reduced to 2 mm (RSPJ-A3), another 35.1% reduction in the initial shear stiffness of the joints was observed. It is also stated that the upright piece fabricated out of thinner material (i.e., RSPJ-A2 and RSPJ-A3) suffered substantial damage compared with the upright piece fabricated using thicker material (i.e., RSPJ-A1).

Comparing the load-deformation curves of the RSPJ-A1 specimen and the RSPJ-A4 specimen in Figure 2, it is evident that the behavior of the rack-to-spine-bracing joints under consideration is sensitive to the thickness of the associated endplate components. Reducing the endplate thickness from 6 mm (RSPJ-A1) to 4 mm (RSPJ-A4) resulted in a 18.8% reduction in the initial stiffness and a 61.0% decrease in the yielding strength of such joints.

5 CONCLUDING REMARKS

This paper summarizes the experimental observations associated with a total of twelve rack-to-spine-bracing joint specimens subjected to lateral load. The three nominally identical joint specimens within the same test group exhibited consistent load-deformation relations. The experimental tests revealed that the design factors including both the upright thickness as well as the endplate thickness had an enormous impact on the behavior of the entire rack-to-spine-bracing joints under consideration. Reducing the upright thickness from 3 mm thick to 2 mm thick would result in a 56.4% reduction in the initial stiffness of such rack-to-spine-bracing joints. The upright piece fabricated out of thinner material evidently suffered much severer damage, which may have an undesired impact on the capacity of the global high-rise storage system. Replacing the 6 mm endplates with 4 mm thick components would lead to a 18.8% reduction in the initial stiffness and a 61.0% decrease in the yielding strength of the considered rack-to-spine-bracing joints. The experimental tests suggested that the rack-to-spine-bracing joints under consideration may fail by the fracture of the connecting bolts, the yielding and tearing of the endplate, and the yielding of the upright flanges.

REFERENCES

Huang Z. & Zhao X., 2019. Elastic buckling analysis of asymmetrically braced steel storage racks, Proceedings of Seventh International Conference on Structural Engineering, Mechanics and Computation (SEMC 2019), Cape Town, South Africa.

Kanyilmaz A., Brambilla G., Chiarelli G. P. & Castiglioni C. A., 2016. Assessment of the seismic behaviour of braced steel storage racking systems by means of full scale push over tests, Thin Walled Structures 107: 138–155.

Teh L. H., Hancock G. J. & Clarke M. J., 2004. Analysis and Design of Double-Sided High-Rise Steel Pallet Rack Frames. Journal of Structural Engineering, ASCE 130(7):1011–1021.

Current Perspectives and New Directions in Mechanics, Modelling and Design of Structural Systems – Zingoni (ed.)
© 2022 Copyright the Author(s), ISBN 978-1-032-39114-4

Experimental and numerical evaluation of various solutions for fire protection of steel structures

M.H. Nguyen, S.E. Ouldboukhitine, S. Durif, V. Saulnier & A. Bouchaïr
Université Clermont Auvergne, Clermont Auvergne INP, CNRS, Institut Pascal, Clermont-Ferrand, France

ABSTRACT: Steel structures are commonly used in construction for their high strength/weight ratio and many other advantages. However, due to their thermal conductivity and the decrease of mechanical performance with high temperature, steel structures need to be protected in fire. Various solutions exist to protect steel structures using materials with appropriate insulation performances (intumescent paints, plasterboards, Spray-applied fire-resistant materials). In the present study, common solutions using intumescent paints and plasterboard are tested. Their behaviour is compared to that of solid wood elements used as insulation materials in fire. Experiments in furnace are performed to obtain the evolution of temperature on the steel profiles surfaces. Thus, thermocouples are installed on the steel profile surface and on different depths of timber elements. In fact, even for intumescent paints, limited data exist on the steel profile heating under fire exposure. These data are useful to develop thermomechanical models of steel structures exposed to fire. A numerical model is also proposed to analyze the heating of fire protected steel profile. The results show that wood can be used for fire protection of steel structures in addition to its advantages provided for the mechanical behaviour of hybrid sections composed of steel and timber. This solution contributes to the development of passive protection of steel structures using bio-based materials.

1 INTRODUCTION

Steel structures are commonly used for their high mechanical performances. However, their thermal properties can be an important drawback in the case of fire mainly for structural elements with high slenderness. Indeed, submitted to high temperatures, steel structures can lose their strength and stiffness significantly. To maintain the load-bearing function during a required time of fire exposure, steel structures usually need to be protected. Steel can be combined with other materials to improve its fire resistance as in composite steel-concrete sections (Ma et al. 2008, Korzen et al. 2010) or steel-timber hybrid sections (Uesugi et al. 2002, Kia & Valipour 2021).

In this paper, the study concerns the thermal behaviour of hybrid steel-timber specimens through experimental tests carried out considering different possibilities to obtain the best solution regarding fire protection. This method of fire protection will be compared with the common fire protection of steel structures using intumescent paints. In parallel, numerical simulations were performed to validate the proposed models and to be able to predict the thermal behaviour under different conditions.

2 DESIGN AND PRODUCTION OF SPECIMENS

The tests are performed on unprotected and protected steel specimens using wood as protection (Table 1).

Table 1. Test samples.

Test	Set-up	Fire exposure configurations
Partially encap-sulated beam	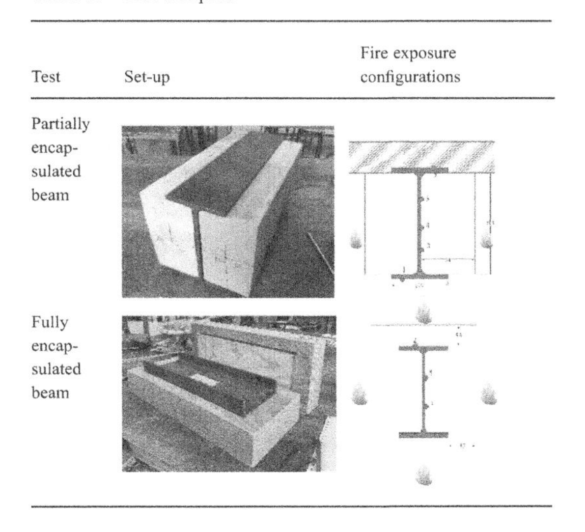	
Fully encap-sulated beam		

DOI: 10.1201/9781003348450-149

The aim is to observe and compare the temperature evolution in different locations in wood, steel and at the interface for a one-hour fire exposure.

3 EXPERIMENTAL RESULTS

Unprotected steel beam shows a very high heating rate that reaches a temperature of 600°C at only 10 minutes (Figure 1). The increase of the width of wood in the partially encapsulated beam allows to significantly increase the heating time (450°C at 60 minutes). The fully encapsulated beam gives the best results as none of the steel parts is exposed.

It can be noticed that the more we add wood for the protection, the more it brings fire load. Thus, the present experimental tests aimed also to observe this influence in the behaviour of each tested specimen. Nevertheless, even for more severe fire conditions due to the charring of wood, those protections brought a significant delay in the steel temperature increase which validate these solutions as suitable for fire protection of steel structures.

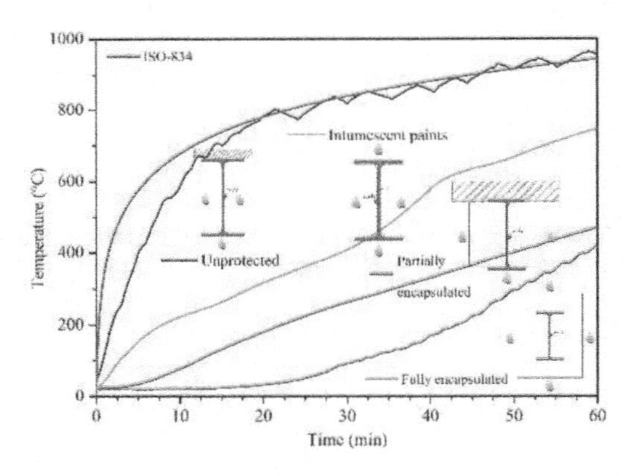

Figure 1. Comparison of temperature on steel for four different fire protections.

4 NUMERICAL MODELLING OF HEAT TRANSFER

The heat transfer in wood and steel is simulated by two-dimensional (2D) finite elements model using Abaqus software. The validation is based on the comparison of temperatures between calculated and measured values. The modelling simulation was verified with a series of fire tests on spruce timber specimens exposed to ISO-fire (König 1999).

The evolution of the temperature obtained by numerical simulation in timber element is also compared to that from tests. The charring rate of wood obtained from the experiment and the model were relatively close to each other with values ranging from 0.59 to 0.62 mm/min. The thickness of the char layer formed after the experiment at 57 minutes was measured again as 35 mm that correspond to a wood charring rate of 0.61 mm/min. This result is compared with the char layer obtained on the model.

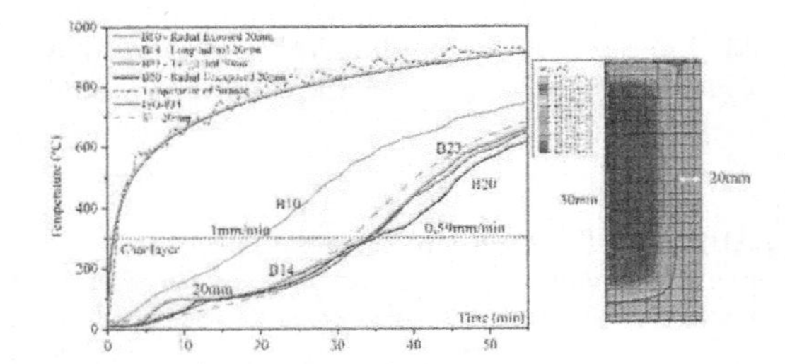

Figure 2. Experimental and numerical temperature-time curve at 20mm and 30mm depth in wood and charring rate.

5 CONCLUSIONS

Timber has an insulating effect that limits the heating of steel beams. However, the evaluation of temperatures in the wood thickness remains a subject for further studies. The tests realized in this study have shown promising results for temperatures on protected steel that reached a critical temperature of 550 °C at 35 to 40 minutes whereas the equivalent unprotected steel reached this value at 5 to 10 minutes. The measured wood average charring rate is reasonable according to Eurocode 5, from 0.59 to 0.62 mm/min. Three measurement methods were evaluated in this study considering the installation of thermocouples in different directions perpendicular, tangential, and parallel to the isotherms. The measurement method with the thermocouple parallel to the isotherm gives the closest results, in comparison with the observation of the char layer after the experiment.

REFERENCES

Kia, L., & Valipour, H. R. 2021. Composite timber-steel encased columns subjected to concentric loading. *Engineering Structures*, *232*(August 2020), 111825.

König, J. 1999. One-dimensional charring of timber exposed to standard and parametric fires in initially unprotected and postprotection situations. *Rapport/ Trätek. NV – 9908029.*

Korzen, M., Rodrigues, J., & Correia, A. (2010). Composite columns made of partially encased steel sections subjected to fire. *Structures in Fire - Proceedings of the Sixth International Conference, SiF'10*, (January 2010), 341–348.

Ma, Z., & Mäkeläinen, P. 2008. Behavior of Composite Slim Floor Structures in Fire. *Journal of Structural Engineering, ASCE*, *20*(2), 284–294.

Uesugi, S., Harada, T., & Namiki, Y. (2002). Fire resistance of sugi covering materials for structural steel. *Journal of Wood Science*, *48*(4), 343–345.

Current Perspectives and New Directions in Mechanics, Modelling and Design of Structural Systems – Zingoni (ed.)
© 2022 Copyright the Author(s), ISBN 978-1-032-39114-4

Microstructure and corrosion resistance of dissimila welded joints

B. Verhoeven & R. Dewil
Department of Chemical Engineering, KU Leuven, Sint-Katelijne-Waver, Belgium

G. Ferraz & B. Karabulut
Department of Civil Engineering, KU Leuven, Sint-Katelijne-Waver, Belgium

S. Sun
University of Oxford, New College, Oxford, UK

B. Rossi
Department of Civil Engineering, KU Leuven, Sint-Katelijne-Waver, Belgium
University of Oxford, New College, Oxford, UK

ABSTRACT: This paper investigates dissimilar weldments between duplex stainless steel and carbon steel in atmospheric corrosive environments. The duplex stainless steel grade 1.4062 was welded to the carbon steel grades S355J2/P460NL2 by means of gas metal arc welding. The fabricated specimens were subjected to 5 months exposure tests under 4 different corrosive atmospheric environments, distributed across 3 different countries: Belgium, England and Portugal. A total of 28 samples were designed, fabricated and sand blasted, in order to satisfy the surface requirements of (ISO 8501-1, 2007). Exposure racks were assembled and exposed to the weather conditions at each test location. The environmental parameters influencing the corrosion of dissimilar welds were monitored on a continuous basis. Finally, prelaminar guidance is given on the necessary coating protection for such weldments in the context of girder bridges.

1 STATE OF THE ART

Dissimilar weldments have been applied in a wide range of engineering components and structures such as bridges, in pursuit of environmental and economic design optimization. The use of dissimilar weldments allows one to combine distinct properties of dissimilar metals, such as strength and corrosion resistance. In a hybrid girder bridge case study by (Yabuki et al., 2017), duplex stainless steel was used as the webs of the bridges, whereas carbon steel was used as the structural stiffeners. This design approach has shown to lead to a relatively lower lifecycle cost in bridge structures.

Corrosion properties of carbon steel and duplex stainless steel were studied in the literature (Silva et al., 2021, Fonna et al., 2021). However, there is limited information about the corrosion behaviours of dissimilar weldments between these two types of steel.

This paper investigates the corrosion resistance of dissimilar weldments between lean duplex grade EN 1.4062 and mild carbon steels (S355 and P460). In particular, the authors evaluate the corrosion resistance of 28 test samples after 5 months exposure tests in four different corrosive environments. Correlations among environmental factors, alloy composition and corrosion resistance are discussed.

2 EXPERIMENTAL SETUP AND EXPOSURE PROGRAM

2.1 *Test samples*

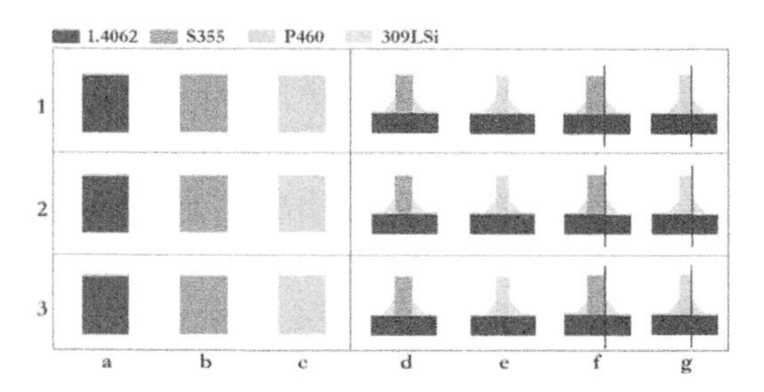

Figure 1. Sample rack used for the exposure tests with a) EN 1.4062 base plate; b) S355JN base plate; c) P460LN2 base plate; d) dissimilar T-joint of 1.4062/S355J2 with 309LSi filler; e) dissimilar T-joint of 1.4062/P460LN2 with 309LSi filler; f) dissimilar joint of 1.4062/309LSi (S355J2 removed); g) dissimilar joint of 1.4062/309LSi (P460NL2 removed).

Seven different types of configurations were investigated in this study, as shown in Figure 1. The first three columns (a, b and c) are attributed to reference

DOI: 10.1201/9781003348450-150

specimens, the original materials without welding (base metals). The consumable used in the dissimilar welded joints is a 309LSi electrode with a diameter of 1 mm. Optimum welding parameters were selected according to (Cools & Staepels, 2021). The last two columns correspond to the joints made of the duplex base plate only and 309LSi filler metal, i.e., where the carbon steels were cut out after welding (f-g). This was done to exclude the corrosion of carbon steel, when comparing to the T-joints. Test sites

In this study, atmospheric corrosion was investigated in four different environments. The aggressivity of these environments can be classified according to a corrosion resistance factor (CRF), a parameter dependent on the distance to the sea or roads with de-icing salts, SO2 concentration and cleaning regime (EN 1993-1-4, 2006):

- Knokke-Heist (KH) (CRF = -7, high risk of exposure): 0.8 km from the coastline, Belgium.
- Sint-Katelijne-Waver (SKW) (CRF = 0, low risk of exposure): 40.9 km from the coastline, Belgium.
- Casalinho da Foz (CDF) (CRF = 0, low risk of exposure): 18.8 km from the coastline, Portugal.
- Oxford (OX) (CRF = 0, low risk of exposure): 96.5 km from the coastline, England.

2.2 *Exposure program*

The test samples were exposure to the weather conditions, in accordance with the ASTM standard (ASTM G50-10, 2015). The duration of the exposure was 5 months. All sites were exposed without any shelter. The environmental parameters influencing the corrosion of dissimilar weldments were monitored on a regular basis at each location (see Figure 2). The devices used to collect the atmospheric information can be categorized as follows:

- Air temperature and relative humidity were collected using an UX100-003 data logger.
- The chloride deposition was collected using a piece of dry gauze (100 mm × 100 mm) that was stretched and held in a Plexiglas frame. Chloride determinations in the test site atmosphere were performed monthly using the 'dry plate' technique (ISO 9225, 2012).

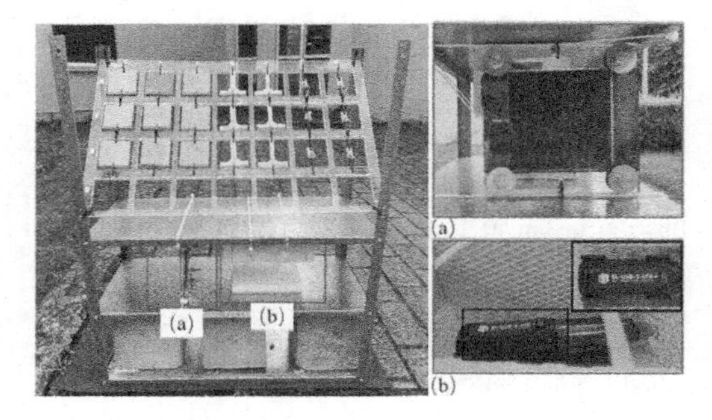

Figure 2. The devices used to collect a) Chloride deposition frame; b) XPS storage box for temperature/humidity sensor.

3 MAIN CONCLUSIONS

From the analysis performed to investigate the influence of corrosion on dissimilar welded joints between duplex stainless steel EN 1.4062 and carbon steel S355J2/P460NL2, the following conclusions are obtained.

- Knokke-Heist is the test site that shows the highest corrosivity to the carbon steel base plates and the dissimilar welded joints. This is somewhat expected because of its geographical location with highest chloride deposition rate. Casalinho da Foz with the highest temperature is more aggressive than Sint-Katelijne-Waver. Oxford which has the greatest distance from the coastline is the least aggressive environment, although it has the highest average relative humidity. Chloride deposition rate and temperature are the key parameters that affect atmospheric corrosion rates.
- The prediction of corrosion rate by weather parameters using the empirical formula based on ISO 9223 is sound, yet conservative.
- The discontinuous interface between the weld metal and the carbon steel is sensitive and susceptible to pitting corrosion, especially when the dissimilar weld is subjected to a chloride-rich environment. At the galvanic zones, the susceptibility is even higher. If the environment has a CRF value of -7 or lower, coating protection on the carbon steel part is recommended.

REFERENCES

ASTM G50–10 2015. Standard Practice for Conducting Atmospheric Corrosion Tests on Metals. ASTM International.

Cools, Y. & Staepels, A. 2021. *Microstructural and mechanical investigations on dissimilar welding.* Master of Science Master's thesis, Ku Leuven.

EN 1993-1-4 2006. Eurocode 3: Design of steel structures - Part 1-4: General rules - Supplementary rules for stainless steels. The British Standards Institution.

Fonna, S., Bin, M. I. I., Gunawarman, Huzni, S., Ikhsan, M. & Thalib, S. 2021. Investigation of corrosion products formed on the surface of carbon steel exposed in Banda Aceh's atmosphere. *Heliyon*, 7: 1–13.

ISO 9225 2012. Corrosion of metals and alloys - Corrosivity of atmospheres - Measurement of environmental parameters affecting corrosivity of atmospheres. International Organization for Standardization.

ISO 8501–1 2007. Preparation of steel substrates before application of paints and related products - Visual assessment of surface cleanliness - Part 1: Rust grades and preparation grades of uncoated steel substrates and of steel substrates after overall removal of previous coatings. International Organization for Standardization.

Current Perspectives and New Directions in Mechanics, Modelling and Design of Structural Systems – Zingoni (ed.)
© 2022 Copyright the Author(s), ISBN 978-1-032-39114-4

3D models of clad steel structures: Assumptions and validation

M.J. Roberts
University of Manchester, UK
SCIT Ltd, Altrincham, UK

J.M. Davies
SCIT Ltd, Altrincham, UK
University of Manchester (Emeritus Professor), UK

ABSTRACT: Cladding systems, on steel-framed buildings, may have a significant effect on the global behavior of the building, particularly under horizontal load. However, cladding is rarely included in the analytical models used for the design of such structures. This paper examines different ways cladding could be included in such models. Validated finite element models of four clad structures, three of which were previously tested and one much larger multi-bay frame, were generated and analyzed using the different methods. It was found that the simplified methods were able to estimate the deflection of the smaller structures but were inaccurate for the larger multi-bay structure. The failure modes of the cladding could not be predicted by the simplified methods. For all structures most of the horizontal load (applied at the eaves level) was resisted by the cladding showing that the deflections and forces used in the design of clad steel structures are largely fictious.

1 INTRODUCTION

It is well known that both secondary steelwork and cladding may have a significant impact on the behavior of single-story steel-framed clad structures (colloquially termed "sheds"), however, both are often conspicuously absent in the Finite Element Models (FEMs) used during the design of such structures.

This paper presents the progress by the authors in developing simple validated FEMs which include sufficient detail to estimate failure modes of real clad structures. FEMs can be used to estimate the behavior of existing buildings, including the effect of remedial measures, design new structures in compliance with national standards or for research in developing simplified methods of appraising buildings.

2 EVALUATION OF MODELLING METHODS

Four structures were selected to demonstrate how parasitic stressed skin action affects the behavior of buildings, FEMs of the structures are shown in Figure 1. A simplified horizontal wind loading of $0.5kN/m^2$ windward and leeward was applied as a point load at the top of the eaves. Each structure was modelled five times using the following different methods:

(UF) – Unclad frame: A representative internal frame with no stiffening effect of cladding. This provides a baseline and indicates how much of an effect the cladding has.

(ECCS) – Bryan method. The cladding panel flexibility and strength is calculated using the ER (ECCS, 1995) and the Bryan Method used to calculate how load is shared between the frames and cladding.

(ETM) – Equivalent Truss Method (Mahendran & Moor, 1999). The cladding stiffness is calculated using the expressions given in the ER. Cross braced, tension only, elements, together with stiff edge members pinned at both ends were used to model the roof and walls.

(CM1) – Component method roof and restrained walls. The roof is modelled in detail using the component method, while the gable and side walls are restrained in plane with nodal restraints

(CM2) – Component method roof and wall. Both the roof and wall are modelled in detail using the component method. Based on comparison with tested results, this method is the best approximation of reality.

3 RESULTS

The results of the analysis are shown in Table 1. The four structures have very different structural forms, and different bracing systems; this dramatically influences the effectiveness of the simplified methods. In general, ECCS and CM1 tended to overestimate the forces resisted by the gables due to the assumption of infinitely stiff gables; this is most prominent in (D) where the internal frames had very little sway stiffness in their own plane. All methods showed a reasonable match for the central deflection of the three smaller frames (A, B, C), however for the larger three-bay structure ECCS/ETM overestimated the central deflection; both methods use the ER expressions for the

DOI: 10.1201/9781003348450-151

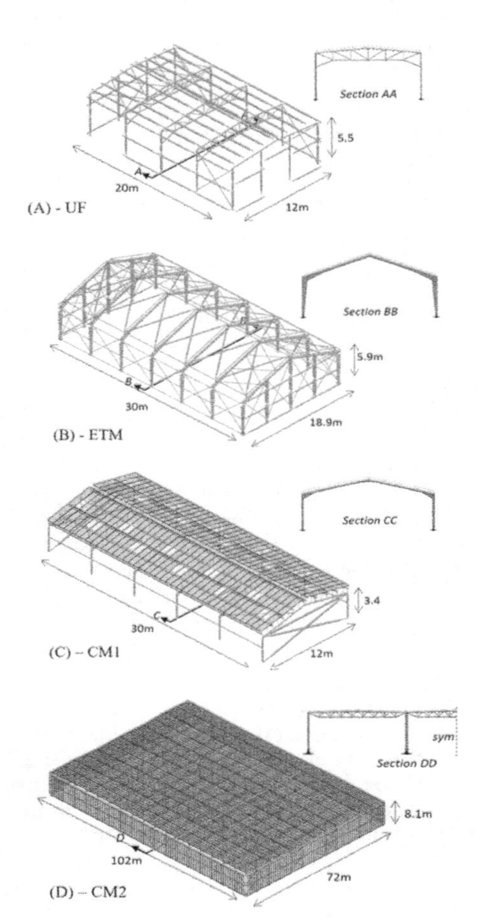

Figure 1. – Tested buildings (A) Trussed rafter building in Poland (UF), (B) Tapered frame in Hungary (ETM), (C) Plastically designed portal in UK (CM1), (D) Three bay warehouse.

flexibility of a panel, which may be inaccurate when the panel is part of a large, multi-bay structure. In addition, neither of these methods give any allowance for interface flexibility thus as the forces in the sheeting build up in the larger building the discrepancy grows. The simplified methods were unable to predict the failure loads of the panels with accuracy in any of the buildings. In general, the central deflection predicted by CM1 was comparable to CM2, but lowered by the deflection at the gable frames, suggesting that a compromise would be to mesh the gables, but restrain the side walls with constraints.

4 CONCLUSIONS

In all four structures the deflections were significantly reduced, and more than 70% of the applied loads at eaves levels (corresponding to around 35%

Table 1. Results of parametric study.

Building	UF	ECCS	ETM	CM1	CM2
		Σ Applied Load / Σ Reaction at Gables			
A	-	0.97	0.92	0.98	0.95
B	-	0.67	0.61	0.79	0.71
C	-	0.92	0.96	0.91	0.90
D	-	0.71	0.76	0.94	0.87
		Deflection at center frame (mm)			
A	122.78†	5.26	11.36	3.94	9.05
B	17.96	9.02	10.28	5.87	7.53
C	62.57	7.34	3.98	2.82	3.50
D	171.51†	31.04	51.71	19.18	27.82
		Deflection at gable frame (mm)			
A	-	0.00	5.50	0.00	5.09
B	-	0.00	2.16	0.00	2.39
C	-	0.00	0.49	0.00	0.42
D	-	0.00	15.89	0.00	7.61
		Roof panel failure load factor			
A	-	1.21(E)	1.34(E)	3.21(G)	3.57(G)
B	-	1.34(E)	1.53(E)	4.96(G)	5.69(G)
C	-	1.30(E)	1.23(E)	6.32(S)	6.93(S)
D	-	0.79(E)	1.05(E)	1.98(G)	1.39(G)

(E) = Edge fasteners, (G) = To gable, (S) = Sheet to purlin
† Whole 3D building analysis including braces and purlins

of the total wind load) was resisted at the gables. Forces in FEMs which do not include the cladding, such as those used for design, may be far less accurate than previously appreciated.

For smaller structures the simplified ECCS and ETM provided a reasonable approximate for the global deflections; the ETM could realistically be incorporated into design models with little extra effort to estimate the influence of parasitic diaphragm action, however, the estimated failure load is very conservative in all cases. For the large structure (D), the simplified methods were far less accurate.

CM2 type models are beyond the capability of typical design offices, it therefore follows that there is no way currently available to (safely) design such structures. This research is ongoing.

REFERENCES

ECCS. (1995). *European recommendations for the application of metal sheeting acting as a diaphragm*. Brussels, Belgium: European Convention for Constructional Steelwork.

Mahendran, M., & Moor, C. (1999). Three-dimensional modelling of steel portal frame buildings. *Journal of Structural Engineering*, *125*(8), 870–878.

Current Perspectives and New Directions in Mechanics, Modelling and Design of Structural Systems – Zingoni (ed.)
© 2022 Copyright the Author(s), ISBN 978-1-032-39114-4

Fatigue resistance of steel arch bridge hanger connection plates due to transverse welding

Ph Van Bogaert
Civil engineering Department, Ghent University, Belgium

ABSTRACT: The connection of hangers to steel tied arch bridges requires the introduction of stiffeners, enabling to transfer the hanger force to the hollow box arch section's webs. However, the hanger force coincides with the arch axis plane, whereas the webs are located in 2 vertical planes at a distance of half of the arch width from the axis. Hence transverse bending of the lower flange will combine with the stress flow in this flange. The research aims at assessing the fatigue resistance for this multiaxial stress condition, first by applying nominal stress evaluation and subsequently by determining the hot spot stress situation. The critical plane method renders the most reliable result and demonstrates the insufficient fatigue resistance for a particular case. The results are generalized in a single diagram.

1 HANGER CONNECTIONS

The connection of hangers to steel tied box section arches essentially is meant to transfer the hanger force as shear to the upper member. This shear force is developed in the arch box webs, located on both sides of the hollow section, as shown in Figure 1. Hence, both these forces are not coplanar, and diaphragms in the arch box section are needed for transferring the central hanger force from one plane to the other. As a consequence, the diaphragms are subjected to bending in a plane perpendicular to the arch plane. It seems logical that the arch box lower flange should be connected to the diaphragms in order to constitute reversed T-section and to develop higher bending resistance. However, all of these details are introducing geometrical discontinuities in the various cross sections and result in stress concentrations, which considerably lower the fatigue resistance. Figure 1 illustrates this situation. The issue is now whether the longitudinal stiffener may or may not be welded to the lower flange as indicated by the arrows.

Built-up cross sections generate a geometric discontinuity of the stress flow and lower the fatigue resistance. The problem is well-known and can be analyzed quite easily through the use of the fatigue classes of EN 1993-1-9. However it should be stressed that these fatigue classes apply to nominal stress only. It is often found that hot spot stresses are compared to the values in the tables, which clearly is not the way to proceed.

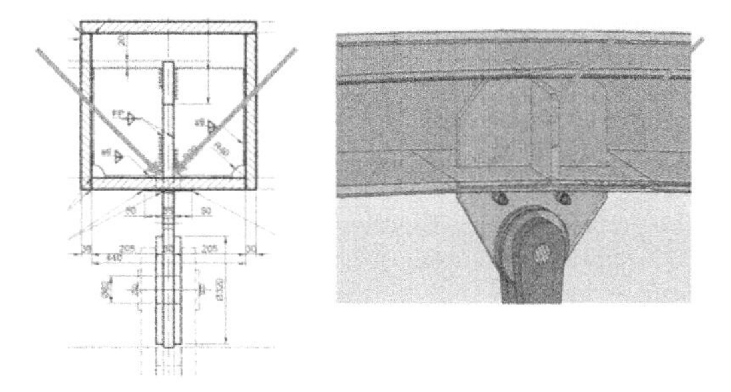

Figure 1. a (left) Cross section connection 1b (right) Connection 3-D view.

2 TRANSVERSE BENDING AND LONGITUDINAL FLANGE COMPRESSION

If the connection of the arrows in Figure 1 exists, obviously, the lower flange plate is subjected to biaxial bending. Various methods, such as nominal stress and hot spot stress have been tested and render diverging results. Figure 2 shows the different nature of hot spot stress due to the hanger force and due to the arch compression and bending.

This illustrates the biaxial character of the nominal stresses. Hence, the alternative of the modified critical plane mode theory (Bäckström 2003) has been considered. This method can be adapted to the two-dimensional problem. If φ is the angle of the perpendicular to the sought plane with respect to the x-axis the normal stress to this plane is found as:

DOI: 10.1201/9781003348450-152

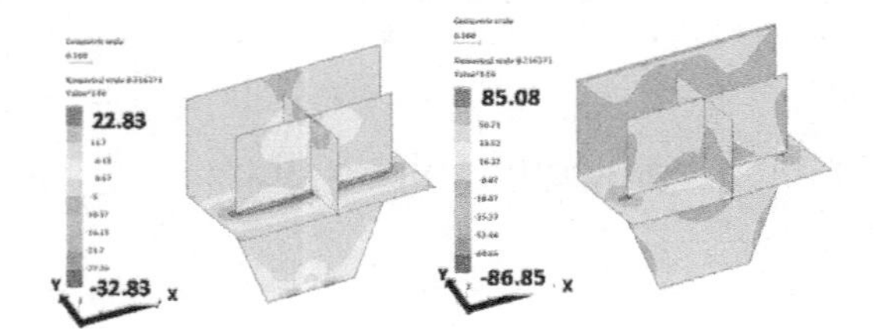

Figure 2. Stress concentrations due to (left) hanger force (right) arch compression and bending.

$$\sigma_x' = \sigma_x \cos^2\varphi + \sigma_y \sin^2\varphi + 2\,\tau_{xy}\sin\varphi\cos\varphi \quad (1)$$

In (1) the x and y axes are identical to the coordinate system. The angle φ is determined to result in maximum stress on the plane. Similar expressions apply to other normal and shear stresses on this plane. If shear is the dominant stress for fatigue failure, the value of (2) should be compared to the damage function.

$$\tau_{hs}' = \Delta\tau_{hs} + 2\,k\,\sigma_{n\,hs}^{max} \quad (2)$$

In (2) $\Delta\tau_{hs}$ is the hot spot variation of shear stress acting on the critical plane and the quantity $\sigma_{n\ hs}^{max}$ is the largest applied hot spot normal stress. The factor k is the material sensitivity to normal stress on a shear plane and has been determined experimentally as equal to 0.3. Finally τ'_{hs} should be compared to (3), verifying that $N_f \geq 2\ 10^6$, since all stress variations were calibrated on this number of cycles.

$$N_f = \left(\frac{\Delta\tau_{hs}'}{\tau_f^*}\right)^{-m} \quad (3)$$

In (3) τ^*_f is the hot spot fatigue category and m is the exponent of the corresponding fatigue SN-curve.

The question than arises as to what should be the magnitude of the hanger force variation to comply with the hot spot fatigue verification. For this, the hanger force has been varied under constant conditions for the arch. Clearly, the hanger force should be limited to 0.861-times its real value, to obtain $N_f = 2\ 10^6$.

Figure 3 shows a diagram which tries to generalize the results. The horizontal axis displays the stress variation parallel to the arch axis, due to compression and bending, whereas the vertical axis renders the normal stress perpendicular to the former and due to the hanger force being introduced. Both values must already include γ_{Mf} and the drawn line corresponds to hot spot fatigue category 90 MPa and thus the number of 2 10^6 loading cycles, which occur simultaneously

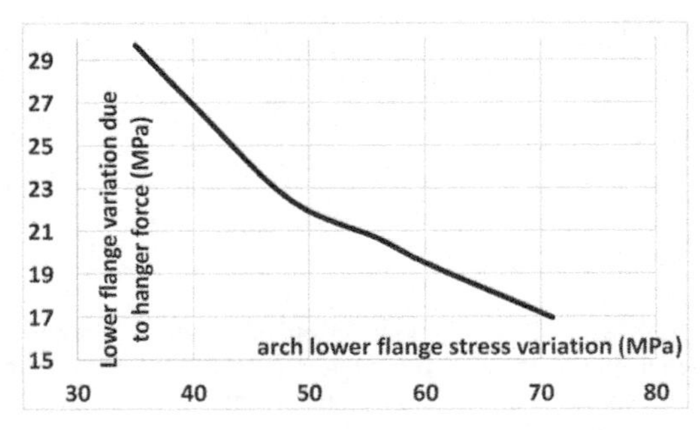

Figure 3. Generalized fatigue failure lower flange stress variations.

3 CONCLUSIONS

The fatigue resistance of the hanger connection to a hollow box section of the arch has been assessed. The main issue of this connection is the two-directional bending of the lower plate of the box section. Because of this complex stress condition, the method of nominal stress does not seem to cover the fatigue requirement. However, this method uses principal stress as a criterion. In this type of joint, the direction of the principal stress does not coincide with a typical fatigue category.

Hence, the fatigue resistance of this type of joint needs to be addressed by the hot spot method and requires to determine the particular stress concentration factors. In most problems, the nominal stress variation method is more conservative. Obviously, this is not the case for the hanger to box section connection, since the use of the modified critical plane method for hot spot stress variations renders a more critical result.

The results have been generalized by summarizing in a single diagram, the stress variation parallel to the arch axis, due to compression and bending, and the normal stress perpendicular to the former and due to the hanger force being introduced. If the stress variations introduced on this curve already include the fatigue partial safety factor they correspond to a fatigue resisting condition.

REFERENCES

Bäckström, M., Multiaxial fatigue life assessment of welds based on nominal and hot spot stresses. PhD thesis Lappeenrata Univ of Techn. VIT Publication, Lappeenrata (Finland) 2003.

Current Perspectives and New Directions in Mechanics, Modelling and Design of Structural Systems – Zingoni (ed.)
© 2022 Copyright the Author(s), ISBN 978-1-032-39114-4

Numerical simulation and design of steel equal-leg angle section beams

B. Behzadi-Sofiani, L. Gardner & M.A. Wadee
Imperial College London, London, UK

1 INTRODUCTION

Comparisons with experimental and numerical results have demonstrated that existing design provisions are generally conservative in predicting the ultimate resistance of steel angles subjected to major-axis bending and often unsafe for minor-axis bending. The aim of the present paper is to develop a new approach to the design of angle section beams, suitable for incorporation into Eurocode 3 (EC3).

2 MECHANICAL BEHAVIOUR

No analytical expression exists to determine the local elastic buckling moment $M_{cr,l,u}$ for angles under major-axis (u-u) bending. Hence an approximate expression was developed on the basis of numerical results and analogies with existing expressions (Bulson, 1969). Lateral-torsional buckling may also be critical for longer angle section beams in major-axis bending.

For equal-leg angle section members subjected to uniform bending about the minor axis (v-v), since both legs are under the same stress distribution, it is reasonable to assume that they will buckle locally simultaneously and hence exhibit simply-supported conditions along the adjoined plate edges, with the other edge being free. Bulson (1969) derived an expression for determining the elastic local buckling stress of plates with such boundary conditions under bending, resulting in the elastic local buckling moment for an equal-leg angle. Under minor-axis bending, both legs have the same shape and direction of buckling (i.e. rigid twisting); hence, the local and torsional buckling modes are essentially the same. This can also be shown by assuming that anticlastic plate bending is restrained due to plate continuity, which leads to identical analytical expressions for the elastic torsional and local buckling stresses for a plate simply-supported along one edge and free along the other (Behzadi-Sofiani et al., 2021a). The general expression for the lateral-torsional buckling moment by Clark & Hill (1960) can be used to show equal-leg angle section beams under minor-axis bending may experience lateral-torsional buckling. An additional consideration for angle section beams is that of flange curling or Brazier flattening. Zhou et al. (2018) developed an analytical expression for predicting the Brazier critical moment $M_{cr,Br}$ in angle section beams.

3 NUMERICAL MODELLING

A parametric study was carried out to investigate the behaviour of steel equal-leg angles. It was found that the numerically obtained buckling reduction factors vary with the $M_{cr,l,u}/M_{cr,LT,u}$ ratio for the same slenderness when local buckling is critical. Therefore, a similar method to that proposed in Behzadi-Sofiani et al. (2021a, b) for fixed-ended equal-leg angle section columns is also proposed herein for equal-leg angle section beams, where the plate buckling and lateral-torsional buckling curves are used as the upper and lower bound limits, respectively.

In the current EC3, no member buckling check is required for beams under minor-axis bending. However, equal-leg angle section beams under minor-axis bending may experience lateral-torsional buckling. In addition, when the tips are in tension, Brazier flattening becomes critical, and reduces the ultimate strength of equal-leg angle section beams, as shown in Figure 3. A new design approach for angle section members in minor-axis bending, giving due consideration to both lateral-torsional buckling and Brazier flattening is therefore presented in Section 4.

4 NEW DESIGN PROPOSALS

In order to address the aforementioned shortcomings in the current EC3 design rules, new design rules are proposed. The proposed design buckling resistance $M_{b,u,Rd}$ of steel equal-leg angle section beams in bending about the major axis is:

$$M_{b,u,Rd} = \frac{\chi W_{pl,u} f_y}{\gamma_{M1}} \tag{1}$$

noting that the plastic section modulus $W_{pl,u}$ is used for all classes of cross-section, but, following the same reasoning as described in Behzadi-Sofiani (2021a, b), the influence of both local and lateral-torsional buckling is nonetheless captured, as shown below. In Equation 1:

$$\chi = \chi_{LT} + \Delta(\chi_l - \chi_{LT}) \tag{2}$$

in which the local buckling reduction factor χ_l is given by:

DOI: 10.1201/9781003348450-153

$$\chi_1 = \frac{\bar{\lambda}_{\max,u} - 0.188}{\bar{\lambda}^2_{\max,u}} \text{ but } \chi_1 \leq 1.0 \quad (3)$$

the lateral-torsional buckling reduction factor χ_{LT} is given by:

$$\chi_{LT} = \frac{1}{\phi_{LT} + \sqrt{\phi^2_{LT} - \bar{\lambda}^2_{\max,u}}} \text{ but } \chi_{LT} \leq 1.0 \quad (4)$$

and Δ is given thus:

$$\Delta = \begin{cases} \left(1 - \frac{M_{cr,l,u}}{M_{cr,LT,u}}\right)^{3.5} \text{ for } \frac{M_{cr,l,u}}{M_{cr,LT,u}} \leq 1 \\ 0 \text{ for } \frac{M_{cr,l,u}}{M_{cr,LT,u}} > 1 \end{cases} \quad (5)$$

with the maximum normalised slenderness $\bar{\lambda}_{\max,u}$ and ϕ_{LT} being given by:

$$\bar{\lambda}_{\max,u} = \sqrt{\frac{W_{pl,u}f_y}{M_{cr}}} \quad (6)$$

$$\phi = 0.5\left[1 + 0.34\left(\bar{\lambda}_{\max,u} - 0.2\right) + \bar{\lambda}^2_{\max,u}\right] \quad (7)$$

where M_{cr} is the minimum of $M_{cr,l,u}$ and $M_{cr,LT,u}$.

4.1 *Bending about the minor axis*

For equal-leg angle section beams in bending about the minor axis, it is proposed that the design buckling resistance $M_{b,v,Rd}$ is obtained thus:

$$M_{b,v,Rd} = \frac{\chi W_{pl,v} f_y}{\gamma_{M1}} \quad (8)$$

where

$$\chi = \frac{1}{\phi + \sqrt{\phi^2 - \bar{\lambda}^2_{max,v}}} \text{ but } \chi \leq 1.0 \quad (9)$$

in which the maximum normalised slenderness $\bar{\lambda}_{\max,v}$ and ϕ are given by:

$$\bar{\lambda}_{\max,v} = \sqrt{\frac{W_{pl,v}f_y}{M_{cr}}} \quad (10)$$

$$\phi = 0.5\left[1 + 0.13\left(\bar{\lambda}_{\max,u} - 0.4\right) + \bar{\lambda}^2_{\max,u}\right] \quad (11)$$

where M_{cr} is the minimum of $M_{cr,LT,v}$ and $M_{cr,Br}$. Comparisons of the collected test and generated FE capacities M_u against the resistance predictions according to the new proposals $M_{b,prop}$ are presented in Figure 1 for angle section beams under bending.

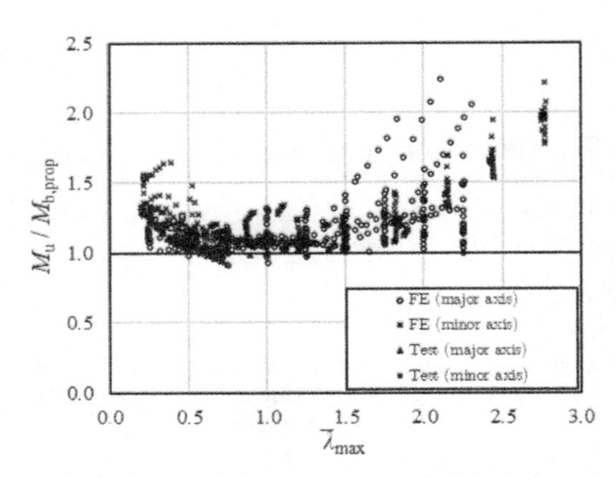

Figure 1. Comparisons of test and FE capacities with resistance predictions obtained using new design proposals.

5 CONCLUSIONS

A comprehensive study into the behaviour and design of steel equal-leg angles subjected to bending has been conducted. For angles under major-axis bending a new design method was proposed, where, as well as the member slenderness, the ratio of the local to lateral-torsional elastic buckling moments are considered. For angles in bending about the minor axis, only a cross-section resistance check is recommended in the current EC3 methodology. However, it has been shown that angles subjected to minor-axis bending can still be susceptible to lateral-torsional buckling. Additionally, Brazier flattening can become important when the tips of the angle section are in tension. A new design approach for angle section beams in minor-axis bending reflecting these findings has been proposed. Overall, the new design proposals have been shown to provide dramatic improvements in the accuracy of strength predictions for steel equal-leg angle section beams.

REFERENCES

Behzadi-Sofiani, B., Gardner, L. & Wadee, M. A. (2021), 'Stability and design of fixed-ended stainless steel equal-leg angle section columns', Eng. Struct. 249, 113281.

Behzadi-Sofiani, B., Gardner, L., Wadee, M. A., Dinis, P. B. (2021), 'Behaviour and design of fixed-ended steel equal-leg angle section columns', J. Constr Steel. Res. 182, 106649.

Brazier, L. G. (1927), 'On the flexure of thin cylindrical shells and other thin sections', Proc. Royal Soc. London. Series A 116(773),104–114.

Bulson, P. S. (1969), The stability of flat plates, Elsevier

Clark, J. W. & Hill, H. N. (1960), 'Lateral buckling of beams', Journal of Structural Division, American Society of Civil Engineers (ASCE) 86, 175–196.

Zhou, Z., Xu, L., Sun, C. & Xue, S. (2018), 'Brazier effect of thin angle-section beams under bending', Sustainability 10(9),1–11.

Current Perspectives and New Directions in Mechanics, Modelling and Design of Structural Systems – Zingoni (ed.)
© 2022 Copyright the Author(s), *ISBN 978-1-032-39114-4*

Behaviour of steel end plate bolted beam-to-column joints

J. Qureshi
School of Architecture, Computing and Engineering, University of East London, London, UK

S. Shrestha
Adept Contracts Ltd, London, UK

ABSTRACT: This paper presents a numerical investigation into moment-rotation behaviour of extended, flush, and partial depth end-plate joints. Two types of steel beam-to-column joints are modelled: I-beam (IPE) to H-column (HEB) and PFC-beam (UPE) to SHS-column. The joints are designed as a single cantilever beam-to-column joint configuration. The connection details are in accordance with Eurocode 3 part 1-8 and SCI guides P358 (Simple joints) and P398 (Moment joints). Three-dimensional finite element modelling of the joints is carried out using ABAQUS. The numerical results are validated against experiments. Both material and geometric nonlinearities are considered. All joints are classified according to their stiffness as per Eurocode 3. The failure modes of different joint configurations are determined and compared with the experiments. The key contribution of this paper comes from stiffness analysis of the joints, characterisation of the joints as per EC3 and modelling nonlinear behaviour of the joints.

1 INTRODUCTION

Joints can be classed as simple or moment-resisting joints. End plate joints are most popular due to ease of fabrication and speed of construction. Joint's behaviour is characterised by its moment-rotation response. The conventional approach is to assume the joint's response nominally pinned or fully rigid. However, in many practical frames the moment-rotation response lies between these two theoretical extremes, introducing semi-rigid action (Qureshi & Mottram 2015). The behaviour of end plate bolted joint is in between pinned and rigid joints.

The aim of this paper is to model end plate steel beam-to-column joints using ABAQUS. Three joint configurations are tried: extended depth, flush or full depth and partial depth end plate joints. Two different member configurations are modelled: IPE beam to HEB column and PFC beam to SHS column. The numerical results are verified against experiments first, followed by a parametric study. Moment-rotation behaviour of joints is studied. Failure patterns are also determined. Joints are also classified, according to their stiffness using Eurocode 3 provisions, as rigid, semi-rigid and pinned.

2 EXPERIMENTAL DATA

The experimental results from previous research (de Lima et al. 2004; Simões Da Silva et al. 2004) are used to validate the finite element models. The data for flush and extended end plate bolted joints is taken from (Simões Da Silva et al. 2004), shown in Figure 1 and (de Lima et al. 2004) in Figure 2, respectively.

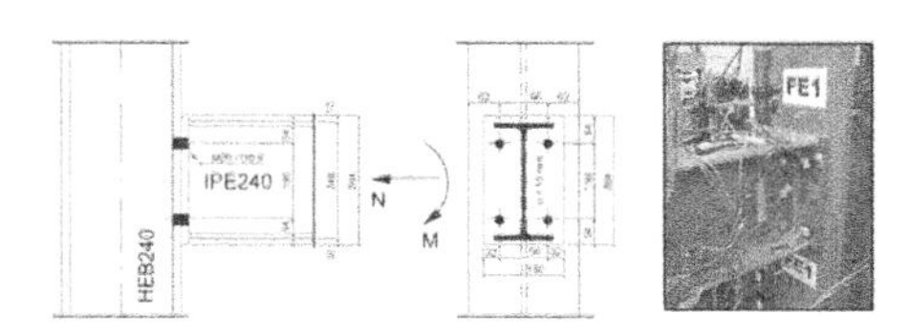

Figure 1. Flush end plate joint (Simões Da Silva et al. 2004).

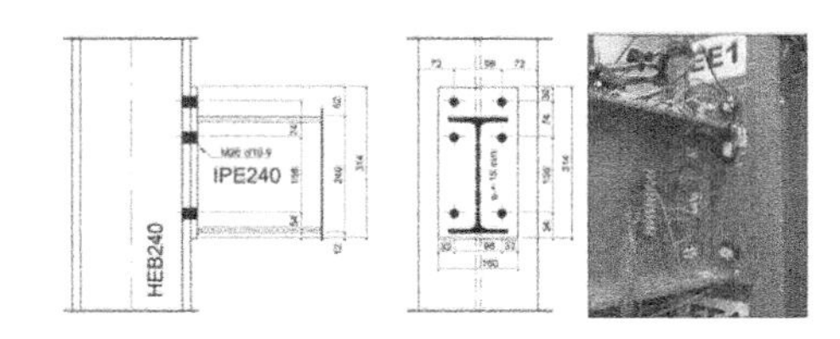

Figure 2. Extended end plate joint (de Lima et al. 2004).

3 FINITE ELEMENT MODEL

Finite element modelling is carried out for steel end plate bolted joints using ABAQUS. The end plate

DOI: 10.1201/9781003348450-154

bolted joints modelled are partial depth, full depth and extended end plate joints as shown in Figure 3.

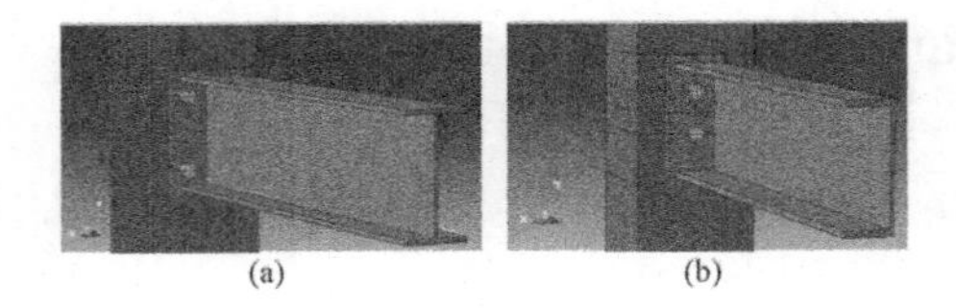

Figure 3. Assembled ABAQUS flush end plate joint models: (a) IPE to HEB joint; (b) PFC to SHS joint.

4 RESULTS AND DISCUSSION

First, the finite element model is verified against the flush and extended end plate joints between IPE and HEB from (Simões Da Silva et al. 2004) and (de Lima et al. 2004), respectively. Failure modes are compared too. The validated model is used to create IPE-HEB partial depth end plate joint and PFC-SHS joints (partial, flush and extended end plate). Figures 4 and 5 show comparison of moment-rotation response and failure modes for flush end plate IPE-HEB joints.

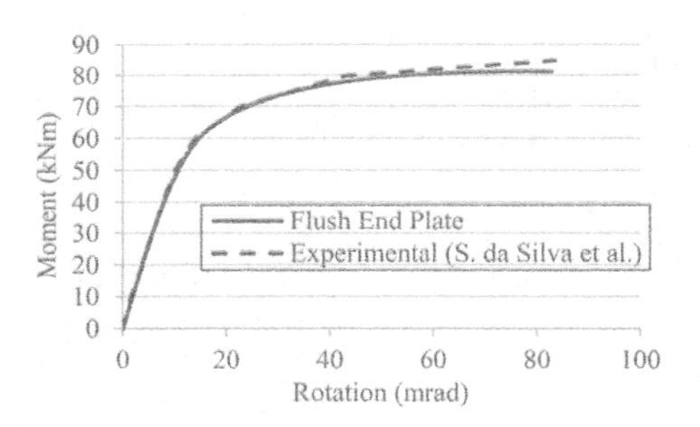

Figure 4. IPE to HEB flush end plate joint FE model comparison with experimental results.

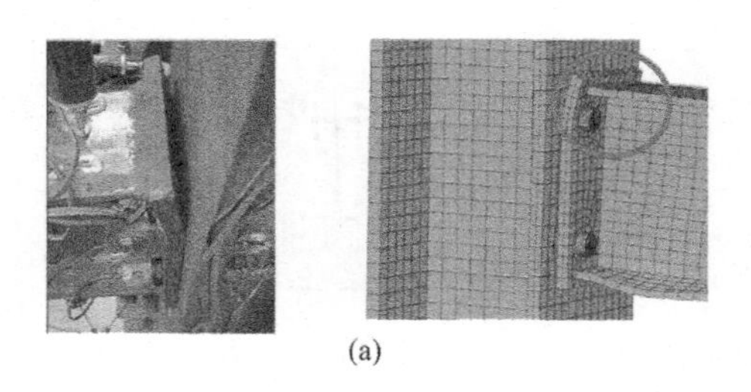

Figure 5. End plate bending in experiment and FE model.

The moment-rotation response of partial, flush and extended depth end plate IPE-HEB joints is presented in Figure 6. The moment resistance of flush end plate and partial depth end plate is about 37% and 79% of the moment in extended end plate joints.

4.1 *Joint classification*

All joints are classified according to their stiffness as per Eurocode 3. Figure 7 presents the joint classification for IPE-HEB joints. Joints with extended and flush end plates are classified as semi-rigid, whereas partial depth end plate joints are classed as pinned.

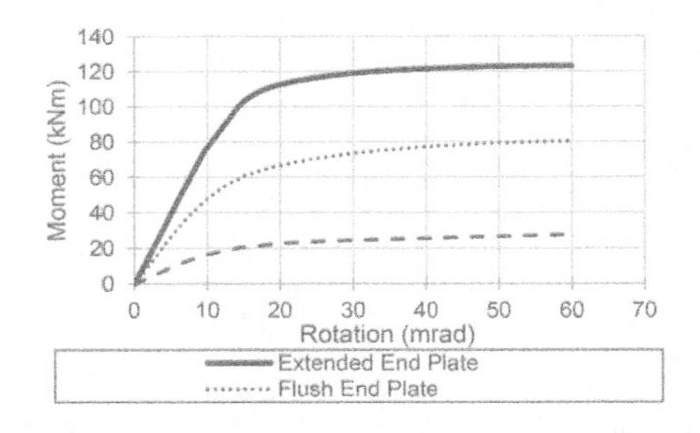

Figure 6. Moment-rotation behaviour of IPE to HEB end plate bolted joints (joint rotation limited to 60 mrad for comparison).

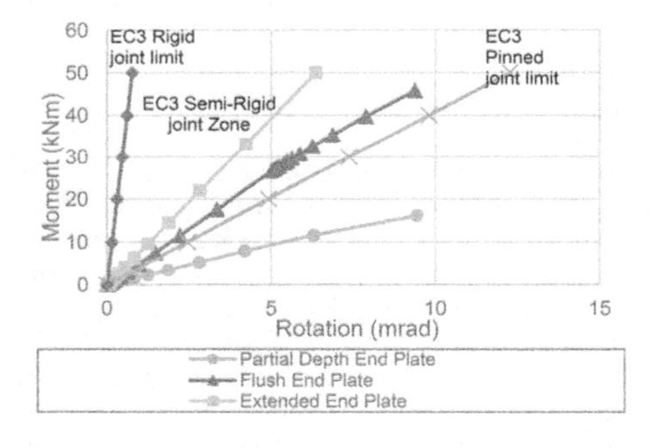

Figure 7. Joint classification as per EC3 for IPE-HEB joints.

5 CONCLUSIONS

- Finite element model matched with experimental moment-rotation response and failure modes.
- IPE-HEB extended depth and flush end plate joints are classified as semi-rigid and partial depth joints as nominally pinned.

REFERENCES

de Lima LRO, Simoes da Silva L, Vellasco PCG d. S, de Andrade SAL. 2004. Experimental evaluation of extended endplate beam-to-column joints subjected to bending and axial force. Eng Struct. 26(10):1333–1347.

Qureshi J, Mottram JT. 2015. Moment-rotation response of nominally pinned beam-to-column joints for frames of pultruded fibre reinforced polymer. Constr Build Mater. 77:396–403.

Simões Da Silva L, De Lima LRO, Pedro PCG, De Andrade SAL. 2004. Behaviour of flush end-plate beam-to-column joints under bending and axial force. Steel Compos Struct. 4(2):77–94.

Current Perspectives and New Directions in Mechanics, Modelling and Design of Structural Systems – Zingoni (ed.)
© 2022 Copyright the Author(s), *ISBN 978-1-032-39114-4*

Compression strength of aged built-up column with vanished lacing bars

T. Miyoshi & T. Nakakita
Department of Civil Engineering, National Inst. of Tech. (KOSEN), Akashi College, Akashi, Hyogo, Japan

K. Iwatsubo
Production System Engineering Crs., National Inst. of Tech. (KOSEN), Kumamoto College, Yatsushiro, Kumamoto, Japan

T. Takai
Department of Civil and Architecture Engineering, Kyushu Inst. of Tech., Kitakyushu, Fukuoka, Japan

K. Tamada
Department of Civil Engineering and Arch., National Inst. of Tech. (KOSEN), Maizuru College, Maizuru, Kyoto, Japan

ABSTRACT: Recently, Japanese aged built-up members have lost their lacing bars due to corrosion. However, remaining strength of the built-up column with vanished lacing bars has not been clarified yet. In addition, residual stress of channel shaped steel which consists built-up column has not made sufficiently clear. The authors measured residual stress distribution of channel shaped steel, which was removed from an aged steel truss bridge. Moreover, the authors numerically calculated ultimate behavior of the built-up column with vanished lacing bars using finite element method. As a result, it is found that residual stress magnitude of the channel shaped steel is small compared to that of welded member, vanishment of lacing bars does not cause significant reduction of the compression strength and column strength curve given in Japan Specifications for Highway Bridges can be used to estimate its remaining strength conservatively.

1 INTRODUCTION

Recently, Japanese aged steel bridges with built-up members have lost their lacing bars due to corrosion. Meanwhile, their remaining strengths and rational methods for estimating the strength have not been clarified yet. In addition, residual stress of channel-shaped steel (hereafter, abbreviated to "CSS"), which is a part of built-up columns has not been made sufficiently clear. This study aims to demonstrate residual stress distribution of CSS and numerically estimate compression strength of aged built-up columns with vanished lacing bars.

In this study, the authors carried out a site investigation of an aged steel bridge with built-up members. According to the result, a vertical compression member lost its lacing bar (hereafter, abbreviated to "LB"). In addition, the authors conducted residual stress measurement of CSSs removed from a steel truss bridge —Morimura Bridge—, which has been passed 115 years since the completion. Furthermore, the authors numerically analyzed the compression strength of built-up columns with vanished lacing bars using the finite element (hereafter abbreviated to "FE") method.

2 RESIDUAL STRESS MEASUREMENT

Built-up members removed from Morimura Bridge consist of two (left and right) CSSs. The authors measured residual stress by the sectioning method. One of the results, Figure 1, shows residual stress measurement points and its distribution of left CSS, where σ_Y is lower yield stress and σ_r is the member axis component of residual stress. In addition, the figure indicates the average values of the exterior and interior and general residual stress distribution models for mild carbon steel welded sections as a reference.

From Figures 1(b) and (c), effects of perforating and riveting on the residual stress distribution are not observed, although rivet holes are positioned in $z/B = \pm 0.45$ of upper and lower flanges. The absolute value of residual stresses is smaller than the magnitude of the residual stress models. From Figure 1(c), as with upper and lower flanges, the residual stress is generally small compared to the models. In addition, the residual stress distribution does not have a regularity as seen in a welded member.

3 NUMERICAL ANALYSIS

The authors analyzed the ultimate behavior of the built-up column (vertical member) with vanished LB using the FE model shown in Figure 2. A 4-node isoparametric shell element was employed to model CSS and end tie plates (hereafter, abbreviated to "ETPs"). The authors adopted a 2-node truss element to model LBs. The authors assumed only initial deflection expressed by a half-sinusoidal curve with

DOI: 10.1201/9781003348450-155

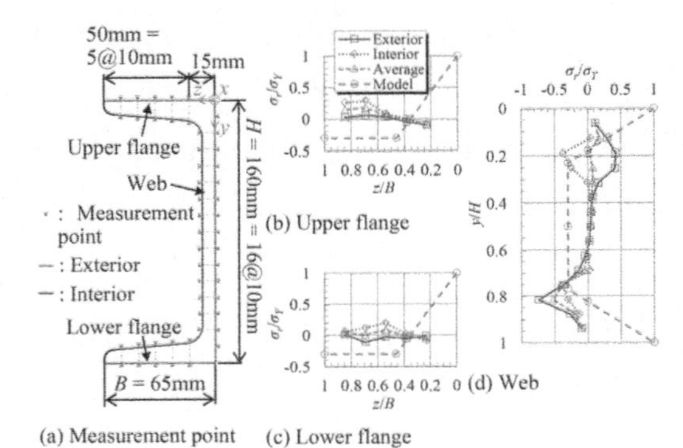

Figure 1. Residual stress distribution of left CSS.

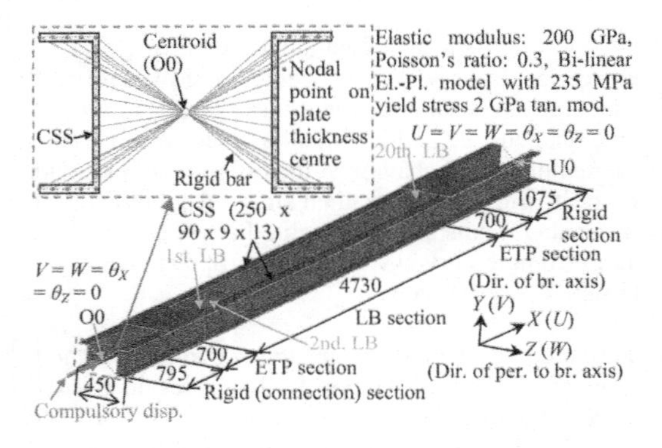

Figure 2. Finite element model of the built-up member.

0.001 times of the column length as the maximum defection concerning initial imperfection. Based on the results of site investigation and axial force of LBs obtained from spatial frame analysis before FE analysis, the authors analyzed four built-up columns shown in Figure 3 to investigate the effect of LBs on the ultimate behavior as a parametric study. The authors show results of the FE analysis using the abbreviations ND, P3, NP4, and P9.

Figure 4 shows load versus axial displacement curves of four FE models compared to those obtained from frame analysis and simple calculation. In Figure 4, LS is a load versus displacement relation based on a simple calculation. FA is a load versus displacement relation obtained from frame analysis. YL is the yield load of cross-sections that consists of two CSSs. UL is the ultimate load by using a column curve specified in Japanese Specifications for Highway Bridges and regarding two CSSs as an effective cross-section.

From Figure 4, There is no significant difference in the maximum load of ND, P3, NP4, and P9. Therefore, even if the LBs such as P3, NP4, and P9 vanished, the load-carrying capacity of the build-up column is hardly reduced. Although the maximum

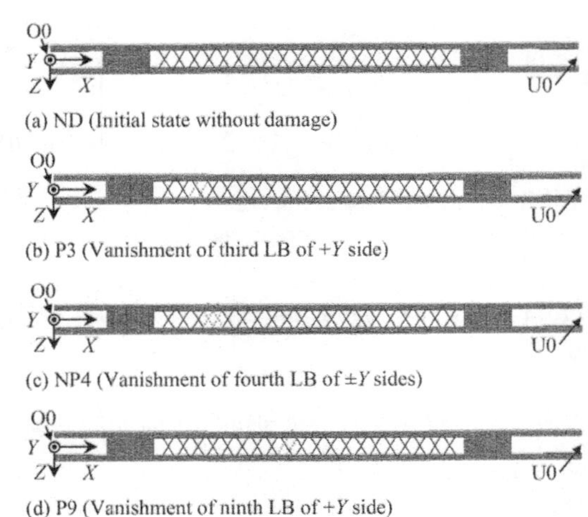

Figure 3. FE models of the parametric study.

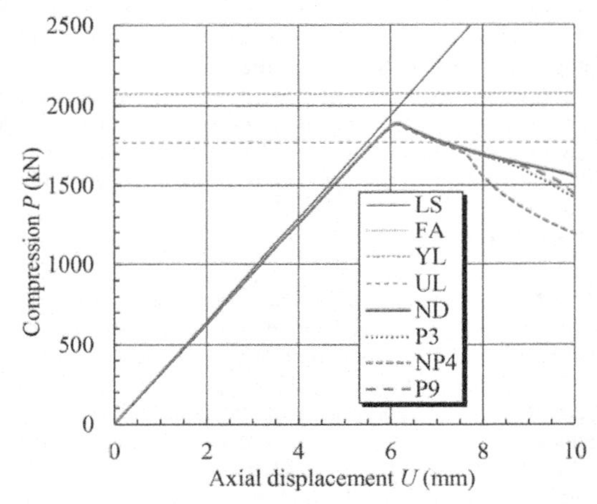

Figure 4. Load versus axial displacement curves.

load of ND, P3, NP4, and P9 does not reach yield load, they exceed UL.

4 CONCLUSIONS

The obtained main results from this study can be summarized as follows:

1) Residual stress of the channel-shaped steel is generally small compared to welded members.
2) Built-up column with a couple of vanished lacing bars has a similar load-carrying capacity to the columns without any damages.

REFERENCES

Japan Road Association 2017. Specifications for Highway Bridges, Part II Steel members and steel bridges (in Japanese)

Current Perspectives and New Directions in Mechanics, Modelling and Design of Structural Systems – Zingoni (ed.)
© 2022 Copyright the Author(s), ISBN 978-1-032-39114-4

Machine learning approach for stress analyses of steel members affected by elastic shear lag

H. Fritz & M. Kraus
Chair of Steel and Hybrid Structures, Bauhaus University of Weimar, Germany

ABSTRACT: In the design of steel members based on beam theory, normal stresses in cross sections are generally determined neglecting shear strain influences (shear lag). However, in cross sections with wide flanges compared to member lengths, shear lag can significantly influence normal stress states. To capture corresponding effects in practical applications, this paper presents a novel stress calculation approach based on machine learning (ML). For implementing the approach, neural networks are employed as supervised ML algorithm using training data generated by finite element calculations (shell models). The performance of the approach is validated using cross sections with wide flanges revealing high accuracy of the neural network. By subsequently interpreting the ML model, influences of different parameters, such as cross section parameters, on the normal stress distributions are quantified, providing deeper understanding of the mechanical problem solved by ML.

1 INTRODUCTION

For simplified calculations of normal stresses in beams, influences of shear distortions are usually neglected going along with the validity of Bernoulli's hypothesis for bending as well as Wagner's hypothesis for torsion. However, this no longer applies to cross sections with wide flanges compared to member lengths, as shear strains associated with bending strongly influence normal stress distributions (shear lag). For capturing this phenomenon, the so-called "effective widths method" has been defined as an approximation, where the nonlinear stress course of wide flanges is substituted by the constant maximum stress, while the flange width is reduced to an "effective width".

In the context of beam theory, corresponding stress distributions may also be determined, for example, with the approach presented by Kraus & Mämpel (2017), provided that the structural system consists of a single-span girder with a distributed load. However, the mechanical formulations do not cover stresses properly in case of discontinuities in the distribution of V_z in longitudinal beam direction. Therefore, and to avoid an approximation by effective widths, machine learning (ML) is applied for such structural systems in the following. This paper investigates a novel regression method based on ML algorithms to autonomously calculate normal stresses of arbitrary thin-walled I cross sections considering shear lag effects.

2 FRAMEWORK

A conceptual framework of the approach presented in this paper is provided in Figure 1.

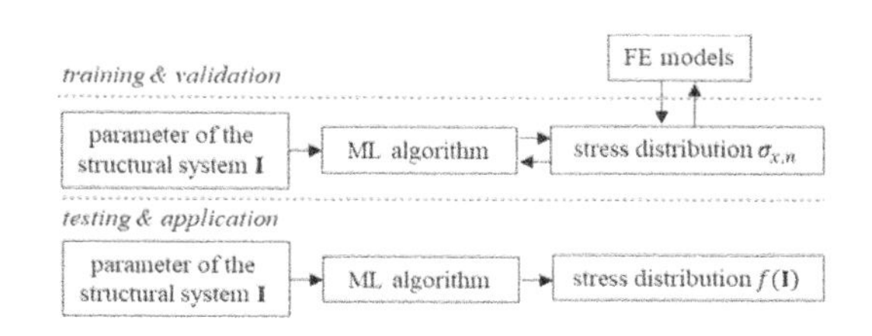

Figure 1. Conceptual framework of calculating stress distributions using ML.

In the approach, first, input data consisting of varying parameters **I** is fed into an ML algorithm. Next to cross section properties, varying input parameters may be load variables and positions or system lengths. In training and validation, the varying parameters **I** and the corresponding pre-determined stresses $\sigma_{x,n}$ in discrete fibers of the cross section n (i. e., the output) are fed into a supervised ML algorithm. Training and validation output data is generated using solutions of finite element (FE) shell models. Based on inherent relationships between the parameters **I** and the corresponding stress distributions $\sigma_{x,n}$, stresses can be approximated in the application process through $f(\mathbf{I}) = \sigma_{x,n}$ by the ML model.

DOI: 10.1201/9781003348450-156

3 IMPLEMENTATION AND VALIDATION

For implementation and validation, a single-span beam with a concentrated vertical load $F = 1000$ kN at mid-span is examined and a singly symmetric I profile is considered, as illustrated in Figure 2. The dataset for training and validating the ML algorithm is defined by varying parameters **I** within predefined limits. The parameters **I** include top flange width b_{tf}, bottom flange width b_{bf}, top flange thickness t_{tf}, bottom flange thickness t_{bf}, web height h_w, web thickness t_w, as well as the structural system length L. For generating the training and validation output data, i.e., the stress distributions $\sigma_{x,n}$ at the upper flange corresponding to the varying input parameters **I**, the structural system is modelled using finite elements and analyzed numerically.

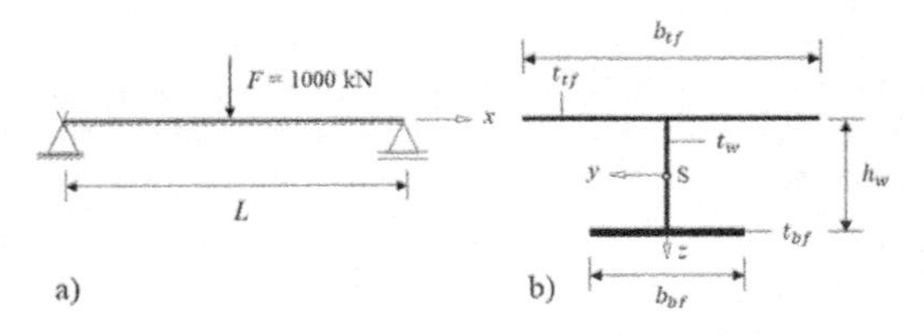

Figure 2. a) Structural system with varying system length and b) varying cross section parameters of the implementation and validation example.

For implementing the approach, a neural network as supervised ML model is developed. The neurons of the input layer correspond to the previously defined input parameters **I**, and the stresses $\sigma_{x,n}$ at discrete points n of the top flange correspond to the neurons of the output layer. During training and validation, a performance of $R^2 = 0.99$ (coefficient of determination) is achieved. The accuracy of the overall performance is analyzed using random test datasets, revealing a good approximation of the neural network with $R^2 = 0.97$. Figure 3 exemplifies a solution of the neural network in comparison to a FE calculation.

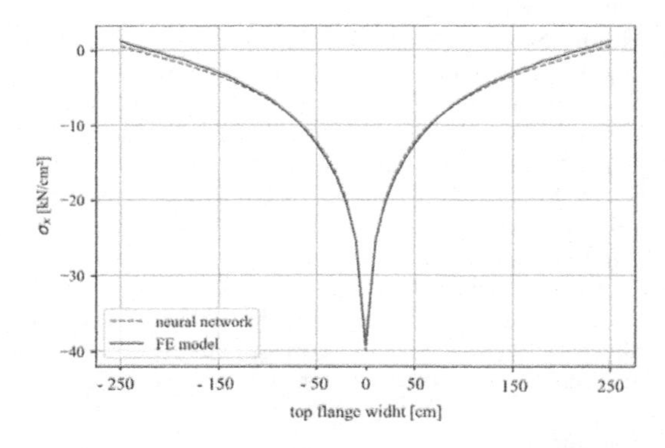

Figure 3. Calculation of top flange stresses by the neural network and a FE model of an I cross section with b_{tf} = 500 cm, b_{bf} = 100 cm, t_{tf} = 2 cm, t_{bf} = 1 cm, h_w = 15 cm, t_w = 1.5 cm, and a system length of L = 450 cm.

4 SENSITIVITY ANALYSIS

In the context of interpretable ML models, influences of input parameters on the output are examined through a variance-based sensitivity analysis using the "total-effect" index (see Saltelli (2002) for details). The total-effect index of the input parameters can be interpreted as the contribution to the output variance, including the impact of all interactions with other input variables. The results demonstrate that the influences of the top flange b_{tf} and the system length L increase towards the edges of the flange ($y = \pm b_{tf}/2$), while the influences of the top flange thickness and the web height decrease, as displayed in Figure 4.

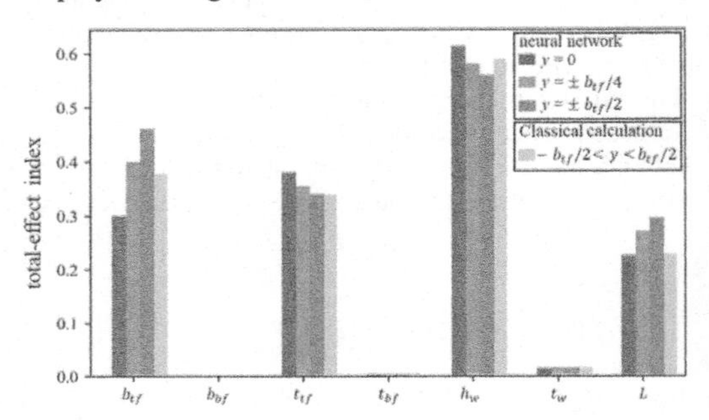

Figure 4. Total-effect indexes, showcasing the influences of varying parameters **I** on the output from the edge ($y = b_{tf}/2$) to the middle ($y = 0$) of the flange in comparison to influences determined based on classical stress calculations ($\sigma_x = M_y/I_y \cdot z$).

5 SUMMARY AND CONCLUSIONS

For calculating stress distributions in steel cross sections affected by shear lag, this paper presents a novel calculation method using machine learning (ML) by means of a neural network. During training, system parameters (e.g., cross section properties) are fed into the ML algorithm as input while providing the corresponding stress output at discrete positions collected from finite element calculations. After training and validation, the neural network reveals a good approximation of stress distributions for singly symmetric I sections with varying cross section parameters and system lengths. Next to the application process, the results are verified through a sensitivity analysis. The method proposed may be used as an automated approximation method with low computational effort.

REFERENCES

Kraus, M., Mämpel, S., 2017. Zur computerorientierten Berechnung schubbeeinflusster Normalspannungen in Stäben. Stahlbau, Vol. 86 (11), pp. 951–960.

Saltelli, A., 2002. Sensitivity analysis for importance assessment. Risk Analysis, Vol. 22 (3), pp. 579–590.

Current Perspectives and New Directions in Mechanics, Modelling and Design of Structural Systems – Zingoni (ed.)
© 2022 Copyright the Author(s), ISBN 978-1-032-39114-4

Behaviour and recommended design method for laterally unsupported monosymmetric steel I-section beams

S. Suman & A. Samanta
Department of Civil and Environmental Engineering, Indian Institute of Technology Patna, India

ABSTRACT: Methods of designing steel monosymmetric I-beams against lateral buckling are not well supported by research. Lateral-torsional buckling (LTB) may predominate the design strength predictions for laterally unsupported beams. The behaviour of monosymmetric I-beams is analyzed for shorter, intermediate, and longer spans with a height-to-width ratio greater than two (h/b > 2). Three regimes are significant in the inelastic buckling resistances of hot-rolled monosymmetric beams under moment gradient, depending on which flange yields first and the end moment ratio. The present manuscript carries out a parametric study using a validated and verified numerical model. The robust computational software ABAQUS is used for simulating a total of 300 elastic and inelastic beam models. The behavior of monosymmetric beam observed under moment gradient and non-linear interaction equation is proposed. Further, the resulted data is also compared with the Eurocode 3 1.1.

1 INTRODUCTION

LTB of monosymmetric I-beams under various load scenarios has been extensively studied in the past (Kitipornchai, Wang, & Trahair, 1986; Kumar & Samanta, 2006; Samanta & Kumar, 2006, 2008; Wang & Kitipornchai, 1986). It was found that the lateral buckling resistance not only limited to pre-buckling deformation but also on section shape and load distribution. It was also established in (Mohri et al., 2010; Suman & Samanta, 2021) that for negative moment gradients all available solutions fail to predict the buckling moment and only numerical solutions are found to be the powerful tool to predict strength of the monosymetric I-beams.

In present study, the behaviour of the beam is investigated under moment gradients. Here, five degrees of monosymmetry of the beam section. The numerical analysis is conducted out using ABAQUS with simply supported boundary conditions.

2 VALIDATION AND NUMERICAL MODELLING

This work investigates full-scale models of monosymmetric steel I-beams. The beams are represented in ABAQUS using C3D8R elements. The Riks approach is utilized to do geometrically and materially nonlinear analysis. The peak amplitude of the imperfection at the beam's mid-span is calculated using Equation 1.

$$y(x) = \frac{l}{1000} \times sin\frac{\pi x}{l} \tag{1}$$

where, l is length of beam span.

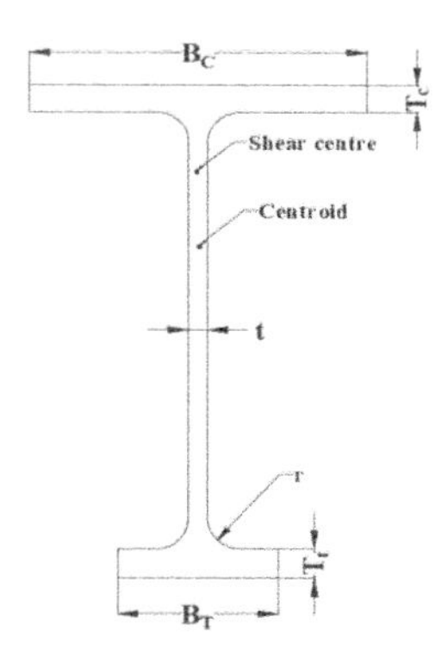

Figure 1. Sectional geometry of beams.

Figure 1 depicts general cross-section geometry. Table 1 shows a description of the geometric characteristics employed in the current work. The degree of monosymmetry of the I-beam cross-section is represented by Equation 2.

$$\rho = \frac{I_{YC}}{I_{YT} + I_{YC}} = \frac{I_{YC}}{I_Y} \tag{2}$$

where, ρ estimates the degree of monosymmetry of I-section beam. To perform the numerous iterations modified Riks method is applied in ABAQUS using very small increment in the analysis. Typical mode shape under reverse curvature bending is shown in (Figure 2d) and corresponding non-linear analysis in Figure 2e. For present analysis yield strength of 250 Mpa and Elastic modulus of 210 Gpa is used.

DOI: 10.1201/9781003348450-157

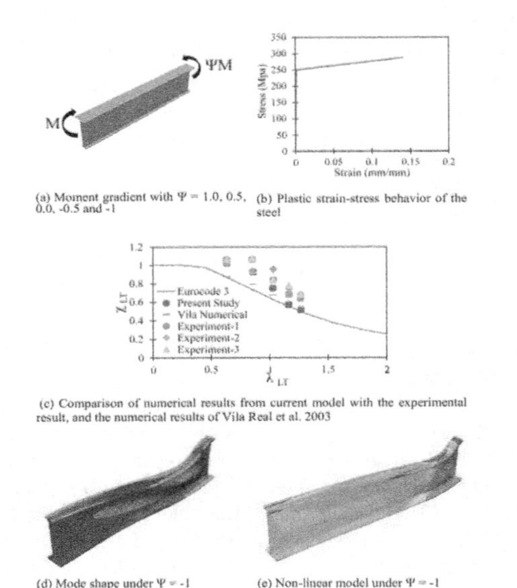

Figure 2. (a) Loading patterns (b) Material plastic properties (c) Validation and verification study and (d) Elastic analysis for imperfection and corresponding non-linear analysis in (e) for ρ = 0.3.

Table 1. ISMB600 monosymmetric beam cross-sectional data.

S. No.	ρ	h	B_C	T_C	B_T	T_T	t	r
1.	0.1	600	100	20	210	20	12	20
2.	0.3	600	150	20	210	20	12	20
3.	0.5	600	210	20	210	20	12	20
4.	0.7	600	210	20	150	20	12	20
5.	0.9	600	210	20	100	20	12	20

* Units: All dimensions are in mm

3 RESULTS AND DISCUSSIONS

3.1 *Interaction curves between degree of mono-symmetry and moment gradient*

The monosymmetric and moment gradient interaction curves are obtained using inelastic moment capacity data using Equation (5) i.e., also used in (Suman & Samanta, 2021). The non-dimensional inelastic buckling moments derived from FEM analysis at ambient temperature (Figure 3) are used to demonstrate the connection between degree of monosymmetry (ρ) and moment gradient parameter (ψ). Equation 5 is used to

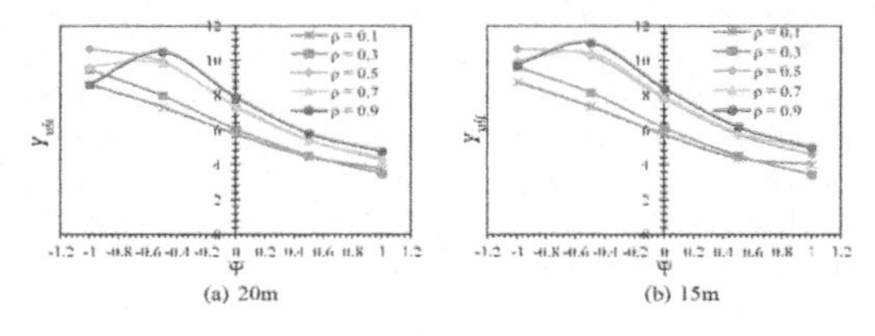

Figure 3. Proposed results for all studied loading.

compute the non-dimensional inelastic critical moment, Y_{ult}, in terms of ultimate inelastic moment, sectional and material parameters.

$$Y_{ult} = \frac{M_{ult} * L}{\sqrt{EGJI_y}} \tag{5}$$

4 CONCLUSION

The collapse of beams with a longer span is not much affected by the non-linear material behaviour of monosymmetric I-beams. Monosymmetric beams with higher width of top flange (i.e., ρ = 0.7 & 0.9) perform exceptionally well despite having same sectional modulus as of sections ρ = 0.1 & 0.3. Under asymmetric bending (ψ = 1), a similar peak moment was obtained due to the same failure mode shape. The inelastic moment-monosymmetry interaction curve showed the constant non-dimensional value of 2.5-3 irrespective of the degree of monosymmetry. Further, the statistical analysis predicted the conservativeness of the current design recommendation in EN 1993-1-1. (2005) under moment gradient pattern (Figure 4).

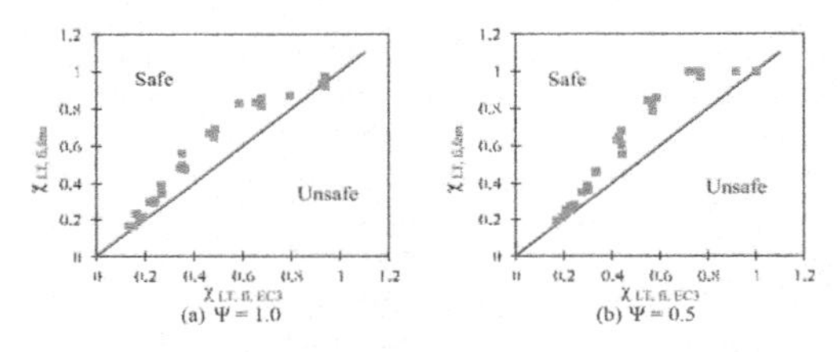

Figure 4. Statistical comparison of FE-result against in EN 1993-1-1. (2005) for all studied loading patterns.

REFERENCES

Kitipornchai, S., Wang, C. M., & Trahair, N. S. (1986). Buckling of Monosymmetric I-Beams under Moment Gradient. *Journal of Structural Engineering*, *112*(4), 781–799. https://doi.org/10.1061/(ASCE)0733-9445(1986)112:4(781)

Kumar, A., & Samanta, A. (2006). Distortional buckling in monosymmetric I-beams: Reverse-curvature bending. *Thin-Walled Structures*, *44*(7), 721–725. https://doi.org/10.1016/j.tws.2006.08.003

Mohri, F., Damil, N., & Potier-Ferry, M. (2010). Linear and non-linear stability analyses of thin-walled beams with monosymmetric I sections. *Thin-Walled Structures*, *48*(4–5), 299–315. https://doi.org/10.1016/j.tws.2009.12.002

Samanta, A., & Kumar, A. (2006). Distortional buckling in monosymmetric I-beams. *Thin-Walled Structures*, *44*(1), 51–56. https://doi.org/10.1016/j.tws.2005.09.007

Samanta, A., & Kumar, A. (2008). Distortional buckling in braced-cantilever I-beams. *Thin-Walled Structures*, *46* (6), 637–645. https://doi.org/10.1016/j.tws.2007.12.004

Suman, S., & Samanta, A. (2021). Behavior of laterally unsupported monosymmetric steel I-section beams at elevated temperature under non-uniform moments. *Structures*, *33*(December2020), 3324–3356. https://doi.org/10.1016/j.istruc.2021.06.070

Wang, C. M., & Kitipornchai, S. (1986). Buckling capacities of monosymmetric I-beams. *Journal of Structural Engineering (United States)*.

Current Perspectives and New Directions in Mechanics, Modelling and Design of Structural Systems – Zingoni (ed.)
© 2022 Copyright the Author(s), *ISBN 978-1-032-39114-4*

Steel beam upstands as a strengthening approach for hot-rolled I-shaped sections

K. Mudenda & A. Zingoni
Department of Civil Engineering, University of Cape Town, Cape Town, South Africa

ABSTRACT: The need to strengthen steel beams in existing structures may become a necessity due to any number of factors such as change in the use of a structure. In this study the use of upstand stiffeners on bi-symmetric steel beams is explored. The upstands convert the section into a monosymmetric section. The effect of monosymmetry leads to unique properties that can be exploited to strengthen an existing beam instead of replacing it with another bi-symmetric section of greater size as obtained in a steel handbook. Elastic and elastic-plastic analysis approaches are used to demonstrate how the use of stiffeners influences the strength and stiffness gain of the existing beam and how the selection process can be implemented. The viability of the approach is demonstrated with examples.

1 INTRODUCTION

Repurposing the use of existing structures may be necessitated by any number of factors. When the changes in use of a structure entail an increase in applied loads then the design engineer may have to consider the possibility of strengthening existing members rather than removal and replacement with new members of greater size or capacity. It may be the case that removal of an existing beam, for instance, and replacement with a new member may be a difficult or even prohibitive undertaking. Access to existing members or limited downtime are some typical examples. In addition, principles of making use of existing members are consistent with the philosophy of sustainable engineering.

Several authors have presented methods of stiffening existing steel beams. Tsavdaridis (2015) highlighted two approaches for reinforcing existing flexural systems as adding new framing to supplement the existing system or reinforcing existing beams, girders and their connections. Reinforcing the existing members was stated as generally being the easy option and possibilities for welding strengthening elements to flanges were presented. Koshmidder-Hatch (2014) discusses some practical methods for strengthening existing steel beams included the use of bolted on members, weld-on plates and concrete encasement. Practical limitations to welding on site were highlighted.

An alternative strengthening technique for hot-rolled I-shaped beams that are laterally unrestrained is presented in the current study. The strengthening method is based on the use of flange upstands so that the initial section, if doubly symmetric, is turned into a monosymmetric cross section. We therefore exploit the beneficial aspects of monosymmetry and use this as the aid to strengthen existing steel beams. The sections of interest here are I-sections, but clearly flange upstands are also applicable to T-sections.

The study looks at the elastic behaviour aspects as well as strength behavior when material yielding together with geometrical imperfections are incorporated.

2 ELASTIC BEHAVIOUR

Laterally unrestrained steel beams bending about their major axis are susceptible to lateral-torsional buckling (LTB) failure. The critical elastic moment is used to determine the load level at which LTB will occur. This provides a basis upon which decisions on selection of beam members can be made. This aspect is considered in this section.

2.1 *Beam model*

The beam model used in the study is a simply supported beam carrying a point load that is applied on the top flange of the beam and thus at a distance ‘a’ above the shear center (SC). The load case employed is that of a point load at midspan.

DOI: 10.1201/9781003348450-158

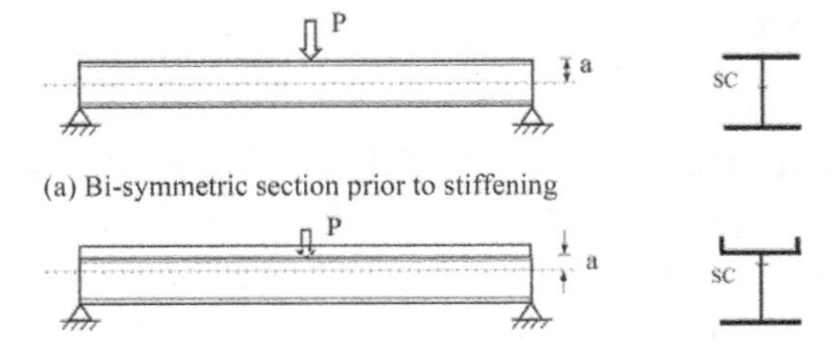

Figure 1. Steel beam stiffening detail.

The thickness of the upstands is taken as being equal to the flange thickness.

2.2 *Member properties and comparison*

For the comparison of using a strengthened member to that of using a bi-symmetric member of greater size the elastic critical moment (or the critical load) needs to be determined. For the strengthened member the elastic critical loads at different values of upstand stiffener heights needed to be determined. The second moment of area was also required for stiffness checks. Unique properties of monosymmetric sections obtained using upstands reported by Mudenda & Zingoni (2018) are utilized. For example, as shown in Figure 2, for a particular section with upstands, by comparing to an 'equivalent doubly/bi-symmetric section' the movement of the shear center with respect to the centroid leads to a region where the gain in critical elastic load at low upstand heights can be attributed to the effect of monosymmetry and not only on an increase in section properties.

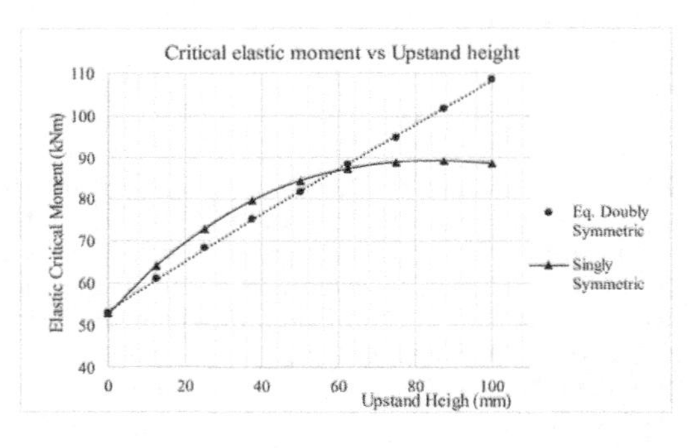

Figure 2. Critical elastic moment variation for beam with stiffeners and 'equivalent doubly symmetric' beam.

Two examples are presented to demonstrate how a stiffened section can have properties similar to or greater than the next higher bi-symmetric section size in the steel handbook. Once the desired upstand height has been determined, based on critical elastic load as well as service limit state requirements, a nonlinear analysis can be carried out so that the comparison could also be made on the basis of member strength.

3 DISCUSSION

From the current study it has been shown that consideration of strengthening existing bi-symmetric beams using flange upstands to create a monosymmetric section is a viable alternative that can be considered by design engineers in lieu of replacing with a new greater sized bi-symmetric member. Beneficial aspects of monosymmetry can be exploited. The application of this approach has been demonstrated with two examples showing its application. For the nonlinear analysis only material plasticity and geometric imperfections were accounted for. Residual stress effects on strength values need to be considered in future studies.

4 CONCLUSION

This study has presented an approach that can be used to consider strengthening existing bi-symmetric beams using flange stiffeners that convert the beam into a monosymmetric section. Using comparison of elastic critical loads as well as actual strength it has been shown how the option of strengthening can be compared to the option of replacing the member with another bi-symmetric section of greater size in the steel handbook. Based on the results of this study, design engineers have a viable option that they can consider. The option seeks to take advantage of beneficial monosymmetry effects.

REFERENCES

Koschmidder-Hatch, D. 2014. *Strengthening Existing Steelwork* [Online]. Newsteelconstruction.Com/Wp/Strengthening-Existing-Steelwork.

Mudenda, K. & Zingoni, A. 2018. Lateral-Torsional Buckling Behavior Of Hot-Rolled Steel Beams With Flange Upstands. *Journal Of Constructional Steel Research*, 144, 53–64.

Tsavdaridis, K. D. 2015. Strengthening Techniques: Code-Deficient Steel Buildings. *In:* Beer, M., Kougioumtzoglou, I. A., Patelli, E. & Au, S.-K. (Eds.) *Encyclopedia Of Earthquake Engineering*. Berlin, Heidelberg: Springer Berlin Heidelberg.

Current Perspectives and New Directions in Mechanics,
Modelling and Design of Structural Systems – Zingoni (ed.)
© 2022 Copyright the Author(s), ISBN 978-1-032-39114-4

Bending capacity of single and double-sided welded I-section girders: Part 1: Experimental investigations

J. Wang, M. Euler & H. Pasternak
Institute of Civil Engineering, Brandenburg University of Technology, Cottbus, Germany

Z. Li
Institute of Civil Engineering, Technical University of Berlin, Berlin, Germany

T. Krausche
IPP Ingenieurgesellschaft Prof. Pasternak & Partner, Cottbus, Germany

B. Launert
Strabag AG, Germany

ABSTRACT: Lateral-torsional buckling tests of four laterally unbraced welded steel girders with thin-walled I-sections are presented, whose flange-to-web junctions were welded from one or two sides in order to evaluate the influence of the manufacturing process, in particular the welding. The investigations include three-point bending tests under combined bending and torsion loading caused by an eccentric vertical single force. The geometric imperfections of the welded test girders were determined by 3D laser scanning. The test results show that the influence of the geometric imperfections and the residual stresses on the lateral-torsional buckling resistance is limited for the tested girders.

1 INTRODUCTION

Numerous lateral-torsional buckling (LTB) tests on I-section girders show that the LTB resistance is dependent on the cross-section, the girder span, the support conditions, the type of loading, the steel properties, the residual stresses and the geometric imperfections (Couto & Real 2019). The last two influences are associated with each other and depend on the fabrication of the girders. Up to now the influence of differently welded flange-to-web junctions of I-section girders on their LTB resistance has not been examined in detail. Therefore, four laterally unrestrained welded steel girders were tested under combined bending and torsion loading in three-point bending tests at Brandenburg University of Technology (BTU) in order to investigate the welding influence on the LTB resistance considering geometric imperfections and residual stresses. The steel girders had welded flange-to-web junctions that were welded from one side or two sides, respectively.

2 LATERAL-TORSIONAL BUCKLING TESTS

2.1 *Test specimens*

The test girders had identical doubly symmetric thin-walled I-sections of cross-sectional class 3 according to EN 1993-1-1 (2005). The flange-to-web junctions were made using Metal Active Gases (MAG) welding. The test girders were not straightened after welding in order not to change the residual stress state due to welding. They had a total length of 6000 mm allowing a span of 5500 mm for the three-point bending tests. Stiffeners were added at the end support and the loading section to avoid local buckling of the directly loaded flange. The webs and flanges of the girders were made each of the same charge of structural steel. The mean yield strength amounts to 375 N/mm² for the flanges and 335 N/mm² for the web. The mean Young's modulus of all girders is about 200 GPa.

2.2 *Geometric imperfections*

The geometric imperfections of the girders were measured before testing with a three-dimensional (3D) laser scanning system using the speckle pattern technique. The scanned imperfect girder geometry was analyzed with the software MATLAB® to separate the global and local imperfections.

The global imperfections, which have a relatively large influence on the LTB resistance, consist of the horizontal imperfection v_0 at the centroid of the web and the initial rotation ϑ_0 around the longitudinal axis as shown in Figure 1. They were used to fit a global sinusoidal bow imperfection function with maximum amplitude e_0 at the midspan for each girder for a numerical simulation (Li et al. 2022).

DOI: 10.1201/9781003348450-159

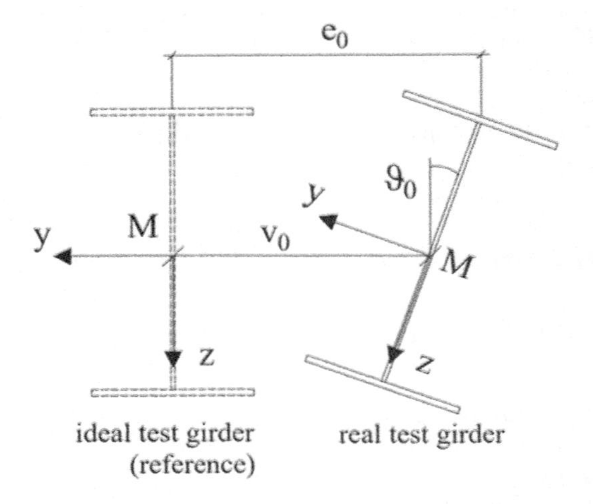

Figure 1. Global imperfections relevant for lateral torsional buckling.

2.3 *Test set-up*

All girders were tested under three-point bending, as shown in Figure 2, with an additional torsional moment caused by an eccentricity of the vertical force in midspan with respect to the girder's shear center. The tested girders were supported with fork-end boundary conditions (denoted as '6' in Figure 2).

The test force was supposed to be vertically directed during the entire test. As LTB inevitably generates horizontal displacements, it was necessary that the vertical hydraulic cylinder (denoted as '5' in Figure 2) also remained upright even if the load application point on the test girder ('8') was laterally displaced. For this purpose, the vertical hydraulic cylinder ('5') was attached to a sleigh ('2'), so that the cylinder could be automatically moved by an additional horizontal hydraulic cylinder ('4') depending on the horizontal displacement of the test girder that was recorded in real-time. All test girders failed under lateral-torsional buckling.

3 SUMMARY AND OUTLOOK

The investigations of the four laterally unbraced girders with thin-walled I-section and one- or two-sided welded flange-to-web connections allow following conclusions:

1. The sinusoidal bow imperfections of all test girders were comparable despite the different welding of the flange-to-web junction.
2. The maximum amplitudes of the geometric imperfection were consistent with or slightly smaller than the maximum value recommended by ECCS (1984).
3. There is no significant difference in the lateral-torsional buckling behavior of the tested one- and two-sided welded girders that were subject to the same eccentricity of the applied vertical single force.

Figure 2. Experimental setup: 1 – test rig, 2 – sleigh, 3 – hinge, 4 – horizontal hydraulic cylinder, 5 – vertical hydraulic cylinder, 6 – fork support, 7 – horizontal displacement transducer, 8 – test girder.

Further investigations, in particular on thick-walled I-sections and different slenderness ratios, seem to be necessary for a general conclusion on the fabrication influence on I-sections welded from one or two sides.

The authors take this opportunity to express their profound gratitude to the ASTRON Building S.A., Luxembourg, in particular to Mr. A. Belica, that supported the investigations by clarifying many helpful questions in advance and producing the test girders.

REFERENCES

Couto, C. Vila Real, P. 2019. Numerical investigation on the influence of imperfections in the lateral-torsional buckling of beams with slender I-shaped welded sections. *Thin-Walled Struct.* 145 106429.

Li, Z. Pasternak, H. Wang, J. Krausche, T. Launert, B. 2022. Part 2 of this paper at this conference.

EN 1993- 1-1. 2005. Eurocode 3: Design of steel structures – Part 1-1: General rules and rules for buildings. CEN. Brussels.

European Convention for Constructional Steelwork (ECCS) – Technical Committee 8, 1984. *Ultimate Limit State Calculation of Sway Frames with Rigid Joints*, ECCS Publication No.33, Brussels, Belgium.

Current Perspectives and New Directions in Mechanics, Modelling and Design of Structural Systems – Zingoni (ed.)
© 2022 Copyright the Author(s), *ISBN 978-1-032-39114-4*

Bending capacity of single and double-sided welded I-section girders: Part 2: Simplified welding simulation and buckling analysis

Z. Li
Institute of Civil Engineering, Technical University of Berlin, Berlin, Germany

H. Pasternak & J. Wang
Institute of Civil Engineering, Brandenburg University of Technology, Cottbus, Germany

B. Launert
Strabag AG, Germany

T. Krausche
IPP Ingenieurgemeinschaft Prof. Pasternak & Partner, Cottbus, Germany

ABSTRACT: For the structural design of welded girders with numerical approach, residual stresses and initial deformations or geometric imperfections need to be considered. Nowadays, with the development of computer technology and finite element theory, the numerical welding simulation approaches have improved significantly. However, the three dimensional welding simulation requires a huge amount of computing resources, which limits its wide application. This paper presents a simplified numerical approach based on a modified, so-called local-global model approach. This approach was developed and calibrated with experimental investigation by means of the single- and double-sided welded girders. By comparing the results of numerical calculations and experimental results, it shows that the simplified welding simulation cannot only calculate the residual stress of welded girders with single- and double-sided welds relatively accurately, but also obtain acceptable results of the deformations caused by the welding process. Besides, the calculated residual stress distribution can be automatically inputted into the numerical model for the subsequently further buckling analysis of the welded girders. The calculation results show that the buckling load obtained by numerical approach are in good agreement with the experimental results and the both welded girders have a relatively smaller different of bending capacity, despite clearly different residual stress distributions is exist.

1 INTRODUCTION

Nowadays, welding is the most commonly used technique for joining and producing structural components in steel structures (Pasternak et al. 2015). However, welding process will inevitably generate residual stresses and corresponding deformations. This residual stresses generating from the fabrication process of welded girders have a significant influence on the stability behavior of structural members. The design recommendations (ECCS 1984) and some recent research results (Schaper et al. 2018) are developed simplified models based on experimentally measured residual stress results. However, experimental results are usually limited to the surface of the welded girders, and it is difficult to obtain the distribution of residual stress in the thickness direction through experiments. The validated numerical method has a huge advantage in studying the residual stress distribution of welded steel girders, especially for heavy sections.

With the development of computer technology and finite element method (FEM) software, the welding simulation can obtain residual stress comparing the experimental results in three dimension. Due to the large dimensional differences in length and height or width of welded structures in steel structures, 3D welding simulation must have a sufficiently fine mesh structure in numerical simulations under the condition of satisfying computational accuracy. Therefore, the traditional welding simulation consumes a lot of computing resources. For welded girders in steel structures which length are often several meters or tens of meters, it is necessary to use a developed simplified two dimensional (2D) welding simulation (Launert 2019).

In this paper, the correctness of the numerical model and method is verified by comparative experiments, including residual stress and bearing capacity of welded girders with lateral torsional buckling. Then, the difference in residual stress distribution of single- and double-sided welded beams is presented

DOI: 10.1201/9781003348450-160

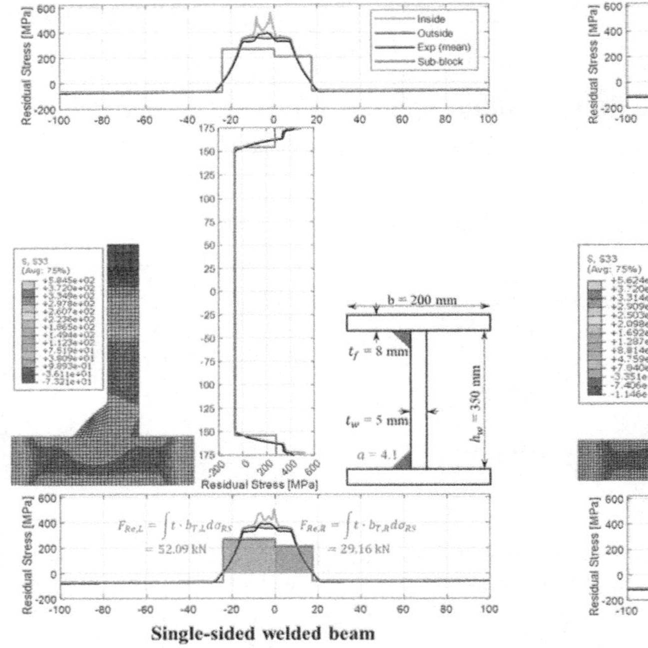

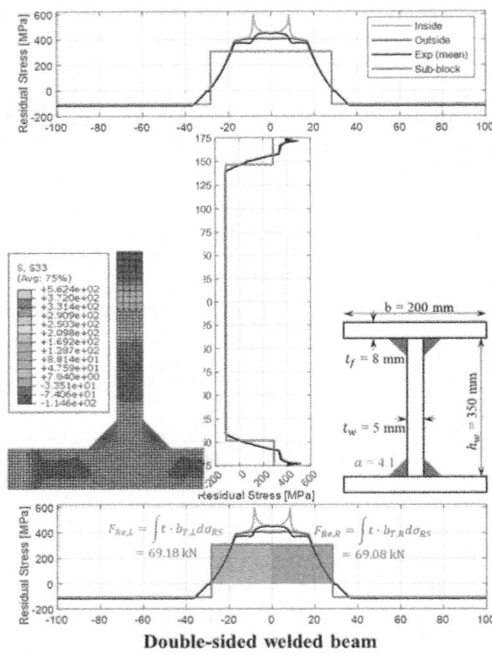

Figure 1. Residual stress distribution for single- and double-sided welded beams.

through extensive parametric studies and the influence of the corresponding distribution on lateral torsional buckling bearing capacity is discussed.

2 NUMERICAL SIMULATION

The distribution of residual stress for the single- and double-side welded beam is presented in Figure 1. It is obviously that residual stress distribution of double-side welded beam is symmetric. However, single-sided welded beams are asymmetric and the tensile stress on weld side is slightly higher.

The load versus displacement diagrams from test results and numerical simulation are shown in Figure 2. It is clear that the calculated bearing capacity is consistent with the experimental results, the max. corresponding different is smaller than 4%. Besides, there is no clear difference in the bearing capacity of beams with lateral torsional buckling for single- and double-sided welded beam, although their residual stress distributions are significantly different.

The residual stress distribution of single- a) and double-sided b) welded girders for different calculated cross-sections is calculated. The results show that single-sided welds inevitably lead to an asymmetric distribution model. Besides, it is worth noting that this sub-block model of residual stress has a strong correlation with the cross-section class. Obviously, a welded beam with cross-section class 1 has generally a larger area of tensile stress or shrink-age force, whether it is a single- or double-sided welds. As the cross-section class increases, the area of tensile stress tends to decrease.

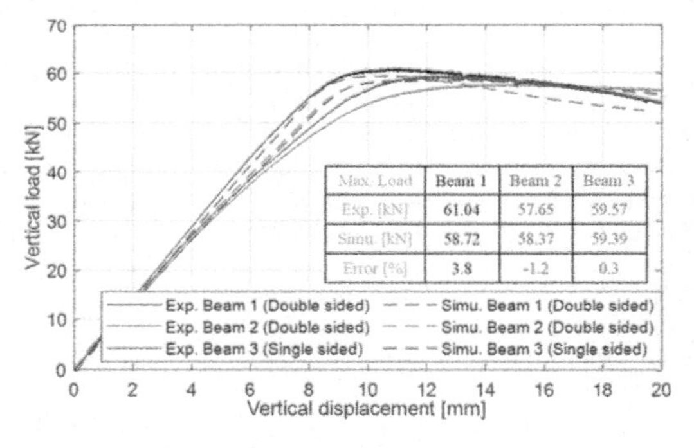

Max. Load	Beam 1	Beam 2	Beam 3
Exp. [kN]	61.04	57.65	59.57
Simu. [kN]	58.72	58.37	59.39
Error [%]	3.8	-1.2	0.3

Figure 2. Bearing capacity with numerical simulation and experiment.

3 CONCLUSION

In this paper, a simplified 2D model for welding simulation is used to study the bearing capacity of single- and double-sided welded girders with lateral torsional buckling. The results show that the single-sided weld leads to an asymmetric residual stress distribution model and level of asymmetry depends on the ratio of the weld size to the web thickness ac-cording to parameter study. Moreover, the ultimate bearing capacities of lateral torsional buckling obtained from numerical calculation with GMNIA and experimental results are comparable with each other and the corresponding average error is just about 1% with real geometric imperfection based on scanned values. However, the bearing capacities of the two different welded beams are almost similar, despite the significant differences in residual stress.

Current Perspectives and New Directions in Mechanics, Modelling and Design of Structural Systems – Zingoni (ed.)
© 2022 Copyright the Author(s), *ISBN 978-1-032-39114-4*

Issues related to the assessment of an existing reinforcement of a lattice telecommunication tower

J. Szafran & J. Telega
Faculty of Civil Engineering, Architecture and Environmental Engineering, Lodz University of Technology, Poland

ABSTRACT: The paper presents complex calculation analyses of a non-standard reinforcement of tower legs (made of L-sections). The construction of a combined cross-section (comprising L-sections and two round solid bars) and the influence of the additional elements on basic geometric characteristics of the elements are shown. Section forces resulting from wind action were compared with buckling resistances of tower legs for three scenarios concerning the way the reinforcement works: (1) as a rigid connection, (2) as a frictional connection in dry conditions, and (3) as a frictional connection in wet conditions. If the actual work of connected elements is not taken into account, the buckling resistance of the legs is significantly overestimated, which leads to wrong conclusions on whether a tower structure can be safely used.

1 INTRODUCTION

Competition in the construction industry and an ongoing focus on optimizing building facilities by reducing the amount of material required to built them clearly limits their capacity to carry additional, unplanned loads. For steel lattice towers used in the telecommunications industry, mobile telephony providers constantly modify the set of antennas to meet consumer expectations and needs (which in most cases involves adding new devices).

When loads exerted by the additional equipment exceed the load-carrying capacity of structural elements of a facility, the latter should be reinforced. In engineering practice, one can find means by which some structures are reinforced but the effect on improving their load-carrying capacity is doubtful or cannot be clearly determined.

This paper focuses on a stress-strength analysis of an existing telecommunications tower with a completed non-standard reinforcement of the tower legs using round bars as the main reinforcing element.

solid bars and metal plates around them. When one surface moves with respect to the other, a friction force is created; its value mainly depends on the surface roughness defined by the coefficient of friction. In dry conditions, the surfaces are in direct contact with each other. In steel building structures, the coefficient of friction is influenced by weather conditions. The effect of introducing a lubricant between the surfaces is to create a band of separation between the upper surface and the lower one. This band reduces the depth of penetration of the upper surface into the lower one, thereby reducing the coefficient of friction.

2 REINFORCEMENT MODELING

Reinforcing elements in segment S-3, the one with the reinforcement at the highest level above ground, was used in the analysis. The length of the leg considered was taken as 1600 m, and spacing of clamps connecting the leg with round bars was assumed as 850 mm.

The model was discretized using 8-node hexahedral finite elements, each having 24 geometric and static properties, using the FEMAP software. The elements include areas with curved contours, which is why isoparametric elements were used in addition to regularly shaped finite elements. Finite elements are 10×10 mm in the base sheet and 5×5 mm in other elements with more complex geometry. The discretized part of the structure was built using 94,762 hexahedral finite elements which were connected with each other by means of 144,591 nodes (Figure 1).

Forces are transferred from the tower structure to the solid bars by means of friction, which is why frictional connections were defined between the solid bars and the plates around them. In order to provide such a connection, dimensionless friction coefficients had to be defined depending on ambient conditions (Table 1).

The structure of the part of the tower with the reinforcing elements was subjected to the linear elastic buckling analysis based on the Lanczos method. Critical forces for three design cases are listed in Table 2.

For comparison purposes, the L-section without the reinforcement was also analyzed. The percentage increase in the buckling resistance of the reinforced arrangement depending on ambient conditions is given in Table 3.

DOI: 10.1201/9781003348450-161

Table 1. Static friction coefficients.

Ambient Conditions	Static friction coefficient
Dry	0.74
Wet	0.16

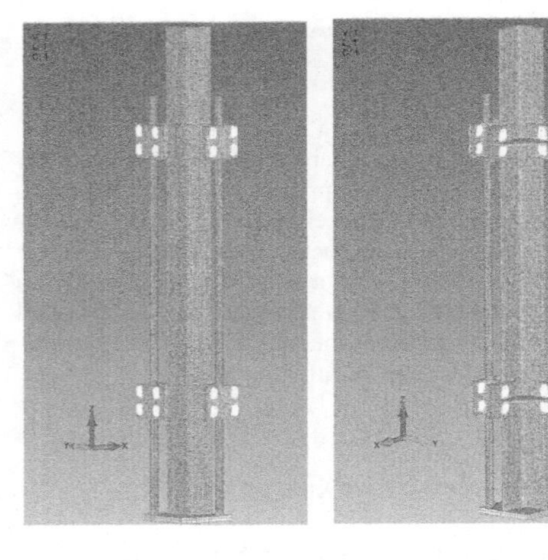

Figure 1. The calculation model of the reinforced leg.

Table 2. The critical forces.

Model type	Environment	Critical force [kN]
without reinforcement	Not applicable	141.72
with reinforcement	Dry	175.05
	Wet	182.74

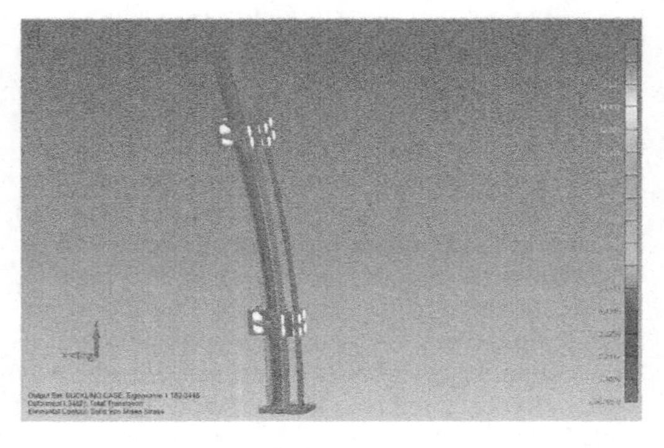

Figure 2. First buckling mode.

Since the structure is exposed to weather conditions, the rise of 26.6% should be considered as a reliable improvement in the buckling resistance of the section of the reinforced leg. This value was taken as the average increase in the resistance of all segments in which the reinforcement was made. The effect of the reinforcement on each segment of the tower is summarized in Table 3 and Table 4.

Table 3. Values of buckling resistance.

Model type	Environment	Buckling resistance [kN]	Buckling improvement [%]
without reinforcement	Not applicable	141.72	-
with reinforcement	Dry	175.05	26.6
	Wet	182.74	31.6

Table 4. Effort of the legs.

	Maximum effort [%]	
Segment	Tower legs unaffected by the reinforcement	Reinforced legs
S-8	138	109
S-7	128	101
S-6	126	100
S-5	137	108
S-4	141	111
S-3	147	116

3 SUMMARY AND CONCLUSION

The paper discusses an engineering problem consisting in a reliable evaluation of the effect of a non-standard reinforcement on the buckling resistance of tower legs.

The key task was to determine to what extent the round solid bars, which reinforce the leg, work with the leg. This was examined in the advanced numerical software FEMAP using the finite element method. The research indicated that the buckling resistance of an L-section with the reinforcement is influenced by the friction coefficient. The percent rise in the buckling resistance was found to be between 26.6% and 31.6% (for two opposite extreme ambient conditions). A scenario assuming fully rigid connections between the reinforcing and the reinforced elements was also analyzed. These three analyses lead to significantly different conclusions.

If the correct behavior (friction) of the connected elements is not taken into account, the buckling resistance of the legs is significantly overestimated. Under no circumstances such a connection should be considered as rigid. A certain effect of ambient conditions on the calculated buckling resistance was also found. The effect is relatively small, up to 5%, but considering its proper value in certain critical conditions may determine whether the results are valid, i.e. whether a structure can be safely used.

Current Perspectives and New Directions in Mechanics, Modelling and Design of Structural Systems – Zingoni (ed.)
© 2022 Copyright the Author(s), *ISBN 978-1-032-39114-4*

Relating stress concentrations in triangular steel bridge piers to simple beam models

Ph. Van Bogaert, G. Van Staen & H. De Backer
Civil Engineering Department, Ghent University, Belgium

ABSTRACT: Triangular piers integrate both support surface for bearings and lateral stiffness to resist horizontal force. It is generally known that in the triangle's corner large stress concentrations may appear. Therefore curved transitions are provided at these locations. To determine these stress concentrations the use of highly dense FE-models and indeed also engineering judgement, is needed. The latter applies to most software packages. On the other hand, one might consider to use simple beam elements, the axes coinciding with the real pier axes. In doing this, the stress concentrations will be missed and unsafe design is the result. The main aim of the research is to compare the results of an equivalent beam model to elaborate shell element models, which include the necessary transition curves. For each type of external force and rounding parameters the stress concentration factor for the undisturbed cross-sections is established. The work is presently in progress and the first results show definitely plausible factors. However, for small stress magnitude, the results are less convincing.

1 INTRODUCTION

Concrete triangular piers became rather popular for long viaducts, from the 90's on, up to date. There are many examples from viaducts. However, the need for rapid construction increases the popularity among designers to consider steel triangular piers. The latter show considerable horizontal resistance to braking and acceleration, as well as lateral loads.

Determining the stresses in triangular piers requires the use of detailed FE-models. Any other type of analysis will not detect the stress concentrations and the design will be obsolete. The use of full FE-models, including mesh refinement to sufficient reliability requires large effort. Therefore it would be interesting and time saving to obtain relevant results from simple beam models. The aim of the research then was to try establishing a reliable relation between the two types of models, including the stress concentrations.

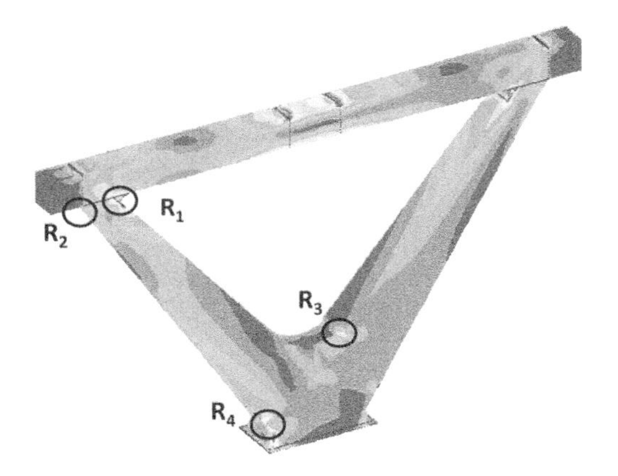

Figure 1. Plate model with stress concentrations.

2 TYPES OF NUMERICAL MODELS

To investigate this, two types of numerical models of a triangular pier for viaduct have been compared. The full FE-model is of the plate type of model and must comply with the real structure. It consists of plate elements, including bending. Figure 1 shows the equivalent (von mises) skin stresses for dead weight, vertical and horizontal loads in 2 directions. The areas of important stress concentrations become clear. Obviously, stresses at the locations of concentrated force can be lowered by local reinforcement of plate sections.

The typical areas of stress concentrations are located at the rounded sharp or obtuse angles of the structure, as well as at the connection with a base plate. All of these locations appear where the main force must change direction and are identified as R_1 to R_4 in Figure 1. This is the main characteristic of triangular structures.

In the plate model, no internal diaphragm stiffeners have been introduced. Hence some of the cross section deformations are due to distortion of the hollow core members. This may give rise to additional stress, which probably cannot be found from simple beam theory.

DOI: 10.1201/9781003348450-162

The second and simpler model consists of 3-D beams with 6 degrees of freedom. As his model should also be representative for the real structure, the element axes have to coincide with the central lines of the tapered sloping members. This also implies that the actual beam cross sections are sometimes trapezoidal instead of rectangular.

Beam models are generally incapable of simulating the rounded transitions of the real structure. This is why stress concentrations are not found. Hence it is not evident to predict which type of model will detect the highest stress levels.

3 COMPARING PLATE AND BEAM MODEL

At first glance, the results from both types of models seem rather disappointing. In particular comparing stresses in cross-sections which are close to the nodes is inefficient, due to stress concentrations (Nagpal 2012). In addition, cross section distortions give rise to additional stress. The latter becomes clear when looking at the deformation of cross sections.

In view of this, direct comparison of stresses from the plate and the beam model does not apply to any cross section at all. The locations for comparing must be chosen outside of the areas of stress concentrations and may even then be troubled by distortion. The result is shown in Table 1 as the ratio of stresses from the plate model divided by the corresponding stress from the beam model.. In a second step, the stress concentration factors are established, which transform the values from plate model at the aforementioned locations to the peak values (SCF's). If this stepwise method is adopted, the results become acceptable.

Table 1. Conversion of stress from beam to plate model.

	R4	R4 oppos	R1	R1 oppos
F_X	0.979	3.087	8.93	0.893
F_Y	1.0786	1.099	1.438	1.438
F_Z	0.956	3.267	2.935	2.935
M_y	0.893	3.109	1.051	3.861
M_z	1.33	0.242	2.786	0.514

Thus stresses found from the beam model can be converted to the locations of the real structure, that are not affected by stress concentrations.

Some of the factors of table 3 may seem incredibly high and inaccurate. This is mostly due to the fact that the beam model renders a low value.

Concerning the SCF's within the plate model, a full set of data is given in the full paper of this contribution.

4 CONCLUSIONS

The aim of this research was to establish a reliable link between results from a simple beam model of a triangular pier and elaborate and refined models from the same pier. In order to be able to address any load combination, unity-loads of 6 types had to be considered.

It became clear that stress concentrations cannot be derived from beam models and hence, the problem should be divided in two. First reliable SCF's for the complete plate model have bene derived. The latter have been found to be constant and accurate. Taking the ratio of the circular rounding of angular transitions to the main dimension of the member as a parameter, SCF's are presented for 3 types of stresses and for the various external force types. Obviously, the data apply to a single type of geometry, sloping angle (55°) of the struts and ratio of stiffness of the upper beam and the struts. The influence of all of these parameters should be explored further.

An important conclusion is that modifying the radius of one of the transitions between elements, does not modify the stress in other areas. Hence, the SCF's are totally independent.

Finding a numerical connection between the plate FE-model and a simple beam model has been less successful. The compared values should not be influenced by the stress concentration effect, nor by distortion of cross sections. This requires to determine relevant locations for cross-sections, still allowing for sufficient bending and shear effect.

The ratio of stresses from plate to beam model, found so far shows insufficient accuracy. The plate model consistently renders higher stress values, often 3-times larger than the beam result. The latter may be important in the present state of the research. If a beam model is used and stress values are multiplied by 3 and further multiplied by the SCF, a rather safe stress value will be found. Obviously, closer location of comparable cross-sections is needed and stresses from the beam model may be supplemented by distortion and warping stresses.

REFERENCES

Nagpal, S., Jain, N.& Sanyal, S.,2012 Stress concentration and its mitigation techniques in flat plate with singularities – a critical review. *Engineering Journal* Vol 16 Issue 1 (2012) http:/www.engj.org DOI:10.4186/EJ.2012.16.1.1.

Current Perspectives and New Directions in Mechanics, Modelling and Design of Structural Systems – Zingoni (ed.)
© 2022 Copyright the Author(s), ISBN 978-1-032-39114-4

Experimental investigation of the rack-to-bracing joints between the high-rise steel storage rack frames and the independent bracing towers

Z. Huang & X. Zhao
Tongji University, Shanghai, China

K.S. Sivakumaran
McMaster University, Hamilton, Ontario, Canada

ABSTRACT: This research experimentally investigated the behavior of the rack-to-bracing joints between the high-rise rack frames and the associated independent spine bracing towers. A total of twelve joint specimens were involved in the experimental program. The analysis of the experimental results indicates that the deformation of the upright components and the endplate components, as well as the bending of the bracing tower post flanges made significant contributions towards the overall deformation of the rack-to-bracing joints under consideration. Among all the design factors considered in these experiments, the initial stiffness of the rack-to-bracing joints is most sensitive to the changes in the thickness of the bracing tower posts. The fracture of the connecting bolts, the yielding of the upright flanges, the yielding and the tearing of the endplates, as well as the yielding of the flanges of the bracing tower posts may cause the failure of these rack-to-bracing joints.

1 INTRODUCTION

High-rise steel storage racking systems provide high storage capacities and are thus widely used in the warehouses. A high-rise steel storage rack needs to be stabilized by vertical spine bracing systems so as to acquire additional stiffness in the down-aisle direction. Unlike conventionally braced steel frames, the spine bracings associated with the high-rise storage racks, however, are commonly located at a short horizontal distance away from the main rack frames so as to avoid obstructing the placement of the pallets. In the rack design practice, the spine bracings made of slender elements (i.e., cables, flats or round bars) should be connected to additional posts not supporting vertical pallet loads, forming an independent spine bracing tower. The joints that connect the main rack frames and the associated spine bracing towers are identified herein as rack-to-bracing joints. A typical rack-to-bracing joint consists of several components, including an upright piece, a bracing tower post stub, a bracket assembly, and a spine bracing connecting plate.

Even though it is well recognized that the rack-to-bracing joints play a dominant role in determining the buckling load (Huang and Zhao 2019) and the sway behavior (Yin et al. 2018) of the high-rise rack systems, the behavior of the rack-to-bracing joints, however, is hitherto unavailable. As such, this research experimentally investigated the behavior of the rack-to-bracing joints between the high-rise racks and the associated spine bracing towers.

2 EXPERIMENTAL INVESTIGATION

The test setup used in the experimental program is shown in Figure 1. Since the rack-to-bracing joints under consideration basically transmit the shear force between the high-rise rack frame and the associated spine bracing tower, the joint specimens were loaded in the horizontal direction in the tests. The specimens were fixed in a rigid frame with the top section and the end section of the upright piece mounted to a rigid beam, respectively. A hydraulic actuator was used to apply the lateral load to the joint specimens. In order to keep the loading direction remain unchanged during the tests, the fixing end of the actuator was mounted to a single-way sliding support. The loading end of the actuator was attached to the connecting plate of the specimen joint through a pin. The lateral load applied by the actuator, F, was recorded by a load cell. The shear deformation of the entire rack-to- bracing joint, Δ, was measured by three transducers.

A total of four groups of the rack-to-bracing joint specimens, namely, specimen groups RSPJ-B1 thru RSPJ-B4, were involved in the experimental program. Each specimen group consisted of three nominally identical joint specimens. Influencing design factor associated with such rack-to-bracing joints, including the upright thickness, the endplate thickness, and the thickness of the bracing tower posts were experimentally considered. The specimen details are summarized in Table 1.

DOI: 10.1201/9781003348450-163

Table 1. The specimen details.

Specimen ID	Upright thickness	Endplate thickness	Bracket thickness	Bracing tower post thickness
	mm	mm	mm	mm
RSPJ-B1	3	6	3	3
RSPJ-B2	3	6	3	2
RSPJ-B3	2	6	3	2
RSPJ-B4	3	4	3	3

3 EXPERIMENTAL RESULTS

The representative load-deformation relations associated with each rack-to-bracing joint specimen group are summarized in Figure 2. It is pointed out that the three nominally identical joint specimens in the same test group exhibited consistent load-deformation relations.

In general, the rack-to-bracing joints under consideration experienced one of four failure modes, including the fracture of the connecting bolts, the yielding of the flanges of the bracing tower posts, the yielding of the flanges of the load-carrying uprights, as well as the yielding and tearing of the endplates. It must be pointed out, however, that the joint specimens in the same specimen group experienced exactly the same failure mode.

4 INFLUENCE OF THE DESIGN FACTORS ON THE INITIAL STIFFNESS OF THE RACK-TO-BRACING JOINTS

Taken as the average of the initial slopes of the corresponding load-deformation curves, the average initial stiffness values of the specimens RSPJ-B1 thru RSPJ-B4 were determined as 838.6 kN/m, 505.4 kN/m, 354.9 kN/m, and 661.7 kN/m, respectively. Comparing the load-deformation relations in Figure 1 we realize that a reduction in the bracing tower post thickness from 3 mm (i.e., RSPJ-B1) to 2 mm (i.e., RSPJ-B2) would cause a 39.7% reduction in the average initial stiffness of such rack-to-bracing joints. Replacing the 3 mm thick upright pieces (i.e., RSPJ-B2) with the 2 mm thick upright pieces (i.e., RSPJ-B3) would lead to a 29.8% reduction in the initial joint stiffness, given they were all connected to 2 mm thick bracing tower posts. When the upright pieces and the bracing tower post stubs were both fabricated using 3 mm thick material, using 4 mm thick endplates (i.e., RSPJ-B4) instead of 6 mm thick plates (i.e., RSPJ-B1) we should expect a 21.1% decrease in the initial stiffness of such rack-to-bracing joints. It is generally stated that the initial stiffness of such rack-to-bracing joints is most sensitive to the changes in the thickness of the bracing tower posts.

5 CONCLUDING REMARKS

The behavior of twelve rack-to-bracing joints connecting the high-rise rack frames and the associated spine bracing towers are experimentally established in this paper. The experimental observations indicated that the deformations of the upright pieces, the endplates and the bracing tower post stubs noticeably contributed to the deformation of the entire rack-to-bracing joints. The initial stiffness of the rack-to-bracing joints under consideration was most sensitive to the changes in the thickness of the bracing tower post components. The fracture of the connecting bolts, the yielding of the upright flanges, the yielding and the tearing of the endplate, as well as the yielding of the bracing tower posts may cause the failure of such rack-to-bracing joints.

Figure 1. The test setup.

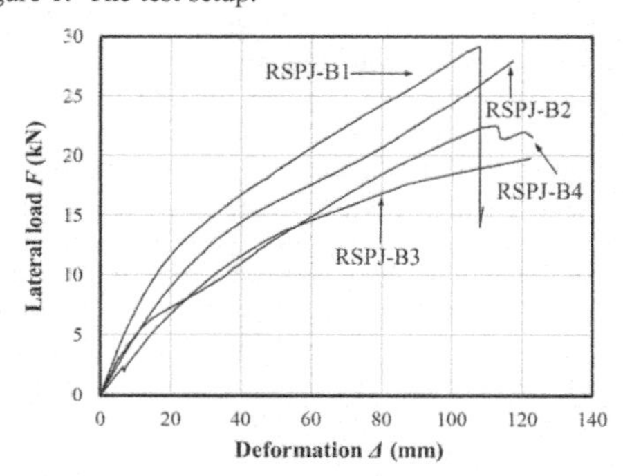

Figure 2. The representative load-deformation relations.

REFERENCES

Huang Z. & Zhao X., 2019. Elastic buckling analysis of asymmetrically braced steel storage racks, Proceedings of Seventh International Conference on Structural Engineering, Mechanics and Computation (SEMC 2019), Cape Town, South Africa.

Yin, L., Tang, G., Li, Z., Zhang, M., & Feng, B. (2018). Responses of cold-formed steel storage racks with spine bracings using speed-lock connections with bolts I: static elastic-plastic pushover analysis. Thin-Walled Structures, 2018, 125:51–62.

Current Perspectives and New Directions in Mechanics, Modelling and Design of Structural Systems – Zingoni (ed.)
© 2022 Copyright the Author(s), *ISBN 978-1-032-39114-4*

Optimum design of hat cold-formed steel members using direct strength method

M.H. Khashaba
Structural Engineering, Faculty of Engineering, Ain shams University, Cairo, Egypt

I.M. El-Aghoury & A.A. El-Serwi
Department of Structural Engineering, Faculty of Engineering, Steel Structures, Ain shams University, Cairo, Egypt

ABSTRACT: Thin-walled steel members are attracting increasing interest due to their economic privilege over equivalent hot-rolled members. However, these slender sections suffer from buckling failure which puts some constraints for the use of these sections. Direct strength method is a recently adopted design approach that requires no effective section calculations. Instead, it depends on performing buckling analysis on members using numerical analysis methods. This research presents a verified analysis program (OpenFSM) developed by the authors that conducts buckling analysis using the finite strip method. OpenFSM offers an alternative analysis tool with an easier interface including features that can facilitate the design process and offer more potential for conducting parametric studies and research projects. OpenFSM is utilized to perform parametric studies to investigate the behavior of the widely used hat cold-formed steel section. Moreover, several observations are deducted and equations regarding critical local buckling stress are presented.

1 INTRODUCTION

1.1 *Motivation*

Light gauge steel is considerably becoming a valuable alternative to hot-rolled steel. Cold-formed steel has many advantages. Not only it offers economic less weighted elements. It is also characterized by its fast production, easy assembly, and high specific strength (strength to weight ratio).

These slender cold-formed members force the engineers to accurately estimate their buckling resistance to different buckling modes (local buckling, distortional buckling, and global buckling). Correct classification of the buckling modes is crucial to the design of such members to safely withstand the applied actions and leverage the extreme efficiencies of these light weighted members.

1.2 *Literature review*

Recently there has been growing interest in the design of cold-formed members. Several design approaches are used to design cold-formed steel members. Since the first introduction of cold formed sections in design specifications, the effective section method was implemented which depends on the reduction of the gross section properties to the effective section dimensions.

The other approach which is the scope of our study is the direct strength method which uses different buckling analysis procedures including (finite element method, finite strip method, generalized beam theory and manual hand predictions) to calculate the critical buckling loads of sections and then uses empirical equations to obtain the ultimate design strength values.

Finite strip analysis is a specialized variant of the finite element method. It was first introduced by (Cheung & Tham 1997). According to their work, Direct strength method was first presented by (Schafer & Peköz 1998).

All these studies encouraged the formal implementation of the direct strength method as the main specification in (American Iron and Steel Institute, 2016).

The behavior of hat-shaped members has received limited attention in the literature. Cold-formed hat omega-section is a commonly used roll formed standard section. Its high torsional stiffness offers an advantage over the similar Z sections also its higher lateral torsional buckling strength allows it to be an adequate alternative especially in short-spanned members.

In this context, the research sheds new light on the buckling behavior of the widely used hat-shaped members utilizing the introduced OpenFSM open-source software used to conduct buckling analysis using finite strip method.

2 OPENFSM SOFTWARE

The programming language used for development of OpenFSM program is C#.

Different operations are carried out to establish the local and global elastic and geometric stiffness matrices (K_e and K_g respectively) then the eigenvalue problem in Equation (1) is solved for the differently presented half-wavelengths (a):

DOI: 10.1201/9781003348450-164

$$K_e \Delta = \lambda K_g \Delta \quad (1)$$

Consequently, all eigen values (λ) and eigen vectors (Δ) are computed and the minimum eigenvalue for each (a) value would be the load factor (critical buckling load/applied load).

Consequently, no error is monitored in the results compared to equivalent case studies in the well-known equivalent software (CUFSM, 2020).

The newly proposed program showed identical results as shown in Figure 1:

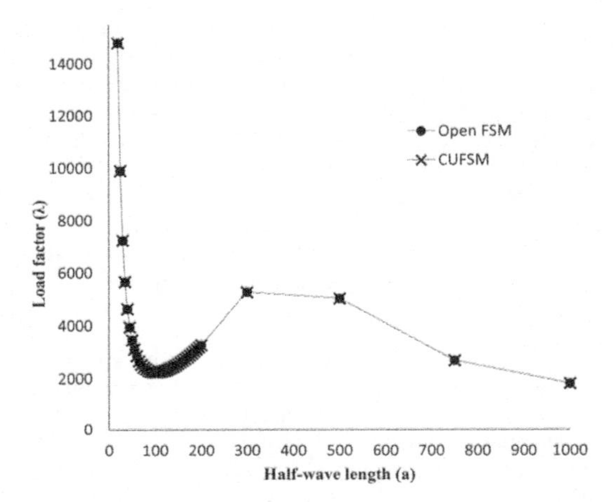

Figure 1. Signature curve for simply supported hat-shaped section subjected to uniform compression using both CUFSM and OpenFSM (h= 100mm, b= 100mm, t= 2mm and d= 30mm).

The software offers an alternative analysis tool with an easier interface including features that can facilitate the design process and offer more potential for conducting parametric studies and research projects. The program is available on the following open-source repository (Link).

3 PARAMETRIC STUDY

Using the parametric study tool incorporated in (OpenFSM), More than 3420 case studies were investigated and buckling analyses were performed on various geometric properties (h/t, h/b, d/h) of the widely used pinned-pinned hat-shaped member under various loading conditions (Figure 2).

The study resulted in developing Equation 2 for computing the local buckling critical stress for hat members subjected to pure compression or bending moment about y-axis, while Equation 3 is presented for hat members subjected to bending moment about x-axis (compression/tension at flange).

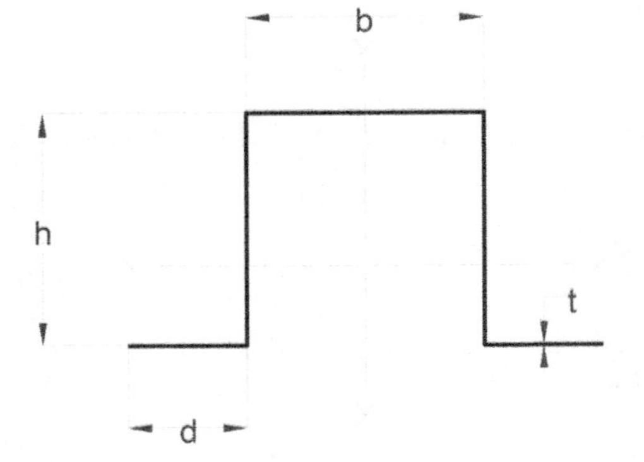

Figure 2. Hat-shaped section.

$$F_{cr} = \beta \times (\frac{b}{t})^{\gamma} \quad (2)$$

$$F_{cr} = \beta \times (\frac{h}{t})^{\gamma} \quad (3)$$

Where β and γ are factors obtained from tables presented in the paper according to the values of h/b and d/h ratios.

4 CONCLUSION

The study utilized the finite strip method to perform buckling analysis of cold formed steel members. A newly proposed open-source software "OpenFSM" is presented to perform buckling analysis using FSM. The program offers an easier and more user-friendly interface for the users and incorporates many features and tools to aid the designers and the researchers. Verification of the program is guaranteed by comparing the results with similar examples from the well-known software CUFSM.

Utilizing OpenFSM, parametric studies are conducted to investigate the behavior of hat-shaped members under various loading conditions. Equations are presented for the critical local buckling stress of hat members.

REFERENCES

American Iron and Steel Institute. (2016). North American specification for the design of cold-formed steel structural members, 2016 edition.

Cheung, Y. K., & Tham, L. G. (1997). Finite Strip Method. https://doi.org/10.4324/9781003068709

Schafer, B. W. (2020). CUFSM v4.05.

Schafer, B. W., & Peköz, T. (1998). Direct strength prediction of cold-formed steel members using numerical elastic buckling solutions. Fourteenth International Specialty Conference on Cold-Formed Steel Structures.

Current Perspectives and New Directions in Mechanics, Modelling and Design of Structural Systems – Zingoni (ed.)
© 2022 Copyright the Author(s), ISBN 978-1-032-39114-4

Performance of light gauge cold-formed steel flexural members subjected to non-uniform fire

R. Singh & A. Samanta
Department of Civil and Environmental Engineering, Indian Institute of Technology Patna, India

ABSTRACT: Light gauge cold-formed steel construction is gaining the interest of structural engineers across the globe. Past studies have shown the need for further research on CFS flexural members at elevated temperatures. From literature, it is found that most of the research in past has been focused on the uniformly elevated temperature. But in real-life scenarios, the CFS members might be subjected to non-uniform temperature distribution due to fire protection. Thus, this research aims to study the CFS flexural members under non-uniform thermal profile. In this study, CFS lipped channel flexural members are used under four-point loading and simply supported boundary condition, and five different thermal profiles. Simulation results show that the critical temperature and member structural behavior of the CFS flexure members are predominantly influenced by the thermal profile of the member.

1 INTRODUCTION

In all of these studies the member was assumed to be exposed to the fire from all sides. However, in real life scenario, the member might be subjected to thermal gradients. In case of forced convection, on the colder side, very high amount of thermal gradient might be established in the member. None of the past study was found to address the issue of extreme thermal gradient on the behaviour of CFS flexural members. Thus, this research aims to study the effects of extreme thermal gradients on the structural performance of thin-walled CFS flexural members.

2 DEVELOPMENT OF FINITE ELEMENT (FE) MODEL

Based on mesh sensitivity analysis, mesh density of 10 mm× 10 mm is selected for the study. FE element type S4R is adopted in the simulations. This element is 4-noded, doubly curved thin shell. These elements use reduced integration approach, and finite membrane strains. Hourglass control was turned on, in order to avoid excessive hourglass deformation in the elements. Four-point loading was used in this study, with simply supported boundary conditions. Load on the member was applied from centre of the top flange of the member at L/3 locations from each end. For the validation of the developed finite element model, the grade of steel selected in the Liu et al. (2002), is selected to be 275 Mpa, following elastic perfect plastic material model. However, in parametric study a detailed stress-strain model based on gradual yielding type was adopted. In parametric study, G450 grade of steel was used, and Ramberg-Osgood stress strain model was adopted following, the Ramberg-Osgood parameters η and β suggested by Kankanamge & Mahendran (2011). Yield strength of the material at ambient temperature was taken as 514.50 MPa, whereas ultimate strength was and 542.5 MPa from Kankanamge & Mahendran (2011). Modulus of elasticity was again taken from literature (Kankanamge & Mahendran 2011) to be 206.33 GPa. Loss of mechanical properties of steel at elevated temperature was accounted by adopting adequate reduction factors for yield strength, ultimate strength and modulus of elasticity.

In order to check the validity of developed model, the results were first compared with the experimental results of Liu et al. (2002), and numerical results of Yin & Wang (2004). As presented in Figure 2. A good agreement between experimental and numerical results available in the literature can be seen with the developed FE model.

3 PARAMETRIC STUDY

Parametric study is conducted in order to evaluate the behaviour of CFS LCBs under gradient thermal profile. 450 Grade of steel is selected, three different geometry of member is selected, having a span of 2.25 m. Three magnitudes of load applied on the member, are 35%, 50% and 65% respectively. Five different thermal profile was selected in this study,

DOI: 10.1201/9781003348450-165

named as T_1-T_5, based on the respective thermal distribution. All the thermal profiles are presented in Figure 4. Failure of the member is reported at a temperature when continuous state of deformation is set in the member. Temperature of the heated flange is reported as the failure temperature of the member.

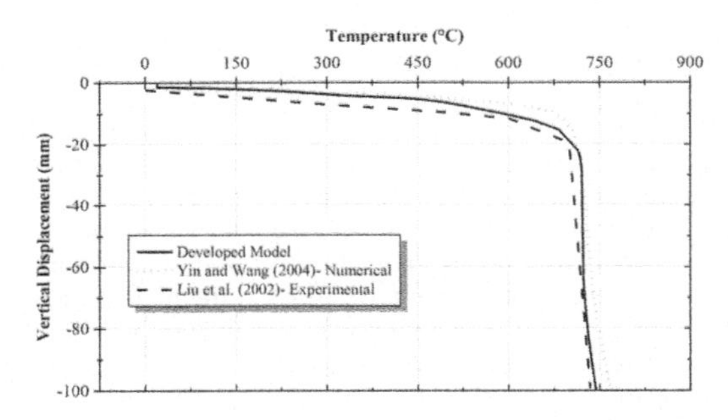

Figure 1. Comparison of developed model with experimental and numerical results.

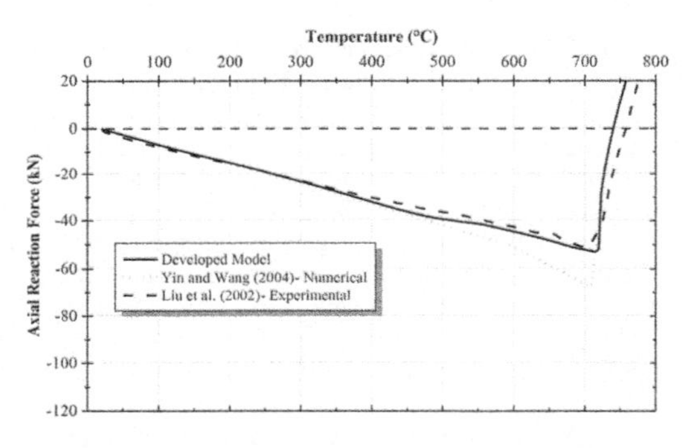

Figure 2. Comparison of developed model with experimental and numerical results.

4 RESULTS AND CONCLUSION

Failure mode obtained after non-linear simulation in case of B_1 beam was found to be lateral-torsional buckling, however in case of deep beams failure is found to be triggered by local buckling of the member. In case of T_4 and T_5 thermal profile due to heated top flange, local buckling of the loaded flange started, which is less significant for the T_3 thermal profile, and vanishes in the T_1 and T_2 thermal profiles. This trigger of the local buckling is due to the heated compression flange of the member, having lower material properties, than the colder part of the member. Out of all three beams, highest critical temperature is obtained for T_1, followed by T_2 thermal profile. This is due to the fact that members having these thermal profiles underwent gradual deformation, and unlike T_3-T_5 thermal profiles, sudden collapse is observed at much higher temperature. For deep members such as B_2 and B_3, lowest critical temperature is found to be governed by fully heated thermal profile (T_3), whereas for more slender sections B_1, it is true for low initial applied loads (0.35 LR) only. However, for higher initial applied load levels (for B_1 beam at 50% and 65% load levels), the lowest critical temperature is obtained for T_5 thermal profile. Critical temperature obtained for different thermal profiles for all three members is presented in Figure 3.

This research clearly highlighted the fact that the critical/ failure temperature of the member is highly dependent on the thermal profile which is exposed on the member.

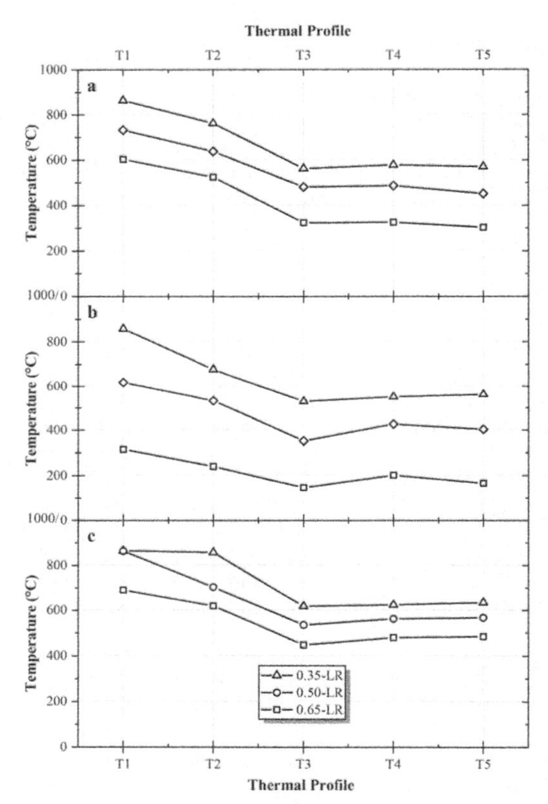

Figure 3. Critical temperature of the all the members under different thermal profiles. a) B_1 beam, b) B_2 beam, and c) B_3 beam.

REFERENCES

Kankanamge, N.D. and Mahendran, M., 2011. Mechanical properties of cold-formed steels at elevated temperatures. *Thin-Walled Structures*, 49 (1), 26–44.

Liu, T.C.H., Fahad, M.K., and Davies, J.M., 2002. Experimental investigation of behaviour of axially restrained steel beams in fire. *Journal of Constructional Steel Research*, 58 (9), 1211–1230.

Yin, Y.Z. and Wang, Y.C., 2004. A numerical study of large deflection behaviour of restrained steel beams at elevated temperatures. *Journal of Constructional Steel Research*, 60 (7), 1029–1047.

Current Perspectives and New Directions in Mechanics, Modelling and Design of Structural Systems – Zingoni (ed.)
© 2022 Copyright the Author(s), *ISBN 978-1-032-39114-4*

Static strength capacity of single-sided fillet welds

T. Skriko
Laboratory of Welding Technology, Lappeenranta-Lahti University of Technology LUT, Lappeenranta Finland

A. Ahola, K. Lipiäinen & T. Björk
Laboratory of Steel Structures, Lappeenranta-Lahti University of Technology LUT, Lappeenranta, Finland

ABSTRACT: The current design standards do not obtain clear recommendations or rules how the secondary bending moment at single-sided fillet weld joint should be considered. However, significantly high tensile stresses can occur at the weld root due to the secondary bending, which can have an impact on the capacity of fillet welds, particularly if such joints are made of high-strength steels. In this regard, the capacity includes both deformation capacity and ultimate load-carrying capacity. In the present work, the static strength capacities of single-sided fillet welds are analytically, experimentally and numerically investigated. From the practical application viewpoints, it is important to understand the mechanisms of secondary moment at the welded details, in which the joint stiffness and type, along with distributions of loads at the joint components, are often more important factors affecting the capacity than the eccentricity effect caused by the single-sided fillet weld. A particular attention should be paid in designing joints made of high-strength steel materials and subjected to the secondary moment causing high tensile stress at the weld root. The presented analytical design method for the single-sided fillet welds and experimental test results showed a reasonable agreement.

1 INTRODUCTION

In accordance with the new preliminary European design standard (prEN 1993-1-8, 2020), the local eccentricity of single-sided fillet welds should be avoided whenever possible. However, the bending moment at welded joint can also be caused by an external bending moment or by a rotation of the joint due to unsymmetric structural behavior of adjacent joint member (Tuominen et al., 2017).

In this study, the effect of secondary bending on the single-sided fillet weld joints was studied by means of experimental tests, along with analytical calculation models and finite element analyses.

2 MATERIALS AND METHODS

A S960MC grade steel with the plate thickness of 10 mm was chosen for the study. Different welding consumables were tested including under-matching and matching filler metals. The mechanical properties of the studied materials are presented in Table 1.

The test specimens are shown in Figure 1. The welding preparation was done with a robotic gas metal arc welding with the given welding consumables. All specimens were quasistatically tested until total rupture of specimen. A digital image correlation system and strain gages were used to monitor the tensile tests.

Table 1. Mechanical properties of the studied materials.

	f_y	f_u	A_5	KV (temp.)
Material ID	MPa	MPa	%	J
Base material				
S960MC*	960	980–1250	7	27 (-40°C)
S960MC**	1053	1175	10	82 (-40°C)
*Filler metals****				
OK Autrod 12.51	430	530	30	–
OK Aristorod 69	715	805	17	–
OK Aristorod 89	920	940	18	47 (-40°C)
Union X96	930	980	14	47 (-50°C)

*nominal values, **material certificate values, ***typical values for filler metals according to the material suppliers.

3 RESULTS

Figure 2 Presents the obtained experimental capacities and theoretical capacities. In addition, different distributions for the bending stress (elastic $k = 6$ and plastic $k = 4$), and strengths (nominal and hardness-corrected) are applied.

Figure 3 Presents the modelled normal stress distributions along the transverse leg length (perpendicular to the loading direction). It can be seen that the plate boundary conditions have major effects on the stress distribution at the leg length.

DOI: 10.1201/9781003348450-166

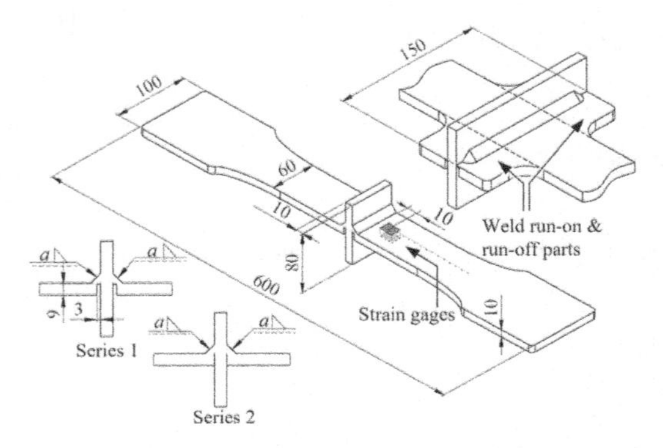

Figure 1. Shape and dimensions of the test specimens (all dimensions in mm).

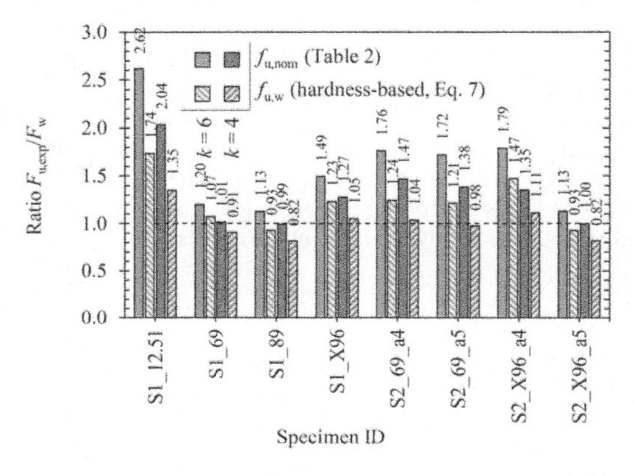

Figure 2. Experimental results compared to the theoretical capacities.

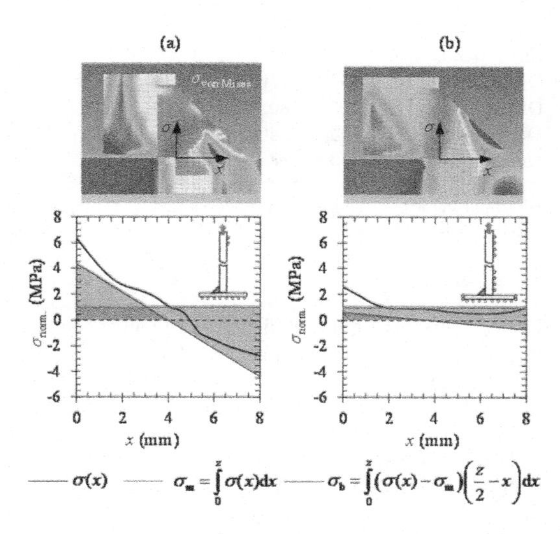

Figure 3. Normalized stress distribution $\sigma_{norm.}(x)$ and stress components along the leg length parallel to the loading direction from the weld root to the weld toe: in the case of (a) free rotation of adjoined plate component (the experimental setup), and (b) prevented rotation.

4 DISCUSSION

The results obtained with high-strength and ultra-high-strength filler metals (OK69, OK89, X96) indicated that the Eurocode 3 proposal for considering elastic stress distribution ($k = 6$) was in line with the obtained ultimate capacities (Figure 2). The plastic model ($k = 4$) in these joints overestimated the weld capacities. For the OK12.51 filler metal, elastic model was overly conservative suggesting that the weld capacity could be obtained using the plastic model ($F_{u,exp}/F_w$ > 1.0, Figure 2). Moreover, a consideration of bending moment direction is important. In this work, only joints with the moment opening the weld root were investigated. If the moment closes the weld root, the highest stress and failure originates at the weld face, and the capacity obtained with the elastic stress distribution model might be too conservative.

5 CONCLUSIONS

Based on the conducted study regarding the static strength capacity of single-sided fillet welds subjected to secondary bending moment, the following conclusions can be drawn:

- Bending moment should be considered using the elastic distribution model ($k = 6$) particularly in high-strength steels. For mild steels and low-strength welding consumables, the plastic stress distribution could be alternatively used to reduce conservatism.
- Weld capacities obtained using the theoretical model showed a reasonable agreement with the experimental results, particularly when considering the actual strength determined by the hardness measurements showing higher strength than the nominal values.
- The failures occurred at the weld metal, slightly inclined from the effective throat thickness section towards loading direction with the shear mode. With different geometrical configuration (weld to plate size and unsymmetric leg lengths), other failure paths could occur.
- Ftprom the experimental testing viewpoint, the determination of bending moment for obtaining the weld capacity is a critical step. The failures did not occur with the maximum moment. In addition, the failure location should be evaluated, and the eccentricity from the plate center should be obtained to calculate bending moment at the critical failure plane.

REFERENCES

prEN 1993-1-8. (2020). *Eurocode 3 - Design of steel structures - Part 1-8: Design of joints.*

Tuominen, N., Björk, T., & Ahola, A. (2017). Effect of bending moment on capacity of fillet weld. In A. Heidarpour & X.-L. Zhao (Eds.), *Tubular Structures XVI: Proceedings of the 16th International Symposium for Tubular Structures* (pp. 675–683).

Current Perspectives and New Directions in Mechanics, Modelling and Design of Structural Systems – Zingoni (ed.)
© 2022 Copyright the Author(s), ISBN 978-1-032-39114-4

Influence of web openings on the load bearing behavior of steel beams

F. Eyben
Institute of Steel Construction, RWTH Aachen, Germany

S. Schaffrath
Institute of Steel Construction, RWTH Aachen, Germany

M. Feldmann
Institute of Steel Construction, RWTH Aachen, Germany

ABSTRACT: A construction should maximize the potential for use through its design. Therefore, flexible use, as with steel beams with web openings e.g. to rout cables through, is important when designing a steel structure. This can also significantly reduce the height of ceiling systems. Until now, beams with web openings were not explicitly considered in the European standard, but a preliminary version of EN 1993-1-13 implements rules for different opening shapes. In order to further develop the design concepts, beams with web openings under bending are therefore to be investigated. For this purpose, first, fundamental factors influencing the load-bearing behavior of girders with web openings under bending load were investigated numerically without taking material damage into account. Various parameter studies were carried out for this purpose. For example, the factors under study were the opening shape and size.

1 INTRODUCTION

In any construction, the maximization of potential for use through the design of the building plays a huge role but on the other hand, limits are given by standards. The design and construction of steel structures is laid down in EN 1993 (Eurocode 3). With the needs of sustainable and resource-saving construction, steel is often the most economical material [1]. Web openings in steel girders can also considerably increase their economic efficiency, e.g. technical installations can be routed through and reduce the ceiling-height [2]. The geometries of the openings can be designed flexibly and thus create a high level of light permeability and a pleasant feeling of space. The goal of this work is to analyze the load bearing behavior of beams with web openings numerically and to check the design recommendations of prEN 1993-1-13 [3].

2 STATE OF THE ART REGARDING WEB OPENINGS IN STEEL GIRDERS

If a steel girder has openings in the web, the load capacity regarding bending, normal stresses and shear forces and also the stiffness is reduced in this area [4]. The remaining cross section can be split in two T-sections. Resulting from the shear stresses, local bending moments, called Vierendeel-moments, occur at the t-sections and lead to higher compression stresses at the corners of the opening [4].

3 NORMATIVE REGULATIONS REGARDING THE DESIGN OF WEB OPENING

In the current version of the EN 1993-1-1 [5] is no explicit design concept for beams with web openings.

The failure of cross-sectionally weakened profiles with opening is only checked in the design of stiffeners with openings acc. to EN 1993-1-5 [6]. This insufficient approach to design also beams with web openings lead a preliminary version of the EN 1993-1-13 [3], in which beams with web openings are explicitly implemented. Here, for example different shapes and maximum dimensions of openings are defined [3].

4 DEVELOPMENT OF THE FEM-MODEL

To investigate beams with web openings, Finite-Element-Method-models (FEM) were developed in

DOI: 10.1201/9781003348450-167

the FEM-software Abaqus. In general, four point-bending tests on standardized sections with an opening in the middle were simulated, so that a constant moment prevails the opening-area. Stiffeners at the supports and the load introduction as well as lateral supports were added. Imperfections where implemented after a buckling analysis. The following investigations were with beams (span length = 5000 mm) made of an IPE 400, and an S355N as well as an S690QL, where the material properties were taken from [7].

5 NUMERICAL SIMULATIONS

First, the influence of the height of a circular opening, in dependence of the dimensions acc. to prEN 1993-1-13 [3] was analyzed.

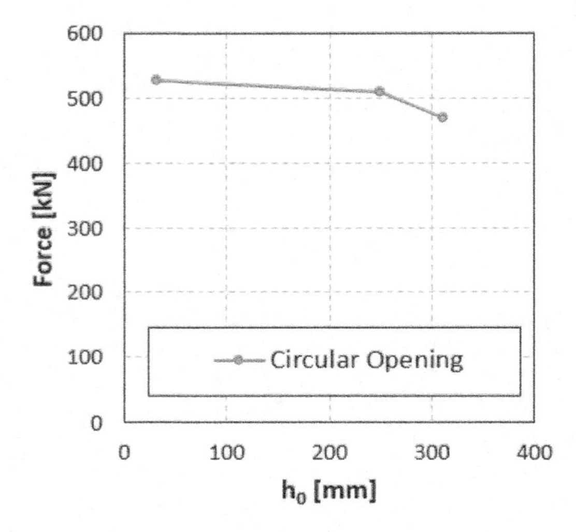

Figure 1. Maximum force in dependence of the height for circular openings for steel grade S690.

The results for steel grade S690 are given in Figure 1. Analogously as for steel grade S355 it can clearly be seen that the influence of a small circular web opening is not significant, while for bigger openings (with a height of around 70 % of the maximum allowed height) a lower load bearing capacity can be found.

The same investigation was made for beams with rectangular openings on beams made of S355 with the cross-section of the IPE 400. The results are given in Figure 2. At constant opening width, the influence of the height on the load bearing capacity is obviously more critical for the rectangular beams than for circular beams. With a height of 30 % of the maximum allowed height, a large loss of capacity is present. This is, because stress hot spots are more present at the corners of rectangular openings under constant moment than for the round circular openings.

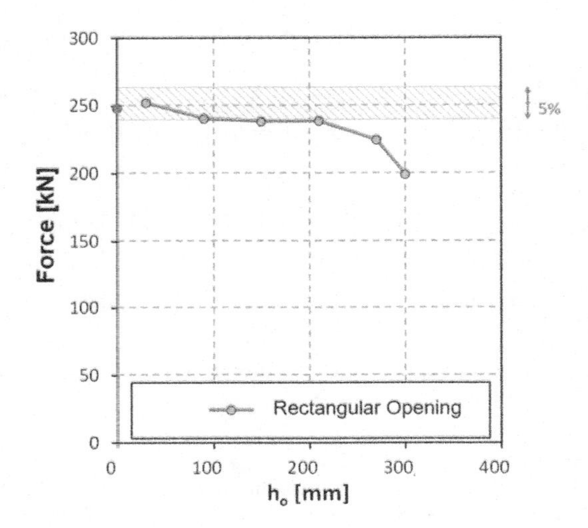

Figure 2. Maximum force in dependence of the height for rectangular openings for steel grade S355.

6 CONCLUSIONS AND OUTLOOK

This paper presents investigations on the influence of web openings of steel beams on the load bearing capacity. Following main conclusions were found:

1. Small web openings have a minor influence on the load bearing capacity.
2. As expected, with increasing heights of the web openings, the reduce on the load bearing capacity gets bigger.
3. Different shapes of openings show different buckling and load bearing behaviors. The circular opening is with regard to the load bearing capacity the best option.
4. The method of equivalent openings should be critically checked and validated with experiments in further research.

REFERENCES

[1] Nachhaltige Argumente für das Bauen mit Stahl, bauforumstahl, Nr. B106, Düsseldorf, 2017.

[2] ACB Lochstegträger, ArcelorMittag Europe - Long Products Sections and Merchant Bars, Luxembourg, 2014.

[3] prEN 1993-1-13, 2020.

[4] K. Tsavdaridis und C. D'Mello, Vierendeel Bending Study of Perforated Steel Beams with Various Novel Web Opening Shapes through Nonlinear Finite-Element Analysis, Journal of structural engineering, University of Brighton, 2012.

[5] EN 1993-1-1: Eurocode 3: Design of steel structures - Part 1-1: General rules and rules for buildings, 2005.

[6] EN 1993-1-5: Eurocode 3 - Design of steel structures - Part 1-5: Plated structural elements, 2006.

[7] M. E. A. Feldmann, Ableitung neuer, verbesserter Festigkeitskriterien für Stahlbauteile", Aachen: Schlussbericht zu IGF-Vorhaben Nr. 17925 N, Forschungsvereinigung Stahlanwendung e.V. (FOSTA), 2017.

Current Perspectives and New Directions in Mechanics, Modelling and Design of Structural Systems – Zingoni (ed.)
© 2022 Copyright the Author(s), ISBN 978-1-032-39114-4

Effects of elevated temperatures on cold-formed stainless steel double shear bolted connections

Yancheng Cai & Ben Young
Department of Civil and Environmental Engineering, The Hong Kong Polytechnic University, Hong Kong, China

ABSTRACT: A total of 35 cold-formed stainless steel double shear bolted connection specimens were tested using transient state test method. The specimens of bolted connections were fabricated by three different grades of stainless steel, namely, EN 1.4301 (AISI 304), EN 1.4571 (AISI 316Ti having small amount of titanium) and EN 1.4162 (AISI S32101). The connection tests were conducted at three different load levels of 0.25, 0.50 and 0.75 of the failure load at room temperature. The stainless steel grade EN 1.4571 generally performed better than the other two stainless steel grades as higher values of critical temperatures were obtained. Two failure modes, namely the bearing and net section tension failures, were observed in the transient state tests.

1 INTRODUCTION

The design rules of cold-formed stainless steel bolted connections in current specifications (e.g., EC3-1.4 2015) are applicable at room temperature only, but not for high temperature conditions. The tests on structural fire resistance are mainly carried out by two methods, namely, steady state test method and transient state test method. For the transient state test method, the test specimen is loaded and maintained at a given load level and then the temperature increased until the failure of specimen.

In this study, a total of 35 cold-formed stainless steel double shear one-bolted and three-bolted connection specimens were conducted using transient state test method. The tests conducted by the authors are detailed in Cai and Young (2015). The transient state test method was adopted with the consideration of 3 load levels of the failure load (P_u), at room temperature (Cai and Young 2014a). The specimens were fabricated by three different grades of cold-formed stainless steel, namely, EN 1.4301 (AISI 304), EN 1.4571 (AISI 316Ti having small amount of titanium) and EN 1.4162 (AISI S32101). For simplicity, these three types of stainless steels, EN 1.4301, EN 1.4571 and EN 1.4162 are labelled as types A, T and L, respectively, in the context of this paper.

2 DOUBLE SHEAR BOLTED CONNECTIONS

The double shear one-bolted and three-bolted connection specimens were designed by varying the number of the bolts, bolt diameter, as well as the type of stainless steel. Each specimen is labelled by four segments, for example specimen "T-D-3-8". The first letter "T" indicates the type of stainless steel of which the bolted connection specimen is assembled. The second letter "D" means the double shear; the third segment of the label is the number of bolt used in the specimen. The fourth part of the label means the nominal diameter of the bolt used in the connection.

The ISO standard fire curve as specified in EC1-1.2 (2002) was adopted in raising the furnace temperature. The applied loads on the double shear bolted connection specimens were chosen as $0.25P_u$, $0.50P_u$, and $0.75P_u$, respectively. The test set-up is shown in Figure 1. Details of the testing procedures are described in Cai and Young (2015).

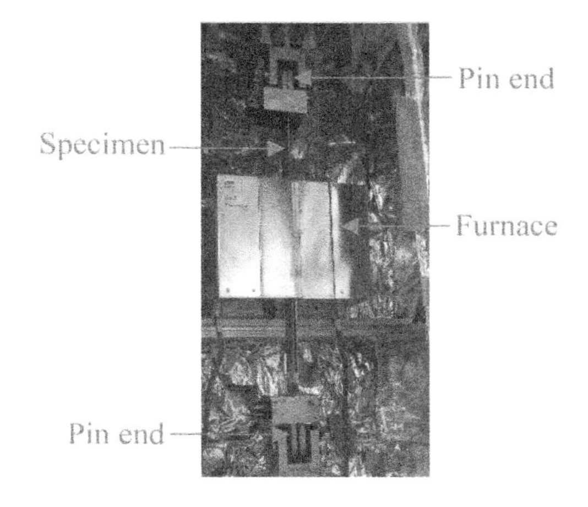

Figure 1. Test set-up of a bolted connection specimen using transient state test method.

DOI: 10.1201/9781003348450-168

3 TEST RESULTS

The critical temperature for the double shear bolted connection can be determined by the load-temperature curve for which the load drops not more than 5% of the pre-selected load level (Cai and Young 2015). The failure modes of bearing and net section tension were found in all the stainless steel double shear one-bolted and three-bolted connections at different load levels. Figure 2 exemplifies the load-temperature curves of specimen series T-D-3-8 at the three different loading levels of $0.25P_u$, $0.50P_u$, and $0.75P_u$.

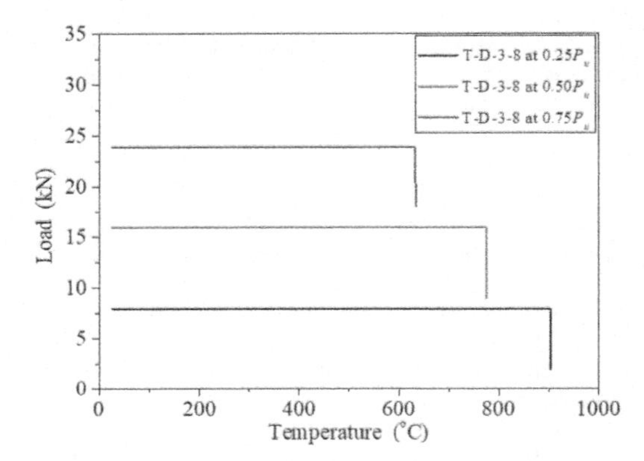

Figure 2. Load-temperature curves of series T-3-8.

4 COMPARISON OF TEST RESULTS

Figure 3 shows the comparison of the transient state test results for the different grades of cold-formed stainless steel double shear one-bolted and three-bolted connections.

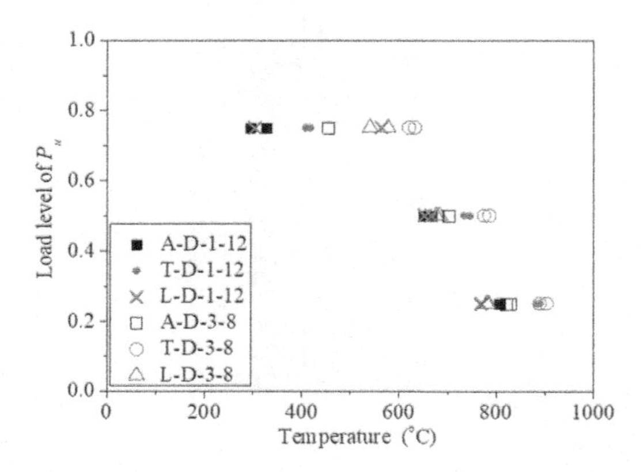

Figure 3. Comparison of critical temperatures at different load levels for different specimen series.

It is clearly shown that the specimens assembled by the stainless steel grade T (AISI 316Ti having small amount of titanium) behave better by providing higher critical temperatures when compared with those assembled by the other two grades A and L under the same load levels. It was also found that the tendency of reduction of the connection strengths performed by transient state tests is similar to those specimens conducted by the steady state tests. These are presented in detail in Cai and Young (2015).

5 CONCLUSIONS

A total of 35 connection specimens were tested at three different load levels of 0.25, 0.50 and 0.75 of the failure load at room temperature. It was found that the austenitic stainless steel grade EN 1.4571 (AISI 316Ti having small amount of titanium) generally performed better than the other two stainless steel grades EN 1.4301 and EN 1.4162, as higher critical temperatures under the same load levels were obtained. The critical temperatures of specimens assembled by stainless steel grades EN 1.4301 and EN 1.4162 are close to each other for the load levels of $0.25P_u$ and $0.50P_u$, but those with EN 1.4162 generally provided higher critical temperatures at the load level of $0.75P_u$. The critical temperatures obtained from the transient state test results are generally slightly higher compared with those deduced from the steady state test results.

ACKNOWLEDGEMENTS

The authors are grateful to STALA Tube Finland for supplying the test specimens. The research work described in this paper was supported by a grant from the Research Grants Council of the Hong Kong Special Administrative Region, China (Project No. HKU718612E). The Start-up Fund for RAPs under Strategic Hiring Scheme by The Hong Kong Polytechnic University is highly appreciated.

REFERENCES

Cai, Y. and Young, B. (2014a). Structural behavior of cold-formed stainless steel bolted connections. *Thin-Walled Structures*, *83*, 147–156.

Cai, Y. and Young, B. (2014b). Transient state tests of cold-formed stainless steel single shear bolted connections. *Engineering Structures*, *81*, 1–9.

Cai, Y. and Young, B. (2015). High temperature tests of cold-formed stainless steel double shear bolted connections. *Journal of Constructional Steel Research*, *104*, 49–63.

EC1-1.2. (2002). Eurocode 1 - actions on structures - Part 1-2: general actions - actions on structures exposed to fire. *European Committee for Standardization*, BS EN 1991-1-2:2002, Brussels.

EC3-1.4. (2015). Eurocode 3. Design of steel structures - Part 1.4: General rules - Supplementary rules for stainless steels. *European Committee for Standardization*, BS EN 1993-1-4:2015, Brussels.

Current Perspectives and New Directions in Mechanics, Modelling and Design of Structural Systems – Zingoni (ed.)
© 2022 Copyright the Author(s), ISBN 978-1-032-39114-4

Simplified approach to calculate welding effect for multi-layer welds of I-girders

T. Krausche
IPP Ingenieurgemeinschaft Prof. Pasternak & Partner, Braunschweig/Cottbus, Germany

H. Pasternak & J. Wang
Chair of Steel and Timber Structures, Brandenburg University of Technology, Cottbus, Germany

Z. Li
Institute of Civil Engineering, Technical University of Berlin, Berlin, Germany

B. Launert
Strabag AG, Germany

ABSTRACT: In steel structures, welding is one of the favorable joining methods. To consider negative effects of the joining technique residual stresses and distortions can be found with different welding simulation tools. The usage of different tools will require know how of the process as well as the input of energy, geometry and the material. In addition, there is a large time amount necessary the build-up the working model with the required boundary conditions. To reduce modelling and welding time a welding tool was developed and will be used to analyses in a quite simple and easy-applicable manner residual stresses. It will show a better direct link between weld manufacturing and capacity calculations. The simplified approach will be compared with earlier fill scale simulation efforts in a case study. The application of such approach for considering welding effects in the component design of welded plate girders is however still quite unusual even in steel research. Especially multi-layer welds, which are found in joining thick plates, require multiple calculation steps and generate the need of simplifications. Therefore, the existing welding tool, which is able to calculate I-girders, is expanded to multi-layer welds. Finally, this approach will have developed and calibrated with experimental investigation by means of a multi-layer welded girders.

1 INTRODUCTION

Welding causes distortions and residual stresses that can change the load bearing behavior to a large extent (Pasternak et al. 2015, Launert et al. 2017). To join thick plates multi-layer welds, have to be considered. To apply this in a practical context (Zebinger et al. 2018) shows a way of calculating an unregularly geometry with two to five layers. With the continuous development of computer technology and finite element method (FEM), realistic weld imperfections can be determined using sequentially coupled thermo-physical and thermo-mechanical analyses, nowadays. However, the application of welding simulation for structural component design of welded girders is still quite unusual even in steel research. Reasons for this is that the numerical simulation requires enormous modelling and computation efforts to perform such analysis on large structural components. Therefore, a simplified approach need to be developed to consider the welding effect for large structural components, which usually need to use the multi-layer welds (Krausche 2021). In recent years, some simplified welding simulation models are made available, which can be also suitable for use on large structural components, such as Analytical Numerical Hybrid Model with shrinkage force model and Local-Global Model Approach (Launert et al. 2019). The existing welding tool, which is able to calculate I- and T-girders are only applicable to single-layer welds and need to be expanded to multi-layer weld butt welds.

2 SIMULATION OF MULTI-LAYER WELD BUTT WELDS

To demonstrate that the simplified approach can be extended to calculate the butt weld with multi-layer welds with the 2D welding simulation. In the following, two steel plates having a thickness of 20 mm and a width of 250 mm are welded with five layer welds. Figure 1 displayed the magnitude of differences in weld pool sizes for different weld layers, where the melting point of steel is defined as 1500 °C. Figure 2 displays the residual stress distribution along the width

DOI: 10.1201/9781003348450-169

direction for the upper and lower edge as well as the mean value for the thickness direction.

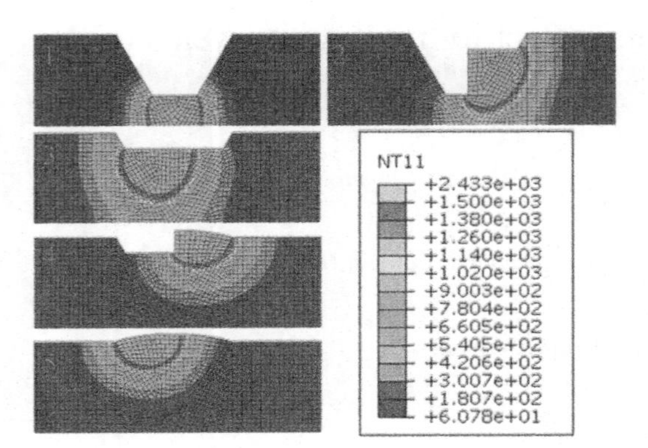

Figure 1. Temperature distribution (in °T) for each weld layer.

It is clear that the residual stress at the multi-layer weld area is significantly larger than the yield strength and the max. value can be exceeding 600 MPa, as noted in experimental research (Launert 2019 & Pasternak 2018).

3 CONCLUSIONS

In this paper, a simplified 2D welding simulation method is proposed to calculate residual stresses for multi-layer welds. Compared with 3D welding simulation, this method has a huge advantage in the required computing resources. Although the 2D welding simulation for multi-layer welds has great potential for practical civil engineering applications, the simplified approach still needs to be verified by experiments. In particular, some key result parameters, such as the magnitude of differences in weld pool sizes, transverse deformations and distribution of Mises residual stresses, need to be studied to modify and improve the 2D welding model. Hence, the future work can focus on the following two ways besides calibrating and verifying the model through experiments. It is necessary to establish relationship functions that can determine the parameters of the heat source model as well as the welding speed and energy through the welding method and the weld size of each layer. Another way is to program an automated 2D welding software for multi-layer welds, so that civil engineering can use this approach with limited finite element and welding simulation knowledge. In summary, it is necessary to apply the benefits of the accurate results of welding simulation to the design of steel structures. This seems particularly necessary in the context of the growing desire on more and more compute-raided designs.

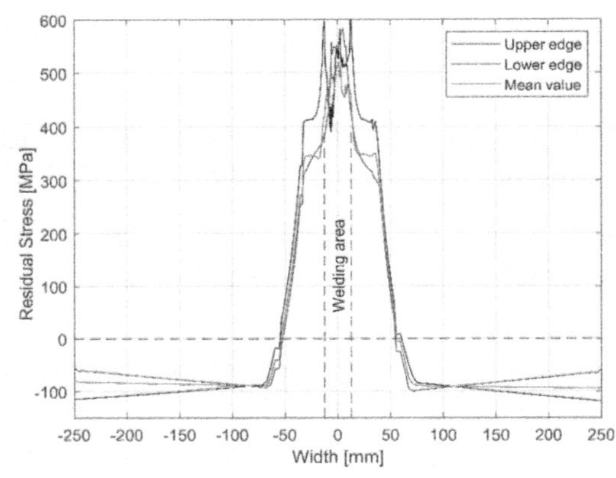

Figure 2. Residual stress distribution in width direction.

REFERENCES

Aichele G. Leistungskennwerte für Schweißen und Schneiden, 2. überarbeitete und erweiterte Auflage. Düsseldorf: DVS-Verlag; 1994.

Goldak J, Chakravarti A, Bibby M. A new finite element model for welding heat sources. Metall Trans B 1984; 15:299–305.

Krausche T. Zum Tragverhalten von geschweißten I-Trägern unter Berücksichtigung des Einflusses von Montagestößen. Brandenburg University of Technology (BTU), 2021

Launert B, Szczerba R, Gajewski M, Rhode M, Pasternak H, Giżejowski M. The buckling resistance of welded plate girders taking into account the influence of post-welding imperfections – Part 1: Parameter study. Mater Test 2017; 59:47–56.

Launert B. Untersuchungen an geschweißten I-Trägern aus normal- und hochfestem Baustahl: Beitrag zur Erweiterung der Tragfähigkeitsnachweise durch Einsatz der Schweißsimulation. Brandenburg University of Technology (BTU), 2019.

Launert B, Li Z, Pasternak H. Development of a new method for the direct numerical consideration of welding effects in the component design of welded plate girders. Adv. Eng. Mater. Struct. Syst. Innov. Mech. Appl., 2019, p. 1143–7.

Pasternak H, Launert B, Krausche T. Welding of girders with thick plates - Fabrication, measurement and simulation. J Constr Steel Res 2015; 115:407–16.

Pasternak H, Launert B, Rhode M, Kannengiesser T. Erhöhung der Tragfähigkeit geschweißter I-Träger aus hochfestem Baustahl durch verbesserte Ansätze zur Berücksichtigung von Eigenspannungen, 2018.

Peil U, Wichers M. Schweißen unter Betriebsbeanspruchung - Werkstoffkennwerte für einen S355 J2G3 unter Temperaturen bis 1200°C. Stahlbau 2004;73:400–16.

Voß O. Untersuchung relevanter Einflußgrößen auf die numerische Schweißsimulation. Technische Universität Braunschweig, 2001.

Zebinger M, Csendes M, Gierat M, Pasternak H, Krausche T. Das Seilnetzdach der Schierker Feuerstein Arena im Harz Teil 2: Herstellung, Montage und Schweißsimulation. Bauingenieur 2018; 93:383–91.

Current Perspectives and New Directions in Mechanics, Modelling and Design of Structural Systems – Zingoni (ed.)
© 2022 Copyright the Author(s), ISBN 978-1-032-39114-4

Resistance of axially and eccentrically loaded steel column at high temperature: A simple expression

G. Somma
Polytechnic Department of Engineering and Architecture, University of Udine, Italy

ABSTRACT: In this paper a rather simple expression for predicting the resistance of steel column subjected to both axial and eccentrical load under high temperature is proposed. This expression considers the dependence of the steel column resistance on the relative slenderness at the collapse temperature. The steel column resistance values at high temperature obtained with the here proposed expression has been compared with those outcoming from other researchers and Code expressions, with respect to the test results found in the literature. The proposed expression has found to lead not only the most accurate but also the most uniform prediction of the column resistance under fire loads. It is also important to point out that the here proposed expression is very simple and easily applicable also by engineers in the practical design.

1 INTRODUCTION

In this paper a rather simple expression for predicting the resistance of steel column subjected to both axial and eccentrical load under high temperature is proposed, based on the model proposed by Rankine in 1866 and modified by Merchant 1954 for columns at ambient temperature: they proposed an empirical expression based on the correlation between the critical elastic load, P_e, and the plastic load, P_p, determining the buckling and plastic collapse, respectively.

Only in the last twenty years some authors have proposed models considering the Rankine-Merchant expression extended to high temperature. Among them Tang et al. 2001 modified the Rankine approach simply introducing some corrective coefficient with the purpose to obtain accurate previsions of the experimental results, but they neglect the eventual eccentricity, and hence their expression is not suitable for column eccentrically loaded. The eccentric load has been considered by Toh et al. 2000, who proposed a simple expression for calculating the column resistance under fire conditions, assuming a temperature constant in the section and in the length, no initial curvature, no local buckling, neglecting little deformations. Despite these simplifying assumptions, their expression has found to provide an accurate prediction of the experimental results.

The expression proposed in this paper, based on the model proposed by Toh et al. 2000, has found to lead not only the most accurate but also the most uniform prediction of the column resistance under fire loads in comparison with the Toh et al. and Eurocode 3 expressions, with respect to 66 test results found in the literature. The reliability of the proposed expression has been verified for both axially and eccentrically loaded columns.

2 PROPOSED MODEL

In the here proposed model, the same simplifying hypothesis done by Toh et al. are considered. The Rankine-Merchant model, considering the influence of temperature, leads to the following expression

$$\frac{1}{P_c(T)} = \frac{1}{P_p(T)} + \frac{1}{P_e(T)} \tag{1}$$

where $P_c(T)$ is the column total resistance, $P_e(T) = (\pi^2 E(T)\, J)/l_0^2$, and $P_p(T) = \mu\, P_y(T)$, with $P_y(T) = A\, f_y(T)$. The modulus of elasticity and the yielding strength are in function of the temperature T: $E(T) = k_E(T)\, E$, and $f_y(T) = k_y(T)\, f_y$, where $k_E(T)$ and $k_y(T)$ are the reduction factor of E and f_y, respectively. The factor μ is a reduction coefficient of the yielding resistance and it depends on the eventually present load eccentricity, e, and on the section factor, $\xi = A/W_p$, where W_p is the plastic modulus of resistance of the section. It has different expression depending on the buckling axis.

In this manuscript the ultimate total resistance, $P_c(T)$, of a steel column is considered depending also to relative slenderness ratio, $\Lambda_T = \lambda/\lambda_T$, where λ is the column slenderness, and $\lambda_p(T)$ is the slenderness of proportionality, depending on T.

DOI: 10.1201/9781003348450-170

The following expression providing the total resistance, $P_c(T)$, is obtained from Equations 1 and considering that it should depend on Λ_T:

$$P_c(T) = c_0 \left[f(\Lambda_T) \left(\frac{1}{P_p(T)} + \frac{1}{P_e(T)} \right) \right]^{-1} \tag{2}$$

where the coefficient c_0 and the function $f(\Lambda_T)$ have the following different values and expressions for axial load (e = 0), load with little eccentricity (e ≤ 50 mm), and load with great eccentricity (e > 50 mm):

- for e = 0: c_0 = 2,28 and

$$f(\Lambda_T) = 2,5\Lambda_T^{-0,1} \tag{3}$$

- for e ≤ 50 mm: c_0 = 1,66 and

$$f(\Lambda_T) = 1,3\Lambda_T^{-0,3} \tag{4}$$

- for e > 50 mm: c_0 = 2,23 and

$$f(\Lambda_T) = 3,2 \tag{5}$$

The here proposed expression (Equation 2) is simple and widely appliable, as it can be used for every load eccentricity, and takes into consideration not only the influence of the temperature, but also the dependence to Λ_T in the resistance calculation.

3 MODEL RELIABILITY

To check the reliability of the here proposed model with respect to those previously provided and here considered, the column critical loads, P_u, obtained in the collected tests on 66 steel columns at the maximum temperature T_u have been compared with the resistance values, $P_c(T)$, obtained applying the proposed final expression (Equation 2), the expression of Toh et al. 2000, and the design formula given by Eurocode 3. From this comparison the following values of the average, AVG, and coefficient of variation, COV, of the experimental to calculated ultimate load ratios have been obtained:

- Toh et al. 2000: AVG = 0,97, COV = 0,41;
- Eurocode 3: AVG = 0,52, COV = 0,62;
- Proposed: AVG = 1,00, COV = 0,33.

It can be deduced that the here proposed model leads to a 92% more accurate prevision than Eurocode 3. As regards the COV comparison the here proposed model provides COV = 0,33, 20% lower than that obtained with the other author expression, and even 47% than the Code formula.

In Figure 1 the ratios between the experimental and calculated resistance values are plotted versus T_u, applying both the proposed expression (Equation 9) and the Eurocode formula (Equation 14). From this graph it is evident that the here proposed model provides a more accurate prevision of the column resistance of a steel column under fire exposure: the dots are more concentrated along the line of the perfect equality between experimental and calculated resistance.

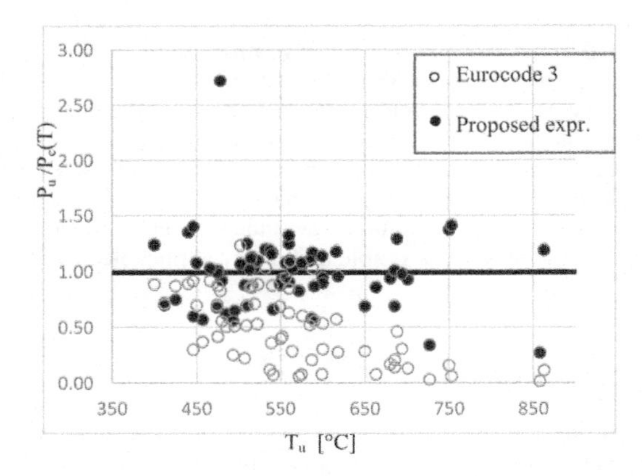

Figure 1. $P_u/P_c(T)$ in function of T_u considering both Eurocode 3 and the proposed expression.

4 CONCLUSIONS

In this paper a new expression for calculating the resistance of column under fire exposition has been proposed. The intention of the author is to provide not only a generic expression valid for column both axially and eccentrically loaded, but also a formula which is easy to apply and therefore widely usable by engineers. The here proposed expression has found to be consistent with the experimental results, providing an AVG = 1,00, indicating the mean perfect equality between experimental and calculated values, and COV = 0,33, the lowest among the expressions considered.

REFERENCES

European Committee for Standardisation EN 1993-1-2. 2005. Eurocode 3: Design of steel structures: Part1.2: General Rules Structural Fire Design. Brussels.

Merchant, W. 1994. The failure load of rigid jointed frameworks as influenced by stability. *Structural Engineering*. Vol.32, pp.185–190.

Tang, C.Y., Tan, K.H., and Ting, S.K. 2001. Basis and application of a simple interaction formula for steel columns under fire conditions. *Journal of Structural Engineering, ASCE*. Vol.127, No.10, pp.1206–1213.

Toh, W.S., Tan, K.H., and Fung, T.C. 2000. Compressive resistance of steel columns in fire: the Rankine approach. *Journal of Structural Engineering, ASCE*. Vol.126, No.3, pp.398–405.

Current Perspectives and New Directions in Mechanics, Modelling and Design of Structural Systems – Zingoni (ed.)
© 2022 Copyright the Author(s), *ISBN 978-1-032-39114-4*

Large-scale fire testing of an innovative cellular beam and composite flooring structural system

Jaleel Claasen, Richard Walls & Antonio Cicione
Department of Civil Engineering, University of the Stellenbosch, Stellenbosch, South Africa

ABSTRACT: This paper presents a summary of experimental and numerical findings on the thermal and structural behavior of a novel modular office steel building system at elevated temperatures. The structural framing of the system consists of horizontal cellular beam sections and vertical columns, with a sandwiched profiled steel decking flooring system attached to the bottom of the horizontal sections. The primary aim of the study and experimental program was to determine the fire resistance of the structural system. A large-scale 1-hour standard fire test was carried out on a 5.66 m × 3.8 m experimental steel frame with the sandwiched decking system attached. Novel temperature and deflection data for the system were obtained from the large-scale experiment conducted, which are used to validate numerical finite element analysis models. The numerical and experimental data are compared and show suitable correlation. The work highlights how the sandwich decking flooring system is pivotal in protecting the steelwork, and any failure of it would result in subsequent failure of the steel system. Also, the high flexibility of the structure means that limited internal axial forces are generated.

1 INTRODUCTION

Steel framed structures have become more competitive in comparison to reinforced concrete buildings as the popularity of long-span cellular steel beams has grown. Cellular beams are widely used in multi-storey buildings, commercial and industrial buildings, warehouses, and portal frames, in the UK and Europe. Furthermore, assessing the performance of structures in the event of a fire has also grown in importance over time. This piqued interest in the behavior of cellular beams at elevated temperatures and how to effectively evaluate their design (Vassart, 2009). However, only a few large-scale (i.e. structural assembly) composite cellular beam fire tests have been conducted (Nadjai *et al.*, 2011).

This paper provides a summary that brings together technical results regarding the behaviour of a large-scale non-composite cellular beam system in fire. This paper forms part of larger study documented across multiple papers and theses, which helps in understanding the fire response of the system in an holistic manner.

2 CELLULAR BEAM STRUCTURAL (CBS) SYSTEM

The experimental frame tested in this study is based on the design of a single module of the cellular beam modular office building system proposed by the Southern African Institute of Steel Construction, and is labelled as the Cellular Beam Structural (CBS) system. The steelwork is protected by the sandwich decking (SD) flooring system suspended from the underside of the beams and acts as the load bearing component of the floor. This allows for the implementation of a false floor on top of the SD system where services can be housed.

The work in this paper was carried out to investigate the behaviour of the CBS system and quantify the fire resistance. The focus of this paper is on the thermal behaviour of the SD system and the structural behaviour of the experimental frame at elevated temperatures. A thermal response model of the SD system and thermal-stress models of the experimental frame have been developed and the results compared to the experimental data.

2.1 *Largescale standard fire test*

The large-scale test was conducted with a newly built furnace at the Ignis Testing facility (Ignis Testing, 2021) located in Cape Town, South Africa. The furnace's inner dimensions are 6 m × 4 m × 1.185 m (L × W × H). The test was done according to SANS 10177-2 (SABS, 2005), which is based on the ISO 834 requirements and time-temperature curve. The experimental frame tested consisted of ten main structural elements (See Figure 1 (Top)) made of grade S355JR steel, which included two secondary symmetrical cellular beams, with the SD system attached to the bottom of the horizontal structural elements.

DOI: 10.1201/9781003348450-171

3 NUMERICAL MODELLING

Using the Finite Element Method (FEM) and the results obtained from the large-scale experiment, numerous numerical model analyses were carried out to predict and further investigate the thermal behaviour of the SD system and the thermal-stress response of the structural frame.

A two-part analysis procedure was followed in this work to carry out the thermal response model analyses of the SD system, with the material thermal properties derived from literature and as supplied by the manufacturers. When comparing the experimental and numerical data of the SD system to that of the Fire Limit State insulation limit temperatures, a fire rating of 57 and 42 minutes is estimated, respectively.

Single element models and a single global structural model was developed to analyze the experimental frame, however, only the global structural model results (See Figure 1) will be considered in this paper. A three-step analysis procedure was adopted in this work for all the thermal-stress model analyses as follows: Step-1 (initial step): set boundary conditions, Step-2 apply mechanical loading, Step-3 apply thermal loading as predefined fields using the captured time-temperature curves from the experiment. The error difference between the numerical and experimental results ranges from 1.4% to 11.4% at the end of the analysis.

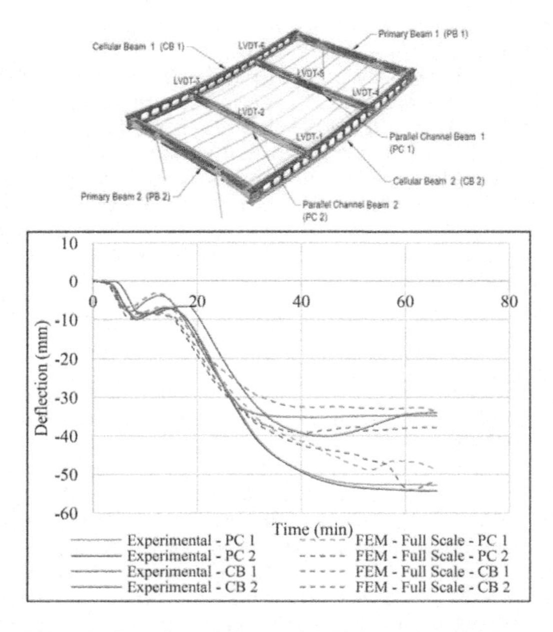

Figure 1. Experimental frame global structural model (Top) and experimental and numerical deflection results (Bottom).

4 CONCLUSION

The key finding of the experiment and numerical model analyses are as follows: (1) The presence of the SD system, which is attached to the bottom of the horizontal structural elements, has resulted in a) the experimental frame undergoing unusual deflection behaviour and b) the structural elements experiencing relatively low elevated temperatures for the entire duration of the test. (2) The deflection behaviour of the structure is governed by three deflection mechanisms, namely, thermal elongation, thermal bowing, and mechanical deflection. (3) No structural failure was experienced by any of the structural components and the deflections recorded for the secondary cellular beams and tertiary channel beams were all within the recommend limit of span/20. (4) The numerical models developed was able to sufficiently predict the thermal behaviour of the SD system and the thermo-mechanical behaviour of the experimental frame.

The deeper implication of this structural system lies within the approach taken to design steel structures in the case of fire. If sufficient evidence is provided that the integrity of the SD system is maintained during a fire, and the steel temperatures remain within the Fire Limit State criteria, the structural elements of the system could potentially only be considered at ambient temperature during the design process. This would negate the complex aspects that need to be considered during the design of steel elements in fire, especially with regards to cellular beam elements. Hence, a novel and light-weight modular system as shown in this work could be used for multi-storey buildings.

REFERENCES

Ignis Testing. (2021), "Ignis Testing - Fire Resistance Testing Laboratory - Cape Town", available at: https://ignis testing.co.za/direction-and-related-issues-for-fire-door-testing/ (accessed 14 October 2021).

Nadjai, A., Bailey, C., Vassart, O., Han, S., Zhao, B., Hawes, M., Franssen, J.M., *et al.* (2011), "Full-scale fire test on a composite floor slab incorporating long span cellular steel beams", *The Structural Engineer*, Vol. 89 No. November, pp. 18–25.

SABS. (2005), *SANS 10177-2:2005 - Fire Testing of Materials, Components and Elements Used in Buildings Part 2: Fire Resistance Test for Building Elements*, SABS Standards Division.

Vassart, O. (2009), *Analytical Model for Cellular Beams Made of Hot Rolled Sections in Case of Fire*, available at: https://www.researchgate.net/publication/279263512_Analytical_model_for_cellular_beams_made_of_hot_rolled_sections_in_case_of_fire.

Current Perspectives and New Directions in Mechanics, Modelling and Design of Structural Systems – Zingoni (ed.)
© 2022 Copyright the Author(s), ISBN 978-1-032-39114-4

Design for de-construction of lightweight infill wall systems

S. Kitayama & O. Iuorio
University of Leeds, Leeds, UK

ABSTRACT: This paper discusses the design for de-construction of lightweight infill wall systems. The paper reports the deconstruction and reassembly processes of a sample light steel frame skeleton. Each of the steel members in the frame skeleton was connected by screws, which were removed by a screwdriver. The observations and findings during the deconstruction and re-assembly processes are presented. The authors observed that the screws in the sample steel frame were safely removed, and the members were re-assembled without causing any damage. The authors compared the stiffness of the frames before and after the deconstruction and observed that there was no change in the stiffness in the frame.

1 BACKGROUND

The United Kingdom Research and Innovation (UKRI) National Interdisciplinary Circular Economy Research (NICER) program was launched on 27 May 2021, which became the largest UKRI investment focusing on Circular Economy (CE) to date.

The authors of this paper belong to the Interdisciplinary Circular Economy Centre for Mineral-based Construction Materials (ICEC-MCM) which is one of the five NICER centres. The authors of this paper form a research group based at the University of Leeds to work on a research project "Design for Deconstruction of Light Steel External Wall Systems" within ICEC-MCM. The research project has collaborated with an industry partner Etex (Headquarters: Brussels, Belgium).

2 DESIGN FOR DECONSTRUCTION OF LIGHT STEEL EXTERNAL WALLS

Design for Deconstruction (DfD) reduces material consumption, embodied carbon, and waste in new constructions by distributing these impacts over a series of circular life cycles (Tingley & Davidson 2012). In DfD, salvaged members from buildings are repurposed directly in new projects, thus eliminating the new material manufacturing and the processes associated with recycling.

Lightweight infill wall for exterior (outdoor) facades is increasingly used on a range of building types and are an economic and efficient method of providing façade walls (SCI 2019). The infill walls are built between the floors of steel or concrete frames and are designed to resist wind loading and support the weight of the cladding. The external infill walls are removed during a refurbishment of the building to adapt spaces to new functions and/or to comply with new energy efficiency or humidity control targets. While a building is used typically for 50 or 60 years before it is demolished, a façade is used only for about 30 years (RICS 2017). A recent study showed that the light steel frame members for infill external walls in multistorey buildings can be used for about 100 years (SCI 2000). Currently, only about 5% of light steels are reused and most of them (93%) are recycled (Sansom & Avery 2013). Thus, deconstruction of light steel members from the infill walls and reuse of these members may increase the reuse rate of light steel.

3 DECONSTRUCTION AND RE-ASSEMBLY OF SAMPLE LIGHT STEEL FRAME SKELETON

This section presents the deconstruction and re-assembly of one of the two sample light steel frame skeletons that was donated by a manufacture of light steel products, EOS (Headquarters: County Durham, UK). The samples of light steel frame skeletons are shown in Figure 1. The two samples are identical specimens. Note that the samples are smaller version of actual sizes of steel frame skeletons used in the real construction of external infill walls. Note that EOS uses S390N/mm^2 proof steel with G275gsm (normal as coated chemical passivation) for their standard products.

In the steel samples, the sections of the head and base tracks are C sections without lip, and the sections of studs are C sections with lip.

Two screws were used in the samples: 5.5x19mm Pan head #2 Lox drive (supplier: Grabber) and 5.5x25mm

DOI: 10.1201/9781003348450-172

Figure 1. Sample lightweight steel frame skeletons from EOS.

Pan head Drywall screw S-MD03ZW (supplier: HILTI). The former screws were used to connect the horizontal steel member that was connected to two studs. The latter screws were used to connect the rest of the members. Note that the authors could not find any screwdrivers that fit the shape of the lox screw head of the former screws. As such, the deconstruction presented in this section is limited to the deconstruction of the latter screws.

Figure 2 presents the steel members after deconstruction. It is observed that the deconstruction of members did not cause any deformation to any of the members and screws. Also, the screw halls were not deformed during the removal of screws.

Figure 2. Steel members after deconstruction.

After the deconstruction of the sample members, the re-assembly of members and screws were conducted. Note that the screws were mixed after the deconstruction, thus some of the screws might be placed in different locations than they were originally located during the re-assembly. One of the challenges during the reassembly was to place the studs and tracks so that the holes in these members align with each other. In some holes, it was difficult to perfectly align one hole perimeter to the other hole perimeter. In such cases, a strong torque from the screwdriver through the author's arm was needed.

To observe the changes in the stiffnesses of sample frames in the lateral, torsional and out-of-plane directions, the sample was tested by the hands of the authors before deconstruction and after the reassembly. The test schemes are illustrated in Figure 3. The authors observed no changes in the stiffnesses in the lateral, torsional and out-of-plane directions. The same tests were conducted for another sample that was not deconstructed. The authors did not recognize any difference in the stiffnesses of these two sample frames. These observations imply that the light steel members in the external infill walls may be deconstructed and re-assembled to reuse for other constructions. To validate the observations presented in this paper, the mechanical connection tests at the laboratory and finite element analysis to simulate the behaviour of these frames need to be conducted.

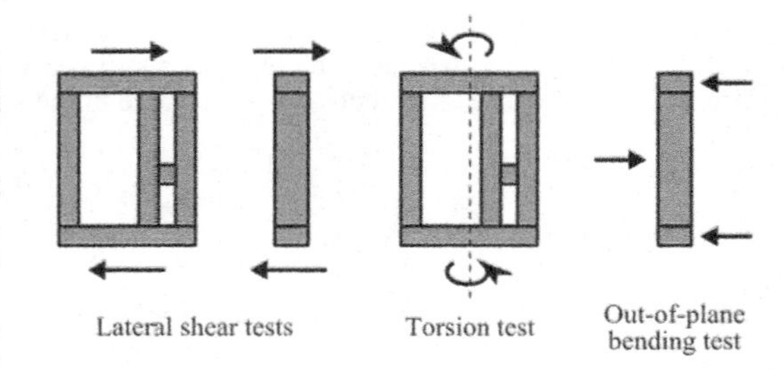

Figure 3. Tests to check stiffness of sample frames.

4 CONCLUSION

This paper demonstrated the deconstruction and re-assembly of a sample lightweight steel frame skeleton to investigate the feasibility of deconstruction and reuse of steel members. The authors observed that: (i) the screws were safely removed, (ii) the members were re-assembled without causing any damage and (iii) no changes were observed in the lateral, torsional and out-of-plane stiffnesses.

REFERENCES

RICS (Royal Institution of Chartered Surveyors). 2017. *Whole life carbon assessment for the built environment.* 1st edition, November. RICS professional standards and guidance. London, UK.

Sansom, M. & Avery N. 2013. Briefing: Reuse and recycling rates of UK steel demolition arisings. *Proceedings of the Institution of Civil Engineers Engineering Sustainability* 167 (ES3): 89–94.

SCI (The Steel Construction Institute). 2000. *Building design using cold formed steel sections - durability of light steel framing in residential building.* SCI Publication. P262. Ascot, UK.

SCI (The Steel Construction Institute). 2019. *Design and installation of light steel external wall systems.* SCI Publication. ED017. Ascot, UK.

Tingley, D.D. & Davison, B. 2012. Developing an LCA methodology to account for the environmental benefits of design for deconstruction. *Building and Environment* 57: 387–395.

Current Perspectives and New Directions in Mechanics, Modelling and Design of Structural Systems – Zingoni (ed.)
© 2022 Copyright the Author(s), ISBN 978-1-032-39114-4

Calibration and validation of slender circular CFDST columns subjected to eccentric loading

T.N. Haas, A.A. Usongo & J.A. Koen
Department of Civil Engineering, Faculty of Engineering, Stellenbosch University, South Africa

ABSTRACT: Concrete-Filled Double-Skin Tubular (CFDST) columns are a novel composite structural column formed by sandwiching concrete between two hollow structural steel tubes. Previous research work predominantly focused on concentric loading of slender circular CFDST columns. Although this work is important, it does not conform to codified requirements, which stipulate that the axial load must be applied eccentrically on the cross section of the column. In the development of Finite Element (FE) models, many investigators do not specify/elaborate how their FE models were validated. In some cases it seems that some investigators also incorrectly assume that calibrating their FE model to a single experimental response automatically implies that the FE model is validated. Another concern observed was that some researchers performed parameter studies where the geometric and material properties are significantly different from the base model, without validating these results. These misconceptions could lead to amplification of errors during sensitivity analyses, resulting in the accuracy of the results being questioned. In this paper, a generic FE model, which predicts the behaviour of circular slender CFDST columns subjected to eccentric axial loading was developed. The generic FE model was calibrated to one experimental response whereafter it was validated against three different CFDST experimental specimen configuration responses, to ensure that it is functional across a wide range of geometrically different circular slender CFDST columns. The investigation revealed that it is incorrect to assume a FE model is validated by simply calibrating it to one experimental response without further validation or performing sensitivity analyses on geometric and material parameters that varies significantly from the initial experimental response.

1 INTRODUCTION

Reinforced concrete and structural steel were extensively studied to developed codified guidelines for column design and construction. With ongoing research, alternative methods of column construction was proposed, which include composite columns such as Concrete Filled Skin Tubes (CFST) and Concrete Filled Double Skin Tubes (CFDST) columns. CFDST columns have seen a seen an increase in research attention due to its numerous advantages; i.e. lighter weight, increased fire resistant properties, higher bending stiffness, greater axial, flexural and torsional strength.

Significant research attention was devoted to studying the behaviour of circular stub and slender CFDST columns subjected to concentric loading. Insignificant research was devoted to studying the behaviour of these columns subjected to eccentric loading. A shortcoming in most published work is that some researchers do not explicitly inform the reader how the FE model was calibrated and validated. In some cases, the calibrated FE model was validated to an experimental response(s) where the geometric and material properties are similar. The effect of using similar CFDST geometric and material properties would not indicate any modelling errors. Some investigators used their calibrated FE models to perform parameter investigations on geometric and materials properties, which are significantly different from the base (calibrated) response, without any validation. These misconceptions can lead to the amplification of errors during a sensitivity analyses.

A Finite Element (FE) analysis was conducted to calibrate and properly validate the FE model using 4 different circular slender CFDST specimens. The aim was to establish whether incorrectly calibrating and properly validating the FE model can affect the behaviour of circular slender CFDST columns.

2 METHODOLOGY AND FE RESULTS

The research was conducted through an experimental investigation on 4 different circular slender CFDST columns subjected to eccentric loads (Koen, 2015).

These results were used in a FE investigation conducted by Usongo (2021). The reader is referred to Koen (2015) and Usongo (2021) for detailed geometric and material properties of the CFDST columns.

DOI: 10.1201/9781003348450-173

The FE model was developed and calibrated to the STK model. During the calibration process it was discovered that the main parameters preventing accurate calibration depended on the curing conditions of the concrete in the CFDST columns and cubes, which were observed as very different. This led to a reduction in concrete strength of 18.5% based on work conducted by Naderi et al. (2009). Also, the structural steel sections are not manufactured perfectly straight. Thus, these imperfections needed to be accounted for.

After implementing the revised concrete strength and the initial lateral imperfection, the FE model accurately predicted the ultimate load of the STK model to within 0.2% of the experimental response.

The modelling techniques were then applied to the STN, LTK and LTN models to validate the FE model. The default Concrete Damage Plasticity (CDP) parameters from the STK model were applied to the STN, LTK and LTN models. After performing the validation simulations on the 3 models, an average error of 2.9% between the FE and experimental results were observed. The error was considered unacceptable, since the FE model would be used in a thorough sensitivity analysis. This led to further investigating the FE model to determine which factors have an influence on the response. This led to observing that the CDP parameters and the confined concrete has a significant influence on the behaviour of the CFDST columns. The CDP parameters were recalculated based on the geometric and material properties.

Since the geometric properties of the columns changes, it has an effect on the concrete confined model. It is therefore important to update the concrete confinement stress vs strain curve when the geometric properties changes. Figure 1 presents the difference in the confined concrete stress vs strain responses for the CFDST columns with a thick "TK" and thin "TN" annulus.

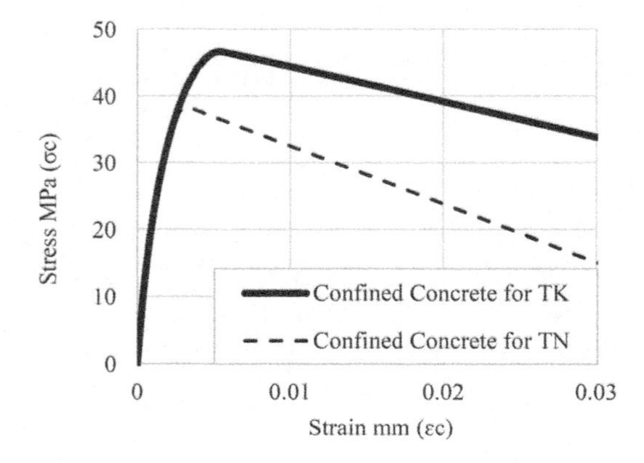

Figure 1. Confined concrete stress vs strain curves for TK and TN models.

After implementation the updated CDP parameter magnitudes and the confined concrete stress strain curves, resulted in a maximum error of 1.8% between the 4 FE and experimental responses as shown in Figure 2. This was considered sufficiently accurate.

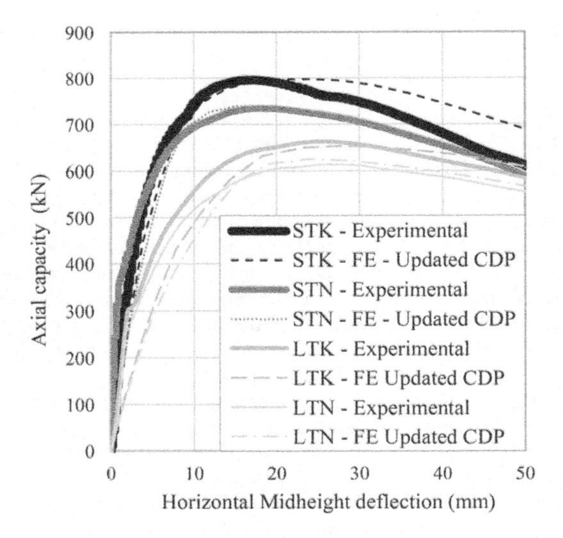

Figure 2. FE vs Experimental horizontal midheight deflection responses for the STK, STN, LTK and STN models.

3 CONCLUSION

The investigation showed that it is important to take all parameters into account during the calibration process and that a proper validation of the FE modelling is required. The FE modelling process was considered validated since the FE and experimental responses differed by 1.8%. It is anticipated that the FE model developed can be used to study the behaviour of circular slender CFDST columns subjected to eccentric loading, for geometric and material properties of double the base state parameters without validation. Beyond this point, experimental research is required to validate the results.

REFERENCES

Koen, J.A. 2015. *An investigation into the axial capacity of eccentrically loaded concrete filled double skin tube columns*. MSc Thesis, Stellenbosch University.

Naderi, M., Sheibani, R. & Shayanfar, M.A. 2009. Comparison of different curing effects on concrete strength. *3rd International conference on concrete & development*. Tehran: Iran.

Usongo, A.A. 2021. *Finite Element Modelling of Eccentrically Loaded Concrete filled Double Skin Tube Columns*. MSc Thesis: Stellenbosch University.

Current Perspectives and New Directions in Mechanics, Modelling and Design of Structural Systems – Zingoni (ed.)
© 2022 Copyright the Author(s), *ISBN 978-1-032-39114-4*

Investigation of CDP parameters and concrete confinement models for use in circular slender CFDST columns subjected to eccentric loading

A.A. Usongo & T.N. Haas
Department of Civil Engineering, Faculty of Engineering, Stellenbosch University, South Africa

ABSTRACT: Circular Concrete-Filled Double-Skin Tubular (CFDST) columns are high performance structural members, formed by sandwiching concrete between two hollow structural steel sections. To date, insignificant research attention was devoted to investigating the effect of eccentric loading on slender circular CFDST columns, thus limiting its implementation in industry, since design codes require columns to be eccentrically loaded to account for construction tolerances. From previous research work on circular CFDST columns, it was observed that geometric and material properties have a significant effect on it's behaviour when subjected to concentric loading. The correct modelling of the concrete confinement and the Concrete Damaged Plasticity (CDP) parameters are critical in obtaining accurate behaviour of CFDST columns in FE simulations. Default CDP parameters were proposed for concentric loading. To date no published work exists that confirm whether the default parameters used for concentric loading are also applicable to eccentric loading. This study investigated the effect of the CDP parameters on 4 different geometric CFDST column specimens subjected to eccentric loading through a FE analysis. Using 4 different CFDST specimens, a generalised overview of the CDP parameters were observed, resulting in adjusted CDP parameters being proposed for circular slender CFDST columns subjected to eccentric loading.

1 INTRODUCTION AND METHODOLOGY

Concrete Filled Double Skin Tubular (CFDST) columns is a relatively new form of composite construction, which is formed by sandwiching concrete between two hollow structural steel sections. Significant research work was conducted on stub and slender CFDST columns subjected to concentric loading to investigate the effect of various geometric parameters on its behavior.

Of particular significance in this study is the Concrete Damage Plasticity (CDP) parameters and the confined stress strain curves for concrete. To date, insignificant published work is available on eccentric loading of circular slender CFDST columns. It is assumed that the default CDP parameters and concrete's stress strain curve is also applicable for eccentric loading, which has not been verified. Thus, the purpose of this research was to determine whether the default CDP and confined concrete parameters used for concentric loading is applicable to eccentric loading when applied to circular slender CFDST columns.

The research was conducted through a Finite Element (FE) analysis investigation using the 4 experimental responses of the circular slender CFDST columns investigated by Koen (2015). The columns were distinguished based on the geometric properties and material properties, i.e. STK, STN, LTK and LTN.

The detailed geometric and material properties can be obtained in Koen (2015) and Usongo (2021). The CDP parameters that were investigated are; **1.** Dilation angle (Ψ); **2.** Flow potential eccentricity (e); **3.** the ratio of compressive strength under biaxial loading to axial compressive strength (f_{bo}/f_c); **4.** the ratio of the second stress invariant on the tensile meridian to that on the compressive meridian (K_c) and **5.** Viscosity parameters (μ).

Also, 2 concrete confinement models used by Hassanein et al (2013) and Pagoulatou et al (2014) were reviewed.

2 PARAMETRIC INVESTIGATION AND RESULTS

The geometric and material properties affect the CDP parameters and the confined concrete model.

A default parameter magnitude between 20° and 30° is generally used for the dilation angle (Ψ). A method to determine the dilation angle based on the actual properties of the CFDST column, resulting in an angle of 19° and 24°, for the TN and TK models was also proposed. The investigation shows that the Ψ changes by an average difference of 1.4% when the parameter is varied between 15° and 30°.

DOI: 10.1201/9781003348450-174

Researchers indicated that the Flow Potential Eccentricity (e) should be a "small positive value". A magnitude of 0.1 was proposed this parameter. The sensitivity analyses on this parameter ranging from 0 to 1 indicated an average difference of 0.012% when the parameter is varied between 0 and 1.

A default magnitude of 1.16 for the strain hardening/softening equation for the ratio of compressive strength under biaxial loading to axial compressive strength (f_{bo}/f_c) was proposed. The sensitivity analyses on this parameter ranging from 0 to 1.16 indicated an average difference of 0.31% when the parameter is varied between 0 and 1.16.

A default magnitude of 0.667 was proposed for the ratio of the second stress invariant on the tensile meridian to that on the compressive meridian, K_c. The sensitivity analyses on this parameter ranging from 0 to 1 indicated an average difference of 0.6% when the parameter is varied between 0 and 1.16.

It can thus be concluded that parameters 1 to 4 has an insignificant effect on the behaviour of slender circular CFDST columns subjected to eccentric loading.

A default value of 0 was proposed to represent the viscosity parameter. Figure 1 presents the results of the sensitivity analysis.

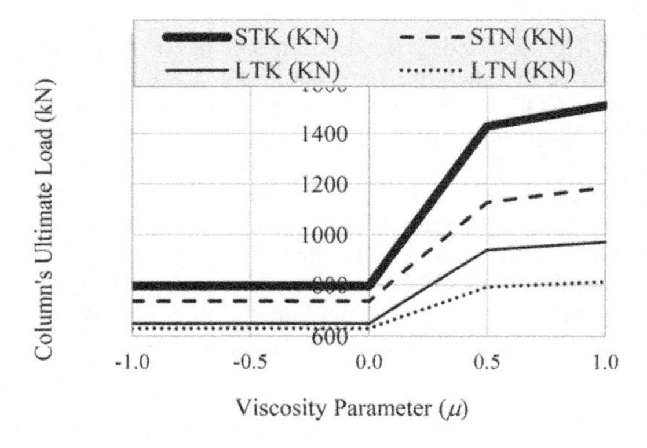

Figure 1. Effect of Viscosity Parameter (μ) on column's ultimate load, (Usongo, 2021).

Figure 1 indicates that the Viscosity Parameter is constant between -1 and 0, then varies by an average of 50.5% between 0 and 0.5, and varying by an average of 4.2% between 0.5 and 1. This is the only CDP parameter, which shows a significant change when the parameter is varied.

Two models commonly used to determine the confined concrete stress-strain curves for the TK model shown in Figure 2. The 2 approaches resulted in a difference of 5.9% in the confined stress. Although not significant, it yielded a 4.1% difference in ultimate load. Thus, either of the approaches can be used in a FE simulation. The 2 approaches resulted in a 5.9% difference in the confined stress. Although insignificant, it yielded a difference of 4.1% in ultimate load. Thus, either of the approaches can be used in a FE simulation.

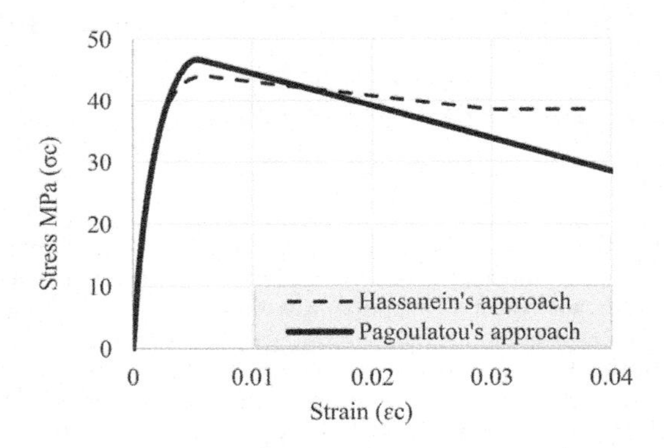

Figure 2. Comparison between Confined Concrete Stress-Strain Curves (Usongo, 2021).

3 CONCLUSIONS

The study investigated the accuracy of the default CDP parameters to model circular slender CFDST columns subjected to eccentric loading. This study used 4 different column configurations to investigate whether the default CDP parameters applied to concentric loading is applicable to eccentric loading. The sensitivity analysis reveal that adjusted CDP parameters can be used for ease of implementation for eccentric loading. The adjusted parameter recommendations for CDP parameters under eccentric loading are;

- A value of unity (1) is proposed to represent e, f_{bo}/f_c and K_C, while a magnitude of 0 is proposed for VP and a magnitude of 25° is proposed for Ψ.

Either of the 2 concrete confined models can be used in the FE simulations as it yields a difference of less than 5%.

REFERENCES

Koen, J.A. 2015. *An investigation into the axial capacity of eccentrically loaded concrete filled double skin tube columns*. Stellenbosch: MSc Thesis: Stellenbosch University.

Usongo, A. (2021). *Finite Element Modelling of Eccentrically Loaded Concrete Filled Double Skin Tube Columns*. Stellenbosch University, Civil Engineering. Stellenbosch University.

Hassanein, M.F., Kharoob, O.F. & Liang, Q.Q. 2013. Circular concrete-filled double skin tubular short columns with external stainless steel tubes under axial compression. Thin-Walled Structures 73: 252–263.

Pagoulatou, M., Sheehan, T., Dai, X.H. & Lam, D. 2014. Finite element analysis on the capacity of circular concrete-filled double-skin steel tubular (CFDST) stub columns. Engineering Structures 72: 102–112.

Current Perspectives and New Directions in Mechanics, Modelling and Design of Structural Systems – Zingoni (ed.)
© 2022 Copyright the Author(s), *ISBN 978-1-032-39114-4*

Evaluation of parameters which influence circular slender CFDST columns subjected to eccentric loading

A.A. Usongo & T.N. Haas
Department of Civil Engineering, Faculty of Engineering, Stellenbosch University, South Africa

ABSTRACT: Significant research was devoted to determine the effect that various parameters have on the ultimate load of circular slender Concrete Filled Double Skin Tubular (CFDST) columns subjected to concentric loading. However, the same research attention has not been devoted to circular/square/rectangular CFDST columns subjected to eccentric loading. This has limited the development of codified requirements for CFDST columns, since design codes require column loads to be applied eccentrically to accommodate construction tolerances. To contribute to the body of knowledge, a Finite Element (FE) analysis investigation was performed to determine the effect that various parameters have on the ultimate load of circular slender CFDST columns subjected to eccentric loading. This investigation was conducted using 4 different column specimens to obtain a holistic overview of the effect of the parameters on the column's response. The parameters that were investigated include; concrete and steel strengths, inner and outer tube thicknesses, column curvature, load eccentricity and column support fixities. The investigation revealed that a change in the concrete and steel strength has a moderate effect on the ultimate load of slender circular CFDST columns subjected to eccentric loading. A change in the inner and outer tube thicknesses as well as an increase in the load eccentricity has a significant effect on the ultimate load. The longitudinal curvature of the column has the least effect on the ultimate load, while the column's support conditions have the greatest effect on the ultimate load. These findings are similar to what was observed for concentric loading.

1 INTRODUCTION AND METHODOLOGY

CFDST columns are constructed using two hollow steel sections with concrete sandwiched between them. Stub and slender circular CFDST columns were extensively studied to determine the behavioural effect of parameters when subjected to concentric loading. Previous research identified several parameters that potentially have a significant effect on the column's behaviour and ultimate load when subjected to concentric loading. These parameters include but not limited to; **1**. Outer tube thickness; **2**. Inner tube thickness; **3**. Concrete strength; **4**. Steel strength; **5**. Concrete thickness; **6**. Column Curvature; **7**. Column fixity and **8**. Load eccentricity. Only parameters 1 through 4 received significant research attention.

Literature shows that limited published work exist investigating the behaviour of circular slender CFDST columns subjected to eccentric loading.

What differentiates this work from others is that a FE sensitivity analyses was performed on 4 geometrically different slender circular CFDST columns (STK, STN, LTK and LTN). This allows for a holistic overview of the column's behaviour and obtain a better understanding of the effect of the parameters.

2 PARAMETRIC STUDY

The reader is referred to Koen (2015) and Usongo (2021) for detailed information on the CFDST columns. A parametric study was performed on parameters 1 through 8 using the FE model described in Haas et al. (2022). A constant load eccentricity of 20 mm was applied to all FE models while the other parameters were individually added, varied and removed, before the next parameter was investigated.

A sensitivity analysis was conducted on concrete strengths between 35 MPa and 55 MPa. As expected, the investigation revealed a corresponding linear increase in ultimate load as the concrete strength increases. The TK and TN model's strength increases by 0.77% and 0.53% per MPa, respectively, for the particular geometric configurations.

Since steel strength varies, a sensitivity analysis was conducted on its strength between 200 MPa and 355 MPa. The TK columns experience an average increase of 27.8% in ultimate load, while the TN columns experience an average increase of 40.8% when the steel strength is increased from 200 MPa to 355 MPa.

Hollow steel sections are manufactured in different thicknesses. Both the inner and outer tube thicknesses were individually investigated between

DOI: 10.1201/9781003348450-175

thicknesses of 3 mm and 6 mm. It was observed that increasing the outer tube thickness leads to a corresponding increase in the ultimate load of approximately 15% per millimetre for the short (S) columns and approximately 17% per millimetre for the long (L) columns. An average increase of approximately 3% per millimetre was observed in the TK columns' response as opposed to an average increase of approximately 8% per millimetre in the TN columns' response. Thus, it is more beneficial to increase the outer tube thickness as opposed to the inner tube thickness. Figure 1 portrays a novel observation not previously observed, where a column with different cross sections, i.e. TN and TK, but with the same length (S or L), the initial strength of the TK models have a greater ultimate load at 3 mm compared to the TN models, however the TN models ultimate load increases beyond the TK models as the inner tube thickness increases.

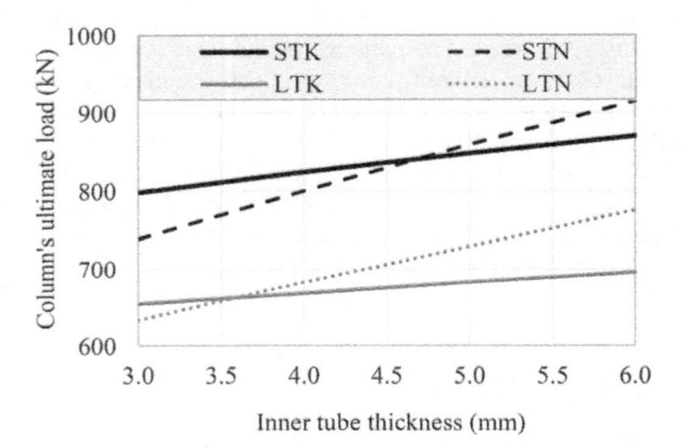

Figure 1. The effect of the inner tube thickness on the ultimate load of the CFDST columns (Usongo, 2021).

Steel tubular sections are not manufactured perfectly straight. This measured imperfection, produced lateral midspan deflections of approximately 2 mm for the 2.5 m outer steel tube. Various codes of practice provide guidelines to deal with this imperfection. The investigation revealed that all the columns exhibit an average decrease of 3.5% when the lateral curvature at midspan increases from 0 mm to 5 mm.

In a design environment, it is assumed that supports are completely rigid or a perfect hinge, which only exist in an ideal environment. The investigation shows that the ultimate load of the rigidly supported columns increase by an average of 85.1% compared with the hinge supported columns. The L columns experiences an average increase of 96.1% compared with the S columns, which only experiences an average increase of 74.2% when the support conditions are changed from hinged to rigid.

It is commonly known that loads are not applied concentrically on the cross section and that some form of eccentricity is inherent in the applied load.

A sensitivity analyses was therefore conducted to investigate the column's behaviour to different load eccentricity magnitudes, as presented in Figure 2.

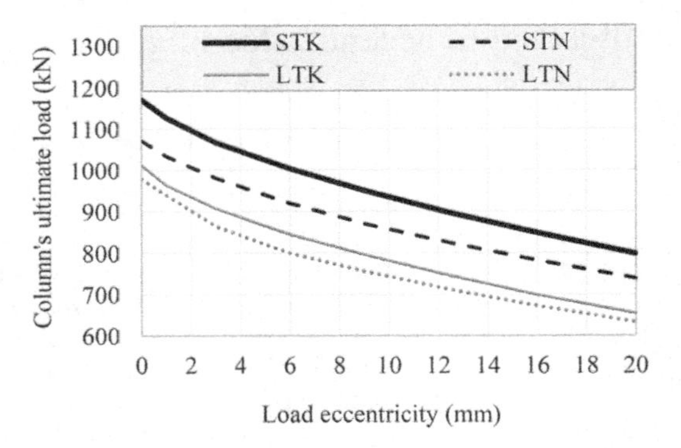

Figure 2. The effect of the load eccentricity on the ultimate load of the CFDST columns (Usongo, 2021).

From Figure 2, it was observed that the ultimate load vs load eccentricity is non-linear for load eccentricities between 0 mm and 9 mm, whereafter the response tends to be linear. The S columns experience an average load reduction of 31.4%, while the L columns experiences an average force reduction of 35.4%.

3 CONCLUSIONS

The investigation revealed that a change in the concrete and steel strength has a moderate effect on the ultimate load of slender circular CFDST columns subjected to eccentric loading. A change in the inner and outer tube thicknesses as well as an increase in the load eccentricity has a significant effect on the ultimate load. The longitudinal curvature of the column has the least effect on the ultimate load, while the column's support conditions have the greatest effect on the ultimate load. These findings are similar to what was observed for concentric loading.

REFERENCES

Koen, J.A. 2015. *An investigation into the axial capacity of eccentrically loaded concrete filled double skin tube columns*. MSc Thesis, Stellenbosch University.

Usongo, A.A. 2021. *Finite Element Modelling of Eccentrically Loaded Concrete filled Double Skin Tube Columns*. MSc Thesis, Stellenbosch University, Stellenbosch.

Haas, T.N., Usongo, A.A. & Koen, J.A. 2022. Calibration and Validation of Circular CFDST Columns Subjected to Eccentric Loading. '*Unpublished results*'.

Current Perspectives and New Directions in Mechanics, Modelling and Design of Structural Systems – Zingoni (ed.)
© 2022 Copyright the Author(s), ISBN 978-1-032-39114-4

Evaluating the effect of load eccentricity on slender circular CFDST columns

J.P. Rust & T.N. Haas
Stellenbosch University, Stellenbosch, Western Cape, South Africa

ABSTRACT: Reinforced concrete and structural steel columns are the most common types of columns used in industry. Recently, there has been development in other forms of columns such as Concrete-Filled Double Skin Tube (CFDST) columns to improve ease of construction, efficiency of structural members and stability during seismic events. The objective of this study was to investigate the effect of load eccentricity on slender circular CFDST columns through the development of a finite element (FE) model to predict column behaviour, investigating the effect of the hollow section ratio and the slenderness ratio as a column performance indicator.

1 INTRODUCTION

Concrete-Filled Double Skin Tube (CFDST) columns consist of two concentric steel tubes with its annulus filled with concrete. The steel tubes can be any combination of square, rectangular or circular hollow sections. The CFDST columns are an improvement of the traditional Concrete-Filled Skin Tube (CFST) columns as it is lighter, has an increased bending stiffness and it is expected to have higher fire resistance (Tao et al. 2004). These columns have high bearing resistance compared to CFST columns, which can be attributed to the double concrete confinement effects of the inner and outer steel tubes (Ci et al. 2021).

Based on previous research work reviewed, it is evident that significant investigation were conducted on CFDST columns. Majority of the work however studied either concentrically loaded or short/stub columns, with minimal work focusing on eccentrically loaded slender CFDST columns. The inclusion of eccentricity in column design is clearly indicated in the codes of practice to account for initial curvature of the column and imperfection of the load on the column cross-section, reducing the ultimate load capacity of compression members.

In this research, the effects of load eccentricity was investigated on slender circular CFDST columns in order to quantify the influence of parameters such as the slenderness ratio and hollow section ratio. An eccentrically loaded CFDST FE model was developed based on the work by Usongo (2021) and calibrated and validated using experimental tests conducted by Koen (2015).

2 METHODOLOGY

2.1 *Column parameters*

The slenderness ratio (λ) of a compression member is the ratio of the effective length to the corresponding radius of gyration (r). The slenderness ratio of structural compression members is determined using Equation 1.

$$\lambda = \frac{L}{r} \tag{1}$$

The hollow section ratio (χ), which provides information pertaining the cavity within the inner section in relation to the full section, is calculated as the ratio between the outer diameter of the inner steel tube (d) and the inner diameter of the outer steel tube ($D - 2t_o$) as expressed in Equation 2, with outer steel tube wall thickness (t_o).

$$\chi = \frac{d}{D - 2t_o} > \tag{2}$$

2.2 *Material properties*

The steel used in the experimental tests had a yield strength, f_{sy} = 300 MPa, and an ultimate strength, f_{su} = 450 MPa. A Young's modulus of steel, E_s = 200 GPa produced a yield strain of ε_{sy} = 0.0015.

The unconfined compressive cylinder strength of concrete was determined as f_c = 34.4 MPa. To simulate the effect of concrete confinement, Richart et al (1928) and Mander et al (1988) recommended that

DOI: 10.1201/9781003348450-176

the compressive strength of concrete be determined using Equation 3.

$$f_{cc} = f_c + k_1 f_{rp} \tag{3}$$

where $k_1 = 4.1$. The expressions for the lateral pressure (f_{rp}) induced by the steel tubes proposed by Hu & Su (2011) was evaluated by Liang (2018), which recommended the use of Equation 4.

$$f_{rp} = 8.525 - 0.166\left(\frac{D}{t_o}\right) - 0.00897\left(\frac{d}{t_i}\right) + 0.00125\left(\frac{D}{t_o}\right)^2 + 0.00246\left(\frac{D}{t_o}\right)\left(\frac{d}{t_i}\right) - 0.0055\left(\frac{d}{t_i}\right)^2 \tag{4}$$

3 ANALYSIS AND RESULTS

3.1 *Effect of load eccentricity on ultimate column capacity*

A sensitivity analysis was conducted to investigate the effect of load eccentricity on the ultimate load capacity of the columns. It was observed that an increase in column length resulted in a reduced ultimate column capacity for CFDST columns with the same cross-section. Also, an increase in the load eccentricity resulted in a decrease in the ultimate load capacity of the columns.

3.2 *Effect of eccentricity and hollow section ratio*

It was observed that an increase in hollow section ratio resulted in a decrease in ultimate column capacity when comparing slender circular CFDST columns with similar outer tube sections. Furthermore, the effect became progressively insignificant as column length increased. It can therefore be concluded, based on the results, that the hollow section ratio cannot be used in isolation as a column behavioural indicator.

3.3 *Effect of eccentricity and slenderness ratio*

For columns with smaller hollow section ratios, the rate of ultimate column capacity reduction increased significantly as slenderness ratio increased as depicted by solid and dashed lines as shown in Figure 1. The reduction gradient decreased as the columns approached the slenderness limit, transitioning from intermediate slender to long slender columns. It was also observed that a maximum reduction value was reached for longer columns. Columns with larger hollow section ratios were more adversely affected by an increase in slenderness ratio, producing higher rates of reduction in ultimate column capacities.

The effect of slenderness ratio was further investigated by varying the column cross-sections, with constant slenderness ratios, represented by crosses in Figure 1. The slenderness ratio yielded a smaller variance in the rate of reduction. It can therefore be concluded that the slenderness ratio is a better indicator of circular slender CFDST column performance compared to hollow section ratio.

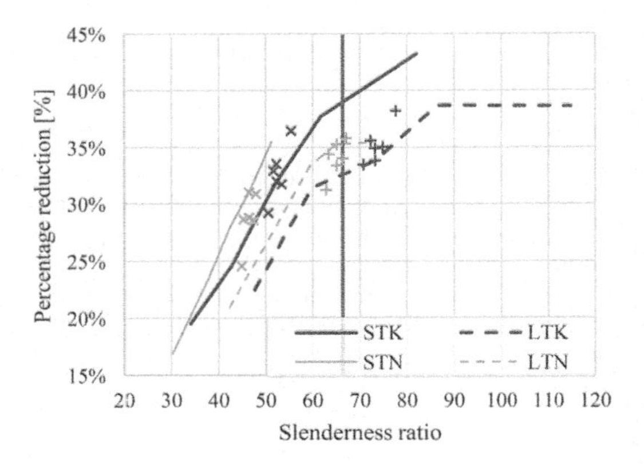

Figure 1. Rate of ultimate column capacity reduction versus slenderness ratio.

4 CONCLUSIONS

Based on literature reviewed, it is evident that research investigating the behaviour of circular slender CFDST columns subjected to eccentric loading are scarce. In this research, FE model was developed to predict the behaviour of circular slender CFDST columns subjected to eccentric loading to investigate the effect of the column hollow section ratio and the slenderness ratio as a column performance indicator. It was concluded, based on the results, that the hollow section ratio cannot be used as a column behaviour indicator in isolation. The slenderness ratio performed better as a column behavioural indicator as opposed to the hollow section ratio.

REFERENCES

Ci, J., Ahmed, M., Jia, H., Chen, S., Zhou, D., & Hou, L. (2021). Testing and strength prediction of eccentrically-loaded circular concrete-filled double steel tubular stub-columns. *Journal of Constructional Steel Research,* 186.

Liang, Q. Q. (2018). Numerical simulation of high strength circular double-skin concrete-filled steel tubular slender columns. *Engineering Structures, 168*, 205–217.

Mander, J. B., Priestley, M. J., & Park, R. (1988). Theoretical stress-strain model for confined concrete. *Journal for Structural Engineering, 114*(8), 1804–1826.

Tao, Z., Han, L., & Zhao, X. (2004). Behaviour of concrete-filled double skin (CHS inner and CHS outer) steel tubular stub columns and beam columns. *Journal of Construction Steel Research, 60*, 1129–1158.

Usongo, A. (2021). *Finite Element Modelling of Eccentrically Loaded Concrete Filled Double Skin Tube Columns.* Stellenbosch University, Civil Engineering. Stellenbosch University

Current Perspectives and New Directions in Mechanics, Modelling and Design of Structural Systems – Zingoni (ed.)
© 2022 Copyright the Author(s), ISBN 978-1-032-39114-4

Shear capacity of the Truedek steel decking system in the formwork

N. Singh, O. Mirza, M. Hosseini, A. Wheeler & D. Allan
School of Engineering, Design and Built Environment, Western Sydney University, Second Avenue, Kingswood, NSW, Australia

ABSTRACT: This paper investigates the vertical shear capacity design of the Truedek steel formwork system. A need for a long span steel decking system to provide the primary structural support and eradicating any secondary support such as frame members to allow retention of the floor space has prompted to expand on experimental testing of the long span steel decking system with various robust design systems. A shear capacity minimises the shear load force and resist a sliding failure of the material along the plane that is parallel to the direction of the shear force. To confirm the viability of vertical shear strength study has been undertaken to provide an adequate data to improve the shear capacity design to increase the durability and shear strength of the system. Shear at end support test were undertaken by varying bearing lengths and different truss depths. Data obtained from the experiment explained the relationship between vertical shear capacity and various bearing lengths with different truss depths. Corroboration of the experimental results attained determines a design guideline relating to the performance characteristics, whereby conforming to a design process in eliminating shear stress and enhancing vertical shear capacity design of the Truedek steel decking system.

1 INTRODUCTION

The Australian construction industry has been going through technological developments for the past fifteen years. In this study vertical shear capacity for innovative Truedek design system was investigated using the experimental tests in accordance to the Australian Standards (AS2327) and the data from the laboratory experiment was analysed and as a result the maximum shear capacity in relation to the required span and bearing length was evaluated.

2 EXPERIMENTAL STUDY

Table 1 outlines the specimen details on the test plan of 2 specimens tested. Load will be applied onto the base plate/web, not directly to the top of steel web profile to avoid any structural deformity of the Truedek design system and evade any ambiguous data results.

Table 1. Specimen details for shear at end support.

Truedek Design System					
Depth Y (mm)	Number of Specimen	Specimen Name	Web Thickness X (mm)	Span Length (mm)	Loading
160	2	ST-H160-T0.7-1	0.7	2000	Static
		ST-H160-T0.7-2			Cyclic

3 RESULTS AND DISCUSSION

Two specimens average slip behaviour was recorded, Specimen ST-H160-T0.7-1 and ST-H160-T0.7-2. as shown in Figures 1 and 2 respectively. At around 12kN it can be seen there is a drop in the load due to sudden buckling of the web both in the North and South side of the specimen. Shear failure contributed to the web crippling of the specimen. For specimen ST-H160-T0.7-2 the ultimate load carrying capacity failure was ascertained at 16.2kN with an average slip of 11.2mm for both the horizontal and vertical shear stress. The difference between the two specimens was that the specimen ST-H160-T0.7-2 went through cyclic loading between 0.65kN and 5.2kN of the ultimate load, then it was finally loaded to failure, with a sudden drop in the load from 16.2kN to 11.8kN where in the North and South sides of the specimen web buckling was observed. Both the Specimen ST-H160-T0.7-1 and ST-H160-T-0.7-2 experienced hogging bending moment in the mid span of the specimen shown in Figures 3 and 4, where the compressive forces act from the bottom and the tensile forces from the top. For specimen ST-H160-T0.7-2 the hogging bending moment was more than the specimen ST-H160-T0.7-1 due to the more observable bending arc.

DOI: 10.1201/9781003348450-177

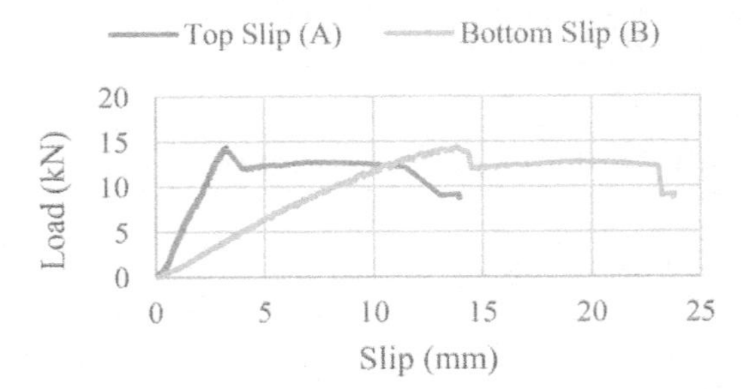

Figure 1. Load versus slip relationship of Specimen ST-H160-T0.7-1.

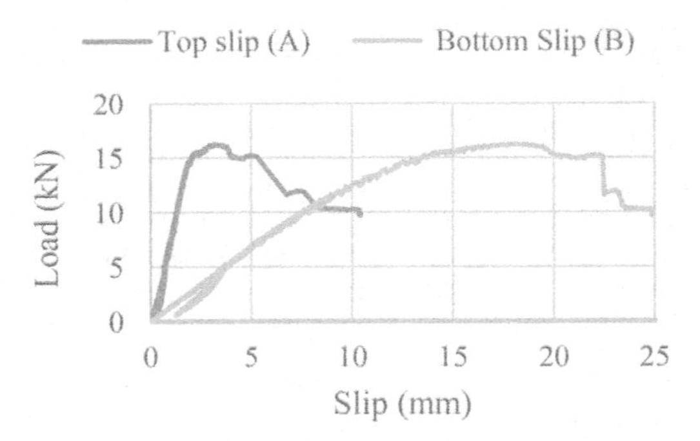

Figure 2. Load versus slip relationship of Specimen ST-H160-T0.7-2.

Figure 3. Bending of specimen ST-H160-T0.7-1.

Figure 4. Bending of specimen ST-H160-T0.7-2.

Figures 5 and 6 shows the shear load graphs supporting the variation in bending moments of the two specimen.

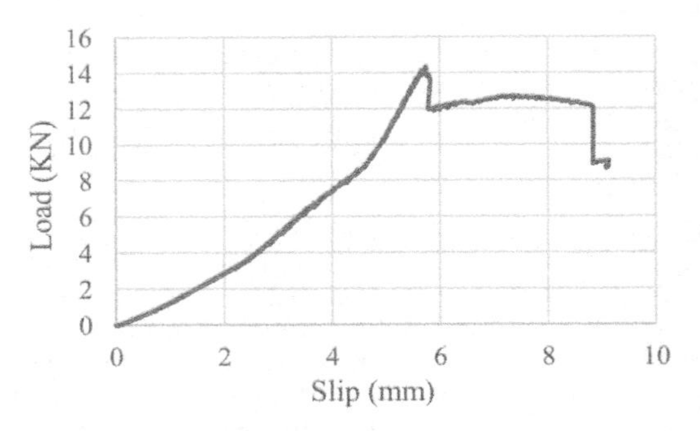

Figure 5. Mid span slip for ST-H160-T0.7-1.

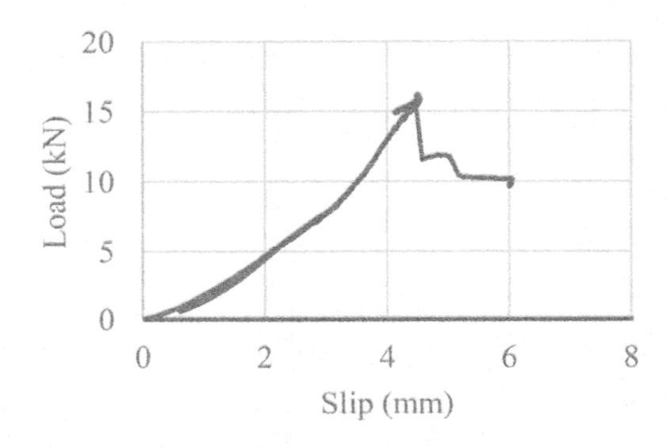

Figure 6. Mid span slip for ST-H160-T0.7-2.

4 CONCLUSION

The following conclusions can be conveyed from the experiment conducted:

- The maximum slip was detected at mid span of the specimen after the ultimate load carrying capacity had been reached.
- A buckling failure experienced by shear failure mechanism can lead to a flexural failure if the load is to be increased.
- A bending arc was observed in the middle span of the specimen due to bending moment.
- Vertical shear was larger than the horizontal shear.

Current Perspectives and New Directions in Mechanics, Modelling and Design of Structural Systems – Zingoni (ed.)
© 2022 Copyright the Author(s), ISBN 978-1-032-39114-4

Numerical modeling for connections of staggered single channel rafters in double-bay cold-formed steel portal frames

J. Kafuko & M. Dundu
Department of Civil Engineering Science, University of Johannesburg, Auckland Park, South Africa

ABSTRACT: Previous work on double-bay cold-formed steel portal frames has shown that it is possible to have internal eaves connection with staggered single channel cold-formed rafters. The rafters and the column are bolted back-to-back at the eaves joint so as to counterbalance the forces in the connected elements. Experiments have shown that sufficient ductility to generate enough plastic rotation at the joint is not achieved. To understand the elastic and inelastic behaviour of this connection, a numerical model is introduced in this study. This numerical model, which has been tested against previous experimental data, gives good agreement with regards to mode and location of the failure, moment-rotation and moment-curvature response of the tested frame. This model facilitates a better understanding of the behaviour of this connection configuration, which will probably result in a wider application of cold-formed steel portal frames.

1 INTRODUCTION

Much of early work on portal frames, constructed from Cold-formed steel (CFS) channels, ensured that the eaves connection was rigid. In later years, there has been a shift to semi-rigid portal frames, because rigid joints are more expensive to fabricate. Most of this work has been on single-bay portal frames.

Tshuma & Dundu (2017) proposed semi-rigid bolted connections for multi-bay CFS portal frames with staggered rafters connected to the column. To extend understanding of the behaviour of this connection, a finite element (FE) model is introduced in this study. This model was created by isolating a portion of the two frames from the eaves joint to points of contraflexure in the rafter and column, respectively, as shown in Figure 1 (Tshuma 2017). Table 1(Tshuma 2017) shows the variables in the frames, with e, f_{yc} and E representing the lever arm, yield strength and elastic modulus of the channels, respectively.

Table 1. Variables in the test frames.

Frame (ECT)	Size (mm)	Channel f_{yc} (MPa)	E (GPa)	Bolts No.	e (m)
1.1	300x75x20x3.0	240.828	207	8, M20	1.08
1.2	300x65x20x3.0	228.666	206	8, M20	1.08
1.3	300x50x20x3.0	255.153	208	8, M20	1.08

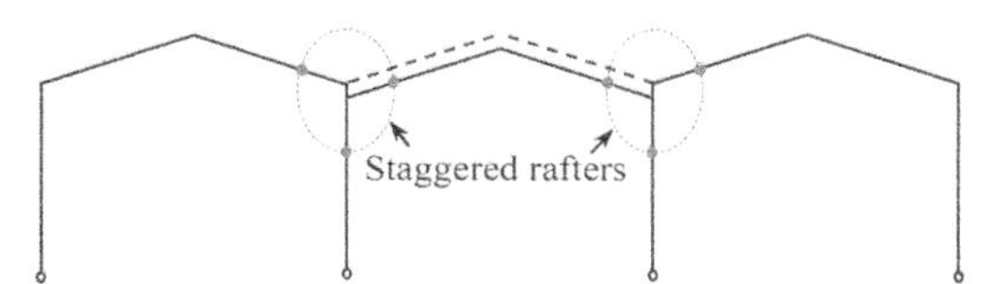

Figure 1. Staggered eaves connection.

Abaqus (2014) was used for the FE model with CFS sections created using 4-noded shell elements (S4R). Element sizes of 2.5 mm x 2.5 mm for high stress zones, around bolt holes, while element sizes of about 30 mm x 30 mm were selected for low stress zones. The bolts were modeled using the 8-noded 3D solid elements (C3D8R) with sizes of 10 mm x 10 mm.

Bilinear stress strain curves were used in the FE model and were derived using the Ramberg-Osgood constitutive model. Initial Geometric Imperfections (GI) were accounted for in the FE model and due to the absence of measured values from experimental data, the cumulative distribution function (CDF), a conservative statistical approach proposed by Schafer & Pekoz (1998) was adopted. To be consistent with the physical model (Tshuma 2017), the base of the column was fixed while the ends of the rafters were pinned. A global general contact interaction was adopted with a penalty behaviour for the tangential direction with a friction coefficient of 0.25.

DOI: 10.1201/9781003348450-178

2 NUMERICAL RESULTS

The local deformation pattern of a typical staggered eaves connection (ECT-1.1), predicted by the FE analysis is shown in Figures 2 and 3. Two failure mechanisms are evident in the FE model, viz; bearing distortion of bolt holes, and local buckling of compression zone of the flange and webs of the channels. The local buckling deformation is located in the compression zone of the flange and webs of the channels of the rafters and column, just outside the connection, due to the large bending moment in the section. The stress concentrations are evident in Figures 2 and 3, around bolt holes and just outside the connection in both rafters and the column. These are the locations in the test frame where the channel sections yielded, aiding in the ductile behaviour of the joint. The bolts did not fail in the frame and the stresses were well below their tensile strength.

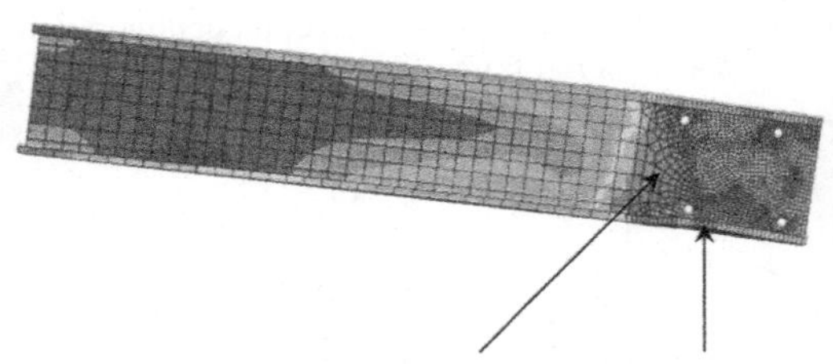

Figure 2. Local deformation in the rafter.

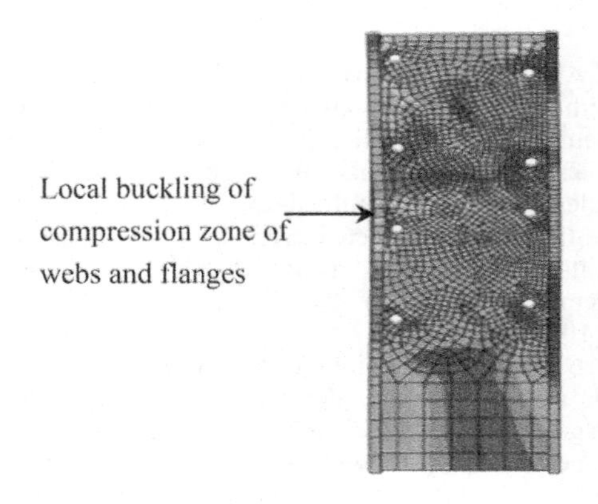

Figure 3. Local deformation in the column.

As shown in Tables 2 and 3, FE model was able to predict for the 3 frames, the maximum moment, rotation, stiffness and curvature within acceptable range. It can also be seen that the change in the flange width of the channel section and yield strength, did not significantly affect the rotation in the connection. However, the sections with larger flanges showed a marginal increase in the stiffness of the connection. The failure modes and the maximum moment were similar and local buckling was the final failure mode at same locations. This limited the moment strength and ductility for the joint proving that the connections were not critical.

Table 2. Average rotation and stiffness results.

	Max. Moment (KNm)		Rotation (rad)		Stiffness (KNm/rad)	
Frame	M_{exp}	M_{fem}	ϕ_{exp}	ϕ_{fem}	K_{exp}	K_{fem}
ECT-1.1	21.80	21.38	0.033	0.030	660.61	712.67
ECT-1.2	20.57	20.20	0.032	0.032	642.81	631.25
ECT-1.3	22.70	20.18	0.033	0.032	687.88	630.63

Table 3. Average curvature results.

	Max. Moment (KNm)		Curvature (1/mm) 10^{-6}	
Frame	M_{exp}	M_{fem}	κ_{exp}	κ_{fem}
ECT-1.1	21.80	21.38	10.41	9.57
ECT-1.2	20.57	20.20	7.16	9.06
ECT-1.3	22.70	20.18	5.28	8.60

3 CONCLUSIONS

The FE model, based on the connection configuration proposed for double-bay portal frames (Tshuma 2017) has been developed using Abaqus (2014), and calibrated against experimental results. Similarly to the experimental investigation, two failure mechanisms were realized using the FE model, viz; bearing distortion of bolt holes, and local buckling of compression zone of the flange and webs of the channels, both in the rafters and the column just outside the connection. Bearing distortion of the bolt holes aided the ductile behaviour of the joint, while local buckling was promoted by the stress concentrations in the inside bolt. It is also shown that the change in the flange width of the channel section and yield strength of the channels, did not significantly affect the rotation in the connection. The FE model was able to predict the moment-rotation and moment-curvature behaviour for this connection configuration. The model as calibrated, is part of ongoing research on CFS moment connections.

REFERENCES

ABAQUS. 2014. Analysis User's Guide, Version 6.14. Dassault Systèmes Simulia USA.

Schafer, B. W. and Pekoz, T. 1998. Computational modelling of cold-formed steel: characterizing geometric imperfections and residual stresses. Journal of Construction Steel Research, Vol. 47: 193–210.

Tshuma, B. and Dundu, M. 2017. Thin-Walled Structures Internal eaves connections of double-bay cold-formed steel portal frames. Thin-Walled Structures,119:760–769.

Current Perspectives and New Directions in Mechanics, Modelling and Design of Structural Systems – Zingoni (ed.)
© 2022 Copyright the Author(s), *ISBN 978-1-032-39114-4*

Numerical modeling of gusseted internal eaves connection of double-bay cold-formed steel portal frames using single channels

J. Kafuko & M. Dundu
Department of Civil Engineering Science, University of Johannesburg, Auckland Park, South Africa

ABSTRACT: It has been demonstrated from previous studies that it is possible to have the internal eaves connection of cold-formed steel rafters, which are connected to the column through a hot-rolled steel gusset plate. This connection configuration eliminates the possibility of unbalanced column moments at the joint, thus allowing for the column to be designed for axial loads only. A numerical model for this connection configuration has been developed in this study and this has been tested against experimental data. The model gives good agreement with regards to the mode and location of the failure, moment-rotation and moment-curvature response of the tested frames. The numerical model provides an opportunity for better understanding of the behaviour of this connection configuration, which could facilitate further development of connecting systems for internal eaves of double-bay cold-formed steel portal frames.

1 INTRODUCTION

Advances have been made over the years in the design of cold-formed steel (CFS) single-bay portal frames including connecting systems. However, there is limited work on internal eaves of CFS double-bay portal frames whose behavior is different to that of the outer eaves connection.

Tshuma & Dundu (2017) proposed a gusseted semi-rigid bolted connection for the internal eaves joints of double-bay CFS single channel portal frames. In the current study, a finite element (FE) model is introduced in order to simulate the behavior of this connection configuration and extend the understanding. The model is based on a portion of two adjoining frames isolated from the eaves internal joint to the adjacent points of contraflexure in the column and the rafters as shown in Figure 1 (Tshuma & Dundu 2017). The variables in the frames are presented in Table 1 with the respective strength of the materials (Tshuma & Dundu 2017), where E, f_y and t_g represent the elastic modulus, yield strength and the thickness of the mild steel gusset plates respectively.

The model was created in Abaqus (2014), adopting 4-noded shell elements (S4R) for the CFS sections and the gusset plates while the bolts were modeled using the 8-noded 3D solid elements (C3D8R). Element sizes of 10 mm x 10 mm were used throughout except in the low stress regions of the rafters and columns which were 20 mm x 20 mm.

The Ramberg-Osgood model was used to generate bilinear stress strain curves which were used in the FE model. Initial Geometric Imperfections (GI) were also included in the FE model and were estimated using the cumulative distribution function proposed by Schafer & Pekoz (1998). To conform with the test model, the column base was fixed while ends of the rafters were pinned. A friction coefficient of 0.25 was used on all contacts in the model.

Table 1. Variables in the test frames.

Frame (ECT)	Section size	Channel f_y (MPa)	E (GPa)	Bolts No.	Gusset Plate t_g (mm)
2.1	300x75x20x3.0	240.828	207	12, M20	6
2.2	300x65x20x3.0	228.666	206	12, M20	6
2.3	300x50x20x3.0	255.153	208	12, M20	6
2.4	300x75x20x3.0	240.828	207	12, M20	8
2.5	300x65x20x3.0	228.666	206	12, M20	8
2.6	300x50x20x3.0	255.153	208	12, M20	8
6 mm	gusset plate	342.754	201		6
8 mm	gusset plate	351.865	200		8

2 NUMERICAL RESULTS

Local buckling in this connection configuration was the final failure mode which occurred after significant rotation of the rafters. Local buckling initiated in both rafters immediately outside of the rafter-to-gusset plate connection. This was more evident in the bottom compression flanges of both rafters. The failure pattern of this gusseted rafter-to-column connection was consistent with other studies (Tshuma & Dundu 2017). In general, two failure mechanisms are evident, namely;

DOI: 10.1201/9781003348450-179

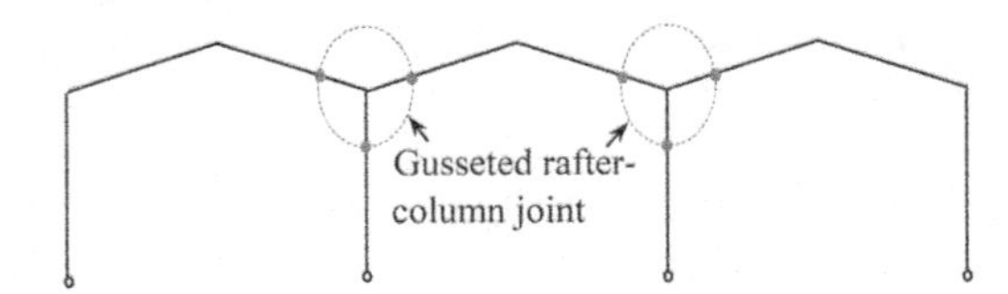

Figure 1. Gusseted rafter-column joints.

bolt hole bearing distortion, and local buckling in the rafters just outside of the rafter-to-gusset plate connection. The column resisted only an axial compression force, as the moments from the rafters balanced at the eaves joint and this was evident by the low stresses in the column as shown in Figure 2. None of the bolts in the joint failed as the stresses were well below their shear strength.

As shown in Tables 2 and 3, the FE model predicted the ultimate moments, rotation, stiffness and curvature with good agreement.

The connections comprising of the 6 mm gusset plates exhibited more ductility than those with the 8 mm gusset plate. These differences are attributed to the material properties and thicknesses of the gusset plate used. However, the curvatures were qualitatively similar in all the frames, largely due to the CFS channel sections failing within the linear-elastic range. The moment strength and ductility for the joints was limited by the local buckling failure experienced in all the frames.

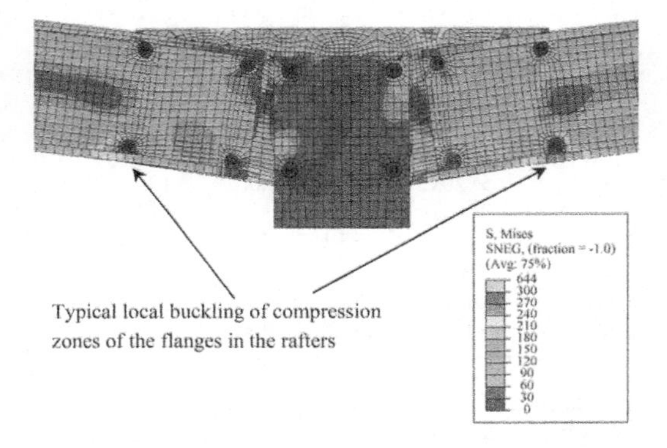

Figure 2. Local buckling deformation of the eaves connection.

Table 2. Summary of average rotation and stiffness results.

Frame	Max. Moment (kNm) M_{exp}	M_{fem}	Rotation (rad) ϕ_{exp}	ϕ_{fem}	Stiffness (kNm/rad) K_{exp}	K_{fem}
ECT-2.1	21.42	21.41	0.016	0.018	1338.75	1189.44
ECT-2.2	21.23	21.23	0.018	0.018	1179.44	1179.44
ECT-2.3	19.51	19.51	0.015	0.017	1300.67	1147.65
ECT-2.4	24.83	22.34	0.012	0.014	2069.17	1595.71
ECT-2.5	23.05	23.04	0.013	0.015	1773.08	1536.00
ECT-2.6	21.63	20.19	0.009	0.011	2403.33	1835.45

Table 3. Summary of average curvature results.

Frame	Max. Moment (kNm) M_{exp}	M_{fem}	Curvature (1/mm) 10^{-6} κ_{exp}	κ_{fem}
ECT-2.1	21.42	21.41	6.35	4.68
ECT-2.2	21.23	21.23	6.26	7.26
ECT-2.3	19.51	19.51	6.18	6.89
ECT-2.4	24.83	22.34	7.67	4.78
ECT-2.5	23.05	23.04	6.52	7.50
ECT-2.6	21.63	20.19	6.33	7.44

3 CONCLUSIONS

A numerical model, based on the connection configuration proposed for double-bay portal frames by Tshuma & Dundu (2017) has been developed and calibrated against test results. Two failure modes were realized using the FE model, and was consistent with tests viz; local buckling in both rafters just outside the connection and bearing distortion of bolt holes. Local buckling was promoted by the stress concentrations around the lower bolts while the bearing distortion of the bolt holes aided the ductile behaviour of the joint. Results also show that the connections with the 6 mm gusset are more ductile than connections with the 8 mm gusset plate. These differences are attributed to the material properties and thicknesses of the gusset plate used. The numerical model predicted the moment-rotation and moment-curvature behaviour for this connection type with good agreement. The calibrated model forms part of ongoing research being conducted on CFS moment connections.

REFERENCES

ABAQUS. 2014. Analysis User's Guide, Version 6.14. Dassault Systèmes Simulia USA.

Schafer, B. W. and Pekoz, T. 1998. Computational modelling of cold-formed steel: characterizing geometric imperfections and residual stresses, J. Constr. Steel Res. Vol. 47:193–210.

Tshuma, B. and Dundu, M. 2017. Internal eaves connections of double-bay cold-formed steel portal frames. Thin Walled Structures,119:760–769.

Current Perspectives and New Directions in Mechanics, Modelling and Design of Structural Systems – Zingoni (ed.)
© 2022 Copyright the Author(s), ISBN 978-1-032-39114-4

Stub concrete-filled double-skin circular tubes in compression

R. Mahlangu & M. Dundu
Department of Civil Engineering Science, University of Johannesburg, Auckland Park, South Africa

ABSTRACT: A total of 24 concrete-filled double skin tubular (CFDST) stub columns were tested to failure to understand their behaviour. The diameter-to-thickness ratio of the outer tubes ranged from 47 to 54, and two concrete grades (normal-to-moderate strength concrete) of self-compacting concrete were used. Failure modes, axial deformations and ductility of the specimens are all investigated, and the results of the experimental investigation revealed that there is some confinement of the concrete core in CFDST columns, and the columns showed significant ductility. The predicted strengths showed great correlation with the experimental strengths.

1 INTRODUCTION

Concrete-filled double skin tubes (CFDSTs) have benefits similar to CFSTs, in that they are economical and quicker to construct compared to conventional reinforced concrete columns since the steel tubes act as formwork (Essopjee & Dundu 2015). The use of an inner hollow section has been found to reduce structural weight, increase bending stiffness, ductility and seismic performance of the CFDST column compared to CFSTs (Tao et al. 2004). Apart from benefiting from the high compressive strength of concrete and the high tensile strength and ductility of the steel in the surrounding skin, CFDSTs have the added advantage in that the inside tubes can be used as downpipes or other services such as electrical wiring, in multi-storey buildings. Ultimately, this improves the aesthetic appearance of the structures.

2 EXPERIMENTAL PROGRAMME

To establish the material properties of steel, tensile tests of coupons were conducted in a 100 kN capacity displacement controlled Instron machine, according to the requirements provided in the British Standard, BS EN ISO 6892-1. The average 0.2% proof stress (f_y), yield strain (ε_y), ultimate strain (ε_u), Young's modulus (E) and the ultimate tensile stress (f_u) are given in Table 1.

Two concrete mixes were designed, and for each mix design, six concrete cubes of 100x100x100 mm size were prepared and cured for 28 days. The average maximum load and compressive strength are given in

Table 2. The modulus of elasticity of the concrete was determined using Table 1 of SANS 10100-1.

Table 1. Average steel properties.

Tubes (mm)	f_y (MPa)	ε_y (%)	f_u (MPa)	ε_u (%)	E_s (GPa)
219.1-1x3.5	417.64	3.88	494.55	15.34	201.99
219.1-2x3.5	435.32	4.71	486.64	16.23	205.64
193.7x3	395.85	5.8	463.19	20.71	202.38
165.1x3	381.85	5.20	439.00	17.47	198.28
152.4x3	460.64	4.39	516.87	12.85	205.54
139.7x3	423.71	4.12	503.43	17.92	195.69
127.0x2	270.13	3.84	322.65	20.69	200.74
76.2-1x2	480.72	5.61	491.86	15.18	201.25
76.2-2x3	324.98	4.38	371.73	10.38	196.59

Table 2. Average concrete properties.

Grade	M (g)	ρ (kg/m^3)	P (kN)	f_{cu} (MPa)	E (GPa)
M30	2430	2429.5	304.2	30.4	28.12
M50	2505	2505.3	530.1	53.0	34.6

A total of 36 concentrically loaded CFDST stub columns were prepared and tested. To ensure that accurate results were obtained, two identical specimens of each CFDST column were fabricated. Two grades of concrete were used in order to study the effect of increasing concrete strength on the behaviour of the columns. A total of eight strain gauges were mounted at mid-height of the inner and outer steel tubes to monitor the longitudinal and transverse strains during loading.

DOI: 10.1201/9781003348450-180

3 EXPERIMENTAL RESULTS AND ANALYSIS

Most of the outer steel tubes failed by outward local buckling from the top to the middle portion of the specimens, while the inner steel tubes experienced inward local buckling. The buckling of the steel tubes was accompanied by crushing of the sandwiched concrete in the buckled regions. The ultimate loads, N_{exp}, obtained from the experimental results and the predicted strengths, N_u, are shown in Table 3. The axial capacities of the CFDST columns were estimated by simply adding together the individual strengths of the steel tubes and the sandwiched concrete.

Table 3. Experimental and predicted capacities.

Outside Tube	Concrete Grade	L_a (mm)	N_u (kN)	N_{exp} (kN)	N_{exp} $/N_u$
127.0x2.0	30.4	382	579.6	498.1	0.86
127.0x2.0	30.4	382	579.6	458.0	0.79
139.7x3.0	30.4	381	974.3	1098	1.13
139.7x3.0	30.4	382	974.3	1117	1.15
152.4x3.0	30.4	420	1172.5	1195.6	1.02
152.4x3.0	30.4	419	1172.5	1188.5	1.01
165.1x3.0	30.4	419	1218.1	1460.0	1.20
165.1x3.0	30.4	418	1218.1	1475.9	1.21
193.7x3.0	30.4	457	1759.6	2045.2	1.16
193.7x3.0	30.4	458	1759.6	2020.5	1.15
219.1x3.5	30.4	458	2545.3	2986.7	1.17
219.1x3.5	30.4	459	2569.0	3012.5	1.17
127.0x2.0	53.0	495	745.0	665.3	0.89
127.0x2.0	53.0	495	745.0	678.6	0.91
139.7x3.0	53.0	496	1188.5	1199.2	1.01
139.7x3.0	53.0	496	1188.5	1194.4	1.00
152.4x3.0	53.0	581	1449.8	1469.8	1.01
152.4x3.0	53.0	581	1449.8	1462.9	1.01
165.1x3.0	53.0	582	1564.3	1731.4	1.11
165.1x3.0	53.0	581	1564.3	1785.4	1.14
193.7x3.0	53.0	658	2201.0	2693.3	1.22
193.7x3.0	53.0	657	2201.0	2657.3	1.21
219.1x3.5	53.0	656	3156.8	3694.7	1.17
219.1x3.5	53.0	657	3133.1	3628.3	1.16

As depicted in Table 3, the experimental and estimated axial capacities are very close, except for the smallest specimen of 127 mm diameter. This can be attributed to improper compaction of the sandwiched concrete during casting. In general, the ratio N_{exp}/N_u increased as the ratio of the outside-to-inside diameter (d_o/d_i) also increased. The largest ratio of $N_{exp}/N_u = 1.22$ suggests that there is confinement of the sandwiched concrete in CFDST stub columns. Increasing the strength of the concrete resulted in an increase in the axial capacity of the CFDST columns.

The load-deformation curves for specimens with normal and moderate strength concrete are shown in Figures 1 and 2, respectively, which shows that M30 is more ductile than M60.

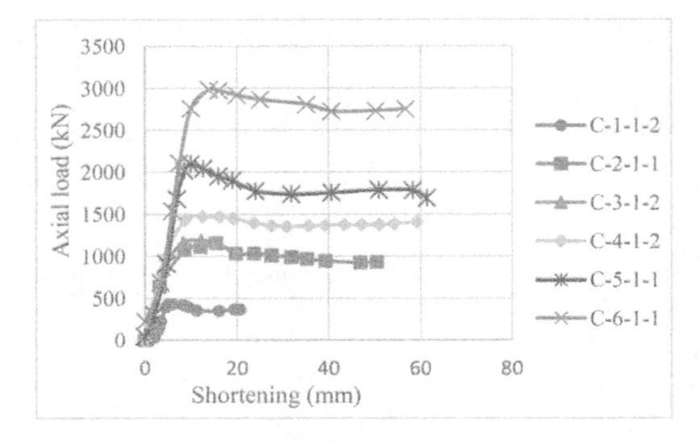

Figure 1. Specimens with M30 concrete.

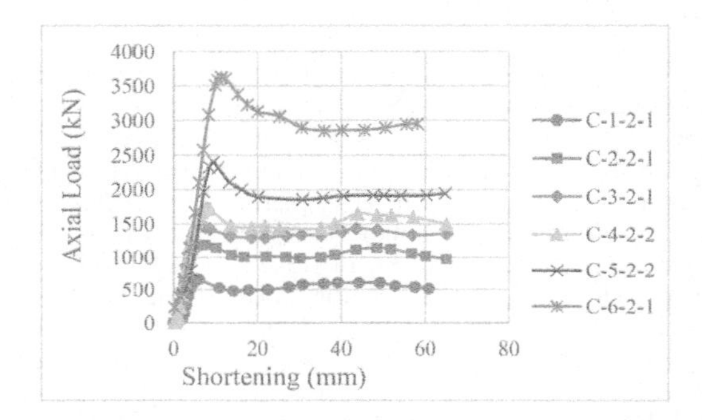

Figure 2. Specimens with M60 concrete.

4 CONCLUSIONS

In most specimens, the outer steel tubes failed by outward local buckling from the top to the middle portion of the specimens, while the inner steel tubes experienced inward local buckling. The buckling of the steel tubes was accompanied by crushing of the sandwiched concrete in the buckled regions. The largest ratio of $N_{exp}/N_u = 1.22$, suggest that there is some confinement of the sandwiched concrete in CFDST stub columns. The CFDST stub columns exhibited a ductile behaviour, however, M30 was more ductile than M60.

REFERENCES

Essopjee, Y. & Dundu, M. 2015. Perfomance of concrete-filled double-skin circular tubes in compression. *Composite Structures* 133:1276–1283.

Tao, Z., Han, L.-H. & Zhao, X.-L. 2004. Behaviour of concrete-filled double skin (CHS inner and CHS outer) steel tubular stub columns and beam-columns. *Journal of Constructional Steel Research* 60: 1129–1158.

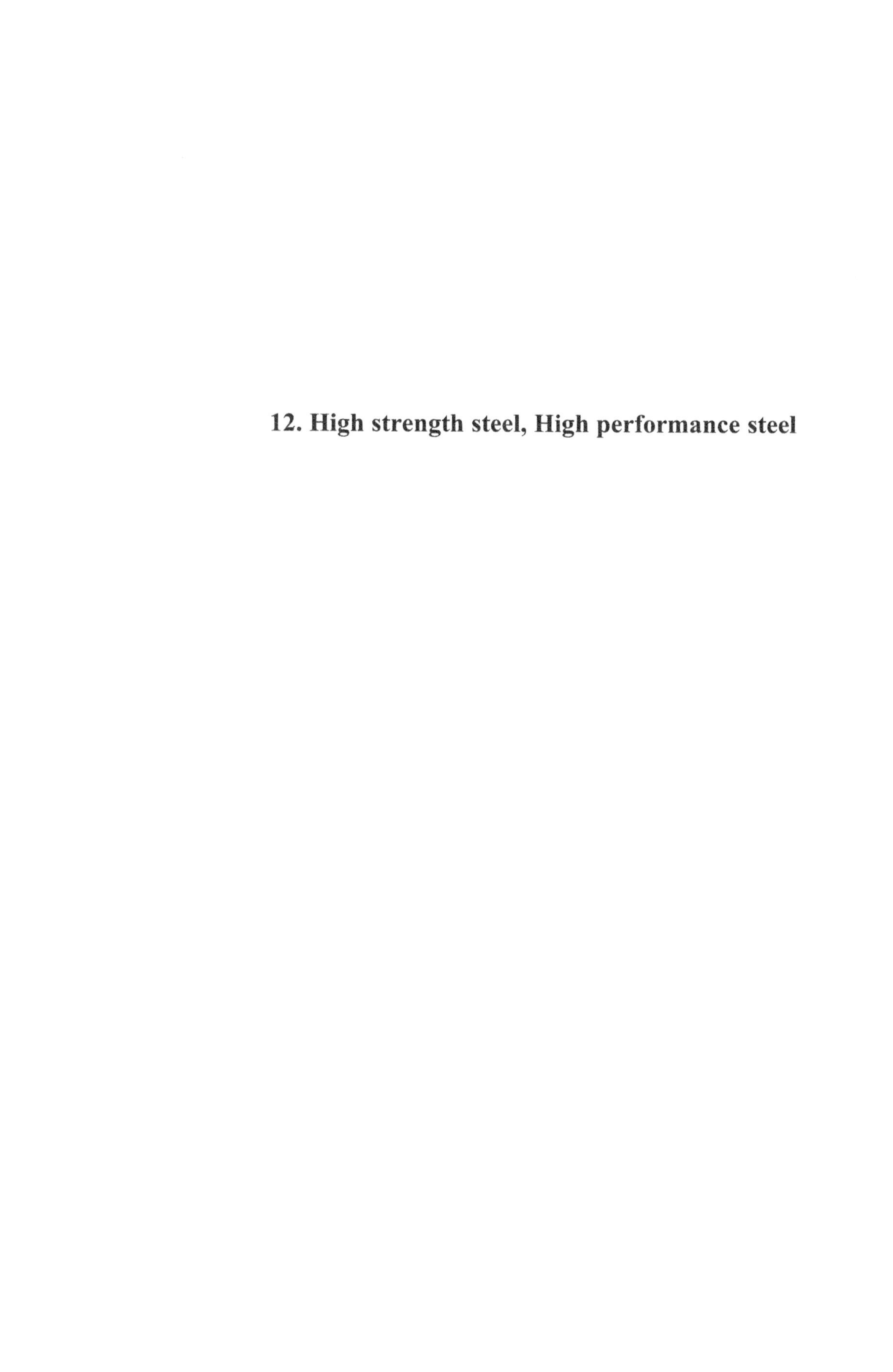

12. High strength steel, High performance steel

Current Perspectives and New Directions in Mechanics, Modelling and Design of Structural Systems – Zingoni (ed.)
© 2022 Copyright the Author(s), ISBN 978-1-032-39114-4

Welded connections of high-strength steels

R. Stroetmann, T. Kästner & B. Rust
Chair of Steel Construction, Technische Universität Dresden, Germany

ABSTRACT: With increasing strength, micro-structural changes in the area of weld seams are becoming more and more important. This is due to the fact that these can negatively influence the load-bearing behavior of welded connections. Particularly in the case of high-strength fine-grain steels with low carbon content, where the strength properties are specifically adjusted by a combination of alloying concepts and heat treatment and, if necessary, rolling at defined temperatures, the importance of the so-called heat-affected zone (HAZ) is increasing. The article gives an overview of current manufacturing processes for high-strength fine-grained structural steels and summarizes current normative and manufacturer's specifications for the welding processing of these. The influence of the cooling time on the mechanical properties of the HAZ and the weld metal is discussed and possible failure modes of welded connections with softening in the HAZ are described. Finally, the future European design rules for welded connections and newly developed models are presented.

1 INTRODUCTION

The use of high-strength steels is becoming increasingly important in the gfield of steel construction. The reasons for this are material savings and weight reductions in structures. Furthermore, the availability of steels and technical regulations for design, construction and execution are drivers of this development. As the strength of steels increases, it becomes more difficult to produce welded connections whose mechanical and technological properties correspond to those of the base material. In steel production the alloy concept, the rolling process and the temperature control are closely coordinated and controlled to optimize the material properties. However, in the production of welded connections, it is the choice of welding process, the filler metals and the temperature control that are important for the connection properties. For a deeper understanding, the following article discusses the production of high-strength steels, the properties of welds and the heat-affected zone as well as the design, failure modes and dimensioning of welded connections.

2 HIGH-STRENGTH STEELS

The specific combination of chemical composition and technical heat treatment, if necessary, in combination with rolling processes, enables the formation of fine-grained microstructures with high strength and toughness properties. For the producer of high-strength steels, quenching with subsequent tempering and thermomechanical rolling are of particular importance. Figure 1 shows schematically the different production processes.

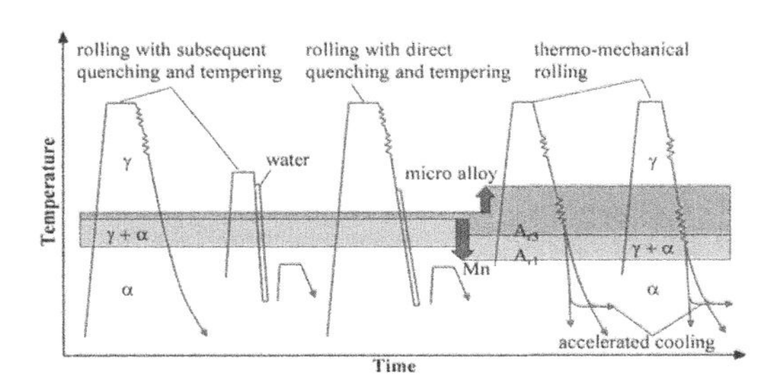

Figure 1. Schematic representation of the manufacturing process of high-strength steels.

3 EXECUTION RULES FOR WELDING CONNECTIONS

The basic principles for the execution of steel structures in Europe are regulated in EN 1090-2. The standard is currently being revised with regard to the use of high-strength steels with yield strengths up to 960 MPa (cf. EN 1090-2/prA1:2021-02). To avoid negative influences on the mechanical properties of the heat-affected zone (HAZ) of welded connections, the cooling times recommended by the steel manufacturers must be followed for steels above S460. An evaluation of manufacturers recommendations for specifying the cooling time $t_{8/5}$ in (Kästner 2022) has shown that this is limited to 20 seconds, and in some cases to 15 seconds, for steels with yield strengths up to 700 MPa. For higher yield strengths, shorter cooling times are generally recommended. In EN 1011-2, cooling times of 10 seconds to 25 seconds are recommended for welding high-strength fine-grained steels. The use of other cooling times is also possible if suitability is verified by welding procedure qualification

DOI: 10.1201/9781003348450-181

tests or a welding test before fabrication. DVS specification 0916:2012, applicable in Germany, recommends a $t_{8/5}$-time interval of 5 seconds to 15 seconds.

4 INFLUENCE OF MANUFACTURING PARAMETERS ON THE PROPERTIES OF WELDED CONNECTIONS

The influence of the manufacturing parameters on the mechanical-technological properties of welded connections of high-strength steels was investigated as part of the AiF-FOSTA research project P1020 in the development of a new design model (Stroetmann 2018, Stroetmann 2021). Thereby, the influence of the peak temperature and the cooling time $t_{8/5}$ on the mechanical strength properties was examined. Figure 2 shows an exemplary representation of the normalized tensile strengths of the steels S700MC and S690QL in the considered cooling time interval of 1.5 seconds to 25 seconds, for the peak temperatures of 1350 °C to 600 °C. The normalized tensile strength is based on the tensile strength of the thermally unaffected base material. For the steel S700MC, softening was observed for all peak temperatures in the cooling time interval considered. In contrast, the S690QL steel exhibits no softening for all peak temperatures up to a cooling time of 15 seconds. For longer cooling times, a decrease in tensile strength below the level of the base material of maximum 5 % is observed at a peak temperature of 800 ° C. The softening of the S700MC steel is in the range of 3 % to 24 %. In addition, the mechanical properties of the weld metal were determined experimentally.

5 FAILURE MODES OF WELDED CONNECTIONS

Even taking into account the manufacturers recommendations on the choice of cooling time, softening in the HAZ region cannot be completely avoided (Figure 2). Therefore, in welded connections of high-strength steels, failure in the HAZ region and mixed forms are possible in addition to weld metal (WM) failure or base material (BM) failure. Figure 3 shows possible failure modes of different welded connections. The course of the fracture lines is significant for the load-bearing resistance of the connection. Different stress states and load-bearing resistances are present. The latter result from the resistances of the zones involved and a possible supporting effect of adjacent zones with higher strength. Within the framework of the AiF-FOSTA research project P1453, the load-bearing behavior of single-sided welded T-connections with local softening in the area of the HAZ is currently being systematically investigated. Within the scope of the research project, recommendations for the welding of one-sided seams and verification methods for HAZ and mixed failure are developed (cf. (Stroetmann 2022)). As there are currently no design regulations in Europe for a failure of the HAZ.

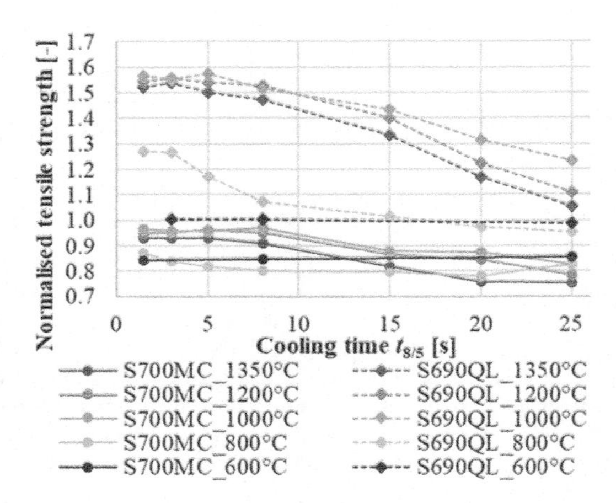

Figure 2. Normalized tensile strengths of S690QL and S700MC steels as a function of peak temperature and cooling time $t_{8/5}$.

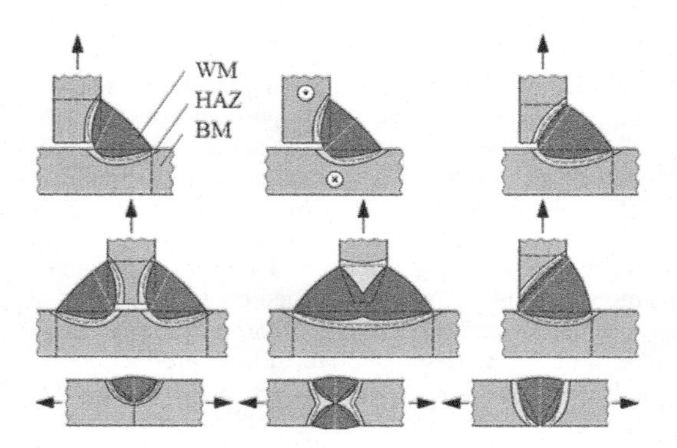

Figure 3. Possible failure modes for welded connections.

REFERENCES

DVS Merkblatt 0916 2012. Metall-Schutzgasschweißen von Feinkornbaustählen, DVS - Deutscher Verband für Schweißen und verwandte Verfahren e. V., Düsseldorf (in German)

EN 1090-2/prA1 2021–02. Execution of steel structures and aluminium structures - Part 2: Technical requirements for steel structures. Amendment A1.

EN 1011–2 2001–05. Welding - Recommendation for welding of metallic materials – Part 2: Arc welding of ferritic steels.

Kästner, T. 2022. Zum Einfluss der Fertigungsparameter auf die mechanischen Eigenschaften von Schweißverbindungen höherfester Stähle. Dissertation, Technische Universität Dresden.

Stroetmann, R.; Kästner, T.; Hälsig, A.; Mayr, P. 2018. Influence of the cooling time on the mechanical properties of welded HSS-connections. *Steel Construction* 11, No. 4. https://doi.org/10.1002/stco.201800019

Stroetmann R.; Kästner, T. 2021. A new design model for welded connections. *Steel Construction* 14, No. 3. https://doi.org/10.1002/stco.202100014

Stroetmann, R.; Kästner, T.; Rust, B.; Schmidt, J. 2022. Welded Connections at high-strength steel hollow section connections. *Steel Construction* 15, No. 2.

Current Perspectives and New Directions in Mechanics, Modelling and Design of Structural Systems – Zingoni (ed.)
© 2022 Copyright the Author(s), *ISBN 978-1-032-39114-4*

A 3D constitutive model of high-strength constructional steels with ductile fracture

Guo-Qiang Li & Yan-Bo Wang
College of Civil Engineering, Tongji University, Shanghai, China

Yuan-Zuo Wang
The Key Laboratory of Urban Security and Disaster Engineering of Ministry of Education, Beijing University of Technology, Beijing, China

Le-Tian Hai
Department of Civil Engineering, Tsinghua University, Beijing, PR China

Wen-Yu Cai
College of Civil Engineering and Architecture, Hainan University, Haikou, China

ABSTRACT: In order to describe and predict cyclic behavior of HSS steels under triaxial stress states, a new cyclic constitutive model with consideration of damage accumulation and influence of triaxial stress states based on ductile fracture model is formulated to simulate the response of HSS steels subjected to a range of loading types in terms of triaxial stress state and loading histories. The model, termed as HSS-3D, is supported by an experimental program consisting of coupon scale monotonic and cyclic tests. The HSS-3D model proposed is able to simulate: (1) the Lode angle dependence through an enhanced yield function which modifies the conventional von Mises yield surface; (2) the deterioration of the elastic region at high accumulated strains through new isotropic softening relationship; (3) the influence of stress states on strength softening behavior of HSS steels and (4) ductile fracture of steels under cyclic loading.

1 INTRODUCTION

According to previous studies by authors [1, 2], it has been demonstrated that the influence of triaxiality should be considered in the yield criterion and ductile fracture model of HSS steels. The next section describes an experimental program developed specifically to accurately characterize key aspects of the cyclic response of HSS steels commonly used in China. These include coupon scale experiments on four types of specimens with different geometries that provide large plastic strains as well as various triaxial stress states. Insights from these experiments are subsequently used to develop a constitutive model that overcomes the limitations of the current models.

2 COUPON SCALE EXPERIMENTS

These HSS steels including Q550, Q690, and Q890 HSS steel plates were produced by Nanjing iron and steel Co., Ltd. To investigate ultra-low cyclic characteristics of HSS steels under various triaxial stress states, monotonic and cyclic tests on specimens with various geometries are designed. The geometry and dimensions of these specimens are shown in **Figure 1** and **Table 1**.

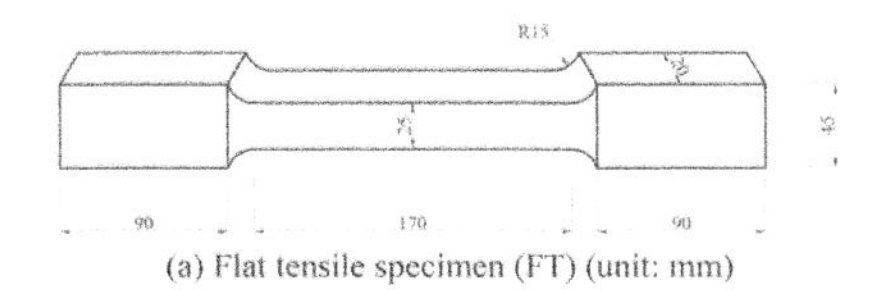

(a) Flat tensile specimen (FT) (unit: mm)

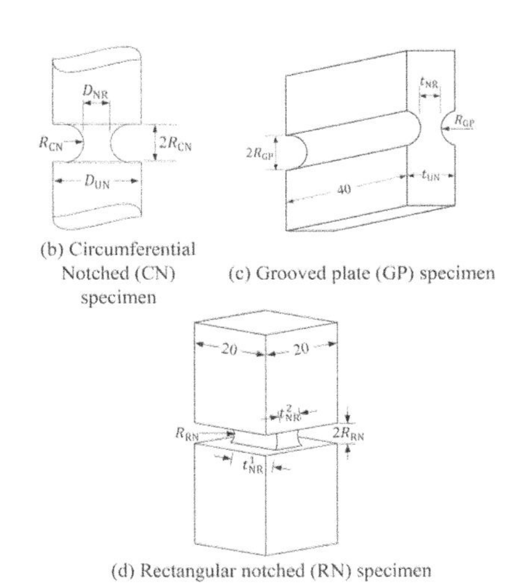

(b) Circumferential Notched (CN) specimen

(c) Grooved plate (GP) specimen

(d) Rectangular notched (RN) specimen

Figure 1. Specimen geometry and dimensions.

3 A 3D CONSTITUTIVE MODEL OF HIGH-STRENGTH CONSTRUCTIONAL STEELS WITH DUCTILE FRACTURE

Based on the observations in [2], a new yield function that takes into account the Lode angle effect is used to describe the plastic behavior of HSS steels, and it is defined as:

DOI: 10.1201/9781003348450-182

Table 1. Dimensions of specimens (unit: mm).

ID	R_{CN}	D_{NR}	D_{UN}
CN-1	1.45	5.20	12.70
CN-1	2.00	6.00	12.70
	R_{GP}	t_{NR}	t_{UN}
GP-1	1.60	2.50	20.00
GP-2	3.20	2.50	20.00
	R_{RN}	t^1_{NR}	t^2_{NR}
RN-1	2.50	15.00	5.00
RN-2	2.50	15.00	10.00

$$F = f(\boldsymbol{\sigma} - \boldsymbol{\alpha}) - \boldsymbol{\sigma}^0 = 0 \tag{1}$$

where $f(\boldsymbol{\sigma} - \boldsymbol{\alpha})$ is the equivalent stress tensor $\boldsymbol{\sigma}$ concerning back-stress tensor $\boldsymbol{\alpha}$, and $\boldsymbol{\sigma}^0$ is the yield stress representing the size of the yield surface given by

$$\begin{aligned} \sigma^0 &= (\sigma_0 + \sigma_{iso}) \cdot g(\theta) \\ g(\theta) &= \left[C_{\frac{\pi}{6}} + \left(C_\theta - C_{\frac{\pi}{6}}\right) \cdot \left|\theta - \frac{\pi}{6}\right| / \frac{\pi}{6}\right] \end{aligned} \tag{2}$$

$$C_\theta = \begin{cases} 1 & \text{for} \quad 0 \le \theta \le \frac{\pi}{6} \\ C_{\frac{\pi}{3}} & \text{for} \quad \frac{\pi}{6} < \theta \le \frac{\pi}{3} \end{cases} \tag{3}$$

where σ_0 is the initial yield stress with zero equivalent plastic strain and the isotropic part σ_{iso} describes expansion or shrinkage of the yield surface.

The isotropic hardening σ_{iso} is expressed as shown below in Eq. (4).

$$\frac{d\sigma_{iso}}{d\bar{\varepsilon}^p} = \begin{cases} k_1 \cdot e^{-b \cdot \bar{\varepsilon}^p} & , \varepsilon^p_{yp} < \bar{\varepsilon}^p \le \bar{\varepsilon}^p_{1,2} \\ k_2 \cdot \dot{D}(\eta, \bar{\theta}) & , \bar{\varepsilon}^p_{1,2} < \bar{\varepsilon}^p \le \bar{\varepsilon}^p_{2,3} \\ k_3 \cdot \dot{D}(\eta, \bar{\theta}) & , \bar{\varepsilon}^p_{2,3} < \bar{\varepsilon}^p \end{cases} \tag{4}$$

where k_i (i=1,2,3) are three material-dependent parameters which have been identified in the previous study [2]. And b is a material-dependent parameter controlling the rate of isotropic softening at the first stage.

Bai-Wierzbicki fracture model (BWM), is selected and calibrated for HSS steels. The formula of this model is given by Eq. (5).

$$\bar{\varepsilon}_f(\eta, \bar{\theta}) = \left[\bar{\varepsilon}^t_f - \bar{\varepsilon}^s_f\right] \cdot \bar{\theta}^2 + \bar{\varepsilon}^s_f \tag{5}$$

where $\bar{\varepsilon}^t_f$ and $\bar{\varepsilon}^s_f$ are functions of loci of two limiting cases corresponding to axial symmetry tension ($\bar{\theta} = 1$) and plane strain ($\bar{\theta} = 0$). For the functions of these two bounds, the influence of stress triaxiality η is considered through the exponent function, given by

$$\bar{\varepsilon}^t_f = F_1 e^{-F_2 \eta} \tag{6}$$

$$\bar{\varepsilon}^s_f = F_3 e^{-F_4 \eta} \tag{7}$$

where F_i (i=1,2,3,4) are four material-dependent parameters. The definition of damage variable D to describe the evolution of the loss of fracture ductility is defined as

$$D(\eta, \bar{\theta}) = \int_0^{\bar{\varepsilon}_f} \dot{D}(\eta, \bar{\theta}) d\bar{\varepsilon} = \int_0^{\bar{\varepsilon}_f} \frac{1}{\bar{\varepsilon}_f(\eta, \bar{\theta})} d\bar{\varepsilon} \tag{8}$$

In Eq. (8), the damage rate index $\dot{D}(\eta, \bar{\theta})$ defining the damage accumulation rate is used to describe the influences of stress states on the strength deterioration of HSS steels under cyclic loading in Eq. (4).

Furthermore, a stress-weighted fracture model is proposed.

$$e^{-\kappa \bar{\varepsilon}_c} \le \sum_{\text{tension}} \int_0^{\bar{\varepsilon}_f} \frac{d\bar{\varepsilon}_t}{\bar{\varepsilon}_f(\eta, \bar{\theta})} - \sum_{\text{compression}} \int_0^{\bar{\varepsilon}_f} \frac{\beta d\bar{\varepsilon}_c}{\bar{\varepsilon}_f(\eta, \bar{\theta})} \tag{9}$$

where $\sum_{\text{tension}} \int_0^{\bar{\varepsilon}_f} \frac{d\bar{\varepsilon}_t}{\bar{\varepsilon}_f(\eta, \bar{\theta})}$ and $\sum_{\text{compression}} \int_0^{\bar{\varepsilon}_f} \frac{\beta d\bar{\varepsilon}_c}{\bar{\varepsilon}_f(\eta, \bar{\theta})}$ represent the decreasing and recovering of the fracture ductility of steels, respectively. $e^{-\kappa \bar{\varepsilon}_c}$ represents the degradation of ductile capacity of steels. β governs the relative rate of decreasing and recovering of ductility during tension loading and compression loading, respectively. κ controls the rate of stress-independent capacity degradation during compressive loading.

4 VALIDATION OF HSS-3D CONSTITUTIVE MODEL

To verify the new damage constitutive model, results of cyclic tests on three types of specimens including CN, GP, and RN in section 3 are used. **Figure** illustrates errors in the peak force determined for different specimens for each cycle.

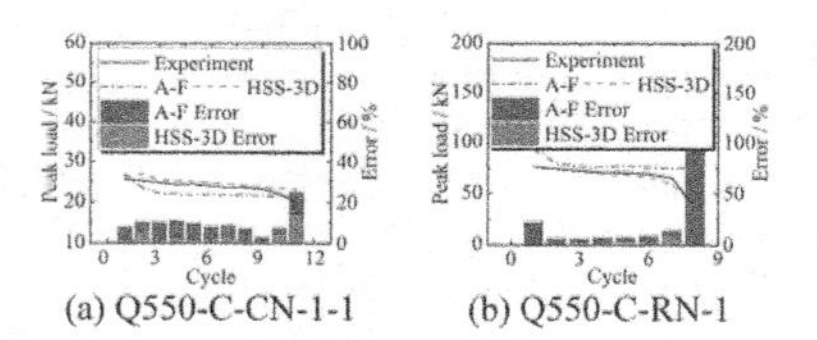

Figure 2. Evolution curves of peak load.

5 CONCLUSION

The HSS-3D model can simulate: (1) the Lode angle dependence through an enhanced yield function which modifies the conventional von Mises yield surface; (2) the deterioration of the elastic region at high accumulated strains through a new isotropic softening relationship; (3) the influence of stress states on strength softening behavior of HSS steels and (4) ductile fracture of steels under cyclic loading.

REFERENCES

[1] Wang Y-Z, Li G-Q, Wang Y-B, Lyu Y-F, Li H. Ductile fracture of high strength steel under multi-axial loading [J]. Engineering Structures, 2020, 210: 110401.

[2] Wang Y-B, Lyu Y-F, Wang Y-Z, LI G-Q, Richard Liew J Y. A reexamination of high strength steel yield criterion [J]. Construction and Building Materials, 2020, 230: 116945.

Current Perspectives and New Directions in Mechanics, Modelling and Design of Structural Systems – Zingoni (ed.)
© 2022 Copyright the Author(s), ISBN 978-1-032-39114-4

Investigation of the effects of an over-elastic preload on the load-bearing behavior of high-strength bolt and nut assemblies

J. Reinheimer & J. Lange
Institute for Steel Construction and Materials Mechanics, TU Darmstadt, Germany

ABSTRACT: In the decisive specifications, for preloaded bolts a subsequent control of the assembly is specified. Among others, preloaded bolts have to be checked for overtightening. If a bolt is classified as overtightened, this can result in the entire bolt group having to be replaced. In addition to the high cost of replacement, this often means that further work has to be temporarily halted. However, there is a lack of knowledge about the degree of utilization of the bolt defined as overtightened. The results of first experimental investigations as well as numerical analyses suggest that over-elastic preload forces only have a significant influence on the load bearing behaviour when exceeding a certain point. The assessment of preloaded bolts with respect to overtightening as given in specifications seems to be too general – and for the bolt lengths so far investigated too conservative.

1 INTRODUCTION

The economic efficiency and sustainability of construction measures are not only determined by the building costs themselves, but also by subsequent improvements and running expenses. This also includes the costs for the replacement of bolts that were not installed in accordance with the requirements given in the respective specifications. In addition to the tightening methods and the desired level of preloading, the inspection of the built-in assemblies is specified. Among others, preloaded bolts are to be checked for overtightening. However, there is a lack of scientific knowledge about the degree of utilization of the bolt defined as overstressed, the influence on the microstructure of the material and the load-bearing behaviour. Therefore, this research project has the aim of gaining in-depth knowledge about the influence of an over-elastic preloading on bolted connections. The effects on the material structure as well as the load-bearing behaviour of high-strength bolts shall be characterised. Therefore, experimental as well as numerical investigations on single bolted connections are conducted. In the future, with the knowledge gained, a limit value of preloading is to be determined for over-elastic preloading.

2 EXPERIMENTAL INVESTIGATIONS

In order to gain an insight into the influence of an over-elastic preload on the load-bearing behaviour of high-strength bolts, experimental tests were carried out. So far, bolts M16 – 10.9 (f_u = 1000 N/mm²) with two different clamping lengths of 68 mm (test series V2) and 88 mm (test series V3) were investigated.

The bolts were tightened referring to the combined method in two steps. In a first step, an initial torque as specified in EN 1090-2 was applied to the bolts using a torque wrench. In a second step, the bolts were tightened to the desired level of preload, which was controlled by using an ultrasonic measurement system. In doing so, the ultrasonic time-of-flight (derived from the maximum achievable change of time-of-flight) acted as a control parameter for the preload and different over-elastic preloading levels could be distinguished.

For both clamping lengths, a clear downward trend with increasing over-elastic prestressing could be recognized. While the load-bearing capacities in test series V3 were all above the minimum tensile strength $F_{m,min}$ acc. ISO 898-1 (European Committee for Standardization 2013), both tests with the highest applied preload in test series V2 fell below this limit. The hypothesis was supported that the effect of exceeding the permissible further rotation angle is more serious on bolts with a shorter clamping length - which is mechanically conclusive, since the resulting strain increases with decreasing initial length for the same applied elongation.

3 NUMERICAL INVESTIGATIONS

In order to extend the experimental results to further connection geometries in future (i.e., bolt diameter

DOI: 10.1201/9781003348450-183

and clamping length), a first approach has been made to numerically simulate a part of the experiments. So far, test series V2 with a clamping length of 68 mm including the realistic geometry of the clamping package has been modelled. For the bolts, a trilinear material model was derived on the basis of tensile tests, while the compliance of the thread pairing was taken into account by an auxiliary body with experimentally derived properties (Figure 1).

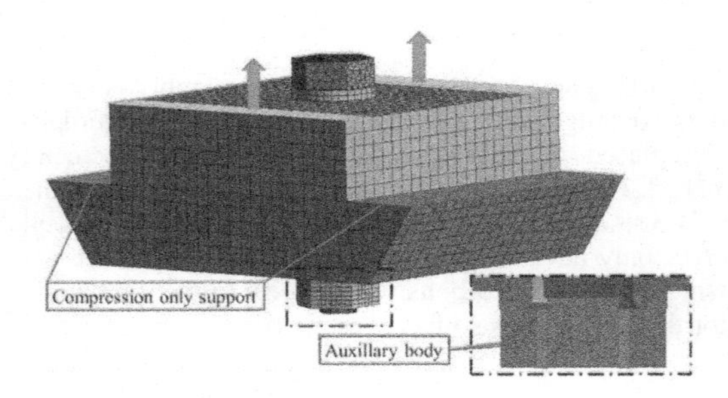

Figure 1. Finite element model.

The preload level was modelled by converting the total applied angle of rotation into a vertical extension of the bolt by using the thread pitch P acc. DIN 13-1 (Deutsches Institut für Normung e.V. 1999).

In total, the numerical analysis consisted of 3 consecutive steps:

1. Simulation of the preload by applying a vertical displacement to the bolt
2. Activating the bonded contact between the auxiliary body and the bolt

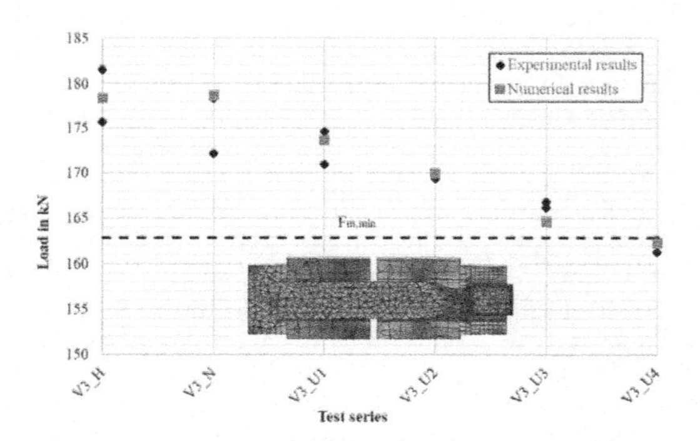

Figure 2. Numerical and experimental results for test series V2.

3. Application of the external load in form of an increasing vertical displacement on the upper plate

Figure 2 shows the comparison of the experimental load-bearing capacities for test series V2 with the results of the numerical simulation. Overall, the model provides good approximations for the load-bearing capacity of over-elastically preloaded bolted joints. The decreasing trend of the ultimate loads is adequately represented by the numerical results.

4 CONCLUSIONS AND OUTLOOK

The experimental results for single bolted connections suggest that increasing over-elastic preloading reduces the load-bearing capacity of the bolts when exceeding a certain degree. However, the minimum tensile strength $F_{m,min}$ was reached or exceeded in all test series except one. Furthermore, it is noticeable that the effect of the same degree of overtightening on bolts of shorter clamping length seems to be more distinct. The results of the numerical simulation show a good agreement with the empirical data, the decreasing trend of the load bearing capacity could be well reproduced.

Future investigation will include additional tests on both, the geometric boundary conditions already presented here as well as on other clamping lengths and bolt diameters. Among others, based on additional experimental investigations, the finite element model will be refined by taking torsional stresses in the bolt resulting of the tightening process into account.

Eventually, with the knowledge gained, a limit value of preloading is to be determined for unplanned over-elastic preloading up to which the load-bearing capacity of the bolts is not decisively affected. The criteria for the classification of an overtightened bolt connection found in specifications will be reviewed.

REFERENCES

Deutsches Institut für Normung e.V. DIN 13–1 1999, *Metrisches ISO-Gewinde allgemeiner Anwendung - Teil 1: Nennmaße für Regelgewinde; Gewinde-Nenndurchmesser von 1 mm bis 68 mm.*

European Committee for Standardization EN ISO 898–1 2013, *Mechanical properties of fasteners made of carbon steel and alloy steel - Part 1: Bolts, screws and studs with specified property classes - Coarse thread and fine pitch thread.*

Current Perspectives and New Directions in Mechanics, Modelling and Design of Structural Systems – Zingoni (ed.)
© 2022 Copyright the Author(s), ISBN 978-1-032-39114-4

Bending effects on the fatigue strength of non-load-carrying transverse attachment joints made of ultra-high-strength steel

A. Ahola & T. Björk
Laboratory of Steel Structures, Lappeenranta-Lahti University of Technology LUT, Lappeenranta, Finland

ABSTRACT: In the present study, the fatigue strength of fillet-welded Non-Load-Carrying Transverse(NLCT) attachment joints made of Ultra-High-Strength Steel (UHSS) under bending loading is experimentally and numerically investigated. Experimental fatigue tests are carried out on UHSS joints in the as-welded condition under four-point bending loading at the applied stress ratios of $R = 0.1$ and $R = 0.5$, together with the relevant residual stress and local weld geometry measurements to identify the key factors influencing fatigue performance. Numerical analyses are carried out to investigate the notch effects employing effective notch stress concept. Compared to the fatigue test data obtained for the similar NLCT joints tested under the axial loading, the bending-loaded joints showed lower fatigue strength capacity and thus, negative bending effect. The numerical analyses revealed that bending causes higher notch stress concentrations than axial loading in the single-sided welded details, and the use of notch stress concepts commeasured the axially-loaded and bending-loaded test results into a single *S-N* curve with relatively low scatter. The tested specimens were small-scale specimens with relatively low residual stresses and consequently, the *R* ratio had a clear effect on the fatigue capacity, that could be considered by employing the 4R method with the mean stress correction in the effective notch stress.

1 INTRODUCTION

Fatigue is a substantial failure mechanism in steel components subjected to the cyclic loads. In welded joints and specifically in those made of high-strength or ultra-high-strength steel (HSS/UHSS), the fatigue strength is amongst the most important design criteria. Experimental verification for the *S-N* curves of the different detail categories in the nominal stress concept, included in design codes and standards, has been principally carried out using axial fatigue testing. Consequently, in many fatigue design standards, the prevailing concept is that stress gradient in the nominal stress at the plate component is neglected. The aim of present study is to investigate bending effect on the fatigue strength in welded UHSS joints. Experimental fatigue tests are carried out on UHSS joints subjected to the load-controlled constant amplitude and uniaxial four-point bending (DOB = 1), and the fatigue test data is compared to the fatigue test data of similar joints tested under the axial load conditions.

2 MATERIALS AND METHODS

The bending fatigue tests are carried out on a S1100 quenched and tempered UHSS grade, similar to the previous axial tests (Ahola et al., 2019). The mechanical properties of the studies materials are presented in Table 1. A total number of 10 non-load-carrying transverse (NLCT) attachment joints were prepared with fillet welds for the bending tests. The shape and dimensions of the test specimens are presented in Figure 1.

Table 1. Mechanical properties of the studied materials. f_y and f_u are the yield and ultimate strength of material, respectively, A_5 is the elongation, and KV is the impact energy.

	f_y	f_u	A_5	KV (temp.)
Material ID	MPa	MPa	%	J
S1100[a]	1100	1200	10	27 (-40°C)
S1100[b]	1157	1188	14	50 (-40°C)
Union X96	930	980	14	50 (+20°C)

[a]Nominal values, [b]Measured

3 RESULTS

Figures 2–4 present the fatigue tests data in the structural HS stress system, the ENS system and using the 4R method. It can be seen that the fatigue

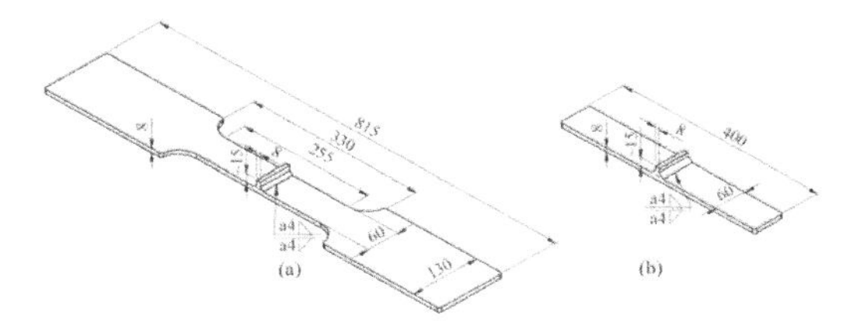

Figure 1. Shape and dimensions of the fatigue test specimens. (a) axial tests and (b) bending tests.

DOI: 10.1201/9781003348450-184

strength of axially-loaded joints is higher than that of the joints tested under bending loading when using structural HS stress system, while for the notch stress-based approaches, axial and bending fatigue test data commeasure into single *S-N* curves.

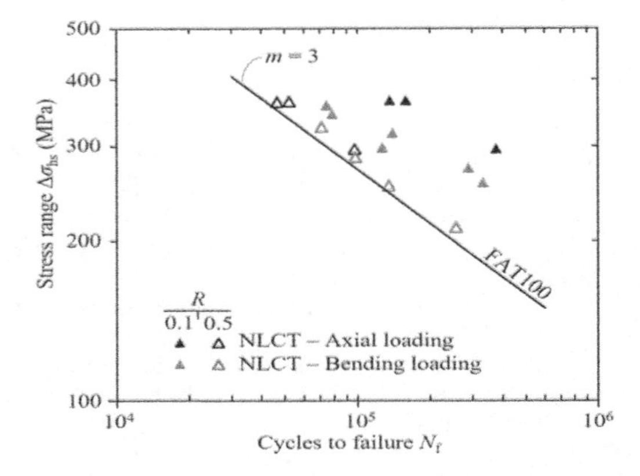

Figure 2. Fatigue test results in the structural HS stress system. *m* is the slope parameter of *S-N* curve.

4 DISCUSSION AND CONCLUSIONS

The results obtained using the structural HS stress system, Figure 1, indicated that bending had a negative effect on the fatigue strength capacity of the NLCT joints. In the investigated joint type with single-sided transverse attachments and weld reinforcement, this was an expected results due to the differences in the notch SCF caused by the axial and bending loading. Due to this fact, an introduction of the notch stress-based approach, namely the ENS concept, in the evaluation of the fatigue test data, Figure 2, showed that the results fit into single scatter band when considering the differences in the applied stress ratio.

As shown in Figures 2 and 3, the applied stress ratio had a clear effect on the fatigue strength capacity although the joints were tested in the as-welded condition. To introduce an approach to understand these differences in the fatigue strength capacities, the 4R method was used to conduct mean stress correction for the effective notch stresses. With the 4R method (Figure 4), all results including both axial and bending fatigue test data and $R = 0.1$ and $R = 0.5$ results could be evaluated using a single *S-N* curve. To summarize the current work and key findings, the following conclusions can be drawn:

- Bending stress causes higher notch stress concentration than axial loading in the single-sided weld reinforcement and thus negative bending effect was experimentally found for the fillet-welded non-load-carrying UHSS joints.

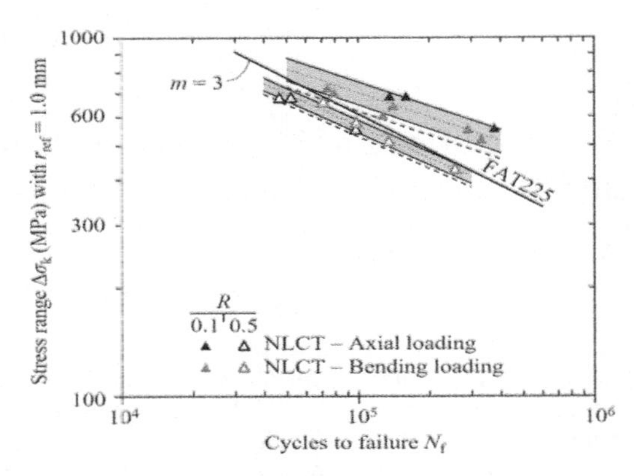

Figure 3. Fatigue test results in the structural ENS stress system.

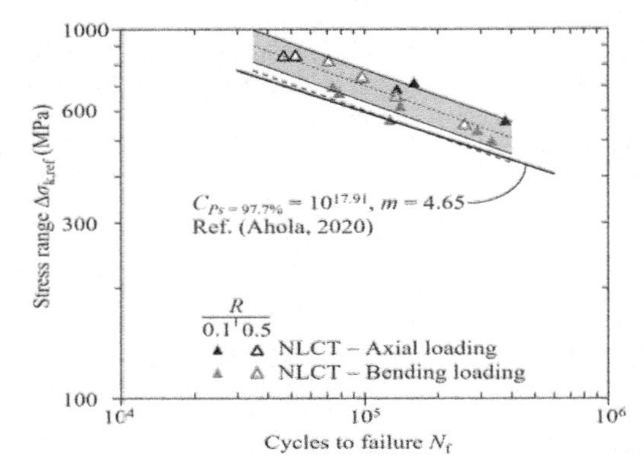

Figure 4. Fatigue test results in the 4R method.

- Reflecting the findings on prior works, careful consideration of joint symmetry and secondary bending stresses is needed since misalignments only cause an increase in bending stress in the joints subjected to axial tension.
- The 4R method was successfully applied to evaluate the effect of applied stress ratio on the fatigue strength capacity. As a result, all results fit into a single scatter band with the scatter index of $T_\sigma = 1.22$.

REFERENCES

Ahola, A., Skriko, T., & Björk, T. (2019). Experimental investigation on the fatigue strength assessment of welded joints made of S1100 ultra-high-strength steel in as-welded and post-weld treated condition. In A. Zingoni (Ed.), *Proceedings of the 7th International Conference on Structural Engineering, Mechanics and Computation (SEMC 2019). Cape Town, South Africa, 2–4 September 2019* (pp. 1254–1259). https://doi.org/10.1201/9780429426506

Current Perspectives and New Directions in Mechanics, Modelling and Design of Structural Systems – Zingoni (ed.)
© 2022 Copyright the Author(s), ISBN 978-1-032-39114-4

Mechanical properties of thin-walled high-strength-steel cold-formed circular hollow sections for crane and scaffolding construction

D. Gubetini & M. Mensinger
Chair of Metal Structures, TUM School of Engineering and Design, Technical University of Munich, Munich, Germany

ABSTRACT: In this paper, the mechanical properties of thin walled, high-induction-frequency welded, cold formed circular hollow sections (CFCHS) suitable for crane and scaffolding construction obtained by experimental results are presented. The effect of different kinds of galvanization (batch galvanization and strip-galvanization) is investigated. The CFCHS for the scaffolding tubes possess a nominal diameter of 48.3 mm and their thicknesses range from 1.8 to 3.0 mm with steel grade classification by the manufacturer of S355, S550, S600, S700 and S960. Tests on the whole tube and on strips, both from the welding seam and on the opposite of the welding seam are performed. For the crane construction tubes with a nominal diameter of 90 mm, experiments on tensile-strips are conducted. Thus, it is possible to quantify the differences of mechanical properties such as $R_{p,0.2}$, R_m and the elongation at failure, depending on the type of galvanization and the testing method.

1 INTRODUCTION

Compressed members in crane and scaffolding construction usually have a circular hollow section, which is traditionally obtained by a cold-forming process. The widely used diameter of D=48.3 mm with the respective thin wall strengths is manufactured by cold rolling and subsequent high-induction-frequency welding (Zhao et al. 2005). Due to the rising application of high-strength steels the idea arises to reduce the wall thicknesses used for scaffolding construction by using high-strength steels. Therefore, this paper conducts experimental research for thin-walled high-strength-steel CFCHS for crane and scaffolding construction

2 EXPERIMENTAL APPROACH

2.1 *Target parameters*

The aim of the presented experiments is the determination of essential mechanical parameters such as yield strength, tensile strength and elongation at failure for the cold formed circular hollow sections (CFCHS). Two methods are applied here according to (DIN German Institute for Standardization 2006) and (DIN German Institute for Standardization 2016). Hereby, also the yielding behavior is inspected, as results with galvanization tend to show a different material behavior than the originally cold-formed member. Three types of galvanization are considered: type A with black/ pickled steel without galvanization, type B as batch-galvanization and type C as strip-galvanization.

2.2 *Experimental matrix*

Following experimental matrix is considered:

Table 1. Experimental matrix.

	diameter D*	thickness t*		
Type	mm	mm	D/t	steel grade*
1	48.3	1.8	26.83	S355
2	48.3	1.8	26.83	S550
3	48.3	1.8	26.83	S960
4	48.3	2.7	17.89	S355
5	48.3	2.7	17.89	S600
6	48.3	3.0	16.10	S700
7	88.9	4.0	22.23	S600
8	90.0	2.7	33.33	S960

* Values relate to the declarations of the manufacturers

Type 1 to Type 6 are all tested with a tensile test on the whole tube. Additionally, for comparison reasons, coupons from Type 2,3,5,6 are tested. For 7 and 8, merely tensile coupons are considered as suitable due to their high load capacity. Contrary to the regulations in the mentioned standard, tensile coupons along the welding seam are extracted for Type 3,5,6 and 8. For the tensile-tests with a thickness lower than 3 mm, the geometry of the tensile coupons is designed for a non-proportional specimen, as due to the low thickness the yielding does not occur proportional to the sample volume.

DOI: 10.1201/9781003348450-185

3 EXPERIMENTAL RESULTS

3.1 *Yielding behavior*

A differentiation in the yielding behavior can be observed depending on the type of galvanization. A gradual yielding process in general occurs in cold-formed steels, whereas the sharp yielding is a characteristic of hot-rolled steel (Yu et al. 2020). The observed yielding types are depicted in Table 3 below:

Table 2. Yielding type according to Figure 1.

	tube test			coupon test*		
Type	A	B	C	A	B	C
1	gradual	sharp	gradual	-	-	-
2	gradual	sharp	gradual	gradual	sharp	gradual
3	gradual	gradual	-	gradual	gradual	-
4	gradual	sharp	gradual	-	-	-
5	gradual	sharp	-	gradual	sharp	-
6	gradual	sharp	gradual	gradual	both**	-
7	-	-	-	-	-	gradual
8	-	-	-	gradual	gradual	-

* results refer to the ordinary coupons
** for one coupon, gradual yielding and for the other sharp yielding was observed

A similar change in the material characteristics was observed in (Najafabadi et al. 2021).

3.2 *Mechanical properties*

As due to the galvanization the yielding might be either sharp either gradual (cf. Figure 1.), in the following diagrams the yield strength $R_{p0,2}$ respectively R_{eH} is referred to as R_{yield}. The mean values for R_{yield} and R_m out of each three samples per type and galvanization and the respective coefficients of variations are evaluated in the figure below:

The same evaluation is performed for the tensile tests of the coupons and shows similar results with a tendency to lower values due to eccentricity moments as a result of the end flattening in the clamping region.

4 CONCLUSIONS

Comparing the results for the galvanization, several conclusions can be drawn. For every type of tested tube, the strip-galvanized tubes showed higher yield stresses and tensile strengths than the non-galvanized. This observation is verified by both the tube and coupon tests. The ductility is reduced due to the strip-galvanization for all tested tubes. The characteristic of the stress-strain-curve is a gradual yielding process as it is typical for cold-formed structures. In comparison, the batch-galvanization leads to other effects: for every type of steel, the yield-strength rises. On the other hand, the tensile strength rises only for the tubes up to S550 whereas

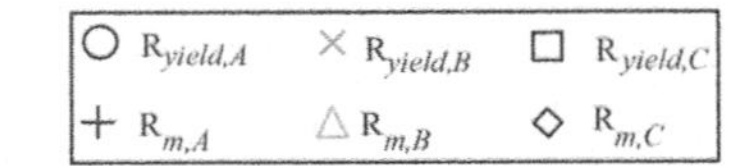

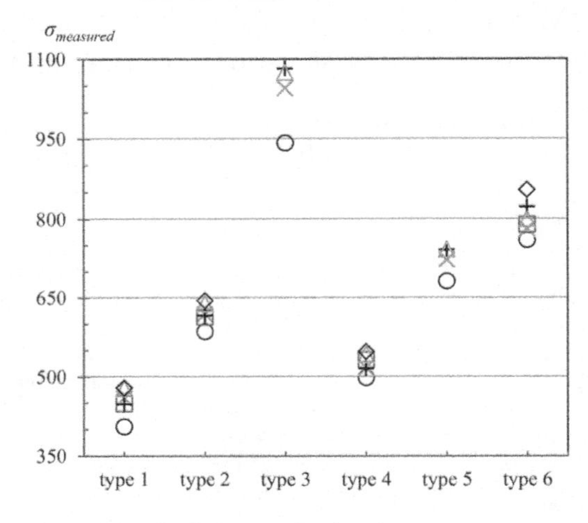

Figure 1. Mechanical properties for tube tests.

the ductility decreases in comparison to the non-galvanized. Vice versa, for the steels from S600 to S960, a slight reduction of the tensile strength accompanied by an increasing ductility could be observed. The characteristic of the curves generally changed from a gradual yielding process to a sharp yielding, except for the S960. Nevertheless, a steeper stress-strain relation in the pre-yielding domain is recorded. As the nonlinearity in this relation for cold-formed structures is accounted for by the residual-stresses, the galvanization leads to a partial reduction of the residual stresses.

The obtained results are statistically evaluated by using at least three specimens for each type of tube and each type of galvanization. For the specimen on the coupons the number of samples varies from two to four. The variational coefficient shows to be lower than 2.2 % for both the tubes and the coupons. This leads to a representative significance of the results.

REFERENCES

DIN German Institute for Standardization. 2006. Cold formed welded structural steel hollow sections – Part 1: Technical delivery conditions; German version EN 10219: 2016. Berlin: Beuth Verlag GmbH

DIN German Institute for Standardization. 2016. Testing of Metallic Materials - Tensile Test Pieces; DIN 50125. Berlin. Beuth Verlag

Najafabadi, E. P. et al. 2021. Hot-dip galvanizing of high and ultra-high strength thin-walled CHS steel tubes: Mechanical performance and coating characteristics. *Thin - Walled Structures* Vol. 164: 107744

Yu, W. et al. 2020. Cold-Formed Steel Design (Fifth Edition). New Jersey: John Wiley & Sons, Inc.

Zhao, X. L. et al. 2005. Cold-formed tubular members and connections: Structural behaviour and design. Oxford: Elsevier

Current Perspectives and New Directions in Mechanics, Modelling and Design of Structural Systems – Zingoni (ed.)
© 2022 Copyright the Author(s), ISBN 978-1-032-39114-4

A basic study on the evaluation of residual axial force of high strength bolts for steel bridges constructed in the same environment at different ages

T. Iida & K. Sugiura
Department of Civil and Earth Resources Engineering, Kyoto University, Kyoto, Japan

T. Yamaguchi
Department of Civil Engineering, Osaka City University, Osaka, Japan

Y. Kitane
Department of Civil and Earth Resources Engineering, Kyoto University, Kyoto, Japan

ABSTRACT: In this study, the specifications of high strength bolts of different types and the friction surface treatment of the connection were different. According to the construction age, high strength bolts were extracted from two in-service two way steel bridges subjected to live loads, temperature changes, and environmental effects under the same environment and bridge type to ascertain whether the residual axial forces of the two bridges were maintained according to the design specifications.

As a result of the investigation, it was found that the residual axial forces of the two bridges were almost all below the design values, except for a few. This decrease in axial force can be attributed; to the influence of the management of the tightening force at the time of construction, the structural type, the bolt type, the friction joint surface treatment, etc.

1 INTRODUCTION

In the maintenance and management of the increasing number of high-strength bolts in the future, it is important to clarify the current issues and problems, as well as to conduct continuous surveys and researches to establish sound evaluation methods, repair and management methods related to these issues. Therefore, the subject of this study is an investigation carried out on two in-service 2 way steel bridges, using the same type and strength of high strength bolts, which are actually subjected to live loads, temperature changes, and the same environmental influences, but with different bridge construction years, types of high strength bolts, and friction surface treatment specifications of the connection. Therefore, it was decided that the high strength bolts be sampled and examined to establish whether the residual axial force was maintained according to the design axial force.

2 OUTLINE OF THE BRIDGE MEASURED

The two bridges to be measured were; 1) a two-span continuous steel two-main box girder bridge with a minimum oblique angle of 56° (below Bridge-A) and 2) a two-span continuous steel two-main box girder bridge with a minimum oblique angle of 56° (below Bridge-B). The bolts were removed from the frictional joints of the high-strength bolts in service in the field, and the residual axial force of the high-strength bolts was measured by laboratory experiments. A summary of the measured bridges is shown in Table 1.

3 RESIDUAL AXIAL FORCE MEASUREMENT METHOD FOR HIGH-STRENGTH BOLTS

To measure the residual axial force of the high strength bolts, triaxle strain gauges were attached to the bolt heads to measure the strain generated when the bolts were removed. Then, using a special calibration device for axial force measurement the load was applied up to the strain at the time of bolt removal in the field, and the resulting load referred to as the residual axial force.

3.1 *Measurement results of Bridge A*

In Bridge A, one joint of the G1 (J3) main girder was targeted for removal of the high strength bolts. A total of 99 bolts (73 from the lower flange of the main girder and 26 from the web) were sampled. The bolts to be measured were M22 (F10T), and the bolt lengths were 70 mm for the main girder web and 85 mm for the lower flange. Table 2(a) shows the measurement results of the residual axial force of Bridge A. Figure 1(a) shows the frequency distribution of the measurement results.

DOI: 10.1201/9781003348450-186

Table 1. Outline of measured bridges.

	Bridge-A	Bridge-B
Bridge Type	2-Span Continuous 2-Box Girder	
Bridge Length	112.0m	
Bridge Span	55.5m+55.5m	
Skew angles	56°(A1), 65°(P1, A2)	
Start year for use	1990 (30years)	2012 (8years)
High-Strength Bolt	M22 (F10T)	M22 (S10T)
Friction Joint Surface	Unpainted	Painted
Distance from the Coastline	30km	
Bridge point	Cross River	

Table 2. Residual axial force measurement results.

(a) Bridge-A

Part	Number of Bolts	Average of Residual Axial Force: N(kN)	Coefficient of Variation: CV (%)
WEB	26	195.0kN (95%)	7.1%
L-FLG	73 [57]	145.1kN (71%) [169.0kN (82%)]	35.0% [12.1%]
Total	99 [83]	158.2kN (77%) [177.2kN (87%)]	31.2% [12.5%]

(b) Bridge-B

Part	Number of Bolts	Average of Residual Axial Force: N(kN)	Coefficient of Variation: CV (%)
WEB	24	163.4kN (80%)	6.8%
L-FLG	32	173.3kN (85%)	9.2%
Total	56	168.9kN (82%)	8.7%

Note: #1: The value in () indicates the percentage of the design axial force of 205kN, #2: The values in [] indicate the values excluding the bolts at the edge of the lower flange of the box girder.

3.2 *Measurement results of Bridge B*

In Bridge B, one joint of the G2 (J5) main girder was targeted for removal of the high strength bolts. A total of 56 bolts (32 from the lower flange of the main girder and 24 from the web) were sampled. The bolts to be measured were M22 (S10T), with a bolt length of 70 mm for both the main girder web and the lower flange. Table 2(b) shows the measurement results of the residual axial force of Bridge B. Figure 1(b) shows the frequency distribution of the measurement results.

4 CONCLUSION

(1) It was confirmed that the residual axial forces of the high strength bolts used in bridges A and B were mostly lower than the design bolt axial forces.
(2) The coefficient of variation due to the difference between the flange and web sections tends to be higher for the lower flange.
(3) Compared to the F10T coefficient of variation (CV):12.1% for Bridge A, the S10T coefficient of variation (CV):8.7% for Bridge B tends to have lower variability.
(4) The estimated residual axial force ratio of the joint with filler after 70,000 hours from the tightening of the high-strength bolts in Bridge D was in general agreement with previous studies[1].
(5) Box girder bridges with a width of about 100 mm at the edge of the lower flange showed extremely low residual axial forces, which was in in agreement with previous studies[2].

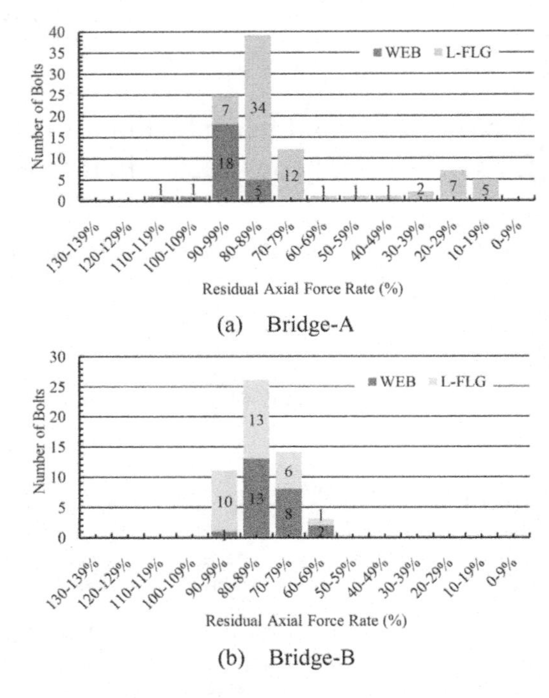

Figure 1. Frequency distribution chart.

REFERENCES

(1) Tetsuya. IIDA, Kunitomo. SUGIURA, Takashi. YAMAGUCHI, Yasuo. SUZUKI, Hiroo. YO-SHIZU, Keiji. MORITA and Hirotaka. ODA, In-vestigation on Residual Axial Force of High-Strength Bolts of Friction Grip Joints in Steel Bridges under Long-Term Service Conditions and Its Tendency Assessment, Journal of JSSC, Vol. 27,No.107, pp.9–21, 2020.

(2) Kuniaki. MINAMI, Hiroshi. TAMURA, Natsuki. YOSHIOKA. Daisuke. UCHIDA, Makoto. Moro and Koki. ANDO, A Study on Initial Bolt Preten-sions of High Strength Bolted Joints Considering Number of Contact Surfaces, Journal of JSCE, Vol.75, No.1, pp.46–57, 2019.

Current Perspectives and New Directions in Mechanics, Modelling and Design of Structural Systems – Zingoni (ed.)
© 2022 Copyright the Author(s), ISBN 978-1-032-39114-4

Design methodology of grade 12.9 bolts considering hydrogen-induced delayed fracture

Y.B. Wang, Z. Sun & B. Tha
College of Civil Engineering, Tongji University, Shanghai, China

G.Q. Li
State Key Laboratory for Disaster Reduction in Civil Engineering, Tongji University, Shanghai, China

X.L. Zhao
School of Mechanical, Electronic and Control Engineering, Beijing Jiaotong University, Beijing, China

ABSTRACT: With the increase demand in high-strength steel members, high-strength steel bolts are needed for high-strength steel structure connections to provide matching tensile strength and shear strength. With the further increasing of bolt steel strength from grade 10.9 to 12.9, hydrogen-induced delayed fracture (HIDF) may become the major concern for the application of grade 12.9 high strength steel bolts in steel constructions. The HIDF resistance of grade 12.9 42CrMo high strength bolt was investigated in comparison with 20MnTiB steel for grade 10.9 bolt by means of constant load test, and analyses of fracture surface morphology and thermal desorption spectroscopy. It is found that the activation energies of the hydrogen traps of grade 12.9 42CrMo steel and grade 10.9 20MnTiB steel are very close to each other, indicating that the hydrogen traps of the two types of steel are of the same type. Because the HIDF resistance of grade 12.9 42CrMo steel is inferior to the grade 10.9 20MnTiB steel, it is not safe to simply extend current steel structure design code to include the design of grade 12.9 bolt connection. To achieve the same service life as grade 10.9 bolts, the prestress reduction factor is recommended for grade 12.9 42CrMo steel bolts.

1 INTRODUCTION

The problem of hydrogen induced delayed fracture (HIDF) becomes more prominent than strength failure, since the susceptibility to HIDF of high-strength steel increases with the tensile strength(Chun et al., 2012, Eliaz et al., 2002). In this study, the HIDF behavior of 42CrMo steel for grade 12.9 bolt is studied in comparison with 20MnTiB steel for grade 10.9 bolt by means of constant load test, and analyses of fracture surface morphology and thermal desorption spectroscopy. A design methodology with considering the effect of HIDF and providing equivalent service life compared to grade 10.9 20MnTiB high strength bolts is proposed.

2 SPECIMEN PREPARATION AND EXPERIMENTAL PROGRAM

2.1 *Materials and specimen preparation*

The steels used in this study were obtained from commonly used grade 12.9 42CrMo high-strength bolts and grade 10.9 20MnTiB high-strength bolts in size of M20×80. Figure 1 shows the circumferentially notched round bar specimen (the stress concentration factor K_t=3.2) for the constant load test.

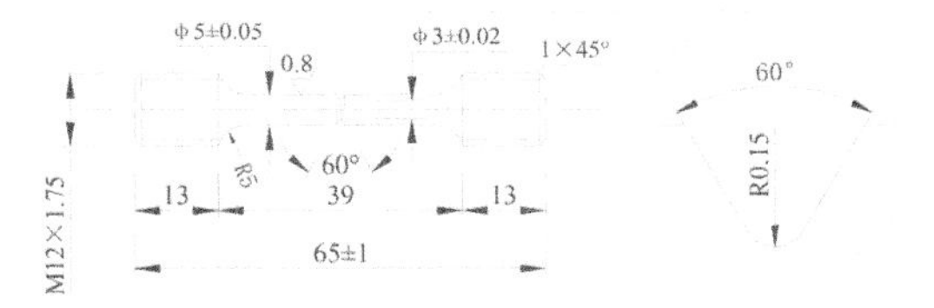

Figure 1. Schematic of notched steel specimen for constant load tensile test (mm).

2.2 *Constant load test*

The constant load test aims to evaluate the critical HIDF stress ratio in an accelerated HIDF process with a specified target fracture time, which is 100 hours in this study. During the constant load test, Walpole solution consisted of sodium acetate and hydrochloric acid aqueous solution with pH value of 3.5, 5.5 and 7.0 was used. Various stress ratios $\beta_i = \sigma_i/\sigma_{u,N}$ were adopted to apply the constant load to specimens, where σ_i is the applied i^{th} tensile stress, $\sigma_{u,N}$ is the fracture stress of the notched specimen tested in air. The critical HIDF stress ratio β_c at 100 h can be obtained by using the following equations.

DOI: 10.1201/9781003348450-187

$$\beta_c = \left(\beta_f + \beta_n\right)/2 \tag{1}$$

$$\beta_f - \beta_n \leq 0.2\beta_c \tag{2}$$

Where β_f is the minimum stress ratio under which the delayed fracture occurs at or before 100hr and β_n is the maximum stress ratio under which the delayed fracture does not occur before 100hr.

3 RESULTS AND DISCUSSIONS

The equilibrium hydrogen concentration under hydrostatic tensile stress are expressed as follows(Li et al., 1966):

$$C_\sigma = C_H \exp(\sigma_h V_H / RT) \tag{3}$$

Where C_σ is the equilibrium hydrogen concentration under hydrostatic tensile stress; C_H is the average hydrogen concentration; σ_h is the hydrostatic tensile stress; V_H is the partial molar volume of hydrogen; R is the gas constant; T is the environmental temperature in Kelvin degree.

Table 1 shows the critical stress ratio of tested steels in Walpole solution with different pH values. Therefore, the applied stress ratio of grade 12.9 42CrMo steel must be smaller than that of grade 10.9 20MnTiB steel in order to achieve the similar HIDF performance, namely the time to fracture.

Table 1. Summaries of the critical stress ratios obtained from constant load tests.

Steel	σ_{NO} (MPa)	pH	σ_c (MPa)	β_c (%)
Grade 12.9 42CrMo	2151	3.5	1452	67.5
		5.5	1570	73.0
		7.0	>1678	>78.0
Grade 10.9 20MnTiB	1929	3.5	1620	84.0
		5.5	1861	96.5
		7.0	≈1929	≈100.0

4 DESIGN METHODOLOGY OF GRADE 12.9 42CRMO BOLT CONNECTION

As specified in EC3 Part 1-8, the preload should be taken as Eq. (4), while γ_{M7} is the safety factor with recommended value of 1.1.

$$P = 0.7 f_u A_e / \gamma_{M7} \tag{4}$$

Similarly, according to GB50017-2017, the pretension force applied to grade 8.8 and grade 10.9 high-strength bolt is determined by Eq. (5), where P is the pre-tension force.

$$P = 0.9 \times 0.9 \times 0.9 f_u A_e / 1.2 \tag{5}$$

The constant load test result shows that the HIDF resistance of grade 12.9 42CrMo high-strength steel bolt is inferior to that of grade 10.9 20MnTiB high-strength steel bolt. To achieve the same time to fracture, the prestress ratio of grade 12.9 42CrMo high-strength steel bolt should be smaller than that of grade 10.9 20MnTiB high-strength steel bolt. The formula for calculating the prestress reduction factor is as follows:

$$\mu_H = \beta_{c12.9} / \beta_{c10.9} \tag{6}$$

where, μ_H is the prestress reduction factor, $\beta_{c12.9}$ is the critical stress ratio of grade 12.9 high strength steel bolt and $\beta_{c10.9}$ is the critical stress ratio of grade 10.9 high strength steel bolt.

Table 2 shows that the prestress reduction factor is not constant under different hydrogen environments. In order to achieve conservative design, the minimum value of the prestress reduction factors is taken, as shown in Eq. (7).

$$\mu_H = min\left(\mu_{H,3.5}, \mu_{H,5.5}, \mu_{H,7.0}\right) = 0.75 \tag{7}$$

Table 2. Pre-tension reduction coefficient under different pH conditions.

pH	Steel	β_c	μ_H
3.5	Grade 12.9 42CrMo	0.675	0.80
	Grade 10.9 20MnTiB	0.840	
5.5	Grade 12.9 42CrMo	0.730	0.75
	Grade 10.9 20MnTiB	0.965	
7.0	Grade 12.9 42CrMo	>0.780	>0.75
	Grade 10.9 20MnTiB	≈1	

5 CONCLUSIONS

the HIDF resistance of grade 12.9 42CrMo steel is inferior to the grade 10.9 20MnTiB steel.

To achieve the same service life as grade 10.9 bolts, the prestress ratio of grade 12.9 42CrMo high-strength steel bolt should be smaller. The prestress reduction factor of 0.75 is recommended for grade 12.9 42CrMo steel bolts.

REFERENCES

Chun, Y. S., Lee, J., Bae, C. M., Park, K.-T. & Lee, C. S. 2012. Caliber-rolled TWIP steel for high-strength wire rods with enhanced hydrogen-delayed fracture resistance. *Scripta Materialia*, 67, 681–684.

Eliaz, N., Shachar, A., Tal, B. & Eliezer, D. 2002. Characteristics of hydrogen embrittlement, stress corrosion cracking and tempered martensite embrittlement in high-strength steels. *Engineering Failure Analysis*, 9, 167–184.

Li, J., Oriani, R. & Darken, L. 1966. The thermodynamics of stressed solids. *Zeitschrift für Physikalische Chemie*, 49, 271–290.

Current Perspectives and New Directions in Mechanics, Modelling and Design of Structural Systems – Zingoni (ed.)
© 2022 Copyright the Author(s), ISBN 978-1-032-39114-4

Influence of high strength steel on the fatigue of welded details

H. Bartsch & M. Feldmann
Institute of Steel Construction, RWTH Aachen University, Germany

ABSTRACT: In addition to static loads, certain steel structures are also exposed to varying stresses, so that verification against fatigue is necessary. Until now, a possible positive influence of high strength steels cannot be considered in the fatigue design of structural steel details in Europe. The presented investigations provide a summary of the state of the art on the influence of the steel grade on the fatigue strength of welded details. Furthermore, an overview of normative regulations in standards is given. Finally, a database evaluation of numerous test data from the literature shows findings on the influence of high strength steel on the fatigue strength of welded details. The assessed welded details comprise apart from load-carrying components, such as the cruciform joint, non-load carrying details as the transverse stiffener for example.

1 STATE OF THE ART REGARDING MATERIAL INFLUENCE ON FATIGUE

The design of fatigue-stressed structures, such as wind turbines, bridges, masts, towers and crane runways, is often governed by the fatigue verification. Currently, there is no optimization by the use of higher-strength steel (HSS) materials in the European standard EN 1993-1-9 [1] because only the fatigue check, which is independent of the strength, dominates the design. This leads, for example, to the fact that conventional-strength steel is still oftentimes used, although higher-strength materials would lead to thinner plate thicknesses and less welding volume.

If the influence of material strength on fatigue strength is to be evaluated, a distinction must be made between welded and non-welded constructions. In welded constructions, the strength of the base material usually has less influence on the fatigue strength in comparison to non-welded details, as the geometric notch resulting from the weld is largely decisive. This relationship has been shown in several investigations.

For components subjected to fatigue, the use of higher-strength steels accordingly does not initially appear to make much sense, as fatigue strength classes are achieved that correspond to those of conventional-strength steels. However, by using suitable post-treatment processes, it is possible to remove the notch sharpness of the weld toe and thus obtain a higher fatigue strength [2].

2 NORMATIVE REGULATIONS REGARDING MATERIAL INFLUENCE ON FATIGUE

As already mentioned before, a material influence on the fatigue of unwelded and welded details is not considered in the European fatigue standard EN 1993-1-9 [1].

The fatigue design recommendations for welded joints and components of the International Institute of Welding [3]allows the use of higher FAT classes, when specific post-treatment in terms of improvement technique is applied.

3 DATABSE EVALUATION

Together with the "University of Stuttgart" and the "Karlsruher Institute of Technology", a structured database was created and published fatigue data was researched [4]. All information relevant to fatigue was gathered. Based on the developed structured database, the information it contains and the coding of corresponding applications, the evaluation of fatigue tests and influencing variables can be easily realized.

In order to determine the influence of the steel grade on the fatigue strength, a statistical evaluation of the test data of each series is carried out in a first step. The individual stress ranges of the experimental test values are then converted to a number of cycles of 2,000,000 cycles using the inverse slope of the

S-N curve of $m = 3$. Thus, the data is normalized and can be presented in comparison to the influencing variable.

DOI: 10.1201/9781003348450-188

The evaluation has been carried out for all data points in the database[4] regarding the load-carrying cruciform joint and the non-load-carrying transverse stiffener, which is presented here in detail. In a second step, the steel grade influence has been evaluated only on the basis of post-treated series. The post-treatment methods include methods for improvement of the weld profile and methods for improvement of residual stress conditions.

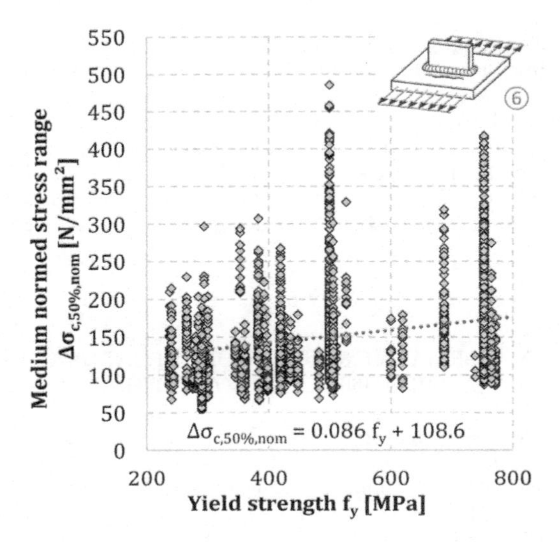

Figure 1. Influence of yield strength on fatigue strength of toe failure at cruciform joint (all test data in database of [4]).

Figure 1 presents the analyzed yield strength influence of the non-load-carrying transverse stiffener based on all data points of the database [4] Here, when looking at all test results, already a slight positive influence of high strength steel on fatigue strength is noticed. When looking only at the post-treated specimen data points, Figure 2, the trend becomes clearer. When calculating the medium fatigue strength with the linear trend function of Figure 2, using an S690 steel leads to a medium fatigue class of 233 MPa, while an S235 leads to 155 MPa. This shows, that the fatigue strength and the lifetime can be increased by 50%.

4 CONCLUSIONS

The following main conclusions can be drawn:

1. The yield strength influence on fatigue strength is in general not significant for welded details.

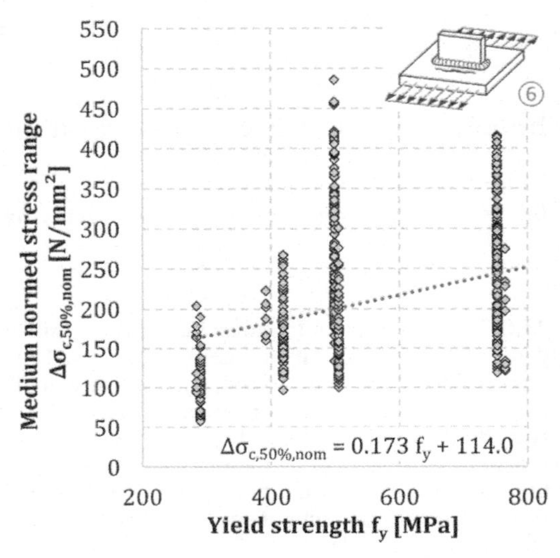

Figure 2. Influence of yield strength on fatigue strength of toe failure at cruciform joint (only post treated test data in database of [4].

2. The yield strength influence can yet be seen more clearly for the non-load-carrying stiffener than for the load-carrying cruciform joints.
3. When applying post-treatment methods, a positive effect of using high strength steel in case of fatigue can be clearly identified.
4. When post-treating the non-load-carrying transverse stiffener, the fatigue strength can be increased by 50% when S690 instead of S235 is utilized.

REFERENCES

[1] EN 1993-1-9: Eurocode 3, Design of steel structures – Part 1-9: Fatigue, 2010.

[2] H. Minner, Schwingfestigkeitserhöhung von Schweißverbindungen aus hochfesten Feinkornbaustählen StE 460 und StE 690 durch Einsatz des WIG-Nachbehandlungsverfahrens, Dissertation, TH Darmstadt, 1981.

[3] A. F. Hobbacher, Recommendations for Fatigue Design of Welded Joints and Components - Second Edition - IIW document IIW-2259-15 ex XIII-2460-13/XV-1440-13, "International Institute of Welding, 2016.

[4] M. Feldmann, H. Bartsch, T. Ummenhofer, B. Seyfried, U. Kuhlmann und K. Drebenstedt, Re-evaluation and enhancement of the detail catalogue in Eurocode 3 for future oriented design of steel construction under high loading, Final Report, 2020.

Current Perspectives and New Directions in Mechanics, Modelling and Design of Structural Systems – Zingoni (ed.)
© 2022 Copyright the Author(s), *ISBN 978-1-032-39114-4*

Influence of residual stresses on innovative composite column sections using high strength steel

R. Röß, M. Schäfers & M. Mensinger
Technical University of Munich, Chair of Metal Structures, Munich, Germany

ABSTRACT: Current developments in steel and steel concrete composite constructions lead to an increasing use of high strength steel grades. Especially in structural members under compressive load, imperfections determine the load bearing behavior significantly. In detail, geometric imperfections can be influenced by the manufacturing process of the columns, whereas structural imperfections such as residual stresses result from the steel's production. In several research projects, innovative composite columns using steels with yield strengths up to 960 MPa with new section designs are being developed. Bundled high strength reinforcement bars are deployed as core within steel tubes and confined with concrete. As there is little knowledge about the residual stress distribution in sections of such steel grades, investigations for each section were carried out. Preliminary results of such examinations will further be used as input parameters for numerical simulations. Comparing calculations with and without consideration of residual stresses of the steel sections are made and their results are presented.

1 NEW TYPES OF COMPOSITE COLUMNS

Bar bundle columns (Figure 1) are a new type of composite column. For this purpose, a bundle of high-strength reinforcing bars with a yield strength of 670 MPa is placed in a steel tube and then grouted with mortar.

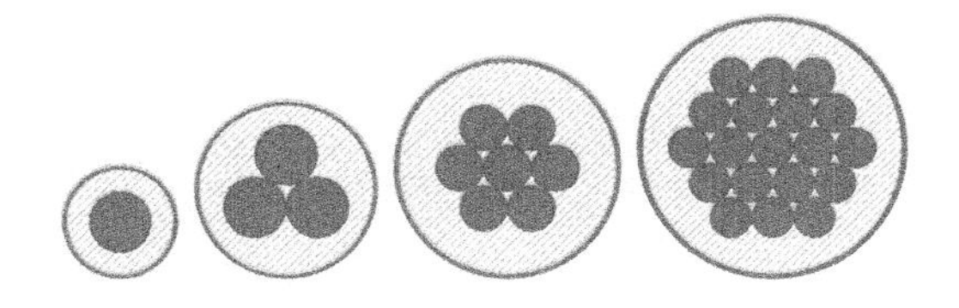

Figure 1. Various bar bundle columns cross sections.

Due to the lower cross-sectional thicknesses of the threaded bars, no strength reduction due to material thickness effects is present. The bundling of the steel bars results in a high steel content in relation to the total cross-section (Mensinger & Röß 2019).

As another approach to tackle high residual stresses due to large steel core sections, innovative composite columns are designed using separate sheet metals with individual shear connections. Limited sheet thicknesses ensure low residual stresses and homogeneous yield strengths throughout the section. Shear connections can be achieved by local longitudinal filled welds on the sheets' edges or local bolting. As a first approach, yield strengths up to 960 MPa are used which result in high load bearing capacities (Figure 2).

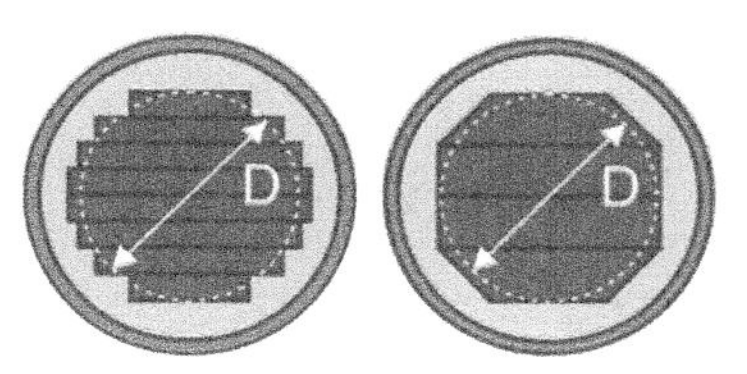

Figure 2. Exemplary sheet metal sections, comparing to circular approximation.

2 RESIDUAL STRESSES

Residual stresses have a considerable influence on the stability behavior of composite columns. Currently, very limited information is available on the residual stress state of high-strength-steel. The residual stresses of the high-strength-steel threaded bars were measured using the X-ray diffraction method. If the residual stress distribution approach from (Roik et al., 1980) is used, the edge stress is as seen in Figure 3. (Ameri & Röß 2021).

Hence, residual stress distributions in the rolling direction of steel sheets were investigated using the sectioning method, a well-established method for measurements of longitudinal residual stresses (Thiébaud & Lebet 2012). Preliminary results show a high correspondence to previous examinations of steel sheets using yield strengths of 355 MPa. Final analysis and a systematic review of the obtained results is still outstanding.

DOI: 10.1201/9781003348450-189

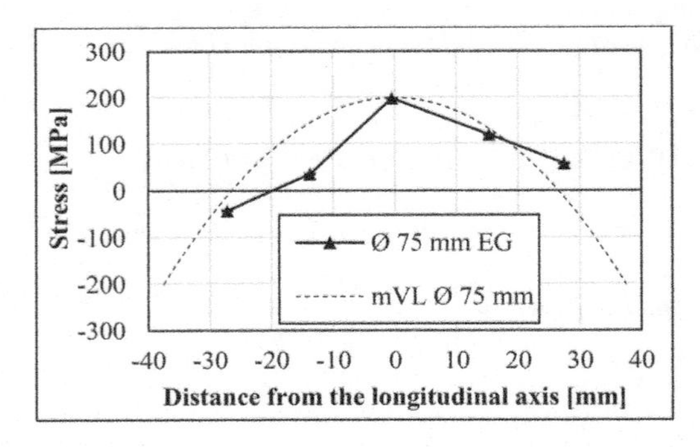

Figure 3. Measurement results (EG) with possible course of residual stress (mVL) according to (Lippes 2008, Chanou 2018, Kleibömer 2018, Roik et al. 1980, Z-26.3-60 2014).

Major influences can rather be explained by the large heat affection of the flame cutting procedure, which result in high compression stresses of the specimens' longitudinal edges (see Figure 4).

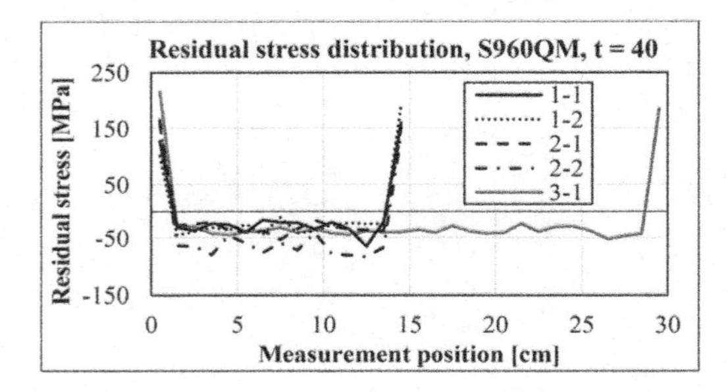

Figure 4. Mean longitudinal residual stresses of S960 steel sheets, obtained by sectioning method (1-1 & 1-2 no post-processing, w = 150 mm; 2-1 & 2-2 shot blasted, w = 150 mm; 3-1 shot blasted, w=300 mm).

3 INFLUENCE OF THE RESIDUAL STRESSES ON THE LOAD-BEARING BEHAVIOR

A numerical model of the column was created using the Abaqus program and calibrated. For the calculation with the measured values with a maximum residual stress value of 260 MPa, a clear deterioration of the values can be seen (Figure 5).

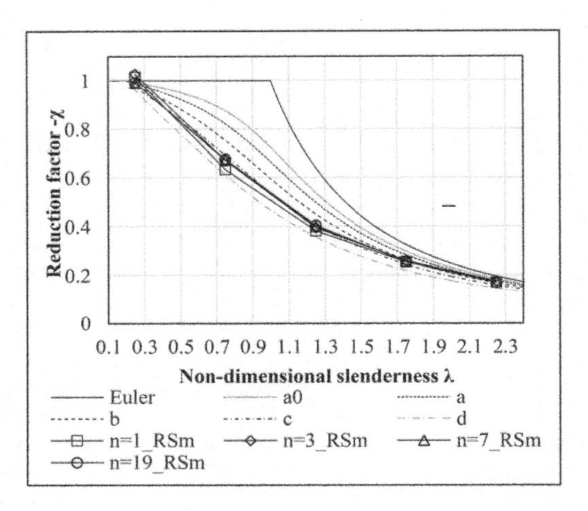

Figure 5. The results from the study with an applied residual stress (solid line corresponds to the number of bars n) at the level of the measured values, the buckling curve a_0 to d and the Euler curve.

The influence of residual stresses is clearly recognizable and therefore cannot be neglected. In the further course of the project, the influence of the residual stress will be investigated together with the yield strength distribution. Once the influence of the two component properties is known, a new parameter study will be carried out resulting in an assignment to the buckling lines.

REFERENCES

Ameri S., Röß R. et al. 2021. *Thermomechanical behavior of bar bundle columns exposed to fire*. Tagungsband - 7th Symposium Structural Fire Engineering – TU Braunschweig.

Mensinger M. and Röß R 2019. *composite column with a bunch of high-strength reinforcement bars*. Ce/papers, 3: 324–330.

Roik, K., Schaumann, P.1980. *Tragverhalten von Vollprofilstützen – Fließgrenzenverteilung an Vollprofilquerschnitten*. Institut für Konstruktiven Ingenieurbau, Ruhr-Universität Bochum.

Thiébaud R., Lebet J.-P. 2012. Experimental study of residual stresses in thick steel plates. Proceedings of the Annual Stability Conference Structural Stability Research Council, Grapevine, Texas.

Current Perspectives and New Directions in Mechanics, Modelling and Design of Structural Systems – Zingoni (ed.)
© 2022 Copyright the Author(s), ISBN 978-1-032-39114-4

Experimental and numerical study of butt welded joints made of high strength steel

Rui Yan, Hagar El Bamby & Milan Veljkovic
Faculty of Civil Engineering and Geosciences, Delft University of Technology, Delft, The Netherlands

ABSTRACT: Welded joints are wildly used in the construction sector for fabrication of steel and aluminium structures. A welded joint is traditionally divided into three regions: the Base Material (BM), the Heat-Affected Zone (HAZ), and the Weld Material (WM). In general, HAZ has a lower material strength compared to BM and WM. Therefore, it is essential to obtain the constitutive model of HAZ to accurately predict the behaviour (strength, stiffness, and ductility) of the HSS welded joint. In this paper, the tensile behaviour of milled and unmilled coupon specimens with a transverse butt weld in the middle are investigated. It is found that the peak deformation would be significantly overestimated if the modified HAZ constitutive model is not used.

1 INTRODUCTION

Welded joints are wildly used in the construction sector. A welded joint consists of three major regions, which are the base material (BM), the heat-affected zone (HAZ), and the weld material (WM). Compared to BM and WM, HAZ has a lower material strength. Therefore, accurately modelling HAZ behaviour plays a key role in the advanced simulation of welded joints.

In this paper, the HAZ stress-strain relationship is calibrated using FEA for two HSS, S500 and S700, based on the tensile tests on the milled coupon specimen with a transverse butt weld in the middle. The modified stress-strain relationship is further verified by the tensile test of the unmilled welded coupon specimen. Finally, the one-quarter square hollow section (SHS) FE model is used to study the tensile behaviour of the welded SHS using the investigated material.

2 EXPERIMENTS

The Cold-formed SHSs made of HSS, S500 and S700, were used as BM in this study. Edges of two hollow section profiles with 200 mm length were prepared with a V groove bevel before welding. The profiles were butt welded using the metal active gas (MAG) welding process. Comparing the strength of BM and the electrode, overmatching and undermatching electrodes were used to weld S500 and S700 SHS profiles, respectively.

The standard coupon specimen for obtaining the BM property was fabricated from the welded tube, as shown in Figure 1. Two welded coupon specimens were fabricated for each steel grade. One welded coupon specimen was milled to a central thickness zone of 3 mm to have a perpendicular HAZ boundary through the thickness before the tensile test. Another welded coupon specimen was directly tested with the complete weld in the middle. A small sample was cut out and prepared for the microstructure observation and the low-force Vickers hardness test (HV 0.5).

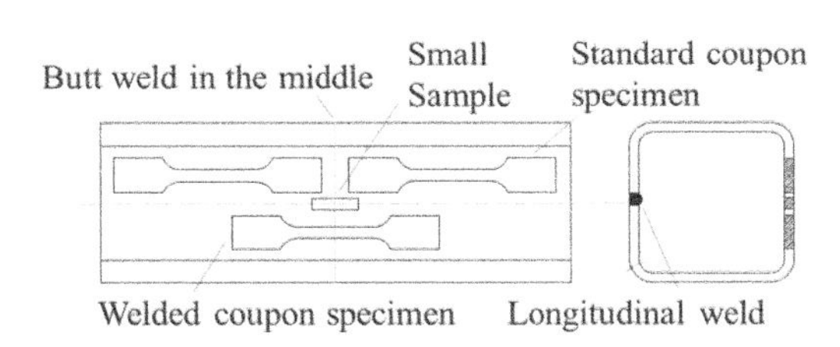

Figure 1. Specimen cutting scheme.

The standard coupon test was conducted in an Instron testing machine with a 100 kN capacity. The loading rate was 0.01 mm/s with displacement control. 3D DIC was employed to measure the deformation during the test.

3 FINITE ELEMENT ANALYSIS (FEA)

The engineering stress-strain relationship of BM is obtained from the tensile test of the standard coupon test. A linear combination of the power law and the linear law is used to generate the BM undamaged true stress-true strain relationship, which is validated

DOI: 10.1201/9781003348450-190

against the standard coupon test following the procedures proposed in [1].

The stress-strain relationship of HAZ and WM is measured from the milled welded coupon specimen. BM and WM imposed a transverse constraint on HAZ during the tensile test of the milled coupon specimen at the plastic stage, indicating that HAZ was under a biaxial tensile stress state instead of a uniaxial tensile stress state. Consequently, the measured stress-strain relationship (the blue dash line in Figure 2) has a higher strength than under the uniaxial stress state (the red dash line in Figure 2) at the same strain level. Therefore, the method proposed by Yan et al. [2] is adapted to modify the measured stress-strain relationship. A linear modification factor is used to reduce the true stress of HAZ between yielding and necking, as shown in Figure 2.

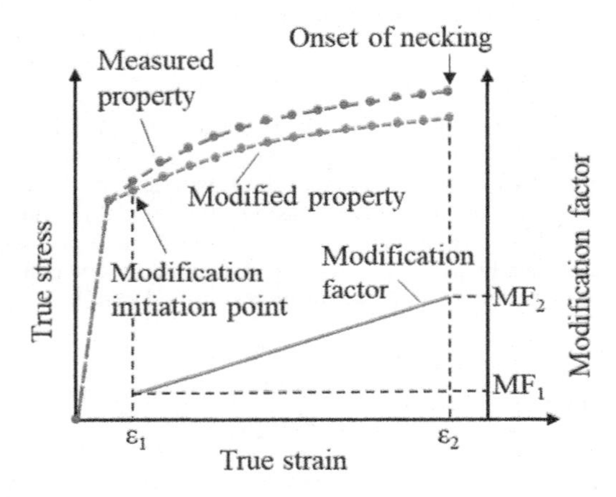

Figure 2. Schematic diagram of the linear stress modification factor.

The ABAQUS:2019 software package is used to conduct FEA. The FE models for the milled and unmilled welded coupon specimens are made based on the measured dimensions. The result of the metallurgical investigation is used to confirm the boundaries of HAZ in the unmilled model. In addition, using the cross-sectional geometry of the unmilled specimen, a FE model of a quarter welded SHS is created for S700 material.

4 RESULTS

Two HAZ constitutive models, the original model measured from DIC (corresponding to FEA-ori) and the modified model based on the linear modification factor (corresponding to FEA-mod), are used in FEA.

The ultimate resistance and the peak deformation of the S500 specimen are well predicted using both the measured original and the modified HAZ stress-strain relationship. For the S700 specimen, using the original HAZ constitutive model, the ultimate resistance is slightly overestimated by 5%, while the peak deformation has a maximum of 30% overestimation. The ultimate resistance and the peak deformation are well predicted with less than 1% deviation using the modified HAZ constitutive model.

In addition to the abovementioned two constitutive models, the BM property is used for all three regions in the quarter SHS model. Compared to the model using the modified HAZ property, the model using the original HAZ property and BM property have a 49% and 34% higher peak deformation, respectively.

5 CONCLUSIONS

S500 and S700 weld joints under different scales are investigated in this paper. The following conclusions are drawn:

1. For S700 welded joints, using the original HAZ property, the predicted peak deformation is maximum of 30% larger than the experimental results, while the deviation is less than 1% using the modified HAZ property.
2. Since no reduction is necessary for S500 HAZ at necking, the prediction by FEA using the original and the modified HAZ stress-strain relationship show a minor difference.
3. Regarding the butt welded S700 SHS joint, the model using the original HAZ property and BM property predict a 49% and 34% larger peak deformation compared to that using the modified HAZ property.

REFERENCES

[1] R. Yan, H. Xin, M. Veljkovic, Ductile fracture simulation of cold-formed high strength steel using GTN damage model, Journal of Constructional Steel Research. 184 (2021) 106832. 10.1016/j.jcsr.2021.106832.

[2] R. Yan, H. Xin, F. Yang, H. El Bamby, M. Veljkovic, K. Mela, A method for determining the constitutive model of the heat-affected zone using digital image correlation (submitted), Construction and Building Materials. (2022).

Current Perspectives and New Directions in Mechanics, Modelling and Design of Structural Systems – Zingoni (ed.)
© 2022 Copyright the Author(s), *ISBN 978-1-032-39114-4*

Behaviour of cold-formed high strength steel tubular X-joints with circular braces and rectangular chords

Madhup Pandey & Ben Young
Department of Civil and Environmental Engineering, The Hong Kong Polytechnic University, Hong Kong, China

ABSTRACT: This paper presents a test program to investigate the static strengths of cold-formed high strength steel X-joints made of circular hollow section braces and square as well as rectangular hollow sections chords. In total, 10 X-joints were tested under compression. The nominal yield strengths of tubular members were 900 and 960 MPa. Two failure modes were observed, namely chord face failure (F) and a combination of chord face and chord side wall failure modes (F+S). The test strengths were compared with nominal strengths predicted from Eurocode 3 and CIDECT. For the range of tests undertaken, it has been demonstrated that the current design rules in these specifications are quite conservative.

1 INTRODUCTION

In recent years, cold-formed high strength steel (CFHSS) tubular members are increasingly used in various engineering applications owing to the exclusive amalgamation of favourable structural efficiency, construction superiorities and aesthetical appearances. For the same design resistance, the evident merits of HSS are in the reduction of the self-weights of structural members and associated handling costs and time. This paper briefly described the experimental investigation on cold-formed S900 and S960 steel grades X-joints made of circular hollow section (CHS) braces and square as well as rectangular hollow sections (SHS and RHS) chords, which has already been comprehensively reported in Pandey and Young (2020). The applicability of existing design rules given in EC3 (2005) and CIDECT (2009) was then examined for cold-formed S900 and S960 steel grades X-joints.

2 EXPERIMENTAL PROGRAM

2.1 *Test specimens and mechanical properties*

A total of 10 X-joints was tested in the experimental program (Pandey & Young, 2020), where brace members were made of CHS and chord members were made of SHS and RHS (hereafter, RHS includes SHS). This joint configuration has been termed as CHS-RHS in this study. The test specimens were fabricated from the identical batch of tubes used in the investigation of CFHSS T-joints (Pandey & Young, 2019a). Therefore, measured static mechanical properties of tubular members can be referred to Pandey & Young (2019a). The measured static mechanical properties of welding filler material were also investigated by Pandey & Young (2019b). In this study, a semi-automatic gas metal arc welding (GMAW) process was used to fabricate test specimens. The welds were designed in accordance with the prequalified tubular joint details given in AWS D1.1 (2020).

2.2 *Test setup and procedure*

A servo-controlled hydraulic testing machine was used to apply an axial compression load through brace members at 0.3 mm/min. A preload between 2 to 4 kN was applied using the load-control mode, which facilitated the self-adjustment of the special ball bearing, and thus, eliminated any possible gaps. The position of the special ball bearing was then locked for the remaining test. During the tests, chord face indentations (u) and chord side wall deformations (v) were measured. After preloading, the test was conducted under displacement control mode.

2.3 *Failure modes*

For test specimens with β<0.88, failure mode was chord face failure (F), where due to the membrane effect of the chord connecting face and strain hardening of the material, no clear peak load was observed. In this study, a combination of chord face and chord side wall failure modes, i.e. combined failure (F+S), occurred when load-deformation curves showed a gentle peak for test specimens with β = 0.88 and 0.89. The definition of joint failure strength (N_f) adopted in this study was in accordance with CIDECT (2009).

DOI: 10.1201/9781003348450-191

3 EXISTING DESIGN RULES

The nominal strengths from EC3 (2005) and CIDECT (2009) were calculated by two methods, first, by duly including the material factor in the design rules, and second, by without including the material factor. When material factor was included in the design rules, the comparisons of static test strengths with reduced nominal strengths ($N_{E,X}$ and $N_{C,X}$) enabled the appropriateness of the current design rules to be checked for the design of CFHSS CHS-RHS X-joints. On the contrary, when material factor was not included in the design rules, the comparisons of static test strengths with normal nominal strengths ($N^*_{E,X}$ and $N^*_{C,X}$) examined the appropriateness of the original design rules, developed for NSS grades, for CFHSS CHS-RHS X-joints. It should be noted that nominal strengths ($N^*_{E,X}$, $N_{E,X}$, $N^*_{C,X}$ and $N_{C,X}$) were calculated using the measured dimensions and material properties. A reliability analysis was performed in this study as per AISI S100 (2016) to evaluate the reliability levels of existing design rules given in EC3 (2005) and CIDECT (2009).

4 COMPARISONS OF TEST STRENGTHS WITH NOMINAL STRENGTHS

The comparisons between static test strengths with their corresponding nominal strengths predicted from EC3 (2005) and CIDECT (2009) are shown in Table 1. From comparisons presented in Table 1, it can be noticed that the existing design rules given in EC3 (2005) and CIDECT (2009), without including the material factor, became slightly unconservative for CHS-RHS X-joint with β = 0.59. On the other hand, current design rules given in EC3 (2005) and CIDECT (2009), with material factor included, became slightly conservative for test specimen with β = 0.59. However, for all remaining CHS-RHS X-joint test specimens with $0.73 \leq \beta \leq 0.89$, the design rules given in EC3 (2005) and CIDECT (2009), with and without including the material factor, were quite conservative.

5 CONCLUSIONS

An experimental investigation looking into the static strengths of cold-formed high strength steel CHS-RHS X-joints made of S900 and S960 steel grades has been presented in this paper. The test specimens were failed by chord face failure (F) and a combination of chord face and chord side wall failure modes, named as combined failure (F+S) mode. The static test strengths were compared with the nominal strengths predicted from EC3 (2005) and CIDECT (2009). The comparisons demonstrated that the existing design rules given in EC3 (2005) and CIDECT (2009) were quite conservative for CHS-RHS X-joints made of S900 and S960 steel grades, except for CHS-RHS X-joint with β<0.6.

Table 1. Comparisons of test strengths with nominal strengths (Pandey & Young 2020).

Specimens X-d_1×t_1-b_0×h_0×t_0	β	Failure mode	Comparisons $\frac{N_f}{N^*_{E,X}}$	$\frac{N_f}{N_{E,X}}$	$\frac{N_f}{N^*_{C,X}}$	$\frac{N_f}{N_{C,X}}$
X-88.9×4-150×150×6	0.59	F	0.83	1.04	0.96	1.07
X-88.9×3-120×60×4	0.74	F	1.17	1.46	1.25	1.39
X-88.9×3-120×60×4-R	0.74	F	1.17	1.46	1.25	1.39
X-88.9×4-120×60×4	0.74	F	1.19	1.49	1.28	1.42
X-88.9×4-120×60×4-R	0.74	F	1.15	1.43	1.22	1.36
X-88.9×4-120×120×6	0.73	F	1.05	1.32	1.15	1.27
X-88.9×4-120×120×6-R	0.73	F	1.06	1.33	1.16	1.29
X-88.9×4-100×60×4	0.89	F+S	1.14	1.43	1.14	1.27
X-88.9×4-100×100×4	0.88	F+S	1.17	1.46	1.24	1.38
X-88.9×4-100×100×4-R	0.88	F+S	1.20	1.50	1.29	1.43
		Mean (P_m)	1.11	1.39	1.19	1.33
		COV (V_p)	0.099	0.099	0.081	0.081
		Resistance factor (ϕ)	1.00	1.00	1.00	1.00
		Reliability index (β_0)	2.09	2.89	2.55	2.94

REFERENCES

AISI. (2016). North American specification for the design of cold-formed steel structural members. *AISI S100-16, American Iron and Steel Institute*, Washington, DC.

AWS D1.1 (2020). Structural Welding Code – Steel, *AWS D1.1/D1.1M*, American Welding Society (AWS), Miami, USA.

CIDECT (2009). Packer, J.A., Wardenier, J., Zhao, X.L., van der GJ, V., Kurobane, Y. Design guide for rectangular hollow section (RHS) joints under predominantly static loading. *Comite' International pour le Developpement et l'Etude de la Construction TuECbulaire* (CIDECT), Design Guide No. 3, 2nd edn., LSS Verlag, Dortmund, Germany.

EC3 (2005). Design of steel structures–Part 1–8: Design of joints, *EN 1993-1-8*, European Committee for Standardization (CEN), Brussels, Belgium.

Pandey, M., & Young, B. (2019a). Tests on cold-formed high strength steel tubular T-joints. *Thin-Walled Struct*, *143*, 106200.

Pandey, M., & Young, B. (2019b). Compression capacities of cold-formed high strength steel tubular T-joints. *J Constr Steel Res*, *162*, 105650.

Pandey, M., & Young, B. (2020). Structural performance of cold-formed high strength steel tubular X-Joints under brace axial compression. *Engineering Structures*, *208*, 109768.

Current Perspectives and New Directions in Mechanics, Modelling and Design of Structural Systems – Zingoni (ed.)
© 2022 Copyright the Author(s), *ISBN 978-1-032-39114-4*

Experimental investigation on cold-formed high strength steel tubular T-joints after fire exposures

Madhup Pandey & Ben Young
Department of Civil and Environmental Engineering, The Hong Kong Polytechnic University, Hong Kong, China

ABSTRACT: An experimental study has been presented to investigate the static strengths of cold-formed S960 steel T-joints after ISO-834 fire exposures. A total of 9 T-joints was tested after cooling down from 300°C, 550°C and 750°C temperatures. Braces and chords were made of square and rectangular hollow sections. Automatic gas metal arc welding process was used to weld brace and chord members. The post-fire residual joint strengths were compared with the nominal strengths predicted from Eurocode 3 and CIDECT using the measured post-fire residual mechanical properties. It is shown that the existing design rules given in Eurocode 3 and CIDECT could not accurately predict the post-fire residual joint strengths of cold-formed steel tubular T-joints made of S960 steel grade.

1 INTRODUCTION

In the design of high rise buildings, trusses, space frames, bridges and many other structures, where steel is often a preferred choice of material for structural elements, the structural fire resistance of steel members is amongst one of the key design considerations due to its sensitivity towards elevated temperatures. The main consideration of the non-occurrence of structural failure at peak fire temperature merely cannot guarantee against its delayed failure during or after the cooling phase of a fire. Thus, a reliable structural evaluation is required to confirm whether a fire exposed structure should be allowed for direct reuse, repaired or completely demolished, thereby highlighting the strategic importance of post-fire assessment. So far, none of the studies were focused on the post-fire behaviour of cold-formed high strength steel (CFHSS) tubular joints, except the investigation carried out by Pandey & Young (2021a). Therefore, this paper briefly described the experimental study by Pandey & Young (2021a) to investigate the post-fire residual static strengths of cold-formed S960 steel grade tubular T-joints.

2 TEST PROGRAM

2.1 *Specimens and material properties*

Three identical sets of tubular T-joints were fabricated and exposed to the predetermined fire exposure temperatures (Ψ) of 300°C, 550°C and 750°C. The brace and chord members of T-joints were made of square and rectangular hollow sections (SHS and RHS) (hereafter, RHS includes SHS). The nominal yield strengths of tubular members were 960 MPa. A total of nine T-joints was tested by applying axial compression loads through braces and by supporting chord ends on rollers (Pandey & Young, 2021a). In this study, the value of brace width (b_1)-to-chord width (b_0) ratio (β) ranged from 0.41 to 1.0, brace thickness (t_1)-to-chord thickness (t_0) ratio (τ) ranged from 0.98 to 1.02 and chord width (b_0)-to-chord thickness (t_0) ratio (2γ) ranged from 30.6 to 35.3. The chord length (L_0) was adopted as h_1+$3h_0$+*width of end bearing plates* as per AISI S100 (2016). In order to avoid the overall buckling of braces, the brace length (L_1) was designed as the maximum of $2b_1$ and $2h_1$. The post-fire residual static material properties of RHS members were investigated by Pandey & Young (2021b). In this study, a fully automatic gas metal arc welding (GMAW) process was used to fabricate the test specimens.

2.2 *Fire exposures*

All test specimens were exposed to fire inside the gas furnace as per ISO-834 standard fire curve (1999). The fire exposure temperatures (i.e. Ψ=300°C, 550°C and 750°C) were carefully selected based on the experimental investigation conducted by Pandey & Young (2021b) on the post-fire mechanical behaviour of S900 and S960 steels.

2.3 *Test setup*

Once the T-joint test specimens cool down to ambient temperature, they were then subjected to brace axial compression load by duly supporting the chord

DOI: 10.1201/9781003348450-192

ends on rollers. During the test, the hydraulic actuator of the test rig was ramped at 0.3 mm/min. A specially designed spherical ball-bearing was pre-bolted on to the top support of the test rig. A preload between 2 to 4 kN was applied at 0.1 kN/sec. Using pre-calibrated transducers (LVDTs), chord face indentation (u) and chord side wall deformation (v) were carefully measured. In order to continue the test in the post-ultimate region, the displacement control mode was used to derive the hydraulic actuator after preloading. The post-fire residual strength of test specimen ($N_{f,\psi}$) was taken as the load corresponding to the first occurrence of ultimate load and ultimate deformation limit load (i.e. $0.03b_0$). Two failure modes were observed in this investigation, namely chord face failure (F) mode for $\beta \neq 1$ and chord side wall failure (S) mode for $\beta=1$.

3 EXISTING DESIGN RULES

Currently, no design rules are available in any code to predict the post-fire residual strengths of tubular joints. Therefore, in this study, the nominal strengths were predicted from CIDECT (2009) and EC3 (2005) using the measured post-fire material properties. The nominal strengths from CIDECT (2009) and EC3 (2005) were obtained by two methods, first, when material factor was incorporated in the design rules, and second, when material factor was not incorporated in the design rules. A reliability analysis was also performed in this study in accordance with the method stipulated in AISI S100 (2016) to evaluate the reliability levels of existing design rules given in CIDECT (2009) and EC3 (2005).

4 COMPARISONS OF TEST STRENGTHS WITH NOMINAL STRENGTHS

The comparisons of post-fire residual joint failure strengths ($N_{f,\psi}$) with the nominal strengths ($N^*_{E,T,\psi}$, $N_{E,T,\psi}$, $N^*_{C,T,\psi}$ and $N_{C,T,\psi}$) are shown in Table 1. From Table 1, it can be noticed that the design rules given in EC3 (2005), with and without including the material factor, were unconservative for CFHSS fire exposed T-joints with $\beta \neq 1$ and became quite conservative when $\beta=1$. However, on comparing the $N_{f,\Psi}$ of fire exposed T-joints with the nominal strengths predicted from CIDECT (2009), with and without including the material factor, a mixed comparison trend can be noticed, wherein CIDECT (2009) predictions varied from marginally unconservative to marginally conservative when $\beta \neq 1$ and became quite conservative when $\beta=1$.

Table 1. Comparisons of test strengths with nominal strengths (Pandey & Young 2021a).

Specimens	Comparisons			
T-$b_1 \times h_1 \times t_1$-$b_0 \times h_0 \times t_0$-$\Psi$	$\frac{N_{f,\psi}}{N^*_{E,T,\psi}}$	$\frac{N_{f,\psi}}{N_{E,T,\psi}}$	$\frac{N_{f,\psi}}{N^*_{C,T,\psi}}$	$\frac{N_{f,\psi}}{N_{C,T,\psi}}$
T-50×100×4-120×120×4-300°C	0.71	0.88	0.89	0.98
T-80×80×4-140×140×4-300°C	0.68	0.85	0.89	0.99
T-140×140×4-140×140×4-300°C	3.76	4.70	3.95	4.38
T-50×100×4-120×120×4-550°C	0.78	0.98	1.07	1.17
T-80×80×4-140×140×4-550°C	0.78	0.97	1.03	1.14
T-140×140×4-140×140×4-550°C	3.52	4.40	3.74	4.14
T-50×100×4-120×120×4-750°C	0.74	0.93	0.97	1.07
T-80×80×4-140×140×4-750°C	0.64	0.79	0.81	0.90
T-140×140×4-140×140×4-750°C	1.82	2.28	1.96	2.16
Mean (P_m)	1.49	1.86	1.70	1.88
COV (V_p)	0.853	0.853	0.742	0.745
Resistance factor (ϕ)	1.00	1.00	1.00	1.00
Reliability index (β_0)	0.82	1.03	1.12	1.22

5 CONCLUSIONS

This paper presents an experimental program to investigate the post-fire residual static strengths of cold-formed S960 steel grade T-joints. The test specimens were failed by chord face failure (F) and chord side wall failure (S) modes. The current design rules given in CIDECT (2009) and EC3 (2005) could not accurately predict the post-fire residual strengths of cold-formed S960 steel grade RHS T-joints using post-fire residual mechanical properties.

REFERENCES

AISI. (2016). North American specification for the design of cold-formed steel structural members. *AISI S100-16, American Iron and Steel Institute*, Washington, DC.

CIDECT (2009). Packer, J.A., Wardenier, J., Zhao, X.L., van der GJ, V., Kurobane, Y. Design guide for rectangular hollow section (RHS) joints under predominantly static loading. *Comite' International pour le Developpement et l'Etude de la Construction TuECbulaire* (CIDECT), *Design Guide No. 3 , 2nd edn.*, LSS Verlag, Dortmund, Germany.

EC3 (2005). Design of steel structures–Part 1–8: Design of joints, *EN 1993-1-8*, European Committee for Standardization (CEN), Brussels, Belgium.

ISO-834 (1999). Fire-resistance tests-Elements of Building Construction-Part 1-General requirements, *ISO 834–1*, International Organization of Standards.

Pandey, M., & Young, B. (2021a). Post-fire behaviour of cold-formed high strength steel tubular T-and X-joints. *J Constr Steel Res*, *186*, 106859.

Pandey, M., & Young, B. (2021b). Post-fire mechanical response of high strength steels. *Thin-Walled Structures*, *164*, 107606.

Current Perspectives and New Directions in Mechanics, Modelling and Design of Structural Systems – Zingoni (ed.)
© 2022 Copyright the Author(s), ISBN 978-1-032-39114-4

Study on mechanical properties of high-strength steel butt-welded joints

W.Y. Cai, Y.B. Wang & G.Q. Li
College of Civil Engineering, Tongji University, Shanghai, China

ABSTRACT: In this research, using the notched tensile specimens, the material properties of hardened heat-affected zone (HHAZ), softened heat-affected zone (SHAZ), welding material zone (WM), and base metal zone (BM) in the butt-welded joints made of Q550 and Q690 steels respectively and their strength-matching welding materials were first experimentally studied. The obtained material properties of each zone in the steel butt-welded joints were used to numerically determine the strength of high-strength steel butt-welded joints and were further validated by the experiment results. According to this research, for a given heat input energy and geometry, the strength of SHAZ in high-strength steel butt-welded joint were deteriorated, compared with that of base metal. Due to the constraining effects from WM and HHAZ, the strength of the high-strength butt-welded joint was lower than that of the base metal but higher than that of SHAZ.)

1 INTRODUCTION

The butt-welded joint is normally formed by joining the ends of flat plates with similar or nearly similar thickness and filling the cavity between the plates by the welding materials [1]. According to the rectangular hardness distribution in Figure 1, four different zones can be simply defined in the high-strength steel butt-welded joint, which is welding material zone (WM), Hardening Heat Affected- Zone (HHAZ), SHAZ, base metal zone (BM).

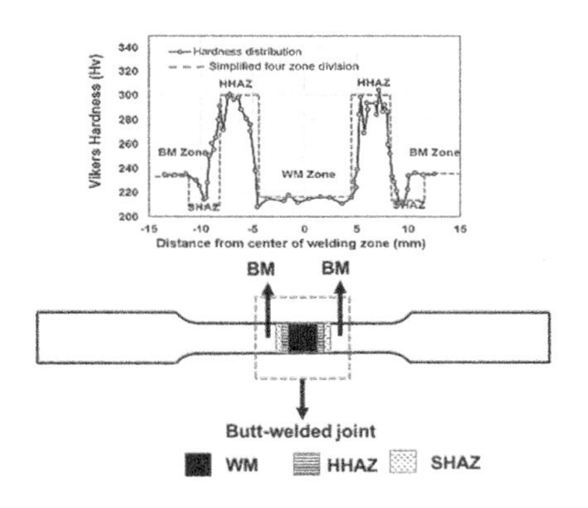

Figure 1. Distribution of each zone.

2 EXPERIMENTAL STUDY ON DIFFERENT ZONES IN HIGH-STRENGTH STEEL BUTT-WELDED JOINT

According to simplified hardness distribution, the notched specimens were designed to make sure that failure occurred in the corresponding zone so that the strength of that zone could be obtained (Figure 2). The tension tests were performed using the MTS machine with a maximum capacity of 500 kN. These experimental results shown in Figure 3 would be used to obtain the stress-strain behavior of each zone in the high-strength steel butt-welded joint in the following part.

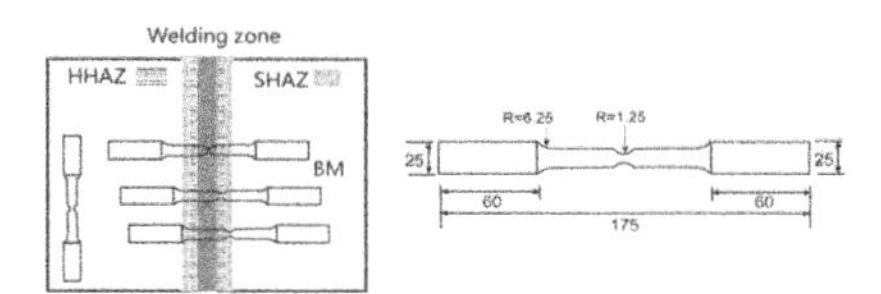

Figure 2. Specimen design and dimensions (Unit: mm).

3 NUMERICAL STUDY ON HIGH-STRENGTH STEEL BUTT WELD JOINT

The true stress-strain behaviors of HHAZ, SHAZ, and WM were subsequently calibrated from test results together with finite element analysis results. The approximate Vickers hardness of HHAZ, SHAZ, and WM and theoretical formula [2] were used to obtain the true stress and strain behavior of each zone. The final true stress-strain model was shown in Figure 4.

The true stress-strain model was further validated by comparing the simulations with the tests on the high-strength steel butt-welded joints reported in Ref [3]. Figure 5 shows the comparison of mechanical behavior of the butt-welded joints made of high-strength steels Q550 with the corresponding strength-matching welding materials, indicating the accuracy of the true stress-strain model used in the simulations.

4 OBSERVATIONS OF THE STRENGTH OF HIGH-STRENGTH STEEL BUTT-WELDED JOINTS WITH DIFFERENT STRENGTH

The load vs. displacement curves for different butt-welded joints and base metal were shown in Figure 6. For both BM specimens and butt-welded joints, they demonstrated an increase of strength with the increase

DOI: 10.1201/9781003348450-193

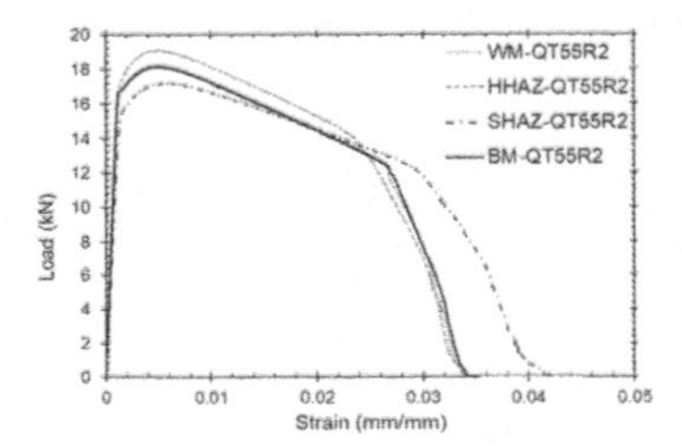

Figure 3. The tension test results of the notched specimens.

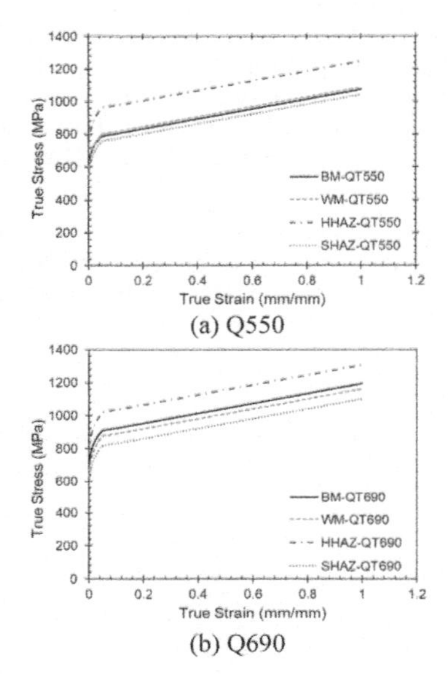

Figure 4. True stress-strain curves of HHAZ, SHAZ, WM, BM for different butt-welded joints. (a) Q550 (b) Q690.

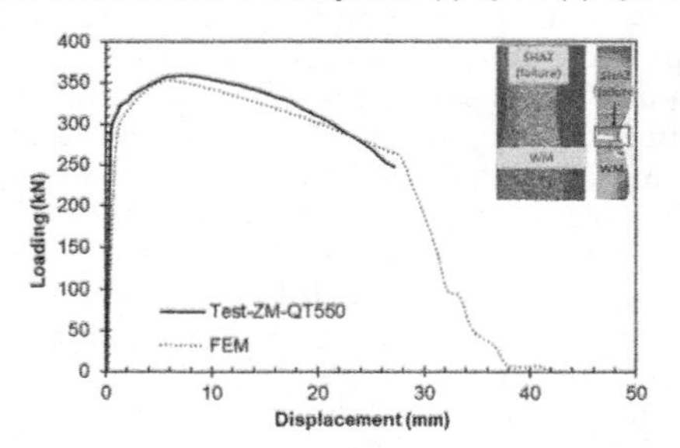

Figure 5. Comparisons of mechanical properties of the high-strength steel butt-welded joints between tests and simulations.

of steel grades. The ratio of the strength of each zone to that of the base material and the ratio of the maximum strength of the butt-welded joint to that of the base material was listed in Table 1.

5 CONCLUSIONS

1) For a given heat input (1.5 kJ/mm), in the Q550 and Q690 steel butt-welded joints, both the true stresses at yielding and at tensile stress demonstrated approximately a 5-17% reduction for SHAZ and a 10-20% increase for HHAZ, compared with the corresponding base metals.
2) Due to the constraining effects from WM and HHAZ, the strength of the high-strength butt-welded joint was lower than that of the base metal but higher than that of SHAZ.
3) This study proposed an approach to experimentally investigate the material behavior of narrow zones in the high-strength butt-welded joints, which can predict the mechanical properties of the high-strength steel butt-welded joints with reasonable accuracy.

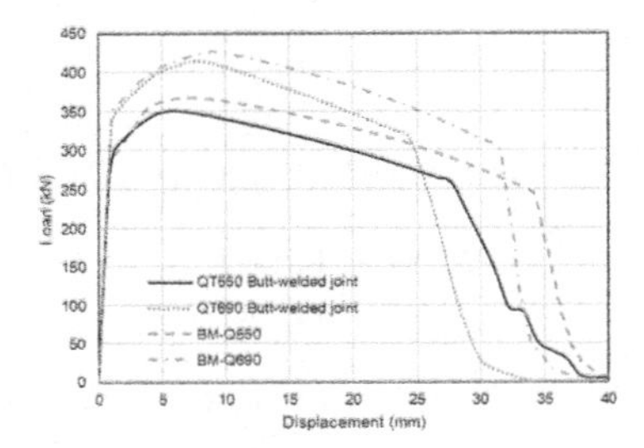

Figure 6. Strength of butt-welded joints and base metals.

Table 1. Strength ratio of each zone and global butt-welded joint to the specimen made of the pure base material.

Steel type	Stress	BM	WM	HHAZ	SHAZ	Butt-Welded joint
Q550	True stress at yielding	590 (1.0)	558 (0.95)	717 (1.22)	515 (0.87)	-
	True tensile stress	789 (1.0)	799 (1.01)	962 (1.22)	756 (0.96)	-
	Engineering Tensile stress	734 (1.0)	760 (1.04)	915 (1.25)	720 (0.90)	720 (0.96)
Q690	True stress at yielding	692 (1.0)	630 (0.91)	775 (1.12)	572 (0.83)	-
	True tensile stress	909 (1.0)	873 (0.97)	1021 (1.13)	814 (0.90)	-
	Engineering Tensile stress	852 (1.0)	822 (0.96)	963 (1.13)	767 (0.90)	794 (0.94)

Note: the values in the bracket are the ratios of the stress to that of BM.

REFERENCES

[1] Salmon C G, Johnson J E & Malhas F A. 2009. *Steel structures – design and behavior (5th ed.)*, Pearson-Prentice Hall, Upper Saddle River (NJ) (2009)

[2] Pavlina E J & Tyne C. 2008. Correlation of yield strength and tensile strength with hardness for steels. *Journal of Materials Engineering & Performance*, 17 (6): 888–893.

[3] Zhao, X. 2020. Study on Mechanical Properties and Design Method of over 500MPa High Strength Steel Butt Welded Connections [D]. *College of Civil Engineering, Tongji University, China*

13. Reinforced concrete structures, Prestressed concrete, Concrete structural elements, Mechanics of concrete

Current Perspectives and New Directions in Mechanics, Modelling and Design of Structural Systems – Zingoni (ed.)
© 2022 Copyright the Author(s), ISBN 978-1-032-39114-4

Using nonlinear finite element modeling for punching shear design of concrete slabs

P.M. Beaulieu, G.J. Milligan & M.A. Polak
University of Waterloo, Waterloo, Ontario, Canada

ABSTRACT: Nonlinear Finite Element Analysis (NLFEA) is currently available in many commercial programs. At the same time computing power has increased to the level that routine use of NLFEA for structural analysis became feasible. Structural concrete shows distinctly nonlinear behaviour and thus the available constitutive models, within FEA packages, include considerations for nonlinear material constitutive behaviour, damage, plasticity and concrete-reinforcement interactions. However, the results of FEA of concrete structures, depend on the choice of numerous modelling parameters making them susceptible to errors in assumptions, and dependent on decisions of the modeler regarding meshing, boundary conditions, and constitutive models for concrete. This paper presents a discussion of how to rationally approach NLFEA of structural concrete with the example of concrete slabs susceptible to punching shear failure. The analyses are done using the concrete damaged plasticity (CDP) model available in ABAQUS. The importance of rational choices of modelling parameters and the calibration of the FEA models based on experimental results is discussed. The paper shows specific examples where NLFEA was used to expand the existing experimental databases, perform detailed parametric studies and propose improvements in the design formulas available in the codes.

1 INTRODUCTION

Application of nonlinear finite element analysis (NLFEA) to the design of concrete structures depends to a great extent on the proper modelling parameters adopted for analyses. In this paper, the key parameters of one of the most popular material models for the NLFEA of reinforced concrete structures, the concrete damaged plasticity (CDP) model, are discussed. The goal of this paper is to provide practitioners with an understanding of the influence of these parameters and guidance for the use of the CDP model. The discussion is presented using NLFEA for concrete slabs with and without shear reinforcement failing in punching shear. To facilitate these discussions the calibration of a nonlinear finite element model (FEM) in ABAQUS based on the experimental work of Adetifa and Polak (2005) is considered.

Concrete flat slabs supported on columns are one of the most popular structural systems for buildings, particularly those where open floor plans are essential. However, due to the lack of additional supporting members a complicated three-dimensional state of stress exists near the slab-column connections. In slabs without shear reinforcement this complicated state of stress can lead to brittle punching shear failures when the applied shear stresses exceed the shear strength provided by the slab concrete. A brittle punching shear failure of one connection can lead to the progressive collapse of an entire structure. As such, shear reinforcement is commonly placed in the vicinity of the slab-column connections to provide additional strength and ductility and mitigate the possibility of punching failures occurring.

Flat concrete slabs have been studied experimentally, but even though the existing experimental database is extensive not all parameters have been adequately addressed due to the significant cost and time required to conduct experimental testing. Thus, NLFEA, calibrated to experimental results, has become a viable alternative to experimental testing. In particular, NLFEA with solid three-dimensional elements can be used to supplement experimental databases and allows for detailed analyses of punching shear behaviour and measurement of quantities which are difficult in the laboratory, such as a detailed strain distribution throughout the slab and reinforcement. However, NLFEA requires an experienced user as the results are dependent on numerous parameters and modelling assumptions.

DOI: 10.1201/9781003348450-194

2 CONCRETE DAMAGED PLASTICITY

The CDP model which has been adopted into ABAQUS was first introduced by Lubliner et al (1989), and was further developed by Lee and Fenves (1998) to decouple the degradation damage from the effective stress determination (Jiang et al, 2021). The formulation of this model results in an evolution of the failure surface dictated by the plastic behaviour of the concrete in tension and compression, as well as the softening behaviour of the tension response as determined by the fracture energy (G_f), and other material parameters such as the dilation angle (ψ). This paper presents the formulation of the CDP model and discusses important parameters for modelling with the CDP model.

3 SPECIMENS USED FOR CALIBRATION

This paper presents several analyses using the CDP model in ABAQUS for specimens SB1, SB2, SB3, and SB4 tested by Adetifa and Polak (2005) at the University of Waterloo. This paper also discusses preliminary modelling of the specimens conducted by Genikmosou (2015). The specimens are a good representation a typical interior slab-column connection that would be susceptible to punching in a concrete flat slab supported on columns system.

4 PARAMETERS TO CALIBRATE

In order to accurately use the CDP model within ABAQUS, many parameters must first be calibrated and inputted into the model for the type of specimen being modelled. Several parameters have default values that can be assumed for the modelling of most specimens. The values for these parameters are discussed and modelling recommendations are presented.

The uniaxial compressive and tensile behaviour of the concrete are critical when modelling using the CDP model and can be described by many models inputted as tabular data within ABAQUS. This paper discusses an approach to calibrating the compressive and tensile behaviour, and also presents a calibration study focused on the fracture energy (G_f) parameter for the SB series specimens.

This paper also presents studies that were conducted for other important modelling parameters, including dilation angle (ψ), element type, mesh size, and boundary conditions. When combined, the calibration of all the discussed parameters provides a rational methodology for calibrating a NLFEA model for a slab-column specimen using the CDP model in ABAQUS.

5 ANALYSIS OF SB SPECIMENS

Finally, this paper presents output of a calibrated NLFEA in ABAQUS for the SB series specimens. A rational method of validating the model output based on the test results of the modelled specimens is presented, including a discussion of load-deflection behaviour, shear reinforcement strains, and post-loading crack patterns.

6 CONCLUSIONS

To summarize, in this paper, the theory of the CDP model in ABAQUS is presented, and the parameters that are influential to the model are identified. The calibration of NLFEA models for slab-column specimens SB1 to SB4 is then presented to illustrate a technique that could be used to calibrate the identified parameters based on a set of test results. Finally, several metrics are presented for specimens SB1 to SB4 with which the accuracy of a calibrated NLFEA model for a slab-column specimen in concentric punching can be assessed. These techniques and results provide a platform for a modeler to understand and use the CDP model to effectively calibrate a NLFEA model which can be used to conduct parametric studies and supplement a testing program for a set of reinforced concrete specimens.

REFERENCES

Adetifa, B., & Polak, M. A. (2005). *Retrofit of Slab Column Interior Connections Using Shear Bolts*. ACI Structural Journal.

Genikomsou, A. S. (2015). Nonlinear Finite Element Analysis of Punching Shear of Reinforced Concrete Slab-Column Connections. *PhD Thesis*. Waterloo, Ontario, Canada: University of Waterloo.

Jiang, L., Orabi, M. A., Jiang, J., & Usmani, A. (2021). *Modelling Concrete Slabs Subjected to Fires using Nonlinear Layered Shell Elements and Concrete Damaged Plasticity Model*. Engineering Structures.

Lee, J., & Fenves, G. L. (1998). *Plastic-Damage Model for Cyclic Loading of Concrete Structures*. Journal of Engineering Mechanics.

Lubliner, J., Oliver, J., Oller, S., & Onate, E. (1989). *A Plastic-Damage Model for Concrete*. Solids Structures.

Current Perspectives and New Directions in Mechanics, Modelling and Design of Structural Systems – Zingoni (ed.)
© 2022 Copyright the Author(s), ISBN 978-1-032-39114-4

Design of precast concrete framing systems against disproportionate collapse using component-based methods

K. Riedel, R. Vollum & B. Izzuddin
Department of Civil and Environmental Engineering, Imperial College, London, UK

G. Rust & D. Scott
Technology & Innovation Group, Laing O'Rourke, Dartford, UK

ABSTRACT: EN 1992-1-1:2004 provides prescriptive rules for determining minimum tying requirements in concrete structures that are not designed for accidental actions. Physically, the tying requirements are intended to localise any accidental damage that might occur by establishing alternative load paths. A major limitation of the EN 1992-1-1:2004 tying rules is that they do not provide any quantifiable level of performance under accidental actions. This paper focuses on the design of precast concrete structures and their connections for prevention of disproportionate collapse under sudden column removal. A component-based method is used in which the structural system is replicated using beam-column elements, rigid links and springs. The proposed approach is demonstrated for the design of a novel D-Frame precast concrete system. Emphasis is placed on the design of a beam-to -beam half-lapped joint for the tying forces resulting from sudden column removal. Means of achieving the required connection strength and ductility are investigated with the latter proving most challenging. Directly connecting the reinforcement ties in adjoining precast beams through couplers is shown to provide greater ductility than connecting beams through top and bottom steel plates bolted to vertical dowel bars embedded in the beams.

1 INTRODUCTION

The D-Frame system, shown in Figure 1, is a demountable precast concrete (PC) product-based building solution (PBBS). It consists of ribbed inverted U-planks sitting on top of inverted T-beams spanning up to 9 m over the columns. The beam-to beam connections are provided by half-lapped joints located at the point of contraflexure. Achieving continuity of tying force at these joints is the main challenge in the design of the D-Frame. This paper overviews the robustness of the D-Frame system by analysing the response of the frame under instantaneous column removal. This is done by using energy balance methods to estimate the dynamic effects through the pseudo-static approach of Izzuddin et al. (2008). An analysis of a typical bay consisting of four 9 m long spans under the middle column removal is carried out.

Two connection types were proposed for the half-lapped joint: plated and bracketed connectors. Both were designed to work under hogging and sagging moments. The plated connection was designed to meet the basic prescriptive tying requirements outlined in Eurocode 2 (BSI 2004). It consists of top and bottom plates with anchored fasteners that join the beams together. The plates were identified early on to fracture at low displacements following column removal. This happens before any significant pseudo-static resistance can be achieved in the system. The bracketed connection provides continuity between the top and bottom 2B40 bars in adjoining beams. The brackets in adjoining beams are connected with bolts onsite. The paper considers the benefit of debonding the B40 bars to either side of the half-lapped joint, by 350 mm and 650 mm, to increase the deformation at failure.

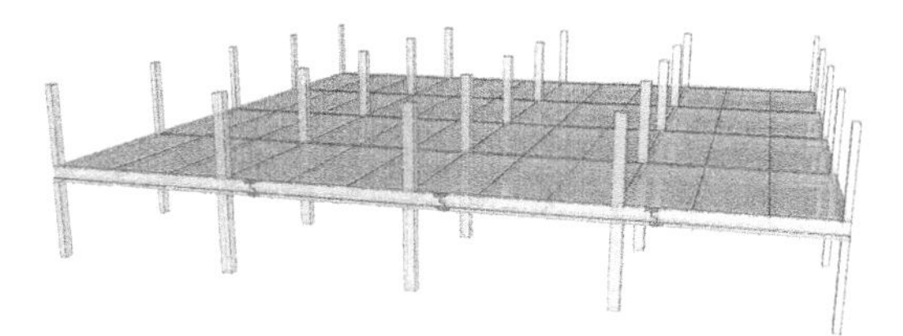

Figure 1. D-Frame system.

2 COMPONENT-BASED MODELLING

The D-frame beams were modelled using a macromodleing approach implemented in the nonlinear finite element analysis (NLFEA) program Adaptic (Izzuddin 1991). The response of the

DOI: 10.1201/9781003348450-195

primary beam in a four-bay frame was investigated under middle column removal. Loads are applied at the rib locations, where planks sit on the beam. The concrete beams were replicated using beam-column elements with separate reinforcement bars, connected together with spring elements. The springs simulate bond-slip properties according to Model Code 2010 (fib 2013). Modelling reinforcement slip is necessary to avoid underestimating beam displacements at reinforcement fracture subsequent to yield.

Component-based modelling was used for half-lapped joints, consisting of a series of beam elements, rigid links and contact springs. A bracketed connection configuration is shown in Figure 2. The response of the plated connection was verified against an Abaqus Dassault Systèmes (2014) micromodel. This was based on ductile fracture theory, and showed good correspondence with the Adaptic results. The modelling of the bracketed connection was validated with laboratory test results (Riedel et al. 2022).

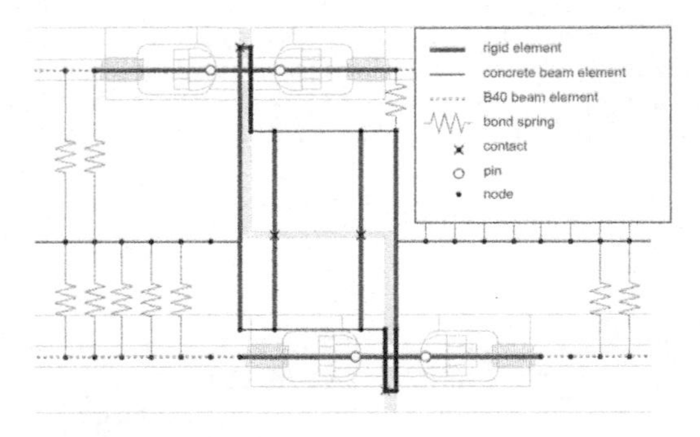

Figure 2. Component-based joint: coupler connection, secondary reinforcement omitted for clarity.

3 RESULTS AND DISCUSSION

The static and pseudo-static responses for all four models are shown in Figure 3 which is expressed in terms of the load level applied at the plank rib locations. The failure in all models was caused by the fracture of the bottom steel assemblies in sagging at the removed column. The plated model failed at the lowest displacements of 137 mm. This was caused by combined tension and localised plate bending within the reduced net area at the first bolt row location. This limited the pseudo-static resistance to 137 kN.

The connector brackets allow significantly greater displacements, and despite having lower yield strength, the long plastic plateau permits higher pseudo-static loads. The B40 bars failed in combined tension and bending about the stirrup closest to the connector. As shown in Figure 3, debonding increases the displacement at which the B40 bars fail resulting in increased pseudo-static resistance. The displacements and pseudo-static loads at failure were

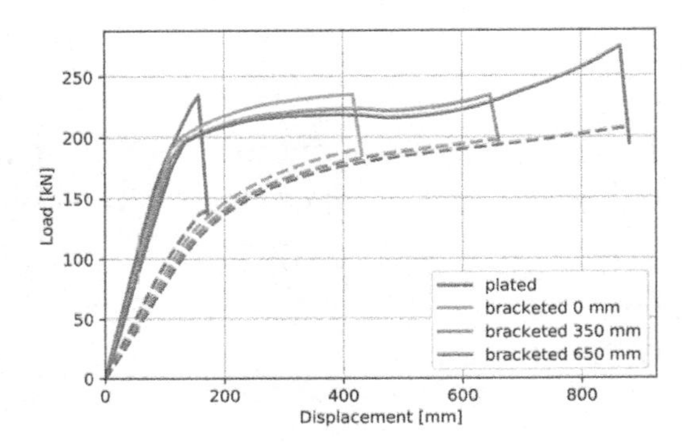

Figure 3. Load-displacement curves (continuous: static curve, dashed: pseudo-static curve).

416 mm and 188 kN for the system with no debonding, 647 mm and 197 kN for the model with 350 mm debonding length and 867 mm and 207 kN for the model with 650 mm debonding.

The horizontal loads in the damaged bay indicated large compressive arching action in all models. Force reversal to tension only developed for the debonded length of 650 mm. This is indicated in Figure 3 by hardening at a displacement of around 750 mm.

The analysis showed, that PBBS precast concrete systems can be designed to achieve high resistance under sudden column removal scenario but designing connections for tying forces alone is insufficient to guarantee the development of catenary action. This is because the PC joints tend to be weaker than the connected beams resulting in localised fracture of ties at connections. In the adopted bracketed connection, premature fracture of ties is mitigated by localised debonding. This increases the system ductility supply and consequently pseudo-static resistance.

REFERENCES

BSI (2004). *BS EN 1992-1-1:2004 Eurocode 2: Design of concrete structures: General rules and rules for buildings.* London, UK: British Standards Institution.

Dassault Systèmes (2014). Abaqus analysis user's guide, version 6.14.

fib (2013). *fib Model Code for Concrete Structures 2010.* Berlin, Germany: The Fédération internationale du béton.

Izzuddin, B. A. (1991). *Nonlinear dynamic analysis of framed structures. Thesis submitted for the degree of doctor of philosophy in the University of London.* Thesis, London, UK.

Izzuddin, B. A., A. G. Vlassis, A. Y. Elghazouli, & D. A. Nethercot (2008). Progressive collapse of multi-storey buildings due to sudden column loss - Part I: Simplified assessment framework. *Engineering Structures 30*, 1308–1318.

Riedel, K., R. Vollum, B. Izzuddin, J. P. Vella, & G. R. D. Scott (2022). Experimental testing of a novel d-frame connection under sudden column removal. In *fib Congress 2022 in Oslo*, Oslo, Norway.

Current Perspectives and New Directions in Mechanics, Modelling and Design of Structural Systems – Zingoni (ed.)
© 2022 Copyright the Author(s), *ISBN 978-1-032-39114-4*

Semi-probabilistic nonlinear assessment of post-tensioned concrete bridge made of KT-24 girders

B. Šplíchal & D. Lehký
Brno University of Technology, Brno, Czech Republic

J. Doležel
Moravia Consult Olomouc, a.s., Olomouc, Czech Republic

ABSTRACT: This paper compares several methods for assessing the design resistance of an existing railway bridge made of KT-24 precast post-tensioned concrete girders. The determination of the load-bearing capacity of the structure is carried out by means of probabilistic nonlinear analysis using finite element method. Load-bearing capacity is determined for the ultimate as well as serviceability limit states. A fully probabilistic approach is compared to selected code-recommended semi-probabilistic methods, which are able to greatly reduce the number of nonlinear calculations needed to estimate the design value of resistance. The results are compared and discussed with respect to accuracy and required computational time, which is a critical issue when performing a global nonlinear analysis of a complex structure.

1 INTRODUCTION

The nonlinear analysis is by its nature a global type of assessment, in which all structural parts interact. Therefore, the safety format suitable for design of concrete structures using nonlinear analysis requires a global approach (Castaldo et al. 2019). This paper compares several methods for assessing the design resistance of an existing bridge made of KT-24 precast post-tensioned concrete girders. The determination of the load-bearing capacity of the structure is carried out by means of probabilistic nonlinear analysis using finite element method. A fully probabilistic (FP) approach is compared to selected code-recommended semi-probabilistic methods, which are able to greatly reduce the number of nonlinear calculations needed to estimate the design value of resistance. Studied methods include the ECoV method according to fib Model Code 2010 (2012), the EN 1992-2 (2004) method and the partial safety factor (PSF) method (EN 1990 2002). These code-supported methods are compared to newly developed semi-probabilistic method based on Taylor series expansion called Eigen ECoV method (Novák & Novák 2021). The results are compared and discussed with respect to accuracy and required computational time, which is a critical issue when performing a global nonlinear analysis of a complex structure.

2 POST-TENSIONED CONCRETE BRIDGE

Post-tensioned concrete railway bridge is made of four 24 m long KT-24 girders, which are made of concrete C35/45. The mild reinforcement is made of V 10 400 (approximately Bst 420) steel and the 17 main prestressing tendons are made of 23 patented wires with diameter 7 mm. The finite element method computational model was created in ATENA 3D software (Červenka et al. 2012). The symmetrical half of the girder was modelled. The girder was loaded with self-weight, secondary dead loads, and a partial continuous load corresponding to the LM71 load model to induce the maximum bending moment (Figure 1).

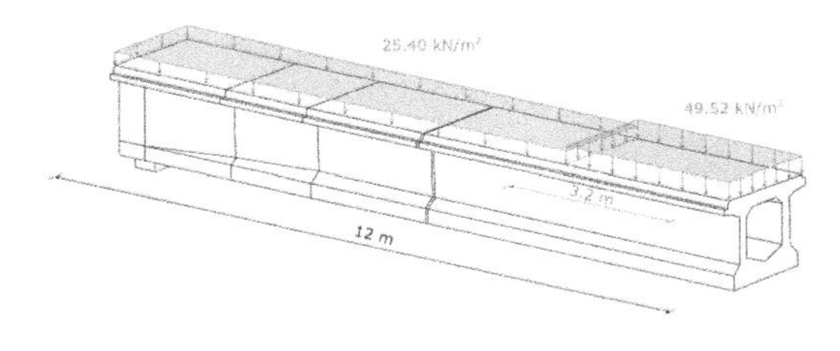

Figure 1. LM71 load model on the symmetrical half of the girder.

DOI: 10.1201/9781003348450-196

3D Nonlinear Cementitious 2 material model was used for concrete. Tendons and mild reinforcement were modelled using bilinear stress–strain law with hardening. Basic material parameters and their statistics were obtained from diagnostic survey and JCSS Probabilistic Model Code (Joint Committee on Structural Safety 2002). In accordance with the standard, the girder was assessed for three limit states – ultimate limit state (ULS, see Figure 2 for model of the structure at the maximum load-bearing capacity, which is accompanied by a rapid increase in deformation, the development of major bending cracks and the progressive collapse of the entire structure), limit state of decompression (LSD), and limitation of compression stress in concrete (LCSC).

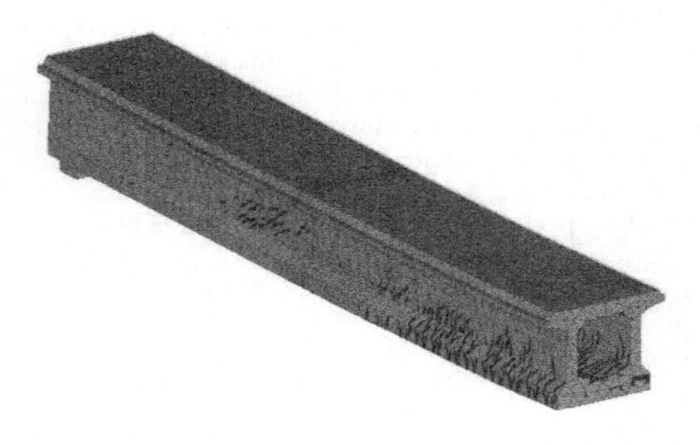

Figure 2. Girder collapse when reaching ULS and bending crack pattern.

3 RESULTS

The design values of the moment resistance were determined for all three limit states and their corresponding reliability levels using above listed methods. In case of the FP method, 32 random realizations of the input variables were generated from the stochastic model. The obtained results are summarized and compared in terms of accuracy, consistency, and time consumption. As an illustration, Figure 3 shows the design moments of resistance for ULS obtained by each method.

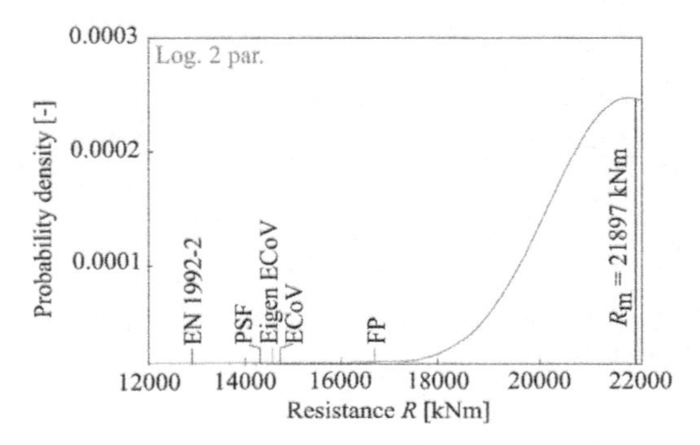

Figure 3. Comparison of design moment resistance values for ULS.

4 CONCLUSIONS

Nonlinear fully probabilistic finite element analysis is currently the most accurate approach for determining the design resistance of structures. However, dozens of simulations are required for this approach, making it time consuming. For this reason, simplified reliability methods that require only a few simulations are often used in engineering practice. The disadvantage of these methods is the inconsistency of the design values obtained for different structures, different limit states and failure modes, and the required reliability values. One of the main objectives of this paper was to compare the available semi-probabilistic methods together with a fully probabilistic approach in the analysis of different limit states on a post-tensioned concrete bridge.

The results show that the ECoV method, the Eigen ECoV method and the PSF method led to similar values of design resistance, and for the ULS and LCSC limit states these are approximately 8% lower than those obtained by the FP approach. The EN 1992-2 method can only be used for ULS, and its results here proved to be very conservative (about 80% of FP). If the FP method cannot be used due to time constraints, the ECoV and Eigen ECoV methods can be recommended based on the obtained results. For the analyzed structure, these methods gave consistent slightly conservative results.

ACKNOWLEDGMENT

This work was supported by the project No. 22-00774S, awarded by the Czech Science Foundation (GACR) and the specific university research project No. FAST-J-22-7914 granted by Brno University of Technology.

REFERENCES

Castaldo, P., Gino, D. & Mancini, G. 2019. Safety formats for non-linear finite element analysis of reinforced concrete structures: discussion, comparison, and proposals. *Engineering Structures* 193: 136–153.

Červenka, V., Jendele, L. & Červenka, J. 2012. *ATENA Program Documentation – Part 1: Theory.* Cervenka Consulting, Prague, Czech Republic.

EN 1990. 2002. *Eurocode: basis of structural design.* Brussels, Belgium.

EN 1992- 1-1. 2004. *Eurocode 2: Design of concrete structures – Part 1-1: General rules and rules for buildings.* Brussels, Belgium.

fib Bulletins 65 & 66. 2012. *Model Code 2010.* Fédération Internationale du Béton (fib), Lausanne, Switzerland.

Joint Committee on Structural Safety. 2002. *Probabilistic Model Code.* Available from: https://www.jcss-lc.org/jcss-probabilistic-model-code.

Novák, L. & Novák, D. 2021. Estimation of coefficient of variation for structural analysis: The correlation interval approach. *Structural Safety* 92: 102101.

Current Perspectives and New Directions in Mechanics, Modelling and Design of Structural Systems – Zingoni (ed.)
© 2022 Copyright the Author(s), ISBN 978-1-032-39114-4

Punching shear behavior of continuous flat slabs: A new test setup incorporating system influences

Matthias Kalus*, Jan Ungermann & Josef Hegger
Institute of Structural Concrete, RWTH Aachen University, Aachen, Germany

ABSTRACT: The punching shear behavior of flat slabs has been investigated in the past mainly on flat slab cutouts since experiments on continuous flat slab systems are expensive and complex. However, these cutouts represent only the hogging moment area around the slab-column connection and thus neglect system influences such as moment redistribution between sagging and hogging moment as well as compressive membrane action (CMA). Nevertheless, a few existing experiments on continuous slabs and several numerical investigations have indicated a beneficial impact of these influences on the punching shear behavior. For an experimentally verified and consistent extension of existing punching shear models as well as the derivation of more progressive design provisions, new systematic experiments with realistic depiction of these influences are needed. To investigate the punching shear behavior of continuous flat slab systems, a new test setup was developed in which isolated flat slab specimens are loaded with varying, load-dependent edge conditions. A numerical model with measured deformations as input parameters is hereby used for real-time simulation of the edge conditions such as CMA and moment redistribution based on an equivalent flat slab system. First test results and numerical calculations are very promising and confirm the positive impact of moment redistribution and CMA on the punching shear behavior due to significant higher load bearing capacities and smaller slab deformations.

1 INTRODUCTION

The punching shear behavior of slabs has been in the focus of research for more than 100 years. Most of the investigations were conducted on isolated slab cutouts which represent just the circular hogging moment area of a slab-column connection. With this chosen geometry for the specimens, most of the relevant influences on the punching shear behavior can be investigated. However, aspects such as moment redistribution and compressive membrane action (CMA) which occur in continuous flat slab systems are excluded. A few experimental and numerical investigations [1–3] indicate a beneficial impact of these aspects on the punching shear capacity and the deformation behavior of the slab.

2 TEST SETUP AND PROGRAM

To quantify the described system influences in more detail, a novel test setup which is an extension to existing conventional punching shear test setups (isolated specimens loaded along a perimeter with radius $r_{q,el}$ = 0.22L), was developed (Figure 1). This test setup enables the application of varying, load-dependent edge conditions on isolated specimens simulating the system influences occurring in continuous flat slabs. These decisive load-dependent edge conditions such as compressive membrane action and moment redistribution are calculated by simultaneous calculations based on a numerical model. Both boundary conditions can be controlled and introduced fully independently of each other.

The core of the test setup consists of 12 hollow piston cylinders for the introduction of the shear force equidistantly positioned in a circle on top of the slab. Threaded rods anchored underneath a rigid floor and on top of the cylinders are used to introduce the shear force to the slab. The influence of moment redistribution is accounted for by introducing a positive bending moment at the edges of the slab. Therefore, the resulting moment in the slab changes and the line of contraflexure shifts closer to the column. The positive bending moment is introduced to the slab by means of four triangular steel frames bonded at the corners to the slab. The diagonal rods are pulled together by two hollow piston cylinders resulting in lifting edges and thus a vertical force introduction at the corners to the slab generating a positive bending moment. The idea of this load introduction was proposed by Clément [4]. The influence of compressive membrane action is considered by four bracings around the slab consisting

*Corresponding author: mkalus@imb.rwth-aachen.de

DOI: 10.1201/9781003348450-197

of tie rods and U-girders which allow the introduction of compressive normal forces by four hollow piston cylinders. Wide load plates are used at the side surfaces of the slabs to initiate a uniform compressive stress inside the load perimeter.

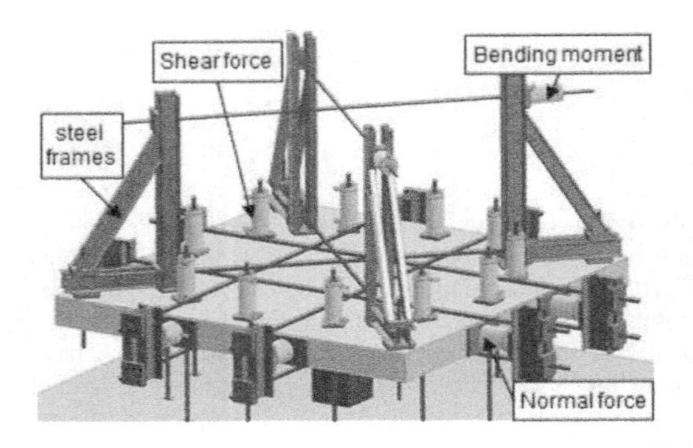

Figure 1. Novel punching shear test setup incorporating moment redistribution and compressive membrane action.

The shear force introduced with the 12 hollow piston cylinders is increased in specific load steps. The boundary forces simulating the system influences are load-dependent and thus need to be calculated for each load step separately. Using the numerical model from Einpaul [2] the load-displacement behavior of an equivalent continuous flat slab system with the span length of L = $r_{q,el}$ /0.22 is calculated. Afterwards, the discrete boundary conditions, i.e. radial normal force n_{sys} and the radial bending moment m_{sys} at the distance of $r_{q,el}$ = 0.22L are derived and introduced to the isolated specimen. By doing so the conventional punching shear test setup is extended by system influences. Further adjustments are planned to incorporate actual measurements such as curvature and dilation in the calculation process in order to update the relationships in real-time during the experiments.

With the new developed test setup, comparative experimental investigations on the punching shear behavior of flat slab cutouts and continuous flat slab systems are possible. A comprehensive test program consisting of 10 specimens (5 isolated specimens, 5 isolated specimens with boundary conditions) is conducted. In the first test series the influence of the reinforcement ratio is analyzed with three different reinforcement ratios (ρ = 0,7 | 1,0 | 1,4 %). The second test series investigates the influence of the concrete compressive strengths with three different strengths (f_c = 20 | 35 | 50 MPa). For each slab configuration two specimens are produced. One of them is tested in conventional way serving as a reference test, and one is loaded with load-dependent boundary conditions simulating slab continuity. By doing so, the influence of slab continuity can be quantified.

3 SUMMARY AND OUTLOOK

A new test setup was developed allowing for the experimental investigation of the punching shear behavior of flat slabs systems by testing a representative flat slab cutout with varying boundary conditions. Discrete load-dependent boundary forces simulating slab continuity are calculated by a numerical model and introduced separately to the slab.

The first ongoing experimental investigations show very promising results which will be presented at the conference and confirm the positive impact of moment redistribution and CMA on the punching shear behavior due to significant higher load bearing capacities and smaller slab deformations. Parallel to the experimental investigations further numerical investigations will be conducted with flat slab systems and flat slab cutouts in order to analyze their different punching shear behavior in more detail.

ACKNOWLEDGMENTS

The presented work is part of the research project 'Method for the investigation of the punching shear behavior of continuous flat slabs under consideration of moment redistributions and membrane actions' at the Institute of Structural Concrete of RWTH Aachen University and has been funded by the Deutsche Forschungsgemeinschaft (DFG, German Research Foundation) – HE2637/36-1 (428862959). The author would like to gratefully acknowledge the support.

REFERENCES

[1] Ladner M, Schaeidt W, Gut S. Experimentelle Untersuchungen an Stahlbeton-Flachdecken. Dübendorf, Schweiz; 1977.

[2] Einpaul J, Fernández Ruiz M, Muttoni A. Influence of moment redistribution and compressive membrane action on punching strength of flat slabs. Engineering Structures 2015;86:43–57.

[3] Kueres D, Schmidt P, Bosbach S, Classen M, Herbrand M, Hegger J. Numerische Untersuchungen zur Durchstanztragfähigkeit von Flachdeckensystemen. Bauingenieur 2018;93(4):141–51.

[4] Clément T, Pinho Ramos A, Fernández Ruiz M, Muttoni A. Influence of prestressing on the punching strength of post-tensioned slabs. Engineering Structures 2014;72:56–69.

Current Perspectives and New Directions in Mechanics, Modelling and Design of Structural Systems – Zingoni (ed.)
© 2022 Copyright the Author(s), ISBN 978-1-032-39114-4

Live load test for pedestrian bridge constructed with innovative concrete bridging system

Hayder H. Alghazali
Civil Engineering, University of Kufa, Al Najaf Governorate, Iraq

John J. Myers*
Missouri University of Science and Technology, Rolla, USA

Kurt E. Bloch
Structural Engineer 2 at Cannon Design, Pacific, USA

ABSTRACT: Changes in deflection and strain during a bridge's life cycle can occur due to time dependent strength losses of the bridge's materials. A live load test can be used to determine bench mark values at early-ages and used to compare them to later-age values to determine any changes that may have occurred throughout a bridge's life cycle. Prestressed precast pedestrian bridge constructed with high strength concrete (HSC) in Rolla, Missouri, USA was utilized in this study. Both mild steel and glass fiber reinforced polymer (GFRP) reinforcement were used in the deck system. A static live load test was utilized to compare the deflection in HSC prestressed precast spandrel beams and precast deck panels. In addition, changes in deflection were compared between deck panels reinforced with mild steel to those reinforced with GFRP. Furthermore, the measured values were compared to theoretical values determined from the basic structural analysis for both simply supported and fixed cases.

1 INTRODUCTION

Throughout the course of history, advancements in the materials used by civil engineers in the design and construction of bridges, buildings, and roads have made improvements to the infrastructure of the nation. Current advancements to bridge construction have lowered costs, reduced construction time, and increased the service life of the structures [1] [2]. One such advancement has been the use of high-strength concrete (HSC) in prestressed bridges. By using HSC, large sustainable bridge structures were built with relatively compact sections. With the improved service life of bridges and reduced concrete in construction, the use of HSC allows for economic savings [3]. Another advancement that has been applied to increasing the durability of bridge structures is substituting mild steel rebar with glass fiber reinforced polymer (GFRP) bars in deck panels. GFRP bars have the positive attribute of being non-corrosive within the concrete where steel reinforcement would normally result in corrosion and cause cracking and spalling of the concrete. By combining the attributes of HSC and GFRP, improved durability in slab sections is possible. By using HSC, which typically will not crack at service loads due to its high strength, any shrinkage cracks that might have resulted will not be as large of a concern because the GFRP will not corrode due to moisture in the concrete.

2 BRIDGE DESCRIPTION

The HSC bridge spans a length of 14.6 m (48 ft) and has a width of 3.0 m (10 ft). The bridge implemented prestressed "L" spandrel beams to function as the structural support of the bridge and the handrails for the pedestrians and has two precast deck panels to form the bridge deck. One precast deck panel was reinforced with mild steel and the other was reinforced with GFRP.

The precast decks of each bridge utilized the same mixture proportion as the beams. Two different types of reinforcements were added to each bridge to monitor the differences in behavior between mild steel and glass fiber reinforced polymer (GFRP) bars in the HSC deck panels. Each deck panel was 8-in. (200 mm) thick and 119-in. (3,000 mm) wide. The beams were placed and welded to embed plates in the abutments.

3 LOAD TEST PROGRAM

A static live load test was conducted on the HSC bridge to obtain a better understanding of the differences in deflection and strain of the concrete bridge. The deflection measurements were determined utilizing laser-based precise surveying. Temperature data

*Corresponding author: jmyers@mst.edu

DOI: 10.1201/9781003348450-198

were monitored with the internal VWSGs with built-in thermistors within the spandrel beams and precast deck panels. In addition, basic structural analysis techniques were utilized to predict the simply supported and fixed behavior of both bridges assuming an uncracked cross-section and comparing them to the measured values. A Toyota Model 7FGU25 forklift truck weighing 8,840 lb. (39.3 kN) was used to load the bridge for the static load test.

Figure 1. Toyota 7FGU25 forklift on the bridge.

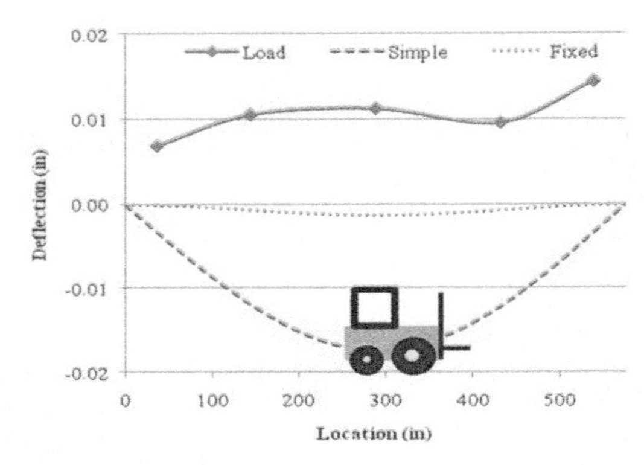

Conversion: 1-in. = 25.4 mm

Figure 2. HSC spandrel beam deflection.

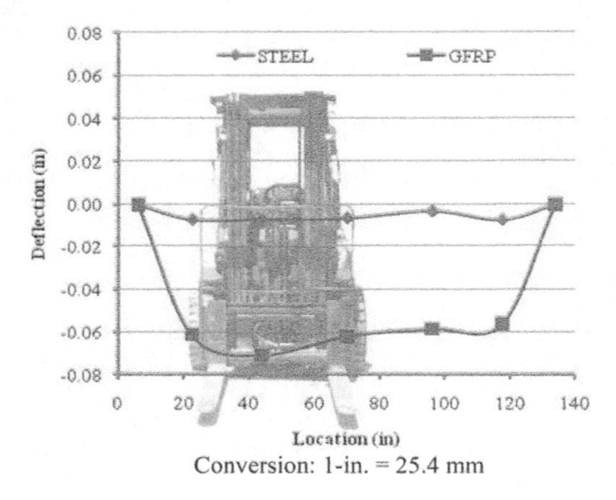

Conversion: 1-in. = 25.4 mm

Figure 3. HSC deck panel with steel vs. GFRP deflection – load case A and C.

4 RESULTS AND DISCUSSION

A comparison of the deflection results to theoretical deflection results are illustrated in Figure 2 for the beam experiencing load case B. From Figure 2 presented below, the HSC deflection appeared to behave more simply supported than fixed. This is expected due to the welded angle connection utilized to attach the beams to the abutments and the deck panels to the beams. A comparison of the deflection between deck panels reinforced with mild steel and those reinforced with GFRP are shown in Figure 3. The deflections of the deck panels were normalized by removing the deflection due to the beams. The maximum deflection was approximately 8 times higher with deck panels reinforced with GFRP than with mild steel. Due to the lower modulus of elasticity of the GFRP, higher deflections were expected in the deck panel with such reinforcement.

5 CONCLUSIONS

The load test utilizing a forklift and precision surveying system was completed on an HSC pedestrian bridge. The HSC bridge displayed deflection results that appeared to be similar to the predicted simply supported deflection values. The deflection of HSC deck panels reinforced GFRP was found to be higher than that reinforced with mild steel.

REFERENCES

[1] Gross, S.P. (1999); "Field Performance of Prestressed High Performance Concrete Highway Bridges in Texas;" Dissertation; University of Texas at Austin.

[2] Myers, J.J.; Yang, Yumin (2005); "High Performance Concrete for Bridge A6130-Route 412 Pemiscot County, MO;" UTC R39.

[3] Alghazali, H. H. and Myers, J.J. (2017). "Evaluation of Strains and Stresses of Prestressed Girders for Bridge A7957, MO, USA (Field Study)," Fourth conference on smart monitoring, assessment and rehabilitation of civil engineering (SMAR), Zurich, Switzerland.

Current Perspectives and New Directions in Mechanics, Modelling and Design of Structural Systems – Zingoni (ed.)
© 2022 Copyright the Author(s), *ISBN 978-1-032-39114-4*

Estimating post-punching capacity and progressive collapse resistance of RC flat plates using shell-based FEA

A.B. Bahnamiri & T.D. Hrynyk
University of Waterloo, Waterloo, Canada

C.Y. Goh
Defence Science and Technology Agency, Singapore

ABSTRACT: Reinforced concrete flat plates are commonly employed due to their structural efficiency; however, have been shown to be susceptible to progressive collapse as evidenced by structural failures stemming from slab punching. Consequently, many building codes and design guides now require progressive collapse specific analyses and design requirements. This paper is focused on demonstrating the suitability of an alternative nonlinear finite element analysis procedure based on the formulations of the Disturbed Stress Field Model and employing low-cost thick-shell finite element modelling procedures for assessing the post-punching performance of flat plate specimens and slab systems. Isolated slab-column connection assemblies, as well as multi-bay substructures, were modelled using static and dynamic analysis procedures and the numerical results are compared with experimental data presented in the literature.

1 INTRODUCTION

Reinforced concrete (RC) flat plates are susceptible to punching failures that can propagate progressive collapse. To reduce the likelihood of propagating failures, slab-column connections must provide a minimum level of shear resistance after a punching event has occurred, permitting the redistribution of loads to neighboring regions of the structure.

Direct design approaches are available to explicitly evaluate a structure's resistance to progressive collapse. Methods involve the notional removal of a column, representing a failed connection, and the design of the remainder of the structure to carry loads in its as-damaged condition. Analyses may be performed using either linear elastic or nonlinear material modelling, in combination with either static or dynamic loading. However, there are codified limitations in terms of the types of structures that may be designed using linear elastic material modelling combined with static loading scenarios.

Procedures employing nonlinear finite element analyses (NLFEA) generally have no limitations for use with acceptance criteria; however, are admittedly more complicated and more expensive. Thus, in assessing flat plate structure response and designing structures to mitigate progressive collapse, there is a need for low-cost NLFEA tools to assess shear critical RC structures under varied loads.

In this paper, the suitability of a low-cost shell-based NLFEA to estimate the post-punching response of RC flat plate slab-column connections is examined. Analyses involving isolated slab-column connections under punching-governed static loading conditions and multi-bay flat plates under dynamic column removal scenarios are considered.

2 NLFEA MODELLING APPROACH

The analyses presented in this paper utilize a 9-node, 42-degree-of-freedom, layered thick-shell finite element. By way of the layered approach employed, the though-thickness composition of the RC slab elements and material response characteristics are considered explicitly. Relevant to post-punching slab response modelling, classical geometric nonlinearity formulations are used to incorporate load and structure reorientation due to slab deformations.

Cracked RC material modelling is done in accordance with the Disturbed Stress Field Model (DSFM) (Vecchio 2000), treating cracked RC as a smeared continuum and using a hybrid rotating-/fixed-crack approach. The constitutive models comprising the DSFM inherently account for mechanisms that have been shown to control the response of cracked RC elements: tension stiffening, compression softening, aggregate interlock, and the influence of variable and changing concrete crack widths.

Slab-column connections comprising RC flat plates are highly disturbed regions where transverse confinement and direct strut action are known to

DOI: 10.1201/9781003348450-199

occur, and can aid in suppressing the development of tension-governed punching shear failures. To account for these beneficial confining effects, localized shear strain suppression, as described in Goh and Hrynyk (2020), is employed.

3 STATIC ANALYSIS OF ISOLATED SLAB-COLUMN CONNECTIONS

An experimental program focused on examining the influence of reinforcement details on the post-punching response of isolated RC slab-column connections was presented in Mirzaei (2010). Slabs were 1.5-m square, 125-mm thick, and were designed with tensile reinforcement ratios that ranged from 0.25 to 1.4 % in the hogging moment regions of the slab comprising the column connection. Eleven test specimens constructed without out-of-plane shear reinforcement were analyzed.

A single finite element mesh, representing a one-quarter slab model, was created and used for all Mirzaei slabs. The mesh consisted of a 5 x 5 grid of layered shell elements, each of which was subdivided into 15 equal thickness concrete layers. Four additional layers were used to represent smeared steel components comprising the top and bottom mats of reinforcement. Material properties used in the modelling of the Mirzaei slabs were defined according to reported data, and previously defined default material models and analysis parameters were employed.

The model was shown to capture the initial stiffnesses, punching capacities, and provided reasonable slab response estimates for post-punching events such as the initial drop in shear resistance following punching and the subsequent increasing slab shear resistance. A mean computed-to-reported punching shear strength ratio of 1.00 with a coefficient of variation of 8 %, and a mean computed-to-reported post-punching shear capacity ratio of 1.24 with a coefficient of variation of 28 % were obtained.

4 DYNAMIC ANALYSIS OF MULTI-BAY SLAB UNDER COLUMN REMOVALS

Peng et al. (2017) presented experimental results pertaining to a 2/5-scale RC flat plate subjected to dynamic column removal loading. The continuous slab consisted of four slab panels in a 2 x 2 configuration, had planar dimensions measuring 6.1 by 5.6 m, and a nominal thickness of 89 mm. It was constructed with varying levels of in-plane reinforcement throughout, but did not contain through-thickness reinforcement. The test specimen was subjected to three consecutive edge column removal events, with the same column removed in each test.

A one-half specimen model consisting of 720 layered thick-shell finite elements, each of which was subdivided into 15 concrete layers, was created. Four additional smeared steel reinforced layers were provided to account for the top and bottom mats of steel. A truss bar finite element was used to simulate the removable edge column. The column removal and subsequent replacement was done by changing the stiffness of the truss bar element from effectively rigid to near-zero. Default material models and analysis parameters were employed, and all mechanical properties were defined according to reported data.

From the analysis results for column removal Tests 1 and 2, the procedure captured both the maximum slab displacement amplitude and the period of vibration well; however, underestimated damping. Under the final impact event (Test 3), the computed response provided good agreement with reported results. In agreement with that reported, the simulation estimated a punching shear failure at the interior column of the multi-bay slab followed by a flexure-induced punching shear failure at the corner column.

5 CONCLUSIONS

The suitability of using a low-cost thick-shell NLFEA procedure to estimate the post-punching response and progressive collapse resistance of RC flat plates was examined. Using previously defined material models and analysis parameters:

- DSFM-based concrete material modelling estimated the different phases of response and governing modes of failure with reasonable accuracy. Reported punching failures were estimated to occur in all of the slabs analyzed.
- Classical geometric nonlinearity formulations in combination with well-established cracked RC material modelling procedures provided adequate post-punching estimates for idealized RC flat plates.
- Dynamic analyses captured peak displacement amplitudes, periods of vibrations, but post column removal event damping was underestimated.

In summary, when combined with appropriate material modelling and disturbed region accommodations, thick-shell finite element modelling can be used to cost-effectively provide RC flat plate response estimates under punching-controlled loading.

REFERENCES

Goh, C.Y.M. & Hrynyk, T.D. 2020. Nonlinear finite element analysis of reinforced concrete flat plate punching using a thick-shell modelling approach. *Engineering Structures*, Elsevier Ltd., 224, 16 pp.

Mirzaei, Y. 2010. Post-punching behavior of reinforced concrete slabs. PhD Dissertation, EPFL.

Peng, Z., Orton, S.L., Liu, J., & Tian, Y. 2017. Experimental Study of Dynamic Progressive Collapse in Flat-Plate Buildings Subjected to Exterior Column Removal. *Journal of Structural Engineering*, ASCE, 143(9): 04017125.

Vecchio, F.J. (2000). Disturbed Stress Field Model for reinforced concrete: formulation. *Journal of Structural Engineering*, ASCE, 126(9): 1070–1077.

Current Perspectives and New Directions in Mechanics, Modelling and Design of Structural Systems – Zingoni (ed.)
© 2022 Copyright the Author(s), *ISBN 978-1-032-39114-4*

Long-term deformation of segmented prestress bridges under harsh weather conditions: Case study of Exit 23 Interchange at Riyadh Ring Road

H.H. Abbas
Department of Structural Engineering, Al Azhar University, Cairo, Egypt

M.M. Hassan
Structural Engineering Department, Faculty of Engineering, Cairo University, Cairo, Egypt
Architectural Engineering Department, University of Prince Mugrin, Madinah, Saudi Arabia

ABSTRACT: Long term deformations greatly influence the design of large span structures. Bridges are usually designed for extended service life reaching over 100 years. During such long periods, they are expected to perform with a certain acceptable level of movement. For segmented prestressed concrete bridges, the concrete creep and shrinkage values determine the level of bridge performance in terms of safety and serviceability. Unexpected bridge movements affect the performance of a bridge and may shed concerns regarding its safety. In this study, field measurements are used to assess the accuracy of the formulas used to predict movements at half bridge joints for an example case study: Exit 23 Interchange at Riyadh Ring Road. After 18 years of service, excessive deformations were observed. The considered case study was investigated to assess the reason behind the large horizontal shortening which caused problems for both the sliding bearing and the expansion joints. The measured values are meant to supplement the available database regarding the longitudinal movements of large span structures.

1 INTRODUCTION

Design of concrete bridges involves different aspects including check on stress values and serviceability limit states. Large-span bridges are usually designed for extended service life reaching over 100 years. For prestressed concrete bridges, several studies have reported the low performance of the long-term movement predictions due to the complexity of the time-dependent interactions between concrete creep, shrinkage, steel relaxation, and deterioration processes [Cakmak et al. (2022) and Bažant et al. (2011a)]. Bažant et al. (2011b) showed through extensive comparisons between available creep and shrinkage models that they cannot provide proper predictions for the long-term movements. Several researchers question the predictions provided by these models due to the inadequate assumptions in modeling and representation of data, Bažant et al. (2013). In-situ measurements provide great source of data in order to supplement and verify the available theoretical-based models, Wendner et al. (2015).

The current study focuses on for the problems observed at half-joints of Exit 23 Interchange at Riyadh Ring Road including excessive shrinkage at several joints. A study was initiated to assess the excessive longitudinal shrinkage.

2 BRIDGE DESCRIPTION

The studied prestressed composite box-girder bridge in the current study is located in Saudi Arabia, Figure 1. Figure 2 shows the detail of the half joint used in the studied bridge.

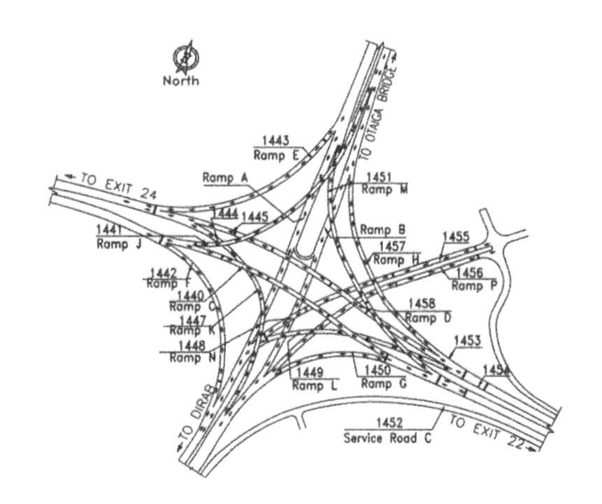

Figure 1. Plan of Interchange - Exit 23.

DOI: 10.1201/9781003348450-200

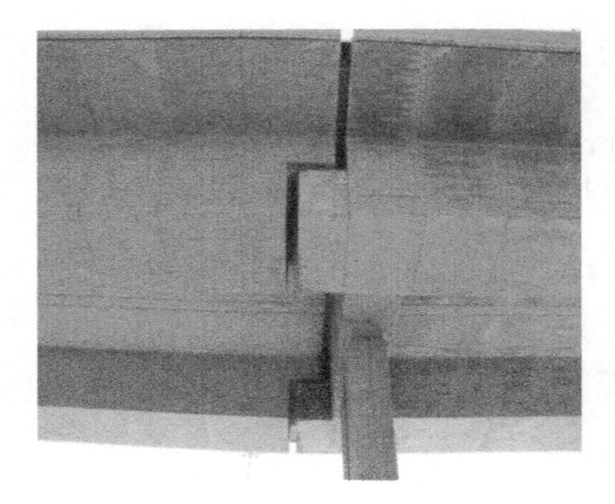

Figure 2. Detail of Bridge Half Joint.

3 OBSERVED STRUCTURAL DEFECTS

Large horizontal shortening of the different platform parts was observed resulting in problems at the different bridge bearings.

Figure 3. Large Movement at Bridge Half Joint.

4 SOURCES OF PLATFORM MOVEMENT

Different design codes require bearings and end gap details to be account for movements due to applied loads, thermal changes, creep, and moisture changes. As per ACI 209R-92, these include thermal effects, creep, and shrinkage. However, other sources can also lead to platform movements including the exact value of pre-tensioning of cables, misalignment of piers or bearings during the construction phase, the effect of the weight component at the sloping parts of the bridge, and the effect of the braking force.

5 VERIFICATION AGAINST FIELD MEASUREMENTS

The movement at each axis was measured at the different bearing locations. Figure 4 shows sample measured versus calculated values. It is observed that the differences vary at the different locations with similar loading conditions and spans which indicate that the movements are not simply due to creep, shrinkage, and temperature.

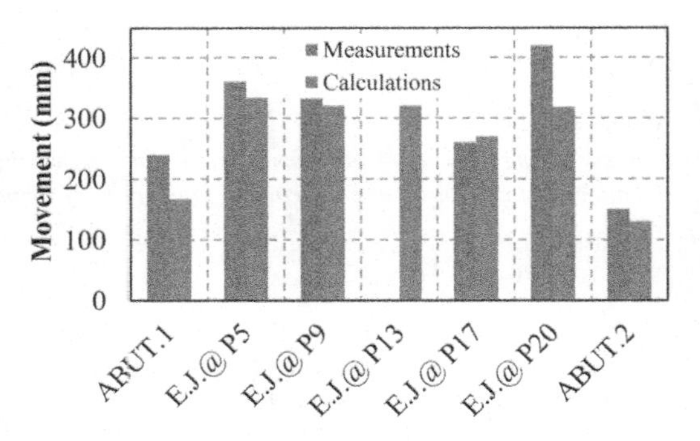

Figure 4. Measured vs. Calculated values for Platform No. 1440(6).

6 CONCLUSIONS

Creep and shrinkage predictions proposed by ACI 209R-92 has been reported to underestimate the expected bridge movement. This usually causes excessive deformations for many prestressed concrete box girder bridges after many years in service. The current study explores an example bridge that suffered from excessive deformations problem and the applied repair procedure.

REFERENCES

Bažant, Z. P., Hubler, M. H., & Jirásek, M. 2013. Improved Estimation of Long-term Relaxation Function from Compliance Function of Aging Concrete. *Journal of Engineering Mechanics*, 139(2): 146–152.

Bažant, Z. P., Hubler, M. H., & Yu, Q. 2011a. Pervasiveness of Excessive Segmental Bridge Deflections: Wake-Up Call for Creep. *ACI Structural Journal*, 108 (6): 766–774.

Bažant, Z. P., Hubler, M. H., & Yu, Q. 2011b. Excessive creep deflections: An awakening. *Concrete International*, 33(8): 44–46.

Cakmak, F., Menkulasi, F. and Eamon, C., 2022. Time-Dependent Flexural Deformations in Composite Prestressed Concrete and Steel Beams. I: Prediction Methodology. *Journal of Bridge Engineering*, 27(5): 04022016.

Prediction of Creep, Shrinkage, and Temperature Effects in Concrete Structures (ACI 209R-92). 1997. American Concrete Institute, Detroit, Michigan.

Wendner, R., Hubler, M.H., & Bažant, Z.P. 2015. Statistical Justification of Model B4 for Multi-Decade Concrete Creep and Comparisons to Other Models Using Laboratory and Bridge Databases. *J Materials and Structures*, 48(4): 815–833.

Current Perspectives and New Directions in Mechanics, Modelling and Design of Structural Systems – Zingoni (ed.)
© 2022 Copyright the Author(s), ISBN 978-1-032-39114-4

Shear capacity of concrete elements with reinforcement made of steel and FRP

S. Görtz & R. Haack
Kiel University of Applied Sciences, Kiel, Germany

ABSTRACT: Based on the approach of the ACI-DAfStb shear databases for structural concrete members with steel reinforcement a database including all internationally published tests on FRP-reinforced concrete beams is being created. The Database currently includes 868 tests. With the conclusions of the experimental program and database evaluations, a model for calculating the shear capacity of concrete members with steel reinforcement is transferred to components with FRP reinforcement.

1 INTRODUCTION

In (Görtz 2004), (Görtz 2020) altogether 32 shear experiments have been carried out with girders made out of normal and high-performance concrete. Both, conventional steel- and FRP- type rebars were used for reinforcement. In addition, analytical and numerical calculations were carried out.

To confirm the results from our own investigations, evaluations using a large database, which contains all published shear tests worldwide, were done. The shear database on steel reinforcement provided by the ACI group 445-D (see DAfStb 2017) was used and based on this, a database for FRP reinforcement with 868 experiments was created.

2 MODEL FOR SHEAR DESIGN

2.1 *Girders without shear reinforcement*

For components, without shear reinforcement, the main part of the shear force in the ultimate limit state is taken by the uncracked compression zone. Regarding this behavior, a design approach for rectangular cross-sections with shear slenderness a/d ≥ 3 was derived in (Zink 1999). The approach was extended to profiled beams and girders with a/d < 3 in (Görtz 2004):

$$V_{Rm,c} = \beta \cdot \frac{2}{3} \cdot f_{ctm} \cdot b_{s,eff} \cdot x \cdot k_1 \cdot k_2 \tag{1}$$

with:

$$\begin{aligned} \beta &= 3/(a/d)\ 1,0 \\ x &= \text{compression zone} \\ &= \frac{2}{1+\sqrt{1+2/\left(\rho_1 \frac{E_s}{E_{cm}}\right)}} \cdot d \end{aligned} \tag{2}$$

$$\begin{aligned} b_{s,\,eff} &= \text{effective width for shear} \\ &= b_w + 0.3 \cdot \sum_i h_{f,i} \end{aligned} \tag{3}$$

$$\begin{aligned} k_1 &= \text{factor, depends on } a/d \\ &= (4 \cdot d/a)^{1/4} \end{aligned} \tag{4}$$

$$\begin{aligned} k_2 &= \text{factor, depends on characteristic length } l_{ch} \\ &= (5 \cdot l_{ch}/d)^{1/4} \end{aligned} \tag{5}$$

ρ_l = longitudinal reinforcement ratio; E_s, E_{cm} = modulus of elasticity of reinforcement and concrete; h_f = height of flanges,

When using FRP reinforcement, the lower modulus of elasticity must be considered. The bending moment leads to a higher strain in the tensile zone and following crack propagation resulting in a reduction of the concrete compression zone.

In the mechanically developed model by Zink the height of the compression zone, respectively the modulus of elasticity of the longitudinal reinforcement is one of the input parameters. Therefore, this model can be applied to concrete girders with FRP reinforcement without any modification.

2.2 *Girders with steel shear reinforcement*

The most effective increase of the shear capacity is reached by shear reinforcement. Assuming a smooth transition between girders without and with shear reinforcement, the shear capacity can be described as follows:

Failure of the shear reinforcement

$$V_{Rd,s} = a_{sw} \cdot f_{yd} \cdot z \cdot \cot\theta + \kappa \cdot V_{Rm,c} \tag{6}$$

Failure of the concrete struts

$$V_{Rd,max} = \nu \cdot f_{cd} \cdot b_w \cdot z/(\tan\theta + \cot\theta) \tag{7}$$

As described, the shear capacity of beams without shear reinforcement $V_{Rm,c}$ corresponds to the shear capacity of the uncracked compression zone. This very "soft" load-bearing system requires large shear cracks and a high reduction in stiffness to reach ultimate limit state. With increasing shear reinforcement ratio, the stiffness of the truss model increases, and the

DOI: 10.1201/9781003348450-201

concrete contribution $V_{Rm,c}$ can only partially be activated. The increase of shear transfer to the concrete contribution starts occurring when shear reinforcement reaches the yielding plateau. However, the concrete contribution can't be activated completely, because a secondary failure of the highly-stressed concrete struts of the truss model occurs beforehand.

Based on investigations from (Görtz 2004), (Hegger et. al 2004), the unknown parameters in eq. (6) – (7) had been derived as follows:

$$\cot\theta = 1 + 0.15 / \omega_{w,ct} - 0.18 \cdot \sigma_c / f_{ctm} \quad (8)$$

$$\omega_{w,ct} = \rho_w \cdot f_y / f_{ctm} \quad (9)$$

with:

$$\kappa = 1 - \omega_{w,ct}/3 \ (0) \quad (10)$$

$$v = 0.75 \quad (11)$$

2.3 *Girders with FRP shear reinforcement*

In the case of FRP reinforcement, the following points must be observed:

(1) The lower modulus of elasticity result in larger strains in the stirrups. This has the following negative effects on the shear capacity:

- The effect of aggregate interlock is reduced. This leads to a steeper inclination of the concrete struts θ in the truss model.
- High stirrup strains favor the critical crack propagation into the uncracked compression zone. Therefore, in some cases, the flexural shear failure due to critical crack propagation occurs before the tensile strength of the stirrup is reached
- High tensile strains reduce the compressive strength of the concrete struts in the truss model.

(2) In general the load-bearing capacity of an FRP stirrup is significantly lower than those of a straight FRP bar. The reasons are:

- Due to the manufacturing process, it is difficult to bend the fibers of the stirrups around corners, this can lead to buckling of the inner fibers.
- Due to the tensile stress in the two stirrup legs, there is a transverse compressive stress on the bar in the area of the bend, see Figure 1 (left). Because in the transverse direction only the polymer matrix acts, the load-bearing capacity is very limited here.

Therefore, the eq. (6) – (11) for FRP shear reinforcement change as follows:

Failure of the concrete struts, taken from (Kurth 2012):

$$V_{Rd,max} = 1.25 \cdot f_{cd}^{2/3} \cdot b_w \cdot z / (\tan\theta + \cot\theta) \quad (12)$$

Failure of the shear reinforcement

$$V_{Rd,s} = a_{sw} \cdot \sigma_{FVK,w,max} \cdot z \cdot \cot\theta + \kappa \cdot V_{Rm,c} \quad (13)$$

with:

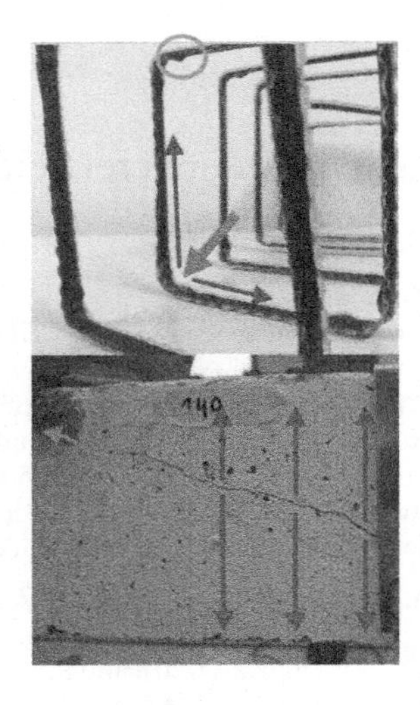

Figure 1. Improper fabrication and transverse pressure in the area of the bend of the stirrups (left) and propagation of a critical shear crack in the uncracked compression zone caused by high stirrup strains (right).

$$\cot\theta = 1 + (0.15/\omega_{w,ct}) \cdot E_{FVK}/E_{steel} - 0.18\sigma_c/f_{ctm} \quad (14)$$

$$\sigma_{FVK,w,max} = \min\begin{cases} f_{FVK,eff} \\ \varepsilon_{FVK,w,max} \cdot E_{FVK} \end{cases} \quad (15)$$

with

$f_{FVK,eff}$ = maximum stress taken by the shear reinforcement, taking into account the anchorage and the bending shape; must be determined experimentally

$\varepsilon_{FVK,w,max}$ = maximum strain of the shear reinforcement to avoid a flexural shear failure

Based on the provided data the design equations were checked with a total of 1921 shear tests. The determination of the strain limitation has not been finally completed yet.

REFERENCES

DAfStb 2017: ACI-DAfStb databases 2015 with shear tests for evaluating relationships for the shear design of structural concrete members without and with stirrups. Berlin 2017.

Görtz, S. 2004: Shear cracking behavior of prestressed and nonprestressed beams made of normal and high performance concrete (in German). Dissertation RWTH Aachen, 2004.

Görtz 2020: Shear capacity of railway sleepers with reinforcement made of basalt fibre reinforced polymer (in German). Research project University of Applied Science Kiel.

Hegger, J. & Sherif, A. & Görtz, S.: Analysis of Pre- and Post-cracking Shear Behavior of Prestressed Concrete Beams Using Innovative Measuring Techniques. ACI Structural Journal, March-April 2004

Kurth, M. 2012: Zum Querkrafttragverhalten von Betonbauteilen mit Faserverbundkunststoff-Bewehrung. Dissertation, RWTH Aachen, 2012

Zink, M. 1999: Zum Biegeschubversagen schlanker Bauteile aus Hochleistungsbeton mit und ohne Vorspannung. Dissertation University of Leipzig, 1999

Current Perspectives and New Directions in Mechanics, Modelling and Design of Structural Systems – Zingoni (ed.)
© 2022 Copyright the Author(s), *ISBN 978-1-032-39114-4*

Modelling temperature and stress development in large concrete elements under sequential construction conditions: Experience with the Msikaba Bridge foundations

Y. Ballim
School of Civil & Environmental Engineering, University of the Witwatersrand, Johannesburg, South Africa

G. Harli
Jones and Wagener Engineering and Environmental Consultants, South Africa

ABSTRACT: This paper describes the development and use of a finite-difference model for predicting the time-temperature profiles in the large concrete foundations for the Msikaba cable stayed bridge on the N2 motorway in South Africa. The important features of the temperature prediction model were the need to account for sequential construction in incremental layers as well as for the effects of solar radiation on the top surface of the concrete layers. The predicted temperature profiles were then used to estimate the likely thermal strains in the concrete and the related risk of cracking. The primary objective of the temperature and stress predictions was to determine a suitable combination of construction layer thickness and time interval between placement of concrete layers to minimize the risk of cracking. The modelling indicated that a concrete layer thickness of 2 m placed at 5 to 7 day intervals would be acceptable for this project.

1 INTRODUCTION

The Msikaba bridge is planned as a cable stay bridge across the Msikaba gorge on the N2 motorway in the Eastern Cape province of South Africa. The bridge spans 580 m and the stay cables are supported on two 127 m high pylons at either end of the gorge. The reinforced concrete pylon foundations measure 14 m x 12 m in plan and are between 7 m to 8 m deep. An important consideration for the construction of these foundation was the vertical concrete layer thickness and time interval between sequential layer placements in order to minimise the thermal stresses and risk of consequent cracking of the concrete that may be caused by the heat of hydration of the cement binder in such a large concrete element.

This paper describes the development and use of a finite-difference model for predicting the time-temperature profiles in the concrete foundation under sequential layer construction conditions. The model was based on a time-step solution to the Fourier heat equation and as input, incorporates the rate of heat of hydration of the actual concrete used on the construction project, measured in a laboratory adiabatic calorimeter. The model incorporated the effects of solar radiation, including exit radiation at night as one of the boundary conditions on the top surface. The model was calibrated against measured temperatures in the 2 m concrete trial cube cast at the construction site, using the concrete intended to be used for the actual foundations.

2 MODEL DEVELOPMENT

The Wits temperature prediction model developed by Ballim (2004) was further developed to allow for prediction under sequential construction conditions. The model uses a two-dimensional, finite difference-approach to solving the Fourier equation for heat flow in order to predict the time-temperature profiles in the concrete. An important feature of this further development of the model was to include the effects of solar radiation in the boundary conditions of the top surface of the concrete element. This was done using measured net solar radiation (incoming minus outgoing radiation) over a 24-hour period during February (summer)in Johannesburg. Equation 1 was used to model the net solar radiation at any time during the day (t_d):

For $6 < t_d < 18$:

$$R_n = 135.62 + 436.18 \cos (0.34.t_d - 4.08) \quad (1)$$

For $t_d < 6$ and $t_d > 18$:

$$R_n = \ - 80W/m^2$$

2.1 *Structure and parameters used for the model*

The model was structured to predict the temperature of a concrete element placed in three layers of equal thickness (h), using the same concrete mixture design and with the same delay period (t_d) between placing of the first and second layers and between the second and third layers. This structure, is shown in Figure 1.

DOI: 10.1201/9781003348450-202

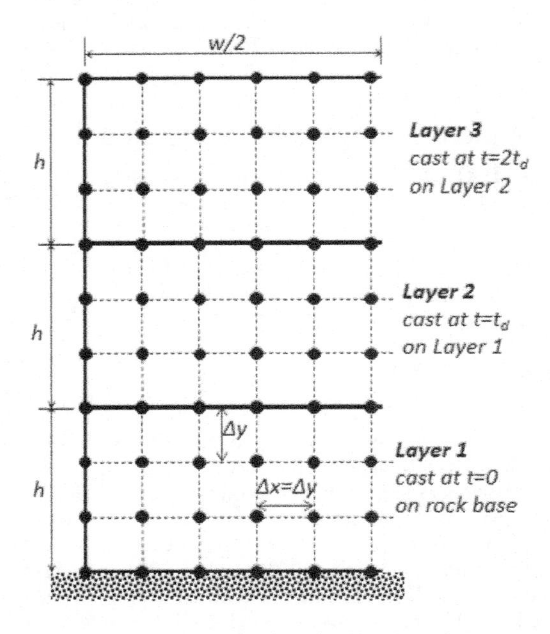

Figure 1. Arrangement for the 3-layer sequential construction used in the model, indicating node points for the finite difference analysis.

The main cable-support pylons at the abutment ends of the bridge are founded on bases 7 m deep and 12 m x 12 m in plan. The cable anchor blocks behind each of the pylons measure 8 m deep and 12 m x 60 m in plan. All the foundation elements are quite heavily reinforced, at a reinforcement ratio of 0.46%. The concrete mixture was designed to be pumped into the formwork and to have a 28-day characteristic compressive strength of 40 MPa.

3 MODEL RESULTS

The model was calibrated against the temperatures measured in a 2 m trial concrete cube that was cast on site. The results showed acceptable reliability of the model in predicting the actual temperature ant this allowed the temperature prediction in the actual foundation elements to proceed. Figures 2 and 3 show the model results for three 2 m thick layers of concrete, cast sequentially in 5 and 7-day intervals between castings.

The model results showed that, for layer thickness of 2 m, the maximum temperature requirement of 70 °C was satisfied but the maximum temperature differential of 22 °C was exceeded for both time delay periods. However, an assessment of the cracking potential using the 'Comprehensive Method' suggested by Bamforth (2007), indicated a theoretical early-age crack width of 0,10 mm, spaced at approximately 1.9 m, which was considered as acceptable.

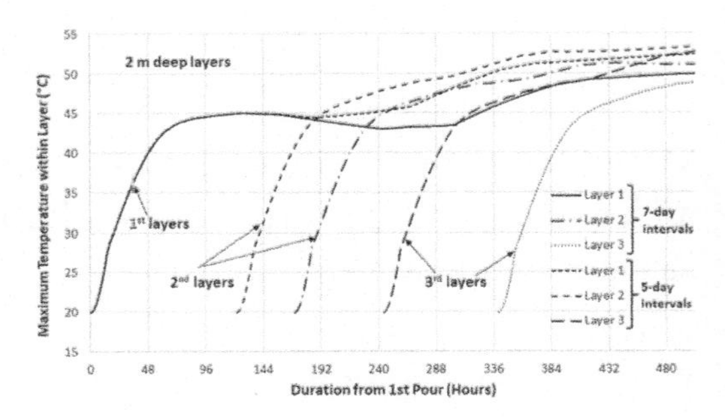

Figure 2. Development of the maximum temperature at the centre of 2 m thick concrete layers, sequentially placed at 5- and 7-day intervals between layers.

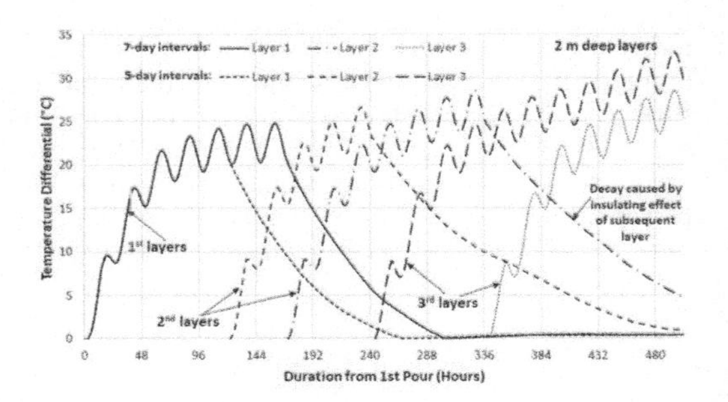

Figure 3. Differential temperature between the centre of 2 m thick concrete layers and the near-top surface, for layers sequentially placed at 5- and 7-day intervals.

4 CONCLUSIONS

The proposed approach for accounting for the effects of solar radiation good approximation of the near-surface temperatures of the concrete

Under sequential construction conditions, the estimates of cracking strain and likely crack widths indicate that temperature differentials of around 30 °C within the concrete element can safely be tolerated in large concrete elements with significant steel reinforcing, using 2 m deep layers at 5 to 7 day intervals.

ACKNOWLEDGEMENTS

The authors express their gratitude to the South African National Roads Agency Ltd. (SANRAL) for their permission to publish this paper.

REFERENCES

Ballim, Y. 2004. A numerical model and associated calorimeter for predicting temperature profiles in mass concrete. *Cement and Concrete Composites* 26: 695–703.

Bamforth, P.B. 2007. *Early-age thermal crack control in concrete*. Report C660. CIRIA, London.

Current Perspectives and New Directions in Mechanics, Modelling and Design of Structural Systems – Zingoni (ed.)
© 2022 Copyright the Author(s), *ISBN 978-1-032-39114-4*

Characterization of major reinforcing bars for concrete works in Botswana construction industry

A.P. Adewuyi, G.B. Eric & O.J. Kanyeto
Department of Civil Engineering, University of Botswana, Gaborone, Botswana

ABSTRACT: The serviceability, functionality, structural reliability and durability of reinforced concrete structure depends primarily on the quality of steel reinforcement. Influx of defective and/or sub-standard construction materials in the market, especially reinforcing steel bars, account for many premature defects, structural failures and frequent collapse of RC constructed facilities during construction and in service. Characterization of steel rebars from two major sources were conducted as a basis for quality assurance and control of reinforcing bars in Botswana construction industry. The qualities were assessed as a function of the physical, chemical, mechanical and morphological properties or individual reinforcing bars and structural behaviour and durability when embedded as reinforcement in concrete. The findings showed that the studied steel bar types, though satisfied the physical and geometric properties, further investigations showed that there were varying degrees of impurities in the rebar products resulting in brittleness with little ductility and plasticity. The comparative flexural behaviour of beams reinforced with the two bar types under normal and induced corrosive environment revealed disparity in bending capacity due to the reduced relatice rib areas, low interfacial bonding, low ductility and plasticity. It can be recommended that relevant stakeholders in the construction industry and the engineering practice as well as construction materials regulatory agencies should intensify to effort to control the influx of defective and substandard materials in Botswanaa construction industry.

1 INTRODUCTION

Reinforced concrete (RC) structures account for majority of the constructed facilities globally and their performance is greatly influenced by the quality of the steel reinforcement in the industry (Adewuyi et al., 2015). Accurate information of the properties of construction materials is fundamental at the design and construction stages, and during the service life of constructed RC infrastructure (Basu et al., 2004). Influx of defective and/or sub-standard construction materials in the market, especially reinforcing steel bars, account for many premature defects, structural failures and ultimate collapse of RC constructed facilities during construction and in service. Defective or substandard rebars in the industry as a result of lack of strict quality control mechanism on the basic material used in the billet production process, rolling process and post-rolling process. Botswana is import dependent for all structural steel and reinforcing bars in construction industry.

However, the influx of reinforcing bars of unverified quality from different sources is a potential threat to the safety and durability of the extensive infrastructural developments in the fast developing country.

Hence, the primary purpose of this paper was to assess the quality of two leading major steel reinforcement products supplied into Botswana both independently and when embedded as reinforcement in RC beams under normal and corrosion induced conditions.

2 MATERIALS AND METHODS

The properties of major ribbed or deformed reinforcing steel bars steel bars imported into Botswana from two leading manufacturers designated as M1 and M2 were investigated in terms of the physical, chemical, microstructural and tensile properties of bare steel rebars and the structural performance when embedded in RC beams.

2.1 *Physical properties of steel reinforcing bars*

Three hundred measurements of ribbed or deformed steel reinforcing bars from M1 and M2 sources were randomly sourced from the warehouses of major suppliers and at on-going construction sites in the Greater Gaborone and the environs to determine the geometric properties as a measure of the actual diameter, rib spacing, rib heights and the gaps.

DOI: 10.1201/9781003348450-203

2.2 *Physical properties of steel reinforcing bars*

Twelve steel rebar samples each from M1 and M2 sources were tested for the elemental composition using S8 Tiger X-ray fluorescence (XRF) spectroscopy machine. XRF spectroscopy coupled with Spectra Plus Bruker analyser was an excellent technology for qualitative and quantitative analysis of the composition of materials in terms of the average elemental percentage by weight (wt. %).

2.3 *Mechanical properties of steel rebars*

Sixty samples each of M1 and M2 deformed steel bars were randomly purchased from suppliers in Gaborone and tested for tensile strength properties using a 600 kN capacity MTS according to ASTM A370 (2008) and E8 (2008). The tensile properties were extracted from the stress-strain curves of each specimen in terms of the yield and ultimate tensile strength, percentage elongation and the modulus of elasticity.

2.4 *Flexural behaviour of RC beams*

Six simply supported RC beams each of dimensions 150 × 200 × 3000 mm of 28th day concrete compressive strength of 35 N/mm^2 reinforced with M1 and M2 steel reinforcements were cast for the experimental study as shown in Figure 1.

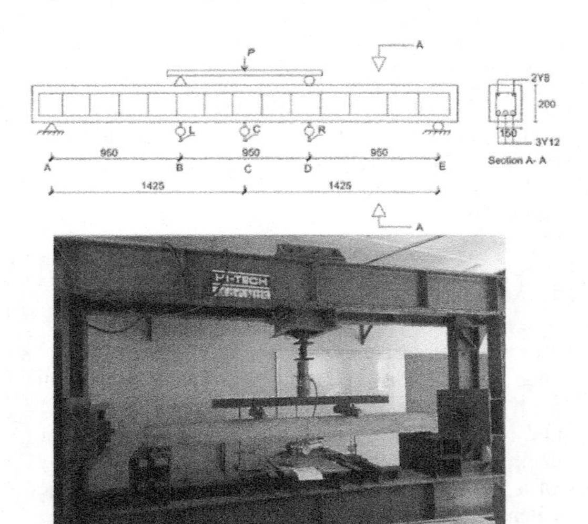

Figure 1. Schematic and experimental setup of third-point loading flexural test.

Three RC beam specimens each were subjected to normal third-point flexural testing procedure, while the remaining three beam specimens separately reinforced with M1 and M2 were subjected to accelerated corrosion of the embedded three tension reinforcing bars by a constant current of 0.5 A via a DC power supply for period of study. A pond of size 150 × 700 × 50 mm centered on top of the beam and containing 5% sodium chloride (NaCl) solution and a stainless-steel cathode was fixed on top of the beam over the length of 700 mm.

The lateral displacements of the RC beams were measured by three dial gauges with accuracy to 0.01 mm and displacement transducers at the mid-span and middle third points (that is 950 mm) to either support under sustained (15 percent of the ultimate load) for 72 hours and later progressive loadings to failure. Crack width was measured on the concrete surface using a crack microscope reading to 0.02 mm, while crack length were measured using crack metre as load was applied in uniform increment.

3 CONCLUSIONS

This paper reports a study on characterization of major steel rebars in Botswana based on the physical, chemical, mechanical and morphological properties of steel bars and the performance of concrete beams reinforced with such beams under normal and corrosion induced conditions. It can be concluded that the rebar types met the physical and geometric requirements of major international standards, but the tensile properties revealed a brittle material with low plasticity and ductility. The morphological assessment revealed defective materials of substandard quality in flexure due to micro-inclusions, internal cracks and impurities. Rebars M2 proved to be better than M1 in flexure under normal and corrosive exposure as well as other physical, chemical and mechanical properties with lesser degree of uncertainty.

REFERENCES

Adewuyi, A. P., Otukoya, A. A., Olaniyi, O. A., & Olafusi, O. S. (2015). Comparative studies of steel, bamboo and rattan as reinforcing bars in concrete: tensile and flexural characteristics. *Open Journal of Civil Engineering*, 228–238.

Darwin, D., & Zuo, J. (2000). Bond Slip of High Relative Rib Area Bars under Cyclic Loading. *ACI Structural Journal*, 331–334.

Current Perspectives and New Directions in Mechanics, Modelling and Design of Structural Systems – Zingoni (ed.)
© 2022 Copyright the Author(s), ISBN 978-1-032-39114-4

Experimental assessment of crack width estimations in international design codes

I. Ridley, M. Shehzad, J. Forth & N. Nikitas
School of Civil Engineering, University of Leeds, UK

A. Elwakeel, K. Elkhoury, R. Vollum & B. Izzuddin
Department of Civil and Environmental Engineering, Imperial College London, UK

ABSTRACT: Cracking due to restraint of early age thermal and long-term shrinkage strains frequently causes serviceability problems in edge restrained reinforced concrete structures, such as liquid retaining and nuclear structures. Consequences of current design recommendations for the design of crack control reinforcement are reported contradictorily in the literature, leading to both excessive crack widths and overly conservative reinforcement designs in practice. This research aims to experimentally investigate the underlying assumptions and parameters affecting the cracking of edge restrained elements. In this work, the results of 8 tests on edge restrained reinforced concrete walls are reported. The crack widths, spacing and profiles obtained from the tests are compared with those estimated using different design codes.

1 INTRODUCTION

Generation of tensile stresses in reinforced concrete members, as a consequence of restraint against imposed thermal and shrinkage strains, often leads to structural cracking. To manage such cracks, horizontal, or transverse, reinforcement is provided, to keep crack widths within the allowable limits (Bamforth, 2018). Depending on serviceability and durability requirements, the allowable maximum crack width is specified, and the reinforcement is designed accordingly. In Europe, detailed guidance for the design of reinforcement to control restraint induced cracking is provided by BSEN1992-3 (BSI, 2006) and CIRIA C766 (Bamforth, 2018). Refer to CIRIA C766 and BSEN1992-3 for the current standardised guidance for crack width calculation. Instances of observed crack widths exceeding those predicted by the guidance have been shown to occur. The uncertainty as to whether this occurs as a result of incomplete/inaccurate design guidance or as a result of assumptions made during the calculation stages merits a re-evaluation of the code standards. This paper is aimed at evaluating the predictive quality of the current design guidance, through a comparison with experimentally observed restraint-induced cracking.

2 EXPERIMENTAL DETAILS

This experimental research is currently being undertaken at University of Leeds (UoL) and Imperial College London (ICL). Results from 8 edge restraint tests are discussed in this paper. The tests have been identified with the following notations; UoL1,2,3,4 and ICL1,2,3,4. Figure 1 shows the cross section of test wall UoL2, subjected to restraint by the mature base slab. The dimensions and reinforcement details in each test were varied to ascertain their influence and the experimental procedure was selected to amplify the restraint present (Stoffers, 1978, Shehzad, 2018). Contributory parameters for theoretically calculating the crack widths for each wall are given in Table 1, including the measured thermal drop (T_1) for each test. The crack widths were visually monitored and measured using a portable microscope of 40x magnification and a precision level of 0.02mm.

Table 1. Details of test parameters.

	c	φ	ρ	f_{ctm}	f_{cm}	E_{cm}	T_1
Wall	mm	mm	mm	MPa	MPa	GPa	°C
UoL1	30	10	0.34	3.39	49.4	31.4	27.7
UoL2	30	10	0.34	1.74	30.9	29.9	24.3
UoL3	30	10	0.34	2.16	35.3	22.1	35.5
UoL4	30	0	0.00	2.57	35.9	24.2	25.7
ICL1	40	10	0.35	3.05	51.8	35.7	34.9
ICL2	25	16	0.89	2.25	42.6	30.9	34.4
ICL3	25	16	0.89	1.98	42.5	31.9	40.4
ICL4	30	10	0.21	1.81	41.3	32.8	30.5

f_{ctm} = concrete mean tensile strength (28 days)
f_{cm} = concrete mean compressive strength (28 days)
E_{cm} = concrete elastic modulus (28 days)

3 RESULTS AND DISCUSSION

Two outputs for the maximum crack width (MCW) have been calculated from each set of guidance. The first used the experimentally obtained concrete properties and environmental data and the second used

DOI: 10.1201/9781003348450-204

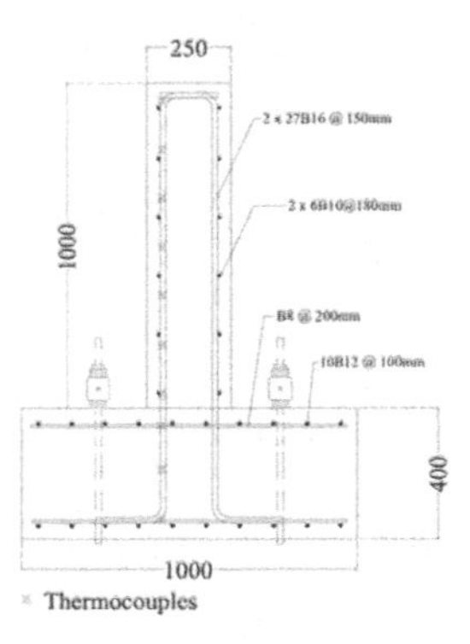

Figure 1. Cross section showing details of UoL2.

parameter values that could be obtained via the design guidance. These calculations have been referred to as 'experimental' and 'design' calculations.

3.1 *Observed crack development*

Cracking did not occur in ICL1 or UoL1-2. The first cracks observed in the remaining 5 walls formed between 7 and 50 days after casting. Within one week of the formation of the first cracks in each wall, cracks increased in length, width and number. The maximum increase in crack width between 3 days (early age, EA) and end of monitoring (long term, LT) observed was 71% for UoL3. According to BSEN1992-3, cracking is not predicted to occur close to the ends of the wall length, for a solid wall restrained along one edge, however UoL4 recorded cracks at a distance of only 380mm from the wall end.

3.2 *Location of maximum crack width*

In UoL3, 4 and ICL3 the MCW occurred between 8-12% of the wall length above the joint. This agrees reasonably well with the literature (Bamforth, 2018). In ICL2 and ICL4, the MCW occurred significantly closer to the joint, at a height of only 5% of the wall length.

3.3 *Comparison of experimentally-observed and code-predicted cracking behaviour – the 'experimental' calculations*

Although cracking was not experimentally observed in UoL1-2 and ICL1, both CIRIA and BSEN1992-3 predict cracking to occur, with BSEN1992-3 calculating the largest cracking risk (ratio of restrained strain to tensile strain capacity) of 186% to be present in the EA for UoL2, compared to a prediction of 39% by CIRIA. The observed crack widths occurring in UoL3 and ICL4 are overestimated by BSEN1992-3 by 50% and 225% respectively. CIRIA over predicts the crack widths for ICL4 by 100%. From these observations, CIRIA appears to predict the observed behaviour better than BSEN1992-3, yet still generates an over estimation in crack width of 100% for ICL4. This indicates the cracking behaviour is not completely understood by CIRIA and parameters that influence the edge restraint behaviour are not correctly incorporated into the guidance. The over estimations in crack widths discussed above cannot be evaluated to be as a result of conservative design codes, as both CIRIA and BSEN1992-3 underestimate the crack widths occurring in walls UoL3-4 and ICL2-3. An underestimation of the crack widths is considered more concerning (BSI, 2006). The inconsistencies in crack width predictions produced using both sets of guidance means that neither can be assumed to safely, accurately, or reliably describe the behaviour of edge restrained cracking.

3.4 *Evaluation of the use of parameter values available from the design guidance – the 'design' calculations*

The variation between expected and measured concrete material properties can be significant. The impact of these variations has been evaluated by replacing the experimentally obtained input parameters with those provided in the design guidance (Bamforth, 2018, BSI, 2005). In all cases, the 'design' crack width predictions are larger than those calculated using the experimental parameters. ICL4 observed the largest increase (58%) in crack width prediction, explained by the largest observed difference between the measured and predicted value for T_1. ICL2 observed the largest variation between the measured and 'design' concrete compressive strength – 9.6MPa, resulting in a 20% increase in the 'design' calculated crack widths using CIRIA, and a 25% increase when using BSEN1992-3. This demonstrates the impact of using more accurate input parameters in crack width prediction calculations.

In the cases where the crack widths had been under predicted by CIRIA and BSEN1992-3 using the experimental input parameters, the MCW predictions using the 'design' calculations, although larger, still did not exceed the observed cracks experimentally.

3.5 *Conclusions*

Due to the magnitude and number of cases in which the crack widths have been underestimated, it is concluded that the guidance provided by both CIRIA C766 and BSEN1992-3 cannot be relied upon to safely or accurately describe the cracking which occurs as a result of edge restraint of imposed strains. Of the two guidance documents considered in this research, as a result of the more comprehensive inclusion of related parameters, it is CIRIA C766 which better predicts the crack widths. Financial support from the EPSRC UK for this work is acknowledged.

REFERENCES

Bamforth, P. 2018. *Control of cracking caused by restrained deformation in concrete*, CIRIA.

BSI 2005. Eurocode 2: design of concrete structures-part 1–1: general rules and rules for buildings.

BSI 2006. 3: Eurocode 2: Design of Concrete Structures–Part 3 Liquid Retaining and Containment Structures.

Shehzad, M. K. 2018. *Influence of vertical steel reinforcement on the behaviour of edge restrained reinforced concrete walls*. University of Leeds.

Stoffers, H. 1978. Cracking due to shrinkage and temperature variation in walls.

Current Perspectives and New Directions in Mechanics, Modelling and Design of Structural Systems – Zingoni (ed.)
© 2022 Copyright the Author(s), ISBN 978-1-032-39114-4

RC flooring system recycling plastic bottles: New innovative RC waffle slab

C. El Mankabady, N. Salama, H. Tolba, K. Nassar, M.N. AbouZeid & E.Y. Sayed-Ahmed
Department of Construction Engineering, The American University in Cairo, Egypt

ABSTRACT: A considerable percentage of plastic bottles (80%+) is not recycled and ends up in landfills; each may take up to 1000 years to decompose. Millions of plastic bottles are not recycled every day in the USA alone; the size of such dilemma is increasing every year. This paper aims to present a kick-off/proof of concept for the development of a new RC flooring system that incorporates plastic bottles. The bottles will reduce the RC slab weight and, of course, the proposed system will contribute to solving the persistent plastic bottle environmental problem. The structural system of the proposed bottled RC slab is presented and explained. This slab flooring system imitates the structural behavior of waffle and/or bubble slabs in some of its aspects. The system uses plastic bottles as a mold substituting the concrete; thus, recycling the plastic bottles which are not acting as a structural element. In this way, the proposed system is more cost-efficient than both waffle or bubble slabs, and it saves the environment from non-biodegraded plastic bottles. Two full-size slabs (2.0m × 2.0m) were casted and tested under monotonically increasing flexure load to failure to prove the concept. The experimental failure load was compared to what is expected from similar ribbed slabs and proved that the concept could efficiently be further improved and used as a flooring RC system. This research work is only the starting point in developing such a system that is expected to be structurally efficient, economically advantageous, and sentimentally friendly.

Keywords: flexural testing, flooring system, reinforced concrete slab, plastic bottle, waffle slab

1 INTRODUCTION

Recycling plastic bottles typically embed additional energy which means an additional carbon footprint. As such, up-cycling is a more effective way to re-use plastic bottles than recycling. This paper presents a proof of concept for the development of a new RC flooring system that upcycles plastic bottles. This new system imitates the structural behavior of waffle or bubble slabs in some of its aspects as it uses plastic bottles as a mold substituting the concrete and reducing the slab weight. Two (2.0m×2.0m) RC slabs were cast and tested under monotonically increasing flexure load to failure to prove the concept.

The idea of incorporating the plastic bottles to create a waffle slab was abstracted from the bubble deck slab invention which was invented by Jorgen Bruenig in 1990 in Denmark (The Constructor, 2020). The proposed slab system imitates the bubble deck slab by replacing the plastic bubble with plastic bottles in a certain configuration and arrangement.

2 THE PROPOSED SLAB SYSTEM

Two RC slabs (2.0m×2.0m) were prepared to be tested in flexure. Each slab is supported on four (390mm×210mm) edge beams and includes four central ribs (100mm×210mm). The beams and ribs have the same slab thickness. The ribs are spaced at 340 mm. The beams were reinforced by 4 no 16 and 2 no 12 bottom and top reinforcement, respectively. No. 8 stirrups spaced at 100 mm were adopted as web reinforcement for the beams. The ribs were reinforced by 2 no. 12 bottom reinforcement with no 8 open stirrups spaced at 200 mm (Figure 1). A 28-day concrete compressive strength of 25 MPa was adopted for the slabs.

The spacing between the ribs is occupied by plastic bottles. This means that the plastic bottles are used as a mold substituting the concrete or the foam molds and are not acting as a structural element. The bottles were integrated into the slabs with a total of 144 empty plastic bottles, 1.5 liters each were used. These bottles were wrapped with 300mm wide cellophane.

DOI: 10.1201/9781003348450-205

Finding the best plastic bottles configuration that could stand the load coming from the slab without deforming was one of the challenges of this project. After trial and error testing by applying a load of around 20 kg on each configuration, the one shown in Figure 2 was selected as it experienced a minimum deformation (~ 1mm) when subjected to the said load.

Figure 1. The proposed RC ribbed slab with plastic bottles.

3 TESTING THE PROPOSED SLABS

Two full-size ribbed slabs with plastic bottles were tested under monotonically increasing flexure load to failure in order to prove the concept. The load was applied at four points at the intersection of the ribs using a hydraulic actuator (Figure 2). A distributor steel beams system was used to transfer the actuator loads to the four points on the slabs. The beam rested on four 100mm×100mm neoprene pads on the slab to assure load uniformity. A load cell was used to record the applied load on the slab four linear variable differential transducers (LVDT) were allocated in different locations to record the slab displacement; three LVDTs were mounted on top of the slab with one LVDT at the bottom. Moreover, the edge beams rested on steel rollers along with the four corners of the slab.

Preliminary design calculations for the slab capacity yielded 520 kN for flexure capacity and 534 kN for shear capacity with the previously mentioned slab configuration and test set-up.

The experimental investigation yielded a failure load of 524 kN with a 26.6 mm deflection for the first slab and 522 kN with a 24.1 mm deflection for the second slab.

Figure 2. The test set up of the ribbed plastic bottle slabs.

4 CONCLUSION

A new ribbed slab system that includes embedded plastic bottles is developed and tested in the work. The pilot experimental investigation of this paper proved that this system can effectively work and yield comparable results to the known waffled slab.

The plastic bottles reduce the slab weight and contribute to solving the persistent plastic bottle environmental problem. The plastic bottles serve as a mold substituting the concrete.

Two RC slabs (2.0m × 2.0m) with the said plastic bottles were tested under monotonically increasing flexure load. The resulting failure load is found to be comparable to what is expected from similar ribbed slabs and proved that the concept could efficiently be further improved and used as a flooring RC system.

ACKNOWLEDGMENT

The work presented in this paper summarizes the work performed by the first three authors as their capstone graduation project under the supervision of the last two authors. The help and support provided by the Department of Construction Engineering of the American University in Cairo and the structural engineering laboratory's technical staff are highly appreciated.

REFERENCES

The constructor, 2020. Bubble Deck Slab - Types, Material Specification, Installation, and Advantages. Retrieved September 2020, https://theconstructor.org/structural-engg/bubble-deck-slab-types-material-advantages/8341/

Current Perspectives and New Directions in Mechanics, Modelling and Design of Structural Systems – Zingoni (ed.)
© 2022 Copyright the Author(s), *ISBN 978-1-032-39114-4*

Investigations on shear transfer by aggregate interlock with unique test setup (TorAx)

Sven Bosbach*, Maximilian Schmidt, Henrik Becks, Martin Claßen & Josef Hegger
Institute of Structural Concrete, RWTH Aachen University, Aachen, Germany

1 INTRODUCTION AND PRELIMINARY INVESTIGATIONS

The structural behaviour of concrete is controlled through shear and normal stress transfer in cracked and uncracked concrete. In the past, different test setups have been used to assess shear and normal stress transfer actions in cracked and uncracked concrete. For example, in [1–3] push-off tests with either variable or constant crack width have been conducted to assess aggregate interlock (mode II). These test methods are partly inconsistent and testing of identical materials may give significantly deviating shear resistances when tests are performed in different test setups.

In test methods like push-off tests with finite shear length, the shear stress distribution is not constant along the interface. The maximum transferred shear strength due to aggregate interlock is only reached at the edges of the ligament length, while the minimum is in the middle of the shear length. Thus, the average shear stress, which is usually calculated by dividing the measured shear force trough the shear length, is biased by the shear length. For higher specimens, smaller average stresses are calculated. In contrast, due to continuous shear length a constant shear stress in all investigated material points is reached [4].

Claßen et al. showed the general application of a novel testing procedure for the investigation of aggregate interlock in a first preliminary study [4]. For this study, plain concrete pipes with a circumferential notch made of a fine grain high strength concrete with a maximum aggregate size $D_{\max}$ of 4 mm and dimensions of $h = 400$ mm, $d_i = 200$ mm, $d_e = 246$ mm have been casted. During the tests, an axial load was first applied for pre-cracking of the specimens with a predefined crack width. Afterwards, the torsional moment was applied deformation-controlled at a constant crack width. During the last step, the slip was measured analogous to the crack width with LVDTs.

The general behaviour of normal and shear stresses for increasing slip are comparable with established test results (e.g. [1; 3]). Otherwise, the obtained maximum stresses are significantly smaller compared to the investigations of e.g. [3]. It should be noted, that the tests in [3] have been carried out with a maximum aggregate size of a least 16 mm. Thus, the results can be classified as qualitatively correct.

2 LARGE SCALE TORAX TEST SETUP

Based on the concept presented before, a novel large-scale test setup for the investigation of e.g. aggregate interlock was developed. In Figure 1 a model of the stiff testing machine is depicted. The total dimensions of the new large-scale test setup for testing of circular concrete specimen with torsionaland axial forces (TorAx) are 10.7 m/4.4 m/ 2.5 m (length/width/height).

On the left-hand side of the TorAx, a tension-compression testing cylinder is installed which can apply axial forces up to 2 MN. On the right-hand side, you will find the torsional unit of the testing machine. Here, a testing cylinder with an eccentricity of 0.5 m is installed. Thus, torsional moments up to 1 MNm can be applied. In the middle section of the testing machine, two massive round flanges (diameter 1.35 m) are installed. Between these two flanges the concrete specimens, e.g. plain concrete pipes for investigation of aggregate interlock, will be installed. For aggregate interlock investigations under pure mode II, the concrete pipes are precracked along a predetermined line (notch) and then twisted while imposing the desired crack widths.

Transmission of loads and moments between TorAx and test specimen are realized by load transmission rings (LTR), which have an outer diameter of 120 cm and are embedded in concrete at both sides of the specimen. Axial forces are transferred into the specimen via circumferential protruding threaded bars which are cast into the specimen and screwed into the inner part of the LTR, while torsional moments are transferred via shear keys which are also screwed into the inner part of the ring.

*Corresponding author: sbosbach@imb.rwth-aachen.de

DOI: 10.1201/9781003348450-206

Figure 1. Large-scale torsional and axial load testing machine (TorAx).

The test specimens have a height of 100 cm and an inner and outer diameter of 66 and 80 cm, respectively. For the casting of the plain concrete pipes with a circumferential notch, reusable and modular steel formwork is utilized. The cylindrical formwork is sectioned into four inner and two outer segments to retract the segments after hardening of concrete. The inner formwork has to be detached from the specimen by moving it towards the inside of the member. Thus, a small gap is left between two parts of the inner formwork to enable retracting. This gap is closed with a sealant to prevent the concrete from leaking through this gap. The width of the gap is specified with spacers between welded flanges to achieve a defined radius of the concrete specimen. At both inner and outer formwork, small rings are assembled, which define a circumferential notch in the specimens where the crack will develop. To adjust the component thickness in the test area for different aggregate sizes, these rings are fixed modular to the formwork. Thus, different ring thicknesses can be installed on the formwork and different component thicknesses can be adjusted. Concreting of the specimens is realized by two small pockets inside the top LTR, which can be re-moved for pouring concrete inside the formwork and ventilation, respectively. After casting, these two parts are assembled back into the ring to ensure a full-surface frictional connection within the LTR.

3 CONCLUSION AND OUTLOOK

Based on the analysis of previous test methods, a new torsion test setup and first preliminary test results on aggregate interlock were presented. The testing procedure can overcome the shortcomings of classic test methods with finite shear length. Furthermore, it allows imposing mode II kinematics by applying tension and torsion to either pre-cracked or non-cracked concrete specimens. For investigation of aggregate interlock, the test setup allows for a continuous shear length at specimens to avoid edge and size effects.

On the basis of the promising test results, a new large-scale test setup was developed in which axial forces up to 2 MN and torsional moments up to 1 MNm can be applied. The unreinforced thin-walled concrete pipes for investigating aggregate interlock can be cast with a new modular steel formwork, while the load is transferred via load transmission rings. With this newly developed testing machine, mode I, mode II (including aggregate interlock), mode III, as well as mixed-mode fracture of concrete can be investigated.

The construction of TorAx and casting of the first test specimens are in progress. At the conference, first results of aggregate interlock tests and experiences with the novel test setup will be presented.

ACKNOWLEDGMENTS

Funded by the Deutsche Forschungsgemeinschaft (DFG, German Research Foundation) – SFB/TRR 280. Projekt-ID: 417002380. The authors would like to thank the DFG for supporting the research project.

REFERENCES

[1] Paulay, T.; Loeber, P. J. (1974) *Shear Transfer By Aggregate Interlock* in: American Concrete Institute [Ed.] *SP-042: Shear in Reinforced Concrete*. Detroit: American Concrete Institute, ACI, pp. 1–15.

[2] Taylor, H. P.J. (1970) *Investigation of the forces carried across cracks in reinforced concrete beams in shear by interlock of aggregate*. No. TR 42.447 Tech. Rpt.

[3] Walraven, J. C. (1981) *Fundamental Analysis of Aggregate Interlock* in: ASCE Journal of the Structural Division 107, ST11, pp. 2245–2270.

[4] Classen, M.; Adam, V.; Hillebrand, M. (2019) *Torsion Test Setup to Investigate Aggregate Interlock and Mixed Mode Fracture of Monolithic and 3D-Printed Concrete* in: Derkowski, W. et al. [Eds.] *fib Symposium 2019*. Krakow, Poland. International Federation for Structural Concrete (fib), pp. 521–528.

*Current Perspectives and New Directions in Mechanics,
Modelling and Design of Structural Systems – Zingoni (ed.)*
© 2022 Copyright the Author(s), ISBN 978-1-032-39114-4

The continuous spalling of reinforced concrete structures of the three bridges on the Lagos lagoon in Nigeria

U.T. Igba, J.O. Akinyele & A. Adetiloye
Department of Civil Engineering, Federal University of Agriculture, Abeokuta, Nigeria

J.O. Labiran
Department of Civil Engineering, University of Ibadan, Nigeria

ABSTRACT: The exposure of reinforced concrete structures to sulphate and chloride salts normally hasten the deterioration of such elements. This research investigated the spalling of the concrete elements of the three bridge structures on the Lagos lagoon in Nigeria and their deterioration as a result of exposure to salt attack within a year. Modelled samples of reinforced concrete beams and cubes were submerged in the lagoon water that contains high salts contents while control samples were submerged in freshwater. Flexural, compressive and split tensile tests were carried out on the concrete elements at 28, 90, 220 and 365 days. The microstructures of the concrete elements were investigated on each day to determine the rate of micro-spalling within the concrete elements due to salt attack. ANSYS software was used to simulate the reactions of the beam elements to loading on the 28th and 365th day for both samples. The results from these tests revealed the impact of sulphate salts on the spalling activities of concrete. The work concluded by developing a mathematical equation for the estimation of concrete deterioration within a year.

1 INTRODUCTION

The continuous deterioration or spalling of the concrete elements of the three major bridges on the Lagos lagoon, which is an offshoot of the Atlantic Ocean has been giving stakeholders in the construction industry serious concerned for some times now.

The durability of concrete depends on many factors which include its physical and chemical properties, the service environment and design life.

The first and major visible manifestations of sulphate attack on cementitious materials that were visible to the naked eyes included macro-cracking, spalling, declamation and loss of cohesion. Common reactions types such as recrystallization, desorption, dissolution or precipitation and adsorption were the consequences of complex chemical reactions and processes between the components of hydrated cement and sulphate compounds (Akinyele *et.al.*, 2021). This research aim at investigating the current state and the continuous spalling of the concrete pile caps of the three major bridges that linked the Lagos Island to the Mainland. This study involved testing for the strength and simulation of the concrete elements.

Figure 1. Pile cap of third mainland bridge.

2 EXPERIMENTAL SET UP

2.1 *Analysis of Lagoon and freshwater*

Parameters such as pH, salinity and the composition of ions such as Cl^-, $SO_4{}^{2-}$, Fe^{2+} in the sample of seawater and their amount and percentage present were determined through chemical analysis.

2.2 *Materials*

Grade 30 concrete was used in this research in line with BS 12390-3 (2009), the sample buried in the lagoon water was designated as BL_{30} while the sample cured in freshwater which is the control sample was designated as $BF_{30.}$ The cube and beam dimensions were 150 x 150 x 150 mm and 150 x 150 x 600 mm respectively.

2.3 *Compressive strength test*

The crushing of the concrete cube was done in the Civil Engineering concrete and soil mechanics laboratory of the University of Lagos with the use of the Compression Testing Machine (the testing machine used had a capacity of 1500KN for the tests and capable of applying the load at the specified rate) following BS EN 12390-3 (2009).

The formula to predict the deterioration in reinforced concrete strength for the future and past years (±) as a result of spalling is:

$$Pf = Pc[1 \pm r]^{\wedge}n \quad (1)$$

Where:

DOI: 10.1201/9781003348450-207

Pf = future year or past year value. ***Pc***= Value at 28 days. ***r*** = One year deterioration rate. ***n*** = number of forecast years after base year (n>1).

2.4 *Flexural strength test*

A three-point loading test was done following BS EN 12390 -5 (2009) to determine the Flexural strength of the beam. Tests were performed at Civil Engineering Laboratory of the Federal University of Agriculture Abeokuta, Ogun State, Nigeria using a 600 kN capacity Zhejiang Tugong Universal Testing Machine.

2.5 *Simulation of beams using ANSYS software*

The beam was modelled using a finite element programme ANSYS Version 19. The number of elements and nodes that subdivided the plates were 16999 elements and 18391 nodes.

3 RESULTS

3.1 *Result of water analysis*

The pH of 6.9 was obtained by the end of one year of curing for both fresh and lagoon water, this showed that both water were neutral. The chloride content in the lagoon and freshwater was 260 and 16 mg/l respectively, Lagoon water had a sulphate content of 20 mg/l, while that of freshwater was 6mg/l at the end of curing days.

3.2 *Compressive strength results*

The compressive strength test results for all the concrete samples are shown in Figure 2.

3.3 *Flexural strength results*

The use of flexural test enabled failure under tension along with the flexural member which was more sensitive to damage such as cracking, shear and deflection. These results are shown in Figure 3.

3.4 *Beam simulation results*

The one-year simulation results are shown in Figures 4 and 5. These showed that the reinforced concrete beam that was cured in freshwater had a damage factor of 0.92857 while the deformation was 4.1314 mm at 365 days. For the sample in lagoon water the damage factor was 0.9525, while the deformation was 4.7886 mm at 365 days.

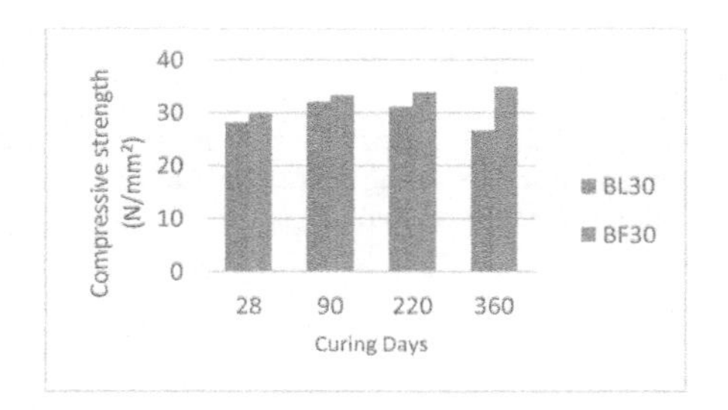

Figure 2. Compressive strength of concrete.

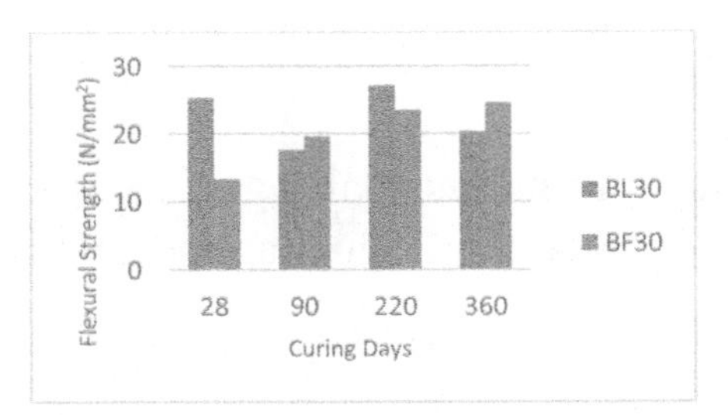

Figure 3. Flexural strength of concrete.

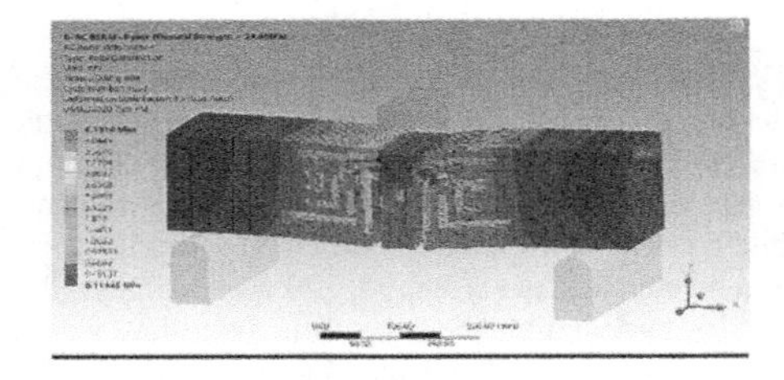

Figure 4. BF_{30} deformation at 365 days.

Figure 5. BL_{30} deformation at 365 days.

4 CONCLUSION

This work has investigated the rate of deterioration and spalling of concrete elements buried in the Lagos lagoon. The simulation of the bridge elements and the deterioration equation developed in this study has been used to estimate the current strength of the concrete elements of the bridge structures which is estimated to be 2.695 N/mm^2, from the initially estimated design strength of 30 N/mm^2 forty years ago. The spalling surfaces observed on all the bridge pile caps was instigated by weak bonds formed in the microstructures of the concrete, this was caused by the penetration of sulphates ions through the concrete porous medium and this led to the reduction in the initial design strength of the concrete elements

REFERENCES

Akinyele J.O., Igba U.T., Alayaki F.M. & Kuye S.I. (2021). Modelling the deformation of steel-bars in reinforced concrete beams submerged in lagoon. Nigerian Journal of Technological development. 18 (3) 219–228.

BS EN 12390–3 (2009) Testing Hardened Concrete. Compressive Strength of Test Specimens. British Standard Institution, London.

BS EN 12390–5 (2009) Testing Hardened Concrete: Flexural Strength of Test Specimens, British Standard Institution, London.

Current Perspectives and New Directions in Mechanics, Modelling and Design of Structural Systems – Zingoni (ed.)
© 2022 Copyright the Author(s), ISBN 978-1-032-39114-4

Theoretical analysis and experimental test results calibrated from concrete beams

G. Sossou
Kwame Nkrumah University of Science and Technology, Kumasi, Ghana

ABSTRACT: This paper relates to pre-stressed cracked beams, subjected to static long-term service loads. This study is based on the nonlinear differential equations of the concrete matrix creep theory which reflects the correlation between the matrix stress and strain by its modulus of elasticity. For crack predictions, virtual work principles are used to estimate (a) transient strains due to the matrix creep and shrinkage, (b) the resulting stress redistribution, as well as (c) the displacement variations in the beams and (d) the pre-stressing losses in the high yield tendons. A series of laboratory experiments are used to support the numerical evaluations. In addition, we (e) calibrate the pre-stressing losses, and (f) predict the beams' stiffness and strength by determining their flexural and nonlinear creep capacity, and hence (g) devising a definition for structural durability and reliability, with regards to the concrete stress-strain relationship under service loads.

Keywords: concrete matrix creep, shrinkage, pre-stressed concrete cracked beams, strength, reliability, durability, steel tendons, pre-stressing losses

1 INTRODUCTION

The main research object in this paper relates to pre-stressed concrete beams of I and II categories of resistance to cracking, not only as the most spread used prefabricated structural elements, but also like the most convenient models for theoretical and experimental studies, which can permit to widely and reliably reveal their positive effect and generalize it to more complicated type of structural members.

This analytical simplified procedure has been aimed to predict the quality, stiffness, strength, reliability, durability, fatigue, life safety, the stability, the non-linear creep and the progressive cracking behavior of the beams at the planning phase. The extended interests include the study of the influence of the material non-linearity, fatigue, creep, integrity and the progressive cracking characteristics under long-term static and quasi-static loads.

2 THEORETICAL ANALYSIS – MAIN DESIGN CONCEPTS

The numerical part of this study is based on the well-known concepts by Yasinskiy, Timoshenko, Barashikov, Yatsenko, Baykov, Vlasov, Drozdov, Juravskiy, … etc and many other Authors.

Similarly to the case of non-cracked beams (details published in previous papers), have been evaluated new geometric parameters of the cracked sections and by using the finite difference method, have been determined the exponents of limit states of cracked members.

For the determination of the new geometric parameters of the cracked beam's section (Figure 1), it is important to evaluate the height of the principal effective concrete layer by using z_T - distance from the beam axis to the crack mouth.

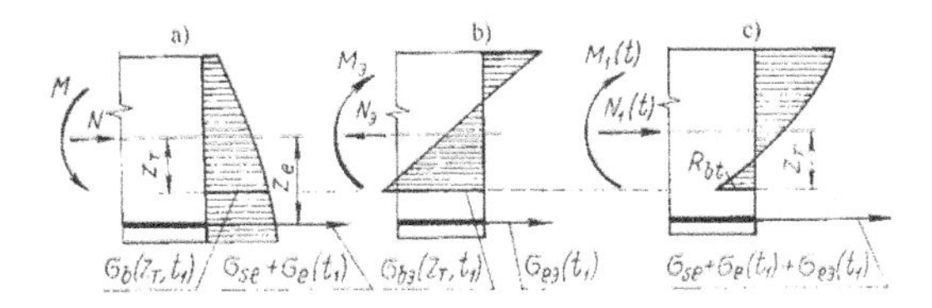

Figure 1. Stress state of the beam element.
a) just before the long-term service loading;
b) with short-term cyclic loading, from the initial value of the service loading;
c) with the total loading as in the case of "a)" included the service loading at the moment t_1.

DOI: 10.1201/9781003348450-208

3 EXPERIMENTAL PROGRAM AND SOME TEST RESULTS

The experimental program consisted of casting twelve 1800x180x100 mm pre-tensioned concrete beams with material properties of 40 MPa for the concrete matrix and E_{SP} = 195 GPa for the high-yield steel tendons with nominal diameters of 14 mm and 18 mm, which will be contained in frames made of 6 mm bent mild steel mesh with E_S = 200 GPa (Figure 2). At the concrete age of 9 days with natural curing, all of the beams, after the transfer of stresses from the steel tendons to the concrete matrix, will be compressed out by the initial pre-stressing forces and then will be taken away from the pre-stressing beds together with their deflection strain gauges, after cutting the steel tendons. 6 beams cast with diameter 14 mm tendons were marked B1 and 6 other cast with diameter 18 mm - B2. Two of B1 and two of B2 respectively marked B_1^{sh} and B_2^{sh}, after the transfer of pre-stress were maintained in the laboratory without long-term loads. By them have been calibrated the variations of deformations both (i) in the steel tendons and (ii) in the concrete, deformations due to the initial pre-stressing forces and also due to the concrete shrinkage.

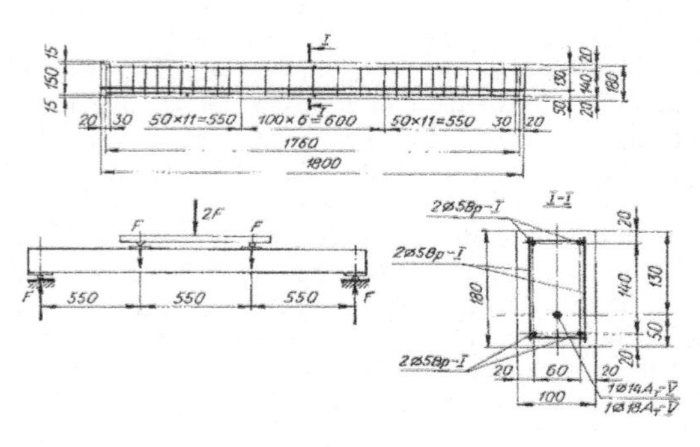

Figure 2. Constructive experimental model of the pre-stressed concrete beam.

4 CONCLUSIONS

(i) Theoretical values of cracking moments M_{crc} of beams B_1^{24} exceeded the same values of beams B_1^{sh} by 14.6 %, but for the experimental data these values have reached 30.9 %. Also the same comparison with beams B_2^{24} and B_2^{sh} shows the exceeding of theoretical data by 15.7 % and the experimental data by 21.8 %. Concerning the comparison of failure moments Mu, so, beams reinforced with tendons of 14 mm diameter have shown almost the same bearing capacity. But the beams reinforced with tendons of 18 mm diameter, loaded under long-term static forces have shown bearing capacity more higher by 2.2-5.6 % than these beams maintained in the laboratory without long-term loads.

(ii) These comparisons have shown that the nonlinear theory described in this thesis, is quite adequate and can be served for practical use. The consideration of this proposed design and experimental model, and these said time-dependent effects can be able to ensure interdependence of design and construction for economies of the steel tendons up to 5 % - 15 %.

(iii) Experimental results have indicated that this proposed numerical analysis model can be used to reach these said objectives and the realistic feasibility of this proposed method has been successfully verified. Theoretical and experimental research results have shown that the finite-difference method based on the displacement formulation is suitable and effective to solve systems of nonlinear equilibrium differential equations. The nonlinear concrete creep is able to contribute to the redistribution of stresses between concrete and steel reinforcement. This said redistribution is able to reduce pre-stressing losses and, like consequence, able to increase the exponents of beams' limit states, to increase the safety and durability, so finally able to contribute for the quality of these said pre-stressed concrete beams.

(iv) Nonlinear concrete creep is able to contribute to the redistribution of stresses between concrete and steel tendons present with long duration forces and service loads influences on the beams. This said redistribution is able to provoke the accumulation of natural initial stresses in reinforcement and concrete, which are equivalent to artificial initial stresses due to pre-stressing of the tendons, and which can reduce losses and, like consequence, able to increase the exponents of beams' limit states, and all of these positive effects are able to contribute for steel tendons' economies for up to 5-15%.

(v) The results of these theoretical and experimental studies have a practical use and have been oriented to current review for their integration into next changes of Codes of reinforced and pre-stressed concrete structures, used by scientists and engineers.

REFERENCES

Backhouse J. K., Houldsworth S.P.T., Cooper B E. D., Horril P.J.F. 1992 *Pure Mathematics, Book 2*, Third Edition, Longman Group Limited.

Edwin H. Gaylord, Jr, Charles N. Gaylord and James E. Stall-meyer 1997, *Structural Engineering Handbook*, Fourth Edition, McGraw-Hill.

Sossou G. 1991 *Influence of Concrete Creep on Limit States of Reinforced and Pre-stressed Concrete Beams*, Ph.D. thesis, Dniepropetrovsk Civil Engineering Institute, Ukraine, 163 pp.

Current Perspectives and New Directions in Mechanics, Modelling and Design of Structural Systems – Zingoni (ed.)
© 2022 Copyright the Author(s), ISBN 978-1-032-39114-4

Behaviour of reinforced concrete beams enhanced with polymer modified ferrocement

N.F.O. Evbuomwan
Accenture Inc, Irving, Texas, USA

ABSTRACT: Dramatic advances in the use of ultra-high-strength microdefect-free materials in the form of extruded cement pastes that achieve high tensile and torsional strength inspired the work reported in this paper. This involved the development of a high-strength polymer modified mortar that can provide quantitative repair enhancement as well as extending the life of deteriorated concrete members. This paper reports on the development of the repair mortar, its application in the repair of 1.0m long concrete beams repaired employing various bonding agents and the structural behaviour under flexural loading. Ultimate strength experimental results are compared with theoretical values based on three (Hognestad et al, BS 8110 Stress block and BS 8110 Simplified rectangular block) approaches to determining ultimate capacity of the beams. Benefits derived from the application of the reinforced repair material and its impact on the ultimate and cracking loads, cracking characteristics, efficiency of bonding systems and improved ductility and stiffness are discussed.

1 INTRODUCTION

The work involved the development a high strength polymer-modified mortar that can be used to provide quantitative repair enhancement as well as extending the life of deteriorated concrete members.

2 DEVELOPMENT OF THE REPAIR MATERIAL

The resulting overall optimum results are shown in Table 1. The mix proportions for the control mix is given by 1:2.13:1/3:0.0533:1/3 representing the ratio of OPC: Sand: PFA: PVA: Water, respectively.

Table 1. Results of materials evaluation of repair mortar.

Mix Proportion	Curing Method Moist + Air	Flexural Strength MPa	Compressive Strength MPa
Control	7M + 21A	13.95	58.00
Control + Superplasticizer (SP)	7M + 21A	13.51	
Control + 5% Microsilica (MC)	7M + 21A	14.67	63.80
Control + 2%SP + 5%MC	7M + 21A	15.69	68.13

3 ENHANCED MODEL BEAMS

The mix of Control + 2% SP + 5% MC) was chosen for the repair application of the test beams. This paper discusses two key series of tests on 1.0m reinforced concrete beams repaired with the mortar.

3.1 *Materials and specimen preparation*

For each series of tests discussed in this paper (Series 2 and Series 3), ten reinforced concrete beams were prepared using Ordinary Portland Cement, 5mm fine sand and 13mm coarse (gravel) aggregate in the proportions of 1:2.70:4.16 with a water/cement ratio of 0.58 and 1:3.86:5.94 with a water/cement ratio of 0.58, respectively. Each beam was reinforced in tension at the bottom with 2M-5.9Ø bars, while 2Y-2.5Ø bars were provided at the top of the beam as nominal reinforcement. Y2.5Ø were provided as links at 60 mm c/c spacing. The loading and support conditions are shown in Figure 1.

3.2 *Repair procedure*

The enhanced beams were demoulded after 24 hours and then covered with damp hessian for a further 6 days, after which the beams were uncovered, coated with a curing membrane, and left to cure in the laboratory environment of about 20 ± 2 °C and 50 ± 5% RH for 21 days before they were tested.

4 TESTING PROCEDURES

Flexural strength tests were performed on the beams using a Denison testing machine. The beams were supported over a span of 914 mm. The applied load readings in kilonewtons were read from a digital display on the Denison testing machine, while deflection readings were read from a dial gauge placed at the bottom of the mid-span position of the beam. Deflection readings were taken throughout the loading regime. During loading, readings for crack widths, spacing and height were taken.

5 RESULTS AND DISCUSSION

The experimental flexural results of all the beams tested in Series 2 and 3 are shown in Tables 2 and 3 below.

DOI: 10.1201/9781003348450-209

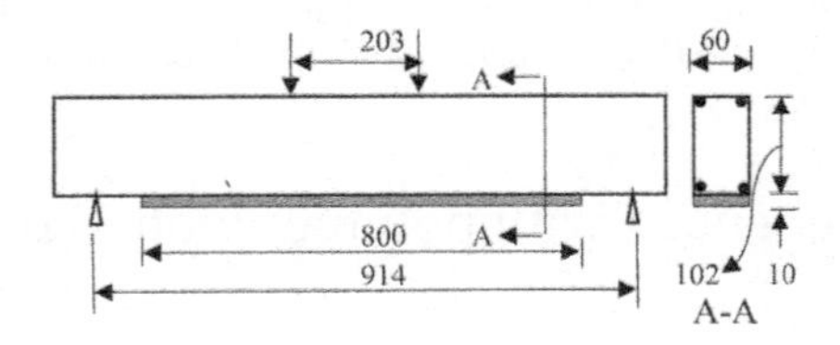

Figure 1. Details of Enhanced RC Beams (All dimensions in mm).

Table 2. Ultimate load and deflection results.

Beam No.	F_{cu} (MPa)	Bond Aid	Cracking Load (kN)	Failure load P_{ult} (kN)	Deflection Δ_{ult} (mm)
S2-A1	55.33	Control	5.00	12.46	13.00
S2-A2	55.33	None	8.00	19.04	7.00
S2-B1	56.00	SBR	9.00	19.03	7.00
S2-B2	56.00	SK 501	10.00	19.05	4.51
S2-C1	55.33	Epoxy A	10.00	19.08	4.00
S2-C2	55.33	Epoxy B	12.00	18.54	6.20
S2-D1	55.57	Epoxy C	14.00	20.20	5.00
S2-D2	55.57	Acrylic	8.00	19.65	6.40
S2-E1	56.90	Epoxy D	18.00	18.54	6.00
S2-E2	56.90	Epoxy E	14.00	18.14	6.50

5.1 *Ultimate load & moment capacity*

It is evident that all the enhanced beams tested in both Series 2 and 3 exhibited higher ultimate loads than the control beam with increases ranging from 46 to 62% for Series 2 beams and from 27 to 55% for the Series 3 beams.

Table 3. Ultimate load and deflection results.

Beam No.	F_{cu} (MPa)	Bond Aid	Cracking Load (kN)	Failure load P_{ult} (kN)	Deflection Δ_{ult} (mm)
S3-A1	30.00	Control	4.00	11.65	4.41
S3-A2	30.00		8.50	11.70	13.81
S3-B1	34.64		10.00	18.08	3.71
S3-B2	34.64	SBR	9.00	18.34	5.47
S3-C1	29.72	SK 121	10.00	19.46	5.23
S3-C2	29.72	Epoxy C	10.00	18.66	5.89
S3-D1	29.70	SK 501	6.00	18.07	6.35
S3-D2	29.70	SK 501	12.00	20.90	5.94
S3-E1	33.67	SK 501	4.00	14.78	10.85
S3-E2	33.67	SK 501	6.00	14.91	14.33

5.2 *Maximum load deflection*

In Series 2, the enhanced beams showed lower deflections observed at maximum load compared to the control beam with percentage reductions between 45 and 70%.

5.3 *Load deflection curves*

The load deflection curves for Series 2 and 3 are shown in Figures 2 and 3. The enhanced Series 2 beam showed increased stiffness characteristics compared to the control beam.

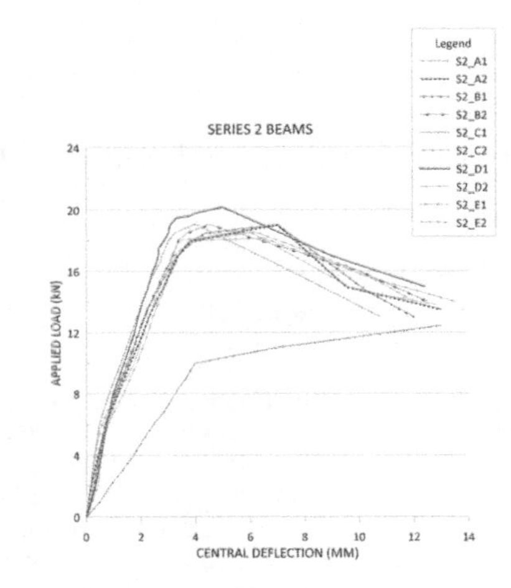

Figure 2. Applied Load vs Central Deflection.

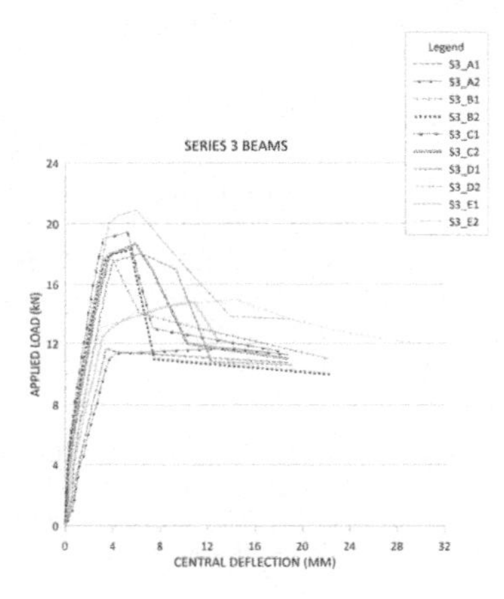

Figure 3. Applied Load vs Central Deflection.

The enhanced beams generally also exhibited higher stiffnesses than the control beam, particularly over the loading regime prior to maximum load.

5.4 *Cracking characteristics*

The enhanced beams from both series tested, exhibited finer and reduced crack widths, with correspondingly higher numbers and closer spacings than the control beam throughout the loading regime.

6 CONCLUSIONS

This paper has clearly demonstrated that the enhancement of reinforced concrete beams with the developed repair mortar can provide quantitative repair to extend the life of downgraded beams. All enhanced beams demonstrated higher cracking and ultimate loads compared with the unenhanced one. Additionally, the enhanced generally demonstrated increased stiffness and ductility. No bond failures were observed or occurred in any of the enhanced beams.

Current Perspectives and New Directions in Mechanics, Modelling and Design of Structural Systems – Zingoni (ed.)
© 2022 Copyright the Author(s), *ISBN 978-1-032-39114-4*

Shear capacity of rib and block slab systems

Y. Essopjee
Department of Civil Engineering Science, University of Johannesburg, Johannesburg, South Africa

ABSTRACT: Rib and block slab systems are commonly used is South Africa for residential and other low-rise building applications. The moment and shear capacity supplied by manufacturers are based on the provisions prescribed in SANS 10100-1. The rib and block can also fail by vertical and horizontal shear. Horizontal shear is typically at the interface of the precast rib and the in-situ concrete. However, the behaviour of rib and block system is different because of its low reinforcement ratio, high yield strength of the reinforcement and non-monolithically cast concrete. The aim of this research was to test ribs to check the validity of SANS 10100-1 shear estimation. A total of 9 rib and block beams were fabricated and tested. The results show that SANS 10100-1 is unnecessarily conservative and that manufacturers shear strength predictions could be revised.

1 INTRODUCTION

A popular type of slab used in South Africa is the rib and block slab system. They are lighter when compared to cast in situ slabs because the blocks are void forming. Their installation is simple and quick. Rib and block slabs consist of a precast concrete beam which has cast in-situ concrete placed on the beam to form a T-section, with the aid of a void former. A typical setup is shown in Figure 1.

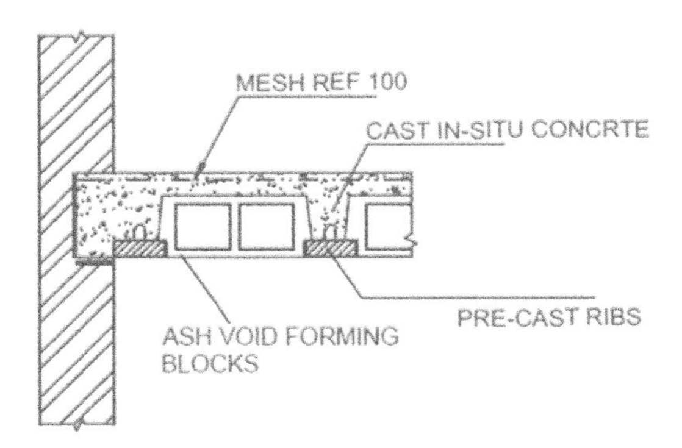

Figure 1. Rib and block slab cross section.

Vertical shear is a possible failure mechanism. While SANS 10100-1 (2000) does cover shear capacity of concrete beams and slabs, the formulae proposed by SANS need to be investigated. This is because the rib and the cast in-situ concrete are not monolithically cast. The aggregate interlock is not the same as for a monolithically cast member.

2 GEOMETRIC PROPERTIES

The width of the ribs were 150mm. The height of the ribs were 78mm inclusive of a 20mm projection. The ribs had between 5 and 8 wire strands each. The final height of the rib beams were selected to be 200 mm. 200 mm rib and block slabs can typically span 5 meters in a single rib format and 6 meters in a double rib format, hence the reason for choosing 200mm.

3 MATERIAL PROPERTIES

The concrete for the ribs was designed as 40 MPa. The diameter of the prestressed wire strands were 4.25mm and they were deformed in the long direction to provide better bond. The wire strands were tested in tension to determine their tensile capacity. The minimum and maximum values found were 1707 and 1812 MPa, respectively. This corresponds to between 1 and 7 % additional strength, hence all the wires were acceptable.

The concrete mix design for the lab mix was designed as a 25 MPa mix. The stone size was 19 mm and a water cement ratio of 0.5 was used. The mix ratio was approximately 1:3:3.5.

4 TEST SETUP

The beams were tested for shear in a 2000kN Instron. A constant rate of 3mm/min was used for all the tests. For the shear tests, two roller supports were placed at the ends. The beam was loaded in the centre with the load actuator to simulate a point load. The load, time and displacement were logged in the Instron data collection system.

DOI: 10.1201/9781003348450-210

5 FAILURE MODE

All the samples failed in a similar manner. A diagonal crack formed from the point of loading through to the support, as shown in Figure 2. This can be classified as a shear-compression failure. The shear span to effective depth ratio ranged between 1 and 1.6 and hence this mode of failure for all the beams was expected. The crack extended all the way through the precast ribs, thus there was no delamination or horizontal shear failure between the precast and cast in-situ concrete surfaces.

Figure 2. Sample 5 with a diagonal crack.

6 TEST RESULTS FOR BEAMS

The test results are shown in Table 1. The results show that on average SANS 10100-1 (2000) underestimated the shear capacity by 3.1 times. The minimum V_{test}/V_{sans} ratio was 2.4 and the maximum was 3.7.

Although SANS 10100-1 (2000) does not consider an increase in the shear capacity when the yield strength of the reinforcement is higher (Tarek et al., 2008) found that when using high strength bars the mode of failure was different and the strength were also higher. Hence, further research is required to establish if this causes an increase in shear capacity.

Table 1. Shear test results.

No	length mm	$100A_s/$ bd	V_{Fail} kN	V MPa	V_{SANS} MPa	V/V_{SANS}
1	600	0.24	115.2	1.92	0.54	3.6
2	600	0.24	127.1	2.12	0.54	3.9
3	600	0.24	125.4	2.01	0.54	3.9
4	600	0.33	102.4	1.71	0.60	2.8
5	600	0.33	87.9	1.47	0.60	2.4
6	650	0.38	116.2	1.94	0.63	3.1
7	840	0.38	100.9	1.68	0.63	2.7
8	720	0.38	138.4	2.31	0.63	3.7

7 CONCLUSIONS

Eight ribs with a depth of 200mm were tested in shear. The rib lengths were designed with an effective span to depth ratio of between 1.6 and 1. This was done so that the ribs would fail in a shear-compression type failure. The results had an average standard deviation of 12 across the three groups and were thus acceptable for shear capacity prediction. The results showed SANS 10100-1 (2000) underestimated the shear strength of the beams between 2.4 and 3.9 times. Hence, SANS 10100-1 (2000) was found to be very conservative. It is proposed that the shear prediction formula of SANS 10100-1 (2000) is revised to include a provision for high tensile wire strands as reinforcement. This should be done by increasing the reinforcement ratio mechanism which would then increase the capacity of the dowel action and aggregate interlock components.

REFERENCES

SANS 10100-1.2000. The structural use of concrete Part 1: Design. Pretoria: *SANS*.

Tarek, K.H., Hatem, M.S., Hazim, D., Sami, H.R. & Paul, Z. 2008. Shear Behavior of Large Concrete Beams Reinforced with High-Strength Steel. *ACI Structural Journal* 105(2):173–179.

14. High strength concrete, High performance concrete, Fibre-reinforced concrete, Fatigue behaviour of HPC and UHPC

Current Perspectives and New Directions in Mechanics, Modelling and Design of Structural Systems – Zingoni (ed.)
© 2022 Copyright the Author(s), ISBN 978-1-032-39114-4

Phase-field modeling for damage in high performance concrete at low cycle fatigue

J. Schröder, M. Pise & D. Brands
Institute of Mechanics, Faculty of Engineering, University Duisburg-Essen, Essen, Germany

G. Gebuhr & S. Anders
Institute for Structural Engineering, Faculty of Architecture and Civil Engineering, Bergische Universität Wuppertal, Wuppertal, Germany

ABSTRACT: In this contribution, a constitutive framework of elasto-plastic phase-field model is presented, which is able to predict nonlinear behavior of high performance concrete (HPC) during low-cycle fatigue. The Drucker-Prager yield criterion is used for the formulation of an elasto-plastic damage model. For modeling of tension-compression behavior of HPC two different data driven degradation functions are interpolated and calibrated. Three-point bending experimental tests at low-cycle using notched beams of pure HPC are performed. For calibration purpose, the load-CMOD (crack mouth opening displacement) is measured. This experimental data were used to calibrate the proposed numerical model. The results show the efficiency of the proposed model to reproduce fatigue behavior of pure HPC as in the experiment.

1 INTRODUCTION

The diverse variety of compositions and wide range of applications of high-performance concretes (HPCs) have attracted a great interest in the construction industry worldwide. However, an extensive research effort needs to be carried out in order to gain deeper understanding of the overall behavior of HPCs during cyclic loading. To achieve this objective, the priority program 2020 (SPP 2020) of the German research foundation (DFG) has been founded which aims at the experimental and numerical analysis of failure of HPC during fatigue. Therein, the authors of this contribution have a joint project.

In this contribution, a constitutive model is developed for the analysis of failure of HPC during cyclic loading. The complex non-linear characteristics of concrete materials due to its elasto-plasticity and distinct behavior in tension and compression are taken into account. An elasto-plastic phase-field model for fracture in HPC using a Drucker-Prager yield criterion is presented, see Storm et al. (2021). Two different data driven degradation functions are interpolated based on the damage parameter, see Gebuhr et al. (2022). Three-point bending beam test during low-cycle for pure HPC is simulated to calibrate the material parameters and the data driven degradation function to reproduced the experimental results.

2 CONSTITUTIVE FRAMEWORK

The free energy function ψ is constructed, cf. Miehe et al. (2016), reads

$$\psi(\varepsilon, \varepsilon^{\mathrm{p}}, \alpha, q, \nabla q) = g(q)\left[\psi_0^{e+} + \psi_0^{\mathrm{p}} - \psi^{c}\right] + \psi_0^{e-} + \psi^{c} + 2\frac{\psi^{c}}{\zeta}\, l\left[\frac{1}{2l}q^2 + \frac{l}{2}||\nabla q||^2\right]. \quad (1)$$

which depends on the phase-field parameter q and its gradient ∇q as well as the equivalent plastic strains α Therein, tensor ε^{e} and ε^{p} denote elastic and plastic part of the total strain tensor ε. Specific critical fracture energy ψ^{c} and a parameter ζ control the crack threshold and shape of the post critical stress softening, respectively. A data driven degradation function $g(q)$ depending on the evolution of phase-field parameter q is considered. The elastic energy function ψ^{e} and plastic energy ψ^{p}, respectively, read

$$\begin{aligned} &\psi^{e}(\varepsilon^{e}, q) = g(q)\psi_0^{e+}(\varepsilon^{e}) + \psi_0^{e-}(\varepsilon^{e}) \\ &\text{with } \psi_0^{e+}(\varepsilon^{e}) = \kappa\langle \mathrm{tr}[\varepsilon^{e}]\rangle_+^2/2 + \mu||\mathrm{dev}\varepsilon^{e}||^2 \\ &\text{and } \psi_0^{e-}(\varepsilon^{e}) = \kappa\langle \mathrm{tr}[\varepsilon^{e}]\rangle_-^2/2, \\ &\psi^{p}(\alpha) = g(q)\,(y_0\alpha + h\alpha^2/2). \end{aligned} \quad (2)$$

Therein, μ and κ are the Lamé coefficients, h is hardening parameter and the operator $\langle\bullet\rangle_{\pm} = 1/2(\bullet \pm |\bullet|)$ denotes Macaulay's notation. Stress tensor σ, reads

$$\boldsymbol{\sigma} = g(q)\left[\kappa\langle \mathrm{tr}\,\varepsilon^{e}\rangle_+\mathbf{I} + 2\mu\,\mathrm{dev}\,\varepsilon^{e}\right] + \left[\kappa\langle \mathrm{tr}\,\varepsilon^{e}\rangle_-\mathbf{I}\right], \quad (3)$$

where $\mathbf{I}$ denotes the identity tensor. Governing equation for phase-field parameter considering a degradation function $g(q) = (1-q)^2$ results in

DOI: 10.1201/9781003348450-211

$$q - l^2 \mathrm{Div}[\nabla q] - (1-q)\mathcal{H} = 0 . \quad (4)$$

To ensure the irreversiblity of the crack evolution a local history field $\mathcal{H} := \max_{s \in [0,t]} \mathcal{H}_0(\mathbf{x}, s) \geq 0$ is considered which depends on the maximum value of a crack driving state function $\mathcal{H}_0$, i.e.,

$$\mathcal{H}_0 = \zeta \left\langle \frac{\psi_0^{e+}(\varepsilon^e)}{\psi^c} + \frac{\psi_0^p(\alpha)}{\psi^c} - 1 \right\rangle . \quad (5)$$

The non-associative Drucker-Prager yield function is

$$\phi_{\mathrm{p}}(\sigma_0, \kappa_{\mathrm{p}}) = ||\mathrm{dev}\, \sigma_0|| / \sqrt{2} + \beta_{\mathrm{p}} \mathrm{tr}\, \sigma_0 - y_0 - h\alpha , \quad (6)$$

and the considered plastic potential function is

$$\phi_n(\sigma_0, \kappa_{\mathrm{p}}) = ||\mathrm{dev}\, \sigma_0|| / \sqrt{2} + \beta_n \mathrm{tr} \sigma_0 , \quad (7)$$

where β_{p} and β_n are the material parameters. To capture an distinct behavior of concrete in tension-compression, two different parameters for the critical fracture energy, i.e. $\psi_{\mathrm{t}}^{\mathrm{c}}$ and $\psi_{\mathrm{c}}^{\mathrm{c}}$, and two data driven degradation functions, i.e. $g^+(q)$ and $g^-(q)$, in tension and in compression respectively, are considered. They are differentiated from each other by sign of the first invariant of the stress tensor

3 NUMERICAL EXAMPLE

The predictive capability of the presented model is checked by simulating the failure of three-point bending beam test of a supported notched beam during low-cycle. Figure 1 shows geometry and the boundary conditions of three-point bending beam test, taken as per the European Standard EN 14651. The material parameters, see Table 1 and support points for degradation functions $g^+(q)$ and $g^-(q)$, see Table 2, are first calibrated using uniaxial tensile and compressive tests, see Storm et al. (2021) and Gebuhr et al. (2022).

In Figure 2, the distribution of stresses in x-direction σ_x, equivalent plastic strain α, phase field parameter q are shown. For the analysis, the experimental and numerical load-CMOD curves are compared in Figure 3a and evolution of degradation function $g^+(q)$ in tension is plotted in Figure 3b.

Table 1. Material parameters used for the simulations.

E	ν	f_{t}	f_{c}	$\psi_{\mathrm{t}}^{\mathrm{c}}$	$\psi_{\mathrm{c}}^{\mathrm{c}}$	y_0	β_{p}	β_n	h	l	ζ
GPa	–	MPa	MPa	MPa	MPa	MPa	–	–	GPa	mm	–
40	0.2	5.7	112	4.2e-4	0.1	5.88	0.45	0.02	6	14	1

Table 2. Calibrated support points for degradation functions for HPC in tension and compression.

q	0	0.1	0.3	0.5	0.6	0.7	0.8	0.9	1
$g^+(q)$	1	1	1	0.85	0.74	0.58	0.35	0.12	0.001
$g^-(q)$	1	0.852	0.585	0.353	0.252	0.164	0.089	0.03	0.001

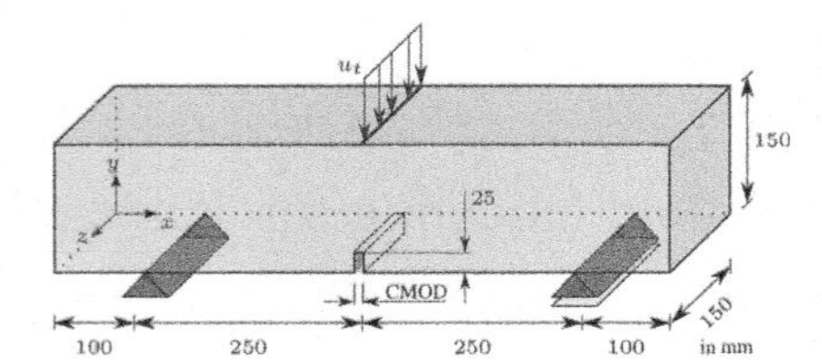

Figure 1. Geometry and boundary value problem for three-point bending beam cyclic test.

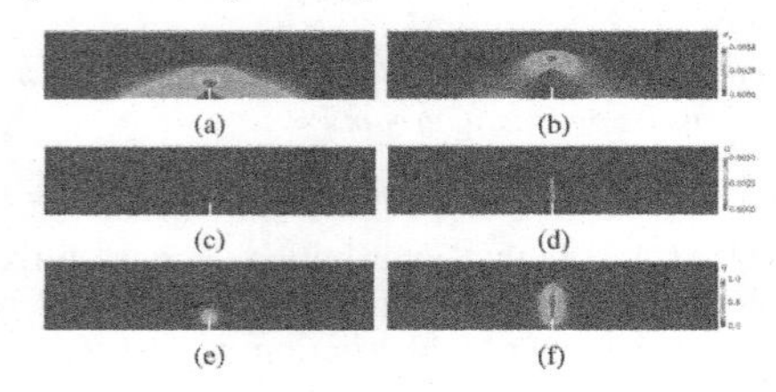

Figure 2. Distribution of stresses in x-direction σ_x (in GPa), the equivalent plastic strain α, the phase field parameter q in (a), (c), (e) at CMOD = 0.022 mm and (b), (d), (f) at CMOD = 0.067 mm, respectively.

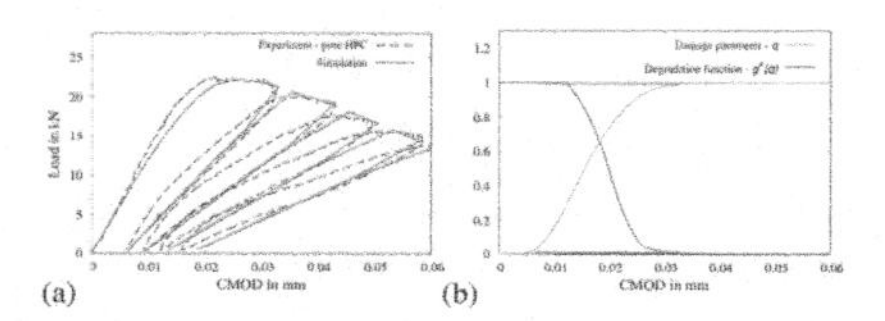

Figure 3. Results: a) load-CMOD diagramm and b) evolution of degradation function $g^+(q)$ in tension with respect to damage parameter q.

4 CONCLUSION

The comparison of load-CMOD curves shows the ability of the presented model to reproduce the congruent response as in the experiment of three-point bending beam test during low-cycle. The concept of data driven degradation functions facilitates the calibration of the presented model.

ACKNOWLEDGEMENT

Funded by the Deutsche Forschungsgemeinschaft (DFG, German research Foundation) - 353513049 (AN1113/2-2, BR5278/2-2, SCHR570/32-2) within the DFG priority program 2020. Computing time provided on the supercomputer magnitUDE, funded by the Deutsche Forschungsgemeinschaft (DFG, German research Foundation), Project number: 263348352 (INST 20876/209-1 FUGG, INST 20876/243-1 FUGG), is gratefully acknowledged.

REFERENCES

Gebuhr, G., M. Pise, S. Anders, D. Brands, & J. Schröder (2022). *Materials 15*(3), 1179. doi.org/10.3390/ma15031179.

Miehe, C., F. Aldakheel, & A. Raina (2016). *International Journal of Plasticity 84*, 1–32.

Storm, J., M. Pise, D. Brands, J. Schröder, & M. Kaliske (2021). *Engineering Fracture Mechanics 243*, 107506. 10.1016/j.engfracmech.2020.107506.

Current Perspectives and New Directions in Mechanics, Modelling and Design of Structural Systems – Zingoni (ed.)
© 2022 Copyright the Author(s), *ISBN 978-1-032-39114-4*

Macroscopic model based on application of representative volume element for steel fiber reinforced high performance concrete

M. Pise, D. Brands & J. Schröder
Institute of Mechanics, Faculty of Engineering, University Duisburg-Essen, Essen, Germany

G. Gebuhr & S. Anders
Institute for Structural Engineering, Faculty of Architecture and Civil Engineering, Bergische Universität Wuppertal, Wuppertal, Germany

ABSTRACT: The reinforcement of steel fibers in HPC show prominent effect on overall material behavior of HPCs in cyclic flexural tests. The complex fiber-matrix interactions at microscale dominate the macroscopic behavior of fiber reinforced high performance concrete (HPC) during failure. To understand the influence of embedded steel fibers, the material behavior of reinforced HPC need to be analyzed. For efficient simulation of the overall material behavior of fiber reinforced HPC a phenomenological material model is developed. Further, a virtual experiment based on an ellipsoidal unit cell (representative volume element, RVE) characterizing the steel fiber reinforced HPC along a preferred fiber direction is simulated. The micro-mechanical model proposed in our recent work is used for the simulation, see Storm et al. (2021). A macroscopic boundary value problem (macroscopic-BVP) is simulated using the presented phenomenological material model. To check the efficiency of the presented numerical model the simulation results of a macroscopic-BVP and a virtual experiment using an ellipsoidal RVE are compared.

1 CONSTITUTIVE FRAMEWORK OF MACROSCOPIC MODEL

The macroscopic stored energy function per unit volume $\bar{\psi}$ is construted using the volume fraction $\mathrm{v}^{\mathrm{HPC}}$ of HPC phase and v^{F} of fiber phase, reads

$$\begin{aligned} \bar{\psi} = & \, \mathrm{v}^{\mathrm{HPC}}\, \bar{\psi}^{\mathrm{HPC}}\left(\bar{\varepsilon}, \bar{\varepsilon}^{\mathrm{p,HPC}}, \bar{q}, \nabla\bar{q}, \alpha^{\mathrm{HPC}}\right) \\ & + \mathrm{v}^{\mathrm{F}}\, \bar{\psi}^{\mathrm{F}}\left(\bar{\varepsilon}, a, e^{\mathrm{p},F}, \alpha^{\mathrm{F}}\right), \\ \text{with } & \mathrm{v}^{\mathrm{HPC}} = 1 - \mathrm{v}^{\mathrm{F}} \text{ and } \bar{\varepsilon}^{\mathrm{e,HPC}} := \bar{\varepsilon} - \bar{\varepsilon}^{\mathrm{p},HPC}. \end{aligned} \tag{1}$$

The energy function $\bar{\psi}^{\mathrm{HPC}}$ of the HPC phase, positive $\bar{\psi}_0^{\mathrm{e+,HPC}}$ and negative $\bar{\psi}_0^{\mathrm{e+,HPC}}$ parts of elastic energy function and plastic energy function $\bar{\psi}_0^{\mathrm{p,HPC}}$ are given

$$\begin{aligned} \bar{\psi}^{\mathrm{HPC}} &= (1-\bar{q})^m \left[\bar{\psi}_0^{\mathrm{e+,HPC}} + \bar{\psi}_0^{\mathrm{p,HPC}} - \psi^{\mathrm{c,HPC}}\right] \\ & \quad + \bar{\psi}_0^{\mathrm{e-,HPC}} + \psi^{\mathrm{c}} + 2\,\tfrac{\psi^{\mathrm{c}} l}{\zeta}\left[\tfrac{\bar{q}^2}{2l} + \tfrac{l}{2}\,||\nabla\bar{q}||^2\right], \\ \bar{\psi}_0^{\mathrm{e+,HPC}} &= \kappa\langle \mathrm{tr}[\bar{\varepsilon}^{\mathrm{e,HPC}}]\rangle_+^2/2 + \mu||\mathrm{dev}\bar{\varepsilon}^{\mathrm{e,HPC}}||^2, \\ \bar{\psi}_0^{\mathrm{e-,HPC}} &= \kappa\langle \mathrm{tr}[\bar{\varepsilon}^{\mathrm{e,HPC}}]\rangle_-^2/2 \quad \text{and} \\ \bar{\psi}^{\mathrm{p,HPC}} &= (1-\bar{q})^m\left[y_0^{\mathrm{HPC}}\alpha^{\mathrm{HPC}} + \tfrac{1}{2}\,h^{\mathrm{HPC}}\alpha^{\mathrm{HPC}^2}\right], \end{aligned} \tag{2}$$

respectively. They depend on the length scale parameter l, a specific critical fracture energy ψ^{c}, the equivalent plastic strains α^{HPC}, a yield stress y_0^{HPC}, a hardening parameter h^{HPC}, the phase-field parameter $\bar{q}$ and its gradient $\nabla\bar{q}$. μ and κ are the Lamc$'$ coefficients of HPC phase and the operator $\langle\bullet\rangle_\pm = 1/2(\,\bullet \pm |\,\bullet\,|)$. The parameters m and ζ control the post critical stress softening during failure. An energy function $\bar{\psi}^{\mathrm{F}}$ for fiber phase using a preferred fiber direction a, cf. Boehler (1987), reads

$$\begin{aligned} \bar{\psi}^{\mathrm{F}} &= \tfrac{1}{2}\,E^{\mathrm{F}}(e^{\mathrm{F}} - e^{\mathrm{p,F}})^2 + y_0^{\mathrm{F}}\alpha^{\mathrm{F}} + \tfrac{1}{2}\,h^{\mathrm{F}}(\alpha^{\mathrm{F}})^2, \\ e^{\mathrm{e,F}} &= e^{\mathrm{F}} - e^{\mathrm{p,F}} \text{ with } e^{\mathrm{F}} = \bar{\varepsilon} : \mathbf{a}\otimes\mathbf{a}, \; ||\mathbf{a}|| = 1, \end{aligned} \tag{3}$$

where the elastic moduli E^{F}, the equivalent plastic strains α^{F}, initial yield stress y_0^{F} and hardening parameter h^{F} are considered for fiber phase. The total macroscopic stress tensor $\bar{\sigma}$ is computed by

$$\begin{aligned} \bar{\boldsymbol{\sigma}} &:= \mathrm{v}^{\mathrm{HPC}}\,\bar{\boldsymbol{\sigma}}^{\mathrm{HPC}} + \mathrm{v}^{\mathrm{F}}\,\bar{\boldsymbol{\sigma}}^{\mathrm{F}} \quad \text{with} \\ \bar{\boldsymbol{\sigma}}^{\mathrm{HPC}} &= g(\bar{q})\left[\kappa\langle tr\,\bar{\varepsilon}^{\mathrm{e,HPC}}\rangle_+\mathbf{I} + 2\mu\;\mathrm{dev}\,\bar{\varepsilon}^{\mathrm{e,HPC}}\right] \\ & \quad + \left[\kappa\langle \mathrm{tr}\,\bar{\varepsilon}^{\mathrm{e,HPC}}\rangle_-\mathbf{I}\right] \quad \text{and} \\ \bar{\boldsymbol{\sigma}}^{\mathrm{F}} &:= \bar{\sigma}^{\mathrm{F}}\mathbf{a}\otimes\mathbf{a} \quad \text{with } \bar{\sigma}^{\mathrm{F}} = E^{\mathrm{F}}(e^{\mathrm{F}} - e^{\mathrm{p},F}). \end{aligned} \tag{4}$$

The evolution equation for phase-field parameter, i.e.

$$\begin{aligned} & \bar{q} - l^2\,\mathrm{Div}(\nabla\bar{q}) - (1-\bar{q})\zeta\mathcal{H}^{\mathrm{HPC}} = 0, \text{ with} \\ & \mathcal{H}^{\mathrm{HPC}} := \max_{s\in[0,t]}\left\langle \tfrac{\mathrm{v}^{\mathrm{HPC}}[\bar{\psi}_0^{\mathrm{e+,HPC}} + \bar{\psi}_0^{\mathrm{p,HPC}}]}{\psi^{\mathrm{c}}} - 1\right\rangle. \end{aligned} \tag{5}$$

To capture the distinct behavior of concrete in tension and in compression, two different parameters for the critical fracture energy in tension $\psi_{\mathrm{t}}^{\mathrm{c,HPC}}$ and in compression $\psi_{\mathrm{c}}^{\mathrm{c,HPC}}$ are considered. The associative Drucker-Prager yield criterion for HPC phase and one-dimensional von Mises yield criterion for fiber phase are considered, respectively as

$$\begin{aligned} \phi^{\mathrm{HPC}} &= \tfrac{1}{\sqrt{2}}\,||dev\,\bar{\boldsymbol{\sigma}}_0^{\mathrm{HPC}}|| - \beta_{\mathrm{p}}\mathrm{tr}\,\bar{\boldsymbol{\sigma}}_0^{\mathrm{HPC}} - \kappa_{\mathrm{p}}^{\mathrm{HPC}} \\ \text{and} \quad & \phi^{\mathrm{F}}\left(\bar{\sigma}^{\mathrm{F}}, \kappa_p^{\mathrm{F}}\right) = |\bar{\sigma}^{\mathrm{F}}| - \kappa_{\mathrm{p}}^{\mathrm{F}}, \end{aligned} \tag{6}$$

DOI: 10.1201/9781003348450-212

where β_p, κ_p^{HPC} and κ_p^F are the material parameters. For detail insight on the presented authors refers the work Storm et al. (2021) and Gebuhr et al. (2022).

2 NUMERICAL SIMULATIONS

The simulation of a virtual experiment is performed by attaching the ellipsoidal RVE to the boundary value problem at microscale, see Figure 1a. A macroscopic homogeneous strain state is applied to the virtual experiment based on the ellipsoidal RVE, see Pise et al. (2020). Moreover, a macroscopic-BVP considering a cuboid is simulated in a single-scale using the phenomenological material model, see Figure 1b.

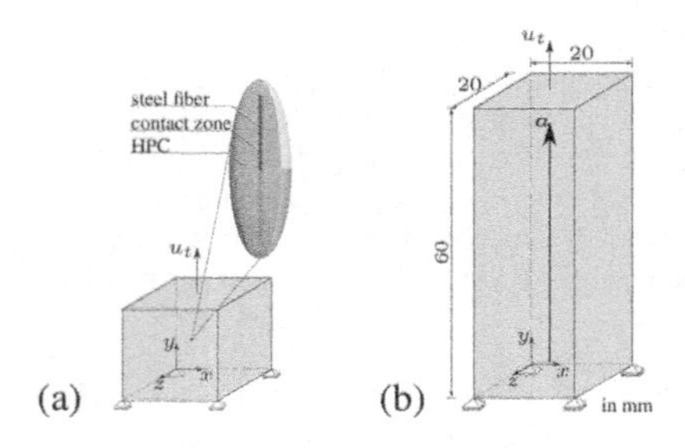

Figure 1. Geometry and boundary conditions for (a) a virtual experiment with attached ellipsodal RVE and (b) a macroscopic-BVP using a cuboid.

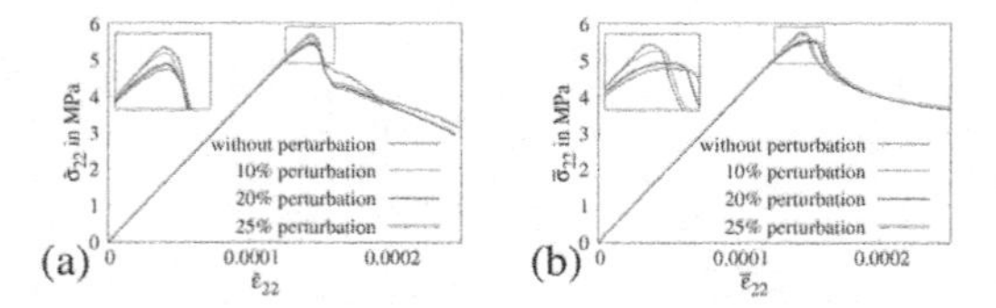

Figure 2. Macroscopic stress-strain characteristic for (a) virtual experiment using ellipsoidal RVE and (b) macroscopic-BVP using a cuboid, without perturbation and with 10%, 20% and 30% perturbation.

The heterogeneity of the real concrete material is considered by the perturbation of the material parameters for HPC matrix randomly, i.e $\kappa = perturb\ \kappa_0$, $\mu = perturb\ \mu_0$ and $\psi^c = perturb\ \psi_0^c$. Therein, perturb is a material parameter which can be computed using a formula perturb = (1+2(pertrand-0.5)maxpert), for details see Pise et al. (2021). The material parameters used for the simulation of macroscopic BVP are taken from Pise et al. (2021). The macroscopic stress-strain characteristic computed from the simulations of virtual experiments using reinforced ellipsoidal RVEs and macroscopic BVP using a cuboid for without perturbation and with 10%, 20% and 25% perturbation are compared in Figure 1a and b, respectively.

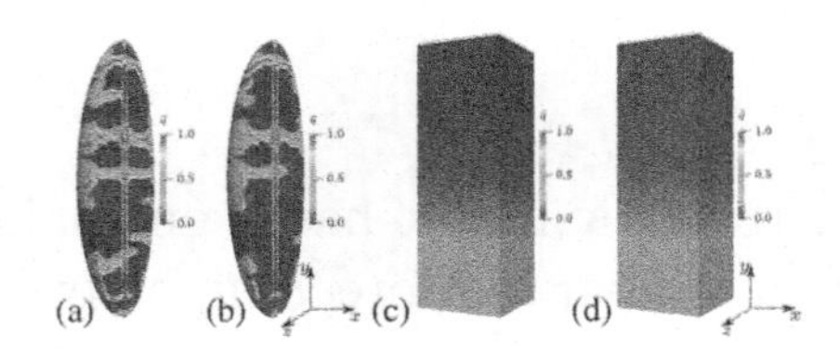

Figure 3. Virtual experiment - uniaxial tensile test without transverse stresses using an ellipsoidal RVE: distribution of microscopic phase-field parameter q (a) with 10% perturbation and (b) with 25% perturbation. Phenomenological material model applied to a macroscopic-BVP: distribution of phase-field parameter $\bar{q}$ (c) with 10% perturbation and (d) with 25% perturbation at macroscopic strains $\bar{\varepsilon}_{22} = 0.000175$.

3 CONCLUSION

In this paper a phenomenological material model for steel fiber reinforced HPC is proposed. To check the efficiency of the presented model, the resulting macroscopic stress-strain curves of virtual experiment based on ellipsoidal RVE and macroscopic-BVP using the different amounts of perturbation are compared. It is observed that the macroscopic the stress-strain characteristic is prominently affected by the amount of heterogeneity. The presented model is efficient to reproduce the similar stress-strain characteristic, specially in post critical stress softening region, as the computed macroscopic responses of virtual experiments. However, in case of the simulations considering perturbation of material parameters the results using presented phenomenological material model differ marginally from the results of virtual experiments.

ACKNOWLEDGEMENT

Funded by the Deutsche Forschungsgemeinschaft (DFG, German research Foundation) - 353513049 (AN1113/2-2, BR5278/2-2, SCHR570/32-2) within the DFG priority program 2020. Computing time provided on the supercomputer magnitUDE, funded by the Deutsche Forschungsgemeinschaft (DFG, German research Foundation), Project number: 263348352 (INST 20876/209-1 FUGG, INST 20876/243-1 FUGG), is gratefully acknowledged.

REFERENCES

Boehler, J. P. (1987). *Applications of tensor functions in solid mechanics*, Volume 292. Vienna: Springer.

Gebuhr, G., M. Pise, S. Anders, D. Brands, & J. Schröder (2022). *Materials 15*(3), 1179. 10.3390/ma15031179.

Pise, M., D. Brands, J. Schröder, G. Gebuhr, & S. Anders (2020). *Proceedings in Applied Mathematics and Mechanics 20*(1).

Pise, M., D. Brands, J. Schröder, G. Gebuhr, & S. Anders (2021). *Proceedings in Applied Mathematics and Mechanics 21*(1).

Storm, J., M. Pise, D. Brands, J. Schröder, & M. Kaliske (2021). *Engineering Fracture Mechanics 243*, 107506. 10.1016/j.engfracmech.2020.107506.

Current Perspectives and New Directions in Mechanics, Modelling and Design of Structural Systems – Zingoni (ed.)
© 2022 Copyright the Author(s),
ISBN 978-1-032-39114-4

Potential of shape memory alloys in fiber reinforced high performance concrete

Bernhard Middendorf & Maximilian Schleiting
Institute for Structural Engineering, Department of Structural Materials and Construction Chemistry, University of Kassel, Kassel, Germany

Ekkehard Fehling
Institute for Structural Engineering, Department of Structural Concrete, University of Kassel, Kassel, Germany

ABSTRACT: Fiber-reinforced concretes have been used in large quantities for the production of precast concrete elements since the 1970s. Fibers of steel, plastic, mineral or other materials are added to the concrete to increase tensile and flexural strength and converting brittle failure into ductile post-cracking behavior. However, based on the rheology, the amount of fiber added is limited. By using shape memory alloys (SMA) as fiber reinforcement, the mechanical properties of ductile fiber-reinforced (ultra) high performance concrete ((U)HPC) can be combined with the good rheological properties of fiber-free concretes.

1 INTRODUCTION

1.1 *Shape Memory Alloys (SMA) in civil engineering*

Even though the shape memory effect has been known since the 1930s, applications in the field of civil/structural engineering only came into the focus of research since the early 1990s. SMA have great potential for civil engineering due to their unique properties and functionality, as shown in a recent review work (Concilio et al. 2021), among others. In principle, three SMA effects are interesting for structural engineering, (1) the one-way effect (also known as pseudoplasticity) with strain restraint, (2) the one-way effect without strain restraint as well as (3) the superelasticity. Due to the inherent damping capacity and reversibility of the transformation, superelasticity is mainly focused by structural engineering (Cladera et al. 2014). In addition to the superelastic properties, the one-way effect, in particular because of the possibility of applying a prestress, is the focus of research, since here a costly elastic prestressing during anchorage can be dispensed with. In this case, unlike the classic one-way effect, the recovery of the SMA is blocked by the anchorage and thus a stress is generated by heating the pre-stretched SMA element, which can be used as a prestress.

1.2 *High- and ultra high-performance concretes in structural engineering*

Ultra high performance concrete (UHPC) has compressive strengths of up to 250 MPa due to a low water/cement ratio, an optimized high packing density and the addition of superplasticizers. However, compared to the high compressive strength, UHPC and other concretes show only low tensile strength as well as a brittle post-cracking behavior. Therefore, concrete is traditionally and standardly reinforced with steel bars. The brittle post-cracking behavior of UHPC also requires the combination of the steel bar with microfibers, e.g. steel fibers or, in perspective, carbon fibers. Steel fibers should primarily lead to a ductile post-cracking behavior of the UHPC, but also have an influence on the load-bearing and deformation behavior of combined reinforced concrete components (Leutbecher & Fehling 2012). In addition, the steel fibers significantly change the processing properties of the fresh concrete. Due to the interaction of the fibers with each other as well as with the aggregate, the flowability and workability of the concrete is very strongly reduced at high fiber volume fractions (> 2.5 - 3.0 vol.%) (Gerland et al. 2019).

2 EXPERIMENTAL INVESTIGATIONS

2.1 *Shape memory alloys with pseudoplastic behavior and the use without strain restraint*

Innovative shape memory alloys (SMA) offer pseudoplastic behavior. They can recapture a previously imprinted shape by thermal activation.

This effect can be applied by thermal curing of the concrete, as occurs in the production of precast concrete elements. Before the transformation, the SMA fibers are converted into a geometry that has a positive effect on the rheology of the fresh concrete; e.g. curls.

This should allow to use a significantly higher fiber volume fraction compared to straight fibers. The reshaping of the fibers in the fresh concrete (see

DOI: 10.1201/9781003348450-213

Figure 1) should result in a significant increase in tensile strength for the hardened concrete as well as greatly improved post-cracking behavior.

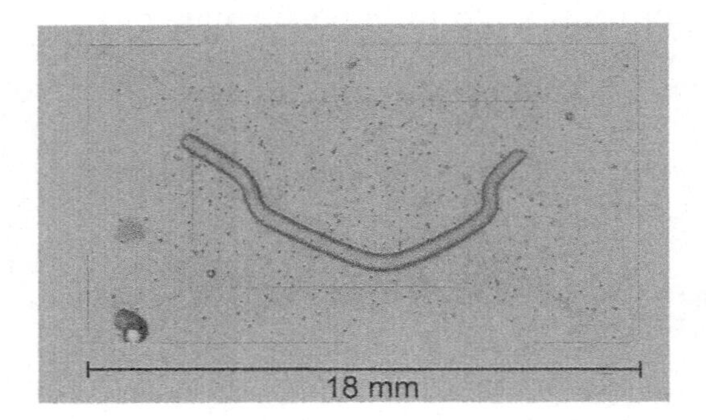

Figure 1. μ-CT image of a single SMA fiber reshaped by a thermal activation in UHPC (Schleiting et al. 2020).

According to own preliminary work (Schleiting et al. 2020; Gerland et al. 2019) that curled fibers lead to a reduction of plastic viscosity and yield point of fresh UHPC by about 30% for the same fiber content, which is accompanied by a significant improvement in workability.

Furthermore, preliminary tests with SMA fibers made of a nickel-titanium alloy (Ni-Ti) have shown that transforming in fresh concrete is possible by thermal activation due to heating in a kiln. The fibers were individually placed in the fresh concrete, which was then heated to 60 °C. After hardening, the transformation of the fibers was visualized with a μ-computer tomograph (μ-CT; see Figure 2).

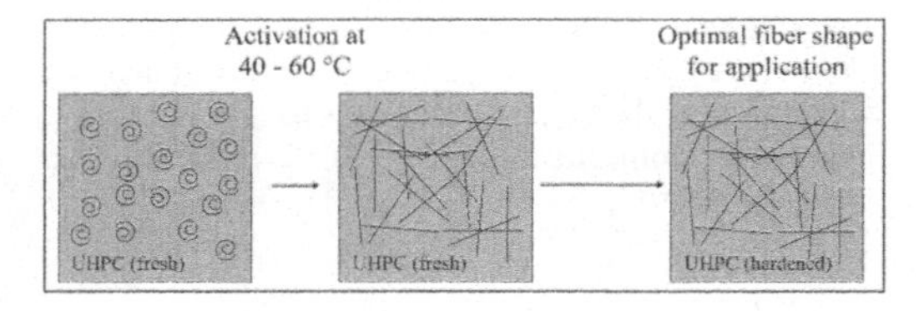

Figure 2. Left: Unfolding of SMA fibers after thermal activation in fresh UHPC. Right: High fiber content in hardened concrete.

2.2 *SMA fibers with pseudoplastic behavior and the use with strain restraint*

The one-way effect with strain restraint of SMA makes it possible to apply prestresses in engineering structures without having to use external actuators such as hydraulic systems. For this use the SMA element has to be stretched prior activation in the construction element. It is important to notice that the activation here is done in the state of hardened concrete.Therefore, due to the impulse of the SMA element to receive its unstretched state, a prestress is transferred on the surrounding UHPC matrix.

3 OPEN SCIENTIFIC QUESTIONS

1. How do SMA fibers of different geometries and increased volume fractions (> 3 vol.%) affect the fresh and hardened concrete properties?
2. How can a certain fiber geometry be used to specifically influence the fiber distribution and orientation?
3. How must the concrete be composed to ensure optimal fiber distribution and not hinder the transforming capacity of the fibers?
4. How does the fresh concrete behave when heated locally around the fiber by induction heating?
5. How do stresses, generated by the shape memory effect, quantitatively transfer on the hardened concrete?
6. How has the interfacial transition zone between fiber and concrete to be strengthen to be able to transfer the occurring stresses?

4 CONCLUSION

By merging high- and ultra-high-performance concretes with SMA, a new, sustainable, innovative and smart generation of construction materials is possible. The use of SMA, due to their shape memory capacity, allows to realize structural components and works with significantly increased safety and service life, while at the same time significantly reducing the CO_2 -intensive material use of construction materials. For example, the reversibility of the SMA effects will make it possible to assign new properties to concrete as a structural material and to realize the prestressing of load-bearing elements permanently without any loss of prestressing force. This opens up completely new possibilities for the design and realization of lightweight and efficient concrete structures. Novel coupled shape memory effects are explored in parallel with new concepts for load-bearing components, their connections and vibration damping design.

REFERENCES

Cladera, A.; Weber, B.; Leinenbach, C.; Czaderski, C.; Shahverdi, M.; Motavalli, M. 2014. Iron-based shape memory alloys for civil engineering structures: An overview. *Constr. Build. Mater.* 63: 281–293.

Concilio, A.; Antonucci, V.; Auricchio, F.; Lecce, L.; Sacco, E. 2021. *Shape memory alloy engineering. For aerospace, structural and biomedical applications*. Elsevier Butterworth-Heinemann, Oxford, Cambridge.

Gerland, F.; Wetzel, A.; Schomberg, T.; Wünsch, O.; Middendorf, B. 2019. A simulation-based approach to eval-uate objective material parameters from concrete rheometer measurements. *Appl. Rheol.* 29 (1): 130–140.

Leutbecher, T. und Fehling, E. 2012. Tensile Behavior of Ultra-High-Performance Concrete Reinforced with Reinforcing Bars and Fibers: Minimizing Fiber Content. *ACI Structural Journal* 109 (2).

Schleiting, M.; Wetzel, A.; Krooß, P.; Thiemicke, J.; Niendorf, T.; Middendorf, B.; Fehling, E. 2020a. Functional microfibre reinforced ultra-high performance concrete (FMF-UHPC). *Cem. Concr. Res.* 130: 105993.

Current Perspectives and New Directions in Mechanics, Modelling and Design of Structural Systems – Zingoni (ed.)
© 2022 Copyright the Author(s), *ISBN 978-1-032-39114-4*

Homogenisation of the material behaviour of UHPFRC under tension

L. Gietz, U. Kowalsky & D. Dinkler
ISD – Institute of Structural Analysis, TU Braunschweig, Germany

J.-P. Lanwer & M. Empelmann
iBMB – Institute of Building Materials, Concrete Construction and Fire Safety, Division of Concrete Construction, TU Braunschweig, Germany

ABSTRACT: Micro steel fibres significantly increase the ductility of ultra-high performance concrete (UHPC) and the post cracking load-bearing behaviour under tensile loading. For an economical material use in fatigue susceptible structures, it is necessary to characterise the crack-bridging load-bearing effect of fibres under tensile loading. The following study describes experimental and numerical investigations of the load-bearing effect of micro steel fibres embedded in UHPC on the macro-level.

1 INTRODUCTION

Ultra-high performance fibre-reinforced concrete (UHPFRC) is characterised by its high compressive strength and, due to the micro steel fibres, by its ductile post-cracking behaviour under tensile loading, which is significantly characterised by the fibre volume content and the fibre type (Wille et al. 2014).

For the development of numerical methods for the computation of realistic engineering structures a homogenisation strategy of the material behaviour, the identification of influencing parameters and processes under loading is essential. The following experimental and numerical investigations illustrate the process of crack formation and activation of the steel fibres on UHPFRC components under tensile loading.

2 EXPERIMENTAL TEST PROGRAMME

The experimental test programme includes quasi-static and cyclic tensile tests on UHPFRC-specimen. Figure 1 shows the tensile stress-displacement curves of two quasi-static reference test series with different fibre contents (VZ1: 1.25 Vol.-%, VZ2: 2.5 Vol.-%). The high fibre content in UHPFRC leads to a very close fibre position. Results of the investigations show that pulling out inclined fibres cause concrete breakout near the exit point of the fibres (Lanwer et al. 2022). It is very likely that in test series VZ2 many fibres are in the breakout and lose their load-bearing effect. This may explain the similar softening behaviour in test series VZ1 and VZ2. The load amplitudes for the cyclic tensile tests are determined from the quasi-static reference tensile strengths. The results of the cyclic load tests suggest that the number of load changes leading to failure is highly scattered for each load amplitude. With the small number of test results, it is currently not possible to create an S/N curve.

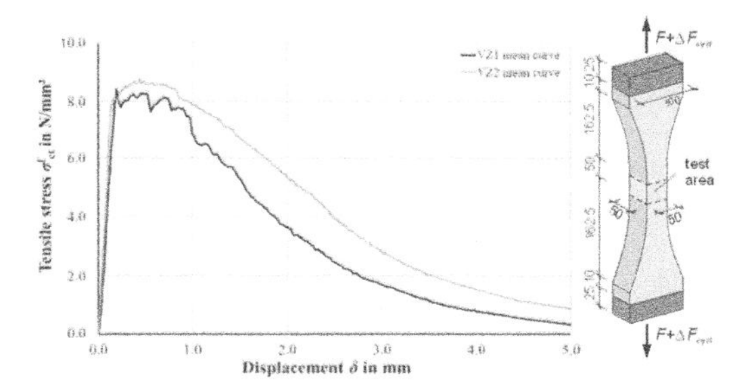

Figure 1. Stress-displacement curves of test series VZ1 and VZ2 as mean values and test specimen.

3 HOMOGENISATION STRATEGY

The basis of numerical methods for the description of the homogenised material behaviour is the strain energy at a virtual representative volume element dV, which contains the components of the concrete matrix and the micro-steel fibres: $\Pi = \Pi_c + \Pi_f$. By means of the volume fractions $dV_c = n_c \cdot dV$ and $dV_f = n_f \cdot dV$ the strain energy densities π_c of the concrete matrix and π_f of the fibres are normalised to the total volume dV and implicitly consider the fibre content in the model equations:

$$\Pi \int_V n_c.\pi_c + n_f \cdot \pi_f dV \quad (1)$$

The material equations of the composite material are developed with the variation of the strain energy density π in a mixed displacement-force variable formulation. It describes the elastic behaviour of the material and is extended by the crack formation with the method of Lagrange multipliers. Here, it is given with

DOI: 10.1201/9781003348450-214

$$\begin{aligned}\pi = n_c\left\{\boldsymbol{\sigma}_c\boldsymbol{\varepsilon} - \frac{1}{2}\boldsymbol{\sigma}_c\boldsymbol{E}_c^{-1}\boldsymbol{\sigma}_c\right\} \\ + n_{\mathrm{f}}\left\{\boldsymbol{\sigma}_{\mathrm{f}}\varepsilon - \frac{1}{2}\boldsymbol{\sigma}_{\mathrm{f}}\boldsymbol{E}_{\mathrm{f}}^{-1}\boldsymbol{\sigma}_{\mathrm{f}}\right\} \\ - n_c\boldsymbol{\varepsilon}_{\mathrm{cr}}\left(\boldsymbol{\sigma}_c - \frac{1}{2}\boldsymbol{E}_{\mathrm{cr}}\boldsymbol{\varepsilon}_{\mathrm{cr}}\right)\end{aligned} \quad (2)$$

The total strain of the concrete matrix is additively divided into an elastic and an inelastic part: $\boldsymbol{\varepsilon}_{c,el} + \boldsymbol{\varepsilon}_{cr}$. The variation according to the crack-strain $\boldsymbol{\varepsilon}_{cr}$ describes the softening behaviour of the concrete matrix during crack formation. The crack-modulus $\boldsymbol{E}_{cr}$ controls the development of the crack-strain.

The discretisation of the strain energy with the FEM is realised with a mixed-hybrid element concept. In addition to the linear approaches for the displacements of an isoparametric volume element (Hex8), linear approaches with generalised free values according to Pian & Chen (1983) are chosen for the variables ($\boldsymbol{\sigma}_c$, $\boldsymbol{\sigma}_f$, $\boldsymbol{\varepsilon}_{cr}$). After integrating the strain energy density of the element, the element stiffness matrix is numerically calculated by static condensation of the elements′ free values.

4 NUMERICAL RESULTS

The 3D cantilever slice in Figure 2 is subjected to bending with an applied load $\sigma_y = 2.0$ N/mm^2 at the end of the slice. Figure 2 shows the numerical results of the distribution of the crack-strain as well as the deformed system. The crack formation begins at the clamped support and develops with increasing load over the longitudinal axis of the structure.

Figure 3 shows the distribution of the total, concrete and fibre stresses over the cross-sectional height at the clamped support of the slice for different load levels. As the load increases, the cross-section begins to crack at the upper edge, so that the tensile zone of the concrete is reduced and the tensile force transmission is only ensured by the fibres. In the compression zone the material behaviour remains elastic.

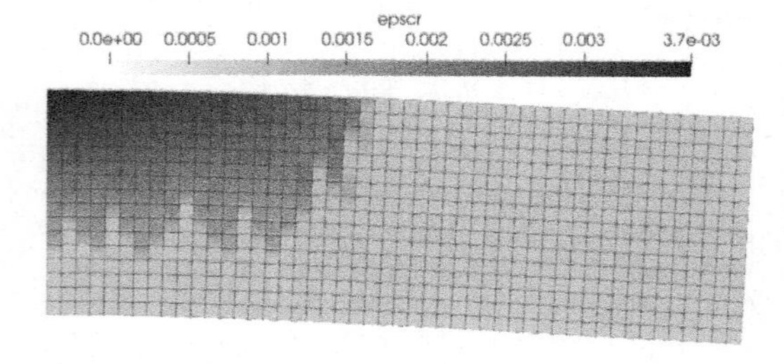

Figure 2. Crack-strain ε_{cr} and deformation at load level σ_y = 1.2 N/mm^2.

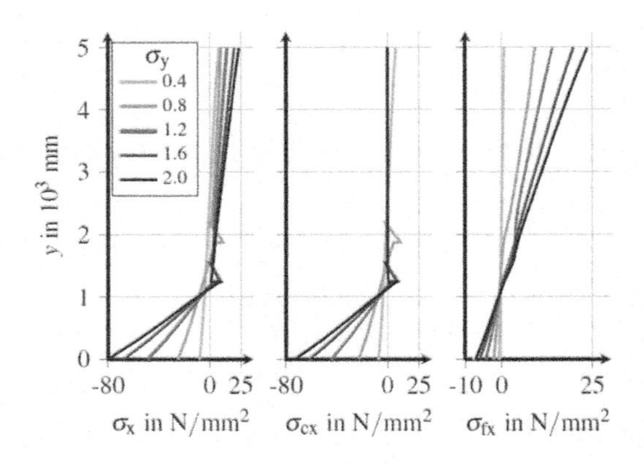

Figure 3. Distribution of the stress components over the cross-sectional height at the clamped support for different load levels.

5 CONCLUSION AND OUTLOOK

The experimental investigations show that the post-crack tensile strength cannot be increased indefinitely by increasing the fibre content, as the fibres interact with each other. The small number of fatigue tensile tests show a high scatter in the load cycles. Currently, no S/N curve can be set up.

Numerical simulations show that the newly developed material model can represent the macro-crack formation in the concrete matrix under tensile stresses and the associated fibre activation. The description of the fibre pull-out, which leads to controlled material softening after exceeding the bond strength between the UHPFRC matrix and the micro-steel fibres, is currently being worked on.

ACKNOWLEDGEMENTS

The investigations presented in this paper are part of the cooperative research project “Damage processes in ultra-high performance fibre-reinforced concrete subjected to cyclic tensile loading” as part of the Priority Programme 2020, funded by the German Research Foundation (DFG).

REFERENCES

Lanwer, J.-P. & Höper, S. & Gietz, L. & Kowalsky, U. & Empelmann, M. & Dinkler, D. 2022. Fundamental Investigations of Bond Behaviour of High-Strength Micro Steel Fibres in Ultra-High Performance Concrete under Cyclic Tensile Loading. *Materials* 15(1): 120.

Pian, T.H.H. & Chen, D. 1983. On the suppression of zero energy deformation modes. *Int. J. Numer. Meth. Eng.* 19: 1741–1752.

Wille, K. & El-Tawil, S. & Naaman, A.E. 2014. Properties of strain hardening ultra high performance fibre reinforced concrete (UHP-FRC) under direct tensile loading. *Cem. Concr. Compos.* 48: 53–66.

Current Perspectives and New Directions in Mechanics, Modelling and Design of Structural Systems – Zingoni (ed.)
© 2022 Copyright the Author(s), ISBN 978-1-032-39114-4

Numerical and experimental analysis of fatigue-induced changes in ultra-high performance concrete

S. Rybczyński & F. Schmidt-Döhl
Institute of Materials, Physics and Chemistry of Buildings, Hamburg University of Technology, Hamburg, Germany

G. Schaan & M. Ritter
Electron Microscopy Unit, Hamburg University of Technology, Hamburg, Germany

M. Dosta
Institute of Solids Process Engineering and Particle Technology, Hamburg University of Technology, Hamburg, Germany

ABSTRACT: This contribution deals with experimental and numerical results of fatigue-induced changes in ultra-high performance concrete (UHPC). For this purpose, a set of experimental investigations has been performed on UHPC and binder samples. To characterize fatigue-induced small scale changes in detail, samples were examined using scanning and transmission electron microscopy (SEM and TEM). A densification of the binder matrix occurs caused by a transformation of nanoscale ettringite. Additionally, the content of unhydrated cement clinker influences the fatigue resistance of UHPC. For numerical investigations of the mechanical behavior of UHPC, the bonded-particle model (BPM) has been used and calibrated with experimental data. A newly derived rheological fatigue formulation for binder and ITZ bonds was developed. Local stress peaks (caused by the difference of stiffness between binder and aggregate) promotes a crack initiation in the ITZ and favors a propagation in the surrounding matrix.

1 EXPERIMENTAL SETUP

1.1 *Mechanical experiments*

Uniaxial static and cyclic compression tests were performed on both UHPC and pure binder specimens. Static compression tests were carried out strain-controlled with a load speed of 0.2 mm/min on three specimens of both materials. The longitudinal strain was measured by three high-precision inductive displacement transducers with a rate of 30 Hz. The mean values of static tests are shown in Table 1.

Table 1. Strength and stiffness of UHPC and binder.

Material	Strength [MPa]	Young's Modulus [GPa]
Binder	172.00 ± 1.63	36.00 ± 1.63
UHPC	193.34 ± 3.77	46.67 ± 0.94

The absent aggregate grains cause a decrease of strength and stiffness in pure binder specimens, while a high strength is still given due to the optimized high packing density.

Cyclic compression tests were carried out force-controlled with a triangular load regime of 1 Hz. The upper and lower load level corresponded to 80% and 5% of the mean value of the static strength and the longitudinal strain was measured by three high-precision inductive displacement transducers with a sampling rate of 30 Hz. However, both materials exhibit the common feature of decreasing stiffness and increasing strain during cyclic loading, Figure 1.

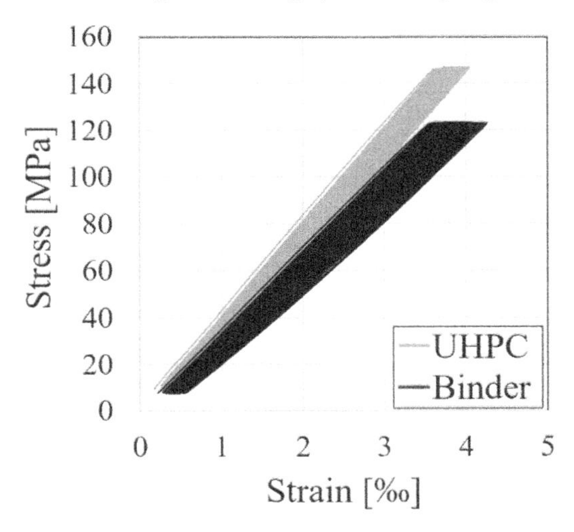

Figure 1. Experimental stress-strain hysteresis of both materials during cyclic loading.

1.2 *Large-area SEM*

To gather statistically robust information on the size, shape, distribution and content of the constituent components in UHPC, we use automated large-area SEM. We

DOI: 10.1201/9781003348450-215

find a positive correlation between the relative content of unhydrated cement clinker (Figure 2) in the binder fraction of a specimen and its eventual lifetime.

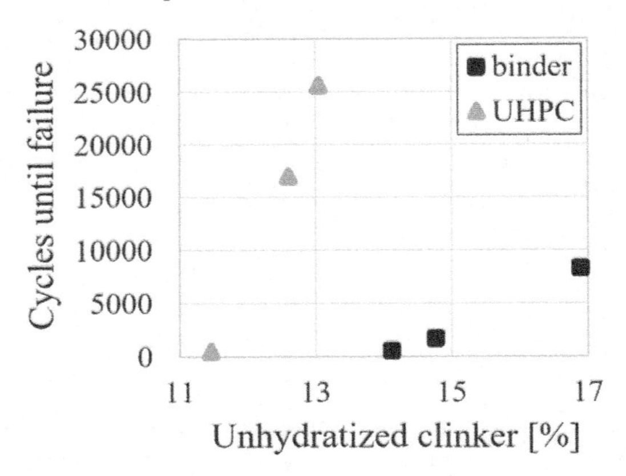

Figure 2. Dependency of specimen lifetime on their relative content of unhydrated cement clinker.

1.3 *Transmission electron microscopy*

Structural changes either shrink or grow in size upon energy uptake from electron irradiation during TEM investigation and preferentially exist in regions of the binder matrix comparatively rich in aluminum and sulfur and with a low silicon content.

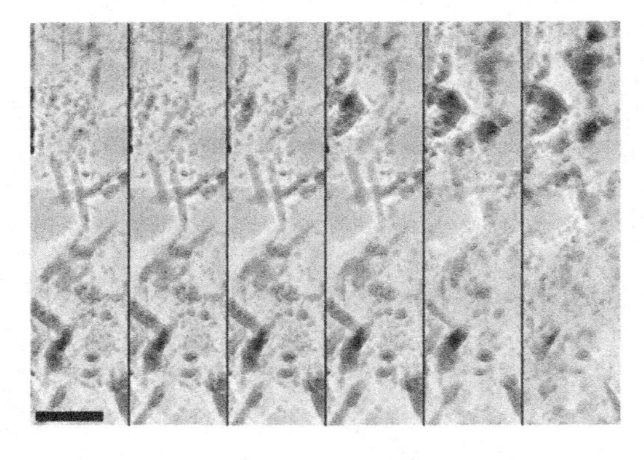

Figure 3. Sections of HAADF-STEM images after increasing durations of electron irradiation showing both shrinking and growing structural changes.

As these structural changes exhibit dark *Z* contrast in HAADF-STEM imaging and readily change size and shape upon energy uptake, we conclude they are connected to locations of a transformation of ettringite.

2 NUMERICAL INVESTIGATION

By comparing Figures 1 and 4, it can be clearly noticed that our fatigue simulations are in good agreement with the experimental stress–strain hysteresis loops. The simulations show the same macroscopic characteristics, e.g., the increase of strain and decrease of stiffness until fatigue breakage.

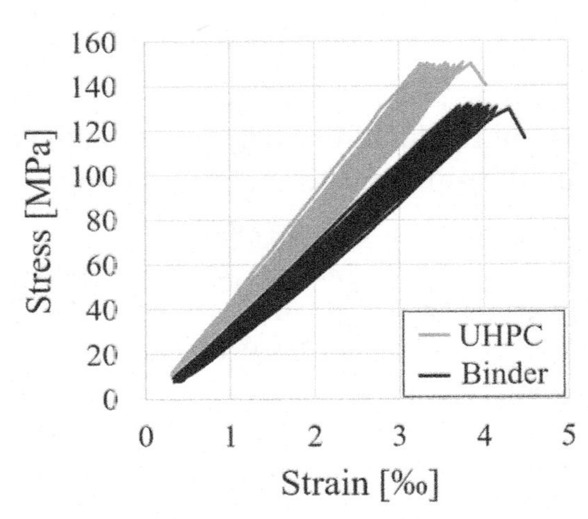

Figure 4. Exemplary macroscopic fatigue simulation hysteresis of UHPC using our proposed bond formulation.

In order to localize the main crack initiation in the investigated UHPC simulations, more numerical evaluations were performed and can be found in the full paper.

3 CONCLUSIONS

The following main conclusions can be drawn from our investigations:

(1) Structural inhomogeneities such as local variations in stiffness or ettringite transformation are possible causes for crack initiation during fatigue.
(2) Deterioration can occur on the meso, micro, and nanoscale. To reproduce the correct mechanisms of degradation during cyclic simulations, nanoscale changes must be taken into account even in the case of mesoscale modeling.
(3) The content of unhydrated cement clinker in the binder matrix results in higher local strength and consequently leads to a longer specimen lifetime during cyclic loading.
(4) Using sulfate-resistant cements containing little or no calcium aluminate could possibly prevent a delayed formation of ettringite and slow down the fatigue process.
(5) Local stress peaks (caused by the difference of stiffness between binder and aggregate) promotes a crack initiation in the ITZ and favors a propagation in the surrounding matrix. Using less stiff and even more fine-grained aggregate could minimize local stress peaks and increase fatigue lifetime.

ACKNOWLEDGEMENTS

We gratefully acknowledge financial support from the German Research Foundation (DFG) within the priority program SPP 2020 "Cyclic deterioration of High-Performance Concrete in an experimental-virtual lab" grant number 353408149.

Current Perspectives and New Directions in Mechanics, Modelling and Design of Structural Systems – Zingoni (ed.)
© 2022 Copyright the Author(s), *ISBN 978-1-032-39114-4*

Damage development of steel fibre reinforced high performance concrete in high cycle fatigue tests

G. Gebuhr & S. Anders
Institute of Structural Engineering, Faculty of Architecture and Civil Engineering, Bergische Universität Wuppertal

M. Pise, D. Brands & J. Schröder
Institute of Mechanics, Faculty of Engineering, Universität Duisburg-Essen

ABSTRACT: In this contribution, damage development in high cycle fatigue of steel fibre reinforced high and ultra-high performance concretes is described and characterized utilizing various damage indicators. For this purpose, high cyclic flexural tests were conducted on one HPC mixture containing hooked-end macro steel fibres and one UHPC mixture with smooth micro steel fibres. Overall, the damage indicators presented, such as the development of residual stiffness, show distinct characteristics depending on the fibre content and compressive strength of concrete.

1 INTRODUCTION

Increasing demands e.g. on the strength of concrete or its fatigue performance lead to various application-specific questions in high-strength structures, especially in areas requiring resistance to dynamic loading, e.g. for bridge deck connections (Graybeal, 2010). High (HPC) and ultra-high performance concrete mixtures (UHPC) are too brittle to be used without fibre reinforcement. Consequently, there is a need to investigate fatigue behavior of steel-fiber reinforced HPC and UHPC and to define suitable parameters to describe deterioration development in fatigue. Since high cycle fatigue tests are very time-consuming, suitable damage indicators must be found that allow transferability of results from short-term low cycle fatigue tests. Attempts to investigate the progression of damage using crack-opening related indicators have been carried out, for example, by (Germano *et al.*, 2016), focusing on low cycle fatigue. In a previous study of low and high cycle fatigue tests in flexure, see (Gebuhr *et al.*, 2019), promising approaches were presented for HPC.

2 DETERIORATION INDICATORS IN LOW AND HIGH CYCLE FATIGUE

Following previous studies in flexural as well as compressive tests, e.g. (Germano *et al.*, 2016), three damage indicators are primarily considered here.

As shown in Figure 1, it is intended to apply the same damage indicators to both, low cycle and high cycle tests.

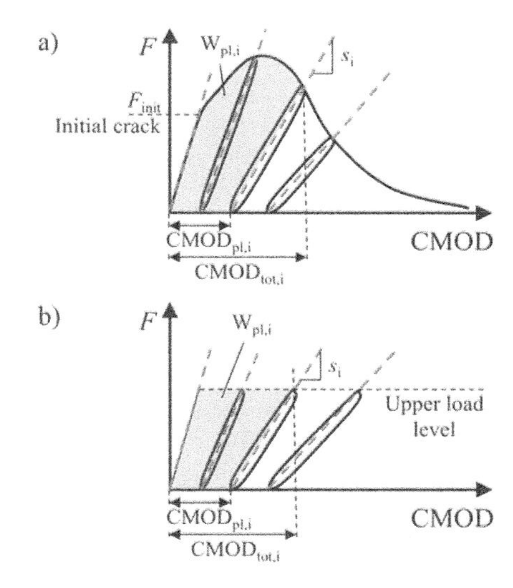

Figure 1. Sketches defining the damage indicators CMOD plastic ($CMOD_{pl,i}$), residual stiffness (s_i) and damage energy ($W_{pl,i}$) in low cycle (a) and high cycle (b) fatigue tests.

Firstly, the residual stiffness s_i is determined, defined as the gradient of the force applied between the lower and upper load levels of each loading cycle in relation to the corresponding CMODs. Secondly, the plastic fraction of the crack mouth opening $CMOD_{pl,i}$ after unloading for each loading cycle is determined. Thirdly, the damage energy $W_{pl,i}$ is calculated as the area between load-deformation curve up to each

DOI: 10.1201/9781003348450-216

individual unloading cycle and corrected by the elastic energy as defined by the residual stiffness s_i.

3 MATERIALS AND TESTING PROGRAM

Two concrete mixes were applied, one HPC and one UHPC. The compressive strengths of the mixes were 102 MPa and 160 MPa, respectively. For the HPC, hooked-end macro steel fibres with single end anchorage were used in quantities of 0/23/57/115 kg/m³. For the UHPC, 0/57/115 kg/m³ straight micro steel fibres were untilized instead, because of the overall smaller aggregate. The tests were performed on standard notched beams to investigate residual flexural strengths in accordance with EN 14651.

Fatigue loading was applied with fixed upper and lower load levels at a frequency of 1 Hz. In literature, e.g. (Lee & Barr, 2004), load levels often refer to the maximum load in static tests. Here, load levels 70/75/80 % refer to F_{init} (Figure 1a), which is the load at occurrence of first crack. The lower load level was kept constant at 5 % for all tests.

4 RESULTS

In this short paper, only the residual stiffness is looked into exemplarily, $CMOD_{pl}$ and damage energy are disussed in the full paper in detail. Generally, mixtures with fibre contents equal to or higher than 57 kg/m³ showed strain hardening in the low cycle reference tests, i.e. transferable forces increase with increasing CMOD due to fibre activation in the crack.

For mixtures without fibres and mixtures containing 23 kg/m³ failure occurred at very small crack openings. Therefore, in the following graphs, both, the behavior for small CMODs up to 0.2 mm and the behavior for crack openings between 0.2 and 2 mm are shown using different scales for better visibility.

Figure 2a depicts the development of residual stiffness for all fibre contents in HPC. Deterioration of stiffness progresses quickly up to a CMOD of approx. 0.05 mm. This is the CMOD where concrete without fibres fails as the first crack is initiated in the matrix. With the initiation of the crack, fibres start to affect the degradation of stiffness from a CMOD of about 0.015 mm onwards. Mixtures with strain hardening behavior generally show higher residual stiffnesses. The curves arrange themselves according to the fibre quantity, which is a generally observable trend for all damage indicators.

Importantly, from the curves for a fibre content of 57 kg/m³ development of stiffness seems to be independent of the upper load level.

a)

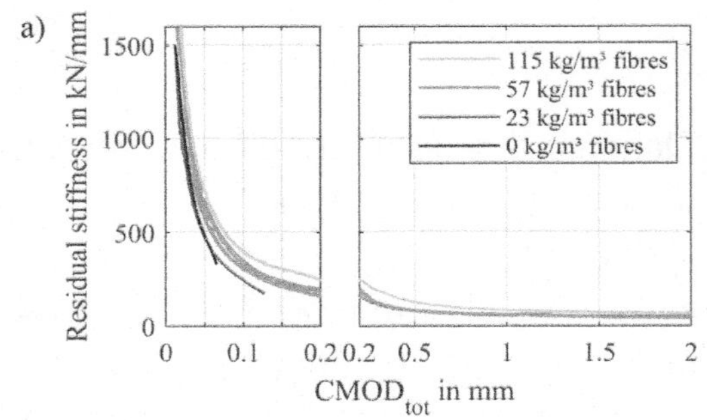

b)

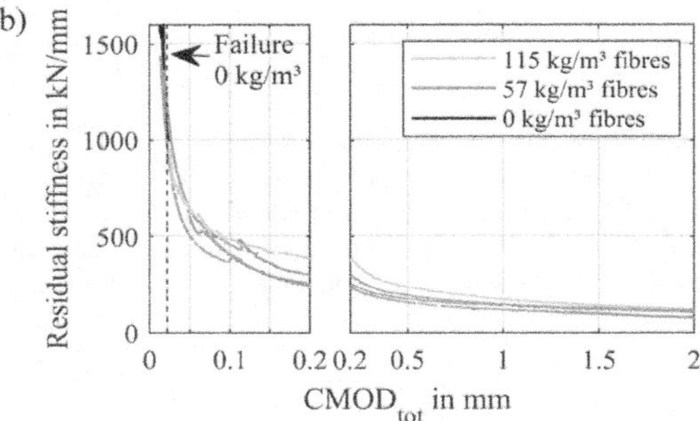

Figure 2. Exemplary development of residual stiffness s with respect to fibre content with a set of curves for 57 kg/m³ fibres in HPC (a), and UHPC (b).

Figure 2b shows the development of residual stiffness for UHPC. Despite the even more brittle material, the trend of the curves is qualitatively comparable to HPC.

Overall, all damage indicators looked into, clearly differentiate damage development depending on the fiber content. The selectivity of the indicators seems to be better for damage energy and residual stiffness. Development of $CMOD_{pl}$ seems to be less selective.

REFERENCES

Gebuhr, G., Anders, S., Pise, M., Sahril, M., Brands, D. & Schröder, J. (2019) Deterioration development of steel fibre reinforced high performance concrete in low-cycle fatigue. In: Zingoni, A. (ed.) *Advances in Engineering Materials, Structures and Systems: Innovations, Mechanics and Applications:* CRC Press. pp. 1444–1449.

Germano, F., Tiberti, G. & Plizzari, G. (2016) Post-peak fatigue performance of steel fiber reinforced concrete under flexure. *Materials and Structures*. [Online] 49 (10), 4229–4245. Available from doi:10.1617/s11527-015-0783-3.

Graybeal, B.A. (2010) *Behavior of field-cast ultra-high performance concrete bridge deck connections under cyclic and static structural loading*. [Online]. Available from: https://rosap.ntl.bts.gov/view/dot/957.

Lee, M.K. & Barr, B. (2004) An overview of the fatigue behaviour of plain and fibre reinforced concrete. *Cement and Concrete Composites*. [Online] 26 (4), 299–305. Available from doi:10.1016/S0958-9465(02)00139-7.

Current Perspectives and New Directions in Mechanics, Modelling and Design of Structural Systems – Zingoni (ed.)
© 2022 Copyright the Author(s), *ISBN 978-1-032-39114-4*

Fatigue behavior and crack opening tests under tensile stress on HPSFRC: Experimental and numerical investigations

Niklas Schäfer*, Vladislav Gudžulić, Rolf Breitenbücher & Günther Meschke
Ruhr-University Bochum, Germany

ABSTRACT: The capability of high-strength steel microfibers to control the degradation in high-strength concrete was experimentally examined and numerically simulated. To this end, notched prismatic high-strength concrete specimens with and without steel microfibers were subjected to static and cyclic tensile tests up to 100,000 cycles. The material fatigue was examined using microscopic analyzes. The increase in strain at the test specimens with high-strength steel microfibers was less than without fibers. And also, the strain stagnates after 10,000 Cycles at high-strength concrete with steel fibers. The microscopic examinations showed that more cracks developed in the microfiber reinforced high-strength concrete than in the unreinforced high-strength concrete. However, these were smaller and shorter, i.e., more finely distributed, and thus had a smaller total crack area than the high-strength concrete without fibers. To investigate the influence of fibers on the behavior of HPSFRC in the cracked state, displacement-controlled crack opening tests, as well as numerical simulations thereof, were carried out. In the finite element model, concrete cracking was simulated using zero-thickness interface elements, and fibers are modeled explicitly, using the Bernoulli beam elements, and bond elements tying microfibers and concrete. Experiments have shown, and the numerical simulations have confirmed, that the inclusion of steel microfibers didn't increase the strength, however, it significantly increased the post-peak carrying capacity, i.e., ductility of the material. Finally, the crack closure model introduced to analyze the unloading/reloading behavior has shown that in this experiment.

1 INTRODUCTION

Increasing requirements in slender design, material savings, and improved durability attract more and more interest in high-performance concretes, often reinforced by fibers to improve material characteristics further. But these developments do not only bring advantages. Slender designs are often more susceptible to fatigue failure. On Behalf of this, systematic investigations on high-strength concrete with and without fibers were carried out to get knowledge about the influence of fibers on the fatigue behavior, e.g. the crack opening behavior of HPSFRC. The investigations were accompanied and mapped by numerical investigations, see [1]. For the test a high-strength concrete with (here named HPC-08-SF) and without fibers (HPC-08) were investigated. HPC-08 and HPC-08-SF had comparably compressive strength of on average 112 MPa.

2 TEST METHODS

2.1 *Static and cyclic tensile tests*

The static tensile tests were performed to determine the maximum static tensile strength. We used prismatic (40 mm × 40 mm × 160 mm) specimens with two notches in the middle for both static and cyclic investigations. The notches define the weakest section, subsequently the breaking point. In displacement-controlled tests, the specimens were loaded with a constant displacement rate until failure of the specimen occurred. In force-controlled fatigue tests, the specimens were subjected to tensile stresses according to the sinusoidal load functions with an upper load level of 70% of the maximum static tensile strength fct, max and a lower level of 35% fct, max.The cyclic loading was stopped after reaching up to 100,000 cycles. After 0, 1, 100, 1000, and 100,000 load cycles, one test specimen was chosen, from which the partial samples for microscopic examinations (amount, length, width, and position of the microcracks) were prepared as thick sections.

2.2 *Crack opening tests*

To investigate the crack closing and re-opening mechanisms, crack opening tests were performed. Therefore, the sample was subjected to a monotonic tensile loading until the first crack appeared. The sample was then unloaded to 10% of the tensile strength, and the difference in strain (Δl) was measured. The displacement was then prescribed on the top end of the specimen until the value of the strain measured at the notch area reached Δl, followed by unloading to 10% of the tensile strength. Subsequently, the procedure of crack opening controlled loading until the crack width of CMOD + 10 Δl was reached, and subsequent force-controlled relieving of the sample was repeated.

*Corresponding author: Niklas.Schaefer@rub.de

DOI: 10.1201/9781003348450-217

3 NUMERICAL MODEL

The investigations carried out can be assumed as a quasi-static process in the numerical simulations. The equilibrium equation reads:

$$\operatorname{div} \sigma + \rho b = 0,$$

with σ being the Cauchy stress tensor, ρ the mass density, and b the specific body force vector. The equation is recast into the weak form and discretized and solved employing the finite element method. The complete description of the numerical model and the boundary conditions can be found in [1].

4 RESULTS AND DISCUSSION

4.1 *Static and cyclic tensile tests*

The maximum tensile strength on average is fct, max = 6.4 MPa for HPC-08 and fct,max = 5.9 MPa for HPC-08-SF (no noticeable influence of the steel fibers). HPC-08 shows a steady increase in strain with a maximum strain of around 70 μm/m until it reaches 100,000 load cycles. For HPC-08-SF, the strain stagnates from approximately 10,000 load cycles and is limited at around 40 μm/m. The rate of strain increase was lower for HPC-08-SF and the strain of HPC-08 grew faster a, the strain of HPC-08-SF only grew marginally. As a result of the microscopic investigations one can note that the cyclic loading lead to the formation of new microcracks and it could rather been observed that the existing ones grow. The HPC-08-SF specimen showed more microcracks after 100,000 load cycles, but these were smaller in size, and when compared to the HPC-08, the crack pattern was finer. The total crack area in the HPC-08-SF specimen was lower. Around 66% of the microcracks have been localized in the interfacial transition zone (ITZ), so that the ITZ is the weakest zone in the case of high-strength concretes. If the fibers in the matrix have sufficient tensile strength and stiffness, they can act as reinforcement delaying macrocrack formation and limiting the opening.

4.2 *Crack opening tests*

The post-cracking behavior of notched prismatic specimens subjected to the loading program described in Section 2.2 is investigated experimentally and numerically. Five specimens (Exp. 1 to Exp. 5) were subjected to cyclic loading. The measured plots of force versus crack opening distance (CMOD) are shown in Figure 1 with dashed lines, and the thick black line shows the response of the numerically simulated fiber-reinforced notched prismatic specimen. The numerical model reproduces well the overall behavior of the short steel fiber reinforced specimen under cyclic loading. The addition of fibers had no influence on the tensile strength and that the numerically determined maximum load (5.63 kN) corresponds to the nominal tensile strength of 7.03 MPa, while the tensile strength of normal concrete was taken as 6.4 MPa in the simulation. It can be seen from the response diagrams in Figure 1 (left) that the presence of fibers has a strong influence on the residual strength and unloading behavior. The residual strength is slightly overestimated in the numerical simulations at later stages of the test. In experiments, the unloading stiffness after a crack is slightly lower than the initial (elastic) stiffness, and as the cycle progresses, the reduction in stiffness in each subsequent cycle is relatively small. The residual stiffness comes primarily from the fibers carrying the load, and that the dissipation manifested in the form of hysteresis is primarily due to the frictional sliding of the fibers in the fiber channels during unloading and reloading.

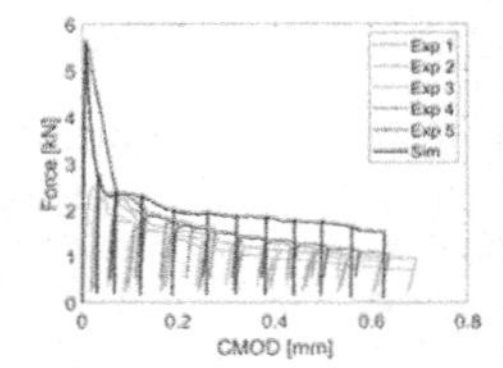

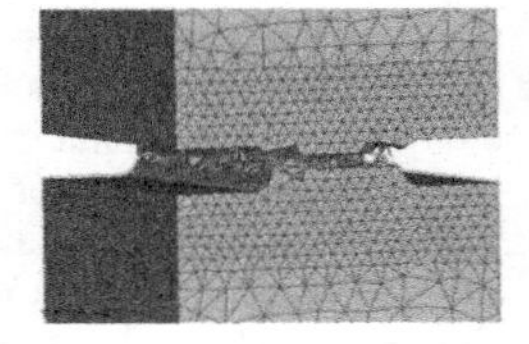

Figure 1. Experimentally and numerically obtained force versus CMOD curves (left); numerically simulated fiber reinforced notched prismatic specimen (right). [1].

5 CONCLUSION

1. Increase in strain of HPC-08-SF is lower of HPC-08 and stagnated after 10,000 cycles.
2. The microscopic evaluation showed that the number of microcracks continuously increased during the load cycles. For the HPC-08, the cracks were wider and longer and the total crack area was larger. HPC-08-SF showed more cracks overall, but they were smaller and shorter, resulting in a lower total crack area. In addition, they were more finely distributed.
3. 66% of the microcracks run or start in the ITZ.
4. Numerical simulations have consistent with the experimental investigations confirmed that the addition of short steel fibers had no influence on the tensile strength, but a significant improvement in the ductility. The unloading stiffness after cracking is slightly lower than the initial (elastic) stiffness; however, it remains stable during the rest of the test, even for large crack openings. Only a small hysteresis loop is formed during the unloading/reloading cycles, which indicates that the fibers carry most of the load. [1]

REFERENCES

[1] Schäfer, N.; Gudžulić, V.; Breitenbücher, R.; Meschke, G.2021. Experimental and Numerical Investigations on High Performance SFRC: Cyclic Tensile Loading and Fatigue. *Materials*, *14*, 7593. https://doi.org/10.3390/ma14247593

Current Perspectives and New Directions in Mechanics, Modelling and Design of Structural Systems – Zingoni (ed.)
© 2022 Copyright the Author(s), *ISBN 978-1-032-39114-4*

Influence of concrete compressive strength on fatigue behaviour under cyclic compressive loading

M. Markert
Materials Testing Institute, University of Stuttgart, Stuttgart, Germany

M. Deutscher
Institute of Concrete Structures, Technische Universität Dresden, Dresden, Germany

ABSTRACT: The development in the building industry is moving towards slimmer, higher and more resource efficient constructions. In reinforced concrete construction, high performance (HPC) and ultra-high performance concretes (UHPC) were developed for these purposes. But the fatigue behaviour of HPC and UHPC is still under research. Therefore, an HPC and a UHPC, were used to determine the extent to which the different concrete strengths have an influence on the number of cycles to failure, the temperature and strain development. For this study, different maximum and minimum stress levels on HPC as well as UHPC will be investigated.

1 INTRODUCTION

Due to research developments, the compressive strength of concretes has greatly increased in recent decades. In the Priority Programme SPP 2020 "Cyclic deterioration of High-Performance Concrete in an experimental-virtual lab", funded by Deutsche Forschungsgemeinschaft (DFG, German Research Foundation), the fatigue behaviour of high-performance concretes is investigated in order to substantially the current state of knowledge. For this purpose, cyclic compressive fatigue tests are carried out on concrete cylinders. In the following, two concretes with different compressive strengths, will be examined in more detail. In addition to comparing the numbers of cycles to failure, the temperature and strain development of the concretes was also compared.

2 EXPERIMENTS

2.1 *Number of cycles to failure*

The results of the fatigue tests of the HPC (99.1 MPa) and the UHPC (188.5 MPa) are given in Figure 1. Therefore, the average number of cycles to failure was plotted in log scale versus the normalised maximum stress S_o. Depending on the concrete, the results are plotted with black quadrats (HPC) and red triangles (UHPC). It can be seen that the number of cycles to failure of the UHPC is lower than that of the HPC at all stress levels. The trend line of the UHPC is to the left of the Model Code 2010 (Fib 2010) while the HPC is to the right. As a result, the Model Code 2010 (Fib 2010) would overestimate the fatigue resistance of the UHPC.

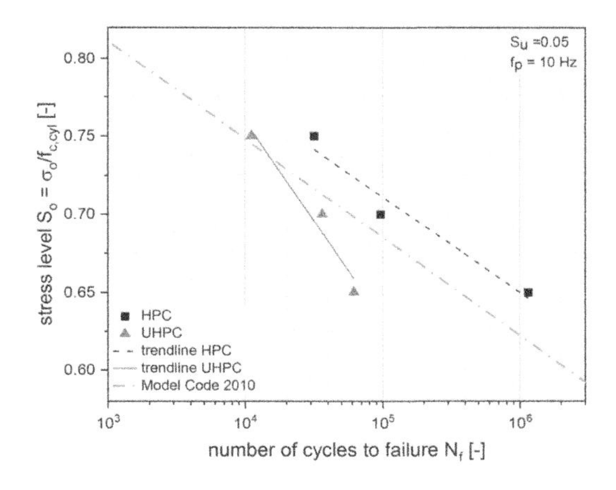

Figure 1. Average values of the number of cycles to failure and the corresponding trend lines.

2.2 *Temperature development*

The fatigue tests were carried out with a test frequency of 10 Hz, which resulted in a significant heating of the specimens. It had been shown that the compressive strength had an influence on the temperature rise as well as on the maximum achievable temperature (Figure 2).

DOI: 10.1201/9781003348450-218

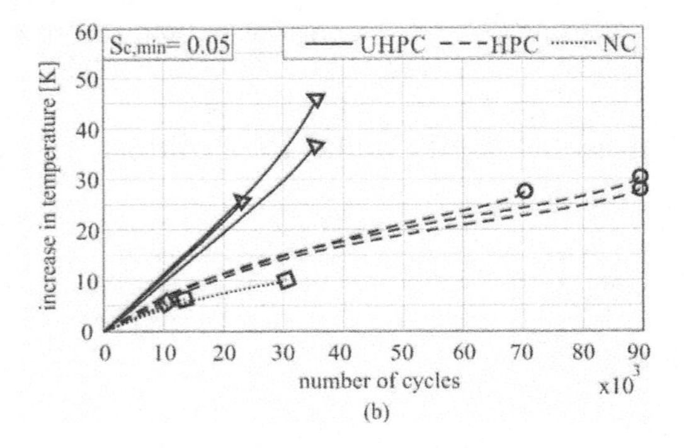

Figure 2. Temperature development on the surface at a maximum stress level of S_o = 0.70 (Deutscher et al. 2020).

2.3 *Strain development as a damage indicator*

For concrete, the strain development showed a characteristic S-shaped curve in general. In order to be able to better compare the different tests with each other, the gradient *m* of the strain in phase II was examined and compared, see Equation (1).

$$m\left(\varepsilon_O^{0.15-0.75}\right) = \left(\frac{\Delta\varepsilon_O^{0.15-0.75}}{\Delta N^{0.15-0.75}}\right) \tag{1}$$

Figure 3 Shows the grades of phase II together with the respective number of cycles to failure on a double logarithmic scale. In addition, to the results of the HPC and the UHPC the results from (Markert et al. 2022) in grey stars.

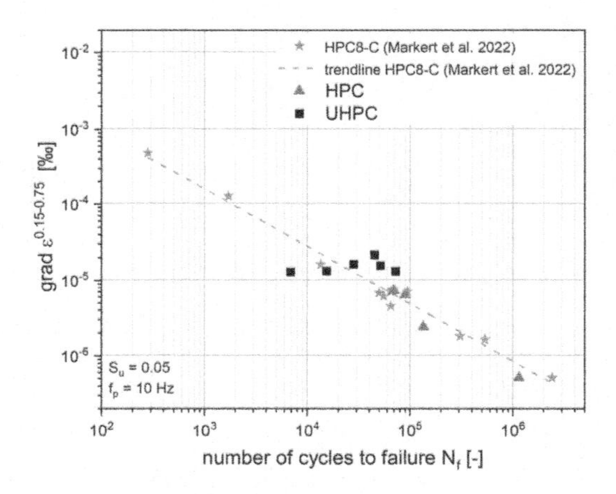

Figure 3. Number of cycles to failure.

The values of the UHPC are plotted in black squares; due to the small number of values, no clear tendency can be identified. Based on the possibility of a parallel straight line to the values of the HPC, the values of the UHPC with a low number of cycles to failure are possible points with a larger deviation. However, it can be seen that the gradient *m* of the UHPC has a different behaviour than that of the HPC. This is a possible indication of material dependence.

3 CONCLUSIONS AND OUTLOOK

An HPC and a UHPC were investigated to study the effect of concrete compressive strength on fatigue behaviour under cyclic compressive loading. It has been shown that the number of cycles to failure at the same stress level was lower for the UHPC compared to the HPC. It can be seen that the Model Code 2010 (Fib 2010) overestimates the fatigue resistance of the UHPC and underestimates that of the HPC. There is also an influence on the temperature development. The UHPC heats up faster and more than the HPC. The heating of the specimen has an effect on the static compressive strength, which is a possible explanation for the fact that the numbers of cycles to failure of the UHPC are lower than those of the HPC. Besides the number of cycles to failure and the temperature development, the evolution of the strain development in damage phase II was analysed. Although the test data is limited, there is a tendency for the strain development to show different behaviour as a function of the compressive strength or the concrete composition. The influence of the temperature on the strain development must be clarified in further investigations.

ACKNOWLEDGEMENTS

Financial support was provided by the Deutsche Forschungsgemeinschaft (German Research Founda-tion, DFG) for the project "Temperature and humidi-ty induced damage processes caused by multi-axial cyclic loading" (project number 353921616) and "Influence of load-included temperature fields on the fatigue behaviour of UHPC subjected to high frequency compression loading" (project number 353981739), both as part of the DFG Priority Programme 2020. This support is gratefully acknowledged.

REFERENCES

Deutscher, M.; Markert, M.; Tran, N.L.; Scheerer, S. Influence of the compressive strength of concrete on the temperature increase due to cyclic loading. Conf. paper, Int. fib Congress, held online, China 22-24.11. 2020, pp. 773–780.

Fib Model Code 2010, fib Model Code for Concrete Structures 2010. International Federation of Structural Concrete (fib), Lausanne, Switzerland, Ernst & Sohn, Berlin, 2013

Markert, M.; Katzmann, J.; Birtel, V.; Garrecht, H.; Steeb, H. Investigation of the Influence of Moisture Content on Fatigue Behaviour of HPC by Using DMA and XRCT. Materials 2022, 15, 91. https://doi.org/10.3390/ma15010091

Current Perspectives and New Directions in Mechanics,
Modelling and Design of Structural Systems – Zingoni (ed.)
© 2022 Copyright the Author(s), ISBN 978-1-032-39114-4

Influence of loading frequency on the compressive fatigue behaviour of high-strength concrete with different moisture contents

M.Abubakar Ali, N. Oneschkow, L. Lohaus & M. Haist
Institute of Building Materials Science, Leibniz University Hannover, Germany

ABSTRACT: The fatigue behaviour of concrete with high moisture content has become an important subject of interest with the expansion of offshore wind energy systems. Investigations in the literature indicate that the numbers of cycles to failure significantly decreases with the increased moisture content in concrete. The damage mechanisms, which are responsible for an accelerated degradation behaviour, are investigated in a joint research project. In this paper, results of compressive fatigue investigations on high-strength concrete subjected to different loading frequencies are presented comparatively for specimens stored in air condition and underwater, i.e. different moisture content of the concrete. The results show that the damage accelerating effect of concrete with high moisture contents is more pronounced for lower load frequencies. Additionally, the results of the stiffness development indicate that different damage mechanisms are acting with increasing the moisture content.

1 INTRODUCTION

The number of fatigue-loaded concrete structures has significantly increased in the last few decades with expansion of offshore wind energy systems. The offshore exposure of those construction leads to a higher moisture content in concrete compared to onshore wind energy systems. Investigations in the literature indicate that the numbers of cycles to failure significantly decreases with the increased moisture content [see e.g. Muguruma et al. (1984)]. Furthermore, Tomann et al (2020) identified the moisture content in concrete - and not the water as an environment - as primarily responsible for the reduction of the fatigue resistance of concrete. Oneschkow et al. (2020) found indication that the influence of loading frequency is higher for specimens with higher moisture contents. Although the impact of the moisture content on the fatigue resistance of concrete has been recognized in the past, there is still a pronounced lack of knowledge, especially concerning the mechanisms responsible for the accelerated degradation behaviour. In a joint project as part of the priority programme SPP 2020 'Cyclic Deterioration of High-Performance Concrete in an Experimental-Virtual Lab', fatigue tests on concrete specimens with different moisture contents have been conducted. In this paper, the numbers of cycles to failure of specimens with two different moisture contents subjected to different loading frequencies are presented. The influence of the loading frequency on the stiffness development is analysed additionally.

2 EXPERIMENTAL INVESTIGATIONS

The experimental investigations were conducted on the reference high-strength concrete RH1 of the priority programme SPP 2020, whose composition is given in Tomann et al. (2019). The mean 28-day compressive strength was determined as $f_{cm,cube}$ = 114 MPa according to DIN EN 12390-2 (2019). The water-saturated specimens (WS) were stored underwater immediately after the concreting process until testing. The specimens of the storage condition/moisture content Cl 65 were stored after the demolding process in a climate chamber under standard climate condition (20°C, 65% RH). Directly before testing, all specimens of the both storage conditions were sealed using epoxy resin. The axial force, axial displacement of the hydraulic cylinder, axial deformations of the specimens, ambient temperature and the temperature at the mid-height of the specimens were measured continuously with a sampling rate between 300 to 2600 Hz. In the compressive fatigue tests, the maximum stress level was kept constant at S_{max} = 0.75 and the minimum stress level at S_{min} = 0.05. The fatigue tests were carried out with loading frequencies of f_t = 0.1, 1.0 and 10 Hz, respectively, for both storage conditions (Cl 65 and WS)

3 RESULTS

The logarithmic numbers of cycles (log N_f) to failure as mean value with respect to the loading frequency are shown in Figure 1. It is obvious from the Figure

DOI: 10.1201/9781003348450-219

that specimens of the storage condition Cl 65 reached a higher mean values than the water-saturated specimens for all loading frequencies. The mean value of log N_f of specimens of the storage condition Cl 65 subjected to the loading frequency of f_t =1 Hz was located quite close to the specimens of f_t = 0.1 Hz. The highest loading frequency f_t = 10 Hz led to the lowest mean value of log N_f. By contrast, specimens of the moisture content WS showed higher values of log N_f at higher loading frequencies. Additionally, the relative reduction of the number of cycles to failure of water-saturated specimens, referred to the value of storage condition Cl 65, decreased for higher loading frequencies. This indicated that lower loading frequencies intensify the moisture induced damage mechanisms, which lead to an accelerated fatigue degradation of concrete.

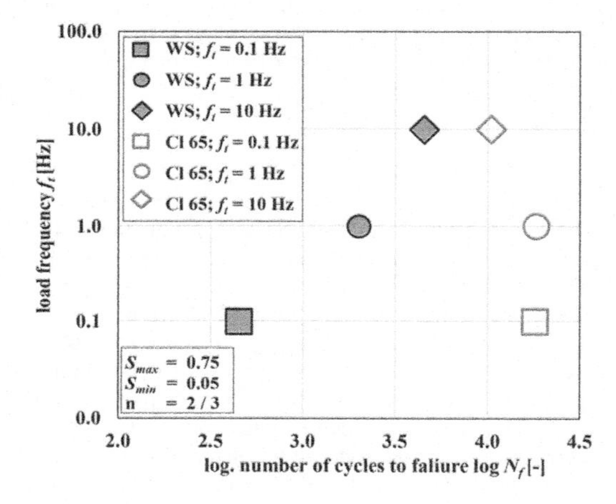

Figure 1. The logarithmic number of cycles to failure with respect to the loading frequency.

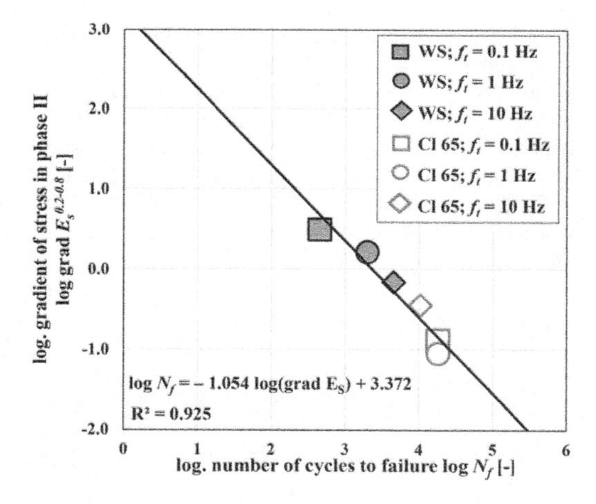

Figure 2. Logarithmic gradient of stiffness in phase II for different loading frequency.

A linear correlation between the log. gradient of stiffness in phase II and the log. numbers of cycles was found (cf. Figure 2). Specimens with the higher moisture content WS show significantly higher mean values of gradient of stiffness compared to the specimens of the storage condition Cl 65. The clear division of the values of the stiffness degradation depending on the moisture content is an indication that different damage mechanisms are acting with increased moisture content in concrete

4 CONCLUSIONS

The first results of the fatigue investigations on high-strength concrete specimens with two different moisture contents subjected to different loading frequencies are presented in this paper. The main results can be summarised as follows:

- Higher moisture contents lead to decreased numbers of cycles to failure and increased gradients of stiffness in phase II.
- The damage accelerating effect of high concrete moisture contents is more pronounced for lower load frequencies.
- A linear correlation exists between the logarithmic gradient stiffness in phase II and the logarithmic number of cycles to failure for both moisture contents. Hereby, a clear division of the values was observed depending on the moisture content indicating different damage mechanisms.

Overall, the results presented reveal the negative effect of high concrete moisture contents on the fatigue behaviour of high-strength concretes with decreasing loading frequencies. Additional damage indicators, such as acoustic emission, will be further analysed, also with respect to different minimum stress levels. For a better understanding of the fatigue damage processes at microscale, microstructural investigations of porosity, pore size distribution and moisture content will be conducted.

REFERENCES

Muguruma, H.; Watanabe, F. On the Low-cycle Compressive Fatigue Behaviour of Concrete under Submerged Condition. In Proceedings of the 27th Japan Congress on Materials Research, Kyoto, Japan, September 1984; pp. 219–224.

Tomann, C.; Oneschkow, N. Influence of moisture content in the microstructure on the fatigue deterioration of high-strength concrete. Struct. Concr. 2019, 20, 1204–1211.

Oneschkow, N.; Hümme, J.; Lohaus, L. Compressive fatigue behaviour of high-strength concrete in a dry and wet environment. Constr. Build. Mater. 2020, 262, 119700.

DIN EN 12390-3:2019-10. German Institute for Standardization. Testing Hardened Concrete—Part 3: Compressive Strength of Test Specimens; German Version of EN 12390-3:2019; Beuth: Berlin, Germany, 2019.

Current Perspectives and New Directions in Mechanics, Modelling and Design of Structural Systems – Zingoni (ed.)
© 2022 Copyright the Author(s), ISBN 978-1-032-39114-4

Influence of the concrete moisture content on the strain development due to cyclic loading

M. Markert, V. Birtel & H. Garrecht
Materials Testing Institute, University of Stuttgart, Stuttgart, Germany

ABSTRACT: Due to research activities on the topic of concrete mixtures during the last decades, concrete compressive strength has increased significantly. At the same time, the lifespan of concrete structures has continuously extended. Experiments of HPC with three different concrete moisture contents were investigated. For this, different types of storage were used to achieve the different moisture levels. The focus is on the strain development. This will be used to investigate in more detail why the fatigue resistance of moist concrete is lower than that of dried concrete. Therefore, the moisture content and the maximum stress level were varied.

1 INTRODUCTION

The fatigue behaviour of HPC is very complex and not fully understood, even after many years of research. So far, mainly the influence of the frequency and amplitude has been studied. The latest research has shown that moisture content of the concrete has a significant influence on the fatigue behaviour (Markert et al. 2019 and Tomann & Oneschkow 2019). The influence on the temperature development and on the fatigue behaviour was discussed in (Markert & Laschewski 2020 and Deutscher et al. 2019), as well as in (Deutscher et al. 2020). In addition to the influence of moisture and temperature, the maximum grain size also seems to have an influence on the fatigue behaviour. In the following, an HPC with three different moisture contents is investigated. The effect of moisture on the number of cycles to failure and temperature development is investigated. Special attention is given to the analysis of the strain development.

2 EXPERIMENTS

2.1 *Material and geometry*

The HPC used in this study was the reference mixture developed for the Priority Programme SPP2020 (Markert et al. 2019 and Tomann & Oneschkow 2019). The resulting concrete moisture can be found in Table 1. The exact configuration for the test setup can be taken from (Markert et al. 2019).

2.2 *Strain development*

For concrete, the strain development showed a characteristic S-shaped. This curve can be divided into three phases (see Figure 1). This damage progression depended on the applied stress levels, as well as the moisture content (Markert et al. 2019). In order to be able to better compare the different tests with each other, the gradient m of the strain in phase II was examined and compared. For this purpose, the strain ($\Delta\varepsilon_o$) within 15% and 75% of the number of cycles to failure (ΔN) was analysed.

Table 1. Storage conditions and concrete moisture of the HPC.

	Series		
	UW	C	D
Moisture Content	5.1%	4.0%	~0.1%

$$m\left(\varepsilon_O^{0.15-0.75}\right) = \left(\frac{\Delta\varepsilon_O^{0.15-0.75}}{\Delta N^{0.15-0.75}}\right) \quad (1)$$

The gradients m of phase II together with the respective number of cycles to failure on a double logarithmic scale is shown in Figure 2. The moisture content seemed to have an influence to some smaller extent. The difference between the series D and the other two series was clearly visible. Between the UW series and the C series, the difference was almost not recognisable.

3 CONCLUSIONS AND OUTLOOK

In order to investigate the influence of moisture content on fatigue behaviour of HPC, the HPC was stored in three different environmental conditions and

DOI: 10.1201/9781003348450-220

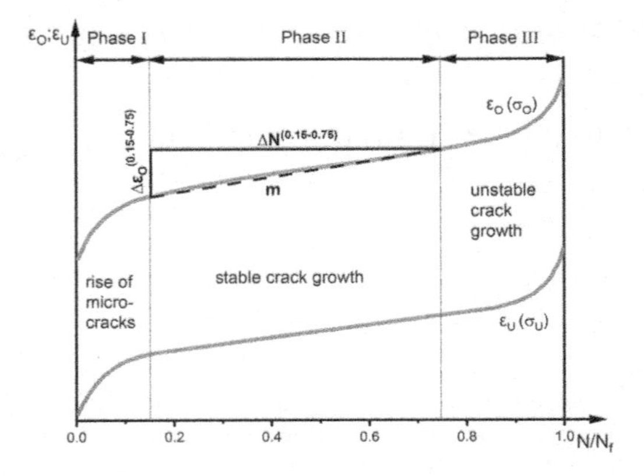

Figure 1. Schematic strain development and division into three phases.

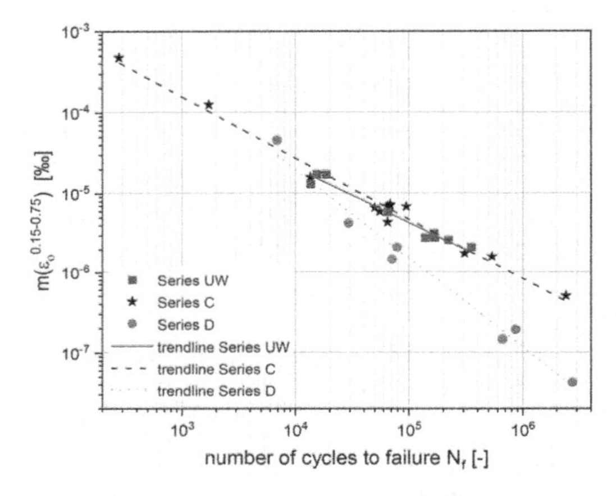

Figure 2. Gradients of strain in damage phase II and the associated trendlines.

then subjected to cyclic loading. All tests were carried out at a constant frequency of 10 Hz and with a variation of the upper stress level. Due to the different storage conditions, three different moisture contents have resulted. The moisture content of the HPC had a negative effect on the fatigue resistance. At the same applied stress level, the fatigue resistance decreased as the moisture of the concrete increased. This influence can be clearly seen in the number of cycles to failure. In addition to the effect on the number of cycles to failure, the moisture content also changes the temperature development of the concrete during the fatigue test. Considering the fact that the compressive strength of concrete is characterised by a temperature dependence, an indirect influence of the compressive strength during the fatigue test can be concluded from the found moisture-temperature coupling. To research this effect in more detail further fatigue tests for instance with external controlled or elevated temperature levels are aspired. By evaluate the development of the strain in the damage phase two of the fatigue experiments an evaluation method was presented that reduces the big scatter range in the plots of failure cycles versus applied load. In summary, it can be mentioned that the moisture content of the concrete has a significant influence on the number of cycles to failure, strain-and temperature evolution This article is an excerpt from the publication in the special issue "Cyclic Deterioration of Concrete" of the journal "materials" (Markert et al. 2022).

ACKNOWLEDGEMENTS

Financial support was provided by the Deutsche Forschungsgemeinschaft (German Research Foundation, DFG) for the project "Temperature and humidity induced damage processes caused by multi-axial cyclic loading" (project number 353921616) as part of the DFG Priority Programme 2020. This support is gratefully acknowledged. We would also like to thank our colleagues and partners in the DFG Priority Programme SPP2020 for great discussion.

REFERENCES

Deutscher, M.; Markert, M.; Tran, N.L.; Scheerer, S. Influence of the compressive strength of concrete on the temperature increase due to cyclic loading. Conf. paper, Int. fib Congress, held online, China 22-24.11. 2020, pp. 773–780.
Deutscher, M.; Markert, M.; Tran, N.L.; Scheerer, S. Influence of the compressive strength of concrete on the temperature increase due to cyclic loading. Conf. paper, Int. fib Congress, held online, China 22-24.11. 2020, pp. 773–780.
Deutscher, M.; Tran, N.L.; Scheerer, S. Experimental Investigations on the Temperature Increase of Ultra-High Performance Concrete under Fatigue Loading. Applied Sciences 2019, 9, 4087.
Markert, M.; Birtel, V.; Garrecht, H. Temperature and humiditiy induced damage progresses in concrete due to pure compressive fatigue loading. In Proceedings of the fib Symposium, Krakow, Poland, 27-29 May 2019; Derkowski, W., Gwozdziewicz, P., Hojdys, Ł., Krajewski, P., Pantak, M., Eds.; FIB: Laussane, Switzerland, 2019; pp. 1928–1935.
Markert, M.; Katzmann, J.; Birtel, V.; Garrecht, H.; Steeb, H. Investigation of the Influence of Moisture Content on Fatigue Behaviour of HPC by Using DMA and XRCT. Materials 2022, 15, 91. https://doi.org/10.3390/ma15010091
Markert, M.; Laschewski, H. Influencing factors on the temperature development in cyclic compressive fatigue tests: An overview. Otto Graf Journal 2020, 19, pp. 147–162.
Tomann, C.; Oneschkow, N. Influence of moisture content in the microstructure on the fatigue deterioration of high-strength concrete. Structural Concrete 2019, 20, 1204–1211.

Current Perspectives and New Directions in Mechanics, Modelling and Design of Structural Systems – Zingoni (ed.)
© 2022 Copyright the Author(s), ISBN 978-1-032-39114-4

Influence of moisture content on strain development of concrete subjected to compressive creep and cyclic loading

B. Kern, N. Oneschkow, M. Haist & L. Lohaus
Institute of Building Materials Science, Leibniz University, Hannover, Germany

ABSTRACT: The expected long-term deformations of concrete structures today are calculated using creep models, assuming constant mechanical loads. Nevertheless, in many structures like bridges constant creep loads are superimposed with cyclic loads of substantial magnitude. The effect of such a cyclic loading onto the concrete deformation behaviour so far has been unkown. An underestimation of the resulting deformation by established creep models might cause serious damage and safety risks for the structure. Before this background, the strain development due to creep and cyclic loading of a normal strength concrete have been comparatively investigated at two different moisture contents (approx. 100 % and 75 %). The results show that viscous strains from cyclic loading are significantly higher than those from purely static loading despite identical mean stress levels. Further, the resulting strains are considerably affected by the moisture content of concrete. The gathered results provide the basis for improving existing creep models.

1 INTRODUCTION

Long-term deformations, measured on existing reinforced and prestressed concrete bridges, can significantly exceed the deformations predicted by creep models. In the literature, the superposition of creep with cyclic loadings is mentioned as a possible reason. However, the effect of cyclic loading onto the deformation behaviour at moderate loadings has been subject of very few experimental investigations (e.g. Mehmel & Kern 1962, Whaley & Neville 1973), where, despite its importance, the humidity was not taken into account. Therefore, fundamental investigations about comparability and differences of creep and cyclic strains were carried out on a normal-strength concrete as part of a collaborative research project. The paper at hand is focused on the influence of moisture with regard to the type of loading. The developments of viscous strain are presented comparatively for creep and cyclic loading with stress levels $S_{max} \leq 0.45$. The characterisation of long-term strain development due to cyclic loading in comparison to creep loading is the basis for the overall aim of ongoing studies to improve approaches describing the cyclic strain development based on creep models.

2 EXPERIMENTAL PROGRAMME

The presented investigations were carried out on a normal strength concrete using cylindrical specimens (h/d = 180/60 mm) at a minimum concrete age of 180 days. The reference compressive strength of the concrete was determined force-controlled immediately before testing in the range of 41.6 to 44.4 MPa. The maximum stress level S_{max} in the cyclic tests was varied between S_{max} = 0.35, S_{max} = 0.45 while the test frequency f_t = 0.1 Hz and the minimum stress level S_{min} = 0.05 were kept constant. Static creep tests were conducted with load levels of $S_{creep} = S_{max}$ and $S_{creep} = S_{mean}$, respectively. All tests were conducted on sealed specimens which were stored before in two different ways to ensure a constant moisture content of approx. 100 % (V100) or 75 % (V75).

The peak strains of the cyclic curves $\varepsilon_{tot,max}$ at the maximum stress level S_{max} were obtained by peak analyses. The elastic strain component $\varepsilon_{el,0}$ was defined as the strain when S_{creep} or S_{max} was reached the first time. The time-dependent viscous strain component $\varepsilon_{v,cr}$ (creep) or $\varepsilon_{v,max}$ (cyclic) presented hereinafter was determined from the total strain ε_{tot} after subtracting the elastic strain component $\varepsilon_{el,0}$.

3 RESULTS AND DISCUSSION

The mean curves of the viscous strains at maximum stress level due to cyclic loading $\varepsilon_{v,max}$ (black) are shown together with the mean curves of viscous strains due to creep loading $\varepsilon_{v,cr}$ at $S_{creep} = S_{mean}$ and $S_{creep} = S_{max}$ (both in grey) in Figures 1 and 2 in relation to the loading duration. The mean viscous strains for a moisture content of approx. 100 % (V100) are presented in Figure 1 and those for

DOI: 10.1201/9781003348450-221

approx. 75 % (V75) in Figure 2 for S_{max} = 0.45 (including the corresponding creep stress levels).

In Figures 1 and 2, it can be seen that the viscous strains due to cyclic loading $\varepsilon_{v,max}$ significantly exceed the strains due to static creep loading $\varepsilon_{v,cr}$ with $S_{creep} = S_{mean}$. The viscous cyclic strains $\varepsilon_{v,max}$ instead were smaller but closer to those caused by static creep loading $\varepsilon_{v,cr}$ at $S_{creep} = S_{max}$. The slope of the viscous cyclic strains $\varepsilon_{v,max}$ and the viscous creep strains $\varepsilon_{v,cr}$ due to $S_{creep} = S_{max}$ was very similar at the beginning of loading. As soon as the curves flatten and become more linear, the differences between the cyclic curves and the creep curves for $S_{creep} = S_{max}$ became evident, which are more or less pronounced depending on the moisture content. While the slope of the cyclic curve is smaller than the slope of the creep curve for $S_{creep} = S_{max}$ for V100 (Figure 1), it can be seen that for V75 (Figure 2), the curve due to cyclic loading at S_{max} = 0.45 flattens less than the creep curve for $S_{creep} = S_{max}$. Therefore, the slope of the cyclic curve was steeper than that of the creep curve for $S_{creep} = S_{max}$ after approx. 2 days (detailed analysis). The steeper slope led to an intersection point of viscous cyclic strain curve and creep curve for $S_{creep} = S_{max}$ at the sixth day of loading for S_{max} = 0.45 and higher strains could be possible afterwards.

For S_{max} = 0.35, the viscous strains due to cyclic loading are also closer to those caused by static creep loading $\varepsilon_{v,cr}$ at $S_{creep} = S_{max}$ compared to those caused by $S_{creep} = S_{mean}$. The viscous cyclic curves proceeded almost parallel to the creep curve for $S_{creep} = S_{max}$ at both moisture contents. As the slopes were very close to each other, there is no indication whether the cyclic curves for V75 would also exceed the creep curves for $S_{creep} = S_{max}$ for loading durations > 91 days.

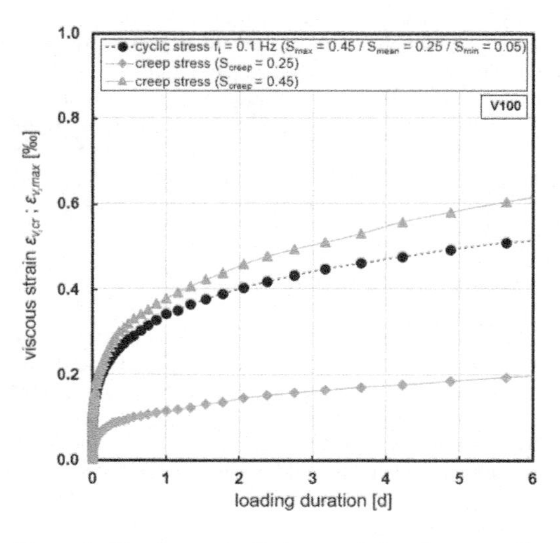

Figure 1. Development of viscous strains due to cyclic loading ($S_{max}/S_{mean}/S_{min}$ = 0.45/0.25/0.05) and comparable creep loading for specimens with a moisture content of approx. 100 % (V100).

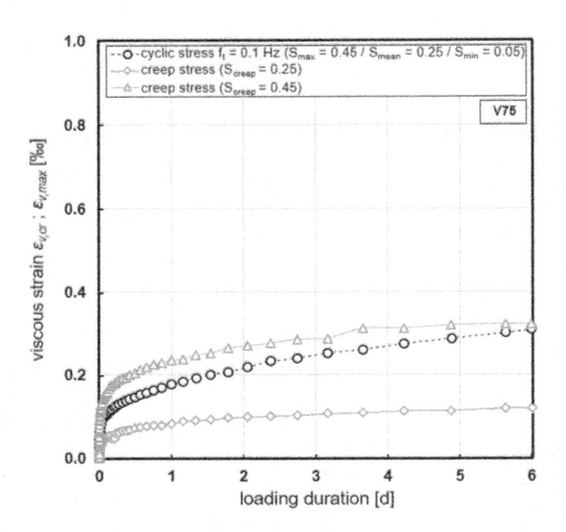

Figure 2. Development of viscous strains due to cyclic loading ($S_{max}/S_{mean}/S_{min}$ = 0.45/0.25/0.05) and comparable creep loading for specimens with a moisture content of approx. 75 % (V75).

4 CONCLUSIONS

The results obtained show that the viscous strains due to cyclic loading are generally higher than those due to creep loading at an equal mean stress level for both moisture contents and maximum stress levels investigated. Instead, the viscous cyclic strains are closer to the strain caused by creep loading with $S_{creep} = S_{max}$. However, there are still differences between those strains, which become more or less pronounced depending on the maximum stress level and the moisture content. While for the moisture content of V100 it is shown that the cyclic viscous strains will be consistently below the creep curve for $S_{creep} = S_{max}$ over the entire loading duration, the investigations on the moisture content V75 show that the cyclic strains may exceed the creep strains for $S_{creep} = S_{max}$. This effect might be further increased by lower moisture contents. As a consequence, cyclic loading, especially at low moisture levels, might lead to serious damage and safety risks of deflection-sensible structures designed with creep formulations, assuming constant loads. Therefore, it is important to investigate the effect of varying loads and different moisture contents on long-term deformations in further detail.

REFERENCES

Mehmel, A. & Kern, E. 1962. Versuche über die Festigkeit und die Verformung von Beton bei Druckschwellbeanspruchung [Investigations on the strength and deformation of concrete under compressive fatigue loading]. *Deutscher Ausschuss für Stahlbeton* 144, Berlin [in German].

Whaley, C.P. & Neville, A.M. 1973. Non-elastic deformation of concrete under cyclic compression. *Mag Concr Res* 25(84): 145–154.

Current Perspectives and New Directions in Mechanics, Modelling and Design of Structural Systems – Zingoni (ed.)
© 2022 Copyright the Author(s), ISBN 978-1-032-39114-4

Behavior of coupled walls with high performance fiber reinforced concrete coupling beams

M. Abdelhafeez
PhD Student, University of Hawai'i at Mānoa, Honolulu, HI, USA
East-West Center Student Affiliate, Assistant Lecturer, Cairo University, Egypt (On Leave)

H. Salem
Professor, Cairo University, Egypt

ABSTRACT: Coupling beams have great effect on the behavior of coupled shear wall systems under lateral loads. Therefore, coupling beams are reinforced by complicated reinforcement, in the form of diagonal bars and confinement reinforcement, however, creates construction difficulties. Because of the shallow angle of diagonal reinforcement, (less than 20) with respect to the beam longitudinal axis, in slender coupling beams where beam's aspect ratio is around 3.0, the effect of this reinforcement is questionable. The purpose of this research is to study the effect of using tensile strain-hardening, high-performance fiber reinforced concrete (HPFRC) in the coupling beams. A nonlinear static analysis was performed using the Applied Element Method investigating the effect of coupling beam with HPFRC. After validating the analysis with previously tested specimens, a parametric study was carried out. The results showed that using of HPFRC can provide higher capacity with enhanced energy dissipation more than traditional reinforced concrete.

1 INTRODUCTION

Individual shear walls can produce larger stiffness and strength if they are connected by a coupling beam. Therefore, coupling beams must be stiff and ductile to dissipate required energy. Numerical studies are nowadays replacing the conventional experimental ones due to their low cost and relatively low time-consuming. In this study, the AEM is used for numerical assessment of multistory reinforced concrete frame structure with high performance fiber reinforced concrete coupling beams (HPFRC) subjected to lateral loading. After validating the AEM, a parametric study was carried out for a mid-rise multistory reinforced concrete frame structure with HPFRC coupling beams.

2 THE APPLIED ELEMENT METHOD (AEM)

The AEM [1] is based on dividing the structure virtually into small elements. Each two elements are connected by one normal and two shear springs at contact points. The AEM is based on discrete crack approach which can track the structural behavior passing through all stages of loading.

Extreme Loading for Structures (ELS) [2] is software based on applied element method AEM [1]. Maekawa compression model is used for concrete modeling under compression [3], and for concrete shear springs.

For reinforcement springs, the model, presented by Ristic et al. [4], used in ELS.

3 VALIDATION OF ELS

The ELS software was validated and had shown good agreement with several cases. For the current study, it's crucial to validate the behavior of the HPFRC coupling beams, so experimental data of coupling beams tested by Setkit [5] was used to validate ELS.

4 CASE STUDY

4.1 *Details of studied structure*

The first reference case is 300×1000 mm coupling beam reinforced with 3T25 as a top and bottom reinforcement, two rows of 2T18 as a shrinkage reinforcement, and two branches of T12-100 as shear stirrups casted in between 450×1000 mm upper and lower concrete blocks reinforced with 4T22 as top and bottom reinforcement, 2T18 as a shrinkage reinforcement and T12-100 as four branches' stirrups.

DOI: 10.1201/9781003348450-222

The second reference case is a twelve-story reinforced concrete coupled shear wall designed to resist 0.15 g. The shear wall is 300×4000 mm reinforced with T16-200, T12-100 distributed vertical and horizonal reinforcement respectively. Also, the wall is reinforced with 9T18 concentrated vertical reinforcement. The Coupling beams are reinforced with the same reinforcement of first case.

4.2 *Material properties*

The materials used in the models are nonlinear. The regular concrete is used for coupled walls, top and bottom blocks, also the HPFRC is used for coupling beams.

HPFRC create several narrower cracks and they begin to fail when the fibers start to pull out from the matrix, therefore the failure occurs because of crack localization. Also, mixture properties and matrix fiber interaction should be considered.

4.3 *Loads*

The flooring, live and wall loads are assumed 1.5, 2 and 1.8 kN/m2 respectively. Quasi-static loading in a displacement-controlled mode consisted of 10 cycles started with 2.4 mm with step of 2.4 mm till it reaches 24.2 mm was assigned to the first reference case. For the second reference case, the coupled walls with HPFRC coupling beams, was subjected to quasi-static loading in a load-controlled mode started with 407 kN with step of 407 kN till it reaches 4884 kN. Loads were distributed to follow triangular distribution. The coupled walls with regular rein-forced concrete coupling beams were subjected to quasi-static loading in a load-controlled mode start-ed with 270 kN with step of 270 kN till it reaches 3240 kN. These cycles were calculated to have the same drift as HPFRC coupling beam case.

4.4 *Structural model*

Figure 1 shows the reference cases as modelled in the current study.

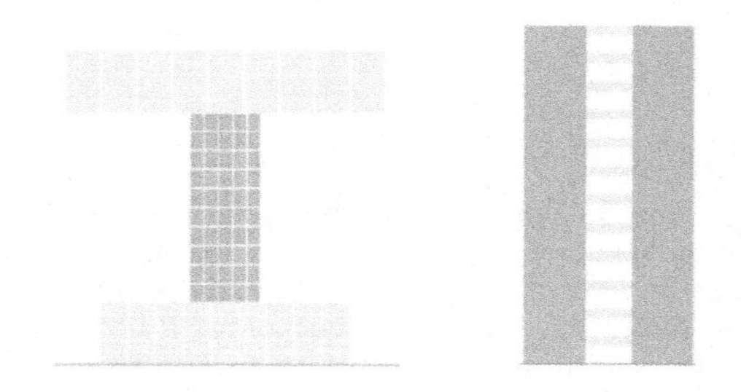

Figure 1. First reference case (left) and second reference case as modeled in ELS.

For the reference case, the coupling beam is 300× 1000 mm with 0.5% reinforcement ratio, f_c'=50 N/mm^2, and f_y= 420 N/mm^2 for reinforcement steel.

4.5 *Studied parameters*

The studied parameters are material type (HPFRC and regular concrete), the longitudinal reinforcement ratio for coupling beams, high performance fiber reinforced concrete embedment inside the coupled walls, presence of diagonal reinforcement with and without confining stirrups, coupling beam's aspect ratio and the fiber's ratio.

5 SUMMARY AND CONCLUSIONS

Based on the numerical results of the case studies, the following conclusions could be obtained.

- Energy dissipated by HPFRC coupling beams reaches as much as 1.3 times energy dissipated by regular RC coupling beams. This is explained by strain hardening behavior and the capability of developing multiple cracks after maximum tensile stress.
- Increasing of longitudinal reinforcement ratio for coupling beams from 0.5% to 0.8%, leads to increase in coupled walls capacity by 1.1 times. This is explained by the ductility enhancement through retarding coupling beam failure.
- As HPFRC embedment inside the coupled walls increases from zero to four coupling beam's depth, coupled walls capacity increases up to 1.14 times. This is explained by the capability of HPFRC to resist higher stresses than regular RC due to strain hardening behavior.
- As beam's aspect ratio decreases from 3.75 to 2.72, coupled walls capacity increases up to 1.15 times. This is explained by increasing depth leads to higher capacity.
- Using two intersected groups of reinforcing bars of 4T22/group lead to an increase in capacity of 40% compared to beams with longitudinal reinforcement.
- Also, using diagonal bars leads to an increase in energy dissipated. This is explained by the capability of diagonal bars to resist diagonal compression generated due to cyclic loading.
- The coupled walls capacity increases up to 1.03 when maximum tensile stress changes from 3.45 MPa to 4.83 MPa. This is explained by enhancing strain hardening behavior.

REFERENCES

[1] Extreme Loading for Structures, Theoretical Manual, 2004–17.

[2] www.extremeloading.com, www.appliedscienceint.com/.

[3] Extreme loading for structures theoretical manual, Appl. Sci. Int. (2013).

[4] D. Ristic, Y. Yamada, H. Iemura, Stress-Strain Based Modeling of Hysteretic Structures under Earthquake Induced Bending and Varying Axial Loads Research report No. 86-ST-01, School of Civil Engineering, Kyoto University, Kyoto, Japan, 1986.

[5] Setkit, M. (2012). Seismic Behavior of Slender Coupling Beams Constructed with High-Performance Fiber-Reinforced Concrete.

15. Safety and reliability, Design philosophy, Bridge technology, Transport infrastructure, Building performance, Engineering education

Current Perspectives and New Directions in Mechanics, Modelling and Design of Structural Systems – Zingoni (ed.)
© 2022 Copyright the Author(s), ISBN 978-1-032-39114-4

Structural reliability estimates by Slepian models

M. Grigoriu
Cornell University, Ithaca, USA

ABSTRACT: Estimates are developed for time-variant reliability problems by using the distribution of the time when a system state leaves a safe set for the first time, e.g., a critical stress or displacement *X(t)* of a dynamical system crosses a level a with positive slope for the first time. The probability density function of this random time, referred to as the first passage time T_a, satisfies an integral equation whose kernel relates to crossings properties of *X(t)*. In contrast to current studies which find this kernel by calculating numerically multidimensional integrals, we obtained it via Slepian models. Two numerical examples dealing with broad and narrow band processes *X(t)* are presented to illustrate the implementation of the proposed method and assess its accuracy.

1 INTRODUCTION

The probability that the state of a physical system satisfies design constraints during a time interval $[0, \tau]$ defines the system reliability $p_s(t)$ in this time interval. For example, if a system state is a real-valued stochastic process $X(t)$, its reliability $p_s(t)$ for the design constraints requiring that $X(t)$ does not leave the safe set $D_s = (-\infty, a]$, $a > 0$, in $[0, t]$ is the probability of the event $\{\sup_{0 \le s \le t}\{X(s)\} \le a\}$.

There are no analytical expressions for the distributions of the extreme event $\sup_{0 \le s \le t}\{X(s)\}$. The mean rate at which $X(t)$ crosses with positive slope the threshold a, referred to as mean a-upcrossing, can be used to construct approximations for the reliability $p_s(t)$ (5) (Chap. 7). These approximations work for broad band processes but can be unsatisfactory for narrow band processes. Two options have been explored to improve mean crossing-based approximations of $p_s(t)$ for narrow band processes. The first uses correction factors which account for the process bandwidth see, e.g., (2) (Sect. 11.7) and (6). The second calculates $p_s(t)$ by using the probability of the event $\{T_a > t\}$, where T_a denotes the first passage time of $X(t)$, i.e., the first time when this process exceeds a. The two definitions of the reliability are equivalent in the sense that the events $\{\sup_{0 \le s \le t} X(s) \le a\}$ and $\{T_a > t\}$ have the same probability.

We follow the second approach which estimates $p_s(t)$ from the distribution of the random variable T_a. The approach requires to construct an integral equation for the density $f_a(t)$ of T_a and subsequently solve of this equation numerically. The construction of this equation uses Slepian models, which characterize the evolution of $X(t)$ following an a-upcrossing of this process. Numerical illustrations include broad and narrow band processes.

2 FIRST PASSAGE TIME DISTRIBUTION

Assume that $X(t)$, $t \ge 0$, is a mean square (m.s.) differentiable process and let $\nu(a;t)$ denote its mean a-upcrossing rate at time t. The density $f_a(t)$ of the first passage time T_a is the solution of

$$\nu(a;t) = f_a(t) + \int_0^t \kappa(t|s) f_a(s)\, ds, \quad t \ge 0, \qquad (1)$$

where $\kappa(t|s)$ denotes the mean a-upcrossing rate of $X(t)$ at time t given that the process exceeds level a at time $s < t$ for the first time (1). To simplify calculations, we approximate $\kappa(t|s)$ by the mean a-upcrossing rate $\tilde{\kappa}(t|s)$ of $X(t)$ at time t given an a-upcrossing at time $s < t$ and calculate it via Slepian models.

3 SLEPIAN MODELS

$X(t)$, $t \ge 0$, be a real-valued stochastic process which has an a-upcrossing at time $s < t$. The Slepian model $S_a(\alpha)$, $\alpha = t - s \ge 0$, describes the evolution of this process following its a-upcrossing at time s. If $X(t)$ is a zero-mean stationary mean square differentiable Gaussian process with correlation function $r(u) = E[X(t)X(t+u)]$, then

$$S_a(\alpha) = a r(\alpha) - \frac{r'(\alpha)}{\lambda_2} Z + \mathcal{K}(\alpha), \quad 0 < s < t, \qquad (2)$$

where $\alpha = t - s$, $\mathcal{K}(\alpha)$ a zero-mean nonstationary Gaussian process and the random slope Z of the

DOI: 10.1201/9781003348450-223

process at the crossing time is independent of $\mathcal{K}(\alpha)$ (3) (Sect. 10.3).

4 NUMERICAL ILLUSTRATIONS

Two numerical examples examined in (4) are presented. The first is a zero-mean, unit-variance, mean square (m.s.) differentiable stationary Gaussian process $X(t)$ with correlation function

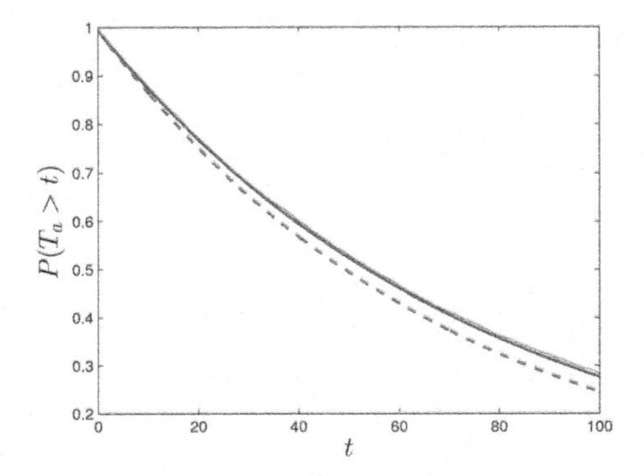

Figure 1. Estimates of $P(T_a > t)$ by the Monte Carlo simulation, integral formulation and mean crossing rate (thin solid, heavy solid and heady dotted lines) for $\lambda = 2$ and $a = 2.5$.

$r(u) = (1 + \lambda|u|) \exp(-\lambda|u|)$, $\lambda > 0$, $u \in \mathbb{R}$, so that it is broad band. The numerical results in Figure 1 are for $\lambda = 2$ and $a = 2.5$. The heavy solid line is an approximation of $P(T_a > t)$ given by the numerical solutions of Equation (1). The thin solid line, which nearly coincides with the heavy solid line, is a Monte Carlo estimate of $P(T_a > t)$ obtained from 10,000 independent samples of $X(t)$. This estimate is viewed as the actual probability $P(T_a > t)$. The dashed line is the approximation of $P(T_a > t)$ based on the mean a-upcrossing rate. The crossing approach provides a reasonable approximation of the first passage time distribution but it is inferior to that by the integral equation.

The second is zero-mean, unit-variance, mean square (m.s.) differentiable narrow band stationary Gaussian process with one-sided spectral density $g(\nu) = 1(\nu_1 < \nu < \nu_2)/(\nu_2 - \nu_1)$, $0 < \nu_1 < \nu_2$. The numerical results in Figure 2 are for $(\nu_1, \nu_2) = (5, 6)$ and $a = 2.5$. The thin solid line is a Monte Carlo estimate of $P(T_a > t)$ obtained by 10,000 samples of $X(t)$. It is viewed as truth. The heavy dashed and solid lines are approximations of $P(T_a > t)$ based on the mean a-upcrossing rate and the integral equation of $\tilde{f}_a(t)$ (Equation 1). The approximations of $P(T_a > t)$ by the solution of the integral equation is superior to that by mean crossings. The difference between these approximations is significant for this narrow band processes.

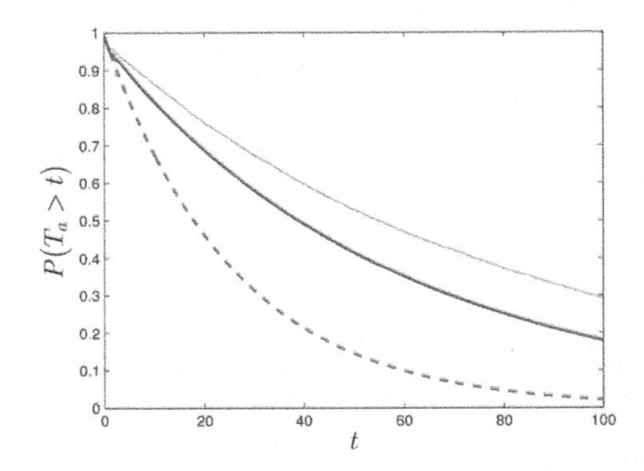

Figure 2. Estimates of $P(T_a > t)$ by the Monte Carlo simulation, integral formulation and mean crossing rate (thin solid, heavy solid and heady dotted lines) for $\nu_1 = 5, \nu_2 = 6$, and $a = 2.5$.

5 CONCLUSIONS

Two numerical examples have been presented to illustrate the construction of Slepian models and the implementation of the integral equations for the distribution of the first passage time T_a. The resulting distributions have been compared with approximations derived from mean crossing rates of $X(t)$ and Monte Carlo simulation for broad- and narrow-based processes. In these examples, the approximations of the distributions of T_a by integral equations are superior to those by crossing rates in the sense that they are closer to Monte Carlo estimates of the distribution of T_a. The difference between these approximations is significant for narrow-band processes.

ACKNOWLEDGEMENTS

The work reported in this paper has been partially supported by the National Science Foundation under the grant CMMI-2013697. This support is gratefully acknowledged.

REFERENCES

[1] M. C. Bernard and J. W. Shipley. The first passage problem for stationary random structural vibration. *Journal of Sound and Vibration*, 24(1):121–132, 1972.

[2] H. Cramer and M. R. Leadbetter. *Stationary and Related Stochastic Processes*. John Wiley & Sons, Inc., New York, 1967.

[3] M. R. Leadbetter, G. Lindgren, and H. Rootzén. *Extremes and Related Properties of Random Sequences and Processes*. Springer-Verlag, New York, 1983.

[4] M. Grigoriu. First passage time for Gaussian processes by Slepian models. *Probabilistic Engineering Mechanics*, 61, 2020.

[5] T. T. Soong and M. Grigoriu. *Random Vibration of Mechanical and Structural Systems*. Prentice Hall, Englewood Cliffs, N.J., 1993.

[6] E. Vanmarcke. On the distribution of the first-passage time for normal stationary random processes. *Journal of Applied Mechanics*, 42:215–220, 1975.

Current Perspectives and New Directions in Mechanics, Modelling and Design of Structural Systems – Zingoni (ed.)
© 2022 Copyright the Author(s), *ISBN 978-1-032-39114-4*

Framing of the design base for load bearing structures against climate change

J.V. Retief
Department of Civil Engineering, Stellenbosch University, South Africa

ABSTRACT: Projected changes to climate extremes due to human activities are regarded as a significant agent of future impact. Due to limited projection skills, changes to climate loads on load bearing structures are to be treated as having deep uncertainty. A framework is provided for incorporating decision making under deep uncertainty into the decision support systems for structural engineering practice to account for changing climate loads.

1 INTRODUCTION

Climate change due to human activities, or anthropogenic climate change (ACC), reached the stage where the observed global warming of 1.1 °C since the pre-industrial era (Table 1) [1] is approaching the critical global warming level (GWL) of 1.5 °C [2]. Attribution of human causes to changes in observed climate extremes is included in the lines of proof of ACC.

Table 1. Main features of human caused climate change.

FEATURE	QUANTITY	COMMENT
Atmospheric CO_2	270 → 410 ppm	Human contribution doubling time 33 y
Climate sensitivity	3[2.5, 4] °C	ΔT per CO_2 doubling
Global warming ΔT (Radiative forcing)	1.1 °C (2.7W/m^2) [1.5, 0.8, -0.5, 0.5] °C	Total (forcing); [GHG, CO_2, SO_2, methane]
Land/Ocean	1.59/0.88 °C	5/91 % of warming
Extremes, regions with changes	Heat [41/0/2] Precipitation [19/0/9]	Numbers – changes [+, -, both]
Frequency/intensity (@ 1.5 °C)	Doubling @ 2 °C, Quadrupling @ 3 °C	High impact events increase 1.5 °C → 2 ° C

The extension of projection skills to climate extremes provides a measure for considering projected changes to climate actions on load bearing structures. Adaptation of the design base for structures to account for the impact of ACC are required, due to the vital socioeconomic role of such structures over a service life that straddles projected changes.

In this survey, the projection skills for climate extremes assessed in [1] are used to rate trends and confidence levels for wind load on structures under South African conditions as an example case. The strategy for accounting for ACC conditions in design conditions proposed in [3] is then extended to a framework for developing the adaptation of the design base for structures across all levels of decision support.

2 CLIMATE EXTREMES AND ACTIONS

Recent inclusion of extreme wind as a class of climate extremes, as related to associated windstorm classes, considered in ACC assessments [1], provides an indication of confidence levels in projecting future wind load conditions.

The overall conclusion of *low confidence* in the observation of changes to climate impact drivers, or of projecting its emergence by the end of the century, derives from limitations of short-term observation of wind extremes and the resolution of simulation models; in addition to uncertainties due to regional downscaling, local influences, resolving parent windstorms such as extratropical and severe convective storms [1].

As a result, wind loads are characterized here to degenerate to *deep uncertainty* for which it is not possible to be treated probabilistically, with an interim stage where *ambiguous uncertainty* is applied to the distribution parameters for historic extreme value probabilistic models for wind loads.

3 DECISIONS UNDER UNCERTAINTY

Uncertainty classification from [4] is adapted as reference scheme to relate future extreme wind load characteristics (uncertainties) to that for design variables for the structural design base, to determine the appropriate management approach (Table 2).

The hierarchy of decision support for the development of the design base for structures can be aligned to the levels of uncertainty and decision approaches (Table 3), integrating the current risk- and reliability-based semi-probabilistic design approach and provision for ACC extreme actions.

4 FRAMING OF ACC DESIGN BASE

The decision support hierarchy shown in Table 3 can be used to launch the immediate implementation of ACC related operational design, supported by guidance and

DOI: 10.1201/9781003348450-224

Table 2. Decision approach for outcome uncertainty level.

UNCERTAINTY LEVEL	DECISION APPROACH
1 NOMINAL Well defined, historical base	Point estimate, sensitivity analysis
2 PROBABILISTIC Outcome	Expected outcome < Acceptable risk
3 SCENARIOS Plausible outcomes, unknown likelihoods	Most favorable policy over all plausible (not predicted) outcomes
4a DEEP Futures many, but bounded	Robust, flexible, preplanning, learning
4b DEEP Futures (Known to be) Unknown	Optimize robustness to *failure* against {known-required} information gap

Table 3. Hierarchy of decision support for design base development relating decision maker to uncertainty level.

LIMIT STATES DESIGN – Structural Engineer
Design performance, optimization using *pre-calibrated reliability* parameters
RISK & RELIABILITY FRAMEWORK – Standards Committee
Acceptable system risk levels and categories, limit states; reliability calibration using *basic variable probability models*
OPTIMAL POLICIES – International Standardization
Guidance on *best practices* for all levels of risk & reliability management
ADVANCEMENT – International Professional Bodies
Unbounded exploration of horizons beyond best practices, *integration* of diverse information sources and extension of *decision systems*

Table 4. Decision support framework for development and implementation of design base for climate extremes.

LIMIT STATES DESIGN – Structural Engineer

Optimize robustness of structure and all subsystems against extreme wind

- Specify strategic function of structure against ACC, including service life
- Stress test against accidental climate extreme wind load, using *pre-calibrated reliability* parameters
- Optimize wind load robustness, utilizing reserve capacities for persistent design situation and assessment approaches for existing structures
- Plan reassessment campaigns over service life, using updated information

RISK & RELIABILITY FRAMEWORK – Standards Committee

Provide guidance to designers on treating of deep uncertainties in wind load

- Include *Design Situation for ACC wind load (Informative)* in standard
- Identify (sub) classes of structures, service life to account for ACC
- Guidance on wind load reserves, using risk & reliability approaches
- Reassess current wind load specifications for refinement
- Reassess design base for acceptable risk levels, categories, limit states, design situation, calibration and probability models

OPTIMAL POLICIES – International Standardization

Guidance on *best practices* for all levels of risk & reliability management

- Extend generic provisions for design base for climate actions to ACC
- Consider all implications of *deep uncertainty* and associated loss of historic and experience base
- Reassess requirements and procedures for climate and general robustness

ADVANCEMENT – International Professional Bodies

(Continued)

Table 4. (Cont.)

Unbounded exploration of horizons beyond best practices, *integration* of diverse information sources and extension of *decision system*

- Advance technical depth of all aspects of decision support for climate actions, trends and (deep) uncertainties
- Keep track of ACC information advances relevant to structures
- Integrate structural practice with issues related to risk and reliability, robustness, resilience, sustainability

development of background information, towards a design base development strategy. The framework for such a broad-based implementation and development program is illustrated by the issues to be considered at each level listed in Table 4.

At all levels of implementation, the decision approach for treating deep uncertainty should be accounted for, the information gap between historic and projected conditions, by providing for maximum flexibility, monitoring, and reassessment, as climate change evolves [4].

5 CONCLUSIONS

Changes to the climate due to human activities have evolved to levels that can be observed at scales from global warming to human influences on individual weather extremes. Climate simulation models, verified by comparison to observations, are used to project future conditions, so as to inform decision-making on mitigation policies and adaptation measures to manage the impact of ACC.

Climate actions on load bearing structures play a significant role in the performance of these structures over a service life related to periods over which significant changes would occur. Assessment of projection skills for climate extremes indicate that actions on structures will deteriorate towards deep uncertainty over the service life for designs in accordance with current practice.

A framework of decision support for integrating the influence of climate change into the design base for structures is proposed, where decision-making can be employed effectively by stakeholders at the respective levels of appropriate uncertainty representation.

REFERENCES

[1] Arias, P. A., N. Bellouin, E. Coppola, et al., 2021. Technical Summary. In: *Climate Change 2021: The Physical Science Basis*. Cambridge University Press. In Press.

[2] IPCC, 2018: Global warming of 1.5°C. An IPCC Special Report. Summary for Policymakers. World Meteorological Organization, Geneva, Switzerland, 32 pp.

[3] Retief JV, Viljoen C, 2021. Provisions for climate change in structural design standards. In Stewart and Rosowsky (Eds.), Engineering for Extremes: Decision-making in an uncertain world. Springer.

[4] Marchau VAWJ, Walker WE, Bloemen PJTM, Popper SW, 2019. Decision Making under deep uncertainty. Springer. https://doi.org/10.1007/978-3-030-05252-2

Current Perspectives and New Directions in Mechanics, Modelling and Design of Structural Systems – Zingoni (ed.)
© 2022 Copyright the Author(s), *ISBN 978-1-032-39114-4*

Mechanical properties of stainless steel coatings formed by build-up spraying

E. Horisawa, K. Sugiura, Y. Kitane & Y. Goi
Graduate School of Engineering, Kyoto University, Kyoto, Japan

T. Tanimoto
Deros Japan Co., Ltd., Ishikawa, Japan

M. Matsumura
CWMD, Kumamoto University, Kumamoto, Japan

ABSTRACT: Thermal spraying is a technology that effectively forms a coating on the surface of a material by spraying molten metal particles. The authors have been studying the use of thick sprayed coatings for repair and reinforcement of steel structures. The purpose of this study is to experimentally understand the mechanical properties of the sprayed coating as basic research on its application to construction. As a result of conventional tensile and compressive tests, it was found that the tensile strength of the sprayed metal was much smaller than that of the base metal. On the other hand, under compressive stress, the strength was equal to or greater than that of the base metal. These results indicate that the coating formed by build-up spraying can be expected to be a load sharing material under compressive stress.

1 INTRODUCTION

Thermal spraying is a technology that effectively forms a coating on the surface of materials and structures by spraying molten metal particles (Ueno 2021). By using a technique called build-up thermal spraying, it is possible to recover from thinning of the base material by forming a coating of up to 5 mm in thickness. Despite the fact that such coatings can be formed, few studies have examined the mechanical properties of thermal sprayed coatings.

The purpose of this study is to investigate the mechanical properties of the coatings formed on stainless steel substrate by build-up spraying. Tensile and compression tests were carried out to clarify the strength characteristics of the sprayed coating.

2 EXPERIMENTAL METHODS

In this study, stainless steel SUS304 and SUS420J2 were used as base material and sprayed metal, respectively. The shape of tensile specimen was determined to be 75 mm in parallel length and 12.5 mm in width. The thickness of the specimen was 6 mm, and the one of the sprayed metal specimen was about 5 mm, which is the maximum thickness of a typical build-up sprayed coating. On the other hand, a steel rod of 25 mm in diameter was used as the compression specimen. The sprayed metal specimens were fabricated by electric arc spraying on SUS304 steel plate. In this process, the surface of the base material was first roughened by grid blasting. Then, about 0.5 mm of Ni-Al alloy was sprayed on the substrate to strengthen the adhesion between the coating and the substrate. The number of test specimens was five for each type of specimen.

Both tensile and compressive tests were carried out using a universal testing machine with a maximum load of 1000 kN. The load and strain were measured by a load cell, and strain gauges attached to the specimen, respectively. The load was manually controlled, and the loading speed was kept below about 10^{-4}/sec. In the tensile test, the load was applied until fracture. In the compression test, the load was applied continuously until the load decreased or the axial strain exceeded 10%.

3 TEST RESULTS

Figures 1 & 2 illustrate examples of stress-strain relationships obtained from tensile and compression tests. The sprayed metal showed an almost linear relationship with a small gradient until failure in the tensile test. On the other hand, the stress-strain relationship in compression of the specimen with the sprayed coating showed nonlinearity from small stress. In addition, the stress of this specimen became larger than that of SUS304, and the load decreased rapidly due to cracking of the coating.

The mechanical properties obtained from the stress-strain curves are shown in Table 1. Where E is Young's modulus, $\sigma_{0.2}$ is 0.2% offset proof stress, σ_t is maximum stress, and ν is the Poisson's ratio. The Young's modulus and tensile strength of the sprayed metal are 70 GPa and 113 MPa, respectively. The Poisson's ratio of the sprayed metal is 0.142. These

DOI: 10.1201/9781003348450-225

properties are attributed to the porous structure of the sprayed coating with micropores inside and the weak bonding between metal particles. Furthermore, the Young's modulus of the specimen with the sprayed coating is 123 GPa in compression. The Poisson's ratio of the specimen is 0.167, which was close to that of the sprayed metal in tension.

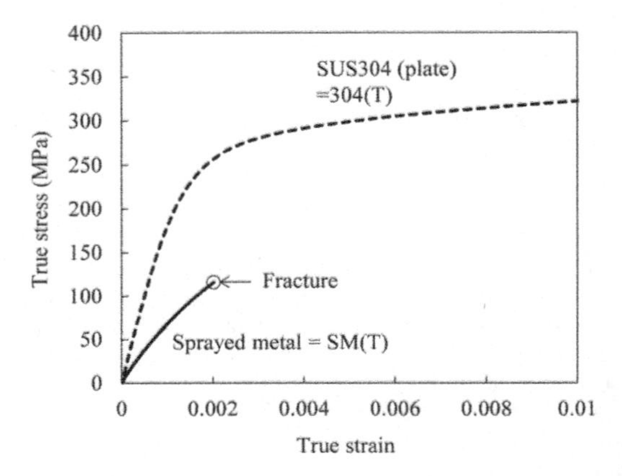

Figure 1. True stress-true strain relationships of tensile specimens up to 1% strain.

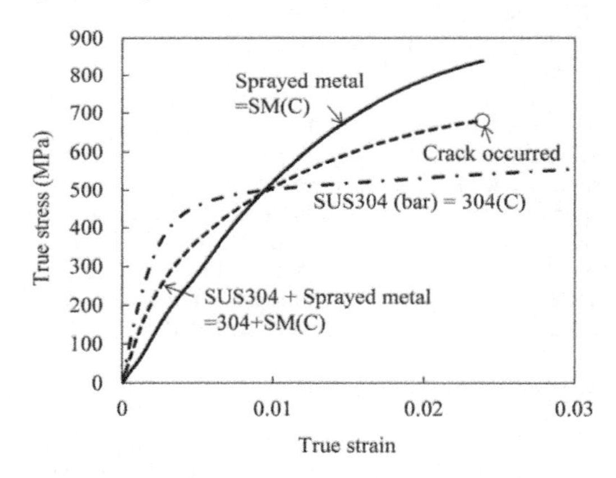

Figure 2. True stress-true strain relationships of compression specimen up to 3% strain.

Figures 3 & 4 show photographs of the specimens after fracture. The sprayed metal tensile specimen showed no elongation or reduction of area after brittle fracture. On the other hand, for the compression specimen, cracks extending in the load axis direction are observed as shown in Figure 4. This suggests that tensile stress in the circumferential direction was generated in the coating when compressive stress was applied.

Using the test results, an attempt was made to derive the stress-strain relationship of the sprayed coating under compression. First, the stress-strain relationship of the stainless steel bars was modeled using equations (CEN 2006). Then, the relationship of sprayed metal was calculated using compound law. An example of the results is shown in Figure 2. It can be confirmed that when the stress of the sprayed metal exceeds 500 MPa, it exceeds the stress of SUS304. In addition, the Young's modulus of the sprayed metal in compression is 65 GPa, which is not much different from that in tension. Moreover, 0.2% proof stress and maximum stress of the coating are larger than those of SUS304.

Table 1. Mechanical properties of sprayed metal under compressive stress.

	E	$\sigma_{0.2}$	σ_u	ν	δ
	GPa	MPa	MPa		%
304(T)	193	286	666	0.267	66.3
SM(T)	70	-	113	0.142	0.3
304(C)	195	434	-	0.256	-
304+SM(C)	123	387	691	0.167	-
SM(C)	65	587	842	-	-

Figure 3. Tensile specimens after fracture.

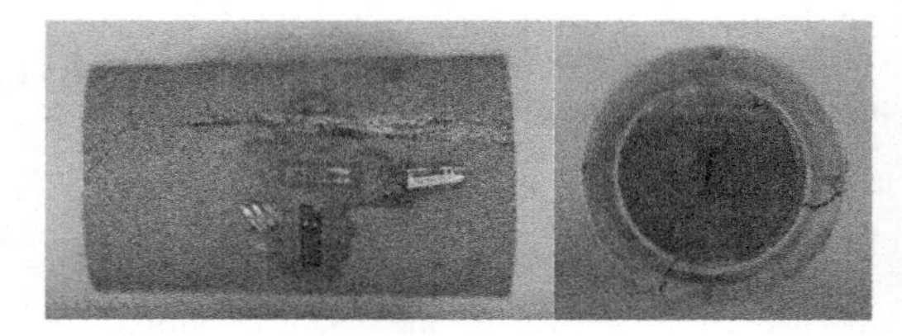

Figure 4. Compression specimen of SUS304 bar with sprayed metal after the test.

4 CONCLUSIONS

In this study, the mechanical properties of the sprayed coating formed by build-up spraying were experimentally investigated. From these results, it can be said that the metal spraying with a thickness of about 5 mm may be applicable to the repair of parts subjected to compression. In the future, the adhesion strength between the build-up sprayed coating and the base metal and the fatigue strength of the coating are going to be investigated.

REFERENCES

European Committee for Standardization 2006. *Eurocode 3 -Design of steel structures - Part 1-4: General rules - Supplementary rules for stainless steels*, EN 1993- 1-4.

Ueno, K. 2021. Historical and prospective aspects of thermally sprayed coatings, *Journal of the society of Materials Science* 70(1), 64–69.

Current Perspectives and New Directions in Mechanics, Modelling and Design of Structural Systems – Zingoni (ed.)
© 2022 Copyright the Author(s), *ISBN 978-1-032-39114-4*

Implication of constructing the new Umhlatuzana River bridge deck monolithically with the existing deck

A.J. Faure
SMEC South Africa (Pty) Ltd, Cape Town, South Africa

ABSTRACT: The paper investigates joining the existing prestressed Umhlatuzana River bridge deck to a new deck monolithically. The new deck will shorten at a higher rate when compared to the existing deck. This is due to differential creep and shrinkage between the new and old decks which will cause restrained differential movement between the two decks and additional stresses.

1 INTRODUCTION

The existing Umhaltuzana river bridges were constructed in 1985 and consist of two 413 m long prestressed box girder decks of 15.1 m width. Figure 1 indicates the final cross-sectional layout which includes the widening.

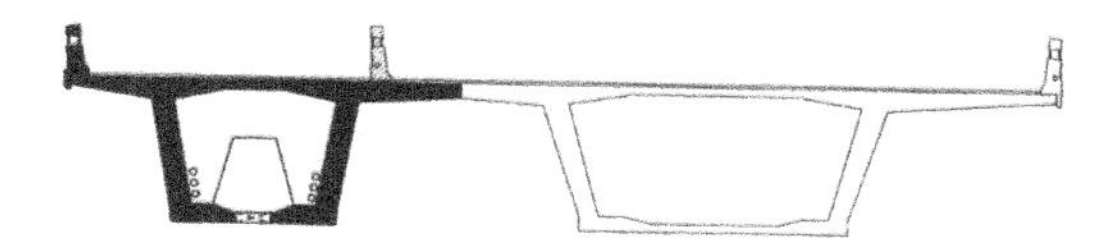

Figure 1. Final cross sections.

The widenings are intended to be constructed between 2023 and 2026 and will be about 40 years younger than the original bridge. Most of the creep and shrinkage of the existing bridges has therefore already occurred. The new bridges have yet to be constructed and creep and shrinkage will be limited to the construction phase. This differential creep and shrinkage will cause the two bridge decks to shorten at different rates. This is problematic when trying to join the two decks together. This paper investigates the possibility of connecting the two decks via a cast in situ concrete slab.

It was expected that the new deck, if unrestrained, would shorten significantly more than the existing deck. The slab would restrain this shortening, developing tensile stresses. The concern was that these tensile stresses would lead to unacceptable cracking and durability issues in the long term. The design process involved calculating the relative shortening from creep and shrinkage and then applying this movement to a FEM model. The stresses in the decks and slab could then be evaluated.

2 ANALYSIS AND DESIGN

Creep and shrinkage curves were calculated for both the new and existing bridges in accordance with the CEB-FIP (1990) Model code. The applied loading was obtained by first calculating the differential strain between the two decks due to creep and shrinkage.

Four months are allowed after the new deck was complete to allow the maximum amount of creep and shrinkage to take place. A total of 250 days was allowed after the average construction of a deck segment and prior to connecting the decks together. Table 1. indicates the equivalent differential creep and shrinkage between the two decks.

Table 1. Differential creep and shrinkage values for Umhlatuzana River bridge widening

Time (days)	Creep factor (Φ)	ΔΦ	Shrinkage strain (με)	Δμε
250	2.77	0.93	-25	-159
36500	3.70		-184	

The creep strain was determined using the following equation: $\varepsilon = \frac{\sigma}{E} x \Delta\Phi$

The resultant creep strain was calculated as 150 με (0.93 x 5 MPa/31 GPa) and the shrinkage strain as 159 με. The resultant effective creep and shrinkage strain from 250 days to 36500 days was therefore 309 με.

A Robot FEM model was set up with girders modelled as beams and with a slab joining them via rigid links. Robot does not have a feature to model the effect of creep and shrinkage. To overcome this shortcoming, an equivalent temperature load was applied.

DOI: 10.1201/9781003348450-226

3 RESULTS AND DISCUSSION

3.1 *Deck movement and forces*

The expected movement of the unrestrained deck due to creep and shrinkage based on an equivalent temperature load is 64 mm. The model shows that the movement in the new deck is 31 mm and the existing deck is 28 mm, a difference of 3 mm. It makes sense that the new deck would shorten more than the existing and its of interest that the existing decks shortening is almost the same as the new. Another point of interest is that if the movements of both decks are added together then the total movement is similar to the unrestrained movement from creep and shrinkage.

The movement indicates that the new deck is restrained by the existing deck. Effectively the new deck induces compression into the existing deck and slab. The compression in the existing deck in turn restrains the new deck, resulting in the new deck undergoing tension. It was therefore important that this effect was considered in the deck design model.

Compression is generally favourable and the additional compression in the existing deck was therefore not considered problematic. The new deck however needed to account for the reduction in axial load. Furthermore, an added benefit from joining the two decks monolithically and the reduced movement was that smaller expansion joints could be used.

3.2 *Deck stresses*

For most of the deck, excluding the ends, the compressive stress in the longitudinal direction is 3.65 MPa with a negligible stress in the transverse direction. This compressive stress is not considered excessive. The deck can accommodate a stress of 0.4 fcu which is a peak stress of 16 MPa. The stresses caused by differential creep and shrinkage are therefore well within the capacity of the slab.

In the last 15 m of the deck, the longitudinal compressive stress increased and peaked at 10 MPa which is within the required limit. The tensile stress' are however above the limit with tensile stresses up to 7 MPa.

To decrease the high tensile stresses, the thickness of the connecting slab is increased, over a length of 15 meters, to a depth of 1 m. The increased depth results in a larger area over which the load is applied, which results in the stress magnitudes decreasing. The deck is also prestressed transversely so that a 5 MPa stress is added to the section. Figure 2 shows the principle stress 1 after prestressing and thickening of the slab.

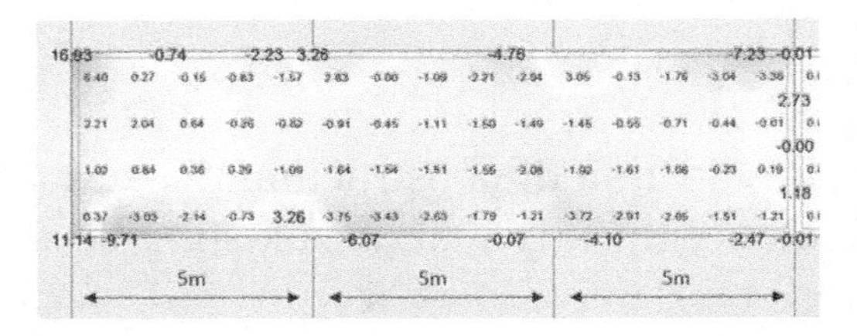

Figure 2. Principle stress 1.

Figure 2 indicates a decrease in tensile stress magnitude throughout the 15m long thickened slab. Most of the thickened slab is in compression with stress values ranging between 0.13 and 3.75 MPa. However, the first 5 m of the slab remains in tension, but the magnitudes of the stresses are below the tensile strength limit of 2.9 MPa.

The compressive stresses have increased in magnitude to values ranging between 3.00 to 5.50 MPa which are within acceptable limits.

4 CONCLUSIONS AND RECOMMENDATIONS

This paper investigated connecting a new box girder deck of 414 m to an existing one. The following key conclusions are made:

- Differential creep and shrinkage plays a major role as the two decks shorten at different rates when the stitch slab is cast;
- Delaying the connection of the decks once the prestressing is installed reduces the differential movement substantially;
- The overall movement of the decks reduced which allowed for smaller expansion joints;
- The analysis indicated that most of the stitch slab is in compression (3-4 MPa) which is easily accommodated by the deck concrete as concrete is strong in compression; and
- High tensile stresses (15 MPa) were evident at the deck ends leading to impractical amounts of reinforcement;
- The previous noted issue was resolved by increasing the thickness of the slab at the deck ends to 1 m depth and prestressing transversely;
- This action reduced the tensile stresses in the stitch slab to below the tensile cracking limit (2.7 MPa) of the concrete. The compressive stress increased but was also within the compressive limit (0.4 fcu) of the concrete.

In summary, the analysis showed that connecting two 413 m decks monolithically is possible.

Current Perspectives and New Directions in Mechanics, Modelling and Design of Structural Systems – Zingoni (ed.)
© 2022 Copyright the Author(s), ISBN 978-1-032-39114-4

Field and desktop survey on failures of rail to concrete connections in South African track on concrete ballastless railway systems

W.M.P. Makwela & J. Mahachi
University of Johannesburg, Auckland Park Campus, South Africa

ABSTRACT: Ballastless track is a railway system that uses concrete instead of ballast and, various systems are used across South Africa. A field survey along with FEA is conducted for the rail to concrete connection. All sites demonstrate that concrete and rails were in good condition, and only the connections failed. The grout pad fails first, and the presence or absence of elastic pads contributes to grout failure. American code specifies 28 MPa grout pad, and most failed grout pads had a strength averaging 40 MPa. Properties affecting durability of grout pad include thickness, strength, and stiffness. Grout pads reduce compressive stresses in concrete. 1:20 inclined rails result in evenly distributed stresses in the connection. Vertical rails distribute loads unevenly. Grout failures are not due to static loads, Dynamic factor of 2 or 4 results in fatigue or instant failure, respectively. A 70 MPa grout with 40-60mm thickness is recommended.

1 INTRODUCTION

Railway track technology is 150 years old. Mainly ballasted track system was used in this period. Over the years, research conducted found that most earlier materials used under the sleepers except ballast were not suitable for modern railway. South Africa has standards for ballasted tracks . In the last 30 years, a new system referred to as ballastless track introduced. The system has shown superior performance, increasing its applications. The two types of ballastless tracks are namely track-on-concrete and track-in-concrete systems. Damage in track-on-concrete connections has been widely reported in experimental studies (Rahmanishamsi, Soroushian, & Maragakis, 2016). In most cases, the grout pad is the first component of the connection system to show failure. The Field survey in this study focussed on the grout pad, which forms part of the widely used direct fixation fastening system in the track-on-concrete rail to concrete connection. The grout pad allows vertical and lateral adjustment of connecting rails to concrete slab/beam, transfers load from rails to concrete by providing a flat continuous support surface, creates adequate height to drain water, and clean particles between the rails. This field survey was conducted to record some of the different types of track-on-concrete systems used in South Africa and access the track-on-concrete system's application, condition, and performance focusing on the grout pad.

2 METHODOLOGY

A field survey across 11 South African rail sidings was conducted in two stages. Firstly a visual information gathering for recording the configuration, performance and application of the system. Secondly a physical tests like rebound hammer along with group pad dimension measurements.

Finite Element Analysis (FEA) is conducted by simulating a matrix of rail to concrete connections with different components, subjected to 20 ton train axle load . The matrix is derived from varying grout pad thickness, 0 and 1:20 vertically inclined rail, Slotted and non-slotted bolt holes, HDPE pad between rail and steel chair present or absent, and varying dynamic factor.

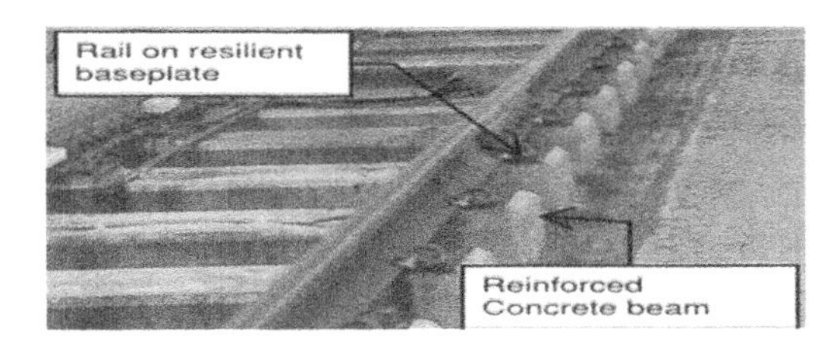

Figure 1. Typical track on concrete setup.

3 FINDINGS

Listed in Table 1 are the 11 visited sidings.

All sidings visited use a cast at grade slab system track on concrete with a direct fixation fastener

DOI: 10.1201/9781003348450-227

Table 1. Case studies.

Case study	Siding
1	Khumani mine stackers and reclaimers line
2	Khumani mine train railway load-out station line
3	Khumani mine train railway weighbridge line
4	Black rock mine train railway load-out station line
5&6	Pretoria & Mamelodi Metrorail commuter station
7	Traxtion traverser pit crane guiding line
8	Glencore mine locomotive inspection pit rail line
9	Minaar train railway load-out station line
10	Sandton Gautrain train railway tunnel line
11	Transnet train railway tunnel line

connection for rail to concrete, except for Sandton station which uses German RHEDA system, and Pretoria station which uses the British PY slab track. The PY slab track was discontinued in Britain due to design fatal flows in the connection. Grout pads used in the rail to concrete connection are 0-100mm thick, few grouts where cracked, and majority are cracked. Two sites with rails directly placed on the concrete experienced concrete local failure.

A 20-ton axle load is mostly used, and most track on concrete systems are less than 10 years. Grout pad is present mostly in direct fixation rail to concrete connections. Most siding lack the 1:20 rail inclination. 5 sidings show that grout pads with strength greater than 55 MPa perform good, cracked ones have strength of 29-58 MPa, and crushed pads have strenght less than 21 MPa.

FEA results show that a thicker grout pad reduces stresses in the concrete slab, making grout a transition layer. 1:20 inclined rails distribute the stresses evenly on both sides of the connection, and vertical rails result in more stresses on the outer side of the connection. The presence of an HDPE pad reduces stresses in underlaying components. An increase in load directly increases the stress in the grout and concrete and can be used on a linear graph to determine which dynamic factor will results in a failing stress in the concrete and grout. A dynamic factor of 4.1 to cause instant damage to a concrete beam or slab with a 25 MPa compressive strength. Case 4 also shows that a repetitive dynamic factor of 2 is required for stresses in concrete to be above 50% of 25 MPa design strength and fatigue to take place

Marais & Perrie (1993) state that fatigue occurs when the applied stress over design stress ratio using greater than 0.5. Using the above findings, calculations can be done to predict resultant stress ratios; hence, designers can limit stresses to be less than 50% for grout and concrete for an expected dynamic factor or load. The presence of a grout pad assists in reducing the stress levels in concrete to be less than 50% of the design stress. Hence a 40 to 60mm grout pad with a minimum of 70 MPa compressive strength is recommended. An elastic pad reduces the stress levels in underlaying components but increases stress in rail, as the rail is subjected to more deflection. The use of circular holes in the steel base plate/chair is found to cause local compressive stresses in grout surrounding the bolt and high pressure in bolt and nut.

4 CONCLUSIONS

Different sites consist of varying configurations of the track-on-concrete system. From field observation confirm that the main weak point of the system is the connection between the rail and the concrete. Primarily the grout pad or concrete fails first. An elastomeric membrane in the connection is needed to reduce the rigidity, vibration, and impact from dynamic loads. The grout pad is for height adjustment and leveling off the rails. The performance of the grout pad can be affected by both extrinsic and intrinsic factors. Measures must be in place for a proper quality control system to construct grout pads, as there is no code of practice to force standards. The elastic pads in the connection reduces stress in the underlying components. The presence of a grout pad acts as a transition layer and reduces high stress in concrete, which can cause failure due to both fatigue and dynamic loading. A dynamic factor of 2 mainly results in fatigue failure and a dynamic factor of 4 results in immediate impact failure. Under normal and perfectly constructed connections, static loading does not cause the failure of the track-on-concrete rail to concrete connection components.

REFERENCES

American Railway Engineering and Maintenance-of-Way Association. (2010). Structures. In *Manual for Railway Engineering* (pp. 8–25). Lanham: American Railway Engineering and Maintenance-of-Way Association.

Marais, L. R., & Perrie, B. D. (1993). *Concrete industrial floors on the ground* (5 ed.). Midrand, Gauteng, South Africa: Portland Cement Institution.

Rahmanishamsi, E., Soroushian, S., & Maragakis, E. M. (2016). Capacity Evaluation of Typical Track-to-Concrete Power-Actuated Fastener Connections in Nonstructural Walls. *ASCE Journal of Structural Engineering, 142* (6). https:// ascelibrary.org/doi/10.1061/(asce)st.1943-541x.0001472

Current Perspectives and New Directions in Mechanics, Modelling and Design of Structural Systems – Zingoni (ed.)
© 2022 Copyright the Author(s), ISBN 978-1-032-39114-4

Long term energy efficiency of non-conventional building systems: Use of Polyblocks in improving thermal performance

D.M. Tshilombo, S.A Alabi & J. Mahachi
Department of Civil Engineering, Faculty of Engineering and Built Environment, University of Johannesburg, Johannesburg, South Africa

ABSTRACT: Buildings' energy efficiency requirements in South Africa are depicted in SANS 204 and SANS 10 400 XA. The thermal performance of buildings is usually assessed by determining the thermal resistance of the building envelope. Although masonry walls have been used for decades, their thermal performance is questionable because they do not meet the minimum requirements. This study investigates the thermal performance of the Polyblock by applying Fourier's one dimensional heat equation to calculate the thermal conductivity of the structure, then establish a connection between the heat and the energy requirements. Results indicated that the thermal resistance of the Polyblock was higher than of the masonry wall, therefore suggesting it offers better thermal comfort.

Keywords: Thermal conductivity, thermal resistance, heat transfer, sustainability, heat flow, energy loss

1 INTRODUCTION

The demand for buildings and infrastructures has grown exponentially over the years, mostly because of the population growth and socio-economic development. Globally, the construction industry accounts for more than 40% of energy usage (International Energy Agency, 2013). In South Africa, fossil fuels is burned to generate energy in form of electricity, and this process releases air-bone pollutants and causes climate change.

The thermal resistance is a property of an element to resist heat flow. Therefore, a building envelope that has a high thermal resistance will be energy efficient.

SANS 204 and SANS 10 400 XA provides effective energy efficient measures that take passive design into account. However, the thermal performance of masonry walls is questionable and often neglected.

2 LITERATURE REVIEW

For decades the construction industry has been dominated by conventional construction materials such as concrete, mortar and bricks. However, a lot of carbon diaxide (CO_2) is release while producting these materials. In addition, construction with conventional materials is usually energy-intensive, and a lot of material is wasted during the construction phase. The construction industry is mostly focused on reducing the project cost when purchasing materials, while the thermal performance of the building is of little interest. But in reality, good thermal performant buildings save more money on the long term. Spisakova and Kozlovska (2013), stated that non-conventional construction would help reduce construction duration and would surely reduce on the CO_2 emissions.

The Polyblock is made by mixing polyester resin with sand and hardeners to create polymer concrete. This is then be molded into a "LEGO-like" brick with polystyrene inside and ready for construction after 24 hours. The Polyblock bricks are then stacked up with tie-rods running inside from bottom to the top of the walls during construction (Mathijsen, 2016).

3 RESEARCH METHODOLOGY

The research commences by determining the average thermal conductivity of Polyblock samples using a FOX 314 heat flow meter test. Then, a classroom built with the Polyblock was heated from the inside in winter to create a significate difference in temperature between the inside and outside. The heat transfer from the inside to outside means that heat could be calculated as a one dimensional equation.

DOI: 10.1201/9781003348450-228

$$\frac{Q}{t} = \frac{kA(T_1 - T_2)}{L} \quad (1)$$

where Q is conductive heat transfer; t is time; k is thermal conductivity; A is area; $(T_1 - T_2)$ = external and internal temperature difference; L is thickness.

4 FINDINGS AND DISCUSION

The fox 314 heat flow meter test helped determine the thermal conductivity of 10 different Polyblock samples. The thermal conductivity of those samples was fairly constant and averaging 0.07645 (W/ m. K).

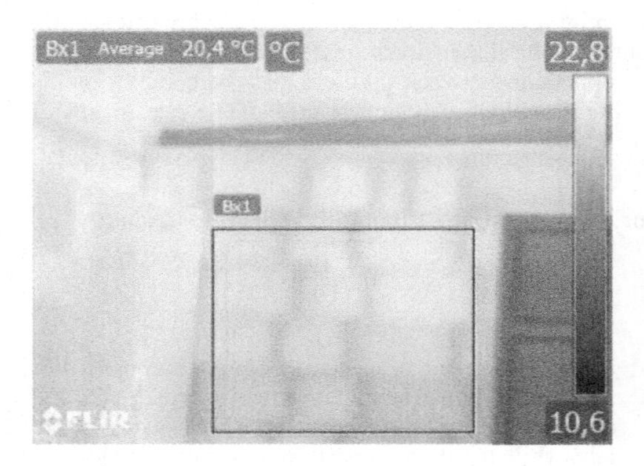

Figure 1. Thermal image from inside of the building.

The whole building had a total area of 184.4 m^2 while the Polyblock walls is estimated to be 38.1% of the total, and of wall thickness of 0.2 meters. Figure 1 provides evidence that the classroom help the heat well with an indoor temperature of 20.4°C and the outdoor temperature of 8.8 °C. Results also showed that the building's total thermal conductivity was 0.187 (W/ m. K).

By applying Fourier's equation to calculate the energy loss, we have:

Q/t = (0.187 x 184.4 x (293.55-281.95))/0.2

Q/t = 2000.002 Watts

The power loss of the building would then be 2000.002 Watts. However, if we calculate the energy required to reach the same temperature without the Polyblock walls, we then have 2 732.709 Watts for the whole building.

This can then be used to illustrate the energy efficiency that the Polyblock over a period of time compared to masonry walls.

It can be seen from Table 1 that the Polyblock offers better thermal performance. Assuming that the municipality electricity rate does not inflate for 10 years, a single building could save more than to R 104 216.3 or 6 950 $.

Table 1. Polyblock vs Conventional masonry wall.

	Polyblock wall	Masonry wall
Area	70.28 m^2	70.28 m^2
Energy used	311.6 W	2 488.6 W
Rate	R 1.64	R 1.64
8 hour daily usage	2.5 kW	19.91 kW
Cost per day	R 4.1	R 32.65
Usage over 1 year	912.5 kW	7 267.15 kW
Cost over 1 year	R 1 496.5	R 11 918.13

5 CONCLUSION

This research focused on thermal performance of the Polyblock compared to conventional construction. The thermal resistance of a masonry wall should not be less than 0.35 (K. m^2/W) in South Africa based on SANS 204 and SANS 10 400 XA. However, the thermal resistance of most masonry walls is 0.2911 (K. m^2/W) or less. The difference is quite large, and unnecessary artificial means are used to achieve thermal comfort by the end-user. Even though the awareness for passive design has increased, the current construction industry has little cognizance for the long term environmental impacts that construction materials extraction have for the planet and the future generation. The schools and housing deficit has worsen over the years, and the South African population rely on the government to provide adequate infrastructure in a short time. The use of the Polyblock in developing countries or in the South African construction industry would help reduce the net energy consumption, reduce on the materials wastage in the sector but also help reduce the housing backlog that Africa is facing in a short period.

REFERENCES

Mathijsen, D. 2016. LEGO-like sand reinforced polyester bricks are set to revolutionize the building world. *Reinforced Plastics*. 60(6): 362–368.

SANS 10400-XA. 2011. *The application of the National Building Regulations Part X: Environmental sustainability. Part XA: Energy usage in buildings*. Pretoria. SABS Standard division.

SANS 204. 2011. *Energy efficiency in buildings*. Pretoria: SABS Standard Division.

Spisakova, M. & Kozlovska, M. 2013. Lean production as an innovative approach to construction. *Journal of Civil Engineering*. 8(1): 87–96

Current Perspectives and New Directions in Mechanics, Modelling and Design of Structural Systems – Zingoni (ed.)
© 2022 Copyright the Author(s), ISBN 978-1-032-39114-4

Cooling materials that help save lives in the context of Covid-19's economic recession

F. Pacheco Torgal
C-TAC Research Centre, University of Minho, Portugal

ABSTRACT: This paper reviews studies concerning the social impacts of heat waves, considered to be the most important and dangerous hazard related to the current climate emergency, combined with the effects of Urban Heat Island. The influence of cooling materials to reduce heat related mortality especially in the context of Covid-19 economic recession is highlighted.

1 INTRODUCTION

According to Watts (2018) in February of last year the temperatures in the Arctic remained 20 °C above the average for longer than a week having increased the melting rate. As a consequence the replacement of ice by water will lead to a higher absorption of solar radiation that makes oceans warmer being responsible for basal ice melting and also for a warmer atmosphere (Ivanov et al., 2016). This constitutes a positive feedback that aggravates the aforementioned problem. Vicious cycles of drought leading to fire leading to more drought constitute another positive feedback further aggravating carbon dioxide emissions and global warming.

2 A DRAMATIC OVERVIEW ON CLIMATE EMERGENCY

The United Nations estimates that by 2030, 700 million people will be forced to leave their homes because of drought (Padma, 2019). Drought and heat waves associated to this climate emergency are responsible for damaging crop yields, deepening farmers' debt burdens and inducing some to commit suicide. Even the discreet and circumspect Joachim Schellnhuber professor of theoretical physics expert in complex systems and nonlinearity, founding director of the Potsdam Institute for Climate Impact Research (1992-2018) and former chair of the German Advisory Council on Global Change lost his frugality views when wrote the foreword of the paper by Spratt and Dunlop (2018) in which he wrote: *"climate change is now reaching the end-game, where very soon humanity must choose between taking unprecedented action, or accepting that it has been left too late and bear the consequences"*. In July of 2018 Professor Bendell authored a dramatic piece warning about the probable social collapse articulating the perspective that it is now too late to stop a future collapse of our societies because of the current climate emergency, and that we must now explore ways in which to reduce harm. He called for a "deep adaptation agenda" that would encompass: "withdrawing from coastlines, shutting down vulnerable industrial facilities, or giving up expectations for certain types of consumption (Bendell, 2018). Some say Bendell has gone too far in his pessimistic views but a Professor of Physics at the University of Oxford wrote in a paper published in Augusts of 2019 the following: *"Let's get this on the table right away, without mincing words. With regard to the climate crisis, yes, it's time to panic"* (Pierrehumbert, R., 2019).

3 HEAT RELATED MORTALITY AND COOLING MATERIALS

According to the IPCC heat waves are the most important and dangerous hazard related to the current climate emergency. Kew et al. (2019) reported that anthropogenic climate change has increased the odds of heat waves at least threefold since 1950 and across the Euro-Mediterranean the likelihood of a heat wave at least as hot as summer 2017 (responsible for temperatures above 40°C in France and the Balkan region and nigh time temperatures above 30°C) is now on the order of 10%. If no adaptation measures are taken this could mean an additional several thousand deaths/year from heat waves (and their synergic effects with air pollution). The consequences due to heat waves prediction do not even take into account the effect associated with Urban Heat Island-UHI. This phenomena is triggered by absorption radiation due to artificial urban materials,

DOI: 10.1201/9781003348450-229

transpiration from buildings and infrastructure, release of anthropogenic heat from inhabitants and appliances, and airflow blocking effect of buildings (Pacheco-Torgal et al., 2020). The use of dark colored surfaces (like dark asphalt pavements) have low reflecting power (or low albedo characteristics) as a consequence they absorb more energy and in Summer can reach almost 60 ºC thus contributing for higher UHI effects. UHI his is probably the most documented phenomenon of the current climate emergency for various geographic areas of the Planet having saw a huge increase on the number of publications since the year 1990. Recent projections shows that in the north west of the UK summer mean temperatures could rise by 5 °C (50% probability, 7 °C top of the range) by the 2080s (Levermore et al., 2018). The expected rise in global temperature is likely to increase the energy needed to cool buildings in the summer. Balaras et al. (2007) mentioned an increase of more than 2000% between 1990 and 2010. A recent review by Santamouris (2019) mentioned a projection of the future mortality of elderly population in Washington State, to increase between 4-22 times by 2045 and heat related mortality in three cities in North-Eastern USA predicted a six to nine times increase by 2080 under the high emission scenario, RCP 8.5. Of course this projections do not account for the economic recession that will be caused by the coronavirus (Michelsen et al., 2020; Leiva-Leon et al., 2020), which in turn will reduce the number of those who can afford air conditioning. Some estimates are so pessimistic (Sraders, 2020) that air conditioning could become a luxury expense. On the positive side Macintyre and Heaviside (2019) concluded that cool roofs could reduce heat related mortality associated with the UHI by ~25% during a heatwave. And this shows how cooling materials can be important in saving lives especially in the context of the coronavirus recession.

4 CONCLUSIOLNS

This paper reviews studies concerning the social impacts of heat waves that show heat related mortality will increase dramatically due to climate emergency. The economic recession brought by Covid-19 will reduce the number of elderly people that can afford air conditioning thus contributing to more heat related deaths. Cooling materials are especially important in this context because they are able to reduce the heat related mortality.

REFERENCES

Bendell, J., 2018. Deep adaptation: a map for navigating climate tragedy. Institute for Leadership and Sustainability (IFLAS) Occasional Papers Volume 2. University of Cumbria, Ambleside, UK. http://insight.cumbria.ac.uk/id/eprint/4166/1/Bendell_DeepAdaptation.pdf

Ivanov, V., Alexeev, V., Koldunov, N.V., Repina, I., Sandø, A.B., Smedsrud, L.H. and Smirnov, A., 2016. Arctic Ocean heat impact on regional ice decay: A suggested positive feedback. Journal of Physical Oceanography, 46(5), 1437–1456.

Kew, S.F., Philip, S.Y., Jan van Oldenborgh, G., van der Schrier, G., Otto, F.E. and Vautard, R., 2019. The exceptional summer heat wave in southern Europe 2017. Bulletin of the American Meteorological Society, 100(1), pp.S49–S53.

Levermore, G., Parkinson, J., Lee, K., Laycock, P., & Lindley, S. 2018. The increasing trend of the urban heat island intensity. Urban climate, 24, 360–368.

Leiva-Leon, D., Pérez-Quirós, G. and Rots, E., 2020. Real-Time Weakness of the Global Economy: A First Assessment of the Coronavirus Crisis.

Michelsen, C., Clemens, M., Hanisch, M., Junker, S., Kholodilin, K.A. and Schlaak, T., 2020. Coronavirus Plunges the German Economy into Recession: DIW Economic Outlook. DIW Weekly Report, 10(12), 184–190.

Pacheco-Torgal, F., Czarnecki, L., Pisello, A. L., Cabeza, L. F., & Goran-Granqvist, C. (Eds.). (2020). Eco-efficient Materials for Reducing Cooling Needs in Buildings and Construction: Design, Properties and Applications.

Padma, T. 2019. African nations push UN to improve drought research. https://www.nature.com/articles/d41586-019-02760-9

Pierrehumbert, R., 2019. There is no Plan B for dealing with the climate crisis. Bulletin of the Atomic Scientists, pp.1–7.

Santamouris, M ed., 2019. Cooling Energy Solutions for Buildings and Cities. World Scientific.

Spratt, D. and Dunlop, I., 2018. What lies beneath: the understatement of existential climate risk. Breakthrough (Nafional Centre for Climate Restorafion). https://climateextremes.org.au/wp-content/uploads/2018/08/What-Lies-Beneath-V3-LR-Blank5b15d.pdf

Sraders, A. 2020 How much will coronavirus hurt the economy? These new estimates are terrifying. https://fortune.com/2020/03/17/how-much-will-coronavirus-hurt-the-economy-these-new-estimates-are-terrifying/ Accessed in April 2 of 2020

Watts, J., 2018. Arctic warming: scientists alarmed by 'crazy' temperature rises, The Guardian, 27.

Current Perspectives and New Directions in Mechanics, Modelling and Design of Structural Systems – Zingoni (ed.)
© 2022 Copyright the Author(s), ISBN 978-1-032-39114-4

Research for achieving high quality air in architecture and urban planning on the example of the Educational and Sports Center in Józefów near Warsaw, Poland

J. Wrana
Professor of the Lublin University of Technology, Faculty of Engineering and Architecture, Lublin, Poland

W. Struzik
Sanitary Engineer, "WAKAD", Lublin, Poland

ABSTRACT: Contemporary research has covered the history of civilization, starting from hunters and gathers, through agrarian civilization related to the sedentary agricultural way of life, through the transitional period of industrial civilization, then the second industrial revolution (electrification, nuclear oil) to enter the "stage of the knowledge society", with the return to attention to renewable sources as a result of the depletion of the natural resources of our land, used since the settlement process. The relationship formulated in this way leads to the need to document the relationship between architecture and urban planning on the stability of renewable energy sources - energy sources, allowing for permanent residence in built-up areas (both in historical and contemporary cities of the future, (a new research area started at the end of the 20th century) - "knowledge-based economy" - in which many researchers conduct a series of synergistic analyzes of new phenomena in this new area: [1], [2], [3], [4], [5], [6]. The simultaneous widespread awareness of the impact of ecology on human life (made the program nature protection a necessary requirement), developing research processes with the implementation of ecological solutions. The direct sourcing of green energy stored in groundwater for HVAC systems to heat and cool buildings and structures is a prime example of the cheapest CO2. utilisation.

Keywords: innovation, new technologies, reduction of operational cost, CO2 utilisation.

1 INTRODUCTION

The issue of ozone depletion was recognised as early as in the 1980s, prompting the governments of most of the world's countries to start reducing the use of chlorofluorocarbons (CFCs) in air-conditioning and cooling systems. Since 1987 we have been changing our questionable habits of old, turning to green technologies, clean air, sustainable energy management and renewable energy sources. One such renewable energy source is groundwater energy obtained using the FCH (Free Cooling and Heating) technology for heating and cooling ventilation air.

It should be remembered that indoor air, its humidity, temperature and cleanliness directly affect the health and well-being of all building's residents and users. Educational buildings, Integration Center of Sport and Recreation in Józefów near Warsaw are public buildings They are not only places of work for lecturers, but primarily places of teaching for young people, who are more sensitive to air pol-lution than adults. Furthermore, educational build-ings feature a considerable variability of heat loads and a large occupancy density, depending on room function. The function of ventilation is the exchange of used-up air from closed rooms and the removal of gases, organic matter, bacteria and fungi.

In this paper we present the enormous untapped potential of groundwater as an energy source, a natural store that is not detrimental to the ozone layer and naturally reduces huge CO2 emissions by not adding to the carbon footprint.

The water deposited under the Earth's sur-face, or groundwater, is a natural store of energy, a free source for heating and cooling. The temperature of groundwater at the depth of about 10 m is constantly about 10 °C, both in summer and in winter, which is an ideal parameter for HVAC systems to heat and cool ventilation air.

DOI: 10.1201/9781003348450-230

2 THE APPLICATION OF EN-VIRONMENTALLY-FRIENDLY LOW-EMISSION TECHNOLOGIES FOR MOD-ERNISING CURRENT WORN-OUT SYSTEMS

This paper presents the concept of green FCH systems to be introduced in the course of modernising the current systems which have been used for 19 years at the Integration Center of Sport and Recreation in Józefów near Warsaw [7] (main. Design. Jan Wrana).

The centre covers an area of 16,400 m^2. which includes Buildings of Middle School with an Auditorium with a usable floor area of 4527.43 m^2, footprint area of 2767.17 m^2, and cubic volume of 22,682.11 m^3. Buildings 3 and 4 of the swimming pool with a wading pool and technological centre, with a usable floor area of 1780.80 m^2, footprint area of 1452.80 m^2, and cubic volume of 13,725.00 m^3. Building 5, Sports Hall, with a usable floor area of 2003.63 m^2, footprint area of 2200.85 mw, and cubic volume of 15,229.2 m^3. Youth hotel with a usable floor area of 603.52 m^2, footprint area of 349 m^2 and cubic volume of 3115.03 m^3, as well as open sports and leisure grounds.

The concept for the complex took into account the functions and Investor's demands, as well as the applicable Technical Conditions.

The concept Middle School with an Auditorium, involves the construction three FCH air handling units 18,000 m3/h on average each with a demand for heat and cooling from the FCH system of 120 kw. Energy will be sourced from 42 bore-holes and FCH units up to 30m.

The swimming pool and wading pool building with a boiler room and technological cen-tre, an as-sumed Vnw of 14, 240 m3/h, two FCH air handling units 7500 m3/h on average each with a demand for heat and cooling from the FCH system of 70 kw. Energy will be sourced from 17 boreholes and FCH units up to 30m.

Sports Hall, an assumed Vnw of 16,024 m3/h, two FCH air handling units 8000 m3/h on average each with a demand for heat and cooling from the FCH system of 90 kw. Energy will be sourced from 20-24 boreholes and FCH units up to 30m.

Youth hotel an assumed Vnw of 5000 m3/h, with a demand for heat and cooling from the FCH system of 90 kw. Energy will be sourced from 12 boreholes and FCH units up to 30m.

Total usable floor area 8910 m2, amount of ventilation air Vnw – 70,000 m3/h,

FCH system power – 560 kW, number of boreholes up to 30 m depth – 90-100. The assumed energy consumption reduction in relation to HVAC up to 50%, reduced heat energy consumption up to 60%, maintaining the operating costs of EUR 4-5 /m^2/year at 2021 prices.

The expected CO_2 emissions reduction. At the implementation stage, the complex should also be equipped with photovoltaic cells.

3 FCH TECHNOLOGY DESCRIPTION

The goal of the FCH technology, which the authors present here along with the study results, is to source heating and cooling energy from water in order to supply it to FCH air handling units, which involve a system of coolers and heaters to cool or heat fresh air to a set temperature.

The described technology provides a way to use the renewable energy of groundwater to reduce CO_2 emissions. The foundation of the FCH technology described in this paper is groundwater temperature. The natural environment provides us with free energy, which has identical parameters as the energy produces in CFC-based systems.

In the described technology, the sourcing of energy from the ground is conduct-ed in several stages: horizontal and vertical PE tubing and vertical tubing made of ac-id-resistant steel in a borehole with a depth of up to 70 m below ground level. Another stage of obtaining the heating and cooling energy is the conversion of energy in special energy-efficient air handling units, which feature a system of intake and exhaust vents allowing them to significantly reduce electrical and heat energy consumption.

Every unit of this technology features a ground heat exchanger with vertical bore-holes to a depth of ca. 20 to 70 m, in which heat/cold is recovered from groundwater. The elements that ensure the desirable flow parameters of the agent are circulator pumps instead of heat pumps.

4 CONCLUSION

The presented solutions and results, paired with the current CO_2 emissions situation, open extensive avenues for research into those environmentally-friendly systems and are an invitation by the authors to other scientists and universities in Poland and around the world. The presented professional results of research, as well as the actual energy consumption reduction demonstrated, display prime examples of environmentally-friendly solutions and confirm the huge potential of groundwater stores.

REFERENCES

[1] Dodge, D.M. 2001-2006 Illustrated history of wind power development, Telos Net Web Development, http://www.telownet.com/wind/idex.html.

[2] Gawlikowska, A.P. 2014. Space and Energy. Managing the social acceptance of change. Przestrzeń i energia. Zarządzanie społeczną akceptacją zmiany. PP. Poznań.

[3] Gomula, S. 2006. *Energetyka wiatrowa*, Kraków: Uczelniane Wydawnictwa Naukowo-Dydaktyczne, AGH.

[4] Harrahin, R. 2007. China building more power, "BBC Environment Analyst, UK", http://news.bbc.co.uk/2/hi/asia-pacfic/67697 43.stm.

[5] Henky, S. 2013. Earth Hour-Global, WWF.

[6] Sternberg, E.M. 2009. *Healingspaces. The science of place, andwell being*, Londyn: The Belknap. Press of Harvard University Press. Cambridge.

[7] Wrana, J., 2011. *Tożsamość miejsca -kryterium w projektowaniu architektonicznym*, Lublin, Politechnika Lubelska, monography(s.139-140, 187) ISBN: 978-83-62596-19-5.

Current Perspectives and New Directions in Mechanics, Modelling and Design of Structural Systems – Zingoni (ed.)
© 2022 Copyright the Author(s), *ISBN 978-1-032-39114-4*

Microorganisms and the healthy built environment

F. Pacheco Torgal
C-TAC Research Centre, University of Minho, Portugal

ABSTRACT: Using the context of the new reality brought by the Covid-19 outbreak this paper focus on the influence of the microorganisms for a healthy built environment. The lacking interactions between the many researchers of the different areas whose work has connections to the built environment like civil engineers, architects, microbiologists and epidemiologists is identified as a worrying bottleneck that hinders the multidisciplinary design of a healthy built environment.

1 INTRODUCTION

The Covid-19 outbreak of zoonotic origin – like MERS (Middle East Respiratory Syndrome) in 2012 and SARS (Severe Acute Respiratory Syndrome) in 2003 - that some believe is the pandemic that Humanity deserves (Harper, 2020; Gills, 2020) has led many (researchers and laymen) to try to understand a little bit more about microorganisms. And thus it also makes sense that this paper can have some basic information about those.

2 THE IMPORTANCE OF VENTILATION TO TACKLE COVID-19 OUTBREAK AND OTHER MICROORGANISMS

Still on the Covid-19 outbreak the plain fact is that this pandemic will continue to be a problem in coming years because according to The Serum Institute, the world largest vaccine manufacturer, there will not be enough doses to inoculate the entire world until 2024 or beyond (Carr, 2020). Not even mentioning the alarming consequences of post-covid syndrome that include exhaustion from small tasks, involving physical activity, cognitive complaints and problems with the autonomic nervous system. Of course that target could be met much sooner if the European Union decide to put enough pressure on big pharma so the vaccines could be produced in hundreds of labs around the world (Pacheco-Torgal, 2021). As it now widely known Also the current rules based on hand washing and maintaining social distance are insufficient to protect populations from respiratory droplets emitted by people infected by Covid-19, as they can remain in the air for long periods and travel long distances while the virus remains infectious, and this problem is aggravated in indoor environments with poor ventilation combined with high levels of occupancy and long periods of exposure. Several studies have shown that airborne transmission is the most likely mechanism to explain the pattern of social infection (Bahl et al., 2020; Van Doremalen et al., 2020), with the highest risk in indoor environments, particularly those with high occupancy levels and insufficient ventilation, which points to an urgent need for the use of protective masks. It is therefore extremely important, that the national authorities acknowledge the reality that the virus spreads through air, and recommend that adequate control measures be implemented to prevent further spread of the SARS-CoV-2 virus.

3 THE DIRE NEED TO FOSTER INTERACTIONS BETWEEN CIVIL ENGINEERS, ARCHITECTS, MICROBIOLOGISTS AND EPIDEMIOLOGISTS

Ironic as it may sound the fact is that the Covid-19 crisis has push science into a reality that is far way of the classic subjects of the researchers that study the built environment where mechanics and structural performance rules. Maybe the main problem of the new approach lies on the dire need to support interactions between researchers of very different areas like civil engineers, architects, microbiologists and epidemiologists (Prussin et al., 2020). Phelan et al (2020) also recognized that the interplay between building energy efficiency, or more generally, building sustainability, and the health and productivity of the building occupants has not received as much attention. He also comment on the fact that the research communities—sustainable buildings and indoor public health—were not often found at the same research conferences. Unfortunately, one of

DOI: 10.1201/9781003348450-231

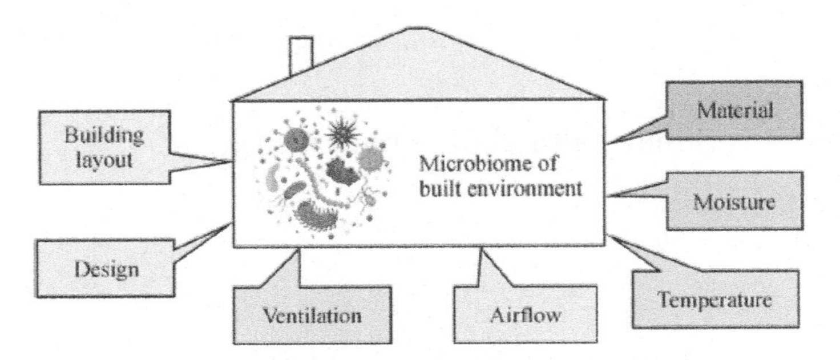

Figure 1. Impacts of built environment attributes on the microbiomes of the built environment (Li et al., 2021).

the problems about the aforementioned lack of multidisciplinary has to do with the fact that usually funding programs are biased against it (Bromham et al. 2016) and that helps to explain why this innovative and most need field is still taking the first steps. And that is why for instance nowadays the professionals of the built environment know very well what green building materials are but have no idea about if those materials are more prone to fungal growth when submitted to high humidity conditions. Buildings of the future should therefore also be designed to promote the presence of beneficial microbes and by reducing exposure to harmful ones by accounting to the several variables as well as the multiple interrelations between them like moisture and temperature that are known to influence mold growth (Figure 1). In order to achieve that goal the most relevant material to measure must be selected, which will be highly dependent on the definition of a healthy microbiome and other factors such as building type. Still it is also unclear which microbial component we need to measure. According to Awda et al. (2021) if we are to achieve healthy buildings, we *must have* a holistic, interdisciplinary research framework through which experts in building science, health, data science, and artificial intelligence collaborate in a coordinated effort. And this book was written having that goal as a guiding compass.

4 CONCLUSIONS

This paper focus on the influence of the microorganisms for a healthy built environment. The importance of ventilation to tackle covid-19 outbreak and other microorganisms was discussed. The paper also includes considerations about the dire need to foster interactions between civil engineers, architects, microbiologists and epidemiologists that hinders the multidisciplinary design of a healthy built environment.

REFERENCES

Awada, M., Becerik-Gerber, B., Hoque, S., O'Neill, Z., Pedrielli, G., Wen, J., & Wu, T. 2021. Ten questions concerning occupant health in buildings during normal operations and extreme events including the COVID-19 pandemic. Building and Environment, 188, 107480.

Bahl, P., Doolan, C., de Silva, C., Chughtai, A.A., Bourouiba, L. and MacIntyre, C.R., 2020. Airborne or droplet precautions for health workers treating COVID-19?. The Journal of infectious diseases.

Bromham, L., Dinnage, R., & Hua, X. 2016). Interdisciplinary research has consistently lower funding success. Nature, 534(7609), 684.

Gills,B. 2020. Deep Restoration: from The Great Implosion to The Great Awakening, Globalizations, 17, 577–579,

Harper, K. (2020) The Coronavirus Is Accelerating History Past the Breaking Point https://foreignpolicy.com/2020/04/06/coronavirus-is-accelerating-history-past-the-breaking-point/

Pacheco-Torgal, F. 2021. Genocidal patents__Intellectual property agains human lives. https://pacheco-torgal.blogspot.com/2021/02/intellectual-property-against-human.html

Phelan, P., Wang, N., Hu, M. and Roberts, J.D., 2020. Sustainable, Healthy Buildings & Communities. Building and Environment, 174, 106806.

Prussin, A. J., Belser, J. A., Bischoff, W., Kelley, S. T., Lin, K., Lindsley, W. G., … & Marr, L. C. 2020. Viruses in the Built Environment (VIBE) meeting report.

Van Doremalen, N., Bushmaker, T., Morris, D.H., Holbrook, M.G., Gamble, A., Williamson, B.N., Tamin, A., Harcourt, J.L., Thornburg, N.J., Gerber, S.I. and Lloyd-Smith, J.O., 2020. Aerosol and surface stability of SARS-CoV-2 as compared with SARS-CoV-1. New England Journal of Medicine, 382(16), pp.1564–1567.

Current Perspectives and New Directions in Mechanics, Modelling and Design of Structural Systems – Zingoni (ed.)
© 2022 Copyright the Author(s), ISBN 978-1-032-39114-4

Do we still need structural engineers?

P.M. Debney
Arup/Oasys, Leeds, West Yorkshire, UK

ABSTRACT: With the increasing power of computers and software, what is the role of the engineer today and tomorrow? Do computer calculations mean that engineers are not needed? Is an engineer just someone who applies maths, or something more? The answer is that we need engineers more, but with a reemphasis on a changing skill set. Computers are better at maths than we are, so let them do that; they are also fast but stupid. Humans are better at imagining and creating, understanding and communicating, so these are the skills that we need to focus on more.

1 INTRODUCTION

Digital automation is superseding many of structural engineer's traditional tasks and AI is threatening to replace more. Does this mean that we might not need structural engineers in the future, or does this instead mean that the engineer's role and skill set needs to change?

The definition "Structural engineers are highly skilled, creative professionals who design the strength and stability of our buildings and bridges". is still true, but the way that engineers work is changing. This means that both the industry and education need to adjust to address the advantages, requirements, and risks of these new working practices.

Today's engineers are familiar with FEA, BIM, and automated section design. There is also a growing emphasis on parametric modelling, digital workflows, and optimisation. Engineering education, on the other hand, focuses heavily on mathematics, and while engineering practice does use both maths and mathematical thinking, engineers are more than just mathematicians. This is especially true today as computers are fundamentally better at maths than humans. Instead, we need to recognise and address our strengths and weaknesses, and those of our digital assistants.

2 DIGITAL ADVANTAGES

Computers do maths very, very quickly; that's it! Computers, or more accurately, computer programs, may seem intelligent but only give us the correct answers when we give them the correct instructions. This means that the responsibility for the accuracy of computer output is ours, whether we are the original programmer or the end user creating a model.

Computers are unlike us in that they work without tiring. This means that they might do one series of calculations, adjust the input parameters a tiny amount, then repeat, and repeat, until it has found the best answer. This means that, when creating a design, we can run optimisations on our structures: constantly adjusting the model until we find the lightest arrangement. A more advanced optimisation, and one that has a greater effect on the result, is optioneering, where we can automatically explore design choices such as column spacing, layout options, and differing materials.

We are also beginning to see the use of AI and machine learning in structural engineering. With machine learning, rather than giving the program a series of rules to derive the answer, we instead give it a lot of relevant data and tell it to derive the answers from that. This can be advantageous when we ask the computer to do something that we can do but cannot describe how, such as how to recognise writing or the objects in a picture.

3 HUMAN SKILLS

Human brains are excellent at spotting patterns in data, filling in gaps, and making predictions from it. But most human of all, along with consciousness, we give meaning and understanding to what we experience in the world.

I am sure that we have all experienced working for a client who thought that they knew what they wanted, guessed the solution from their limited knowledge, then asked you to deliver that. In this situation the wise move is to question the client, to understand why they are asking those questions and to understand what the real need is, then address that. Computers, on the other

DOI: 10.1201/9781003348450-232

hand, only do what they are told, regardless of how well or badly those instructions are given to them. Understanding is a very human skill.

Another human skill is dealing with the unexpected. AI doesn't know that it doesn't know, but that will not stop it giving an answer. Humans can be aware when something is unusual and will act accordingly, possibly creatively.

Humans are different to AI because we are intelligent, though slow.

4 RISKS

Unlike humans, computers are fast but stupid, even the ones running artificial intelligence programs. A machine learning program may be trained to recognise the difference between items, but only because we tell it how it is going to learn from thousands of pictures, all carefully labelled. Computers will not do what we want them to do, only what we instruct them. We must ensure that we get what we require.

As we have seen in the history of finite element-led structural disasters, the fault often lies in the abdication of responsibility of the engineer, accepting the output blindly without ever considering what the programs were not doing: the collapses of the Hartford Civic Center, Sleipner A oil rig, and the CTV building in Christchurch were caused by the engineer not realizing what they were not getting.

5 ADVANTAGES

Optioneering, optimisation, parametric design, automated workflows: all these digital techniques make the engineers and designs of today and tomorrow more efficient. This is essential as, while our duty to safe construction is the same now as it has always been, our duty of protection has expanded to the world. Buildings and the construction industry account for nearly 40% of the worlds energy use and CO_2 emissions. We must both build with less and build less: reduce, reuse, recycle. Engineers must take the lead in driving towards zero carbon construction. But what are the skills needed by structural engineer today and tomorrow to achieve this?

One of the main tasks in future must be for determining how we can minimise the material, energy, and embodied CO_2 in our structures. Optimising section sizes is just the final polish on a structure that might be far from ideal. The key optimisation stage is the conceptual design if we are to eliminate embodied carbon in our work.

Engineers creating analysis models must know what is the question that they are trying to answer, and then how to create a model to give them that answer. It is the engineer who is responsibility for what they include in their model and what they leave out. Models are simplified representations of reality that retain its essential qualities. To simplify the model the engineer must understand what those essential qualities are.

Today programs do not sit in isolation, but instead can work as platforms for your own work. Rather than creating your own analysis and design functions you just need scripts to control the geometry and other parameters, leaving the heavy lifting to the host application.

Computers also free up the engineers to focus more on the nonmathematical aspects. Things like creativity, imagination, innovation, and interpretation. Creativity comes from an appreciation of the wide range of choices available combined with the courage to explore without worrying about all the difficulties at that stage. A good design process first expands to explore the options, before refining and honing back to the final choice that considers the practicalities of analysis, detailing, and construction.

6 CONCLUSION

We come back to the age-old questions: what is an engineer, what do they do, and how should we train them? Undoubtably the skills used in industry are changing, and thus so should the university degrees, but how are we to change them? The core engineering skills are still essential, yet we need more added into already full syllabuses. Engineering education and design needs to embrace more creativity and automation to help us go beyond the old, heavy, inefficient designs of the past to create a world that has a future.

Computing is the key tool in addressing both the shortage of qualified staff and the excess of embodied carbon dioxide in our designs. We can work smarter to produce more efficient designs faster, but also calculate what capacity old structures possess and thus avoiding building at all!

Zero carbon construction is a massive challenge, but one that we can meet by working intelligently with fast but stupid machines.

Current Perspectives and New Directions in Mechanics, Modelling and Design of Structural Systems – Zingoni (ed.)
© 2022 Copyright the Author(s), ISBN 978-1-032-39114-4

Facilitated teambuilding processes to enhance teamwork skills in engineering education

L. Bücking
Institute of Mechanics and Computational Mechanics, Leibniz University, Hannover, Germany

ABSTRACT: Teamwork skills are essential for engineers, as they often collaborate across disciplines and – in increasingly digitalized settings - across cultural backgrounds. This requires universities to teach their engineering students the respective soft skills. Hence, this paper explores how the acquisition of such competences can be integrated into the structure of the courses Engineering Mechanics and Computational Mechanics. The components of the framework enable the students to work in small group settings in which they receive guidance by a facilitator. A facilitator supports the students in engaging in continuous reflective practices. Qualitative and quantitative assessments show that the approach was successful in furthering the students' development of teamwork and communication skills. The components of the framework can easily be adapted to other courses and universities to better prepare the students for their future professional practice.

1 INTRODUCTION

Throughout their careers, engineers contribute to teams facing the modern world's complex challenges, such as climate change. Therefore, teamwork competences are an important asset for engineers entering the labor market. This requires higher education institutions to support the students in developing their respective soft skills, namely teamwork and communication skills. Unfortunately, German universities struggle to successfully integrate such activities into their engineering curricula (VDI, VDMA & Stiftung Mercator GmbH 2016).

Team-building activities in small groups were successfully integrated into existing course structures at the Institute of Mechanics and Computational Mechanics (IBNM) at Leibniz University Hannover (LUH). The framework aims to increase the students' awareness for group processes, their relevant soft skills and their self-efficacy when working in teams. The lecturers and student tutors act as facilitators of the team development processes. They foster the students' competence development by facilitating frequent reflective activities based on the Experiential Learning theory (Kolb 1984).

Hence, this paper describes the facilitated team-building activities and processes which were developed and implemented in the undergraduate courses at IBNM. Further, the paper evaluates the activities using qualitative and quantitative assessment methods.

2 FRAMEWORK

Team-building components have been integrated into the IBNM's existing undergraduate courses Engineering Mechanics A and B (first year) and Computational Mechanics (third year). In the first year, the team-building activities take place in tutorial classes and aim to support the students in building first connections and forming study groups within a cohort of 200 to 300 students. In the third year, the Computational Mechanics students work on group projects over the course of two to three months. These targets require different sets of activities and team development processes.

The tutorial classes are facilitated by student tutors who explain the course content on a peer-to-peer level and encourage collaboration among the first-year students. To catalyze the cohesion of each group, the tutors lead the first-year students through team-building activities. The tutors are second and third-year students who form a team themselves in which they reflect their groups' dynamics and their own soft skill development. At first, the lecturer of the course facilitates the tutors' team meetings using interactive methods. In their second term working as tutors, the tutors facilitate their team meetings themselves to further improve their leadership and facilitation skills.

The course Computational Mechanics offers the context of a project-based learning scenario. While the students work on an engineering task in groups of three to four, they undergo a guided team

DOI: 10.1201/9781003348450-233

development process (Bücking 2021). In the accompanying team-building sessions and reports, they learn to clarify expectations and define common goals (Kickoff), analyze their group dynamics and relate them to their performance in the project (Team Update/ Intermediate Report). After finishing the project, they identify factors for successful collaboration and reflect on their own competence development (Adjourning/ Process Report). Based on their findings, they receive feedback from the facilitator.

The team-building activities were implemented in two subsequent years during the Covid19 pandemic which required a digital or hybrid format. The facilitators and student tutors received training in digital facilitation methods. Based on the Five Stage Model (Salmon 2013), the students were introduced to using digital platforms and tools. This enabled them to interact with their group members, organize their resources in the virtual space and to collaboratively work on documents.

3 EVALUATION AND CONCLUSION

The guided team-building activities and processes were evaluated regarding the students' competence development. Two cohorts of tutors' teams of Engineering Mechanics A and two cohorts engaging in the project-based learning scenario of Computational Mechanics participated in the assessment. The evaluation showed that the students tutors improved their presentation and leadership skills and have become more self-confident and empathetic when working with a group. The third-year students participating in the group project reported that they developed their collaboration and communication skills and derived insights about how to successfully organize their teams. These quantitative findings are supported by a self-assessment conducted in the first cohort of Computational Mechanics students which showed an increased skill level in six out of eight competences relevant for successful teamwork, as shown in Figure 1.

These findings show that it is possible and necessary to include team-building components in the core curriculum of engineering education to enhance the students' teamwork skills. Teachers or student tutors can take on the role of a facilitator guiding the team-building processes to ensure that the students learn from their positive – and possibly negative – experiences. The activities were successfully implemented in online and hybrid teaching scenarios and are easily adaptable to other universities and educational contexts. In the future, they will be transferred to on-campus teaching. Additionally, the framework should be extended to graduate students. This way, the students are prepared to contribute to digital and interdisciplinary teams in their future work environment.

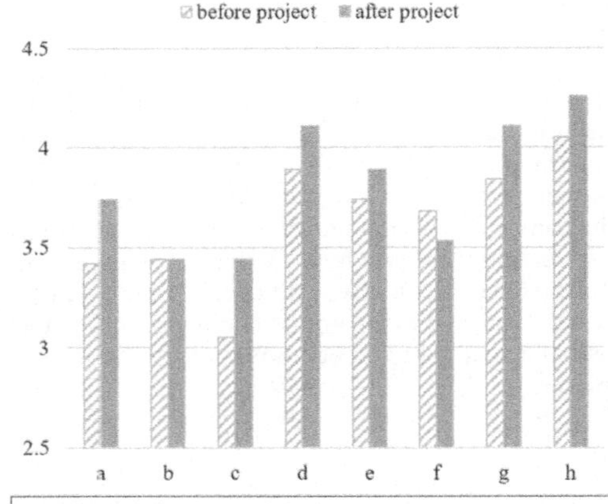

a: asking for help when needed
b: working with people who think differently than oneself
c: clearly expressing thoughts and feeling
d: solving conflicts constructively
e: taking responsibility
f: involving others in what one thinks and does
g: receiving constructive criticism and developing based on it
h: giving constructive criticism, based on which the recipient can improve

Figure 1. Development of teamwork skills during the team project in the course Computational Mechanics (first cohort). Results obtained from students' self-assessment; modified from Bücking (2021).

REFERENCES

Bücking, L. (2021). A guided team-building process in a virtual project-based learning scenario. *ITS21 Conference*: 40-51. Copenhagen, Denmark.

Kolb, D. A. (1984). *Experiential learning: experience as the source of learning and development.* Englewood Cliffs, NJ: Prentice Hall.

Salmon, G. (2013). *E-tivities: The Key to Active Online Learning*. New York: Routledge.

VDI, VDMA, Stiftung Mercator GmbH. (2016). *15 Jahre Bologna-Reform. Quo vadis Ingenieurausbildung?*

Current Perspectives and New Directions in Mechanics, Modelling and Design of Structural Systems – Zingoni (ed.)
© 2022 Copyright the Author(s), ISBN 978-1-032-39114-4

Peer review of teaching/learning paradigms: A new proposal for engineering education

Maram Saudy, Ibrahim Abotaleb, Khaled Nassar & Ezzeldin Sayed-Ahmed
The Department of Construction Engineering, The American University in Cairo, Egypt

ABSTRACT: In higher education, it is common to hear "we have the best engineering education" or "we have talented and experienced instructors who offer the best teaching/learning paradigm," etc. In general, the four main components of higher education are simply students, faculty, facilities, and curriculum. We may have a flawless four components, yet the outcomes may not be as good as expected: an indicative of the crucial need for teaching/learning continuous evaluation. Peer review can be one of the most effective tools for the continuous improvement of higher education offerings. In this paper, the authors form a team to create an effective peer-review process with new and innovative formative and summative functions. A new checklist is being designed and used by two or more reviewers to rate the course syllabus, objectives, outcomes, and activity/assessment tools; this is the summative evaluation phase. Two or more peers used another designed checklist to rate different aspects of the instructor's pedagogy based on a class visit for the formative evaluation. Both lists are shared with the faculty but not used as part of his/her evaluation. A pilot study is performed on the four authors of this paper and four courses of the construction engineering curriculum in the American University in Cairo. The four faculty adopt the process as they believe that what they have excelled in for some time will unequivocally be surpassed by something that is newer and indubitably works better.

1 INTRODUCTION

The necessity for quality instructional practice and assessment in engineering education is of paramount concern (1). However, the fact that the quality of education depends on giving faculty members more control of their practice is seemingly neglected (2). Developments in the scholarship of teaching and learning have changed from an information transmission approach to a quality learning approach. Currently, there is a need to change the traditional method of teaching evaluation to a more collegial design. Peer review and tailored evaluation interventions are increasingly proposed in higher education entities as alternatives to improve the evaluation process and teaching quality (2). Peer review of teaching (PRT) includes observing lectures and tutorials, monitoring online teaching, examining curriculum design, and using student assessments (3). The essence of PRT is about furthering the development of faculty members through expert input based on knowledge and understanding. However, it can be used as part of performance appraisal and tenure portfolios (4). PRT also sharpens individual skills, such as observing and critically reflecting on the dynamics and social context of teaching (5).

Accordingly, the main objective of this paper is to apply the peer review of teaching on an engineering curriculum with new and innovative formative and summative functions.

2 METHODOLOGY

To apply the peer review of teaching on an engineering curriculum, the four authors of this research formed a team to create a pilot study performed on four different courses of the construction engineering curriculum in the American University in Cairo (AUC). First, a summation evaluation was conducted by designing a new checklist to rate the course syllabus, objectives, outcomes, and activity/assessment tools. Then, the formative assessment was conducted by developing another list to rate different aspects of the instructor's pedagogy based on a class visit of each of the four instructors by the other four instructors.

3 SUMMATIVE EVALUATION

The standard way to evaluate teaching for the last half-century has been to collect course-end student rating forms and compile the results. While student ratings have considerable validity, students may not be fully-qualified to evaluate an instructor's offering of the course subject, the currency and accuracy of the course

DOI: 10.1201/9781003348450-234

content, the appropriateness of the level of course rigor, and the teaching and assessment methods used in its delivery, and whether the course content and learning objectives are consistent with the course's intended role in the program curriculum (14); for example, as a prerequisite to another course. Only faculty colleagues may be in a position to make these judgments. Accordingly, a summative checklist is designed, to be used by faculty raters, to evaluate the ability of an instructor to design a course in terms of the course content, objectives, outcomes, availability of the course materials, and the consistency of the assessment tools to the course learning objectives and outcomes.

4 FORMATIVE EVALUATION

As for the formative evaluation, students also have limited ability to provide individual formative feedback to their instructors; only colleagues can freely provide such feedback more realistically and effectively. A formative checklist is designed to rate the instructor's teaching style and his/her students' class engagements.

Ten different assessments have been provided to rate the instructor after a class visit, and the faculty rater should circle his/her level of expectation of the instructor's achievement against the ten different instructional assessments.

5 PILOT STUDY

It is well known that the peer review process may many limitations. As such, a pilot study was carefully designed among four instructors from the construction engineering department in the American University in Cairo to implement this process with trying to avoid and manage those different limitations. First, more than one class visit is recommended for each instructor by the remaining three ones. Second, the opinions of the three observers or raters were compiled after meetings of sharing thoughts and feedback. Finally, all instructors who were reviewed were invited to meet with each other to discuss the evaluation and formulate measures they might take to improve both their teaching and their peer evaluations.

Since the process is still ongoing, the final results are to be compiled by the end of this current semester to provide more comprehensive and representative findings, the final observations cannot be provided now. It is expected to analyze these two different types of data; the summative and the formative. The summative data are going to be analyzed and each of the assessment measures of each instructor will be compared to the average value, among the four members of the study, of this assessment measure to provide an overall summative evaluation. Also, the formative measures are going to be assessed using the same approach to give an overall evaluation of the expected teaching style and the students engagement of each reviewed instructor.

6 CONCLUSIONS

The current pilot study is designed to address many of the peer review evaluation concerns. The analysis of the results are expected to reach one of two findings; if the assessment measure of a reviewed instructor was found to be below the average, of the analysis sample of instructors, this will address the points that need further consideration from the instructor to be improved, on the other hand, if the measures were found to be above average assessment will be considered as the points of strengths of the instructor's teaching that should be shared with other colleagues.

REFERENCES

1. Falchikov, N. and Goldfinch, J., Student peer-evaluation in higher education: a meta-analysis comparing peer and teacher marks. Review of Educational Research, 70, 3, 287–322 (2000).
2. Murray, C. E., & Grant, G. (1998). Teacher peer review: Possibility or pipedream? Contemporary Education, 69, 202–204.
3. Hatzipanagos, S., & Lygo-Baker, S. (2006). Teaching observations: Promoting development through critical reflection. Journal of Further and Higher Education, 30, 421–431.
4. Kohut, G. F., Burnap, C., & Yon, M. G. (2007). Peer observation of teaching: Perceptions of the observer and the observed. College Teaching, 55, 19–25. doi:10.3200/CTCH.55.1.19-25
5. Peel, D. (2005). Peer observation as a transformatory tool? Teaching in Higher Education, 10, 489–504.

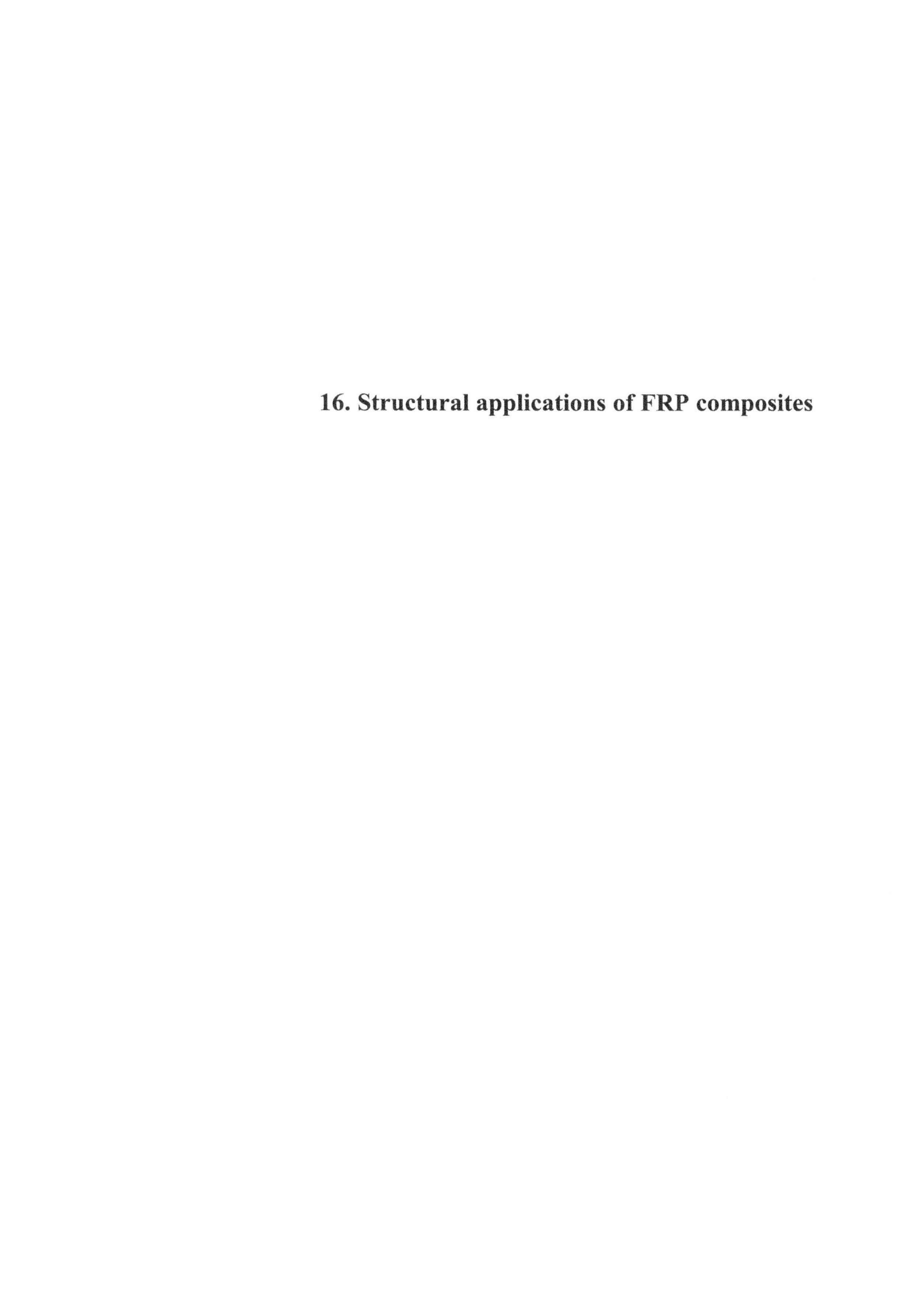

16. Structural applications of FRP composites

Current Perspectives and New Directions in Mechanics, Modelling and Design of Structural Systems – Zingoni (ed.)
© 2022 Copyright the Author(s), *ISBN 978-1-032-39114-4*

Case studies of repurposing FRP wind blades for second-life new infrastructure

L.C. Bank, T.R. Gentry, T. Al-Haddad, A. Alshannaq, Z. Zhang, M. Bermek, Y. Henao, A. McDonald, S. Li, A. Poff, J. Respert & C. Woodham
Georgia Institute of Technology, Atlanta, USA

A. Nagle & P. Leahy
University College Cork, Cork, Ireland

K. Ruane
Munster Technological University, Cork, Ireland

A. Huynh, M. Soutsos, J. McKinley, E. Delaney & C. Graham
Queen's University Belfast, Belfast, UK

ABSTRACT: This paper presents two case studies of the repurposing projects of decommissioned wind turbine blades in architectural and structural engineering applications conducted under a multinational research project is entitled "Re-Wind" (www.re-wind.info) that was funded by the US-Ireland Tripartite program. The group has worked closely together in the Re-Wind Network over the past five years to conduct research on the topic of repurposing of decommissioned FRP wind turbine blades. Repurposing is defined by the Re-Wind team as the reverse engineering, redesigning and remanufacturing of a wind blade that has reached the end of its life on a turbine and taken out of service and then reused as a load-bearing structural element in a new structure (e.g., bridge, transmission pole, sound barrier, sea-wall, shelter). Further repurposing examples are provided in a publicly available Re-Wind Design Catalog. The Re-Wind Network was the first group to develop practical methods and design procedures to make these new "second-life" structures. The Network has developed design and construction details for two full-size prototype demonstration structures – a pedes- trian bridge constructed in Cork, Ireland in January 2022 and a transmission pole to be constructed at the Smoky Hills Wind Farm in Lincoln and Ellsworth Counties, in Kansas, USA in the late 2022. The paper pro- vides details on the planning, design, analysis, testing and construction of these two demonstration projects.

1 INTRODUCTION

With the increase in wind energy over the last few decades, many countries around the world are beginning to divest from traditional nonrenewable energy sources. However, a new issue has arisen in the form of wind turbine blade waste. Turbine blades are made primarily of glass fiber reinforced polymer (GFRP) composite materials which cannot be recycled, and they and are only designed for a service life of 20-25 years. The scale of EOL blade waste makes it an especially concerning issue, as there will be an estimated cumulative total of 43 million tonnes of blade waste worldwide by 2050 if no blades are disposed of in the interim and are stock- piled, with Europe and the United States processing a combined 41% of this waste (Liu and Barlow 2017.

To address the global issue of wind turbine blade waste, the Re-Wind Network (www.re-wind.info) was founded in 2017 as a partnership between the US, UK, and Republic of Ireland. The main goal of this research team is to investigate the use of decommissioned wind turbine blades in second-life structural applications. In early 2022, the Re-Wind team constructed its first ever wind blade pedestrian bridge ('BladeBridge') in Cork, Ireland. In addition to BladeBridge, the team has also been investigating the application of decommissioned blades as a power transmission pole, known as 'BladePole'. A prototype was constructed for load testing in 2021, and a demonstration project is planned for construction in Kansas, USA in late 2022. The Re-Wind Network has proposed a number of additional repurposing solutions as well, all of which can be found in the ReWind 2021 Design Catalogue (McDonald et al. 2021).

DOI: 10.1201/9781003348450-235

2 CASE STUDY 1: BLADEBRIDGE

2.1 *Characterization of a turbine blade for use as a BladeBridge*

Using wind turbine blades in second-life structural applications raises several structural issues. The loads that the blades will experience in these applications differ greatly from those experienced while the blades were in service, and there is the added possibility of adhesive failure with the different air- foil components. In addition, he fact that the blades are EOL products means that each individual blade may have developed unique defects over the course of its service life. These challenges make it especially pertinent to develop a methodology for properly characterizing each unique blade. Further details and results of the materials and structural testing can be found in Ruane et al. (2022).

2.2 *BladeBridge design, fabrication, & installation*

The design of the BladeBridge occurred in two phases – initial renderings were created for use in communicating with various stakeholders and to determine aesthetic details, and a final structural design was then completed by the MTU team for use in construction.

The bridge was constructed by local fabricators and contractors in Cork, Ireland in late 2021 and was installed on the Midleton-Youghal Greenway site in January 2022 (Figure 1).

Figure 1. BladeBridge installed on site. The bridge spans 5m and is 4m wide, with concrete abutments and steel hardware.

3 CASE STUDY 2: BLADEPOLE

3.1 *Characterization of a turbine blade for use as a BladePole*

Detailed analyses have been conducted by the Re- Wind Network to characterize the material and structural properties of a turbine blade for use in a power transmission pole application (Alshannaq et al. 2021a, Alshannaq et al. 2021b). It was found that wind blades are incredibly strong in this application, which can be attributed to the fact that the blades were originally designed to withstand much greater flexural loads on a moving turbine. In addition to the structural properties of the blade, the FRP material of the blade is also non-corrosive and nonconductive, making it a durable material choice for use as a tangent pole.

3.2 *BladePole design & prototype fabrication*

The BladePole team is focused on the repurposing of blades as vertical tangent poles for 230 kiloVolt (kV) transmission lines, which is the basis for structural analysis and comparisons. A rendering of a BladePole in a double circuit 230 kV transmission line is shown in Figure 2.

Figure 2. BladePole rendering as a 230 kV tangent pole.

In 2021, a mockup of the BladePole was constructed on a 6m portion of a GE37 wind blade. The BladePole team has conduct full-scale structural testing of the mockup and will report results in the future.

4 CONCLUSIONS

Wind turbine blades have been proven to be strong and durable structural components, even in their second life after decommissioning. The BladeBridge and BladePole case studies presented in this paper reveal that wind blade structures are becoming increasingly viable in real-life applications around the world.

REFERENCES

Alshannaq, A. A., Bank, L. C., Scott, D. W. & Gentry, R. 2021a. A Decommissioned Wind Blade as a Second-Life Construction Material for a Transmission Pole. *Construction Materials*, 1: 95–104.

Alshannaq, A. A., Bank, L. C., Scott, D. W. & Gentry, T. R. 2021b. Structural Analysis of a Wind Turbine Blade Repurposed as an Electrical Transmission Pole. *Journal of Composites for Construction*, 25: 04021023.

Liu, P. & Barlow, C. Y. 2017. Wind Turbine Blade Waste in 2050. *Waste Management*, 62: 229–240.

McDonald, A., Kiernicki, C., Bermek, M., Zhang, Z., Poff, A., Kakkad, S., Lau, E., Arias, F., Gentry, R. & Bank, L. 2021. *Re-Wind Design Catalog Fall 2021* [Online]. Available: https://static1.squarespace.com/static/5b324c409772ae52fecb6698/t/618e5b5f5c9d244eec10e788/1636719473969/Re-Wind±Design±Catalogue±Fall±2021±Nov±12±2021±reduced±size.pdf [Accessed December 20, 2021].

Ruane, K., Zhang, Z., Nagle, A., Huynh, A., et al. 2022. Material and Structural Characterization of a Wind Turbine Blade for use as a Bridge Girder. *Transportation Research Record*.

Current Perspectives and New Directions in Mechanics, Modelling and Design of Structural Systems – Zingoni (ed.)
© 2022 Copyright the Author(s), ISBN 978-1-032-39114-4

Prediction on stress-strain behavior of FRP-confined concrete with passive confinement-based 3D constitutive model

J.F. Jiang
Tongji University, Shanghai, China

P.D. Li
Shenzhen University, Shenzhen, China

B.B. Li
Nanjing Tech University, Nanjing, China

ABSTRACT: The confinement through fiber reinforced polymer (FRP) wrapping belongs to passive confinement. Since the concrete is a path-dependent material, the passive confinement causes typical stress-strain behavior of concrete which is different from concrete under active confinement. The first author's research group have developed the true-triaxial database associated with the passive confinement by the novel true-triaxial testing system. Accordingly, two types of 3D constitutive models for FRP confined concrete are proposed. They are analysis-oriented and FEM-based constitutive models, which can simulate the compression behavior of concrete square columns with FRP confinement. The well-matched performance demonstrates the improved constitutive models' applicability in simulating the mechanical behavior of concrete with complex passive confinement fields. At last, the arching effect was re-examined for the square section.

1 INTRODUCTION

It is well-known that confinement for concrete can enhance both strength and ductility. FRP confinement, the passive confinement, is characterized by a level of stress state intensity that increases up to failure because of the lateral expansion of concrete.

The local mechanism for concrete under passive confinements still lacks direct test data. Recent studies presented by Jiang et al.(2017a,b) made a notable contribution to developing a true-triaxial testing system specifically related to the passive confinement load path. In this paper, based on the test results reported in Jiang et al. 2017a,b), two types of 3D constitutive models for FRP confined concrete are proposed. The simulation results reveal the characteristic of confinement fields within a square section.

2 TRUE-TRIAXIAL TESTING

2.1 *Experimental framework*

The new true-triaxial testing system (Figure 1) for passive confinement was built at Tongji University by Jiang et al. (2017a). The system enables the testing of concrete mechanical behavior under a non-uniform and explicit confinement field (Jiang et al. 2017a).

The testing system accomplished a series of compression tests on cubes subjected to a simultaneous pressure increase in two mutually orthogonal directions, represented by different confinement stiffness values in the 1st and 2nd directions. The constitutive model has developed accordingly.

Figure 1. True-triaxial testing system (Jiang et al. 2017a).

2.2 *Dilation relationship and stress-strain model*

An explicit function of lateral strain vs. axial strain is proposed according to the feature of the experimental tests adopted herein as references; thus, a bilinear relationship is assumed in Equations (2) and (3):

$$\varepsilon_{l,i} = \nu_0 \varepsilon_c \text{ if } \varepsilon_c \leq \varepsilon'_{cc} \quad (2)$$

$$\varepsilon_{l,i} = \nu_0 \varepsilon'_{cc} + \mu_i(\varepsilon_c - \varepsilon'_{cc}) \text{ if } \varepsilon_c > \varepsilon'_{cc} \quad (3)$$

The typical stress-strain curves for passive confined concrete are modeled by the following equations, presenting post-peak strain-softening or hardening behavior.

$$f(x) = \frac{\sigma_c}{\sigma'_{cc}} = \frac{a \cdot x}{a - 1 + x^{a(x+\delta)^b + c}} \quad (4)$$

$$E_2 = \frac{f_{cu} - \sigma'_{cc}}{\varepsilon_{cu} + \varepsilon'_{cc}} \quad (5)$$

DOI: 10.1201/9781003348450-236

The detailed information on these parameters (a, b, c, ε'_{cc}, σ'_{cc}, E_2) can refer to Jiang et al. (Jiang et al. 2020).

3 NUMERICAL MODELS

3.1 *Analysis-oriented model*

The proposed methodology to derive the global stress-strain behavior of FRP confined concrete square columns follows the flow chart as follows.

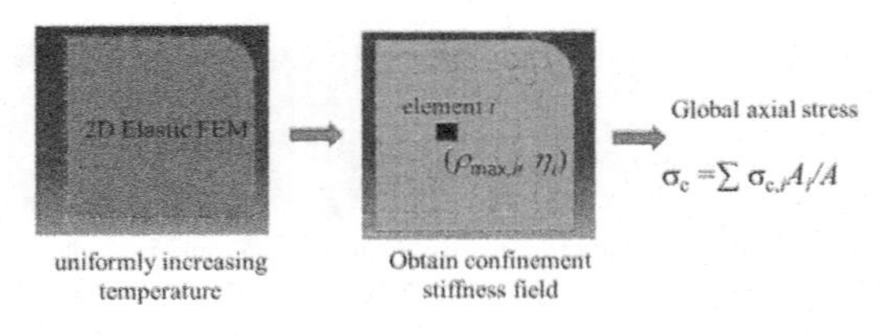

Figure 2. Flow chart of analysis-oriented model.

3.2 *FEM model*

Based on true-triaxial testing, a 3D concrete constitutive model for passively confined concrete is proposed. The characteristic functions for damage variables, the hardening/softening rule, and the dilation angle model are inserted into the constitutive model with the user subroutine USDF. The proposed model was verified by using the commercially available software ABAQUS 6.13-1.

3.3 *Validation and application of proposed models*

The test results reported by Wang and Wu (2008) were selected for verification. The full stress-strain curves were compared between the test and the calculated results (Figure 3). The comparison shows that the proposed method can well capture both strain-hardening and softening behavior. As the analysis-oriented model has simplified the boundary condition, the simulation has a larger deviation in the case of columns with a smaller corner radius than the FEM simulation.

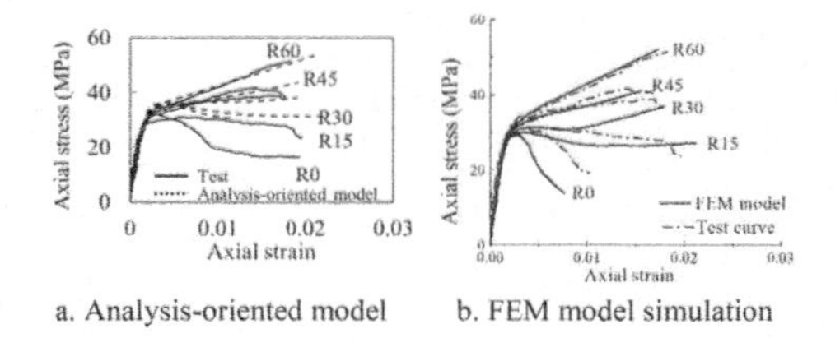

a. Analysis-oriented model b. FEM model simulation

Figure 3. Comparison results of axial stress-strain curves.

4 ARCHING EFFECT FOR THE SQUARE SECTION

Based on the numerical results from the models in Section 3, the axial stress distribution across the section is reported in Figure 4, where the 1-ply CFRP wrap of the C30 section (Wang and Wu 2008) is considered. Based on the arch action theory, the effectively confined zone for a small corner radius is drawn within the boundary lines (dash curves in Figure 4a). The effectively confined zone hardly has a specific clue to the axial stress distribution. Based on the work of Jiang and Wu (20) on the classification of confinement levels (high, moderate and low), the confinement level of each local element can be quantified and illustrated in Figure 4b. It is quite different from the effective confinement zone (area within two solid curves in Figure 4b) derived from the traditional arching action. Assuming the effectively confined area ratio of highly and moderately confined area to the total section area, the value

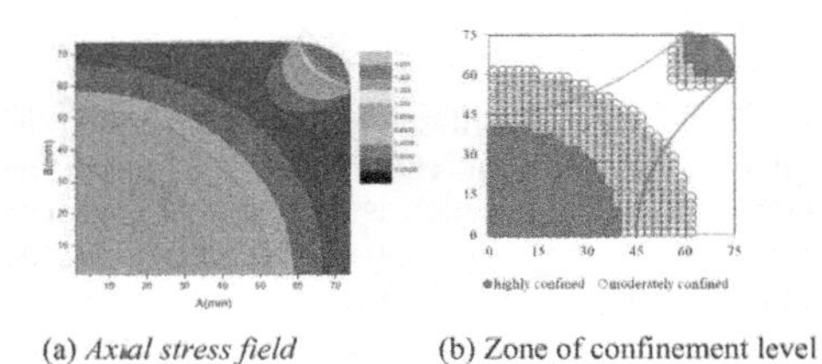

(a) *Axial stress field* (b) Zone of confinement level

Figure 4. Axial stress field and z one of confinement level within the square section (corner radius = 0.2).

depends on the FRP stiffness and normalized corner radius (r/R). Therefore, the limits on the corner radius and the FRP stiffness are essential conditions needed to improve strength and ductility.

5 CONCLUSIONS

This work presents two methodologies to evaluate the effect of FRP confinement in square sections of concrete. The essential part is the constitutive performance of concrete, which is based on the cubic test results through the novel true-triaxial test system associated with passive confinement (Jiang et al. 2017a). The test data help develop the 3D constitutive model for confined concrete, which is incorporated into the FEM model in the platform of ABAQUS software.

The studies on the axial stress distribution of the FRP confined square section also improved arching action theory. A uniform (hydrostatic) condition did not occur in the square confined section but after forcing the subdivision of the homogeneous zones. It allows an essential conclusion for the design that the limits on the corner radius and the FRP stiffness are essential conditions needed to improve strength and ductility.

REFERENCES

Jiang, J. Xiao, P. & Li, B. 2017a. A novel triaxial test system for concrete under passive confinement. J Test Eval 46(3):913–923.

Jiang, J.F. Xiao, P. & Li, B. 2017b. True-triaxial compressive behavior of concrete under passive confinement. Constr Build Mater 156:584–598.

Wang, L.M. & Wu, Y.F. 2008. Effect of corner radius on the performance of CFRP-confined square concrete columns: Test. Eng Struct 30(2):493–505.

Current Perspectives and New Directions in Mechanics, Modelling and Design of Structural Systems – Zingoni (ed.)
© 2022 Copyright the Author(s), ISBN 978-1-032-39114-4

Development and testing of new precast concrete tunnel segments reinforced with GFRP bars and ties

S.M. Hosseini, S. Mousa, H.M. Mohamed & B. Benmokrane
Department of Civil and Building Engineering, Université de Sherbrooke, Sherbrooke, Canada

ABSTRACT: This paper presents the results of an experimental study conducted to investigate the structural performance of GFRP-reinforced PCTL segments. Three full-scale tunnel segment specimens were fabricated and tested under bending load. The investigated parameters were the type of reinforcement and concrete compressive strength. Test results indicated that the replacement of steel reinforcement with GFRP bars enhanced the load carrying capacity and deflection at peak load by 33% and 50%, respectively. In addition, increasing the concrete strength enhanced the cracking load, post-cracking stiffness, and ultimate load by 28, 7 and 17%, respectively.

1 INTRODUCTION

Using non-corroding glass fiber-reinforced polymer (GFRP) bars as a replacement of conventional steel reinforcement is an effective approach to deal with the corrosion problem in PCTL segments (Caratelli et al. 2016). Limited number of GFRP-reinforced PCTL segments were tested in the literature under bending load. In the tested GFRP-reinforced PCTL segments in the literature, tension-controlled failure was reported, which is unfavorable. In addition, based on the results reported in the previous study, the cracking behavior of PCTL segments reinforced with GFRP bars was not satisfactory enough especially for those reinforced with smooth GFRP bars (Caratelli et al. 2016, Caratelli et al. 2017, Spagnuolo et al. 2017). Therefore, before major application of GFRP bars in tunneling projects, it is necessary to do further investigation to evaluate their structural performance.

Three full-scale tunnel segment specimens tested to investigate the effect of reinforcement type and concrete compressive strength are presented herein. The GFRP-reinforced specimens were designed to undergo favorable compression failure mode. In addition, the utilized curvilinear bars were sand-coated and produced through an innovative manufacturing process with improved propertied compared to what was used in the literature. Furthermore, this study pioneers the investigation of the effect of concrete compressive strength on the behavior of GFRP-reinforced PCTL segments.

2 EXPERIMENTAL PROGRAM

2.1 *Test specimens*

Three full-scale tunnel segment specimens were designed, fabricated, and tested under three-pint bending load. Figure 1 shows a typical assembled cage. The test parameters were type of reinforcement (GFRP and steel) and concrete compressive strength (40 MPa and 80 MPa). The test matrix is presented in Table 1. The specimens measured 3100 mm in length, 1500 mm in width, and 250 mm in thickness.

Figure 1. A typical assembled GFRP cage.

DOI: 10.1201/9781003348450-237

Table 1. Specimen details and test matrix.

ID	Reinforcement type	Concrete strength (MPa)	Transverse reinforcement
7G#5	GFRP	40	No. 4 @ 200 mm
7G#5H	GFRP	80	No. 4 @ 200 mm
7S15M	Steel	40	10M @ 200 mm

3 EXPERIMENTAL RESULTS AND DISCUSSION

3.1 *Cracking behavior and failure mode*

Table 2 presents the experimental test results. At the peak load, concrete spalling occurred in 7G#5 and 7G#5H followed by concrete crushing which denotes favorable compression-controlled failure mode. Tension-controlled failure mode was observed in the steel-reinforced specimen. Replacement of steel reinforcement with GFRP bars in PCTL segments enhanced the load-carrying capacity and deflection corresponded to the peak load by 33% and 50%, respectively. Increasing the concrete strength by 100% enhanced the load-carrying capacity and deflection at peak load by 17% and 6%, respectively.

Table 2. Experimental test results.

ID	Cracking load (kN)	Crack width width at service load (mm)	Peak load (kN)	Deflection at peak load(mm)
7G#5	57	0.35	315	66
7G#5H	73	0.4	370	70
7S15M	68	0.2	236	44

3.2 *Load-deflection relationship*

The load versus mid-span deflection of the tested specimens is presented in Figure 2. The specimens experienced linear behavior in pre-cracking stage. After formation of the first crack, the stiffness of the specimens decreased due to transition from gross-section to the effective section. The GFRP-reinforced specimens experienced a nearly linear behavior until the peak load while the steel-reinforced specimen followed the steel yielding plateau. Using HSC enhanced the post-cracking stiffness by 7%. The deflection at service load was 2.47, 2.23, and 3.67 mm in 7G#5, 7G#5H, and 7S15M, respectively (note that the service load was different in the specimens). Therefore, the deflection at service load decreased by 10% when the concrete strength increased by 100%. In addition, while replacement of steel rebars with GFRP bars decreased the post-cracking stiffness by 62%, it decreased the deflection at service load by 33%.

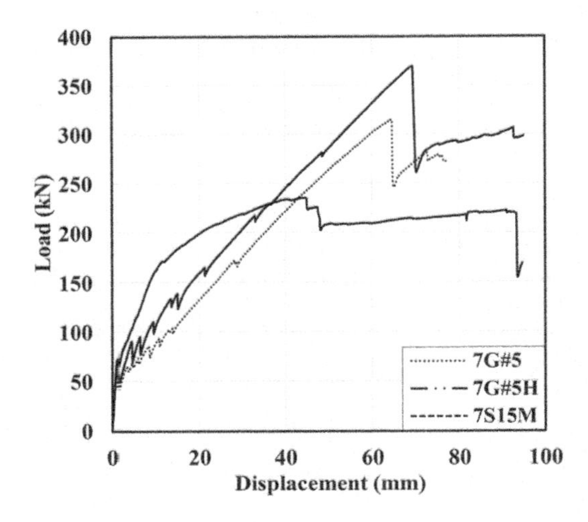

Figure 2. Load-deflection relationship of the specimens.

4 CONCLUSIONS

1. Replacement of steel reinforcement with GFRP bars enhanced the load-carrying capacity and deflection corresponded to the peak load by 33% and 50%, respectively.
2. Increasing the concrete compressive strength by 100% in the GFRP-reinforced PCTL segments enhanced the load-carrying capacity and deflection at peak load by 17% and 6%, respectively, while did not have a noticeable effect on the cracking behavior. In addition, using HSC decreased the deflection at service load by 10%.
3. Replacement of steel reinforcement with GFRP one can secure the durability of PCTL segments while provides satisfying structural performance. The structural behavior of such segments can be improved by increasing the concrete strength.

REFERENCES

Caratelli, A., Meda, A., Rinaldi, Z. and Spagnuolo, S., 2016. Precast tunnel segments with GFRP reinforcement. *Tunnelling and Underground Space Technology*, *60*, pp.10–20.

Caratelli, A., Meda, A., Rinaldi, Z., Spagnuolo, S. and Maddaluno, G., 2017. Optimization of GFRP reinforcement in precast segments for metro tunnel lining. *Composite Structures*, *181*, pp.336–346.

Spagnuolo, S., Meda, A., Rinaldi, Z. and Nanni, A., 2017. Precast concrete tunnel segments with GFRP reinforcement. *Journal of Composites for Construction*, *21*(5), p.04017020.

Current Perspectives and New Directions in Mechanics, Modelling and Design of Structural Systems – Zingoni (ed.)
© 2022 Copyright the Author(s), ISBN 978-1-032-39114-4

An experimental study on the compression behavior of CFRP-Jacketed plastered RC columns

M.N. Yavuzer, S. Kolemenoglu, C. Balcı, M. Ispir & A. Ilki
Department of Civil Engineering, Istanbul Technical University, Istanbul, Turkey

ABSTRACT: A preliminary experimental study is conducted to examine the effectiveness of external FRP jacketing of RC columns without removal of plaster on their surfaces. For this purpose, eight RC columns (D=250 mm, H=500 mm) with plaster on the surface are produced. The test results show that external jacketing with FRPs without removing the plaster can be as effective as traditional ones.

1 INTRODUCTION

Over the last three decades, the axial compressive behavior of fibre-reinforced polymer (FRP)-confined concrete has received significant attention. Many experimental studies have been conducted on this topic (Ilki et al. 2008; Au & Buyukozturk 2005) The majority of studies have focused on effect of; strength of concrete, specimen's cross-sectional sizes and shapes etc. The number of studies examining the effect of plaster is very few. Karantzikis et al. (2005) examined the effectiveness of FRP confinement of cylindrical columns with mortar plastering as a part of the study. In compliance with the results of their study; plastered jacketed RC specimen show nearly same performance as the simple ones.

According to ACI440.2R-17, the plaster must be removed from the surface of the member before the FRP applications due to potential of plaster to reduce the effectiveness of FRP applications. However, removing the plaster from the member surface causes construction workers to work more, increases construction costs, and produces more construction waste. In addition, the concrete surface may be damaged due to workmanship-related reasons during plaster removal. Therefore, this study aims to examine the effectiveness carbon fiber-reinforced polymer (CFRP) jacketing without plaster removal on circular columns. The reason why chosing with circular columns is that the confinement pressure is uniform around the perimeter of the specimen.

2 SPECIMEN AND MATERIALS

In this study, 8 reinforced concrete short columns with circular cross section were produced with a ready-mix concrete. The average 28-day compressive strength of concrete was obtained as 12.2 MPa by testing two unjacketed concrete cylinders. After the short RC column specimens were produced, the specimens were plastered in different thicknesses. As shown in Table 1, the thicknesses of plaster are planned to be 10 and 40 mm and the strength of plaster is planned to be 2.5 MPa. The compressive strength of plaster at the time that short column experimental program start was determined as 2.3 MPa. 12 K Dowaksa Carbon A 49 sheets were used to wrap the specimens.

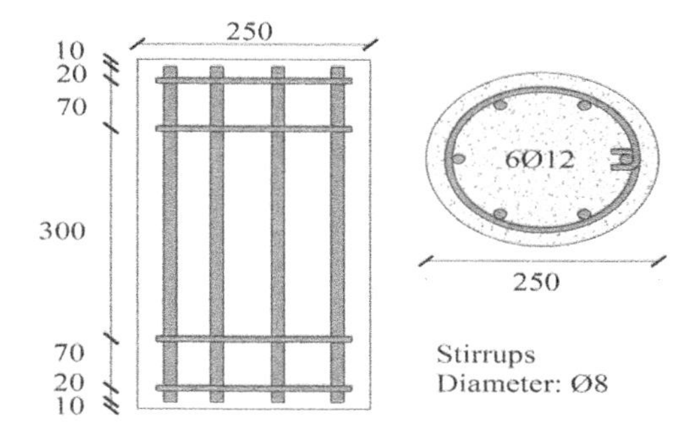

Figure 1. Reinforcement details (mm).

Table 1. Matrix of experimental program.

Specimen Code	D mm	f_{co} MPa	PT mm	PS mm	Plies
C15-PT0-PS0-W0a	250	15	-	-	-
C15-PT0-PS0-W0b	250	15	-	-	-
C15-PT0-PS0-W2a	250	15	-	-	2
C15-PT0-PS0-W2b	250	15	-	-	2
C15-PT1-PS2.5-W2a	250	15	10	2.5	2
C15-PT1-PS2.5-W2b	250	15	10	2.5	2
C15-PT4-PS2.5-W2a	250	15	40	2.5	2
C15-PT4-PS2.5-W2b	250	15	40	2.5	2

D: Diameter of the specimen, f_{co}: cylindrical compressive strength of concrete, PT: plaster thicknesses, PS: plaster strength

DOI: 10.1201/9781003348450-238

3 TEST SETUP

The specimens were tested under monotonic uniaxial compressive loads at a rate of 0.01 mm/sec by using an Instron testing machine with a capacity of 5,000 kN. Before testing, the specimens were pre-loaded/ unloaded to $0.05f_{cj}A_g$. For all column tests, the axial strains were calculated from seven longitudinally oriented linear variable differential transformers (LVDTs) as shown in Figure 2.

4 TEST RESULTS

Table 2 and Figure 3 shows the summary of test results. All specimens failed due to CFRP rupture. The rupture lateral strain of the specimen with 10 mm plaster is nearly same as for the ones without plaster. However, the rupture strain of the specimen with 40 mm is lower than others. Even if both 40 mm and 10 mm plaster layers failed before CFRP reaches its ultimate rupture strain, 10 mm ones crushed without local damage. Therefore, it should be stated that local damage in plaster play an important role on earlier rupture in CFRP.

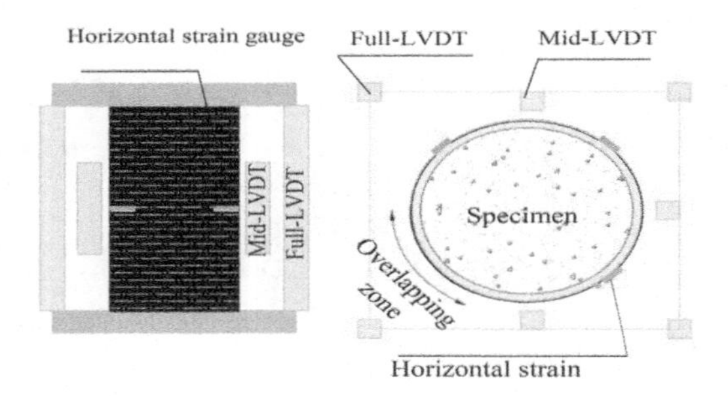

Figure 2. Short RC column test setup and measurement devices locations.

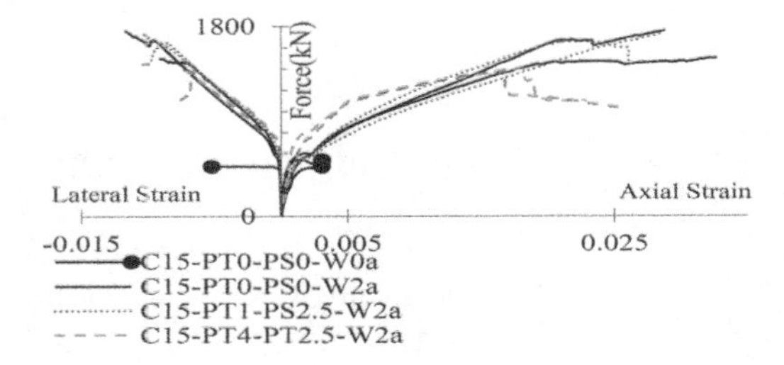

Figure 3. Force-strain relationship of specimens.

5 CONCLUSION

In this study, CFRP external jacketing application without plaster removal was examined with an experimental program. Short RC column specimens with different plaster thicknesses were prepared and tested under monotonic axial load either with or without external FRP jackets. Test results are presented with a focus on both plaster thickness and the presence of plaster. Based on the test results it is possible to conclude that the existence of plaster has not a significant effect on the efficiency of jacketing application. However, as the thickness increases, the ultimate load and axial strain of the specimen reduce.

REFERENCES

ACI Committee 440. 2017. *Guide for the Design and Construction of Externally Bonded FRP Systems for Strengthening Concrete Structures*. Michigan: ACI.

Au, C., A. & Buyukozturk, O. 2005. Effect of fiber orientation and ply mix on fiber reinforced polymer-confined concrete. *J. Compos. Constr.* 9(5): 397–407.

Ilki, A., Peker, O., Karamuk, E., Demir, C. & Kumbasar, N. 2008. FRP retrofit of low and medium strength circular and rectangular reinforced concrete columns. *J. Mater. Civ. Eng.* 20(2): 169–188

Karantzikis, M., Papanicolaou, C.G., Antonopoulos, C.P. & Triantafillou T.C., 2005. Experimental investigation of non-conventional confinement for concrete using FRP. *J. Compos. Constr.* 9: 480–487.

Table 2. Test results.

Specimen Code	F_{co} (kN)	$F_{cc\text{-}wop}$ (kN)	$F_{cc\text{-}wp}$ (kN)	ε_{co}	$\varepsilon_{cc\text{-}wop}$	$\varepsilon_{cc\text{-}wp}$	ε_{ch}	$\frac{F_{cc}}{F_{co}}$	$\frac{F_{cc-wp}}{F_{cc-wop}}$	$\frac{\varepsilon_{cc}}{\varepsilon_{co}}$	$\frac{\varepsilon_{cc-wp}}{\varepsilon_{cc-wop}}$
C15-PT0-PS0-W0a	475	–	–	0.0030	–	–	–	–	–	–	–
C15-PT0-PS0-W0b	541	–	–	0.0031	–	–	–	–	–	–	–
C15-PT0-PS0-W2a	–	1757	–	–	0.0289	–	0.009	3.46	–	9.63	–
C15-PT0-PS0-W2b	–	1500	–	–	0.0322	–	0.012	2.95	–	10.73	–
C15-PT1-PS2.5-W2	–	–	1656	–	–	0.0209	0.009	3.26	1.02	6.96	0.68
C15-PT1-PS2.5-W2	–	–	1731	–	–	0.0286	0.010	3.41	1.06	9.53	0.93
C15-PT4-PS2.5-W2	–	–	1400	–	–	0.0181	0.007	2.76	0.86	6.03	0.59
C15-PT4-PS2.5-W2	–	–	1384	–	–	0.0168	0.007	2.72	0.85	5.60	0.55

F_{co}: ultimate load of unconfined specimen, $F_{cc\text{-}wp}$: ultimate load of FRP jacketed specimen with plaster, $_{cc\text{-}wop}$: ultimate load of FRP jacketed specimen without plaster, ε_{co} strain corresponding to ultimate load of unconfined specimen, $\varepsilon_{cc\text{-}wp}$: strain corresponding to ultimate load of FRP jacketed specimen with plaster, $\varepsilon_{cc\text{-}wop}$: strain corresponding to ultimate load of FRP jacketed specimen without plaster, ε_{ch}: maximum measured transverse strain on CFRP jacket

Current Perspectives and New Directions in Mechanics, Modelling and Design of Structural Systems – Zingoni (ed.)
© 2022 Copyright the Author(s), *ISBN 978-1-032-39114-4*

Parametric study of CFFT systems for small-scale wind turbine towers

Y. Gong & M. Noël
University of Ottawa, Ottawa, Canada

ABSTRACT: This paper studied the feasibility of concrete-filled fibre-reinforced polymer tubes (CFFTs) for small-scale wind turbine towers using the finite element method. Cantilever tower models were developed and tested to study the effect of different parameters. The influence of height-to-diameter ratio was also studied and a conservative preliminary design that can be refined for specific turbine systems and wind conditions was adopted using the results. In conclusion, CFFTs can be an efficient alternative to steel and concrete towers for distributed wind energy systems for remote communities.

1 INTRODUCTION

Over 200,000 individuals in Canada live in one of nearly 300 remote off-grid communities. Typically, they rely on diesel power which is associated with both high emissions and cost. In those areas, wind power can be a good replacement to diesel energy.

Wind turbine towers are often made of steel tubes or concrete, but concrete-filled FRP tubes (CFFTs) is considered as an alternative solution due to its benefits. The lightweight outer FRP tube is a permanent formwork for the concrete and ensures that CFFTs have a good tensile behaviour and are suitable for flexural loading. Tapered tubes can also be used to facilitate segmental construction to simplify transportation and erection of the towers (Watfa et al. 2021).

The finite element (FE) method is useful since large columns and beams are difficult to build and test in the laboratory. Fam and Son (2008) modeled hollow FRP tubes and CFFTs under uniform loading and four-point bending, and they developed a simple design method to predict the optimum concrete filling length (Fam & Son, 2008).

This work contributes to the overall goal of improving the feasibility of wind energy solutions for remote areas.

2 MODEL VALIDATION

2.1 *Finite element models*

One hollow beam model, two CFFT beam models, and two short columns were developed using ABAQUS and validated using experimental results reported by Fam (2000).

The slight asymmetry in the laminate structure in one CFFT beam was ignored to produce a symmetric quarter-model. Other models were developed as full models. The beams were restricted in x and y direction at both sides and were restricted in z direction at the mid of the model. The short columns were fixed at the bottom surface.

For concrete, plastic properties were defined using the concrete damaged plasticity model in ABAQUS. The stress-strain relationship for the concrete in compression was obtained using the Hognestad concrete model. The FRP material was defined as a lamina structure, and Tsai-Wu failure criteria was considered.

The concrete was modeled using a C3D8R element and the FRP tube was modeled using a S4R element. The FRP tube and concrete core were connected using surface-to-surface contact.

2.2 *Results*

The mid-span deflection and longitudinal strains in the maximum moment region outputs were obtained.

The estimated load and deflection results at failure of the three beams agreed with experimental results well, but some discrepancies were observed with respect to ultimate strain values. The behaviour of the column models is acceptable for the purpose of the study, since the performance of wind turbine towers is mostly controlled by bending. Overall, the behaviour of the models was considered as satisfactory for the purpose of this study.

3 PARAMATRIC STUDY

3.1 *Finite element models*

The concrete and FRP were modelled in the same way as the validation study using a concrete compressive strength of 47 MPa. Steel reinforcing bars

DOI: 10.1201/9781003348450-239

were modeled as a perfectly elasto-plastic material while prestressing tendons were modelled with a non-linear stress-strain relationship. The initial temperature load method was used to introduce prestressing.

The concrete core was modeled using C3D8R elements, the FRP tube was modeled using S4R elements, and the steel reinforcing bars and prestressing tendons were modeled using T3D2 elements.

The FRP tube and concrete core were connected using surface-to-surface contact. The steel reinforcing bars and prestressing tendons were connected to the concrete core using embedded region constraints.

3.1.1 *Part A: Geometry and reinforcement*

Ten CFFT wind turbine tower models (D1 to D10) with different parameters including taper ratio, fibre orientation, steel reinforcement ratio, prestressing level, and concrete filling ratio were created. All the models were subjected to a horizontal concentrated load applied at the top of the tower.

Symmetry was enforced along the x-z plane, and the bottom surface of the tower model was fixed.

3.1.2 *Part B: Height/Diameter ratio*

Four models with different h/D ratios were simulated in this section. The taper ratio, concrete filling ratio, concrete strength, steel reinforcement ratio, number of prestressing tendons and prestressing level were 2%, 50%, 47 MPa, 2.3%, 6 and 50%, respectively. Six 45M steel rebars and six seven-wire strands were modeled for each specimen. Table 3.1 shows the geometric details.

Table 3.1. The geometric details of CFFT models.

Model number	L1	L2	L3	L4
Height (mm)	10000	25000	20000	20000
Base diameter (mm)	1000	1500	1500	1000
Top diameter (mm)	800	1000	1000	800

3.2 *Results*

All the models failed due to FRP rupture near the base of the tower. Overall, a bilinear load-deflection response was obtained for CFFT tower models without inner reinforcement. With the increase of steel reinforcement ratio and prestressing ratio, the results became more non-linear.

Increasing the steel reinforcement ratio, prestressing ratio, and fibre oriented in the longitudinal direction can increase the load capacity. The taper ratio only influences the stiffness of CFFT models and does not affect the load capacity. The increase of concrete filling ratio is an effective method to improve the behaviour of CFFT models when it is lower than 50%; however, when over it is over 50%, the improvement in behaviour declines massively.

Figure 3.1 shows the load-deflection relationships of four models. The circular marks represent the service load that was obtained using wind speed and wind turbine tower parameters, and the triangular marks represent the deflection limit that was considered to be 1.25% of the height (Nicholson, 2011). It is clear that model L1 (h/D = 10) and L3 (h/D = 13.3) were over-designed, since the service load was lower than the load at the deflection limit; however, model L2 (h/D = 16.7) and L4 (h/D = 20) were under-designed, as the opposite result was observed.

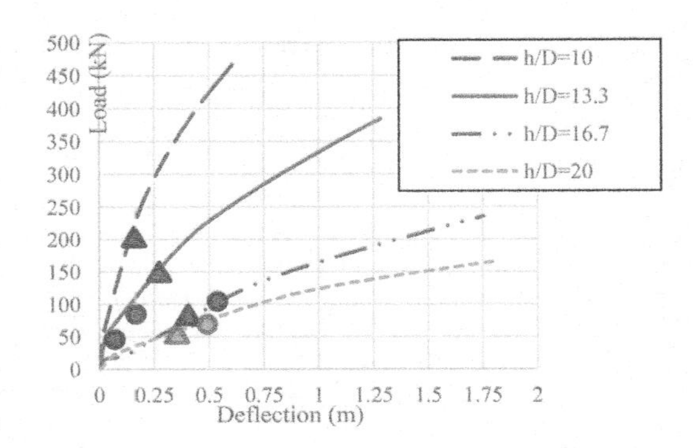

Figure 3.1. Load-deflection relationships of different model L1 to L4 (circle = service load, triangle = deflection limit).

4 CONCLUSION

According to the results, a preliminary design of a CFFT wind turbine tower has been completed using FE modelling. It is recommended to have approximately 15 h/D ratio, 2% taper ratio, 50% concrete filling ratio, 10 mm GFRP tube thickness, and 2% steel reinforcement ratio.

REFERENCES

Fam, A., & Son, J.-K. (2008). Finite element modeling of hollow and concrete-filled fiber composite tubes in flexure: Optimization of partial filling and a design method for poles. *Engineering Structures*, *30*(10), 2667–2676. Scopus.

Fam, A. Z. (2000). *Concrete-filled fiber reinforced polymer tubes for axial and flexural structural members* [Ph.D. thesis]. The University of Manitoba.

Nicholson, J. C. (2011). *Design of wind turbine tower and foundation systems: Optimization approach* [Master of Science, University of Iowa]

Watfa, A., Green, M., Fam, A., Noël, M. (2021). Segmental hollow concrete filled FRP tubes (CFFT) for wind turbine towers. *10th International Conference on FRP composites in Civil Engineering, Dec.* 8–11.

Current Perspectives and New Directions in Mechanics, Modelling and Design of Structural Systems – Zingoni (ed.)
© 2022 Copyright the Author(s), *ISBN 978-1-032-39114-4*

Estimating shear strengths of GFRP reinforced concrete deep beams with indeterminate strut-and-tie method

S. Liu & M.A. Polak
Department of Civil and Environmental Engineering, University of Waterloo, Waterloo, Ontario, Canada

ABSTRACT: This paper discusses the method on how to construct and analyze the indeterminate strut-and-tie models (ISTMs) for glass fibre reinforced polymer (GFRP) reinforced concrete (RC) deep beams. This paper focuses on how to properly model the concrete struts including their sizes, stiffnesses and strengths that relate to the softening factors. The predicted ISTM results are compared to the test results. Conclusions and recom-mendations on the application of ISTM for concrete reinforced with linear elastic brittle reinforcements are provided.

1 INTRODUCTION

GFRP bars are linear elastic and brittle, hence GFRP RC deep beams cannot assume reinforcement yielding and require concrete crushing as the preferred failure mode. Therefore, an indeterminate STM (ISTM) becomes suitable for such beams. This study introduces a detailed way to construct and analyze such models with the beams tested by Krall and Polak (2019).

2 BEAMS TESTED BY KRALL AND POLAK (2019)

This paper focuses on analyzing the GFRP RC deep beams with stirrups tested by Krall and Polak (2019), and the analyses are performed on 6 beams tested by Krall and Polak (2019) listed in Table 1.

3 CONSTRUCTING AND ANALYZING INDETERMEDAITE STRUT-AND-TIE MODELS

3.1 *Proposed model*

The proposed ISTM makes the whole deep beam into fanning regions. Figure 1 shows the proposed ISTM for beams with 220mm stirrup spacings as an example.

3.2 *STM elements*

STMs are composed of ties, struts and nodes. Ties represent reinforcements. Properties of ties are the properties of the reinforcement. Struts represent the concrete elements creating the load path. The widths of the struts are computed from the nodes, and the elastic modulus of the struts are the tangent modulus of elasticity. Node regions are where the struts and ties meet and are sized based on the model geometries.

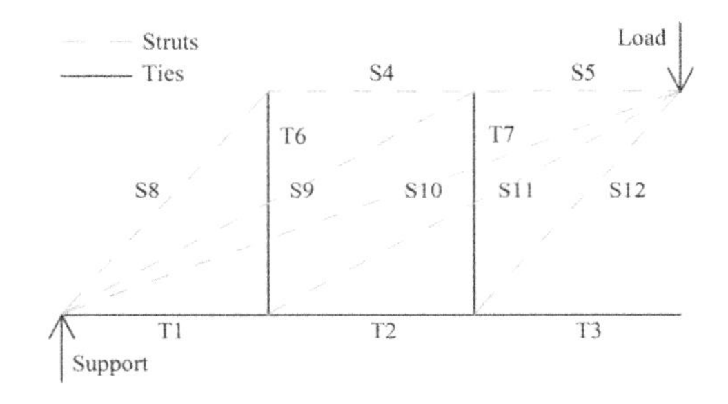

Figure 1. Proposed ISTM for beams with 220mm stirrup spacings.

3.3 *Failure modes*

The desired failure mode of the ISTM is having enough numbers of struts crushed making the model unstable. The failure of a strut occurs when its tangent modulus of elasticity reaches zero. Models fail by horizontal struts are failed in compression; models fail by inclined struts are failed in shear; and models fail by both kinds of struts are failed in both shear and compression. The ISTM might also be failed by rupture of ties or crushing of node regions, but such failures are not desirable.

3.4 *Height of the compression block*

Because strut crushing is the preferred failure mode, the height of the compression block (h_C) is important. Based on a preliminary finite element analysis performed for the specimens, h_C is conservatively determined to be 0.2d.

3.5 *Softening factors*

Softening factors (k) are utilized to reduce the estimated strengths of concrete struts from tested cylinder strengths. Approaches adopted and utilized in the

DOI: 10.1201/9781003348450-240

Table 1. Beam details*.

Specimens	Height (mm)	Width (mm)	Depth (mm)	a/d	Flexural Reinforcement d_{bar} (mm)	f_u (MPa)	E (GPa)	ρ_f (%)	Shear Reinforcement d_{bar} (mm)	$f_{u,bent}$ (MPa)	E (GPa)	$s_{stirrup}$ (mm)
BM12-220	350	200	270	2.5	12	1000	60	2.51	12	700	50	220
BM12-150	350	200	270	2.5	12	1000	60	2.51	12	700	50	150
BM16-220	345	200	270	2.5	16	1000	64	2.23	12	700	50	220
BM16-150	345	200	270	2.5	16	1000	64	2.23	12	700	50	150
BM25-220	330	200	270	2.5	25	1000	60	1.82	12	700	50	220
BM25-150	330	200	270	2.5	25	1000	60	1.82	12	700	50	150

* Data obtained from Krall and Polak (2019).

Table 2. Experimental and predicted results.

			ACI 318-19 (2019)		Nehdi et al. (2008)		CSA S806-12 (R2017)		Proposed Approach	
Specimens	$P_{exp.}$ (kN)	Failure Mode	$P_{predict}$ (kN)	Failure Mode	$P_{predict}$ (kN)	Failure Mode	$P_{predict}$ (kN)	Failure Mode	$P_{predict}$ (kN)	Failure Mode
BM12-220	382	Shear	401	Node	401	Node	223	Shear	353	Shear
BM12-150	405	Shear	359	Compression	359	Compression	307	Shear	363	Combined
BM16-220	309*	Shear	393	Combined**	393	Combined	216	Shear	345	Shear
BM16-150	417	Shear	350	Compression	350	Compression	296	Shear	355	Combined
BM25-220	360	Shear	375	Combined	375	Combined	195	Shear	319	Shear
BM25-150	416	Shear	326	Compression	326	Compression	266	Shear	337	Combined

* Tested with errors, lower than actual strength.
** Combined failure mode is a combination of compression and shear failure modes

analysis include those from ACI 318-19 (2019), Nehdi et al. (2008), CSA S806-12 (R2017) that is based on Modified Compression Field Theory (MCFT) by Vecchio and Collins (1986). A new approach modified from MCFT is also proposed to include the strength increase from having transverse reinforcement.

3.6 *Analysis process*

The analysis is iterative, and the load is increased in small increments till failure of the model.

4 RESULTS AND DISCUSSION

4.1 *Results and discussions*

The results are organized in Table 2. The best results are predicted through the proposed softening factor approach, which are conservative, follow the correct trend and are accurate. The results from the approach by CSA S806-12 (R2017) are also good, as they are also conservative and follow the correct trend.

5 CONCLUSIONS AND RECOMMENDATIONS

GFRP RC deep beams from Krall and Polak (2019) are analyzed based on the proposed ISTM with a constant compression block height of 0.2d and softening factor approaches from ACI 318-19 (2019), Nehdi et al. (2008) and CSA S806-12 (R2017) and the proposed softening factor approach. By comparing the analysis results with the experimental results, the ISTM works best with the proposed softening factor approach. For further studies, the proposed model can be tested on more beams, and a more detailed analysis can be conducted on h_C. Because the approach from CSA S806-12 (2012) also predict reasonable results, it can be further modified to increase the accuracy.

REFERENCES

ACI Committee 318. 2019. *Building Code Requirements for Structural Concrete (ACI 318-19)*. Farmington Hills: American Concrete Institute.

CSA Group. 2019. *A23.3-19 Design of Concrete Structures*. Toronto: CSA Group.

CSA Standard. 2017. *S806-12 Design and Construction of Building Structures with Fibre-Reinforced Polymers*. Mississauga: Canadian Standards Association.

Krall, M. & Polak, M.A. 2019. Concrete beams with different arrangements of GFRP flexural and shear reinforcement. *Engineering Structures* 198.

Nehdi, M., Omeman, Z. & El-Chabib, H. 2008. Optimal efficiency factor in strut-and tie model for FRP-reinforced concrete short beams with (1.5 < a/d < 2.5). *Materials and Structures* 41: 1713–1727.

Vecchio, F.J. & Collins, M.P. 1986. The modified compression-field theory for reinforced concrete elements subjected to shear. *ACI Journal* 83(2): 219–231.

Current Perspectives and New Directions in Mechanics, Modelling and Design of Structural Systems – Zingoni (ed.)
© 2022 Copyright the Author(s), ISBN 978-1-032-39114-4

State of the art in research and field applications of post-tensioned structures strengthened with prestressed composites

R. Kotynia & M. Staśkiewicz
Department of Concrete Structures, Lodz University of Technology, Lodz, Poland

ABSTRACT: The paper focuses on two items: research and field applications of the post-tensioned girders strengthened with prestressed carbon fiber reinforced polymer (CFRP) composites. The research presents a description of the post-tensioned girders with a comparative analysis of the girders differing in several parameters summarised in the database of experimental results. The research of prestressed girders confirmed high efficiency of the CFRP prestressing methods even for low CFRP prestressing levels. The second part of the paper describes several examples of field applications: Lauter River Bridge, Escher River Bridge, Rijeka River Bridge and Pilsia River Bridge. The most important parameter in strengthening with prestressed CFRP materials was the prestressing level. Increase in the prestressing level improved not only the load capacity but the serviceability limit states as well.

1 INTRODUCTION

1.1 *Flexural strengthening techniques with prestressed FRP materials*

A key parameter affecting the quality and effectiveness of strengthening using prestressed carbon fiber reinforced polymer (CFRP) composites is anchorage. Since the beginning of the development of the prestressing systems, numerous concepts of prestressing-anchoring systems have been developed. An overview of the prestressing systems was published by Czaderski & Motavalli (2007) and El-Hacha & Rojob (2016). One of the best solutions is to introduce prestressing to the composite indirectly. This method involves forcing an inverse deflection in the reinforced concrete (RC) members and then bonding the composite reinforcement to the concrete surface in a passive manner. In the next step, the forced deflection is reduced, which in turn causes an increase in deformation in the external reinforcement, resulting in its compression.

The most effective prestressing methods used for external structures are based on the direct prestressing of the CFRP materials with specialized anchorage systems and bonding them to the structure.

This concept was invented at the EMPA Laboratory and published by Deuring (1993). Several prestressing systems require installation on the lateral and end anchores on the strengthened members (Wight et al. 2001), (Franca et al, 2007), (Kim et al. 2008), (You et al. 2012). However, the most popularity gained the prestressing anchorage systems based on the devices that are installed directly to the surface of the strengthened structure.

The most popular systems are prestressing-anchoring systems, which use devices that are installed directly to the surface, usually using anchors installed into the concrete by the mechanical anchoring elements allowing to transfer of the prestressing force directly to the structure. Numerous research on prestressing systems were developed all over the world by Wu (2007). Numerous systems have found widespread recognition and they are successfully used in practice in the structural applications: Sika Leoba CarboDur II, Sika Stress-Head (Sika 2002), Neoxe Prestressing System II, BBR-Stahlton System S&P System, Gradient Method Stöcklin & Meier (2003).

2 RESEARCH ON POST-TENSIONED GIRDERS STRENGTHENED WITH PRESTRESSED CFRP MATERIALS

2.1 *Description of selected research*

The first world's composite application of the anchorless prestressed CFRP system was performed at the Swiss Federal Laboratories for Materials Science and Technology – EMPA in Zurich by Meier's team Stöcklin & Meier (2003). The steel anchors protecting the ends of the CFRP composite from detachment were changed for the gradual reduction of the prestressing force at the end sections of the CFRP laminates. This process was achieved by glueing the subsequent sections of the laminate under progressively lower stress,

DOI: 10.1201/9781003348450-241

which in turn is possible by using the action of increased temperature to significantly accelerate the setting of the adhesive mortar, bonding the CFRP laminate to the concrete surface. This method is named the *gradient system* based on the prestressing devices in the form of tensioning wheels fixed to the surface of the reinforced element on steel frames containing at the same time heating devices and an air cushion to accelerate the setting of the adhesive at the end sections. The research by (Fernandes et al. 2013) confirmed the effectiveness of strengthening large-scale girders using CFRP strips. Despite a very large span of the girder (20 m) and the application of only two passive CFRP laminates, the load capacity increased by 22%. While using the prestressed CFRP strips with pressing to 35 % of the tensile strength indicated an increase in the cracking load by 38% and 30 % increase in the steel yielding.

The implementation of mechanical anchoring in prestressed elements is troublesome due to the presence of cables or prestressing tendons, which cannot be cut through by drilling holes. It is necessary to use the special arrangement of CFRP strengthening using anchorless methods by Czaderski & Motavalli (2007) and (Aram et al. 2008). This *gradient system* was successfully tested on the post-tensioned bridge girders performed in the TULCOEMPA Project referred to the bridge that was reconstructed and strengthened in Poland in 2015 (Kotynia et al. 2015).

2.2 *International field applications*

As a result of numerous scientific studies, many structural strengthening systems have been applied not only in research programmes but also in the field applications of the strengthening of existing buildings and bridge structures.

The world's first use of prestressed CFRP laminates to strengthen a bridge structure took place in Gomadingen, Germany in 1998. The four-span bridge over the Lauter River required strengthening due to excessive cracking of two post-tensioned concrete girders in the support zone, resulting from the improper location of the prestressing cables Andrä & Maier (2000). Sika Leoba CarboDur system was used for this strengthening. The Sika Stress-Head reinforcement system was first used for the structural strengthening of the three-span bridge in Weesen, Switzerland (Berset et al. 2002) that was highly cracked. The Sika Stress-Head system was also used for strengthening the Sung San viaduct in Seoul in 2002, the Hütten bridge in Lucerne in 2003, the Clinton and Hopkins bridges in Ohio in 2003 and the A7 road viaduct at Zaandijk in 2004 (Sika 2004).

2.3 *Analysis of experimental test results of post-tensioned members strengthened with prestressed CFRP materials*

In the literature review a very small number of studies were available on the post-tensioned concrete members. The purpose of developing this database was to analyse the effectiveness of this type of strengthening depending on the tested variable parameters, which differ in the strengthened elements. To determine the effectiveness of the strengthening efficiency, the parameters of the strengthening ratio in failure (η_u) and the strengthening ratio in cracking (η_{cr}) were defined as the relative increase in the load capacity of the tested elements, expressed as the strengthening ratio:

$$\eta_u = (M_u - M_{u0})/M_{u0} \tag{1}$$

$$\eta_{cr} = (M_{cr} - M_{cr0})/M_{cr0} \tag{2}$$

where $M_{u;}$ M_{u0} = ultimate bending moment of the strengthened and reference member and $M_{cr;}$ M_{cr0} = cracking bending moment of the strengthened and reference member.

2.4 *Conclusions*

Based on the collected test data the following conclusions may be drawn:

1. The most important parameter in strengthening with prestressed CFRP materials is the prestressing level. The increase in the prestressing level increases not only the load capacity but improves the serviceability limit states (mainly deflections).
2. The increment of the prestressing level concerning cracking ($\Delta\eta_{cr}$) was even four times greater than the increase in the load-bearing capacity ($\Delta\eta_u$). The increment of the crack loading was dependent on the prestressing level (ε_{fp} from 3.0 ‰ to 8.4 ‰), which corresponded to an increase in the crack loading from 19.3 % to 58.8 %.
3. Test results indicated that the prestressing level did not have a significant effect on the load-bearing capacity of a post-tensioned member, because of the CFRP rupture, which made a big difference between the cracking load and ultimate load.
4. The last important parameter was the type of anchored post-tensioned members with prestressed laminates, which indicated the highest strengthening degree and high utilisation of the tensile strength of the CFRP laminates (η_u = 0.347 ÷ 0.738) for elements that failed due to CFRP rupture.

The most efficient *gradient anchorage system* indicated very high efficiency in the large scale girder with the strengthening ratio of with the strengthening degree of η_u = 0.452 and the CFRP tensile strength utilisation of 82 %.

Current Perspectives and New Directions in Mechanics, Modelling and Design of Structural Systems – Zingoni (ed.)
© 2022 Copyright the Author(s), ISBN 978-1-032-39114-4

Reinforcement made of Basalt Fibre Reinforced Polymer (BFRP): Load-bearing capacity, durability and applications

S. Görtz & K. Lengert
Kiel University of Applied Sciences, Kiel, Germany

D. Glomb, B. Wolf, A. Kustermann & C. Dauberschmidt
Munich University of Applied Sciences, Munich, Germany

ABSTRACT: This paper covers the usability of bars made of basalt fibre reinforced polymer (BFRP). In particular an overview is given on durability, load bearing capacity for bending and shear loads and cracking behaviour. Furthermore, sustainability is evaluated based on the carbon footprint. Finally examples of first applications are given.

1 INTRODUCTION

Because of high costs for repair and maintenance of concrete buildings with steel reinforcement, there is a worldwide increase of research and use of reinforcement made of fibre-reinforced polymer.

Basalt fibres are cost-effective and the tensile strength is sufficient for many applications. Further advantages are that basalt is nearly unlimitedly available and due to the mineral basis, recycling is much easier in case of demolition.

Since 2014, several research projects and various preliminary studies on concrete components with basalt fibre reinforcement have been carried out or are ongoing at the Universities of applied Sciences in Munich (MUAS) and Kiel (see Wolf et al. 2019, Wolf 2018, Glomb 2019, Görtz 2019, Görtz 2020, Kustermann et al. 2021).

2 MATERIAL PROPERTIES

The material properties of the basalt fibres differ depending on the mining regime of the mineral but can be summarized as follow (see Liu, 2007):

- Tensile strengths from 3.0 to 4.5 GPa
- Modulus of elasticity approx. 85 GPa (comparable to glass fibre reinforcement)
- High thermal and chemical resistance
- Low thermal and electrical conductivity
- Low density (with approx. 2.7 g/cm³ comparable to concrete)

3 DURABILITY

Investigations into the alkaline resistance of basalt reinforcement were carried out in artificially produced concrete pore solutions and sodium hydroxide solutions in order to determine alkaline susceptibility and thus the chemical potential for damage.

4 FLEXURAL CAPACITY

The flexural strength can be determined by iterating the strain line of the girder. From an economic point of view, it is necessary to design the girders, so that the maximum tension strength is reached in ultimate limit state.

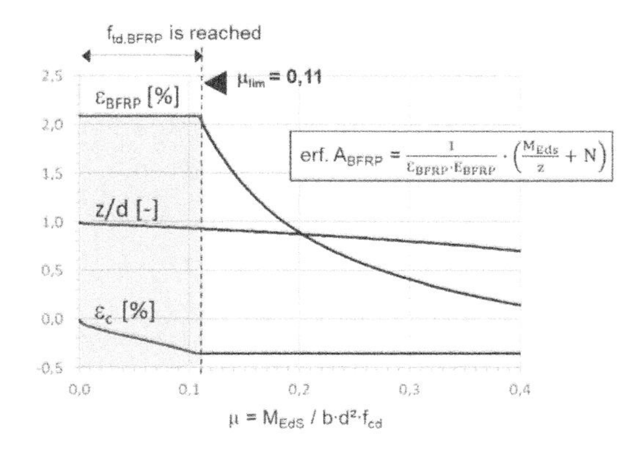

Figure 1. General dimensioning diagram for the BFRP reinforcement T18 from the Deutsche Basalt Stab GmbH.

DOI: 10.1201/9781003348450-242

Based on this, similar diagrams for the design can be derived as for using steel reinforcement. Figure 1 shows an example for the general dimensioning diagram for the BFRP reinforcement T18 from the Deutsche Basalt Stab GmbH.

5 SHEAR CAPACITY

The lower modulus of elasticity of the basalt fibres leeds to a smaller uncracked compression zone and results in a minor flexural shear capacity compared with concrete girders with steel reinforcement.

In the case of girders with shear reinforcement, it is important to note that the load capacity can often not be reached because of problems with the anchoring.

6 CRACK CONTROL/BOND STRENGTH

The bond characteristics (and thus the anchoring lengths) are dependent on the profilation of the rebar, which differs in the currently available types of FRP reinforcement. In case of a high profiled surface it is possible to achieve higher bond strength than with steel reinforcement, so that limiting crack widths under service loads to values between 0.2 - 0.4 mm are possible despite of the low modulus of elasticity.

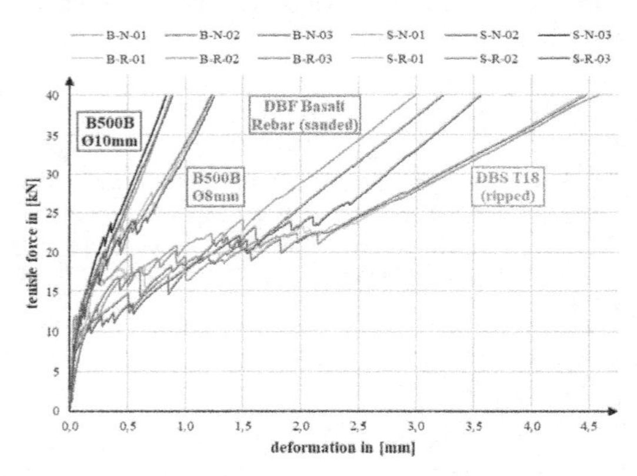

Figure 2. Summary of the deformation curves of the uniaxial tensile tests at MUAS (results for steel Ø10 mm and sanded BFRP see also Wolf et al. 2019 and Glomb 2019).

7 CARBON FOOTPRINT

Investigations show that basalt fibre reinforcement have a better CO2-footprint than steel reinforcement or carbon reinforcement.

The carbon footprint for the production of the BFRP reinforcement is around 2.5 kgCO2/kg BFRP. In contrast, for 1 kg steel reinforcement only approx. 0.68 kg CO2 are required, each as mean values for Germany (see www.oekobaudat.de).

However, if the lower density and the higher strength of the BFRP reinforcement are considered, the result is a CO_2 saving of approx. 45% compared to concrete steel.

$$\Delta \mathrm{CO2} = \left(1 - \frac{CO_{2,BFRP}}{CO_{2,s}} \cdot \frac{\rho_{BFRP}}{\rho_s} \cdot \frac{f_{yk}}{s} / \frac{f_{t,BFRP}}{f_{t,BFRP}}\right) \cdot 100$$
$$= \left(1 - \frac{2.5kgCO2}{0.68kgCO2} \cdot \frac{1,900kg/m}{7,850kg/m} \cdot \frac{500MPa}{1,15} / \frac{930MPa}{1,3}\right) .100$$
$$= 45\% \tag{1}$$

8 FIRST APPLICATIONS

First applications of basalt reinforcement took place in the repair of damaged facade panels: here, basalt reinforcement was installed to replace the existing reinforcement close to the surface. By installing the basalt reinforcement, thin layers of retrofitting mortar could be ensured.

Bridge caps or railroad ties made of reinforcement with basalt fibre are currently being developed. The results so far show that reinforcement made of BFRP is an interesting alternative to reinforcement made of steel or glass and carbon fibre from a technical and also an economic point of view.

REFERENCES

Wolf, B. & Dix, S. & Kustermann, A. & Dauberschmidt. C. & Schuler, C.: Instandsetzung vorgehängter Sichtbetonfassaden durch dünnwandige Fassadenergänzungen aus basaltbewehrtem Beton (Akronym: „FASALT“), Abschlussbericht Forschungsvorhaben FHprofUnt vom Oktober 2019

Görtz 2019: Tastversuche an Betonbauteilen mit Basalt-Bewehrungsstäben T6. Forschungsprojekt FH Kiel

Görtz 2020: Querkrafttragfähigkeit von Bahnschwellen mit Basaltfaserstabbewehrung. Forschungsprojekt FH Kiel

Kustermann, A. & Dauberschmidt, C. & Glomb, D. & Wolf, B. & Görtz, S. & Lengert, K. & Burgard, S. & Schwarzer, M. 2021: Entwicklung ressourcenschonender, dauerhafter und frostbeständiger Brückenkappen auf Grundlage nichtmetallischer Bewehrung und Betonen mit 100% rezyklierter Gesteinskörnung. ZIM-Projekt KK5102701KI0

Wolf, B., Glomb, D. S., Kustermann, A., Dauberschmidt, C.: Untersuchung des Zug- und Verbundverhaltens von Basaltfaserverstärkter Kunststoff-Stabbewehrung in Beton. Beton- und Stahlbetonbau 114 (2019), S. 454–464.

Glomb, D.: Experimentelle und numerische Untersuchungen zum Riss- und Zugtragverhalten von Betonkörpern mit Bewehrung aus Basaltfaserverbundkunststoff. Beitrag im DBV-Heft 45 März 2019, Deutscher Bau- und Bautechnik-Verein, März 2019

Deutsche Basalt Stab GmbH: https://www.deutsche-basalt-stab.com/wp-content/uploads/2021/01/Datenblatt_T18B-1.pdf, January 2021 (last access: February 2022)

Current Perspectives and New Directions in Mechanics, Modelling and Design of Structural Systems – Zingoni (ed.)
© 2022 Copyright the Author(s), ISBN 978-1-032-39114-4

Onsite manufacturing and applications of FRP pipes

M. Ehsani
QuakeWrap, Inc., Tucson, AZ, USA

ABSTRACT: The traditional method of constructing pipes in segments and shipping them to the site for assembly is inefficient. The cost of shipment, especially for larger diameter pipes could be tremendous and the leaks at joints are a constant maintenance concern for the life of the pipeline. This paper introduces a special technique for building FRP pipes that allow construction of the pipe at or near the job site. Case studies demonstrate how pipe segments are manufactured and used to repair deteriorated pipes and culverts. In an extension of this techniques shown in this video, layers of resin-saturated carbon or glass fabric are wrapped around a heated mandrel. The mandrel is heated and within minutes the pipe is cured. The mandrel is partially collapsed, and the finished pipe is removed, leaving a small length of the pipe at the tip of the mandrel. The process will continue by adding more layers of fabric until a pipe of desired length is constructed. This project has been funded through various grants from the US Dept. of Agriculture, and the National Science Foundation. One of the main uses of this technique is for developing countries or remote sites where the lack of manufacturing and or road infrastructure adds significant cost to construction of pipelines.

1 INTRODUCTION

Construction of pipes requires heavy equipment and complex manufacturing facilities. As a result, pipes are traditionally constructed in short segments and shipped to the job site, where they are joined together. The result is a pipeline with joints every 3-10m (10-30 ft) or so. These joints are a potential source of leaks, which can inflict significant loss of revenue as well as harm to the environment. For large diameter pipes, the transportation costs alone from the plant to the job site add significant expense to the project.

To overcome the above shortcomings, a new Fiber Reinforced Polymer (FRP) pipe is proposed. FRP products are comprised of fabrics constructed with carbon or glass fibers that are saturated with epoxy resin. When the resin cures, FRP reaches a tensile strength that is two to three times that of steel. The tensile strength of FRP is derived from the fiber content and fiber orientation.

2 TECHNICAL APPROACH

The StifPipe® described here uses the sandwich construction technique that has been used extensively in the aerospace and ship-building industries. In this approach, to increase the rigidity of the pipe at a low cost, light-weight core or 3D fabric is used as a filler material, like the web of an I-beam. Additional layer(s) of carbon or glass FRP will be used as the outer skin of the pipe (Figure 1).

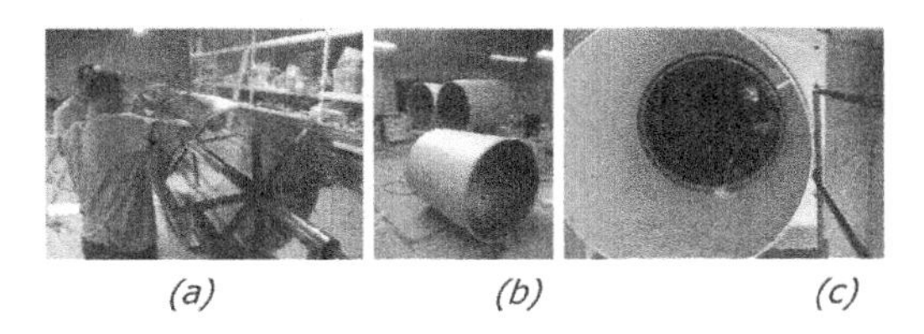

Figure 1. Manufacturing of StifPipe® for slip-lining corroded steel pipes: (a) adjustable diameter mandrel, (b) pipe segments prior to shipment, (c) finished repair showing a snug fit with between the StifPipe® and the host pipe.

Several case studies have been reported where StifPipe® has been used to repair 29 riser pipes in a pressurized system (e.g. in Puerto Rico) and a gravity flow culvert (e.g. in Queensland, Australia). In a record-setting application, 4.88m (16 ft) diameter pipe segments are being manufactured near the job site in Detroit, MI for repair of concrete tunnel.

3 CONTINUOUS MANUFACTURING PROCESS

The relative ease of the manufacturing of StifPipe® lends itself to continuous onsite production of a pipeline of any length, dubbed InfinitPipe®. With recent financial support through several U.S. government grants, this technology has been further tested

DOI: 10.1201/9781003348450-243

and developed a Mobile Manufacturing Unit (MMU) (Figure 2). The video linked on Figure 3 shows how InfinitPipe® is manufactured.

The pipe can be cut to any length and fittings, valves, elbows, flanges, etc. can be added at a later stage. These systems have been tested and it is demonstrated that they meet the pressure rating requirements of water pipes.

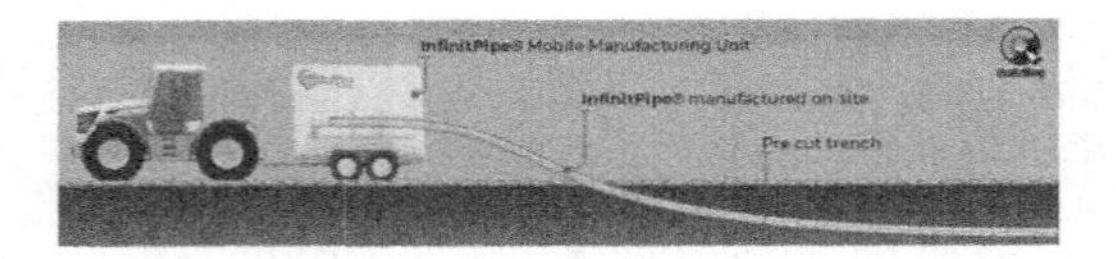

Figure 2. InfinitPipe®: A game-changing technology for onsite manufacturing of pipes.

Figure 3. Semi-automated Mobile Manufacturing Unit (MMU); watch a video of pipe being made at: www.tinyurl.com/InfinitPipe.

4 APPLICATIONS OF INFINITPIPE®

The pipe described above has many other applications beyond the conventional use for conveyance of water. Some of these unique applications are briefly discussed here.

One is for pipe construction in remote sites and developing countries where local villagers can quickly build pipes that meet the tough U.S. standards for potable water pipes. Please watch this video.

Horizontal Directional Drilling (HDD) has become very popular in recent years. InfinitPipe® offers a great solution for such projects where as the pipe is being built, it is pulled into the hole underground. This system reduces the laydown area significantly which is a concern in congested areas.

Ocean Thermal Energy Conversion (OTEC) is a process that can produce electricity, cooling or heating by using the temperature difference between deep cold ocean water and warm tropical surface waters. InfinitPipe® can be manufactured on a barge on the ocean and the pipe is lowered to the seabed as it is being manufactured. This technique of onsite manufacturing can result in significant cost savings. A similar application is for addressing the concerns with coral bleaching that is caused by global warming. In a solution proposed to the National Geographic's Chasing Genius Challenge we offered the use of this technique to address this problem in Australia's Great Barrier Reef. A mechanical pump activated by the waves in the ocean can pump cooler water from the deep portions to the warmer shallow parts where the corals live.

In-Situ Leaching (ISL) is a relatively new and increasingly applied method of mineral recovery, which costs less and has a smaller environmental impact in appropriate hydrogeological circumstances compared with other methods of recovery. ISL projects require hundreds of kilometers of pipe inserted into wells. InfinitPipe® allows the placement of a MMU directly above the well. As the pipe is manufactured, it is lowered into the well. The continuous pipe has no joints to leak and its non-corroding materials ensure a long service life.

Subsea pipelines are frequently used by oil and gas companies. Many of these pipelines are old and require repairs that are very challenging. Using InfinitPipe®, an MMU can be setup on shore and as the pipe is being manufactured it can be pulled with a winch into the host pipe. The new pipe/liner can be designed for a snug fit inside the host pipe with little loss of flow capacity in the pipe after the repairs.

5 SUMMARY AND CONCLUSIONS

This paper describes the sandwich construction technique that can be used to build pipes of any shape and size close to the job site. Examples of such applications are also presented, including a case where the largest diameter pipes are being produced for a project in Michigan with this technology.

The ease of manufacturing lends itself to automation for onsite production of the pipe and such equipment has been presented. The high cost of transportation limits the pipe manufacturers to supply their goods to a fairly limited geographic area around their plants. The technology presented here is game changing in that it allows the pipe manufacturer to participate in such projects globally regardless of the project location. The MMU and the required materials, i.e. rolls of fabric and drums of resin can be shipped to any point in the world and once the pipe is built, the equipment can be returned to its home base for maintenance before deployment to another site.

Several examples where this technology can be superior to the existing pipeline solutions have also been presented.

Current Perspectives and New Directions in Mechanics, Modelling and Design of Structural Systems – Zingoni (ed.)
© 2022 Copyright the Author(s), ISBN 978-1-032-39114-4

Bolted and hybrid beam-column joints between I-shaped FRP profiles

J. Qureshi
School of Architecture, Computing and Engineering, University of East London, London, UK

Y. Nadir & S.K. John
Department of Civil Engineering, College of Engineering Trivandrum, Thiruvananthapuram, Kerala, India

ABSTRACT: Presented are test results from five full-scale pultruded FRP beam-column joints under cyclic loading. The parameters include cleat position, joining method and cleat material. The joints' behaviour is assessed through hysteresis moment-rotation loops, accumulated dissipated energy and failure modes. The hybrid joints with steel cleats showed the best cyclic performance with the dissipated energy 75% higher than bolted joints. The bolted joint with FRP cleats gave the lowest dissipated energy, four times lower than steel cleated joints. Three failure modes were observed: shear-out failure, adhesive debonding and delamination cracking.

1 INTRODUCTION

Fibre reinforced polymer material has corrosion resistance, lower ecological impact, lightweight, and aesthetic possibilities. FRP composites contain carbon or glass fibres embedded in a polyester or vinylester resin matrix (Bank 2006; Qureshi et al. 2020). Limited research exists on cyclic response of FRP beam-column joints (Martins et al. 2021). This paper aims to investigate cyclic response of FRP joints between I-shaped sections. Five full-scale physical tests are performed. The parameters are cleat material (steel or FRP), cleat position (Flange or web cleated) and joining method (bolted or hybrid).

2 TEST CONFIGURATION

Figure 1 shows a single-sided beam-to-column joint sub-assembly test arrangement. The cleat is a - 50×50×6 mm leg-angle. A steel grade of Fe410 with a yield and ultimate strength of 250 MPa and 410 MPa is used for steel parts. Grade 8.8 M12 threaded bolts are used. There are five tests: SFSc2-bolted joint with web and flange steel cleats; SFSc2A-companion hybrid joint with bolting and bonding; SFStc2-bolted joint with steel flange cleats only; SFStc2A-companion hybrid joint and SFFc2- bolted joint with web and flange FRP cleats.

3 RESULTS AND DISCUSSION

3.1 *Moment-rotation hysteresis curves*

Moment-rotation hysteresis curves for steel and FRP flange and web cleated joints are shown in Figures 2-4. These figures also show companion monotonic response from (Qureshi et al. 2020) and cyclic backbone or reference skeleton curves. Backbone curves are produced by linking together peaks of full cycles.

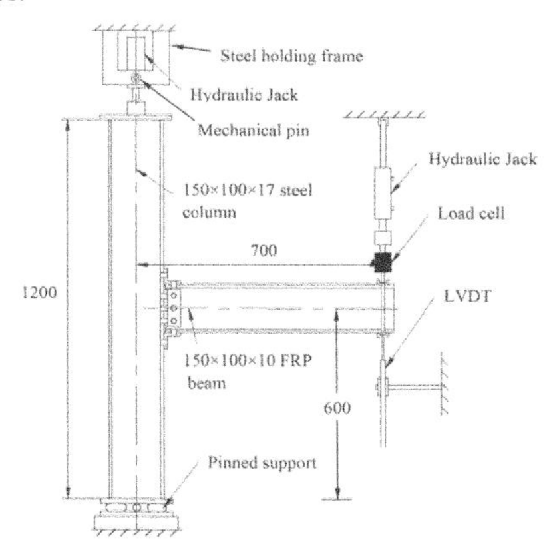

Figure 1. Schematic diagram of test configuration.

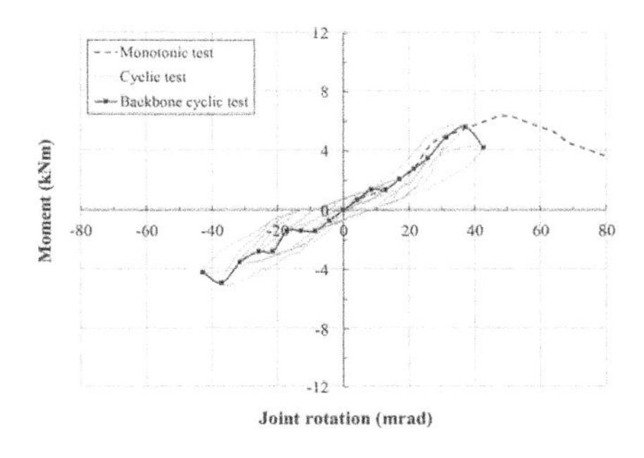

Figure 2. Cyclic moment-rotation hysteresis curves: SFSc2 – bolted joint with double web and flange steel cleats.

DOI: 10.1201/9781003348450-244

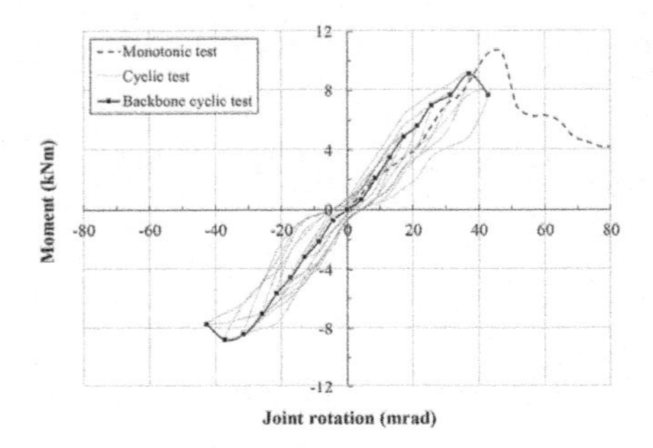

Figure 3. Cyclic moment-rotation hysteresis curves: SFSc2A – hybrid joint with double web and flange steel cleats.

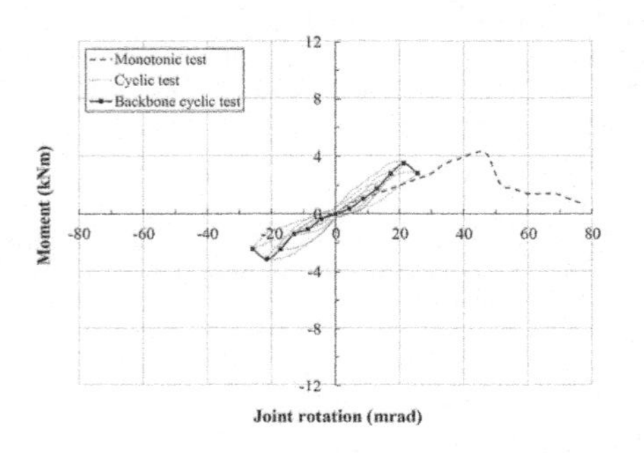

Figure 4. Cyclic moment-rotation hysteresis curves: SFFc2 – bolted joint with double web and flange FRP cleats.

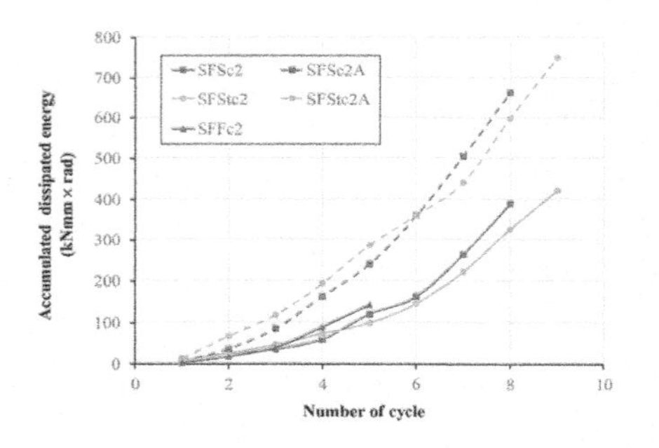

Figure 5. Accumulated dissipated energy of each joint versus number of cycles.

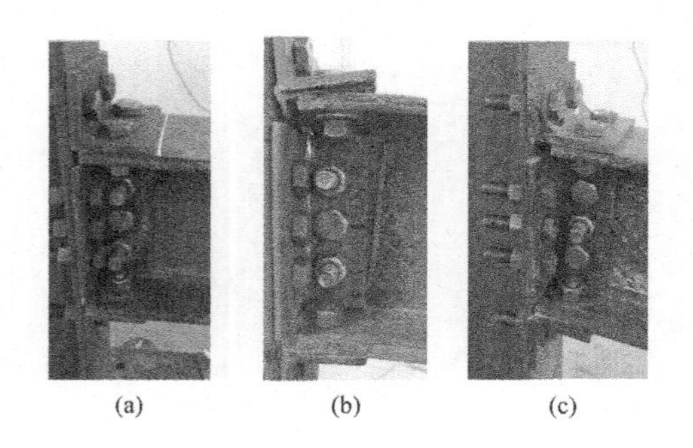

(a) (b) (c)

Figure 6. Failure modes: (a) Shear-out in SFSc2; (b) Adhesive debonding in SFSc2A; (c) Delamination cracking in SFFc2.

3.2 *Cyclic response*

The cyclic performance is evaluated by plotting the accumulated dissipated energy versus number of cycles in Figure 5. The hybrid joints presented the best overall performance in terms of dissipated energy. The final accumulated dissipated energy for hybrid joints was about 75% higher than bolted only joints. FRP cleated joint performed poorly with dissipated energy about half of the joint with steel cleats.

3.3 *Failure patterns*

Figure 6 shows failure modes: shear-out failure, adhesive debonding and delamination cracking. The shear-out failure happens when the distance from bolt centreline to the free end of the beam is short. This failure happened in steel cleated bolted joints, SFSc2 and SFStc2. For hybrid joints SFSc2A and SFStc2A, the failure was due to adhesive debonding followed by shear-out failure of the beam's bolted region. FRP cleated joint SFFc2 showed delamination cracking failure in FRP cleats.

4 CONCLUSIONS

- Shear-out failure, adhesive and delamination cracking failure modes were observed.
- Hybrid joints showed twice as much stiffness as bolted joints, regardless of joint detailing.
- The accumulated dissipated energy of hybrid joints was 75% higher than bolted joints. The worst cyclic performance was shown by FRP cleated joint with dissipated energy only half of the steel cleated FRP joint.

REFERENCES

Bank LC. 2006. Composites for construction - Structural design with FRP materials. NJ, USA: John Wiley & Sons.

Martins D, Gonilha J, Correia JR, Silvestre N. 2021. Exterior beam-to-column bolted connections between GFRP I-shaped pultruded profiles using stainless steel cleats, Part 2: Prediction of initial stiffness and strength. Thin-Walled Struct. 164.

Qureshi J, Nadir Y, John SK. 2020. Bolted and bonded FRP beam-column joints with semi-rigid end conditions. Compos Struct. 247:112500.

Current Perspectives and New Directions in Mechanics, Modelling and Design of Structural Systems – Zingoni (ed.)
© 2022 Copyright the Author(s), ISBN 978-1-032-39114-4

Performance of self-drilling screw connections for structural pultruded fibre reinforced polymer composites

Z. Cai & Y. Bai*
Department of Civil Engineering, Monash University, Clayton, Australia

L.C. Bank
School of Architecture, Georgia Institute of Technology, Atlanta, USA

C. Qiu
Department of Civil Engineering, Monash University, Clayton, Australia

X.L. Zhao
Department of Civil and Environmental Engineering, UNSW Sydney, Australia

ABSTRACT: Self-drilling screw is an efficient and cost-effective connection method for lightweight structural members. In this study, the performance of self-drilling screw connections for fibre reinforced polymer (FRP) structural members was investigated by a series of pull-out tests on connection specimens consisting of FRP plates joined to FRP square hollow sections (SHS) at a right angle. The varied parameters included the pultrusion direction of the FRP plate, screw gauge size, and coarseness of screw threads. The results indicated that the increase in screw gauge size and thread coarseness may improve the ultimate connection strength. A simplified equation was further presented to estimate the pull-out connection strength.

1 INTRODUCTION

Fibre reinforced polymer (FRP) composites manufactured by the pultrusion process are increasingly used in building construction, due to their high strength-to-weight ratio, durability, and availability in various standard section shapes (Bank, 2006). Extensive research has been done on connections for pultruded glass fibre reinforced polymer (GFRP) composites in structural applications (Bank et al., 1996, Qiu et al., 2018, Fang et al., 2019). However, results on the use of self-drilling screws in GFRPs are still scarce in literature.

In this study, screw connections were tested under static uplifting load when used for the components of GFRP SHS and GFRP connection plates. The effects of the pultrusion direction in the GFRP connection plate and the screw parameters on the mechanical behaviour of the screwed connection were discussed and clarified. A simplified analysis of the pull-out connections at the ultimate state was also conducted for strength estimation.

2 EXPERIMENTAL PROGRAME

The specimens were fabricated by joining pultruded GFRP SHS to pultruded GFRP flat plates (referred to as 'FP' in figures and table) at a right angle using self-drilling screw fasteners, as shown in Figure 1. Two types of flat plates were tested, with the pultrusion direction along either the longitudinal or the transverse plate direction. The self-drilling screw employed had a countersunk head and a self-drilling point. Four screw specifications were covered in the experiments with three screw gauges (6-, 8- or 10-gauge) and two thread types (coarse thread or fine thread). Therefore five types of specimens were prepared as shown in Table 1, where three repeating ones were considered for each type.

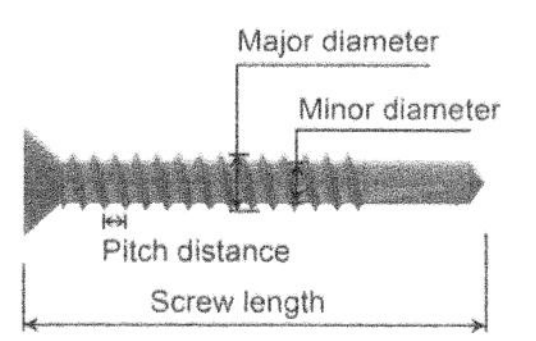

Figure 1. Self-drilling screw and its geometric parameters.

As shown in Figure 2, the specimens were examined using an Instron 50kN testing machine. Through a set of steel fixtures, the flat plate was clamped at the two ends while the SHS was lifted to the failure of the specimen.

3 RESULTS AND DISCUSSION

All the specimens with pultrusion direction oriented longitudinally in the flat plates (L-series) failed

*Corresponding author: yu.bai@monash.edu

DOI: 10.1201/9781003348450-245

Table 1. Configuration of specimens.

Specimens	Specification of screw		Pultrusion direction (GFRP FP)
	Thread type	Screw gauge	
L6C	Coarse	6 gauge	Longitudinal
L8C	Coarse	8 gauge	Longitudinal
L10C	Coarse	10 gauge	Longitudinal
L10F	Fine	10 gauge	Longitudinal
T8C	Coarse	8 gauge	Transverse

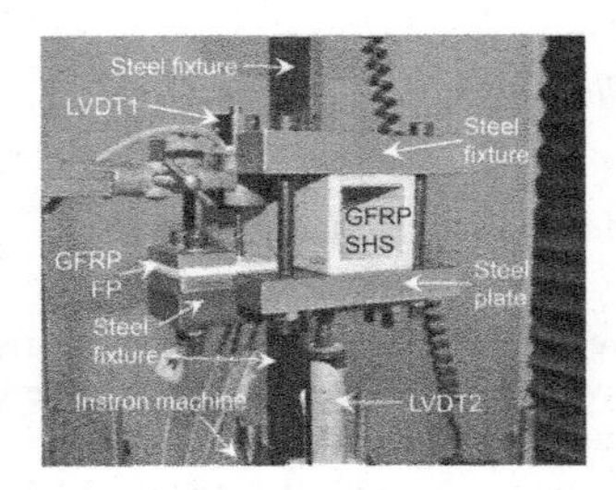

Figure 2. Experimental setup for all specimens.

ultimately at load $P_{u,Exp}$, as the screws were gradually pulled out from the SHS, striping off the GFRP materials of the screw-drilled holes on the SHS. The flat plates remained intact when the ultimate loads were reached (Figure 3a).

For the T-series specimens (T8C), the flat plates were bent upward until the ultimate load $P_{u,Exp}$ where they fractured in the middle section at the SHS side. After $P_{u,Exp}$, the screw heads were indented into the ruptured plates. No pull-out damage of the screw fasteners (from the SHS) was observed (Figure 3b).

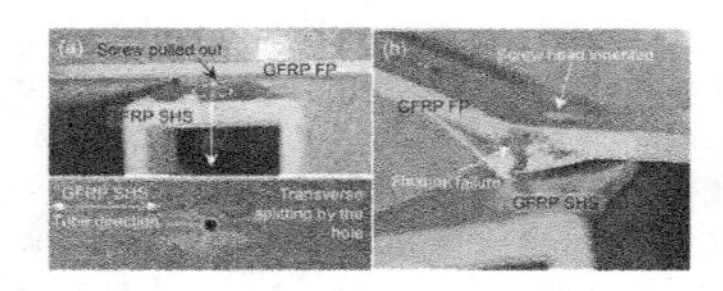

Figure 3. Typical failed specimens of (a) L8C; (b) T8C.

The experimental results of initial stiffness $K_{i,Exp}$, and ultimate load $P_{u,Exp}$ are plotted in Figure 4 for all specimens. Compared to T8C, L8C was 45% higher in $K_{i,Exp}$ and 35% higher in $P_{u,Exp}$. The L-series specimens also showed a less brittle failure mode.

Comparison of the specimens with different screw gauge sizes (i.e., L6C, L8C, and L10C), L10C exhibited the highest ultimate load of 2.403 kN on average. An increase in the gauge size of the screws in the connection enhanced the ultimate strength, while no clear trend was noted for the initial stiffness. Comparing the coarse-thread specimens L10C and the fine-thread specimens L10F, a 19% reduction in $P_{u,Exp}$ was associated with the decrease in pitch distance.

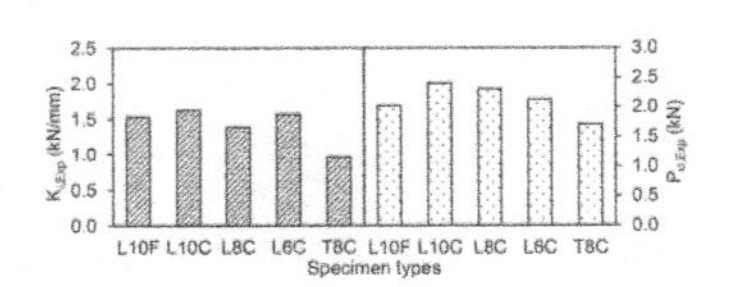

Figure 4. Comparison for experimental results of initial stiffness ($K_{i,Exp}$) and ultimate load ($P_{u,Exp}$).

4 ESTIMATION OF PULL-OUT STRENGTH

Governed by the pull-out of the self-drilling screws, the ultimate load P_u may be estimated assuming that the shear-out of the GFRP materials at screw threads in the SHS occurred simultaneously. The ultimate load $P_{u,Cal}$ is calculated for the L-series specimens by

$$P_{u,Cal} = \tau_f A_{s,eff} \tag{1}$$

where τ_f is the interlaminar shear strength of GFRP SHS (31.9 MPa), $A_{s,eff}$ is the effective shear area in the threaded hole on the SHS.

Satisfactory agreement was found for the specimens connected by coarse-thread screws (L6C, L8C, L10C), with a discrepancy of less than 10% from the experimental results. A larger discrepancy was found for the fine-thread specimens L10F ($P_{u,Cal}/P_{u,Exp}$ = 138%). The discrepancy can be attributed to the assumption that the GFRP materials at screw threads in the effective shear area $A_{s,eff}$ fail simultaneously.

5 CONCLUSIONS

The performance of screwed connections in pultruded GFRPs is investigated in this work through a series of pull-out tests. The specimens with pultrusion direction oriented transversely in the flat plates presented the lowest connection capacity. For the specimens with pultrusion direction oriented longitudinally in the flat plates, the use of screws with larger gauge sizes and coarser threads improved the ultimate connection strength. A simplified equation was used to estimate the strength, with a 38% higher result for the fine-thread L-series specimens while within the 10% difference for the coarse-thread ones.

REFERENCES

Bank, L. C. 2006. Composites for construction: structural design with FRP materials, John Wiley & Sons.

Bank, L. C., Yin, J., Moore, L., Evans, D. J. & Allison, R. W. 1996. Experimental and numerical evaluation of beam-to-column connections for pultruded structures. Journal of reinforced plastics and composites, 15, 1052–1067.

Fang, H., Bai, Y., Liu, W., Qi, Y. & Wang, J. 2019. Connections and structural applications of fibre reinforced polymer composites for civil infrastructure in aggressive environments. Composites Part B: Engineering, 164, 129–143.

Qiu, C., Ding, C., He, X., Zhang, L. & Bai, Y. 2018. Axial performance of steel splice connection for tubular FRP column members. Composite Structures, 189, 498–509.

Current Perspectives and New Directions in Mechanics, Modelling and Design of Structural Systems – Zingoni (ed.)
© 2022 Copyright the Author(s), *ISBN 978-1-032-39114-4*

Sustainable sandwich composites made of recycled plastics

R.A. Kassab
PhD Candidate, Dalhousie University, Canada

P. Sadeghian
Associate Professor and Canada Research Chair in Sustainable Infrastructure, Dalhousie University, Canada

ABSTRACT: This research investigates the use of recycled polyethylene terephthalate (r-PET) plastic—the type used to create water bottles and beverage containers—as a core component of sandwich beams. The facing component of the sandwich beams derives from bio-based polymer resin through waste by-products and is reinforced with continuous PET fibers in a form of fiber-reinforced polymer (FRP) composites. The sandwich beam sets were fabricated in a wet-layup process and tested in four-point bending configuration. The changing parameter within the sandwich sets included the density of the honeycomb core, as well as the thickness of the facing and core components. For the bending test, the load-applied, midpoint-displacement, and strain-change at the top and bottom sandwich facing were monitored and recorded; this occurred until each beam specimen reached ultimate load capacity. The load-deflection relation of the sandwich beams resulted in a nonlinear trend stemming from the beams' thermoplastic components.

1 INTRODUCTION

This study explores the use of recycled PET (r-PET) plastic, and bio-resin reinforced with continuous PET fibers, as the core and facing components of a novel form of sandwich beams, respectively. The recycled plastic is sourced from postconsumer beverage/water bottles, whereas the bio-resin is derived from agricultural waste. The use of recycled plastic and bio-sourced resin yields a sandwich composite with a lower carbon footprint in regard to traditional sandwich panels/beams used structurally.

2 EXPERIMENTAL METHOD

2.1 *Testing matrix*

The changing parameters for sets included the density of core component, which varied between 80 kg/m^3 and 155 kg/m^3, and the thickness of the facing and core components, which varied between 1mm and 2mm, and 7mm and 15mm, respectively. Table 1 provides the test matrix summary; as listed, three identical sandwich specimens were tested from each set. In total, 12 beams were tested under four-point bending condition.

Table 1. Test matrix.

Set No.	Facing Thickness (mm)	Core Thickness (mm)	Core Density (kg/m3)	Total Specimens
1	1	12	100	3
2	1	15	80	3
3	2	15	80	3
4	1	7	155	3
			Total	12

2.2 *Instrumentation and testing setup*

Two strain gauges were mounted on the midpoint top and bottom facing component of each sandwich beam and two linear load potentiometers (LPs) were used to capture the midspan deflection. Figure 1 depicts the four-point bending loading fixture applied on all tested specimens.

3 TEST RESULTS AND DATA ANALYSIS

The outcome data obtained from each bending test was processed and analyzed. Ultimately, the load-deflection, load-strain, and moment-curvature data from all beam sets was computed, plotted, and compared. At peak load, two modes of failure were observed. Firstly, core shear, which occurred for the first, second, and fourth sets. Secondly, debonding of the top face component, which was the failure mode for the third set.

As shown in Figure 2, the midspan deflection of the beams throughout bending was captured and plotted against the applied load. Load-deflection relation for all tested beams followed a nonlinear trend that could be idealized into two linear profiles with slopes shifting from high to low. This nonlinearity stems from

DOI: 10.1201/9781003348450-246

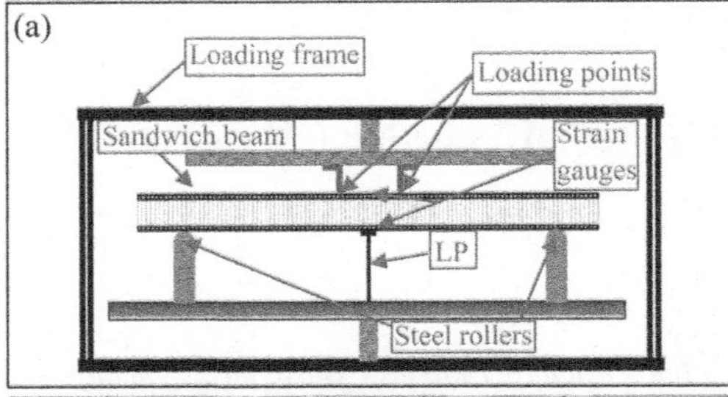

Figure 1. Four-point bending setup: (a) schematic drawing; (b) focused in image of specimen during bending; (c) focused out image depicting loading fixture, DAQ unit, and loading frame.

the thermoplastic core and facing components. The third sandwich set was comprised of beams with the thickest facing components and had the highest overall stiffness. Thus, as the thickness of the facing component increases, the overall stiffness of the beam will also increase accordingly. Additionally, the load-deflection slope changes occur at a significantly higher load for beams comprised of thicker facing component. The second variable that impacted beam stiffness was the thickness of the core component.

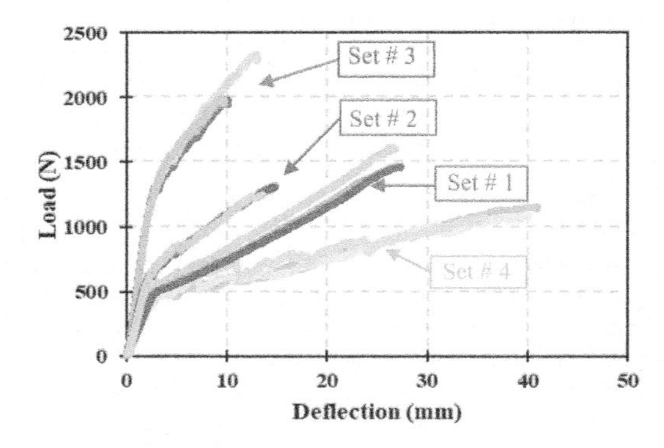

Figure 2. Load vs deflection curves of the tested sandwich beam sets.

As the thickness of the core component decreased, the stiffness of the beam progressively decreased. Despite the decrease in honeycomb density from 100 kg/m^3 for Set 1 to 80 kg/m^3 for Set 2, the stiffness was higher for beams in Set 2 due to an increased core thickness. As the density of the honeycomb increased, the area of the individual hexagonal cells decreased.

Despite possessing lower stiffness, sandwich beams from Set 1 reached a higher ultimate load due to the strong bond between the facing and core components compared to the Set 2 beams. Regardless of its highest core density, sandwich beams in Set 4 had the lowest stiffness and ultimate load capacity because of its thin facing and core components.

The load-strain relation of all tested beams followed a nonlinear trend. Sandwich beams with thicker facing component (2mm) had lower strain at a certain load compared to beams comprised of thin facing component (1mm). In addition, as the thickness of the core component among the different sandwich sets decreased, the strain corresponding to a given applied load increased.

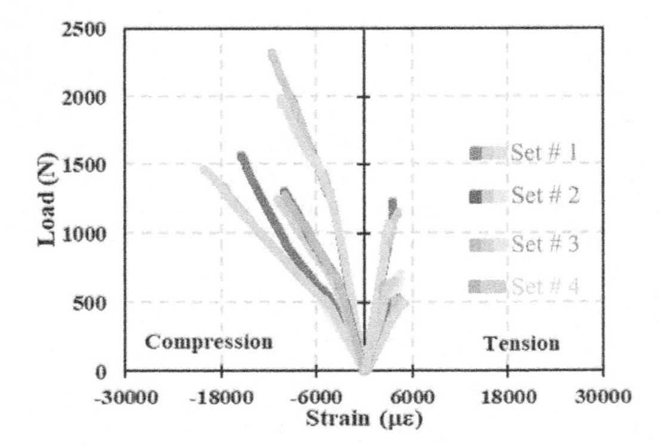

Figure 3. Load vs strain curves of the tested sandwich beam sets.

Thickness of the core component had a direct impact on the degree of curvature of the beams during bending. As the thickness of the beams increased, the degree at which the beam bended at a given load decreased.

4 CONCLUSIONS

In this study, four sets of sandwich beams were fabricated and tested in four-point bending. The difference among the sets were the thickness and density of core component, as well as the thickness of facing component. Recycled PET core and bio-resin reinforced with continuous PET fibers were the materials used in the core and facing components, respectively. The main research findings are as follows:

- The stiffness of the facing component had the greatest impact on the overall stiffness of the sandwich beams with a directly proportional relationship.
- As the thickness of the core component increased, the stiffness of the overall beam increased.
- As the thickness of the core component decreased, the strain capacity of the facing component increased under a given load.
- The degree of curvature increased as the core thickness increased under a given moment.

Current Perspectives and New Directions in Mechanics, Modelling and Design of Structural Systems – Zingoni (ed.)
© 2022 Copyright the Author(s), ISBN 978-1-032-39114-4

Flexural characteristics of bio-based sandwich beams made of paper honeycomb cores and flax FRP skins

Yuchen Fu & Pedram Sadeghian
Department of Civil and Resource Engineering, Dalhousie University, Halifax

ABSTRACT: In this study, a total of 18 large-scale bio-based sandwich beams made of flax FRP skins and two types of paper honeycomb core were studied. The parameters of the tests were skin thickness (1,2 and 3 layers of flax FRP) and core types (namely, hollow and foam-filled). Each specimen was 1200 mm long, 100 mm wide and approximately 80 mm thick and was tested by three-point bending. The failure modes were observed, and the test data were collected and processed. The test results were shown by Load-Deflection diagrams, and Moment-Curvature diagrams. Overall, the bio-based sandwich structures have potential to be used for building applications with much less environmental footprints in comparison with other synthetic counterparts.

1 INTRODUCTION

Sandwich structure is a composite material consisting of a thick core and two thin skins, and it is a very effective system with high performance and minimum weight. The core is normally lightweight and separates the strong skin materials apart to provide higher moment of inertia under flexure, and it also provide shear resistance and insulation (water and sound) for the sandwich system. The two skins are very stiff and strong to resist tension and compression force that is resulting from the bending moment [1]. Nowadays, sandwich composites made by fiber-reinforced polymer (FRP) skins and core have been practices for couple decades. With the growing understanding of the impact on the environment caused by human activity, people are finding the ways to make building construction more sustainable by using bio-based material. In this study, flax was used for manufacturing the FRP skin. Paper honeycomb (hollow and foam-filled) was selected to be the core material. The purpose of this study is to show that this kind of sandwich structure has required strength on building applications.

2 TEST MATRIX, MATERIAL, AND SETUP

A total of 18 flax FRP and paper honeycomb core sandwich beams which were 1200 mm long, 100 mm width, and 80 mm thickness, were tested. The main parameters were the facing thickness and the type of paper honeycomb core (namely, hollow and foam-filled). Three skin thickness were studied: one layer, two layers, and three layers of flan FRP on each side.

All the sandwich panels were fabricated by flax FRP skins and paper honeycomb cores. The density of fully stretched hollow core were measured to be 19.12 kg/m^3. The density of the spray foam that used to fill the hollow core was measured to be 27.8 kg/m^3. The density of the foam-filled paper honeycomb core was measured to be 43.87 kg/m^3. The flax FRP was fabricated by a bidirectional flax fabric with a density of 410 g/m^2 and a bio-based epoxy resin with a bio-content of 21% after mixing. The flax FRP that fabricated by the same material was tested by Betts [2] in previous research. The tensile modulus, strength, and ultimate strain of this flax FRP were tested to be 7.51 ± 0.69 GPa, 45.4 ± 1.8 MPa and 0.0083 ± 0.0009 mm/mm, respectively.

Each specimen was tested under three-point bending using a 1 MN actuator. The load was applied to the specimen at a rate of 2 mm/min through a -150x150x275 mm Hollow Structure Section (HSS). A strain gauge was installed at the center of each specimen face to measure the longitudinal strain. A strain pot was applied at the center bottom of the specimen to measure deflection. One support was a roller, and the other one is hinge. A data acquisition system was used to record the force, deflection, and strains at a rate of 10 samples per second. The test setup is shown in Figure 1.

3 TEST RESULTS AND DISCUSSION

3.1 *Failure modes*

There are couple failure modes of sandwich beams. However, in this study, only two failure modes were obtained: core shear and debonding. The failure modes of each group of specimens are shown in Table 1. Specimen 1FL-F and 1FL-H with one layer flax FRP were failed by core shear. No surprisingly, when the thickness of facing is increasing, the rest of specimens would also be failed by core shear. On the other hand, looking at specimens with debonding failure mode, all of them were specimens with foam-filled honeycomb core. This was caused by low fabrication quality.

DOI: 10.1201/9781003348450-247

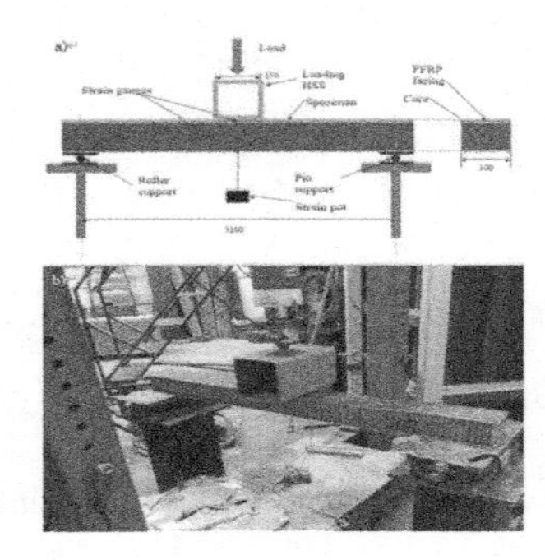

Figure 1. Test setup and instrumentation: a) schematic drawing (dimensions in mm); b) test setup photo.

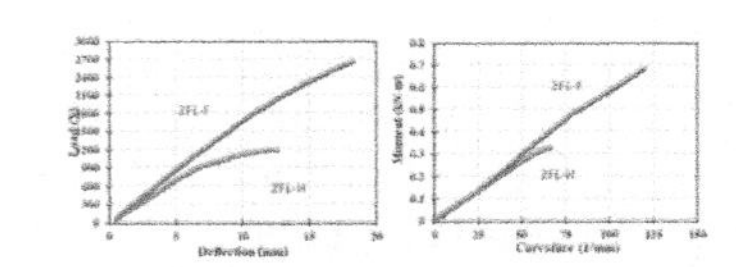

Figure 2. Effect of honeycomb core type on load-deflection and moment-curvature diagrams.

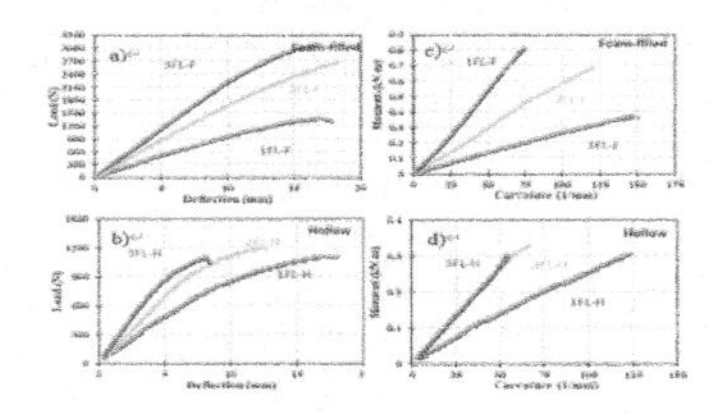

Figure 3. Effect of facing thickness on load-deflection and moment-curvature diagrams.

3.2 *Effect of honeycomb core types*

As shown in Figure 2, for example, looking at the difference between 2FL-F and 2FL-H, the initial stiffness and the initial flexural rigidity almost stay the same. However, the change in core type had a major impact on the peak load and ultimate moment at the peak load.

3.3 *Effect of facing thickness*

As shown in Figure 3, the change of facing thickness had a major impact on initial stiffness and initial flexural rigidity for each honeycomb core type. For the specimens with foam-filled core, their peak load and corresponding moment were increased by adding more flax FRP layers. On the other hand, for the specimens with hollow core, the increasing of facing thickness did not affect their peak load and ultimate moment.

4 CONCLUSION

In this study, 18 sandwich beams were fabricated using FRPs made by bidirectional flax fabrics and bio-based epoxy resin for the facings and paper honeycomb cores (hollow and foam-filled). All the specimens were tested under three-point bending. Based on the test results, it can be concluded that: by filling hollow paper honeycomb core with foam, the peak load and corresponding moment of specimens were significantly increased, but the initial stiffness and the initial flexural rigidity were not. By increasing the thickness of facing, the specimens with foam-filled core obtained higher peak load, corresponding moment, initial stiffness, and initial flexural rigidity. However, for the specimens with hollow core, higher initial stiffness and initial flexural rigidity were obtained, but the peak load and ultimate moment were not obviously changed.

REFERENCES

[1] Sadeghian, P., Hristozov, D., & Wroblewski, L. (2018). Experimental and analytical behavior of sandwich composite beams: Comparison of natural and synthetic materials. The Journal of Sandwich Structures & Materials, 20(3), 287–307.

[2] Betts, D., Sadeghian, P., & Fam, A. (2018). Experimental Behavior and Design-Oriented Analysis of Sandwich Beams with Bio-Based Composite Facings and Foam Cores. Journal of Composites for Construction, 22(4), 4018020.

Table 1. Summary of test results.

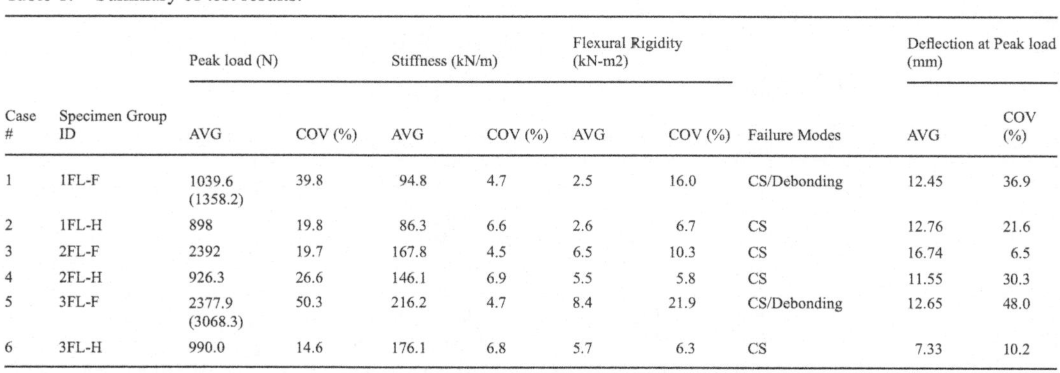

		Peak load (N)		Stiffness (kN/m)		Flexural Rigidity (kN-m2)			Deflection at Peak load (mm)	
Case #	Specimen Group ID	AVG	COV (%)	AVG	COV (%)	AVG	COV (%)	Failure Modes	AVG	COV (%)
1	1FL-F	1039.6 (1358.2)	39.8	94.8	4.7	2.5	16.0	CS/Debonding	12.45	36.9
2	1FL-H	898	19.8	86.3	6.6	2.6	6.7	CS	12.76	21.6
3	2FL-F	2392	19.7	167.8	4.5	6.5	10.3	CS	16.74	6.5
4	2FL-H	926.3	26.6	146.1	6.9	5.5	5.8	CS	11.55	30.3
5	3FL-F	2377.9 (3068.3)	50.3	216.2	4.7	8.4	21.9	CS/Debonding	12.65	48.0
6	3FL-H	990.0	14.6	176.1	6.8	5.7	6.3	CS	7.33	10.2

Note: a specimen identification with format of XFL-Y is used to identify each specimen. X is the number of flax FRP layers, and FL represents "Flax Layers". Y indicates the core type, where F is foam-filled, and H is hollow. The number in the brackets is the average peak load ignoring the specimens with debonding failure.

Current Perspectives and New Directions in Mechanics, Modelling and Design of Structural Systems – Zingoni (ed.)
© 2022 Copyright the Author(s),
ISBN 978-1-032-39114-4

Behavior of NSM FRP flexurally-strengthened RC beams with embedded FRP anchors

Y. Ke, S.S. Zhang* & X.F. Nie
Huazhong University of Science and Technology, Wuhan, China

ABSTRACT: Flexural strengthening of RC beams by applying the near-surface mounted (NSM) fiber-reinforced polymer (FRP) into the bottom concrete cover of the beam has been proved to be efficient by abundant previous studies. However, it was found from these studies that NSM FRP-strengthened RC beams frequently failed by FRP premature debonding, which limited the strengthening efficiency of the NSM FRP. In order to fully exploit the strengthening efficiency of the NSM FRP, anchorage devices are normally needed to mitigate the debonding failure. However, existing studies on the anchorages for NSM FRP-strengthened beams are still limited. A novel embedded FRP anchor (EFA) is therefore proposed by the authors for the anchoring of NSM FRP. In the present paper, the concept, manufacture and installation process of the proposed EFA are first presented. Then, the effectiveness of the EFA in NSM FRP-strengthened beams is verified through a test program consisting of four full-scale RC beams. The test results showed that the EFA is able to mitigate NSM FRP end debonding and therefore improve the load-carrying and deformation capacities of NSM FRP-strengthened RC beams.

1 INTRODUCTION

The near-surface mounted (NSM) fiber-reinforced polymer (FRP) strengthening technique is a promising alternative to the externally bonded (EB) FRP strengthening method for reinforced concrete (RC) members, on account of its various advantages. Flexural strengthening of RC beams is one of the mainstream applications of NSM FRP that has been proved to be efficient by numerous experimental studies. However, premature FRP debonding frequently happened in NSM-strengthened beams, which limited not only the behavior of NSM-strengthened beams but also the utilization efficiency of FRP strength. To prevent or delay the debonding failures of NSM-strengthened beams, some end anchorage measures have been used in previous studies. Although these anchorages have been proved to be effective, some deficiencies still exist in these anchorages.

Against the above background, this paper proposes a novel embedded FRP anchor (referred to as *EFA* hereafter) for NSM FRP flexurally-strengthened RC beams. The concept, manufacture and installation process of the EFA are first introduced. Then, an experimental program consisting of four full-scale RC beams is presented to ascertain the effectiveness of the EFA.

2 CONCEPT, MANUFACTURE AND INSTALL-ATION OF EMBEDDED FRP ANCHOR (EFA)

The NSM-strengthened beam installed with EFAs and the close-up view of EFA are respectively shown in Figures 1a and b. The EFA is composed of sleeve and spike components, and made of a rectangular fiber sheet through a wet lay-up process.

The manufacture of an EFA and the installation process of EFAs onto the NSM strip are introduced in detail in the full paper.

3 EXPERIMENTAL PROGRAM

The experimental program consisted of four full-scale RC beams. One of these beams was not strengthened and served as un-strengthened control beam (referred to as *CB*); one was strengthened in flexure with a compound NSM strip (referred to as *SB*); the other two beams were not only strengthened with NSM FRP strip but also anchored with vertical and inclined EFAs respectively. These beams were tested under four-point bending.

*Corresponding author: shishun@hust.edu.cn

DOI: 10.1201/9781003348450-248

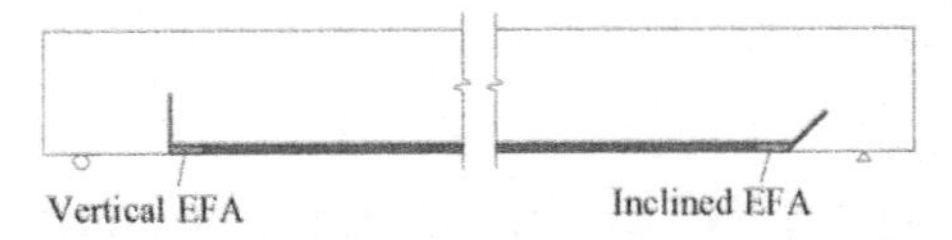

(a) NSM-strengthened beam installed with EFAs

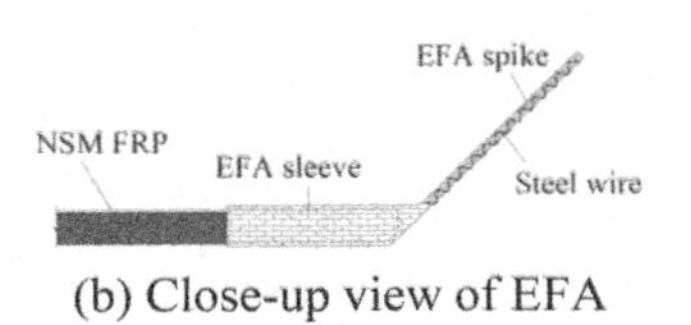

(b) Close-up view of EFA

Figure 1. Schematic representation of embedded FRP anchor (EFA).

4 TEST RESULTS AND DISCUSSION

4.1 *Failure modes*

The un-strengthened control beam (Specimen CB) failed by yield of tension steel bars followed by concrete compressive crushing (referred to as *TC*). The strengthened control beam (Specimen SB) failed by concrete cover separation at the end of NSM strip (referred to as *CCS*). The beam anchored with vertical EFAs (Specimen EB-V) failed by sleeve-concrete interfacial debonding followed by concrete cover separation (referred to as *SLD+CCS*), as shown in Figure 2. The beam anchored with inclined EFAs (Specimen EB-I) failed by delamination of EFA followed by concrete cover separation (referred to as *DS+CCS*), as shown in Figure 3.

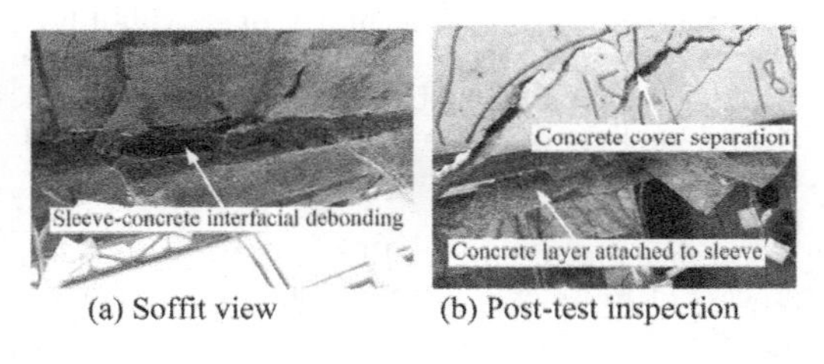

(a) Soffit view (b) Post-test inspection

Figure 2. Failure mode of Specimen EB-V (SLD+CCS).

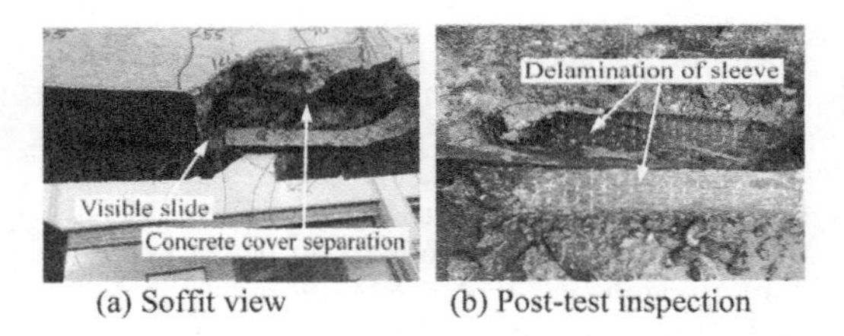

(a) Soffit view (b) Post-test inspection

Figure 3. Failure mode of Specimen EB-I (DS+CCS).

4.2 *Load-deflection responses*

The load versus mid-span deflection responses of all specimens are shown in Figure 4. The ultimate load of NSM-strengthened control beam (Specimen SB) is higher than the un-strengthened control beam (Specimen CB) by 21%. With the additional anchoring by EFAs, the ultimate load and corresponding mid-span deflection of NSM-strengthened beam are further improved by more than 21% and 54% respectively, compared to the strengthened control beam SB.

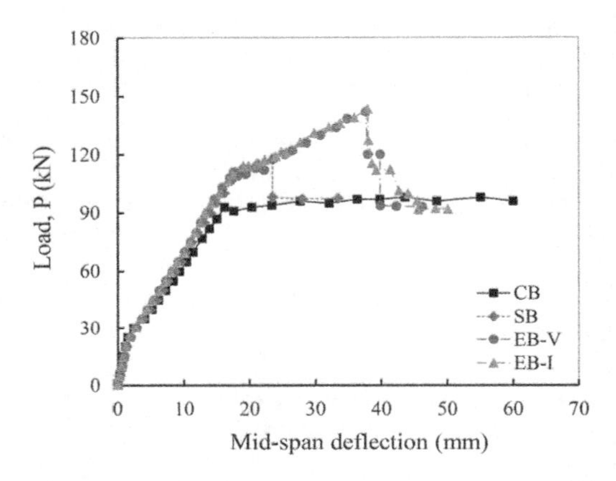

Figure 4. Load-deflection responses of all specimens.

5 CONCLUSIONS

This paper proposes a novel embedded FRP anchor (EFA) for mitigating the debonding failures of NSM FRP flexurally-strengthened beams. The concept, manufacture and installation processes of the EFA are first presented. The advantages of the EFA over the anchorages in previous studies are also clarified. Then, the effectiveness of the EFA in mitigating the debonding failures and improving the performance of NSM-strengthened beams is verified through a test program of four full-scale RC beams. The test results not only prove that the proposed EFA is capable of delaying debonding failures and therefore improving the performance of NSM-strengthened beams, but also provide a solid basis for future studies.

Current Perspectives and New Directions in Mechanics, Modelling and Design of Structural Systems – Zingoni (ed.)
© 2022 Copyright the Author(s), *ISBN 978-1-032-39114-4*

Standardization, guide development and long-term durability of fiber reinforced polymers (FRP): In situ field results from FRP RC bridge decks after 15+ years of service exposure

John J. Myers*
Professor of Civil, Architectural, and Environmental Engineering, Missouri University of Science and Technology, Rolla, Missouri, USA

Ali F. Al-Khafaji
Structural Engineer, IMEG Corp., Portland, Oregon, USA

ABSTRACT: This paper presents information about one of the most recent large-scale field durability studies in the United States to autopsy FRP reinforcing bars from eleven bridges constructed with FRP between 15 and 20 years ago. The work provides insight on how the FRP bars have been performed over the last two decades under various filed service conditions through physical and microscopic examination. To assess the performance of bridges' concrete that surrounds the GFRP bars, pH, carbonation depth, and chlorides content were conducted to see if any chemical changes and/or microcracks occurred in the concrete. Scanning electron microscopy (SEM) was performed to observe any microstructural degradations that might take place in both the GFRP and the surrounding concrete including the interfacial transition zone (ITZ). Energy dispersive spectroscopy (EDS) was applied to see if there were any elemental changes. Short bar shear, glass transition temperature, and fiber content were conducted on the GFRP samples to evaluate the GFRP's state. This study adds new evidence to the validation of the long-term durability of GFRP bars as concrete reinforcing used in filed applications.

1 BRIDGES INFORMATION

Eleven bridges with almost twenty years of in-field service are included in this study. The bridges investigated were: Bettendorf Bridge in Iowa, Cuyahoga Country Bridge and Salem Ave. Bridge in Ohio, Gills Creek Bridge in Virginia, McKinleyville Bridge in West Virginia, Sierrita de la Cruz Creek Bridge in Texas, Walker Box Culvert Bridge and Southview Bridge in Missouri, and Thayer Road Bridge in Indiana.

2 SPECIMENS EXTRACTION, PREPARATION, AND CONDITIONING

To extract the GFRP bars from the concrete bridge decks, concrete cores of 102 mm (4 in.) diameter were taken out from the decks. Multiple cores were removed from each of these bridges. After the extraction was made, a fast-curing cementitious grout was used to fill the holes in the decks.

3 GFRP AND CONCRETE TESTS

3.1 *Scanning Electron Microscopy (SEM)*

SEM examination was utilized to check for any microstructural degradations in-terms of cracks of fibers/resin and debonding issues. There were no significant signs of microcracking or debonding.

3.2 *Energy Dispersive Spectroscopy (EDS)*

EDS examination was utilized to see if there were any chemical changes in the GFRP bars resulted from the long-term in-situ exposure to harsh environment. The same apparatus used in SEM test was implemented to perform the EDS test. A 10 to 20 KeV electron was applied on the samples. For almost all the bridges no signs of chemical changes were noticed.

3.3 *Short Bar Shear (SBS)*

SBS test was performed to check for the horizontal shear capacity after all these years of service. The

*Corresponding author

DOI: 10.1201/9781003348450-249

test was made following the recommendations of ASTM D4475 (ASTM-D4475, 2016).

The results ranged from 30 to 45 MPa (4316 to 6809 psi), which indicated a retention of strength from 72% to 92%. Bars fell at the weaker end of the spectrum were noticed to be at the shorter end of the limit of span-to-diameter ratio.

3.4 *Fiber content*

This test is also named Burn-off test and was implemented to detect the ignition loss of cured resin. This test was conducted following the recommendations of ASTM D2584 (ASTM-D2584, 2018). From the eleven bridges tested, the results met or exceeded the 70% requirement of the ASTM standard with only one exception which is one of the bridges exhibited a fiber content of 69.2%. It is recommended that an enhanced test procedure to accommodate for the effect of residual filler materials stuck to the glass fibers.

3.5 *Glass transition temperature*

Glass transition temperature can be defined as the temperature range where the behavior of the resin changes from being in its solid state to rubbery state. In this study, DSC technique was used to determine the glass transition temperature following the recommendations of ASTM E1356 (ASTM E1356 – 2014a). The results showed that the values of the temperature were around 115° C (239° F).

3.6 *pH*

Concrete pH test was utilized to evaluate the alkalinity levels as highly alkaline medium negatively affects the resin matrix of the GFRP bar. The normal range of concrete is around 11 or 12. Concrete pH test results were recorded for all bridges. Most of the bridges exhibited a pH of more than 11. Two bridges among the eleven ones showed a pH value of 10, which can be used as an indication of corrosion initiation of steel reinforcement.

3.7 *Carbonation depth*

In this study, carbonation was noticed in most concrete cores. In addition, one of the bridges showed carbonation depth larger than 38 mm (1.5 in.), which can indicate that carbonation reached to the GFRP bars. Figure 7 shows a carbonation depth test result of one of the bridges with no signs of carbonation owning to the change in color of the entire surface to purple.

3.8 *Chloride content*

This test was carried out to examine for the chlorides levels in concrete as chlorides is considered one of the main reasons for corrosion of mild reinforcement in RC concrete. In this study, the water-soluble technique was utilized to measure the chlorides that can damage the oxide film that covers the reinforcement in concrete. In most of the bridges, there were no signs of chlorides attack. In one of the bridges, some moderate signs of chlorides were noticed and it was attributed to the use of deicing salts.

REFERENCES

[1] Andrew, R. M. (2018) 'Global CO2 emissions from cement production', Earth System Science Data, pp. 195–217. 2018.

[2] ASTM-D4475 (2016) 'Standard Test Method for Apparent Horizontal Shear Strength of Pultruded Reinforced Plastic Rods By the Short-Beam Method', ASTM Standard, pp. 3–5. 2016

[3] D2584 (2018) 'Standard Test Method for Ignition Loss of Glass Strands and Fabrics', ASTM Standard, pp. 2–5. 2005.

[4] Gooranorimi, O. et al. (2017) 'GFRP Reinforcement in Concrete after 15 Years of Service', Journal of Composites for Construction, pp.1–9. 2017.

[5] Grubb, J., Limaye, H., and Kakade, A. (2007) 'Testing pH of concrete: Need for a standard procedure', Concr. Int., 29 (4), pp. 78–83. 2007.

[6] Micelli, F. and Nanni, A. (2004) 'Durability of FRP rods for concrete structures', Construction and Building Materials, 18(7), pp. 491–503. 2007.

[7] Mufti, A. et al. (2005) 'Durability of GFRP reinforced concrete in field structures', Proceedings of the 7th International Symposium on Fiber Reinforced Polymer Reinforcement for Concrete Structures - FRPRCS-7, pp. 889–895. 2014.

[8] RILEM and Materials, T. R. for the T. and U. of C. (1994) 'RILEM 7-II-128. RC6: Bond Test for Reinforcing Steel - Pullout Test'.1994.

[9] Sagues, A. (1997) 'Carbonation in concrete and effect on steel corrosion', Final Rep., No. 99700-3530-119, Dept of Civil and Environmental Engineering, Univ. of South Florida, Tampa, FL.

[10] Transportation, M. D. of (2012) 'Self-Consolidating Concrete (SCC) for Infrastructure Elements Report E – Hardened Mechanical Properties and Durability Performance. Rolla.

Current Perspectives and New Directions in Mechanics, Modelling and Design of Structural Systems – Zingoni (ed.)
© 2022 Copyright the Author(s), *ISBN 978-1-032-39114-4*

Strength behaviour of fibre reinforced polymer concrete beams under flexural loading

J.T. Senosha & S.D. Ngidi
Department of Civil Engineering Technology, University of Johannesburg, Gauteng, South Africa

ABSTRACT: The paper presents investigation on the flexural behaviour of the reinforced concrete beams repaired and strengthened with Fibre Reinforced Polymer on its soffit. Six beams including one control beam were tested under two-point static loading. The control beam was tested to failure without preloading, whereas four beams were preloaded at 60% (serviceability load) and 85% (ultimate load) and were further repaired with 50 and 100 mm wide and 1.2 mm thick FRP. One beam was strengthened with 100 mm FRP without preloading. The flexural strength capacity of FRP strengthened and repaired beams increased by more than 100% as compared to the control beam, and the deflection decreased immensely with the increase in width-to-thickness ratio of FRP. The experimental loads of all the beams did not reach the SANS 10100-1:2000 predictions.

Keywords: Preloading, Fiber Reinforced Polymer, Repairing, Flexural Behaviour, Deflections

1 INTRODUCTION

Reinforced concrete (RC) beams tends to have flexural failure due to the tension steel reinforcement reaching its yielding stress at an early stage, and not allowing the beam to have adequate bending when subjected to static loads. It is then mandatory to repair the structure to prevent extended deterioration. Repairing reinforced beams with Fibre Reinforced Polymer (FRP) for flexural strength enhancement is a method of repairing reinforced concrete beams' flexural load resistance capacity and controlling the deflection. The main objective of this research was to investigate the effect of adhesively FRP repairing of preloaded RC beams on the flexural load capacity and structural integrity.

Previous studies by Triantafillou and Plevris (1992) and Benjeddou et al. (2006) gave a practical exhibition and explanation that, FRP can be used as a repairing or strengthening construction material of concrete elements, nonetheless there are gaps involved in the studies. The studies lacked the extensive documentation on the effect of width to thickness ratio of FRP on concrete beams' strength capacity.

2 EXPERIMENTAL PROGRAMME

2.1 *Details of test beams*

The concrete beams for the study were casted using the PPC Pronto ready-mix concrete with a predicted average compression strength of 20 MPa at 28 days. The material properties of ready-mix concrete were determined by casting three 100 mm x 100 mm x 100 mm cubes which were cured for 28 days. The cubes were casted in three layers as per SANS 5863:2006-2with a rounded tamping rod to get the air bubbles out of the concrete. The investigation of the study used concrete beams with cross section of 250 mm wide x 450 mm deep x 4800 mm long and spanning 4500 mm. The beams were reinforced with internal longitudinal reinforcement steel of high-yield 12 mm diameter (Y12), two top compression reinforcement, two bottom tension reinforcement of yield-strength (f_y = 450 MPa) and 25 high-yield steel stirrups of 8 mm diameter of yield-strength (f_y = 350 MPa) spaced at 200 mm.

2.2 *Strengthening and repairing techniques*

Four beams were preloaded first, and one beam was strengthened without being preloaded first. Two beams preloaded at serviceability loads (60%) load were repaired with 50 and 100 mm FRP, and other two preloaded close to ultimate loads (85%) were also strengthened with 50 and 100 mm FRP, with one beam preloaded to 0% strengthened with 100 mm FRP. The adhesive Sikadur 30 epoxy resin was applied to all the Sika S512 (50 mm) and S1012 (100 mm) FRP plates and were attached to the surface of the beams and were gently pressurised with a steel roller to eliminate any excessive adhesive epoxy.

DOI: 10.1201/9781003348450-250

Table 1. Experimental and theoretical results.

Beam Type	Width-to-thickness	Preloading loads kN	Test Max Load kN	Theoretical loads kN	Preloading Deflection mm	Test Max Deflection mm	Failure Modes*
CB	-	-	34.74	43.06	-	96.99	SY
B-100	83.33	-	73.53	175.77	-	32.11	FED
BS1-50	41.67	22.35	51.97	116.58	14.92	25.87	D
BS2-100	83.33	23.64	56.95	175.77	7.85	33.70	FED
BU1-50	41.67	29.72	53.09	116.58	38.79	20.91	RP
BU2-100	83.33	30.91	86.17	175.77	27.95	31.95	FED

* SY: Steel Yielding; FED: FRP-End-Debonding; D: Delamination; RP: Rapture, FM: Failure Mode

3 EXPERIMENTAL AND THEORETICAL RESULTS

3.1 *Flexural strength results*

The experimental and theoretically analysed results by SANS 10100-1:2000 for both control beam (CB), repaired beams (BS1-50, BS2-100, BU1-50, BU2-100), strengthen beam (B-100) are summarised in Table 1. The preloaded and strengthened beams that were further repaired with FRP achieved the maximum full strength which is more than twice the strength of the control beam. However, all the FRP repaired and strengthened beams did not achieve the full strength as per the SANS 10100-1:2000 predictions. For beams preloaded at ultimate load (85%) of the control beam, the results show a significant high strength capacity as compared to the beams preloaded serviceability (60%) of the control beam. This gives an indication that preloading degree does not necessarily affect the load capacity resistance of repaired beams. This can also be proven by comparing B-100, which was strengthened with 100 mm FRP with 0% preloading and BU2-100 which was preloaded at ultimate load (85%) of the control beam which is closest to failure and was also repaired with 100 mm FRP, but still achieved highest full strength as compared to all repaired and the strengthened beams.

3.2 *Load-Deflection*

The control beam (CB) shows an experimental maximum deflection of 96.99 mm deflection. At this position the beam's steel reinforcement reached its maximum load capacity of 34.47 kN. From the deflection of 11.98 mm at 17.70 kN, the load-deflection curve of the control beam started showing a uniform horizontal increase in deflection, between loads slightly below 20 kN and a slightly above 30 kN. This indicates a zone where the tension steel reinforcement of the beam was yielding until it failed. The deflections of repaired and strengthened beams, however indicates the decrease in deflection which is three times lesser than the deflection of the control beam, while the flexural strength capacity increased. The increase in width of FRP plate, increased the load capacity but did not have an impact on the deflection.

3.3 *Conclusion*

The flexural strength capacity of FRP strengthened and repaired beams significantly increased with more than twice the strength capacity of the control beam. Therefore, this method is effective in restoring the structural integrity of the beams.

The study indicated an outstanding decrease in deflection for both strengthened and repaired beams compared to the control beam. The deflection decreased immensely with the increase in width-to-thickness ratio of FRPs, while the flexural strength was increased. This gives an idea that this repairing technique can increase the stiffness of the beams and improve the cracking.

REFERENCES

ACI Committee, 2017. Guide for the design and construction of externally bonded FRP systems for strengthening concrete structures (ACI 440.2 R-17). *American Concrete Institute, Farmington Hills, MI.*

Benjeddou, O., Ouezdou, M.B. and Bedday, A., 2007. Damaged RC beams repaired by bonding of CFRP laminates. *Construction and building materials*, *21*(6), pp.1301–1310.

South African National Standard, 2000. SANS 10100-1: 2000. The Structural Use of Concrete, Part 1: Design.

Current Perspectives and New Directions in Mechanics, Modelling and Design of Structural Systems – Zingoni (ed.)
© 2022 Copyright the Author(s), ISBN 978-1-032-39114-4

Properties of a 37 m long FRP wind turbine blade after 11 years in service

A.A. Alshannaq, J.A. Respert, L.C. Bank & T.R. Gentry
Georgia Institute of Technology, Atlanta, USA

D.W. Scott
Georgia Southern University, Statesboro, USA

ABSTRACT: The waste coming from decommissioned wind turbine blades poses an environmental problem since they are made primarily of Fiber Reinforced Polymers (FRPs) which are non-biodegradable materials. Millions of tons of these materials are expected to be disposed of in the coming years. Thus, responsible end-of-life (EOL) solutions are needed. The use of the blades for civil engineering infrastructure (e.g. power transmission poles, pedestrian bridge girders, highway sound barriers and roofing materials) has been proposed. In order to perform a structural and stress analysis for any of these design concepts, the as-received mechanical and physical properties of the specific type of wind blade to be used are needed. This paper presents an extensive testing program of the GFRP (Glass Fiber Reinforced Polymer) material in the spar cap of a 1.5 MW GE 37 decommissioned wind turbine blade that had been used for 11 years at a wind farm in Langford, Texas. Procedures for specimen preparation and testing are highlighted, since the blade parts have varying thicknesses (50 mm or more in some parts), layups and curvatures. As-received tensile, compressive, and shear strengths and stiffnesses were determined and compared to results from static testing of fatigued wind blades from the literature. Bolt testing was performed to obtain bearing and pull-through properties of the spar cap materials for second-life (i.e. EOL) applications where bolt performance will be a major design aspect.

1 INTRODUCTION

A sustainability problem already exists for decommissioned wind turbine blades (primarily made of Glass-FRP materials). Thousands (currently) and millions (in future years) of tons of composites now need, and will continue to need, disposal, this poses a significant threat to the environment (Liu & Barlow 2017). These materials are non-biodegradable and are currently being disposed of using landfilling and incineration. Mechanical or thermal and chemical methods are being studied for recycling. All these methods have drawbacks - in being environmentally harmful options (for the case of landfilling or incineration, which are also expected to have legislative restrictions soon) or in providing recycled fibers with reduced properties compared to composites made of virgin glass fiber (for the case of mechanical or thermal and chemical recycling of glass fiber), which eventually makes them less attractive options. Furthermore, the relative short service lives of these structures in their first life as wind blades may allow for viable structural repurposing, defined here as using the entire wind blade or large cuts in load-bearing structural applications (e.g. bridges, poles) or consumer products (e.g. furniture, urban architecture) (Jensen & Skelton 2018).

The Re-Wind Network (www.re-wind.info) is aimed at providing greener and sustainable ways of recycling, reusing, and repurposing decommissioned wind turbine blades. The project covers various aspects regarding the end-of-life applications. The main objective of the research is to propose a methodology for dealing with end-of-life applications which will pave the way for implementation in the field by relevant stakeholders.

This paper summarizes the testing of tensile, compressive, shear, bearing, and pull-through properties at the coupon-scale of the spar cap of a decommissioned 1.5 MW GE37 wind turbine blade that has a total length of 37 m and has been in service for 11 years in a wind site in Langford, Texas.

2 EXPERIMENTAL PROGRAM

Any cross-section of a wind blade consists of three major parts forming a multicellular cross-section, except the root (i.e. the end of the wind blade that connects to the hub) which is a thin-walled circular cross-section. The three main parts are the spar cap (main load-carrying part), the aerodynamic shell, and the web (or webs).

2.1 *Materials*

The materials tested were cut from the spar cap of a 1.5 MW decommissioned GE37 wind blade that has a total length of 37 m. The goal is to assess the as-

DOI: 10.1201/9781003348450-251

received mechanical properties of this wind blade after 11 years in a wind farm in Langford, Texas for possible second-life applications. Spar cap samples were cut from the wind blade in the root-transition region (i.e. where the cross-section changes from circular to an aerodynamic shape) where the thickness reaches 50 mm.

2.2 *Specimens and Equipment*

For tensile, compressive, bearing, and pull-through testing, a 250-kN 810 MTS testing machine using 69 MPa hydraulic grip pressure, while for shear testing, a 100-kN 810 MTS testing machine was used. A constant crosshead displacement of 1.27 mm/min was used according to relevant ASTM standards. Extensometers and strain gages were used for strain acquisition. NI cDAQ-9178 was used to acquire simultaneous load, displacement, and strain data. Note that longitudinal and transverse testing was performed for all tested properties except transverse interlaminar shear properties.

2.3 *Results*

The results and number of specimens are summarized in Table 1.

Table 1. Summary of results.

Property	*N*	Strength (MPa)	Modulus (GPa)	Strain at Failure (%)
Tension (L)	53	597±54.4	36.8±1.95	1.94±0.15
Tension (T)	13	33.6±3.58	10.7±0.45	0.29±0.02
Compression (L)	74	504±39.8	42.7±2.54	1.22±0.05
Compression (T)	11	114±2.05	-	-
Open-Hole (L)	11	584±29.5	-	-
Open-Hole (T)	16	22.4±2.33	-	0.24±0.02
V-Notch Shear (L)	26	60.8±2.26	4.57±0.12	-
V-Notch Shear (T)	10	27.9±2.47	-	-
Short-Beam Shear (L)	14	55.0±3.57	-	-
Bearing – 19.1 mm pin (L)	10	302±17.4	-	-
Bearing – 19.1 mm pin (T)	10	212±9.50	-	-
Bearing – 25.4 mm pin (L)	10	296±32.2	-	-
Bearing – 25.4 mm pin (T)	10	246±10.3	-	-
Pull-though – Thin BB	10	65.6±6.50	-	-
Pull-though – Thick BB	10	84.5±2.62	-	-
Pull-though – Thin RS	10	130±6.48	-	-
Pull-though – Thick RS	10	132±5.90	-	-

3 DISCUSSION

Table 2 shows a comparison between the obtained results and published data from the literature for fatigued 100-kW wind blades produced by other manufacturers with the same fiber type. The GE37 results provide promising strength and stiffness values for potential second-life applications by outperforming the data of 100 kW wind blades. It is important to mention that larger capacity wind blade testing (i.e. in the range of 1.0-2.0 MW) is not available in the literature, to the authors knowledge, and thus 100 kW wind blade's data are presented for comparison purposes.

Table 2. Comparison of test results to data from the literature.

Property	Spar Cap of GE37	Sayer et al. (2013)	Ahmed et al. (2021)
Tensile Strength - L (MPa)	597	477	350
Compressive Strength - L (MPa)	504	447	225
Shear Strength - SBS (MPa)	55.0	32.3	-
Tensile Modulus - L (GPa)	36.8	26.7	15.6
Compressive Modulus - L (GPa)	42.7	26.2	-

4 CONCLUSIONS

This paper presents results of tests of the GFRP material spar cap of a 1.5 MW decommissioned GE37 wind turbine blade for tensile, compressive, shear, bearing, and pull-through properties. The following conclusions can be drawn from the results:

- Strength and stiffness properties of the GE37 after 11 years in a wind farm provide values which outperform the data published previously on 100 kW fatigued wind blades. More testing with age and capacity is still needed to judge strength and stiffness retention levels.
- Data presented herein can be used for a structural analysis and design, and prepare for reliability-based strength reduction factors for second-life applications.
- Bolts through the thick spar cap of a 1.5 MW GE37 wind blade can achieve full capacity of the bolt itself when blind bolting techniques are employed.

REFERENCES

Ahmed, M. M. Z., Alzahrani, B., Jouini, N., Hessien, M. M. & Ataya, S. 2021. The Role of Orientation and Temperature on the Mechanical Properties of a 20 Years Old Wind Turbine Blade GFR Composite. *Polymers*, 13: 1144.

Jensen, J. P. & Skelton, K. 2018. Wind Turbine Blade Recycling: Experiences, Challenges and Possibilities in a Circular Economy. *Renewable and Sustainable Energy Reviews*, 97: 165–176.

Liu, P. & Barlow, C. Y. 2017. Wind Turbine Blade Waste in 2050. *Waste Management*, 62: 229–240.

Sayer, F., Bürkner, F., Buchholz, B., Strobel, M., van Wingerde, A. M., Busmann, H.-G. & Seifert, H. 2013. Influence of a Wind Turbine Service Life on the Mechanical Properties of the Material and the Blade. *Wind Energy*, 16: 163–174.

Current Perspectives and New Directions in Mechanics, Modelling and Design of Structural Systems – Zingoni (ed.)

© 2022 Copyright the Author(s), ISBN 978-1-032-39114-4

The axial behavior of the low and extremely low strength concrete confined with large-rupture-strain PET-FRPs made of production wastes

Ali Gurkan Genc*, Omer Faruk Eskicumali & Neslihan Aslan
Department of Civil Engineering, Istanbul Technical University, Turkey

Tuluhan Ergin & Ugur Alparslan
Kordsa Teknik Tekstil A.Ş., Turkey

Medine Ispir & Alper Ilki
Department of Civil Engineering, Istanbul Technical University, Turkey

ABSTRACT: This study aims to examine the axial behavior of low and extremely low strength concrete externally confined with fiber reinforced polymers (FRPs) having large rupture strain capacity. In order to achieve this aim, as a preliminary study, twelve standard cylinder specimens are produced to be tested. The large-rupture-strain FRP (PET-FRP) sheets used in this study are produced by wastes of tire manufacturing. Compressive strength, axial and lateral deformation capacities of the specimens have been obtained under monotonic axial compression loading. The monotonic compression test results show that the axial compressive strength and axial strain capacity of LSC and ELCS increase significantly through external confinement with PET-FRPs and the strength enhancement ratios obtained for ELCS are higher than those for LCS whereas enhancement ratio for ELCS specimens in terms of axial strain is smaller than LCS.

1 INTRODUCTION

As a retrofitting technique, external confinement of concrete members with FRP composites has been a widely examined topic for the past few decades (Ilki & Kumbasar, 2003; Rousakis et al., 2012). Although the deformation and ductility capacities of the members externally confined with conventional carbon, glass and aramid fiber polymer composites increase considerably, the sudden failure mechanism cannot be prevented. To retard the sudden failure remarkably, the composites with much higher deformation capacities like PET-FRP composites can be used (e.g., Anggawidjaja et al., 2006a; Anggawidjaja et al., 2006b; Ispir, 2015).

In this preliminary study, unlike the studies exist in the literature, LCS or ELCS have been considered and their performance has been examined experimentally after external jacketing with PET-FRP sheets.

2 EXPERIMENTAL PROGRAM

2.1 *Specimen design*

In this study, 12 specimens have been tested under monotonic axial compression loading, 8 of them have been externally jacketed and 4 of them have been remained unconfined as reference specimens. All specimens have a height of 300 mm and a diameter of 150 mm. Two identical specimens were prepared for each test parameter. Within the scope of the study, two different variables have been considered, namely concrete strength (low and extremely low) and the number of PET-FRP (600 gr/m^2) layers (one and two). Specimens are denoted as ELCS-n-m and LCS-n-m, for which ELCS and LCS refer to extremely low strength concrete and low strength concrete, respectively. The letters of n and m show the number of FRP layer (i.e. 1 and 2) and the identical specimen number, respectively. For example, ELCS-1-2 indicates that the specimen is produced with extremely low concrete strength and confined with one-layer PET-FRP and the specimen number is two.

2.2 *Materials*

The mix-proportions of ELCS and LCS are given in Table 1. The average 28-day unconfined concrete compressive strengths are obtained as 3.0 MPa for ELCS and 6.6 MPa for LCS. Mechanical properties of PET-FRP are given in Table 1.

Table 1. Mechanical Properties of PET-FRP.

f_{fu} (MPa)	E_1 (MPa)	E_2 (MPa)	ε_{fu} (%)	t (mm)
160	5200	1800	7.01	1.59

2.3 *Specimen preparation*

PET-FRP composite sheets have been determined to confine the concrete specimens in one or two layers and have an overlap length of approximately 260 mm. The overlap length has been kept long to eliminate the risk of slipping.

*Corresponding author: gencali1@itu.edu.tr

DOI: 10.1201/9781003348450-252

2.4 *Instrumentation and testing*

The displacement-controlled test has been carried out at a constant rate of 0.6 mm/min. Axial strains were measured with strain gages (60 mm gage length), linear variable displacement transducers (LVDTs) as shown in Figure 1.

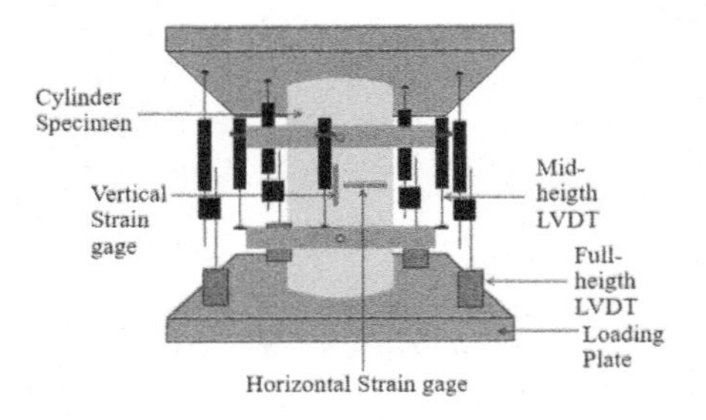

Figure 1. Test Setup.

3 TEST RESULTS

The average unconfined compressive strengths of two identical specimens are obtained as 3.2 MPa for ELCS and 6.8 MPa for LCS. Some specific points obtained from the test results are listed in Table 2 for the confined specimens. Owing to potential problems during confinement applications, ELCS-2-2 and LCS-2-1 specimens did not exhibit proper behavior in tests. Therefore, these specimens were not considered in calculating the average strength and strains (Table 2). It should also be noted that no separation or slipping of PET-FRP jackets has been observed in the overlap zones of the specimens. At the end of the study, using the models proposed by Ilki & Kumbasar (2003) and Rousakis et al. (2012), the values of confined strength and corresponding axial strain are predicted (Table 3).

4 CONCLUSIONS

In this preliminary study, the compression behavior of LCS and ELCS specimens confined with PET-FRP sheets, which were obtained from waste materials in tire manufacturing process, is investigated under monotonic axial compression loading. It was shown that significant improvement was observed in the strength and axial strain capacities of low and extremely low strength concretes. It should be noted that the strength enhancement ratios obtained for ELCSs are higher than those for LCSs while the axial strain enhancement ratios have a reverse situation. Consequently, practically no significant difference was observed between recycled PET-FRP and normal PET-FRP.

Table 3. Comparison of Test Results with Model Results.

Specimen		ELCS-1-ply	ELCS-2-ply	LCS-1-ply	LCS-2-ply
Ilki & Kumbasar (2003)	$f_{cc}(P)/f_{cc}(T)$	0.57	0.44	0.58	0.61
	$\varepsilon_{cc}(P)/\varepsilon_{cc}(T)$	1.08	1.62	0.51	0.71
Rousakis et al. (2012)	$f_{cc}(P)/f_{cc}(T)$	0.36	0.30	0.42	0.42
	$\varepsilon_{cc}(P)/\varepsilon_{cc}(T)$	1.10	1.75	0.43	0.64

$f_{cc}(P)$: predicted confined strength, $f_{cc}(T)$: confined strength obtained experimentally, $\varepsilon_{cc}(P)$: predicted ultimate strain, $\varepsilon_{cc}(T)$: ultimate strain obtained experimentally

REFERENCES

Ilki, A., & Kumbasar, N. 2003. Compressive behaviour of carbon fibre composite jacketed concrete with circular and non-circular cross-sections. J. Earthquake Eng., 7(3), 381–40. https://doi.org/10.1080/13632460309350455.

Rousakis, T. C. & Karabinis, A. I. 2012. Adequately FRP confined reinforced concrete columns under axial compressive monotonic or cyclic loading. Mater. Struct. 45 (7): 957–975. https://doi.org/10.1617/s11527-011-9810-1.

Anggawidjaja, D., Ueda, T., Dai, J., and Nakai, H. 2006a. Deformation capacity of RC piers wrapped by new fiber-reinforced polymer with large fracture strain. Cem. Concr. Compos., 28(10), 914–927.

Anggawidjaja, D., Ueda, T., Dai, J., and Nakai, H. 2006b. Shear Deformation of Pier Wrapped by High Fracturing Strain Fiber. ABSE Symposium, Budapest, Hungary, 31-38. https://doi.org/10.2749/222137806796185030.

Ispir, M. 2015. Monotonic and cyclic compression tests on concrete confined with PET-FRP. J.Compos.Constr. 19 (1): 04014034. https://doi.org/10.1061/(ASCE) CC.19435614.00009

Table 2. Test results of confined specimens.

Specimen	f_{cc} (MPa)	ε_{cc} (mid-height LVDTs)	ε_{cc} (mid-height strain gages)	f_{cc}/f_{co}	$\varepsilon_{cc}/\varepsilon_{co}$ (mid-height LVDTs)	$\varepsilon_{cc}/\varepsilon_{co}$ (mid-height strain gages)	$\varepsilon_{rup,hoop}$ (strain gages)	$\varepsilon_{rup,hoop}/\varepsilon_{f,u}$ (k_ε)
ELCS-1-1	20.9	0.0909	0.0477	6.0	17.5	15.0	0.0628	0.90
ELCS-1-2	17.3	-	0.0454					
ELCS-2-1	38.9	0.1000	0.0522	12.2	19.2	16.8	0.0618	0.88
ELCS-2-2	22.6	-	0.0660					
LCS-1-1	27.8	0.0600	0.0345	3.9	22.7	19.6	0.0548	0.78
LCS-1-2	24.6	0.0900	0.0398					
LCS-2-1	-	-	-	5.4	24.8	22.5	0.0543	0.77
LCS-2-2	36.5	0.0819	0.0428					

f_{cc}: strength of the specimen, ε_{cc}: strain corresponding maximum strength of the specimen, $\varepsilon_{rup,hoop}$: maximum hoop strain, $\varepsilon_{f,u}$: tensile strain capacity of PET-FRP coupon

Current Perspectives and New Directions in Mechanics, Modelling and Design of Structural Systems – Zingoni (ed.)
© 2022 Copyright the Author(s), ISBN 978-1-032-39114-4

Unique FRP solutions for structural repair of piles, seawalls and decks

M. Ehsani
QuakeWrap, Inc., Tucson, USA

ABSTRACT: The dry-wet cycle in marine environments causes rapid corrosion of structures such as piles and seawalls. This paper introduces two new Fiber Reinforced Polymer (FRP) products developed by the author. These products have been tested by several agencies and are being used by the US Military worldwide. One is a very thin cured FRP laminate about 1200 mm wide that can be wrapped around a pile of any shape and size above water to create a shell. The shell is lowered into water, and additional segments can be added to create a shell of desired height. The bottom of the shell is sealed and the annular space between the shell and the pile is filled with grout and optional reinforcing bars. The other is a sandwich construction FRP panel that is used as a stay-in-place form to repair seawalls and deteriorated beams and decks. The panels are connected in the field and secured to the existing structure, creating an annular space between the panels and the host structure that is filled with concrete or grout. The impervious FRP creates a moisture barrier that keeps water and oxygen away from the host structure, thus significantly lowering the corrosion rate.

1 INTRODUCTION

Repair of submerged structures requires diving teams and that adds significant cost to the project. There is a big demand for techniques that can reduce this cost and increase the service life of these structure.

The applications in this paper are for repair or strengthening of corrosion-damaged structures such as timber, concrete or steel piles, seawalls and decks. All these solutions provide an impervious barrier around the host structural element that will keep moisture and oxygen away, thus significantly reducing the corrosion rate. FRP materials do not corrode and offer a very long service life. They are typically stronger than steel in tension and in some applications, this combination of strength and durability can result in unique cost-effective solutions that can extend the service life of the structure.

2 REPAIR OF PILES

The product used for repair of piles is a special type of FRP laminate developed by the author over a decade ago (Ehsani 2010) commonly known as PileMedic®. The system is covered by multiple patents (Ehsani 2018 & 2020). The laminates are typically supplied in rolls that are 1200mm (4 ft) wide by hundreds of feet long. The repair steps require attachment of spacers around the pile and insertion of reinforcing bars. Next, the laminate is cut to a length approximately twice the perimeter of the pile; the second half of the laminate is coated with an epoxy paste and is wrapped around the spacers to create a two-ply shell (Figure 1). A primary advantage of this system is that it eliminates the need for ordering jackets of different size and shape for each project. The special adhesive cures underwater, eliminating the need for coffer dams. Additional 1200mm shells can be similarly built to cover the desired repair height. Lastly the annular space is sealed at the bottom and it is filled with an underwater grout.

Figure 1. Repair of deteriorated timber piles.

The solution has been used in many projects worldwide and videos of sample projects are available for concrete piles: www.tinyurl.com/ASR-Pile and timber piles: www.tinyurl.com/PLM-Perdue

When repairing steel piles, there is not sufficient bond between the surface of the steel and the ring of concrete to transfer the loads. We

DOI: 10.1201/9781003348450-253

have developed special devices called ShearWrap™ and ShearClamp™ for attachment to circular and H-Piles, respectively (Figure 2). These devices are attached to the host pile by predetermined torque levels that ensures proper transfer of loads from the ring of concrete to the host pile by friction. This provides a continuous load path that bypasses the damaged zone of the pile.

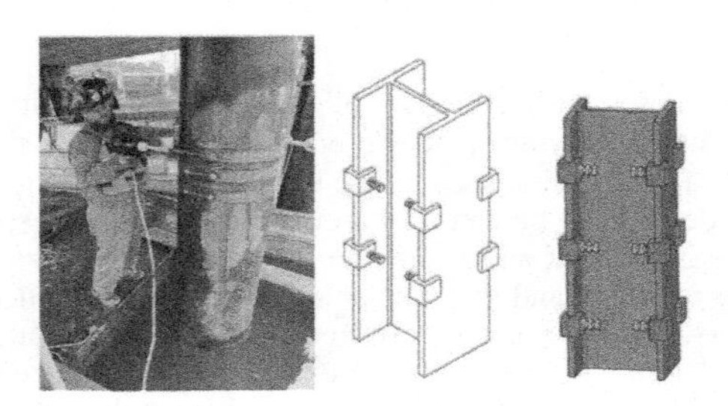

Figure 2. ShearWrap™ being installed on cylindrical steel pile and ShearClamp™ for use on H piles.

3 REPAIR OF SEAWALLS AND DECKS

The patented Sheet Pile Repair (SPiRe®) system is comprised of rigid FRP panels that are custom made for each project using the sandwich construction technique (Ehsani 2021). A lightweight core around 6-8mm (0.2-0.3 in.) in thickness is used. On the exterior surfaces, one or more layers of unidirectional or biaxial glass or carbon fabric will be used. The type and strength of these fabric layers is a function of the project requirements and will be designed for each application. The overall thickness of the product is typically less than 12 mm (0.5 in.).

The most common panels are flat 1200mm (4 ft) wide x 6m (20 ft) long. In the field, the panels get secured with J-bolts to the deteriorated seawall (Figure 3). The edges of the panels are epoxied together to create a seamless impervious stay-in-place form. If desired, reinforcing bars can also be placed in the annular space before it is filled with grout or concrete. Videos of two sample projects for an industrial client www.TinyURL.com/SPiRe-WestRock and the US Coast Guard www.TinyURL.com/SPiRe-USCG are available at the links provided.

Figure 3. Repair of corroded seawall with flat Sheet Pile Repair (SPiRe®) panels.

The decks in piers are often very close to the water surface, leaving little space for the workers to perform repairs under the deck. The same SPiRe® system can be manufactured in various shaped shells as shown in Figure 4. These can be lifted from below and anchored to the deck and the beams while concrete is being placed. Additional reinforcing bars can also be placed in these shells. This will reduce the construction time significantly. The shells will stay in place permanently, providing future protection for the deck against corrosion. More details of this system are presented in this video: www.tinyurl.com/SPiRe-Pier

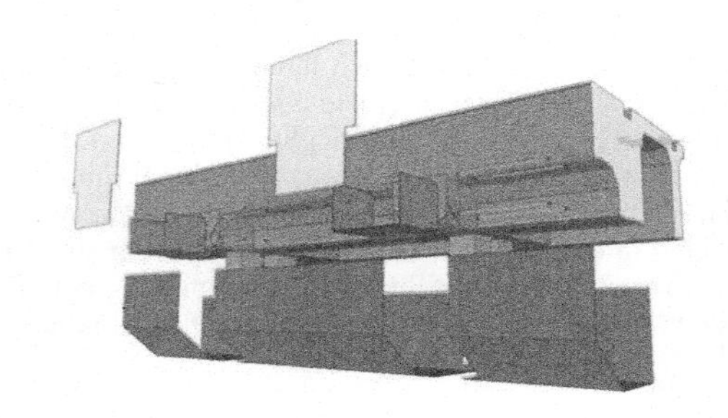

Figure 4. Sheet pile repair FRP panels can be custom made to any shape.

4 SUMMARY AND CONCLUSIONS

Corrosion of submerged piles and seawalls is a major concern in many coastal structures. Techniques for repair and strengthening of these structures using non-corroding FRP products have been introduced. A novel way for improving the transfer of bond stresses in steel piles has also been presented.

Case studies from challenging projects discussed demonstrate the efficacy of these solutions. When substantial strengthening, speed of construction or limited access restrict the use of conventional repair techniques, the solutions offered here can be of great value to the design engineer. In all applications, the FRP system provides a permanent impervious barrier that keeps the oxygen and moisture away from the host structure, i.e. pile or bulkhead. This brings the corrosion rate to extremely low levels, and prolongs the service life of the repaired structural element.

REFERENCES

Ehsani, M. 2010. "FRP Super Laminates: Transforming the Future of Repair and Retrofit with FRP," *Concrete International*, American Concrete Institute 32(3): 49–53.

Ehsani, M. 2018. "Reinforcement and Repair of Structural Columns," U.S. Patent #9,890,546.

Ehsani, M. 2020. "Spacers for Repair of Columns and Piles," U.S. Patent #10,908,412.

Ehsani, M. 2021. "Structural Reinforcement Partial Shell," U.S. Patent #10,968,631.

Current Perspectives and New Directions in Mechanics, Modelling and Design of Structural Systems – Zingoni (ed.)
© 2022 Copyright the Author(s), ISBN 978-1-032-39114-4

Evaluation of using FRP bond equations in alternative types of advanced composite externally bonded to concrete

Zuhair Al-Jaberi
Department of Civil Engineering, College of Engineering/Al-Nahrain University, Al-Jadriya, Baghdad, Iraq

Rana Mahdi
State Commission for Dams and Reservoirs, Studies Department, Baghdad, Iraq

John J. Myers*
Department of Civil, Architectural & Environmental Engineering, Missouri University of Science and Technology, Rolla, USA

ABSTRACT: The use of advanced composite materials has become common for increasing the structural element capacity. Most of the structural buildings are in need of strengthening to extend their service life as a result of changing their function or by steel reinforcement deterioration in the form of corrosion. The bond performance of fiber reinforced polymer (FRP) in both techniques (NSM or externally bonded) has been extensively studied in the past. As a result, several equations have been derived and proposed to predict both the maximum load and the effective bond length for FRP-concrete joints. On the other hand, there is no standard code or specific proposed model to predict the bond performance of the new generation of advanced composite (steel reinforced polymer or SRP). This paper is an attempt to evaluate the possibility of using FRP bond equations in predicting SRP bond behavior based on wide range of database collected from literature. The reason behind this attempt is that both FRP and SRP systems have the same concept of using organic material as a paste material. As a conclusion, the selected equations used in FRP composite system underestimated in term of SRP behavior prediction.

1 INRODUCTION

ch15_504The bond of advanced composites to a concrete or masonry surface for the purpose of strengthening or rehabilitation is a common topic in civil engineering [1,2]. Advanced composites in both forms (NSM or EB) are usually attached to the concrete or masonry surface, either by using polymeric or cementitious material to ensure efficient performance during the service life of structural element. The interfacial bond between the fiber and the structural element substrate plays the main and important role in achieving the action of composite.

The bond between FRP and substrate has been studied extensively, consequently, mathematical models and equations have been proposed to predict the maximum load capacity and the fiber effective length. In order to overcome the drawbacks of FRP that's related to stiffness and ductility, steel reinforced polymer (SRP) was introduced. Determining the effective bond length to be adopted in the SRB system is still a concern for researchers, as well as knowing the amount of force transmitted between concrete and SRP composite.

The aim of this paper is to investigate the SRP bond behavior by evaluating the possibility of using different FRP standards and specifications. A comparison has been made between seven equations used to calculate the effective bond length in various specifications, codes and design guidelines. Also, the common semi-empirical formula and modified formula that used to predict the maximum debonding load were evaluated based on collected database.

2 TEST SET-UP

Different experimental set-ups were employed to evaluate and study the adhesion mechanism of advanced composite strips externally bonded to concrete. Such as single shear lap test, double shear lap test, direct tension pull-off test and bending test. Figure 1 show different test set-up.

3 EFFECTIVE BOND LENGTH

The tensile forces generated in concrete are transmitted to the SRP through the shear stresses in the adhesive material for a short length close to the point of applied load. When the applied load increases, the

*Corresponding author

DOI: 10.1201/9781003348450-254

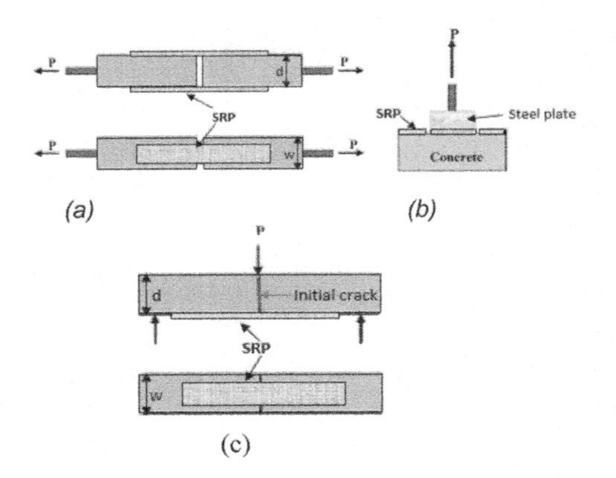

Figure 1. (a) Double shear lap (b) Direct tension (c) Bending test.

active bond zone shifts down away from the point of applied load. The effective part of bonded length that is responsible for transmitting the load is called effective bond length. The efforts of researchers in previous studies focused on investigating the bond between the advanced composite and concrete. Based on literature, there is no highly acceptable analytical model for the required effective bond length. The relationship between the effective length and SRP stiffness of is as shown in Figure 2.

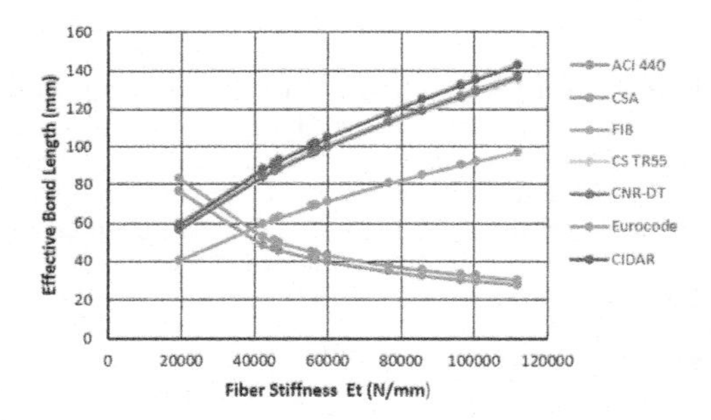

Figure 2. Effective bond length vs. SRP stiffness.

It is noted that some standards or codes have an increasing effective length with the increase in the stiffness of the fiber (E_f t_f), meaning that it is a proportional relation, while other codes have an inverse proportion between these parameters.

Based on these results, it can be concluded that the mechanism of the advanced composites bonding concrete has not been fully understood, and we also conclude that it is not possible to rely on these codes and guidelines in determining the effective bond length for the SRP composite.

4 SRP- CONCRETE BOND CAPACITY

In order to evaluate the existing analytical models that are used in predicting the bond strength transmitted between the advanced composite and the substrate, a wide range of data were collected from various previous experimental studies. Due to the lack of design-guidelines and codes for the SRP system, the models used in the FRP system will be studied to find out the possibility of use for the SRP system. The common semi-empirical formula that used to predict the maximum debonding load was proposed by Chen and Teng.

Based on the results of data base, this model fails to accurately predict the adhesive strength since the average experimental-to-proposed bond capacity is 1.44, but remains within the limits of the underestimated values. The expected reason for these results is that the model did not take into account the properties of the adhesive material that is used to bond the fiber on the concrete surface, which plays a key role in the process of continuing the load transfer before debonding failure.

Diab and Omer modified the Chen and Teng model by considering type of adhesive material. Therefore, the β_a coefficient, which represents the properties of the bonding material, will be considered in the equation of effective bond length and maximum debonding load that proposed by Chen and Teng. It can be noticed that the ratio of experimental-to-proposed bond capacity has improved after adopting the properties of the adhesive material in the proposed equations and its 1.15.

5 CONCLUSIONS

This paper presents a bond characteristic between SRP and the concrete. According to this research, the following conclusions can be drawn:

The effective adhesive length adopted from a number of specifications and codes is inconsistent due to the trend difference between the parameters on which this length depends. The common semi-empirical model (Chen and Teng model) that used to predict the FRP maximum debonding load fails to accurately predict the SRP debonding load and its under estimation by 44% as a result of neglecting the effect of the properties of the adhesive material. The modified model, which took into account the effect of the properties of the bonding material, contributed to an increase in the accuracy of SRP debonding load compared to the common model of Chen and Teng. The average experimental-to-proposed bond capacity for this model is 1.15.

REFERENCES

[1] Al-Jaberi, Z., J. J. Myers, and K. Chandrashekhara. "Direct Service Temperature Exposure Effect on Bond Behaviour between NSM Composites and Concrete Masonry Units." Advanced Composites in Construction, ACIC 2019 - Proceedings of the 9th Biennial Conference on Advanced Composites in Construction 2019, pp. 149–154

[2] Zou, Xingxing, and Lesley H. Sneed. "Bond behavior between steel fiber reinforced polymer (SRP) and concrete." International Journal of Concrete Structures and Materials 14.1 (2020): 1–17.

Current Perspectives and New Directions in Mechanics, Modelling and Design of Structural Systems – Zingoni (ed.)
© 2022 Copyright the Author(s), ISBN 978-1-032-39114-4

Sustainable and corrosion-free bridge structures

A. Belarbi, D. Vecchio & D. Stefaniuk
Department of Civil and Environmental Engineering, University of Houston, Houston, USA

1 TOWARDS CORROSION-FREE BRIDGE STRUCTURES

Prestressing techniques are a solid and trustable way to induce a state of pretension in a concrete member to enhance its capacity under serviceability levels. Nevertheless, its efficiency, either due to the nature of concrete itself or to the type of external environment, small and wider cracks may appear and propagate, leaving the reinforcement under no protection and exposing it to corrosion. Recently, about 27% of all the highway bridges in the United States are prestressed concrete bridges. However, maintenance and rehabilitation of those concrete bridge elements to maintain their serviceability and safety are costly. Therefore, the use of non-corrosive composite materials has increased within the last two decades. Fiber-Reinforced Polymers (FRP) made of glass (GFRP), carbon (CFRP), or basalt (BFRP) fibers are becoming an alternative material to typical construction materials with a broader range of civil engineering applications. Also, high-strength stainless steel (HSSS) strands were introduced to the construction industry as another corrosion-resistant and potentially viable alternative to conventional carbon steel (CS) strands in corrosive environments. Accordingly, this paper presents an overview of three projects conducted within the National Cooperative Highway Research Program (NCHRP) to develop design guidelines specifications according to American Association of State Highway and Transportation Officials (AASHTO), Load and Resistance Factor Design (LFRD) format, for prestressed concrete girders reinforced with non-corrosive materials. A summary of the three projects can be seen in Figure 1.

1.1 *NCHRP 12-97*

The objective of this project was to develop design and material guide specifications in the AASHTO LRFD format for concrete beams prestressed with CFRP longitudinal reinforcement, using either pretensioning or post-tensioning. After an initial phase of literature review and understanding of the state-of-art, analytical and numerical models were assessed to understand the mode of failure at ultimate load, prestress losses, transfer lengths as well as strength reduction due to harping. The experimental campaign included the design, construction, testing, and analysis of 12 full-scale AASHTO Type I CFRP prestressed concrete beams. The project ended with flexural design formulations, covering bonded and unbonded prestressing CFRP, minimum reinforcement, and design examples.

1.2 *NCHRP 12-120*

Differently from CFRP, HSSS exhibits an unclear yielding point, a short plastic deformation beyond the elastic range, and a relatively lower strength. Parametric studies were analyzed for AASHTO Type I and extended to AASHTO Type II and a comparative design procedure was proposed. Advanced numerical techniques were used to model harping, including material properties and simulation stages, and geometry of seven-wire strands. Currently, the project is in the experimental campaign which includes both small and full-scale tests, including wobble and friction losses, bond characteristics, transfer and development length, and immediate and long-term deflections, as well as elastic shortening, friction losses, camber measurement, flexural and shear behavior and minimum reinforcement.

1.3 *NCHRP 12-121*

The objective of this project is to cover all aspects of FRP (BFRP and GFRP) auxiliary reinforcement which have not been established in the previous guidelines with CFRP strands. Similarly, this project is based on an initial phase of literature review and collection of the state-of-art to address the knowledge gaps. Also, potential modifications to the AASHTO LRFD design specifications were proposed. Currently, the second phase is ongoing and both analytical and numerical simulations were already performed to understand the differences against CS reinforced beams, in terms of ultimate capacity and failure modes. An experimental database with FRP reinforced concrete beams was collected, following the idea of the ACI-DAfStb Group "Shear Database" published in 2015.

DOI: 10.1201/9781003348450-255

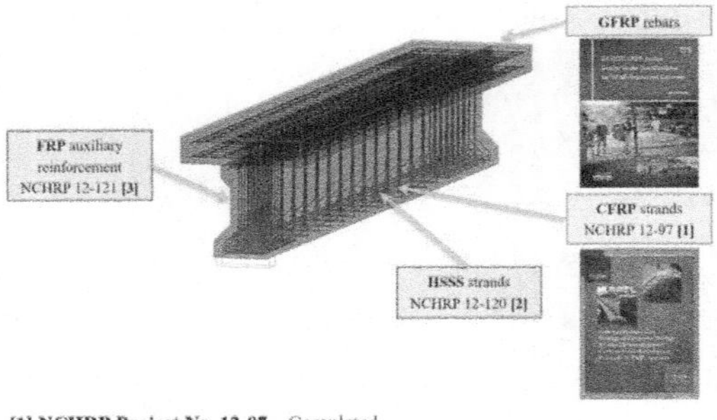

[1] **NCHRP Project No. 12-97** – Completed
*Guide Specification for the Design of Concrete Bridge Beams Prestressed with **CFRP** Systems*

[2] **NCHRP Project No. 12-120** – Ongoing
***Stainless Steel** Strands for Prestressed Concrete Bridge Elements*

[3] **NCHRP Project No. 12-121** – Ongoing
*Guidelines for the Design of Prestressed Concrete Bridge Girders Using **FRP** Auxiliary Reinforcement*

Figure 1. Summary of NCHRP projects conducted at University of Houston, TX, USA.

2 NUMERICAL AND EXPERIMENTAL PROGRAM

2.1 *Harping properties*

Harping of CS is a common practice. However, for CFRP and HSSS, harping can be not feasible due to the substantial losses of jacking capacity due to their brittle nature and low ultimate strain. The results of CFRP performed in NCHRP 12-97 project showed that CFRP can substantially lose its capacity due to harping, and as a result, should not be performed (e.g. for 25.4 mm (1 in.) deviator, the loss of more than 50% of capacity has been predicted).

2.2 *Transfer length*

The force from the prestressed strand is transferred to the concrete over a certain distance, called the transfer length. To analyze the transfer length of CFRP and HSSS, a 3D non-linear model was created using the concrete damage plasticity model in Abaqus software.

2.3 *Flexural behavior*

Flexural behavior was analyzed using the geometry of the AASHTO Type I beam with deck slab, using four-point bending static scheme. The load-deflection curves are shown in Figure 2. As it can be seen, smaller deflections were obtained at failure for HSSS and CFRP. It should also be noted that the lowest cracking and failure loads were achieved in the case of HSSS due to its low ultimate strength.

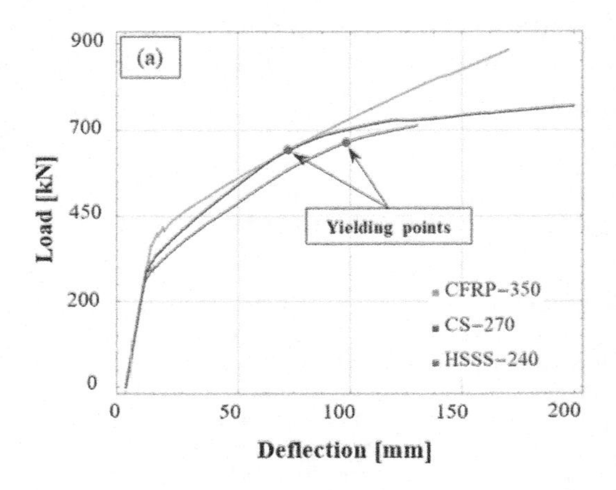

Figure 2. Comparison between HSSS, CS, and FRP strands: (a) the load-deflection curves with marked yielding points.

2.4 *Shear behavior*

The shear strength is widely accepted as the sum of concrete and stirrups contribution, represented by $V_n = V_c + V_s$. In prestressed beams, an additional term is provided by the vertical component of prestressing force in the case of draped strands (V_p). A database, which contains 156 test setups with FRPs stirrups (CFRP, BFRP, GFRP, and in a few cases AFRP) and 237 concrete beams reinforced with FRPs longitudinal rebars, was used as a reference for the analytical computations given by existing shear models.

3 CONCLUDING REMARKS

Conventional carbon steel is not sustainable in an aggressive environment due to corrosion. The alternative is to use non-corrosive reinforcement such as HSSS or FRP. Guidelines and specifications have to be extended or developed. For this purpose, comprehensive analytical and numerical analyses, as well as experimental tests, have been performed (and are still in progress) in order to validate or modify existing design recommendations that already exist for CS.

17. Construction technology, Construction projects, Construction materials, Properties of concrete

Current Perspectives and New Directions in Mechanics, Modelling and Design of Structural Systems – Zingoni (ed.)
© 2022 Copyright the Author(s), *ISBN 978-1-032-39114-4*

Full-scale experiments on hybrid tunnel lining segments

D.N. Petraroia & P. Mark
Institute of Concrete Structures, Ruhr University Bochum, Germany

ABSTRACT: The design of tunnels constructed by mechanized means is usually defined by local effects at the longitudinal joints of the segments employed. This paper presents hybrid segments manufactured with alternative fabrication techniques, so that their longitudinal joints are reinforced with high performance fiber reinforced concrete. Experimental testing is performed on a testing device able to retrieve real boundary conditions of tunnels. The results show higher bearing capacities when compared to a conventional reference. Additionally, the failure is shifted to the middle part of the segment, avoiding brittle failure. Altogether, it allows material savings by thickness reduction of the lining.

1 INTRODUCTION

Mechanized tunnelling has increased over the last decades. Therefore, the optimization of its construction in terms of time and use of material is essential. In the final serviceability state, the contact regions of the lining segments installed by the tunnel boring machine are generally decisive for the structural capacity (Liu et al., 2015), which leads to thicker designs. The goal of this study is to show the improved performance of segments manufactured with strengthened longitudinal joints by means of tests that retrieve the boundary conditions present in real tunnels.

2 FABRICATION AND EXPERIMENTS

2.1 *Fabrication*

The segments were manufactured with a vertical formwork which was conveniently adapted to isolate the regions of the longitudinal joints from the rest of the body (Figure 1). To this end, steel sheets were employed. They counted with vertical hollows to make them compatible with the longitudinal reinforcement of the segments. High-performance steel fiber reinforced concrete with dry consistency was cast in the longitudinal joints, whereas normal concrete was used for the middle region, saving the use of fibers in the specimen. The casting was performed gradually and homogeneously with the two concretes at the same time, so that after removing the steel sheets, both mixtures were in contact wet-on-wet (Plückelmann & Breitenbücher, 2019).

Figure 1. Detail of device employed for the separation of concretes.

2.2 *Specimens and experimental setup*

Three segments were manufactured. Firstly, a reference specimen (R) was cast employing only one concrete for the whole body and conventional reinforcement, that is, stirrups arranged in parallel direction to the load introduction. Secondly, two hybrid specimens (H1 and H2) were cast using the method descrubed in Section 2.1. In H1, the longitudinal joint counted with HPSFRC only, whereas in H2, it additionally counted with welded steel rebars. The HPSFRC counted with a fiber cocktail of 60 kg/m^3 of 1.5-hook ended macrofibers and 60 kg/m^3 of straight microfibers. The mechanical parameters are listed in Table 1.

The tests were performed on a testing device that allows the introduction of radial loads on the segment extrados (V) and reproduces the effects of partial area

DOI: 10.1201/9781003348450-256

loading at the longitudinal joints (Figure 2), where a horizontal load is introduced. Further description can be found at Petraroia & Mark (2021).

Table 1. Material parameters of the test specimens (MPa).

Concrete	*fcm*	E_C	*fct*
Reference	42.8	29658.7	3.7
H1 (NC)	30.7	28425.0	2.6*
H1 (HPSFRC)	77.9	41608.0	-
H2 (NC)	34.6	29886.0	3.0
H2 (HPSFRC)	79.0	42325.7	-

* Result out of one specimen.

Figure 2. Test setup.

2.3 *Results and Discussion*

Figure 3 shows the load-displacements attained.

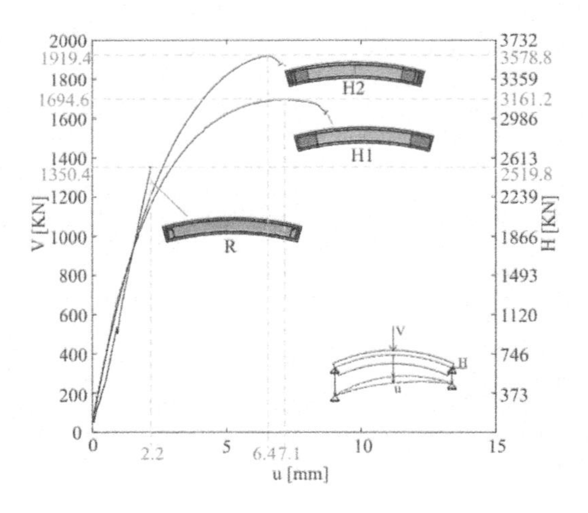

Figure 3. Load-deformation curves of the test specimens.

The segment R reached V = 1350.4 kN for a deformation of 2.2 mm, which was followed by a brittle failure at one of the longitudinal joints due to partial area loading effects. H1 and H2 reached V = 1694.6 kN and V = 1919.4 kN with deformations of 7.1 mm and 6.4 mm, respectively, with failure in the middle region of the segments due to compression with low eccentricity.

If the ratio between maximal vertical load and compressive strength of the NC is computed for each hybrid segment, it results approximately 1694.6/30.7 ≈ 1919.4/34.6 ≈ 55. If this factor is extended to a concrete as the one used for the segment R, an increase in the global capacity of 74.3% with respect to the one achieved by R can be obtained.

This failure mode offers prior notice, which enables to shorten security margins in the design. This, together with the higher bearing capacity, offers the possibility to considerably reduce thickness of the tunnel structure.

3 CONCLUSIONS

In this work, tunnel lining segments manufactured by a non-traditional method were presented that offer a considerable increase in the bearing capacity due to the strengthening of their longitudinal joints with high performance steel fiber reinforced concrete. This, combined with a more favorable failure mode, offers potential for reduction in the tunnel thickness, and with it, of the excavation diameter, reducing excavation times and use of material.

ACKNOWLEDGMENTS

The authors thank the German Research Foundation (DFG) for the financial support within the Collaborative Research Center SFB 837 “Interaction modeling in mechanized tunneling” (Project B1) - project number 77309832.

REFERENCES

Liu, X., Bai, Y., Yuan, Y., & Mang, H. A. 2015. Experimental investigation of the ultimate bearing capacity of continuously jointed segmental tunnel linings. *Structure and Infrastructure Engineering*, 12(10): 1364–1379. https://doi.org/10.1080/15732479.2015.1117115

Petraroia, D. N., & Mark, P. 2021. Variable, full-scale tester for tunnel linings. *Structural Concrete*, 22(6): 3353–3367. https://doi.org/10.1002/suco.202000806

Plückelmann, S., & Breitenbücher, R. 2019. Hybrid lining segments – bearing and fracture behavior of longitudinal joints. *Tunnels and Underground Cities: Engineering and Innovation Meet Archaeology, Architecture and Art Proceedings of the WTC 2019 ITA-AITES World Tunnel Congress*: 2881–2890. https://doi.org/10.1201/9780429424441-305

Current Perspectives and New Directions in Mechanics, Modelling and Design of Structural Systems – Zingoni (ed.)
© 2022 Copyright the Author(s), ISBN 978-1-032-39114-4

Environmental inefficiency of the world construction industry

Magdalena Kapelko
Wroclaw University of Economics and Business, Wroclaw, Poland

ABSTRACT: The measurement of environmental inefficiency that assesses economic sectors' performance integrating the production of both marketed (good) outputs and the negative environmental externalities (bad outputs) is an increasingly important focus of recent economic research. This study assesses environmental inefficiency of the world construction industry over the period 2000-2014. The study's contributions are twofold. In differentiation to previous research on the environmental inefficiency in the construction industry, this study relies on the latest so called by-production model that in the inefficiency measurement considers two separate and parallel technologies: a standard technology generating good outputs, and a polluting technology for the by-production of bad outputs. Moreover, the focus on the construction industry in forty two world developed and developing countries is the second contribution of this study. The results indicate more inefficiency in the generation of bad outputs than in the production of good outputs for the vast majority of countries.

1 INTRODUCTION

Construction industry plays an important role in economic growth and urban development in developed and developing countries, contributing to 13% of the world's GDP (Barbosa et al., 2017). At the same time, construction sector consumes a large amount of resources creating environmental problems such as carbon dioxide (CO_2) emissions and waste generation. In particular, construction industry (together with operation of buildings) was responsible for 38% of global energy-related CO_2 emission in 2015 (United Nations Environment Programme, 2021). CO_2 emission is a major reason responsible for climate change, which is now a critical challenge elsewhere (Hu & Liu, 2015). Therefore, the analysis of environmental performance of the construction industry is of particular importance.

The present paper contributes by assessing environmental inefficiency of the world construction industry. In particular, we focus on construction industry in forty two world developed and developing countries over the period 2000–2014.

2 METHODOLOGY

The by-production model introduced by Murty et al. (2012) is one of the preferred options to the measurement of environmental inefficiency in the recent literature. The by-production approach builds on the idea that production systems are made up of several independent processes. In this model, the technology is an intersection of two sets of sub-technologies: one for the production of good outputs (T_1) and one for the generation of bad outputs (T_2).

Following Aparicio et al. (2021) for the assessment of environmental inefficiency we apply the by-production output-oriented directional distance function using Data Envelopment Analysis (DEA), measuring inefficiency separately for T_1 and T_2. We then calculate an overall measure of inefficiency.-

3 DATASET

This paper uses data on construction industry in forty two developed and developing countries for 2000-2014. The source of standard input-output variables is World Input-Output (WIOD) database (Timmer et al., 2015), while environmental data is derived from the dataset of Joint Research Centre of European Commission (Corsatea et al., 2019). We use the following variables: one good output, gross value added; one bad output, the CO_2 emissions; two non-pollution causing inputs, number of employees and gross fixed capital formation; and one pollution-causing input, emission relevant energy use.

4 RESULTS

We analyze the inefficiencies of construction industry for all years together: for good output technology (IE_{T1}) and for bad output technology (IE_{T2}), and overall inefficiency (OIE), as summarized in Table 1. The findings indicate that on average for 2000-2014 period the more inefficiency is found with regard to bad outputs than in the production of good outputs (0.739 of inefficiency versus 0.421 of inefficiency). Nevertheless, some differences can be observed between countries as some of them are obtaining better performance outcomes for bad output than for good output.

DOI: 10.1201/9781003348450-257

Table 1. Inefficiencies for T_1 and T_2 and overall per country, 2000-2014 (average values reported).

Country	IE_{T1}	IE_{T2}	OIE
Australia	0.075	0.866	0.465
Austria	0.056	0.881	0.464
Belgium	0.021	0.908	0.464
Bulgaria	0.601	0.828	0.589
Brazil	0.710	0.833	0.615
Canada	0.027	0.840	0.433
Switzerland	0.073	0.888	0.476
China	0.453	0.336	0.298
Cyprus	0.107	0.756	0.413
Czech Republic	0.302	0.852	0.541
Germany	0.111	0.790	0.442
Denmark	0.400	0.850	0.567
Spain	0.068	0.478	0.270
Estonia	1.083	0.727	0.622
Finland	0.368	0.843	0.554
France	0.154	0.875	0.503
UK	0.004	0.937	0.471
Greece	0.349	0.832	0.537
Croatia	0.601	0.875	0.621
Hungary	0.893	0.895	0.672
Indonesia	1.226	0.836	0.691
India	0.097	0.198	0.138
Ireland	0.386	0.806	0.537
Italy	0.254	0.906	0.553
Japan	0.081	0.000	0.033
Korea	0.605	0.869	0.620
Lithuania	0.113	0.614	0.352
Luxembourg	0.244	0.669	0.431
Latvia	0.691	0.858	0.624
Mexico	0.415	0.839	0.564
Malta	0.000	0.000	0.000
Netherlands	0.174	0.860	0.502
Norway	0.449	0.814	0.560
Poland	0.064	0.798	0.423
Portugal	0.982	0.922	0.701
Romania	0.771	0.831	0.613
Russia	0.881	0.721	0.575
Slovakia	0.108	0.981	0.534
Slovenia	0.770	0.686	0.557
Sweden	0.262	0.852	0.526
Turkey	2.642	0.869	0.786
USA	0.000	0.014	0.007
Average	0.421	0.739	0.484
Significance betwen inefficiencies	a, b, c		

* a denotes significant differences between IE_{T1} and IE_{T2} at 1%
b denotes significant differences between IE_{T1} and OIE at1%
c denotes significant differences between IE_{T2} and OIE at1%

5 CONCLUSIONS

This paper contributed by assessing environmental inefficiency of the construction industry in forty two developed and developing countries over the period 2000-2014. Moreover, this study's value added relied on the application of the recently developed a by-production model that in the measurement of efficiency considers two separate and parallel technologies: for a production of good outputs and for the generation of bad outputs.

The results indicated the potential for construction industry to improve its competitiveness by enhancing the production of good output, as well as to lower a harmful environmental impact. On average for 2000-2014 construction industry was more inefficient with regard to bad output of CO_2 emissions than in good output of gross value added.

ACKNOWLEDGEMENTS

Financial support for this article from the National Science Centre in Poland (grant no. 2016/23/B/HS4/03398) is gratefully acknowledged. This paper is also financed by the Ministry of Science and Higher Education in Poland under the programme "Regional Initiative of Excellence" 2019 - 2022 project number 015/RID/2018/19 total funding amount 10 721 040,00 PLN. The calculations for the adapted Li test were made at the Wroclaw Centre for Networking and Supercomputing (www.wcss.wroc.pl), grant no. 286.

REFERENCES

Aparicio, J., Kapelko, M. & Ortiz, L. 2021. Modelling environmental inefficiency under a quota system. *Operational Research* 21(2): 1097–1124.

Barbosa, F., Woetzel, J., Mischke, J., Ribeirinho, M.J., Sridhar, M., Parsons, M., Bertram, N. & Brown, S. 2017. Reinventing construction: A route to higher productivity. McKinsey Global Institute. Available at: https://www.mckinsey.com/business-functions/operations/our-insights/reinventing-construction-through-a-productivity-revolution

Corsatea. T.D., Lindner. S., Arto, I., Román, M.V., Rueda-Cantuche. J.M., Velázquez Afonso. A., Amores, A.F. & Neuwahl, F. 2019. World Input-Output Database Environmental Accounts. Update 2000-2016, Publications Office of the European Union, Luxembourg.

Hu, X. & Liu, C. 2015. Managing undesirable outputs in the Australian construction industry using Data Envelopment Analysis models. *Journal of Cleaner Production* 101: 148–157.

Murty, S., Russell, R.R. & Levkoff, S.B. 2012. On modeling pollution-generating technologies. *Journal of Environmental Economics and Management* 64: 117–135.

Timmer, M.P., Dietzenbacher, E., Los, B., Stehrer, R. & de Vries, G.J. 2015. An illustrated user guide to the World Input–Output database: The case of global automotive production. *Review of International Economics* 23: 575–605.

United Nations Environment Programme. 2021. Global status report for buildings and construction: Towards a zero-emission, efficient and resilient buildings and construction sector. Available at: https://globalabc.org/sites/default/files/2021-10/GABC_Buildings-GSR-2021_BOOK.pdf

Current Perspectives and New Directions in Mechanics, Modelling and Design of Structural Systems – Zingoni (ed.)
© 2022 Copyright the Author(s), *ISBN 978-1-032-39114-4*

Natural language processing as work support in project tendering

L. Cusumano, R. Rempling, R. Jockwer & R. Saraiva
Chalmers University of Technology

M. Granath
University of Gothenburg

N. Olsson
NCC Sweden AB

S. Okazawa
Savantic AB

ABSTRACT: When producing a tender, contractors manually analyze client requirements contained within many different text documents. The combination of requirements lead to crucial design decisions and every decision is related to costs and risks. This study explores the possibility of making the client requirement analysis in design-bid contracts automated to reduce the risk of conceptual design mistakes. The research approach chosen includes developing a work support tool based on natural language processing and evaluating its usefulness through a combination of surveys and a workshop for tendering specialist. The results show that applying digitalized working methods and using artificial intelligence in the tender phase can enable data-informed decision making and generate benchmarking and risk management opportunities.

1 INTRODUCTION

In the tender phase of a construction project, contractors invest time and money to find technical solutions optimizing costs, quality, production time, and climate impact. Traditionally, the process of analyzing client requirements and choosing conceptual solutions consists of a lot of manual and time-consuming work (Laryea et al. 2011), and the body of knowledge available for the analysis is the combined experience of the individuals in the tendering team. As a result, little data-informed knowledge is transferred from previous tendering and production projects, other than the specific team members experiences.

Serval factors affect the decision-making in the bid process (Egemen & Mohamed, 2007) but few studies show how the data representing those factors are obtained, processed, and analyzed. With increased digitalization within the construction industry, this high-risk labor-intensive activity could potentially be data-informed, enabling more accurate offers, data-driven bid/no-bid decisions, and automating the client requirement analysis process.

Besides drawings, a lot of information analyzed in the tender phase is text, which makes the area of computer-supported text analysis, Natural Language Processing (NLP), particularly interesting (Otter et al. 2020).

This study explores the possibility of using natural language processing to make the tender procurement analysis more automated and serve as a tool for making data-driven decisions. The aim was to identify opportunities and obstacles for using artificial intelligence (AI) as a tool for tendering specialists.

2 METHOD

The study was divided into the following steps:

- Step 1: Adoption of a proprietary tender phase AI-tool prototype
- Step 2: Pre-demonstration survey that assessed tendering specialists' optimism towards the suggested tool concept.
- Step 3: AI-tool demonstration for tendering specialists
- Step 4: Post-demonstration survey that assessed changes in optimism towards the proposed tool concept.
- Step 5: Data analysis and conclusions

The purpose of introducing an AI-tool to tendering specialists was to give an example of how AI can be used as work support and inspire the survey participants to reflect on the possibilities it might provide within their field of specialization.

The AI-tool is Python-based, using Word2Vec for text processing and cosine similarity to compare text. The AI-tool was given a web-based user interface, giving its users the ability to:

- Extract client requirements from PDF documents.

DOI: 10.1201/9781003348450-258

- Find similarities between the requirements of a new project and previous projects already inside a database.
- Compare different project requirements

The intended usage of the AI-tool is in the initial phase of the tender project when analyzing the client requirements, see Figure 1.

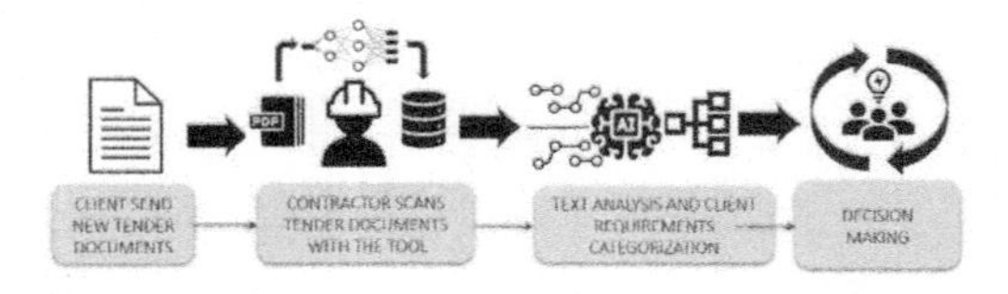

Figure 1. Intended use of AI-tool developed in the study.

The AI-tool was demonstrated to a group of tendering specialists within structural engineering, geotechnology, cost-estimations, energy calculations, life cycle cost calculations, sustainability classification systems, and building service systems. The specialists evaluated the AI-tools potential and identified opportunities and challenges with implementing such a tool through a pre- and post-demonstration survey. The first survey was sent to 59 specialists and had an answer frequency of 64%. The second survey was sent to persons who answered the first survey and attended the demonstration and had an answer frequency of 66%.

3 RESULTS

The survey results were thematically divided into challenges, opportunities, future features, and implementation needs. Then, the answers were analyzed and divided into subthemes. Figure 2 shows the frequency of responses corresponding to each subtheme.

There were three subthemes of challenges identified: the tool's reliability, the quality of the data extracted, and how such a tool can cause misunderstandings and misinterpretations among the humans using it.

The most significant opportunity area was in the early stages, such as the tender stage or early involvement projects. The other two subthemes within opportunities were the tool's potential for estimating material quantities and time and benchmarking purposes related to key figures for data and prizing.

The two most prominent desired future features were found in risk management, bid/no bid decisions, and benchmarking by creating specialist-specific templates.

Before such an AI tool can be implemented, the need for some actions was identified in the surveys. One frequent answer was gained knowledge among potential users regarding how the tool works, its precision, and its limitations. The other two subthemes were the need to create and maintain accurate databases and a need for changing work methods and working more standardized.

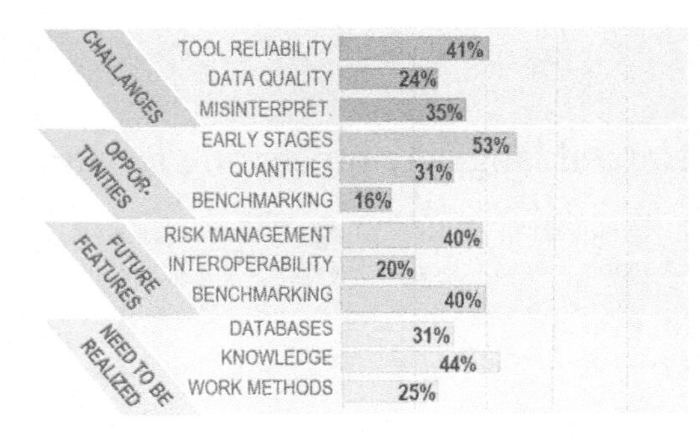

Figure 2. Themes and response rate of subthemes presented as a summarized result from both surveys.

4 CONCLUSION

In this study, the possibility of using natural language processing as a tool for analyzing client requirements has been expslored by demonstrating an AI-tool to tendering specialists and assessing its potential by a pre- and post-demonstration survey. The results show that such a tool could be helpful support for tendering specialists, particularly for project benchmarking and risk management purposes.

NLP applied in the tendering process allows analyzing large datasets and extracting meaningful insights to support decision-making. Using such tools could reduce the manual work invested in client requirement analysis. Still, more importantly, it can increase the tender accuracy by enabling benchmarking of project key figures and access to historical data.

However, it requires creating and maintaining databases and higher standardization levels regarding collecting, storing, and comparing data. In addition, there must be a systematic way of logging both client requirements and actual project output. If data from projects won and executed can be collected, it can serve as an essential input for new tender projects, allowing knowledge transfer from production to early stages.

REFERENCES

Egemen, M. & Mohamed, A.N. 2007. A framework for contractors to reach strategical correct bid/no bid and mark-up size decisions. *Building and Environment*, 42 (3), 1373–1385.

Laryea, S & Hughes, W. 2011. Risk and price in the bidding process of contractors. *Journal of Construction Management and Engineering* 137(4): 248–258. https://doi.org/10.1061/(ASCE)CO.1943-7862.0000293

Otter, D. W., Medina, J. R. & Kalita, J. K. 2020. A survey of the usages of deep learning for natural language processing. *IEEE Transactions on Neural Networks and Learning Systems* 32(2): 604–624.

Current Perspectives and New Directions in Mechanics, Modelling and Design of Structural Systems – Zingoni (ed.)
© 2022 Copyright the Author(s), ISBN 978-1-032-39114-4

Improvement of mechanical performance of cement-based composites using a new type of low-cost low-energy demand graphene

T.D. Nguyen & M. Su
Department of Mechanical, Aerospace and Civil Engineering, The University of Manchester, Manchester, UK

M. Watson
First Graphene (U.K.) Ltd, Sedgefield, County Durham, UK

ABSTRACT: Graphene is employed as an additive to enhance the performance of Portland cement composites. The presence of 2-D nano-size graphene improves the nucleation of cement hydrates and the prevention of crack propagation within cement composites. However, the high cost and problems with homogeneous dispersion limit the use of graphene in a cement matrix. To seek low-cost high-performance cement composites, the use of hydrated graphene (HG), a low-cost product with high water dispersibility, was investigated. Cement composites with different HG contents were examined and compared with the corresponding samples with ordinary graphene nanoplates (i.e., dried graphene (DG)) and control sample (without any graphene). The results indicated that HG performed better dispersion and stability than DG. Accordingly, samples using HG gained highest compressive strength and reached a 3-day compressive strength of 61 MPa, which was 42% higher than the control sample (43 MPa).

1 INTRODUCTION

Ordinary Portland cement (OPC.), with low-cost and high compressive strength, is widely used and has a considerable impact on modern construction industry. Graphene, a single-layer sheet of carbon atoms arranged into a 2D honeycomb lattice pattern, is employed as an additive to enhance the performance of OPC composites. The improvement in hydration and microstructure of OPC mortar samples using 0.07% graphene resulted in an increase in the compressive and tensile strength by 34 and 27%, respectively (Ho et al. 2020). Furthermore, the inclusion of 2.5% graphene decreased the chloride diffusion coefficient and chloride migration coefficients of OPC mortar samples by 70 and 31%, respectively (Du et al. (2015)).

Although cement-based composites incorporating graphene exhibits outstanding performance, they have not been widely used because of: (i) the high cost of graphene, and (ii) the difficulty of dispersing graphene due to its hydrophobic nature, high surface energy and strong Van der Waals interactions. To overcome the agglomeration and high cost of graphene, this study introduces the use of hydrated graphene in the OPC composite. The hydrated graphene (HD) is an intermediate graphene product, and it is supplied as a paste without drying. Therefore, its price is less than that of dried graphene (DG) which has been used in the previous research. Furthermore, the hydrated graphene in the paste form can eliminate potential agglomeration so that it requires less energy for dispersion than the dried graphene in the powder form. The cement-based samples with different HG contents were examined and compared with the corresponding samples with DG and control sample (without any graphene).

2 MATERIALS AND METHODOLOGY

2.1 *Materials*

In this study, The OPC used was CEM1 52.5N (Hanson-UK). The HD and DG graphenes were provided by First Graphene (UK). A superplasticizer (SP) (ADVA 650, UK) was used to disperse and improve the stabilization of dried and hydrated graphene in the mixing water.

2.2 *Sample preparation and methodology*

In addition to the control sample (CS), two sets of samples (i.e. DG and HG) were designed. Each set included graphene contents of 0.005, 0.01 and 0.017% of the mass of cement. A ratio of OPC:sand:water:SP was 1:1.94:0.36:0.09. The sedimentation of graphene aqueous solution was obtained by using UV-Vis Camspec - M550. The compressive strength of the mortar sample was measured by servo-controlled hydraulic testing machine. FTIR was performed by using a Bunker Vertex 80v instrument.

DOI: 10.1201/9781003348450-259

3 RESULTS

The observations of the sedimentation of graphenes in aqueous solution with times based on the absorbance of these solutions measured by UV–vis are shown in Figure 1. A solution with a higher absorbance with time indicates for better dispersion of graphene aqueous solution with time. The reduction in absorbance of the HG suspension was slower with time than that of DG suspension, indicating its better stablity and disperssion than DG. Since the stability and dispersion of graphen with time is the importance factor to assure the acceleration in hydration of OPC until final setting, the use of HG should offer better accelated hydration of OPC than the use of DG.

Figure 2 presents the infrared transmittance spectrums of C-S-H phases of paste samples at band of ~440, 490, ~670 and 740cm^{-1} after 3 days of hydration (Zhu et al. 2020). The samples including DG or HG demonstrated higher intensities of C-S-H phases than the CS.

Figure 3 presents the compressive strength development of OPC mortar samples over 28 days of curing. In line with the FTIR results, samples including DH or HG presented higher compressive strength than the CS, especially at early ages. The higher compressive strength of DH and HG samples than CS was attributed to the role of nucleation sites of DH an HG in promoting the hydration of OPC to form a denser microstructure of OPC mortars. Generally, samples using HG produced higher compressive strength than samples using DG with the same graphene content. The compressive strength of CS slowly gained 43 MPa after 3 days. The use of DG led to 35% improvement in 3-day compressive strength compared to the CS to reach 58 MPa. The improvement in compressive strength was more obvious in samples including HG (~42%), resulting in a 3-day compressive strength of 58 to 61 MPa. Afterwards, the compressive strength of CS fastly developed to gain 58 and 73 MPa after 7 and 28 days of hydration, respectively. The inclusion of DG enabled the OPC mortar samples reached up to 69 and 78 MPa after 7 and 28 days of hydration, respectively. Samples using HG demonstrated the highest compressive strengths as they reached up to 78 and 82 MPa after 7 and 28 days of hydration, respectively.

4 CONCLUSIONS

The results indicated that the inclusion of HG stimulated the formation of hydration products and increased the compressive strength of OPC mortar by 42% and 12% after 3 and 28 days of hydration, respectively. HG performed better dispersion and stability than DG during the hardening of OPC thus HG mortar samples presented better performance than DG mortar samples. Therefore, HG with cost-effective and well-dispersed performance is considered a perfect alternative to DG in OPC-based materials.

ACKNOWLEDGEMENT

This work was supported by the Engineering and Physical Sciences Research Council [grant number: EP/T021748/1].

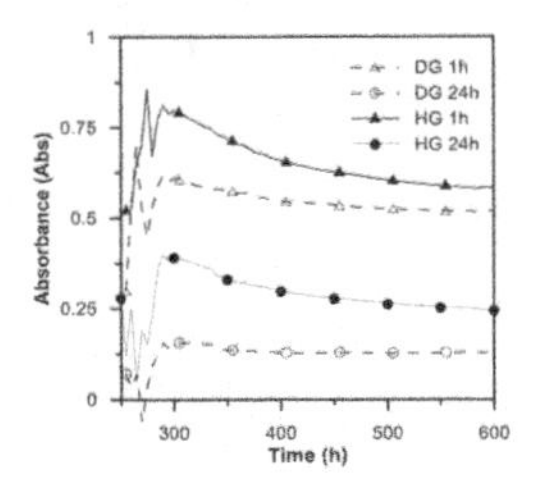

Figure 1. UV-vis test of graphene suspensions.

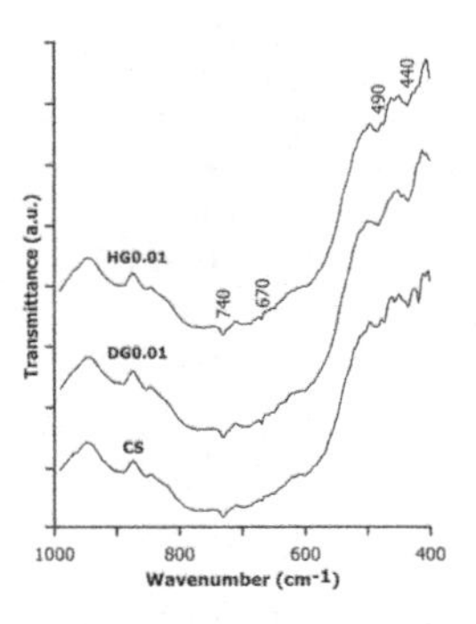

Figure 2. FTIR spectra of paste samples.

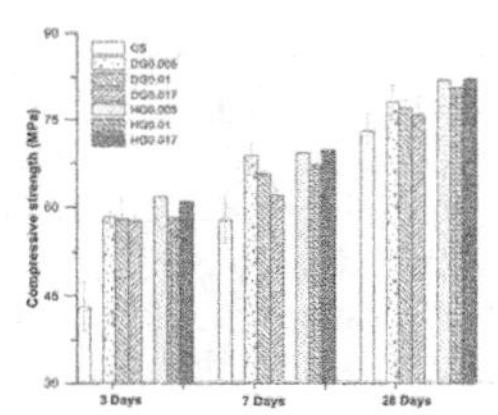

Figure 3. Compressive strength of mortar samples.

REFERENCES

Du, H. & S. D. Pang. 2015. "Enhancement of barrier properties of cement mortar with graphene nanoplatelet." *Cement and Concrete Research* 76: 10–19.

Ho, V. D., et al. 2020. "Electrochemically produced graphene with ultra large particles enhances mechanical properties of Portland cement mortar." *Construction and Building Materials* 234: 117403.

Zhu, X., et al. 2020. "Experimental study on the stability of C-S-H nanostructures with varying bulk CaO/SiO2 ratios under cryogenic attack." *Cement and Concrete Research* 135: 106114.

Current Perspectives and New Directions in Mechanics, Modelling and Design of Structural Systems – Zingoni (ed.)
© 2022 Copyright the Author(s), ISBN 978-1-032-39114-4

Relationship between flexural strength and compressive strength in concrete and ice

Alessandro P. Fantilli, Barbara Frigo & Farmehr M. Dehkordi
Politecnico di Torino, Torino, Italy

ABSTRACT: Ice is a locally available material in cold regions, where it is used in temporary constructions and permanent hydraulic structures. Thus, it is of practical interest the introduction of a model capable of computing the flexural strength of ice as function of its compressive strength. Accordingly, three point bending test and compression test have been performed for the first time on the same ice prism (40 × 40 × 160 mm^3). In accordance with the testing procedure suggested by UNI EN 196 for cement-based mortars, compression loads are applied on the two halves of the specimens previously broken in bending. In this way, the ratio between the modulus of rupture and the compressive strength of ice can be measured. As a result, although the specific strengths of cement-based materials are higher than those of ice, the flexural/compressive strength ratio of ice is larger than that obtained in normal strength mortar.

1 INTRODUCTION

1.1 *State of the art*

In recent years, interest in ice has increased not only as a scientific research topic, but also as a promising material for solving real problems in Civil Engineering, regarding temporary ice construction and permanent hydraulic structures (Vasiliev Pronk, Shatalina Janssen Houben. 2015).

Snow and ice have been used as construction material since ancient times (Timco. 1989), especially in countries with a polar environment.

Ice is a natural material locally available. However, there are several issues with the structural application of ice, such as weak strength, brittle failure, etc. This limits the application of ice and snow structures to some extent.

Because of the low strength, a considerable shell thickness is required, resulting in a comparatively long construction time. The layer thickness could be drastically reduced by using fibers within the ice structure. Moreover, the effect of temperature on mechanical characteristics needs to be studied at the molecular level.

On the other hand, concrete is a well-known material for structural application. Concrete is the second-most-used substance in the world after water (Colin. 2014). Concrete has relatively high compressive strength, but significantly lower flexural strength. The ratio between flexural strength and compressive strength is an important material property of concrete.

Considering the complexity, cost, and time-consuming nature of performing flexural tests, many researchers and building guidelines are interested to predict the flexural strength from compressive strength therefore their relationship is really useful for practical application

1.2 *Research significance*

This study introduces a relationship between the ratio of flexural strength to compressive strength (f_f/f_c) of plain ice specimens, and the results are compared with f_f/f_c ratio of cement-based materials.

2 EXPERIMENTAL ANALYSIS

An experimental campaign has been performed at Politecnico di Torino (Italy) on cement-based mortars and plain ice which will be discussed in the next part.

2.1 *Material and specimens*

The procedure suggested by UNI EN 196-1 (2005) was adopted to cast mortar specimens, three cement-based mortars were prepared. The 3 prisms were demolded one day after casting and stored at 20°C (90% RH). After that the prisms were left for 28 days to reach their full strength.

Comparing to concrete samples, the main issue of ice specimens is the preparation of the samples. To tackle this issue, a new mold was made with resin. Due to the flexibility of the mold, ice specimens were easily detached without any distortion.

DOI: 10.1201/9781003348450-260

2.2 *Experimental procedure for cement-based mortar*

28 days after casting, all of the cement-based prisms were tested in three-point bending by applying the load through a loading cell. The applied force P and the mid-span deflection η were determined and reported. Having the result of flexural tests we can evaluate the flexural strength using the following equation.

$$f_f = \frac{3FL}{2bd^2} \tag{1}$$

As a results of the three point bending test, each specimen is divided into two parts, on which compression tests are performed. The values of compressive strength f_c measured on the mortar.

2.3 *Experimental procedure for ice*

Similarly to mortar specimens, ice prisms 40 × 40 × 160 mm^3 were obtained and tested both in tension and in compression according to UNI EN 196-1 (2005). During the test, the melting of ice did not occur significantly, and the weight of the specimens remains more or less the same

As expected, cement-based materials have higher tensile and compressive strengths than ice. On average the flexural strength of concrete is around 5 times larger than ice while the compressive strength is around 18 times greater than ice.

3 ANALYSIS OF THE RESULTS

3.1 *Comparing the data with literature*

Table 1 reports the values of compressive strength of ice as measured by Wu et al. (2020). The strength increases as the freezing point decreases, and the compressive strength measured at −5°C (i.e., 1.78 MPa) is more or less similar to the average value obtained in the current tests. This is a valid standard approach to measure the mechanical properties of both cement-based mortars and ice.

3.2 *Ratio of tensile to compressive strength*

Since the evaluation of the compressive strength is generally easier, the relationship flexural strength *vs.* compressive strength is used to avoid flexural tests.

Table 1. Ice compressive strength evaluated in the article for different temperatures.

Fiber Content	Temperature	f_c
	°C	*MPa*
0	−5	1.78
0	−10	2.59
0	−15	3.08

To obtain flexural strength from compressive strength for ice materials, the approach suggested by Euro-code 2 can be adopted.

$$f_{ctm,fl} = \max\left\{\left(1.6 - \frac{h}{1000}\right) \times f_{ctm}; f_{ctm}\right\} \tag{2}$$

Where, h=40 mm and the average direct tensile strength f_{ctm} has to be calculated with

$$f_{ctm} = \alpha \times f_{ck}^{(2/3)} \tag{3}$$

where α is a coefficient depending on the type of the material. The evaluated alpha for our cement-based mortar is 0.32. If the same formulae are considered in the case of ice, due to the higher flexural to compressive ratio, the value of α must be higher. In particular for our case, α needs to be 0.42, which is higher than that of the cement-based mortar.

4 CONCLUSION

Even though cement-based materials have much higher flexural and compressive strengths than ice, the flexural/compressive strength ratio of ice is greater than that of normal strength mortar.

On the other hand to obtain the flexural strength from the compressive strength, the traditional approach suggested by Euro-code 2 for concrete can be easily extended to ice.

FURTHER RESEARCH

In the next phase of our investigation, it is necessary to consider the effect of the temperature and also the addition of fibers to our ice specimens.

REFERENCES

N. K. Vasiliev, A.D.C. Pronk, I.N. Shatalina, F.H.M. E. Janssen, R.W.G. Houben, 2015. A review on the development of reinforced ice for use as a building material in cold regions. Cold Regions Science and Technology, 115, 56–63.

Timco, G.W. (Editor) (1989) International Association for Hydraulic Research (IAHR) Working Group on Ice Forces– 4th State-of-the-Art Report. US Army Cold Regions Research and Engineering Laboratory Special Report 89–5, 385 pp.

Gagg, Colin R. (1 May 2014). "Cement and concrete as an engineering material: An historic appraisal and case study analysis". Engineering Failure Analysis. 40: 114–140

UNI EN 196–1 (2005) Methods of testing cement - Part 1: Determination of strength. European Committee for Standardization, Brussels.

Yue Wu. Xiaonan Lou. Xiuming Liu. Arno Pronk(2020) The property of fiber reinforced ice under uniaxial Compression, Materials and Structures

Current Perspectives and New Directions in Mechanics, Modelling and Design of Structural Systems – Zingoni (ed.)
© 2022 Copyright the Author(s), *ISBN 978-1-032-39114-4*

Comparison and statistical evaluation of Marshall stability and stiffness modulus for asphalt mixtures

J. Valentin, P. Vacková & M. Belhaj
Department of Road Structures, Faculty of Civil Engineering, Czech Technical University in Prague, Prague, Czech Republic

ABSTRACT: The Marshall test has been used for decades as a basic approach to optimize asphalt mix design. It has gradually been replaced by new approaches over the last two decades, yet it can serve as a quick indication for predicting the deformation behavior of an asphalt mix. The stiffness of the asphalt mixture expressed in terms of its modulus, which is determined by different test methods usually at a temperature of 10-20 °C, represents one of the key factors for the mechanistic design of the asphalt pavement structure. In contrast to the Marshall test, the determination of the stiffness modulus is always based on a non-destructive principle, which also uses measurements in the linear viscoelastic region of the bituminous binder. The negative aspect of the non-destructive test for determining the stiffness of the asphalt mix is the higher demand on laboratory equipment and at the same time the use of appropriate software for the actual stiffness calculation. This is a limiting for many commercial laboratories and limits the tests to universities or research centres. For this reason, in recent years, attention has been given to verifying the possibility of an indicative prediction of the sufficient stiffness of an asphalt mix based on the Marshall test characteristics, which can be performed by essentially any road laboratory. Measurements have been made on more than 250 different types of asphalt mixtures in order to find correlations between the stiffness determined by repeated indirect tensile loading and the Marshall stability, which is the defining characteristic of the Marshall test. The results are presented in the paper.

1 INTRODUCTION

The Marshall test determined according to EN 12697-34, was used in the past as one of the basic tests for asphalt mixtures design and characterization (at least in the Czech Republic). It is still used in some parts of the world, both for basic characterization of asphalt mixtures and for optimization of mix design. The Marshall stability evaluates the ability of the asphalt mixture to withstand applied loads and can be first and easy indicator of other asphalt mix properties.

The determination of Marshall stability is advised to be used in the Czech Republic as the first stiffness indicator for asphalt mixtures with high stiffness modulus (HMAC). If the HMAC meets the minimum stability limit of 14 kN and achieves strain between 20-50 dmm (or up to 60 dmm for modified bituminous binders). Furthermore there is some probability that it will also meet the minimum stiffness modulus limit of 9 000 or 9 500 MPa at 15°C (depending on the used non-destructive test method either performed as 2-point bending test or indirect tensile stress test).

The paper summarizes the results of more than 250 asphalt mixtures analyzed between 2018 and 2022, for which air voids content, the stiffness modulus at 15 °C (by the IT-CY method according to EN 12697-26) and the Marshall test were determined.

2 TEST RESULTS

A certain correlation was found between the Marshall stability determined at 60 °C and the stiffness modulus determined at 15 °C when all asphalt mixtures were included (coefficient of determination R^2=0.52). When the asphalt mixes were divided into "surface" (approx. 85) and "binder+base" course (approx. 165), a quite good correlation was found for the "binder+base" course group, but the correlation was very low for the asphalt mixes for the surface courses (R^2=0.15). The larger dispersion of the measurements may be caused by a greater variety of asphalt mixes, a higher use of asphalt mixes with PMB or, conversely, a smaller statistical sample.

2.1 *Asphalt mixtures for binder and base layers*

The asphalt mixes for binder and base layers include conventional asphalt concretes AC_{binder} and AC_{base}, or RBL mixes (known as *rich bottom layers*). Another

DOI: 10.1201/9781003348450-261

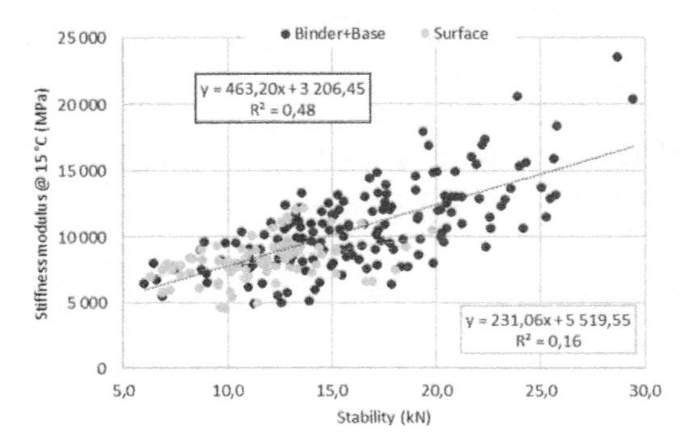

Figure 1. Correlation between Marshall stability and the stiffness modulus – according to use in course.

representatives are SMA_{binder} mixes intended for binder layers and high modulus asphalt concretes (HMAC) eminent for their higher stiffness values. All types of asphalt mixes have representatives with a maximum grain size of 16 mm as well as 22 mm.

From general knowledge, there was a concern about a higher heterogeneity of results in case of higher grain size, but this concern was not fulfilled according to the obtained results. If the asphalt mixes with a maximum grain size of 16 mm and 22 mm were compared in terms of the correlation between stiffness and stability, better results would have been obtained for the asphalt mixes with a maximum grain size of 22 mm.

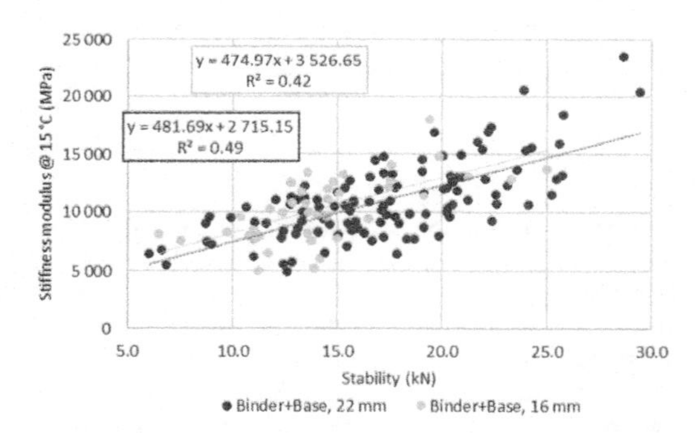

Figure 2. Comparison of asphalt mixtures for binder and base layers – different maximum grain sizes.

Among the asphalt mixtures for binder and base layers, the highest correlation of stiffness modulus and Marshall stability is undoubtedly obtained for the HMAC mixtures, although the coefficient of determination of 0.53 still does not represent a statistically very significant dependence between these variables. The largest linear regression directive is also observed for the set of HMAC mixes, i.e. the increase in stiffness modulus occurs "faster" than for conventional asphalt concretes. For these, the correlation between stiffness modulus and Marshall stability is also significantly lower.

2.2 *High modulus asphalt concretes*

In the case of the 15 mixes tested, it is evident that the asphalt mixes here did not meet the stiffness modulus requirement, even though they exceeded the 14 kN stability threshold. This represents approximately 20% of all cases. The remaining 7 variants achieved stability below 14 kN, with only one variant achieving a stiffness modulus higher than 9 500 MPa.

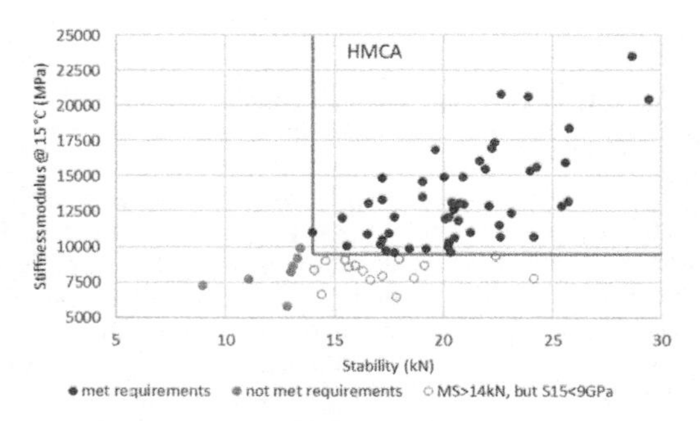

Figure 3. Comparison of HMAC mixes.

2.3 *Different types of bituminous binder*

The results show that asphalt mixes with hard binders (in this case, exclusively HMAC asphalt mixes) achieved the highest stiffness moduli and Marshall stability. The slope of the regression function directive is also higher for this type of binder - the increase in characteristics occurs faster than, for example, for conventional paving grade binders – unmodified binders. The lowest increase is then for asphalt mixes with PMB binders, which tend to achieve lower stiffness moduli at higher stabilities.

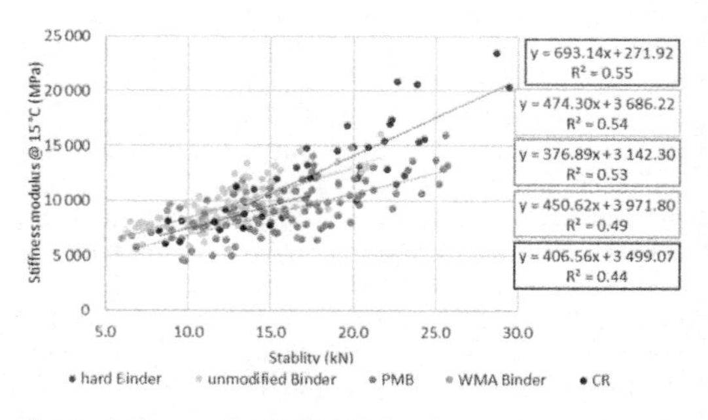

Figure 4. Comparison of asphalt mixtures in terms of the asphalt binder used.

ACKNOWLEDGEMENTS

This paper was written within the activities of the project TE01020168 of the Centers of Competence program of the TAČR.

Current Perspectives and New Directions in Mechanics, Modelling and Design of Structural Systems – Zingoni (ed.)
© 2022 Copyright the Author(s), ISBN 978-1-032-39114-4

Using recycled concrete aggregate from bombed building in new concrete

Salhin Alaud
Assistant professor of Civil Engineering, Garaboulli Engineering faculty, Elmergib University, Libya

Nouri Ali Droughi
Professor of Radiation Protection and Radiation Measurement, Nuclear Research Center, Tajoura, Libya

Salem Alaud
Radiation Protection Supervisor

ABSTRACT: The natural level of radioactivity in building materials is one of the major precautionary measures for the buildings that were bombarded by the NATO forces in Libya. For the purpose of this research project number of samples were taken from one building to verify that components of these buildings are free from any natural (especially depleted uranium) or man-made radioactive materials. To reuse these building materials, it has to be measured using highly sensitive gamma-ray detectors to specific activity concentrations of ^{226}Ra, ^{232}Th, and ^{4o}K in these building materials from one big building in Tripoli-Libya. The concentrations of the natural radionuclides and the radium equivalent activity in studied samples were compared with the corresponding results of different countries. The analysis found that these materials were free from any contamination of depleted uranium and therefore may be safely reused as construction materials and do not pose any significant radiation hazards. After verifying that the building was free from radiation, the properties of concrete containing recycled aggregate from the selected building were studied. It has been proposed to use this building that is destroyed as aggregate in new concrete in weight ratios (0%, 25%, 50%, 75%, 100%) of natural aggregate as an alternative part of the weight of coarse aggregate. Several tests have been carried out for recycled aggregate as well as fresh and hardened concrete. The results showed that an increase of recycled aggregate replacement decreases the workability and strength of new concrete.

1 INTRODUCTION

The aggregate obtained from the concrete wastes can be re-used instead of the natural aggregate in part or in whole to produce new concrete. The use of recycled concrete aggregate in construction work started when many structures were demolished by bombing after the Second World War (Ogar 2017). The largest construction and demolition waste (CDW) obtained from the construction waste at present is concrete (Symonds 1999). Basically, most research is based on tests of mechanical properties of recycled aggregate concrete. However, the challenge is that this research point requires checking the amount of radiation especially depleted uranium that was supposedly used by NATO forces in a bombing large number of buildings and compounds in Waster part of Libya some of the buildings were hit by ordinary bombs and rockets. For these reasons to reuse the leftover of these damaged buildings, it has to be verified that it is clear from any natural or depleted uranium and to verify that it is environmentally suitable and free from any types of radioactivity. For this reason, three samples were taken from one building that was targeted by NATO forces during the appraisal of the February revolution and measured using gamma-ray spectrometry.

2 METHODOLOGY AND RESULTS

The presence of radiation in the demolished building is examined firstly in this paper. After verifying the absence of radioactivity, the proper replacement of recycled aggregate is determined according to the appropriate application of concrete.

2.1 *The radioactivity in the building*

Three samples were taken from different places in the building to examine the radiology that may have accompanied the bombs dropped on the building. Using gamma-ray spectrometry the activity concentration due to the total natural radioactivity namely depleted uranium in the studied building materials was estimated using High Purity Germanium detector at the Tajura Nuclear Center using the Equation 1 Activity:

$$B_q = \left[\frac{CPS \times 100 \times 100}{BI \times E}\right] \pm \left[\frac{CPS_{error} \times 100 \times 100}{BI \times E}\right] \quad (1)$$

Where *CPS* is the net count rate per second, *BI* is the Branching Intensity and *E* is the Efficiency of the

DOI: 10.1201/9781003348450-262

detector. The results are illustrated in Table 1. In the present study, arithmetic means concentrations are derived.

The values fall within the lowest range and are three to six lower than the reported world average values in the UNSCEAR 2000 Report (UNSCEAR 2000).

Table 1. Activity concentration of U-238, Th-232 and K-40 for three building material samples.

S. No.	Activity Concentration (Bq/Kg)		
	U-238	Th-232	K-40
1	80 ± 5	35±1.8	370±15
2	61 ± 2.0	42±2.5	270±14
3	50 ± 4.0	41±3.7	280±10

2.2 *Recycling the building as aggregate*

Since the amount of radiation in the building is free from any contamination of depleted uranium, it is now can be reusing the building debris as aggregates in new concrete after some tests. Ordinary Portland cement, Fine and coarse aggregates were used.

Preliminary physical and mechanical tests on the natural and recycled aggregate including specific weight, adsorption, and Impact value was conducted.

Cubes (15 cm) for compressive and Prisms with (10x10x50) cm to find the bending resistance of concrete mix has been designed. Mixture (C0) as a reference and four other mixes have been designed with replacing the recycled aggregate by 25%, 50%, 75%, and 100% of concrete denoted by RC25, RC50, RC75, and RC100 respectively.

2.3 *Results and dissection of recycled concrete*

2.3.1 *The workability of recycled concrete*

The slump test results for the mixes are illustrated in Figure 1. From the plot shown results, the workability decreases with an increase in the recycled aggregate (RA) replacement although the w/c content are the same in all mixes.

2.3.2 *Compressive strength*

The average of the results of the compressive strength test on the 7th day and on the 28th day are illustrated in Figure 2. It can be observed from the plot that the compressive strength generally at the 7th and 28th day slightly reduces with an increase of more than 25% of RA replacement ratio, However, in the literature, some researchers indicate.

2.3.3 *Flexural strength*

Figure 3 Shows the influence of the RA replacement ratio on the flexural strength of RAC. It is evident from the plot that with an increase in RA ratio, the flexural strength of RAC slightly decreases.

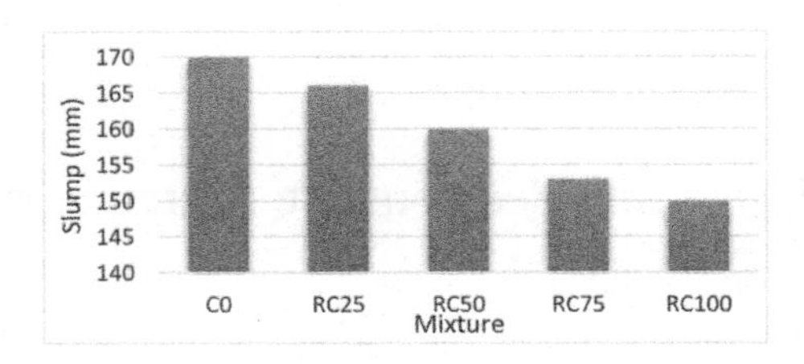

Figure 1. Slump test value for all mixes.

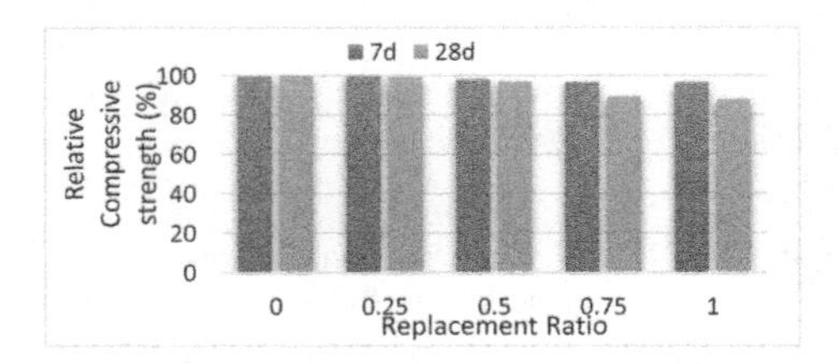

Figure 2. Relative compressive strength of RAC for different.

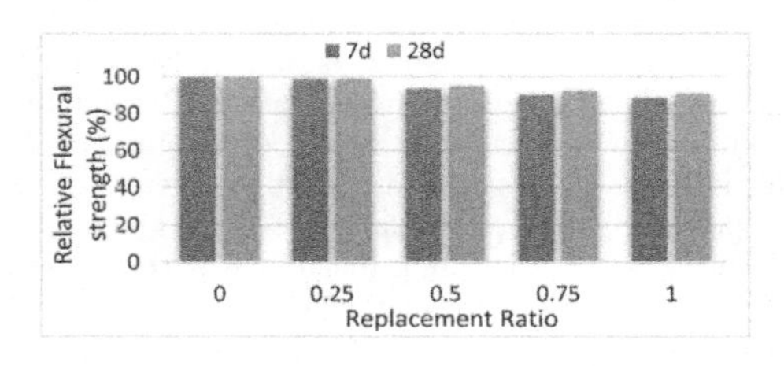

Figure 3. Relative flexural strength of RAC for different.

3 CONCLUSION AND RECOMMENDATIONS

High-resolution γ-ray spectrometry was used to determine the elemental concentrations of the radioactive radionuclides of thorium, uranium namely depleted uranium, and potassium in three samples collected from the designated demolished building in the suburb of Tripoli city have shown a normal range of radioactivity of the specified radionuclides. The workability of recycled aggregate concrete RAC decreases with increased recycled aggregate (RA) replacement. The replacement ratio of RA has a significant effect on the compressive and flexural strength of RAC. It is generally accepted that replaced RA by up to 25% of NA without significant changes in the mechanical properties of concrete.

REFERENCES

Ogar, Idagu F. 2017. "The Effects of Recycled Aggregates on Compressive Strength of Concrete." *(IJAERS)* 4 (1): 250–258.

Symonds. 1999. *European Commission. Construction and Demolition Waste Management Practices*. Report to DGXI, European Commission.

UNSCEAR. 2000. *Sources and effects of Ionizing Radiation*. Report to the General Assembly, with Scientific Annexes, New York: United Nations Scientific Committee on the Effects of Atomic Radiation

Current Perspectives and New Directions in Mechanics, Modelling and Design of Structural Systems – Zingoni (ed.)
© 2022 Copyright the Author(s), ISBN 978-1-032-39114-4

Using waste from paint manufacturing in concrete

H. Muller & R. Combrinck
University of Stellenbosch, Stellenbosch, South Africa

ABSTRACT: Cement is the main binder material of the most used construction material known as concrete. However, to produce cement, a large amount of CO_2 is produced, which drives the search for alternative fillers and binders in concrete. This has led to the investigation into the possibility of using waste from the paint industry in concrete. The waste was used as both cement and sand replacement in concrete to increase the possibility of finding an application for the waste material. The paint waste material used in this study is referred to as "filter cake" and can be collected from the settlement of the wash water after cleaning the mixing tanks used for a production run of a certain paint. Experiments were conducted to determine the influence of the paint waste material on the fresh and hardened properties of a conventional concrete. A process of milling was used to control particle size. Compressive, tensile and oxygen permeability tests were implemented to determine the influence on hardened state properties; slump and setting time test for influence on fresh state properties. The results show a negative influence on the fresh properties as well as hardened state properties of concrete, especially at high replacement percentages. An exception was the permeability, which decreased with the addition of the paint waste, indicating a possible improvement in durability. At low replacement percentages of up to 5%, the results show that the paint waste can be used in concrete without any significant negative influence on the fresh and hardened properties of the concrete.

Keywords: Paint waste, concrete, replacement material, fresh properties, hardened properties

1 INTRODUCTION

Concrete is considered to be one of the most used and popular construction materials in current times (Naik and Moriconi, 2005). For concrete to achieve its well-known physical properties a binder is needed, and the main binder used in concrete is cement, specifically Ordinary Portland Cement (OPC). In order to produce OPC, manufacturing processes that leads to significant pollution, are the norm. In line with that, the production process of the paint industry produces solid waste which has to be discarded on landfills, as there is currently no use for the material. This waste material is the coagulated leftover paint collected from the wash water after cleaning paint mixers from various paint production streams, called "filter cake". No previous research could be found that uses this specific waste material in concrete. Research found, focussed on using leftover or old unused paint in concrete. Assaad (2016) conducted a study on the effects of adding waste latex paint to concrete, finding an increase in flowability and an increase in compressive strength, at low replacement percentages. Almesfer et al. (2012) attempted to use waste latex paint as a replacement for other chemical admixtures in concrete, finding an increase in workability, compressive and tensile strengths at low addition percentages. Feng et al. (2018) conducted a study on the use of paint sludge from the automotive industry in concrete, resulting in an increase in flexural and compressive strength, up to 10% of replacement content.

The aim of this study was to investigate the influence of adding the paint waste material, or filter cake, on the fresh and hardened properties of concrete. This is done not only to consider a more environmentally friendly manner of discarding the paint waste, but also to find suitable waste materials for supplementing some of the constituents used in concrete. If the results of this study show that the waste can be used in concrete as a viable replacement or filler, be it for cement or fine aggregates, the environmental impact of the paint and concrete industry can be reduced.

2 EXPERIMENTAL FRAMEWORK

Two batches of paint waste were implemented for this study and utilized in different forms. The first batch was spread out to dry in the sun, after which it was milled down to a powder form. The second batch was

DOI: 10.1201/9781003348450-263

mixed and implemented in a wet state. The waste material was considered as a replacement for both cement and sand in comparative concrete mixes. The influence of the waste on the fresh state properties of concrete was determined by conducting a slump and slump flow test as well as setting time tests for both initial and final setting times. The influence of the waste material on the hardened state was determined by comparing test results for compressive strength, tensile strength, and durability of concrete mixes with varying replacement percentages of both sand and cement.

3 RESULTS

The replacement of concrete mix constituents with powdered waste material resulted in a decrease in slump and initial setting times with the exception of 30% cement replacement and 10% sand replacement resulting in an increase in initial setting times. An increase in final setting times was noted for all replacements. Using the waste material in liquid form resulted in a self-compacting concrete like mix.

Comparing hardened state test results, an overall decrease in compressive strength was noted, with the exception of 5% cement replacement's late strength gain surpassing the reference mix. Similar results were found for tensile strength comparisons. A decrease in permeability was seen for all considered mix designs.

4 CONCLUSIONS

The objective of this study was to determine the possibility of utilising waste from the paint industry in concrete as both cement and sand replacement. The investigation was done by considering the influences that the waste material has on concrete properties, both in the fresh and hardened state. The following main conclusions can be drawn from this study:

Fresh state influences:

- The waste material decreased the workability of the concrete when it was added as a replacement material, but significantly increased the workability when added as an additive without compensating for the water content.
- The waste material contributed to an unpleasant working environment, as the waste material added an unpleasant smell to the concrete, especially during mixing, but subsided after curing.
- Concrete setting times were not significantly influenced when using the waste material at low replacement percentages, 5% and 15%, but at 30% cement replacement and all wet waste replacements the setting times increased.

Hardened state influences:

- At a low cement replacement of 5%, the compressive and tensile strength was not noticeably influenced, however at higher replacement percentages the strength reduced as the replacement quantity increased.
- When using the dry waste material, a decrease in permeability of all considered samples were found. The 15% cement replacement resulted in the biggest decrease in permeability of all the considered samples.

Final remarks are that the paint waste did not show enough positive influences on concrete properties to be considered as a viable cement replacement. However, the material can still be used in low dosages to improve durability of the concrete and reduce the carbon footprint of both the concrete and paint industries. Using the wet paint waste would be recommended instead of using the dried-out waste as the drying and milling process adds to cost and emissions.

REFERENCES

Almesfer, N., Haigh, C., Ingham, J., 2012. Waste paint as an admixture in concrete. Cement and Concrete Composites 34, 627–633. https://doi.org/10.1016/j.cemconcomp.2012.02.001

Assaad, J.J., 2016. Disposing waste latex paints in cement-based materials – Effect on flow and rheological properties. Journal of Building Engineering 6, 75–85. https://doi.org/10.1016/j.jobe.2016.02.009

Feng, E., Sun, J., Feng, L., 2018. Regeneration of paint sludge and reuse in cement concrete. E3S Web of Conferences 38, 6. https://doi.org/10.1051/e3sconf/20183802021

Naik, T.R., Moriconi, G., 2005. Environmental-friendly durable concrete made with recycled materials for sustainable concrete construction, in: Center for By-Products Utilization. Presented at the CANMET/ACI International Symposium on Sustainable Development of Cement and Concrete, Toronto, Canada, p. 22.

Current Perspectives and New Directions in Mechanics, Modelling and Design of Structural Systems – Zingoni (ed.)
© 2022 Copyright the Author(s), *ISBN 978-1-032-39114-4*

Considerations regarding the toxicity of construction and building materials

F. Pacheco Torgal
C-TAC Research Centre, University of Minho, Portugal

ABSTRACT: Toxicity aspects have been a field outside the boundaries of the construction industry practitioners belonging to the realm of health professionals. And that helps to explain why architects, civil engineers and other professionals involved in the construction industry have so little knowledge about this area. However, since Covid-19 outbreak has brought a new focus to health issues then it is important that the professionals connected built environment also be more concerned with such health implications. Therefore, this paper briefly reviews indoor pollutants and also toxic materials used in the built environment in order to highlight the importance of its absence when designing a healthy built environment.

1 INTRODUCTION

The general population in North America and Europe already spent on average almost 90% of their time indoors (Phelan et al, 2020). Since recent studies show that a healthy environment is crucial to strength the immune system (Haahtela, 2019) which in turn is crucial to develop anti-bodies to tackle infections. No wonder then that traditional non healthy built environments and the potential toxic effects of building materials on health has gain a new importance in the context of the Covid-19 health crisis in in which more people spent more time indoors.

2 SOME CONSIDERATIONS ON YHE TOXICITY OF CONSTRUCTION AND BUILDING MATERIALS

Its worth remember that toxicity aspects have been a field outside the boundaries of the construction industry practitioners belonging to the realm of health professionals. That's why architects, civil engineers and other professionals involved in the construction industry have so little knowledge about this area. Even despite the fact that many buildings suffer from many problems that can cause several health related problems such as: Asthma; itchiness; burning eyes, skin irritations or rashes, nose and throat irritation; nausea; headaches; dizziness; fatigue; reproductive impairment; disruption of the endocrine system; cancer; impaired child development and birth defects; immune system suppression and even cancer. Indoor pollutants may include volatile organic compounds (VOC) such as formaldehyde, benzene, toluene, xylene, styrene, acetaldehyde, naphthalene, limonene, and hexanes (Liu and Little, 2012). Some major sources, such as adhesives, sealants, paints, solvents, wood stain, wallboard, treated wood, urethane coatings, pressed-wood products, and floorings. They also include semi-volatile organic compounds (SVOC), such flame retardant), and phthalate are also important toxic chemicals that can be found in the indoor environment (Liu and Little, 2012a). Also important in the context of the toxicity of building materials concerns the use of flammable building materials that can lead to catastrophic consequences that were seen in the Grenfell Tower tragedy, when a fire erupted on a block of flats in London on 14 June of 1917 that have resulted in 72 casualties and 70 physically injured. McKenna et al (2019) showed that for instance the use of polyethylene-aluminium composites used in the Grenfell Tower showed 55x greater peak heat release rates and 70x greater total heat release, when compared to the least flammable panels on the market. A different problem is related to the fact that the use of flame retardant chemicals, like organ halogens, are associated with adverse health effects such as diminished immune function, endocrine disruption, and cancer (Charbonnet et al., 2020).

3 A HEALTHY BUILT ENVIRONMENT AS A PRECONDITION OF RESILIENCE

In 2018 Professor Bendell authored a dramatic piece warning that it is now too late to stop a future collapse of our societies because of the current climate emergency, and that we must now explore ways in which to reduce harm (Bendell, 2018). A healthy built environments contributes to the first R and it is therefore an important goal to be pursued. More recently Gills (2020) called for a deep restoration of

DOI: 10.1201/9781003348450-264

the awareness of the necessity for maintaining ecological balance within the context of earth system dynamics. A deep restoration of the ethics of harmony with the web of life, including not only all species of creatures but also with the water, the oceans, the forests, and the soils of the earth, in which we are deeply embedded and mutually interdependent. Recently Carmichael et al. (2020) also referred to the need of considering health impact in future building regulations. Still Horve et al. (2019) recalled the anecdotally fact that, when architects describe "healthy building" principles, they routinely speak of access to daylight and outside air, and this is supported by the prioritization of daylight in building performance rating systems. The United Nations' Sustainable Development Goals (SDGs) offer an overarching framework for improving the environment and health in cities and the SDGs indicator 11.1.1 explicitly refers to the need for adequate housing standards. Worden et al. (2020) recognized that green building were focused on environmental sustainability topics like energy use and water utilization, so health issues had not been given enough importance. Still theses authors claim that in what concerns the LEED v4 rating system many of health related credits require practitioners to possess some degree of public health expertise in order to recognize and implement the credits in a health-oriented manner. Be there as it may Knoll (2020) recall that the Code of Ethics and Professional Conduct of the American Institute of Architects demands those professionals to "select and use building materials to minimize exposure to toxins and pollutants in the environment to promote environmental and human health…". A different issue however is the growing gap between expert knowledge on design of green and healthy buildings and lay knowledge of house builders and house holders, as design and construction of housing becomes more sophisticated knowledge gaps widen between different groups. In terms of health impact a more holistic approach to both policy and skills training could lead to more considerate building practices where there is greater knowledge of the links between building design and health. Also improved access to education materials on green and healthy buildings and how to use them could help (Carmichael et al. (2020).

4 CONCLUSIONS

This paper briefly reviewed indoor pollutants and also toxic materials used in the built environment in order to highlight the importance of its absence when designing a healthy built environment. The paper also highlights the importance of a healthy built environment in the context of the resilience condition of the Deep Adaptation Agenda.

5 REFERENCES

Bendell, J., 2018. Deep adaptation: a map for navigating climate tragedy. Institute for Leadership and Sustainability (IFLAS) Occasional Papers Volume 2. University of Cumbria, Ambleside, UK. http://insight.cumbria.ac.uk/id/eprint/4166/1/Bendell_DeepAdaptation.pdf

Carmichael, L., Prestwood, E., Marsh, R., Ige, J., Williams, B., Pilkington, P., Eaton, E. and Michalec, A., 2020. Healthy buildings for a healthy city: Is the public health evidence base informing current building policies?. Science of the Total Environment, 719, p.137146.

Charbonnet, J.A., Weber, R. and Blum, A., 2020. Flammability standards for furniture, building insulation and electronics: Benefit and risk. Emerging Contaminants.

Gills, B. 2020. Deep Restoration: from The Great Implosion to The Great Awakening, Globalizations, 17:4, 577–579,

Haahtela, T. 2019. A biodiversity hypothesis. Allergy, 74, 1445–1456.

Horve, P.F., Lloyd, S., Mhuireach, G.A., Dietz, L., Fretz, M., MacCrone, G., Van Den Wymelenberg, K. and Ishaq, S.L., 2019. Building upon current knowledge and techniques of indoor microbiology to construct the next era of theory into microorganisms, health, and the built environment. Journal of Exposure Science & Environmental Epidemiology, 1–17.

McKenna, S.T., Jones, N., Peck, G., Dickens, K., Pawelec, W., Oradei, S., Harris, S., Stec, A.A. and Hull, T.R., 2019. Fire behaviour of modern façade materials–Understanding the Grenfell Tower fire. Journal of hazardous materials, 368, 115–123.

Pacheco-Torgal, F. 2013. Toxicity Issues: Radon. In Nearly Zero Energy Building Refurbishment, 361–380. Springer, London.

Phelan, P., Wang, N., Hu, M. and Roberts, J.D., 2020. Sustainable, Healthy Buildings & Communities. Building and Environment, 174, 106806.

Current Perspectives and New Directions in Mechanics, Modelling and Design of Structural Systems – Zingoni (ed.)
© 2022 Copyright the Author(s), ISBN 978-1-032-39114-4

Fire performance of metakolin concrete and mortar blends using Nigerian kaolinite clays

J.O. Labiran & B.U. Ezea
Department of Civil Engineering, University of Ibadan

U.T. Igba & J.O. Akinyele
Civil Engineering Department, Federal University of Agriculture Abeokuta

ABSTRACT: Concrete Spalling at fire loads, depends on the expansion of residual $Ca(OH)_2$ in the hydrated matrix. The modified Chappelle's test was carried out on Nigerian calcined kaolin from 4 locations, samples with Metakaolin (MK) blends were subjected to fire loads. Concrete cylinders specimens were made, with the MK to cement replacement of 0, 2.5, 5, 7.5, and 10%, subjected to fire loads range from 200-900°C sustained 4 hours exposure. The optimum calcining temperature at 655 +/- 15°C, fixed 1731mg of $Ca(OH)_2$ /mg of metakaolin. Strength gain at 57 days MK replacement levels of 0%, 2.5% and 10% were 25%, 46.45% and 59% respectively. Compressive strength recorded an increase at 200°C for 28-day maturity, with strength gains for 0, 2.5, 5 and 7.5% were 85.1%, 96.5% 122% and 40% respectively. Mass loss ranges from 0.46% to 11%. Blended samples had improved strength development and better residual strength than the control.

1 INTRODUCTION

Reduction goals for carbon dioxide emissions in concrete production have demonstrated clear sustainability in clinker substitution. The paper identifies the kaolin samples sourced from South west Nigeria, measuring reactivity levels of at various calcination temperatures using the modified Chappelle's test. Optimum replacement percentages of metakaolin in concrete was identified, reviewing the relationship of residual strength with performance to fire loading at elevated temperatures.

2 FIRE RESISTANCE

According to Hertz (2005) with the concrete exposure less than 300°C, the specimen can absorb the moisture from the air and recover. At temperatures between 400°C -600°C, the residual compressive strength drops to 20% of actual unfired strength. The strength enhancement is be due to the large surface area of MK which fills the pores, the acceleration of cement hydration due to large surface area, and the pozzolanic reaction of MK with calcium hydroxide (Wild, Khatib & Jones, 1996). Spalling which is the disintegration of concrete matrix is usually due to the degradation of portlandite $Ca(OH)_2$ contained in the matrix.

3 METHODS

3.1 *Modified chappelle test*

The modified Chappelle assesses the consumption of calcium hydroxide by a test material in a dilute heated suspension as a measure of pozzolanicity. 1g of a supplementary cementitious material reacts with 2g of calcium oxide slaked in 250ml of distilled water at 85±5°C for 16hrs. The unreacted lime is then analyzed by a chemical titration to determine the quantity in mg of $Ca(OH)_2$ fixed for every 1g of the pozzolan (AFNOR-NF P 18-513, 2010).

3.2 *Water absorption*

The addition of MK is known to refine the pore system of concrete. The relative effectiveness of MK at performing these primarily results from the smaller particle size, higher degree of reactivity, and chemical composition. Hanif, Syuhaili, & Rafikullah (2020) prepared a composite mix of nylon fibers and metakaolin of different dosages. Their findings as shown in table 2.2 indicate that with the addition of metakaolin comes a decrease in the water absorption property of the mix. Concrete containing 10% metakaolin is the least water-absorbing on the 7th day and 28th day. This is due to the characteristics of the metakaolin itself, which acts as a filler to fill the empty pores in the concrete thus reducing the percentage of water absorption. Ashok & Arun (2015) assessed the effect of high temperatures on the compressive strength of M30, M25 & M20 concrete. After heating between temperatures 200°C to 800°C, the test samples were cooled at room temperature before a crushing test is done. The results revealed up to 500°C, fairly robust performance, with strength coming down slightly. at 650°C, the fall in concrete strength would be a cause for concern. Beyond 650°C, concrete stood completely decimated (Castillo & Durrani, 1990).

DOI: 10.1201/9781003348450-265

4 RESULTS

Reactivity levels of metakaolin at various calcination temperatures were measured. The results are:

Table 1. Modified Chappelle test on Sample A.

Temperature (°C)	Volume of HCL required	Amount of $Ca(OH)_2$ fixed
400	6.97	251.7
500	4.43	1121.21
600	4.07	1247.06
700	5.27	835.19
800	5.47	766.54

Table 2. Comp. strength for various % metakaolin at 27°C.

	Cylinder(N/mm^2)	compressive	strength
MK (%)	7 days	28 days	57 days
0	9.16	10.18	12.73
2.5	12.83	14.25	20.87
5.0	11.45	13.75	11.71
7.5	11.71	15.27	20.87
10.0	12.22	15.78	17.82

Table 3. Residual compressive strength at 28 days curing.

	Heat Load (0C) /Compressive Strength at 28 days (N/mm2)				
MK (%)	27	200	400	600	800
0	10.18	18.84	13.24	5.60	2.04
2.5	14.25	28.00	14.25	8.15	5.09
5.0	13.75	30.55	17.82	12.73	2.04
7.5	15.27	21.89	14.76	12.22	4.07
10	15.78	12.73	10.69	8.15	7.13

Table 4. Residual compressive strength at 57 days curing.

	Heat Load (°C) vs. Compressive Strength at 57 days (N/mm^2)				
MK (%)	27	200	400	600	800
0	12.73	12.73	17.82	10.69	3.56
2.5	20.87	20.36	10.18	11.71	5.6
5.0	11.71	10.69	16.80	9.16	5.09
7.5	20.87	12.73	12.22	12.73	6.62
10	17.82	15.78	14.76	15.78	6.62

5 DISCUSSION AND CONCLUSIONS

The following conclusions was drawn from experimental results on the concrete specimens.

1. Kaolin is reactive between calcination temperature of 610-650°C.
2. CEM II /B-L recorded 56 and 72% of predicted mix design strengths using CEMI.. There was a clear strength gain of up to 35% higher than control in compression.
3. The test results showed that the compressive strength increases up to 200°C and then decreases as the exposure temperature increases up to 800°C.
4. Metakaolin is effective at enhancing the early concrete strength gain, alters the pore structure of concrete paste to improve resistance to water absorption.
5. There is an optimum percentage addition level is 10% by weight of cement above which strength drop.
6. Explosive spalling was observed in the MK concrete specimens particularly between 400 and 800°C. The rate of spalling increased with the higher MK content.
7. The experimental results indicate that the compressive strength of concrete before heat treatment (F_c) has no significant effect on depletion to the residual percentage reductions in compressive strength (F_{ct}). As temperature increases, the stress-strain curve the concrete is becoming ductile, calcination duration stabilizes at two hours and flash cooling is recommended.

REFERENCES

AFNOR-NF P 18–513. 2010. "Pozzolanic addition for concrete – Metakaolin – Definitions, Specifications and Conformity Criteria." Association Française de Normalisation - Standard 19.

Hager, I. 2013. "Behaviour of cement concrete at high temperature." Bulletin of polish academy of science technical sciences.

Hanif, M.I, M.N Syuhaili, and D Rafikullah. 2020. "Strength and water absorption of concrete containing metakaolin and nylon fiber." International Journal of Sustainable Construction Engineering and Technology, 11 (1) 230–242.

Newman, John B, and Seng Choo Ban. 2003. Advanced Concrete Technology: Concrete Properties. Britain: Butterworth-Heinemann An imprint of Elsevier Linacre House, Jordan Hill, Oxford OX2 8DP 200 Wheeler Road, Burlington MA 01803

Current Perspectives and New Directions in Mechanics, Modelling and Design of Structural Systems – Zingoni (ed.)
© 2022 Copyright the Author(s), ISBN 978-1-032-39114-4

Rheological and strength characterisation of limestone calcined clay cement 3D printed concrete

K.A. Ibrahim, G.P.A.G. van Zijl & A.J. Babafemi
Division for Structural Engineering and Civil Engineering Informatics, Stellenbosch University, Stellenbosch, South Africa

ABSTRACT: The aim of this paper is to investigate the effect of mix composition such as the percentages of limestone powder and calcined clay (LC^2) on the fresh/rheological and mechanical properties of 3D printed concrete (3DPC) reinforced with micro synthetic polypropylene fibre. The material mix design satisfying 3DPC requirements for early strength and stiffness, shrinkage cracking resistance and mechanical properties, including interfacial bond, was developed. The physical properties and the particle size distribution of both the binders and aggregate in compliance with Fuller-Thompson's ideal curve/theory produce a high-performance concrete mix without compromising the rheological design approach for 3DPC. Finally, the high performance is demonstrated by buildability and shape retention fresh state 3DPC results and characterised hardened mechanical test results on printed and cast samples.

Keywords: 3DPC, LC^3, rheological characterisation, buildability, mechanical performance

1 INTRODUCTION

Digital fabrication of composite materials has attracted attention in recent years and has grown as a promising technique in the construction industry. However, the bond between the deposited filament layers is weak, leading to a reduction in strength and durability performance of 3D printed objects and hindering the implementation by mass industries (Kruger, 2019). Therefore, the rheological properties of binders should be actively designed for at various stages throughout the printing process, and a fundamental understanding of all the printing parameters is necessary.

Furthermore, as the research and development (R&D) of the technology advances, the high cement content required in the mix design of digital concrete must be lowered. With the availability of clay and limestone worldwide, LC^2 can substitute cement, making LC^3 and enhance sustainability. Chen et al. (2019) have investigated the extrudability of several mixtures based on ternary blends using a ram extruder. However, much remains unknown on the behaviours of limestone and calcined clay-based cementitious materials for extrusion-based 3DPC. Hence, this paper seeks to evaluate the effects and performance of LC^2 inclusion in cement for 3DPC production.

2 MATERIALS AND METHODS

Primary cementitious (conforming to BS EN 197-1: 2000) and other materials selected for the study are depicted in Table 1. X-ray fluorescence spectrometry (XRF) was used in carrying out the chemical analysis of the Portland cement and other raw materials.

Table 1. Materials mix constituent proportioning (kg/m^3).

Mixture	CEM I 42.5R	LC^2	Sand	Water	SP	VMA	PP
LC^3 3DP-FRC	420.0	343.7	1229	343.7	6.1	2.3	9.0

Mixing was done mechanically in a Hobart concrete mixer. Fresh/rheological properties, including buildability performance were measured immediately. Samples for hardened properties were cast and compacted using a standard vibration table. The printing technique adopted is extrusion-based using a gantry-type 3D printer of three degrees of freedom with a build volume of 1 m^3 (Kruger, 2019). The printing and curing for the cast and printed samples were made in a climate-controlled room at 23°C (± 2°C) and 65% (± 5%) RH until the test days.

The strength evolution characterisation for compressive (f_{cu} & f_c), flexural (f_f) and tensile (f_t) strengths on the cast and printed (cored and

DOI: 10.1201/9781003348450-266

cut-sawn in different orientations, i.e., D1, D2, and D3) samples were executed after 7 and 28 days of curing conforming to ASTM and NEN-EN 196-1: 2016.

3 RESULTS AND DISCUSSION

The rheological characterisations of freshly mixed LC^3 3DPC are listed in Table 2, indicating appropriate workability and buildability, and suitability for printing.

Table 2. Results of the rheological parameters, slump, and slump flow of the mixes.

Mixture	$\tau_{S,i}$ (Pa)	$\tau_{D,i}$ (Pa)	R_{thix} ($Pa.s^{-1}$)	A_{thix} ($Pa.s^{-1}$)	μ (Pa.s)	Slump (mm)	Slump flow (mm)
LC^3 3DPC	1211.2	675.6	1.27	0.33	7.6	23.0	171.0
LC^3 3DPC-FRC	2051.7	1333.7	7.12	1.06	10.4	7.0	150.3

The mechanical performance of the cast and printed samples for compressive, tensile, and flexural strength test results are summarised and illustrated in Figures 1 and 2.

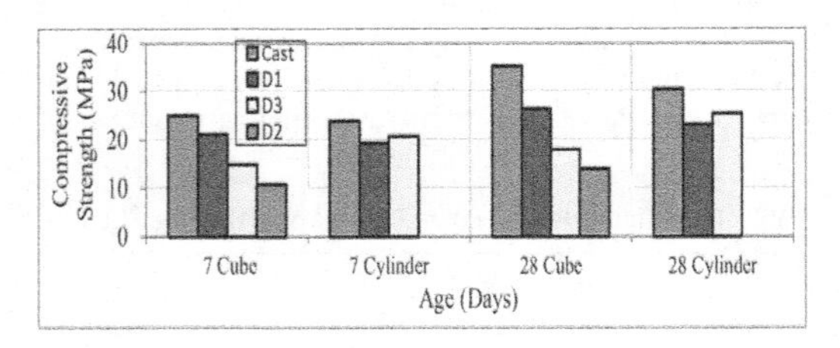

Figure 1. Compressive strength of cast and printed samples.

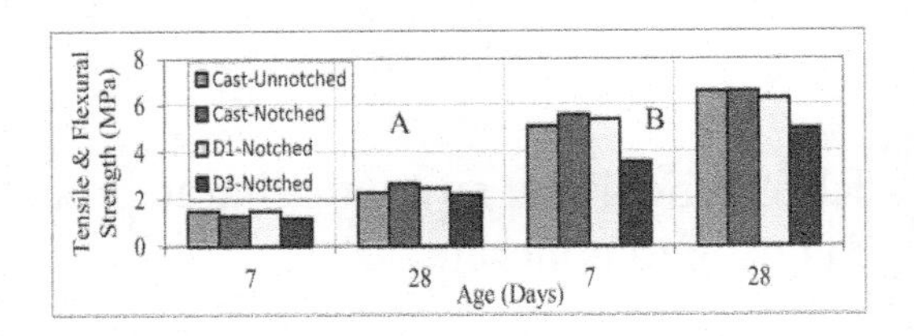

Figure 2. Comparison of 7- and 28-days A) tensile and B) flexural strength of cast and printed prism samples.

4 CONCLUSIONS

Based on the study above, the slump and slump flow table values recorded were within the recommended limits for suitable 3D printability. The buildability performance, shape retention ability and yield shear stress evolution were enhanced in LC^3 3D printable mixture. Lastly, the anisotropic behaviour is exhibited for printed samples containing LC^3 in compression, tension, and flexure. The highest strengths were recorded in the D1 (samples parallel to the loading direction) and D3 (samples perpendicular to the loading direction) orientations for cubes and cylindrical samples, respectively, and D1 (intralayer) for both the direct tension and flexure.

REFERENCES

BS EN 197-1. 2000. British Standard Institution; Cement-Composition, specification and conformity criteria for common cements, London.

Chen, Y., Li, Z., Figueiredo, S.C. Copuroglu, O. Veer, F., Schlangen, E. 2019. Limestone and calcined clay-based sustainable cementitious materials for 3D concrete printing: a fundamental study of extrudability and early-age strength development, Ap.Sc-9, https://doi.org/10.3390/app9091809.

Kruger, P.J. 2019. "*Rheo-mechanics modelling of 3D concrete printing constructability*", PhD Dissertation in Engineering, Stellenbosch University, South Africa, 223 pp. https://scholar.sun.ac.za.

Current Perspectives and New Directions in Mechanics, Modelling and Design of Structural Systems – Zingoni (ed.)
© 2022 Copyright the Author(s), *ISBN 978-1-032-39114-4*

Fresh properties and strength evolution of slag modified fibre-reinforced metakaolin-based geopolymer composite for 3D concrete printing application

M.B. Jaji, G.P.A.G Van Zijl & A.J. Babafemi
A Division for Structural Engineering and Civil Engineering Informatics, Department of Civil Engineering, Stellenbosch University, Stellenbosch, South Africa

ABSTRACT: This paper develops mixes for slag modified fibre-reinforced metakaolin-based geopolymer composite (GPC) to address slow setting time and efflorescence. The fresh properties of the mixes showed that slump and slump flow reduced with increasing slag content. There is an increase in compressive and flexural strength as the slag content increases. The visual observation on the hardened specimens shows no efflorescence at the age of 28-day. The control sample (Mix 1) and Mix 2 are 3D printable. The mix with slag content shows better structural build-up.

Keywords: Metakaolin, slag, alkaline activator, 3D printed geopolymer, fresh and mechanical properties

1 INTRODUCTION

The Geopolymerisation process largely depends on the alkaline activator and the aluminosilicate content in the geopolymer composite. User-friendly alkaline activators have slow geopolymerisation reactions, hence a slow setting time, and consequently must be cured at autoclave temperature. On the other hand, user-hostile alkaline activators, have a quick geopolymerisation reaction; hence, short setting time, and no need for curing at elevated temperature. This research proposes to partially replace metakaolin in a two-part geopolymer (GPC) with slag up to 15% by mass to hasten the setting, avoid autoclave curing, prevent efflorescence, and apply the mixes in 3D concrete printing.

2 MATERIALS AND METHODS

The matrix mix proportion is shown in Table 1. The materials were batch mixed in a rotary mixer for in 5 minutes and cast in a 50 mm cube metal mould for compression test, and 40 mm × 40 mm × 160 mould for flexural test. The specimens were stripped after 1 day and subsequently cured in a climate-controlled room (23 ± 2 °C and 65 ± 5% relative humidity) and tested at 7-day and 28-day curing ages.

Table 1. Mix composition (kg/m^3).

Materials	Mix 1	Mix 2	Mix3	Mix 4
MK_{60}	680.47	646.45	612.42	578.4
Slag	-	34.02	68.5	102.07
Sand	1088.76	1088.76	1088.76	1088.76
AA	496.75	496.75	496.75	496.75
Water	34.02	34.02	34.02	34.02
Fiber	4.60	4.60	4.60	4.60

The slump, slump flow and initial and final setting time were determined using mini-slump cone, flow table and Vicat apparatus. The mechanical tests were conducted using a Zwick Z250 Universal Material testing Machine conforming to ASTM C39. The Mix 1 and 2 were assessed for their application in 3D concrete printing.

3 RESULTS AND DISCUSSION

The The mini-slump test results of the mixes are presented in Figures 1 and 2. Slag inclusion reduced the slump of the mixes and workability. The final setting time of the control mix (Mix 1) shown in Figure 3 is high, but as the slag content increases, the initial and

DOI: 10.1201/9781003348450-267

final setting time reduces in line with Huo et al. (2014). The average compressive (f_{cu}) and flexural (f_t) strength values increased with increasing slag inclusion as shown in Figures 4 and 5.

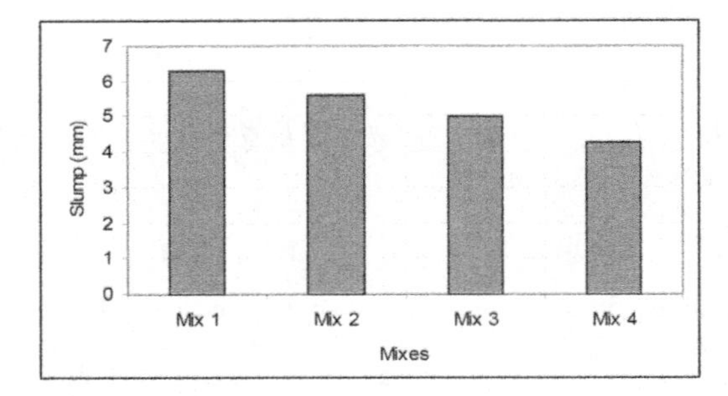

Figure 1. Slump of mixes.

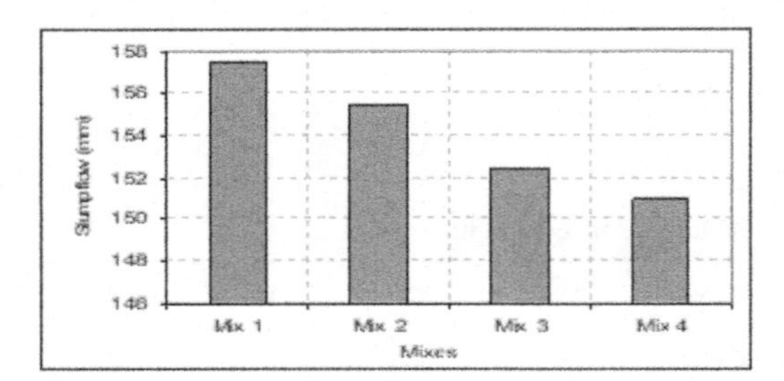

Figure 2. Slump flow of mixes.

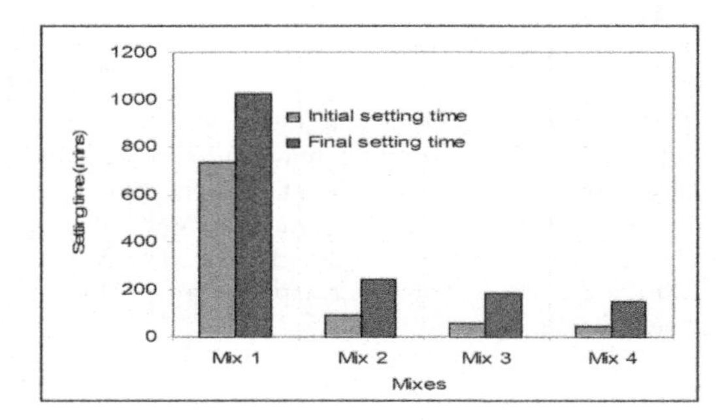

Figure 3. Setting time of mixes.

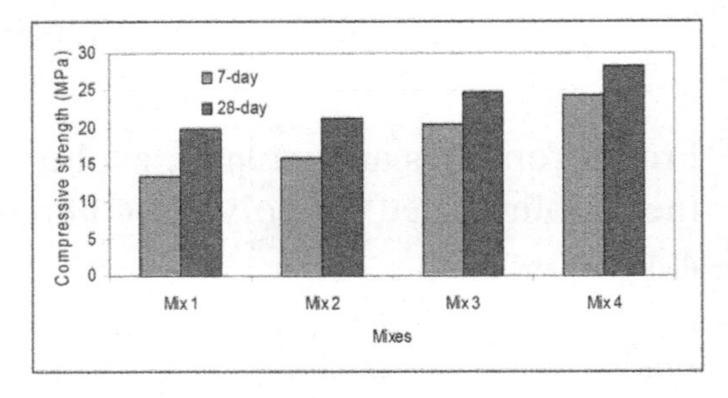

Figure 4. Compressive strength results.

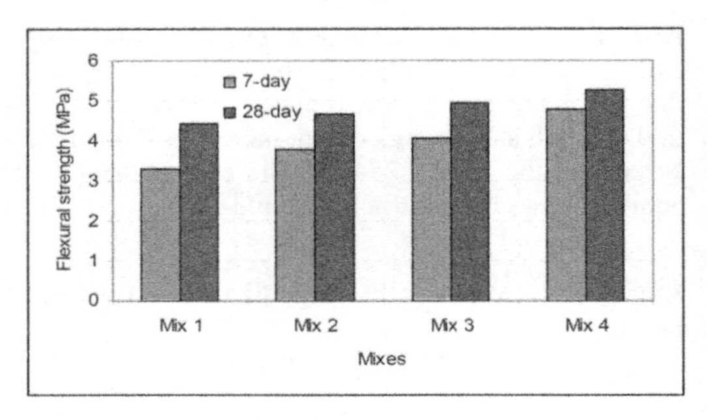

Figure 5. Flexural strength results.

4 CONCLUSIONS

It was concluded that slag inclusion reduced workability, decreased setting time, and improved strength. This could be due to the high CaO content in the slag. Efflorescence was not observed on 28-day, this could be due to the low NaOH in the activator. The structural buildup of the printed Mix 2 was enhanced because of slag inclusion.

REFERENCES

ASTM C39/C39M-18. 2018. Standard Test Method for Compressive Strength of Cylindrical Concrete Specimens; ASTM International: West Conshohocken, PA, USA.

Huo, W., Zhu, Z., Zhang, J., Kang, Z., Pu, S., & Wan, Y. 2021. Utilization of OPC and FA to enhance reclaimed lime-fly ash macadam based geopolymers cured at ambient temperature. *Construction and Building Materials, 303 (March) 124378.* https://doi.org/10.1016/j.conbuildmat.2021.124378.

Current Perspectives and New Directions in Mechanics, Modelling and Design of Structural Systems – Zingoni (ed.)
© 2022 Copyright the Author(s), ISBN 978-1-032-39114-4

Expression for calculating the compressive strength of concrete containing Rice Husk Ash

G. Somma
Polytechnic Department of Engineering and Architecture, University of Udine, Italy

ABSTRACT: Rice Husk Ash (RHA) is the residue of completely incinerated rice husk, that is the covering part of the grain of rice. Rice husk is abundant in many parts of the world, and is classified as a highly reactive pozzolana, so it can be used as a partial replacement of cement. During the years, researchers have studied the mechanical properties and the durability of the concrete incorporating different percentage of RHA. The objective of the study presented in this paper is not only to find the "perfect percentage" of RHA replacement, but also to find a new expression for calculating the concrete compressive strength, considering the percentage of RHA replacement.

1 INTRODUCTION

Rice is one of the major food crops in the world, and its production generates an equally great amount of waste, named Rice Husk (RH). By the burning of rice husk, Rice Husk Ash (RHA) is obtained, and being it mostly composed of silica (80-95%) (Singh et al. 2014), it is considered for its pozzolanic activity in partial substitution of cement present in the concrete mix. Several studies conducted by various authors have demonstrated that the RHA partial replacement in cement can also improve the mechanical properties of concrete, such as the compressive strength and the splitting tensile strength (Singh et al. 2014, Abalaka 2013).

By analysing the 97 test data done by other authors from 1996 to 2015 on the compression strength of concrete samples without RHA and with different percentage of RHA replacement, in this paper an expression for calculating the compressive concrete strength also in presence of RHA is proposed, with the intention to stimulate the future use of RHA concrete (RHAC) in structural applications.

2 EXPERIMENTAL ANALYSIS

In this manuscript the author intention is to consider a great number of test data done by other authors from 1996 to 2015 and found in the literature on concrete samples with different percentage of RHA, ρ_{RHA}, varying from 0% (ordinary concrete) to 30% as partial replacement of cement. The objective of this work is to investigate the influence of rice husk ash on the compressive strength of concrete, and to find the "perfect percentage" of RHA replacement providing the higher compressive strength.

2.1 *Proposed compressive strength expression*

Based on the observation of the concrete compressive strength values obtained in the experiments, f'_c, for RHA percentage, ρ_{RHA}, variable from 0% (ordinary concrete, OC) to 30%, the average value of ρ_{RHA} which meanly provides the greatest resistance values is 12%. It is observed that f'_c tends to increase with ρ_{RHA} until an average value of 12%, and then decreases. Hence, a second order polynomial expression in function of ρ_{RHA} has been thought to better fit the experimental f'_c results:

$$f'_{c,RHA} = A\ \rho^2_{RHA} + B\ \rho_{RHA} \tag{1}$$

where two of the three constant coefficients have been found by imposing the following boarding conditions:

$$f'_{c,RHA}(\rho_{RHA} = 0) = f'_{c,OC} \tag{2}$$

$$\frac{df'_{c,RHA}}{d\rho_{RHA}}(\rho_{RHA} = 0, 12) = 0 \tag{3}$$

where $f'_{c,OC}$ is the compressive strength of the "control sample": in every experimental campaign done by various author and considered in this paper the tested samples mixtures differ only in ρ_{RHA}, and the

DOI: 10.1201/9781003348450-268

"control sample" is made by ordinary concrete, with $\rho_{RHA} = 0$.

Considering Equations 2 and 3, Equation 1 becomes:

$$f'_{c,RHA} = A\left(\rho^2_{RHA} - 0,24\rho_{RHA}\right) + f'_{c,OC} \quad (4)$$

Observing that the concrete strength values obtained in the tests have a maximum meanly for $\rho_{RHA} = 12\%$, the imposing of the following condition

$$\frac{d^2 f'_{c,RHA}}{d\rho^2_{RHA}}\left(\rho_{RHA} = 0,12\right) < 0 \quad (5)$$

provides A<0. The value of the coefficient A has been determined by imposing that the coefficient of variation, COV, of the ratios between the tested compressive strength, f'_c, and the strength obtained with Equation 4, $f'_{c,RHA}$ is minimum, hence minimizing the scattering of the predicted results. Accordingly con this procedure the value A= - 203,17 has been determined, and Equation 4 becomes

$$f'_{c,RHA} = 203{,}17\rho_{RHA}\left(0,24 - \rho_{RHA}\right) + f'_{c,OC} \quad (6)$$

This expression can be generally used, as it provides the concrete compressive strength both in case of absence and in presence of RHA as replacement of cement.

2.2 *Model reliability*

To check the reliability of the here proposed model the concrete compressive strength values, $f'_{c,RHA}$, of 97 specimens have been calculated applying the here proposed expression (Equation 6), together with the compressive strength calculated without considering the presence of RHA, $f'_{c,OC}$. The values of the corresponding tested-to-calculated ratios, $f'_c/f'_{c,RHA}$ and $f'_c/f'_{c,OC}$, have been calculated too.

The following AVG and COV values of the experimental to calculated compressive strengths have been obtained: AVG = 0,98, COV = 0,07 with Equation 6, and AVG = 1,02, COV = 0,11 without considering the presence of RHA. It is evident that the here proposed model leads to an equivalently accurate but more uniform prediction of the concrete compressive strength also in presence of RHA, providing a 36% reduction in the COV.

This result is also evident from the graph in Figure 1, where the $f'_{c,RHA}$ values calculated by means of Equation 6 are plotted versus the experimental values, f'_c, outcome from the tests considered. In Figure 1, in fact, the dots relative to the proposed model are concentrated all along the bisector line, representing the perfect equality between theoretical and experimental values, indicating a good uniformity and accuracy in the strength prediction.

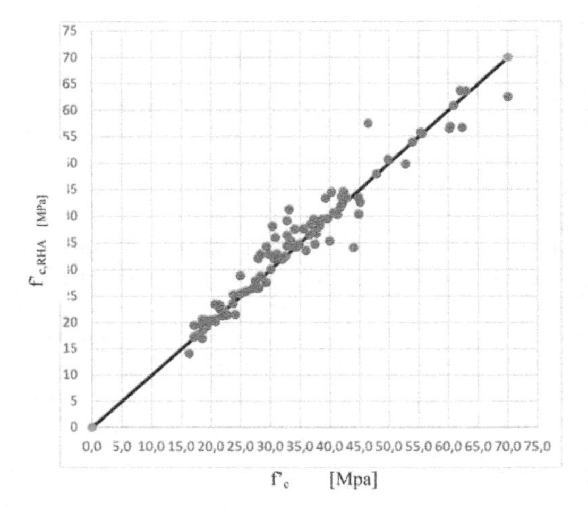

Figure 1. Correlation between the experimental compressive strength, f'_c, and that calculated by the here proposed expression, $f'_{c,RHA}$.

3 CONCLUSIONS

From this study it is concluded that the optimum percentage of RHA replacement has found to be meanly equal to 12%. An expression for calculating the compressive strength of concrete also considering the presence of RHA is provided in this paper, and it has found to be consistent with the experimental results: with the proposed expression there is a 36% more reliable prediction of the real strength.

Hence, the use of RHA in concrete mixture is encouraged not only for ecological reasons reported here and in several precedent manuscripts, but also because, in limited percentage, it increases the compressive strength, consistently calculated by the here proposed expression.

REFERENCES

Abalaka, A.E. 2013. Strength and Some Durability Propoerties of Concrete Containing Rice Husk Ash Produced in a Charcoal Incinerator at Low Specific Surface. *International Journal of Concrete Structures and Materials*. Vol.7, No.4, pp.287–293.

Singh, P., Lallotra, B., and Patyal, V. 2014. Effect on Strength of Concrete after Adding Rice Husk Ash. *International Journal of Engineering Research and Applications. National Conference on Advances in Engineering and Technology*. pp.21–25.

Current Perspectives and New Directions in Mechanics, Modelling and Design of Structural Systems – Zingoni (ed.)
© 2022 Copyright the Author(s), ISBN 978-1-032-39114-4

Compatibility issues between dehydrated calcium sulphate cement and plasticiser/superplasticiser

Johandré M.H. Bessinger, Marnu Meyer & Riaan Combrinck
Stellenbosch University, Stellenbosch, South Africa

ABSTRACT: In the production of cement, the milling of clinker and calcium sulphate causes temperature peaks at localized areas within the mill that can result in the dehydration of calcium sulphate dihydrate to hemihydrate. The dehydration of calcium sulphate influences the solubility and dissolution rate of the calcium sulphate in cementitious materials. Ultimately, the type of calcium sulphate influences the hydration kinetics of the cement, in particular the setting time and the susceptibility to false set, or flash set. Chemical admixtures such as plasticiser/superplasticiser have an affinity towards tricalcium aluminates that can result in compatibility issues caused by the variation in the type of calcium sulphate present in the cement. This study aims to establish the influence of the interaction between calcium sulphate and plasticiser/superplasticiser on concrete performance in the fresh state. The results show that the presence of plasticiser/superplasticiser increases the solubility of calcium sulphate hemihydrate and to a lesser extent, - dihydrate. The presence of hemihydrate in cement is found to cause a decrease in the workability while increasing the rate of concrete stiffening.

1 INTRODUCTION

The hydration process within cementitious materials start instantaneously with the addition of water. It is due the addition of calcium sulphate that concrete initially maintains its workability, allowing time for construction processes. This is referred to as the dormant period. Calcium sulphate in the form of gypsum is added to clinker during the milling phase. The temperature at milling determines the type of calcium sulphate present in cement i.e. dihydrate (gypsum), hemihydrate (basanite) or anhydrite (Chakkamalayath et al., 2011). Hemihydrate (HH) and anhydrite are a result of gypsum dehydration at different temperatures (Camarini & De Milito, 2011).

The dehydration of dihydrate to hemihydrate, and to a lesser extent anhydrite, occurs naturally during the milling process of clinker due to spikes in temperature at localized areas within the mill. It has been found that the main influence of the temperature during milling on cement quality can be related to the dehydration of dihydrate to hemihydrate (Suzakawa & Kobayashi, 1966). This dehydration of dihydrate to hemihydrate particularly influences the setting time and susceptibility to false set of cementitious materials (Suzakawa & Kobayashi, 1966). The normal, false, quick or flash set of a cementitious materials due to the hydration of tricalcium aluminates (C_3A) is dependent on both the reactivity of the C_3A and availability of soluble sulphate (Rößler et al., 2007).

Plasticiser/superplasticiser modifies the hydration of C_3A (Plank et al., 2010). When C_3A hydrates in the presence of both calcium sulphate and superplasticiser, the reaction of superplasticiser with C_3A is heavily dependent on the concentration of sulphate in the mix solution. The concentration of dissolved sulphate being dependent on the type of calcium sulphate present in cement.

2 EXPERIMENTAL INVESTIGATION

Variables to the mixes included the type of calcium sulphate (Ordinary Portland cement partially substituted by either calcium sulphate DH or HH powder), water-reducing chemical admixture and high-range water-reducing chemical admixture present in cement.

2.1 *Materials*

Ordinary Portland cement (OPC) and blended cements, by replacing a portion of the cement with calcium sulphate powder, were used. Lignosulphonate (LS), Sulphonate naphthalene formaldehyde (SNF) and Polycarboxylic ether (PCE) type plasticisers/superplastizers were used.

DOI: 10.1201/9781003348450-269

2.2 *Solubility test*

The solubility test methodology of Bock (1961) was adopted. Solubility cells were filled with water-plasticiser/superplasticiser solutions. The solutions were oversaturated with a pre-weighed quantity of calcium sulphate dihydrate or hemihydrate and agitated. After agitation the undissolved solids are removed from the solution and the solubility of each solution calculated.

2.3 *Sump/Slump flow test*

Slump or slump flow tests, depending on the mix fluidity, were performed in accordance with BS EN 12350-2:2019 (BS EN, 2019) and BS EN 12350-8:2019 (BS EN, 2019), respectively.

2.4 *ICAR Rheometer test*

The Germann Instruments ICAR Rheometer was used to determine rheological parameters. Two tests can be performed using the ICAR Rheometer i.e. the flow curve and stress growth test.

The flow curve test was performed to obtain the plastic viscosity.

The stress growth (SG) test was performed to obtain the static (SYS) and dynamic yield stress (DYS) as well as characterize the potential thixotropic behavior of the specimen. To determine the rate of structuration (A_{thix}) of a specimen, various stress growth tests were performed at incremental periods of rest i.e. zero shear. Periods of rest consisting of 5, 10, 20, 40 and 60 minutes were used in between tests. Kruger's bi-linear thixotropic was adopted to determine A_{thix} (Kruger et al., 2019).

3 SUMMARY OF EXPERIMENTAL FINDINGS

Based on the results of this experimental investigation the following conclusions were drawn:

1. The nature of the calcium sulphate in cement influences both initial fresh state properties and hydration of concrete. Hemihydrate increases the water demand as well as the hydration rate of cementitious materials. The use of hemihydrate, as a source of calcium sulphate, can be beneficial for applications where rapid stiffening of concrete is required.
2. It was found that the presence of hemihydrate significantly reduces the initial fluidity induced by the PCE type superplasticisers. However, the nature of the calcium sulphate present in cement does not influence the rate at which concrete stiffens, while the dosage have a significant influence on the rate of structuration.
3. SNF type superplasticisers maintain their efficiency as a fluidising agent irrespective of the type of calcium sulphate present in cement. However, the workability period is significantly reduced when hemihydrate is present in cement as it causes accelerated structuration i.e. stiffening of mixes containing SNF type superplasticiser. Hemihydrate, as a source of calcium sulphate thus counters the delayed hydration associated with SNF type superplasticiser without reducing the initial workability.
4. The performance of LS type plasticisers are dominated by the type of calcium sulphate present in cement. The presence of hemihydrate, in place of dihydrate, eliminates the ability of a LS type plasticiser to induce fluidity. A significant increase in initial stiffness was observed with the use of LS type plasticiser. Furthermore, after 10 min of resting time, the mix exhibited false set.

It can be concluded that the temperature at which clinker and calcium sulphate (dehydrate) is milled together to produce cement, should be carefully monitored as high milling temperatures could result in the dehydration of dihydrate to hemihydrate, and to a lesser extent anhydrite. From this study it is seen that the type of calcium sulphate present in cement influences the fresh state properties of concrete. Furthermore, the calcium sulphate to plasticiser/superplasticiser relation, and the influence thereof on the performance of concrete, is unique to each calcium sulphate and chemical admixture combination. Improper combinations could result in undesirable concrete performance.

REFERENCES

Bock, E. (1961). Solubility of Anhydrous Calcium Sulphate and of Gypsum in Concentrated Solutions of Sodium Chloride. Can. J. Chem., 39(1061).

BS EN. (2019a). BS EN 12350-2: 2019.

BS EN. (2019b). BS EN 12350-8: 2019.

Camarini, G., & De Milito, J. A. (2011). Gypsum hemihydrate-cement blends to improve renderings durability. Construction and Building Materials, 25(11), 4121–4125. https://doi.org/10.1016/j.conbuildmat.2011.04.048

Chakkamalayath et al. (2011). Cement-superplasticiser compatibility - Issues and challenges. Indian Concrete Journal. https://www.researchgate.net/publication/286714849

Germann Instruments A/S. (2016). ICAR Rheometer Manual.

Kruger, J., et al. (2019). An ab initio approach for thixotropic characterisation of (nanoparticle- infused) 3D printable concrete. 1–45.

Plank, J., et al. (2010). Fundamental mechanisms for polycarboxylate intercalation into C_3A hydrate phases and the role of sulfate present in cement. Cement and Concrete Research, 40(1),45–57. https://doi.org/10.1016/j.cemconres.2009.08.013

Rößler, et al. (2007). Influence of superplasticisers on C3A hydration and ettringite growth in cement paste. https://www.researchgate.net/publication/305770868

Suzakawa, Y., & Kobayashi, W. (1966). Relations between the Grinding Temperature and Quality of Cement. 82(82).

18. Timber structures, Timber technology, Properties of wood

Current Perspectives and New Directions in Mechanics, Modelling and Design of Structural Systems – Zingoni (ed.)
© 2022 Copyright the Author(s), ISBN 978-1-032-39114-4

Reliability of statically indeterminate timber structures: Modelling approaches and sensitivity study

D. Caprio, R. Jockwer & M. Al-Emrani
Division of Structural Engineering, Department of Architecture and Civil Engineering (ACE). Chalmers University of Technology, Göteborg, Sweden

ABSTRACT: In statically indeterminate timber structures, the stiffness of the members and the semi-rigid behaviour of joints influence the force distribution in the structure, affecting its performance and reliability. This paper aims at identifying the impact of connections on the reliability of such structures by performing a study of a simple statically indeterminate system, i.e. a simple beam with rotational springs at its supports. Using UQlab, a Matlab-based toolbox, a parametric study varying the stiffness and the level of ductility of connections was performed. Different methods such as Subset, Importance Sampling and Adaptive Kriging with Subset simulation were applied and compared in terms of efficiency and computational time. The results showed that an increment in reliability is gained increasing the level of ductility of joints of intermediate and high stiffness.

1 INTRODUCTION

In statically indeterminate systems too simplistic assumptions on the behaviour of joints might lead to uneconomic or unsafe design of a structural system.

Different methodologies exist to compute the reliability of the structural system. In this work Adaptive Kriging Subset (APCK-Subset), that combines Subset sample simulation with PC Kriging as metamodel was tested together with Monte Carlo Simulation (MCS), Subset Simulation (Subset) and importance sampling (IS). The methods were compared in terms of computational time and accuracy. Moreover, a parametric study on a beam supported by semi-rigid joints is performed varying the stiffness and ductility of joints and computing the variation of the probability of failure.

2 THE CASE STUDY

A typical example of a statically indeterminate structure is a beam restrained by semi-rigid joints. If the beam is less strong than the acting moment at the joint and at midspan and in case the joints are still in the elastic domain, then brittle failure occurs.

The failure domain of the brittle failure mode is:

$$BR = (\{g_{b1} \leq 0\} \cup \{g_{b2} \leq 0\}) \cap \{g_{j1} \leq 0\} \quad (1)$$

Where $g_{b1} \leq 0$ and $g_{b2} \leq 0$ represent the failure domain of the beam at the joints location and at midspan, respectively, while $g_{j1} \leq 0$ defines the elastic domain of the joints.

If the bending capacity corresponding to the yielding point is reached in the joints, the ductile failure domain of the structural system can be defined as:

$$DU_1 = (\{g_{b3} \leq 0\} \cup \{g_{b4} \leq 0\}) \cap \{g_{j2} \leq 0\} \quad (2)$$

$g_{b3} \leq 0$ and $g_{b4} \leq 0$ define the failure domain of the beam at the joint and at mid span, respectively. $g_{j2} \leq$ represents the plastic domain of the joints.

Finally, if the joint reaches the ultimate rotation after certain plasticization, the failure domain of the structural system is defined by:

$$DU_2 = \{g_{b5} \leq 0\} \cap \{g_{j3} \leq 0\} \quad (3)$$

$g_{b5} \leq 0$ represents the failure domain for bending failure of the beam at mid span after loss of rotational stiffness of the joints and $g_{j3} \leq 0$ defines failure domain for the joint exceeding the ultimate rotation capacity.

Based on these different failure domains the probability of failure of the entire structural system is defined as:

$$P_f = P[BR \cup DU_1 \cup DU_2] \quad (4)$$

3 ANALYSIS AND RESULTS

Different methodologies were applied based on the input variable reported Table 1. The probabilistic analysis was performed with the UQLab toolbox (Marelli & Sudret

DOI: 10.1201/9781003348450-270

2014). The relative deformation-based definition of ductility (Jockwer et al. 2021) was used within this paper for the evaluation of the results. D_u was varied from 1 to 20.

Table 1. Distribution parameters of the considered random variables.

Input	Unit	Distribution	Mean V.	Char. V.	CV
q	$\frac{kN}{m}$	Gumbel	-	4.6	0.4
f_m	$\frac{N}{mm^2}$	Lognormal	37	24	0.25
K_{el}	$\frac{kNm}{rad}$	Lognormal	1000	-	0.3
D_u	-	Lognormal	1 to 20	-	0.1
$M_{R,j}$	kNm	Lognormal	18.4	-	0.2
E	$\frac{N}{mm^2}$	Lognormal	11,000	-	0.13
W_{el}	m^3	Constant	0.001	-	
L	m	Constant	6.5	-	-

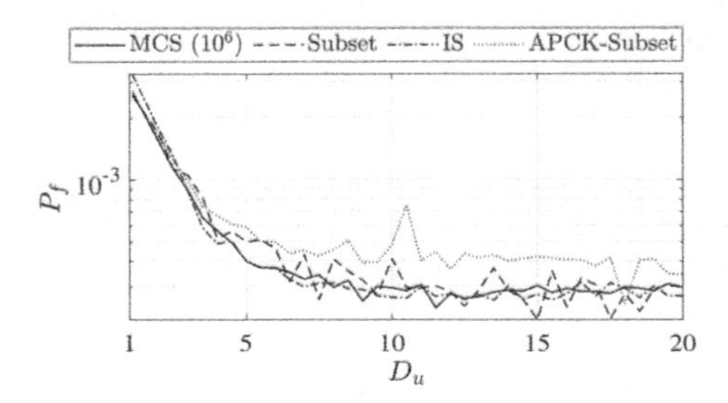

Figure 1. Comparison of the convergence of different modelling techniques.

The resulting probabilities of failure are shown in Figure 1 and summarized in Table 2 for the case of a ductility $D_u = 20$.

Table 2. Results for $D_u = 20$.

Method	Evaluations	P_f	CV	Time
MCS	10^7	$2.85e^{-4}$	0.02	89.44 sec
MCS	10^6	$2.76e^{-4}$	0.06	2.18 sec
Subset	36998	$2.56e^{-4}$	0.09	0.17 sec
IS	1196	$2.94e^{-4}$	0.07	0.32 sec
APCK-Subset	162	$4.02e^{-4}$	0.02	1211 sec

MCS returned the probability of failure with good accuracy and at relatively low computational time, despite the number of simulations required (10^6). When the maximum sample size of MCS was set to 10^7 to decrease the CV, the computational time for MCS increased by a factor of ≈ 45. Subset and IS required lower number of evaluations to return a similar estimation of probability, however the accuracy was less compared to MCS with maximum number of simulation set to 10^7. APCK-Subset did not return a good estimation of the probability of failure, in particular for greater values of D_u and the algorithm showed rather poor convergence for the majority of cases.

4 PARAMETER AND SENSITIVITY STUDY

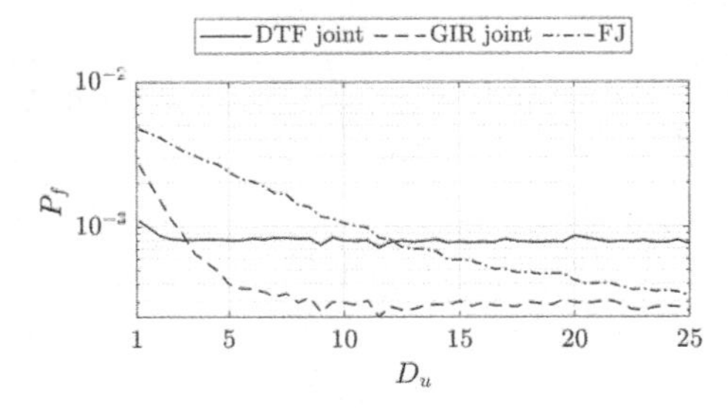

Figure 2. Parametric study results.

In order to study the effect of ductility on the reliability level of the beam, a parametric study was carried out by varying the mean value of the ductility D_u from 1 to 25 for each of the three mean values of joint stiffness K_{el} corresponding to joints with dowel-type fasteners (DTF, $K_{el} = 300 \frac{kNm}{rad}$), with glued-in rods (GIR, $K_{el} = 1000 \frac{kNm}{rad}$) and with large finger joints (FJ, $K_{el} = 3800 \frac{kNm}{rad}$). MCS with maximum number of simulations set to 10^6 was selected as method to carry out the analysis. In the case of DTF joint, the ductility had none or little effect on the probability of failure of the system. When D_u ranges from 0 to 6 the probability of failure drops for the beam with GIR joints. Regarding FJ joint, the ductility range with a significant variation in probability of failure is larger (from $D_u = 0$ to $D_u = 25$), however the variation of P_f is less sensitive to an increase in ductility if compared to the case with GIR joint. If GIR joint or FJ provides sufficient ductility, the effect of the joint can be eliminated from the failure probability of the beam system. In fact, after a certain value of D_u, P_f stabilizes around a value of 10^{-4}.

CONCLUSIONS

Increasing the ductility in a limited range in the case of GIR joint and FJ, had an impact on the the reliability of the structural system. Regarding the different methodologies, it was observed that the large number of evaluation needed by MCS requires great computational efforts if the model is not fast to evaluate. Advanced methods, such as IS, Subset or APCK-Subset may serve as more powerful alternatives, however metamodel-based adaptive method might be affected by approximations errors.

REFERENCES

(2014). *UQLab: A Framework for Uncertainty Quantification in MATLAB.*

Jockwer, R., D. Caprio, & A. Jorissen (2021). Evaluation of parameters influencing the load-deformation behaviour of connections with laterally loaded dowel-type fasteners. *Wood Material Science and Engineering*, 1–14.

Current Perspectives and New Directions in Mechanics, Modelling and Design of Structural Systems – Zingoni (ed.)
© 2022 Copyright the Author(s), ISBN 978-1-032-39114-4

Future perspectives about timber-hybrid systems: The role of connections

G. Di Nunzio, L. Corti & G. Muciaccia
Politecnico di Milano, Milan, Italy

ABSTRACT: Timber is characterized by a very low carbon footprint, and it is a true alternative to reduce the construction impact on the environment. The potential of using timber as structural material is even increased if a combination with concrete or steel elements is considered in the design (i.e. timber-hybrid system) In a hybrid structure, components are arranged in sub-systems to resist gravity and lateral loads respectively. It is intuitive that detailing is crucial as the design of the elements themselves. Within this context, a literature review is presented focusing on the role of connections.

1 INTRODUCTION

Modern society is facing new challenges as never seen before in human history. The continuous growth of population and economy is increasing the pressure on our planet resources. The effects are there for all to see: the climate change is revealing in increasing catastrophic events. Construction industry is one of the main responsible of carbon dioxide emissions, particularly concerning the construction stage. The engineering and the scientific community should then promote initiatives with the aim of reducing the impact of the built environment. Timber is characterized by a very low carbon footprint, and it is a true alternative to reduce the construction impact. The potential of using timber as structural material is even increased if a combination with concrete or steel elements is considered in the design. Such a system is defined as "timber-hybrid" and may be adopted to overcome some issues related to full timber structures. In a structure, components are arranged in sub-systems to resist gravity and lateral loads respectively. It is intuitive that detailing is crucial as the design of the elements themselves. The efficiency and the robustness of a structural system is, in fact, directly linked to the performances of the connections.

2 DESIGN CODES AND GUIDELINES

For brevity, only European codes are revised. The use of timber-concrete composites (TCC) is regulated by EN 1995-1-1 for slabs, and by EN 1995-2 for bridge decks. In order to evaluate the ultimate load-bearing capacity, the so-called "γ method" is adopted, where the γ parameter ranges between 1, for a fully rigid connection, and 0, for an ineffective connection. Deflections of TCC tend to increase over time due to the different rheology of the materials, namely shrinking/swelling for timber and creep for concrete. In Eurocodes, long-term and environmental effects on TCC floor are covered by the approximation proposed by Ceccotti in 1995. The only standard available for the assessment of joints made with mechanical fasteners in timber structures is EN 26871, which regulates the so-called "push-out tests".

3 LITERATURE RIVIEW

In this paper, a literature review is presented focusing on (i) TCC slabs and (ii) timber-hybrid structural systems with emphasis on multi-stories buildings. Timber structural elements can be connected by using one, or the combination, of (i) direct contact between the elements, (ii) mechanical fasteners, (iii) adhesive bonding.

A "hybrid joint" is every fastening solution in which at least two of the abovementioned technologies are combined (Schober & Tannert 2016). Strong and stiff connections seem associated to hybrid joints; however, the mechanical description is complicated by the possible mismatch of stiffness.

3.1 *Timber-concrete composite slabs*

The concept of coupling a timber element, either a panel or a beam, and a concrete slab is an elegant and effective structural solution for members in bending. In fact, timber provides the required tensile strength while concrete is mainly stressed in compression (Yeoh et al. 2011). The two parts are coupled by means of special designed fasteners either nails, screws, dowels, notches cut or

DOI: 10.1201/9781003348450-271

continuous glued connectors. The connectors are usually spaced according to the shear force diagram, with fasteners closely spaced in the proximity of the supports. In a recent publication (Mirdad et al. 2021), an analytical study to minimize the carbon footprint of TCC slab was presented.

3.2 *Timber-hybrid structures*

In few years, several studies have been dedicated to review the future challenges for multi-stories timber building (Kuzmanovska et al. 2018, Stepinac et al. 2020, Leskovar & Premrov 2021). Mid to high-rise buildings are of major interest in structural timber application. Although high-rise buildings may push to the limit a structural system, however, it can be argued that sometimes the existing studies were not consistent with the expected behavior under different lateral loads. In the current design approaches, wind load exponentially increases above a certain height, while seismic forces are almost linear with the height. Therefore, depending on the floor mass and on the location, a tall building could be more prone to wind load than to earthquake.

4 TOWARDS INNOVATIVE SOLUTIONS

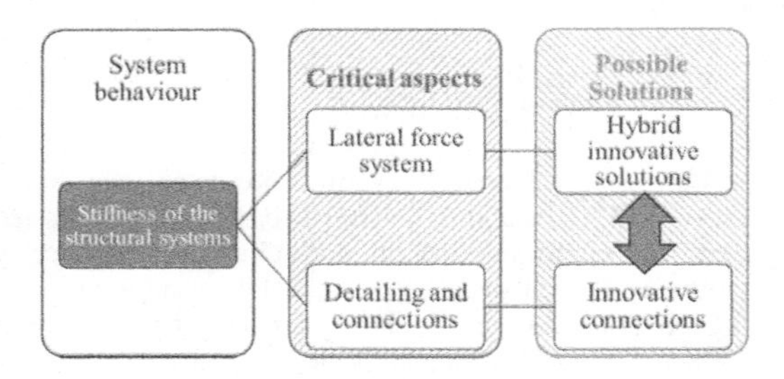

Figure 1. Flow char for innovative timber-hybrid structures.

A flow chart for the development of innovative multi-stories timber buildings is shown in Figure 1. For construction in seismic zone, a comparison with a traditional reinforced concrete structure can be carried out in qualitative terms as shown in Table 1. It is worth noticing how the connections are of paramount importance in the energy dissipation.

Table 1. Timber vs concrete: qualitative comparison.

Parameter	Timber pros	Timber cons
Self-weight	Lighter material	-
Construction	Prefabrication	-
Dissipation	Connections	q-factor

On this basis, case studies were grouped into three categories as: (i) timber-hybrid elements, (ii) existing timber-hybrid structures, and (iii) Proposed timber-hybrid structures.

5 MAIN OUTCOMES

In this paper, a literature review about timber-hybrid systems was presented. Scientific publications were analyzed, where a hybrid structural system was identified as every structure in which mass timber products are used in combination with other elements as concrete cores, steel frames or bracings. The following conclusions can be drawn:

- Hundreds of papers have been published in the last decades for promoting the structural use of mass timber products.
- Knowledge about TCC for floors seems to be more consolidated than for other applications.
- Several innovations have been recently proposed to improve the structural performances and to promote the use of timber-hybrid components, as well as full structural systems for multi-stories buildings. Such solutions demonstrate the existence of ready to use solutions.
- Solutions have been proposed to connect timber to other elements creating an effective hybrid system. The variety of the connectors, as well as their use in the structure, raises questions on their structural response. For this reason, further research is essential in order to provide effective solutions and reliable design methods.

ACKNOWLEDGMENTS

The European Consortium of Anchors Producers (ECAP) is thanked for partially supporting the work.

REFERENCES

Kuzmanovska, I. et al. 2018. Tall timber buildings: Emerging trends and typologies. In *WCTE 2018; Proc. conf., Seoul, South Korea.*

Leskovar, Z.V. & Premrov, M. 2021. A Review of Architectural and Structural Design Typologies of Multi-Storey Timber Buildings in Europe. *Forests* 12 (6): 757.

Mirdad, M.A.H. et al. 2021. Sustainability Design Considerations for Timber-Concrete Composite Floor Systems. *Advances in Civil Engineering* 2021 (6688076).

Schober, K.U. & Tannert, T. 2016. Hybrid connections for timber structures. *European Journal of Wood and Wood Products* 74 (3): 369–377.

Stepinac, M. et al. 2020. Seismic design of timber buildings: Highlighted challenges and future trends. *Applied Sciences* 10 (4).

Yeoh, D. et al. 2011 State of the Art on Timber-Concrete Composite Structures: Literature Review. *Journal of Structural Engineering* 137 (10): 1085–1095.

Current Perspectives and New Directions in Mechanics, Modelling and Design of Structural Systems – Zingoni (ed.)
© 2022 Copyright the Author(s), *ISBN 978-1-032-39114-4*

Recent developments in timber-concrete composite construction

K. Holschemacher
Structural Concrete Institute, HTWK Leipzig, Leipzig, Germany

ABSTRACT: Composite members combining the beneficial properties of different building materials become increasingly important in civil engineering. Whereas steel-concrete composite has been an often-applied method in construction practice for many decades, timber-concrete composite (TCC) was a less preferred technology in past. Nowadays, situation is changing because of the obvious advantages that TCC offers in respect of environmental issues. Especially, TCC slabs are a recognized construction method, meanwhile. The paper reports the state-of-the-art and recent developments in TCC. Advantages of the application of steel fibre reinforced concrete (SFRC) for the construction of the concrete slab are discussed. Furthermore, the construction process for strengthening of existing timber beam ceilings with TCC and application of precast TCC members is reported. An increasing application of TCC in construction practice may be expected because of their lower environmental impact in comparison to other usual floor constructions. This tendency is supported by future normative regulations enabling the easy calculation of mechanical properties of shear connectors.

Basically, TCC members consist of timber beams connected with a concrete slab by special shear connectors. Main application of TCC are floor constructions, in which the timber beams are arranged with spacing or without any spacing between the neighboured beams. There may be arranged an interlayer, e.g. a boarding, between the timber beams and the concrete slab. Usually, the concrete slab is located at the top of the timber beams, but in few cases also between them (Figure 1).

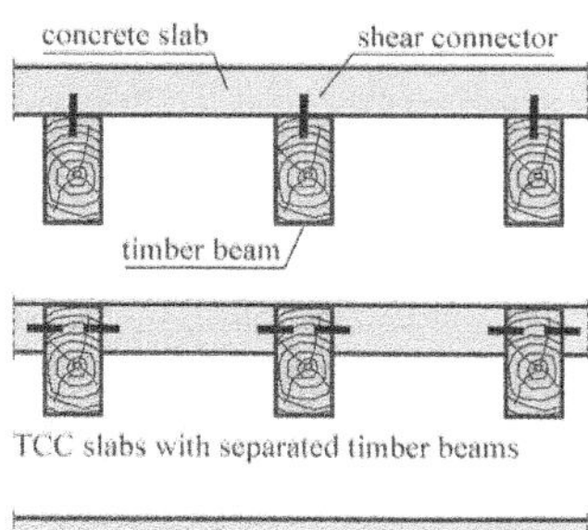

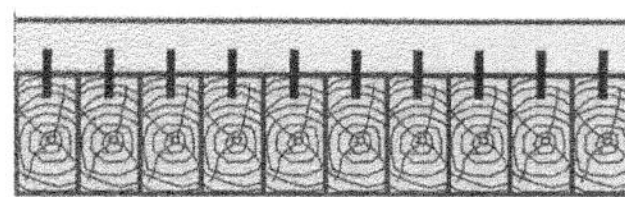

Figure 1. Typical TCC sections.

The interaction and the influence of shear connectors' stiffness, mechanical properties and dimensions of the concrete slab as well as the timber beams on the flexural stiffness of a TCC member is exemplary demonstrated in Figure 2. The flexural stiffness was calculated according to EN 1995-1-1 (2004), Annex B. Modulus of Elasticity of concrete E_c, height of the concrete slab h_c, height of the timber beams h_t, and stiffness of the shear connectors K_{con} were varied in ranges, representative for most practical applications. It is obvious that the height of the concrete slab and the timber beams are the predominant influence parameter. But, also stiffness of shear connectors is of importance. If there is no bond between timber beams and concrete slab, the resulting flexural stiffness is only 1.93 MNm^2, respectively 70% of the value according to initial situation in Figure 2.

The tensile stress σ_{ct} in plain concrete must be limited in the ultimate limit state (ULS):

$$\sigma_{ct} \leq f_{ctd} \tag{1}$$

where f_{ctd} = design value of tensile strength of concrete, considering the partial safety factor for plain concrete. In many practical TCC applications it is not possible to satisfy the demand according to Eq. (1). However, the height of the tensile zone normally is relatively low in usual TCC, e.g. few millimetres or centimetres.

Therefore, steel bar or mesh reinforcement is often not active in this situation because the tensile zone is not larger than the concrete cover and reinforcement is located in or near to compression zone. In this case,

DOI: 10.1201/9781003348450-272

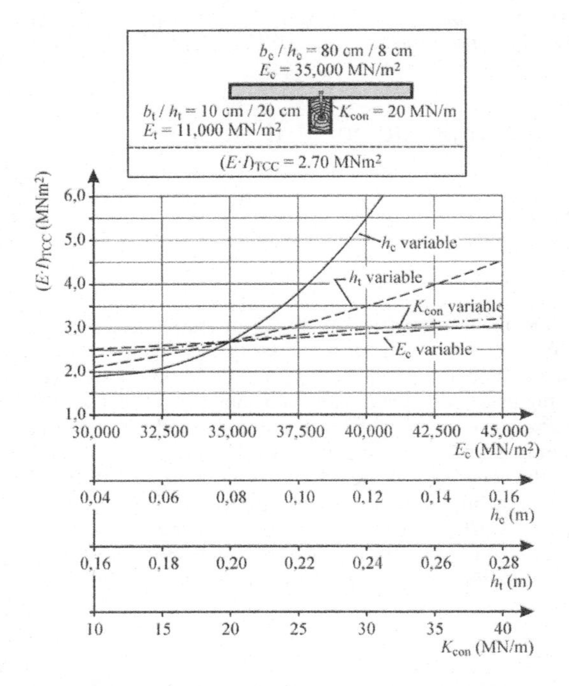

Figure 2. Influence of various parameters on flexural strength of TCC.

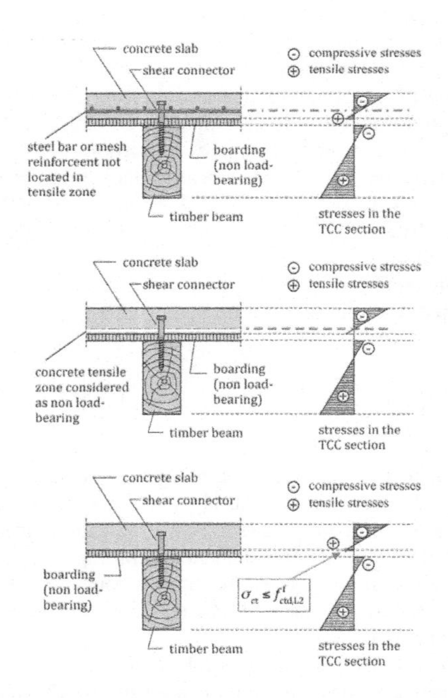

Figure 3. Tensile stresses in the concrete slab when applying reinforced concrete, plain concrete, and SFRC.

the consideration of the tensile zone as not load-bearing interlayer is a suitable design solution. Another possibility is the application of steel fibre-reinforced concrete (SFRC) in the slab. SFRC without any additional steel bar or mesh reinforcement does not need a concrete cover and is able to bear tensile stresses over the full height of the tensile zone in the slab (Figure 3).

In last years, the application of TCC in construction of new buildings has become more important. One reason for this tendency is the lower environmental impact of TCC in comparison to traditional floor constructions. Recent investigations proved that global warming potential of TCC slabs is 30% lower than for reinforced concrete slabs under same conditions (Holschemacher, Hoffmann & Heiden 2021).

After years of stagnation, there is a revival of TCC recently. TCC is increasingly applied for strengthening of existing structures and construction of new buildings. One of the main reasons is the lower environmental impact of TCC, an issue gaining high importance in today's civil engineering. Future normative regulations will contribute to facilitated application of TCC in construction practice.

REFERENCES

Holschemacher, K. & Kieslich, H. 2021. Timber-concrete composite (in German). In K. Bergmeister, F. Fingerloos & J.-D. Wörner (eds.), *Betonkalender 2021*, Berlin, Ernst & Sohn. https://doi.org/10.1002/9783433610206.ch4.

Dias, A., Schänzlin, J., & Dietsch, P. 2018. *Design of timber-concrete composite structures*, State-of-the-art report by COST Action FP1402/WG 4. Shaker, Germany.

EN 1995-1-1 (2004). *Eurocode 5: Design of Timber Structures – Part 1-1: General – Common rules and rules for buildings*, European Committee for Standardization.

EN 1992-1-1 (2004). *Eurocode 2: Design of Concrete Structures – Part 1-1: General – Common rules and rules for buildings*, European Committee for Standardization.

CEN/TS 19103:2021 (2021). *Eurocode 5: Design of Timber Structures – Structural design of timber-concrete composite structures – Common rules and rules for buildings*, European Committee for Standardization.

Holschemacher, K. 2021. Timber-concrete composite – a high-efficient ad sustainable construction method. *Proceedings of International Structural Engineering and Construction*, 8 (2021), 1. doi:10.14455/ISEC.2021.8(1).STR-59.

Holschemacher, K. & Quapp, U. 2021. Timber-concrete composite – a contribution to sustainability in architecture and civil engineering. *Proceedings of the 8th International Conference on Architecture and Built Environment*, Rome, Italy, 329–338. ISBN 978-3-9820758-7–7.

Holschemacher, K., Hoffmann, L. & Heiden, B. 2021. Environmental impact of timber-concrete composite slabs in comparison to other floor systems. *Proceedings of the World Conference on Timber Engineering (WCTE 2021)*, Santiago de Chile, Chile. ISBN 978-1-713-84097–8.

Current Perspectives and New Directions in Mechanics, Modelling and Design of Structural Systems – Zingoni (ed.)
© 2022 Copyright the Author(s), ISBN 978-1-032-39114-4

Improving the modelling of tall timber buildings

O. Flamand
CSTB, centre scientifique et technique du bâtiment, Nantes, France

M. Manthey
CSTB, centre scientifique et technique du bâtiment, Marne la vallée, France

ABSTRACT: Verticalization of housing is one answer to densification of cities. The use of timber materials for all or part of the structural elements offers light structures. Beyond resistance criteria, service comfort criteria (vibrations) is a significant challenge for lightweight constructions intended for housing, since the expectations are more severe than for office use. Lighter means more excitable: the structural damping of building materials can make a difference but there are still many unknowns that complicate the task of designers.

As part of the European research project Dyna-TTB the dynamic behavior of high-rise timber towers under the effect of wind has been studied. The objectives are as follows: to experimentally quantify the structural damping in as built high-rise timber buildings; to make FEM models more reliable (behavior law of connections, effect of non-structural elements on stiffness, damping and the wind-induced dynamic response); to establish a guide of recommendations for in-situ tests and digital modeling of tall timber buildings.

Characterization of the dynamic behavior of timber towers under imposed loads (heavy mechanical exciters) was carried out. In France, the Hyperion (16 floors, 56m high) and Treed It (11 floors, 36m high) towers were instrumented and tested at various stages of construction (before/after the installation of the interior partitions).

The FE models were made using Ansys software, with material properties derived from the Eurocodes; particular attention was paid to the modelling of the foundations and joints.

1 INTRODUCTION

The dynamic response of a tall timber building (9+ levels) using timber element for the use of structural construction (e.g. floors, beams, columns, etc.) still has several uncertainties.

Use of timber for structural use in tall construction is relatively new, comparing to steel or concrete solutions, timber structures do have another mass/rigidity ratio which can affect their dynamic behavior.

Timber-based structure, being relatively lightweight and flexible could induce vibration issue. The topic of dynamic serviceability of TTB deserves a better understanding. Unexpected vibration/disturbance level in low-frequency range are experienced by the occupants, especially in the range of service loading (fluctuation of the wind/strong wind).

A lack of understanding of the dynamic behavior of TTBS may lead to larger drift (both total and inter-story drift ratio). As always with dynamic behavior, three kinds of parameters are of importance: masses, stiffnesses and damping. The first part of this paper deals with experimental techniques used in this research program to assess the structural damping of the timber buildings.

In the second part of this paper initial FEM, used to prepare the experimental campaign, will be described and updated regarding experimental results. Considerations on masses and stiffnesses will be widely reported.

2 EXPERIMENTAL ASSESSMENT OF STRUCTURAL DAMPING AND FREQUENCIES

Structural damping of buildings is often amplitude dependent, what pushes the experimenter to shake the existing towers with large amplitude, with respect to their integrity. In the DynaTTB research program an agreement was found with the buildings' owners providing assurance to them that a careful FE modelling will be used for calculation of stresses and relative displacements resulting from a strong excitation. The shaker used by CSTB was first closely calibrated in laboratory, then the same force signal was introduced in the first FE model of the timber building and stresses were computed and shared with the building owner.

On the TreedIt tower a series of tests was performed when all the main structure was achieved but before the inner walls have been put in place. A second

DOI: 10.1201/9781003348450-273

campaign was done after the full completion with inner walls in place, to check if these internal partitions add only weight, of stiffness, or damping.

A strong excitation up to 9000N was made and released to measure the decay of oscillation at the frequency of excitation as illustrated in Figure 1. Calculating the damping at each oscillation during the decay provides the evolution of modal damping with amplitude. The process showed to be very repeatable and the variation of damping with amplitude was great, ranging from 3.5% of critical for medium amplitudes to 1% of critical for low amplitudes. A reduction of damping was observed for very large amplitudes, what was unattended and should be explained by modelling.

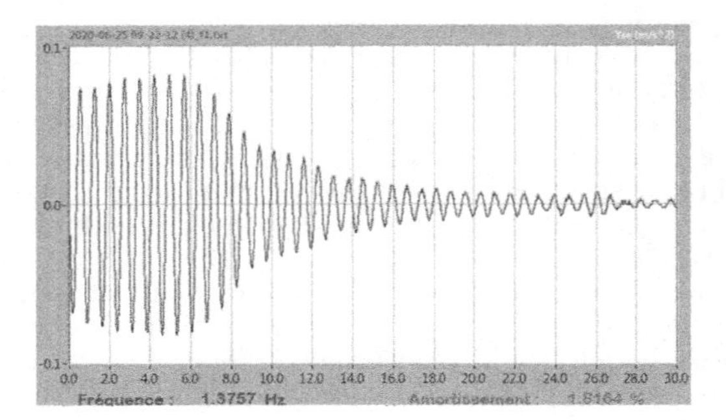

Figure 1. Shutdown test on TreedIt building with excitation of the first mode. Accelerations in m/s², time in s.

Accelerations measured at three levels of both buildings with 3 sensors in the horizontal plane at each level can be compared to the mode shapes calculated by FE models, using the MAC matrix.

In TreedIt tower displacement sensors were attached between timber piles and beams, to measure local deformation by mean of LVDT sensors. Deformations of some micrometers have been recorded, to be compared with the ones calculated in the FE modelling.

3 FE MODELLING

As described previously, 2 towers were studied, Hyperion and the Treed It. As both towers are representative of timber-concrete hybrid tower as made in France, only the modelling on Treed It tower will be presented and discussed hereafter.Treed It is a timber-concrete hybrid tower. The tower includes a concrete core, timber-concrete floors, piles foundations and glue-laminated columns and beams.

Finite Element modelling has been realised with various level of knowledge about the tower. A very first model, named "basic model" in this study, was made initially by the designers. Three FEM modelling made during this study were done. They represent the various stages of construction of the tower corresponding to the performed experimental campaign.

The computer program ANSYS Mechanical was used to develop FE models for the buildings. The program can perform linear and non-linear, static, and dynamic analyses.

Following table sums up the main results from FEM regarding frequencies modelled.

Table 1. Frequencies obtained with on-site testing and with FEM modelling.

FEM-/test- results	1stmode (Hz)	2ndmode (Hz)	3rdmode (Hz)
Basic model	0,45	0,81	0,88
1st on site testing (no partition wall in place)	1,33	1,57	1,70
1st FEM (structural elements only)	1,30	1,54	1,70
2nd on site testing (partition wall in place)	1.39	1.49	1.62
2nd FEM (partition's mass added)	1,04	1,22	1,35
3rd FEM (partition as shell element)	1,29	1,57	1,71

Good correlation was reached regarding the frequencies of the 1st experimental campaign and 1st FEM. 2nd and 3rd model tends to demonstrate that inner partition should be modelled not only as masses but also as shell contributing to the lateral stiffness of the tower.

4 CONCLUSION

A good correlation was obtained between a predictive FE model based on best engineering judgment and results from an experimental campaign on a tall timber building without non-structural element. After the non structural elements (internal partitions) have been added to the timber building, the FE model was no more accurately describing its dynamics. Improvement of the FE model was necessary with a focus on the way to model internal walls and their connexion to the main structural bearing elements.

The DynaTTB research program will issue recommendations to structural designer to help them model non-structural element effect on stiffness.

In situ measurement of the damping showed to be helpful to refine FEM, especially with temporal analysis/harmonic response.

REFERENCES

Jeary A P (1997). Damping in structures. Journal of Wind Engineering and Industrial Aerodynamics 72

Current Perspectives and New Directions in Mechanics, Modelling and Design of Structural Systems – Zingoni (ed.)
© 2022 Copyright the Author(s), *ISBN 978-1-032-39114-4*

Influence of initial crack width in Mode I fracture tests on timber and adhesive timber bonds

S.A. Rahman, M. Ashraf, M. Subhani & J. Reiner
Deakin University Geelong, School of Engineering, Waurn Ponds, Australia

ABSTRACT: We present the experimental quantification of fracture energy release rates in timber and timber adhesive bonds subjected to Mode-I double cantilever beam (DCB) tests with different pre-crack geometries to establish the optimal pre-crack width of the test samples. Additionally, crack resistance curves for both timber and timber adhesive bonds are calculated and compared. From the analysis of experimental results, maximum fracture energy release rates of a pure bond delamination are achieved from the minimum possible width of the initial crack. Thin (0.1 mm) polyethylene duct tape was used in the initial crack zone to maintain a minimal gap between timber joints. For DCB tests on pure timber, 0.5 mm was the lowest achievable crack width found to yield more stable fracture propagation compared to test samples with 1 mm crack widths.

Keywords: Fracture process zone, double cantilever, fracture energy, glue line failure, crack resistance curve

1 INTRODUCTION

Timber and timber composites are becoming popular in the construction industry due to their environmentally friendly and renewable characteristics. Most of the timber products used in construction is either engineered timber product (ETP) or sawn timber sourced from sustainable fast grown plantations. Timber from sustainable resources has a range of defects resulting from smaller diameter of logs than those natively grown in forests. With the help of adhesive technologies, the relative low-quality timber logs could be turned into versatile structural members such as cross-laminated timber, glue-laminated timber, and laminated veneer lumber (Li et al., 2020).

Although a significant amount of progress has been achieved in timber adhesive technologies, extensive investigations are still required to characterise bond strength, fracture energy evolution and crack propagation in the bond. Most recent research investigations have employed random geometric and initial crack width conditions which ultimately leads to a wide range of fracture energy values.

To understand the glue delamination of a softwood timber joint, three different initial crack width conditions of 3, 1 and 0.1mm in DCB test orientation have been investigated. A separate DCB test on sawn timber has also been conducted to distinguish between glue failure and timber failure.

2 MATERIALS AND METHODS

All testing samples were prepared from home grown Australian radiata pine (Pinus radiata). The moisture content was tested for all samples before the experiment and found to be 9.6 % on average.

Polyurethane based adhesive HB S309 PURBOND was used for timber adhesive bond joints. A schematic overview of DCB test is shown in Figure 1. The test specimen represented a cross-section of 300 × 30 ×60 mm, and the load was applied by using two pins as shown in Figure 1.

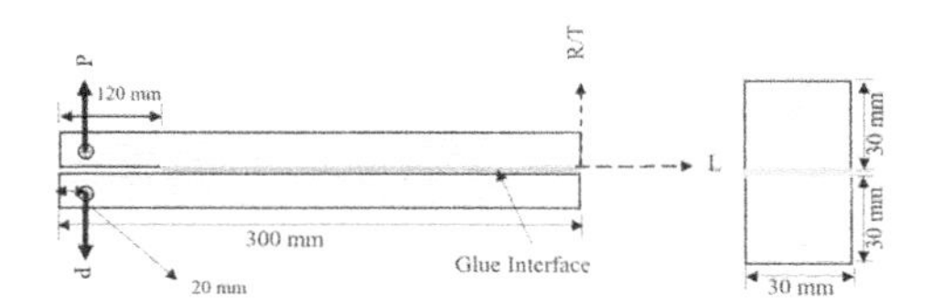

Figure 1. Double Cantilever Beam (DCB) specimen geometry.

3 RESULTS AND DISCUSSION

Load-displacement curves from DCB tests with 3 mm, 1 mm and 0.1 mm notches are illustrated in Figure 2. The average maximum strength is 909 N from 0.1 mm notch which is considerably higher than in the two other configurations. According to the experiments, it has been observed that introducing a notch at the sample by sawn blade, damages the surrounding timber fibre and creates some hairline cracks. During the test those faults easily connect with pre-existing gaps along the timber fibre direction. As a result, it reduces the stiffness, strength and overall load carrying capacity of the specimen. 3 mm and 1 mm notch samples were

DOI: 10.1201/9781003348450-274

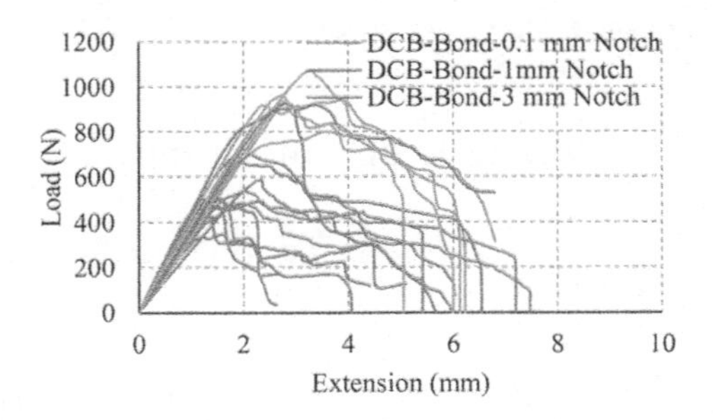

Figure 2. Load vs Extension of DCB with 0.1,1- and 3-mm notch width.

not suitable for pure bond delamination. Hence results from those tests were not considered in analysing fracture energy release rate of bond delamination. Only results from 0.1 mm notches are considered herein as pure delamination and were analysed to achieve strain energy release rate for DCB-Bond test.

A separate DCB fracture test on sawn timber was also carried out to understand differences between timber bond delamination and timber fracture. DCB fracture tests on sawn timber were only performed for 1 mm and 0.5 mm of initial crack width. Load-displacement of DCB tests on sawn timber is shown in Figure 3. The average failure strength for the considered cases were 548 N and 527 N. However, 0.5 mm initial crack width produced a consistent load-displacement response and relatively smooth stable crack propagation compared to 1 mm width notches. Therefore, results from 0.50 mm wide notch samples were selected to analyse the fracture energy release rates.

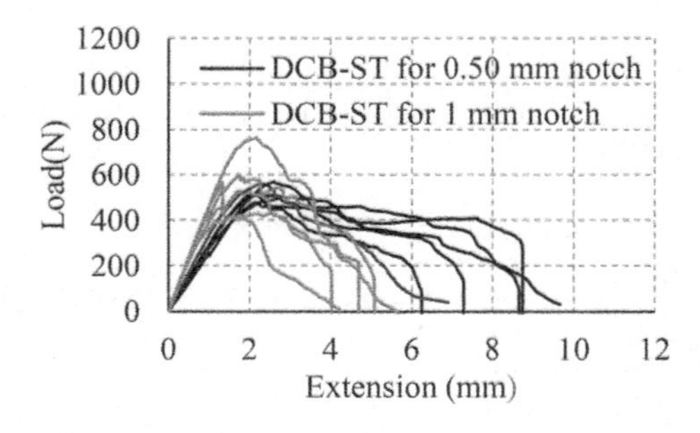

Figure 3. Load vs Extension of DCB test on sawn timber.

Fracture resistance curve-R-curves for timber and timber bond delamination using only Compliance calibration CC method for all specimens are presented in Figures 4 and 5. Timber bond is stronger than timber itself and hence crack propagation in timber requires less load than in bond. At the edge, glue delamination bridges with timber failure, which reduced the load and fracture energy. On the other hand, R-curves for sawn timber in Figure 5 are

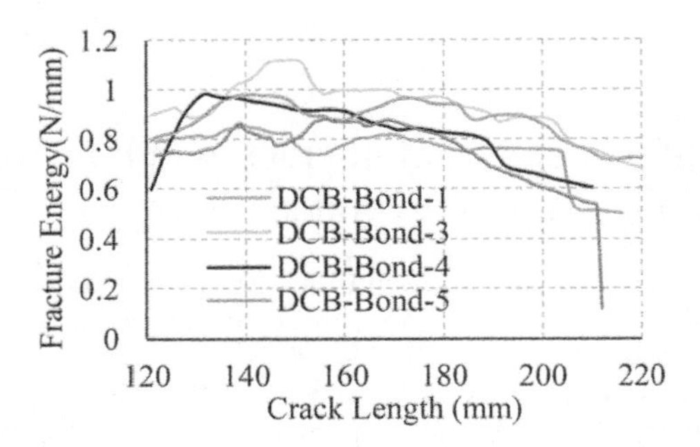

Figure 4. Resistance R-curve for timber bond delamination.

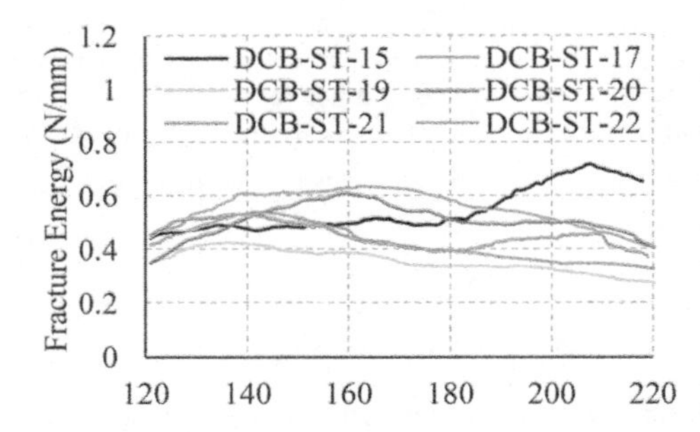

Figure 5. Resistance R-curve for fracture of swan timber.

somewhat followed a constant rate. Crack propagation in timber deals with one material but bond delamination involves both glue and timber. Softwood timber bond delamination always takes part in a combined failure of bond and soft timber. As a result, higher rate at the beginning and comparatively lower rates at the end of the test can be observed in fracture energy evaluations of bond delamination tests.

4 CONCLUSIONS

This study has successfully differentiated between timber and bond failure. The strength and load caring carrying capacity of the bond fracture is slightly higher than fracture in timber. Initial crack width has a profound influence on timber bond delamination. The lowest possible crack width ensures smooth and stable crack propagation. Cutting a notch along the glue line is inappropriate for the glue delamination test. It damages the timber and generates fault in timber to channel the crack inside timber. However, fracture energy rates for sawn timber has an insignificant influence on initial crack width. Lower initial crack width only contributes to stable and consistent crack propagation.

REFERENCES

Li, X., Ashraf, M., Subhani, M., Kremer, P., Kafle, B. & Ghabraie, K. 2020. Experimental and numerical study on bending properties of heterogeneous lamella layups in cross laminated timber using Australian Radiata Pine. *Construction and Building Materials*, 247, 118525.

Current Perspectives and New Directions in Mechanics, Modelling and Design of Structural Systems – Zingoni (ed.)
© 2022 Copyright the Author(s), *ISBN 978-1-032-39114-4*

Laminating effect in South African pine glue laminated timber beams

F.J. Pretorius & C. Roth
University of Pretoria, Pretoria, South Africa

ABSTRACT: This work presents a methodology for analysing glue laminated beams (GLB). More specifically, it concerns an investigation of the raw material (South African Pinus) used to manufacture the GLB's. The GLB's were tested through non-destructive and destructive tests to determine the bending strength or Modulus of Rupture (MOR). The values were then compared to a mathematical model that was developed and refined for South African soft wood. The approach used in the simulation model PROLAM analyses for strength using a transformed section method and analyses for stiffness using a complementary virtual work technique. The aim was to develop a better understanding of the structural behaviour of a naturally grown building material and to develop confidence for structural engineers to work with a renewable resource. The laminating effects found ranged from 2.2 to 2.8, with an average of 2.5.

1 INTRODUCTION

Structural timber is generally divided into three construction material groups: (1) timber in its raw material form, e.g. logs and poles; (2) timber as a processed material, e.g. planks and boards; and (3) Engineered Wood Products (EWP), e.g. glue laminated beams (GLB), cross laminated timber (CLT), plywood and laminated veneer lumber (LVL). These different types of timber products can be distinguished by the method utilised in their manufacturing process and their final application.

GLB's are manufactured by gluing together a number of smaller timber boards known as laminates, with the grain of the laminates parallel to the longitudinal axis of the beam. Among the advantages of GLB's are the range of sizes and cross-sectional shapes that can be achieved; as well as the efficient use of timber.

An important characteristic of glulam manufacture is that the bonding of laminates can result in beams of higher strength than the strength of the single laminates from which they are constructed. This laminating effect is the increase in strength of lumber laminates when bonded in a glulam beam compared with their strength when tested individually.

This paper looks at GLB's manufactured from South African Pine. The objective was to quantify the laminating effect for such timber, as existing literature only covers European and North American timber. The effect of using laminates of varying strength, rather than a consistent strength, in a beam was a secondary objective.

2 LAMINATING EFFECT

Falk & Colling (1995) investigated the laminating effect for European and North American GLB's. They defined the effect numerically as:

$$\lambda = \frac{f_{b,gl}}{f_{b,lum}} \quad (1)$$

where λ = laminating effect; $f_{b,gl}$ = mean bending strength of a population of GLB's; and $f_{b,gl}$ = mean tensile strength of a population of lamination lumber.

They found that the laminating effect ranged from 1.06 to 1.59 for European GLB's and from 0.95 to 2.51 for North American GLB's. In general, they found that the lower the quality and strength of the timber, the higher the potential laminating effect.

They proposed several reasons for this effect.

3 EXPERIMENTAL PROGRAM

3.1 *Test samples*

48 Pieces of 38 × 114 mm South African Pine were selected from the York Timbers Driekop sawmill near Graskop in Mpumalanga.

3.2 *Laminate testing*

A four-point bending test to SANS 6122 (2014) was carried out on each of the boards at the University of Pretoria as shown in Figure 2. The test was a non-

DOI: 10.1201/9781003348450-275

destructive test to determine the modulus of elasticity (MOE). The results were used to assign a specific MOE to each of the 48 boards. The MOE values ranged from 5.47 to 15.3 GPa.

3.3 *GLB manufacturing and testing*

Five different beams were designed, each using 8 of the 48 laminates. The beams were designed to use the higher stiffness laminates in the outer layers of the beam where the stresses are highest, and lower stiffness laminates near the neutral axis of the beam.

The faces of the laminates were pre-planed to a thickness of 33.3 mm and screened for any surface defects that might influence the bonding. Thereafter the adhesive, melamine formaldehyde resin, was applied, and the beams were pressed and planed to final dimensions of 276 mm × 100 mm.

The MOE and modulus of rupture (MOR) of the beams were then determined through a four-point destructive bending test as per SANS 6122 (2014). Results are in Table 1.

Table 1. Experimental program results.

	Failure load	MOR	MOE
Beam	kN	MPa	GPa
1	124	52	9.9
2	138	58	10.9
3	104	44	9.7
4	109	46	10.1
5	125	53	10.4

4 ANALYSIS

4.1 *Model details*

An approach consistent with that used by PROLAM was used to predict the strength of the GLB's based on the strength of the individual laminates. The transformed section method was used to analyse the assembled beam. The transformed section method analyses a glulam beam by transforming the widths of each laminate in the composite cross section so simple elastic flexural formulas can be applied.

4.2 *MOE-tension relationship*

The above approach requires an estimate of the tensile strength of each laminate. The MOE of a timber section can be determined with little effort through a non-destructive static 4-point bending test, while the tensile strength can only be determined through destructive testing.

Dowse (2010) completed a study of local South African timber and found a correlation between static MOE and tensile strength. He found the following equation with a R^2 correlation of 0.63:

$$T = -1.8998 + 1.6 \times MOE \qquad (4)$$

where T = tensile strength in MPa, and MOE is in GPa.

As the Dowse equation was based on a substantial testing program on timber similar to that used here, it was selected as the most realistic equation.

5 RESULTS

The comparison of the predicted and tested MOR values for each beam is shown in Table 2, along with the calculated laminating effects.

It is clear that the laminating effect is substantial, with an average of 2.5.

Table 2. Laminating effect results.

	Predicted MOR	Tested MOR	
Beam	MPa	MPa	Laminating effect
1	19	52	2.7
2	21	58	2.8
3	20	44	2.2
4	19	46	2.4
5	21	53	2.5

6 CONCLUSIONS

The values for the laminating effect here are generally higher than those found by Falk & Colling (1995). This agrees with their finding that the lower the quality and strength of the timber, the higher the potential laminating effect. The mean tensile strength of the pine tested here was lower than all of the European and North American timber they tested.

It should be noted that the values calculated are dependent on the MOE vs tensile strength relationship selected. It is possible that the relationship used here underestimates the tensile strength of the laminates, resulting in a higher laminating effect. Further work is needed in this area for South African pine.

REFERENCES

Dowse, G. 2010. Selected mechanical properties and the structural grading of young Pinus patula sawn timber. Stellenbosch: University of Stellenbosch

Falk, R. & Colling, F. 1995. Laminating Effects in Glue Laminated Timber Beams. *Journal of Structural Engineering*, 121(12): 1857–1863.

Current Perspectives and New Directions in Mechanics, Modelling and Design of Structural Systems – Zingoni (ed.)
© 2022 Copyright the Author(s), *ISBN 978-1-032-39114-4*

Development of cross-laminated timber composite panels from C16 timber

E. McAllister & D. McPolin
Queens University Belfast, Northern Ireland

ABSTRACT: This research investigates the potential of using timber of grade C16 to manufacture cross laminated timber (CLT) panels with comparable mechanical properties to those manufactured from C24 timber. The use of locally sourced C16 timber will mean the CLT panels will be more sustainably attractive with reduced transportation costs. The effect of how drying and conditioning of the raw timber affects the panel strength characteristics has been considered. The study also investigates how the addition of a strengthening material, GRP mesh will affect the panels performance in bending and shear tests. A theoretical model has been created to determine strength characteristics of CLT panels with and without the additional material.

1 INTRODUCTION

The use of natural resources as a reputable building material is rapidly developing, particularly with timber. This paper will solely focus on cross laminated timber (CLT), which is becoming an increasingly favorable building material within Europe.

This study investigates the feasibility of the use of C16 timber instead of C24 as a source material for the production of CLT. The mechanical and physical performance will be determined using a series of tests and the future of CLT manufactured within the United Kingdom will be assessed.

2 DETERMINING THE PERFORMANCE OF CLT

In past research, the bending capacity of CLT panels has been found using the stiffness values of each individual length of timber in the CLT panel (Sikora et al. 2016). Using this method, the stiffness of any 'timber-only' CLT panel can be calculated at a predictive level. Eqs. (1) and (2) based on BS EN 310, can be used to obtain the modulus of elasticity and bending strength of CLT panels.

$$E_m = \frac{L_1^3(F2 - F1)}{4bt^3(a2 - a1)} \quad (1)$$

$$f_m = \frac{3F_{max}L_1}{2bt^2} \quad (2)$$

where L_1 is the distance between the support centres; b is the width of the test piece (mm); t is the thickness of the test piece (mm); $F2 - F1$ is the increment of load on the straight-line portion of the load-deflection curve; $a2 - a1$ is the increment of deflection at the mid-length of the test piece (corresponding to F2 - F1); and F_{max} is the max load applied (N).

A method of assessing CLT panel strength has been established which uses the Young's modulus of the material in each layer of the panel and transforms it into a section using modular ratios. From this, the parallel axis theorem can then be used to calculate the moment of inertia for the cross section of the panel and the bending stresses on positions of the panel can simply be calculated for various loads. However, it is important to note that this modelling is only suitable for the calculation of strength characteristics up to the moment of failure of the outermost component in the CLT panel. Post elastic behaviour, most notably in the outmost compression zone is not yet considered in the model. This theoretical modelling technique can also be used to calculate the same characteristics when the panel has an existing strengthening layer.

Figure 1 below shows how various panels deflected during physical experimental testing or how they would theoretically deflect when a load of 1.5kN/m^2 is applied. This would be a typical domestic floor loading. Only one panel was physically tested (span of 2000mm) and the other results in Figure 1 are theoretical results from the modelling programme.

DOI: 10.1201/9781003348450-276

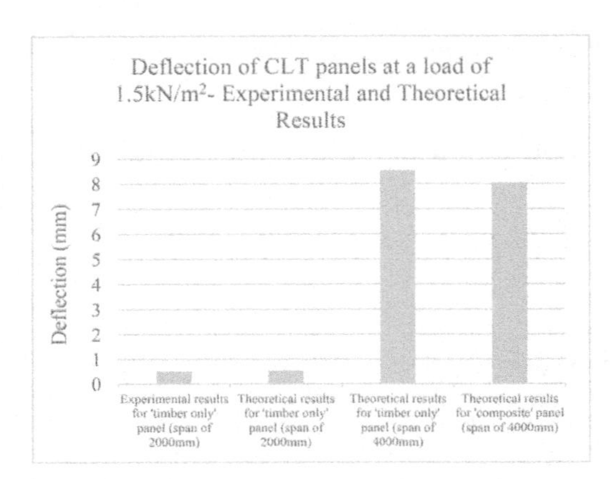

Figure 1. The deflection of various panels at a load of $1.5kN/m^2$.

3 IMPROVING THE PERFORMANCE OF CLT

3.1 *Additional layer*

As determined from the predicted values from the theoretical modelling programme, the addition of a layer of GRP mesh improves the strength of the CLT panel. With the same load of $1.5kN/m^2$ applied each time, the presence of the additional layer resulted in a deflection reduction of 5.8% on average. This additional layer is required when using the C16 timber as a source material for the CLT panels. The panel strength is improved as the tensile forces acting on the bottom of the panel are not only resisted by the overlapping pattern of the CLT but also by the additional material which has a high tensile force resistance. The use of GRP mesh can be substituted using any layering material and the modelling programme can calculate the strength parameters when the value for the Young's modulus for the additional material is imputed. This programme allows us to determine the combination of materials which produces panels with the highest performance.

3.2 *Drying and conditioning*

A set of CLT panels were manufactured in the laboratory under replicated conditions as in a factory setting. The timber was allowed to air dry for 3-4 days in an area of high ventilation and temperature of 18-20 °C. The moisture content of the timber was checked daily using a moisture probe and the panels were manufactured once the timber had a moisture content reading of 12-15%. After some time, gaps started appearing in the panels as illustrated in Figure 2. These gaps resulted in insufficient bonding between the layers, leading to a reduction in panel strength. The average modulus of elasticity of the five panels manufactured in the laboratory are shown in Table 1 below and have be compared with industrially manufactured CLT panels.

Table 1. Data reproduced from Metsawood Leno and KLH CLT technical brochures.

Source of Data	Average Modulus of Elasticity (N/mm^2)	Moisture Content before assembly (%)
Metsawood (Leno)	10,590	10% (+/- 2%)
KLH	12,000	12% (+/- 2%)
CLT manufactured in the laboratory	5,496	12% - 15%

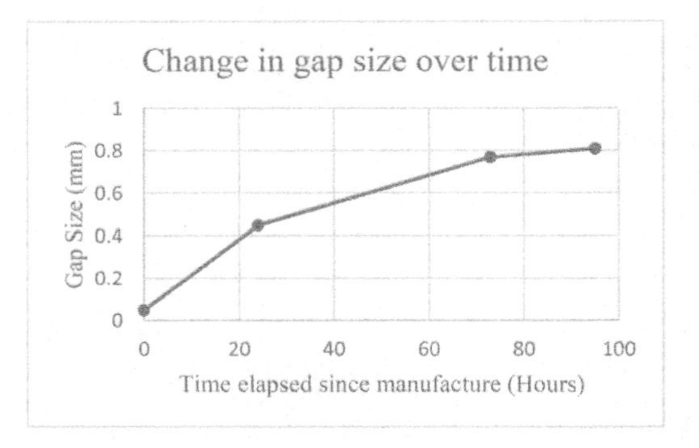

Figure 2. The change in gap size (mm) over some time after manufacture of the panel.

4 CONCLUSIONS

From the work undertaken to date, it is confirmed that the addition of a layer of GRP mesh would increase the overall panel strength properties. Through the theoretical calculations, it is confirmed that with the additional material, the panel deflection during a loading of $1.5kN/m^2$ is reduced by 5.8% on average. It is also confirmed that the theoretical calculations over-predict the performance of the panel deflection when compared to experimental results.

REFERENCES

BSI, British Standards Institutions. (1993). BS EN 310 - Wood-based panels — Determination of modulus of elasticity in bending and of bending strength.

Sikora, K. S., McPolin, D. O., & Harte, A. M. (2016). "Effects of the thickness of cross-laminated timber (CLT) panels made from Irish Sitka spruce on mechanical performance in bending and shear" [online] Available at: https://www.sciencedirect.com/science/article/pii/S0950061816307000 (Accessed Date: 10 January 2021).

Current Perspectives and New Directions in Mechanics, Modelling and Design of Structural Systems – Zingoni (ed.)
© 2022 Copyright the Author(s), *ISBN 978-1-032-39114-4*

Influence of knots and density distribution on compressive strength of wooden foundation piles

G. Pagella, M. Mirra & G.J.P. Ravenshorst
Delft University of Technology, Delft, The Netherlands

J.W.G. van de Kuilen
Technical University of Munich, Munich, Germany & Delft University of Technology, Delft, The Netherlands

ABSTRACT: This work investigated the influence of knots on the compression strength of wooden foundation piles. The study involved 110 pile segments sawn from 18 spruce and 9 pine piles with a mean diameter of approximately 200 mm, and moisture contents above fiber saturation. The mechanical properties were determined performing both full-scale compression tests on pile segments, and small-scale experiments on discs sawn from selected segments, considering samples with and without knots. A knot ratio (KR) was defined analysing the knots layout of each wooden pile, and evaluating how the compressive strength was influenced by size, number and layout of knots. As final step, a prediction model was implemented based on the dry density and KR of wooden piles, to estimate the influence of knots on their compressive strength.

1 INTRODUCTION

This paper investigates the influence of knots on the compressive strength along wooden piles, since the presence of knots can be governing for the capacity of the pile. This is because the low stiffness of the cross section with knots in the axial direction of a pile, due to fiber deviation around a knot, implies that only a very small amount of the load is taken up by knots. Thus, the influence of knots as a strength reducing factor on the load bearing capacity of timber piles has been studied along the piles.

2 MATERIALS AND METHODS

This study investigated: 18 spruce (*picea abies*) piles from the Netherlands; 9 pine (*pinus sylvestris*) piles from Germany. All the piles were cut into 110 segments with a length of approximately six times the average diameter. For most piles, three segments (sawn from the head, middle and tip part) were tested according to the procedure described in section 2. For six piles, eight segments over the length of the piles were tested to gain insight into a more precise distribution of knots and strength along the pile. For three segments that were tested, two discs of 100 mm were sawn: one disc without knots and one with a branch whorl. Mechanical testing was performed to determine the compression strength $f_{c,0,meas,wet}$ of water-saturated pile segments. The influence of knots on the stress distribution of areas where fibers are not parallel to the axial load was considered calculating both the diameter of the knot (Φ_1) and the diameter around the knot where fiber deviations are visible ($\Phi_1 + \alpha\ \Phi_1$). An equivalent knot diameter Φ_2 (Eq. 1) was defined as the diameter to which no strength is accounted for the prediction models described in section 4.

$$\Phi_2 = \Phi_1 + \alpha \beta \Phi_1 \tag{1}$$

β describes how much of $\alpha\Phi_1$ should be taken into account to obtain a value for Φ_2 as an equivalent size to which no strength is assigned. The influence of knots on $f_{c,0,meas,wet}$ was evaluated with a knot ratio (KR), defined as the sum of the diameters $\Phi_{2,i}$ of each knot in a branch whorl within a section of a pile and the circumference O of that pile (Eq. 2).

$$\mathrm{K}R = \frac{\sum_{i=1}^{n} \Phi_{2.i}}{O} \tag{2}$$

The global analysis was validated by testing in compression also three 100-mm-thick wooden discs sawn from a section with and without knots, i.e. a section with maximum KR and a clear wood section, respectively, of the same pile segments. A value of $f_{c,0,KR,meas,wet}$ and $f_{c,0,CW,meas,wet}$ was defined. Finally, the mechanical properties of six full-length piles, divided in eight sub-parts, were determined, by testing in compression each sub-part.

3 RESULTS

A coefficient α = 0.95 was determined for the majority of knots with $20 \leq \Phi_1 \leq 30$. However, larger knots contributed less to fiber deviation, resulting in a lower α, while smaller knots caused higher fiber deviation, resulting in a higher α. Table 1 shows the mechanical properties and KRs of 110 pile segments. The results

DOI: 10.1201/9781003348450-277

Table 1. Average mechanical properties of the tested 110 pile segments. In brackets the standard deviation is reported.

	Avg. ρ_{dry}	Avg. m. c.	Avg. KR Φ_2	Avg. $f_{c,0,meas,wet}$
Segments	kg/m^3	%	-	MPa
Head	400 (43)	87 (14)	0.14 (0.07)	16.7 (3.4)
Middle	395 (24)	78 (10)	0.21 (0.05)	16.3 (1.7)
Tip	390 (25)	80 (10)	0.28 (0.08)	14.9 (1.9)
All	395 (33)	82 (12)	0.22 (0.09)	16.0 (2.5)

showed an average decrease of $f_{c,0,meas,wet}$ from the head to the tip. This can be attributed to higher KRs and to a larger presence of juvenile wood in tips. The compressive strength values obtained from discs with and without knots (although on a limited amount of tests), were similar to the strength values of segments with and without knots.

4 ANALYSIS

Based on the comparison between $f_{c,0,meas,wet}$ and KR, the size of Φ_1 is not sufficient to describe a zero strength zone caused by a knot. Thus, Φ_2 was calculated according to Equation 1, with a factor $\beta = 0.462$. The equivalent zero strength zone of a knot which is assumed to take up no load, is defined by Φ_1 plus approximately half of Φ_1. The discs sawn from a clear wood part of the pile, showed $f_{c,0,CW,meas,wet}$ values which could be adopted, in combination with KR (Φ_2), to predict the failure at any position of a full-length pile. Thus, for the parts without knots it is assumed that this basic strength is related to the dry density, based on the research of Ravenshorst (2015). On this basis, a linear regression analysis was conducted to determine $f_{c,0,CW,mod,wet}$ (Eq. 4) with reference to dry density (for a m.c. of 0%).

$$f_{c,0CW,mod,wet} = C_1\, \rho_{dry} + C_{k,1} + \varepsilon_{k,1} \tag{4}$$

where C_1 = experimental factor; ρ_{dry} = dry density; $C_{k,1}$ = experimental factor; $\varepsilon_{k,1}$ = error term in regression. The correlation between mechanical properties and knots of 110 pile segments was used to implement a prediction model including the knot ratio KR (Φ_2). Equation 4 was extended with the term ($1 - C_2$ KR) that takes into account the reduction of the active cross section (Eq. 5).

$$f_{c,0KR,mod,wet} = (C_1\, \rho_{dry} + C_{k,1})(1 - -C_2 KR) + \varepsilon_{k,2} \tag{5}$$

where C_2 = experimental factor; $\varepsilon_{k,2}$ = error term in regression. The prediction model (Eq. 4-5), was

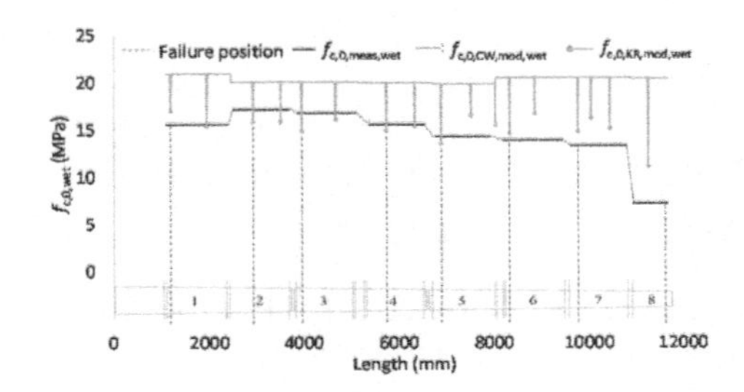

Figure 1. Profile of $f_{c,0,CW,mod,wet}$ calculated from dry density (m.c. = 0%) and $f_{c,0,KR,mod,wet}$ from KR (Φ_2), compared with failure positions of 8 segments along the full pile. The grey lines drawn on the pile represent the cutting positions of the segments.

calibrated on five specimens without knots, resulting in $C_1 = 0.028$, $C_{k,1} = 9.62$, $\varepsilon_{k,1} = 3.6$ MPa (Eq. 4); $C_2 = 1$, $\varepsilon_{k,2} = 3.7$ MPa (Eq. 5). This resulted in Equation 6, representing $f_{c,0,KR,meas,wet}$ (Figure.7).

$$f_{c,0KR,mod,wet} = (0.027\rho_{dry} + 10.1)(1 - -1KR) + \varepsilon_{k,2} \tag{6}$$

The formulated prediction model was applied on a full-length pile divided in eight segments (Figure 1), where $f_{c,0,CW,mod,wet}$ and $f_{c,0,KR,mod,wet}$ were compared with $f_{c,0,meas,wet}$. The values of $f_{c,0,KR,mod,wet}$ were calculated based the respective KR (Φ_2) values along the pile.

5 CONCLUSIONS

In this work, a correlation was found between $f_{c,0,meas,wet}$ and the area where fibers were deviated by knots, by calculating a knot ratio KR, expanding on the research of van de Kuilen (1994). It was concluded that the equivalent size of the influence of a knot is about 1.5 its diameter, with the assumption that the equivalent area of knots do not contribute to the strength of the pile. The $f_{c,0,meas,wet}$ of segments was comparable to that of the relative discs. Based on this, it was possible to reliably predict the strength of pile segments from discs with maximum KR, confirming the assumption that the knot area governs the resistance to axial loading. This allows to identify the critical sections along a pile by using KR. A prediction model was implemented to estimate the decrease in compressive strength based on the dry density and KR of a pile. The developed model can have in-situ future implications to predict the failure positions of wooden foundation piles in use, by determining their critical sections, opening up the opportunity of a more optimal use of these structural elements.

REFERENCES

Ravenshorst G.J.P. 2015. Species independent strength grading of structural timber, Doctoral Thesis.

Van de Kuilen J.W.G. 1994. Bepaling van de karakteristieke druksterkte van houten heipalen. TNO-report 94-CON-R0271, 1994 (in Dutch).

Current Perspectives and New Directions in Mechanics, Modelling and Design of Structural Systems – Zingoni (ed.)
© 2022 Copyright the Author(s), *ISBN 978-1-032-39114-4*

The effect of wood microstructure on the mechanical properties of some selected tropical hardwood species used in construction

J.O. Akinyele, A.B. Folorunsho & U.T. Igba
Department of Civil Engineering, Federal University of Agriculture, Abeokuta, Nigeria

P.O. Omotainse
Department of Agricultural Engineering, Federal University of Agriculture, Abeokuta, Nigeria

J.O. Labiran
Department of Civil Engineering, University of Ibadan, Nigeria

ABSTRACT: This work aims to research the effect of wood microstructures on the mechanical properties of some commonly used tropical hardwood in the construction industry. Five kinds of wood were selected, which included: African Mahogany (*Afzelia africana*), West African Albizia (*Albizia zygia*), African Birch (*Anogeissus leiocarpus*), Beech Wood (*Gmelina arborea*), Salt and Oil (*Cleistopholis patens*). None of these woods was listed among the tropical hardwoods in BS 5268, hence the need to study these species. Both the mechanical (Compressive, Tensile, Static bending and Brinell hardness) and physical (Moisture content and density) properties were determined, while Scanning Electron Microscopy (SEM) analysis was used to study the wood microstructures. The microstructural analysis revealed that the size and bonds between the vessels and lignin of the individual wood contributed to the variation in mechanical properties.

1 INTRODUCTION

Wood is one of the few natural and renewable construction materials that exists but has its limitations in general use for construction, carpentry and upholstery. It is a complex building material owing to its heterogeneity and species diversity. Wood does not have consistent, predictable, reproducible and uniform properties as the properties vary with species, age, soil and environmental conditions (Kliger, 2016).

There are different species of both the hardwood and softwood spread all over the tropical regions of the world. The mechanical properties of these wood are mostly responsible for their strength and use.

One of the very important properties of engineering material is its microstructures. The mechanical and microstructural properties of hardwood were investigated by Wikete *et al.* (2010), the microstructure-stiffness relationship for hardwood was observed through macroscopic mechanical experiments. Particularly, the large cylindrical vessels that run through the stem direction, the work was able to classify hardwood into 'ring-porous' and diffuse-porous' as a result of what was observed in the experiment.

This research aims to investigate the effect of wood microstructures on the mechanical properties of some selected tropical hardwood. Since several of these tropical hardwood species was not been listed in Table 14 of BS 5268-2 (2002). The need to study these five tropical species arose because they are readily available and commonly used for construction and furniture purposes, but with very little information on their mechanical and microstructural properties to the best of our knowledge.

2 EXPERIMENTAL SET UP

Five samples of mature hardwood species were obtained from Malaaka village, in Abeokuta, Ogun State Nigeria. The local and botanical names for each tree species are African Mahogany (*Afzelia africana*), West African Albizia (*Albizia zygia*), Salt and Oil wood (*Cleistopholis patens*), African Birch (*Anogeissus leiocarpus*) and Beechwood (*Gmelina arborea*).

2.1 *Sawing and seasoning*

The sawn wood was arranged in an open environment protected from rain and the ground. It was arranged in such a way that air freely circulate it and was left in this state for 2 months (air seasoning).

2.2 *Moisture contents*

The sample was weighed after the period of seasoning and then dried in an oven at a temperature of 103 ± 2 °C (217 ± 4 °F) for 3 days until the weight was constant. The loss in weight expressed as a percentage of the final oven-dry weight was taken as the moisture content of the specimens

2.3 *Mechanical properties*

The determination of various mechanical properties was carried out following BS 5268-2 (2002). The strength tests (Compressive, tensile, Bending and Brinell hardness) were carried out with the aid of a Universal Testing Machine

DOI: 10.1201/9781003348450-278

(Testometric material testing machines) with a maximum load capacity of 100 kN at the National Centre for Agricultural Mechanization (NCAM) Ilorin, Nigeria.

Figure 1. Bending strength test.

2.4 *Scanning Electron Microscopy analysis (SEM).*

The samples were cut from each wood species (along the longitudinal direction or parallel to the grains) at 5 mm thickness and about 10 mm x 10 mm width and breadth; the fractured small samples were, firstly treated with carbon before mounting on the SEM stubs. Images from polished surfaces of these wood species fragments allow evidence of the physical arrangement of the wood fiber.

3 RESULTS AND DISCUSSIONS

3.1 *Mechanical and physical test results*

Pure tension tests in the grain direction showed that the force-elongation relationship was not linear up to failure for all the wood samples (Figure 2), while other mechanical and physical test results are shown in Table 1.

3.2 *SEM analysis results*

The microstructure of wood is generally known to consist of lignin and fibers. (Figures 3) shows the microstructures of the different woods at a magnification of (682 x) for each specimen. The figures showed the spongy shape grains, cut paralleled to each other to reveal the fiber structures of the specimens.

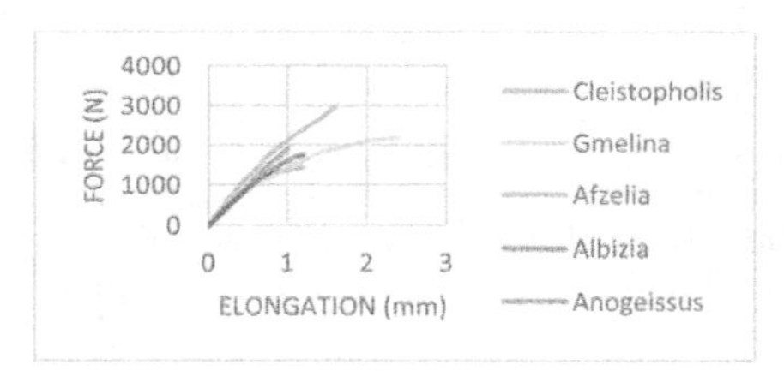

Figure 2. Tension parallel to grain.

The results obtained for *Anogeissus leiocarpus* showed that it is the best hardwood species from this experiment and the results are almost in agreement with the findings of Bello and Jimoh (2018) who obtained a compressive strength parallel to grains at 57. 15 MPa, tensile strength of 128.08 MPa, Modulus of rupture (MOR) 108.96 MPa, but a density of 1150 kg/m^3 at a moisture content of 10.1%. The difference in mechanical and physical properties may be attributed to the difference in moisture content. However, Bello and Jimoh (2018) did not study the microstructural properties.

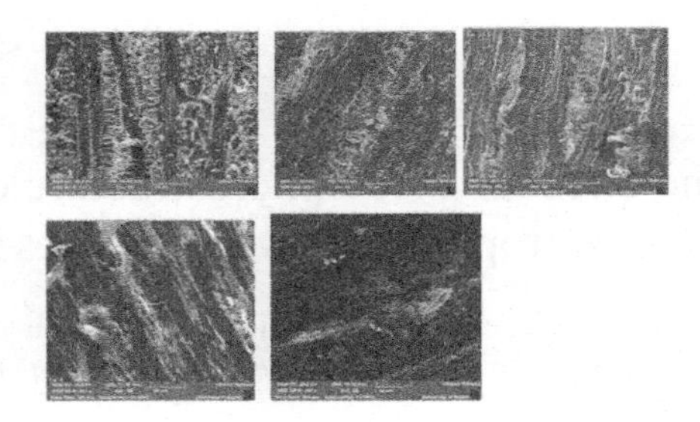

Figure 3. Parallel fiber of (*Afzelia Africana, Cleistopholis patens, Albizia zygia, Gmelina arborea and Anogeissus leiocarpus).*

Table 1. Mechanical and physical properties of wood samples.

Wood Type	Comp. stress MPa	Tens. stress MPa	Hardness kg/m^2	Density kg/m^3	MOR MPa	Moisture content (%)
Afzelia africana	41.46	165.54	1984.72	821.67	90.27	16.47
Albizia zygia	47.82	98.89	2942.16	1033.3	79.03	22.07
Anogeissus leiocarpus	53.77	106.37	4308.67	1030.3	107.3	20.79
Cleisto-pholis patens	32.36	79.28	2089.73	784.17	76.49	18.90
Gmelina arborea	27.44	122.78	1374.05	783.33	37.57	27.58

4 CONCLUSION

This work has investigated the microstructures of some selected tropical hardwoods that were not listed by BS 5268; this investigation has shown that the arrangement and orientation of the wood microstructures contributed to both the physical and mechanical properties of wood. The microstructure, through the SEM images, has revealed the importance of wood lignin and vessel/pore sizes to the properties of each wood species.

REFERENCES

Bello A.A. and Jimoh A.A. 2018. Some Physical and Mechanical Properties of African Birch Timber. *Journal of Applied Science and Environmental Management*, 2018; 22 (1) 79–84.

BS 5268 – 2. Mechanical use of Wood: Code of practice for permissible stress design, materials and workmanship. British Standard Institution, London. 2002.

Kliger R. (2016) Introduction to design and design process: Design of Wood structures; (1) 8–25. Ed. Eric Borgstrom: Swedish Forest industries federation, Swedish wood. Stockholm.

Wikete C. Bader T.K, Jäger A, Hofstetter K., and Eberhardsteiner J. (2010). Mechanical properties and microstructural characteristics of hardwood. 27th DANUBIA-ADRIA SYMPOSIUM on Advances in Experimental Mechanics, Vienna University of Tech.

Current Perspectives and New Directions in Mechanics, Modelling and Design of Structural Systems – Zingoni (ed.)
© 2022 Copyright the Author(s), ISBN 978-1-032-39114-4

Mechano-sorptive behaviour on crack propagation of notched beams of Okume

M. Asseko Ella
Clermont Auvergne Université, CNRS, Clermont Auvergne INP, Institut Pascal, Clermont-Ferrand, France

G. Goli
University of Florence, DAGRI-Department of Agriculture, Food, Environment and Forestry, Firenze, Italia

J. Gril
Clermont Auvergne Université, CNRS, Clermont Auvergne INP, Institut Pascal, Clermont-Ferrand, France
Université Clermont Auvergne, INRA, PIAF, Clermont Ferrand, France

R. Moutou Pitti
Clermont Auvergne Université, CNRS, Clermont Auvergne INP, Institut Pascal, Clermont-Ferrand, France
CENAREST, IRT, BP, Libreville, Gabon

ABSTRACT: The paper presents the experimental results of the mechano-sorptive induced cracking process on notched beams of Okume. The beams tested have dimensions equal to 160x12x60mm and a notch length of 20mm. The tests were carried out in a temperature and humidity-controlled room. These specimens were tested in 3-point bending for 7 days under an initial stress rate of 80% of the failure stress. At the end of the 7 days, further small stress increments were added to the previous one to accelerate the cracking process. The tests were carried out under a sorption cycle of 45-75% relative humidity and at constant temperature of 20°C. Monitoring of cracking parameters, including crack propagation and crack opening, was carried out using a USB microscope. The moisture content of the specimens was monitored by mass monitoring of control specimens of the same size as those tested. The results show that crack propagation is influenced by viscoelastic and mechano-sorptive effects on the one hand and by mechanical loading effects on the other hand. Crack propagation is greater when mechano-sorptive effects occur, especially during the drying process.

1 INTRODUCTION

The current economic and environmental context is a major advantage for timber constructions, which are otherwise dominated by steel and concrete constructions. Unfortunately, in their long-term or short-term structural use, the interactions between mechanical and environmental stresses on these timber structures can strongly modify their mechanical behaviour. This makes their implementation more complex and can eventually lead to a shortened service life. The objective of this work is therefore to study the influence of mechano-soptive and viscoelastic effects on wood crack to better understand the contribution of moisture content on wood crack. To this end, we carried out creep and crack tests on Okume specimens under controlled relative humidity cycles and at a constant temperature of 20°C.

2 MATERIAL AND METHODS

2.1 *Viscoelastic and mechano-sorption tests*

We performed viscoelastic and mechano-sorptive creep test in 3-point bending on the beams. (Figure 1). Each beam is equipped with 3 transducers, 2 at the sides and in the center (Figure 2). The transducers at the sides (T_L and T_R) are used to estimate the shrinkage and swelling on the one hand and to correct the deflection at the center on the other.

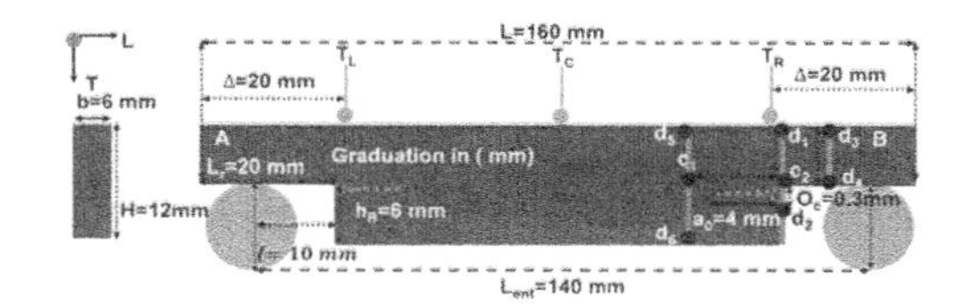

Figure 1. Geometry of the tested beams.

The central transducer (T_C) is used to estimate the displacements in the center of the beam. At the two ends of the beams (A) and (B), we made notches of length (Lr) of 20mm and height (h_R) of 6mm corresponding to half the height of the beam. The tests are carried out in a plastic box covered with a glass pane, this box is placed in a climatic room (Figure 2a) where the temperature and relative humidity are kept constant at 75% and 20°C as shown by the external control system

DOI: 10.1201/9781003348450-279

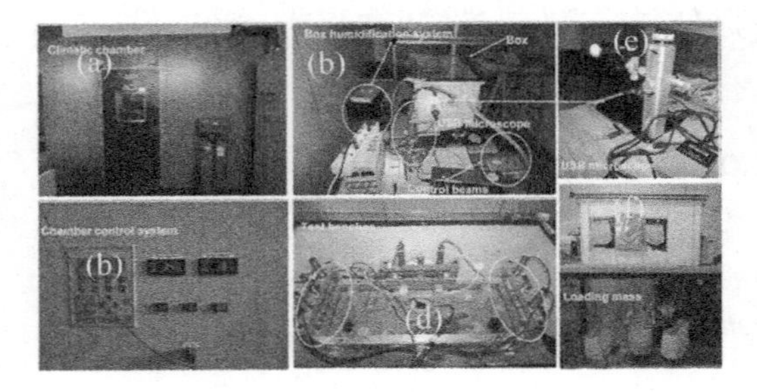

Figure 2. Experimental set up; (a) Climatic chamber; (b) chamber control system; (c) experiment set up; (d) test benches; (e) USB microscope; (f) loading for creep test.

(Figure 2b). The box (Figure 2c) is equipped with 3 mini-3-point bending benches with 3 transducers (Figure 2d). The box is humidity controlled by a humidity pump. The moisture content of the specimens is monitored by successive weighing of the control specimens. The monitoring of the crack parameters, in particular the length and opening of the crack, was done by taking images with an electronic USB microscope (Figure 2e). The test specimens were loaded in the wet state under constant stress corresponding to 80% of the stress at failure under a first hydric cycle of 45 to 75% relative humidity (RH). After this first cycle, if no crack was observed, additional loads were added to the initial loading to enhance the crack process and beam failure. After this overloading we continue with a second moisture cycle from 45 to 75% RH.

3 RESULTS

3.1 *Effects of mechano-sorptive viscoelastic behaviour on crack propagation*

In a first step we show the strain due to the mechano-sorptive and viscoelastic behaviour effect on the crack propagation of the tested beams until the rupture of the test (Figure 3a). In parallel, we present the evolution of the experimentally measured moisture content (MC_Exp) and the one we estimated (MC_Int) in (Figure 3b). Thus, these figures show the propagation of the crack increment under the effects of viscoelastic and mechano-soprtive strain of the Okume specimens during the first cycle and the beginning of the second sorption cycle. The viscoelastic and mechano-sorptive strain were estimated by the following Equation 2.

$$\varepsilon = \frac{2 \cdot 3 \cdot H}{1 + 7 \cdot \lambda^3} \frac{y_c - y_{RL}}{L_{ent}^2} \tag{1}$$

With $L\lambda = 2l/L_{ent}$ y_{RT} is the average of the deflections measured by the transducers on the T_R and T_L sides and the deflection at the center measured by the T_C transducer. In these figures it can also be seen that the instantaneous loading of the beams leads to crack initiation. The crack propagation tends to develop on one side of the beam while it remains almost constant on the opposite side, this is the case in (Figure 3a). In general, it appears that drying favours crack propagation in contrast to humidification where there is no crack. This observation confirms the results of Pambou et al. (2019).

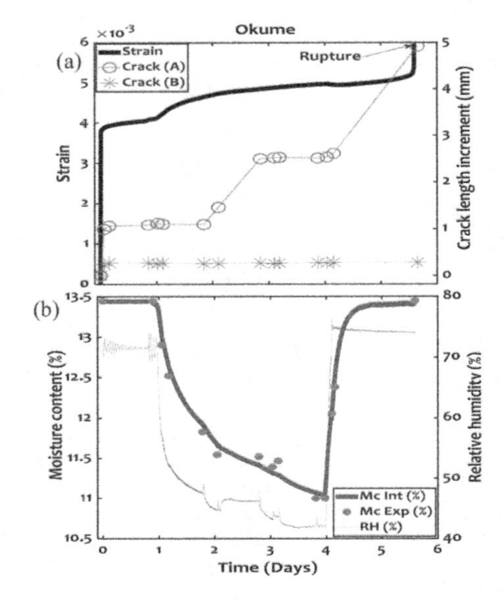

Figure 3. Impact of crack propagation on the strain; (a) Strain; (b) Moisture content.

4 CONCLUSION

This paper presents the study of the effects of mechano-sorptive and viscoelastic creep on wood crack. The study is done on notched specimens of Okume. These specimens were tested in the wet state and initially loaded to 80% of the rupture force for 6 days under a sorption cycle of (45% and 75% RH) before being loaded to 100% of the rupture force in a second stage. The monitoring of the crack parameters was done with a USB microscope. The results show that the mechano-sorptive effects accentuate the propagation of cracks and that the coupling of the two effects, namely crack and the mechano-sorptive effect, accelerates the strain of the wood until it breaks. Regarding the effect of drying on crack propagation, our observations agree with those of the literature. On the effect of drying on strain, we have however noticed that drying accentuates strain in the case of a mechano-symmetric creep test, contrary to a creep test in an uncontrolled environment where the effect of drying on strain is reversed.

REFERENCES

Pambou Nziengui, C. F. *et al.* (2019) 'Notched-beam creep of Douglas fir and white fir in outdoor conditions: Experimental study', *Construction and Building Materials*. Elsevier Ltd, 196, pp. 659–671. doi: 10.1016/j.conbuildmat.2018.11.139.

Current Perspectives and New Directions in Mechanics, Modelling and Design of Structural Systems – Zingoni (ed.)
© 2022 Copyright the Author(s), ISBN 978-1-032-39114-4

Link between growth strategies and physical-mechanical properties of wood of tropical species from Gabon

E. Nkene Mezui
CIRAD, UPR BioWooEB, University Montpellier, Montpellier
Université Clermont Auvergne, Clermont Auvergne INP, CNRS, Institut Pascal, Clermont Ferrand

L. Brancheriau & D. Guibal
CIRAD, UPR BioWooEB, University Montpellier, Montpellier

R. Moutou Pitti
Université Clermont Auvergne, Clermont Auvergne INP, CNRS, Institut Pascal, Clermont Ferrand

ABSTRACT: Gabon is a tropical country that still has vast forest areas, covering more than 80% of its territory. These forest areas contain a great diversity of tree species that are still little studied today, especially with regard to the different growth strategies of the species in relation to their properties. Indeed, current studies on temperate zone species seem to indicate that there are significant differences between the different ecological temperaments of trees with regard to their properties. The objective of our study was to highlight these differences by analyzing some physical-mechanical properties of a panel of 49 tropical species from Gabon from the CIRAD database. The results show significant differences according to ecological temperament. Sciaphilous species show better mechanical properties but higher hydric deformations. These results would be related to variations in the anatomical and chemical characteristics of the wood.

1 INTRODUCTION

During their growth, trees adopt different strategies necessary for their survival depending on their environment and their competitors. As highly variable living organisms immersed in sometimes constraining environments, their choice of strategies will have an impact on the quality of the wood they produce (Larson 1962). The quality of wood can be determined by analyzing its physical, mechanical and chemical properties, for example. Tree growth strategies include shade-tolerant (sciaphilous), hemi-tolerant (semi-heliophilous) and shade-intolerant (heliophilous) species (Blanc et al., 2003). Nowadays, only a few studies have been conducted, but the majority of them concern temperate tree species, a handful concern a few tropical species and only one study to our knowledge concerns two tropical species from Gabon (Ondo et al., 2021). Gabon has a high diversity of species and would benefit from more studies in this area. For this reason, we selected 49 tree species of Gabonese origin divided into heliophilous, semi-heliophilous and sciaphilous temperaments. The aim was to highlight any differences between these ecological profiles by analyzing the physical and mechanical properties of their wood.

2 MATERIALS AND METHODS

The plant material used in our study consists of 49 tropical species of Gabonese origin and divided into three groups of balanced numbers: heliophilous (18 species), semi-heliophilous (15 species) and sciaphilous (16 species). The properties studied are taken from the CIRAD wood database (Gerard et al., 2017). They were chosen for their technological interest (some properties are rarely studied) and because they are related to anatomical, structural, hydraulic and chemical traits of wood. These traits are characteristics of different trees growth strategies. The tests used to determine these properties are normalized (Sallenave, 1955). The tests are carried out on wood stabilized in a climatic chamber at a temperature of 20 °C ± 2 °C and a relative air humidity of 65% ± 5%.

3 RESULTS AND DISCUSSION

We present the boxplot results of three significant properties: density, shrinkage ratio and tensile strength. The discussion will be between the Sciaphilous and Heliophilous groups with the exception of density.

Figure 1 above shows the boxplots of the density variable ρ (in kg/m^3) which is related to the porosity of the material (proportion of fibers in the wood) for each ecological temperament. The means are 714.7, 658.65 and 729.15 for the heliophilous, semi-heliophilous and sciaphilous respectively. It can be observed that the highest average is for the sciaphilous species group. This means that the higher the density, the more sciaphilous rather than semi-heliophilous the species are. From the literature, we know that fast-growing (heliophilous) species are less

DOI: 10.1201/9781003348450-280

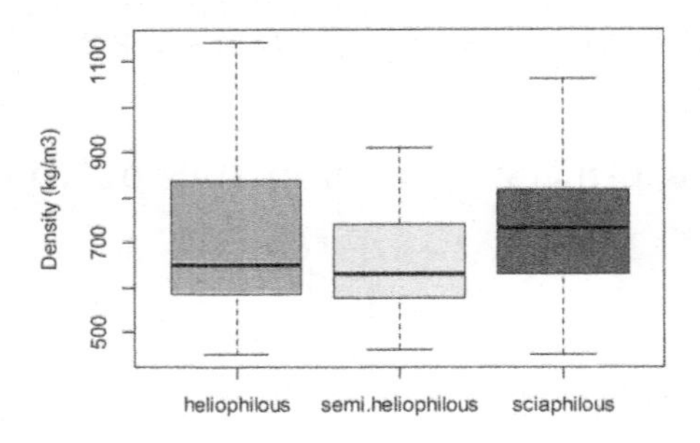

Figure 1. Density distribution according to ecological temperament.

dense than slow-growing (sciaphilous) species (Poorter et al., 2012). Since semi-heliophilous species are intermediate in temperament, it is logical that they are less dense than sciaphilous species and therefore less hard.

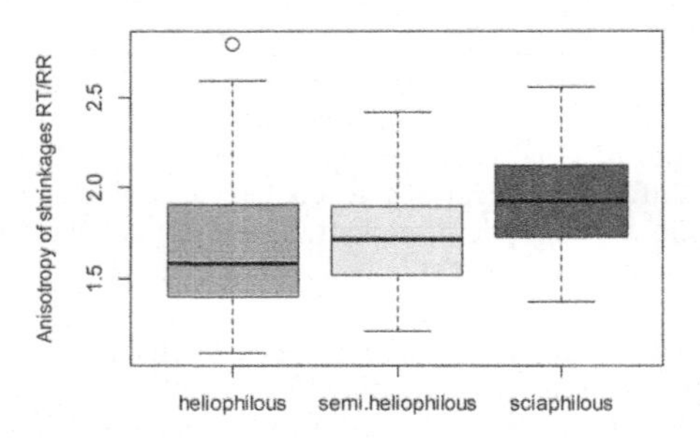

Figure 2. Distribution of the quotient R_T/R_R by ecological temperament.

Figures 2 above shows the results of the R_T/R_R shrinkages ratio for each temperament. This ratio quantifies the deformation of a piece of wood during moisture variations. The closer this ratio is to 1, the more stable the wood is. If, on the other hand, it tends towards a value greater than or equal to 2, this means that a species is sensitive to deformations. The results of the averages of this ratio are respectively 1.68, 1.72 and 1.93, all close to 2. These values close to 2 reflect a very pronounced anisotropy of shrinkage. Sciaphilous species seem to be more sensitive to hydric deformations (Figure 2) and the results of the previous total tangential shrinkage are in line with this, considering the low total radial shrinkage R_R of about 4%.

Figure 3 present the perpendicular tensile strength as a function of ecological temperament. These is one property rarely discussed in the literature. The tensile strength perpendicular to the fibers is the force that tend to separate the fibers from each other. This stress should generally be avoided. The boxplots in Figure 3 indicate that sciaphilous species would be more resistant to perpendicular traction with average values of around 2 MPa consistent with the literature for the three profiles, indicating however a very low strength (Guibal et al., 2015). An analysis of the fibers and the lignin content would allow a better interpretation of these results.

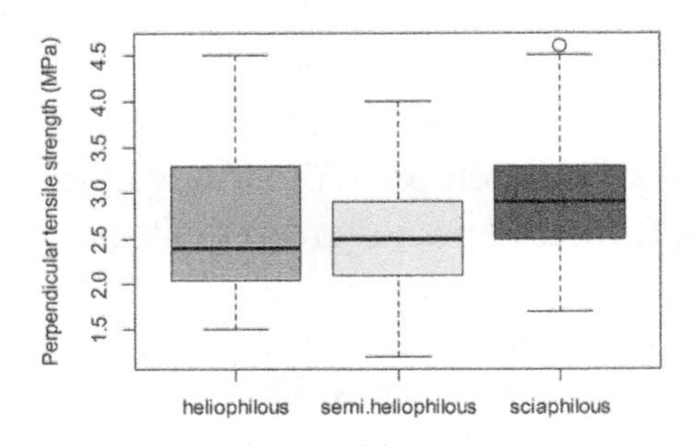

Figures 3. Perpendicular tensile strength as a function of ecological temperament.

4 CONCLUSION

The objective of our study was to highlight the differences between the ecological profiles of the trees, by analyzing some physical and mechanical properties of their wood. The shade-tolerant (sciaphilous) species seem to have better tensile strength than the intolerant (heliophilous) species. Although they also seem to be more vulnerable to hydric deformations. This would be related to the characteristics of the fibers and the lignin content. Additional anatomical and chemical tests must be carried out to allow a better interpretation of our results.

ACKNOWLEDGEMENTS

Our sincere thanks to the Association Academic of Civil Engineering (AUGC) for the financial support given to us to take part in this congress.

REFERENCES

Blanc, L., Florès, O., Molino, J.F., Gourlet-Fleury, S. & Sabatier, D. 2003. Specific diversity and grouping of tree species in the Guiana Forest. French Forestry Review, *55* (SPEC.ISS.), 131–146.

Gérard, J., Guibal, D., Cerre, J.C. & Paradis, S. 2017. *Tropical timber atlas: Technological characteristics and uses*. Editions Quae.

Guibal, D., Langbour, P. & Gérard, J. 2015. Physical and mechanical properties of wood (in French). In: Memento of Forestier tropical. Mille Gilles (ed.), Louppe Dominique (ed.). Versailles: Ed. Quae, 873–884. ISBN 978-2-7592-2340-4.

Larson, P.R. 1962. A biological approach to wood quality. *Tappi*, *45*, 443–448.

Ondo, J.L.Z., Ruelle, J., Dlouhá, J. & Fournier, M. 2021. Characterization and variability of physical and structural properties of kevazingo, *Guibourtia tessmannii*, and okume, *Aucoumea klaineana*, wood from the natural forests of Gabon (in French). WOODS & FORESTS OF THE TROPICS, *347*, 43–59.

Poorter, L., Lianes, E., Moreno-de las Heras, M. & Zavala, M.A. 2012. Architecture of Iberian canopy tree species in relation to wood density, shade tolerance and climate. *Plant Ecology*, *213*(5), 707–722.

Sallenave, P. 1955. Physical and mechanical properties of French Union tropical woods (in French). Nogent-sur-Marne: CTFT, 128 p. (Publication of the Tropical Forestry Technical Center, 8).

Current Perspectives and New Directions in Mechanics, Modelling and Design of Structural Systems – Zingoni (ed.)
© 2022 Copyright the Author(s), ISBN 978-1-032-39114-4

High performance light timber shear walls and dissipative anchors for damage limitation of wooden buildings in seismic areas

V. Wilden, G. Balaskas, B. Hoffmeister & L. Rauber
Institute of Steel Construction, RWTH Aachen University, Germany

B. Walter
Walter Reif Ingenieurgesellschaft, Germany

ABSTRACT: This contribution describes the development and the experimental investigation of innovative enhanced light weight timber frame buildings, which can be a very efficient structural solution for wooden multistory buildings. This development concepts comprise high-performance timber frame walls, also named "powerwalls" with variable arrangement of sheathing planes and dissipative anchorings at the base of the timber buildings, which can limit damages of the timber structure and can be replaced after a significant earthquake. In order to testify the improved performance of the developed elements in the context of a real building, a full scale two storey building core was designed, erected (in the laboratory) and tested experimentally for its seismic performance. The results of the tests are presented. Observations during the test also reveal some aspects of the design and detailing which need consideration in the application of the timber walls in the context of the complete building.

1 INTRODUCTION

Light-weight timber frame elements represent an efficient structural solution for wooden buildings. The tendency towards multi-storey structures leads to challenges regarding the stiffness, strength and ductility of the buildings and the accurate predictability of them. As the number of storeys grows, the requirements not only for fire protection, but also for stiffness increase, especially for structures exposed to lateral actions, such as wind or earthquake loads. In order to construct buildings entirely in wood, high performance timber walls are required for structural and seismic performance. The special challenge in developing the new high-performance wall, is to increase the resistance and stiffness on the one hand and simultaneously take advantage of ductile and dissipative properties of wooden light frame elements. To reach this aim, the development and investigation of high-performance wall elements and their crucial parts of timber frame elements were conducted. Tests on connections (Wilden & Hoffmeister 2020), tests on anchoring systems (Balaskas et al. 2021), on wall elements and tests on one full scale two storey building core were performed, to identify the influence on different scale.

2 WALL ELEMENTS

Basically, the light-weight timber element consists of a solid timber frame stiffened by a sheathing, which provides diaphragm action in wall and floors. The conceptional idea of the enhanced wall element was to introduce an additional shear layer which acts independently (no common fasteners to multiple sheathing) and increases its stiffness, resistance and ductility. An end rabbet at the inner edge on one side of the frame is cut and the additional shear layer is arranged.

The test set-up was developed in order to allow for testing of wall elements under monotonic or cyclic loading and extended to include the simultaneous application of gravity loads. In total 13 different tests on conventional and enhanced wall elements were conducted and evaluated regarding strength, stiffness and ductility. The test on wall elements led to the following conclusion:

- Initial stiffness of conventional walls is remarkably higher than estimated by (Blaß et al. 2004) – this might lead to an underestimation of the spectral seismic acceleration.
- The enhancement by the third OSB-layer increased the resistance by 37%.
- The results for ductility depend very much on the evaluation method, the performance observed in the tests, however, indicate high ductility. Low cyclic degradation was observed, due to irreversible damage of the connection between sheathing and frame of changing load directions.
- The comparison between conventional and enhanced wall elements shows no significant different in the equivalent viscous damping values.

DOI: 10.1201/9781003348450-281

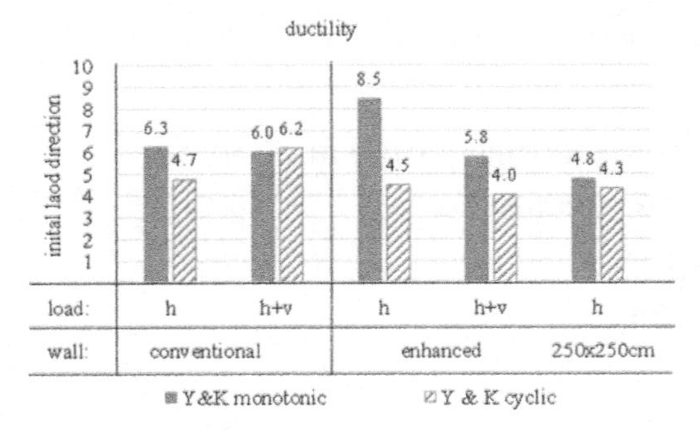

Figure 1. Ductility as results of experimental tests.

3 ANCHORING SYSTEMS

Another arrow in the quiver to increase the ductility and dissipation capability of the enhanced wall element is to arrange an appropriate novel ductile and replaceable anchoring system. Therefore, investigation on a new anchoring system was conducted, (Balaskas et al. 2021). The supporting and anchoring detail is made of steel, behaves nonlinear in a ductile manner under severe seismic loads and provides additional energy dissipation capacity. The anchoring device is designed to transfer only vertical loads. Shear forces should be resisted by additional shear keys or angle profiles. It consists of a steel rectangular hollow section (RHS) and a steel plate, welded to the upper flange of the hollow section. Energy dissipation occurs only under tensile loading of the anchorage in case of uplift. If designed appropriate, the anchors serve as initial dissipating elements and allow until a certain seismic load level limitation of plastic deformations to the anchors only. The anchors act as a dissipative mechanism also limiting the forces transferred into the timber building. Consequently, the timber members remain undamaged for moderate earthquakes and slightly damaged in case of strong seismic events. The anchors can be exchanged after a significant earthquake

4 FULL SCALE TWO STOREY BUILDUNG

In order to prove the effective performance of the developed elements in the context of a real building, a full scale two storey building core was designed and erected in the laboratory. The core consisted of two storeys. To exclude a soft story failure during the tests, the timber walls in the main direction of the ground floor were designed with the new enhanced wall elements, while the walls on the first floor were designed with conventional wall elements. The connections between floor element, made of CLT, and the walls were overdesigned, in order to reassure elastic response during the load transfer between the floors. Dynamic tests to determine the dynamic characteristics and push-over tests to determine the initial stiffness, the strength and the ductility of the core were conducted with different anchoring systems.

From the observations during tests, the following conclusions can be drawn:

- The measured frequencies, determined by dynamic tests, were greater than the predicted values. A higher initial stiffness of the core was observed and the assumption, that the additional secondary wall elements, which were connected to the wall elements in main direction, contributes to the initial stiffness.
- The ultimate resistance in push-over-test was lower than the determined resistance based in test on wall elements. The reason was premature buckling and detachment of initial sheathing plates due to contact induce restraint of the rotational movement of the OSB-sheathing.

5 CONCLUSION

All tests confirmed the very good performance of light-weight-timber buildings under seismic loads. It was, however, observed, that the numerical predictions of strength according to EC 5 lead to a significant underestimation of the overstrength of of the full-size walls. Test on full scale two storey building underline the need of detailing according to the assumptions for the design of the individual walls.

The authors are grateful for the financial support by BMWI, which funded the ZIM research project and for cooperation with the project partners Walter & Adams.

REFERENCES

Balaskas, G. & Wilden, V. & Hoffmeister, B. 2021. Experimental numerical investigation of innovative ductile and replaceable anchoring systems for wood shear walls under seismic loads. *COMPDYN 2021 - Computational Methods in Structural Dynamics and Earthquake Engineering, Greece Athen, 27-30 June 2021*, Greece.

Blaß, H. J. & Ehlbeck, J. & Kreuzinger, H. & Steck, G. 2004. Erläuterungen zu DIN 1052: 2004-08. Entwurf, Berechnung und Bemessung von Holzbauwerken. *Hg. v. DGfH Innovations- und Service GmbH*. München.

Wilden, V. & Hoffmeister, B. 2020. Experimental analyses of innovative wood-shear walls under seismic loads. *IABSE Congress - Resilient Technologies for sustainable infra-structure, Christchurch, 2-4 September 2020*, New Zealand.

Current Perspectives and New Directions in Mechanics, Modelling and Design of Structural Systems – Zingoni (ed.)
© 2022 Copyright the Author(s), *ISBN 978-1-032-39114-4*

Creep tests on notched beams of silver fir wood (*Abies alba*)

A. Bontemps, G. Godi & E. Fournely
Université Clermont Auvergne, Clermont Auvergne INP, CNRS, Institut Pascal, Clermont-Ferrand, France

J. Gril
Université Clermont Auvergne, CNRS, Institut Pascal, Clermont-Ferrand, France
Université Clermont Auvergne, INRAE, PIAF, Clermont Ferrand, France

R. Moutou Pitti
Université Clermont Auvergne, CNRS, Institut Pascal, Clermont-Ferrand, France

ABSTRACT: Silver fir is an important species in the wood french industry. Many works have been done to find new economic opportunities and to overcome the problems encountered in the use of this species. This paper presents the results of creep tests made on partially dried silver fir notched beams. The central deflection and the crack length are measured. The results show two types of creep evolution where neither the initial mechanical characteristics nor the crack propagation are to blame. The current hypothesis is sun exposure.

1 INTRODUCTION

Silver fir (*Abies alba*) is an emblematic European species, accounting for 8% of the French forest and is appreciated by carpenters for its mechanical properties. However, drying this type of wood is a sensitive issue due to the presence of water pockets in heartwood and a high average initial moisture content. Understanding and forecasting the mechanical evolution of green silver fir wood beams under outdoor sheltered conditions would help at developing new building applications, using poorly dried silver fir. Agricultural buildings with good ventilation, for instance, might avoid some of the constraint imposed on civil buildings.

This paper presents the results of 4-points bending creep tests on notched beams of silver fir wood that were conducted during summer 2021. These experiences are scheduled in a project studying the long-term mechanical behaviour of silver fir wood at different moisture content. As a consequence, the beams studied were partially dried or even still in a green state.

2 MATERIALS AND METHODS

The 4-points bending tests conditions comply with (AFNOR). Six notched beams made of silver fir wood were loaded on the test benches, these are beams S03, S04, S07, S14, S12 and S21. The dimensions of the notched beams and the conditions of the 4-points bending creep tests are shown in Figure 1.

The initial properties of the beams were measured, including basic density, moisture content and specific modulus in dynamic bending. The initial loads applied to the creep tests are 345kg concrete beams. The maximum force F_c, i.e. the force at which the first crack occur, was measured on two other beam of the same dimensions (S13 and S20). This allows to compute the load ratio Φ of each beam. This load ratio was calculated assuming proportionality of F_c to the basic density of the beam. The results of the initial characterisation of the beams are shown in Table 1, the moisture content MC was measured just before the creep loading.

The test benches are in a semi-sheltered outdoor environment. Semi-sheltered means that they are sheltered from rain but partially from solar radiation. Meteorological parameters are measured during the entire test period. For this purpose, a weather station is installed near the test benches, which measures, among other things, temperature, relative air humidity and solar radiation intensity. In order to control the duration of the tests, it was decided to increment the loading gradually. Each month, between 15 and 30kg is added to beams S03, S14 and S21. Throughout the tests, the central deflection was measured on all beams. Moreover a marker tracking method using ArUco markers (Garrido-Jurado et al. 2014) was developed for these tests. This method allows to measure displacements from photos of the beams to which

DOI: 10.1201/9781003348450-282

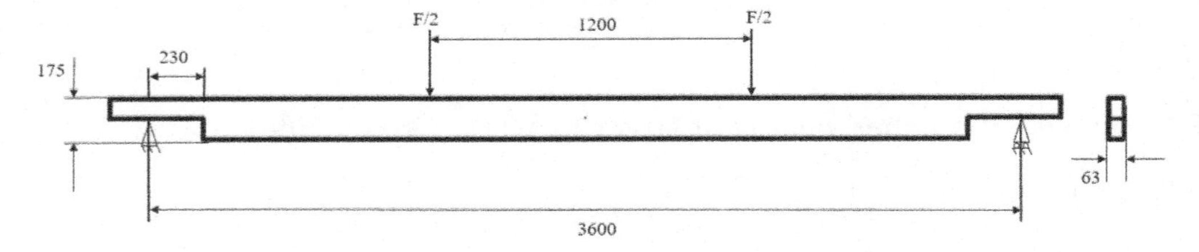

Figure 1. Dimensions of notched beams and conditions for the 4-point bending creep tests. All numerical values are in mm.

Table 1. Summary of initial properties.

	ρ_B [∅]	E_S^d $[GPa]$	MC [%]	Φ [%]
S03	0.37	28.7	21.6	90.3
S04	0.4	30.5	27.1	82.8
S07	0.34	28.8	49.5	97.4
S14	0.4	32.6	31.8	82.8
S12	0.38	39	34.1	86.6
S21	0.39	34.4	36.4	85.7
S13	0.42	28.1	16.6	/
S20	0.41	29.4	19	/

the markers are glued. The crack opening and crack length of the beams were measured with this method.

3 RESULTS AND DISCUSSIONS

3.1 *Measurements of instant central deflections*

Beams S04 and S07 broke during loading. Therefore, no interesting data could be extracted from their test. In Figure 2 the temporal evolution of the compliance is plotted. The compliance is computed using the equation derived by (Pambou Nziengui 2019):

$$J(\alpha, U_c(t), F(t)) = \frac{U_c(t)}{A \times f(\alpha) \times F(t)} \quad (1)$$

Where $U_c(t)$ is the central deflection at time t; $F(t)$ is the applied load at time t and $f(\alpha)$ is a function depending on the notch level α. This allows to compare the beams independently of crack propagation and load ratio. Two types of creep can be distinguished: much more deflection for S03 and S14 than for S21 and S12, especially in the first 60 days. It is worth noted that beam S14 have quite the same initial mechanical characteristics than S12 and S21, hence this is not the cause of this difference. One explanation may come from the fact that beams S03 and S14 were sometimes exposed to the sun while S12 and S21 remained in the shade for the entire duration of the tests. The solar radiation hitting S03 and S14 directly during a part of the day caused local drying, activating a mechano-sorptive effect.

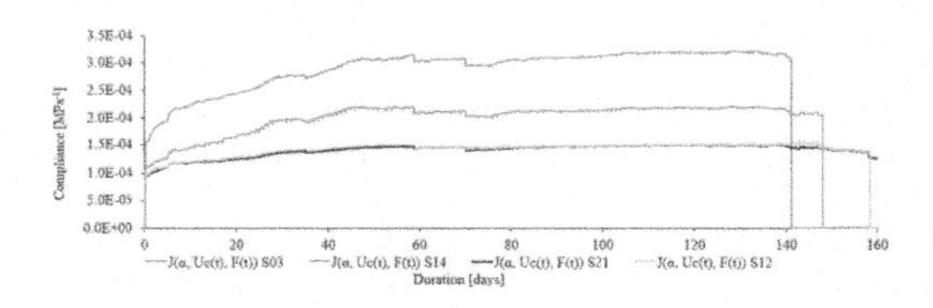

Figure 2. Temporal evolution of longitudinal compliance.

4 CONCLUSION

Six beams were experimented in 4-points bending creep tests during summer 2021. Two beams broke instantly. The temporal evolution of the compliance J was plotted for the four remaining beams. The calculation takes into account crack length and multi-stage loading. Nevertheless, the four beams left show two types of creep behaviour. Two beams, S03 and S14, were subjected to solar radiation and have undergone a significant higher creep than the two beams in the shade, S12 and S21. One explanation, that have to be demonstrated, is that the solar radiation accelerate a non linear mechano-sorptive effect on S03 and S14.

ACKNOWLEDGEMENTS

The authors would like to thank the Auvergne Rhône-Alpes region for the PhD scholarship as well as the fundings of the Hub-Innovergne program of the Clermont Auvergne University.

REFERENCES

AFNOR. NF EN 408+A1 - Structures en bois. Bois de structure et bois lamellé-collé.

Garrido-Jurado, S., R. Muñoz-Salinas, F. J. Madrid-Cuevas, & M. J. Marín-Jiménez (2014, June). Automatic generation and detection of highly reliable fiducial markers under occlusion. *Pattern Recognition 47*(6), 2280–2292.

Pambou Nziengui, C. F. (2019, July). *Fissuration du bois en climat variable sous charges de longues durées : applications aux essences européennes et gabonaises*. These de doctorat, Clermont Auvergne.

19. Structural health monitoring, Damage detection, System identification, Maintenance, Durability, Long-term performance

Current Perspectives and New Directions in Mechanics, Modelling and Design of Structural Systems – Zingoni (ed.)
© 2022 Copyright the Author(s), ISBN 978-1-032-39114-4

Structural Health Monitoring in civil engineering: Status and trends

G. De Roeck, D. Anastasopoulos & E. Reynders
Department of Civil Engineering, KU Leuven, Leuven, Belgium

ABSTRACT: For Structural Health Monitoring (SHM) many methods have been developed that use features extracted from vibrations under operational conditions. Most frequently derived features from modal analysis are natural frequencies, damping factors, mode shapes and possibly modal strains. Applications, SHM system design and recent trends in sensor technology will be highlighted. Finally, results are presented of a long term monitoring campaign on a steel bowstring bridge, illuminating the performance of optical strain sensors.

1 INTRODUCTION

Many civil engineering structures (e.g. bridges, dams, offshore platforms …) are reaching or exceeding their design life time. Although the annual maintenance cost is relatively small compared to the investment cost (less than 1%), the sum of the maintenance cost over the service lifetime is of the same order of magnitude as the investment cost. Early damage detection and localization allows maintenance and repair works to be properly programmed, minimizing the annual costs of repair and avoiding a long out of service time.

Vibration-based Structural Health Monitoring (VBSHM) has distinct advantages over other inspection methods, like e.g. periodic visual checks: the state of the structure can be continuously, even remotely, tracked. Moreover, internal damage, otherwise invisible, can be assessed. VBSHM is based on the principle that modal characteristics of a structure are like a signature of structural mechanics behavior. Special attention is being paid to techniques making use of operational data (ambient vibration testing).

2 EXCITATION SOURCES

For commissioning of new bridges usually static and increasingly also dynamic structural monitoring campaigns are prescribed. The deformation under a series of trucks with known weight is used to check with static design calculations. From dynamic measurements modal characteristics are extracted that are compared with their calculated counterparts.

A crucial aspect of VBSHM is the scope of the delivered modal information, which is strongly related to the excitation that produces the vibrations. Ambient vibration testing can be used for periodic inspection but also for continuous monitoring. On a bridge, traffic has not to be interrupted, so reducing not only the cost of the test itself, but also the economic impact of closing the bridge. The inherently present ambient excitation can be supplemented with a relatively small artificial excitation, like an impact from a hammer or a drop weight or a force generated by a pneumatic mechanical muscle. In the subsequent system identification, the combined (un-known) stochastic and (measured) deterministic force inputs are taken into account (Reynders & all., 2008). Big advantages are the extension of the suitable frequency interval and the possibility to obtain absolute scaling of those modes that are excited by the small artificial forces.

3 APPLICATIONS

A SHM system will deliver an experimental baseline model that can be used to validate and to calibrate a computational (FE-) model of the measured structure. When combined with a calibrated numerical model, the measured responses can also predict responses at possibly critical locations, like strains close to welds or forces in bolts (virtual sensing).

From measurements a digital, load independent signature can be obtained, like a set of modal characteristics. A change in this signature can be used to detect, localize and quantify damage. A well-designed SHM system can minimize the overall maintenance costs of a structure by detecting damage at an early age. The SHM system can also be used to evaluate the effectiveness of retrofit and repair works.

4 DAMAGE ASSESSMENT

The most widely used features for damage assessment are the modal characteristics. They can easily

DOI: 10.1201/9781003348450-283

be obtained from ambient vibration measurements. However, one of the main obstacles to extract damage from modal features is that, apart from sensitive to damage, they are also dependent on changing environmental conditions such as temperature variations. These disturbing effects are well known for natural frequencies but less observed and understood for mode shapes and modal strains.

Non-model based methods try to extract damage sensitive (modal and other) features from the pure response measurements. Alternatively, there is the family of methods that try to assess damage by means of a model representing the structure before and after damage. By an optimization process, differences between measured and calculated modal characteristics are minimized, taking as updating variables parameterized damage patterns that are modelling the stiffness reductions.

5 SENSORS

Usually uniaxial/triaxial accelerometers or velocity meters are used to record vibrations in the SHM system. However, accelerometers do not provide a direct measurement of displacements.

On the other hand, strains are directly related to stresses. Dynamic strain measurements can deliver natural frequencies, modal damping ratios and modal strains. The introduction of fiber-optic sensors that can accurately measure dynamic strains and offer ease of installation, resistance in harsh environment and long-term stability, contributed to an increased interest in adopting these sensors for VBSHM. With Fiber Bragg grating (FBG) sensors sub-microstrain accuracy can be attained. FBGs share the important advantages of other FOS, but additionally they are easy to multiplex. Other sensor types that receive increasing interest are pattern interferometry, laser Doppler interferometry, electromagnetic wave interferometry and photogrammetry.

6 DESIGN OF A SHM SYSTEM

Unfortunately, in general the development of the SHM system is not yet integrated from the beginning as part of the project management process. Rather, in most cases, it is mainly implemented to fulfill specific requirements by the relevant authorities, like e.g. maximum bridge accelerations under service loads. Guidelines will be given for the design of a SHM system to fully exploit its inherent potentials.

7 CASE STUDY: KW51 BRIDGE

A case will be presented illustrating the long term monitoring of detailed strain mode shapes in operational conditions. The KW51 railway bridge (Figure 1) is a 117 m long steel bowstring bridge. The bridge deck consists of a steel orthotropic (i.e., rib-stiffened) steel plate that is supported by two longitudinal steel girders and thirty-three steel transverse beams. The bridge deck is suspended from the arch with thirty-two inclined braces.

Dynamic macrostrains of the bridge have been monitored with four chains of FBG sensors. Each fiber contains 20 FBG sensors (Figure 1). They are attached to the flanges using custom-made small clamping blocks (Anastasopoulos & all., 2019, 2021). Each FBG sensor measures the average strain between two clamping blocks. The strain acquisition is conducted with a FAZT-I4 interrogator. Throughout the monitoring period, the temperature of the bridge has also been measured. During the monitoring period, the bridge underwent a retrofit: all connections of the hangers between the arches and the bridge deck were strengthened. The influence of this retrofitting on the natural frequencies and strain mode shapes is investigated and compared with the temperature influence.

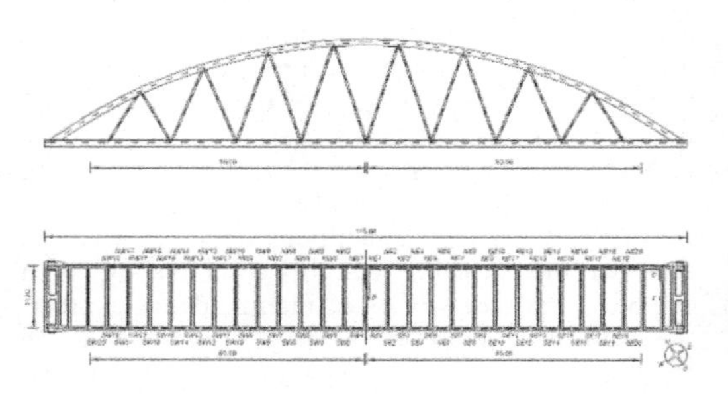

Figure 1. Bridge KW51 with the layout of the FBGs.

REFERENCES

Anastasopoulos, D., De Roeck, G. & Reynders, E.P.B. 2019. Influence of damage versus temperature on modal strains and neutral axis positions of beam-like structures. *Mechanical Systems and Signal Processing*, 134:106311.

Anastasopoulos, D., De Roeck, G. & Reynders, E.P.B. 2021. One-year operational modal analysis of a steel bridge from high-resolution macrostrain monitoring: influence of temperature vs. retrofitting. *Mechanical Systems and Signal Processing* 161(107951).

Reynders, E. & De Roeck, G. 2008. Reference-based combined deterministic-stochastic subspace identification for experimental and operational modal analysis. *Mechanical Systems and Signal Processing* 22(3):617–637.

Current Perspectives and New Directions in Mechanics, Modelling and Design of Structural Systems – Zingoni (ed.)
© 2022 Copyright the Author(s), *ISBN 978-1-032-39114-4*

Development of vibration-based early scour warning system for railway bridge piers

C.W. Kim, J. Qi & D. Kawabe
Department of Civil and Earth Resources Engineering, Graduate School of Engineering, Kyoto University, Kyoto, Japan

ABSTRACT: This study proposes a real-time scour warning system of railway bridges by means of a remote ambient vibration monitoring as an alternative method for the impact test. The real-time scour warning system consists of an edge computing system, wireless communication and cloud computing. The monitored ambient vibration signals are processed and the target frequency is identified in the edge computing system, and the identified frequencies are sent to a cloud computing system via Wi-Fi. In the cloud computing, possibility of scour is estimated and send out warnings as needed. A stochastic approach is investigated for the real-time scour warning because of relatively poor signal to noise of the ambient vibration signals comparing to that from the impact test. It is also investigated what kind of probability distribution best represents the observed frequencies. The stable distribution showed the best fit to the observed frequency distributions. Observations demonstrated an extremely low probability of scour of the target bridge during swollen river water period.

1 INTRODUCTION

This study aims to propose a real-time scour assessment utilizing ambient vibration data of railway bridge piers. A remote monitoring system with edge computing algorithm is designed and deployed to a bridge under operation. To estimate the natural frequency of the target bridge, accelerometers were installed on the top of the bridge pier and the bridge girders for the impact test, and two sensors on the pier top were left for scour monitoring.

The estimated posterior distribution of the target frequency is used for the real time scour detection. Two approaches are investigated. One is the method utilizing an index from the ratio of newly identified frequency to that of healthy state. Possibility of the scour is estimated by comparing the index with a scour assessment scale specified in the Japanese guideline (Ministry of Land, Infrastructure, Transport and Tourism, Railway Bureau 2007). The other is the method utilizing probability distributions of the identified frequency during normal river water period (hereafter "normal period") and swollen river water period (hereafter "swollen period").

2 MONITORING SYSYTEM AND BRIDGE

The monitoring bridge is a steel plate girder railway bridge with a span length of 22.5 m, the pier height of 9 m and width of 3 m, designed for single railway track. 13 triaxial sensors were installed on the top of the pier and connecting girders during the impact test, while all sensors except two sensors installed upstream and downstream of the pier top were removed after the impact test. Figure 1 shows sensor deploying map. Two sensors left on the pier top are used for the long-term ambient vibration monitoring. The sampling frequency of the measurement is 200Hz. The monitored ambient vibration signals are processed and the target frequency is identified in the edge computing system, and the identified frequencies are sent to a cloud computing system via Wi-Fi. In the cloud computing, possibility of scour is estimated and send out warnings as needed.

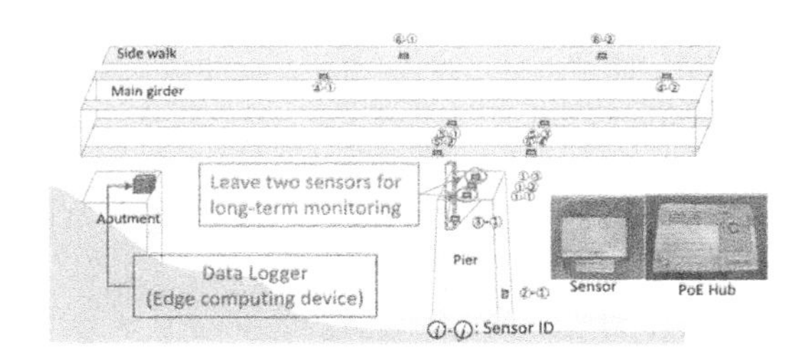

Figure 1. Sensor deploying map.

DOI: 10.1201/9781003348450-284

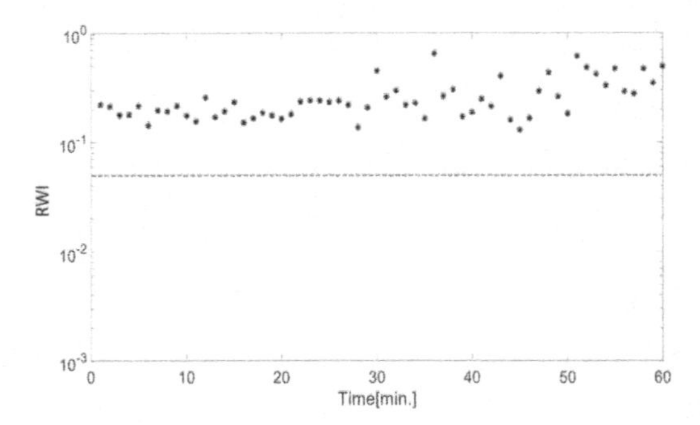

Figure 2. Time series of observed real time stochastic warning index by *Method-C*.

Table 1. Relationship between soundness index and stochastic warning index.

Freq.	SoundnessIndex	Category	SWI (Ψ_c)
$X \leq 6.4$Hz	$\kappa \leq 0.70$	A1	$\Psi_c \leq 0.0016$
6.4Hz$<X \leq$ 7.8Hz	$0.70<\kappa \leq 0.85$	A2	$0.0016<\Psi_c \leq 0.0053$
7.8Hz$<X \leq$ 9.1Hz	$0.85<\kappa \leq 1.00$	B	$0.0053<\Psi_c \leq 0.46$
9.1Hz$<X$	$1.00<\kappa$	S	$0.46<\Psi_c$

3 REAL-TIME SCOUR ASSESSMENT

The PDF of the stable distribution (Adler et al. 1996) of the frequency during the normal river water period is estimated. Three real-time stochastic approaches were investigated following assumption of stable distribution or normal distribution. A stochastic warning index (SWI), real-time stochastic warning index (RWI), and a concept of marginal indicator of stochastic warning index (MISWI) are proposed to assess scour occurrence. If RWI during swollen river water period reaches the MISWI, it indicates possible occurrence of scour. The MISWI can be defined considering relationships with the soundness index in Table 1. The soundness indices 0.70, 0.85 and 1.00 correspond to frequency of 6.4Hz, 7.7Hz and 9.1Hz respectively if the frequency in healthy condition is 9.1Hz.

The relationship between the soundness index κ and the stochastic warning index Ψ_c are summarized in Table 1. When the soundness index κ is 0.70, the corresponding frequency $Z_m = 6.4$Hz and the corresponding SWI is $\Psi_c = 0.0016$. Indeed, category A1 shows extremely low probability as SWI = 0.0016, and the stochastic approach can link to risk analysis.

Figure 2 Plots RWI defined as Equation 1.

$$\Psi_c = \int_{-\infty}^{\infty} L_F(X) \int_{-\infty}^{X} S_N dZ\, dX \tag{1}$$

where, $L_F(X) = S_F(\alpha_F, \beta_F, \gamma_F, X_t, X)$, S_N and S_F denote the PDF of the stable distribution under normal river water and swollen river water conditions, respectively. X_t is the identified frequency from a data sample measured at a time span. Parameters of the PDF of a stable distribution are identified by the maximum likelihood estimation (MLE).

4 CONCLUSIONS

This study investigates a way of scour detection by means of ambient vibration monitoring of the bridge pier during swollen river water period. A remote monitoring system was proposed. The natural frequency of the bridge pier was identified with high accuracy from the ambient vibration during the swollen river water period. Reliability-based scour assessment was proposed in which parameters of the PDF of the stable distribution are identified by means of the maximum likelihood method using the frequency of the pier during normal and swollen river water periods, and the real-time stochastic warning index is proposed and data collected from real swollen river water period is used as a case study. Observations demonstrated that that it is possible to assess probabilistic scour occurrence by using the ambient vibration data during swollen river water period. Moreover, feasibility of the real-time stochastic scour assessment was observed.

ACKNOWLEDGMENTS

This work is supported by the Grant-in-Aid for JSPS (Grant No: 19H0225). The financial support is greatly acknowledged.

REFERENCES

Adler, R., Feldman, R & Taqqu, M.S. 1996. A Practical guide to heavy tails. Statistical Techniques and Applications, pp313–314.

Au, S.K. 2011. Fast Bayesian FFT method for ambient modal identification with separated modes, *J Eng Mech ASCE* 137(3), 214–226.

Ministry of Land, Infrastructure, Transport and Tourism, Railway Bureau. 2007. Railway structures maintenance management standard and commentary (structural edition), pp.169–170. (in Japanese)

Current Perspectives and New Directions in Mechanics, Modelling and Design of Structural Systems – Zingoni (ed.)
© 2022 Copyright the Author(s), ISBN 978-1-032-39114-4

Vibration-based Bayesian anomaly detection of PC bridges

D. Kawabe, C.W. Kim & K. Takemura
Graduate School of Engineering, Kyoto University, Nishikyo-ku, Kyoto, Japan

K. Takase
Graduate School of Global Environment Studies, Kyoto University, Sakyoku, Kyoto, Japan

ABSTRACT: This study aims to investigate an effective damage detection for PC-box girders. A Bayesian anomaly detection is thus proposed. The Bayesian anomaly detection consists of two steps: one is to extract damage sensitive features from autoregressive model, and the other is to detect changes in these features by means of Bayesian hypothesis testing. Three half-scaled PC-box girders were tested to verify validity of the proposed anomaly detection method: one was used for the reference girder and remaining two were used for damaged girders which have unfilled grout zone around some of the post-tensioned tendons. In order to obtain the adoptable vibration data, ten accelerometers have been installed and impact tests have been conducted at the intervals of static loading steps. Observations showed that the proposed method detect changes in the damage sensitive feature when initial cracks propagate in the PC box girder. It also showed the possibility of detecting tendon-cut events.

1 INTRODUCTION

Innovative inspection approaches utilizing sensing technologies to detect bridge damages have been examined for over a decade. The vibration-based SHM, which is one of the innovative inspection method, measures structural dynamic response by sensors like accelerometers, and formulates it as mathematical model. Then, modal characteristics such as modal frequencies, modal damping ratio as well as mode shapes have been considered as damage indicators showing the structural state. Kim et al (2013) and Goi & Kim (2017) focused on autoregressive parameters, and proposed an anomaly index whose parameters are extracted from an autoregressive model derived from the acceleration data for the purpose of improving the sensitiveness.

This study is intended to discuss the feasibility of a practical vibration-based SHM through a loading experiment on PC box girders whose damage in the post-tensioned tendons were introduced in the laboratory (Luna Vera et al. 2017). The sensitivity of the proposed anomaly index against damages in the experiment is investigated even by comparing with cracks length of the PC box girders.

2 BAYESIAN ANOMALY DETECTION

The proposed Bayesian anomaly detection (BAYAD) method consists of learning and anomaly detection steps. In the learning step, the posterior distribution of vibration characteristics from observed data at healthy condition, which is called reference data, is identified by Bayesian inference, and damage sensitive features are extracted. In the anomaly detection step, statistical changes between features estimated from the target data and reference data, are evaluated.

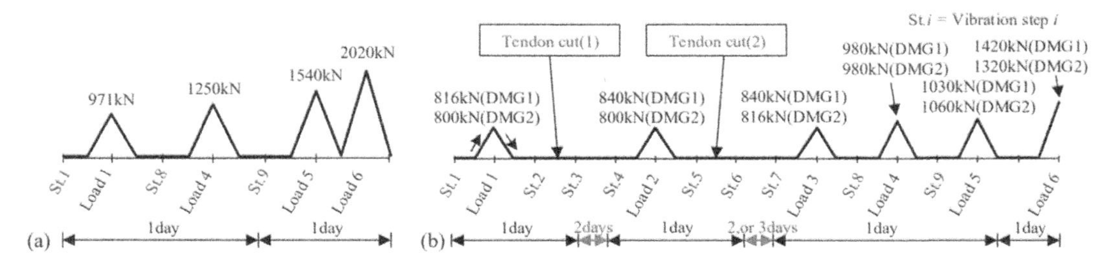

Figure 1. Loading and vibration test process: (a) G0, (b) G1 and G2.

DOI: 10.1201/9781003348450-285

3 LABORATORY EXPERIMENT

3.1 *Experiment information*

Three PC-box girders (named as G0, G1 and G2, respectively) were tested to assess their bending performance, from which G0 is the reference girder because no damage was introduced in any of its post-tensioned tendons. Then, the girder G1 was used for the introduction of damage in the post-tensioned tendons located in one of its webs and the girder G2 introduces the same type of damage, but this time located in both webs of the girder.

The loading and vibration tests, which hit the plate with an impact hammer, history of the PC-box girder G0 is shown in Figure 2a. Four loading steps with increasing magnitude were performed, from which three vibration tests were carried out before and after the first two loads. All the experiment events for girders G1 and G2 are summarized in Figure 2b.

3.2 *Application Bayesian anomaly detection to G2 focusing on certain hitting points*

Since influences of the impact points to BAYAD are observed at every stage in G1 and G2, only the vibration test data when the plate directly above the two accelerometers in the center of the span is hit is used, and the result in G2 is shown in Figures 2. All of both 2*ln*B at St2 show positive values, and it means initial cracks were detected by the BAYAD. In regard with tendon-cut, clear differences of anomaly indices due to tendon-cut are observed in Figure 2.

Focusing on one day apart event at the interval of St3-4 and St6-7, anomaly indices in St3-4 in G2 increase intensively. Although it might be considered that the gradual redistribution of residual prestress would be progressed over time after tendon-cut, the reduction of prestress in the overall system might be considered to have an effect on the changes in vibration characteristics.

4 CONCLUSIONS

Bayesian anomaly detection (BAYAD) was applied to static loading test and vibration test of PC box girders for damage detection, and it is observed that there is a possibility of detecting initial cracks of PC-box girder because the slight but gradual increasing trends due to each damage events were observed.

Since influence of impact location to damage detection was observed at every stage in G2, only the vibration test data when the plate directly above the two accelerometers in the center of the span is hit is used for the feasibility investigation of the BAYAD. Observations demonstrated that it clearly detects the initial cracks in both girders. The effects of the redistribution of residual prestress due to tendon-cut could be revealed, but the relations between the sensitivity of anomaly indices and the number of tendon-cut should be discussed.

ACKNOWLEDGEMENTS

This work is supported by the Grant-in-Aid for JSPS (Grant No: 19H0225). The financial support is greatly acknowledged.

REFERENCES

Kim, C.W., Kawatani, M. & Hao, J. 2012. Modal parameter identification of short span bridges under a moving vehicle by means of multivariate AR model, *Structure and Infrastructure Engineering* 8(5):459–472, 2012.

Kim, C.W., Isemoto, R., Sugiura, K. & Kawatani, M.: Structural fault detection of bridges based on linear system parameter and MTS method, *J. of JSCE*, JSCE, 1(1):32–43, 2013.

Luna Vera, O. S., Kim, C. W., & Oshima, Y. 2017: Energy capacity influence on modal parameters of prestressed concrete box girders, *Proc. of VIII ECCOMAS Thematic Conference on Smart Structures and Materials SMART 201*: 785–795, 2017.

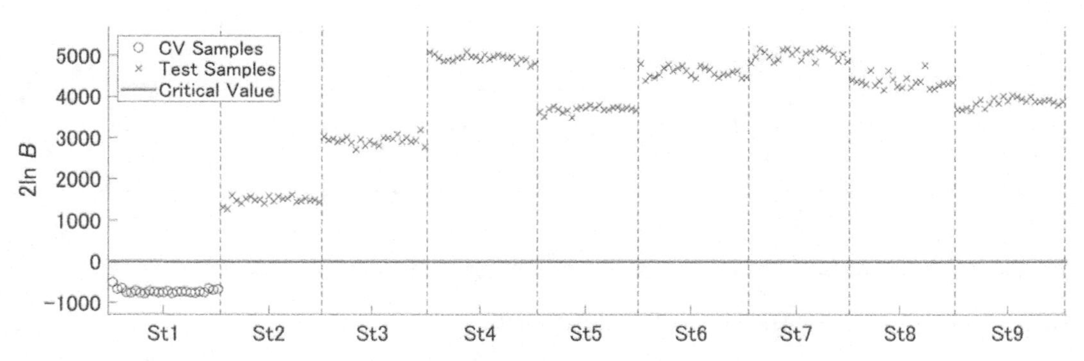

Figure 2. BAYAD results in G2 girder focusing on middle span hitting data. Tendon cut events are at the interval of St2-3 and St5-6. St3-4 and St6-7 have one day apart.

Current Perspectives and New Directions in Mechanics, Modelling and Design of Structural Systems – Zingoni (ed.)
© 2022 Copyright the Author(s), ISBN 978-1-032-39114-4

Vibration-based structural health monitoring of bridges based on a new unsupervised machine learning technique under varying environmental conditions

M. Salar
Department of Civil and Environmental Engineering, Politecnico di Milano, Milano, Italy

A. Entezami
Department of Civil and Environmental Engineering, Politecnico di Milano, Milano, Italy
Department of Civil Engineering, Faculty of Engineering, Ferdowsi University of Mashhad, Mashhad, Iran

H. Sarmadi
Department of Civil Engineering, Faculty of Engineering, Ferdowsi University of Mashhad, Mashhad, Iran

B. Behkamal
Department of Civil and Environmental Engineering, Politecnico di Milano, Milano, Italy
Department of Computer Engineering, Faculty of Engineering, Ferdowsi University of Mashhad, Mashhad, Iran

C. De Michele & L. Martinelli
Department of Civil and Environmental Engineering, Politecnico di Milano, Milano, Italy

ABSTRACT: In most real-world applications, adverse influences caused by multiple sources of environmental variability conditions can mask extracted features and may lead to false indications of structural damage. Hence, it is thus fundamentally significant to investigate the effects of these variations on the damage-related features and damage detection procedure. This article proposes a new unsupervised machine learning technique for early damage detection of bridge structures, which are always exposed to environmental variability conditions. The proposed method is based on a data dependent dissimilarity measure. At last, the effectiveness and robustness of the proposed approach are assessed and verified through the well-known Tianjin-Yonghe Bridge; additionally, the proposed unsupervised machine learning methodology succeeds in early detecting damage under variability of environmental conditions.

1 INTRODUCTION

The main goal of structural health monitoring (SHM) is to identify any potential damage in civil structures. In the context of SHM, novelty or anomaly detection on the basis of unsupervised learning is an influential method for feature normalization and classification to avoid yielding erroneous results of damage detection owing to the influences of environmental and/or operational variability (EOV) conditions. Accord-ingly, the main objective of this article is to propose a novel data dependent dissimilarity measure for using *k*NN anomaly detection method in long-term SHM of bridge structures under strong environmental variability conditions (Diez et al., 2016, Sarmadi and Karamodin, 2019).

2 DATA ANOMALY DETECTION FOR SHM

Anomaly detection is of paramount importance for a wide range of SHM applications due to high costs caused by human and economic losses linked to misunderstanding anomalies than those of other main damage-related features (Farrar and Worden, 2013, Pimentel et al., 2014). In SHM context, the process of anomaly detection typically consists of two phases, i.e. training and monitoring.

3 PROPOSED MASS-BASED *k*NN ANOMALY DETECTION METHOD FOR FEATURE NORMALIZATION

The proposed method within this study is based on a data dependent dissimilarity measure for using

DOI: 10.1201/9781003348450-286

kNN anomaly detection method instead of standard distance- and statistics-based measures. The central idea behind the proposed mass-based anomaly detection method is to improve the performance of kNN anomaly detector employing distance-based function (Hoang et al., 2021, Pimentel et al., 2014).

4 APPLICATION TO A CABLE STAYED BRIDGE

Model-based feature datasets extracted from acceleration time histories of the Tianjin-Yonghe Bridge is utilized to validate the accuracy of the proposed method. This bridge is one of the earliest cable-stayed bridges constructed in Mainland China as shown in Figure 1.

Figure 1. The Tianjin-Yonghe Bridge.

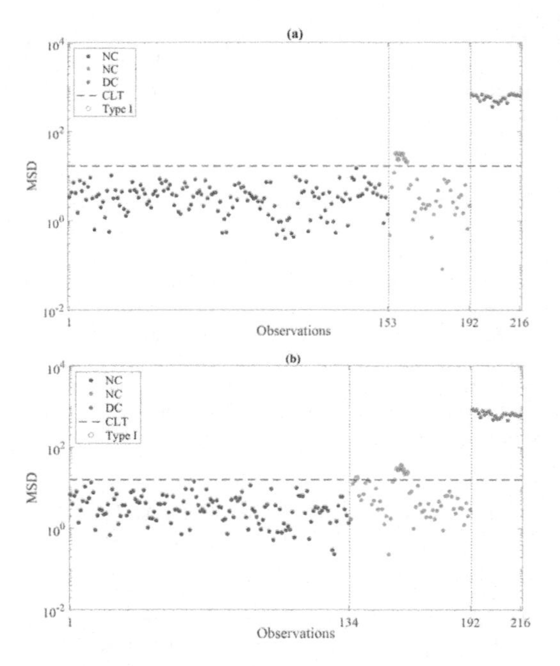

Figure 2. Early damage detection in the Tianjin-Yonghe Bridge by the proposed mass-based kNN anomaly detection method with different training sample sizes: (a) 80%, (b) 70%.

Figure 2 illustrates the results of early damage detection in the Tianjin-Yonghe Bridge by the proposed unsupervised learning method. In Figure 2(a) and (b), 80% and 70% of the observations of the normal condition of the bridge are considered, respectively. In this figure, the observations 1-153 and 1-134 belong to the normal condition (*NC*) and the observations 192-216 are related to the damaged condition (*DC*) of the bridge.

It is apparent from this figure that all the obtained values regarding the damaged state of the bridge exceed the CLT threshold limit, which means that the proposed mass-based kNN anomaly detection in conjunction with the MSD method is capable of detecting damage and distinguishing the damaged condition from the normal state even under varying environmental conditions as well as changing training data samples. However, it is seen that there are a few violations in the validation datasets that exceed the threshold limit, which imply a few false alarm or Type I errors in damage detection process.

5 CONCLUSIONS

This article proposed an unsupervised learning method based on two main steps consisting of feature normalization by a mass-based kNN anomaly detection method using data dependent dissimilarity measure and feature classification via the Mahalanobis distance metric. Modal frequencies extracted from vibration measurements regarding a large-scale cable-stayed bridge, the Tianjin-Yonghe Bridge, were utilized to demonstrate the performance and effectiveness of the proposed method. The results illustrated that the proposed unsupervised learning technique is capable of highly suppressing the variations caused by the environmental conditions and accurately detecting damage in the bridge by using the normalized features obtained from mass-based dissimilarity measure and Mahalanobis distance metric.

REFERENCES

Diez, A., Khoa, N. L. D., Makki Alamdari, M., Wang, Y., Chen, F. & Runcie, P. (2016) A clustering approach for structural health monitoring on bridges. *Journal of Civil Structural Health Monitoring*, 6, 429–445.

Farrar, C. R. & Worden, K. (2013) *Structural Health Monitoring: A Machine Learning Perspective*, John Wiley & Sons Ltd.

Hoang, A., Mau, T. N., Vo, D.-V. & Huynh, V.-N. (2021) A mass-based approach for local outlier detection. *IEEE Access*, 9, 16448–16466.

Pimentel, M. A. F., Clifton, D. A., Clifton, L. & Tarassenko, L. (2014) A review of novelty detection. *Signal Processing*, 99, 215–249.

Sarmadi, H. & Karamodin, A. (2019) A novel anomaly detection method based on adaptive Mahalanobis-squared distance and one-class kNN rule for structural health monitoring under environmental effects. *Mechanical Systems and Signal Processing*, 140, 1–24.

Current Perspectives and New Directions in Mechanics, Modelling and Design of Structural Systems – Zingoni (ed.)
© 2022 Copyright the Author(s), ISBN 978-1-032-39114-4

An embedded physics-based modeling concept for wireless structural health monitoring

K. Dragos & K. Smarsly
Institute of Digital and Autonomous Construction, Hamburg University of Technology, Germany

ABSTRACT: In structural health monitoring (SHM), the inherent processing capabilities of wireless sensor nodes have enabled data-driven models to be embedded into the processing units of wireless sensor nodes. The merits of data-driven modeling in SHM have been known for years, particularly with respect to models being independent from assumptions on structural properties, thus reducing epistemic uncertainty. However, current SHM has been associated with civil structures of increasing complexity, the detailed assessment of which cannot rely exclusively on the limited information obtained through data-driven modeling. This paper presents an embedded physics-based modeling concept for wireless SHM systems, exploiting the processing capabilities of wireless sensor nodes and the analysis capabilities of physics-based models in structural engineering to obtain estimates of structural conditions that cannot be derived from data-driven models. The proposed concept is presented in detail, and future work on implementing the concept in wireless SHM systems is briefly discussed.

1 INTRODUCTION

The culture of devising structural health monitoring (SHM) strategies for assessing the structural condition of critical infrastructure has been well-established, as evidenced by the numerous structures equipped with long-term SHM systems, nowadays (Nagarajaiah & Erazo 2016). In recent years, SHM practice has turned to wireless technologies, motivated by the rapid advances in informatics. Despite the efficiency in installation and the considerable cost reduction, wireless sensor nodes have inherent limitations, mainly regarding the limited power autonomy and the unreliability of wireless communication. To overcome the limitations of wireless SHM systems, embedded models have been devised, facilitating on-board data analysis, aiming to avoid the wireless transmission of structural response data. State-of-the-art embedded models for wireless SHM are predominantly data-driven, which do not account for the physical principles governing the structural behavior and may provide limited information on the behavior of complex structures.

This paper presents an embedded physics-based modeling (PBM) concept for wireless SHM. The embedded PBM concept leverages the descriptive as well as the predictive capabilities of physics-based models to enhance the performance of smart wireless sensor nodes. To make the best use of the structural response data collected by the sensor nodes, the embedded PBM concept builds upon associating the structural response data directly with structural properties. In what follows, the embedded PBM concept is presented, and a brief discussion on future work for implementing the proposed concept into wireless SHM systems is provided.

2 EMBEDDED PHYSICS-BASED MODELS FOR WIRELESS SHM

The embedded physics-based modeling concept is based on segmenting the monitored structure into N "substructures". In the jth substructure, structural response data from L specific locations $\mathbf{S}_j = [S_{j1}, S_{j2},\dots,S_{jL}]$ are associated with r_j structural properties $\boldsymbol{\theta}_j = [\theta_{j1}, \theta_{j2},\dots,\theta_{jr}]$ of the substructure, such as stiffness k_j and mass m_j.

Associating structural response data with structural properties using classical mechanics may be impractical for SHM due to (i) the ill-posed nature of inverting dynamic equilibrium equations, which is a result of several structural parameters in these equations being unknown, (ii) the coupling between dynamic equilibrium equations, which would require intensive wireless communication between wireless sensor nodes for obtaining solutions and, eventually, hinder the autonomy of the sensor nodes, and (iii) the lack of information on external forces, which would allow only approximative and/or stochastic solutions to dynamic equilibrium equations.

To address the aforementioned limitations, machine learning (ML) methods incorporating physical principles into, referred to as "physics-informed

DOI: 10.1201/9781003348450-287

neural networks" (Raissi et al. 2019), are adopted for the embedded PBM concept. Specifically, an initial physics-based model is created and updated using preliminary structural response data, as shown in Figure 1. Next, the model is segmented into substructures, and a ML model, e.g. an artificial neural network (ANN), is trained to "learn" relationships between structural response data and structural properties of interest $\mathbf{\theta}_j$ for substructure j.

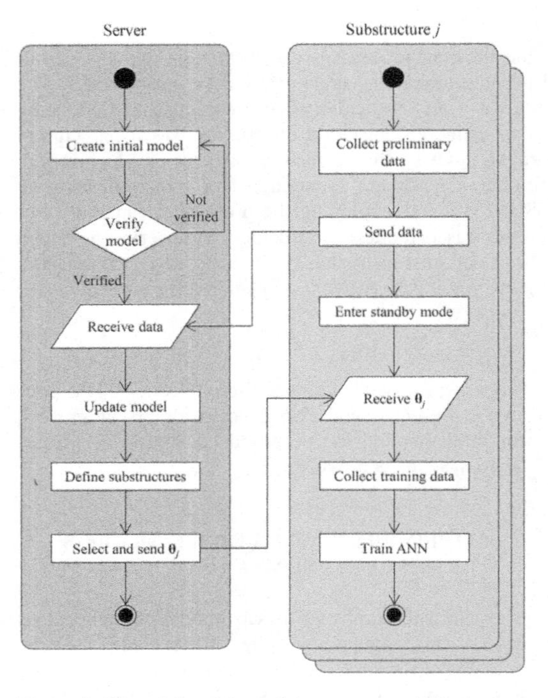

Figure 1. Flowchart of the training process within the proposed embedded PBM approach.

The first step of the embedded PBM approach is to create a physics-based model of the structure on the server and update the model using preliminary structural response data. The next step involves defining the substructures and sending the structural properties of interest to the respective wireless sensor nodes. For defining the substructures, the geometry of the structure needs to be taken into account, and the interfaces between the substructures need to be clear. Finally, the structural properties of interest are sent to the wireless sensor nodes, along with functions parametrized with the structural properties of interest, the values of which are used as target output. Once the structural properties of interest and the functions are received, the wireless sensor nodes use structural response data as input and function values as output to train the ML model. The schematic description of the embedded physics-based concept is depicted in Figure 2.

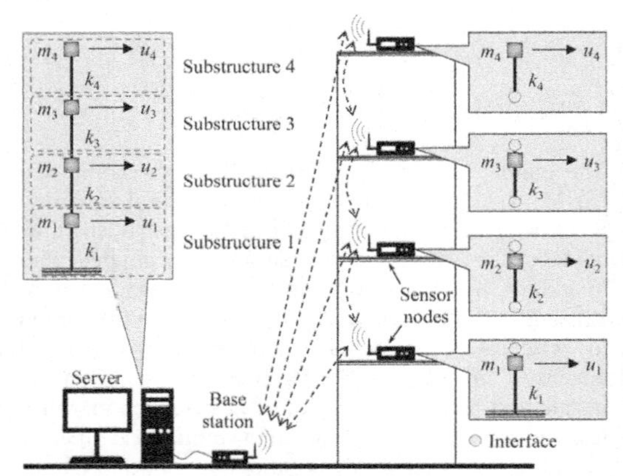

Figure 2. Schematic description of the embedded physics-based concept applied to the shear-frame structure.

3 FUTURE WORK

The procedure for implementing the concept in wireless SHM, which will be part of future work, is illuminated in this section. The experimental setup considered for the future work comprises a simple four-story shear-frame structure, as shown in Figure 2. The wireless SHM system installed on the structure consists of one wireless sensor node placed at each story. The wireless sensor nodes communicate via a gateway node, denoted as "base station", with a computer, operating as a server.

REFERENCES

Nagarajaiah, S. & Erazo, K. 2016. Structural monitoring and identification of civil infrastructure in the United States. Smart Monitoring and Maintenance 3(1): 51–69.

Raissi, M., Perdikaris P. and Karniadakis, G.E. 2019. Physics-informed neural networks: A deep learning framework for solving forward and inverse problems involving nonlinear partial differential equations. Journal of Computational Physics 378: 686–707.

Current Perspectives and New Directions in Mechanics, Modelling and Design of Structural Systems – Zingoni (ed.)
© 2022 Copyright the Author(s), ISBN 978-1-032-39114-4

Damage evaluation and tracking using kriging approaches on wave-guide dispersion curves

O.A. Bareille, L. Nechak & M. Ichchou
Ecole Centrale de Lyon – LTDS, Ecully, France

A. Zine
ICJ – Institut Camille Jordan, Ecully, France

ABSTRACT: The calculation of the wave numbers is sometimes a prerequisite as well as a potential indicator of the evolution of the structure's state. On the one hand, when addressing medium- and high-frequency monitoring signals, even Wave Finite Elements Method's calculations can take a long time to obtain the dispersion curves. The present work proposes the kriging based metamodeling for both the prediction and the measurement of dispersion curves in waveguides. The main idea is to predict the dispersive curves by using a small number of WFEM based simulations. This study hence analyses the reliability and validity of the kriging based metamodeling when used for the prediction and the evaluation of the dispersion curves for a waveguide structure.

1 PROPERTIES OF WAVEGUIDES FOR STRUCTURAL HEALTH MONITORING

The importance of determining properties of guided waves such as the dispersion curves is well established in numerous fields relating in particular to Structural Health Monitoring (SHM). Indeed, for a given structure, the dispersion curves permit the characterizing of the velocity-frequency relationship for all the modes which may propagate in a given structure, which, in the defects and damage detection with non-destructive approaches, is necessary to obtain insight about the efficiency of the wave transmission. Then, predicting the dispersion curves has been addressed few decades ago and numerous approaches have been addressed in literature. The most popular classification of these approaches is that dividing them into two main groups namely analytical or exact methods and numerical methods

When based on propagative approaches, the health monitoring techniques require that the behavior of the structure shall be somehow understood in order to get an accurate evaluation of the damage level. The calculation of the wave numbers is sometimes a prerequisite as well as a potential indicator of the evolution of the structure's state. Indeed, it permits to getting insight about how wavemodes are propagating and interacting in structures.

On the one hand, when addressing medium- and high-frequency monitoring signals, even Wave Finite Elements Method's calculations can take a long time to obtain the dispersion curves. On the other hand, when time has come to obtain the experimental dispersion curves to identify the occurrence of a damage, the question arises about the selection of the measurement points. The kriging tools are then suited for this.

The present work proposes the kriging based metamodeling for both the prediction and the measurement of dispersion curves in waveguides. The main idea is to predict the dispersive curves by using a small number of WFEM based simulations. This way is expected to drastically reduce the above-mentioned numerical difficulties and thus enhance the efficiency. The distribution and the number of the measurement points can also be derived.

This study hence analyses the reliability and validity of the kriging based metamodeling when used for the prediction and the evaluation of the dispersion curves for a waveguide structure. Different parameters of kriging metamodels are hence considered. The main objective is to search for the kriging parameters that maximize the accuracy of the kriging based predictions of the dispersive curves while ensuring an acceptable computing time.

2 WAVE FINITE ELEMENT METHOD FOR PERIODIC WAVEGUIDE

The Wave Finite Element Method (WFEM) is widely used to study free and forced vibration in periodic structures. Basically, this method helps for the describing of waves travelling in left and right directions in a periodic structure by using a finite

DOI: 10.1201/9781003348450-288

element model of only a single cell with a small number of degrees of freedom, instead of considering the whole periodic structure.

For wave guides with damping (namely dissipative waveguide), the associated wavenumbers are complex and are related to decaying waves. The dispersion curve can be determined from the relationship between wavenumbers and frequency like in Figure 2.

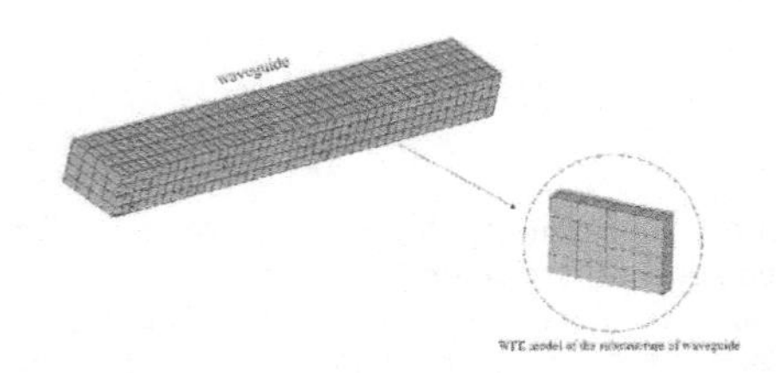

Figure 1. FE model of a substructure.

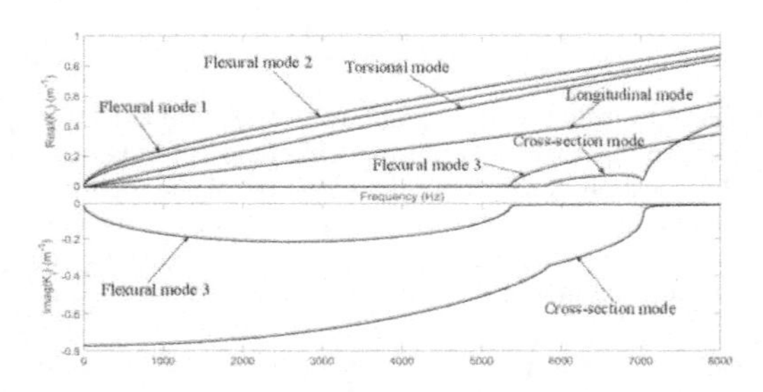

Figure 2. Dispersion curve of a substructure.

3 KRIGING BASED META-MODELLING FOR THE PREDICTION OF THE DISPERSION CURVE

3.1 *Advantage of the kriging based metamodelling*

The key idea and advantage in kriging based metamodelling is related to the using of only a small size data set of suitable input/output samples in order to mimic and predict the behavior of an unknown function. To acquire this mimicry capacity, a kriging metamodel exploits information about spatial correlations between sample points by considering the associated distance. This is performed by using predefined correlation models. The latter should well represent the most likely relationships between samples. In other words, models are built such as two infinitely distant points be not correlated while two identical points have a unitary correlation. In the here-reported woks, the main principles are used to build a kriging metamodel in order to approximate and predict the dispersion curves.

4 APPLICATION AND RESULTS

A substructure of the beam-like waveguide is first described by a finite element model which is then exploited through the WFEM in order to obtain the dispersion curves. These are considered as the reference which will be used to evaluate the accuracy of the kriging based predictions. This accuracy is analyzed with respect to numerous parameters of the kriging predictor namely the regression order, the correlation function and the size of the learning data set. The main objective is to determine the parameters of kriging predictor that ensure suitable robustness properties for the accuracy of the predictions with respect to the complexity of the dispersion curves. The Exponential and especially the Gauss correlation models already offer with these few samples a suitable overview about the function.

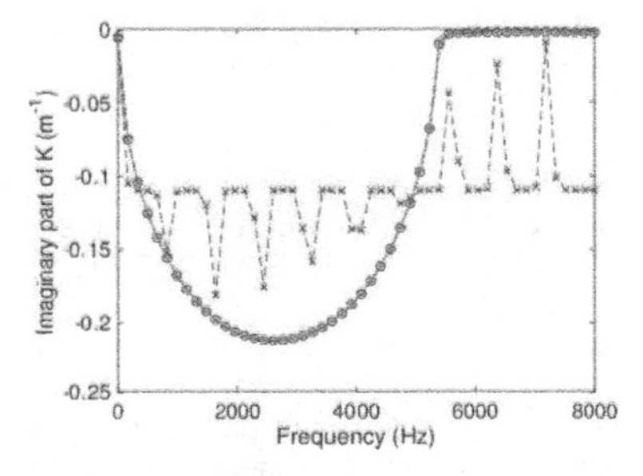

Figure 3. The 3rd flexural mode predicted by a zero-order regression based kriging metamodels with spline correlation function and different sizes.

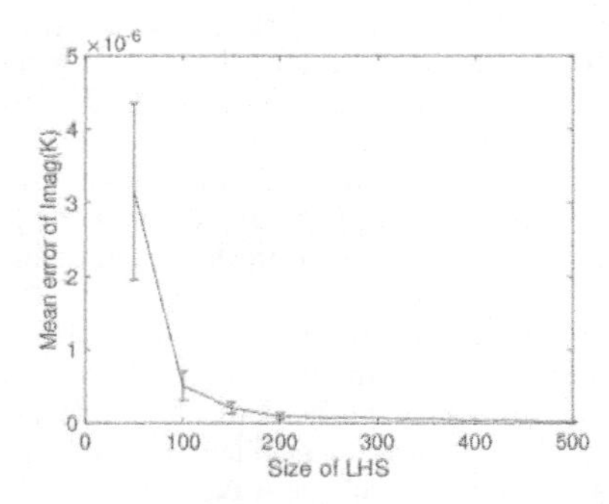

Figure 4. Evolution of the mean square error and the standard deviation versus the size of the LHS design on imaginary part of wavenumbers.

It can be observed that the shape of the curve is correctly approximated by all the kriging predictors from N = 50. The difference between the different predictors is located in particular at the zone of occurrence of the singularity (near 5400Hz). It appeared that the predictor with the Gauss correlation model has difficulties.

Current Perspectives and New Directions in Mechanics, Modelling and Design of Structural Systems – Zingoni (ed.)
© 2022 Copyright the Author(s), ISBN 978-1-032-39114-4

Prompt modal identification with quantified uncertainty of modal properties

Y. Goi
Department of Civil and Earth Resources Engineering, Kyoto University, Kyoto, Japan

ABSTRACT: This study proposes an efficient modal identification method for bridges under operation. The noisy condition caused by the traffic loadings is one of the difficulties involved in the operational modal identification. To cope with the problem, this study quantifies uncertainty involved in the modal properties utilizing Bayesian statistics. A vector autoregressive model, which is one of classical models for modal analysis, is adopted for prompt computation. Bayesian inference produces posterior distribution related to modal properties involved in the vector autoregressive model. This study adopts variance of the posterior distribution to quantify the uncertainties of the estimated modal properties. The quantified uncertainty enables to extract the stably estimated modal properties. The proposed method is applied to traffic induced vibration measured from an actual single-span truss bridge. Eighteen modes are extracted from the vibration measured with eight uniaxial accelerometers. Eight stably-estimated modes well correspond to peaks in power spectral density curves. The estimated mode shapes describe that five of them are bending modes and the other three are torsional modes. The quantified uncertainty also indicated ten faint modes.

1 INTRODUCTION

Vibration based structural health monitoring provides a research field for nondestructive evaluation based on physical measurements and computer analyses to complement existing visual inspection methods. In actual bridges, however, the noisy condition caused by the traffic loadings is one of the difficulties involved in the operational modal identification. That is because the noisy condition induces spurious estimators of modal properties. In existing methods, it is troublesome to distinguish the physically meaningful estimators from the spurious estimators.

To cope with the problem, this study simplifies the uncertainty quantification process using a vector autoregressive (VAR) model, which is one of classical models for modal analysis. As proposed in an Authors' previous study [1], Bayesian inference produce the posterior distribution of regressive parameters of VAR model without numerical iteration. The posterior distribution is converted to the distribution of mode vectors and poles of the linear system. This study adopts coefficient of variance (C.V.) of the posterior distribution to quantify the uncertainties of the estimated modal properties. The quantified uncertainty indicates the physically meaningful modal properties among the estimators. The proposed method is applied to traffic induced vibration measured from an actual truss bridge.

2 METHODOLOGY

Letting $\boldsymbol{y}_k \in \mathbb{R}^{m\times 1}$ $(k = 1 \ldots n)$ denote a column vector at the k-th time step of the time series with components corresponding to the measurement locations, then the following VAR model with sufficient model order p is known to approximate the time series obtained from a linear structural system excited by white noise [16].

$$\boldsymbol{y}_k = \sum\nolimits_{i=1}^{p} \boldsymbol{\alpha}_i \boldsymbol{y}_{k-i} + \boldsymbol{e}_k$$

Letting $\mathbf{W} = [\boldsymbol{\alpha}_1, \ldots, \boldsymbol{\alpha}_p] \in \mathbb{R}^{m\times mp}$ and $\boldsymbol{\phi}_k = [\boldsymbol{y}_{k-1}; \ldots; \boldsymbol{y}_{k-p}] \in \mathbb{R}^{mp\times 1}$ for simplicity, the AR model can be rewritten as

Figure 1. Target bridge.

$$\boldsymbol{y}_k = \mathbf{W}\boldsymbol{\phi}_k + \boldsymbol{e}_k$$

Let $\mathbf{Y}_i = [\boldsymbol{y}_{p+1}^{(i)}; \ldots; \boldsymbol{y}_{n_i}^{(i)}]$ denote the i the time series and $\sum$ denote the covariance matrix of $\boldsymbol{e}_k$. The posterior distribution is given as the following inverce

DOI: 10.1201/9781003348450-289

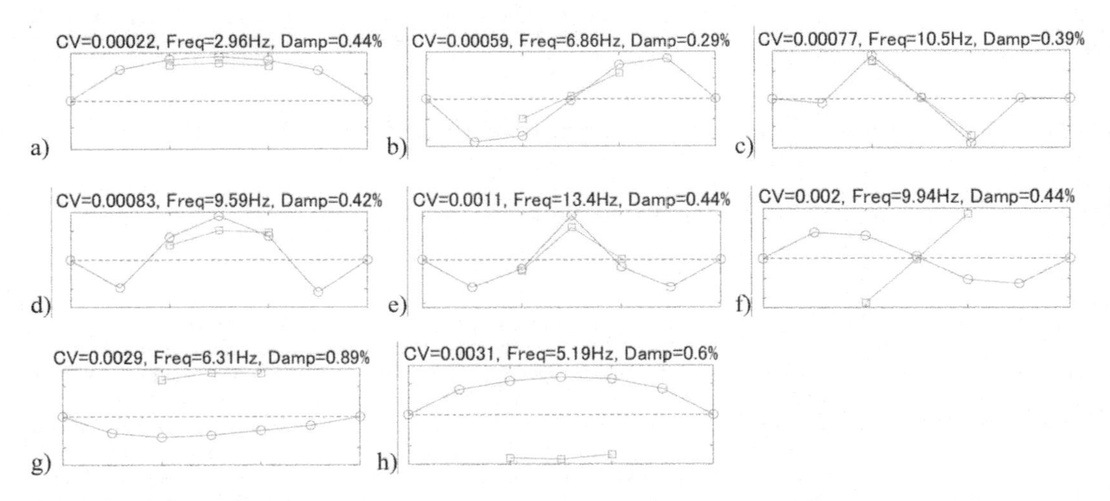

Figure 2. Identified modes.

$$\mathrm{p}\left(\mathbf{W}, \sum | \mathbf{Y}_1, \ldots, \mathbf{Y}_l\right) = \mathcal{MN}\left(\mathbf{W}|\mathbf{M}, \sum, \mathbf{L}^{-1}\right)\mathcal{IW}\left(\sum|\psi, \nu\right)$$

Let v_i and λ_i respectively represent mode vector and the pole, and let p_i represent the corresponding eigenvector and of the system. The following relation is given.

$$\mathbf{W}_i = \boldsymbol{v}_i\lambda_i$$

This study proposes uncertainty quantification of the modal properties assessing the variance of posterior distribution of $\boldsymbol{v}_i\lambda_i$. using C.V.

3 CASE STUDY

3.1 *Preliminary investigation*

This study provides a case study for a simply supported bridge to assess the feasibility of the proposed method through comparison to an earlier investigation [2]. Field experiments were conducted to the bridge with a moving vehicle. The bridge is a single-lane through-type steel Warren truss bridge as presented in Figure 1. Eight uniaxial accelerometers were installed on the deck of the bridge to measure vertical vibrations.

The proposed method extract certainly estimated modes by sorting the calculated C.V. Figure 2 shows the mode shapes of the certain estimators in ascending order of the C.V. The diagrams also includes the values of the C.V., frequencies and damping ratios on caption. In the diagrams, the horizontal axis represents relative location of the sensor along the longitudinal direction on the bridge, and the vertical axis represents relative modal deformation. The mode shapes in Fig. 6a), b), c), d), and e) seem to correspond to bending modes because the modal deformations on both edges are deforms simultaneously. These results correspond to the previous study by Chang and Kim [2]. Contrarily, it is reasonable to regard the mode shapes in Fig. 6f), g) and h) as torsional modes because the both edges deforms alternately.

FUNDING

This study was partly sponsored by a Japanese Society for Promotion of Science (JSPS) Grant-in-Aid for Scientific Research (B) under Project No. 16H04398 and for Early-Career Scientists under Project No. 19K15072. That financial support is gratefully acknowledged.

REFERENCES

[1] Y. Goi, C.W. Kim, Bayesian outlier detection for health monitoring of bridges, *Proceedia Eng.* 199, 2120–2125, 2017.

[2] K.C. Chang, C.W. Kim, Modal-parameter identification and vibration-based damage detection of a damaged steel truss bridge, *Eng. Struct.*, 122, 156–173, 2016.

Current Perspectives and New Directions in Mechanics, Modelling and Design of Structural Systems – Zingoni (ed.)
© 2022 Copyright the Author(s), ISBN 978-1-032-39114-4

Wavelet-based transmissibility for structural damage detection

Kajetan Dziedziech, Krzysztof Mendrok, Tadeusz Uhl & Wiesław J. Staszewski
Department of Robotics and Mechatronics, AGH University of Science and Technology, Krakow, Poland

ABSTRACT: Short-time, abrupt events – such as earthquakes and other shock loadings – often lead to damage that is difficult to detect in structures using output-only vibration measurements. The time-variant transmissibility is proposed to tackle this problem. The approach is based on two-dimensional wavelet power spectra. The time-frequency transmissibility and relevant coherence function are used for structural damage detection in structural elements in buildings. Numerical simulations and experimental tests are used in these investigations. The results are compared with the classical transmissibility and time-variant input-output wavelet approach. The paper shows that output-only measurements and the wavelet-based transmissibility can be used to monitor abrupt damage-related changes to structural dynamics.

1 INTRODUCTION

Vibration characteristics are often used for structural damage detection. Methods based on excitation and response measurements have been utilised to obtain natural frequencies, damping, deflection/mode shapes, curvatures, flexibilities, Frequency Response Functions (FRF) and many other parameters and characteristics based on modal analysis, as reviewed in [1]. When excitation measurement are not available operational modal analysis is employed [2]. In addition, transmissibility – often defined as the ratio of two response spectra [3] has been also used for structural damage detection [4].

The paper demonstrates the application of wavelet-based transmissibility for structural damage detection. Simulated vibration responses from a simple building model 3 are used to detect abrupt changes of stiffness.

The structure of the paper is as follows. Section 2 provides the theoretical background of the wavelet-based transmissibility. The application of the method is illustrated in Section 3. A simple time-variant building model is described, and damage detection results – based on the input-output and output-only analysis - presented. Finally, the paper is concluded in Section 4.

2 WAVELET-BASED TRANSMISSIBILITY

For two independent measured responses x_i and x_j the wavelet-based transmissibility and coherence can be defined as

$$T(a,b) = \frac{G_{ij}(a,b)}{G_{jj}(a,b)} \tag{1}$$

$$\gamma^2(a,b) = \frac{\left|\widehat{G}_{ij}(a,b)\right|^2}{\widehat{G}_{ii}(a,b)\widehat{G}_{jj}(a,b)} \tag{2}$$

where G_{jj} and G_{ij} is the wavelet-based auto-spectrum and cross-spectra, respectively. Since b is a translation operator indicating locality in time, a is a scale operator indicating locality in frequency, the above characteristic is defined in the combine time-frequency domain.

The wavelet-based transmissibility and coherence were implemented numerically. The continuous wavelet transform defines as

$$W_x(a,b) = \frac{1}{\sqrt{a}}\int_{-\infty}^{+\infty} x(t)\psi^*\left(\frac{t-b}{a}\right)dt \tag{3}$$

Was used in this implementation. Here, b is a translation operator indicating locality in time, a is a scale operator indicating locality in frequency, $\psi(t)$ is an analysing wavelet and superscript "$*$" indicates a complex conjugate. The complex Morlet wavelet function was used in the calculation of the continuous wavelet transform.

3 STRUCTURAL DAMAGE DETECTION - NUMERICAL SIMULATIONS

A simple Linear Time-Variant (LTV) building model is simulated using a 3-DOF system presented in

DOI: 10.1201/9781003348450-290

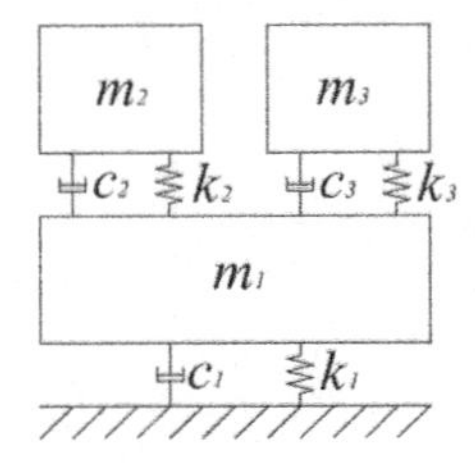

Figure 1. Frame structural element and experimental damage detection arrangements.

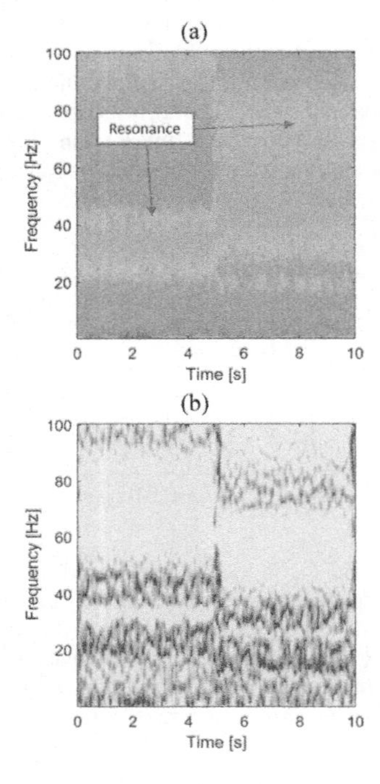

Figure 2. Wavelet-based analysis for the simulated LTV system: (a) wavelet-based FRF in logarithmic scale; (b) wavelet-based FRF coherence; (c) wavelet-based transmittance; (d) wavelet-based transmittance coherence.

Figure 1. The values of constant physical parameters were used as: $k_1 = 800\ kN/m$, $k_2 = 400\ kN/m$, $c_1 = 50\ N/(m/s)$, $c_2 = 50\ N/(m/s)$, $c_3 = 100\ N/(m/s)$, $m_1 = 7\ kg$, $m_2 = 7\ kg$, $m_3 = 10\ kg$. In contrast, the stiffness k_3 was assumed to be time-variant and was abruptly decreased from $k_3 = 960\ kN/m$ to $k_3 = 200\ kN/m$ in 5 seconds.

The LTV 3-DOF system was simulated with stiffness k_3 decreasing in time as explain above. Relevant functions presented in Section 2 are calculated and shown in Figure 2.

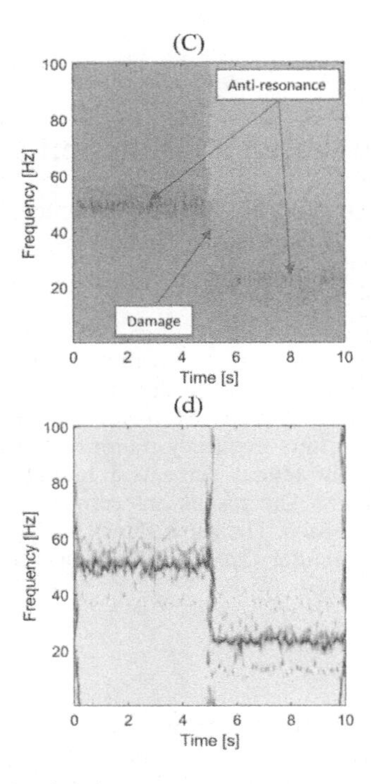

Figure 2. Continued.

4 CONCLUSIONS

The method was illustrated using a simple 3-DOF model of building structure exhibiting sudden reduction of stiffness. The paper shows that the wavelet-based transmissibility is capable to detect abrupt changes in dynamics. Major advantage relates to the fact that only two output measurements are needed to detect damage.

REFERENCES

1. S.W. Doebling, C.R. Farrar and M.B. Prime. 1998. A summary review of vibration-based damage identification methods. *Shock Vib Digest* **30**(2) 91–105.
2. B. Peeters and G. De Roeck. 1999. Reference-based stochastic subspace identification for output-only modal analysis. *Mechanical Systems and Signal Processing* **13** (6) 855–878.
3. W. Liu and D.J. Ewins. 1998. Transmissibility properties of MDOF systems. Proc. of the 16th International Modal Analysis Conference (IMAC) 83–90, Santa Barbara, California.
4. K.Worden, Structural fault detection using a novelty measure1997. *Journal of Sound and Vibration*$\mathbf{201}$(1) 85–101.

Current Perspectives and New Directions in Mechanics, Modelling and Design of Structural Systems – Zingoni (ed.)
© 2022 Copyright the Author(s), ISBN 978-1-032-39114-4

Explainable framework for Lamb wave-based damage diagnosis

L. Lomazzi, M. Giglio & F. Cadini
Dipartimento di Meccanica, Politecnico di Milano, Milan, Italy

ABSTRACT: Structural health monitoring (SHM) has been widely employed in several engineering fields as support tool for condition-based maintenance policies aimed at increasing structural safety levels of thin-walled structures. In this field, tomographic algorithms are typically used to process ultrasonic guided waves and generate damage probability maps. More recently, machine learning-based frameworks, specifically neural networks, have also been employed to perform damage diagnosis, successfully overcoming some intrinsic limitations of classic methods. However, the black box-like nature of neural networks has built mistrust in such tools, thus slowing down their employment outside the academic world. This work aims at contributing to build trust in SHM frameworks based on neural networks by presenting an explainable machine learning framework for ultrasonic guided wave-based damage detection using a convolutional neural network for classification. The capabilities of the framework are demonstrated by means of a realistic numerical case study involving crack-like damage affecting a metal plate and the behavior of the convolutional neural network is explained through the layer-wise relevance propagation algorithm.

1 INTRODUCTION

Structural health monitoring has been widely employed in several engineering fields as support tool for condition-based maintenance policies aimed at increasing structural safety levels of thin-walled structures. Recently, machine learning-based methods, mainly employing deep learning tools such as convolutional neural networks (CNNs), have been proposed to process Lamb waves (LWs) and have been proven to outperform classic tomographic reconstruction-based approaches (Rautela and Gopalakrishnan, 2021). However, despite the satisfactory capabilities of deep learning-based methods in diagnosing damage, the black box-like nature of neural networks has built mistrust in such algorithms, thus limiting their application outside the academic world. While the aforementioned issue appears to be still unsolved in the SHM field, solutions have been proposed in the literature to open up those black boxes in other fields, thus giving rise to explainable artificial intelligence (XAI) algorithms (Thomas *et al.*, 2019).

In this work, an explainable framework for Lamb wave-based damage diagnosis using a CNN for classification to detect potential damage is presented. The LRP algorithm is integrated in the framework to explain the CNN behavior with the purpose of identifying the ultrasonic guided waves features learned by the machine learning tool and comparing them to the intuition of human experts, thus eventually building trust in the network.

2 METHODOLOGY

During each acquisition, LWs are gathered through a network of piezoelectric (PZT) sensors installed on the structure. The healthy baseline is subtracted from the acquired LWs, which are then corrupted with noise to enhance the database and collected into grayscale images (GSIs). Note that the term acquisition herein refers to the process of exciting LWs from one PZT device at a time, while the others serve as sensors for the elastic waves. GSIs are labeled using a binary variable, such that 0 corresponds to healthy conditions and 1 to damage conditions, and are used to train and test a CNN for classification performing damage detection. After training, the CNN is used to analyze new GSIs and each prediction is backpropagated through the CNN architecture by employing the layer-wise relevance propagation (LRP) framework, thus allowing assigning relevance values to each piece of input data (Bach *et al.*, 2015). Positive relevance values support the presence of a learned structure. Moreover, in this work input relevance values are aggregated into path relevance scores PR_l to rank the couples of PZT devices, as shown in Equation 1:

$$PR_l = \sum_{i \in S_l} R_i \tag{1}$$

where S_l is the set of samples of the LW corresponding to the l-th couple of PZT devices and R_i is the input relevance value of the i-th piece of input data.

DOI: 10.1201/9781003348450-291

3 CASE STUDY

An experimentally validated case study involving crack-like damage affecting a metal plate (Sbarufatti, Manson and Worden, 2014) is replicated to test the capabilities of the framework described in Section 2. Signals are acquired considering 43 damage positions and 12 crack lengths equally spaced in the range [5 mm, 60 mm]. Training is performed resorting to the Adam optimizer; the trained CNN labels the GSIs in the test set with accuracy 98.3%. Then, CNN predictions are backpropagated through the network architecture using a standalone Python implementation of the LRP algorithm (Lapuschkin *et al.*, 2016), according to the workflow described in Section 2. Finally, path relevance scores PR_l are computed and the four couples of PZT devices driving the CNN prediction, i.e., the four paths described by the highest PR_l values, are identified for each GSI. The results of the explainability approach are shown in Figure 1 for two reference crack positions.

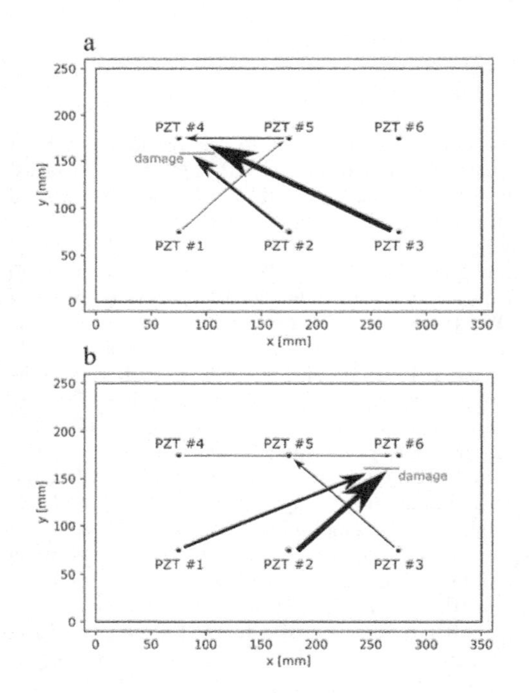

Figure 1. Couples of PZT devices driving the CNN prediction.

4 CONCLUSIONS

In this work an explainable framework for Lamb wave-based damage diagnosis has been proposed. The framework involves (i) a Lamb waves acquisition stage, (ii) a database generation stage, (iii) a prediction stage and (iv) an explainability stage. The performances of the proposed method have been demonstrated by means of a realistic numerical case study involving crack-like damage affecting a thin metal plate. Results have shown that the neural network bases its prediction on those paths crossing or passing closer to the damage. This result agrees with the physics governing Lamb waves propagation and with the intuition of human experts, according to which the closer Lamb waves pass to the damage, the more perturbed is their propagation. Hence, since the convolutional neural network behavior is compatible with that of human experts, trust may be built in such a tool.

Further work is ongoing to apply the same explainability approach to CNNs for regression performing damage localization and quantification. Moreover, further postprocessing the results of the explainability algorithm may allow underlying possible damage-related hidden physics in the propagation of Lamb waves.

REFERENCES

Bach, S. *et al.* (2015) 'On Pixel-Wise Explanations for Non-Linear Classifier Decisions by Layer-Wise Relevance Propagation', *PLOS ONE*, 10(7), p. e0130140. doi: 10.1371/JOURNAL.PONE.0130140.

Lapuschkin, S. *et al.* (2016) 'The LRP Toolbox for Artificial Neural Networks', *Journal of Machine Learning Research*, 17, pp. 1–5. Available at: https://github.com/happynear/caffe-windows (Accessed: 5 October 2021).

Rautela, M. and Gopalakrishnan, S. (2021) 'Ultrasonic guided wave based structural damage detection and localization using model assisted convolutional and recurrent neural networks', *Expert Systems with Applications*, 167, p. 114189. doi: 10.1016/J.ESWA.2020.114189.

Sbarufatti, C., Manson, G. and Worden, K. (2014) 'A numerically-enhanced machine learning approach to damage diagnosis using a Lamb wave sensing network', *Journal of Sound and Vibration*, 333(19), pp. 4499–4525. doi: 10.1016/j.jsv.2014.04.059.

Thomas, A. W. *et al.* (2019) 'Analyzing Neuroimaging Data Through Recurrent Deep Learning Models', *Frontiers in Neuroscience*, 0, p. 1321. doi: 10.3389/FNINS.2019.01321.

Current Perspectives and New Directions in Mechanics, Modelling and Design of Structural Systems – Zingoni (ed.)
© 2022 Copyright the Author(s), *ISBN 978-1-032-39114-4*

Application development of distributed fibre optic sensors for monitoring existing bridges

B. Novák, F. Stein & A. Dudonu
Institute for Lightweight Structures and Conceptual Design, University of Stuttgart, Germany

J. Reinhard
Schoemig-Plan Ingenieurgesellschaft mbH, Kleinostheim/Stuttgart, Germany

ABSTRACT: Concerning the ageing infrastructure and increasing traffic load, the importance of monitoring systems for structural health monitoring of existing bridges and early damage detection is constantly increasing. Due to progress in monitoring methods, recent developments offer new opportunities to monitor bridge structures over the entire geometrical length using distributed fibre optic sensors (DFOS), however, with little experience for practical use on structures outside the laboratory environment. To develop a monitoring system applicable under environmental conditions, different concrete surface applications of DFOS for crack detection and permanent monitoring of existing bridges were evaluated in laboratory tests. In subsequent field tests on a highway bridge, the potential of crack detection and monitoring traffic-induced crack developments using DFOS was successfully demonstrated when exposed to temperature, humidity, and vibrations from traffic.

1 INTRODUCTION

The ageing infrastructure and steady increase of heavy traffic loads impose rising demands on sophisticated monitoring systems for bridge assessment and maintenance management. Current developments of distributed fibre optic sensing (DFOS) based on Rayleigh scattering show new potentials for structural health monitoring. Distributed measurements provide new possibilities to detect and localise non-visible local developments of deteriorations at concrete structures. To study the applicability of DFOS as a permanent structural health monitoring on existing bridges, different concrete surface applications were analysed in laboratory tests and subsequently examined in field tests on a prestressed highway bridge under environmental and traffic impacts. The series of laboratory and field tests summarized in this paper is presented in detail in (Novák *et al.*, 2021).

2 LABORATORY TESTS

2.1 *Objectives and monitoring requirements*

For the application of distributed fibre optic sensors as a permanent monitoring of existing bridges, the ease of installation outside the laboratory environment is to be considered in addition to the required performance of the distributed fibre optic strain gauge cable. Distributed fibre optic sensors are commercially available with various coating materials and layers for protecting the fibre from fracture. The strain in the fibre depends on the strain transfer from the concrete surface to the adhesive and coating layers to the fibre core and is mitigated by the shear and bond deformations of the adhesive and coating layers (Ansari & Libo, 1998; Bassil *et al.*, 2020). To protect the fibre from mechanical fracture a sufficiently soft coating, a flexible adhesive or a combination of both can be used (Barrias *et al.*, 2019; Bassil *et al.*, 2020). A soft multi-layer coated fibre combined with a stiff adhesive (A1) and a single-layer coated fibre combined with an elastic adhesive (A2) were chosen for tests on concrete surfaces, see Figure 1. Concerning structural health monitoring of existing bridges, it was previously defined that the fibre must not fail up to a crack mouth opening of 0.6 mm. To protect the fibre from vandalism and drop-off due to undesired bond loss, the fibre was embedded in a groove and tested for different fibre-adhesive combinations.

2.2 *Tensile and bending tests*

The surface applications were examined on concrete specimens in tensile, three- and four-point bending tests. The results can be summarized as follows:

- DFOS provide a high strain sensitivity at a micrometre resolution, thus allowing cracks to be detected with their formation.

DOI: 10.1201/9781003348450-292

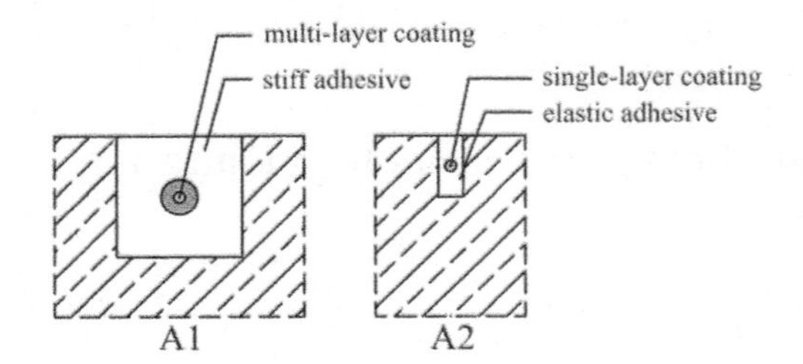

Figure 1. Studied surface applications in a groove.

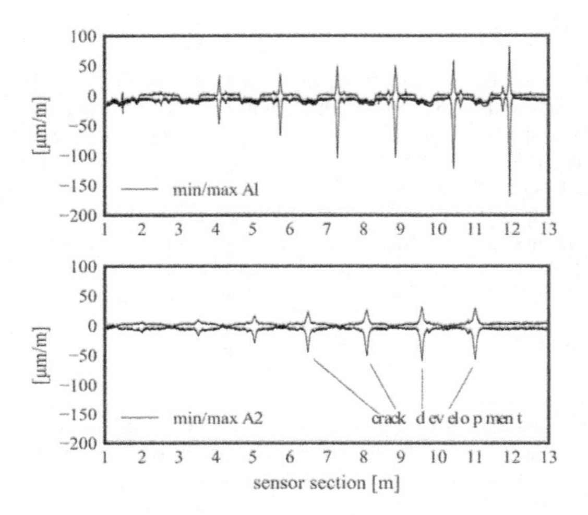

Figure 2. Strain curve of crack development in the coupling joint area resulting from traffic load.

- In the event of crack formation, a distinctive strain curve with a characteristic strain transfer length develops for each surface application. No significant changes for the specific strain transfer lengths could be observed with an increasing crack mouth opening up to the defined limit.
- A linear strain behaviour of both surface applications for an increasing crack mouth opening can be determined.
- No fibre fracture occurred for the defined limit. At strain higher 1.6% a bond deterioration of application A1 can be detected.

3 FIELD TESTS

3.1 *Objectives and setup*

Four-week field tests were realized on a prestressed highway bridge to confirm the applicability of the previously studied surface applications and measuring technology under environmental conditions and traffic-induced vibration impacts. A coupling joint with known crack formation from a presently installed permanent monitoring system was chosen as a sampling point. Displacement transducers (LVDT) and temperature sensors were installed distributed over the height on both sides of the web to verify the fibre-optic strain measurements.

3.2 *Field test results*

Figure 2 shows a detail of the fibre optic measurements at the coupling joint on the outer web. The crack displacement resulting from heavy load traffic can be identified from the developing characteristic strain curves of the respective surface application. The experimental results of the field test can be summarized as follows:

- Crack localisation and crack monitoring is both possible under environmental and traffic impacts. The potential of DFOS for crack monitoring of minor crack developments for the application on existing structures can be confirmed.
- Vibrations from traffic cause no significant errors in the measurement.
- Comparable results can be achieved for LVDT and fibre optic measurements of crack developments resulting from traffic loads
- Temperature compensation is mandatory for permanent monitoring due to thermal sensitivity of DFOS based on Rayleigh scattering.

4 CONCLUSIONS

The series of laboratory and field tests demonstrates the suitability of surface applied DFOS for structural assessment even under environmental impacts and shows the significant potential for permanent monitoring of bridges or as complement support for structural inspections of existing structures.

REFERENCES

Ansari, F. & Libo, Y. (1998) Mechanics of Bond and Interface Shear Transfer in Optical Fiber Sensors. *Journal of Engineering Mechanics*, 124(4), 385–394. Available from: https://doi.org/10.1061/(ASCE)0733-9399(1998)124:4(385).

Barrias, A., Casas, J.R. & Villalba, S. (2019) Distributed optical fibre sensors in concrete structures: Performance of bonding adhesives and influence of spatial resolution. *Structural Control and Health Monitoring*, 26(3), e2310. Available from: https://doi.org/10.1002/stc.2310.

Bassil, A., Chapeleau, X., Leduc, D. & Abraham, O. (2020) Concrete Crack Monitoring Using a Novel Strain Transfer Model for Distributed Fiber Optics Sensors. *Sensors 2020*, Volume 20(Issue 8). Available from: https://doi.org/10.3390/s20082220.

Novák, B., Stein, F., Reinhard, J. & Dudonu, A. (2021) Einsatz kontinuierlicher faseroptischer Sensoren zum Monitoring von Bestandsbrücken. *Beton- und Stahlbetonbau*. Available from: https://doi.org/10.1002/best.202100070.

Current Perspectives and New Directions in Mechanics, Modelling and Design of Structural Systems – Zingoni (ed.)
© 2022 Copyright the Author(s), ISBN 978-1-032-39114-4

Real-time health monitoring of civil structures by online hybrid learning techniques using remote sensing and small displacement data

A. Entezami & C. De Michele
Department of Civil and Environmental Engineering, Politecnico di Milano, Milano, Italy

A. Nadir Arslan
Finnish Meteorological Institute (FMI), Helsinki, Finland

ABSTRACT: Structural Health Monitoring (SHM) has become essential in modern societies due to critical importance of economic and human losses caused by occurring damage. To avoid any catastrophic event, many data-based methods under machine learning concepts have been proposed. On the other hand, SHM via Synthetic Aperture Radar (SAR) images has become popular among civil engineers. However, an SAR-based SHM strategy itself contains major challenges such as considering insufficient displacement data from SAR images and direct data analysis. To deal with these challenges, this article proposes online hybrid learning methods to detect damage via small displacement data from SAR images. Small displacement samples extracted from satellite images is used to assess the accuracy of proposed online hybrid methods. Results show that the proposed methods are successful in pre-collapse prediction of the bridge even under strong EOV conditions.

1 INTRODUCTION

Structural health monitoring (SHM) is a practical technology for assessing the safety and stability of civil structures (Rizzo and Enshaeian, 2021). Recently, SHM based on synthetic aperture radar (SAR) images have been popular among researchers (Biondi et al., 2020, Bakon et al., 2020, Amoroso et al., 2020). However, the underlying limitations of a SAR-based SHM strategy are the problem of environmental and/or operational variability conditions, the availability of small data, and directly analysis such data (Selvakumaran et al., 2018, Qin et al., 2017, Milillo et al., 2016, Milillo et al., 2019).

Thus, this article aims to propose automated online hybrid learning methods compatible with an unsupervised learning class on the basis of an online deep transfer learning (ODTL) algorithm using sequential auto-associative neural networks (AANNs).

2 PROPOSED ONLINE HYBRID LEARNING METHODS

2.1 *Deep transfer learning*

The process of feature normalization by the proposed ODTL algorithm is the heart of the proposed methods. In this step, the main objective is to learn a deep network of sequential AANNs (Kramer, 1992). The deep transfer learning (DTL) is intended to cope with one of the major challenges in machine learning regarding the problem of hyperparameter optimization (e.g., the neuron sizes of the hidden layers).

2.2 *Online deep transfer learning*

In this section, the improved version of DTL under an online learning algorithm is proposed. The proposed ODTL algorithm consists of two stages. The first stage is an online learning strategy until the stopping condition terminates the learning process. In this case, the algorithm makes a decision that the new data no longer belongs to the undamaged or normal condition and it is most likely concerned with a damaged state. To put it another way, the algorithm alarms the occurrence of damage. Hence, the second stage starts by learning a one-step-ahead sequential AANN.

2.3 *Feature classification by EMSD*

In order to make the final decision on the current state of the structure, the proposed online learning methods do not explicitly use an alarming threshold to compare novelty score with this limit, but instead, a scalar index is defined to detect damage in an online manner. This index is the Euclidean norm of the MSD values of the residual samples at each iteration step of the ODTL algorithm. It should be noted that when the second stage of the ODTL algorithm starts, this means that the structure suffered from damage.

3 APPLICATION: THE TADCASTER BRIDGE

The Tadcaster Bridge is a historic bridge over the River Wharfe in Tadcaster, United Kingdom. This bridge partially collapsed on December 29th, 2015. Figure 1 illustrates the Tadcaster Bridge.

DOI: 10.1201/9781003348450-293

Figure 1. The Tadcaster Bridge (Selvakumaran et al., 2018).

Figure 2 illustrates the outputs of the proposed methods for damage detection. As can be seen, the EMSD quantities on 15th November, 2015 regarding the image #44 is larger than the previous images. The comparison between the HMC-ODTL-EMSD and SLS-ODTL-EMSD methods demonstrates that

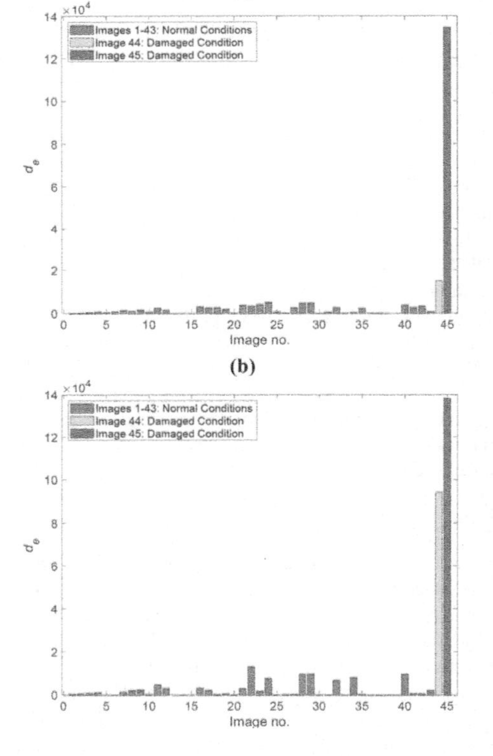

Figure 2. Decision-making for early damage detection: (a) HMC-ODTL-EMSD, (b) SLS-ODTL-EMSD.

the latter provides better and larger output and decision-making than the former regarding the image #44. The other important conclusion is that the EMSD values on 26th November, 2015 associated with the image #45 suddenly increase so that these are the largest distance quantities among the other values. This clearly indicates the growth of damage or extend the damage severity. Therefore, one can state that the proposed methods could quantitatively estimate the severity of damage. However, it can be concluded that the proposed SLS-ODTL-EMSD method can provide more appropriate results of early damage detection compared to the HMC-ODTL-EMSD technique.

4 CONCLUSIONS

This article intended to propose new online hybrid learning methods HMC-ODTL-EMSD and SLS-ODTL-EMSD for early damage detection and pre-collapse prediction of civil structures. The small displacement samples of the Tadcaster Bridge obtained from some satellite images of TerraSAR-X were applied to validate the proposed methods. The main conclusions of this study can be summarized as follows: (i) the proposed idea for the augmentation of the small displacement data through the HMC and SLS provided better observations of structural responses, structural behavior, and EOV conditions, (ii) the HMC outperformed the SLS in this evaluation, (iii) the online data normalization and online damage detection strategies accurately detected the damaged state of the bridge.

REFERENCES

Amoroso, N., Cilli, R., Bellantuono, L., Massimi, V., Monaco, A., Nitti, D. O., Nutricato, R., Samarelli, S., Taggio, N., Tangaro, S., Tateo, A., Guerriero, L. & Bellotti, R. (2020) PSI Clustering for the Assessment of Underground Infrastructure Deterioration. *Remote Sensing*, 12, 3681.

Bakon, M., Czikhardt, R., Papco, J., Barlak, J., Rovnak, M., Adamisin, P. & Perissin, D. (2020) remotIO: A Sentinel-1 Multi-Temporal InSAR Infrastructure Monitoring Service with Automatic Updates and Data Mining Capabilities. *Remote Sensing*, 12, 1892.

Biondi, F., Addabbo, P., Ullo, S. L., Clemente, C. & Orlando, D. (2020) Perspectives on the Structural Health Monitoring of Bridges by Synthetic Aperture Radar. *Remote Sensing*, 12, 3852.

Kramer, M. A. (1992) Autoassociative neural networks. *Computers & Chemical Engineering*, 16, 313–328.

Milillo, P., Giardina, G., Perissin, D., Milillo, G., Coletta, A. & Terranova, C. (2019) Pre-collapse space geodetic observations of critical infrastructure: the Morandi Bridge, Genoa, Italy. *Remote Sensing*, 11, 1403.

Milillo, P., Perissin, D., Salzer, J. T., Lundgren, P., Lacava, G., Milillo, G. & Serio, C. (2016) Monitoring dam structural health from space: Insights from novel InSAR techniques and multi-parametric modeling applied to the Pertusillo dam Basilicata, Italy. *International Journal of Applied Earth Observation and Geoinformation*, 52, 221–229.

Qin, X., Liao, M., Yang, M. & Zhang, L. (2017) Monitoring structure health of urban bridges with advanced multi-temporal InSAR analysis. *Annals of GIS*, 23, 293–302.

Rizzo, P. & Enshaeian, A. (2021) Challenges in Bridge Health Monitoring: A Review. *Sensors*, 21, 4336.

Selvakumaran, S., Plank, S., Geiß, C., Rossi, C. & Middleton, C. (2018) Remote monitoring to predict bridge scour failure using Interferometric Synthetic Aperture Radar (InSAR) stacking techniques. *International Journal of Applied Earth Observation and Geoinformation*, 73, 463–470.

Current Perspectives and New Directions in Mechanics, Modelling and Design of Structural Systems – Zingoni (ed.)
© 2022 Copyright the Author(s), *ISBN 978-1-032-39114-4*

Permanent tunnel lining monitoring system for the purpose of further design optimization

M. Jonáš & J. Zatloukal
Experimental centre, Faculty of Civil engineering, Czech Technical University in Prague, Czech Republic

ABSTRACT: Thermal actions considered during the design of tunnel structures in the Czech Republic are not sufficiently defined by standards, therefore permanent monitoring system was designed and prepared for purpose of thermal and strain monitoring of tunnel linings. This system is installed into three tunnel structures, its design is adjusted to ensure its similarity in all the structures. Monitored spots of the system are situated in central and boundary segments and the system measures temperature profiles and strain in crown and wall of tunnel segments. Measurement should provide sufficient basis for statistical evaluation of temperature changes reached in tunnel linings during the extended period of time. Result of tunnel lining monitoring will be an adjusted design temperature profile for purpose of tunnel lining design optimization, which should lead to the reduction of final costs of newly constructed tunnels.

1 INTRODUCTION

In the Czech Republic, consideration of thermal actions on tunnel linings is not clearly defined by standards and the technical conditions issued by the infrastructure management organizations. The basic standard defining loads on tunnel linings is the ČSN 73 7501, however, this standard does not describe thermal actions on tunnel linings. In case of railway tunnels, TKP ČD prescript is used. Document prescribes temperature values in certain spots, resulting in vertical temperature difference of 10 °C across the thickness of lining. Usually, this difference is applied only to the secondary lining, as the secondary lining is considered load bearing element of whole tunnel lining.

This situation in tunnel structures design led to installation of measurement systems into more tunnels, which were constructed in last 20 years. Specifically, measurements by strain gauges with integrated thermometers for purpose of temperature monitoring were performed in the Klimkovice tunnel (Šourek et al., 2008).

According to these measurements, usually used values from TKP ČD were unrealistically high. These measurements confirmed the anticipated significantly lower reached vertical temperature differences across the thickness of lining. Measured values of temperature difference did not exceed 3 °C during the summer period, in winter period were differences even lower (up to 1.5 °C). (Ďuriš et al., 2013) However, for proper consideration of vertical temperature difference on tunnel lining, more measurements should be performed and results should be statistically processed to provide sufficient basis for lining analysis.

2 PERMANENT MONITORING SYSTEM

In the aim of obtaining quantity of temperature loading data, permanent monitoring system was designed and built into the linings of three tunnels. Monitoring system measures strain and temperature in defined spots of lining in chosen cross-sections of tunnels.

System is designed with maximal similarity in each structure for economy. System can be modified for specific purposes and allows wide usage for monitoring measurements at various structures.

2.1 *Sensor deployment*

Sensors installed in each segment tunnel lining are deployed according to one constant scheme. Basic division of cross-sections is made to obtain data for spots next to portal and for zones more far from portal in central parts of the tunnel. Therefore, first cross-section is usually situated right next to the portal. Second cross-section is situated in inner part of the tunnel, at least about 150 m from the nearest portal.

Basic deployment on cross-section is the same in each measured segment. Temperatures are measured in two profiles, one is situated in wall, about 1.5-2 m from the bottom of lining, second one is situated in the crown of tunnel, if possible, in the highest point of tunnel cross-section. Each temperature profile consists of 7 or 5 (according to lining thickness) thermometers divided according to usual vertical temperature division in concrete structures.

DOI: 10.1201/9781003348450-294

Strain is measured in three spots in the cross section. In each spot, one pair of strain gauges was installed, each strain gauge was fixated to reinforcement bars. Two spots are situated in wall, in the same height as thermometers, third spot in the top of the vault, next to the thermometers. Position of sensors is shown in Figure 1.

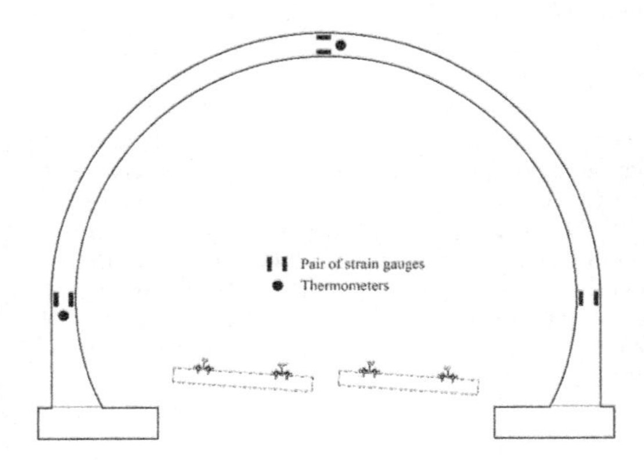

Figure 1. Deployment of sensors in the cross-section of the tunnel.

2.2 *Preparation, installation and data acquisition system*

To ease installation, secure right position and protect sensors from damage, thermometers were pre-concreted into tiny casing beams. Dimensions of these casing beams were given by the reinforcement grid and height was given by lining thickness.

Prior to sensor encasement in casing beams, sensors were covered by specific type of high-performance concrete shell made of MSK 2 substance (Bažantová et al., 2016) to ensure insulation between metallic parts of reinforcement and sensor itself, therefore measurement should not be influenced by local thermal changes caused by higher thermal conductivity of steel. Measurement will be processed by the data acquisition system designed for purpose of relatively simple monitoring systems, where higher number of individual systems have to be used and costs have to be reduced.

Core of data acquisition system is Raspberry Pi computer, which processes its whole operation. Resistive sensor measurement will be performed by LucidControl I/O RTD input units. Vibrating wire strain gauges measurement will be operated by VibWire-108-485 units. Portal system will be connected via wireless LTE modem to the internet and will send measured data to the server regularly.

3 RESULT EVALUATION & FURTHER PROGRESS

Measurements should be processed for extended period of time; basic period is at least one year to cover all the weather seasons. However, to obtain extensive data amount, the monitoring system should be in operation for few years, ideally ten years.

During this period, extreme thermal action situations should be evaluated, typically uniform temperature changes and vertical temperature differences. After specific period, vertical temperature differences should be processed statistically as the 98% quantile, to provide more detailed and secure data of thermal actions on tunnel linings. Strain measurements will be evaluated with combination of computation data about lining loading and deformation to verify real versus expected states of lining.

4 CONCLUSION

Installation of permanent monitoring systems into tunnel linings is the initial step for obtaining wider knowledge about thermal actions on tunnel linings. Relatively simple and inexpensive measurement systems will be operated long-term. Unfortunately, works and processes progress on construction sites did not allow start of measurement operation. Therefore, no measurement results could be presented in the paper.

According to previous performed measurements in Klimkovice tunnel (Ďuriš et al., 2013), expected data will likely not exceed design assumed actions. Adjusted thermal action effects applied in design of new structured might lead to notable cost and material reduction, what would be suitable, especially today, in the times of environmental impact reduction. Therefore, proper monitoring by measurement systems is suitable on any structure type, where design optimization can be made.

REFERENCES

Bažantová et al. (2016). Multifunkční silikátový kompozit programovatelných vlastností nejen pro rychlé opravy cementobetonových konstrukcí. *Materiály pro stavbu*, 34–37.

Ďuriš et al. (2013). Thermal loading action on final linings of underground structures. *Tunel*, 44–52.

Šourek et al. (2008). Measurement fo deformations amd temperatures on final tunnel liners. *Tunel*, 70–76.

Current Perspectives and New Directions in Mechanics, Modelling and Design of Structural Systems – Zingoni (ed.)
© 2022 Copyright the Author(s), ISBN 978-1-032-39114-4

On the data-driven damage detection of offshore structures using statistical and clustering techniques under various uncertainty sources: An experimental study

M. Salar
Department of Civil and Environmental Engineering, Politecnico di Milano, Milan, Italy

A. Entezami
Department of Civil and Environmental Engineering, Politecnico di Milano, Milan, Italy
Department of Civil Engineering, Faculty of Engineering, Ferdowsi University of Mashhad, Mashhad, Iran

H. Sarmadi
Department of Civil Engineering, Faculty of Engineering, Ferdowsi University of Mashhad, Mashhad, Iran

C. De Michele & L. Martinelli
Department of Civil and Environmental Engineering, Politecnico di Milano, Milan, Italy

ABSTRACT: Damage detection procedure of offshore structures based on data-driven techniques is of paramount importance to ensure their safety and integrity, especially in real-world applications. Therefore, the main aim of this article is to propose an improved clustering-based method for data-driven damage detection with the aid of a distance scaling technique. The feasibility and reliability of the method presented in this study is exemplified by application in a laboratory jacket-type offshore platform under different damage scenarios along with several comparative studies. Results are demonstrated to be effective and successful in detecting early damage of the offshore structure in the presence of various uncertainty sources.

1 INTRODUCTION

The offshore structure industry has experienced fast development during past recent decades, and there are a number of offshore structures all over the world, which have already outlived their intended service life. Hence, there is a great attention in applying Structural Health Monitoring (SHM) in offshore structures in order to ensure their safety and integrity and to prevent potential catastrophic events (Vestli et al. 2017). As a result, some techniques have been proposed and developed to detect, localize and estimate the severity of damage in offshore structures.

2 DAMAGE DETECTION TECHNIQUES

2.1 *The k-means clustering*

Suppose that $\mathbf{X} = [\mathbf{x}_1, \ldots, \mathbf{x}_n] \in \mathbb{R}^{n\times r}$ is a matrix containing n observations and r variables. The k-means clustering technique divides the data set $\mathbf{X}$ into k clusters, which the number of clusters (k) is known as *a priori*. The objective function of k-means clustering is defined as follows:

$$Q_C(\mathbf{X}) = \sum \min D(\mathbf{c}_i, \mathbf{x}_j) \tag{1}$$

2.2 *Proposed clustering method*

The proposed clustering-based technique is on the basis of the k-means algorithm. This technique also introduces a new damage indicator based on an $L_{p,r}$-distance metric to increase damage detectability so that p and r are scalar values implying the powers of the $L_{p,r}$-distance. Assume the two arbitrary vectors $\mathbf{x}$ and $\mathbf{w}$, the $L_{p,r}$-distance measure is defined as follows (Deza and Deza 2016):

$$L_{p,r} = \left(\sum |\mathbf{x} - \mathbf{w}|^p\right)^{\frac{1}{r}} \tag{2}$$

Therefore, a new damage indicator according to the definition of the $L_{p,r}$-distance by employing the feature vector of the testing data and the cluster means can be defined as follow:

$$DI = \min\left(L_{p,r}(\mathbf{w} - \mathbf{c}_1), \ldots, L_{p,r}(\mathbf{w} - \mathbf{c}_k)\right) \tag{3}$$

where

$$L_{p,r}(\mathbf{w},\mathbf{c}_j) = \left(\sum |\mathbf{w} - \mathbf{c}_j|^p\right)^{\frac{1}{r}} \tag{4}$$

2.3 *Damage threshold estimation method*

In this study, the measurement data of the undamaged state of the offshore jacket platform is assumed

DOI: 10.1201/9781003348450-295

Table 1. The undamaged and damaged conditions of the experimental offshore jacket platform.

Case no.	Structural Condition	Description
1	Undamaged	Baseline
2	Undamaged	Baseline + Added 40 kg on deck
3	Undamaged	Baseline + Added 80 kg on deck
4	Damaged	Removing horizontal member in 3rd level + Added 80 kg on deck
5	Damaged	Inducing 5% damage in diagonal member of 5th level and 45% damage in diagonal member of 3rd level + Added 80 kg on deck

Figure 1. The overall view of the experimental offshore jacket platform.

Table 2. Numbers and percentages of Type I, Type II, and total errors in detecting damage for experimental offshore jacket.

Case no.	Type I	Type II	Total
4	537(5.37%)	656(6.52%)	(5.965%)
5	537(5.37%)	1276(12.76%)	(9.065%)

to be normally distributed, and the threshold is defined as the lower bound of 95% Confidence Interval (CI), which is expressed as:

$$T = \mu \pm \sigma f_{(1-\alpha)} \quad (5)$$

3 APPLICATION TO THE EXPERIMENTAL OFFSHORE JACKET PLATFORM

The tested structure is a laboratory jacket-type offshore platform. The platform including four main legs, horizontal and diagonal bracing members, as shown in Figure 1. Several damage scenarios are simulated in the tests as listed in Table 1.

Figure 2 depicts the results of early damage detection in the experimental offshore jacket platform using the Chebychev distance metric for damage scenarios 4 and 5, where the horizontal dashed line is indicative of the threshold limit gained by the

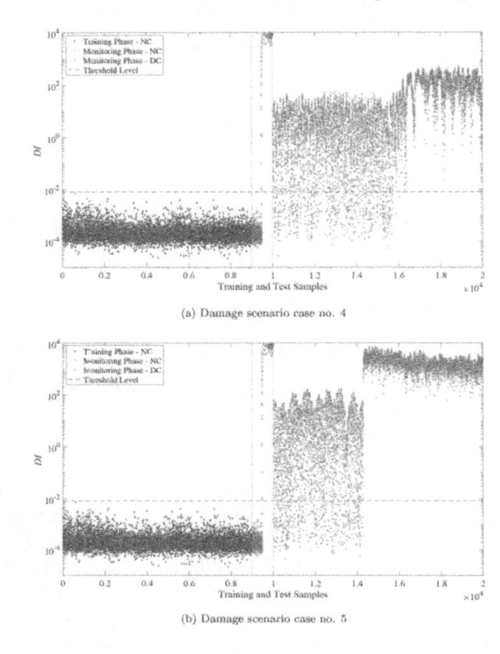

Figure 2. Damage detection using the proposed method.

standard CI method. In addition, the number and percentages of Type I, Type II, and total errors for experimental offshore jacket structure are shown in Table 2.

4 CONCLUSIONS

In this article, the new clustering method was developed based on the k-means algorithm with the aid of a novel approach to choose an appropriate number of cluster for dealing with the effects of environmental variations. Moreover, the application of the $L_{p,r}$-distance metric to the k-means clustering algorithm was presented in order to increase the detectability of damage. The results show that the proposed method is remarkably effective for identifying damage in the presence of environmental changes.

REFERENCES

Deza, M. M. & E. Deza (2016). Distances in graph theory. In *Encyclopedia of distances*, pp. 277–311. Springer.

Vestli, H., H. G. Lemu, B. T. Svendsen, O. Gabrielsen, & S. C. Siriwardane (2017). Case studies on structural health monitoring of offshore bottom-fixed steel structures. In *The 27th International Ocean and Polar Engineering Conference*. OnePetro.

Current Perspectives and New Directions in Mechanics, Modelling and Design of Structural Systems – Zingoni (ed.)
© 2022 Copyright the Author(s), *ISBN 978-1-032-39114-4*

Sustainable structural health monitoring using e-waste and recycled materials

P. Peralta & K. Smarsly
Institute of Digital and Autonomous Construction, Hamburg University of Technology, Germany

ABSTRACT: With the apparent effects of climate change, sustainability has come to the forefront in civil engineering, aiming for reliable systems and structures of low environmental impact. However, the market for electronic devices is growing rapidly, generating a non-negligible amount of electronic waste (e-waste). Components in e-waste, such as sensors from disposed electronic devices, may be recycled and reused in structural health monitoring (SHM) systems. In this paper, the design and implementation of a sustainable wireless SHM system composed of e-waste and recycled materials is presented. The sustainable SHM system is validated in field tests on a pedestrian bridge using a conventional wireless SHM system as benchmark. As a result, it is demonstrated that e-waste components and recycled materials are viable alternatives to implement sustainable SHM systems. With sustainable SHM systems, a contribution to minimize the environmental impact of the devices may be achieved.

1 INTRODUCTION

Fueled by the digital transformation, the market for electronic devices is growing rapidly worldwide, generating substantial amounts of electronic waste (e-waste). The problem also becomes apparent in the architecture, engineering, and construction (AEC) industry, which is shifting towards circular economy approaches (i.e. closed-loop production cycles) accelerated by Internet of Things (IoT) solutions (United Nations Environment Programme 2020). In particular, structural health monitoring (SHM) systems are characterized by dense arrays of sensors installed in the structures. The sensors are attached to sensor nodes, also installed in the structures, that are devised to collect, process, and analyze the sensor data as a basis to assess the structural condition (Luckey et al. 2022). SHM systems enable damage detection to facilitate maintenance and sustainable retrofitting of structures, albeit the environmental impact of the SHM systems themselves is usually neglected. Hence, it is necessary to develop sustainable approaches towards designing sensor nodes to reduce the environmental impact of SHM systems.

Based on the circular energy framework presented in (Ingemarsdotter et al. 2019), concepts that reuse and recycle resources, such as reusing sensors in SHM systems, have not yet received enough attention. Taking advantage of e-waste and the reusability of microcomputers, an additional service life may be given to e-waste sensors (i.e. sensors retrieved from disposed or outdated electronic devices) by deploying the e-waste sensors in sensor nodes for SHM systems, reducing the waste generated by SHM systems. To reuse e-waste sensors in SHM systems, special care must be given to select e-waste sensors that are compatible with microcomputers and that are capable of measuring the parameters relevant to SHM, such as vibration (e.g. accelerometers), strain and force (e.g. load cells), and temperature (e.g. thermistors).

In this paper, the design and implementation of a sustainable SHM system is presented. The sustainable SHM system is composed of sensor nodes assembled from e-waste sensors and recycled materials. Sensors found in smartphones and disposed refrigerators are connected to microcomputers to measure acceleration and temperature. The sustainable SHM system is validated in a field test on a pedestrian bridge and compared with a conventional SHM system serving as a benchmark system.

2 A SUSTAINBLE SHM SYSTEM

The sustainable SHM system applies circle economy concepts to reuse and to recycle resources for hardware in SHM systems, while maintaining accuracy. The software design provides real-time sensing, embedded computing, and IoT connectivity (Figure 1). The sustainable SHM system is composed of two sensor nodes, (i) an outdated smartphone and (ii) a sensor node based on Raspberry Pi microcomputers, and a cloud server. The sensor nodes measure acceleration and temperature, which

DOI: 10.1201/9781003348450-296

are essential to monitor the condition of civil engineering structures.

Using the outdated smartphone as a sensor node, the data of the accelerometer embedded into the main chip board of the smartphone can be accessed easily and transmitted via Bluetooth to the Raspberry-Pi-based sensor node. Using the Raspberry-Pi-based sensor node, an NTC thermistor from a disposed refrigerator is attached to the Raspberry Pi microcomputer, which is capable of reading and processing the sensor output signals. Algorithms are embedded into the Raspberry-Pi-based sensor node to perform on-board data analysis of the acceleration data using operational modal analysis methods, such as fast Fourier transforms (FFT) and peak picking.

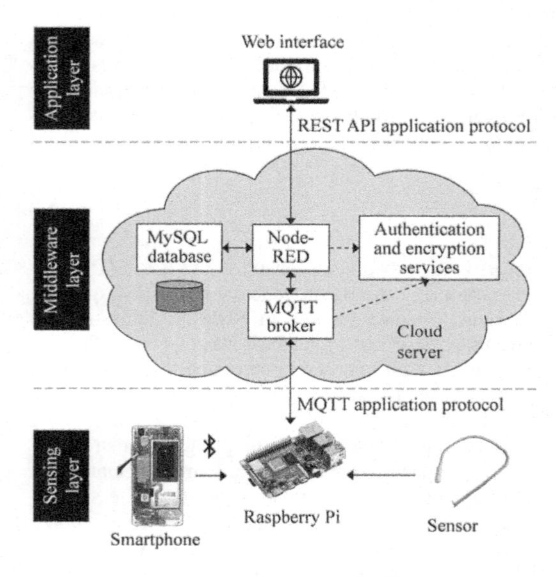

Figure 1. Architecture of the sustainable SHM system.

3 VALIDATION OF THE SUSTAINABLE SHM SYSTEM

The sustainable SHM system is validated in a field test on a pedestrian bridge over the Lotse canal at Harburg Inland Port, Hamburg, Germany. The validation aims to determine the performance of the e-waste sensors by comparing the sensor data recorded by the sustainable SHM system with sensor data recorded by a benchmark SHM system. Moreover, the data analyses performed by the algorithms embedded into the sensor nodes of the sustainable SHM system are used for validation.

The structural response of the bridge (i.e. vibration) under pedestrian traffic is monitored to perform an operational modal analysis, including environmental conditions. As shown in Figure 2, the bridge is instrumented with the sustainable SHM system and a benchmark SHM system for comparison.

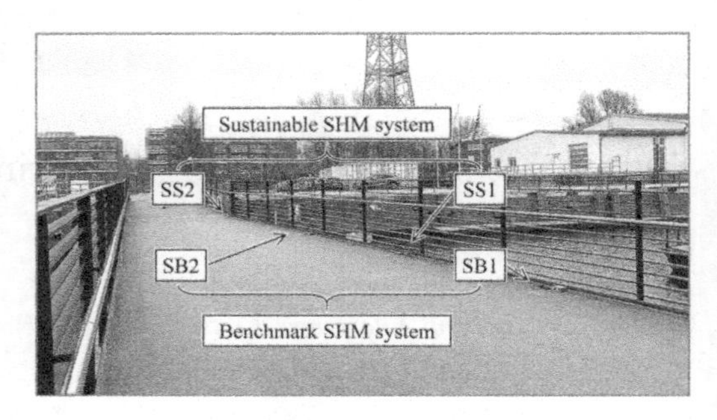

Figure 2. Setup of the field test.

The accuracy of the sustainable SHM system is evaluated by determining the frequencies of the vibration modes, which are in accordance with the benchmark SHM system with a difference of 0.049 Hz for the first vibration mode (f_1) and of 0.147 Hz for the second vibration mode (f_2). Furthermore, the temperature data is compared using the cosine similarity coefficient, yielding a similarity coefficient of 0.999 with a maximum difference of 0.77 °C, indicating high similarity between the two SHM systems.

4 SUMMARY AND CONCLUSIONS

A concept to reduce the waste generated by SHM systems has been proposed, addressing the environmental impact of SHM systems. Developing sensor nodes using e-waste components and recycled materials is a viable alternative to implement sustainable SHM systems. The reusability of e-waste components is limited by the accessibility, the functioning principles, and the compatibility with microcomputers (e.g. regarding input voltage, amperage, and output signals). By reusing e-waste sensors, the service life of sensors and devices may be extended, reducing waste and contributing to minimize the environmental impact of SHM systems.

REFERENCES

Ingemarsdotter, E., Jamsin, E., Kortuem, G., & Balkenende, R. (2019). Circular Strategies Enabled by the Internet of Things – A Framework and Analysis of Current Practice. Sustainability, 11(20), 5689.

Luckey, D., Fritz, H., Legatiuk, D., Dragos, K., & Smarsly, K. (2020). Artificial intelligence techniques for smart city applications. In: Proceedings of the International ICCCBE and CIB W78 Joint Conference on Computing in Civil and Building Engineering 2020. Sao Paolo, Brazil, 08/ 18/2020.

United Nations Environment Programme (2020). 2020 Global Status Report for Buildings and Construction: Towards a Zero-emission, Efficient and Resilient Buildings and Construction Sector.

Current Perspectives and New Directions in Mechanics, Modelling and Design of Structural Systems – Zingoni (ed.)
© 2022 Copyright the Author(s), ISBN 978-1-032-39114-4

Corrosion of fasteners in concrete: Literature review and discussion of current test methods

M. Cervio & G. Muciaccia
Department of Civil and Environmental Engineering, Politecnico di Milano, Milan, Italy

ABSTRACT: The aim of this paper is to present the existing knowledge on fasteners' corrosion behavior in concrete. The paper investigates the available literature and discusses the most significant test methods currently in use for the evaluation of fastener's corrosion performance in concrete. In this context, design standards and product norms often provide inconsistent requirements, while assessment procedures of non-zinc coatings are left to case-by-case implementation. Moreover, the study of existing test methods revealed that, beside the hydrogen embrittlement test for concrete screws, none of the other considered test methods are specifically conceived to investigate the fastener's corrosion behavior at the steel-to-concrete connection scale, nor they account for service conditions. Therefore, the significance for both designers and manufacturers of the remarks pointed out in this paper.

1 INTRODUCTION

Durability is an essential characteristic of fasteners for use in concrete. Hence, corrosion damage impairing the functioning and the load-bearing behavior of concrete fasteners during their design service life must be avoided.

In this context, the current European Assessment Documents (EADs) in case of corrosion of fasteners for use in concrete provide a conservative approach. However, it might be questionable that such design provisions may duplicate and/or override the one given by the related design standard of fastenings for use in concrete EN 1992-4:2018 (2018).

The literature review presented in this paper tries to clarify the background of current specifications as well as to highlight possible applicative issues. Finally, the study of existing test methods is of fundamental importance in pointing out that, beside the case of hydrogen embrittlement test for concrete screws, none of the other considered test methods are specifically conceived to investigate the fastener's corrosion behavior at the steel-to-concrete connection scale, nor they account for service conditions.

2 FUNDAMENTALS OF CORROSION

2.1 *General vocabulary*

ISO 8044:2020 (2020) associates the term "corrosion" to the process, that is a physicochemical interaction between a metallic material and its environment. For construction products i.e. fasteners for use in concrete, this interaction is of electrochemical nature and it generally leads to significant impairment of the metal.

2.2 *Basics of corrosion*

In case of wet corrosion of construction products, the anodic reaction is an oxidation process in which the metal loses one or more electrons and corrosion products are produced. In this context, the corrosion current density is by convention the partial current due to metal oxidation per anodic unit area. The corrosion current density is equivalent to the corrosion rate according to Faraday's law. The relative sizes and locations of anodic and cathodic areas are important variables affecting corrosion rates.

3 CORROSION FORMS IN CONCRETE FASTENERS

To the scope of this paper, corrosion forms in concrete fasteners may be grouped into four main types: (i) uniform, (ii) localized, (iii) selective, (iv) environmental induced. Fasteners in real conditions may experience a single form of corrosion or a combination of them.

Uniform corrosion is the most common form of corrosion since it may affect exposed parts of fasteners to the atmosphere and/or to carbonated concrete. Localized corrosion is located at discrete sites of the fastener's metal surface exposed to the corrosive environment. Typical forms of localized corrosion affecting fasteners in concrete are: pitting, crevice

DOI: 10.1201/9781003348450-297

and galvanic corrosion. Selective corrosion comprises dealloying and intergranular corrosion forms. Only the latter form is generally relevant for fasteners in concrete, as it may affects some stainless steels when subjected to quenching. Environmental induced corrosion forms refer to the conjoint action of different corrosive agents and environments. Among environmental induced corrosion forms the ones of interest for fasteners in concrete are: stress corrosion hydrogen embrittlement corrosion. consists in a decrease of ductility of a metal due to absorption of hydrogen.

4 CORROSIVITY IN CONCRETE AND PROTECTION OF FASTENERS

4.1 *Corrosivity in concrete*

Durability of fasteners in concrete is discussed in an informative annex of EN 1992-4:2018 (2018). The mentioned standard provides guidance by simplifying and adapting exposure classes originally developed for corrosion of reinforcement in concrete structures.

4.2 *Corrosion protection methods*

Among corrosion resistant materials, stainless steels are the preferred choice, while concerning protective coatings, zinc is particularly useful to sacrificially protect fastener's parts exposed to the atmosphere or parts into carbonated concrete.

Electroplating is the most used method for fastener coating deposition. Hot-dip galvanizing is a typical process to deposit zinc coatings by immerging the fasteners into a bath of molten zinc. Mechanical plating is a process in which metallic coatings are deposited onto small metallic parts, such as fasteners, using kinetic energy at room temperature. Zinc flake coating is a non-electrolytically applied metallic coating. Sherardizing coatings consist of Zn-Fe alloy layers obtained by sherardizing. Finally, there are nonmetallic coatings which are generally categorized into two groups: (i) organic, e.g. paints; (ii) inorganic, e.g. phosphates.

5 CURRENT TESTING METHODS

5.1 *Hydrogen embrittlement*

In Europe, hydrogen embrittlement testing is required for product qualification of concrete screws and the method is detailed in Section 2.2.1.3 of EAD 330232-00-0601 (2016). The outcome of the test is a reduction factor obtained by comparing the residual tensile capacity of the anchor after 100 h of exposure to the above-mentioned conditions with respect to results obtained in a reference test series.

5.2 *Salt spray*

Salt spray tests are performed by placing samples in test cabinets that have been designed and operated in accordance with a specific standard. The period of testing shall be as designated by the specification covering the material or product being tested.

At international level, i.e. ISO 9227:2017 (2017), the neutral salt spray (NSS) is the oldest and most widely used accelerated corrosion test. The NSS equipment allows for spraying a solution with (50 ± 5) g/l of sodium chloride. Recommended periods of exposure are within a range from 2 h to 1008 h.

5.3 *Sulfur dioxide*

Exposure of metals to moist air containing sulfur dioxide leads to a corrosion form resembling that occurring in industrial environments. This method is detailed in ISO 6988:1995 (1995). The appropriate evaluation criterion and test duration are usually provided by the product standard related to the tested object. In conclusion, as for other accelerated tests, results should not be taken as a direct guide to corrosion resistance in service.

6 CONCLUSIONS

This paper presents an initial literature review of corrosion of fasteners in concrete.

The analysis of the current state of the art highlighted that the definition of environment corrosivity and the consequent selection of a suitable corrosion protection method are key factors for both manufacturers and designers. However, current product and design standards lack of consistent and specific provisions for fasteners for use in concrete.

The study of such existing test methods revealed that, the current approach does not account for the real complexity of the corrosion system at the structural scale. Therefore, the need to propose modifications to existing assessment methods to account for the mentioned drawbacks.

REFERENCES

EAD 330232-00-0601 2016. Mechanical fasteners for use in concrete. Brussels:EOTA.

EN 1992–4:2018 2018. Eurocode 2 – Design of concrete structures – Part 4: Design of fastenings for use in concrete. Brussels: CEN.

ISO 6988:1995 1995. Metallic and other non-organic coatings – Sulfur dioxide test with general condensation of moisture.

ISO 8044:2020 2020. Corrosion of metals and alloys – Vocabulary.

ISO 9227:2017 2017. Corrosion tests in artificial atmospheres – Salt spray tests.

Current Perspectives and New Directions in Mechanics, Modelling and Design of Structural Systems – Zingoni (ed.)
© 2022 Copyright the Author(s), ISBN 978-1-032-39114-4

Plastic strain localization behavior of corroded steel plate under tensile loading

Y. Kitane
Department of Civil and Earth Resources Engineering, Kyoto University

N.K. Gathimba
Department of Civil, Construction and Environmental Engineering, Jomo Kenyatta University of Agriculture and Technology

K. Sugiura
Department of Civil and Earth Resources Engineering, Kyoto University

ABSTRACT: Recent studies have shown that ductility of steel members under tension can be reduced due to the presence of corrosion surface roughness. This reduction in ductility of corroded steel members is caused by the stress concentration and plastic strain localization phenomena, resulting from the unevenness of the thickness. This paper presents results from the investigation on ductility reduction of corroded steel plates under tensile loading for different degrees of corrosion surface roughness, and finally proposes a relationship between the corrosion surface roughness and a degree of plastic strain localization. To identify a region of strain localization, the digital image correlation method is used on the images taken during tensile tests of corroded steel specimens, and area of large strain is identified at the two different times during the tensile test: maximum load and just before fracture. As a result of this study, it is found that the reduction in ductility of corroded steel under tension can be explained quantitatively by the reduction of area of large plastic strain region and that the reduction of area of large plastic strain region can be estimated by surface roughness parameters.

1 INTRODUCTION

Although there are several researches (Kariya et al. 2005, Ahmmad & Sumi 2010) regarding a reduction in ductility of corroded steel members due to corrosion surface roughness, there is a lack of detailed investigations on the ductility performance of corroded steel members with regard to surface configuration. To understand the mechanism of a reduction in ductility under tensile loading and to propose a relationship between the ductility reduction and surface roughness parameters, the authors have conducted a series of numerical and experimental studies (Gathimba & Kitane 2021a, 2021b).

In these previous studies, the authors proposed the mechanism of ductility reduction of corroded steel members as follows. The surface roughness and thickness unevenness cause stress concentration under loading, which prevents uniform yielding of the section. Non-uniform yielding causes plastic strain localization, resulting in a reduction in ductility of the corroded steel specimen.

The objective of this study is to fully understand this ductility reduction mechanism and to propose a relationship between the corrosion surface roughness and a degree of plastic strain localization based on results from the previous experimental investigations on ductility reduction of corroded steel plates under tensile loading. To identify a region of strain localization, the digital image correlation method is used on the images taken during tensile tests of corroded steel specimens, and area of large strain is identified at the two different times during the tensile test: maximum load and just before fracture.

2 TENSILE TEST OF CORRODED SPECIMENS FROM PREVIOUS STUDIES

Based on tensile test results of corroded steel specimens from the previous studies (Gathimba & Kitane 2021a, 2021b), the normalized total ductility $\bar{\mu}_f$ is plotted against average surface height ratio ζ_{Sa} in Figure 1, where $\bar{\mu}_f$ is a ratio of total plastic ductility of a corroded specimen to that of a control specimen, and ζ_{Sa} is a ratio of an average surface height S_a to an average thickness t_{ave}. As can be seen in the figure, there is a clear trend that total plastic ductility decreases as the average surface height ratio increases.

3 IDENTIFICATION OF PLASTIC STRAIN LOCALIZATION

Figure 2 shows strain distributions identified by DIC just before the final fracture for bridge girder specimens with different degrees of surface roughness. As can be seen in the figure, although about 12% of the gauge length

DOI: 10.1201/9781003348450-298

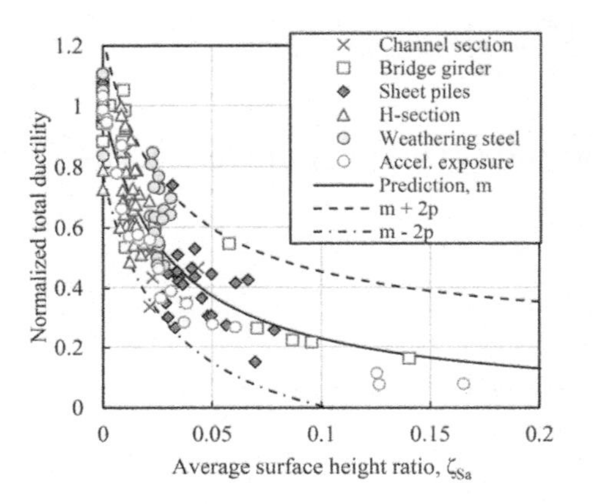

Figure 1. Normalized total ductility versus average surface height ratio.

area experiences a strain of over 40% for the control specimen, this area reduces to 5.6% and 4.7% for the specimens with ζ_{Sa}=0.024 and 0.14, respectively, implying that the greater the surface roughness, the smaller the gauge length area that experiences a significant plastic strain is. Therefore, the reason why the ductility decreases as the surface roughness increases may be explained by the plastic strain localization.

To quantitatively evaluate this plastic strain localization, a region of specimen that experiences strain greater than a threshold value (30%, 35%, or 40%) was identified for each specimen at the two different times during the tensile test: maximum load and just before fracture. A size of the plastic strain localized region was expressed as an average length of specimen over the width.

Figure 3 shows a relationship of normalized total ductility $\bar{\mu}_f$ and normalized average length of plastic strain localized region before the final fracture when a threshold strain value is 35%. The normalized average length of plastic strain localized region is defined as a ratio of an average length of plastic strain localized region before the final fracture $L_{ave,f}$ to an average length of plastic strain localized region before the final fracture for the control specimen $L_{ave,f0}$. As can be seen in the figure, there is a clear correlation between $\bar{L}_{ave,f}$ and $\bar{\mu}_f$. Therefore, it can be concluded that ductility of a corroded specimen decreases due to

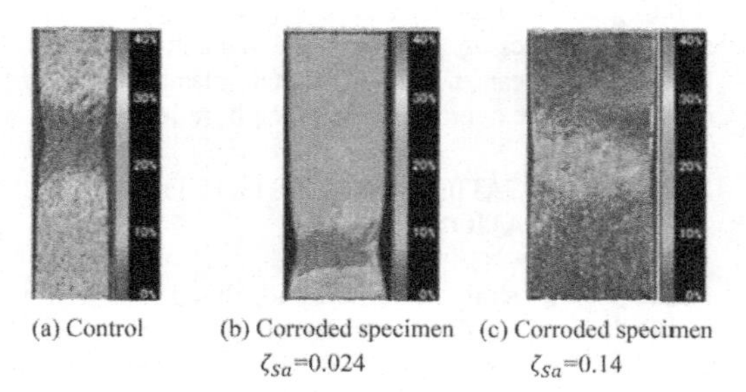

(a) Control (b) Corroded specimen ζ_{Sa}=0.024 (c) Corroded specimen ζ_{Sa}=0.14

Figure 2. Examples of strain localization behavior of specimens from bridge girders with different degrees of surface roughness before the final fracture.

plastic strain localization. When $\bar{L}_{ave,f}$ is plotted against the average surface height ratio ζ_{Sa} as shown in Figure 4, there is a decreasing trend of an average length of plastic strain localized region as a surface roughness parameter increases.

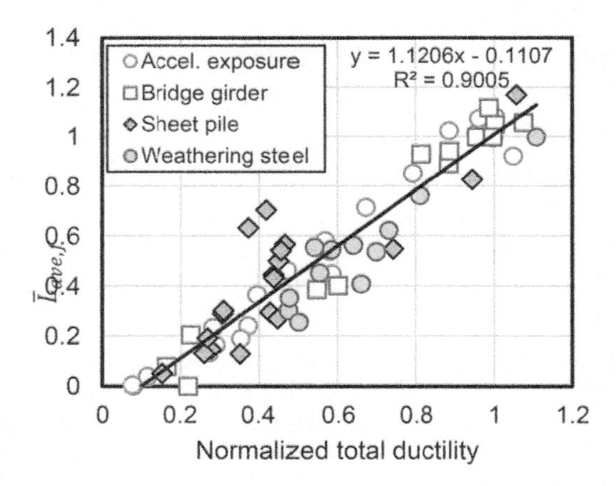

Figure 3. Relationship between normalized average length of plastic strain localized region and normalized total ductility.

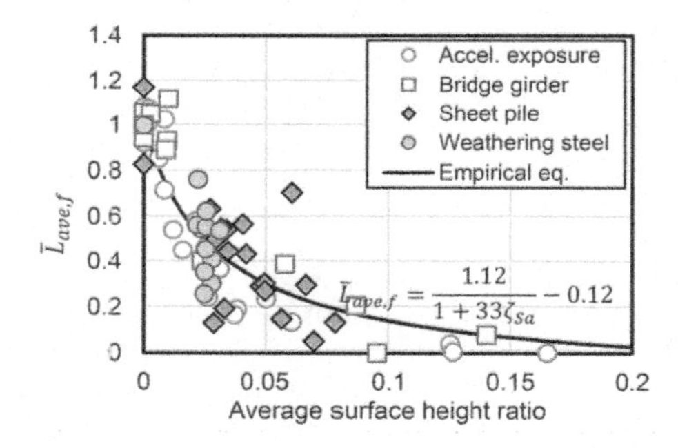

Figure 4. Normalized average length of plastic strain localized region before the final fracture versus average surface height ratio.

REFERENCES

Ahmmad, M.M. & Sumi, Y. 2010. Strength and deformability of corroded steel plates under quasi-static tensile load. *Journal of Marine Science and Technology* 15(1): 1–15.

Gathimba, N. & Kitane, Y. 2021a. Surface roughness characteristics and ductility of steel plates corroded by underwater accelerated exposure test. *Journal of Constructional Steel Research* 176: Article 106392.

Gathimba, N., & Kitane, Y. 2021b. Estimating remaining ductility of corroded steel by using surface roughness characteristics based on tensile coupon test results. In Wang C.M., Dao V., Kitipornchai S. (eds), *Proc. of the 16th East Asia-Pacific Conference on Structural Engineering & Construction (EASEC16), Brisbane, Australia, December 3- 6,2019 , Lecture Notes in Civil Engineering* 101: 1177–1187.

Kariya, A., Tagaya, K., Kaita, T., & Fujii, K. 2005. Mechanical strength of corroded plates under tensile force. *Collaboration and Harmonization in Creative Systems*: 105–110. Taylor and Francis Group.

Current Perspectives and New Directions in Mechanics, Modelling and Design of Structural Systems – Zingoni (ed.)
© 2022 Copyright the Author(s), ISBN 978-1-032-39114-4

Fundamental study on crack detection method for steel members by thermal image processing

Shinya Watanabe, Yoshinao Goi, Yasuo Kitane, Kazuo Takase & Kunitomo Sugiura
Kyoto University, Kyoto, Japan

ABSTRACT: Fatigue cracks of steel structures, which may bring significant risk on bridge safety, occur at near welded lines and at the edge of sole plate. Early detection and repair of them can be an important key and research have eagerly been conducted to develop easy detection techniques of them. Magnetic particle flaw detection tests, penetrant flaw detection tests, ultrasonic flaw detection tests, etc. have been used as conventional detection methods for fatigue cracks in steel structures, but these crack detection methods require a removal of the coating film, so that the efficient inspection cannot be performed. To achieve an easy and efficient inspection, infrared thermography images are used in this research as a method for detecting fatigue cracks in steel structures without the need to remove the coating film.

1 INTRODUCTION

Steel is one of the major construction materials for such as civil engineering structures. In urban highways, many viaducts have been made of steels over the major intersections. However, most of them were constructed during the period of high economic growth, and the aged structures over 40 years old are going to be up to about 60%. Especially, fatigue damage of bridge due to traffic load had been reported from around 1980s. Such fatigue cracks may cause brittle fracture so as to have a significant impact on the safety of bridges. Therefore, an early detection and its repair are important. However, these methods require the removal of the coating film, which does not allow for efficient inspection. However, these methods require the removal of the coating, which makes it impossible to perform efficient inspection. Therefore, research has been conducted to remotely detect and monitor fatigue cracks in steel structures in a nondestructive manner without the need for the removal of the coating. In this study, we aim to establish a practical method that can be applied not only to the inspection of steel deck plates but also to welding points and cross-sectional changes of gussets and other members, and that can be applied to various field environments without limitation of the observation position of the object to be measured. The objective of this study is to identify cracks and their shapes by using the temperature distribution near cracks in steady-state steel, focusing on active thermal loading instead of passive thermal loading, and using various crack shapes and steel sections.

2 PROPERTY VERIFICATION BY FE-ANALYSIS

In general, heat conduction in a three-dimensional object can be solved by solving the following three-dimensional unsteady heat conduction equation. However, the theoretical equations are applicable to very simple geometries, and while they can be used to determine trends in structural properties, they are not applicable to structural members with cracks or defects, as proposed in this study. Therefore, a finite element analysis is conducted to study its application.

2.1 *Analysis model and method*

In the numerical experiments, a crack shown in Figure 2. was introduced in the center of a 400 mm × 800 mm × 12 mm steel plate shown in Figure 1, where heat dissipates sufficiently, and an analytical model was set up with the parameters of crack length l, crack depth h_c, and crack opening width w. Thermo-elastic finite element analysis, in which a steel plate with a surface crack is thermally loaded, was used to verify the temperature transition characteristics of each analytical model. 8-node linear integral solid elements were used to model the steel plate. In this method, the thermal loading temperature is higher than the ambient temperature, and there is no restriction on the application of the method. However, the thermal boundary condition of the analytical model is the same as that of the laboratory environment used in the verification experiment (room temperature: 298 K (25°C)), and the initial temperature of

DOI: 10.1201/9781003348450-299

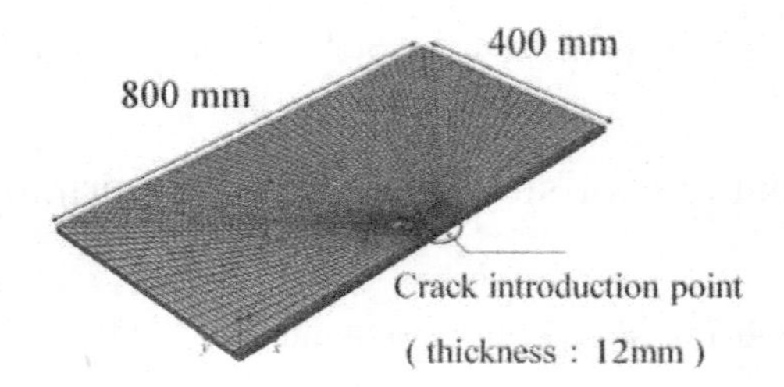

Figure 1. Numerical analysis model.

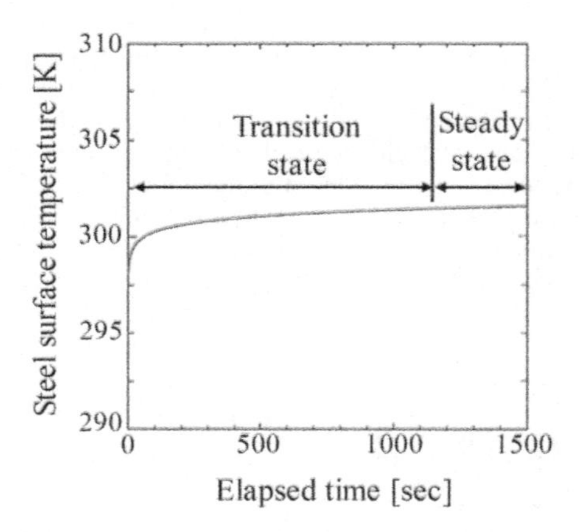

Figure 3. Relationship between Steel surface temperature in the vicinity of the crack and elapsed time.

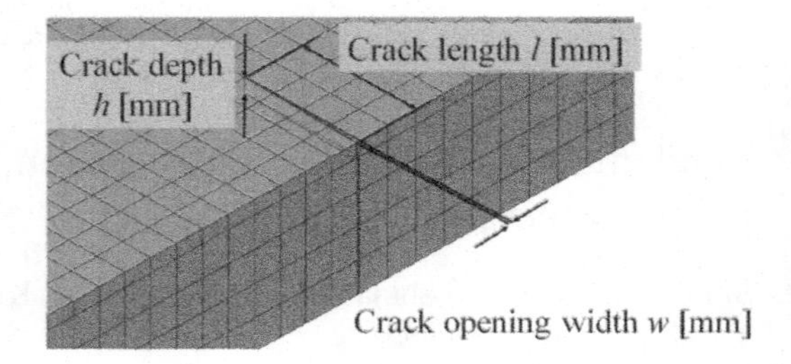

Figure 2. Enlarged view of the vicinity of a crack.

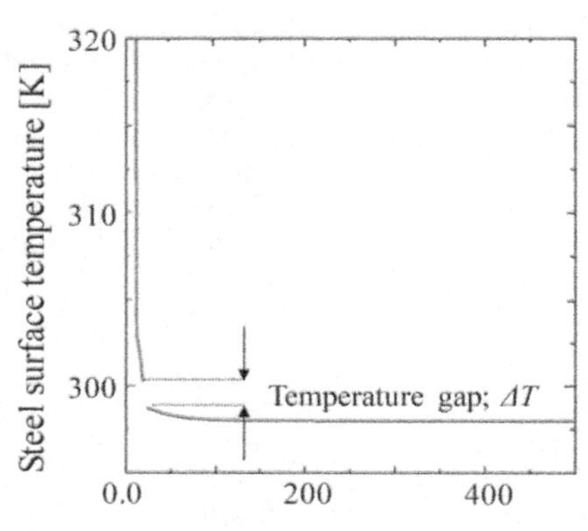

Figure 4. Temperature gradient curve near crack.

the steel plate is uniformly set at 298 K (25°C). For crack detection under nondestructive conditions, heat of 353 K (80°C), which is the heat-resistant temperature of the coating film, is applied, referring to Table 1. As shown in Figure 3, when heat is continuously applied to the steel plate from the heat source, a steady state is reached in which the temperature does not change with time. The temperature gradient curve of the steel becomes dis continuous near the crack, as shown in Figure 4, when the surface temperature of the steel and the distance from the heat source are considered. By focusing on the temperature gap ΔT at this discontinuity and the crack depth h_c, crack width w, crack length l, and plate thickness t, the following relationship equations were obtained.

$$h_c = \frac{t \cdot \Delta T \times 2.6 \times 10^{-2}}{(1-1.21e^{-5.45l})e^{-37.0d/1000}\sin\theta}$$

3 CONCLUSION

In this study, as an efficient and simple detection method for fatigue cracks in steel structures, we focused on expanding the application of nondestructive inspection methods using infrared thermography and investigated a method that focuses on the temperature gradient around a crack by applying heat from the outside to a steady state. Specifically, he effects of both the crack and the heat source are clarified by heat conduction analysis, and by considering the relationship between each parameter, it is found that three-dimensional crack depth can be expressed as a relational equation from two-dimensional image information.

REFERENCES

Kinki Regional Development Bureau, Ministry of Land, Infrastructure, Transport and Tourism, Current status and deterioration of bridges Estimation of the number of buildings, *Kinki Regional Development Bureau, Ministry of Land, Infrastructure, Transport and Tourism*, 2013.

T. Sakagami, T. Nishimura, S. Kubo, T. Nozaki, K. Ishino, Development of a Self-reference Lock-in Thermography for Remote Nondestructive Testing of Fatigue Crack: 1st Report, Fundamental Study Using Welded Steel Samples, *The Japan Society of Mechanical Engineers Journal*, A73(724), pp.1860–1867, 2006.

Y. Nishina, D. Imanishi, Y. Yoshinaga, K. Yokoyama, W. Maeda, Crack Detection Technology by Infrared Thermography, *Materia Japan*, Vol. 50 (2), pp.79–81, 2011.

Y. Mizogami, A. Koganemaru, S. Kusano, C. Morita, Optical full-field inspection method of steel fatigue cracking due to thermal load without removing coating film, *Journal of Japan Society of Civil Engineers*, Vol. 72(2), pp. 643–652, 2016.

Current Perspectives and New Directions in Mechanics, Modelling and Design of Structural Systems – Zingoni (ed.)
© 2022 Copyright the Author(s), ISBN 978-1-032-39114-4

Deflection condition monitoring of a steel bridge via remote sensing techniques

M.S. Miah & W. Lienhart
Institute of Engineering Geodesy and Measurement Systems, Graz University of Technology, Graz, Austria

ABSTRACT: The monitoring task of bridges are obtaining consideration due to their importance to the community. And to perform the overall structural health monitoring task various type of sensors have been adopted over the last few decades. Herein, a simple steel bridge has been considered to perform tests using remote sensing techniques, like video streaming via camera, high dynamic reflector-less distance measurements, and digitalized total station measurements. The bridge has been tested in a resting position (e.g. preventing the self-weight sagging using clamps) and in a released condition after removing the clamps. Both the static and dynamic loads are applied to the bridge and the following loading scenarios are tested: (i) no-load case, (ii) manual impulse, (iii) putting extra masses on the bridge, and (iv) under various dynamic loads. The tests data have been post-processed and analysed by employing MATLAB. The observed displacement results of aforementioned scenarios are evaluated and presented herein. In a nutshell, the obtained results show the changes of the displacements throughout the bridge for the aforementioned loading conditions. The test results of different sensors scenario have been investigated and the results are inter-compared. It can be summarized that remote sensing techniques are effective as the pre-warning can be deliver and identified before any change (e.g. man-made or natural) takes place. This study not only assist to visualize the variation of the deformation contour but also the video streaming of whole structure, as a result, the overall condition monitoring task becomes straightforward.

1 INTRODUCTION

The remote-sensing approach is one of the most recent development in the area of structural health monitoring (SHM). There are many advantages of utilizing such approach and among them one of the most important one is that it does not require to visit the site often. Due to underlying advantages of the aforementioned approach, it is also attracting the researchers in for applications in civil structures as well as infrastructures. As for example, (Soldovieri, Ponzo, Ditommaso, & Cuomo 2021) reported the applicability of the multi-modal sensing strategy for critical infrastructures and built environment for structural monitoring purpose.

The remote-sensing techniques have been investigated by many and also implemented into real-life problems for monitoring purpose. For instance, remote-sensing approaches have been adopted to monitor bridges by employing various sensing technologies (Biondi, Addabbo, Ullo, Clemente, & Orlando 2020, Chen, Liu, Dai, Bian, & Hauser 2011, Hopkins, Beckham, & Sun 2007, Zou, Sato, Tosti, & Alani 2020). Step further, high-resolution satellite remote-sensing technologies are employed to monitor bridge (Gagliardi, Ciampoli, D'Amico, Alani, Tosti, Battagliere, & Benedetto 2020). Further study in (Cusson, Ghuman, & McCardle 2011, Cusson, Ghuman, Gara, & Mccardle 2012) have investigated the satellite sensing method for bridges as well as for civil infrastructures to perform the monitoring task. And (LaFlamme, Turkan, & Tan 2015) has adopted 3D imaging to evaluate the structural condition of bridge. It is always a concern of data accuracy, hence, a novel image-assisted total station for accurate measurement was investigated in (Ehrhart & Lienhart 2017).

In order to apply the early discussed strategy, it is also important that one needs to be aware of the related steps of remote-sensing method. Besides the general parameters of a monitoring sensor (e.g. measurement frequency, resolution), the remote sensing approach consists of several specific components like (i) the instrument itself, (ii) the stability of the setup location, (iii) the atmospheric conditions along the measurement path and the (iv) properties of the target.

Due to many early mentioned advantages, herein, we focused to investigate the possibility of remote-sensing method by employing different sensors. To do this end, a prototype beam is selected as toy example where both static and dynamic loads are applied and the responses are evaluated.

The rest of the paper structured as follows; the immediate section contains the description of the experimental setup, the next section comes with

DOI: 10.1201/9781003348450-300

results and discussion, and the last section has summarized the outcome of the study.

2 DESCRIPTION OF THE PROBLEM

A simply supported steel bridge of 2m span in the laboratory setup has been utilized for the tests. The tests are performed by employing different state-of-the art techniques like high frequent 1D laser distance measurements with a laser triangulation sensor (LTS), tracking of a prism with a robotic total station (RTS), image based monitoring based on a video captured with a single lens reflex (SLR) camera and point cloud data acquired by a terrestrial laser scanner (TLS). An overview of the setup is given in Figure 1. Hereafter the RTS sensor has been labelled as TS15 in the rest of the paper. All tests have been made in the temperature and humidity controlled geodetic measurement laboratory and therefore atmospheric effects did not occur.

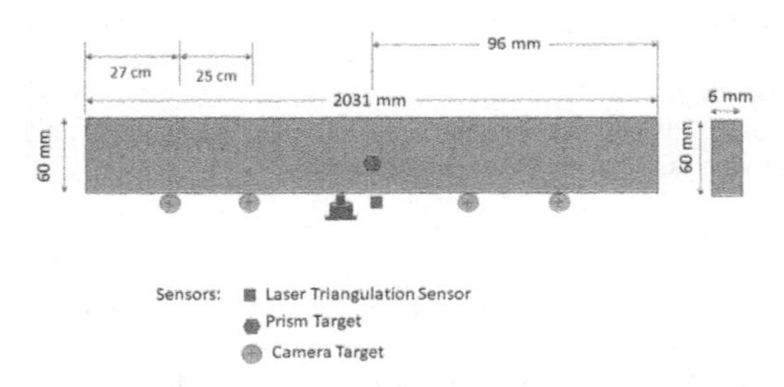

Figure 1. The sketch of the experimental setup of the tested bridge.

3 CONCLUSIONS

Herein, the deformation of a steel bridge is monitored by employing remote-sensing approach under various loading conditions. To attain the goal, the state-of-the-art laser scanning technology, robotic total station, and video streaming technologies have been adopted. The displacement time-history of the bridge is captured considering no extra weight placed on the bridge. Further, the displacements are measured placing small and heavy extra weight on the bridge. All the experimentally measured data of the bridge have been post-processed by using MATLAB. A detail comparison of the results are performed and a quite good match has been noticed among different sensors. It is noteworthy to mention that even tough different sensors have different sampling rate but interestingly all of the sensors have managed to capture the overall characteristics of the bridge. However, it is also observed that the employed remote-sensing techniques can be very sensitive to the dynamic loads. Last but not least, the evaluated results of the studied multi-sensors scenario provides better understanding of the dynamical phenomena of the structure. Hence such strategy can be beneficial for the SHM to make comprehensive judgement of the investigated structures. In real world scenarios additional atmospheric effects may degrade the overall performance and may introduces additional noise and systematic effects. Furthermore, it has to be noted that the measurement noise in metric units does increase for angular based measurements and therefore the high sensitivity achieved at the short distance in the lab will degrade for longer distances. Finally, remote sensing techniques rely on stable setups of the instrument. In order to verify the stability and if necessary apply numerical corrections, internal tilt readings or stable reference targets can be used.

REFERENCES

Biondi, F., P. Addabbo, S. L. Ullo, C. Clemente, & D. Orlando (2020). Perspectives on the structural health monitoring of bridges by synthetic aperture radar. *Remote Sensing 12*(23), 1–25.

Chen, S.-E., W. Liu, K. Dai, H. Bian, & E. Hauser (2011). Remote Sensing for Bridge Monitoring. In *Condition, Reliability, and Resilience Assessment of Tunnels and Bridges*, Volume 47625, Reston, VA, pp. 118–125. American Society of Civil Engineers.

Cusson, D., P. Ghuman, M. Gara, & A. Mccardle (2012). Remote Monitoring of Bridges From Space. *Anais do 54 Congreso Brasileiro do Concreto* (1), 25.

Cusson, D., P. Ghuman, & A. McCardle (2011). Satellite sensing technology to monitor bridges and other civil infrastructures. *SHMII-5 2011-5th International Conference on Structural Health Monitoring of Intelligent Infrastructure.*

Ehrhart, M. & W. Lienhart (2017). Accurate Measurements with Image-Assisted Total Stations and Their Prerequisites. *Journal of Surveying Engineering 143*(2), 04016024.

Gagliardi, V., L. B. Ciampoli, F. D'Amico, A. M. Alani, F. Tosti, M. L. Battagliere, & A. Benedetto (2020). Bridge monitoring and assessment by high-resolution satellite remote sensing technologies. In M. Kimata, J. A. Shaw, and C. R. Valenta (Eds.), *SPIE Future Sensing Technologies*, Volume 11525, pp. 17–26. International Society for Optics and Photonics: SPIE.

Hopkins, T. C., T. L. Beckham, & C. Sun (2007). Implementation of Remote Sensing Technology on the I-64 Bridge over US 60. Technical report.

LaFlamme, S., Y. Turkan, & L. Tan (2015). Bridge Structural Condition Assessment using 3D Imaging. *2015 Conference on Autonomous and Robotic Construction of Infrastructure*, 159–170.

Soldovieri, F., F. C. Ponzo, R. Ditommaso, & V. Cuomo (2021). Multimodal sensing for sustainable structural health monitoring of critical infrastructures and built environment. In E. Stella (Ed.), *Multimodal Sensing and Artificial Intelligence: Technologies and Applications II*, Volume 11785, pp. 31–39. International Society for Optics and Photonics: SPIE.

Zou, L., M. Sato, F. Tosti, & A. M. Alani (2020). Advanced Bridge Monitoring Strategies by Polarimetric GB-SAR. In *EGU General Assembly 2020*, pp. 10–11.

Current Perspectives and New Directions in Mechanics, Modelling and Design of Structural Systems – Zingoni (ed.)
© 2022 Copyright the Author(s), *ISBN 978-1-032-39114-4*

Condition rating for maintenance of existing reinforced concrete bridges in Soweto, South Africa

R.K. Maphosa, J Mahachi & S.O. Ekolu
University of Johannesburg, Auckland Park, South Africa

ABSTRACT: Reinforced concrete (RC) bridges are key highway structures of national importance to any country's economy. In developing countries such as South Africa, these structures are not adequately maintained. The purpose of this paper was to obtain inventory and inspection data from the various road agencies and analyze the relationships that exist on the inspected bridges. The data was obtained from bridges in Soweto, South Africa. The analysis involved the following: determining the predominant defects and presenting their relationships with inventory data such as bridge age. STATA logit models were used to investigate whether the relationships were statistically significant. Furthermore, the condition indices (average structure condition index and priority condition index) were also investigated to assess the condition of the bridge structures. Cracking and spalling were the predominant defects and were distributed mostly on abutments and deck slabs of RC bridges in Soweto. There were no statistically significant relationships identified between the predominant defects and age. However, there was a general decrease in condition indices with an increase in age. The average structure condition index of the RC bridges indicated that their conditions were fair to good. The data collected from the various road agencies on bridges inspected in Soweto should have indicated that inspections were done every five years. However this was not the case and therefore limited the findings from the analysis. Ultimately, a delay to conduct timeous bridge inspections or repairs may lead to its premature replacement or permanent closure before reaching its service life. Further studies should be undertaken on the severity of these defects.

Keywords: Bridge management system (BMS), Condition indices, Logistic regression, Defects

Reinforced concrete (RC) bridges are key highway structures of national importance to any country's economy. However, in developing countries such as South Africa, these structures are not adequately maintained.

The bridge inspection procedure commonly implemented in South Africa for maintenance and rehabilitation of bridges is the Bridge Management System (BMS-SA) (Nordengen and Fleuriot, 1998). The data contained on such BMSs provides useful data related to the defects on bridges and the conditions of bridges. This data could also be used to provide understanding into predominant defects and their relationships with inventory data such as age. According to (Mbanjwa, 2014), there is no evidence indicating that relationships exist between the defects and the above-mentioned inventory data. Furthermore, the condition of bridges in relation to the specified inventory data is unknown. This paper was then conducted to provide a better comprehension of the condition of the bridges in the Soweto.

The purpose of this paper was to obtain inventory and inspection data from the various road agencies such as Johannesburg Roads Agency (JRA) and South African National Roads Agency (SANRAL) to analyze the relationships that exist on the bridge data within Soweto, a suburb of Johannesburg. The analysis involved the following: determining the predominant defects and presenting their relationships with inventory data such as bridge age. STATA logit models were used to investigate whether the relationships were statistically significant. Furthermore, the condition indices (Average Structure Condition Index (ASCI) and Priority Condition Index (PCI)) were also investigated to assess the condition of the bridge structures.

Three binary matrices containing the defects and the inventory data for the identified RC bridges based on Johannesburg Roads Agency (JRA) and South African National Roads Agency (SANRAL) past field inspection were computed. JRA had provided inspection data for 2014 and one bridge inspected in 2019. While, SANRAL had provided inspection data for 2005 and 2017. The defects are the response/dependent variable which has only two values, such as 0 and 1 or the RC structure either has the defect or does not have the defect.

Using the binary matrices, the frequencies of the defects were established and the most frequently occurring defects were identified (i.e., most predominant

DOI: 10.1201/9781003348450-301

defect). Subsequently, relationship between the response variable and the predictor variable (structure age) were investigated using Logistic Regression Analysis (LRA) models in STATA.

The highest percentage of RC bridges with cracking was observed on the JRA inspections (30.8 %) followed by the SANRAL 2017 inspections (23.1 %). The highest percentage of RC bridges with spalling and honey combing was also observed on the JRA inspections (30.8 %) followed by the SANRAL 2005 inspections (23.1 %).

The main defects of cracking and spalling were grouped as they were considered related and typical of the deterioration mechanism of possibly carbonation. The warm, semi-arid climate of this area favours carbonation. The main effects of corrosion include cracking and spalling. Corrosion is a time dependant defect which is expected to spread with time.

The data suggest that most of the cracking (45.5 %) was observed on the abutments and most of the spalling (28.6 %) on the abutments, deck slab, piers and columns.

The continuous predictor variable, age, was categorised as indicated in Table 1 and Table 2 and the categorical variables investigated for the main defects in relation to the base category 20 to 39 years.

Table 1. LRA results between cracking and ages groups for RC bridges.

	20 – 39 years	40 – 59 years
Odds Ratio	-.7050121	-2.467542
Std. Err.	.6220479	2.654587
z	-1.13	-0.93
P>z	0.257	0.353
[95% Conf.	-1.924204	-7.670438
Interval]	.5141794	2.735353

Table 2. LRA results between spalling and ages groups for RC bridges.

	20 – 39 years	40 – 59 years
Odds Ratio	-.2069485	.4725518
Std. Err.	.2718398	.9455386
z	-0.76	0.50
P>z	0.446	0.617
[95% Conf.	-.7397448	.3258478
Interval]	-1.38067	2.325773

None of the RC bridge age groups were found statistically significant with regards to any of the defects as the p-values obtained were all higher than 0.05. In other words, there is no relationship between defects and age groups being studied.

The ASCIs and corresponding average PCIs for the increasing age categories showed a general decrease with increasing age. The overall ASCIs and average PCIs were 86.3 and 75.1 for all age groups respectively. The ASCIs for the age group of 20 to 39 years, which contained the most RC bridges was 97.8. Moreover, all the RC bridge ASCIs were greater than 70 implying that they were generally in good condition. The ASCI and average PCI computed for JRA bridges in Soweto was 81.8 and 60.5, respectively. There was one bridge inspected in 2019 and owned by JRA with an ASCI less than 70, suggesting it had some visible deteriorations and would require preventative maintenance or renewal of isolated areas.

All SANRAL bridges in Soweto inspected in 2005 and 2014 showed ASCIs greater than 70, suggesting that most of the bridges are almost new, and no issues are expected and that some of the bridges are in a condition that requires routine maintenance to sustain its condition. The ASCI and PCI computed for SANRAL bridges in Soweto based on 2017 inspections was 98.4 and 96.4, respectively, suggesting that the SANRAL bridges were generally in better condition than the JRA bridges.

Cracking and spalling were the predominant defects and were distributed mostly on abutments and deck slabs of RC bridges in Soweto. There were no statistically significant relationships identified between the predominant defects and age. However, the condition indices showed a general decrease with an increase in age. The average structure condition index of the RC bridges indicated that their conditions were fair to good. The data collected from the various road agencies on bridges inspected in Soweto should have indicated that inspections were done every five years, however this was not the case, therefore limiting the findings from the analysis. Ultimately, a delay to conduct timeous bridge inspections or repairs may lead to its premature replacement or permanent closure before reaching its service life. Further studies should be undertaken on the severity of these defects. To improve this study culverts should be taken into consideration and more bridges from different regions in Johannesburg would have been ideal to analyze, compare and provide accurate results. The computation of the Struman BMS condition indices for bridges is independent of the bridge age. Attempts should be made to investigate condition index indicator limits that would take bridge age into consideration.

REFERENCES

Mbanjwa, T. (2014). *An Investigation of the Relationships between Inventory and Inspection Data of RC Bridges and RC Culverts in the Western Cape Province*. University of Cape Town.

Nordengen, P., & Fleuriot, E. D. (1998). The Development and Implementation of a Bridge Management System for South African Road and Rail Authorities. *Investing in Transport*. Sydney: ARRB Group Ltd.

Current Perspectives and New Directions in Mechanics, Modelling and Design of Structural Systems – Zingoni (ed.)
© 2022 Copyright the Author(s), ISBN 978-1-032-39114-4

Bridge collapse prediction by small displacement data from satellite images under long-term monitoring

A. Entezami & C. De Michele
Department of Civil and Environmental Engineering, Politecnico di Milano, Milano, Italy

A. Nadir Arslan
Finnish Meteorological Institute (FMI), Helsinki, Finland

ABSTRACT: Thanks to developments of satellite sensors and synthetic aperture radar (SAR), it has been possible to exploit their benefits for long-term structural health monitoring (SHM) via satellite images. However, the major challenge in most of the long-term SHM projects is related to variations caused by environmental and/or operational variability (EOV). Therefore, the main objective of this study is to propose effective and efficient multi-stage unsupervised learning methods for addressing this limitation. Displacement samples extracted from a few satellite images of TerraSar-X regarding a long-term monitoring scheme of the Tadcaster Bridge is applied to validate the proposed method. Results show that the proposed methods effectively deal with the major challenges in the SAR-based SHM and provide practical tools for real applications.

1 INTRODUCTION

Despite appropriate researches on SHM via non-contact sensors (Sony et al., 2019), it is may be difficult to monitor large civil structures such as dams and cross-sea in wide areas. This issue may be more difficult for long-term SHM under EOV. To deal with these limitations, the use of SAR images makes a new technology in SHM and provides an excellent opportunity for safety assessment of civil structures.

The main objective of this article is to propose two hybrid machine learning methods, i.e., called here MCMC-AANN-MSD and MCMC-TSL-MSD, by (i) augmenting the small set of displacement samples by Markov Chain Monte Carlo (MCMC) (Van Ravenzwaaij et al., 2018) and Hamiltonian Monte Carlo (HMC) sampler (Neal, 2011), (ii) removing or reducing the EOV conditions through an auto-associative neural network (AANN) and a teacher-student learning (TSL) algorithm.

2 PROPOSED METHODS

Suppose that $\mathbf{X} \in \mathbb{R}^{p\times n}$ and $\mathbf{Z} \in \mathbb{R}^{p\times m}$ and Z are the augmented training and test datasets from their small sets $\mathbf{X}_0 \in \mathbb{R}^{p\times n_0}$ and $\mathbf{Z}_0 \in \mathbb{R}^{p\times m_0}$, where $n>n_0$ and $m>m_0$. On this basis, n_0 and m_0 denote the number of actual displacement samples, while n and m refer to their augmented samples. Moreover, p is the number of target points for extracting the displacement samples. The augmented datasets are determined by using the HMC sampler, which is the first part of both proposed methods. The second part is to remove the effects of the EOV. Hence, the augmented training data is used to learn two feature normalization models, i.e., AANN and TSL. Subsequently, both the augmented training and test matrices are applied to these models to extract the residuals between the inputs and outputs. For the training and test datasets, the residual matrices are defined here $\mathbf{E}_\mathbf{x} \in \mathbb{R}^{p\times n}$ and $\mathbf{E}_\mathbf{z} \in \mathbb{R}^{p\times m}$, respectively. Finally, in the third part of the proposed methods, the MSD is used to determine damage indices for decision-making (feature classification).

3 PERFORMANCE EVALUATION BY THE TADCASTER BRIDGE

The Tadcaster Bridge is a historic nine-arch masonry bridge in Tadcaster, United Kingdom. This bridge partially collapsed on 29 December 2015 after flooding that followed Storm Eva. Figure 1 illustrates the damaged area of the Tadcaster Bridge.

DOI: 10.1201/9781003348450-302

Figure 1. The Tadcaster Bridge and its collapsed area (Selvakumaran et al., 2018).

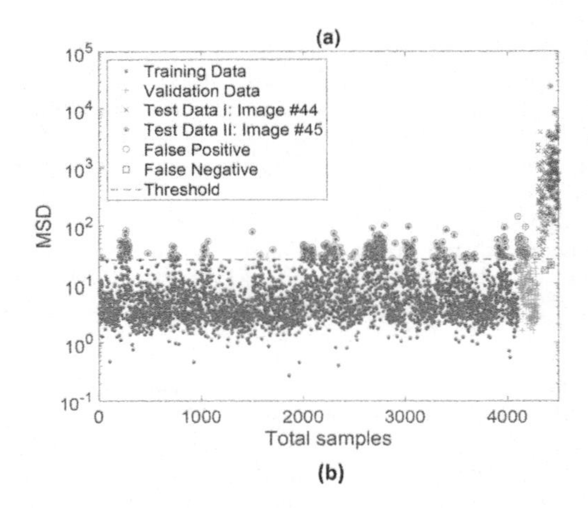

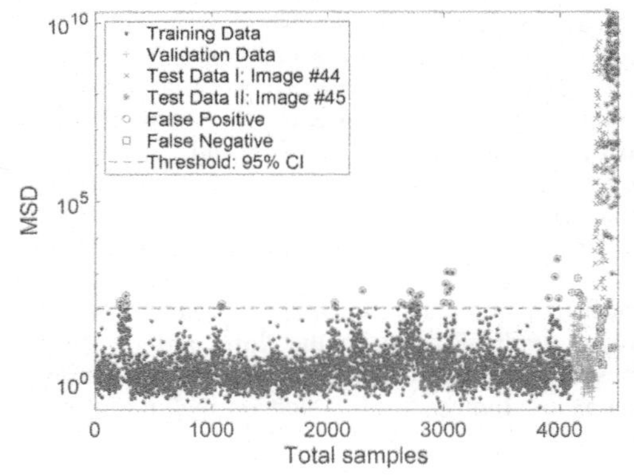

Figure 2. Damage detection in the Tadcaster Bridge (a) MCMC-AANN-MSD, (b) MCMC-TSL-MSD.

To evaluate the performances of the proposed methods for the collapse prediction or early damage detection, the augmented displacement samples are divided into the training, validation, and test samples. For this aim, the first 41 satellite images of the Tadcaster Bridge are used to make the training data. Having considered that the last two images are related to the damaged state, each of the validation and test (damaged) data include 200 augmented samples.

For the final decision-making, Figure 2 shows the results of damage detection (partial collapse prediction) in the Tadcaster Bridge, where the horizontal lines are indicative of the threshold limits obtained from the 95% confidence interval of the novelty scores of the augmented training samples. Despite proper reducing the EOV, it is observed in Figure 2(a) regarding the proposed MCMC-AANN-MSD method that some false alarms are available in the training and validation samples. Nonetheless, MCMC-AANN-MSD could properly discriminate the damaged conditions regarding the images 44 and 45 from the normal conditions. From Figure 2(b), it can be deduced that the use of the proposed TSL approach significantly reduces the rates of false positive and false negative errors.

4 CONCLUSIONS

This article proposed two hybrid unsupervised learning methods, MCMC-AANN-MSD and MCMC-TSL-MSD. Both methods intended to augment small data by HMC sampler, remove the EOV conditions via an AANN and a TSL algorithm, and determine novelty scores by the MSD for decision-making. The small displacement samples of the Tadcaster Bridge obtained from some satellite images of TerraSAR-X were applied to validate the proposed methods. The main conclusions of this study can be summarized as follows: (i) the proposed strategy for augmenting the small data enabled us to better visualize the EOV conditions and any unknown variability sources, (ii) both proposed methods could deal with the problem of the EOV conditions by obtaining discriminate novelty scores, and (iii) the MCMC-TSL-MSD outperformed the MCMC-AANN-MSD technique.

REFERENCES

Neal, R. M. (2011) MCMC using Hamiltonian dynamics. *Handbook of Markov Chain Monte Carlo*. Boca Raton, FL, USA, CRC Press.

Selvakumaran, S., Plank, S., Geiß, C., Rossi, C. & Middleton, C. (2018) Remote monitoring to predict bridge scour failure using Interferometric Synthetic Aperture Radar (InSAR) stacking techniques. *International Journal of Applied Earth Observation and Geoinformation*, 73, 463–470.

Sony, S., Laventure, S. & Sadhu, A. (2019) A literature review of next-generation smart sensing technology in structural health monitoring. *Structural Control and Health Monitoring*, 26, e2321.

Van Ravenzwaaij, D., Cassey, P. & Brown, S. D. (2018) A simple introduction to Markov Chain Monte–Carlo sampling. *Psychonomic Bulletin & Review*, 25, 143–154.

Current Perspectives and New Directions in Mechanics, Modelling and Design of Structural Systems – Zingoni (ed.)
© 2022 Copyright the Author(s), *ISBN 978-1-032-39114-4*

Structural damage identification using optimization-based FE model updating

K. Lamperová
Slovak University of Technology in Bratislava, Slovakia

D. Lehký & O. Slowik
Brno University of Technology, Brno, Czech Republic

ABSTRACT: Early detection of incipient damage of aging bridges provides the opportunity for early maintenance and can ensure their reliability and extend their service life. Identification of structural damage is usually performed using non-destructive vibration tests combined with a mathematical procedure called model updating. This modifies the original model of the undamaged structure to achieve a good match with the damaged structure. This paper describes the use of a metaheuristic optimization method aimed multilevel sampling to efficiently search the design parameter space to achieve the best match between the deformed structure and its model. The presented methodology is applied to damage identification of single- and double-span steel trusses. The effect of the damage rate and location on the identification speed and the accuracy of the solution is investigated. In all studied cases, the locations and extent of damage were correctly identified with acceptable accuracy.

1 INTRODUCTION

Structural damage commonly occurs due to various factors. Early detection of incipient damage provides the opportunity for early maintenance of the structure and can ensure its reliability and extend its service life. The identification of structural damage is usually performed using non-destructive vibration experiments to determine the static and dynamic characteristics of the structure (modal properties, displacements etc.). Changes in these characteristics compared to an intact structure then indicate structural damage. The finite element (FE) model updating process is a mathematical procedure that modifies the original model of an intact structure to achieve a good agreement with a damaged structure. Once a match is reached, a local modification of the FE model indicates damage.

In this paper, model updating is approached as an optimization problem in which the physical parameters of the FE model are updated to achieve a relative match between the model and the damaged structure. Due to the usually high computational and time requirements for structural analyses, it is necessary to employ a suitable optimization method that will be able to find the optimal solution to the problem using a relatively limited number of numerical simulations. Therefore, the metaheuristic optimization method Aimed multilevel sampling (AMS) is used in combination with the stratified small sample simulation method (Lehký et al. 2018). The basic idea is to divide the optimization process into several levels and to perform efficient sampling at each level within a defined design space. This is suitably localized and reduced at each level to achieve the best convergence.

2 DAMAGE IDENTIFICATION

2.1 *FE model updating*

The term FE model updating refers to a procedure aimed at calibrating the FE model to match the experimental and numerical static and dynamic properties of the structure. It is naturally defined as an inverse problem that can be solved by optimization. The goal is to determine the optimal value of the structural parameter vector **d** that minimizes the objective function $\phi(\mathbf{d})$ defined as

$$\phi(\mathbf{d}) = \sum_{i=1}^{n} w_i \left[\frac{e_i - m_i(\mathbf{d})}{e_i} \right], \qquad (1)$$

where e_i denotes ith experimental static or dynamic property (structural response parameter) to be matched, $m_i(\mathbf{d})$ is the same response parameter produced by the FE model. The parameters w_i are the weight coefficients of the response parameters. The weight coefficients were determined by sensitivity analysis. These coefficients expressed the degree of

DOI: 10.1201/9781003348450-303

influence of response parameters on the value of $\phi(\mathbf{d})$. Four response parameters were used for the FE model updating: natural frequencies, mode shapes, maximum displacements, and displacement vectors of multiple points of the structure. To compare the response of the models, a static and dynamic analysis were performed. The calculated response of the damaged structure can later be replaced by experimental data.

2.2 *Aimed multilevel sampling*

The basic idea of the AMS method (Lehký et al. 2018) is to divide the optimization process into several levels. At each level, the simulation is then run within a defined space using stratified simulation method (Novák et al. 2014). The sample with the best properties with respect to the definition of the optimization problem is selected and used as a vector of mean values of random variables for simulation at the next level. The sampling space around the best sample is "compressed". The next simulation is then run in this reduced space. This results in an increasingly detailed search of the regions around the samples with the best properties.

3 APPLICATIONS

The presented methodology is applied to damage identification of single- and double-span steel trusses. All numerical models consisted of several sections (Figure 1). The original (undamaged) models were made of steel with the Young's modulus $E = 210$ GPa. The damage was modeled by reducing E of selected beam sections. For each section, a separate material model was created, while the original modulus of elasticity E was multiplied by the coefficient k, which expressed how severe is a damage of the section. The coefficients k represent random variables for the FE model updating.

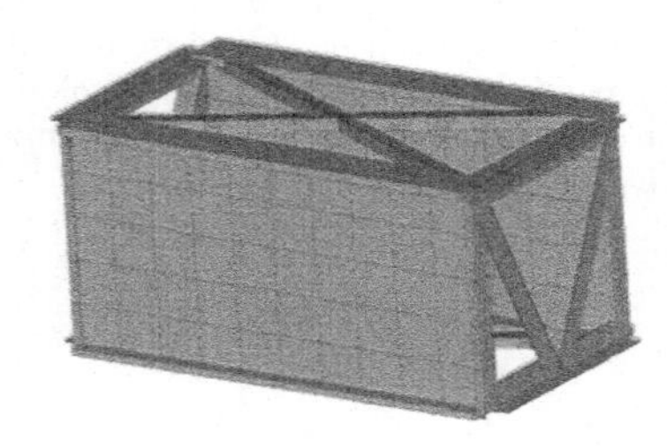

Figure 1. A section of the FE model.

In this short paper, one of the solved cases is presented: a double-span steel truss that consists of twelve sections (the first span has eight sections and the second span has four sections). In this case, the damage was modeled in sections 3, 8 and 9 (in the first span and above the support on both sides, see Figure 2a). The value of E was reduced by 30% (i.e., the value of the coefficient k_3, k_8 and k_9 was 0.70). The interim results of the optimization are shown in Figure 2c. The values of the objective function $\phi(\mathbf{d})$ approached zero (Figure 2b). The location of damage was identified and even the resulting damage rate (at the last level) is in good agreement with the modeled damage (Figure 2c).

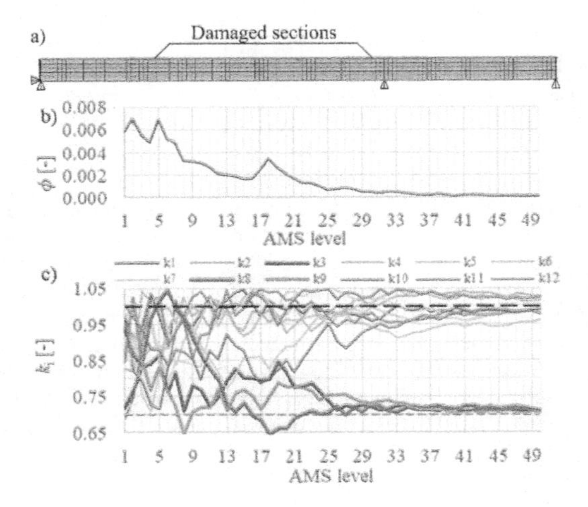

Figure 2. A double-span truss: a) indication of the damaged sections, b) evolution of the objective function, c) evolution of the damage level in individual sections.

4 CONCLUSIONS

In principle, FE model updating methods are suitable tools to address the problem of damage identification. In this paper, a metaheuristic FE model updating method was presented. This method effectively combines AMS optimization method with a stratified small sample simulation method. Numerical examples of steel trusses show that the proposed method can find an optimal solution to the FE model updating problem by using a relatively limited number of numerical simulations.

ACKNOWLEDGMENT

This work was supported by the project No. 20-01734S, awarded by the Czech Science Foundation (GACR).

REFERENCES

Lehký, D., Slowik, O. & Novák, D. 2018. Reliability-based design: Artificial neural networks and double-loop reliability-based optimization approaches. *Advances in Engineering Software* 117: 123–135.

Novák, D., Vořechovský, M. & Teplý, B. 2014. FReET: Software for the statistical and reliability analysis of engineering problems and FReET-D: Degradation module. *Advances in Engineering Software* 72: 179–192.

Current Perspectives and New Directions in Mechanics, Modelling and Design of Structural Systems – Zingoni (ed.)
© 2022 Copyright the Author(s), ISBN 978-1-032-39114-4

The implications of climate change effects on the response parameters of concrete arch dams with respect to anomaly detection

T. Tshireletso & P. Moyo
University of Cape Town

ABSTRACT: Anomaly detection is of great significance in dam safety management. There is an abundance of literature looking at anomaly detection, but they have not accounted for the effects of climate change. The aim of this article was to predict the deformations (response parameter) of Roode Elsberg dam deformations using the risk-based framework, with an account to daily operational behaviour of the dam and with respect to the South African climate change scenarios by mid-century, 2053 using nonparametric machine learning model and look at the implications of the results with respect to anomaly detection of arch dams. SGBM was chosen as the best model for making prediction at root mean square error (RMSE) of 0.0019. Roode Elsberg radial deformations are expected to move upstream during winter as a subset of the three water level scenarios. The precipitation of Roode Elsberg catchment is expected to decrease therefore the highest probability of occurrence of seasonal rainfalls matching the LSS. Warmer (<3°C) and drier conditions will cause Roode Elsberg dam to move upstream during winter and spring between 2012-2017 and 2048-2053, while hotter (>3°C) and drier conditions will cause the dam to move upstream during winter and spring with slight upstream movement in autumn.

1 INTRODUCTION

Lombardi (2004) founded objectives of dam and foundation monitoring. These objectives pose four questions to be answered in relation to the behavior of a dam, for dam operations to be ensured, and potential anomalous performance to be detected as early as possible to avoid serious malfunctioning or failure:

- Does the dam behave as expected/predicted?
- Does the dam behave as in the past?
- Does any trend exist which could impair its safety in the future?
- Was any anomaly in the behavior of the dam detected?

In the short term, the measurements of some devices used in dam monitoring are compared to reference values, which corresponds to the dam responses to the occurring loads which are normally termed as 'normal' or 'safe'. These reference values and associated prediction intervals above and below them are obtained from some behavioral model which accounts for the actual value of the acting loads. Those measurements outside the cited intervals are considered as potential symptoms of anomalous behavior. For the medium to long term, behavioral models and observed data are analyzed to draw conclusions on the overall dam performance. In particular, the association between each load and output is observed, and the evolution over time is evaluated.

2 MATERIAL AND METHODS

The study focuses on Roode Elsberg dam, which is a double curvature arch dam located in the Sanddriftkloof Hex Valley, near the town of De Doorns, Western Cape, South Africa. The reservoir capacity is 8.21 million m3 and it was built for irrigations of the vineyards in De Doorns and surrounding areas. Two studies by Tshireletso and Moyo were referenced, these studies were based on the prediction of water temperatures of concrete arch dams with respect to a changing climate, and the effects climate change will have on the natural frequencies of concrete arch dams.

Figure 1. Roode Elsberg dam.

DOI: 10.1201/9781003348450-304

3 ROODE ELSBERG DAM MONITORING SYSTEM

To understand the behaviour of Roode Elsberg dam, two monitoring systems were installed on the dam crest. These are the continuously monitored GPS system at four survey beacons in 2010 and the dynamic monitoring system at the dam crest in 2013. The environmental and operational conditions were also measured.

Figure 2. Roode Elsberg dam GPS monitoring system.

4 NON-PARAMETRIC MACHINE LEARNING MODELS

Non-parametric Machine Learning models were used for the study, this is because their data distribution cannot be defined by a finite set of parameters, but by assuming an infinite dimensional θ, which is usually a function. The amount of information that θ captures about the data can grow as the amount of the data grows, making non-parametric models more flexible (Amberg, 25-29 May 2009), therefore it is this that makes them suitable for this study.

5 DEFORMATION PREDICTIONS

Three water level cases (wet season steady state; dry season steady state; and a scenario where recorder water levels were used combined with their respective ambient temperature) were considered. The deformations were predicted for the period 2012-2017 and 2048-2053 and the differences compared. The machine model that was used was the stochastic gradient boosted machines.

6 RESULTS AND DISCUSSION

The South African climate scenarios studied in this article shows that between 2012-2017 and 2048-2053, Roode Elsberg dam will move upstream during winter, as a subset of the three water level scenarios (see Table 1).

Table 1. Roode Elsberg dam seasonal deformations location.

water level	autumn	summer	winter	spring
RWL	+	-	-	-
FSS	-	-	-	-
LSS	+	-	-	-

When the RWL was used with warmer model only winter upstream movements were seen between 2012-2048 and when the hotter model was used, win-ter upstream movements were seen with a slight up-stream movement in autumn.

7 CONCLUSIONS

The following conclusions can be made with respect to anomaly detection of concrete arch dams in a changing climate.

- Response variables and loads will be affected differently by climate change, this will be with respect to different climatic zones.
- The effects of climate change will be greater in other seasons compared to others; therefore, it is important that this change is applied whenever anomaly detection models are made.

REFERENCES

Amberg, F., 25–29 May 2009. *Interpretative models for concrete dam displacement*. Brasilia, Brazil, s.n.

Lombardi, G., 2004. *Advanced data interpretation for diagnosis of concrete dams*, s.l.: CISM.

Tshireletso, T., Moyo, P. & Kabani, M., 2021. Predicting the Effects of Climate Change on Water Temperatures of Roode Elsberg Dam Using Nonparametric Machine Learning Models. *Infrastructures*.

Tshireletso, T., Moyo, P. & Kabani, M., 2021. Predicting the effects of climate chnage on water temperatures of Roode Elsberg dam using non-parametric machine learning models. *Infrastructures*, 6(14, https://doi.org/10.3390/infrastructures6020014).

Current Perspectives and New Directions in Mechanics, Modelling and Design of Structural Systems – Zingoni (ed.)
© 2022 Copyright the Author(s), ISBN 978-1-032-39114-4

Evaluation of changes in flexural rigidity of cracked concrete railway bridges under high-speed train passages

K. Matsuoka
Railway Technical Research Institute, Tokyo, Japan

ABSTRACT: Crack breathing under train passages is an essential aspect of managing the existing concrete bridges since it can cause bridge stiffness to decrease or fluctuate. On the other hand, nonlinear modal identification methods corresponding to the rapid system frequency fluctuation under forced vibration state, on the other hand, are required to estimate bridge responses. The Bayesian TV-ARX method, which is a time-varying system identification method that takes external force characteristics into account, is suggested in this study and applied to the recorded displacement responses of a high-speed railway concrete girder when train passages. The identified time-varying bridge frequencies are compared to the results of some wavelet analysis. As a result, it was determined that the suggested Bayesian TV-ARX approach can achieve the instantaneous decrease in bridge frequency that cannot be estimated by the continuous wavelet transform.

1 INTRODUCTION

Cracking is an important index for evaluating irregularities, defects, and decreased bridge stiffness in prestressed concrete (PC) and partial prestressed concrete (PPC) (Chikami et al. 2018). The excitation frequency of trains traveling at high speed tends to be close to the bridge frequency on high-speed railway (HSR) bridges (Matsuoka et al. 2019). As a result, the decrease in bridge frequency associated with the decrease in bridge stiffness may result in resonance between the traveling train and the PC and PPC bridges. Therefore, it is an important to evaluate the decrease in flexural stiffness (and frequency) of PC and PPC bridges due to cracks.

The author created a Time-Varying Auto-Regressive with eXogenious (TV-ARX) model and its hierarchical Bayesian estimation, which considers the modal-externalized traveling train load as an exogenous variable of the time-varying time series model (Matsuoka et al. 2020, Matsuoka et al. 2021). This study examines a PPC bridge with 30 m span for HSR where there is a concern that the bridge stiffness may decrease as a result of crack breathing beneath train passages. The time-varying frequency under train passage is discovered and compared using the Bayesian TV-ARX method and CWT.

2 RESULTS

2.1 *Measurement results*

Figure 1 illustrates the displacement response of the bridge's mid-span measured when a train passes. The train travels at a speed of 220 km/h (61.1 m/s).

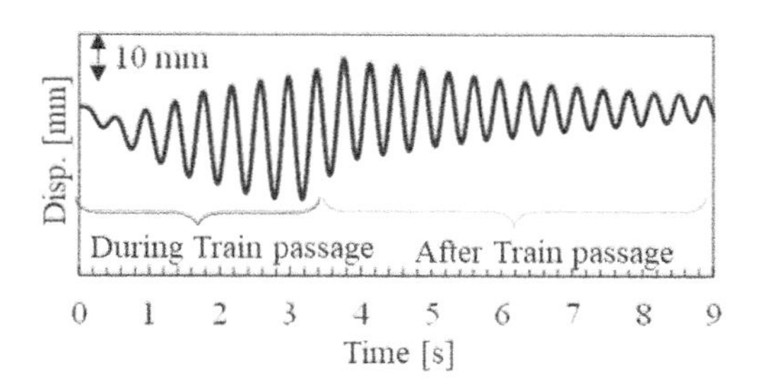

Figure 1. Measured displacement response.

2.2 *Time-varying frequency estimation results*

Figure 2 illustrates the instantaneous frequency and PSD approximated using the Bayesian TV-ARX method and CWT from the measured displacement response.

DOI: 10.1201/9781003348450-305

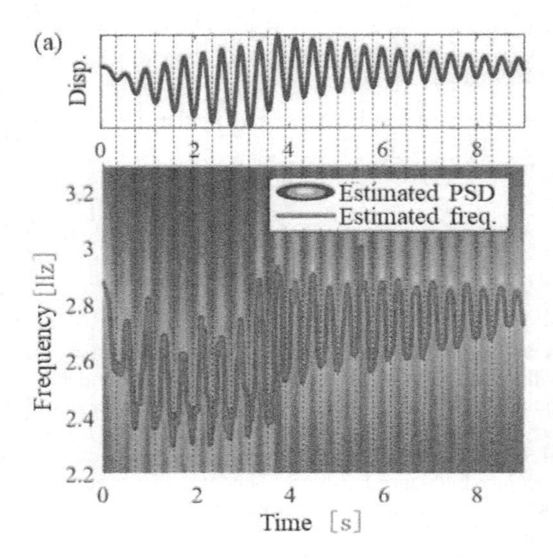

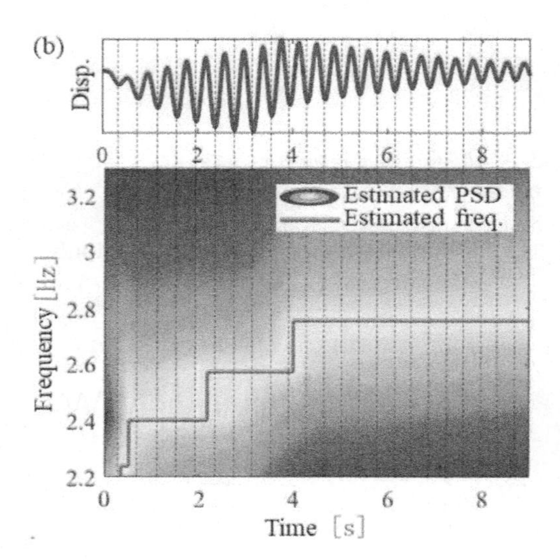

Figure 2. Estimated PSD and frequency: (a) Bayesian TV-ARX, (b) CWT.

Figure 2 also illustrates the displacement response for the purpose of analyzing the relationship with the estimated instantaneous frequency variations. Additionally, a dotted line is drawn at the point at which the measured displacement becomes the minimum.

Compared to the results of each method in Figure 2, CWT was unable to determine the bridge frequency's instantaneous fluctuation during and after the train passes. On the other hand, the Bayesian TV-ARX method's estimation results indicate that the bridge frequency repeats rapid fluctuations. Even with the train passing, it is possible to confirm that the bridge frequency estimated by the Bayesian TV-ARX method fluctuates almost in agreement with the vibration cycle of the measured bridge displacement. Additionally, the momentary decrease in bridge frequency tends to be large at the point of the dotted line where the displacement amplitude becomes the downward maximum. As a result of the foregoing, it was clarified that the tested bridge had a momentary fluctuation in the frequency depending on the amplitude, which was difficult to estimate by the CWT when the train passed.

3 CONCLUSIONS

In this study, the Bayesian TV-ARX method and CWT were used to estimate the instantaneous decrease in the flexural stiffness of a PPC bridge caused by cracks opening when passing a train, and the factors causing the decrease were analyzed experimentally. The obtained results are summarized below.

- The instantaneous fluctuation of the cracked PPC bridge frequency under a railway passage, which is difficult to quantify using CWT, can be quantified using the Bayesian TV-ARX method by separating from the external force.
- The Bayesian TV-ARX method was used to measure the measured HSR bridge displacement response during a train passage and revealed that the bridge frequency decreased momentarily when the train passed.

REFERENCES

Chikami, K., Sugito, K. & Tsukishima, D. 2018. Cause of resonance in PPC girder when Shinkansen passing and external cable reinforcement. *40th IABSE Symp., Nantes*, 19-21 September 2018. 56: 19–28.

Mustafo, A.E., Mizutani, T. & Nagayama, T. 2018. Numerical analysis of large amplitude nonlinear vibration of high-speed train PPC bridge. *40th IABSE Symp., Nantes*, 19-21 September 2018. 520: 19–26.

Matsuoka, K., Collina, A., Somaschini, C. & Sogabe, M. 2019. Influence of local deck vibrations on the evaluation of the maximum acceleration of a steel-concrete composite bridge for a high-speed railway. *Engineering Structures* 200: 109736.

Matsuoka, K., Kaito, K. & Sogabe, M. 2020. Bayesian time–frequency analysis of the vehicle–bridge dynamic interaction effect on simple-supported resonant railway bridges. *Mechanical Systems and Signal Processing* 135: 106373.

Matsuoka, K., Tokunaga, M. & Kaito, K. 2021. Bayesian estimation of instantaneous frequency reduction on cracked concrete railway bridges under high-speed train passage. *Mechanical Systems and Signal Processing* 161: 107944.

Current Perspectives and New Directions in Mechanics, Modelling and Design of Structural Systems – Zingoni (ed.)
© 2022 Copyright the Author(s), ISBN 978-1-032-39114-4

Applying automated damage classification during digital inspection of structures

J. Flotzinger
Institute for Structural Engineering, University of the Bundeswehr Munich, Germany

P.J. Rösch & F. Deuser
Institute for Distributed Intelligent Systems, University of the Bundeswehr Munich, Germany

T. Braml
Institute for Structural Engineering, University of the Bundeswehr Munich, Germany

S. Reim
MoBaP research project
Ilp Ingenieure GmbH & Co. KG

B. Maradni
MoBaP research project
Khoch GmbH, Germany

ABSTRACT: With an ever-growing building stock that is showing signs of age-related damage, the inspection of structures is more important than ever. Digitized inspections can help speed up and facilitate the current analog process. The generation of digital twins from building information models with detected and located damage further accelerates the process. Recognizing thereby depicts the classification and localization which is made possible by the deployment of convolutional neural networks (CNNs). This work focuses on the development and deployment of CNNs for live image classification of damage with an Android application in the context of digital inspections.

1 INTRODUCTION

In recent years many bridges reach a critical age at which damage occur more often due to the disproportionate increase in heavy traffic, fatiguing and their exposition (ARTBA 2022). To ensure the stability, traffic safety and durability of the buildings, they have to be frequently inspected to prevent sudden blocking as well as weight limits. Authorities are struggling to accomplish inspections of all structures in time due to staff shortages and limited monetary resources. Simultaneously, traffic obstructions are to be reduced to a minimum which further complicates the organization and execution. In order to overcome these challenges, this work proposes a digitalized process for accelerating and facilitating the assessment of built structures. The workflow for a digital inspection applying mobile devices can be divided into three main components according to Figure 1: (*i*) developing a building information model (BIM), (*ii*) automatically classifying, localizing and evaluating damage under human oversight, (*iii*) fusing damage metadata from the previous step with the BIM to generate a digital twin and a report according to the country-specific standard. One crucial component during digital inspections is the automated recognition of damage using CNNs. In this way, damage that are situated on the area

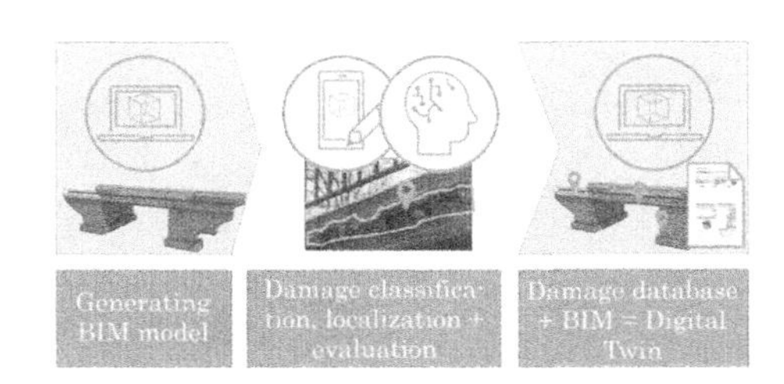

Figure 1. Digital Inspection Process using a mobile device.

the inspector's facing can be predicted or at least preselected. This enables a faster surveying of the building by an inspector, who is therefore directly confronted with neuralgic areas and a reduced number of damage classes.

In this work, the authors will show the deployment and application of state-of-the-art CNNs for live damage classification on Android devices by introducing the *dacl-app*.

DOI: 10.1201/9781003348450-306

2 DACL-APP

The *dacl-app*, which is the acronym for "damage classification-application", performs live image multi-target classification of six damage classes with two state-of-the-art CNNs, shown in Figure 2. Live image classification means that a specific amount of video frames of the live camera output is constantly fed into the deployed CNN. The app uses two pre-trained DL models from *Building Inspection Toolkit* (*bikit*) (Rösch & Flotzinger 2022): ResNet50 (RN50) which achieves the best exact match ratio (EMR) of 73.73 % and MobileNetV3-Large (MNV3) with an EMR of 69.46 % (Flotzinger et al. 2022).

The *dacl-app* includes one page showing the current camera output and all damage classes with the corresponding probability between zero and one hundred percent. Additionally, predicted classes which exceed the threshold of 50.0, are highlighted. The class *No Damage* is marked in green color as soon as the threshold is reached while the others are marked in red. This allows the user to evaluate constantly the *correctness* and *confidence* of the model's predictions on the underlying image data. Furthermore, metrics to evaluate the computational performance are displayed above the list of classes which include the average inference time over the last minute, the amount of frames that can be predicted in one second and the time that the last inference took. In addition, the flashlight can be switched on in case of adverse light conditions with the toggle at the top of the screen.

Figure 2. The *dacl-app's* predictions for a surface showing spalling, exposed rebars and rust.

As shown in Table 1, the most FPS can be classified by utilizing MNV3 on the Google Pixel. This can be attributed to the lower number of model parameters of MNV3 (5.4 m) compared to RN50 (24.0 m) and the computational superiority of this device.

During on-site testing, the smartphone was placed over significant damage and healthy parts of a reinforced concrete plate. The first tests showed that the recognition of the class *No Damage* and *Spalling* is nearly always correct.

Table 1. Average frames per seconds (FPS) for inference of the models MobileNetV3-Large (MNV3) and ResNet50 (RN50) on two different smartphones with the CPU clock signal frequency in GHz (CR) and the number of cores (C).

			FPS	
Smartphone	CR	C	MNV3	RN50
Umidigi A3X	2.0	4	3.7	1.0
Google Pixel 4a 5G	1.8-2.4	8	12.5	2.2

In case of multiple damage on one live frame the appropriate classification of all labels appears to be less reliable and stable. Figure 2 shows two examples for the app's predictions inside a screen cast. Classes the model often appears to predict incorrectly are *Bars Exposed* and *Rust*. The incorrect classifications may be caused by a mismatch between the training data from CODEBRIM dataset and the actual video images captured in the field (e.g., image quality, distance of the object from the lens). Mundt et al. generated their dataset with multiple cameras, which had a large output resolution, mounted on an UAV. Moreover, on the slab on which the app was tested, the texture of concrete and the general appearance of damage in CODEBRIM is not the same, which is another reason for the lack of performance.

3 CONCLUSIONS AND FUTURE WORK

The *dacl-app* sets an impulse for the deployment of CNNs in order to automatically recognize bridge damage in a real-world environment. This component plays an essential role during the digital inspection of structures and thus advances research in this field. Nevertheless, improvements need to be made, in particular regarding the DL models, to ensure a more reliable and supportive damage recognition. To enhance the performance of the models further, collecting of data, especially from many different structures is essential because the appearance of each building varies – when dealing with reinforced concrete structures – depending on multiple factors (e.g. local weather conditions, concrete recipe, concrete quality). Furthermore, capturing relevant data to form and display a digital twin while only making use of one device constitutes the research team's ultimate goal.

REFERENCES

ARTBA (2022). Bridge report. Technical report, American Road and Transportation Builders Association, Washington D.C., US.

Flotzinger, J., P. J. Rösch, N. Oswald, & T. Braml (2022). Building inspection toolkit: Unified evaluation and strong baselines for damage recognition.

Mundt, M., S. Majumder, S. Murali, P. Panetsos, & V. Ramesh (2019, June). Meta-learning convolutional neural architectures for multi-target concrete defect classification with the concrete defect bridge image dataset. In *Proceedings of the IEEE/CVF Conference on Computer Vision and Pattern Recognition (CVPR)*.

Rösch, P. J. & J. Flotzinger (2022). Building inspection toolkit. https://github.com/phiyodr/building-inspection-toolkit.

Current Perspectives and New Directions in Mechanics, Modelling and Design of Structural Systems – Zingoni (ed.)
© 2022 Copyright the Author(s), ISBN 978-1-032-39114-4

Crack intensity closed form solutions by frequency measurements on damaged beams

S. Caddemi, I. Caliò, F. Cannizzaro & N. Impollonia
Dipartimento di Ingegneria Civile ed Ambientale, Università di Catania, Catania, Italy

ABSTRACT: Based on Sherman and Morrison formula, the present work proposes an explicit formulation of the natural frequencies of multi-cracked beams in terms of crack intensity to avoid solution of the non linear frequency equation. Among the problems that can be tackled with an explicit formulation of the multi-cracked beam natural frequencies, in this work the crack intensity identification problem by frequency measurements is addressed. Precisely, closed form solutions of the crack intensity in terms of frequency measurements are proposed and employed for the identification in single and multi-cracked beams.

1 INTRODUCTION

The presence of damage in beam-like structures is a subject of considerable interest in the structural engineering literature in view of the influence on the structural safety. The presence of damage is often not detectable by means of simple visual inspection, hence, for this reason a number of procedures have been devised to identify damage by measurements of natural frequencies (Morassi and Rollo, 2001). When the damaged area is very limited, damage can be considered concentrated and it is usually referred to as a crack. First, in this work the natural frequencies of beams in presence of multiple cracks are explicitly formulated by making use of the Sherman and Morrison formula (Sherman and Morrison, 1949), already exploited for structural problems (Rao and Berke, 1997; Impollonia, 2006), by imposing that the exact frequency values are reproduced for reference crack intensity values and otherwise provides approximate frequency values for multi-cracked beams over a specified range of crack intensity values. Then, this work shows how the proposed explicit expressions of natural frequencies in terms of crack intensities can be easily inverted to provide closed form expressions of the crack intensity in terms of natural frequencies. The latter closed form expressions are employed for the identification in single and multi-cracked beams when frequency measurements are available. Applications to single and multi-cracked beams are presented to show the accuracy of the presented closed form expressions and how the latter can be improved by an iterative procedure.

2 NATURAL FREQUENCIES APPROXIMATION OF MULTI-CRACKES BEAMS

In this section a procedure to formulate explicit expressions of the natural frequencies of beams with multiple cracks at fixed cross section, in terms of crack intensity parameters, is proposed.

Let us consider a beam with n cracks, the i-th crack intensity parameter, denoted with λ_i, is considered as a function of a continuous variable ε_i representing the deviation from a reference value $\lambda_{o,i}$ as follows: $\lambda_i = \lambda_{o,i} + \varepsilon_i$. Moreover, the upper and lower bounds of the crack severities λ_i identified as $\underline{\lambda}_i = \lambda_{o,i} - \Delta\varepsilon_i$ and $\overline{\lambda}_i = \lambda_{o,i} + \Delta\varepsilon_i$, respectively, are introduced such that $-\Delta\varepsilon_i \leq \varepsilon_i \leq \Delta\varepsilon_i$ where $2\Delta\varepsilon_i$ is the range of variation of the variable ε_i.

In order to provide an estimation of n_f frequency parameters α_r^4 as functions of the damage severities variation parameters ε_i, the following interpolating Sherman and Morrison formula is adopted:

$$\alpha_r^4(\varepsilon_i) \cong \alpha_{o,r}^4 + \sum_{i=1}^{n} \frac{\varepsilon_i a_{r,i}}{1 + \varepsilon_i b_{r,i}}, \quad r = 1, \ldots, n_f \quad (1)$$

where $\alpha_{o,r}^4$ is the r-th frequency parameter associated to the reference crack severity distribution $\lambda_o = \left[\lambda_{o,1}, \lambda_{o,2}, \ldots, \lambda_{o,n}\right]$, whereas $a_{r,i}, b_{r,i}$ represent n_f appropriate sets of $2n$ coefficients. The latter are to be evaluated by enforcing the correspondence between exact values and approximate expression of the frequency parameters at the following $2n$ significant damage configurations:

DOI: 10.1201/9781003348450-307

$$\bar{\lambda}_i = \left[\lambda_{o,1}, \lambda_{o,2}, \ldots, \bar{\lambda}_i = \lambda_{o,i} + \Delta\varepsilon_i, \ldots \lambda_{o,n}\right]$$
$$\underline{\lambda}_i = \left[\lambda_{o,1}, \lambda_{o,2}, \ldots, \underline{\lambda}_i = \lambda_{o,i} - \Delta\varepsilon_i, \ldots \lambda_{o,n}\right], \quad i = 1, \ldots, n \qquad (2)$$

3 CRACK INTENSITY IDENTIFICATION BY FREQUENCY MEASURMENTS

The proposed explicit expressions of the frequency parameters of cracked beams can be exploited to address the inverse problem aiming at the estimation of the damage magnitude of cracks whose positions are provisionally detected by means of appropriate procedures. For the simple case of a beam with a single crack $n = 1$, Equation (1), once a generic r-th frequency parameter appearing on the left-hand side is experimentally measured (named $\tilde{\alpha}_r^4$), is written as:

$$\tilde{\alpha}_r^4 \cong \alpha_{o,r}^4 + \frac{\varepsilon_1 a_{r,1}}{1 + \varepsilon_1 b_{r,1}} \qquad (3)$$

The parameter ε_1 is explicitly provided by solving Equation (3) as follows:

$$\varepsilon_1 = \frac{\tilde{\alpha}_r^4 - \alpha_{o,r}^4}{a_{r,1} - b_{r,1}\left(\tilde{\alpha}_r^4 - \alpha_{o,r}^4\right)} \qquad (4)$$

where the coefficients $a_{r,1}, b_{r,1}$ are given as mentioned in the previous section and expressed as follows:

$$a_{r,1} = \frac{2}{\Delta\varepsilon_1} \frac{\left[\bar{\alpha}_r^4 - \alpha_{o,r}^4\right]\left[\alpha_{o,r}^4 - \underline{\alpha}_r^4\right]}{\bar{\alpha}_r^4 - \underline{\alpha}_r^4}, \quad b_{r,1} = \frac{1}{\Delta\varepsilon_1} \frac{2\alpha_{o,r}^4 - \underline{\alpha}_r^4 - \bar{\alpha}_r^4}{\bar{\alpha}_r^4 - \underline{\alpha}_r^4} \qquad (5)$$

Equations (4), (5) provide explicitly the deviation of the single crack severity with respect to the chosen reference value $\lambda_{o,1}$. Equations (4), (5) require the knowledge of the measured r-th frequency parameter $\tilde{\alpha}_r^4$, provided the frequency values $\underline{\alpha}_r^4, \bar{\alpha}_r^4$ correspondent to the damage magnitudes $\underline{\lambda}_1 = \lambda_{o,1} - \Delta\varepsilon_1$, $\underline{\lambda}_2 = \lambda_{o,2} - \Delta\varepsilon_2$, respectively, as well as a suitable choice of the range of variation of $\Delta\varepsilon_1$. The identified crack severity parameter is finally given by $\lambda_1 = \lambda_{o,1} + \varepsilon_1$.

It is worth to note that, contrarily to many crack identification procedures based on the comparison between the healthy and damaged structures, the presented solution does not require the preliminary measurement of the undamaged beam frequency.

In general, for a n-cracked beam, the system in Equation (1) written for n measured frequencies $\tilde{\alpha}_r^4$, $r = 1, \ldots, n$, can be linearized, under the assumption of small cracks and small severity deviations ε_i, $i = 1, \ldots, n$ from the reference values, as follows:

$$\tilde{\alpha}_r^4 - \alpha_{o,r}^4 = \sum_{i=1}^{n} \varepsilon_i \left[a_{r,i} - b_{r,i}\left(\tilde{\alpha}_r^4 - \alpha_{o,r}^4\right)\right]; \quad r = 1, \ldots, n \qquad (6)$$

4 APPLICATION

For the case of a simply supported beam with $n = 2$ cracks at normalized positions $\xi_1 = 0.4$, $\xi_2 = 0.7$ with actual values $\lambda_1 = 0.25$, $\lambda_2 = 0.15$, the linear system proposed in Equation (6) is used to evaluate the damage severity deviation parameters $\varepsilon_1, \varepsilon_2$ and in turn the damage severities $\lambda_1 = \lambda_{o,1} + \varepsilon_1$, $\lambda_2 = \lambda_{o,2} + \varepsilon_2$, starting from chosen reference severity values $\lambda_{o,1} = 0.2$, $\lambda_{o,2} = 0.1$. The results, also in terms of relative errors, are reported in Table 1 together with an improvement due to an iterative update of the reference severity values.

Table 1. Identification of damage severities of a double cracked beam with an iterative procedure.

Iter	$\lambda_{o,1}$	$\Delta\varepsilon_1$	$\lambda_{o,2}$	$\Delta\varepsilon_2$	λ_1	E[1] %	λ_2	E[2] %
1	0.2	0.2	0.1	0.1	0.2415	3.416	0.1621	8.094
2	0.2415	0.1	0.1621	0.05	0.2504	0.147	0.1495	0.342
3	0.2504	0.05	0.1495	0.025	0.2500	0.00019	0.1500	0.00049

5 CONCLUSIONS

This work intended to provide a contribution towards crack intensity identification by frequency measurements by means of closed form expressions. In particular, first, explicit expressions of natural frequencies of multi-cracked beams have been presented to avoid the solution of the complex trascendental frequency determinatal equations. Then, it has been shown how the proposed explicit solution can be employed to identify the crack intensities of multi-cracked beam based on frequency measurements, by simply solving a linear system.

REFERENCES

Impollonia N., 2006. A method to derive approximate explicit solutions for structural mechanics problems. *Int. J. Solids Struct.* 43, pp. 7082–7098.

Morassi A., Rollo M., 2001. Identification of two cracks in a simply supported beam from minimal frequency measurements. *J. Vib. Control* 7 (5), pp. 729–739.

Rao S.S., Berke L., 1997. Analysis of uncertain structural systems using interval analysis. *AIAA J.* 35 (4), pp. 727–735.

Sherman J., Morrison W.J., 1949. Adjustment of an inverse matrix corresponding to changes in the elements of a given column or a given row of the original matrix (abstract). *Ann. Math. Stat.* 20, 621, doi: 10.1214/aoms/1177729959.

Current Perspectives and New Directions in Mechanics, Modelling and Design of Structural Systems – Zingoni (ed.)
© 2022 Copyright the Author(s), *ISBN 978-1-032-39114-4*

Effect of 2-directional chloride ingress on concrete resistivity and corrosion rate of steel bars at orthogonal edges

Z.G. Zakka
Department of Civil Engineering, Walter Sisulu University, Ibika Campus, Eastern Cape, South Africa
School of Civil and Environmental Engineering, University of the Witwatersrand, Johannesburg

M. Otieno
School of Civil and Environmental Engineering, University of the Witwatersrand, Johannesburg

ABSTRACT: Concrete resistivity is one of the factors that control the corrosion rate of steel reinforcement bars in concrete. In chloride-induced corrosion, the presence of chloride ions and moisture in the concrete improves its ionic conductivity by allowing easy flow of ions between the corroding (anode) and noncorroding (cathode) parts of the steel bar. In this paper, the corrosion rate of steel bars at orthogonal edges of cracked fly ash and slag concrete beams that were exposed to 2-directional chloride ingress increased exponentially although there is no significant decrease in the concrete resistivity. It is concluded that apart from the concrete resistivity, the penetration of oxygen chloride ion from adjacent faces of the concrete element increased the availability of corrosion species thereby resulting in an increase in the steel corrosion rate.

Keywords: Steel corrosion rate, Concrete resistivity, Cracked concrete, Uncracked concrete, 1D chloride penetration, 2D chloride penetration

1 INTRODUCTION

Corrosion of embedded steel is the most common durability problem in steel reinforced concrete (RC) structures (Mehta, 2003). In general, there is a reduction in the load carrying capacity of structures due to loss of the steel cross-sectional area (CSA) and reduction in bond between the steel and concrete due to accumulation of corrosion products (rust) at the steel-concrete interface (SCI) result in the development of corrosion-induced cover cracks (Broomfield, 2007; Marcotte & Hansson, 2007). Cover cracks may allow further penetration of corrosion agents into the concrete thereby increasing the rate of steel corrosion (Pease, 2010). Concrete resistivity which can easily be measured on most sites has been identified as parameter that can be used to predict the risk of corrosion in RC elements.

In this paper, the relationship between corrosion rate and concrete resistivity in cracked and uncracked concrete prisms exposed to either 1D or 2D chloride ingress are studied. It is assumed that the corrosion rate of the steel bars will be influenced by their location (i.e. at an orthogonal corner exposed to 2D chloride ingress or at a location that is farther away from the concrete edge that is exposed to 1D chloride ingress) within the concrete. The steel bars that are located at square (orthogonal) edges of vertical and horizontal RC elements serve as tie points for stirrups, hence the prediction of their corrosion with the aid of concrete resistivity measurement will be beneficial in setting planned repair works to avert collapse of RC structures that are in aggressive environments. The corrosion of steel bars at orthogonal edges that are exposed to 2D chloride ingress causes delamination at the concrete corners.

2 EXPERIMENTAL APPROACH

2.1 *Concrete specimens and chloride exposure*

A total of 24 prisms of size 150 × 150 × 625 mm were used in this study. Using a w/b = 0.40, two types of concrete specimens, Portland cement (PC; CEM I 52.5 R), fly ash (F; class F), and ground granulated blast furnace slag (S) in the proportion 70/30 PC/F and 50/50 PC/S were cast. A cross-section of the prism depicting the position of a single (10 mm diameter by 620 mm long) steel bar relative to the chloride ingress exposure is shown in Figure 1. The cracked beams were placed back-to-back onto a 3-point load frame and the load adjusted until a single crack (≈ 0.30 mm width) via a notch was induced while no load was applied to the uncracked specimens.

DOI: 10.1201/9781003348450-308

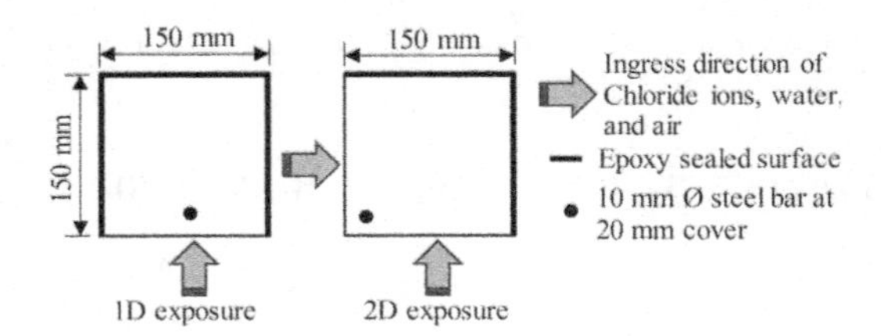

Figure 1. Specimens exposure to 1D, 2D ingress of corrosion agents.

The specimens were subjected to a 2-week wetting (in a 5% NaCl solution) and drying cycle (in ambient laboratory conditions for a period of 110 weeks. The corrosion rate was measured using a device that operates based on the coulostatic technique while the concrete resistivity was measured with a 4-point Wenner resistivity meter. Both measurements were taken approximately 24 hours after each wetting cycle.

3 RESULTS AND DISCUSSION

Each plot point in Figure 2 is a 3-point average obtained from 3 readings. The results indicate that a weak inverse C-R relationship exists between the corrosion rate and resistivity in both the uncracked and cracked concrete, The cracked specimens yielded significantly higher corrosion rates.

While a linear C-R relationship was observed in the cracked (C) fly ash (F), and slag (S) prisms that where exposed to 1D chloride ingress, the corrosion rate increased exponentially in the specimens that were exposed to 2D chloride ingress with steel bars at the edge of the prism. This result was obtained despite the fact there was no significant change in the concrete resistivity. Other factors such as oxygen and chloride penetration from adjacent faces of the concrete edge may have been responsible for the higher corrosion rate in the specimens exposed to 2D ingress. The ductility of steel bars in the plastic deformation phase was lower in steel bars exposed to 2D chloride ingress. This could result in sudden collapse of RC structures as against a desired ductile failure (Otieno & Zakka, 2020).

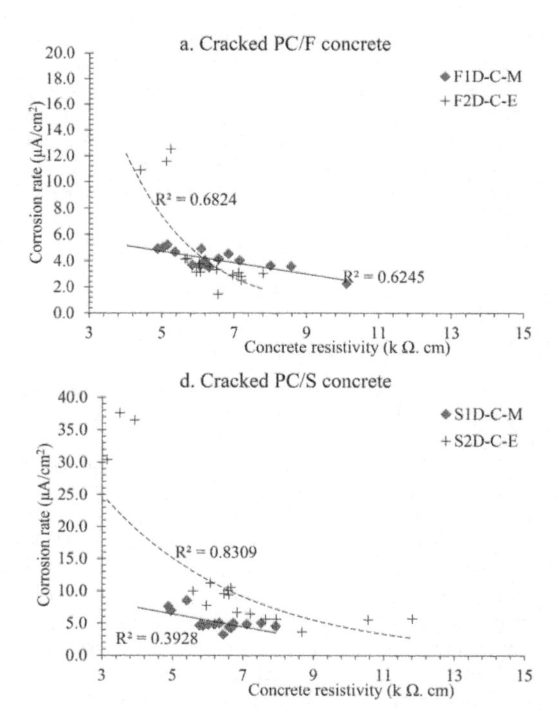

Figure 2. Relationship between corrosion rate and concrete resistivity.

4 CONCLUSIONS

1. There is a weak linear C-R relationship in the uncracked concrete prisms
2. An exponential C-R relationship was observed in prisms that were exposed to 2D chloride ingress
3. The presence of cover cracks, location of the steel bar, chloride ion concentration at depth of steel bar, and the prisms exposure to chlorides influenced C-R relationship
4. The use of 1D models is not sufficient to predict the corrosion of steel bars at orthogonal edges of concrete elements that are exposed to 2D chloride ingress.

ACKNOWLEDGEMENT

The authors acknowledge the support that was received from the South African National Research Foundation (NRF), PPC Ltd., AfriSam South Africa (Pty) Ltd, Galvaglow (Pty) Ltd., Sika South Africa (Pty) Ltd, and ABE (Pty) Ltd.

REFERENCES

Broomfield, J.P. 2007. *Corrosion of Steel in Concrete* Understanding, *Investigation and Repair.* 2nd Ed.

Marcotte, T.D. & Hansson, C.M. 2007. "Corrosion products that form on steel within cement paste," *Materials and Structures*, 40, pp. 325–340.

Mehta, P.K. 2003. *Concrete in Marine Environment.* Edited by Taylor & Francis Books, Inc.

Otieno, M. & Zakka, Z. 2020. "Strength and ductility performance of corroded steel bars in concrete exposed to 2D chloride ingress," *Materials Research Society*, pp. 2817–2825.

Pease, B.J. 2010. *Influence of concrete cracking on ingress and reinforcement corrosion.*

Current Perspectives and New Directions in Mechanics, Modelling and Design of Structural Systems – Zingoni (ed.)
© 2022 Copyright the Author(s), ISBN 978-1-032-39114-4

Measurement and simulation of pipeline attached to bridge for vibration-based SHM

D. Kobayashi, T. Nakanishi, Y. Sakurada & A. Aratake
NTT Access Network Service Systems Laboratories, Tsukuba, Ibaraki, Japan

ABSTRACT: There has been very little research on the vibration-based structural health monitoring (SHM) of pipes attached to bridges compared to that on bridges with independent pipelines. Therefore, in this work we utilize a theoretical model to measure and verify the vibration of pipes attached to a bridge and propose a method to identify deterioration based on the theoretical model. We set up ten accelerometers for an 18-m- long bridge and the pipe and performed eigenmode and frequency response analysis by finite element method (FEM). The results showed that the pipe vibrates with a mixture of natural vibrations near 25 Hz and bridge-derived vibrations below 20 Hz. We also found through the theoretical simulation that it is effective to use the natural vibration with large intensity and unique mode shape above 80 Hz as a marker of deterioration occurrence.

1 INTRODUCTION

Pipelines are important facilities for social life, and it is desirable that vibration-based SHM be implemented for efficient and high-quality maintenance. While there has been much research on vibrationbased SHM for bridges and independent pipelines, very few works have examined pipelines attached to bridges (Kobayashi et al., 2021). This is primarily due to the complex structures of such pipelines, which contain a mixture of bridge and pipe vibrations. In this work, we measure and verify the vibration of pipes attached to a bridge with a theoretical model and then propose a method to identify deterioration based on the theoretical model for the eventual implementation of vibration-based SHM.

2 METHOD

The target of this study was a rigid polyvinyl chloride (PVC) communication pipeline in Japan measuring about 18 m in length (Figure 1). Seven pipes, fixed their edges with mortar, were them by beams placed at intervals of 2.2 to 2.3 m. The bridge accommodated a 2-lane road made of pre-stressed concrete. Accelerometers were installed at three points on the bridge, three points on the pipe, and four points on the support beams, and the sampling rate was set to 500 Hz. The measurement time was 300 seconds, during which 15 vehicles passed by.
The weather during the time of measurement was sunny with a light wind.

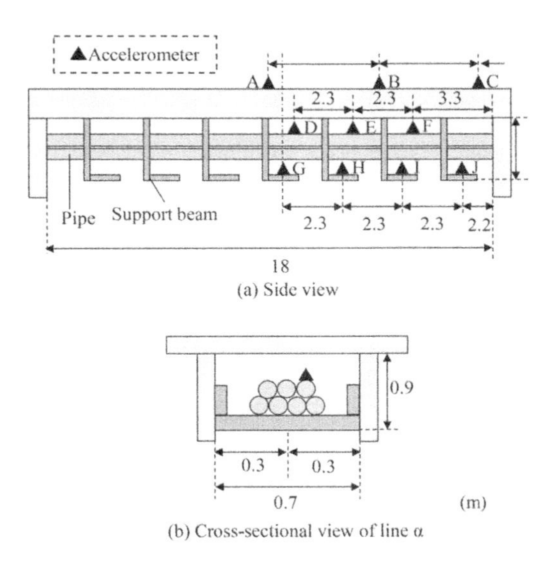

Figure 1. Diagram of bridge, pipes, and sensors placement.

In basic experimental modal analysis, the frequency response function (FRF) is calculated from the excitation force and response acceleration. However, for large structures, it is difficult to measure the excitation force. We, therefore, analyzed the FRF by using operational modal analysis (OMA) to obtain the pseudo excitation force (Schwarz & Richardson, 2007).

We also performed operating deflection shapes (ODS) analysis to clarify the relative deformations

DOI: 10.1201/9781003348450-309

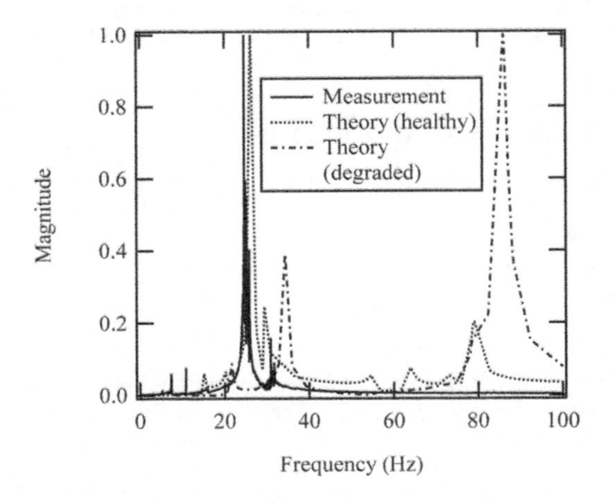

Figure 2. FRF of measurements and theoretical model. Measurement is FRF of pipe obtained by OMA. Degraded means model of loss at the center.

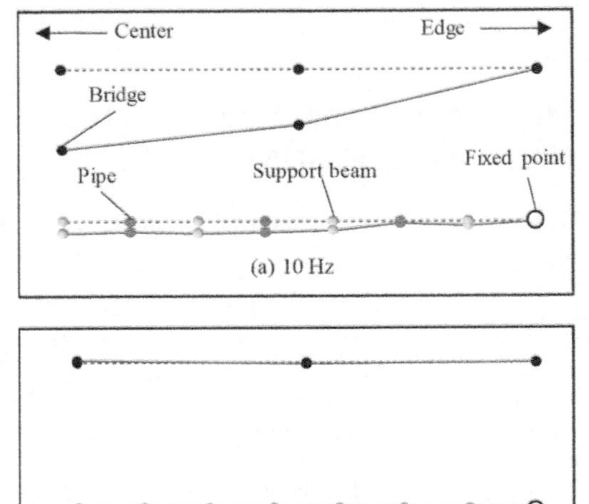

Figure 3. ODS shape at 10 and 25 Hz. Solid lines indicate deformation, dashed lines indicate reference

of the bridge, support beams, and pipes in the frequency domain based on the magnitude and phase of the Fourier spectrum.

Eigenmode analysis and frequency response analysis were performed by the FEM of beam elements. The pipe was fixed with mortar at both ends, and the support points between them were fixed with thin beams. Therefore, the boundary conditions were set to a fixed support at both ends, and the support points between them were set to hinge support. The Young's modulus was set to 2400 MPa, which is assumed for PVC.

In the model reproducing the deterioration, loss of parts was reproduced by setting the target part as free support. In this paper, a model with loss of the central support beam was shown.

3 RESULTS AND DISCUSSION

As shown in Figure 2, the measured and theoretical models (healthy) showed a strong magnitude at about 25 Hz and 30 Hz. The ODS analysis result (Figure 3) at 10 Hz, which was only present in the measurement, shows that the pipe followed the deformation of the bridge. On the other hand, only the pipe vibrated at 25 Hz, which means that the pipe was forced to displace by the eigenmode of the bridge at 10 Hz, and 25 Hz was the natural frequency of the pipe. The peak of 79 Hz on the theoretical model was not detected by the measurement because it was a second-order mode deformation, and the acceleration position was just at the node.

Our comparison of the healthy and degraded models shows that the degradation had its strongest peak at 86 Hz, and the mode shape at 86 Hz showed a unique deformation at the section where the support point was lost.

On the basis of these findings, we propose using the magnitude of the second-order mode peak and the change in mode shape as markers for SHM. To observe these, the accelerometer should be placed in the quadrant between the support points.

4 CONCLUSION

We clarified the dynamic behavior of a pipeline attached to a bridge through measurements and theoretical simulations. We also investigated the markers of deterioration that should be used for vibration-based SHM. Ten accelerometers were attached to the bridge, support beams, and pipe, and measurements were taken under vehicle traffic. The theoretical simulation was analyzed by FEM with fixed supports and hinge supports containing beam elements.

The results showed that the pipe vibrated with a mixture of the eigen vibration of the pipe at 25 Hz and the eigen vibration of the bridge at 10 Hz. From the theoretical simulation, we found that it is effective to use the natural vibration with large intensity and unique mode shape above 80 Hz as a marker of deterioration occurrence.

REFERENCES

Kobayashi, D., Ikeguchi, Y., Nakagawa, M. & Aratake, A. (2021) Vibration-Based Detection of Loosened Bolts on Pipes Attached to Bridges. Materials Research Proceedings.

Schwarz, B. & Richardson, M. (2007) Using a De-Convolution Window for Operating Modal Analysis. IOMAC 2007.

Current Perspectives and New Directions in Mechanics, Modelling and Design of Structural Systems – Zingoni (ed.)
© 2022 Copyright the Author(s), ISBN 978-1-032-39114-4

Investigation of modulation transfer due to nonlinear shear wave interaction with local source: Numerical and theoretical approach

R. Radecki, A. Ziaja-Sujdak, M. Osika & W.J. Staszewski*
Department of Robotics and Mechatronics, AGH University of Science and Technology, Krakow, Poland

ABSTRACT: Presented paper examines the shear wave modulation transfer due to the wave interaction with the nonlinear contact interfaces. The analytical results based on the Harmonic Balance expansion, are compared with a numerical solution based on the Local Interaction Simulation Approach (LISA). To analyze the modulation transfer, two sources of the shear wave excitations are considered. First, a single frequency probing wave, and a second, referred as pumping wave which is a sine wave modulated by a one-order lower frequency signal. A local through thickness defect is modelled in a linear elastic medium. The propagating waves interact with the nonlinear frictional interfaces of a crack, causing a modulation transfer through the shear stick-slip movement of crack surfaces. The mechanism is numerically implemented as Coulomb friction formulation of crack surfaces behavior. The pumping and probing waves are excited simultaneously in the structure by a displacement uniformly distributed over the thickness of the plate. It is shown, that despite the symmetric characteristic of considered source of nonlinearity, the transfer of all orders of sidebands is observed.

1 INTRODUCTION

1.1 *Scope of the paper*

As early detection allows cost-effective maintenance planning, the effort is made to understand an incorporate the knowledge about nonlinear defect-related ultrasonic signal features. Higher order harmonic generation, hysteresis, slow dynamics, or modulation phenomenon are examples of nonlinear effects that can be related to microcracks or plastic zone in material under inspection.

The crack-induced nonlinearity involves multiple mechanisms responsible for the higher harmonic generation. For instance, under certain levels of tension and compression, crack can open and close. This closing/opening motion is often associated with clapping, kissing, friction, adhesion, thermo–elasticity, slow–dynamic behaviour and various nonclassical wave-interaction effects. Various modelling techniques have been developed such as bi-linear stiffness simplification and stress-strain hysteresis (Broda et al. 2014).

In this paper, the modulation transfer phenomenon is scrutinized. In particular, the propagation of the Shear Horizontal (SH) waves in a medium with a local nonlinearity is considered. To observe the modulation transfer, modulated pumping wave and a monoharmonic probing waves are excited simultaneously in the structure. The nonlinearities in the structure result in the modulation transfer to the probing wave. To analyse this effect, numerical simulations using Local Interaction Simulation Approach (LISA) (Delsanto et al. 1994), and a semi-analytical approach based on Harmonic Balance method (HBM) (Radecki et al. 2021) are employed.

1.2 *Harmonic balance method*

The Harmonic Balance method is a suitable tool for analysis of nonlinear systems. It imposes a multi–frequency plane wave solution upon the considered wave propagation phenomenon. The method was successfully employed to obtain the wave dispersion and propagation characteristics in nonlinear complex and periodic media. In the presented study the HBM is used for solving the nonlinear contact problems.

1.3 *Local interaction simulation approach*

LISA is an example of a numerical tool that can be successfully applied for nonlinear ultrasound (Osika et al. 2020). The method was firstly proposed for wave propagation in physics for modelling complex media (i.e. media with sharp interfaces and inclusions of different material properties) (Delsanto et al. 1994). Next, it has been used extensively for linear

*Corresponding author

DOI: 10.1201/9781003348450-310

guided waves. LISA relies on a particular application of the Finite Difference formulas in the discrete approximation of the space derivatives in the governing differential equations.

2 NUMERICAL SIMULATIONS

2.1 *Numerical model (LISA)*

A two-dimensional model of a semi-infinite 1000 mm plate of 2 mm thickness was simulated. A linear homogenous material property corresponding to aluminium are assumed (i.e., density ρ = 2700 kg/m^3, Poisson's ratio of 0.33, Young modulus E=70 GPa). The model grid size is set to $\Delta x = \Delta y = 0.1$ mm, leading to 20 elements through the thickness of the considered plate. The time step for the simulations is set as $\Delta t = 0.01$ μs. The adopted mesh size and time step ensures sufficient convergence to the analytical characteristics and the numerical stability.

2.2 *Coulomb friction model*

To facilitate the comparison of LISA and HBM, the crack is modelled as a through-thickness crack localized in the middle of the plate length. Due to the adopted z-invariance, only the shear movement of crack surfaces (perpendicular to the SH wave propagation direction) is modelled. In this case, the faces of the crack interact mechanically by the friction force, which results from the contact between asperities under a normal force (pressure).

In the LISA models, the Coulomb friction formulation with a stick-slip behavior is implemented with the additional normal pressure acting on crack surfaces. The details of the implemented model are given in (Osika et al. 2020).

Excitation is employed by a displacement uniformly distributed over the thickness of the plate. The pumping and probing waves frequencies are set the same as in the HMB (i.e., amplitude modulated wave with f_1 = 490 kHz and f_m = 21 kHz and the probing wave with frequency f_2 = 85 kHz).

3 RESULTS AND DISCUSSION

The results from the semi-analytical HMB and the numerical simulation using LISA method are presented in Figure 1.

Using both method of investigation it can be observed that the modulations excited in the vicinity of the pumping wave frequency f1 indeed have transfer in a particular percentage to

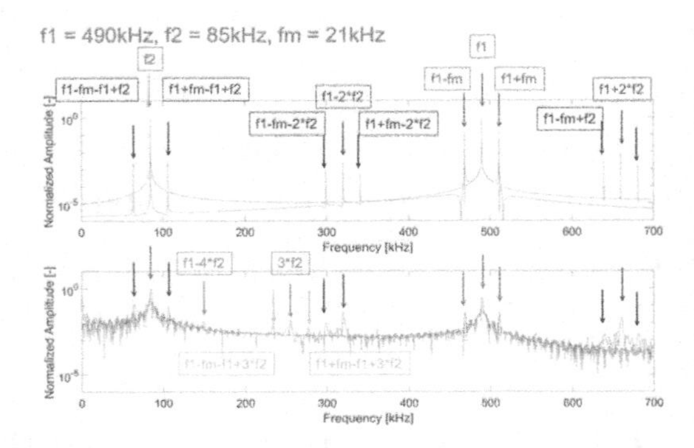

Figure 1. Numerical results obtained from HB (top) and SH-LISA (bottom) in frequency domain for intact and damaged structures.

the vicinity of the probing wave frequency f2 as a result of the frictional interaction between the contact interfaces.

Apart from the aforementioned phenomenon, it is also possible to observe additional frequency components, which are the sum and/or difference of the frequencies of the excited waves.

4 SUMMARY AND CONCLUSIONS

In the presented research paper, the shear horizontal wave interaction with local nonlinearity is addressed using a semi-analytical and numerical approaches. The results show that the modulation transfer from the high frequency pumping wave to the lower frequency probing wave is present for the considered type of defect.

ACKNOWLEDGMENTS

The work presented in this paper was performed within the scope of the research grant UMO-2018/30/Q/ST8/00571 financed by the Polish National Science Centre. This grant is part of the Polish-Chinese collaborative research SHENG initiative.

REFERENCES

Jhang, K.Y. 2008. Int. J. Precis. Eng. Manuf. 10(1), 123–135.

Broda, D., et al. 2014. Journal of Sound and Vibration, 333(4), 1097–1118.

Delsanto, P., et al. 1994 Wave Motion 20(4), 295–314.

Osika, M., et al. 2020 In European Workshop on Structural Health Monitoring (pp. 200–209). Springer, Cham.

Radecki, R., et al. 2021 Nonlinear Dynamics, 103(1), 541–556.

20. Structural assessment, Historic structures, Masonry structures, Repair, Strengthening, Retrofitting

Current Perspectives and New Directions in Mechanics, Modelling and Design of Structural Systems – Zingoni (ed.)
© 2022 Copyright the Author(s), *ISBN 978-1-032-39114-4*

Intermediate Isolation System for the seismic retrofit of existing masonry buildings

D. Faiella, M. Argenziano, G. Brandonisio, F. Esposito & E. Mele
Department of Structures for Engineering and Architecture, University of Naples Federico II, Naples, Italy

ABSTRACT: This paper deals with the potentials of Intermediate Isolation Systems (IIS) for retrofitting masonry buildings by means of a vertical extension, isolated at its base, and realized on the rooftop of the existing structure. By carefully defining the dynamic properties of the superstructure and isolation system, in previous papers the authors have demonstrated that this passive technique is able to reduce a priori the seismic demand on the existing building. For this reason, the non-linear behavior has been only assumed for the isolation system, with the masonry structure in the elastic field. This result has been validated by carrying out spectrum-compatible time history analyses on the 3D FE models of some case studies, and design criteria have been also derived. However, ground motions of intensity larger than the design level cannot be excluded, and the hysteretic behavior of the masonry structure should be taken into account. For this aim, Pivot-type hysteretic models are examined for assessing the representativeness of masonry behavior. By utilizing a case study, 3D FE models are developed, and non-linear time history analyses are carried out for assessing the effect of masonry nonlinearities on the retrofit solution designed according to linearity assumptions.

1 INTRODUCTION

In densely urban areas characterized by medium/high seismic hazard, an Intermediate Isolation System (IIS) can be used for vertical extensions without generating additional forces in the existing structure. The isolation layer can be designed to convert the vertical addition into a huge mass damper. Actual applications of this approach mainly refer to r.c. buildings, as in (Dutta et al. 2009). This idea has been already proposed by the authors for masonry structures, by examining some case studies (Faiella et al. 2020, Argenziano et al. 2021) and assuming a non-linear behavior only in the isolation system.

In this paper, the effect of the masonry nonlinearities is examined. Once selected as case study, linear parametric analyses are firstly carried out on lumped mass models to choose the configurations of the isolated vertical addition that minimize the global seismic response of the overall structure. Then, the seismic behavior of the retrofit solutions is assessed through 3D FE models, accounting the masonry structure's hysteretic behavior with Pivot-type models (Pasticier et al. 2008).

2 CASE STUDY

The case study (Figure 1) is an aggregate of four masonry buildings located in Pozzuoli, which is ascribed in a wide plan to reconstruct the rooftop volumes, demolished in the 1980s. The mechanical properties assumed for the tuff masonry material are: average weight of 16 kN/m^3, Young modulus of 1080 MPa, shear modulus of 360 MPa, and compression strength of 3.0 MPa (NTC 2018).

The idea is to rebuild the demolished floors (Figure 1a) by means of the IIS technique, by erecting an isolated vertical addition in structural steelwork (S275) on the roof of the existing building. For creating a unique level of the isolation system, a filling steel structure has been preliminarily added on the roof of the existing building on the left, and designed for resisting to gravity loads only and to not modify the dynamics of the as is configuration.

The 3D FE model of the structure is developed in SAP2000. The masonry building capacity is assessed with push over analyses, while its seismic behavior

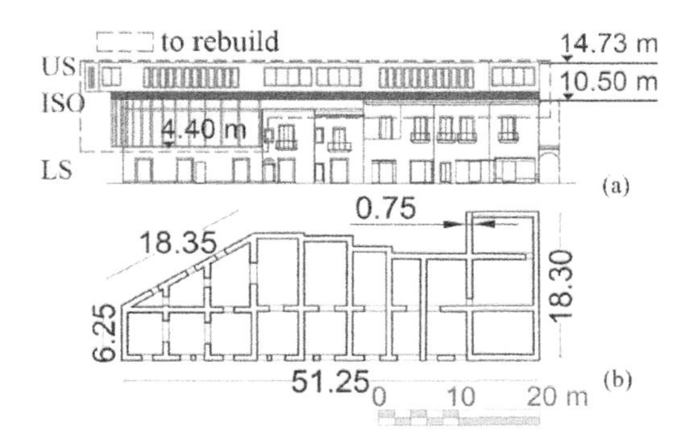

Figure 1. Case study building: (a) front view and (b) first floor plan.

DOI: 10.1201/9781003348450-311

with non-linear time history analyses (section 3) in which the hysteretic behavior is described by the Pivot-type model ($\alpha = \beta = 0.45$, $\eta = 0$).

3 IIS EFFECTIVENESS AND CONCLUSIONS

An IIS system can be preliminary described through a three-degree-of-freedom model (3DOF IIS, Figure 2a), in which the DOFs refer to the lower structure (LS), isolation system (ISO), and upper structure (US). The dynamic properties of the isolated extension can be described by some design parameters, i.e.: the mass ratio $\mu = (m_{ISO}+m_{US})/m_{LS} = 0.25$; the stiffness ratio $K = k_{US}/k_{LS} \in [0.1, 2]$; the isolation ratio $I = T_{ISO}/T_{US} \in [0.1, 10]$. Response spectrum analyses are carried out by adopting the design spectrum prescribed by the Italian seismic code for the case study site (NTC 2018). By considering the ratio $d = d_{1,IIS}/d_{1,LS}$ between the first DOF's displacement of the IIS and LS configurations, different design solutions (with K = 1.85) are selected in the shaded area of Figure 2b. The conventional elevation (LS+US), without isolation system, is also considered for comparison purpose. Then, non-linear spectrum-compatible time history analyses are carried out on the 3D FE LS, LS+US and IIS models of the design solutions through SAP2000, in which the masonry nonlinearities are accounted through the Pivot-type model while the isolated vertical addition is assumed elastic. The main results are here discussed for the weakest (Y) direction. The Figure 3a depicts the peaks' average displacements obtained from the spectrum-compatible input waves; Figures 3b shows the plastic hinge's loops for the most stressed pier in the LS model and the IIS model with T_{ISO} = 5 s, by adopting one record of the spectrum-compatible set; for the same input and models, Figures 4 provides the cyclic response of the base shear vs. the displacement at the top of the existing building. Reductions of LS displacements are obtained by designing isolation systems with long T_{ISO} (Figure 3a). The presence of the isolated vertical addition provides a smaller inelastic engagement of the masonry building (Figure 3b) that globally shows a smaller dissipated energy (Figure 4). The disconnection between the vertical addition and the existing building by means of the isolation layer improves the response of the original structure

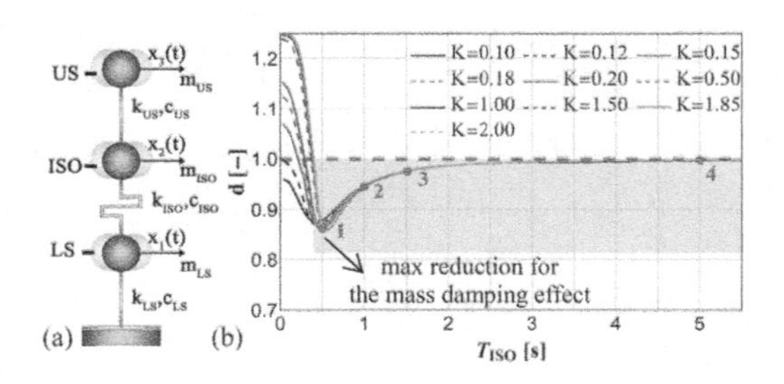

Figure 2. (a) 3DOF IIS model and (b) displacement ratio d as a function of the isolation period T_{ISO}.

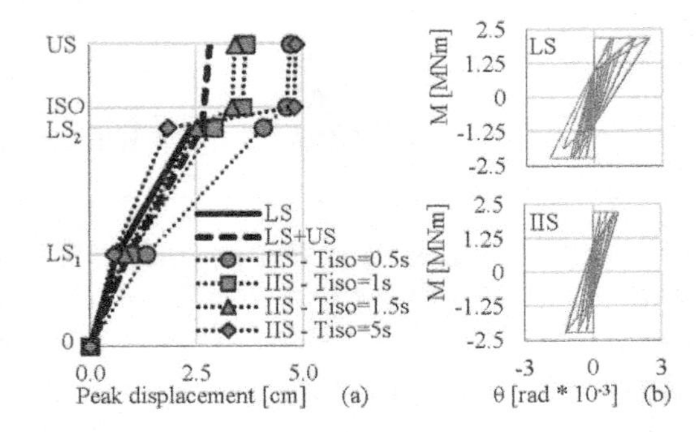

Figure 3. (a) Peak story displacements and (b) plastic hinge's loops for the most stressed pier in the LS and IIS models for one record.

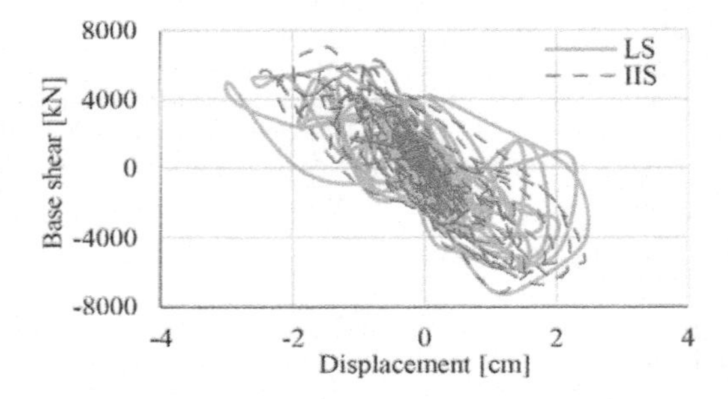

Figure 4. Hysteresis loops in the 3D LS and IIS models for one record.

compared to the case of conventional extension. The actual behavior of the masonry can be further improved under seismic action, by adopting retrofit interventions for which the existing building works in the elastic field.

REFERENCES

Argenziano, M., Faiella, D., Bruni, F., De Angelis, C., Fraldi, M., Mele, E. 2021. Upwards - Vertical extensions of masonry built heritage for sustainable and antifragile urban densification. *J. Build. Eng.* 44, 102885.

Dutta, A., Sumnicht, J.F., Mayes, R.L., Hamburger, R. O. 2009. An innovative application of base isolation technology. *Proceedings of 18th Analysis and Computation Specialty Conference-structures Congress* 841–854.

Faiella, D., Calderoni, B., Mele, E. 2020. Seismic retrofit of existing masonry buildings through Inter-story Isolation System: a case study and general design criteria. *J. Earthq. Eng.* 26(4): 2051–2087.

NTC. 2018. Consiglio Superiore Lavori Pubblici, Aggiornamento delle Norme Tecniche per le Costruzioni. Gazzetta Ufficiale della Repubblica Italiana 2018; 42. (in Italian).

Pasticier, L., Amadio, C., Fragiacomo, M. 2008. Non-linear seismic analysis and vulnerability evaluation of a masonry building by means of the SAP2000 V.10 code. *Earthq. Engng Struct. Dyn* 37: 467–485.

Current Perspectives and New Directions in Mechanics, Modelling and Design of Structural Systems – Zingoni (ed.)
© 2022 Copyright the Author(s), *ISBN 978-1-032-39114-4*

A seismic retrofitting design approach for activating dissipative behavior of timber diaphragms in existing unreinforced masonry buildings

M. Mirra & G. Ravenshorst
Bio-Based Structures and Materials, Delft University of Technology, Delft, The Netherlands

ABSTRACT: The region of Groningen (NL) has experienced increasing human-induced seismicity caused by gas extraction in the last decades. The local building stock, not designed for seismic loads, consists for more than 50% of unreinforced masonry buildings with timber diaphragms. In this context, this work presents a design approach for creating strengthened dissipative timber diaphragms, and maximizing the seismic capacity of existing masonry buildings through this retrofitting method. The results from the performed numerical analyses prove that the proposed design approach for timber floors can increase the energy dissipation capacity of masonry buildings, while improving the box behavior at both damage and near-collapse limit state.

1 INTRODUCTION

In the northern part of the Netherlands, within the Province of Groningen, human-induced earthquakes caused by gas extraction have occurred in the last decades. Because these events were absent until recently, the local building stock was not designed or realized accounting for seismic loads. These buildings consist for more than 50% of unreinforced masonry (URM) structure with slender walls and poorly connected, flexible timber floors. The focus of this work is on the retrofitting of these diaphragms, designed to improve their strength and stiffness, and to activate additional energy dissipation. The strengthening method consisted of an overlay of plywood panels screwed along their perimeter to the existing sheathing, and the performed experimental tests (Mirra et al. 2020) confirmed the potential of this technique. Starting from these results, both analytical (Mirra et al. 2021a, b) and numerical (Mirra et al. 2021c, Mirra & Ravenshorst 2021) modeling strategies were developed for an advanced simulation of their nonlinear in-plane response. These formulations open up the opportunity of designing the diaphragms in such a way that they can fully transfer the expected seismic shear forces, while activating a beneficial energy dissipation.

2 DESIGN METHODOLOGY

The design approach presented in this work refers to URM buildings sufficiently regular in plan and elevation (EN 1998). Once that a preliminary (on-site) investigation on geometrical and material properties of the building is conducted, the first step consists of estimating its seismic capacity in terms of base shear. To this end, a nonlinear static (pushover) analysis can be conducted, preliminarily considering rigid diaphragms. After determining the base shear F_b in the weakest direction, the corresponding seismic forces F_i at each floor level are calculated with the lateral force method. The shear forces F_i constitute the design seismic loads to be transferred by each retrofitted diaphragm without causing an out-of-plane collapse of masonry walls. In other words, these forces represent the minimum value of the strength of the diaphragms, according to which a sufficient number of fasteners for applying the plywood panels overlay can be designed.

In order for the energy dissipation to be activated, the diaphragms should display in-plane deflection capacity, but without causing the out-of-plane collapse of masonry walls. Therefore, besides the definition of the strength of the retrofitted floors according to F_i, also a displacement limit to avoid an excessive deflection is set. In the present work, the maximum in-plane deflection $\delta_{max,i}$ of each retrofitted diaphragms was limited by the minimum value between half of the wall thickness and 2.0% of its height.

Once that the in-plane strength and limit deflection are known for the diaphragm, the retrofitting intervention is designed by considering the floor as a timber shear wall. In this way, starting from the load-slip response of the single fastener connecting plywood panels and planks, by means of equilibrium relations the backbone curve and pinching cycles of the whole diaphragm can be determined. It should be noticed that the in-plane strength and stiffness of the retrofitted floor is governed by the number and

DOI: 10.1201/9781003348450-312

diameter of fasteners, but also by the dimensions of the plywood panels. Therefore, in agreement with the specific design choices, it is possible to create stiffer or more flexible floors, and to predict their energy dissipation and pinching cycles accordingly.

In numerical analyses, the designed retrofitted floors can then be conveniently modeled following a macro-element approach (Mirra et al. 2021c). A floor is subdivided in a mesh of quadrilaterals composed of rigid truss elements, with two diagonal truss elements containing the constitutive law simulating the in-plane response of the diaphragm.

3 APPLICATION OF THE DISSIPATIVE FLOORS IN URM BUILDINGS

The presented design and modeling approach was applied in three case-study buildings, analyzed considering three configurations: an as-built one, with existing flexible floors; one featuring rigid diaphragms realized with the casting of a concrete slab on the existing floors; one in which the retrofitting solution activating energy dissipation was applied. In order to realistically capture the seismic response of the case-study buildings, each configuration was subjected to seven seismic accelerograms applied in both plan directions, and nonlinear behaviour in masonry and diaphragms was considered.

The results in terms of peak ground acceleration (PGA) at collapse show that the impact of dissipative diaphragms is very beneficial for the case-study buildings, because of the additional energy dissipation and damping effect on masonry walls: on average, the highest values of PGA at collapse are retrieved (Figure 1). It should be noticed that a major improvement in the seismic response and box behavior of the analyzed buildings is already retrieved with rigid concrete floors, but the additional dissipation of the designed solution results in an increase of the PGA at collapse of 30% on average.

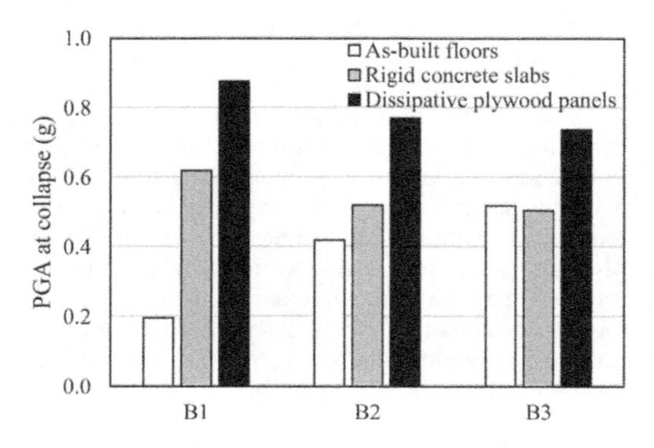

Figure 1. Average PGA at collapse of the studied configurations.

Besides, it was proved that the additional dissipation induced by the floors designed following the proposed approach is a relevant part of the hysteretic energy activated in the whole building. This confirms that well-designed, dissipative diaphragms, able to retrieve the box behavior in the building, can be a better alternative than rigid diaphragms, concentrating the dissipation only in masonry walls.

Besides, although the retrofitted dissipative floors are deformable in plane, the out-of-plane displacement of masonry at damage limit state never overcomes the prescribed drift limits. The presented design approach can thus be suitably adopted for a retrofitting of floors balancing strength, stiffness, and energy dissipation at both near-collapse and damage limit state.

4 CONCLUSIONS AND OUTLOOK

In this work, a design approach for retrofitting timber diaphragms in existing URM buildings has been shown. This method allows to activate additional energy dissipation in the floors, which can greatly improve the seismic performance of the buildings. Besides showing a design example, the results of numerical time-history analyses on three case-study URM buildings have been discussed. It has been proved that diaphragms retrofitted with the proposed design approach are indeed able to activate large hysteretic energy dissipation. For future research, it is recommended to further validate these results, and to evaluate the effect of such retrofitted diaphragms in terms of behavior factor, also by means of experimental tests.

REFERENCES

EN 1998, Eurocode 8. Design of Structures for Earthquake Resistance.

Mirra, M. et al. 2020. Experimental and analytical evaluation of the in-plane behaviour of as-built and strengthened traditional wooden floors. *Engineering Structures* 211

Mirra, et al. 2021a. An analytical model describing the in-plane behaviour of timber diaphragms strengthened with plywood panels. *Engineering Structures* 235.

Mirra, M. et al. 2021b. Comparing In-Plane Equivalent Shear Stiffness of Timber Diaphragms Retrofitted with Light and Reversible Wood-Based Techniques. *Practice Periodical on Structural Design and Construction* 26.

Mirra, M et al. 2021c. Analytical and numerical modelling of the in-plane response of timber diaphragms strengthened with plywood panels. *8th ECCOMAS Thematic Conference on Computational Methods in Structural Dynamics and Earthquakes Engineering*, Athens, Greece.

Mirra, M. & Ravenshorst, G.J.P. 2021. Optimizing Seismic Capacity of Existing Masonry Buildings by Retrofitting Timber Floors: Wood-Based Solutions as a Dissipative Alternative to Rigid Concrete Diaphragms. *Buildings* 11.

Current Perspectives and New Directions in Mechanics, Modelling and Design of Structural Systems – Zingoni (ed.)
© 2022 Copyright the Author(s), ISBN 978-1-032-39114-4

Lifetime assessment for historical cast steel bridge bearings

D. Siebert & M. Mensinger
Technical University of Munich, Chair of Metal Structures, Munich, Germany

ABSTRACT: At the end of the 19th century different types of cast steel bearings were used for railroad bridges. These have to be evaluated concerning their risk of brittle fracture and fatigue safety to guarantee a safe use in the future despite their advanced age. In this paper, the fracture mechanical methods used for the verification against brittle fracture and adequate fatigue safety are presented. Non-destructive testing on bridge bearings, numerical simulations, and analytical formulas for the determination of stress intensities at an assumed crack tip are used as a basis. Crack propagation calculations are conducted for determining the remaining lifetime by considering the interaction of forces due to train crossings by friction. A concept for the assessment of the usability of historical bridge bearings, even after a long period of use, is provided by which a safe and sustainable use of the considered bearings can be ensured.

1 INTRODUCTION

For historical cast steel bridge bearings, it is essential to evaluate the residual life as well as the risk of brittle fracture for assessing further use. The assessment of existing cast steel bearings is not covered normatively by the Eurocodes. By means of appropriate fracture mechanical calculations, further use can be assessed and thus frequently made possible.

For fracture mechanical investigations, an existing defect in the form of an initial crack is generally assumed. To define possible positions and sizes of defects, numerical casting simulations can be used in addition to the results from non-destructive testing (UT & MT). Based on the defined initial crack and the applied stress, determined by using finite element analysis (FEA), the stress intensity at the crack tip can be determined. A comparison with the fracture toughness of the material allows a statement on the risk of brittle failure. In addition, crack growth until failure due to the cyclic loading from train crossings is possible. This crack growth is described by means of crack propagation calculations so that the period of safe further use can be determined.

The fracture mechanical background as well as the applied nondestructive testing methods are summarized in the full paper. This short paper briefly describes the procedure for conducting the verification against brittle fracture and fatigue safety of historical cast steel components. This is illustrated by means of a historical bridge bearing. Finally, the concept for the assessment and the results are discussed and a conclusion is drawn.

2 VERIFICATION AGAINST BRITTLE FRACTURE AND FATIGUE SAFETY

2.1 *Verification against brittle fracture*

In this study, the verification against brittle failure is conducted in a simplified manner by comparing the stress intensity factor at the crack tip K_I with the resisting material fracture toughness K_{IC}. K_I is calculated analytically in dependence on initial crack size, applied stress, and dimensions of the structural component which is approximated by the respective crack model in a plate.

Based on assumed quality grade 2 of the cast bearings, the maximum permissible defect indication according to EN 12680 (DIN German Institute for Standardization, 2003a, 2003b) and the unfavorable aspect ratio of the crack geometry of depth to width a/c = 0.4 (Erhard & Otremba, 2014) are used to determine the crack dimensions of the semi-elliptical surface crack. This results in a crack depth of a = 5.6 mm and a crack width of 2c = 28.2 mm.

The crack is placed conservatively perpendicular to the maximum main tensile stress. Stresses are calculated by applying bearing reactions for the extraordinary load event with the help of FEA.

2.2 *Crack propagation calculations*

Crack propagation is calculated stepwise by integrating the Paris law. Every step the resulting stress intensity range at the crack tip is assessed by means of the Failure Assessment Diagram (FAD) until failure is predicted. The initial crack size is assumed analogously to section 2.1 even if a smaller size for crack propagation

DOI: 10.1201/9781003348450-313

simulation is reasonable as well. The applied stress ranges from train crossings are determined with the help of influence lines, main tensile stresses from the uniform load cases, superposition of acting stresses, and rainflow-counting. For details, the reader is referred to the full paper.

Interacting bearing reactions might be challenging so that they need to be examined in detail. For example, normal forces and transverse forces are interacting at the arch crown and arch base bearings. Straightforward superposition of the respective forces multiplied by the uniform load cases results in highly conservative results, as in reality normal forces contribute to the transfer of transverse forces. Thereby the corresponding stresses in the uniform load case of isolated transverse force are exaggerated. For this reason, a simplified consideration of the interaction is proposed by incorporating friction. To be specific, the transverse force time series is reduced by a fraction removed by friction.

In this study, a coefficient of kinetic friction of μ_k = 0.2 is applied. For a deeper understanding of the influence of the assumed coefficient of kinetic friction, a parametric study is undertaken, see Figure 1.

A superposition that is incorporating friction is especially beneficial if high transverse stresses are present in the uniform load case. For the considered position, a crack propagation calculation without reduction of the transverse forces by friction results in 0.55 years of residual life, whereas applying μ_k = 0.2 results in at least 80 years. Starting from μ_k = 0.17 the transverse forces are completely dissipated through the acting normal forces. Most notably are the huge differences in lifetime between the smaller assumed values of kinetic friction coefficients as high stress ranges are the crack driving force.

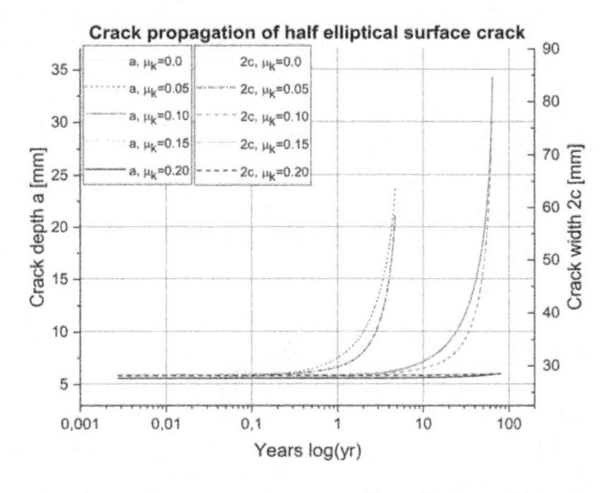

Figure 1. Influence of the assumed coefficient of kinetic friction on the crack propagation in the directions of a and c, exemplarily shown for one position with high stresses in the uniform load case for shear force of the cast arch crown bearing.

3 CONCLUSION

In this paper, the assessment of historical cast steel bridge bearings based on fracture mechanical calculations is presented. The assumed initial crack sizes are determined dependent on the results of nondestructive testing (MT, UT) and on additional casting simulations which allow the assessment of possible shrinkage cavities due to the casting process. The adopted crack size as well as acting main tensile stresses are used to determine the stress intensities at the crack tip. A comparison with the fracture toughness, which is conservatively reduced by the maximum reduction factor for possible plastifications, predicts no risk of brittle fracture for the considered arch crown bearing. Crack propagation calculations based on the Paris-law are presented based on the assumed initial crack size. As abort criterion brittle fracture as well as yielding are implemented according to the FAD. Stress ranges for the crack propagation calculations are calculated by multiplying the forces of train crossings by ultimate load cases for normal force and transverse force. As the acting normal forces are contributing to the dissipation of transverse forces at the considered arch crown bearing, a reduction by friction is applied. For this purpose, a coefficient of kinetic friction is assumed. The influence of the assumed kinetic friction coefficient on the calculated crack propagation is shown and discussed. With the help of the crack propagation calculations, the residual lifetime of the considered cast steel bearing due to crack propagation is determined. It has to be respected that the results of fracture mechanical calculations are quite sensitive to the assumed input parameters. For this reason, assumptions have to be drawn carefully and the period of prediction should be limited to e.g. 80 years as this is already a prediction far in the future. Inspections are recommended for updating the applied models in the future if needed. Even longer periods of use than calculated can be assured if no cracks are detected during inspections.

Altogether, the presented assessment concept based on fracture mechanical considerations enables a further safe use of historical cast steel bridge bearings.

REFERENCES

DIN German Institute for Standardization. 2003a. Founding - Ultrasonic examination - Part 1: Steel castings for general purposes; German version EN 12680-1:2003. Berlin: Beuth Verlag GmbH, (DIN EN 12680-1:2003–06).

DIN German Institute for Standardization. 2003b. Founding - Ultrasonic examination - Part 2: Steel castings for highly stressed components; German version EN 12680-2:2003. Berlin: Beuth Verlag GmbH, (DIN EN 12680-2:2003–06).

Erhard, A., & Otremba, F. 2014. Nondestructive Testing and Fracture Mechanics. In 5th International CANDU In-Service Inspection Workshop in conjunction with the NDT in Canada 2014 Conference.

Current Perspectives and New Directions in Mechanics, Modelling and Design of Structural Systems – Zingoni (ed.)
© 2022 Copyright the Author(s), ISBN 978-1-032-39114-4

Strengthening of existing infrastructure with concrete screws as post-installed reinforcement

J. Feix
Unit of Concrete Structures and Bridge Design, University of Innsbruck, Austria

J. Lechner
Feix Ingenieure GmbH, Munich, Germany

ABSTRACT: In many parts of the world, infrastructure has reached a critical age. Due to enviromental influences, higher loads or changes of code regulations many structures in this field show a lack of resistance concerning shear, punching and bending. During the last decade a new strengthening system for existing structures based on the use of so called concrete screws was developed. These special concrete screws are installed in predrilled holes in the concrete structrues and act as post-installed reinforcement. The big advantages of this system are a significant rise of the resistance of the existing structures, the easy and fast installation without disturbance of the traffic and the low costs of the measures. The contribution deals with the development of the strengthening method especially for shear and punching loads, the laboratory tests and numerical investigations. Also a design approach based on Eurocode provisions is presented. These calculation rules and installation principles are embedded in a technical approval order to allow the application for all practical engineers. Finally some best practice projects of bridges, tunnels and parkdecks strengthened with this method will be presented. Especially when we look at the upcoming questions regarding sustainability, minimizing CO2-emissions and the efficient use of natural resources the extended use of existing structures by strengthening these structures will gain a huge importance in the future.

1 INTRODUCTION

The majority of existing bridge structures in Central European countries were built between 1960 and 1990. That means more than 45 % of the bridge structures on German federal roads are between 60 and 40 years old. An examination of the use of materials used for these bridges by the German Federal Highway Research Institute shows that the majority - almost 90 % of the structures - were built using reinforced or prestressed concrete. In addition to the aging bridge infrastructure in Central Europe, the increase in traffic volume, especially heavy goods traffic, also plays a significant role in the deterioration of the bridge infrastructure. The example of the Brenner motorway built in the 1970s in Austria shows very clearly that the loads on the infrastructure have increased massively since it was put into operation. Also the code provisions regarding shear strength have changed significantly compared in the last decades.

2 CONCRETE SCREWS AS POST-INSTALLED REINFORCEMENT

The basic idea of the new system is to use socalled concrete screw anchors with nominal diameters of $d_0 = 16mm$ and $d_0 = 22mm$ as post-installed reinforcement for concrete structures. These easily installable anchors, known from the use as anchor in concrete constructions, can fulfil the requirements of an efficient strengthening system, such as installation during ongoing use, installation from only one side of the structure and robust and immediate load bearing characteristic due to their load-transfer mechanism on basis of interlock. The concrete screws for post-installed reinforcement are screwed into a previously created drill hole. In this way, they create a force transmission in the front area via the thread, which can carry the acting forces. At the rear end of the screw, reanchoring elements are attached to the screw at the outside of the structure.

3 SCIENTIFIC INVESTIGATIONS

The basic idea for the use of bonded anchor bolts as post-installed reinforcement was developed in 2010. In a first step, the general suitability of the bonded anchor bolts as a post-installed reinforcement element was successfully demonstrated in several tests for both shear force reinforcement and punching

DOI: 10.1201/9781003348450-314

shear reinforcement. Based on this findings of the first tests, a test programme was developed and carried out, which finally led to technical approvals for the system.

4 DESIGN PROVISIONS

Through these extensive tests and numerical investigations carried out in parallel with the experimental tests, two technical approvals by the German Institute for Structual Engineering were obtained for the new method. These technical approvals specify not only the field of application and the design guidelines, but also a assessment model on basis of the known equations of Eurocode 2 for the planning engineer. These assessment models are derived from the achieved test results.

5 PILOT-PROJECTS

On the basis of the technical approval and the design concept contained therein, numerous reinforcement projects for building structures and bridge structures have been carried out in recent years.

For example a railway bridge carries a two-lane road under a railway line, whereby the supporting structure is located in the entrance area of a railway station and thus a switch is arranged on the structure. The bridge was built in the form of a single-span slab bridge with a span of 11.3 m and a width of approx. 18 m. Due to the switch on the bridge no measures on top of the structure were possible. The applied reinforcement was installed entirely from below from the two walkways. Thus, the reinforcement did not interfere with the traffic under the bridge and could be carried out within a few days.

In the course of a recalculation of a highly skewed, three-span slab bridge in Brandenburg, a deficit in punching shear capacity was detected. The bridge is supported at the intermediate supports on 6 and 7 columns respectively. The reinforcement is installed exclusively from the underside of the structure from a bridge inspection device. Therefore only minimal disruptions of the use on the bridge were given.

A railway crossing over the German motorway A70 was built with tendons out of steel which is known for its tendency to stress corrosion. To raise the shear load capacity concrete screw anchors were drilled through the hollow boxes from underneath the bridge. To increase the flexural bending failure load steel plates were attached as external reinforcement at both sides of the girder. The whole installation time of the reinforcing system was four weeks and no disruption of the railroad traffic was necessary.

6 SUSTAINABILITY OF REINFORCEMENT

An investigation of an existing railway bridge over a federal motorway with regard to the environmental impacts in comparison of a retrofit, a new construction with demolition of the existing structure and the closure of a structure for a certain period of time. The effects shown on the greenhouse effect, acidification and the non-renewable cumulative energy input are related in each case to the temporal closure of the supporting structure.

7 CONCLUSIONS

A significant number of concrete structures lack of reinforcement. Therefore, several research projects aim to develop a new strengthening system to increase their strength. Such systems aim at easy installation without disruption of the use of the structure.

In several pilot projects the easy and fast installation of the new retrofltting system was proved. Serveral projects regarding shear and punching shear reinforcement were carried out with great success. In all projects the application under ongoing use without disruption of the traffic on the bridge was confirmed. Also the fast installation was one of the main advantages of this system compared to other reinforcement measures. Summarising, it can be seen that the usage of concrete screw anchors as post-installed reinforcement can raise the load capacity of existing structures signiflcantly. Because of the easy and fast installation of the system, it is very well-suited for the strengthening of structures under ongoing use. Using the new system, the lifetime of existing structures can be increased signiflcantly and disruptions to traffic by road closures or rerouting can be avoided or reduced.

In an investigation regarding the sustainability of reinforcing it was shown, that strengthening of existing structures with minimal road-closures on and under the bridge reduces the impact on the global warming potential, the acidiflcation and on non-renewable culmulative energy input. Therefore, efforts should be made to keep the existing infrastructure in operation longer through reinforcement measures in order to reduce the impact on the environment and global warming.

Current Perspectives and New Directions in Mechanics, Modelling and Design of Structural Systems – Zingoni (ed.)
© 2022 Copyright the Author(s), *ISBN 978-1-032-39114-4*

Analytical formulation for plain and retrofitted masonry wall under out-of-plane loading

J.A. Dauda
School of Architecture, Built and Natural Environments, University of Wales Trinity Saint David, UK

O. Iuorio
School of Civil Engineering, Faculty of Engineering and Physical Sciences, University of Leeds, UK

F.P. Portioli
Department of Structures for Engineering and Architecture, University of Naples Federico II, Italy

ABSTRACT: This paper presents an analytical formulation to complement considerable experimental and numerical programs conducted to propose a new timber-based retrofit technique for masonry walls. The formulation is based on the static analysis of an idealised structural scheme adopted in previous experimental tests, and it obtains the bending moment equation which predicts the maximum failure load and its location on both plain and retrofitted masonry walls. Thereafter, the expected failure load in the masonry wall is analytically estimated from the moment of resistance of both the plain and retrofitted walls using ultimate section analysis. The results show that the analytical formulation correctly evaluated the capacity of both the plain and retrofitted masonry wall to within 5% variation. It thus concluded that the developed analytical model can be extended into more complex models and thus fit for a parametric analysis to analyse further the efficiency of the proposed timber masonry retrofit technique.

1 INTRODUCTION

Analytical models and calculations are basic skills employed by engineers as a useful complement or alternative to experimental tests and numerical models. This is because full experimental investigation on structural elements, particularly masonry is always tedious and expensive. Therefore, this paper presents an analytical formulation to complement considerable experimental and numerical programs conducted to propose a new timber-based retrofit technique for masonry walls (Dauda et.al, 2021; Iuorio et.al, 2021). Masonry is a very complex, heterogenous and composite structure that possess non-linear mechanical properties. The non-linearity in the mechanical behaviour of masonry causes high uncertainty in analysing its structural response (Lourenco and Silva, 2020; Ismail and Ingham, 2016). As such, the analytical prediction of masonry strength and responses is a very complex subject area that is attracting interest from various researchers in the field.

Therefore, this paper focuses on an analytical formulation for plain and retrofitted masonry walls under out-of-plane loading. The formulation is based on the static analysis of the idealised structural schemeadopted in the experimental test setup. The formulation obtains the bending moment equation which predicts the maximum failure load and its location on both plain and retrofitted masonry walls. The results of the analytical formulation were compared against the experimental results obtained from Iuorio et.al (2021).

The paper was articulated as follows. In section 2, the analytical formulation for both plain and retrofitted masonry walls was presented. Section 3 presents a brief comparison of the analytical result with the experimental results. Finally, the conclusion to the study is presented in section 4.

2 ANALYTICAL PREDICTION OF MASONRY WALL STRENGTH

An analytical model in form of a simplified structural scheme (Figure 1) which accounts for a horizontal restraint due to the friction at the base of the four-point bending experimental test setup (Figure 2) was used in this study to obtain the strength of masonry wall as subsequently described.

Referring to Figure 1, the support reaction at the top and bottom of the masonry wall (R_A and R_B) are estimated from static equilibrium equations following equations 1 to 4.

$$R_A + R_B = P + F_\mu \qquad (1)$$

DOI: 10.1201/9781003348450-315

$$R_A = \frac{P}{2} - \left(F_\mu * \frac{X_3}{X_1 + 2X_2} \right) \quad (2)$$

$$R_B = \frac{P}{2} + \left(F_\mu * \frac{X_1 + 2X_2 + X_3}{X_1 + 2X_2} \right) \quad (3)$$

$$F_\mu = \mu \times \gamma_{mw} \quad (4)$$

Where:
F_μ: is the resultant horizontal force due to friction at the base of the specimen.
μ: coefficient of friction between steel roller and plate at the base of the wall (0.8)
γ_{mw}: unitary weight of wall calculated using the density of brick unit (2200kg/m^3)

Meanwhile, in order to estimate the bending moment equation which will predict the maximum failure load at any location (X_i) within the inner bearing of the masonry wall, a section Y-Y is considered, and the moment equation is derived using equation 5 or 6.

$$M = \frac{P}{2}(X_i) - R_A(X_2 + X_i) \quad (5)$$

$$M = \frac{P}{2}(X_2) - \left(F_\mu * \frac{(X_2 + X_i) * X_3}{X_1 + 2X_2} \right) \quad (6)$$

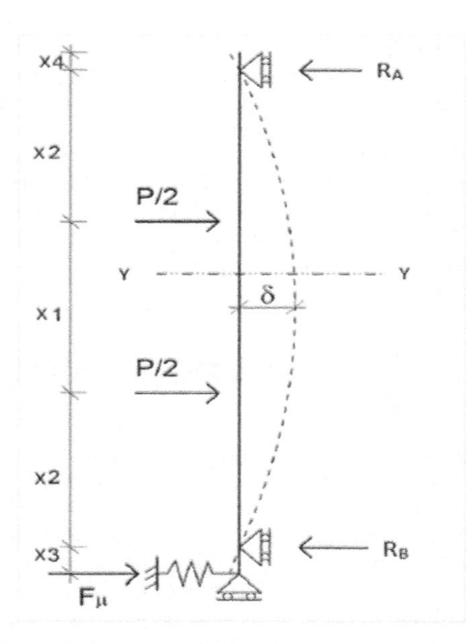

Figure 1. Idealised structure scheme for four point bending test.

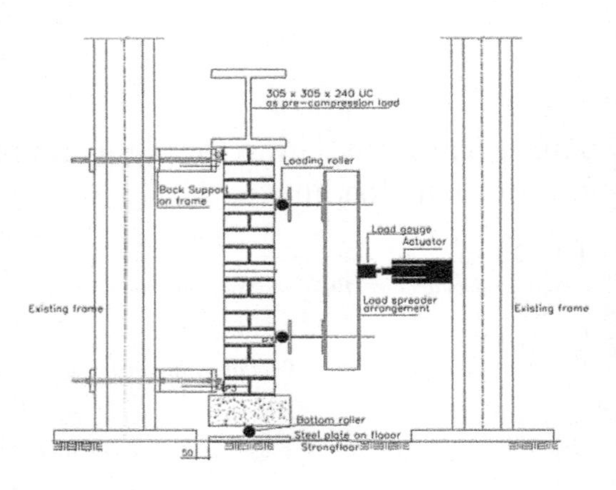

Figure 2. Test set up for four point bending test.

Thereafter, the expected failure load in the masonry wall was estimated analytically by considering the moment of resistance of both the plain and retrofitted walls using ultimate section analysis in section 2.1 and section 2.2.

3 CONCLUSIONS

The study presented in this paper has demonstrated that the results of experimental tests conducted on four bending tests (Iuorio et.al, 2021) can be predicted analytically. The analytical formulation was based on using the stress state of the generic cross-section of masonry specimen. The formulation assumed the combined effect of the axial force, due to the masonry self-weight, and the bending action induced by the two horizontal loads. The model predicted the peak load and the corresponding failure to within less than 5% of the average results obtained from the test.

Current Perspectives and New Directions in Mechanics, Modelling and Design of Structural Systems – Zingoni (ed.)
© 2022 Copyright the Author(s), ISBN 978-1-032-39114-4

Structural behavior of masonry walls with soft-layers: An overview of experimental work

N. Mojsilović
Institute of Structural Engineering, ETH Zurich, Switzerland

ABSTRACT: A numerous series of static-cyclic shear tests on masonry elements (triplets), wallettes and full-scale unreinforced masonry (URM) walls with soft layer membrane placed in the bed joint to induce sliding have been performed during last ten years at ETH Zurich within several research projects on the seismic behavior of unreinforced masonry walls with such layers. Specimens were constructed using typical perforated Swiss clay and calcium-silicate blocks and standard cement mortar. As a soft layer type, the six types were considered (rubber granulate, cork, cork-rubber granulate, extruded elastomer, bitumen and polyvinylchloride.) In order to develop the most suitable joint in the first phase the specimens were tested placing the soft layer in the mortar joint or between the mortar and the blocks. Based on the results, the so-called multi-layer bottom bed joint, which comprises a core soft layer protected by two layers of extruded elastomer and placed in the middle of mortar joint, was developed. As a core soft layer the four layer types were considered: rubber granulate, cork, cork-rubber granulate and bitumen. The following testing phase was aimed at choosing the most suitable core soft layer – rubber granulate. The final testing phase comprised five tests on story-high URM walls with rubber granulate core soft layers performed to investigate the influence of the size, the pre-compression level and the aspect ratio on the seismic behavior of URM walls with a multi-layer bed joint. This paper presents an overview of the mentioned experimental work.

1 INTRODUCTION

The central idea of the presented research is to modify the seismic response of individual structural masonry walls by changing their response mechanism to control the horizontal shear force they receive. This can be achieved by placing engineered deformable soft layers at the bottom of masonry wall. The soft layers are designed so, that their presence will allow the wall to slide, albeit with some friction, thus controlling shear force in the wall. Building on the fundamental findings of previous research the concept of using soft layers in individual masonry walls is developed by: 1) investigating the mechanical properties of soft layers required to achieve the desired performance of the structural wall; 2) conducting the series of proof-of-concept and parameter exploration tests; and 3) developing the mechanical models to describe the seismic response of structural masonry walls with incorporated soft layers and optimization of mechanical properties of soft layer to achieve the desired seismic performance of structural masonry walls. In the present paper we report and discuss the results of the first two (experimental) tasks.

The idea of controlling the shear behavior of masonry walls using soft layers was initially explored during the author's academic/research visits to University of Newcastle, Australia. The main goal of the joint research project by the University of Newcastle, Australia, and ETH Zurich, was to investigate the influence of the bed joint placed damp-proof course membrane on the structural behavior of unreinforced masonry under cyclic shear loading.

The findings of the joint research project laid the fundaments for the long-term research project at ETH Zurich, which will be presented in the following chapters.

2 TESTS ON SMALL SPECIMENS (TRIPLETS)

Experimental investigations presented in this chapter mainly concentrated on the assessment of the shear parameters and overall performance of joints containing soft layers (made of different materials) by conducting monotonic and static-cyclic experiments on masonry triplets. Based on the obtained results, it can be concluded that shear force can be transmitted through the joints containing soft layer membranes and that due to sliding failure along the soft layer a considerable energy dissipation and quasi ductile behavior could be expected for masonry with soft

DOI: 10.1201/9781003348450-316

layers. Initially, the tests with single soft layers placed in bed joint have been performed and it was found that elastomer-based membrane was the most durable soft layer material and was not subjected to the degradation induced by sliding motion, especially under repeated (cyclic) load reversals. Thus, it was decided to develop the so called multi-layer bed joint. Such joints consist of a core soft layer protected with two thin extruded elastomer membranes, which in turn are placed in a bed mortar joint. The extruded elastomer membranes are employed to prevent and/or limit the deterioration of the core soft layer during the cycling action, which has been observed in previous investigations.

3 WALLETTE TESTS

As a next step in the research, four series of static-cyclic test on clay block masonry wallettes were performed. These wallettes had two different types of soft layers: rubber granulate and extruded elastomer. Two of the four series were performed on I-shaped wallettes.

Soft layers were beneficial when the tested wallettes failed in sliding. Such sliding failure was observed for wallettes with rubber granulate soft layers placed in the first bed joint. This resulted in quasi-ductile behavior of the wallettes, with a displacement and energy dissipation capacities significantly larger than those of the conventional unreinforced masonry control specimens. The progressive deterioration of the rubber granulate layers lead to a decreasing shear strength at larger horizontal displacements. Further, the test results show that the flanges considerably increased the horizontal force resistance and the initial stiffness of the I-shaped wallettes compared to rectangular wallettes.

4 WALL TESTS

A test series on five large, story-high walls with multi-layer bed joint concluded the experimental investigation. Prior to this test series, a four test on smaller walls has been conducted in order to determine the most suitable type of the multi-layer bed joint, i.e. the core soft layer material. Four different types of (core) soft layer, namely rubber granulate, cork, bitumen and cork-rubber granulate have been used and placed in the bed joint at the bottom of the masonry wall. Based on the results of this preliminary phase the rubber granulate has been chosen as the material for the core soft layer for test series on large walls.

The main testing phase comprised five tests on story-high specimens with rubber granulate core soft layers performed to investigate the influence of the size, the pre-compression level and the aspect ratio on the seismic behavior of unreinforced masonry walls with a multi-layer bed joint.

The response of all specimens was initially concentrated at the bottom bed joint and followed by sliding at the interface between the core soft layer and the elastomer. Such sliding motion caused lateral tensile stresses at the bottom brick course of the specimens, which led to (vertical) tensile cracks in the head joints. Simultaneously with the sliding motion, diagonal shear cracks formed along the bed and head joints, starting from the middle of the wall. The tensile cracks in the bottom brick course spread to the upper courses of the wall as the tests progressed and caused a reduction of the wall effective area.

The level of the reduction of the wall effective area, which strongly influenced the response of tested walls, depended on the level of pre-compression, aspect ratio, position of the tensile cracks and whether or not they subsequently merged with the formed diagonal shear cracks. Nevertheless, shear was finally the common failure mode of all specimens, but for two specimens in which the reduction of the effective area was not significant, and they could slide theoretically to infinity.

The response of the wall specimens can be interpreted using a multilinear horizontal force-displacement response idealization, whose parameters can be estimated using the developed equations. In addition, a model, which is capable of predicting the position of the previously described vertical cracks has been developed and validated with the experimental findings.

5 CONCLUSIONS

From presented experimental investigations, it can be concluded that structural masonry walls with engineered soft layers, including multi-layer bed joint, will enable the masonry structure to achieve the following performance goals (in regions of low and moderate seismicity):

- Experience essentially no damage in frequent earthquakes and under high wind loads when the structure is expected to remain elastic.
- Experience controlled damage while preserving the gravity load-carrying capacity in design-basis earthquakes through controlled lateral sliding deformation and an elongation of the structural response period due to low stiffness in the sliding regime.
- Prevent collapse in beyond-design-basis earthquakes through engagement of the full shear strength of the structural masonry walls.

Current Perspectives and New Directions in Mechanics, Modelling and Design of Structural Systems – Zingoni (ed.)
© 2022 Copyright the Author(s), *ISBN 978-1-032-39114-4*

Investigating the causes, impact, and repair methods of popping bricks in house construction

K. Nongwane, S.A. Alabi & J. Mahachi
Department of Civil Engineering, Faculty of Engineering and Built Environment, University of Johannesburg, Johannesburg, South Africa

ABSTRACT: Popping bricks are commonly used in areas with extreme temperatures, particularly those with repeated freeze-thaw cycles. The study presents the causes and impacts of popping bricks in house construction and repair methods. Site observation and experimental tests such as water absorption and compressive strength were carried out on the existing popping bricks. The outcome of the water absorption test for the popping bricks is at 8%, indicating that the bricks were freeze/thaw resistant and reliable for housing construction, implying that water infiltration was not the reason for the bricks popping out. The average compressive strength of the popping bricks was 16.35 MPa, higher than the 3.50 MPa recommended by the standards. Therefore, the bricks were suitable for construction.

Keyword: Pop-outs, popping, water, brick, pop-out repair, freeze/thaw cycles, free lime

1 INTRODUCTION

Brick is a highly durable and effective insulator, which is why it has been used for centuries. Pop-outs are caused by the expansion of porous aggregate particles with a high absorption rate. With increased awareness of the causes of popping bricks in houses and their effects on construction, contractors and owners must keep up with changes in climatic conditions, which are the primary triggers of popping bricks in houses. Hence, the goal of this study is to investigate and increase literature to the body of knowledge of popping bricks in house construction and what measures need to be adopted to prevent this matter. Furthermore, this study will identify the impacts of popping bricks in houses and its repair methods that need to be applied to improve the service life of brick structures.

2 METHODOLOGY

A non-random sample technique was used in this study, including purposeful sampling. The following data were acquired using two simple tests that are commonly performed on typical cement bricks in quality control laboratories on building sites: a) water absorption test according to the ASTM C34-13 standard. b) compressive strength testing by SANS 227:2007. c) on-site observation of the samples that have been damaged.

3 EXPERIMENTATION AND RESEARCH RESULTS

Water absorption tests on bricks are used to assess long-term durability features such as degree of burning, quality, and weathering behavior. Compression strength tests are performed using a compression testing machine to assess the load-carrying ability of bricks under compression.

3.1 *Water absorption test*

The water absorption test for the popping bricks results is approximately at 8% water absorption, indicating that the bricks were resistant enough for the freeze/thaw cycle and reliable for house construction. This indicates that water infiltration was not the cause of the brick pop-outs.

3.2 *Compressive strength test results*

The compressive strength of bricks used in house construction should not be less than 3.5 N/mm2. The average compressive strength test results of the bursting bricks are 16.35 MPa.

3.3 *Site observation*

The bricks used in the test contained free lime, which was added to the aggregates during the manufacturing process. Figure 1 shows the lime pitting that caused the surface of the bricks to pop and resulted in a defect.

DOI: 10.1201/9781003348450-317

Table 1. Water absorption of popping bricks results.

Brick number	Mass of bricks (g)					
	Initial mass	After 24 hours mass	2 hours	2 hours (W_0)	Water mass (W_1)	% percentage absorption
1	2890.00	2888.00	2860.8	2850.81	3152.07	10.57
2	2984.07	2969.07	2954.07	2937.77	3219.48	9.59
3	2839.94	2824.24	2814.94	2799.24	3062.41	9.401
4	2794.90	2776.16	2764.2	2757.42	3068.74	11.29
5	2933.46	2914.88	2904.93	2897.98	3180.18	9.74
6	3015.24	3005.39	2985.24	2975.39	3270.23	9.91

Figure 1. Lime pitting in popping bricks samples.

4 CONCLUSIONS AND RECOMMENDATIONS

This study aimed to look into the causes and effects of popping bricks in houses and repair methods. Water absorption tests have shown that popping bricks are freeze/thaw resistant and reliable for home construction. Free lime pitting was observed in the bricks, which was the primary source of pop-outs.

According to water absorption test findings on the popping bricks, the water absorption was 8.22%. Pop-outs were discovered to be repairable by chipping away at the remnant aggregate in the surface cavity, cleaning it, and then filling it with a proprietary repair material such as dry pack mortar and epoxy mortar. (1–10)

REFERENCES

1 Courard, L., Degée, H. & Darimont, A., 2014. Effects of the presence of free lime nodules into concrete: Experimentation and modelling. Cement and Concrete Research, III (64), pp. 73–88.

2 CTI, 2019. Cemen Tech Momentum. [Online] Available at: https://cti-ia.net/what-are-aggregate-popouts-how-can-i-avoid-them/ [Accessed 28 August 2021].

3 Fay, K. F. V., 2015. Guide to Concrete Repair. 2nd ed. America: Technical Service Center.

4 Friedman, D., 2007. InspectApedia.com. [Online] Available at: https://inspectapedia.com/structure/Brick_Wall_Repairs.php [Accessed 06 July 2021].

5 Lee, M.-H. & Lee, J.-C., 2009. Study on the cause of pop-out defects on the concrete wall and repair method. Construction and Building Materials, 23(1), pp. 482–490.

6 Lewis, R., 2015. Spalling Bricks [Interview] (5 September 2015).

7 Malo, R. C., 2021. Howtocivil. [Online] Available at: https://www.howtocivil.com/spalling-bricks-cause-and-repair/ [Accessed 22 July 2021].

8 Netinger, I., Vračević, M., Ranogajec, J. & Vučetić, S., 2014. Evaluation of brick resistance to freeze/thaw cycles according to indirect procedures. GRAĐEVINAR, III (66), pp. 197–209.

9 Starmark, 2018. Starmark Home Inspections. [Online] Available at: https://www.charlottehomeinspector.com/brick-spalling-explained/ [Accessed 22 July 2021].

10 Thompson, J., 2013. Stlbrickrepair. [Online] Available at: https://stlbrickrepaircom/2013/05/20/what-causes-brick-faces-to-pop-off-or-deteriorate-why-is-this-happening/ [Accessed 22 July 2021].

Current Perspectives and New Directions in Mechanics, Modelling and Design of Structural Systems – Zingoni (ed.)
© 2022 Copyright the Author(s), ISBN 978-1-032-39114-4

Experimental investigation of the influence of embedment depth on the sustained load performance of adhesive anchors

Andrea Carolina Oña Vera, Sahand Sartipi, Wouter Botte & Roman Wan-Wendner
Ghent University, Ghent, Belgium

ABSTRACT: Adhesive anchors are fastening solutions widely used in new and old structures for their strengthening and rehabilitation. Their flexible nature and ease of use facilitate the installation and attachment of new load bearing components to old concrete members. Nowadays, sustainability represents an important requirement in the design and rehabilitation of structures and their connections. This requires understanding their long-term behavior, since they are affected by sustained loads that, due to the gradual increase in deformations, ultimately lead to failure. Despite their advantages due to different influencing factors such as concrete aging, non-linear viscosity, and post-curing of the adhesive and anchor geometry, the load-bearing behavior of adhesive anchors under sustained loading is still not very well understood. This paper presents the summary of the results of an extensive experimental test campaign, in particular isolating the potential effect of the embedment depth on tertiary creep rupture of a vinyl-ester based commercially available anchor system. This was achieved by means of an experimental campaign consisting of a series of tensile tests in a confined configuration. Each configuration consists of a set of short-term pull-out tests to determine the ultimate capacity followed by sustained load tests at various load levels. The ultimate load is used to define the load levels for long-term tests which are applied to a limited number of specimens. The time to failure is recorded for every sustained-load test and the so called time-to-failure curves are constructed by fitting the resulting data points based on different regression models. Finally, the different models are analyzed according to the long-term behavior prediction of the adhesive anchors.

1 INTRODUCTION

Adhesive anchors, or chemically bonded anchors, are post-installed fasteners installed in drilled holes in concrete or masonry. They are commonly used in the strengthening and rehabilitation of existing structures as connections between the load bearing elements and concrete. Loads are transferred by means of bond stress between the anchor and the adhesive and between the adhesive and concrete.

The long-term behavior of adhesive anchors has been the subject of several investigations in the past. The results of these investigations have been reflected in current standards in form of prediction laws based on the total displacement of anchors throughout their lifespan. ACI 318 and EAD 330499-01-0601 propose methods in which the assessment of the long-term behavior is extrapolated based on the results of sustained load tests using the Findley method (Findley and Khosla) In this method, first, a series of short-term pull-out tests are conducted on the desired geometry and the maximum capacity is recorded. A series of sustained-load tests are then conducted using fractions of the maximum capacity. The load levels are chosen in a way to ensure the anchors will not fail for a determined period of time under the sustained load. The displacements are measured during the tests. The displacement for the 50-year service life is then predicted based on the results of the sustained load tests. The estimated displacement must be compared against the critical displacement of the adhesive obtained from the short-term pull-out tests (EAD). Although simple and conservative, this method fails to estimate a failure time for a given load level. In addition, it is sensitive to the chosen function for the fitting (Cook et al.) and the data basis (Wan-Wendner and Podroužek, Podroužek and Wan-Wendner).

2 EXPERIMENTAL INVESTIGATION

This experimental campaign consisted of tensile testing of adhesive anchors installed in concrete. The results are presented in the form of time-to-failure (TTF) curves where the relative load level is plotted against the time to failure. The data is fitted according to the current models to assess the long-term behavior of adhesive anchors.

2.1 *Test series and setup*

A total of 3 anchor configurations were performed in each set, each with a different embedment depth. A configuration consists of a set of six short-term tests to failure to determine the mean ultimate load. followed by several long-term tests at different relative load levels, from which the time to failure (TTF) is determined.

Loading of the specimens was performed by applying a tensile load to the anchor by means of a pressure-controlled hydraulic system. Hydraulic jacks with 22 Ton capacity were used and the target load was actively kept constant throughout the sustained load test with

DOI: 10.1201/9781003348450-318

a variation smaller than ±5% in accordance with EAD 330499-01-0601. The loading rate was 10 kN/s, to minimize creep damage during loading.

The tests were performed on a confined configuration to prevent concrete-driven failure. Additionally, a spherical coupling connecting the threaded bar and hydraulic cylinder was provided to prevent transfer of bending moments.

The pressure and the time were monitored throughout the test. Finally, the time-to-failure (TTF) was determined as the difference between the time at which the target load is fully applied and the time at which the anchor fails. This concept is shown in Fig.2.

Displacements in the load direction were measured by a potentiometer placed at the top of the bar. The used linear potentiometers have nominal strokes of 10 mm with a linearity of ±0.3%. The displacements were recorded at a rate of 50 Hz for the initial loading. In order to optimize the processing of the data, for tests lasting longer than 2 hours the following rate was set as 1 Hz. The pressure was continuously recorded by means of a 700 bar pressure sensor at a rate of 20 Hz.

2.2 *Materials*

The adhesive product used in this investigation is a vinyl ester. Concrete class selection, installation, and testing of anchors were performed according to EAD and ETAG 001. The concrete mix was designed for a concrete class C25/30. The anchors were installed following the manufacturer's instructions in pre-drilled holes in uncracked-dry concrete slabs. The concrete slabs were cast at a controlled temperature of approximately 18 °C. Characterization specimens were cast and stored at 20oC temperature and relative humidity of 95% at 28 days of concrete age. The concrete's mechanical properties are presented in Table 2

Table 2. Concrete's mechanical properties at 28 day.

Age (days)	f_{cu} (N/mm2)	f_{cy} (N/mm2)	f_t (N/mm2)	G_F (N/mm)
28	34.8 *CoV* 4%	32.2 *CoV* 4%	2.9 *CoV* 7%	0.08 *CoV* 8%

3 EXPERIMENTAL RESULTS AND DISCUSSION

Figure 1 illustrates the TTF curve for configuration V40 and the fit functions. It was observed that the logarithmic model (LM), power-law model (PLM) and creep rate model (CRM) are not able to reproduce the upper asymptote, while the sigmoid function (SFM) does. The lower asymptotes according to CRM take place at 16%, 41%, and 10% for V40, V75 and V110 respectively. According to SFM, it takes place at 0% for V40 and V110 and at 36% for V75. Additionally, the 50-year prediction according to CRM and SFM does not differ significantly The predicted load level for V40 is 0.61 according to CRM and 0.60 according to SFM. Both models predict values of 0.66 and 0.60 for V75 and V110 respectively.

The accuracy of the models is measured by comparing the residual sum of squares reported in Table 4. No difference is observed in the case of V40 (0.007 for all models) and V75 (0.025 for all models) while for V110 the difference is less than 0.1%, 0.030 for LM and SFM, 0.033 for PLM and 0.031 for CRM. The models with the lowest errors are the logarithmic model and the SFM.

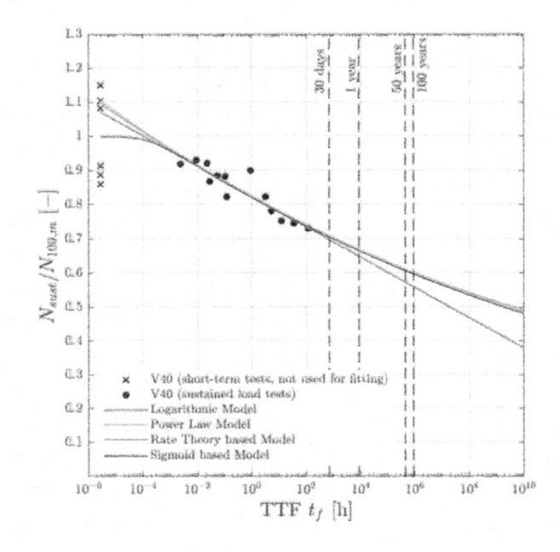

Figure 1. Comparison of the four fitting models for configuration V40 with h_{eff} = 40 mm

4 CONCLUSIONS

This contribution represents a systematic experimental investigation on the sustained load behavior of adhesive anchors for different curing times of two adhesive products. The main conclusions of this investigation are:

1. The current models adopted for the prediction of the time to failure of bonded anchors under sustained load, i.e. logarithmic and power-law models in stress vs. logarithmic time to failure plots, cannot accurately predict the lower asymptotes, i.e. the load level that reflects the transition from linear to non-linear viscoelasticity where the damage due to creep starts to develop.
2. The presented models result in equal or very similar values of RSS.
3. Although similar to the SFM, the CRM is not able to reproduce the upper asymptote at which the relative load level reaches 1 s.

REFERENCES

ACI 318-11 Building Code Requirements for Structural Concrete. Farmington Hills, MI, USA: American Concrete Institute:.

European Organization for Technical Approvals. 2018. EAD 330499-01-0601: Bonded fasteners for use in concrete. Brussels.

COOK, R. A., DOUGLAS, E. P., DAVIS, T. M. & LIU, C. 2013. *Long-Term Performance of Epoxy Adhesive Anchor Systems*.

FINDLEY, W. & KHOSLA, G. 1956. An equation for tension creep of three unfilled thermoplastics. *SPE Journal*, 12, 20–25.

Current Perspectives and New Directions in Mechanics, Modelling and Design of Structural Systems – Zingoni (ed.)
© 2022 Copyright the Author(s), ISBN 978-1-032-39114-4

The review of diagonal compression tests of URM panels strengthened with NSM steel bars

S.D. Ngidi
University of Johannesburg, Johannesburg, Gauteng, South Africa

ABSTRACT: Unreinforced masonry structures often crack due to different factors which include differential movement of structures, temperature variation, wind loads and seismic waves. Retrofitting is introduced to ensure that these cracks do not expand to a point of making the wall fail. The study focused on the in-plane shear behaviour of unreinforced masonry (URM) wall strengthened using near surface mounted (NSM) high strength steel bars. The objective of this work was to use previous experimental work done by different authors to check the efficiency and effectiveness of the NSM steel bar technique to strengthen and stabilize masonry walls.

Keywords: Wall panels, Diagonal compression test, NSM

1 INTRODUCTION

Masonry structures constitute a large percentage of the building infrastructure in South Africa and in most countries around the world. Many of these buildings were designed and constructed before the building codes were updated for seismic resistance. In fact, recent earthquakes worldwide have highlighted the vulnerability of URM buildings to significant damage in the event of a large earthquake eruption.

The advantage of using the NSM steel insertion technique is the minimal disturbance of occupants' business activities in the building during its application and there is no loss of space inside the structure. Furthermore, the dynamic properties of the building remain fixed and are not changed due to little addition of the weight and stiffness.

The in-plane performance of masonry walls with NSM retrofit solution has been investigated by several researchers (Ismail et al. 2011, Casacci et al. 2019, Li et al. 2005, Mahmood et al. 2011 & Sondoval et al. 2021). Most of the recent experimental studies have been on the use of FRP bars / strips while only limited recent research is on stainless steel reinforcing bars. The stainless-steel reinforcement bars are commercially available and are cheaper compared to FRP materials. The objective of this work was to investigate the performance of diagonally compression tests on masonry wall panels with NSM steel retrofit solutions.

The test setup consisted of a hydraulic actuator along with a load cell, high strength steel rods to induce compression, load cell, backing channel, linear variable differential transducers (LVDTs) and a data acquisition system. Figure 1 shows the setup used for diagonal shear testing. Displacement controlled loading was applied along the diagonal of the test wall at a rate of approximately 0.1 mm/s.

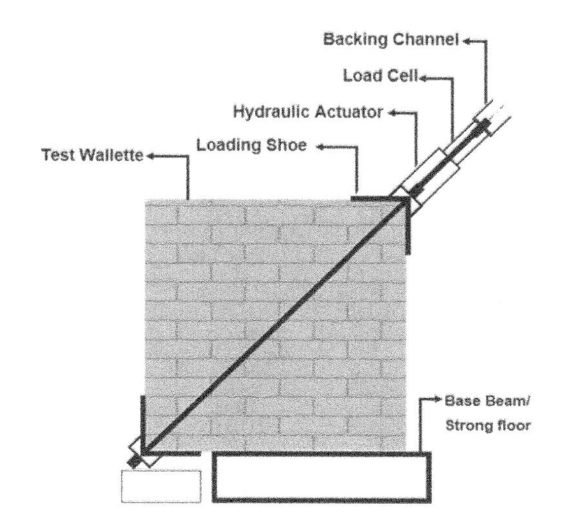

Figure 1. Test setup.

DOI: 10.1201/9781003348450-319

2 RESULTS ANALYSIS

Table 1. Experimental results.

Designation	(H/B)	ρ (%)	P_{max} (kN)	v_{max} (MPA)	$\frac{v}{v_{max}}$ (%)
COW1	1	URM	76.5	0.22	-
COW6	1	0.029	89.9	0.26	118
COW7	1	0.029	155.2	0.44	200
WS-1	1	URM	39.0	0.11	-
WS-7	1	0.020	88.4	0.22	200
W1C-1/2	1	URM	143.0	0.77	-
W1S-7	1	0.380	105.0	0.56	73
W1S-8	1	0.380	124.0	0.66	87
W2C-3	1	URM	51.0	0.14	-
W2S-14	1	0.11	71.6	0.19	140
NRW	1	URM	195	0.54	-
RW3H	1	0.092	145	0.40	74
SW-L_0	1.17	URM	27.5	0.16	-
SW-L_S/3-3	1.17	0.151	24.3	0.14	86
SW-L_S/2-3	1.17	0.151	24.8	0.14	86
TW-L_0	1.17	URM	146.0	0.25	-
TW-L_S/2-3	1.17	0.068	90.0	0.16	64
TW-L_S/2-6	1.17	0.068	101.3	0.18	72
TW-L_D/2-3	1.17	0.137	155.9	0.27	108

H/B = aspect ratio, ρ =reinforcement ratio, P_{max} = maximum applied diagonal force, v_{max} = maximum shear stress, v = shear stress of the as-built wallet

The response of the specimens is presented in terms of failure modes and shear strain response. Test results from the retrofitted specimens were compared to their counterpart from non-retrofitted specimens. Based on the evaluation of specimen COW1, COW6 and COW7, it can be observed that the technique of strengthening with NSM steel bars is very efficient in improving shear capacity of the URM wall regardless of its distribution on one or both faces of the wall. The maximum increment in shear capacity ratio of 200% shows that the technique is effective in significantly increasing the in-plane strength and ductility. Increase in shear strength of retrofitted specimens for Ismail & Ingham (2010) was a ratio of 200%. Increasing panel thickness on the study by Sandoval et al (2021) had a positive effect on shear strength. On average, shear strength of triple-thickness wall specimens was significantly higher than that of single-thickness specimens. Reinforcement had a negligible effect on shear strength, regardless of wall specimen thickness. However, on reinforced triple-thickness wall specimens, those reinforced on both faces exhibited significantly higher shear strength compared to those that had been reinforced on a single face. The effect of increasing the reinforcement depth from 30 to 60 mm in triple-unit-thick panels, with a single reinforced face, led to an increase in maximum shear strength of 12.5% for wall panels.

Test unit WS-1 diagonally cracked along the mortar step joint and upon further increase of the applied load the upper portion of the unit started to slide in the horizontal direction. The force displacement response exhibited a peak and sudden strength degradation once the elastic limit was reached. WS-7 failed with a diagonal crack As-built tested URM wall panels W1C-1 and W1C-2 behaved almost linearly up to peak load and then failed suddenly at small deformations with a diagonal cracking failure mode. The horizontal reinforcement scheme was effective in bridging diagonal cracks which formed close to peak load and allowed the wall panel to deform further until ultimate failure occurred. The helical profile of the stainless-steel reinforcing bars resulted in excellent mechanical bond between the reinforcement and masonry over short, bonded lengths.

3 CONCLUSIONS

The in-plane shear behaviour of URM wall panels strengthened using NSM high strength twisted stainless steel reinforcement bars was investigated. The experimental results showed that the NSM reinforcing steel bars are effective in improving the structural integrity of the URM elements subjected to shear loads, such as those expected in a seismic event. Improved post peak behavior characterized by larger ultimate shear strain were evident in reinforced panels, with smaller crack openings observed on their reinforced faces.

REFERENCES

Ismail, N., Petersen, R.B., Masia, M.J., Ingham, J.M. 2011. Diagonal shear behaviour of unreinforced masonry wallettes strengthened using twisted steel bars, *Construction Building Materials*. 25 (12): 4386–4393.

Soti, R. & Barbosa, R.B. 2019. Experimental and applied element modeling of masonry walls retrofitted with near surface mounted (NSM) reinforcing steel bars. *Bulletin Earthquake Engineering*. 17 (7): 4081–4114.

Ismail, N. & Ingham, J.M. 2010. Structural testing of unreinforced masonry assemblages retrofitted using twisted steel inserts. 8th International Masonry Conference., Dresden.

Current Perspectives and New Directions in Mechanics, Modelling and Design of Structural Systems – Zingoni (ed.)
© 2022 Copyright the Author(s), ISBN 978-1-032-39114-4

Rehabilitation applications of roadway structures

H.H. Abbas
Department of Structural Engineering, Al Azhar University, Cairo, Egypt

M.M. Hassan
Structural Engineering Department, Faculty of Engineering, Cairo University, Cairo, Egypt
Architectural Engineering Department, University of Prince Mugrin, Madinah, Saudi Arabia

ABSTRACT: Deterioration of existing bridge structures in many countries imposes technical and economic problems. Hence, bridge rehabilitation plays an important role in extending the service life of the existing bridges in addition to increasing their capacity to meet new traffic loads. Lack of rehabilitation might lead to sudden complete or partial failure of the concerned structure. In many cases, rehabilitation of an existing bridge is more difficult than constructing a new one depending on the condition of the rehabilitated structure. This paper explores the different rehabilitation practices performed for a group of roadway bridges within the province of Cairo Governate in Egypt. Many of these bridges were built during the last two centuries with some of them exceeding their service life. The work was part of an inspection campaign targeting the bridge stock across the country. The study lists the different stages starting by assessment through site visits and investigation of available design documents. This is followed by pinpointing critical conditions and defects and proposing the most suitable rehabilitation procedure. Reported problems include cracks, unsafe members under increased loads, permeant deformations, spalling, asphalt creep...etc. The study is meant to focus on condition of major structures and explore applied rehabilitation procedures.

1 INTRODUCTION

Maintaining the acceptable levels of performance of bridges requires periodic assessment of structural conditions during service life. Conditions of different bridges vary significantly based on time of construction, year of last repair, and surrounding environmental conditions. Many researchers have tackled this aspect through detailed studies [Hassan et al. (2017), Abbas and Hassan (2021), and Tang (2021)]. Efforts to document conditions of bridges in different countries is considered one of the strategic projects as it sets priorities for maintenance.

The current paper shows assessment efforts for major roadway bridges in Egypt during the past years as per a comprehensive assessment study for bridges across the whole country.

2 DESCRIPTION OF CONSIDERED BRIDGES

The study will focus on selective roadway bridges to show the applied inspection procedures and the observed defects. These include 6th October Bridge, Sayeda Eisha Bridge, Fardous tunnel, Galaa bridge, Abo Hasheesh overpass, and Helmiat Zaytoun bridge.

3 INSPECTION & ASSESSMENT PROCEDURE

Assessment of existing structures starts by gathering data concerning their current condition and precise review of available documents. Assessment includes different steps which can be summarized as visual inspection works, testing, structural analysis, fatigue assessment, and finally rehabilitation recommendations.

4 OBSERVED DEFECTS

Inspection of the considered structures was based upon a request of Cairo Governorate as part of the efforts to elevate the quality of transportation network in Egypt. The study started by several visits to the concerned structure. Accordingly, extensive investigation was initiated depending on visual inspection observations in addition to tests to pinpoint defects. In many cases, inspection teams had to measure different dimensions due to lack of data and as-built drawings. Upon inspection, several defects were observed as listed in the following figures.

DOI: 10.1201/9781003348450-320

Figure 1. Observed defects for 6th October Bridge: Corrosion of steel and crushing of concrete at expansion joint.

Figure 2. Observed defects for Sayeda Eisha Bridge: Corrosion at bottom parts of column.

Figure 3. Observed defects for Fardous Tunnel: Cracks in lateral beam.

Figure 4. Observed defects for Galaa bridge: Misalignment of guardrail.

Figure 5. Observed defects for Abo Hasheesh Overpass: Signs of friction with passing vehicles.

5 REHABILITATION PROCEDURES

Upon inspection and analysis, rehabilitation procedures are considered based on suitable approach for each case. Decision in strengthening applications requires detailed studies with consideration o several aspects including cost, traffic interruptions, know how availability…etc, Venturi et al. (2021).

6 CONCLUSIONS

This paper explored the defects observed during inspection of sample bridges. The assessment of such structures starts by exploring available documents and perform preliminary visits to understand the structural system of any considered structure. This is followed by pinpointing critical conditions and defects and proposing the most suitable rehabilitation procedure. Reported problems include cracks, unsafe members under increased loads, permeant deformations, spalling, asphalt creep…etc. The current study explored different structures showing the different observed defects. The considered structures represented different types of structures including reinforced concrete box girders, steel orthotropic decks, steel girders, reinforced concrete girders…etc. The study showed that the inspected structures suffered from many defects including corrosion, excessive deformations, damping effects, inadequate levels, and loss of expansion joints and loss of asphalt layer.

REFERENCES

Abbas, H.H. and Hassan, M.M., 2021. REHABILITATION STUDY OF AN OLD METALLIC ORTHOTROPIC DECK BRIDGE. *CSCE Annual Conference Inspired by Nature, Virtual, Canada.*

Hassan, M.M., Elsawaf, S.A. and Abbas, H.H., 2017. Existing metallic bridges in Egypt: current conditions and problems. *Journal of Civil Structural Health Monitoring*, 7(5) 669–687.

Tang, L., 2021. Maintenance and Inspection of Fiber-Reinforced Polymer (FRP) Bridges: A Review of Methods. Materials, 14(24): 7826.

Venturi, G., Simonsson, P. and Collin, P., 2021. Strengthening old steel railway bridges: a review. *In IABSE Congress*, Ghent, Belgium, September 22- 24.

Current Perspectives and New Directions in Mechanics, Modelling and Design of Structural Systems – Zingoni (ed.)
© 2022 Copyright the Author(s), ISBN 978-1-032-39114-4

Heritage protection and safety requirements: A difficult relationship

U. Quapp & K. Holschemacher
Faculty of Civil Engineering, HTWK Leipzig University of Applied Sciences, Germany

ABSTRACT: The domination of today's construction processes by safety regulations for structures contributes to the prevention of personal injury, property loss, and environmental damage. Nevertheless, while revaluation of historic buildings, current safety requirements may collide with heritage protection interests. The purpose of the study is to show that treating each historic building as an individual case and balancing aspects of usage, efficiency, and aesthetics can help to preserve the beauty and the uniqueness of cultural heritage for posterity. During their research, the authors analyzed current German safety regulations for reconstruction of historic buildings with the focus on load bearing capacity of structural systems. Using literature studies and own experimental investigations, they provide solutions how to balance the interests of construction safety authorities, owners, and authorities for the protection of historic buildings. The conclusion is that for planning and performing construction works in existing protected structures, architects and civil engineers need specialized high-level competence. Knowledge about historic building techniques as well as about today's safety requirements is necessary for acting in compliance with current safety regulations and building heritage interests. It should be in the interest of owners, planners, construction safety authorities as well as engineers to find individual solutions for protected historic structures to support cultural heritage protection.

1 INTRODUCTION

High requirements on structural stability and durability of buildings prevent collapse of structures and, therefore, save life and limb of humans and animals property and environment. However, in old buildings, today's structural requirements often cannot be met without affecting the characteristic historic appearance of the building. In order to preserve the beauty and the uniqueness of cultural heritage for posterity, individual solutions are needed to balance safety and building heritage interests.

2 SAFETY REQUIREMENTS

Since the 19th century, industrialization required more dynamically stressed constructions and reliable calculations of structural stability. Calculation of structural stability is a demanding and comprehensive task. Various regulations contain safety requirements for building structures. On the European level, Regulation (EU) No 305/2011 of the European Parliament and of the Council (2011) sets out basic requirements for construction works in Annex I.

In Germany, requirements for construction works can be found at national level and at Federal States' level.

The requirements are applicable to all structures, whether they are new or still existing. If owners start construction works in an existing structure, they must comply with the current safety regulations. However, not only building law must be obeyed, but also requirements from heritage protection.

3 HERITAGE PROTECTION

Currently, in Germany, more than one million protected buildings exist, of which one third urgently needs revaluation.

The German Federal State enacted heritage protection legislation in order to protect and maintain historic structures by monitoring them and preventing them from danger. During renovation of protected buildings, requirements laid down in the heritage protection law, must be met.

4 CONFLICT OF INTERESTS

While construction works in a protected building, the owner has to comply with safety and heritage protection requirements. However, often, safety regulations collide with preservation interests. The structural status, historic building materials as well as the intended usage of building heritage (e.g. as theatres, opera houses, museums) therefore requires individualized and innovative safety solutions.

DOI: 10.1201/9781003348450-321

5 EXAMPLES FOR SOLUTION

The authors analyzed two methods for solving the conflict of interests: Timber-Concrete Composite (TCC) and strengthening of timber beams with fiber reinforced polymer systems.

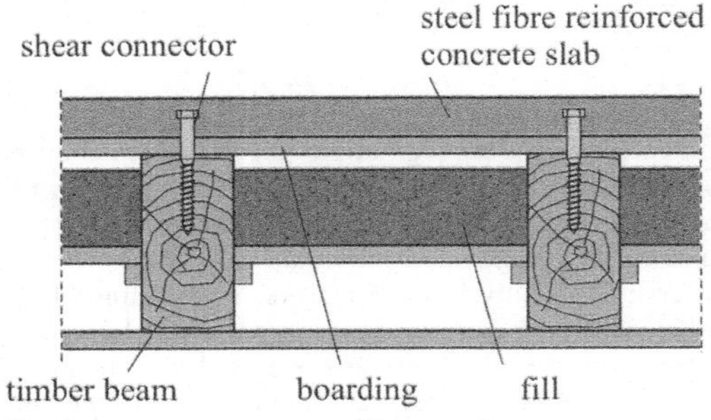

Figure 1. Existing timber beam ceiling strengthened by TCC.

A TCC member is the result of the arrangement of a concrete slab at the top site of an existing timber beam ceiling. Special shear connectors realize the connection between the timber and the concrete part of the composite section. Under the precondition that the existing timber beams are not damaged, the boarding can be left in place. Without removal of the boarding, holes can be drilled through it and, subsequently, the shear connectors can be assembled.

Using this method saves construction time and money. Usually, in the concrete slab steel mesh or bar reinforcement is arranged for bearing tensile stresses. An alternative possibility is the application of steel fiber reinforced concrete without any mesh reinforcement. In this case, the concrete slab can be constructed with low height because there is no need for a concrete cover for the steel reinforcement (Kieslich, & Holschemacher 2016). A typical TCC section with respect to strengthening of timber beam ceilings is presented in Figure 1.

Besides the increased flexural stiffness and load-bearing capacity of the strengthened timber beam ceilings, the TCC technique results in strongly improved building-physical properties.

Revaluation of existing timber beams with TCC allows compliance with all structural requirements and may contribute to preservation of historic buildings if the bottom side of the strengthened existing timber beam ceiling stays unchanged when applying TCC technique.

A possible technique to strengthen timber beams without significantly changing their visible surface is the adhesion of fiber reinforced polymer sheets or laminates in the tensile zone of the beam (Rescalvo et al. 2017), see Figure 2. In general, the strengthening sheets and laminates are produced on the basis of glass or carbon fibers. Both fiber types are characterized by a very high tensile strength and modulus of elasticity. By adhesive bonding of FRP sheets or laminates in a notch along the length of the beam, there is only a low visible intervention in the existing timber structure. Consequently, this solution is the preferred one for strengthening of timber beams in historic buildings because it also results in a considerable increase of flexural load-bearing capacity.

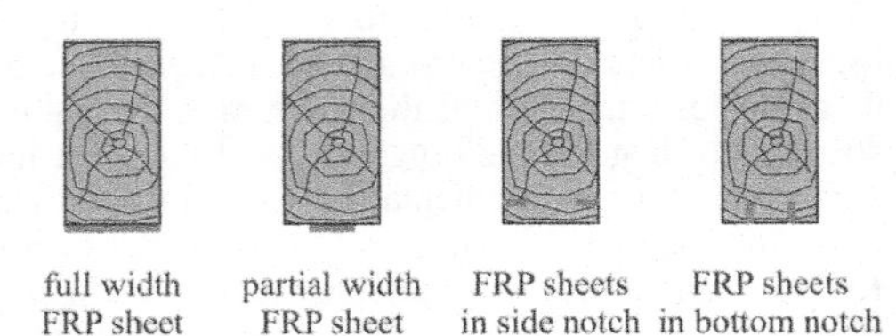

Figure 2. Strengthening of timber beams with FRP sheets and laminates.

6 CONCLUSION

For planning and performing construction works in existing protected structures, architects and civil engineers need specialized high-level competence. Knowledge about historic building techniques as well as about today's safety requirements is necessary. Each one of the protected structures must be seen as an individual case, where aspects of usage, efficiency and aesthetics must be balanced, under compliance of current safety regulations. The aforementioned methods for revaluation of historic timber beams show that meeting stability requirements, economic considerations, and heritage protection do not have to be contradictions.

REFERENCES

European Parliament and the Council 2011. Regulation (EU) No 305/2011 of the European Parliament and of the Council of 9 March 2011 laying down harmonized conditions for the marketing of construction products and repealing council directive 89/106/EEC. OJ 2011 L 88: 5–43.

Kieslich, H. & Holschemacher, K. 2016. Transversal load sharing in timber-concrete floors – experimental and numerical investigations. *Proc. World Conference on Timber Engineering, WCTE 2016: 2355–2364*. Vienna: TU Verlag Wien.

Rescalvo, F. J. et al. 2017. Experimental comparison of different fiber composites in reinforcement layouts of wooden beams of historical buildings. *Materials 2017, vol. 10: 1–22, doi:10.3390/ma10101113*. Basel: MDPI.

Current Perspectives and New Directions in Mechanics, Modelling and Design of Structural Systems – Zingoni (ed.)
© 2022 Copyright the Author(s), ISBN 978-1-032-39114-4

Reconnaissance survey of historic monumental unreinforced masonry building by visual inspection: A case study of senate hall building of Allahabad University, India

A. Kumar
Civil Engineering Department, Motilal Nehru National Institute of Technology Allahabad, Prayagraj, India

K. Pallav
Department of Civil Engineering and Surveying, Cape Peninsula University of Technology, Cape Town, South Africa

ABSTRACT: The Indo-Saracenic style of architecture reconnaissance survey is presented in the present work. The Senate Hall building is 106 years old unreinforced masonry of Allahabad University, India, built-in 1915. It was designed from Hindu (Stone Columns, Windows), Muslim (Domes, arches) architecture with British plans (inner rooms) like large halls, wide openings, porches, and facades. The building has been survived two major earthquakes, *viz.*, Bihar Nepal (1934) and Nepal (2015). Masonry walls were constructed from cellular wall patterns ranging from 1.07m to 0.61m (thickness) from the ground floor to the first-floor level. Most buildings have observed major cracks and damages in masonry walls, connections, arches, and porches. The construction materials (binder and plaster) have continuously deteriorated due to environmental factors like atmospheric conditions, seepage, and moisture. This work is focused on developing a digital model of the SH building using visual inspection for future preservation.

Keywords: Historical structure, masonry buildings, visual inspection, structural geometry, damages

1 INTRODUCTION

Masonry constructions are one of the important practices in most ancient buildings around the world. The construction of such a building has been started around 1000BC and was made up of stone or brick masonry. It is a simple combination of bricks/stone, mortar, and joint interface units. The bricks are mainly made up of unfired, dried mud (adobe) to fired clay, baked brick, whereas the stone material is commonly in the ordinary form such as rubble/fieldstone and handmade artificial shape. It was mainly the preferred material in constructing different structures in ancient times. The material has a unique quality for construction aesthetics, permanence, workability, and accessibility. Rounded stones have been designed the heavy structures as these stones are used for filling the gap between different stones. They are most durable in all weather and were widely used for old buildings and now in the new buildings of the rural areas.

Allahabad city is rich in old buildings constructed during the 16th - 19th century, and the Senate hall building is one of the oldest masonry structures of Allahabad University. It was constructed in the physics department of old Katra, Allahabad. The architecture of the SH building is one of the unique examples of the Indo-Saracenic style of architecture.

This work has been focused on developing a digital model for preserved 106 years old SH building. This study calculates the covered area (floors, masonry walls, openings) for future reference if the building is damaged/collapsed by earthquakes and other atmospheric conditions.

2 CONDITION SURVEY

The current condition of the old unreinforced masonry SH building has been performed to assess its state viz., construction techniques, architecture style, structural materials, cracking patterns, material deterioration, damages and failures. Shows the actual spalling of materials, damages, and failure of its components of the SH building. The geometrical plan and elevation drawing is prepared during the building's continuous survey. The cracks, deterioration and damages of the building are noted and marked.

DOI: 10.1201/9781003348450-322

3 GENERAL CRITERIA

The typology of the structure should be identified (churches, buildings, towers, bridges, temples, etc.) and whether the building is isolated or a combination of masonry and stone. The building is a combination of stone and brick masonry in the present case. These details provide a base level of knowledge for the structure.

4 HISTORIC CONDITION

A detailed survey is shown for all the components in the subsequent section of the old SH building during Visual Obtained on: (1) Walls: (a) External wall conditions: (b) Internal wall conditions: (c) Baradari: (2) Ceiling: (3) Flooring: (4) Parapet walls and baradari: (5) Doors/window/ventilators: (a) Ground Floor (GF): (b) First Floor (FF): (c) Second Floor (SF): (6) Stone shades: (7) Tile roof.

5 COVERED AREA OF THE BUILDING

The building has symmetrical patterns on the ground and first floor levels. Three major halls have been constructed on ground floor level height going up to the second floor. In the building, eleven towers have construed from the combination of brick masonry and stone in different locations. In the visual inspection of the building, entire dimensions have been measured for each structural element and room, masonry walls, openings on each respected floor level. This calculation may be beneficial in reconstructing the structural elements and each portion of the senate hall building for future damages by the earthquake, seepage, and atmospheric conditions.

6 EXPLANATORY RESULTS

The destructive 2015 Nepal earthquake has made engineers and researchers to focus on the structural behaviour of the senate hall building. The building actual conditions, architectural style, geometrical assessment, a structural element (Towers, Arches, Domes), construction materials (Bricks, Stones, Timbers) and mechanical properties, cracks and damages, collapse and failure, strengthen and restoration, retrofitting and restoration are prime importance to our heritage/Old building. Due to numerous atmospheric conditions like seepage, moisture, temperature effects, rainwater, the construction and decorating material have continuously deteriorated overall building elements such as masonry walls, arches, towers, and roofs. The entire building dimensions, geometrical plans, and actual covered area of each specific component of the structures (floors, openings, and masonry walls) have recorded for future perspective. The 3D model of the entire building has been built using the FEM tool, which can be used in the restoration work for future references.

7 CONCLUSIONS

The study has been focused on the detailed measurement of the building for developed geometrical plans after visual inspection. The covered area (Masonry walls, floors, openings) has been recorded on each building level during visual inspection. It is beneficial for the renovation and rehabilitation of the building after future consequences occur.

The study is constructive for retrofitting and renovating the building due to major cracks and damages clearly visible during the survey after the 2015 Nepal earthquake. As exemplified in the senate hall building, the general damage and failure patterns may be summarized a) Huge relative displacement led to vertical cracking between the main central hall walls. The excess of shear or tension in the construction material may lead to such failures. b) The second-floor wall is observed to witness overturning and cracking. Also, the upper part of the wall may face vertical cracks generated due to bending about its vertical axis. c) Dislocation of the corners. The shear and tearing stress may lead to cracks in the vertical corners. These cracks may lead to initial failure. Also, the wall panels get displaced due to a deficiency of link between the wall and corner of the tower. The top portion of the boundary wall may fail due to excessive displacement. d) The roof structure has witnessed deformation and displacement. The wall panels may face cracks in the horizontal direction. e) Roof tiles are falling and slipping. f) Failure at the corners and openings are observed. g) The masonry wall of the SH buildings may also fail due to the lack of earthquake-resistant attributes. In addition, the use of new material during construction is also a severe threat to the health of a structure. It is further recommended that the 3D model of the building along with its floor area could be related to its mechanical properties.

ACKNOWLEDGEMENTS

Both the authors gratefully acknowledge the cooperation of the engineer in charge of Mr. Naveen Kumar Allahabad university to provide the information about the senate hall structure.

REFERENCES

[1] Kumar, A. & Pallav, K. 2017. Static and Dynamic Analysis of Masonry Tower of Allahabad University, India, *VIII ECCOMAS Thematic Conference on Smart Structures and Materials (SMART 2017)*, pp 821–830.

[2] Kumar, A., Pallav, K. & Singh, D.K. 2017. Static and Dynamic analysis of Stone arch of Allahabad University, *Proceeding of International Conference on Recent Innovation in Engineering and Technology, Jaipur, India,* 18th -19th Feb 2017, *ISBN: 978-93-86291-63-9*.

[3] INTECH. 2011. *Survey reports of full assessment of Allahabad University was conducted by INTECH India Private Limited.*

Current Perspectives and New Directions in Mechanics, Modelling and Design of Structural Systems – Zingoni (ed.)
© 2022 Copyright the Author(s), *ISBN 978-1-032-39114-4*

On the characterization of materials and masonry walls of historical buildings: Use of optical system to obtain displacement maps in double-flat jack tests

M. Acito & E. Magrinelli
Department of Architecture, Built Environment and Construction Engineering, Politecnico di Milano, Milano, Italy

C. Tiraboschi & M. Cucchi
Material Testing Laboratory, Politecnico di Milano, Milano, Italy

ABSTRACT: Among the testing techniques aiming at the mechanical characterization of masonry, the double flat-jack testing method is widely adopted to identify the local value of significant parameters needed to perform structural analyses, such as elastic modulus, Poisson's ratio and compressive strength. The experience gained from many applications has allowed not only to collect experimental data concerning different types of masonry, but also to highlight the difficulty in the interpretation of the results and the limitations of both single and double flat-jack tests. Although the accuracy of the flat-jack technique in detecting strength and deformability behavior of masonry is still debated in the technical literature and practical activities, changes in the testing procedure aiming at ascertaining the validity of the test results have not been formally defined yet. After a brief description of the standard test procedure and its uncertainties, the present paper proposes an upgrade of the test procedure for improving the level of reliability of the test results. In particular, an experimental case study related to a historical brick masonry building located in Italy is presented to point out the additional information necessary to validate the results of the testing process.

1 INTRODUCTION

An adequate knowledge of the mechanical properties of masonry is essential to carry out accurate numerical analyses of existing masonry buildings. Among the most performed in-situ tests, the flat-jack test allows to characterize the acting state of stress as well as the deformability properties of masonry. Being only slightly destructive, this testing technique is suitable also for buildings of historical value. Even though the flat-jack method is widely used, it involves some uncertainties which may affect the validity of the test results. Several experienced authors have discussed about the difficulties in the interpretation of the test results due to the anomalies commonly found in the outcomes as well as the lack of a strong theoretical framework to correlate experimental data with actual mechanical properties. As a consequence, the definition of objective criteria for ensuring the reliability of the test results is essential. Over the years, test results have been indirectly validated using back analysis and relatively sophisticated numerical models. Two distinct approaches for the numerical analysis of masonry are commonly used, namely the micro- and macro-modeling. The micro-modeling is particularly apt to investigate local mechanisms, such as cracking phenomena, while the global behavior of masonry structures is usually analyzed by macro-modeling, where mean values of stress and strain are considered. The present work is intended to provide practical suggestions for upgrading the double flat-jack test procedure and improving the reliabilty of the outcomes for a better interpretation of the masonry behavior. To this purpose, a complementary optical-based displacement monitoring method is proposed in addition to the standard displacement measurement through LVDTs. The system allows for the real-time monitoring of the stress-strain evolution in the wall specimen as well as in the flat-jacks in an easy and detailed way.

2 INSIGHT INTO THE FLAT-JACK TESTS

Interpreting double flat-jack test results is a challenging step of the test procedure not only when clear anomalies affect the outcomes but also when they are apparently highly reliable. The uncertainties concerning the validity of the test results may impair the definition of the correspondence between the mechanical properties determined by flat-jack tests and the actual ones. The outcomes are highly dependent on the type of the tested masonry, which may be affected by

DOI: 10.1201/9781003348450-323

hidden irregularities. The main aspects which have great influence on the validity of the results of the double flat-jack tests are: 1. the state of stress in the masonry above the top flat-jack, which may be significantly lower than the pressure applied by the flat-jack; 2. the improper choice of the slot positions and the arrangement of the reference points; 3. the irregular composition of the wall, whose deformability properties may be significantly variable; 4. the masonry texture and the type of mortar. The execution of sonic tests by trasparency prior to the flat-jack ones is useful to detect masonry irregularities and, therefore, to choose the most suitable position for the flat-jack tests.

3 STRESS-STRAIN RELATIONSHIP

The double flat-jack test is not comparable to the classical uniaxial test and the results may be unreliable for the following reasons. Firstly, since the double flat-jack test is usually performed on masonry walls thicker than flat-jack dimensions, the flat-jacks transfer an eccentric load to the WS. Secondly, being the WS confined, the experimental $\sigma - \varepsilon$ relationship comes from a multiaxial stress state and its correlation to uniaxial state could be considered improper. Thirdly, the $\sigma - \varepsilon$ relationship is determined by measuring the WS deformations, without considering that they can be affected by geometrical effects due to flat-jack displacements. During the test, indeed, the relative distance between the flat-jacks can increase or decrease depending on the stress-strain scenarios in the masonry surrounding the WS. When the distance between the flat-jacks increases, the $\sigma - \varepsilon$ relationship of the WS could be more rigid than the actual one. On the contrary, when the distance between the flat-jacks decreases, the relationship of the WS could be less rigid than the actual one. Therefore, three different scenarios are herein considered:

- case 1: the WS and the surrounding masonry are characterized by the same stress-strain behavior;
- case 2: the WS is less rigid than the wall volumes above and under the tested area;
- case 3: the WS is more rigid than the wall volumes above and under the tested area.

It can be stated that the σ-ε relationship obtained from a double flat-jack test well fits the actual one only when the flat-jacks do not move with respect to each other during the test, otherwise the measured σ-ε relationship can be considered only representative of the real one. Therefore, monitoring the displacements of the flat-jacks and the strain diffusion is necessary to verify and validate the test results. This paper focuses on the optical system for the acquisition of the displacement field of the tested area and the results of its application during the double flat-jack tests performed on the masonry walls of Sommariva Palace located in Lodi, Italy, are provided.

4 EXPERIMENTAL CAMPAIGN ON THE MASONRY WALLS OF SOMMARIVA PALACE IN LODI, ITALY

Within the rehabilitation intervention of Sommariva Palace in Lodi, Italy, sonic tests, single and double flat-jack tests were performed on the masonry walls of the ground floor by the Material Testing Laboratory at Politecnico di Milano. Flat-jack tests were carried out on two different masonry typologies, in the J1 and J2 positions. The J1 double flat-jack test was interrupted at an applied stress equal to 1.4 MPa, when the opening of cracks occurred in the WS. The J2 double flat-jack test was clearly affected by anomalies because of the poor masonry quality and the insufficient confinement effect provided by the volume above the WS. During the double flat-jack tests, the measurement of the displacements was carried out not only by traditional LVDTs, but also by arranging an optical system which allowed to fix absolute coordinates and to acquire information like displacement vectors of selected points. The displacement trend measured during the J1 test was quite regular: according to paragraph 3, the experimental behavior detected by the J1 double flat-jack test can be considered well representative of the actual behavior of the WS. The displacement trend measured during the J2 double flat-jack test was quite irregular. Since a general displacement directed upward can be observed on the monitored points, it can be deduced that the WS was pushed upward because of an insufficient confinement effect provided by the masonry volume above the WS. According to paragraph 3, the experimental behavior detected by the J2 double flat-jack test can be considered not representative of the actual behavior of the WS.

5 CONCLUSIONS

Experimental flat-jack tests are valid tools to detect the local state of stress (single flat-jack test) and the stress-strain behavior (double flat-jack test) in masonry walls. The reliability of the results obtained from double flat-jack tests has been debated in the literature since their very first experimental applications. This work proposes an upgrade of the procedure aiming at ascertaining the reliability of the results objectively. The upgrade requires monitoring the relative displacements which the flat-jacks may undergo during the test because of the peculiarities of the tested masonry. Indeed, the control of the reliability of the test results allows for a more precise definition of the stress-strain behavior of the masonry wall specimen included between the two flat-jacks. The preliminary experimental-analytical considerations developed in this work allow to highlight that monitoring the displacements of the WS, as well as of the surrounding masonry and flat-jacks, is important to validate the test results.

Current Perspectives and New Directions in Mechanics, Modelling and Design of Structural Systems – Zingoni (ed.)
© 2022 Copyright the Author(s), *ISBN 978-1-032-39114-4*

A computational procedure interacting with the Italian "CARTIS" online database to derive residential building portfolios for large scale seismic assessments

A. Basaglia, G. Brando, G. Cianchino, G. Cocco, D. Rapone, M. Terrenzi & E. Spacone
Department of Engineering and Geology, University "G. d'Annunzio" of Chieti-Pescara

ABSTRACT: Acquiring detailed information on the building vulnerability for seismic risk assessments at the large scale is arduous, given the high heterogeneity of the built environment. For this reason, it is convenient to derive vulnerability information from existing databases. A project funded by the Italian Civil Protection Department started collecting data on residential buildings in Italy since 2014 through the so-called CARTIS form. The form, developed by the Italian Plinius Study Centre, is based on interviews with local experts and collected data are stored in an online database. In this paper, a recently proposed computational procedure interacting with the CARTIS database is adopted to derive building portfolios at the large scales. A case study application in Central Italy is presented, using an automated tool that that is freely downloadable.

1 INTRODUCTION

Defining the seismic vulnerability is a challenging task for large scale assessments, due to the high number of structures and the high heterogeneity of the built environment. With respect to residential buildings, the seismic vulnerability may be defined using survey forms. The CARTIS form (Zuccaro et al., 2015) was developed in 2014 by the Italian Plinius Study Centre. Data collected through this form may be used to define building portfolios for large scale seismic assessments. To this aim, a semi-automated procedure interacting with the CARTIS database was proposed (Basaglia et al., 2021). This procedure uses a MATLAB® script (available at: https://github.com/albertobasaglia-code/CARTIS-Automated-Subtypologies-Definition/).

2 THE CARTIS FORM AND DATABASE

Data collection through the CARTIS form-(Zuccaro et al., 2015) is carried out through an interview with a local expert of notable experience and knowledge.

The CARTIS form assumes that a municipality can be divided into "compartments", that are areas characterized by similarities in terms of buildings peculiarities. Each compartment may comprise up to four Unreinforced Masonry (URM) and four Reinforced Concrete (RC) building typologies. Data collected through the form are uploaded in the open source pgAdmin platform (https://www.pgadmin.com).

3 SEMI-AUTOMATED PROCEDURE FOR DEFINING BUILDING TYPOLOGIE

For each municipality, two different "comma-separated values" (.csv) files are downloaded - one for URM and the other for RC buildings. These files include the geometrical and structural parameters for each compartment. At first, the data accuracy is verified. In case of errors, compartments are discarded. Then, compartments are "disaggregated" by equally distributing values of parameters according to their relative percentage. The disaggregation can be performed for all parameters or only a selection of them. To reduce the number of resulting sub-typologies, parameters may be "aggregated" assuming that they do not alter the building seismic behavior. As an example, roofs made of bricks or in RC may be grouped into a "heavy" material category. Finally, sub-typologies with low relevance (e.g., frequency<1%) are discarded and only those relevant are included in the building portfolio. Yet, discarded sub-typologies may be "assimilated" to those relevant if they are "similar".

4 CASE STUDY APPLICATION: THE MARCHE REGION (CENTRAL ITALY)

As of early 2022, 25 municipalities of the Marche Region (Central Italy) have been surveyed through the CARTIS form. Obtained sub-typologies are reported in Table 1 and Table 2. They represent the 69.3% and 81.8% of all URM and RC buildings, respectively.

DOI: 10.1201/9781003348450-324

Table 1. URM sub-typologies with frequency higher than 1% for the Marche Region based on CARTIS data.

No. of Buildings	Masonry Type	Number of Stories	Construction Period	Ring Beams and Tie Rods	Slab Type	Roof Material	Relevance (%)
7374	Blocks	3 stories	1861-1919	Present	Rigid	Light	12.82
7357	Blocks	2 stories	1861-1919	Present	Rigid	Light	12.79
4029	Irregular masonry - w/o courses	3 stories	< 1860	Present	Flexible	Light	7.01
4029	Irregular masonry - w/o courses	4 stories	< 1860	Present	Flexible	Light	7.01
3162	Blocks	3 stories	1861-1919	Absent	Rigid	Light	5.5
3154	Blocks	2 stories	1861-1919	Absent	Rigid	Light	5.48
2731	Blocks	3 stories	1861-1919	Present	Flexible	Light	4.75
2684	Blocks	2 stories	1861-1919	Present	Flexible	Light	4.67
1167	Blocks	3 stories	1861-1919	Absent	Flexible	Light	2.03
1147	Blocks	2 stories	1861-1919	Absent	Flexible	Light	1.99
940	Blocks	3 stories	1945-1981	Absent	Semi-rigid	Heavy	1.63
911	Blocks	3 stories	1919-1945	Absent	Semi-rigid	Heavy	1.58
578	Irregular masonry - w/o courses	3 stories	< 1860	Absent	Flexible	Light	1.01
578	Irregular masonry - w/o courses	4 stories	< 1860	Absent	Flexible	Light	1.01

Table 2. RC sub-typologies with frequency higher than 1% for the Marche Region based on CARTIS data.

No. of Buildings	Main Resisting System	Number of Stories	Construction Period	Design Type	Unidirectional Frames	Ground Floor Infill Walls	Relevance (%)
21229	Frames with inner shear walls	3 Stories	Before WWII	Seismic	No	Regular config.	12.38
21119	Frames with sturdy infill walls	3 Stories	After WWII	Seismic	No	Regular config.	12.31
21119	Frames with sturdy infill walls	4 Stories	After WWII	Seismic	No	Regular config.	12.31
11639	Frames with sturdy infill walls	3 Stories	DM 92 or 96	Seismic	No	Regular config.	6.79
8271	Frames with sturdy infill walls	3 Stories	DM 81 or 85	Seismic	No	Regular config.	4.82
6773	Frames with sturdy infill walls	4 Stories	DM 92 or 96	Seismic	No	Regular config.	3.95
6767	Frames with sturdy infill walls	4 Stories	DM 81 or 85	Seismic	No	Regular config.	3.95
4866	Frames with sturdy infill walls	2 Stories	DM 92 or 96	Seismic	No	Regular config.	2.84
3785	Frames with sturdy infill walls	3 Stories	After WWII	Seismic	Yes	Regular config.	2.21
3785	Frames with sturdy infill walls	4 Stories	After WWII	Seismic	Yes	Regular config.	2.21
3384	Frames with sturdy infill walls	2 Stories	DM 72	Seismic	No	Absent	1.97
3384	Frames with sturdy infill walls	2 Stories	DM 81 or 85	Seismic	No	Absent	1.97
3384	Frames with sturdy infill walls	3 Stories	DM 72	Seismic	No	Absent	1.97
3384	Frames with sturdy infill walls	3 Stories	DM 81 or 85	Seismic	No	Absent	1.97
3384	Shear walls	2 Stories	DM 92 or 96	Seismic	No	Absent	1.97
3384	Shear walls	3 Stories	DM 92 or 96	Seismic	No	Absent	1.97
3362	Frames with sturdy infill walls	2 Stories	Post NTC2008	Seismic	No	Regular config.	1.96
3362	Frames with sturdy infill walls	3 Stories	Post NTC2008	Seismic	No	Regular config.	1.96
1982	Frames with sturdy infill walls	3 Stories	After WWII	Gravity	Yes	Regular config.	1.16
1982	Frames with sturdy infill walls	4 Stories	After WWII	Gravity	Yes	Regular config.	1.16

5 SUMMARY AND CONCLUSIONS

This paper employed a recently proposed procedure interacting with the CARTIS database to derive building portfolios at the large scale. The procedure is applied to the Marche Region, Central Italy. 14 URM and 20 RC building sub-typologies were derived, representing most of the built environment. Future developments include the application of the proposed procedure to other Italian Regions (alone or combined).

REFERENCES

Basaglia, A., Cionchino, G., Cocco, G., Rapone, D., Terrenzi, M., Spacone, E., & Brando, G. 2021. An automatic procedure for deriving building portfolios using the Italian “CARTIS” online da-tabase. Structures, 34: 2974–2986. https://doi.org/10.1016/j.istruc.2021.09.054.

Zuccaro, G., Dolce, M., De Gregorio, D., Speranza, E., & Moroni, C. 2015. La scheda CARTIS per la caratterizzazione tipologico-strutturale dei com-parti urbani costituiti da edifici ordinari. Valuta-zione dell’esposizione in analisi di rischio sismi-co. Proceedings of the 34th GNGTS.

Current Perspectives and New Directions in Mechanics, Modelling and Design of Structural Systems – Zingoni (ed.)
© 2022 Copyright the Author(s), *ISBN 978-1-032-39114-4*

Seismic assessment of a colonial adobe building in Cusco, Peru

A. Tancredi, G. Cocco, E. Spacone & G. Brando
University "G. d'Annunzio" of Chieti-Pescara, Italy

ABSTRACT: Peruvian territory, in particular in the area of Cusco, is characterized by a high seismic risk. The city of Cusco, whose historic center has been recognized as UNESCO site since 1983, has a long story of telluric events and the study of the seismic vulnerability of its buildings is a duty of great concern. Furthermore, protection and preservation of its monumental and residential assets is an essential issue, especially for those buildings made of adobe, which are an important legacy testifying the life and the development of past civilizations. Based on these premises, the proposed study aims to assess the seismic behavior of a colonial adobe building located in the historic center of Cusco. This category of buildings represents the so-called "casa cusqueña", i.e. the typical residential building realized by Spanish colonists, often characterized by architectural and structural elements coming from the ancient Inca population. Squared plan with internal courtyard, stone foundations and adobe walls are some of the main structural characteristics of this type of building. After the description of the reconnaissance campaign, a careful historical analysis of the building is presented. Then, the global seismic behavior of the structure is analyzed by means a finite element model implemented in ABAQUS. A step-by-step linear dynamic analysis is performed and cracks evolutions as the ground acceleration increases is described.

1 INTRODUCTION

In the Peruvian Andes, many heritage buildings are made of adobe masonry. Over time, these structures have shown low resistance to earthquakes (Chácara et al., 2019). For this reason, the Seismic Vulnerability Assessment of adobe masonry buildings is a crucial topic, especially in those areas that are most prone to earthquake actions, such as the city of Cusco in Peru.

In this paper, seismic vulnerability assessment of a typical colonial adobe building located in the historic center of Cusco is dealt with. After a careful investigation campaign presented in Section 2, the finite element modeling in Abaqus is carried out, and a multi-step dynamic linear analysis is performed to analyze the structure's behavior stressed by increasing ground accelerations (Section 3). Cracks evolution and kinematic mechanisms activation following the collapse were evaluated in Section 4.

2 THE SURVEY CAMPAIGN

The analyzed building is located in one of the oldest areas of the historic center of Cusco. After the 1950 Cusco earthquake, it suffered several damages and was rebuilt in 1988. The structural configuration of this residential building reflects the typical Spanish colonial style with a square plan and an internal courtyard. Stone foundations one-meter height from the street level, adobe masonry thickness ranging from 90 to 120cm, and wooden floors are the main structural features common to all residential buildings in the historic downtown area (Figure 1a, 1b, and 1c). In addition, stone arches with a balcony are present in the inner courtyard, as shown in Figure 1d. The roof is composed of wooden trusses with bamboo canes, a layer of earth and straw as thermal insulation, and clay tiles.

3 NUMERICAL MODELLING

A step-by-step linear dynamic analysis and crack evolutions as the ground acceleration increases are described to analyze the seismic behavior of this typical adobe colonial building. The finite element model of the structure has been developed in Abaqus. Adobe walls were modeled using quadrilateral four-node reduced integration shell elements (S4R), with a mesh size of 200 mm. Two-node linear beam elements (B31) were used to model lintels and roof beams with rectangular and circular sections. Stone foundations were not modeled, but their presence was assumed as rigid constrains at the base of the adobe walls.

Furthermore, the interactions between orthogonal walls were simulated through tie constraints. The elastic material properties used in the model are reported in Table 1, based on current studies in the literature (Silveira et al., 2012), (Miccoli et al., 2017).

By scaling the elastic response spectrum of Cusco (Ministerio de Vivienda, Construcción y Saneamiento, 2018), incremental values of ground accelerations have

DOI: 10.1201/9781003348450-325

a) 

b) 

c) 

d) 

Figure 1. Structural features of the analyzed colonial adobe building in Cusco. a), b) Two main façades; c) Typical stone foundation and c) Stone arches in the internal courtyard.

Table 1. Elastic material properties.

	E	γ	υ	f_c	f_t
	[MPa]	[KN/m^3]	[-]	[MPa]	[MPa]
Adobe	200	20	0.20	1.32	0.17
Wood	10000	6.87	0.15		
Stone	1230	20	0.20		

been considered. Linear dynamic analyses have been performed for each ground acceleration, modifying the model, in each step, by removing those elements for which the maximum tensile stress exceeded the maximum corresponding failure stress f_t, equal to 0.17 MPa. This procedure allowed to identify the main mechanisms' activation up to the collapse.

4 SUMMARY AND CONCLUSIONS

This paper deals with the seismic vulnerability assessment of a typical colonial building in Cusco, Peru. The structure was analyzed by a sophisticated numerical model, implementing a multi-step dynamic linear analysis applied on variable geometry of the building in order to define the state of damage for different seismic demand. This allowed to identify the activation of the main kinematic mechanisms up to the collapse.

REFERENCES

Miccoli, L., Silva, R. A., Garofano, A., & Oliveira, D. V. (2017). In-plane behaviour of earthen materials: A numerical comparison between adobe masonry, rammed earth and COB. *COMPDYN 2017 - Proceedings of the 6th International Conference on Computational Methods in Structural Dynamics and Earthquake Engineering*, *1* (February 2019), 2478–2504. https://doi.org/10.7712/120117.5583.17606

Silveira, D., Varum, H., Costa, A., Martins, T., Pereira, H., & Almeida, J. (2012). Mechanical properties of adobe bricks in ancient constructions. *Construction and Building Materials*, *28*(1), 36–44. https://doi.org/10.1016/j.conbuildmat.2011.08.046

Ministerio de Vivienda, Construcción y Saneamiento, P. (2018). *Norma Técnica E.030 Diseño Sismorresistente*.

Chácara, C., Pantò, B., & Aguilar, R. (2019). Evaluation of the seismic response of a historical earthen structure based on a discrete macro-element modelling approach. *COMPDYN Proceedings*, *1*, 1391–1400. https://doi.org/10.7712/120119.7006.19198

21. Soil-structure interaction, Foundations, Underground structures, Geotechnical engineering, Rock mechanics

Current Perspectives and New Directions in Mechanics, Modelling and Design of Structural Systems – Zingoni (ed.)
© 2022 Copyright the Author(s), ISBN 978-1-032-39114-4

Interaction between the deepest foundation piles in Germany and the superstructure

R. Katzenbach, A. Werner & S. Fischer
Ingenieursozietät Professor Dr.-Ing. Katzenbach GmbH, Frankfurt, Germany

ABSTRACT: In northern Germany a huge skyscraper is in planning. The new skyscraper will get an over all height of 245 m. Due to the little bearing capacity of the ground a deep foundation is needed. To found the building Germanys deepest foundation piles, with length up to 75 m, will be executed. In advance, to determine the bearing capacity of the surrounding soil in detail, four testpiles, with lengths up to 110 m, the deepest testpiles in Germany so far, have been executed. Numerous numerical simulations were carried out to recalculate the behavior of the testpiles and also for the prediction of the foundation behaviour.

1 DESCRIPTION OF THE BUILDING

The skyscraper will be built in the so called area "Hafencity", which is located in the middle of Hamburg directly on the river Elbe. Figure 1 shows the famous building "Elbphilharmonie" and the planned new skyscraper.

Figure 1. Elbphilharmonie, left and Elbtower, right.

2 SOIL CONDITIONS

In the area of the construction site an intersection of two glacial channels was suspected, so extensive soil investigations were carried out to explore the subsoil to a depth of approx. 200 m. These geotechnical site investigations consisted of:

- Exploration drillings
- Crosshole measurements
- Cone pressuremeter testing
- Laboratory testing

Based on these extensive soil investigations the following simplified soil model (Figure 2) has been elaborated:

- Layer 1 : sandy filling, partly rubble
- Layer 2 : clay and mud, cohesive filling, partly rubble
- Layer 3 : sand with silty deposits
- Layer 4 : sand middle dense to dense
- Layer 5 : sand dense to very dense
- Layer 6a/b: clay/silt with thin sandy layers
- Layer 7 : sand partly silty with silty layers

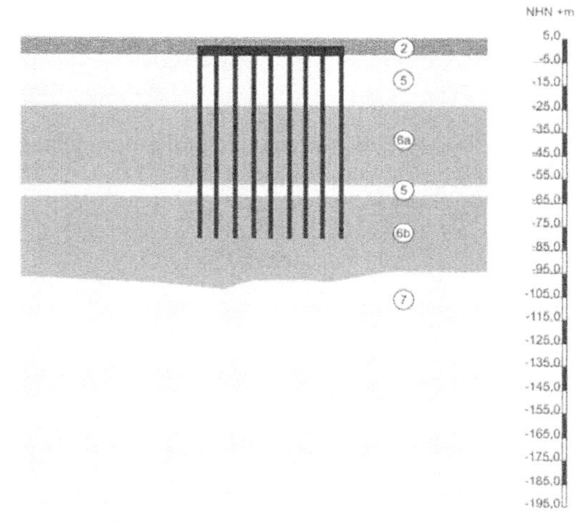

Figure 2. Simplified soil model of the construction site.

As a result of the soil investigations only low bearing capacities were identified in most layers. For this reason a deep foundation, in form of a combined pile raft foundation (CPRF) is needed. To safely dimension the deep foundation, testpiles had to been executed.

DOI: 10.1201/9781003348450-326

3 PILE LOAD TESTS

To determine the bearing capacity, 4 pile load tests have been carried out (Figure 3). The executed test-piles are the deepest piles in the context of a skyscraper in Germany so far. To be able to make a prediction for all affected soil layers, piles with different lengths have been executed. The most important layer for the load transfer is layer 7. For this reason 2 piles reaches into this layer. The piles were loaded by O-cells (England 2009, Zhussupbekov & Oamrov 2019), which are placed approx. 6 m below pile feet.

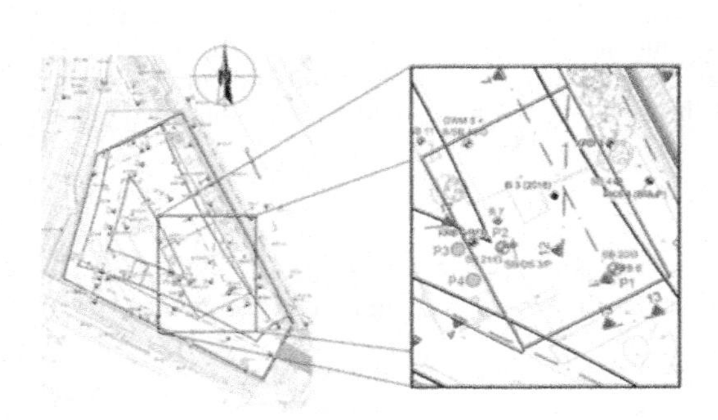

Figure 3. Location of the testpiles P1 to P4.

One of the main results of a pile load test is the load-displacement curve. On the basis of the measured distribution, soil characteristics can be recalculated e.g. by using the finite-element method.

Figure 4 shows the measured and the recalculated load-displacement curves of the testpile 1.

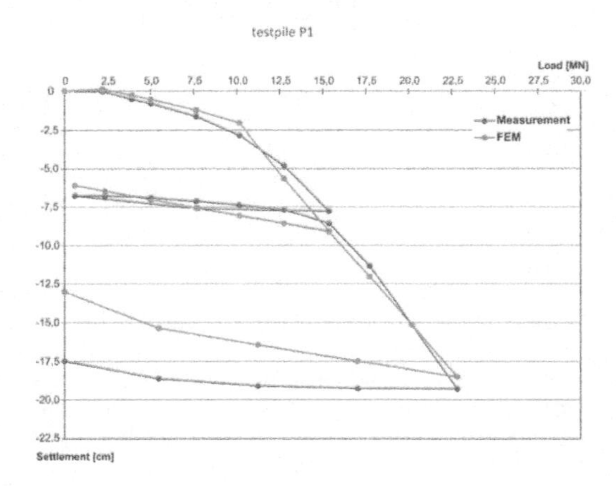

Figure 4. Load-displacement of testpile 1.

4 DEEP FOUNDATION

An overview on the design of piled rafts and its behavior for example is given in Reul (2000). For the prediction of the settlements and especially to proof the acceptable inclination of the skyscraper, various three dimensional numerical calculations were executed taking into consideration scientific works like Katzenbach et al. (1994), Randolph (1994), etc. Figure 5 shows the dimensions of the numerical model.

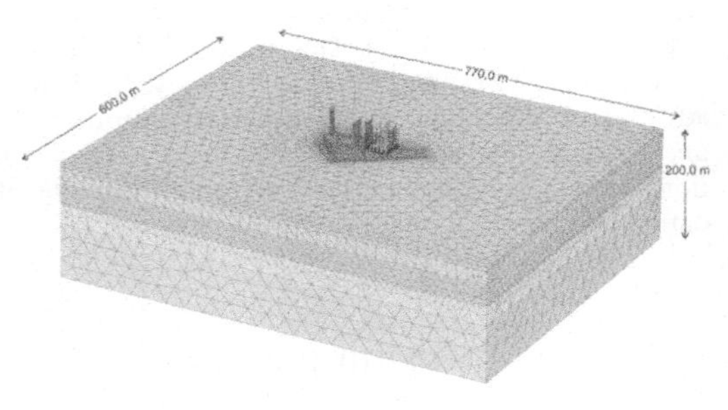

Figure 5. Dimension of the numerical model for the prediction of the settlement.

In the result of the calculations, including several sensitivity studies, the final layout of the CPRF with 66 large caliber piles with diameter d = 2.0 m and length l = 75 m is, in accordance to the CPRF-Guideline, a safe and economical design of the foundation for the third highest high-rise-building in Germany.

REFERENCES

England M Review of methods of analysis of test results from bi-directional static load tests 2009 Deep Foundations on Bored and Auger piles, pp 235–239, Taylor & Francis Group, England.

Katzenbach R, Arslan U and Gutwald J 7-9 September 1994 A numerical study on pile foundation of the 300 m high Commerzbank Tower in Frankfurt am Main Proc. 3rd Europ. Conf. on Num. Methods in Geomech., Manchester, UK, pp 271–277.

Randolph M F Design Methods for pile groups and piles rafts 1994 Proceedings 13th ICSNFE, New Dehli, Vol. 5, pp 61–82, Rotterdam: Balkema.

Reul O In-situ-Messungen und numerische Studien zum Tragverhalten der Komninierten Pfahl-Plattengründung 2000 Mitteilungen des Institutes und der Versuchsanstalt für Geotechnik der Technischen Universität Darmstadt, Heft 53.

Zhussupbekov A, Omarov *A Complex of static loading tests of bored piles* 2019 International Journal of GEOMATE, Vol. 16, pp 8–13.

Current Perspectives and New Directions in Mechanics, Modelling and Design of Structural Systems – Zingoni (ed.)
© 2022 Copyright the Author(s), ISBN 978-1-032-39114-4

Derivation of cyclic p-y curves for the design of monopiles in sand

M. Achmus & J. Song
Institute for Geotechnical Engineering, Leibniz University Hannover, Hannover, Germany

ABSTRACT: As foundation for offshore wind energy converters, the monopile has proven to be an economic and robust solution. In design practice, usually the p-y method is applied in the calculation of the monopile behavior. Here, the effect of cyclic loading with varying number of load cycles must be considered. A simple-to-use approach for the derivation of cyclic p-y curves would be highly desirable, which just modifies a chosen static p-y curve depending on the number of load cycles and other relevant parameters. A calculation approach termed "Stiffness Degradation Method (SDM)" has been applied to calculate pile deflection lines and bedding resistances of monopiles embedded in sand dependent on the actual number of cycles of a given load and with that also p-y curves for a given number of load cycles. The obtained results were utilized for the development of an approach to derive cyclic p-y curves for a given load cycle number based on a chosen static p-y model (cyclic p-y overlay model). The results show that a y-multiplier approach is suitable to account for the effect of cyclic loading. The proposed calculation approach gives the y-multiplier dependent on number of load cycles, pile geometry, load eccentricity and considered depth.

1 INTRODUCTION

In recent researches, the efficient approaches to minimize the costs for offshore wind energy exploitation has become a principal topic, which can be enabled by optimizing the foundation elements of offshore wind converters.

In practice, a special subgrade reaction method (p-y method) is usually applied in calculations to determine the bearing behavior of laterally loaded monopiles. Here, the dependence of the resultant soil reaction force p and the horizontal deflection y at each pile depth is described by depth-dependent and non-linear functions ("p-y curves").

A method utilizing the tanh-function to derive p-y curves for flexible piles in sand soils is proposed in the offshore guideline API (2014). This method also accounts for cyclic loading by adaptation of the static p-y curves, but does not take the actual number of cycles into account. However, the accurate estimation of deformation accumulation with the actual load cycle number is necessary in design to ensure the serviceability of the structure. A simple-to-use approach for the derivation of cyclic p-y curves would thus be highly desirable, which just modifies a chosen static p-y curve by p- or y-multipliers depending on the actual number of load cycles and other relevant parameters.

2 CYCLIC OVERLAY MODEL

Numerical simulations for cyclically horizontal loaded piles with large-diameter have been conducted by utilizing the stiffness degradation method (SDM) (details in Kuo (2008) and Achmus et al. (2009)) in ABAQUS (Simulia 2018) to derive the cyclic overlay model. The SDM used here accounts for cyclic effects by reducing the stiffness of the soil around the pile. according to the results of cyclic triaxial tests.

First, the "basic p-y curves" were derived for constant prescribed horizontal deflections of a monopile. The pile diameter was varied between D = 3 m and D = 8 m in 1 m-increments. The pile length was set to L = 25 m in all calculations. By comparison of the basic p-y curves for load cycle numbers greater than 1 with the basic p-y curve for N=1, the basic overlay model accounting for different relative densities (cf. Figure 1) was derived, i.e. equations were developed which describe the change of the basic p-y curves with number of load cycles considered. These equations form the "basic overlay model", in which it seems reasonable to account for cyclic effects in the p-y approach by applying the determined y-multipliers (y_N/y_1-values) of the lateral displacement for the results depicted in Figure 1. The derived regression curves are also presented.

Subsequently, the correction functions for the basic overlay model were derived by applying a horizontal load as well as a bending moment at the point of embedment in the same numerical model with varying load eccentricities (e/L = 0, 0.4, 0.6 and 1.0) and normalized pile length (L/D = 5, 6, 7 and 8), for which the bending moment is the product of horizontal force multiplied by the investigated load eccentricity. The results of a parametric study give depth-dependent p-y curves accounting for the actual load conditions and deformation modes. These p-y curves were systematically compared to the basic p-y curves. The distributions of the correction factor (cf. Figure 2) depending on number of load cycles over the whole pile depth were evaluated through these comparisons.

DOI: 10.1201/9781003348450-327

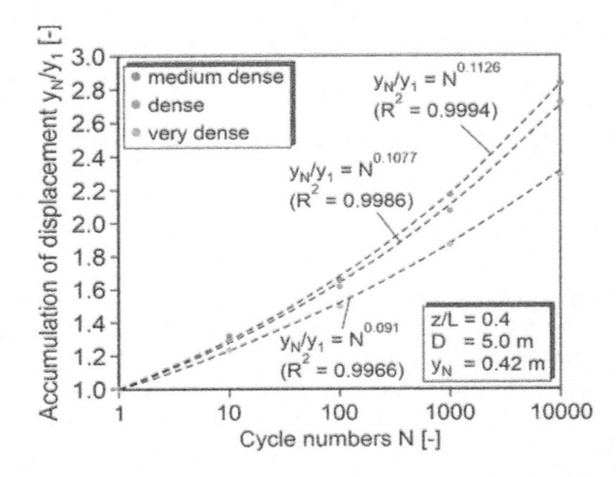

Figure 1. Y-multipliers for a monopile with variable relative densities.

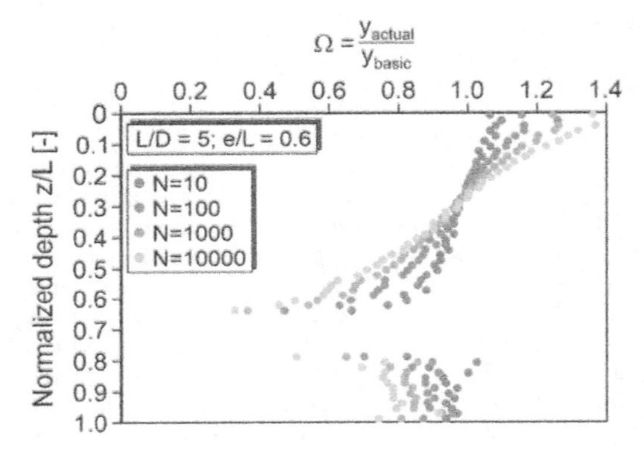

Figure 2. Distribution of the correction factor (D=5m, L=25m).

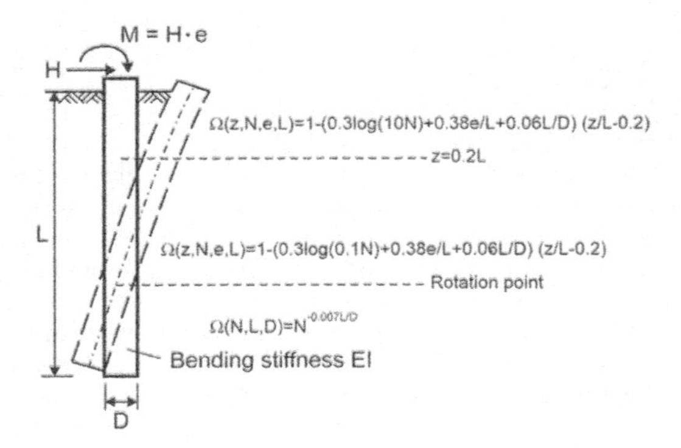

Figure 3. Conception and equation of correction factors for D=5m.

3 CONCLUSION AND OUTLOOK

New static p-y approaches applicable to large-diameter monopiles were developed recently. However, a general approach to consider cyclic load effects on p-y curves dependent on the number of load cycles is missing. Such an approach would be highly desirable for practical design in particular of monopiles for offshore wind energy converters.

The "cyclic overlay model" presented here bases on the comparison of numerical simulations of monopile behavior in sand soil once under monotonic and once under cyclic loading with a defined number of load cycles (utilizing the SDM). Applying the derived equations for y-multipliers, arbitrary p-y curves for monopile behavior under monotonic loading can be adapted to a given number of load cycles. The y-multipliers are formulated as a product of a term ($y_N/y_1=N^A$) (cf. Figure 1) valid for constant horizontal deflection of a pile and a correction term Ω (cf. Figure 3) accounting for the actual deflection line and loading conditions. Hence, the actual cyclic p-y curves can be approximated by applying the correction factors additionally to the y-multipliers as:

$$y_N/y_1 = N^A \cdot \Omega.$$

The new cyclic overlay model was proven to resemble the numerical simulation results with good accuracy. An extension of the Ω-approach given in Figure 3 for other pile diameters needs to be investigated in the next step of the work. An experimental validation of the model is yet also missing, but this is planned by cyclic large scale pile load tests in an ongoing research project.

ACKNOWLEDGEMENTS

This study was carried out in the scope of the research project "Ho-Pile" funded by Bundesministerium für Wirtschaft und Energie (Germany). Project No. 0324331A. The authors sincerely acknowledge BMWi support.

REFERENCES

Simulia (2018): Abaqus 2018 Standard User's Manual. Version 6.14. Dassault Systèmes Simulia Corp.

Achmus, M., Kuo, Y.-S. and Abdel-Rahman,K. (2009). "Behavior of monopile foundations under cyclic lateral load," Computers and Geotechnics, Vol. 36, No. 5, pp.725–735

API (2014): Recommended Practice 2GEO-Geotechnical and Foundation Design Considerations. American Petroleum Institute, Version October 2014.

Kuo, Y.-S. (2008): On the behavior of large-diameter piles under cyclic lateral load. Ph.D. thesis, Leibniz Universität Hannover, Hannover, Institute for Geotechnical Eng., Vol. 65.

Current Perspectives and New Directions in Mechanics, Modelling and Design of Structural Systems – Zingoni (ed.)
© 2022 Copyright the Author(s), ISBN 978-1-032-39114-4

Climate-smarter design of soil-steel composite bridges using set-based design

J. Lagerkvist
Swedish Transport Administration, Chalmers University of Technology, Sweden

C.G. Berrocal
Thomas Concrete Group, Chalmers University of Technology, Sweden

R. Rempling
Chalmers University of Technology, Sweden

ABSTRACT: Soil-Steel Composite Bridges (SSC-Bridges) is one of the most built bridge types in Sweden. This paper investigates the potential climate and economic savings that could be done by implementing a novel design method called Set-Based Design (SBD). The results shows that there is great potential to reduce the climate and material cost by up to 20%. Considering the large number of SSC-Bridges that is built, this is an efficient way to reduce the climate impact when building these bridges in the future.

1 INTRODUCTION

Since 2010 and up to the first half of 2021, 522 Soil-Steel Composite Bridges (SSC-Bridges) has been built in Sweden. SSC-Bridges are often considered as an attractive structural solution compared to the slab frame bridge made of reinforced concrete. SSC-Bridges are traditionally designed with Point Based Design (PBD), which means that one single design is proposed and as long as that single design fulfils the design criteria, that is the final design (Rempling *et al.*, 2019). With PBD we do not know whether we get the most suitable solution when a SSC-Bridge is bult, we just know that we have one solution that fulfils the design criteria. Because of the large number of SSC-Bridges that are built, there could be large potential savings with regard to cost and CO_2-equivalents by investigating more solutions for every SSC-Bridge that is built.

2 OBJECTIVE AND LIMITATIONS

The aim of this study was to investigate the potential economic and environmental savings that could be achieved for SSC-Bridges by adopting an alternative design approach.

This research has studied closed SSC-Bridge profiles and it has been carried out under the criteria of how these bridges are designed in Sweden. For the optimization, cost and climate impact, in terms of CO_2-equivalents, from the materials have been taken into account.

3 STATE OF THE ART

For optimization of SSC-Bridges some studies have been carried out, looking into how the optimal shape would look like if it was possible to design it in any shape, how different shapes perform against each other and how different parameters affects the optimal design.

To the best of the authors knowledge, none of the-previously found research about optimization of these structures has studied the potential economic and environmental savings that could be done by implementing a Set-Based Design approach.

4 SET-BASED DESIGN METHOD

The applicability and potential of the proposed method were assessed by implementing an automated Set-Based Design (SBD) for six different profiles of closed SSC-Bridges. Every profile consists of between 8 and 80 different sizes, which gives us a total of 233 different geometries and sizes.

Five existing bridges were chosen as case studies that we could compare our results with. Those bridges were designed and built before this research was initiated.

For the SBD method we need an initial set of alternatives. These sets are generated from selected geometrical design parameters and their corresponding ranges of values. The different parameters, range of values and their contribution to the number of alternatives are presented in Table 1.

DOI: 10.1201/9781003348450-328

Verification of the design was performed for ULS, SLS, FAT and for loads during construction and they follow section 5 in (Pettersson and Sundquist, 2014).

Table 1. Parameters and variations to produce alternatives.

Parameter	Values	Step	Unit	Number of alternatives
Bridge width	6.0	-	[m]	1
Cover height	0.5 – 4.0	0.05	[m]	71
Plate thickness	3 - 8	0.5	[mm]	11
Backfill material	Sub-base Base-course	-	-	2

5 RESULTS

The reductions were analyzed in different steps. In the first step we looked at a pipe with the same profile and with the same size but changed the cover height which also affect the plate thickness of the pipe. In the second step we also changed the backfill material to see if it was possible to make even more reductions. In the third step we looked at the criteria, in terms of a free space rectangle inside the pipe, that the bridge should fulfil to see if we could find a smaller size within the same shape that still fulfils this rectangle. The final step was to look at all the geometries and sizes that fulfills the required free space rectangle to see if it was possible to find a better solution.

The results for the final step for built bridge 5 is presented in Figure 1. The results in the figure are normalized with the value 1,0 for the built bridge.

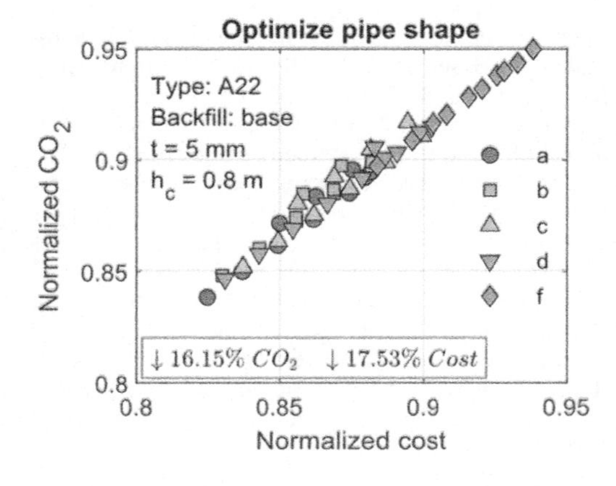

Figure 1. Final optimization with SBD for built bridge 5.

6 DISCUSSION

For the five existing bridges that were used to evaluate the method we could see that it was possible to make reductions in every step except when the bridges were already designed with base course as the backfill material, which gave the optimal solution.

For the final step, when we also varied the different profiles and sizes to find the most optimal solution that fulfil the free space rectangle that was needed for that certain bridge, we saw that we had a lot of potential. One thing that could be mentioned is that we do not know if there were any other parameters that has been important for the choice of profile for the different built bridges. It is also worth mentioning that the design of the bolted connections of the bridges was not trivial and was often critical for our design.

7 CONCLUSION

The purpose of this research was to investigate the potential cost and CO_2 reductions that could be done by implementing a Set-Based Design approach for Soil-Steel Composite Bridges. In this research project SCC-Bridges were designed using SBD where parameters described in chapter 4 varied which generated a lot of possible solutions for each profile and size.

In our research, we have showed that there is a large potential to design Soil-Steel Composite Bridges in a more optimized way and we have verified our results with five different existing bridges. The reduction of CO_2-equivalents varied between 1.7% - 20.8% and the reductions of cost was between 1.8% - 19% for the five existing bridges.

Considering the large number of Soil-Steel Composite Bridges that are built in Sweden every year, this method could be readily implemented to further reduce the climate impact from the Swedish Transport Administration projects.

ACKNOWLEDGEMENTS

This project is part of an Industrial PhD project that is financed by the Swedish Transport Administration.

REFERENCES

Pettersson, L. and Sundquist, H. (2014) *Design of soil steel composite bridges, Report 112, 5th Edition*. Available at: www.byv.kth.se (Accessed: 30 October 2021).

Rempling, R. *et al.* (2019) 'Automatic structural design by a set-based parametric design method', *Automation in Construction*, 108, p. 102936. doi: 10.1016/j.autcon.2019.102936.

Current Perspectives and New Directions in Mechanics, Modelling and Design of Structural Systems – Zingoni (ed.)
© 2022 Copyright the Author(s), ISBN 978-1-032-39114-4

Parametric modelling of integral bridge spring reactions

N.R. Featherston
Ovum Corporation (Pty) Ltd, South Africa

C. Viljoen
Department of Civil Engineering, Stellenbosch University, South Africa

ABSTRACT: In integral type bridges, the soil's interaction with the abutment and the piles (the structure) is usually modelled in practice using soil springs. The characteristics of these soil springs ultimately determine the structure's response in terms of the applied bending moments, deflections and shears in the deck, abutment and piles (similarly for a bridge on footings). In this study, the influence on the soil-structure interaction (SSI) spring reactions in integral bridges was investigated. To this end, a series of parametrically varied 2D and 3D bridge models with soil springs were created and subsequently analysed. The parameters of span length, abutment height and soil condition for different percentages of live load and thermal expansion/contraction/gradient were considered and their influence on the SSI reactions was revealed through a series of parametric model testing. Both a piled foundation type and a footing foundation type were considered in the analysis work that was undertaken, and simplified models by authors such as Hambly (1991), Lehane (1999) and O'Brien and Keogh (1999) were assessed. The model test results showed that the maximum spring reactions (for abutments, piles and footings) followed either a linear relationship with increasing span or tended towards more non-linear type relationships. The results showed distinct and significant differences between the maximum spring reaction vs span relationships for the abutments and the piles (for the same bridge types). Further interpretation of the test results also showed that as spans increase (irrespective of the abutment height), the maximum spring reaction ratio (abutment/pile reaction ratio) tends towards unity.

1 INTRODUCTION

The integral bridge is a bridge structure in which the deck, piers and abutments are all made integral without the use of expansion joints. The joints would normally allow the expansion and contraction of the bridge deck, however now the flexibility inherent in the piers and abutments is utilized. The main design issue in an integral bridge is the effect of the temperature fluctuations and the corresponding movements in the bridge deck.

By its nature, the integral bridge has ends that interact with the embankment soil, as the bridge is subjected to cycles of movement that result in changes in soil pressure from the active to the passive pressure state. Forces inherent in conventional bridge systems also act longitudinally against the backfill soil, however the magnitude of the soil-structure interaction is negligible since the cyclic forces that are created are minor in comparison with those acting on integral bridges. This makes the integral bridge design both a structural and a geotechnical problem of interest.

Finite Element (FE) approaches to Soil-structure interaction (SSI) usually fall into approaches that characterize the soil using continuum elements and those that represent the soil through springs. In the spring method of analysis, the resistance of the soil lying adjacent to the piles and to the abutment is represented by springs which can be linear, compression only or non-linear in character. The theories used for the calculation of the spring stiffnesses are quite different between the abutments and the piles.

In this study, the influence on SSI reactions of pertinent bridge geometry and load parameters was investigated through the use of a series of parametrically varied 2D and 3D bridge models with soil springs.

The parameters of span length, abutment height and soil condition for different percentages of live load and thermal expansion/contraction/gradient were investigated and their influence on the spring reactions was shown through the series of model testing. Pile lateral loading vs deflection relationships tend to be non-linear, however the spring reaction relationship for increasing span and load (once a pile is incorporated into an integral bridge system) was unknown prior to the analysis work captured in this study.

2 MODEL SETUP AND LOADCASES

The following parameter variations were studied, with the focus on examining the effects of the maximum spring reactions in each instance:

DOI: 10.1201/9781003348450-329

- Live load variability (for constant span)
- Span variability (for constant live load)
- Temperature variance effects (effective temperature and temperature gradient)
- Abutment height variability (for constant span)
- Ground condition variability (for constant span)

Note that ground condition relates to the footing or pile founding condition, not the backfill condition, which remains constant in general.

The basic loading scenarios that were investigated in this set of models were as follows:

- Live load (20% DL, 40% DL, 60% DL, 80% DL)
- Temperature Max (+15°C rise, from 20°C ambient)
- Temperature Min (-15°C drop, from 20°C ambient)
- Temperature Gradient (in accordance with TMH7 Parts 1&2 - Section 4.5.5, Figure 22 and Table 16).

Both 3D and 2D models were investigated in this study (an example is shown below in Figure 1). The expansion and contraction effects were examined and various maximum spring reaction graphs were produced. The effects of soil spring stiffness were also studied.

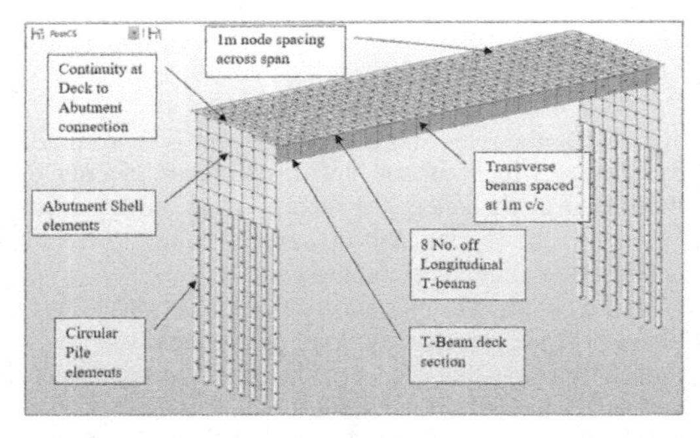

Figure 1. Typical 3D Grillage model for integral abutment modelling.

3 RESULTS AND OBSERVATIONS

Note that the results that were recorded in the various graphs are for the highest spring reaction value found in each load or scenario case.

Of note are the results from Figure 2, which shows the distinct difference in the Abutment/ Pile spring reaction ratio for the 9m high abutment vs the 3m high abutment. For a normal higher (9m) abutment (with no soil), one would expect less force due to the increased flexibility, however since there is an increased abutment soil-structure interaction, there is more area and more passive spring force, therefore the force is in fact increased, not reduced. The inference of these diagrams is that for short spans, the 9m high abutment generates significantly more reaction forces in the abutment, with smaller forces and bending moments being generated in the piles.

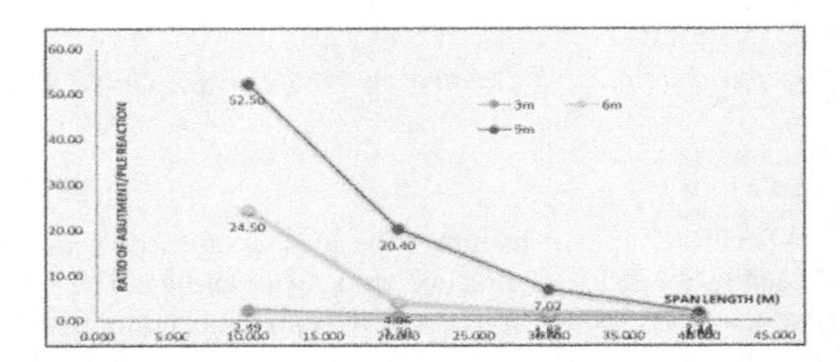

Figure 2. (Abutment/Pile) Spring Reaction ratios for variable abutment height vs Span length (founding in stiff clay, 40% Live load, 3D Grillage model).

4 CONCLUSION

- For the span range 10m - 20m, the effects of foundation soil type made very little difference to the abutment spring reactions. After the 20m span is exceeded, the effects of the foundation soil type made more of a difference in the spring reaction results.
- The model testing showed that for the smaller spans, the higher abutment heights generated significantly more spring reaction than in the piles – this would imply that the abutments could possibly be made larger in dimension than the piles.
- An unexpected result was that the model testing showed that as the span lengths increase, the abutment and pile spring maximum loads tend to converge and become the same, ie. they match each other in magnitude.
- A lack of good correlation was shown in this study between spring stiffness and spring reaction as evidenced in various graphs that were produced.

REFERENCES

Elson W.K., (1984) CIRIA Report 103 - Design of laterally-loaded piles. Construction Industry Research & Information Association (CIRIA).

Hambly, E.C. (1991) Bridge deck behaviour – Second Edition. Taylor & Francis, Bath.

Lehane, B. (1999) Predicting the restraint to integral bridge deck expansion, in Proceedings of 12th European Conference on Soil Mechanics and Geotechnical Engineering, Amsterdam, Balkema, Rotterdam.

Current Perspectives and New Directions in Mechanics, Modelling and Design of Structural Systems – Zingoni (ed.)
© 2022 Copyright the Author(s), ISBN 978-1-032-39114-4

A parametric study of the vibration of beams resting on elastic foundations with nonlinear cubic stiffness

E. Feulefack Songong & A. Zingoni
Department of Civil Engineering, University of Cape Town, Rondebosch, Cape Town, South Africa

ABSTRACT: This paper presents a parametric study of beams resting on elastic foundations having nonlinear cubic stiffness properties. The work is achieved through an analytical study using the expansion method and a numerical study using Finite Element Analysis (FEA) in Abaqus. Both free and forced vibration responses of the beam are investigated. The study's objective is to understand how the beam frequencies are influenced by the cubic stiffness nonlinearity of the foundation and explore the external excitation amplitude effect on the beam amplitude of vibration. As expected, the beam's natural frequencies increase as the stiffness parameter increases, and the beam's amplitude of vibration reduces in the presence of nonlinearity. Finally, numerical results are compared to analytical results for validation.

1 INTRODUCTION

Beams on elastic foundations found their applications in civil and mechanical engineering. Some examples include vibrating machines on elastic foundations, railroad tracks, bridges, buildings, etc. (Hetényi 1971, Akour 2010). The concept of beams on elastic foundations was first introduced by Hetenyi, who studied beams on elastic supports (Hetényi 1971). Among the available studies, the analysis of beams on the Winkler foundation is the most popular.

The paper investigates analytical and numerical studies of the simply supported beam resting on an elastic foundation with nonlinear cubic stiffness properties and subjected to external excitation. The analytical research uses the straightforward expansion method, whereas the numerical study employs FEM in Abaqus/Standard. The aim is to understand how the natural frequencies of the nonlinear system are modified due to the cubic nonlinearity. That will provide new results and interpretations related to nonlinear cubic elastic foundation systems. Moreover, the load effects on the beam are highlighted to picture the variation of the beam amplitude as time increases.

2 PRESENTATION AND SOLUTION OF THE EQUATION OF MOTION OF THE SYSTEM

The system in Figure 1, adopted from (Hetényi 1971) shows a beam simply supported at both ends, resting on elastic springs, and a rigid base supporting the foundation. All the springs and the rigid base form the elastic foundation on which the beam rests.

The equation of motion of the system presented in Figure 1 is given by

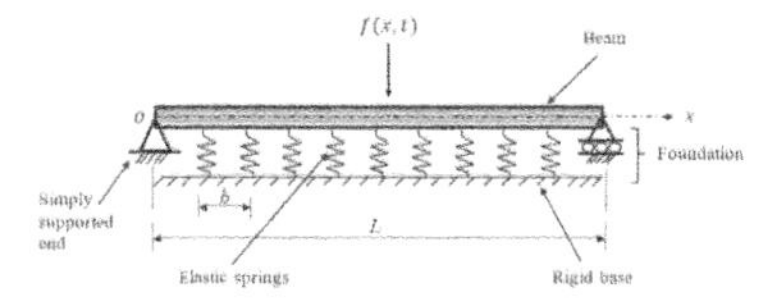

Figure 1. Simply supported beam resting on an elastic foundation.

$$\rho S \frac{\partial^2 W(x,t)}{\partial t^2} + EI \frac{\partial^4 W(x,t)}{\partial x^4} + k_1 W(x,t) + k_3 W(x,t)^3 = f(x,t), \quad (1)$$

where $W(x,t)$ is the beam transversal displacement, k_1 and k_3 are respectively the linear and nonlinear cubic stiffness coefficients. The right-hand side is given by $f(x,t) = F_0 \cos(\Omega t)\delta(x - x_0)$, where F_0 is the initial amplitude, Ω the frequency of excitation, and x_0 the point of application of the load.

The solution of Eq. (1) is assumed to be

$$W(x,t) = \sum_{n=0}^{\infty} Q_n(t) \sin\left(\frac{n\pi}{L}x\right), \quad (2)$$

where $Q_n(t)$ is the temporal part and $\sin\left(\frac{n\pi}{L}x\right)$ is the spatial part of the solution. By replacing Eq. (2) into Eq. (1) and applying orthogonality to the resulting equation and simplifying it, one obtains

$$\ddot{Q}_n + \omega_n^2 Q_n + \beta Q_n^3 = F_n \cos(\Omega t). \quad (3)$$

where $\omega_n^2 = \frac{EI}{\rho S}\left(\frac{n\pi}{L}\right)^4 + \frac{k_1}{\rho S}$; $\beta = \frac{3}{4}\frac{k_3}{\rho S}$; and $F_n = \frac{2F_0}{\rho SL} \sin \left(\frac{n\pi}{L}x_0\right)$.

In Eq. (3), $\ddot{Q}_n$ is the second derivative of Q_n concerning the time t, whereas Q_n^3 is the cubic term that brings the nonlinearity to the equation. Using the straightforward expansion method (He 2004, Nayfeh & Mook 1995), a small parameter ε is used. Then, the solution of the homogeneous part of Eq. (3) is assumed to be

$$Q_n(t) = Q_1 + \varepsilon^2 Q_2 + \varepsilon^3 Q_3 + \cdots + \varepsilon^n Q_n, \quad (4)$$

where $\varepsilon << 1$. Moreover, a new time $\tau = \Omega_c t$ is set so that the second derivative $\ddot{Q}_n$ is with the new time τ. Next, Ω_c is also written as follows.

$$\Omega_c = \omega_n + \varepsilon\, \Omega_2 + \varepsilon^2 \Omega_3 + \cdots + \varepsilon^n \Omega_n, \quad (5)$$

where ω_n is the linear natural frequency. By replacing Equations (4) and (5) into Equation (3), the overall solution of the equation without the right-hand side is found as

$$W(x,t) = \sum_{n=0}^{\infty} \sin\left(\frac{n\pi}{L}x\right)\{C_0 \cos(\Omega_c t + \phi_0) -$$

DOI: 10.1201/9781003348450-330

$$\frac{\varepsilon\beta}{32\omega_n^2}C_0^3[\cos(\Omega_c t) - \cos(3\Omega_c t)]\Big\}, \quad (6)$$

and the nonlinear natural frequency Ω_c is deduced as

$$\Omega_c = \omega_n\left[1 + \frac{3\varepsilon\beta}{8\omega_n^2}C_0^2\right]. \quad (7)$$

3 PRESENTATION OF THE RESULTS

3.1 *Comparison between analytical and numerical results from linear and nonlinear analyses*

The following table shows values used for analytical and numerical analyses. The parameters were chosen arbitrarily.

Table 2 reveals linear and nonlinear natural frequencies computed from both analytical and numerical analyses. The errors are higher at lower modes when comparing linear to nonlinear results. On the other hand, nonlinear errors are minimal (less than 5%), which confirm the agreement between the two methods. Then, the results from Abaqus are an excellent benchmark for the analytical results.

Table 1. Properties of the steel beam and other parameters to be used for analytical calculations and numerical analysis.

Item	Notation	Value	Others parameters	
Young's modulus	E	200 GPa	k_1	$1.69 \times 10^5 N/m^2$
Length	L	2 m	k_3	$1.3 \times 10^7 N/m^4$
width	a	0.05 m	ε	0.01
height	h	0.03 m	F_0	$1000N$
Poisson's ratio	ν	0.3	x_0	$\frac{L}{2}$
mass density	ρ	8050 kg/m^3	Ω	100π (rad/s)

Table 2. Comparison between linear and nonlinear natural frequencies obtained analytically and in Abaqus.

	Analytical			Abaqus			
Mode	Linear	Nonlinear	Error (%)	Linear	Nonlinear	Error (%)	Nonlinear Error (%)
1	159.1	302.3	47.7	284.9	288.7	1.33	4.5
2	442.1	493.7	10.5	500.8	502.96	0.43	1.8
3	965.8	989.4	2.4	991.5	992.6	0.11	0.32
4	1708.2	1721.6	0.78	1715.4	1716.1	0.04	0.32
5	2665.4	2673.9	0.32	2653.6	2654.0	0.02	0.75
6	3836.2	3842.1	0.15	3797.6	3797.9	0.008	1.2
7	5220.3	5224.7	0.084	5141.1	5141.3	0.004	1.6
8	6817.6	6821.0	0.05	6678.1	6678.3	0.003	2.1
9	8628.1	8630.7	0.03	8402.6	8402.8	0.003	2.7
10	10652	10654	0.02	10305	10305	0.0	3.4

Concerning the plots presented in Figure 2, the paths followed by the analytical and numerical curves are similar. That confirms the agreement between the two methods.

3.2 *Effects of the external excitation*

To solve Equation (3) with the right-hand side, the fourth-order Runge-Kutta method (Cortell 1993) is used in MATLAB. The following plots were obtained.

Figure 3, reveals that the beam amplitude is minimal for both cases when the foundation is nonlinear. For mode 1, the linear amplitude is constant while the nonlinear amplitude decreases with time. As for mode 3, both linear and nonlinear curves decrease as time increases, but the nonlinear amplitude is still minimal.

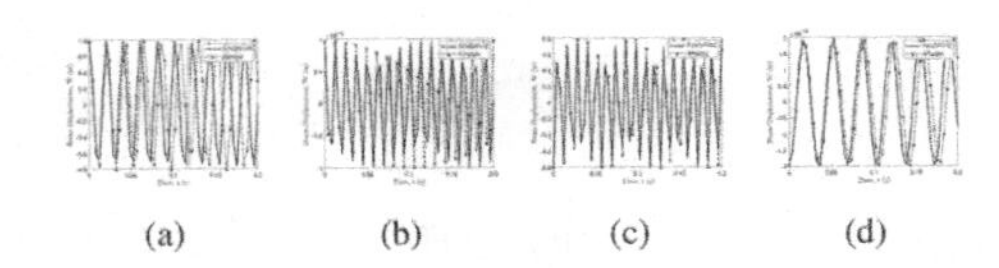

(a) (b) (c) (d)

Figure 2. Temporal displacement comparison between analytical and numerical analyses: (a) to (d) are respectively modes 1 to 4.

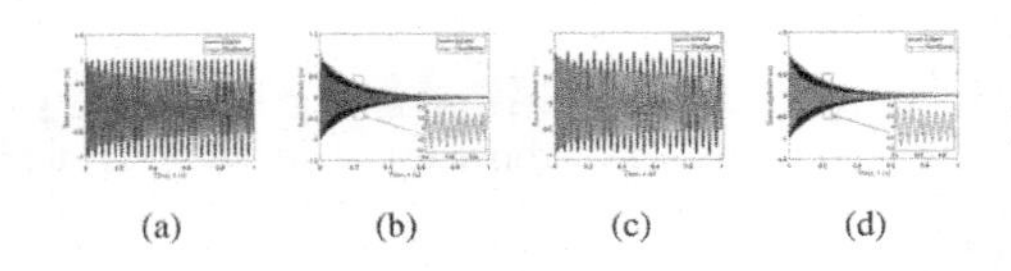

(a) (b) (c) (d)

Figure 3. Temporal amplitude comparison between linear and nonlinear analyses for forced vibration of the beam on a nonlinear elastic foundation: (a) and (b) are modes 1 and 3 for $F^0 = 1000N$, (c) and (d) are modes 1 and 3 for $F_0 = 10,000N$.

4 CONCLUSION

This paper studied analytical and numerical analyses of the linear beam resting on a nonlinear cubic elastic foundation. It is found that the nonlinear natural frequencies are more significant than the linear natural frequencies. The errors between linear and nonlinear analyses proved that the linear system might fail to represent the real situation. The minimal nonlinear errors proved the agreement between analytical and numerical methods. Also, the nonlinear foundation reduced the beam amplitude for forced vibrations.

ACKNOWLEDGEMENT

This work was supported by The Organization for Women in Science for the Developing World (OWSD) Fellowships, the Swedish International Development Corporation Agency (SIDA), and the International Students' Scholarship from the University of Cape Town.

REFERENCES

Akour, S. N. (2010). Dynamics of nonlinear beam on elastic foundation. In *Proceedings of the World Congress on Engineering*, Volume 2, pp. 1427–1433. Citeseer.

Cortell, R. (1993). Application of the fourth-order runge-kutta method for the solution of high-order general initial value problems. *Computers & structures 49*(5), 897–900.

He, J.-H. (2004). Comparison of homotopy perturbation method and homotopy analysis method. *Applied Mathematics and Computation 156*(2), 527–539.

Hetényi, M. (1971). *Beams on elastic foundation: theory with applications in the fields of civil and mechanical engineering*. University of Michigan.

Nayfeh, A. & D. Mook (1995). *Nonlinear oscillations*. Wiley Classics Library. Wiley-VCH.

Current Perspectives and New Directions in Mechanics, Modelling and Design of Structural Systems – Zingoni (ed.)
© 2022 Copyright the Author(s), *ISBN 978-1-032-39114-4*

Degradation of axial friction resistance on buried district heating pipes

T. Gerlach & M. Achmus
Institute for Geotechnical Engineering, Leibniz University Hannover, Hanover, Germany

ABSTRACT: The axial soil resistance is a crucial component within the design process of buried district heating pipelines and is known to degrade during the service life time due to axial movement of the pipeline under cyclic temperature loading. In current design methods, a large bandwidth of the coefficient of friction μ between pipe and soil is considered to account for this effect. However, a more accurate consideration of cyclic effects on the axial resistance could lead to a more economic design of district heating pipelines. Experimental investigations were carried out to identify the main factors influencing the friction degradation. It was found that beside stress and relative density of the surrounding soil, the amount of resistance degradation and also its rate are strongly affected by the magnitude of cyclic pipe movement and the number of load cycles. It could also be observed that the degradation can partly be healed dependent of the order of load packages, i.e. subsequent cycles with constant displacement amplitude.

1 STATE OF THE ART

1.1 *Axial friction resistance*

In common design practice in Germany [AGFW 2007], the maximum friction resistance is estimated from:

$$F_{max} = \mu \cdot (F'_R + F'_G) \tag{1}$$

Herein F_{max} denotes the maximum resistance force per meter length, μ is the friction coefficient and F'_R and F'_G are the integrated normal stresses around the pipes circumference and the sum weight force of the filled pipeline, respectively. To incorporate the effects resulting from changes of operating temperature, the coefficient of contact friction is taken to $\mu = 0.4$ for increasing temperature conditions. During unloading (decrease of temperature), a value of $\mu = 0.2$ shall be applied [AGFW 2007]. However, the coefficient of contact friction is a material specific constant.

1.2 *Dependency on cyclic movement*

During the operating life of a district heating network, the pipeline moves back and forth several times. These cyclic displacements lead to changes of the surrounding soil state and thus the mobilizable friction resistance. Since this affects all system components involved, the description of these cyclic effects is of great importance.

With respect to mobilizable shear stresses, it is well known in soil mechanics that cyclic shear loading of noncohesive soil leads to an increase in relative density at constant imposed load. In the context of the axially displaced pipelines, this means that cyclic displacements result in compaction in the pipe-soil contact zone. This leads to a decreasing contact normal stress and thus to a lower mobilizable frictional resistance. First experimental investigations on the decrease of the maximum frictional resistance were carried out by Weidlich [Weidlich 2008]. In a testing box, pipelines of nominal sizes DN 40, DN 65 and DN 80 were covered with sand and then axially displaced for several cycles. The displacement amplitude was selected so that the full frictional resistance was mobilized at each reversal point of the displacement. Already after a few cycles, a residual value F_{res} is reached, which corresponds to about half of the value reached during the initial displacement F_0.

2 INSIGHTS FROM EXPERIMENTAL INVESTIGATIONS

Within the same test box used by Weidlich, further experimental investigations were carried out. At first, the magnitude of cyclic displacement was varied. Figure 1 shows that this has a significant impact on the evolution of friction resistance degradation. While for large magnitudes, e.g. relative displacements between 0mm and 11mm, a constant value of D_F is reached after about 10 cycles as observed by Weidlich, for small magnitudes the number of cycles required is tending towards N=200.

In the following, tests were carried out with subsequent load packages of different displacement

DOI: 10.1201/9781003348450-331

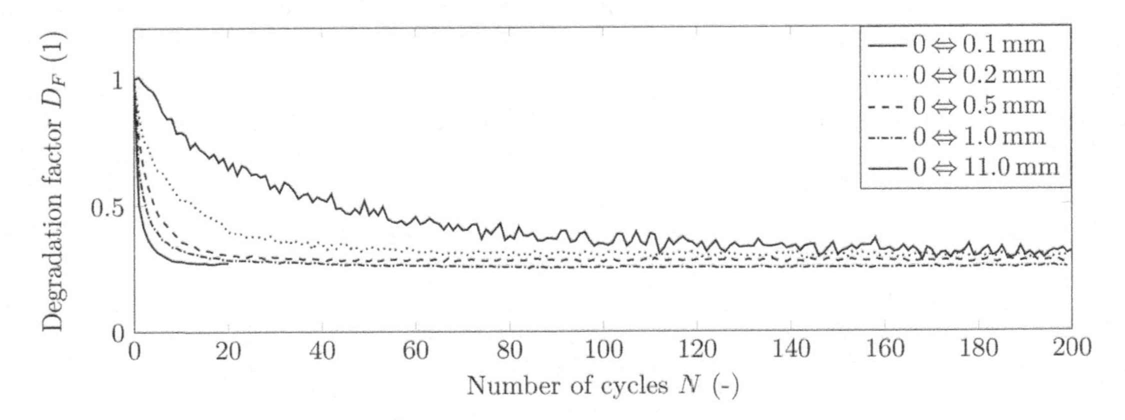

Figure 1. Degradation factor D_F for different displacement magnitudes.

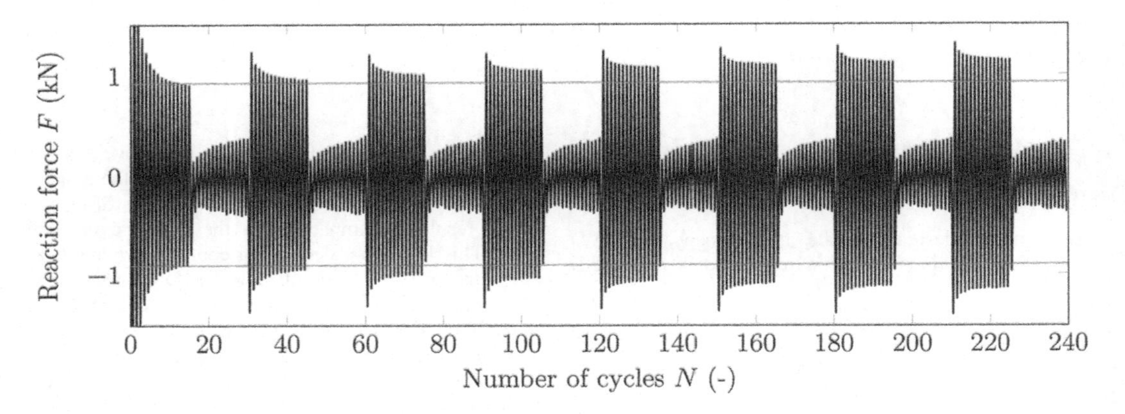

Figure 2. Reaction force F for alternating displacement amplitudes.

magnitude. The pipeline was a DN80 with an outer diameter of 16cm covered with 80cm dense sand ($I_D = 0.8$). Within the test shown at Figure 2, 14 cycles were performed with a displacement range between 0 mm and 11 mm followed by 15 cycles with a displacement range between 0 mm and 0.2 mm. This sequence was repeated eight times. While for the large magnitude the friction resistance is constantly decreasing towards a constant value, within the following load package with the small magnitude the resistance is increasing for a certain number of cycles. Secondly, if the large magnitude packages are compared to each other, an increase of the residual value can be observed. For the sake of clarity, the red lines in Figure 2 indicate the residual values of friction from the first load package. This suggests that the degradation impacted by large magnitude cycles can partly be healed by small magnitude cycles.

3 CONCLUSION AND PERSPECTIVE

Within the paper at hand, the state of the art concerning axial friction resistance of axially displaced pipelines is summarized briefly. In addition, results from additional experimental investigation are presented. While for large amplitude cycles the degradation process is finished after 10-15 cycles, the necessary number of cycles increases with decreasing magnitude. Also, it was shown that small magnitude cycles subsequent to a full resistance degradation due to large magnitude cycles can partly "heal" previous degradation. Therefore, further research is needed and it is planned to extend the scope of experiments. It should then be possible to predict the friction resistance for arbitrary situations that can occur during the service life.

REFERENCES

AGFW (2007). Arbeitsblatt FW 401 Teil 10: Verlegung und Statik von Kunststoffmantelrohren (KMR) für Fernwärmenetze: Statische Auslegung, Grundlagen der Spannungsermittlung.

Weidlich, I. (2008). Untersuchung zur Reibung an zyklisch axial verschobenen erdverlegten Rohren. Ph. D. thesis, Leibniz Universität Hannover.

Current Perspectives and New Directions in Mechanics, Modelling and Design of Structural Systems – Zingoni (ed.)
© 2022 Copyright the Author(s), ISBN 978-1-032-39114-4

A hybrid structure to protect infrastructures from high energy rockfall impacts

M. Marchelli & V. De Biagi
Politecnico di Torino, Torino, Italy

ABSTRACT: Among natural phenomena, rockfall probably is the most hazardous due to its abruptness and high energies involved. Rock blocks can impact a structure even with a kinetic energy of 50 MJ, involving consistent damages. To protect inhabited areas and infrastructures, mitigation measures are installed along slopes. Focusing on protective measures, several solutions can be adopted, i.e., net fences, embankments, etc. Such structures intercept and stop the falling block by dissipating its kinetic energy through permanent large deformations, needing large downslope free area, which can constitute a strong geometric constraint. Nevertheless, RC walls alone are not diffused as they cannot withstand rockfall impact forces. We propose a compelling solution for protecting infrastructures with a hybrid structure made of multiple vertical layers: a high deformable downhill earth face coupled with a RC wall. This solution ensures energy dissipation and reduced deformability of the downslope face of the structure and can be installed close to roads. A simplified design procedure and an example are proposed with reference to a real case study. The effectiveness of the rockfall risk mitigation measures is discussed and a cost-energy capacity design chart is presented.

1 INTRODUCTION

Rockfall events can represent one of the most hazardous among the landslide phenomena. Due to high possible damages, rockfall protection structures are well diffused along transportation routes to protect vehicles/trains, etc. or in proximity of urbanized areas. Net fences and rockfall protection embankments (RPE) are the most adopted solutions.

Net fences, which are usually preferred since they are easy to install and have low visual impact, can sustain impacts up to 10000 kJ and dissipate energy through large displacements, requiring the need of a minimum distance from the element at risk. RPE have been considered a suitable solution against events involving very high kinetic energies, up to 50 MJ, together with their ability to sustain repeated impacts before collapse (Lambert & Kister 2017). Different types of embankments have been developed in the past. Among them, the reinforced earth RPE, i.e., a system made up of overlaid layers of compacted soil, each wrapped in a tensile resistant element, represents a valuable solution, allowing a side inclination up to 70°. Nevertheless, the massive size of RPE needs a suitable topographical configuration in terms of space and inclination of the site.

In the present work, a compelling solution is proposed for protecting infrastructures with a hybrid structure made of two vertical layers: a high deformable uphill face made of reinforced earth coupled with a downhill reinforced concrete (RC) wall. A design scheme to support impact loads, based on the on the results of experiments and studies on rockfall protection tunnels, is proposed. Parametric analyses and a discussion about the economic advantages in installing such coupled rockfall protection structure are reported.

2 THE HYBRID ROCKFALL PROTECTION STRUCTURE

The proposed rockfall protection structure consists in an earth reinforced embankment laterally supported by a RC structure, i.e., a wall (Figure 1)

Differently from a simple reinforced earth RPE, the proposed solution presents limited site constraints since the overall cross-section is reduced. In addition, no downhill displacements occur, thus, it can be installed closer to the road infrastructure.

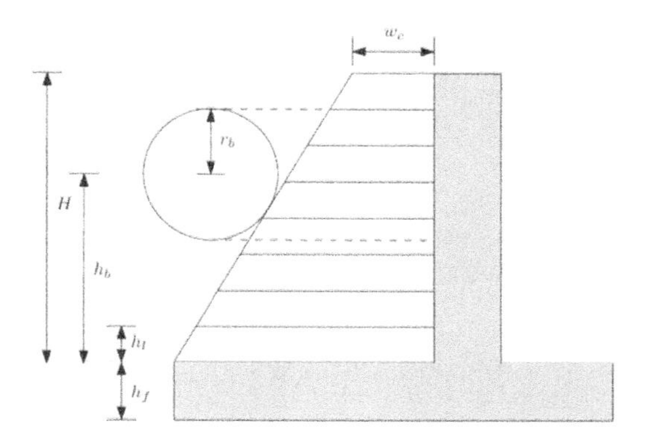

Figure 1. Sketch of the hybrid rockfall protection structure.

DOI: 10.1201/9781003348450-332

3 METHODOLOGY

The proposed design method derives from the procedures adopted for rockfall protection tunnel (Astra 12 006 2008), in which the reinforced concrete structure is covered by a granular soil stratum. The hybrid structure herein proposed can thus be considered as a system composed by two vertical layers, similar to rockfall protection tunnels. In the proposed structure, block kinetic energy is dissipated in the earth layer (only through compaction, as the block penetrates the earth layer) and the resulting forces are turned into pressures on the concrete wall.

3.1 *Reinforced earth layer*

Following the findings of experimental studies on reinforced earth protection embankments (Peila et al. 2007, Ronco et al., 2009), the stress geometrical diffusion (the zone disturbed by the impact) is evaluated and the equivalent load acting on the RC wall is computed. The maximum force $F_{E,max}$ is evaluated as for rockfall protection tunnels:

$$F_{E,max} = 2.8t^{-0.5} r_b^{0.7} M_E^{0.4} tan\phi E_k^{0.6} \tag{1}$$

where t is the thickness (in meters) of the earth layer in the point of the impact (considering the center of mass of the block), r_b the block radius (in meters), M_E the elastic modulus of the soil (in kPa), ϕ the internal friction angle, and E_k the kinetic energy of the impacting block (in kJ).

Layer deformation δ_E, i.e., compaction of the soil and crater formation, can thus be evaluated as the ratio of twice the kinetic energy and the correspondent resistant force, assuming that a linear compaction force-displacement relationship holds until the maximum force $F_{E,max}$ is reached. From the crater, a diffusion of the stresses is considered.

3.2 *Reinforced concrete wall*

The pressure acting over the upslope face of the RC wall is assumed as unform over the area on the earth-concrete interface influenced by the impact and is equal to:

$$q_{RC} = \frac{F_{E,max}}{A_{RC}} \tag{2}$$

Considering the wall as a cantilever, the maximum forces (bending moment and shear) are recorded at the base.

4 PARAMETRIC ANALYSES

The kinetic energy of the block E_k is the key parameter in rockfall engineering, as well as the height of the impacting block. The parametric analysis for defining the capacity of the system to mitigate rockfall hazard considers variable block size and block impacting velocity. Following the design rule proposed in Marchelli & Deangeli (under review), the height of the impacting block h_b was considered as:

$$h_b = H - r_b - h_l \tag{3}$$

where H is the total height of the structure, while h_l is the height of each layer of compacted soil.

A set of simulations was performed in order to evaluate the q_{RC} for different impact energies, varying the block radius from 0.5 m to 1.5 m, and its velocity from 5 m/s to 30 m/s. The obtained values of q_{RC} span from 7 kPa to 427 kPa. The resulting forces on the RC wall were computed according to the geometric diffusion resulting in bending moment and shear at the base. Following the impact forces, then obtained minimum thicknesses of the RC wall range from 0.15 m (theoretical) to 1.2 m.

The parametric analysis herein presented reveals that the system can support an impact of maximum kinetic energy smaller of 17 MJ with an area occupation of about 5 m wide. To assess the economic sustainability of the proposed solutions, construction costs of the most recurrent rockfall protections structures are compared (Figure 2).

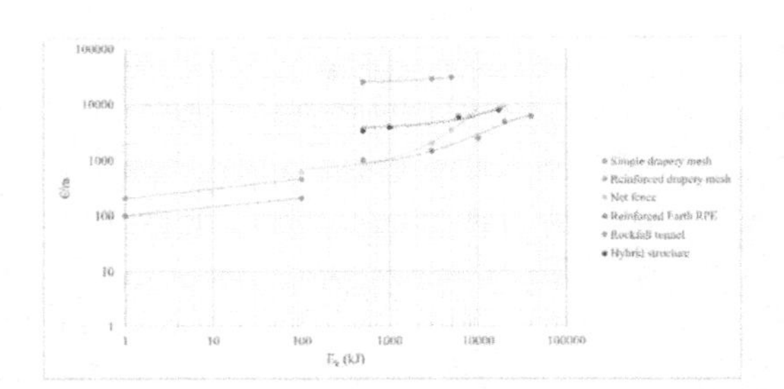

Figure 2. Prices of the most common rockfall mitigation measures, considering a mean height of 5 m.

REFERENCES

ASTRA 12 006 2008: *Actions de chutes de pierres sur les galeries de protection.* Office Fédéral des routes OFROU, Berne, Switzerland.

Lambert, S., and Kister, B., 2017. Analysis of Existing Rockfall Embankments of Switzerland (AERES); Part A: State of Knowledge. Federal Office for the Environment, Bern. 1–55.

Marchelli, M. & Deangeli, C. (under review). Towards a codified design procedure for rockfall reinforced earth embankments. *GEAM*

Peila, D., Oggeri, C., & Castiglia, C. 2007. Ground reinforced embankments for rockfall protection: design and evaluation of full scale tests. *Landslides*, 4(3), 255–265.

Ronco, C., Oggeri, C., & Peila, D. 2009. Design of reinforced ground embankments used for rockfall protection. *Natural Hazards and Earth System Sciences*, 9(4), 1189–1199.

Current Perspectives and New Directions in Mechanics, Modelling and Design of Structural Systems – Zingoni (ed.)
© 2022 Copyright the Author(s), ISBN 978-1-032-39114-4

Author Index

Abbas, H.H. 425, 679
Abdelhafeez, M. 471
Abotaleb, I. 497
AbouZeid, M.N. 435
Abubakar Ali, M. 465
Achmus, M. 695, 703
Acito, M. 197, 685
Adali, S. 173
Adam, J.M. 179, 187
Adetiloye, A. 439
Adewuyi, A.P. 431
Adriaenssens, S. 261
Aegerter, D.W. 75
Afoutou, J. 213
Agüero, A. 61
Ahola, A. 353, 391
Akinyele, J.O. 439, 563, 591
Alabi, S.A 485, 673
Alaud, S. 557
Al-Emrani, M. 575
Alghazali, H.H. 421
Al-Haddad, T. 501
Al-Hadi, H. 69
Al-Jaberi, Z. 539
Al-Jabri, K. 43
Al-Khafaji, A.F. 529
Allan, D. 375
Al-Nuaimi, A. 69
Alparslan, U. 535
Alqahtani, H. 217
Alshannaq, A. 501
Alshannaq, A.A. 533
Ameijeiras, M.P. 233
Anan, Y. 143
Anashpaul, S. 173
Anastasopoulos, D. 603
Anders, S. 449, 451, 459
Andersen, L.V. 19, 37, 39
Andersson, A. 269
Annoura, K. 63
Aratake, A. 655
Argenziano, M. 23, 661
Arrè, L. 135
Arslan, A.N. 623
Ashcroft, I. 115
Ashraf, M. 583
Aslan, N. 535
Asseko Ella, M. 593
Azzara, R.M. 73

Babafemi, A.J. 565, 567
Babawarun, T. 49
Bahnamiri, A.B. 423
Bai, Y. 521
Baitinger, M. 295
Balaskas, G. 597
Balcı, C. 507
Baldassino, N. 309
Ballim, Y. 429
Bamby, H.E. 403
Bank, L.C. 501, 521, 533
Bareille, O.A. 613
Baric, M. 293
Barretta, R. 109
Barszcz, A.M. 245, 247
Bartsch, H. 399
Basaglia, A. 687
Bastek, M. 207
Battini, J-M. 29
Bauer, A. 53
Beaulieu, P.M. 413
Becks, H. 437
Behdinan, K. 175
Behkamal, B. 609
Behzadi-Sofiani, B. 327
Bekele, A. 231
Belarbi, A. 541
Belhaj, M. 555
Belis, J. 289
Benfratello, S. 83
Benmokrane, B. 505
Bergmann, T. 157
Bermek, M. 501
Bernardi, E. 59, 79
Bernardi, M. 309
Bernardini, L. 25
Berrocal, C.G. 697
Bessinger, J.M.H. 571
Bhuta, W.H. 43
Bhuwal, A.S. 115
Birtel, V. 467
Bissing, H. 209
Biswas, S. 77, 205
Björk, T. 353, 391
Bloch, K.E. 421
Bluhm, J. 95
Bondsman, B. 21
Bontemps, A. 599
Bosbach, S. 437
Botte, W. 675
Bouchaïr, A. 319
Bradley, R.A. 277
Braml, T. 649
Brancheriau, L. 595
Brando, G. 687, 689
Brandonisio, G. 661
Brands, D. 89, 449, 451, 459
Breitenbücher, R. 461
Bristogianni, T. 283, 287
Brunesi, E. 185
Bücking, L. 495
Buitrago, M. 179, 187
Bukasa, G.M. 253, 255
Bukieda, P. 297

Caddemi, S. 651
Cadini, F. 619
Cai, W.Y. 409
Cai, W-Y. 387
Cai, Y. 357
Cai, Z. 521
Calabrese, A. 111
Caliò, I. 651
Calvo-Gallego, J.L. 107
Cannizzaro, F. 651
Caprio, D. 575
Cardinali, V. 73
Cardone, D. 111
Caspeele, R. 289
Ceresara, M. 59
Cervio, M. 631
Chen, Y. 35
Chesi, C. 197
Chiaia, B. 195
Chisari, C. 55

Cianchino, G. 687
Cicione, A. 141, 363
Claasen, J. 363
Clarke, S.D. 181
Claßen, M. 437
Classen, M. 133
Cocco, G. 687, 689
Collina, A. 25
Colmenares, D. 269
Combrinck, R. 559, 571
Corti, L. 577
Costa, G. 269
Costanzi, C.B. 127
Cucchi, M. 685
Curry, R.J. 181
Cusumano, L. 549
Czichos, R. 157

da Porto, F. 59, 79
Damen, W. 287
Danielsson, H. 21
Dauberschmidt, C. 515
Dauda, J.A. 669
Davies, J.M. 323
De Backer, H. 345
De Biagi, V. 195, 705
De Cicco, S. 105
De Marchi, N. 97
De Matteis, G. 55
De Michele, C. 609, 623, 627, 641
De Roeck, G. 603
De Silva, K.K.V. 103
Debney, P.M. 493
Deckner, F. 29
Dehkordi, F.M. 553
Del Giudice, L. 129
del Pozo, G. 269
Delaney, E. 501
Desai, D. 215
Detournay, E. 165
Deuser, F. 649
Deutscher, M. 463
Dewil, R. 321
di Gioia, A. 309
Di Nunzio, G. 577
Di Sarno, L. 251
Dinka, M.O. 225
Dinkler, D. 455
Dittmann, J. 113
Doležel, J. 417
Donà, M. 59, 79
Dosta, M. 457
Dragos, K. 611
Drosopoulos, G.A. 173
Droughi, N.A. 557
Duan, X. 175
Dubois, F. 213
Dudonu, A. 621
Dundu, M. 253, 255, 377, 379, 381
Durif, S. 319
Dziedziech, K. 617

Edri, I.E. 199
Ehsani, M. 517, 537
Ekolu, S.O. 639
El Anwar, H.H. 9
El-Aghoury, I.M. 349
El-Hussain, I. 43
Elkhoury, K. 433
Elnashai, A.S. 9
El-Serwi, A.A. 349
Elwakeel, A. 433
Empelmann, M. 455
Engel, A. 169
Engel, T. 127
Engelhardt, M.D. 183
Entezami, A. 609, 623, 627, 641
Ergin, T. 535
Eric, G.B. 431
Erven, M. 131
Eskicumali, O.F. 535
Esposito, F. 661
Essopjee, Y. 445
Euler, M. 339
Evbuomwan, N.F.O. 443
Eyben, F. 355
Ezea, B.U. 563

Faiella, D. 661
Fan, L. 35
Fantilli, A.P. 273, 553
Faure, A.J. 481
Featherston, N.R. 699
Fehling, E. 53, 453
Feix, J. 667
Feldmann, M. 355, 399
Ferraz, G. 321
Ferronato, M. 97
Feulefack Songong, E. 701
Fischer, S. 693
Flamand, O. 581
Flodén, O. 19, 21, 37, 39
Flora, A. 111
Flotzinger, J. 649
Folorunsho, A.B. 591
Forman, P. 137
Forth, J. 433
Foti, F. 27
Fournely, E. 599
François, S. 29
Frigo, B. 273, 553
Fritz, H. 155, 333
Fu, F. 193
Fu, Y. 525

Gabriel, S. 181, 201
Gajewski, J. 221, 223
Galvanetto, U. 161
Gardner, L. 3, 327
Garrecht, H. 467
Gasparini, G. 135
Gathimba, N.K. 633
Gebuhr, G. 449, 451, 459
Genc, A.G. 535
Gentry, T.R. 501, 533
Georgiadis, M. 275
Georgiou, N. 147
Gerlach, T. 703
Giese-Hinz, J. 295
Gietz, L. 455
Giglio, M. 619
Giordano, P.F. 11
Girardi, M. 73
Gizejowski, M.A. 237, 245, 247
Glomb, D. 515
Godi, G. 599
Godoy, L.A. 233
Gogolewski, D. 139, 145, 149
Goh, C.Y. 423
Gohnert, M. 277
Goi, Y. 211, 479, 615
Goi, Y. 635
Goli, G. 593
Gong, Y. 509
Görtz, S. 427, 515
Graham, C. 501
Granath, M. 549
Gribanov, I. 99
Grigoriu, M. 475
Gril, J. 593, 599
Guan, H. 227
Gubetini, D. 393
Gudžulić, V. 5, 461
Guest, S.D. 13
Guibal, D. 595
Gullapalli, V. 45

Haack, R. 427
Haas, T.N. 367, 369, 371, 373
Hadjioannou, M. 183
Haese, A. 291
Hai, L-T. 387
Haist, M. 465, 469
Hajnys, J. 151
Hamed, E. 159
Hamilton, S. 203
Hänig, J. 299
Harli, G. 429
Hashimoto, K. 189
Hassan, M.M. 425, 679
Hassan, M. 167
He, X. 31
Hegger, J. 419, 437
Hellebrand, S. 89
Henao, Y. 501
Ho, W.H. 49
Hoffmann, H. 295
Hoffmeister, B. 597
Holschemacher, K. 579, 681
Horisawa, E. 479
Hosseini, M. 375
Hosseini, S.M. 505
Hrynyk, T.D. 423
Huang, Z. 317, 347
Huang, Z-W. 65
Hudert, M.M. 19, 37
Huynh, A. 501

Ibrahim, K.A. 565
Ichchou, M. 613
Igba, U.T. 439, 563, 591
Iida, T. 395
Ilki, A. 507, 535
Ilunga, D.T. 225
Impollonia, N. 651
Inoue, S. 143
Iskhakov, T. 5
Ispir, M. 507, 535
Iuorio, O. 365, 669
Iwatsubo, K. 331
Izzuddin, B. 415, 433

Jaji, M.B. 567
Jiang, J.F. 503
Joachim, A. 301
Jockwer, R. 549, 575
John, S.K. 519
Jonáš, M. 625
Juraschitz, T. 305

Kafuko, J. 377, 379
Kaluba, C. 249
Kalus, M. 419
Kanyeto, O.J. 431
Kapelko, M. 547
Karabulut, B. 321
Karoumi, R. 269
Karpiński, R. 223
Kassab, R.A. 523
Kästner, T. 385
Katzenbach, R. 693
Kawabe, D. 605, 607
Ke, Y. 527
Kern, B. 469
Khashaba, M.H. 349
Kiakojouri, F. 195
Kiama, Y 47
Kiefer, B. 53
Kim, C.W. 605, 607
Kitane, Y. 189, 211, 395, 479, 633, 635
Kitayama, S. 365
Knaack, U. 127
Knobloch, M. 209
Kobayashi, D. 655
Koen, J.A. 367
Kolemenoglu, S. 507
Koniari, A.M. 287
Koopman, D. 287
Kotynia, R. 513
Kowalsky, U. 455
Kozior, T. 139, 145, 149
Kozlov, D. 263
Kraus, M. 155, 333
Krausche, T. 339, 341, 359
Krpec, P. 151
Kruger, P.J. 141
Kruis, J. 91
Kulshreshtha, S. 205
Kumar, A. 683
Kumar, R.R. 45
Kustermann, A. 515

Labiran, J.O. 439, 563, 591
Lagerkvist, J. 697
Laghi, V. 135
Lamperová, K. 643
Langdon, G.S. 181, 201
Lange, J. 127, 131, 169, 171, 389
Lanwer, J.-P. 455
Launert, B. 339, 341, 359
Leahy, P. 501
Lebea, L. 121, 215
Lechner, J. 667
Lehký, D. 417, 643
Lengert, K. 515
Leppla, S. 167
Lewis, W.J. 259
Li, B.B. 503
Li, C. 313
Li, G.Q. 397, 409
Li, G-Q. 387
Li, P.D. 503
Li, Q-Y. 311
Li, S. 41, 501
Li, Y. 227
Li, Z. 339, 341, 359
Lienhart, W. 637
Limongelli, M.P. 11
Lipiäinen, K. 353
Liu, S. 511
Liu, T. 115
Lohaus, L. 465, 469
Lomazzi, L. 619
Lombaert, G. 29
Louter, C. 305
Lu, C. 35
Lu, X.Z. 227

Mabuda, I. 119, 123
Madyira, D.M. 225
Magnucki, K. 239
Magrinelli, E. 197, 685
Mahachi, J. 47, 483, 485, 639, 673
Mahdi, R. 539
Mahlangu, R. 381
Maier, N. 243
Makoond, N. 179, 187
Makwela, W.M.P. 483
Mangliar, L. 19, 37
Mangliár, L. 39
Manthey, M. 581
Maphosa, R.K. 639
Maradni, B. 649
Marchelli, M. 705
Marelli, S. 129
Marini, F. 73
Mark, P. 137, 545
Markert, M. 463, 467
Marotti de Sciarra, F. 109
Martinelli, L. 27, 609, 627
Masouros, S.D. 107
Matsumura, M. 63, 143, 479
Matsuoka, K. 647
Mazza, F. 67

McAllister, E. 587
McDonald, A. 501
McKinley, J. 501
McPolin, D. 587
Mele, E. 23, 661
Mendrok, K. 617
Mensinger, M. 243, 393, 401, 665
Merkl, L. 207
Meschke, G. 5, 461
Mesicek, J. 151
Metrikine, A.V. 33
Meyer, M. 571
Mezui, E.N. 595
Miah, M.S. 637
Middendorf, B.
Middendorf, B. 53, 453
Middendorf, P. 93, 113, 157, 207
Milligan, G.J. 413
Mirra, M. 589, 663
Mirza, O. 375
Miyoshi, T. 331
Möckel, M. 305
Modungwa, D. 117, 121, 203
Mohamed, H.M. 505
Mojsilović, N. 671
Molenkamp, T. 33
Molkens, T. 133, 191
Moriyama, H. 63, 143
Mostert, J.P. 141
Mousa, S. 505
Moutou Pitti, R. 599
Moyo, P. 645
Mozzon, S. 59
Msibi, M. 119, 123
Muciaccia, G. 577, 631
Mudenda, K. 337
Mulas, M.G. 27
Muller, H. 559
Müller, J. 127
Myers, J.J. 421, 529, 539
Myśliński, A. 163

Nadir, Y. 519
Nagatani, H. 189
Nagle, A. 501
Nakakita, T. 331
Nakanishi, T. 655
Nakatsuka, S. 315
Napier, J.A.L. 165
Nassar, K. 435, 497
Ndogmo, J. 243
Nechak, L. 613
Nemavhola, F. 117, 119, 121, 123, 215
Netshivhulana, A.G. 49
Neu, G. 5
Ngidi, S.D. 531, 677
Nguyen, M.H. 319
Nguyen, T.D. 551
Ngwangwa, H.M. 49, 117, 119, 121, 123
Ngwangwa, H.M. 215
Nicklisch, F. 295
Nie, X.F. 527
Niendorf, T. 53
Nikitas, N. 433
Noël, M. 509
Nongwane, K. 673
Novák, B. 621

Offereins, D. 285
Oikonomopoulou, F. 283, 287
Okazawa, S. 549
Olsson, N. 549
Omotainse, P.O. 591
Oneschkow, N. 465, 469
Oertzen, V.v. 53
Osika, M. 657
Osterminski, K. 219
Otieno, M. 653
Ouldboukhitine, S.E. 319

Pacheco Torgal, F. 487, 491, 561
Pacoste, C. 29
Padovani, C. 73
Pagac, M. 151
Pagella, G. 589
Palermo, M. 135
Palizzolo, L. 83
Pallarés, F.J. 61
Pallarés, L. 61
Pallav, K. 683
Palmeri, A. 23
Pandelani, T. 49, 107, 117, 119, 121, 203
Pandey, M. 405, 407
Pang, Y. 115
Papangelis, J. 241
Parisi, F. 185
Parisi, M.A. 197
Pasternak, H. 339, 341, 359
Pauli, A. 293
Pejatovic, M. 289
Pellegrini, D. 73
Peplow, A. 39
Peralta, P. 629
Perotti, F. 27
Persson, P. 19, 21, 37, 39
Petraroia, D.N. 545
Phillips, A.T.M. 231
Phocas, M.C. 147
Pickering, E.G. 181
Pinnola, F.P. 109
Pise, M. 449, 451, 459
Poff, A. 501
Polak, M.A. 413, 511
Portioli, F.P. 669
Pourbehi, M.S. 275
Pretorius, F.J. 585
Pugliese, F. 251

Qi, J. 605
Qiu, C. 521
Quapp, U. 681
Qureshi, J. 329, 519

Rabinovitch, O. 199
Radecki, R. 657
Rahman, S.A. 583
Rajapakse, N. 103
Ranzi, G. 235
Rapone, D. 687
Rapp, D. 87
Rappl, S. 219
Rauber, L. 597
Rauch, M. 209
Ravenshorst, G. 663
Ravenshorst, G.J.P. 589
Ray, A. 217
Rębielak, J. 267
Reddy, G.R. 45
Reddy, G. 81
Reichert, J. 295
Reim, S. 649
Reinecke, J.D. 203
Reiner, J. 583
Reinhard, J. 621
Reinheimer, J. 389
Reinold, J. 5
Rempling, R. 549, 697
Respert, J. 501
Respert, J.A. 533
Retief, J.V. 477
Reynders, E. 603
Rheinschmidt, F. 101, 281

Ridley, I. 433
Riedel, K. 415
Rigby, S.E. 181
Ritter, M. 457
Roberts, M.J. 323
Robertson, I.N. 57, 75
Rogala, M. 221, 223
Rößß, R. 401
Rösch, P.J. 649
Rosendahl, P.L. 101, 281
Rossi, B. 133, 321
Roth, C. 585
Rouhi, J. 55
Ruane, K. 501
Rumpf, A. 303
Rust, B. 385
Rust, G. 415
Rust, J.P. 373
Rybczyński, S. 457

Sadeghian, P. 523, 525
Sajadi, F. 87
Sakurada, Y. 655
Salama, N. 435
Salar, M. 609, 627
Salem, H. 471
Saler, E. 79
Salomoni, V.A. 97
Samanta, A. 335, 351
Sapsathiarn, Y. 103
Saraiva, R. 549
Sarmadi, H. 609, 627
Sartipi, S. 675
Sato, A. 211, 315
Sato, T. 71
Saudy, M. 497
Saulnier, V. 319
Sauvat, N. 213
Sayed-Ahmed, E.Y. 435
Sayed-Ahmed, E. 497
Scabbia, F. 161
Scalvenzi, M. 185
Schaan, G. 457
Schäfer, N. 461
Schäfers, M. 401
Schaffrath, S. 355
Scheunemann, L. 89
Schleiting, M. 453
Schmauder, S. 87
Schmidt, M. 437
Schmidt-Döhl, F. 457
Schneider, J. 281
Schröder, J. 89, 95, 449, 451, 459
Schwarz, A. 95
Scott, D. 415
Scott, D.W. 533
Sebastiao, L. 241
Seif, P. 113
Seip, M. 167
Semakane, L. 117
Sengupta, P. 77, 205
Senosha, J.T. 531
Serrano, E. 21
Shangguan, G.H. 271
Shehzad, M. 433
Shen, S. 65
Shkundalova, O. 133
Shrestha, S. 329
Shrikhande, M. 81
Siebert, B. 291
Siebert, D. 665
Siebert, G. 285, 293
Singh, N. 375
Singh, R. 351
Sivakumaran, K.S. 317, 347
Skriko, T. 353
Slowik, O. 643
Smarsly, K. 611, 629
Soldavini, G. 25
Somma, G. 361, 569
Sommer, D. 93
Song, J. 695
Sossou, G. 441
Soutsos, M. 501
Šplíchal, B. 417
Spacone, E. 687, 689
Starossek, U. 17
Staśkiewicz, M. 513
Staszewski, W.J. 617, 657
Stefanaki, I.M. 287
Stefaniuk, D. 541
Stein, F. 621
Steineck, S. 171
Stindt, J. 137
Strasheim, J. 275
Stroetmann, R. 385
Struzik, W. 489
Su, M. 551
Subhani, M. 583
Sugiura, K. 189, 211, 395, 479, 633
Sugiura, K. 635
Suman, S. 335
Sun, S. 321
Sun, W. 115
Sun, Y. 35
Sun, Z. 397
Sunten, J. 95
Szafran, J. 343

Tahir, M.N. 159
Takai, T. 331
Takarli, M. 213
Takase, K. 607, 635
Takemura, K. 607
Tamada, K. 331
Tan, P. 59
Tancredi, A. 689
Tanganelli, M. 73
Tanimoto, T. 479
Taylor, R.S. 99
Telega, J. 343
Terrenzi, M. 687
Tha, B. 397
Theland, F. 29
Thiemicke, J. 53
Tinmitonde, S. 31
Tiraboschi, C. 685
Tolba, H. 435
Troff, B. 93
Trombetti, T. 135
Tshilombo, D.M. 485
Tshireletso, T. 645
Tsouvalas, A. 33
Tyas, A. 181

Uhl, T. 617
Ungermann, J. 419
Usongo, A.A. 367, 369, 371

Vaccaro, M.S. 109
Vacková, P. 555
Valentin, J. 555
Van Bogaert, Ph. 325, 345
van de Kuilen, J.W.G. 589
Van Staen, G. 345
van Zijl, G.P.A.G. 565, 567
Vaserchuk, Y. 263
Vassiliou, M.F. 129
Vazzano, S. 83
Vecchio, D. 541
Veljkovic, M. 403
Vera, A.C.O. 675
Verhoeven, B. 321
Verma, G.K. 77
Viljoen, C. 699
Vollmer, M. 53
Vollum, R. 415, 433
von Klemperer, C.J. 181, 201

Vorel, J. 91
Vozáb, J. 91

Wadee, M.A. 231, 327
Waldschmitt, B. 127
Walls, R. 141
Walls, R. 363
Walter, B. 597
Walther, C. 155
Wang, J. 265, 271, 339, 341, 359
Wang, Y.B. 397, 409
Wang, Y-B. 387
Wang, Y-Z. 387
Wan-Wendner, R. 675
Waris, M.B. 43, 69
Watanabe, S. 635
Watson, M. 551
Weißgraeber, P. 101
Weller, B. 295, 297, 299, 301, 303
Werner, A. 693
Wheeler, A. 375
Wiedro, P. 247
Wijesinghe, R.A.R. 103
Wilden, V. 597
Williamson, E.B. 183
Wittwer, J. 303
Wolf, B. 515
Woodham, C. 501
Wrana, J. 489
Wu, B-L. 65
Wünsch, J. 303

Xotta, G. 97
Xu, C. 87
Xu, X. 87

Yamaguchi, T. 395
Yan, L. 31
Yan, R. 403
Yang, Z. 227
Yankelevsky, D.Z. 199
Yavuzer, M.N. 507
Yildirim, R. 209
Yoshimoto, D. 189
Young, B. 311, 357, 405, 407
Yuen, S.C.K. 181

Zaccariotto, M. 161
Zakka, Z.G. 653
Zandonini, R. 309
Zatloukal, J. 625
Zhang, A.L. 265, 271, 313
Zhang, Y.X. 265, 271
Zhang, Y-X. 65
Zhang, Z. 501
Zhang, S.S. 527
Zhao, X. 265, 317, 347
Zhao, X.L. 7, 397, 521
Ziaja-Sujdak, A. 657
Zijl, G.V. 141
Zine, A. 613
Zingoni, A. 249, 337, 701
Zizi, M. 55
Zmarzły, P. 139, 145, 149
Zonta, A. 59, 79
Zou, M. 271